Parabolic Geometries I

Background and General Theory

Mathematical
Surveys
and
Monographs

Volume 154

Parabolic Geometries I

Background and General Theory

Andreas Čap
Jan Slovák

American Mathematical Society
Providence, Rhode Island

2000 *Mathematics Subject Classification.* Primary 53C15, 53A40, 53B15, 53C05, 58A32; Secondary 53A55, 53C10, 53C30, 53D10, 58J70.

The first author was supported during his work on this book at different times by projects P15747–N05 and P19500–N13 of the Fonds zur Förderung der wissenschaftlichen Forschung (FWF).

The second author was supported during his work on this book by the grant MSM0021622409, "Mathematical structures and their physical applications" of the Czech Republic Research Intents scheme.

Essential steps in the preparation of this book were made during the authors' visit to the "Mathematisches Forschungsinstitut Oberwolfach" (MFO) in the framework of the Research in Pairs program.

For additional information and updates on this book, visit
www.ams.org/bookpages/surv-154

Library of Congress Cataloging-in-Publication Data

Čap, Andreas, 1965–
Parabolic geometries / Andreas Čap, Jan Slovák,
v. cm. — (Mathematical surveys and monographs ; v. 154)
Includes bibliographical references and index.
Contents: 1. Background and general theory
ISBN 978-0-8218-2681-2 (alk. paper)
1. Partial differential operators. 2. Conformal geometry. 3. Geometry, Projective. I. Slovák, Jan, 1960– II. Title.

QA329.42.C37 2009
515′.7242–dc22 2009009335

∞ The paper used in this book is acid-free and falls within the guidelines
established to ensure permanence and durability.
Visit the AMS home page at http://www.ams.org/

10 9 8 7 6 5 4 3 2 1 14 13 12 11 10 09

Contents

Preface

The roots of the project to write this book originated in the early nineteen–nineties, when several streams of mathematical ideas met after being developed more or less separately for several decades. This resulted in an amazing interaction between active groups of mathematicians working in several directions. One of these directions was the study of conformally invariant differential operators related, on the one hand, to Penrose's twistor program (M.G. Eastwood, T.N. Bailey, R.J. Baston, C.R. Graham, A.R. Gover, and others) and, on the other hand, to hypercomplex analysis (V. Souček, F. Sommen, and others). It turned out that, via the canonical Cartan connection or equivalent data, conformal geometry is just one instance of a much more general picture. Over the years we noticed that the foundations for this general picture had already been developed in the pioneering work of N. Tanaka. His work was set in the language of the equivalence problem and of differential systems, but independent of the much better disseminated developments of the theory of differential systems linked to names like S.S. Chern, R. Bryant, and M. Kuranishi. While Tanaka's work did not become widely known, it was further developed, in particular, by K. Yamaguchi and T. Morimoto, who put it in the setting of filtered manifolds and applied it to the geometric study of systems of PDE's. A lot of input and stimulus also came from Ch. Fefferman's work in complex analysis and geometric function theory, in particular, his parabolic invariant theory program and the relation between CR–structures and conformal structures. Finally, via the homogeneous models, all of these studies have close relations to various parts of representation theory of semisimple Lie groups and Lie algebras, developed for example by T. Branson and B. Ørsted.

Enjoying the opportunities offered by the newly emerging International Erwin Schrödinger Institute for Mathematical Physics (ESI) in Vienna as well as the long lasting tradition of the international Winter Schools "Geometry and Physics" held every year in Srní, Czech Republic, the authors of this book started a long and fruitful collaboration with most of the above mentioned people. Step by step, all of the general concepts and problems were traced back to old masters like Schouten, Veblen, Thomas, and Cartan, and a broad research program led to a conceptual understanding of the common background of the various approaches developed more recently. In the late nineties, the general version of the invariant calculus for a vast class of geometrical structures extended the tools for geometric analysis on homogeneous vector bundles and the direct applications of representation theory expanded in this way to situations involving curvatures. In particular, the celebrated Bernstein–Gelfand–Gelfand resolutions were recovered in the realm of general parabolic geometries by V. Souček and the authors and the cohomological substance of all these constructions was clarified.

The book was written with two goals in mind. On one hand, we want to provide a (relatively) gentle introduction into this fascinating world which blends algebra and geometry, Lie theory and geometric analysis, geometric intuition and categorical thinking. At the same time, preparing the first treatment of the general theory and collection of the main results on the subject, gave us the ambition to provide the standard reference for the experts in the area. These two goals are reflected in the overall structure of the book which finally developed into two volumes. The second volume will be called "Parabolic Geometries II: Invariant Differential Operators and Applications".

Contents of the book. The basic theme of this book is canonical Cartan connections associated to certain types of geometric structures, which immediately causes peculiarities.

Cartan connections and, more generally, various types of absolute parallelisms certainly played a central role in Cartan's work on differential geometry and they still belong to the basic tools in several geometric approaches to differential equations. However, in the efforts in the second half of the last century to put Cartan's ideas into the conceptual framework of fiber bundles, the main part was taken over by principal connections. The isolated treatments in the books by Kobayashi and Sharpe stopped at alternative presentations of the quite well–known conformal Riemannian and projective structures. As a consequence, even basic facts on Cartan connections are neither well represented in the standard literature on differential geometry nor well known among many people working in the field. For this book, we have collected some general facts about Cartan connections on principal fiber bundles in Section 1.5. While this is located in the "Background" part of the book because of its nature, quite a bit of this material will probably be new to most readers.

The other peculiar consequence of the approach is that the geometric structures we study display strong similarities in the picture of Cartan connections, while they are extremely diverse in their original descriptions. Therefore, in large parts of the book we will take the point of view that the Cartan picture is the "true" description of the structures in question, while the original description is obtained as an underlying structure. This point of view is justified by the results on equivalences of categories between Cartan geometries and underlying structures in Section 3.1, which are among the main goals of this volume. This point of view will also be taken in the second volume, in which the Cartan connection (or some equivalent data) will be simply considered as an input.

Let us describe the contents of the first volume in more detail. The technical core of the book is Chapter 3. In Section 3.1 we develop the basic theory of parabolic geometries as Cartan geometries and prove the equivalence to underlying structures in the categorical sense. This is done in the setting of $|k|$–gradings of semisimple Lie algebras, thus avoiding the use of structure theory and representation theory. The structure theory is brought into play in Section 3.2 to get more detailed information on the applicability of the methods developed before. Section 3.3 contains an exposition and a complete proof of Kostant's version of the Bott–Borel–Weil Theorem, which is needed to verify the cohomological conditions that occur in several places in the theory, and proves to be extremely useful later, too.

In Chapter 4, the general results of Chapter 3 are turned into explicit descriptions of a wide variety of examples of geometries covered by our methods.

In particular, we thoroughly discuss the geometries corresponding to $|1|$–gradings (which can be described as classical first order G–structures) and the parabolic contact geometries, which have an underlying contact structure. In Sections 4.4 and 4.5, we discuss two general constructions relating geometries of different types, the construction of correspondence spaces and twistor spaces, and analogs of the Fefferman construction.

The developments in Chapter 5 admit two interpretations. On the one hand, via the notion of Weyl structures, we associate to any parabolic geometry a class of distinguished connections and we define classes of distinguished curves. On the other hand, the data associated to a Weyl structure offer an equivalent description of the canonical Cartan connection in terms of objects associated to the underlying structure. In this way, one also obtains a more explicit description of the canonical Cartan connections. Throughout Sections 5.2 and 5.3 we also discuss various applications of the theory developed in the book.

The first part of the book (Chapters 1 and 2) provides necessary background and motivation. Chapter 1 is general and rather elementary and should be digestible and enjoyable even for newcomers. Here Cartan's concept of "curved analogs" of Klein's homogeneous spaces and also the related general calculi are explained using the effective general language of Lie groups and Lie algebras but no structure theory. As mentioned before, some of the material presented in this discussion is not easy to find in the literature. Section 1.6 contains an explicit and elementary treatment of conformal (pseudo)–Riemannian structures. Apart from motivating further developments, this also indicates clearly that a deep understanding of the algebraic structure of the algebras and groups of symmetries in question is the key to further progress. This naturally leads to Chapter 2, which contains background material on semisimple Lie algebras and Lie groups. While the material we cover in this chapter is certainly available in book form in many places, there are some unusual aspects. The main point is that, apart from the complex theory, we also discuss the structure theory and representation theory of real semisimple Lie algebras. The real theory is typically scattered in the textbooks among the advanced topics and hence rather difficult to learn quickly elsewhere. In this way, the first part of the book makes the whole project more or less self–contained. In addition, it should be of separate interest as well. As an important counterpart to the theory developed in Chapter 2 we provide tables containing the central structural information on semisimple Lie algebras in Appendix B.

The second volume will be devoted to invariant differential operators for parabolic geometries, in particular, the technique of BGG–sequences, and several applications. While the links of the Cartan geometry to the more easily visible and understandable underlying structures are among the main targets of the first volume, the second one will treat the Cartan connections as given abstract data. This will further underline the algebraic and cohomological character of the available tools and methods.

Suggestions for reading. We have tried to design the book in a way which allows fruitful reading for people with different interests. Readers interested in one or a few specific examples of the geometries covered by the general theory could start reading the parts of the fourth and fifth chapters devoted to the structures in question, and return to the earlier chapters to get background or general results and concepts as needed. Apart from the well–known conformal and projective structures, the book contains extensive material on almost Grassmannian structures, almost quaternionic geometries, CR–geometries and Lagrangean contact geometries, quaternionic contact geometries, low–dimensional distributions, path geometries, and many others. This includes part of their twistor theory, correspondence spaces and further functorial constructions.

Readers familiar with differential geometry and Lie theory, who are interested in the general approach to parabolic geometries might prefer to look at the "Prologue" 1.1 in order to get a sense for the typical examples of the structures in question, inspect briefly the generalities on Cartan connections in 1.5 and then begin a serious reading straight from Chapter 3. If necessary, it should be possible to find background material, concepts and technicalities quickly in Chapters 1 and 2.

Finally, both Chapters 1 and 2 are also intended to be useful as a broader introduction to the subject and have been used successfully as underlying material for various graduate courses, both in Vienna and Brno. Thus we also believe that readers at the graduate student level may enjoy reading the book in the order it is written, with possible glimpses backward and forward for illustrations of the general considerations as they go.

Acknowledgement. Although we heard that delays are common with mathematical books, the work on this one certainly has taken much longer than anyone expected, and we are grateful to the publisher for his patience and support during all of these years. We would also like to mention here many of our collaborators and colleagues who contributed countless suggestions and comments during the ten years of the development of this project. Our particular thanks are due to those colleagues and students of ours who attended our seminars in Vienna and Brno, where most of the contents of this first volume of our book was read, presented and discussed in detail — the project could hardly have been accomplished without their efforts.

Vienna and Brno, August 20, 2008

Part 1

Background

CHAPTER 1

Cartan geometries

The concept of a "generalized space" was introduced by Élie Cartan in order to build a bridge between geometry in the sense of Felix Klein's Erlangen program and differential geometry. In the Erlangen program, a geometry is given by a manifold endowed with a transitive action of a Lie group, and thus (up to the choice of a base point) by a homogeneous space G/H of a Lie group G. Cartan's idea was to associate to such a homogeneous space a differential geometric structure, whose objects may be thought of as curved analogs of the homogeneous space G/H, just like n–dimensional Riemannian manifolds may be thought of as curved analogs of Euclidean space, viewed as a homogeneous space of the group of Euclidean motions. In modern terminology, such structures are called Cartan geometries, and they are defined as principal bundles endowed with Cartan connections.

This chapter provides a quick introduction to Cartan geometries as well as a sketch of the necessary background from differential geometry. A more detailed explanation of the relation to the Erlangen program and a comprehensive study of several basic examples of Cartan geometries can be found in the book [**Sh97**].

There are several results on general Cartan geometries, which are highly non-trivial even for special cases like Riemannian or conformal structures. Among these are, for example, the fact that the automorphisms of a Cartan geometry always form a Lie group, or that local automorphisms of the homogeneous model always uniquely extend to global automorphisms. Apart from these aspects, in which the Cartan geometry is viewed as an input, there is also a second important point of view: The most interesting examples of Cartan geometries are those, in which the Cartan geometry is actually determined by some underlying geometric structure, so that results on the Cartan geometry directly give results on the underlying structure.

Of course, one may also turn this around and ask about the structures underlying a Cartan geometry of given type and whether (assuming a suitable normalization condition) the Cartan geometry is determined by these underlying structures. This point of view will also be important in several parts of this chapter. The underlying structures can already be seen on the homogeneous model G/H of a Cartan geometry, so this question boils down to understanding G–invariant geometric structures on G/H.

Let us briefly describe the contents of the individual sections: Section 1.1 should be viewed as a prologue. We consider several simple actions of Lie groups on spheres and projective spaces and show how to find geometric structures which are invariant under these actions. On the one hand, this shows the diversity of structures showing up already in simple situations. On the other hand, except for the Riemannian round sphere, all the examples we consider are actually the homogeneous models of some parabolic geometry, so this provides the first overview of the structures available in that way.

Sections 1.2 and 1.3 provide background from differential geometry needed in the further development. Section 1.2 collects standard material on smooth manifolds and basic Lie theory. Section 1.3 provides an overview on various concepts of connections, which ranges from linear connections on vector bundles, via connections on general fiber bundles to principal connections and connections on G–structures.

In Section 1.4 the question of invariant geometric structures on homogeneous spaces is taken up. Apart from the study of invariant sections of homogeneous vector bundles, the main emphasis is on the question of existence of invariant connections, which has a decisive influence on the nature of the associated Cartan geometry.

The basic general theory of Cartan geometries is developed in Section 1.5, in which we also prove the general results on Cartan geometries mentioned above. We develop a differential calculus for Cartan geometries and prove the basic Ricci and Bianchi identities. These proofs also provide a quick way to the classical versions of these identities on G–structures. We offer a first glimpse at constructions of invariant (natural) differential operators on Cartan geometries, which will be a central topic of volume two. We introduce the general construction of correspondence spaces and characterize geometries obtained in this way. Then we show how a homogeneous Cartan geometry on a homogeneous space extends to a functor relating Cartan geometries of different types. Finally, we discuss automorphisms and infinitesimal automorphism of Cartan geometries, as well as distinguished curves. Some of the material in Section 1.5 is (to our knowledge) new, at least in this generality.

Finally, Section 1.6 is devoted to simple instances of the question of whether a Cartan geometry is determined by an underlying geometric structure. We first clarify this question completely in the case of affine extensions of G–structures. Next, we give an elementary construction of the canonical normal Cartan geometry associated to a pseudo–Riemannian conformal structure. This will be kept in mind as a basic example throughout the rest of the book.

1.1. Prologue — a few examples of homogeneous spaces

We start by considering several examples of viewing spheres and projective spaces as homogeneous spaces of various Lie groups. In any of these examples, there is a geometric structure which is easily seen to be invariant under the action of the group in question. Much more subtle arguments show that in all examples the group can be characterized as the group of those diffeomorphisms which preserve that geometric structure.

The structures showing up in this way can then be defined on (almost) arbitrary smooth manifolds. While they are very different, they have some common features:

- They are interesting from a local point of view, i.e. there are nontrivial local invariants.
- Any automorphism of such a structure is determined by a finite jet (i.e. its value and the value of finitely many derivatives) in a single point.
- The automorphism group of such a structure on any smooth manifold is finite dimensional (with the dimension being bounded from above by the dimension of the group showing up in our examples).

- In the case of the homogeneous model, any local automorphism of the structure uniquely extends to a global automorphism, i.e. to the action of an element of the group in question.
- In general, such structures are very far from being homogeneous. Indeed, generically they do not admit any nontrivial local automorphism.

As we shall see later on, all these properties (together with many others) are consequences of the fact that the structures can be described as Cartan geometries. This means that in spite of the last point of this list, they admit a description which is quite similar to a homogeneous space.

The main purpose of these examples is, on one hand, to show that in spite of the simple nature of the actions, which all are based on the standard representation of some classical simple Lie group, a broad variety of geometric structures is obtained. On the other hand, many of the constructions that we will present in the book are easily visible in the homogeneous models, so we will regularly refer back to these examples.

This section is intended as a prologue only, so we partly sketch proofs or completely omit them. Also, we use many notions which are more formally introduced in Sections 1.2 and 1.3 below, so some readers may prefer to first go through those two sections and return to the prologue later on.

1.1.1. The Riemannian round sphere. Consider the sphere S^n as the unit sphere $\{x : |x| = 1\}$ in $\mathbb{R}^{n+1}$. The Lie group $G := O(n+1)$ of orthogonal $(n+1) \times (n+1)$–matrices acts on $\mathbb{R}^{n+1}$ by norm preserving maps, so restricting to unit vectors, we get a smooth left action $O(n+1) \times S^n \to S^n$. Since the first vector e_1 in the standard basis of $\mathbb{R}^n$ can be mapped to any other unit vector by an orthogonal transformation, this action is transitive. The isotropy group of e_1 is immediately seen to be

$$H = \left\{ \begin{pmatrix} 1 & 0 \\ 0 & A \end{pmatrix} \right\} \subset G.$$

A matrix of the latter form is orthogonal if and only if the $n \times n$–matrix A is orthogonal. Thus, $H \cong O(n)$ and our left action induces a diffeomorphism of the sphere S^n with the homogeneous space $O(n+1)/O(n)$.

To find the geometric structure, which is invariant under this action, we just have to analyze the tangent spaces of S^n in this picture. In terms of the standard inner product $\langle\ ,\ \rangle$ on $\mathbb{R}^{n+1}$ we have $S^n = \{x \in \mathbb{R}^{n+1} : \langle x, x\rangle = 1\}$ and hence $T_xS^n = \{y \in \mathbb{R}^{n+1} : \langle x, y\rangle = 0\}$. The standard inner product on $\mathbb{R}^{n+1}$ induces a positive definite inner product on each tangent space of S^n, which obviously depends smoothly on the footpoint, so we get a *Riemannian metric* g on S^n. This is usually called the *standard metric* or the *round metric*.

For a matrix $A \in O(n+1)$, a point $x \in S^n$ and an element $y \in T_xS^n$, orthogonality of A immediately implies that $Ay \in T_{Ax}S^n$, and since $x \mapsto Ax$ is the restriction to S^n of a linear map, the tangent map $T_xS^n \to T_{Ax}S^n$ of the action by A is given by $y \mapsto Ay$. Since the Riemannian metric is induced from the standard inner product on $\mathbb{R}^{n+1}$, this tangent map is orthogonal, so $O(n+1)$ is contained in the group $\mathrm{Isom}(S^n, g)$ of isometries of the round sphere.

There are at least two ways to see that these two groups coincide, i.e. any isometry of S^n is induced by an orthogonal transformation of $\mathbb{R}^{n+1}$. Since analogs of these two lines of argument will be important for us in the sequel, we outline both of them.

The first line of argument is based on reducing to fixed jets. Let $\phi \in \mathrm{Isom}(S^n, g)$ be any isometry, and consider $\phi(e_1) \in S^n$. Since the action of $O(n+1)$ on S^n is transitive, we find a matrix $A \in O(n+1)$ such that $Ae_1 = \phi(e_1)$, so $A^{-1} \circ \phi$ lies in the subgroup $\mathrm{Isom}_{e_1}(S^n, g)$ of isometries fixing e_1. Clearly, $\phi \in O(n+1)$ if and only if $A^{-1} \circ \phi$ lies in the isotropy subgroup H. Hence, it suffices to show that any isometry which fixes e_1 actually lies in H.

But if ϕ is such an isometry, its tangent map in e_1 is an orthogonal automorphism of $T_{e_1}S^n$, and thus we can find a (unique) matrix in the isotropy subgroup H which induces the same orthogonal map on $T_{e_1}S^n$. As above, this reduces the problem of showing that $\mathrm{Isom}(S^n, g) = O(n+1)$ to the case of those isometries, whose one–jet in e_1 coincides with the one–jet of the identity. Hence suppose that $\phi \in \mathrm{Isom}(S^n, g)$ satisfies $\phi(e_1) = e_1$ and $T_{e_1}\phi = \mathrm{id}$. Then we can conclude from elementary Riemannian geometry that ϕ must be the identity map as follows:

Since ϕ is an isometry, it must map geodesics to geodesics. More precisely, if $I \subset \mathbb{R}$ is an interval containing zero and $c : I \to S^n$ is a geodesic starting at $c(0)$ in direction $c'(0)$, then $\phi \circ c$ is the geodesic with starting point $\phi(c(0))$ and initial direction $T_{c(0)}\phi \cdot c'(0)$. Since any point x_0 has an open neighborhood whose points can be reached by geodesics starting at x_0, the set on which the one–jet of ϕ coincides with the one–jet of the identity is open and nonempty, and clearly its complement is open as well. Since S^n is connected, we conclude that $\phi = \mathrm{id}_{S^n}$.

The second line of argument is based on the analysis of the Maurer–Cartan form on G. It is less elementary, but easier to generalize to the other cases of interest. We have already observed above that $A \mapsto Ae_1$ induces a diffeomorphism $O(n+1)/H \to S^n$, where $H \cong O(n)$ is the isotropy subgroup of e_1. In particular, the resulting projection $p : O(n+1) \to S^n$ is a principal fiber bundle with structure group $O(n)$. Let $\mathfrak{g}$ be the Lie algebra of $G = O(n+1)$, i.e. the Lie algebra of all skew symmetric $(n+1) \times (n+1)$–matrices. Splitting matrices into blocks of sizes one and n, we see that

$$\mathfrak{g} = \left\{ \begin{pmatrix} 0 & -v^t \\ v & A \end{pmatrix} : v \in \mathbb{R}^n, A \in \mathfrak{o}(n) \right\}.$$

Evidently, the subalgebra $\mathfrak{h} \subset \mathfrak{g}$ corresponding to H is given by those matrices, for which $v = 0$. From the block form, it follows immediately that the vector subspace $\mathfrak{n}$ consisting of the matrices with $A = 0$ is complementary to $\mathfrak{h}$. Hence, $\mathfrak{g} = \mathfrak{h} \oplus \mathfrak{n}$ as a vector space, and $\mathfrak{n} \cong \mathbb{R}^n$.

The adjoint action restricted to the subgroup H makes $\mathfrak{g}$ into a representation of the group $H \cong O(n)$. Since the adjoint action for matrix groups is given by conjugation, one immediately sees that the decomposition $\mathfrak{g} = \mathfrak{h} \oplus \mathfrak{n}$ is H–invariant, and $\mathfrak{n}$ is (isomorphic to) the standard representation of $O(n)$ on $\mathbb{R}^n$.

Now consider the Maurer–Cartan form $\omega \in \Omega^1(G, \mathfrak{g})$ on the group G, i.e. the trivialization of the tangent bundle of G by left translations, viewed as a differential form; cf. 1.2.4. Then ω is G–invariant and thus, in particular, H–invariant. Moreover, the H–invariant decomposition $\mathfrak{g} = \mathfrak{h} \oplus \mathfrak{n}$ splits ω into a sum $\omega = \omega_{\mathfrak{h}} \oplus \omega_{\mathfrak{n}}$ of H–invariant forms with values in $\mathfrak{h} \cong \mathfrak{o}(n)$ and $\mathfrak{n} \cong \mathbb{R}^n$. For any point $A \in G$ the kernel of $\omega_{\mathfrak{n}}(A)$ is exactly the vertical tangent space V_AG in A of the principal fiber bundle $G \to G/H$. Hence, $\omega_{\mathfrak{n}}(A)$ induces a linear isomorphism $T_AG/V_AG \cong T_{p(A)}S^n \to \mathbb{R}^n$. Since in addition $\omega_{\mathfrak{n}}$ is H–invariant, we can view $p : G \to S^n$ as a reduction to the group $O(n)$ of the frame bundle of S^n and this is exactly the orthonormal frame bundle corresponding to the Riemannian metric g. In particular, any isometry $\phi : S^n \to S^n$ lifts to an automorphism $\tilde{\phi}$ of the H–principal bundle $p : G \to S^n$, which in addition satisfies $\tilde{\phi}^*\omega_{\mathfrak{n}} = \omega_{\mathfrak{n}}$.

On the other hand, since the component $\omega_{\mathfrak{h}} \in \Omega^1(G, \mathfrak{h})$ of ω is H–invariant, it follows immediately from the properties of the Maurer–Cartan form, that $\omega_{\mathfrak{h}}$ defines a principal connection on the principal bundle $p : G \to S^n$. This principal connection induces a linear connection on the tangent bundle of S^n which preserves the Riemannian metric g and which is torsion free by the Maurer–Cartan equation. Thus, $\omega_{\mathfrak{h}}$ is the principal connection corresponding to the Levi–Civita connection on S^n. In particular, this implies that for any isometry ϕ of S^n, the induced diffeomorphism $\tilde{\phi} : G \to G$ satisfies $\tilde{\phi}^*\omega = \omega$. But now there is a rather simple general result (cf. Theorem 1.2.4 below), which says that any diffeomorphism with this property is the left multiplication by some element of G. In particular, this implies that $\text{Isom}(S^n, g) = O(n + 1)$.

The curved analogs of S^n viewed as a homogeneous space of $O(n + 1)$ in the sense of É. Cartan's generalized spaces are n–dimensional Riemannian manifolds. Usually, one would rather take the Euclidean space E^n, viewed as a homogeneous space of the group $\text{Euc}(n)$ of Euclidean motions, as the homogeneous model of Riemannian n–manifolds. As we shall see in 1.1.2 below, this makes only a difference in the notion of curvature, but not in the notion of a curved analog itself. For Riemannian manifolds, the properties listed in the beginning of 1.1 are well known, and can be proved without using Cartan connections, which, however, offer a very conceptual approach.

1.1.2. Some variations on the Riemannian sphere. There are several variations of the Riemannian sphere, in which the overall picture looks very similar. The first possible variation is to replace the group $O(n + 1)$ by a group with the same Lie algebra, with $SO(n+1)$ and $Spin(n+1)$ being the most important choices. Clearly, $SO(n + 1)$ acts on the sphere S^n by orientation preserving isometries and precisely as in 1.1.1 one shows that any orientation preserving isometry of S^n is obtained in that way. Moreover, the projection $SO(n + 1) \to S^n$ is exactly the oriented orthonormal frame bundle of S^n with structure group $SO(n)$. The Cartan geometries corresponding to this homogeneous space are oriented n–dimensional Riemannian manifolds. In the case of $Spin(n+1)$, we get isotropy subgroup $Spin(n)$ and the action of $Spin(n + 1)$ on $\mathbb{R}^{n+1}$ and thus the action on S^n factors through the covering homomorphism $Spin(n + 1) \to SO(n + 1)$. This means that the resulting $Spin(n)$–principal bundle $Spin(n+1) \to S^n$ is an extension of the oriented orthonormal frame bundle of S^n and thus defines a *spin structure* on the oriented Riemannian manifold S^n. The group $Spin(n + 1)$ is then exactly the group of all automorphisms of this spin structure, and the curved analogs of this homogeneous space are n–dimensional Riemannian spin–manifolds.

The second possible variation is to replace $O(n + 1)$ by the orthogonal group $O(p + 1, q)$ of an indefinite inner product of signature $(p + 1, q)$. Then one has to replace S^n by the (noncompact) unit sphere of that inner product. Choosing the inner product in such a way that the first vector in the standard basis of $\mathbb{R}^{p+1+q}$ has norm one, one proceeds similarly as in 1.1.1 to show that the action on the unit sphere is transitive and the isotropy subgroup of e_1 is isomorphic to $O(p, q)$. Visibly, $O(p + 1, q)$ acts by isometries of the pseudo–Riemannian metric (with signature (p, q)) on the unit sphere induced from the inner product on $\mathbb{R}^{p+1+q}$, and similarly as in 1.1.1 one shows that all isometries are obtained in this way. The curved analogs of this homogeneous space are exactly $(p+q)$–dimensional pseudo–Riemannian manifolds with metrics of signature (p, q). Clearly, one can pass to other groups with Lie algebra $\mathfrak{o}(p+1, q)$, in particular $SO(p+1, q)$, $SO_0(p+1, q)$ (the connected component of the identity in $SO(p+1, q)$), and $Spin(p+1, q)$, which leads to oriented pseudo–Riemannian manifolds, space– and time–oriented pseudo–Riemannian manifolds, and pseudo–Riemannian spin–manifolds of fixed dimension and signature, respectively.

The next variation we discuss is the passage to Euclidean space. At first sight, this does not look like a variation, sincc the group $\text{Euc}(n)$ of Euclidean motions in n–dimensions is very different from $O(n+1)$. Nevertheless, we obtain the same notion of a curved analog as for the sphere. To get a nice presentation of the Euclidean group $\text{Euc}(n)$, we consider the n–dimensional Euclidean space E^n as the affine hyperplane $\{x_1 = 1\}$ in $\mathbb{R}^{n+1}$. Then the usual Euclidean metric on E^n is induced from the standard inner product on $\mathbb{R}^{n+1}$. The group $\text{Euc}(n)$ can be viewed as the group of all linear automorphisms of $\mathbb{R}^{n+1}$, which map the hyperplane $\{x_1 = 1\}$ to itself and restrict to an isometry of that hyperplane. A simple computation shows that

$$\text{Euc}(n) = \left\{ \begin{pmatrix} 1 & 0 \\ v & A \end{pmatrix} : A \in O(n), v \in \mathbb{R}^n \right\}$$

and the isotropy group $H \cong O(n)$ of the first vector e_1 in the standard basis of $\mathbb{R}^{n+1}$ consists exactly of those elements of $\text{Euc}(n)$, for which $v = 0$. On the level of Lie algebras, we have

$$\mathfrak{euc}(n) = \left\{ \begin{pmatrix} 0 & 0 \\ v & A \end{pmatrix} : A \in \mathfrak{o}(n), v \in \mathbb{R}^n \right\}$$

which as above gives a decomposition $\mathfrak{euc}(n) = \mathfrak{h} \oplus \mathfrak{n}$ as a vector space. Using again the fact that the adjoint action on a matrix group is given by conjugation, we see that this actually is an H–invariant decomposition and $\mathfrak{n}$ is isomorphic to the standard representation $\mathbb{R}^n$ of $H \cong O(n)$.

This shows that while $\text{Euc}(n)$ and $O(n+1)$ are very different Lie groups, they both contain $O(n)$ as a Lie subgroup and via the adjoint actions, the Lie algebras $\mathfrak{euc}(n)$ and $\mathfrak{o}(n+1)$ are isomorphic representations of $O(n)$. Since in the definition of a Cartan geometry corresponding to a homogeneous space G/H only the subgroup H and the H–module $\mathfrak{g}$ (and *not* the Lie algebra $\mathfrak{g}$) enters, the homogeneous spaces $E^n = \text{Euc}(n)/O(n)$ and $S^n \cong O(n+1)/O(n)$ lead to the same generalized spaces, namely n–dimensional Riemannian manifolds. The only point in which the Lie algebra structure of $\mathfrak{g}$ enters is the definition of curvature, which differs in the two settings.

There is a third realization of Riemannian manifolds as curved analogs of a homogeneous space. Namely, we can consider $\mathbb{R}^{n+1}$ endowed with an inner product of Lorentzian signature $(n, 1)$. Let $\mathcal{H}^n$ be the set of all vectors in $\mathbb{R}^{n+1}$ of length -1. For a point $v \in \mathcal{H}^n$, the tangent space $T_v\mathcal{H}^n$ is the orthocomplement of v, so the restriction of the inner product to this subspace is positive definite. Hence, we get a Riemannian metric on $\mathcal{H}^n$, and the resulting Riemannian manifold is called *hyperbolic n–space*. The orthogonal group $O(n, 1)$ clearly acts by isometries on $\mathcal{H}^n$ and the action is immediately seen to be transitive. The isotropy groups of this action are isomorphic to $O(n)$. Choosing the inner product in such a way that the first vector in the standard basis has length -1, we can realize the Lie algebra

$\mathfrak{g} = \mathfrak{o}(n,1)$ as

$$\left\{ \begin{pmatrix} 0 & v^t \\ v & A \end{pmatrix} : v \in \mathbb{R}^n, A \in \mathfrak{o}(n) \right\}.$$

Hence, we again have $\mathfrak{g} = \mathfrak{o}(n) \oplus \mathbb{R}^n$ as a representation of the subgroup $O(n)$.

The fact that the three homogeneous spaces $O(n+1)/O(n)$, $\mathrm{Euc}(n)/O(n)$ and $O(n,1)/O(n)$ lead to the same curved analogs is an instance of *model mutation* as discussed in 4, §3 of [**Sh97**].

1.1.3. Projective space and the projective sphere. The natural interpretation of projective space as a homogeneous space of $SL(n+1,\mathbb{R})$ is our first example of a parabolic geometry. As usual, we view the real projective space $\mathbb{R}P^n$ as the space of all lines through the origin in $\mathbb{R}^{n+1}$. Consider the open subset $C := \mathbb{R}^{n+1} \setminus 0 \subset \mathbb{R}^{n+1}$ and the equivalence relation $x \sim y$ if and only if there exists a number $t \in \mathbb{R}$ such that $y = tx$. Of course, $C/\sim \, = \mathbb{R}P^{n+1}$. Note that the action of $\mathbb{R}^* = \mathbb{R} \setminus \{0\}$ on $\mathbb{R}^{n+1}$ by multiplication is clearly smooth, and restricts to a free action on C. Thus, the natural projection $\pi : C \to \mathbb{R}P^n$ is a principal bundle with structure group $\mathbb{R}^*$.

The standard action of $SL(n+1,\mathbb{R})$ on $\mathbb{R}^{n+1}$ restricts to an action on C, and since the action is linear, it induces an action on $\mathbb{R}P^n$, which is obviously transitive. The resulting diffeomorphisms of $\mathbb{R}P^n$ are exactly the projective transformations, which are the basis for classical projective geometry. The isotropy group of the line through the first vector in the standard basis of $\mathbb{R}^{n+1}$ is given by

$$P := \left\{ \begin{pmatrix} \det(A)^{-1} & \phi \\ 0 & A \end{pmatrix},\ A \in GL(n,\mathbb{R}),\ \phi \in \mathbb{R}^{n*} \right\}.$$

This is an example of a *parabolic subgroup* of the simple Lie group $SL(n+1,\mathbb{R})$. The subgroup P is not connected, with the two connected components distinguished by the sign of the determinant of A. If one only considers the connected component P_0 of the identity, then one obtains the stabilizer of the ray spanned by the first basis vector. The corresponding homogeneous space is the space of rays in $\mathbb{R}^{n+1}$ which can be evidently identified with S^n.

To see the geometric structure on $\mathbb{R}P^n$ which is invariant under the action of $SL(n+1,\mathbb{R})$, we first observe that any projective transformation maps projective lines to projective lines. Indeed, the projective lines are exactly the images under π of the intersections of planes through the origin with C. From this description it is obvious that the action of $SL(n+1,\mathbb{R})$ on $\mathbb{R}P^n$ maps projective lines (viewed as unparameterized curves) to projective lines. The fundamental theorem of projective geometry (see e.g. [**Je00**]) says that any bijection of $\mathbb{R}P^n$ which maps projective lines to projective lines is a projective transformation. Hence, the appropriate geometric structure is given by the family of all projective lines in $\mathbb{R}P^n$. On the sphere S^n, one similarly obtains the family of great circles in S^n, and the diffeomorphisms of S^n which map great circles to great circles are exactly the actions of elements of $SL(n+1,\mathbb{R})$.

This structure can be generalized to arbitrary manifolds by looking at appropriate smooth families of unparameterized curves. A simpler picture is available, however. The great circles on S^n are the geodesics of the round metric on S^n. Via the two–fold covering $S^n \to \mathbb{R}P^n$, we see that also the projective lines in $\mathbb{R}P^n$ can be realized as geodesics of a connection. Such geodesics come with a preferred parametrization, while we are concerned with unparameterized curves. To describe our family of unparameterized curves we should therefore not look at the individual connections (which are not invariant under the action of $SL(n+1,\mathbb{R})$). Instead, we have to consider the class of all connections on the tangent bundle whose geodesics are the curves in the family (with some parametrization).

Recall that from any connection one can construct a torsion free connection by symmetrization. This process does not change the geodesics, so we restrict to torsion free connections. It is a classical result that two torsion free linear connections ∇ and $\hat{\nabla}$ on the tangent bundle of a smooth manifold M have the same geodesics up to parametrization if and only if there is a one–form $\Upsilon \in \Omega^1(M)$ such that

$$\hat{\nabla}_\xi \eta = \nabla_\xi \eta + \Upsilon(\xi)\eta + \Upsilon(\eta)\xi$$

for all vector fields $\xi, \eta \in \mathfrak{X}(M)$. If this is the case, then ∇ and $\hat{\nabla}$ are called *projectively equivalent.* From the definition it follows immediately that projectively equivalent connections have the same torsion.

Now let $[\nabla]$ be the projective equivalence class of the Levi–Civita connection of the round metric on S^n. We claim that a diffeomorphism $f : S^n \to S^n$ preserves this class if and only if it preserves the family of great circles. From the definitions one immediately verifies that c is a geodesic for $f^*\nabla$ if and only if $f \circ c$ is a geodesic for ∇.

Suppose first that for one (or equivalently any) connection ∇ in the projective class $f^*\nabla$ also lies in the projective class. Then any great circle can be parameterized in such a way that it becomes a geodesic c for $f^*\nabla$. Then $f \circ c$ is a geodesic for ∇, i.e. a parametrization of a great circle. Thus, f maps great circles to great circles.

Conversely, if f maps great circles to great circles, then take a connection ∇ from the projective class, and consider $f^*\nabla$. Taking a great circle, we can choose a parametrization c for which $f \circ c$ is a geodesic for ∇. But then c is a geodesic of $f^*\nabla$, so this connection has exactly the great circles as its geodesics. Thus, it must be projectively equivalent to ∇. Of course, this argument applies for $\mathbb{R}P^n$ as well.

The curved analogs of the homogeneous spaces $\mathbb{R}P^n \cong SL(n+1,\mathbb{R})/P$ are n–dimensional projective manifolds, i.e. n–manifolds endowed with a projective equivalence class of torsion free connections. For $S^n \cong SL(n+1,\mathbb{R})/P_0$, the curved analogs are oriented projective manifolds of dimension n.

1.1.4. The contact projective sphere. The idea to get a transitive action on $\mathbb{R}P^n$ (respectively S^n) by viewing it as the space of lines (respectively rays) in $\mathbb{R}^{n+1}$ clearly generalizes to certain subgroups of $SL(n+1,\mathbb{R})$. If we take the sphere and an orthogonal group, then we return to the setting of 1.1.1, since then the embedding of S^n as the unit sphere is equivariant, so we can view it as a subspace as well. Hence, in the realm of simple groups the main remaining possibility is to consider the group $Sp(2n,\mathbb{R})$ of all linear automorphisms of $\mathbb{R}^{2n}$ which preserve a nondegenerate skew symmetric bilinear form ω. This bilinear map also induces the standard symplectic form on $\mathbb{R}^{2n}$.

As in 1.1.3 above consider S^{2n-1} as the space of rays and $\mathbb{R}P^{2n-1}$ as the space of lines in $\mathbb{R}^{2n}$. The linear action of $Sp(2n,\mathbb{R})$ on $\mathbb{R}^{2n}$ induces transitive actions on S^{2n-1} and $\mathbb{R}P^{2n-1}$. The isotropy group P of the line through e_1 again turns out to be a parabolic subgroup of $Sp(2n,\mathbb{R})$. This subgroup has two connected components and restricting to P_0, we get the stabilizer of a ray rather than a line.

In the following exposition, we will focus on the sphere rather than on projective space, which can be treated similarly.

The geometric structure corresponding to this transitive action is significantly more complicated than a projective structure, since it involves a nontrivial filtration on the tangent bundle. More precisely, this is an example of a so–called parabolic contact structure, i.e. a parabolic geometry which has an underlying contact structure.

Put $C := \mathbb{R}^{2n} \setminus 0$ and let $\pi : C \to S^{2n-1}$ be the natural projection. This is a principal fiber bundle with structure group $\mathbb{R}_+$. For any point $x \in C$, the tangent map $T_x\pi$ induces a bijection $\mathbb{R}^{2n}/\mathbb{R}x \to T_{\pi(x)}S^{2n-1}$. Since ω is non–degenerate, $\{y \in \mathbb{R}^{2n} : \omega(x,y) = 0\}$ is a codimension one subspace, and by skew symmetry it contains the line $\mathbb{R}x$. Passing to a nonzero multiple of x, this subspace remains unchanged, so we obtain a well–defined codimension one subspace $H_{\pi(x)} \subset T_{\pi(x)}S^{2n-1}$. By construction, these subspaces fit together to define a smooth subbundle $H \subset TS^{2n-1}$ of rank $2n-2$.

To understand this subbundle, let us temporarily view S^{2n-1} as the unit sphere in $\mathbb{R}^{2n}$. Let E be the Euler vector field on $\mathbb{R}^{2n}$, i.e. $E(x) = x \in T_x\mathbb{R}^{2n}$. Denoting by i_E the insertion operator, we have the one–form $i_E\omega \in \Omega^1(\mathbb{R}^{2n})$. We can restrict this to a one–form α on S^{2n-1}. By definition, $\alpha(x)(\xi) = \omega(x,\xi)$, so the kernel of $\alpha(x)$ is exactly the subspace $H_x \subset T_xS^{2n-1}$.

To compute $d\alpha$, we first note that the one–parameter group of diffeomorphisms corresponding to the vector field E is multiplication by e^t, i.e. $\mathrm{Fl}^E_t(x) = e^t x$. The derivative of this is also multiplication by e^t, and since ω is constant we get $(\mathrm{Fl}^E_t)^*\omega = e^{2t}\omega$. This implies that for the Lie derivative we get $\mathcal{L}_E\omega = 2\omega$. By Cartan's formula $\mathcal{L}_E\omega = di_E\omega + i_E d\omega$ and the well–known fact that $d\omega = 0$, we conclude that $di_E\omega = 2\omega$.

Restricting this to S^{2n-1}, we obtain $d\alpha$, and nondegeneracy of ω implies that $d\alpha$ is nondegenerate on H_x. Otherwise put, $\alpha \wedge (d\alpha)^{n-1}$ is a volume form on S^{2n-1}, so α is a contact form. This means that $H = \ker(\alpha)$ is a contact structure on S^{2n-1}, i.e. maximally non–integrable.

From the construction it is clear that the contact structure $H \subset TS^{2n-1}$ is invariant under the action of $Sp(2n,\mathbb{R})$. However, it is well known that the group of contact diffeomorphisms of a contact manifold is infinite dimensional, so this contact structure is by far not enough to characterize the diffeomorphisms coming from the action of $Sp(2n,\mathbb{R})$. A simple way to get additional input is to just observe that $Sp(2n,\mathbb{R}) \subset SL(2n,\mathbb{R})$, so the projective structure from 1.1.3 on $\mathbb{R}P^{2n-1}$ is preserved by the action of $Sp(2n,\mathbb{R})$. As we have seen in 1.1.3 the diffeomorphisms preserving this projective structure are exactly the actions of elements of $SL(2n,\mathbb{R})$. From this one easily concludes that the actions of elements $Sp(2n,\mathbb{R})$ can be characterized as those diffeomorphisms, which preserve both the contact structure and the canonical projective class.

As we shall see (much) later in 5.3.11, a similar interpretation is possible for the curved analogs of this situation, but this needs quite a lot of theory. As a starting point, this characterization is not satisfactory, since the contact structure and the projective structure are not well compatible. This is particularly transparent if one looks at the associated projective class of connections. None of the connections in this class preserves the subbundle H, and indeed any linear connection that preserves H necessarily has nonzero torsion.

To obtain the "right" characterization, we have to look at the relation between projective lines and the contact structure. The great circles in S^{2n-1} are exactly the intersections of S^{2n-1} with planes through the origin in $\mathbb{R}^{2n}$. If x is a point in such an intersection, then the intersection is tangent to H_x if and only if for any vector ξ in the plane which is perpendicular to x, we have $\omega(x, \xi) = 0$. But this is equivalent to the fact that the restriction of ω to the plane is identically zero, which in turn implies that the intersection with S^{2n-1} is always tangent to H. Hence, a great circle which is tangent to H in one point remains tangent to H in all its points. Of course, the family of great circles tangent to H is invariant under the action of $Sp(2n, \mathbb{R})$. It turns out that this family of great circles is exactly the structure we need.

Similarly, as discussed for projective structures in 1.1.3, this family of unparameterized curves admits an equivalent description related to connections. However, since we are only interested in curves tangent to H, we do not need a connection on the whole tangent bundle TS^{2n-1}, but on the contact subbundle H. But even this is too much, since we only have to differentiate in contact directions in order to have geodesics tangent to the contact subbundle.

The right notion, therefore, is a *partial contact connection*, i.e. a bilinear operator $\nabla^H : \Gamma(H) \times \Gamma(H) \to \Gamma(H)$ written as $(\xi, \eta) \mapsto \nabla^H_\xi \eta$, which is linear over smooth functions in the first variable and satisfies a Leibniz rule in the second variable. For curves tangent to the contact distribution, one can then write down the usual geodesic equation. As in the classical case, one gets existence and uniqueness of solutions, and so on.

To obtain such a connection on S^{2n-1} let us again view it as the unit sphere in $\mathbb{R}^{2n}$. Associated to the contact form α constructed above, one has the *Reeb vector field* which is defined as the unique vector field $r \in \mathfrak{X}(S^{2n-1})$ such that $\alpha(r) = 1$ and $i_r(d\alpha) = 0$. With the help of r, we define $p : TS^{2n-1} \to H$ by $p(\xi) := \xi - \alpha(\xi)r$. Evidently, $\alpha(p(\xi)) = 0$, so this indeed has values in H and for $\xi \in H$ we get $p(\xi) = \xi$, so we have defined a projection onto the subbundle H. Let ∇ be the Levi–Civita connection of the round metric on S^{2n-1}. For $\xi, \eta \in \Gamma(H)$ define

$$\nabla^H_\xi \eta := p(\nabla_\xi \eta).$$

One easily verifies that this indeed is a partial contact connection. From the definition, it follows immediately that great circles tangent to H are geodesics for ∇^H.

There is one more point that we have not touched yet. It turns out that to a partial contact connection one may intrinsically associate the so–called *contact torsion*, which is a skew symmetric bilinear bundle map $H \times H \to H$ (cf. 4.2.6). In contrast to the case of projective structures, one cannot eliminate this contact torsion without changing the geodesics (in contact directions). For the connection ∇^H constructed above, this contact torsion vanishes identically.

Similarly, as for ordinary connections, one can characterize when two partial contact connections have the same geodesics up to parametrization. The result is a formula which is similar to the one in 1.1.3 but a bit more complicated and it involves a section of the dual bundle H^* rather than a one–form; see 4.2.6 and 4.2.7. This formula is then used to define projective equivalence for partial contact connections. It turns out that connections which are equivalent in this sense always have the same contact torsion.

The curved analogs of S^{2n-1}, viewed as a homogeneous space of $Sp(2n,\mathbb{R})$, are given by contact manifolds endowed with a projective equivalence class of partial contact connections with vanishing contact torsion. These structures are called *contact projective structures.*

It should be remarked at this point that projective structures and contact projective structures are rather untypical examples of parabolic geometries, since they involve the choice of a class of connections. In fact, essentially these are the only two examples in which such a choice is required, while all other parabolic geometries are determined by a first order G–structure or even a weakening of such a structure.

1.1.5. The conformal sphere. Our next example is certainly the best studied example of a parabolic geometry, namely conformal structures. This example will be kept in mind as a guiding line throughout the book.

Consider $\mathbb{R}^{n+2}$ endowed with an inner product $\langle\ ,\ \rangle$ of signature $(n+1,1)$, and the "light cone" C of nonzero null–vectors, i.e. $C = \{v \in \mathbb{R}^{n+2} : v \neq 0, \langle v, v\rangle = 0\}$. Taking the standard inner product of this signature, C consists of all nonzero solutions of the equation $x_{n+2}^2 = x_1^2 + \cdots + x_{n+1}^2$. In particular, for any element of C the last coordinate is nonzero. Similarly to 1.1.3, consider the space of lines in C, i.e. the quotient of $C/{\sim}$, where two points in C are equivalent if and only if they are nonzero multiples of each other. Again, the natural projection $\pi : C \to C/{\sim}$ is a principal bundle with structure group $\mathbb{R}^* = \mathbb{R} \setminus \{0\}$.

Viewing S^n as the unit sphere in the subspace $\mathbb{R}^{n+1} \subset \mathbb{R}^{n+2}$, the map $x \mapsto (x, 1)$ includes S^n into C and thus induces a smooth map to the quotient space $C/{\sim}$. Conversely, for $v = (v_1, \dots, v_{n+2}) \in C$ we get $v_{n+2} \neq 0$ and $\frac{1}{v_{n+2}}(v_1, \dots, v_{n+1})$ lies in $S^n \subset \mathbb{R}^{n+1}$. This defines a smooth map $C \to S^n$ which visibly factors to an inverse $C/{\sim} \to S^n$ of the map $S^n \to C/{\sim}$ from above. Hence, we may view S^n as the space of null lines in $(\mathbb{R}^{n+2}, \langle\ ,\ \rangle)$.

But this immediately suggests yet another way to make S^n into a homogeneous space. Namely, consider the orthogonal group $O(n+1,1)$ of the inner product $\langle\ ,\ \rangle$. This acts on $\mathbb{R}^{n+2}$ and by definition leaves the null–cone C invariant. Since it acts linearly, this action factors to a smooth action on $S^n \cong C/{\sim}$. Elementary arguments show that $O(n+1,1)$ acts transitively on C and thus we get a diffeomorphism $S^n \cong O(n+1,1)/P$, where P is the isotropy subgroup of some point of S^n. The subgroup P is usually called the *Poincaré conformal group.* We will describe this group in more detail in 1.6.3. For now, let us just note that P is the semidirect product of the conformal group $CO(n)$ with an n–dimensional vector group and that P is a parabolic subgroup of $O(n+1,1)$.

To analyze the geometric meaning of this transitive action, let us determine the tangent spaces of S^n. Denote by $\pi : C \to S^n$ the natural projection. For $v \in C$, the tangent space to C at v clearly is given by $T_vC = \{w \in \mathbb{R}^{n+2} : \langle v, w\rangle = 0\}$. Since v by definition is a null vector, the line $\ell_v = \{tv : t \in \mathbb{R}\}$ is contained in T_vC. The projection π induces a linear isomorphism $T_vC/\ker(T_v\pi) \to T_{\pi(v)}S^n$, and clearly the kernel $\ker(T_v\pi)$ is exactly $\ell_v \subset T_vC$. Otherwise put, given $v \in C$ we obtain a linear isomorphism $\ell_v^\perp/\ell_v \to T_{\pi(v)}S^n$. Since v is null, the bilinear form $\langle\ ,\ \rangle$ on $\mathbb{R}^{n+2}$ induces a positive definite inner product on $\ell_v^\perp/\ell_v$, which we can carry over to $T_{\pi(v)}S^n$. Replacing v by λv for $\lambda \in \mathbb{R}^*$, we of course have $\ell_{\lambda v} = \ell_v$ and the isomorphism to $T_{\pi(v)}S^n$ changes by a nonzero multiple. Hence, the induced inner product on $T_{\pi(v)}S^n$ changes by multiplication by λ^2. Consequently, on each tangent space we get a positive definite inner product up to positive multiples, and thus a conformal structure on S^n. By construction, this structure is invariant under the action of $O(n+1,1)$.

Both lines of argument outlined in 1.1.1 can be used to show that a diffeomorphism preserving the standard conformal structure on S^n is given by the action of an element of $O(n+1,1)$, and we will prove this later as a consequence of the existence of a unique normal Cartan connection on a conformal manifold. Clearly, one may also replace $O(n+1,1)$ by $SO(n+1,1)$ or $Spin(n+1,1)$ or the respective groups in signature $(p+1,q+1)$ to get similar variations as in 1.1.2. However, there is no analog to the flat Euclidean space, i.e. conformal geometry does not allow a homogeneous model isomorphic to $\mathbb{R}^n$. On $\mathbb{R}^n$ there are local conformal transformations which do not globalize, and this is impossible for the homogeneous model of a Cartan geometry; see 1.5.2.

1.1.6. The CR–sphere. Our final example of viewing spheres as homogeneous spaces is the strictly pseudoconvex CR–structure on odd dimensional spheres, which certainly is the most complicated of the examples we discuss. The basis for this example is a Hermitian analog of the construction of the conformal sphere in 1.1.5. As the projective contact sphere from 1.1.4, the CR–sphere is an example of a parabolic contact structure.

Consider $\mathbb{C}^{n+2}$ with a Hermitian inner product $\langle\ ,\ \rangle$ of signature $(n+1,1)$. Let C be the cone of nonzero null vectors, which is a submanifold of real codimension one, and consider the quotient $C/\sim$ by the action of $\mathbb{C}^* = \mathbb{C}\setminus 0$. Hence, $\pi : C \to C/\sim$ is a principal bundle with structure group $\mathbb{C}^*$. Similarly, as in 1.1.5 above, one easily verifies directly that $C/\sim$ is diffeomorphic to the sphere S^{2n+1}, and we will return to this below. The standard action of $SU(n+1,1)$ on $\mathbb{C}^{n+2}$ leaves the null cone C invariant and thus factors to an action on S^{2n+1}, which can be easily shown directly to be transitive. The stabilizer of a point in S^{2n+1} is a parabolic subgroup P of $SU(n+1,1)$.

While the analysis of tangent spaces is analogous to the conformal case, there are some significant differences. For $v \in C$, the tangent space T_vC is given by $\{w \in \mathbb{C}^{n+2} : \mathrm{re}(\langle v,w\rangle) = 0\}$, so this is no more independent of the choice of a point v lying over a fixed point $x \in S^{2n+1}$. The projection $\pi : C \to S^{2n+1}$ induces a linear isomorphism $T_vC/\ell_v \to T_{\pi(v)}S^{2n+1}$, where ℓ_v is the complex line $\mathbb{C}v$. The orthogonal complement $\ell_v^\perp$ is a complex subspace of $\mathbb{C}^{n+2}$, which sits inside of T_vC as a subspace of real codimension one. The image of this subspace in $T_{\pi(v)}S^{2n+1}$ is a subspace $H_{\pi(v)}$ of real dimension $2n$ and we can carry over the complex structure from $\ell_v^\perp/\ell_v$ to $H_{\pi(v)}$.

It is easy to see that both the subspace $H_{\pi(v)}$ and its complex structure remain unchanged if one replaces v by λv for some $\lambda \in \mathbb{C}^*$. Moreover, both these data depend smoothly on $\pi(v)$ so we get a subbundle $H \subset TS^{2n+1}$ of real rank $2n$. This bundle can be naturally viewed as a complex vector bundle, and we denote by $J : H \to H$ the bundle map given by multiplication by $\sqrt{-1}$. This means that the subbundle H defines an *almost CR–structure* (of CR–dimension n and codimension one) on S^{2n+1}. By construction, this structure is invariant under the action of $SU(n+1,1)$.

Explicitly, the diffeomorphism $S^{2n+1} \to C/\sim$ is given by $z \mapsto \pi(z,1)$ for a choice of Hermitian form analogous to the bilinear form in 1.1.5. The tangent map of this is of course $\xi \mapsto T\pi\cdot(\xi, 0)$. This implies that, viewing S^{2n+1} as the unit sphere for the standard Hermitian inner product on $\mathbb{C}^{n+1}$, the subbundle $H_z \subset T_z S^{2n+1}$ is again the complex orthocomplement of z sitting inside its real orthocomplement. But identifying $\mathbb{C}^{n+1}$ with $\mathbb{R}^{2n+2}$, the real part of the Hermitian inner product is the standard inner product, while its imaginary part is a nondegenerate skew symmetric bilinear form. Hence, we obtain exactly the same subbundle H as in 1.1.4, so in particular we again have obtained a contact structure on S^{2n+1}.

Moreover, as in 1.1.4 we obtain a contact form α, such that the restriction of $d\alpha$ to H is just the imaginary part of a positive definite Hermitian form. In the language of almost CR–structures, this means that the structure is strictly pseudoconvex and partially integrable. In fact, viewing S^{2n+1} as the unit sphere in $\mathbb{C}^{n+1}$, the subspace H_x is the maximal complex subspace of $T_x S^{2n+1} \subset \mathbb{C}^{n+1}$. Hence, the almost CR–structure is embeddable and thus integrable, so it is a CR–structure. This, however, is not important for our purposes.

As in the previous examples, it follows from the existence of a canonical Cartan connection for CR–structures that the actions of elements of $SU(n+1, 1)$ are exactly the CR–diffeomorphisms of the CR–structure on S^{2n+1}. Here a CR–diffeomorphism is a diffeomorphism which preserves the contact structure H and further has the property that the restriction of the tangent map to the contact subbundle is complex linear. The curved analogs of this homogeneous space are strictly pseudoconvex partially integrable almost CR–manifolds of dimension $2n + 1$. These are smooth manifolds M of dimension $2n + 1$ endowed with a rank n complex subbundle $H \subset TM$ which defines a contact structure on M. In addition, one has to require existence of local contact forms α such that the restriction of $d\alpha$ to $H \times H$ is the imaginary part of a definite Hermitian form.

1.2. Some background from differential geometry

In this section, we review some basic facts on differential geometry and analysis on manifolds which will be necessary for further development. Our main purpose here is to fix the notation and conventions used in the sequel, as well as to give a more detailed collection of prerequisites for the further text. The basic reference for this section is [**KMS**]. At the same time, we stress the basic concepts of frames, natural bundles, and the role of the symmetry groups in the properties of geometric objects. This will remain one of the main features of our exposition in the entire book.

1.2.1. Smooth manifolds. Unless otherwise stated, all manifolds we consider are finite dimensional and second countable and we assume that all connected components have the same dimension. Any manifold comes equipped with a maximal *atlas*, i.e. a maximal collection of open subsets $U_\alpha \subset M$ together with homeomorphisms $u_\alpha : U_\alpha \to u_\alpha(U_\alpha)$ onto open subsets of $\mathbb{R}^n$, such that for all α, β with $U_{\alpha\beta} := U_\alpha \cap U_\beta \neq \emptyset$ the subset $u_\alpha(U_{\alpha\beta})$ is open and the composition $u_{\alpha\beta} := u_\alpha \circ u_\beta^{-1} : u_\beta(U_{\alpha\beta}) \to u_\alpha(U_{\alpha\beta})$ is smooth (C^∞). A *chart* on M is any element (U_α, u_α) of this maximal atlas. Such a chart gives rise to local coordinates $u^i_\alpha : U_\alpha \to \mathbb{R}$ on M.

A map $f : M \to N$ between smooth manifolds is smooth, if and only if its expression in one (or equivalently any) local coordinate system around any point in M is smooth. A *diffeomorphism* is a bijective smooth map, whose inverse is smooth, too. A *local diffeomorphism* $f : M \to N$ is a smooth map such that for each point $x \in M$ there is an open neighborhood U of x in M such that $f(U) \subset N$ is open and the restriction $f|_U : U \to f(U)$ is a diffeomorphism.

The space $C^\infty(M, \mathbb{R})$ of smooth real–valued functions on M forms an algebra under the pointwise operations. For $f \in C^\infty(M, \mathbb{R})$ the *support* $\operatorname{supp}(f)$ of f is the closure of the set of all $x \in M$ such that $f(x) \neq 0$. The concept of support generalizes in an obvious way to smooth functions with values in a vector space and to smooth sections of vector bundles; see 1.2.6 below.

A fundamental result is that any open covering of a smooth manifold M admits a subordinate *partition of unity.* This means that if $\{V_i : i \in I\}$ is a family of open subsets of M such that $M = \bigcup_{i \in I} V_i$, then there exists a family $\{f_\alpha\}$ of smooth functions on M with values in $[0,1] \subset \mathbb{R}$ such that for any α there exists an $i \in I$ with $\operatorname{supp}(f_\alpha) \subset V_i$, any point $x \in M$ has a neighborhood which meets only finitely many of the sets $\operatorname{supp}(f_\alpha)$, and $\sum_\alpha f_\alpha(x) = 1$ for all $x \in M$.

For a point $x \in M$, the *tangent space* T_xM to M in the point x is defined to be the space of all linear maps $\xi_x : C^\infty(M, \mathbb{R}) \to \mathbb{R}$ which are derivations at x, i.e. which satisfy the Leibniz rule $\xi_x(\phi\psi) = \xi_x(\phi)\psi(x) + \phi(x)\xi_x(\psi)$. These derivations form a vector space whose dimension equals the dimension of the manifold. Let $c : I \to M$ be a smoothly parametrized curve defined on an open interval $I \subset \mathbb{R}$. Then for $t \in I$ and $x := c(t)$, one obtains a tangent vector in T_xM by mapping $f \in C^\infty(M, \mathbb{R})$ to $(f \circ c)'(t) \in \mathbb{R}$. This tangent vector will be denoted by $c'(t)$. It turns out that any tangent vector can be obtained in this way.

If $f : M \to N$ is a smooth map between smooth manifolds, then for $x \in M$ and $\xi_x \in T_xM$, we define $T_xf \cdot \xi_x : C^\infty(N, \mathbb{R}) \to \mathbb{R}$ by $(T_xf \cdot \xi_x)(\phi) := \xi_x(\phi \circ f)$. One immediately verifies that $T_xf \cdot \xi_x \in T_{f(x)}N$ and this defines a linear map $T_xf : T_xM \to T_{f(x)}N$, the *tangent map* of f at x. The tangent map T_xf is bijective if and only if there is an open neighborhood U of x in M such that $f(U) \subset N$ is open and f restricts to a diffeomorphism from U to $f(U)$. In particular, $f : M \to N$ is a local diffeomorphism if and only if all tangent maps T_xf are linear isomorphisms. A smooth map f is called an *immersion* if all of its tangent maps are injective and a *submersion* if all of its tangent maps are surjective. The images of injective immersions are called *immersed submanifolds.*

A k–dimensional *submanifold* $N \subset M$ in a smooth manifold M of dimension n is a subset such that for each $x \in N$ there is a chart (U, u) for M with $x \in U$ such that $u(U \cap N) = u(U) \cap \mathbb{R}^k \subset \mathbb{R}^n$. Such a chart is called a *submanifold chart.* Restricting submanifold charts to N and their images to $\mathbb{R}^k$ one obtains an atlas for N, so N itself is a smooth manifold. The inclusion of N into M is not only an injective immersion but also an embedding, i.e. a homeomorphism onto its image. In view of this fact the name *embedded submanifold* is also used in this situation.

The union TM of all tangent spaces is called the *tangent bundle* of the manifold M. For each smooth map $f : M \to N$ we get the *tangent map* $Tf : TM \to TN$ of f by putting together the tangent maps at the individual points of M. The tangent bundle is endowed with the unique smooth structure such that the obvious projection $p : TM \to M$ and all tangent maps Tf become smooth maps. In this picture, the chain rule just states that T is a covariant functor on the category of smooth manifolds, i.e. $T(g \circ f) = Tg \circ Tf$. The individual tangent spaces T_xM are vector spaces and each point $x \in M$ has an open neighborhood U in M such that $p^{-1}(U) \subset TM$ is diffeomorphic to $U \times \mathbb{R}^m$ in a way compatible with the natural projections to U. Thus, TM is naturally a vector bundle over M and the tangent map to any smooth mapping is a vector bundle homomorphism; see 1.2.6 below for the terminology of bundles.

A *vector field* is a smooth section $\xi : M \to TM$ of the projection p, i.e. a smooth map such that $\xi(x) \in T_xM$ for all $x \in M$. The space of all smooth vector fields on M will be denoted by $\mathfrak{X}(M)$. It is a vector space and a module over $C^\infty(M,\mathbb{R})$ under pointwise operations. For $\xi \in \mathfrak{X}(M)$ and $f \in C^\infty(M,\mathbb{R})$, one defines a function $\xi \cdot f : M \to \mathbb{R}$ by $(\xi \cdot f)(x) = (\xi(x))(f)$. Smoothness of ξ easily implies that $\xi \cdot f \in C^\infty(M,\mathbb{R})$, while the fact that any $\xi(x)$ is a derivation at x immediately implies that $\xi \cdot (fg) = (\xi \cdot f)g + f(\xi \cdot g)$. Thus, ξ defines a derivation $C^\infty(M,\mathbb{R}) \to C^\infty(M,\mathbb{R})$ and one shows that this induces a bijection between $\mathfrak{X}(M)$ and the space of all derivations. Since the commutator of two derivations is again a derivation, we may associate to two vector fields $\xi, \eta \in \mathfrak{X}(M)$ a vector field $[\xi,\eta] \in \mathfrak{X}(M)$, which is called the *Lie bracket* of ξ and η and characterized by $[\xi,\eta] \cdot f = \xi \cdot (\eta \cdot f) - \eta \cdot (\xi \cdot f)$.

Any local diffeomorphism $f : M \to N$ induces a *pullback* operator $f^* : \mathfrak{X}(N) \to \mathfrak{X}(M)$ defined by $f^*\xi(x) = (T_xf)^{-1}(\xi(f(x)))$. This pullback is compatible with the Lie bracket, i.e. for $\xi, \eta \in \mathfrak{X}(N)$ we obtain $f^*[\xi,\eta] = [f^*\xi, f^*\eta]$. If P is another manifold and $g : N \to P$ another local diffeomorphism, then $(g \circ f)^* = f^* \circ g^* : \mathfrak{X}(P) \to \mathfrak{X}(M)$. For a diffeomorphism $f : M \to N$, there is also a covariant action on vector fields, i.e. an operator $f_* : \mathfrak{X}(M) \to \mathfrak{X}(N)$, which may be simply defined by $f_* := (f^{-1})^*$.

Given a vector field ξ on M, we may ask for integral curves, i.e. smooth curves $c : I \to M$ defined on open intervals in $\mathbb{R}$ such that $c'(t) = \xi(c(t))$ for all $t \in I$. The theorem on existence and uniqueness of solutions of ordinary differential equations implies that for each point $x \in M$ there are a unique maximal interval $I_x \subset \mathbb{R}$ and maximal integral curve $c_x : I_x \to M$ such that $c_x(0) = x$. A slightly finer analysis also using the smooth dependence of the solution on the initial conditions implies that the union of all I_x forms an open neighborhood $\mathcal{D}(\xi)$ of $\{0\} \times M$ in $\mathbb{R} \times M$, and $\mathrm{Fl}^\xi_t(x) := c_x(t)$ defines a smooth mapping $\mathrm{Fl}^\xi : \mathcal{D}(\xi) \to M$ called the *flow* of the vector field ξ. For $t, s \in \mathbb{R}$ and $x \in M$ one has the basic equation $\mathrm{Fl}^\xi_t(\mathrm{Fl}^\xi_s(x)) = \mathrm{Fl}^\xi_{t+s}(x)$, which is usually referred to as the flow property or the one–parameter group property. It is also known that under additional assumptions existence of one side of the equation implies existence of the other side. More precisely, if (s,x) and $(t, \mathrm{Fl}^\xi_s(x))$ lie in $\mathcal{D}(\xi)$, then $(t+s,x) \in \mathcal{D}(\xi)$ and the opposite implication also holds provided that t and s have the same sign. Finally, it turns out that for any point $x \in M$ and any $t_0 \in I_x$ there is a neighborhood U of x in M such that the restriction of Fl^ξ_t to U is a diffeomorphism onto its image for all $0 \le t \le t_0$. A vector field is called *complete* if $\mathcal{D}(\xi) = \mathbb{R} \times M$, i.e. if its flow is defined for all times. On a compact manifold, any vector field is automatically complete. Let us notice the obvious relation between flows and pullbacks, namely for a local diffeomorphism $f : M \to N$ and $\xi \in \mathfrak{X}(N)$ we have $\mathrm{Fl}^\xi_t \circ f = f \circ \mathrm{Fl}^{f^*\xi}_t$.

The dual bundle to $TM \to M$ is the *cotangent bundle* $T^*M \to M$, so for $x \in M$ the *cotangent space* T^*_xM is the space of all linear maps $T_xM \to \mathbb{R}$. In contrast to the tangent functor, T^* only has functorial properties for local diffeomorphism, and it can be viewed either as a contravariant or as a covariant functor. The smooth sections of the cotangent bundle are called *one–forms*. We write $\Omega^1(M) = \Omega^1(M,\mathbb{R})$ for the space of all smooth real one–forms on M. The pointwise operations make $\Omega^1(M)$ into a vector space and a module over $C^\infty(M,\mathbb{R})$ and for any smooth map

$f : M \to N$ we obtain the *pullback* $f^* : \Omega^1(N) \to \Omega^1(M)$ defined by

$$(f^*\phi)(x)(\xi_x) = \phi(f(x))(T_x f \cdot \xi_x).$$

A simple example of a one–form is the differential df of a real–valued function f, defined by $df(x)(\xi_x) := \xi_x(f)$. One may easily generalize this and consider one–forms with values in a (finite–dimensional) vector space V. These are smooth maps ϕ which associate to each point $x \in M$ a linear map $T_xM \to V$. The space of all such forms is denoted by $\Omega^1(M, V)$. For $f \in C^\infty(M, V)$ one obtains as above $df \in \Omega^1(M, V)$.

The antisymmetric k–linear maps $\alpha : \Lambda^k T_xM \to \mathbb{R}$ are the elements of the kth exterior power $\Lambda^k T^*M$ of the cotangent bundle and the sections of this bundle are called *k-forms* on M. The space of all k–forms on M, which again is a vector space and $C^\infty(M, \mathbb{R})$–module under pointwise operations, is denoted by $\Omega^k(M)$. By convention, $\Omega^0(M) = C^\infty(M, \mathbb{R})$. Differential forms can be pulled back along arbitrary smooth functions, by defining

$$(f^*\alpha)(x)(\xi_1, \dots, \xi_k) = \alpha(f(x))(T_x f \cdot \xi_1, \dots, T_x f \cdot \xi_k).$$

In particular, for $h \in \Omega^0(M) = C^\infty(M, \mathbb{R})$ one has $f^*h = h \circ f$. Inserting the values of vector fields $\xi_1, \dots, \xi_k$ into a k–form α one obtains a smooth function $\alpha(\xi_1, \dots, \xi_k) \in C^\infty(M, \mathbb{R})$, so α gives rise to a k–linear, alternating map $\mathfrak{X}(M)^k \to C^\infty(M, \mathbb{R})$. One shows that such a mapping is induced by a k–form if and only if it is linear over $C^\infty(M, \mathbb{R})$ in one (and thus any) variable.

The differential of functions, $d : C^\infty(M, \mathbb{R}) \to \Omega^1(M)$, is a special case of the *exterior derivative*. In general, the exterior derivative $d : \Omega^k(M) \to \Omega^{k+1}(M)$ is given by the formula

$$\begin{aligned} d\omega(\xi_0, \dots, \xi_k) = &\sum_{i=0}^{k} (-1)^i \xi_i \cdot (\omega(\xi_0, \dots, \widehat{\xi_i}, \dots, \xi_k)) \\ &+ \sum_{i<j} (-1)^{i+j} \omega([\xi_i, \xi_j], \xi_0, \dots, \widehat{\xi_i}, \dots, \widehat{\xi_j} \dots, \xi_k) \end{aligned}$$

for all $\xi_i \in \mathfrak{X}(M)$, where the hats denote omission of an argument. The same formula applies for differential forms with values in any finite–dimensional vector space V.

The exterior derivative d is the only linear differential operator which is invariantly defined on all manifolds; see [**KMS**, Theorem 34.2]. Here invariance means commuting with the action of local diffeomorphisms, i.e. $\phi^*(d\omega) = d(\phi^*\omega)$. One of the goals of this book is to develop general tools for the study of such basic operators in the realm of more specific geometric structures on manifolds.

1.2.2. Distributions and foliations. A *distribution* $\mathcal{D}$ on a manifold M is a subset $\mathcal{D} \subset TM$ such that for each $x \in M$ the subset $\mathcal{D}_x = \mathcal{D} \cap T_xM$ is a vector subspace in T_xM. By elementary linear algebra, each distribution can be defined as the kernel of a (not necessarily continuous) one–form ω with values in a suitable vector space V. The dimension of V is at least $\dim M - \max_{x\in M}\{\dim \mathcal{D}_x\}$. The distribution $\mathcal{D}$ is said to be of *constant rank* k if $\dim \mathcal{D}_x = k$ is constant, and $\mathcal{D}$ is *smooth* if it can be defined by a smooth form ω. Equivalently, locally there must be smooth vector fields which span the distribution in each point. A distribution is called *regular*, if it is of constant rank and is smooth. An *integral manifold* N of a distribution $\mathcal{D}$ is an immersed submanifold such that $T_xN \subset \mathcal{D}_x$ for all $x \in N$. A *maximal integral submanifold* is an integral manifold N whose dimension is equal to the rank of $\mathcal{D}$ in all points $x \in N$. A distribution $\mathcal{D}$ is called *integrable* if there is a maximal integral submanifold through each point $x \in M$.

Each vector field ξ on M spans a one–dimensional distribution whose maximal integral submanifolds are the (unparametrized) flow lines $\mathrm{Fl}_t^\xi(x)$. In general, there are no maximal integral submanifolds and the obstruction to their existence is given by the Lie bracket of vector fields lying in $\mathcal{D}$. The distribution $\mathcal{D}$ is called *involutive* if for all vector fields ξ and $\eta \in \mathfrak{X}(M)$ with $\xi(x) \in \mathcal{D}$, $\eta(x) \in \mathcal{D}$ for all $x \in M$, also $[\xi, \eta](x) \in \mathcal{D}$ for all $x \in M$. The following theorem (called the *Frobenius theorem*) appears in all standard textbooks on differential geometry. For a general version for distributions of nonconstant rank see e.g. [**KMS**, section 3].

THEOREM 1.2.2. *A regular distribution $\mathcal{D} \subset TM$ is integrable if and only if it is involutive.*

Given an integrable regular distribution of rank k on M, the maximal integral submanifolds decompose M into a union of k–dimensional immersed submanifolds. This is called the *regular foliation* on M defined by the distribution and the maximal integral submanifolds are referred to as the *leaves* of this foliation. On the other hand, smooth foliations define the associated distributions (defined by the tangent spaces to the leaves) which are integrable by construction.

An immediate and very useful consequence of the last Theorem and the above formula for the exterior differential reformulates the involutivity of $\mathcal{D}$ in terms of the defining one–form ω.

COROLLARY 1.2.2. *Let $\omega \in \Omega^1(M, V)$ be a smooth V–valued one–form and assume that the dimension of* $\ker \omega(x)$ *is constant for all $x \in M$. Then the distribution $\mathcal{D} = \ker \omega$ is integrable if and only if $d\omega(x)(X, Y) = 0$ for all X, $Y \in \mathcal{D}_x$.*

Equivalently, the condition of the corollary can be stated as follows: Representing $\mathcal{D}$ as the intersection of the kernels of $\dim V$ many one–forms $\omega_i \in \Omega^1(M)$, the exterior derivatives $d\omega_i$ belong to the ideal in $\Omega^*(M)$ generated by the forms ω_i.

1.2.3. Lie groups and their Lie algebras. A *Lie group* G is a smooth manifold endowed with a smooth mapping $\mu : G \times G \to G$, the *multiplication*, which defines a group structure on G. Using the implicit function theorem one then shows that the inversion mapping $\nu : G \to G$ is smooth, too. Given an element $g \in G$, we define the *left translation* $\lambda_g : G \to G$ by $\lambda_g(h) = \mu(g, h) = gh$, and the *right translation* $\rho^g : G \to G$ by $\rho^g(h) = hg$. Both λ_g and ρ^g are diffeomorphisms of G with inverses $\lambda_{g^{-1}}$ and $\rho^{g^{-1}}$, respectively. Further, one clearly has $\lambda_g \circ \lambda_h = \lambda_{gh}$ and $\rho^g \circ \rho^h = \rho^{hg}$, which also explains the use of subscripts and superscripts.

Let G be a Lie group and let $\xi \in \mathfrak{X}(G)$ be a smooth vector field on G. Then ξ is called *left invariant*, if and only if $(\lambda_g)^*\xi = \xi$ for all $g \in G$, or equivalently $\xi(gh) = T_h\lambda_g \cdot \xi(h)$ for all $g, h \in G$. The latter equation shows that any left invariant vector field ξ is uniquely determined by its value $\xi(e) \in T_eG$ in the unit element e of G. Conversely, it is easy to see that any $X \in T_eG$ extends to a left invariant vector field L_X on G. Consequently, there is a linear isomorphism between the space $\mathfrak{X}_L(G)$ of left invariant vector fields on G and the tangent space T_eG of G at the unit element. Since the pullback along a diffeomorphism is compatible with the Lie bracket of vector fields, the subspace $\mathfrak{X}_L(G) \subset \mathfrak{X}(G)$ is a Lie subalgebra. Via the linear isomorphism, this gives rise to a Lie bracket on the tangent space T_eG, which is explicitly given by $[X,Y]=[L_X,L_Y](e)$ for $X,Y\in T_eG$. The space T_eG together with this Lie bracket is called the *Lie algebra* $\mathfrak{g}$ of the Lie group G.

Similarly, as for left invariant vector fields, one has the Lie subalgebra $\mathfrak{X}_R(G)$ of *right invariant* vector fields on G. Any right invariant vector field on G is uniquely determined by its value in e, and any $X\in\mathfrak{g}$ extends uniquely to $R_X\in\mathfrak{X}_R(G)$. It is easy to see that $R_X=\nu^*(L_{-X})$, where ν is the inversion on G, which, in particular, implies that $[R_X,R_Y]=R_{-[X,Y]}$. Another basic result is that right invariant vector fields commute with left invariant vector fields, i.e. $[L_X,R_Y]=0$ for all $X,Y\in\mathfrak{g}$.

It follows from the construction of the Lie algebra of a Lie group that if G and H are Lie groups with Lie algebras $\mathfrak{g}$ and $\mathfrak{h}$, and $\phi:G\to H$ is a *homomorphism*, i.e. a smooth map compatible with the group structures, then the tangent map $\phi'=T_e\phi:\mathfrak{g}\to\mathfrak{h}$ is a homomorphism of Lie algebras. For example, if G is a Lie group and $g\in G$ is any element, then the conjugation $h\mapsto ghg^{-1}$ by g defines an automorphism of G, so the tangent map at zero defines an automorphism $\mathrm{Ad}(g)$ of the Lie algebra $\mathfrak{g}$. This is called the *adjoint action* of $g\in G$ on $\mathfrak{g}$.

The simplest example of a Lie group is the group $GL(n,\mathbb{K})$ of linear automorphisms of $\mathbb{K}^n$, where $\mathbb{K}$ is $\mathbb{R}$ or $\mathbb{C}$, which may be also viewed as the group of invertible $n\times n$–matrices with entries from $\mathbb{K}$. This is a smooth manifold, since it is an open subset of the vector space $M_n(\mathbb{K})$ of all $n\times n$–matrices with entries from $\mathbb{K}$, and clearly matrix multiplication is smooth. From this it follows that the Lie algebra $\mathfrak{gl}(n,\mathbb{K})$ of $GL(n,\mathbb{K})$ equals $M_n(\mathbb{K})$, and one easily shows that the adjoint representation is given by the conjugation of matrices, while the Lie bracket is given by the commutator of matrices.

If G is an arbitrary Lie group, then a real or complex (finite–dimensional) *representation* of G is a homomorphism ϕ from G to some $GL(n,\mathbb{R})$, respectively $GL(n,\mathbb{C})$. For such a representation ϕ, the tangent map at $e\in G$ is a homomorphism $\phi':\mathfrak{g}\to\mathfrak{gl}(n,\mathbb{K})$ of Lie algebras, i.e. a representation of $\mathfrak{g}$, called the *infinitesimal representation* corresponding to ϕ. Equivalently, one may describe a representation of G as a smooth map $\hat{\phi}:G\times\mathbb{K}^n\to\mathbb{K}^n$ which is linear in the second argument and has the property that $\hat{\phi}(gh,v)=\hat{\phi}(g,\hat{\phi}(h,v))$ for all $g,h\in G$ and $v\in\mathbb{K}^n$. Slightly more generally, one may consider representations $G\to GL(V)$ for any finite–dimensional real or complex vector space V. The corresponding infinitesimal representation then has values in the space $L(V,V)$ of all linear mappings. For both group and Lie algebra representations, if there is no risk of confusion, we will often omit the name of the representation and simply use the notation $(g,v)\mapsto g\cdot v$ or gv. The adjoint action associates to any element $g\in G$ an automorphism $\mathrm{Ad}(g):\mathfrak{g}\to\mathfrak{g}$, which, in particular, is a linear isomorphism, so this defines a map $\mathrm{Ad}:G\to GL(\mathfrak{g})$. Since mapping g to the conjugation by g is a homomorphism, Ad is a group homomorphism and it is easy to see that it is smooth, so it defines a representation of G, the *adjoint representation*. The infinitesimal representation $\mathrm{ad}=\mathrm{Ad}':\mathfrak{g}\to L(\mathfrak{g},\mathfrak{g})$ turns out to be given by $\mathrm{ad}(X)(Y)=[X,Y]$ for $X,Y\in\mathfrak{g}$.

For each element X in the Lie algebra $\mathfrak{g}$ of a Lie group G, we have the corresponding left invariant vector field L_X on G. The invariance of L_X easily implies that the vector field L_X is complete, i.e. that its flow $\mathrm{Fl}^{L_X}_t$ is defined for all times t. In particular, we can define the *exponential mapping* $\exp:\mathfrak{g}\to G$ by

$$\exp(X):=\mathrm{Fl}^{L_X}_1(e).$$

One readily verifies that $\exp : \mathfrak{g} \to G$ is a smooth map, whose tangent map at $0 \in \mathfrak{g}$ is the identity, so exp restricts to a diffeomorphism from an open neighborhood of $0 \in \mathfrak{g}$ to an open neighborhood of $e \in G$. Further, the flows of left invariant vector fields are given by $\mathrm{Fl}_t^{L_X}(g) = g\exp(tX)$ and for right invariant vector fields one gets $\mathrm{Fl}_t^{R_X}(g) = \exp(tX)g$. In particular, the flows through e of L_X and R_X coincide. By the flow property, this flow defines a *one–parameter subgroup* of G, i.e. a smooth homomorphism from the additive group $\mathbb{R}$ to G. Conversely, any such one–parameter subgroup of G is determined by its derivative at e, so $t \mapsto \exp(tX)$ is the unique curve $\alpha : \mathbb{R} \to G$ such that $\alpha(t+s) = \alpha(t)\alpha(s)$ and such that $\alpha'(0) = X$. For the group $GL(n, \mathbb{K})$ of invertible $n \times n$–matrices, the exponential mapping is given by the usual exponential of matrices, i.e. $\exp(A) = e^A = \sum_{n=0}^{\infty} \frac{1}{n!} A^n$.

If $\phi : G \to H$ is a homomorphism of Lie groups with Lie algebras $\mathfrak{g}$ and $\mathfrak{h}$ and exponential mappings $\exp^G$ and $\exp^H$, then the description of the exponential map as the solution of an ordinary differential equation easily implies that $\phi \circ \exp^G = \exp^H \circ \phi'$. In particular, the values of ϕ on the image of $\exp^G$ are completely determined by the Lie algebra homomorphism ϕ'. From the fact that the image of $\exp^G$ contains an open neighborhood of the unit $e \in G$, one next concludes that the subgroup generated by this image is exactly the connected component G_0 of e in G, so the restriction of ϕ to G_0 is determined by ϕ'. In particular, if G is connected, then any homomorphism from G to some Lie group is determined by its tangent map at e. We may apply this to representations of G. For any representation $\phi : G \to GL(n, \mathbb{K})$ and any $X \in \mathfrak{g}$, we get

$$\phi(\exp(X)) = e^{\phi'(X)} = \sum_{n=0}^{\infty} \frac{1}{n!} \phi'(X)^n.$$

In particular, $\phi'(X) = \frac{d}{dt}|_0 \phi(\exp(tX))$, and if G is connected, then any representation is determined by the corresponding infinitesimal representation.

1.2.4. The Maurer–Cartan form. The left invariant vector fields lead to a trivialization of the tangent bundle of any Lie group. More precisely, the mapping $G \times \mathfrak{g} \to TG$ which is given by $(g, X) \mapsto L_X(g)$ is an isomorphism of vector bundles. The inverse of this isomorphism can be conveniently encoded as a one–form $\omega \in \Omega^1(G, \mathfrak{g})$ on G with values in the Lie algebra $\mathfrak{g}$, which is defined by $\omega(g)(\xi) := T_g\lambda_{g^{-1}} \cdot \xi$. This one–form is called the (left) *Maurer–Cartan form* on G. From the definition of ω it is obvious that $\omega(L_X) = X$, $\lambda_g^*\omega = \omega$, and for each $g \in G$ the map $\omega(g) : T_gG \to \mathfrak{g}$ is a linear isomorphism. Moreover, by definition, $(\rho^g)^*\omega(h)(\xi) = \omega(hg)(T\rho^g \cdot \xi) = T\lambda_{g^{-1}h^{-1}} T\rho^g \cdot \xi$. Since left multiplications commute with right multiplications and $\lambda_{g^{-1}} \circ \rho^g$ is the conjugation with g^{-1}, we conclude that this equals $\mathrm{Ad}(g^{-1})(\omega(h)(\xi))$, and thus we get $(\rho^g)^*\omega = \mathrm{Ad}(g^{-1}) \circ \omega$. Finally, consider the exterior derivative $d\omega$ of the Maurer–Cartan form. Since ω is constant on left invariant vector fields, the standard formula for the exterior derivative (see 1.2.1) implies $d\omega(L_X, L_Y) = -\omega([L_X, L_Y])$ for $X, Y \in \mathfrak{g}$, which equals $-[X, Y]$ by definition of the Lie bracket on $\mathfrak{g}$. Since in each point the values of left invariant vector fields exhaust the whole tangent space, this implies the *Maurer–Cartan equation* $0 = d\omega(\xi, \eta) + [\omega(\xi), \omega(\eta)]$ for all ξ, $\eta \in TG$.

The Maurer–Cartan form leads to a notion of differentiation of functions with values in a Lie group. Indeed, if G is a Lie group with Lie algebra $\mathfrak{g}$, M is an arbitrary smooth manifold, and $f : M \to G$ is an arbitrary smooth map, then we define the *left logarithmic derivative* $\delta f : TM \to \mathfrak{g}$ by $\delta f(\xi_x) = \omega(f(x))(T_x f \cdot \xi_x)$. This means that δf is obtained by composing the tangent map $Tf : TM \to TG$ with the trivialization of TG provided by ω and projecting out the second component. Another way to express this is $\delta f(x) = f^*\omega(x) = T\lambda_{f(x)^{-1}} \circ T_x f$ and, in particular, the Maurer–Cartan form itself equals $\omega = \delta\, \mathrm{id}_G$.

Let us look at some examples. If G is the additive real line $G = (\mathbb{R}, +)$ with standard coordinate x, then $\omega_G = dx \in \Omega^1(G, \mathbb{R})$ and the left logarithmic derivative is just the usual differential. The isomorphic example with the multiplicative positive real line $G = (\mathbb{R}^+, \cdot)$ leads to $\omega = \frac{1}{x}dx$ and the usual logarithmic derivative of real functions, $f \mapsto \frac{f'}{f}dx$. For the general linear group G we easily compute $\omega_G \in \Omega^1(GL(n,\mathbb{R}), \mathfrak{gl}(n,\mathbb{R}))$ in the usual matrix component coordinates $g = (x_{ij}) \in GL(n,\mathbb{R})$, $\omega_G(g) = g^{-1}(dx_{ij})$. For example, in dimension two,

$$\omega_{GL(2,R)} = \begin{pmatrix} x_{22}/\det(g) & -x_{12}/\det(g) \\ -x_{21}/\det(g) & x_{11}/\det(g) \end{pmatrix} \begin{pmatrix} dx_{11} & dx_{12} \\ dx_{21} & dx_{22} \end{pmatrix}.$$

The Maurer–Cartan form provides the infinitesimal information on the multiplication while the Maurer–Cartan equation gives the only local obstruction to its integrability. The explicit local formulation is contained in the following theorem; see [**Sh97**, Chapter 3] for more details.

THEOREM 1.2.4. *Let G be a Lie group with Lie algebra $\mathfrak{g}$ and Maurer–Cartan form ω_G. Let M be a smooth manifold, and let $\omega \in \Omega^1(M, \mathfrak{g})$ be a $\mathfrak{g}$–valued one-form. Then for any $x \in M$ there exist an open neighborhood U of x in M and a function $f : U \to G$ such that $\delta f = f^*\omega_G = \omega$ if and only if ω satisfies $d\omega(\xi, \eta) + [\omega(\xi), \omega(\eta)] = 0$ for all $\xi, \eta \in \mathfrak{X}(M)$.*

If M is connected and $f_1, f_2 : M \to G$ have the property that $\delta f_1 = \delta f_2$, then there is a unique element $c \in G$ (integration constant) such that $f_2(x) = c \cdot f_1(x)$ for all $x \in M$.

PROOF. A straightforward computation establishes the formulae for the actions of the multiplication μ and inversion ν on the Maurer–Cartan form ω_G; cf. [**Sh97**, page 113]. For $g, h \in G$, $\xi \in T_g G$ and $\eta \in T_h G$ one has

$$(\mu^*\omega_G)(\xi, \eta) = \mathrm{Ad}(h^{-1})(\omega_G(\xi)) + \omega_G(\eta),$$
$$(\nu^*\omega_G)(\xi) = -\mathrm{Ad}(g)(\omega_G(\xi)).$$

We start by proving the last statement of the theorem. Consider two functions $f_1, f_2 : M \to G$ such that $f_1^*\omega_G = f_2^*\omega_G$ and define $h : M \to G$ by $h(x) = f_2(x)f_1(x)^{-1}$. We have to show that h is constant, for which it suffices to show that $h^*\omega_G = 0$, since this implies that $\omega_G \circ Th$ and thus Th is identically zero. But by definition, we have $h = \mu \circ (\mathrm{id}, \nu) \circ (f_2, f_1) \circ \Delta$, where $\Delta(x) = (x, x)$. Using the above formulae we thus compute that for $\xi \in T_x M$ we have

$$\begin{aligned}(h^*\omega_G)(\xi) &= (\mu^*\omega_G)(T_x f_2 \cdot \xi, T_x(\nu \circ f_1) \cdot \xi) \\ &= \mathrm{Ad}(f_1(x))(\omega_G(T_x f_2 \cdot \xi)) + (\nu^*\omega_G)(T_x f_1 \cdot \xi) \\ &= \mathrm{Ad}(f_1(x))(\delta f_2(x)(\xi) - \delta f_1(x)(\xi)) = 0.\end{aligned}$$

Concerning existence, $d(f^*\omega_G)(\xi, \eta) + [f^*\omega_G(\xi), f^*\omega_G(\eta)]$ clearly vanishes because of the Maurer–Cartan equation. Thus, it suffices to prove that for $\omega \in \Omega^1(M, \mathfrak{g})$ satisfying the equation we can find a function $f : M \to G$ such that $\omega = \delta f$. To do this, we (locally) construct the graph of f as a leaf of an integrable distribution on $M \times G$. Consider $\Omega = \pi_M^*\omega - \pi_G^*\omega_G$, where $\pi_M : M \times G \to M$ and $\pi_G : M\times G \to G$ are the natural projections. Identifying $T(M\times G)$ with $TM\times TG$, the kernel of Ω is given by the set of all (ξ,η) such that $\omega(\xi) = \omega_G(\eta)$. Since ω_G restricts to a linear isomorphism in each tangent space, there is a unique solution η for this equation for any chosen tangent vector ξ, so the distribution $\ker(\Omega)$ is regular, its rank equals the dimension of M, and $T\pi_M$ restricts to a linear isomorphism on each fiber of $\ker(\Omega)$. To check involutivity, we note that by construction

$$\begin{aligned} d\Omega((\xi_1,\eta_1),(\xi_2,\eta_2)) &= d\omega(\xi_1,\xi_2) - d\omega_G(\eta_1,\eta_2) \\ &= -[\omega(\xi_1),\omega(\xi_2)] + [\omega_G(\eta_1),\omega_G(\eta_2)], \end{aligned}$$

and this obviously vanishes if both (ξ_i,η_i) lie in the kernel of Ω.

By Corollary 1.2.2 this implies integrability of the distribution $\ker(\Omega)$. Given $x \in M$ and $g \in G$, there is a submanifold $N \subset M \times G$ containing (x,g) whose tangent spaces are the fibers of $\ker(\Omega)$. We have observed above that $T\pi_M$ restricts to a linear isomorphism on each of these spaces, so $\pi_M : N \to M$ is a local diffeomorphism. Hence, we can find a neighborhood U of $x \in M$ and a local inverse $j : U \to N$ of this projection. Defining $f := \pi_G \circ j$, we obtain a smooth function $f : U \to G$, and for $\xi \in T_xM$ we get $T_xj \cdot \xi = (\xi, \omega_G^{-1}(\omega(\xi)))$, and thus $\omega_G(T_xf\cdot\xi) = \omega(\xi)$, which means $\omega = f^*\omega_G$. □

If we add the requirement that $\omega : T_xM \to \mathfrak{g}$ is a linear isomorphism, then the theorem implies that there is a unique group structure locally around $x \in M$ which is locally isomorphic to that of G via f, and has Maurer–Cartan form $\omega = f^*\omega_G$. A global version of this theorem works for connected and simply connected manifolds M, or under suitable conditions on the so–called monodromy representation; see again [**Sh97**, Chapter 3] for more details.

From this theorem we may easily conclude that Lie algebra homomorphisms integrate to local group homomorphisms. If G and H are Lie groups with Lie algebras $\mathfrak{g}$ and $\mathfrak{h}$ and $\phi : \mathfrak{g} \to \mathfrak{h}$ is a Lie algebra homomorphism, then consider $\omega := \phi\circ\omega_G \in \Omega^1(G,\mathfrak{h})$. Clearly, we get $d\omega = \phi\circ d\omega_G$, which together with the Maurer–Cartan equation for ω_G and the fact that ϕ is a homomorphism immediately implies that $d\omega(\xi,\eta) + [\omega(\xi),\omega(\eta)] = 0$ for arbitrary tangent vectors ξ and η. By the theorem, we find an open neighborhood U of e in G and a smooth function $f : U \to H$ such that $f(e) = e$ and $\omega = f^*\omega_H$. But then $\omega_H(T_ef\cdot X) = \omega(X) = \phi(X)$, so $\phi = T_ef$. Moreover, we claim that f is a local group homomorphism. For $g_0 \in U$ consider the function $f\circ\lambda_{g_0}$, which is defined locally around e. One immediately verifies that this function also pulls back ω_H to $\omega = \phi\circ\omega_G$, which implies that it coincides with f up to a left multiplication by a fixed element of H. Looking at the values in e, we see that we must have $f = \lambda_{f(g_0)^{-1}}\circ f\circ\lambda_{g_0}$, which implies that $f(g_0g) = f(g_0)f(g)$ if g and g_0g lie in U. The global version of the theorem, in particular, implies that if G is simply connected, then there is a globally defined homomorphism $f : G \to H$ such that $T_ef = \phi$.

Note that in the special case that M is an interval in $\mathbb{R}$, all two–forms on M are automatically zero, so the Maurer–Cartan equation is satisfied for any $\omega \in \Omega^1(M,\mathfrak{g})$. In this case it is also straightforward to deduce global existence of f from local existence. In the special case $G = (\mathbb{R},+)$, one obtains the theorem on existence of an antiderivative of any smooth function and uniqueness up to an additive constant. Therefore, the whole theorem is referred to as the fundamental theorem of calculus in [**Sh97**].

1.2.5. Lie subgroups, homogeneous spaces, and actions. An embedded *Lie subgroup* H of a Lie group G is a submanifold, which at the same time is a subgroup. We shall omit the adjective embedded in the sequel. A Lie subgroup is automatically a closed subset of G and, conversely, it can be shown (see [**KMS**, Theorem 5.5]) that any closed subgroup of a Lie group G is an (embedded) Lie subgroup. For a Lie subgroup $H \subset G$, the tangent space $\mathfrak{h} = T_eH \subset T_eG = \mathfrak{g}$ is a Lie subalgebra, i.e. the Lie bracket of two elements of $\mathfrak{h}$ again lies in $\mathfrak{h}$. A connected Lie subgroup $H \subset G$ is normal if and only if the Lie algebra $\mathfrak{h} \subset \mathfrak{g}$ is an *ideal*, i.e. $[X, Y] \in \mathfrak{h}$ for any $X \in \mathfrak{h}$ and $Y \in \mathfrak{g}$.

Given a Lie group G with Lie algebra $\mathfrak{g}$, there may exist Lie subalgebras $\mathfrak{h} \subset \mathfrak{g}$ for which there is no Lie subgroup $H \subset G$ having $\mathfrak{h}$ as the tangent space at the identity. This can be seen from the case of the real torus $\mathbb{T}^2 = S^1 \times S^1$. The Lie algebra of this is $\mathbb{R}^2$ with the trivial Lie bracket, so a subalgebra is just a linear subspace. Now if one takes a line of irrational slope, then it is easy to see that any subgroup $H \subset G$, which contains a small submanifold around the unit element that is tangent to the given line, must be dense in G, so it cannot be a Lie subgroup. This generalizes to one–parametric subgroups $\exp tX : \mathbb{R} \to G$, for $X \in \mathfrak{g}$ in general Lie groups. Either the image is topologically a circle or it is a line. A circle is a (embedded) subgroup, while lines are only immersed in general.

To avoid this problem, one defines a *virtual Lie subgroup* of a Lie group G to be a Lie group H together with a homomorphism $i : H \to G$ which is an injective immersion. The derivative $i' : \mathfrak{h} \to \mathfrak{g}$ is then the inclusion of a Lie subalgebra. Using the global version of the Frobenius theorem, one shows that for any Lie subalgebra $\mathfrak{h} \leq \mathfrak{g}$, there is a virtual Lie subgroup $i : H \to G$ with Lie algebra $\mathfrak{h}$; see [**KMS**, Theorem 5.2]. The latter result can also be used to prove that for any finite–dimensional Lie algebra $\mathfrak{g}$ there is a Lie group G with Lie algebra $\mathfrak{g}$. For this one uses the theorem of Ado that asserts that $\mathfrak{g}$ admits a finite–dimensional faithful representation and hence is isomorphic to a Lie subalgebra of $\mathfrak{gl}(N, \mathbb{R})$ for some N. For a short proof of the Ado theorem see [**Ne03**].

For any subgroup H in a Lie group G, one may consider the set G/H of cosets gH with $g \in G$. In order that the topology of G/H induced by the canonical projection $p : G \to G/H$ is Hausdorff, it is necessary that H is a closed subgroup, so from above we know that H is even a Lie subgroup of G. In this case, one shows that G/H is a smooth manifold and the structure is uniquely determined by requiring that p is a smooth surjective submersion; see [**KMS**, 5.11]. In particular, for any manifold M smooth maps from G/H to M are exactly the smooth maps from G to M which are constant on each coset.

Lie groups appear often as the symmetry groups on some manifolds, i.e. as groups of transformations on these manifolds. More explicitly, a *left action* of a Lie group G on a manifold M is a smooth mapping $\ell : G \times M \to M$, such that $\ell(e, x) = x$ and $\ell(g, \ell(h, x)) = \ell(gh, x)$. If there is no risk of confusion, we simply write $(g, x) \mapsto g \cdot x$ for the action, so that the defining properties become $e \cdot x = x$ and $g \cdot (h \cdot x) = (gh) \cdot x$. Otherwise put, the action associates to $g \in G$ a smooth map $\ell_g : M \to M$, defined by $\ell_g(x) = \ell(g, x)$ such that $\ell_e = \mathrm{id}_M$ and $\ell_g \circ \ell_h = \ell_{gh}$. In particular, each ℓ_g is a diffeomorphism with inverse $\ell_{g^{-1}}$, so we can view the action as a homomorphism from G into the group of diffeomorphisms of M. In the special case of a finite–dimensional vector space V, a representation of G on V as defined in 1.2.3 is exactly a left action such that all the maps ℓ_g are linear.

Similarly, a right action is a smooth map $r : M \times G \to M$ such that $r(x, e) = x$ and $r(r(x, g), h) = r(x, gh)$, or writing the action as a dot, $x \cdot e = x$ and $x \cdot (gh) = (x \cdot g) \cdot h$. As above, this can be interpreted as associating to any $g \in G$ a diffeomorphism r^g of M such that $r^{gh} = r^h \circ r^g$, so we get an anti–homomorphism from G to the diffeomorphism group.

Given a left (or right) action of G on M and a point $x \in M$ there are two canonical objects associated to x. First, there is the *orbit* $G \cdot x = \{g \cdot x : g \in G\}$ through x, and second there is the *isotropy subgroup* $G_x = \{g \in G : g \cdot x = x\}$, also called the *stabilizer* of x. By definition, G_x is a closed subgroup and thus a Lie subgroup in G. The map $\ell^x : G \to M$, $\ell^x(g) := \ell(g, x)$ then induces a smooth bijection $G/G_x \to G \cdot x$, so any orbit looks like a coset space. Clearly, two orbits are either disjoint or equal, so M is the disjoint union of all G–orbits. The set of all orbits is denoted by M/G. Note that for $y = g \cdot x \in G \cdot x$, one has $G_y = \{ghg^{-1} : h \in G_x\}$, so along an orbit all isotropy subgroups are conjugate.

An action is called *transitive* if there is just one orbit, or equivalently if for arbitrary elements $x, y \in M$ there is an element $g \in G$ such that $g \cdot x = y$. An action is called *effective* if only the neutral element $e \in G$ acts as the identity of M, or equivalently if the intersection of all isotropy subgroups consists of e only. If all the isotropy subgroups are trivial, then the action is called *free*.

The coset spaces G/H are related to actions in two ways. First, right multiplication by elements of $H \subset G$ defines a free right action of H on G, and by definition G/H is exactly the space of orbits for this action. On the other hand, the left multiplication of G on itself induces a smooth left action of G on G/H defined by $g \cdot (g'H) = (gg')H$. Clearly, this action is transitive. In view of this, the coset space G/H is called the *homogeneous space* of G corresponding to the subgroup H.

Each left action ℓ of a Lie group G on a manifold M defines the so–called *fundamental vector fields* by the formula $\zeta_X(x) = \frac{d}{dt}|_0 \ell_{\exp tX}(x)$, for all $x \in M$ and X in the Lie algebra $\mathfrak{g}$ of G. Similarly, we obtain fundamental vector fields for right actions. These fundamental vector fields provide infinitesimal versions of the Lie group actions. In particular, the left–invariant vector fields on the Lie group G itself are obtained as the fundamental vector fields with respect to the right multiplication by elements in G. The fundamental field mapping for right actions yields a Lie algebra homomorphism $\mathfrak{g} \to \mathfrak{X}(M)$, while $\zeta_{[X,Y]} = -[\zeta_X, \zeta_Y]$ for a left action. A simple computation yields $T_x\ell_g \cdot \zeta_X(x) = \zeta_{\mathrm{Ad}(g)X}(g \cdot x)$, i.e. $\ell_g^*\zeta_X = \zeta_{\mathrm{Ad}(g^{-1})\cdot X}$, for left actions. For each point $x \in M$, we also define the isotropy subalgebra $\mathfrak{g}_x \subset \mathfrak{g}$ of elements X with $\zeta_X(x) = 0$. This isotropy Lie algebra by construction is exactly the Lie algebra of the isotropy subgroup G_x.

1.2.6. Fiber bundles, vector bundles and principal bundles. A *fibered manifold* is a surjective submersion $p : Y \to M$, a trivial fibered manifold with fiber S is $\pi_M : M \times S \to M$. A *section* of a fibered manifold $p : Y \to M$ is a smooth map $\sigma : M \to Y$ such that $p \circ \sigma = \mathrm{id}_M$. The space of all smooth sections is denoted by $\Gamma(Y)$. *Fibered morphisms* $\phi : Y \to Y'$ are smooth mappings between fibered manifolds which cover a smooth mapping $\phi_0 : M \to M'$ between the base manifolds, i.e. $p' \circ \phi = \phi_0 \circ p$. A *fiber bundle* with base M and standard fiber S is a fibered manifold $Y \to M$ which is locally isomorphic (via fibered morphisms) to a trivial fibered manifold. Otherwise put, one must have a fiber bundle atlas $\{(U_\alpha, \phi_\alpha)\}$, i.e. an open covering $\{U_\alpha\}$ of M and diffeomorphisms

$\phi_\alpha : p^{-1}(U_\alpha) \to U_\alpha \times S$ which are fibered morphisms. Each of the pairs (U_α, ϕ_α) is called a fiber bundle chart. For two fiber bundle charts (U_α, ϕ_α) and (U_β, ϕ_β) such that $U_{\alpha\beta} := U_\alpha \cap U_\beta \neq \emptyset$, one has the *transition function* $\phi_{\alpha\beta} : U_{\alpha\beta} \times S \to S$ defined by $\phi_\alpha(\phi_\beta^{-1}(x, y)) = (x, \phi_{\alpha\beta}(x, y))$.

Assume next, that we have given a fiber bundle $p : Y \to M$ whose standard fiber is a finite–dimensional vector space V. Then two fiber bundle charts are called *compatible* if the corresponding transition function is linear in the second variable. A fiber bundle atlas consisting of pairwise compatible fiber bundle charts is then called a *vector bundle atlas.* There is an obvious notion of equivalence of vector bundle atlases and a *vector bundle* is a fiber bundle with standard fiber a vector space endowed with an equivalence class of vector bundle atlases. In this case, each of the fibers $p^{-1}(x)$ is canonically a vector space, and one may interpret the transition functions as smooth functions $\phi_{\alpha\beta} : U_{\alpha\beta} \to GL(V)$. A *vector bundle homomorphism* is a fibered morphism between two vector bundles such that the restriction to each fiber is a linear map.

The simplest example of a vector bundle is the tangent bundle TM. Its sections are the smooth vector fields and for a smooth map $f : M \to N$ the tangent map $Tf : TM \to TN$ is a vector bundle homomorphism. Given a vector bundle $E \to M$, there is the notion of an E–valued differential form. An E–valued k–form ϕ is a smooth function which associates to each $x \in M$ a k–linear alternating map $(T_xM)^k \to E_x$, where $E_x = p^{-1}(x)$ is the fiber of E over x. The space of E–valued k–forms is denoted by $\Omega^k(M, E)$. For $k = 0$, one obtains the space $\Gamma(E)$ of smooth sections of E.

One of the motivating examples for a principal fiber bundle is the projection $p : G \to G/H$ onto a homogeneous space. Since this is a surjective submersion, it admits local smooth sections, so any point $x \in G/H$ admits an open neighborhood U such that there is a smooth function $\sigma : U \to G$ with $\sigma(y)H = y$ for all $y \in U$. Such a section immediately gives rise to a fiber bundle chart $p^{-1}(U) \to U \times H$ by mapping $g \in p^{-1}(U)$ to $(p(g), \sigma(p(g))^{-1}g)$ with inverse given by $(x, h) \mapsto \sigma(x)h$. The corresponding transition functions are given by $(x, h) \mapsto \sigma_{\alpha\beta}(x)h$, where $\sigma_{\alpha\beta} : U_{\alpha\beta} \to H$ is given by $\sigma_{\alpha\beta}(x) = \sigma_\alpha(x)^{-1}\sigma_\beta(x)$.

Given a general fiber bundle $p : \mathcal{P} \to M$ with standard fiber a Lie group H, we define a *principal bundle atlas* to consist of charts which are compatible in the sense that the transition functions are given by $(x, h) \mapsto \phi_{\alpha\beta}(x)h$ for smooth functions $\phi_{\alpha\beta} : U_{\alpha\beta} \to H$. A *principal bundle* is then defined as a fiber bundle $p : \mathcal{P} \to M$ with standard fiber a Lie group H which is endowed with an equivalence class (in the obvious sense) of principal bundle atlases. The group H is referred to as the *structure group* of the principal bundle and principal bundles with structure group H are also called principal H–bundles. Multiplication from the right in charts defines a smooth right action of the structure group H on the total space $\mathcal{P}$ of the principal bundle. This is called the *principal right action.* It is by construction free, and its orbits are exactly the fibers of $p : \mathcal{P} \to M$. Conversely, given a smooth map $p : \mathcal{P} \to M$ and a right H–action on $\mathcal{P}$ which is free and transitive on each fiber, then this is a principal H–bundle if and only if p admits local smooth sections. A *morphism of principal bundles* is a fibred morphism commuting with the principal actions, i.e. $\phi(u \cdot h) = \phi(u) \cdot h$. There is a more general notion of morphisms between principal bundles with different structure groups. Fixing a homomorphism ψ between the structure groups, one imposes the equivariancy condition $\phi(u \cdot h) = \phi(u) \cdot \psi(h)$.

As in the case of homogeneous spaces, a local section $\sigma : U \to \mathcal{P}$ of a principal bundle $p : \mathcal{P} \to M$ defines a principal bundle chart $\phi : p^{-1}(U) \to U \times H$ whose inverse is given by $(x, h) \mapsto \sigma(x) \cdot h$. In particular, a principal H–bundle is trivial, i.e. isomorphic to $M \times H$, if and only if it admits a global smooth section.

Consider a principal bundle atlas $\{(U_\alpha, \phi_\alpha)\}$ for a principal H–bundle $p : \mathcal{P} \to M$ with transition functions $\phi_{\alpha\beta} : U_{\alpha\beta} \to H$. Then clearly $\phi_{\beta\alpha}(x) = \phi_{\alpha\beta}(x)^{-1}$ for all $x \in U_{\alpha\beta}$ and for three indices α, β, γ such that $U_{\alpha\beta\gamma} := U_\alpha \cap U_\beta \cap U_\gamma \neq \emptyset$, one has the *cocycle identity* $\phi_{\alpha\beta}(x)\phi_{\beta\gamma}(x) = \phi_{\alpha\gamma}(x)$ for all $x \in U_{\alpha\beta\gamma}$. Conversely, given an open covering $\{U_\alpha\}$ of M, a family $\phi_{\alpha\beta} : U_{\alpha\beta} \to H$ of smooth functions satisfying these two conditions is called a *cocycle of transition functions*. From such a family, one constructs a principal H–bundle as an appropriate quotient of the disjoint union of the sets $U_\alpha \times H$, which has the given cocycle as transition functions. It is easy to see that the transition functions determine the bundle up to isomorphism.

An important special case of the general concept of morphisms of principal bundles is provided by reductions of structure group. A *reduction* of the principal H–bundle $p : \mathcal{P} \to M$ to the structure group K, where $K \subset H$ is a Lie subgroup, is given by a principal bundle $\mathcal{R} \to M$ with structure group K, together with a principal bundle morphism $\iota : \mathcal{R} \to \mathcal{P}$ with respect to the inclusion $i : K \to H$, which covers the identity on M. The question of whether given $\mathcal{P}$ and K there exists a reduction of structure group is difficult in general, but it can be reduced to the question of existence of a smooth section of a certain bundle. Indeed, restricting the principal action to K, we obtain a free right action of K on $\mathcal{P}$, and one easily shows that the space $\mathcal{P}/K$ of orbits of this action is a smooth manifold and a fiber bundle over M with fiber H/K. A simple argument based on the cocycles of transition functions shows the following fact:

LEMMA 1.2.6. *Let $\mathcal{P} \to M$ be a principal bundle with structure group H and let $K \subset H$ be a subgroup. Then reductions of $\mathcal{P}$ to the structure group K are in bijective correspondence with the set of global smooth sections of the fiber bundle $\mathcal{P}/K \to M$.*

In contrast to the case of vector bundles, the individual fibers of a principal bundle $p : \mathcal{P} \to M$ do *not* carry the structure of a Lie group, since left multiplications are not group homomorphisms. The fibers should rather be thought of as the Lie group analog of affine spaces. Indeed, the simplest example of a principal bundle $\mathcal{P} \to \mathrm{pt}$ (with a one–point base manifold) is the space of all bases of an m–dimensional vector space V, which in turn may be identified with the set of all linear isomorphisms between $\mathbb{R}^m$ and V. Clearly, once we fix one basis (or one isomorphism), we may identify $\mathcal{P}$ with $GL(m, \mathbb{R})$, but there is no canonical choice like that.

An important example of a principal bundle is the *linear frame bundle* $\mathcal{P}^1 M \to M$ of a smooth manifold M. Its fiber over $x \in M$ is the set of all bases of the tangent space $T_x M$. The structure group is $GL(n, \mathbb{R})$ where n is the dimension of M. We may equivalently view the fiber $\mathcal{P}^1_x M$ over $x \in M$ as the space of all linear isomorphisms between $\mathbb{R}^n$ and $T_x M$. In analogy to this example, we shall often call all elements in principal bundles *frames*. More generally, there is the linear frame bundle $\mathcal{GL}(\mathcal{V})$ of a vector bundle $\mathcal{V} \to M$ with structure group $GL(V)$, where V is the standard fiber of $\mathcal{V}$.

1.2.7. Associated bundles. Recovering a vector bundle from its linear frame bundle is a special case of forming an associated bundle. The idea of this process is rather simple to understand in the toy example of all bases of a vector space V viewed as a principal bundle over a point. Any element $v \in V$ can be described by its coordinates in any of the bases of V, and since none of the bases is preferred, we should view v as an equivalence class of basis–coordinates–pairs. The correct equivalence relation is easily seen in the picture of isomorphisms $\phi : \mathbb{R}^m \to V$. The coordinate vector of v in the basis corresponding to ϕ is $\phi^{-1}(v)$, and the principal right action of $A \in GL(m, \mathbb{R})$ is given by composition from the right. This implies that the pair (ϕ, x) with $x \in \mathbb{R}^n$ must be considered as equivalent to $(\phi \cdot A, A^{-1}(x))$.

Now assume that $p : \mathcal{P} \to M$ is an H–principal bundle and S is a smooth manifold endowed with a left action $H \times S \to S$. Then we define a right action of H on the product $\mathcal{P} \times S$ by $(u, s) \cdot h := (u \cdot h, h^{-1} \cdot s)$. The space $\mathcal{P} \times_H S := (\mathcal{P} \times S)/H$ is called the *associated bundle* to the principal bundle $\mathcal{P}$ with standard fiber S. From a principal bundle atlas for $\mathcal{P}$ one constructs a fiber bundle atlas for $\mathcal{P} \times_H S$ showing that it is indeed a fiber bundle with standard fiber S. Moreover, the obvious projection $\mathcal{P} \times S \to \mathcal{P} \times_H S$ is an H–principal bundle. For $(u, s) \in \mathcal{P} \times S$, we write $[\![u, s]\!] \in \mathcal{P} \times_H S$ for the orbit of (u, s). We will sometimes write $\mathcal{P} \times_\ell S$ to emphasize the role of the left action ℓ. If the left action is a linear representation of the structure group on a vector space V, then the associated bundle $\mathcal{P} \times_H V$ is canonically a vector bundle.

Each principal bundle morphism $\phi : \mathcal{P} \to \mathcal{P}'$ between bundles with structure group H defines the fibered morphisms $\mathcal{P} \times_H S \to \mathcal{P}' \times_H S$ between associated bundles, which is characterized by $[\![u, s]\!] \mapsto [\![\phi(u), s]\!]$. Thus, the construction of associated bundles corresponding to a fixed left action is functorial. Of course, for linear actions this functorial construction has values in vector bundles and vector bundle homomorphisms. On the other hand, any smooth mapping $f : S \to S'$ commuting with given left actions defines the fibered morphism $\mathcal{P} \times_H S \to \mathcal{P} \times_H S'$, given by $[\![u, s]\!] \mapsto [\![u, f(s)]\!]$, which covers the identity on the base manifold M.

Let us consider a few examples in the case of the frame bundle $\mathcal{P}^1 M \to M$ of an n–dimensional smooth manifold M. The trivial representation $\mathbb{R}$ of $H = GL(n, \mathbb{R})$ provides the trivial associated bundle $\mathcal{P}^1 M \times_H \mathbb{R} = M \times \mathbb{R}$, whose sections are the smooth functions on M. For the standard representation of H on $\mathbb{R}^n$ we see from above that the associated bundle $\mathcal{P}^1 M \times_H \mathbb{R}^n$ may be identified with the tangent bundle TM by mapping $[\![u, x]\!]$ to the tangent vector with coordinates x in the frame u. Forming associated vector bundles is compatible with natural constructions on vector spaces. In particular, for the dual $\mathbb{R}^{n*}$ of the standard representation the associated bundle is the cotangent bundle T^*M, forming tensor powers of the standard representation and its dual, one obtains all tensor bundles, and so on.

PROPOSITION 1.2.7. *Let $p : \mathcal{P} \to M$ be a principal H–bundle, and S a smooth manifold endowed with a left H–action. Then there is a natural bijective correspondence between the set $\Gamma(\mathcal{P} \times_H S)$ of all smooth sections s of the associated bundle and the set $C^\infty(\mathcal{P}, S)^H$ of all smooth maps $f : \mathcal{P} \to S$, which are H–equivariant, i.e. satisfy $f(u \cdot h) = h^{-1} \cdot f(u)$. Explicitly, the correspondence is given by $s(p(u)) = [\![u, f(u)]\!]$.*

PROOF. Starting from an equivariant smooth function f, equivariancy implies that $[\![u, f(u)]\!]$ depends only on $p(u)$, so we can use this expression to define $s : M \to \mathcal{P} \times_H S$. Choosing a local smooth section σ of $\mathcal{P}$, we get $s(x) = [\![\sigma(x), f(\sigma(x))]\!]$, which immediately implies smoothness of s.

Conversely, any element in the fiber over $p(u)$ may be uniquely written in the form $[\![u, y]\!]$, so given s, the equation $s(p(u)) = [\![u, f(u)]\!]$ can be used to define f. Smoothness of f follows easily by writing this in terms of a local smooth section of $\mathcal{P}$, while equivariancy is an immediate consequence of $[\![u \cdot h, f(u \cdot h)]\!] = [\![u, f(u)]\!]$. □

In the case of associated vector bundles, we can generalize this result to a description of differential forms with values in an associated bundle. Given a fibered manifold $p : Y \to M$ and a point $y \in Y$ a tangent vector $\xi \in T_yY$ is called *vertical* if $T_yp \cdot \xi = 0$. The vertical tangent vectors form a subbundle $VY \subset TY$, called the *vertical tangent bundle.* This leads to the notion of a vertical vector field on Y. Now if ϕ is a differential form on Y (with values in $\mathbb{R}$, a vector space, or a vector bundle), then ϕ is called *horizontal* if it vanishes upon insertion of one vertical vector field. Suppose further that $\mathcal{P} \to M$ is a principal bundle with structure group H. Then for any $h \in H$ we have the principal right action $r^h : \mathcal{P} \to \mathcal{P}$, and we can use this to pull back differential forms with values in $\mathbb{R}$ or a vector space.

COROLLARY 1.2.7. *Let $p : \mathcal{P} \to M$ be a principal fiber bundle with structure group H and let $\rho : H \to GL(V)$ be a representation of H on a vector space V. Then for each k, the space $\Omega^k(M, \mathcal{P} \times_H V)$ of k–forms with values in the associated bundle is in bijective correspondence with the space of all $\phi \in \Omega^k(\mathcal{P}, V)$ which are horizontal and equivariant in the sense that $(r^h)^*\phi = \rho(h^{-1}) \circ \phi$ for all $h \in H$.*

PROOF. Consider a form $\alpha \in \Omega^k(M, \mathcal{P} \times_H V)$. For $u \in \mathcal{P}$, $x = p(u)$, and tangent vectors $\xi_1, \dots, \xi_k \in T_u\mathcal{P}$, there is a unique element $\phi(u)(\xi_1, \dots, \xi_k) \in V$ such that

$$\alpha(p(u))(T_up \cdot \xi_1, \dots, T_up \cdot \xi_k) = [\![u, \phi(u)(\xi_1, \dots, \xi_k)]\!]. \tag{1.1}$$

This defines a k–linear alternating map $\phi(u) : (T_u\mathcal{P})^k \to V$, which evidently vanishes if one entry is a vertical tangent vector. One easily verifies that $\phi(u)$ depends smoothly on u, so we have constructed a horizontal V–valued k–form on $\mathcal{P}$. For $h \in H$ we get $p \circ r^h = p$, and hence $Tp \cdot Tr^h \cdot \xi_i = Tp \cdot \xi_i$ for each i. This shows that

$$[\![u, \phi(u)(\xi_1, \dots, \xi_k)]\!] = [\![u \cdot h, \phi(u \cdot h)(Tr^h \cdot \xi_1, \dots, Tr^h \cdot \xi_k)]\!],$$

from which equivariancy follows immediately.

Conversely, suppose we have given a horizontal, equivariant form ϕ. Then for each $x \in M$ we can choose $u \in \mathcal{P}$ such that $p(u) = x$, and any tangent vector at x can be written as $T_up \cdot \xi$. Fixing u we can use equation (1.1) to define $\alpha(x)$. This does not depend on the choice of the lifts of the tangent vectors since ϕ is horizontal. It does not depend on the choice of u either by equivariancy of ϕ. Finally, it is easily verified that smoothness of ϕ implies smoothness of α. □

1.2.8. Natural bundles and jets. A *natural bundle* F on the category $\mathcal{M}_n$ of n–dimensional manifolds and local diffeomorphisms is a functor assigning to any n–manifold N a fiber bundle $p_N : F(N) \to N$ and to any local diffeomorphism $f : N_1 \to N_2$ a bundle map $F(f) : F(N_1) \to F(N_2)$ with base map f, i.e. such that $p_{N_2} \circ F(f) = f \circ p_{N_1}$. Furthermore, F has to be regular, i.e. if M is any smooth manifold and $f : M \times N_1 \to N_2$ is smooth and such that for each $x \in P$

the map $f_x : N_1 \to N_2$ defined by $f_x(y) := f(x,y)$ is a local diffeomorphism, then we assume that the map $M \times F(N_1) \to F(N_2)$ defined by $(x,a) \mapsto F(f_x)(a)$ is smooth, too. Thus, regularity means that smoothly parametrized families of local diffeomorphisms are transformed into smoothly parametrized families. Usually, one assumes in addition that F is local, i.e. that for any inclusion $i : U \hookrightarrow N$ of an open subset, $F(i)$ is the inclusion $p_N^{-1}(U) \hookrightarrow F(N)$.

Developing the general theory of natural bundles has been one of the main aims of [**KMS**]. It turns out that regularity follows from functoriality and locality, and one obtains an explicit description of all natural bundles as associated bundles. In order to state the result, we have to recall another basic concept of differential geometry.

Let M, N be smooth manifolds, and $x \in M$ a point. Two smooth mappings f, $g : M \to N$ are said to have the same *jet of order* r (briefly r–jet) at x if $f(x) = g(x)$ and their partial derivatives at x up to order r in some local charts around x and $f(x)$ coincide. (Then this is true in all charts around these points by the chain rule.) This defines an equivalence relation, whose classes are called r–jets at x and denoted by $j^r_x f$. The point x is called the *source* while $f(x)$ is called the *target* of the jet $j^r_x f$. The space of all r–jets with source in M and target in N is denoted by $J^r(M,N)$. Similarly, we write $J^r_x(M,N)$, $J^r(M,N)_y$, or $J^r_x(M,N)_y$ if the source, target or both are fixed. For $s < r$, we may send an r–jet to the underlying s–jet, thus obtaining a canonical map $\pi^r_s : J^r(M,N) \to J^s(M,N)$. Putting $s = 0$, the source and target map define $\pi^r_0 : J^r(M,N) \to M \times N$.

The chain rule immediately implies that the composition of jets in $J^r(M,N)_y$ and $J^r_y(N,Q)$ is well defined by the formula $(j^r_y g) \circ (j^r_x f) = j^r_x(g \circ f)$. A jet $j^r_x f \in J^r_x(M,M)$ is called invertible if there is a jet $j^r_{f(x)} g \in J^r_{f(x)}(M,M)$ such that $j^r_x(g \circ f) = j^r_x \mathrm{id}_M$ and $j^r_{f(x)}(f \circ g) = j^r_{f(x)} \mathrm{id}_M$.

Using the canonical charts on $\mathbb{R}^n$ and $\mathbb{R}^m$ and the translations, we obtain an identification $J^r(\mathbb{R}^m, \mathbb{R}^n) \cong \mathbb{R}^m \times \mathbb{R}^n \times J^r_0(\mathbb{R}^m, \mathbb{R}^n)_0$, and the Taylor coefficients yield canonical coordinates on $J^r_0(\mathbb{R}^m, \mathbb{R}^n)_0$. This, of course, works similarly for open subsets in $\mathbb{R}^m$ and $\mathbb{R}^n$. The construction of the jet spaces $J^r(M,N)$ is functorial in both arguments, and via this, arbitrary charts on M and N give rise to charts on $J^r(M,N)$. Hence, each $J^r(M,N)$ is a smooth manifold and by construction the natural maps π^r_s for $0 \le s < r$ from above are smooth.

For any fibered manifold $p : E \to M$ we write $J^r(E \to M)$, or briefly $J^r(E)$, for the subset of $J^r(M,E)$ consisting of all jets of local sections of p. This turns out to be a smooth submanifold and a fibered manifold over M. Clearly, $J^r(\ )$ is a functor acting on locally invertible fibered morphisms, the *jet prolongation* functor. There is a universal rth order differential operator j^r which maps sections of $E \to M$ to sections of $J^r(E) \to M$ and is defined by $s \mapsto (x \mapsto j^r_x s)$. For every operator $D : \Gamma(E \to M) \to \Gamma(E' \to M)$, which is an rth order differential operator in local coordinates, there exists a fibered morphism $\tilde{D} : J^r(E) \to E'$ such that $D = \tilde{D} \circ j^r$.

Jets lead to alternative descriptions of many of the basic geometric objects on smooth manifolds. For example, the tangent bundle TM can be naturally identified with the space $J^1_0(\mathbb{R}, M)$ of first order jets of curves in M at $0 \in \mathbb{R}$. Similarly, the cotangent bundle T^*M can be naturally identified with the space $J^1(M,\mathbb{R})_0$. Let us note that in this kinematic approach one naturally obtains a vector bundle structure on the cotangent bundle. Consider the set $J^1_0(\mathbb{R}^n, \mathbb{R}^n)^{\mathrm{inv}}_0$ of invertible one–jets on $\mathbb{R}^n$ with source and target zero. Of course, this is a group under jet composition and it can be naturally identified with the group $GL(n,\mathbb{R})$. This leads to an interpretation of the linear frame bundle $\mathcal{P}^1M$ of an n–dimensional manifold M as $J^1_0(\mathbb{R}^n,M)^{\text{inv}}$ of invertible one–jets from $\mathbb{R}^n$ to M with source 0. In this picture, the principal action of $GL(n,\mathbb{R})\cong J^1_0(\mathbb{R}^n,\mathbb{R}^n)^{\text{inv}}_0$ is given by jet composition from the right.

The jet interpretation leads to higher order generalizations of all the above bundles. The *rth order frame bundle* $\mathcal{P}^rM$ is defined as $J^r_0(\mathbb{R}^n,M)^{\text{inv}}$, so it consists of r–jets of local charts. The structure group of this bundle is the rth *differential group* $G^r_n := J^r_0(\mathbb{R}^n,\mathbb{R}^n)^{\text{inv}}_0$ and the principal action is given by the jet composition from the right. The rth order analog of the tangent bundle is the bundle $T^r_1M = J^r_0(\mathbb{R},M)$. This is the associated bundle to $\mathcal{P}^rM$ with respect to the obvious left action of G^r_n on $S=J^r_0(\mathbb{R},\mathbb{R}^n)$. Now the description of natural bundles is as follows.

THEOREM 1.2.8 ([**KMS**, Theorem 22.1]). *Any local natural bundle on n–dimensional manifolds can be obtained as an associated bundle to some $\mathcal{P}^rM$ with respect to a left action of the differential group G^r_n on a finite–dimensional manifold.*

The lowest possible choice for r in the theorem is called the *order of the natural bundle.* Notice that the composition of the jet prolongation functor with a kth order natural bundle F is the $(k+r)$th order natural bundle J^rF. The theorem was first proved assuming regularity by Palais and Terng (see [**PT77**]), and then in full generality by Epstein and Thurston (see [**ET79**]). Sharp estimates for the orders depending on the dimensions of the base and fiber were obtained by Zajtz. See [**KMS**] for further results and more bibliographic details.

1.2.9. Lie derivatives. Natural bundles also provide the right framework for defining Lie derivatives. Let us first observe that local diffeomorphisms act on the sections of any natural bundle F. For a section $s\in\Gamma(FN)$ and a local diffeomorphism $f:M\to N$, one obtains $f^*s\in\Gamma(FM)$ locally as $F\phi\circ s\circ f$, where ϕ is a local inverse to f. The section f^*s is called the *pullback* of s along f. In particular, this can be applied to the local flow of a vector field $\xi\in\mathfrak{X}(M)$. Fixing $x\in M$, we obtain a curve $t\mapsto(\mathrm{Fl}^\xi_t)^*s(x)$ in the fiber F_xM of FM over x, which is defined for sufficiently small t. Regularity of F implies that this curve is smooth, so we may consider its derivative at $t=0$. If F is a natural vector bundle, then this derivative may be interpreted as an element of the fiber F_xM itself, while for a general natural fiber bundle it has to be viewed as an element of the vertical tangent space $V_{s(x)}FM$. Regularity of F again implies that this element depends smoothly on x, so we obtain the *Lie derivative* $\mathcal{L}_\xi s$ of s along ξ defined by

$$\mathcal{L}_\xi s = \tfrac{d}{dt}|_0(\mathrm{Fl}^\xi_t)^*s = \tfrac{d}{dt}|_0 F(\mathrm{Fl}^\xi_{-t})\circ s\circ \mathrm{Fl}^\xi_t .$$

This is a smooth section of FM in the case of a natural vector bundle and a smooth section of the vertical tangent bundle VFM in the case of an arbitrary natural fiber bundle.

An alternative way to view the Lie derivative is the following: Fixing a point $x\in M$, the flow Fl^ξ_t is defined locally around x for sufficiently small t. Hence, $F(\mathrm{Fl}^\xi_t)$ is a family of locally defined diffeomorphisms of FM, which depends smoothly on t by regularity. Differentiating at $t=0$, one obtains a vector field $\mathcal{F}\xi$ on FM, which is p_M–related to ξ, and $\mathcal{L}_\xi s = Ts\circ\xi-\mathcal{F}\xi\circ s$.

In the special case of the tangent bundle TM we obtain the standard Lie bracket, i.e. $\mathcal{L}_\xi\eta=[\xi,\eta]$ and for all tensor bundles one recovers the classical approach. In particular, it is easy to deduce the Leibniz rule for general tensor products of natural vector bundles

$$\mathcal{L}_\xi(s_1 \otimes s_2) = (\mathcal{L}_\xi s_1) \otimes s_2 + s_1 \otimes (\mathcal{L}_\xi s_2)$$

and compatibility with contractions. For example, for a k–times covariant tensor field τ and $\xi, \eta_1, \dots, \eta_k \in \mathfrak{X}(M)$, we obtain

$$(\mathcal{L}_\xi \tau)(\eta_1, \dots, \eta_k) = \mathcal{L}_\xi(\tau(\eta_1, \dots, \eta_k)) - \sum_{i=1}^{k} \tau(\eta_1, \dots, \mathcal{L}_\xi(\eta_i), \dots, \eta_k).$$

In the case of k forms, i.e. antisymmetric k–times covariant tensor–fields, this formula easily leads to the formula $\mathcal{L}_\xi = i_\xi \circ d + d \circ i_\xi$ for the Lie derivative in terms of the exterior derivative d (see 1.2.1) and the insertion operator i_ξ defined by $(i_\xi \tau)(\eta_2, \dots, \eta_k) = \tau(\xi, \eta_2, \dots, \eta_k)$.

The Lie derivative $\mathcal{L}_\xi$ depends on derivatives of the vector field ξ up to the order of the natural bundle. Thus, the only case in which the Lie derivative is tensorial in the direction ξ are natural bundles of order zero, which are always trivial.

1.2.10. Complex manifolds and complex differential geometry. We conclude this section with a brief discussion of holomorphic aspects of differential geometry. More basic information on these issues can be found, for example, in [**KoNo69**].

A *complex manifold* M is defined similarly to the real case discussed in 1.2.1, but with charts having values in $\mathbb{C}^n$ and holomorphic transition functions. The number n is called the *complex dimension* of M, and of course we can view M also as a manifold of real dimension $2n$. For functions between complex manifolds (and in particular for functions with values in $\mathbb{C}^m$) one defines holomorphicity by requiring holomorphicity in some (or equivalently any) chart. A holomorphic diffeomorphism whose inverse is holomorphic, too, is called a *biholomorphism.* Two complex manifolds are called *biholomorphic* if there is a biholomorphism between them. It happens often that complex manifolds are diffeomorphic without being biholomorphic.

Of course, the product of two complex manifolds is canonically a complex manifold. On the one hand, this implies that there is a well–defined notion of a *complex Lie group* as a complex manifold endowed with a holomorphic group structure. In particular, the group $GL(n, \mathbb{C})$ is a complex Lie group and thus for any complex Lie group one can talk about *holomorphic representations* on complex vector spaces. On the other hand, given complex manifolds M and S, one can define *holomorphic fiber bundles* over M with standard fiber S similarly as in 1.2.6. One just has to require the total space to be a complex manifold and the fiber bundle charts to be holomorphic. In particular, one has the subclass of *holomorphic vector bundles* among complex vector bundles. For principal bundles with structure group a complex Lie group, there is the subclass of holomorphic principal bundles. Given any holomorphic fiber bundle, there is a natural notion of holomorphicity for sections via holomorphicity in some (or equivalently any) fiber bundle chart. One can then consider jets of holomorphic sections similarly as in 1.2.8, and so on.

Since functions between open subsets of $\mathbb{C}^n$ are holomorphic if and only if they have complex linear derivatives, one can use the charts of a complex atlas to make any tangent space of a complex manifold M into a complex vector space. In this way, the tangent bundle TM becomes a complex (and even a holomorphic) vector bundle. The most convenient way to encode this is to consider the linear maps on the tangent spaces of M given by multiplication by $i = \sqrt{-1}$. These fit together to define a smooth bundle map $J : TM \to TM$, which clearly satisfies $J^2 = J \circ J = -\mathrm{id}$. If M and $\tilde{M}$ are complex manifolds and J and $\tilde{J}$ are the corresponding bundle maps, then a smooth map $f : M \to \tilde{M}$ is holomorphic if and only if its tangent map is complex linear, i.e. $Tf \circ J = \tilde{J} \circ Tf$. In particular, the underlying real manifold M and the bundle map J encode the structure of a complex manifold on M, since one may characterize holomorphic charts $u : U \to \mathbb{C}^n$ using J.

Now one can turn the game around, starting with a real manifold M and a bundle map $J : TM \to TM$ such that $J^2 = -\mathrm{id}$. Such a bundle map is called an *almost complex structure* on M, and existence of such a structure implies that the dimension of M is even. A manifold M endowed with an almost complex structure J is called an *almost complex manifold.* The classical *Newlander–Nirenberg theorem* characterizes those almost structures which are integrable, i.e. which come from the structure of a complex manifold on M: For vector fields $\xi, \eta \in \mathfrak{X}(M)$ consider the expression

$$[\xi, \eta] - [J\xi, J\eta] + J([J\xi, \eta] + [\xi, J\eta]).$$

One immediately verifies that this expression is bilinear over smooth functions on M, so it defines a tensor field $N = N_J : TM \times TM \to TM$, called the *Nijenhuis tensor* of the almost complex structure J. From the definition one immediately verifies that N_J is skew symmetric and conjugate linear in both variables, i.e. $N_J(J\xi, \eta) = -J(N_J(\xi, \eta))$, and likewise in the other variable. In particular, N_J always vanishes if M is of real dimension two, since then a basis of each tangent space is given by $\{\xi, J\xi\}$ for some nonzero tangent vector ξ. The Newlander–Nirenberg theorem (see [**NeNi57**]) states that J is induced by a complex structure on M if and only if $N_J = 0$. It should be remarked that this theorem is not too difficult in the case that M and J are assumed to be real analytic; see [**KoNo69**]. The hard part is to show that vanishing of the Nijenhuis tensor implies real analyticity.

One of the basic features of almost complex and complex manifolds is a natural decomposition of the spaces of complex valued differential forms. Let (M, J) be an almost complex manifold and V a complex vector space, and for some $0 \leq k \leq \dim_{\mathbb{R}}(M)$ consider the space $\Omega^k(M, V)$ of real k–forms with values in V. For $\phi \in \Omega^k(M, V)$ and $x \in M$, the value $\phi(x)$ is a map $(T_xM)^k \to V$, which is $\mathbb{R}$–linear in each entry. The finer decomposition has the form $\Omega^k(M, V) = \bigoplus_{p+q=k} \Omega^{p,q}(M, V)$ with $p, q \geq 0$, and is often referred to as the decomposition into (p, q)–types. There are two ways to describe this decomposition: On the one hand, one can look at the action of multiplication by nonzero complex numbers in the real picture. From this point of view, the subspace $\Omega^{p,q}(M, V)$ is formed by all ϕ such that for each nonzero complex number λ and all vector fields $\xi_1, \dots, \xi_k$ on M, one has

$$\phi(\lambda\xi_1, \dots, \lambda\xi_k) = \lambda^p \bar{\lambda}^q \phi(\xi_1, \dots, \xi_k).$$

In particular, for $k = 1$, the subspaces $\Omega^{1,0}(M, V)$ and $\Omega^{0,1}(M, V)$ consist of those ϕ for which each of the maps $\phi(x) : T_xM \to V$ is complex linear, respectively, conjugate linear. For $k = 2$, the subspaces $\Omega^{2,0}(M, V)$ and $\Omega^{0,2}(M, V)$ consist of those forms whose values are complex linear, respectively, conjugate linear in both arguments. The subspace $\Omega^{1,1}(M, V)$ consists of forms ϕ whose values are totally real, i.e. such that $\phi(J\xi, J\eta) = \phi(\xi, \eta)$ for all vector fields ξ and η on M.

The second approach to obtain the decomposition into (p,q)–types is via complexification. Let (M,J) be an almost complex manifold and consider the complexified tangent bundle $T_{\mathbb{C}}M := TM \otimes_{\mathbb{R}} \mathbb{C}$. The bundle map $J : TM \to TM$ extends to a complex linear bundle map $J_{\mathbb{C}} : T_{\mathbb{C}}M \to T_{\mathbb{C}}M$ such that $J_{\mathbb{C}}^2 = -\mathrm{id}$. Since each tangent space is a complex vector space, it splits into the direct sum of eigenspaces for $J_{\mathbb{C}}$ with eigenvalues $\pm i$, and of course, this depends smoothly on the base point. Thus, one obtains a splitting $T_{\mathbb{C}}M = T^{1,0}M \oplus T^{0,1}M$ into a direct sum of smooth subbundles. The eigenspaces $T^{1,0}_xM$ (respectively $T^{0,1}_xM$) are spanned by all vectors of the form $\xi - iJ\xi$ (respectively $\xi + iJ\xi$) with $\xi \in T_xM$. Vanishing of the Nijenhuis tensor is easily seen to be equivalent to thc fact that the sections of either (or equivalently both) these bundles are closed under the Lie bracket.

The dual space $T^*M \otimes \mathbb{C}$ of $T_{\mathbb{C}}M$ splits accordingly as $T^*_{1,0}M \oplus T^*_{0,1}M$, where $T^*_{1,0}$ is the annihilator of $T^{0,1}$. Next, we get

$$\Lambda^k(T^*M \otimes \mathbb{C}) = \bigoplus_{p+q=k} \Lambda^p T^*_{1,0}M \otimes \Lambda^q T^*_{0,1}M.$$

For a complex vector space V, one may then identify $L_{\mathbb{R}}(\Lambda^k T^*M, V)$ with $\Lambda^k(T^*M \otimes \mathbb{C}) \otimes_{\mathbb{C}} V$, which leads to the decomposition into (p,q)–types.

Let us next specialize to the case $V = \mathbb{C}$ of complex–valued forms. Then it is natural to look at the compatibility of the exterior derivative d (which is defined exactly as the real counterpart) with the decomposition into (p,q)–types. For a general almost complex structure, one can only show that for $\phi \in \Omega^{p,q}(M) := \Omega^{p,q}(M,\mathbb{C})$ the exterior derivative $d\phi$ can have nontrivial components only in bidegrees $(p+2,q-1)$, $(p+1,q)$, $(p,q+1)$, and $(p-1,q+2)$. In the case of a complex structure, the situation is much nicer, since in that case the nontrivial components can only lie in degrees $(p+1,q)$ and $(p,q+1)$. Decomposing $d\phi$ into these two parts leads to two natural first order operators $\partial : \Omega^{p,q}(M) \to \Omega^{p+1,q}(M)$ and $\bar\partial : \Omega^{p,q}(M) \to \Omega^{p,q+1}(M)$ such that $d = \partial + \bar\partial$. The fact that $d^2 = 0$ immediately implies that $\partial^2 = 0$, $\bar\partial^2 = 0$ and $\bar\partial \circ \partial = -\partial \circ \bar\partial$.

In the case of complex structures, the splitting into (p,q)–types and the operators ∂ and $\bar\partial$ can be nicely described in local coordinates. Suppose that one has local $\mathbb{C}^n$–values coordinates $(x^1 + iy^1, \dots, x^n + iy^n)$. Then from above we conclude that we obtain local frames $\{\frac{\partial}{\partial z^j} : j = 1, \dots, n\}$ for $T^{1,0}M$ and $\{\frac{\partial}{\partial \bar z^j} : j = 1, \dots, n\}$ for $T^{0,1}M$ by defining

$$\tfrac{\partial}{\partial z^j} := \tfrac{1}{2}\left(\tfrac{\partial}{\partial x^j} - i\tfrac{\partial}{\partial y^j}\right) \qquad \tfrac{\partial}{\partial \bar z^j} := \tfrac{1}{2}\left(\tfrac{\partial}{\partial x^j} + i\tfrac{\partial}{\partial y^j}\right).$$

(Observe that J maps $\frac{\partial}{\partial x^j}$ to $\frac{\partial}{\partial y^j}$.) The dual frames for $T^*_{1,0}M$ and $T^*_{0,1}M$ consist of the elements $dz^j = dx^j + idy^j$, respectively, $d\bar z^j = dx^j - idy^j$. Hence, the forms $dz^{a_1} \wedge \dots \wedge dz^{a_p} \wedge d\bar z^{b_1} \wedge \dots \wedge d\bar z^{b_q}$ with $p + q = k$, $a_1 < \dots < a_p$ and $b_1 < \dots < b_q$ form a local frame for $\Omega^k(M,\mathbb{C})$, and the splitting into (p,q)–types corresponds to the number of unbarred and barred factors. Using the sum convention, we get for $\phi = \phi_{a_1\dots a_p b_1 \dots b_q} dz^{a_1} \wedge \dots \wedge d\bar z^{b_q}$ the formulae

$$\partial(\phi) = \frac{\partial \phi_{a_1\dots a_p b_1\dots b_q}}{\partial z^j} dz^j \wedge dz^{a_1} \wedge \dots \wedge d\bar z^{b_q},$$

$$\bar\partial(\phi) = \frac{\partial \phi_{a_1\dots a_p b_1\dots b_q}}{\partial \bar z^j} d\bar z^j \wedge dz^{a_1} \wedge \dots \wedge d\bar z^{b_q}.$$

Finally, observe that by the Cauchy–Riemann equations a complex–valued function $f : M \to \mathbb{C}$ is holomorphic if and only if $\frac{\partial f}{\partial \bar{z}^j} = 0$ for all $j = 1, \dots, n$. This implies that a (p,q)–form $\phi \in \Omega^{p,q}(M)$ is holomorphic if and only if $\bar{\partial}\phi = 0$, i.e. if and only if $d\phi \in \Omega^{p+1,q}(M)$. This works in the same way for forms with values in any finite–dimensional complex vector space.

1.3. A survey on connections

This section provides background on various versions of connections. We start from the simple idea of linear connections on vector bundles, then pass to general connections on fiber bundles, and specialize to principal and induced connections. Finally, we discuss affine connections and, more generally, connections on G–structures, which are the simplest special cases of Cartan connections.

1.3.1. Linear connections. The general idea of a connection is to provide a notion of directional derivatives for sections of bundles or fibered manifolds. The directions are given by vector fields on the base, and for real–valued smooth functions, the action of vector fields provides a natural operation of this type. We have seen two further instances of such operations already.

First, for smooth functions with values in a Lie group G, the trivialization of the tangent bundle defined by the Maurer–Cartan form allows us to define a derivative, which has values in the tangent space at the unit element. This is the left logarithmic derivative defined in 1.2.4. We shall see below how to link this derivative to the trivial principal connection on the trivial principal bundle $M \times G \to M$.

Second, on all natural bundles, there is the notion of Lie derivative of sections along vector fields; cf. 1.2.9. This is, however, not an analogue of a directional derivative, since the value of $\mathcal{L}_\xi s$ in a point does not only depend on the value of ξ in that point but on a higher jet of the vector field. Hence, a different concept is needed.

Let us consider an arbitrary vector bundle $\mathcal{V} \to M$ with standard fiber V. We wish to have derivatives of sections $s \in \Gamma(\mathcal{V})$ in the direction of a vector field $\xi \in \mathfrak{X}(M)$ which are tensorial in ξ. In the other variable, one requires a Leibniz rule with respect to the multiplication by smooth functions. Such an operation is usually called a "connection" or a *covariant derivative* on the vector bundle $\mathcal{V}$. It is well known (see e.g. [**KMS**, Lemma 7.3]) that the requirement to be tensorial in ξ can be equivalently formulated as being linear over $C^\infty(M)$. Thus, a *linear connection* on a vector bundle $\mathcal{V} \to M$ is often defined as a bilinear operator $\nabla : \mathfrak{X}(M) \times \Gamma(\mathcal{V}) \to \Gamma(\mathcal{V})$, written as $(\xi, s) \mapsto \nabla_\xi s$, such that for all $\xi \in \mathfrak{X}(M)$, $s \in \Gamma(\mathcal{V})$ and $f \in C^\infty(M, \mathbb{R})$ one has

$$\begin{aligned} \nabla_\xi f s &= (\mathcal{L}_\xi f)s + f\nabla_\xi s, \\ \nabla_{f\xi} s &= f\nabla_\xi s. \end{aligned} \tag{1.2}$$

Choosing local charts $\mathbb{R}^n \times \mathbb{R}^k \to \mathcal{V}$ for M and $\mathcal{V}$ and using the usual summation convention, we may write the vector field ξ as $\xi = \xi^i \frac{\partial}{\partial x^i}$, where x is the coordinate on $\mathbb{R}^n$, and the section s as $s^p e_p$, where $\{e_1, \dots, e_k\}$ is the the local frame of $\mathcal{V}$ corresponding to the chosen chart. From the defining properties (1.2) of ∇ we get

$$\nabla_\xi s = \xi^i \nabla_{\frac{\partial}{\partial x^i}} (s^p e_p) = \xi^i \frac{\partial s^p}{\partial x^i} e_p + B^p_{qi} s^q \xi^i e_p,$$

for smooth functions B^p_{qi} characterized by $\nabla_{\frac{\partial}{\partial x^i}} e_q = B^p_{qi} e_p$. This formula shows that $\nabla_\xi s(x)$ depends only on $j^1_x s$. Moreover, for each fixed target $y = s(x)$ there is a unique one–jet $j^1_x s$ such that $\nabla_{\xi_x} s = 0$ for all $\xi_x \in T_x M$. Conversely, knowing that one–jet is equivalent to knowing the functions B^p_{qi} and thus to knowing the linear connection ∇.

The last observation directly leads to the definition of the *horizontal distribution* $\mathcal{H} \subset T\mathcal{V}$ associated to ∇. Given $x \in M$ and $y \in \mathcal{V}_x$, we can find a smooth section $s \in \Gamma(V)$ such that $s(x) = y$ and $\nabla_\xi s(x) = 0$ for all $\xi \in \mathfrak{X}(M)$ and the one–jet $j^1_x s$ is uniquely determined by this condition. In particular, $T_x s$ is unambiguously defined, and we put $\mathcal{H}_y := T_x s(T_x M)$. Of course, $T_y p$ is inverse to $T_x s$, so it restricts to a linear isomorphism between $\mathcal{H}_y$ and $T_x M$. Hence, for each tangent vector ξ_x at x and each point y over x, we can find a unique lift of ξ_x in $\mathcal{H}_y$. This defines the *horizontal lift* of tangent vectors on M.

In local coordinates as above, the distinguished one–jet $j^1_x s$ is characterized by $\frac{\partial s^p}{\partial x^i}(x) = -B^p_{qi}(x) s^q(x)$. In particular, in a point x_0 the distinguished jet for $s(x_0) = y$ can be represented by the map $y - B^p_{qi}(x_0) y^q (x^i - x^i_0)$. This immediately shows that the horizontal lift of $\xi^i(x)\frac{\partial}{\partial x^i}$ is given by

$$\xi^i(x)\frac{\partial}{\partial x^i} - \xi^i(x) B^p_{qi}(x) y^q \frac{\partial}{\partial y^p}.$$

Starting from a smooth vector field $\xi \in \mathfrak{X}(M)$, the horizontal lifts fit together to define a smooth vector field $\xi^{\mathrm{hor}} \in \mathfrak{X}(\mathcal{V})$, called the *horizontal lift* of ξ. It is the unique projectable vector field over ξ whose value in each point lies in the horizontal subspace. Since such horizontal lifts span the horizontal subspaces, we see that the horizontal distribution is smooth.

Let us put this into a broader perspective. Working in local coordinates as above, the fact that the horizontal lift in a point is a linear map implies that we may write

$$\xi^{\mathrm{hor}}(x, y) = \xi^i(x)\frac{\partial}{\partial x^i} + \gamma^p_i(x, y)\xi^i(x)\frac{\partial}{\partial y^p},$$

for some functions γ^p_i. We have seen that in our case $\gamma^p_i(x, y) = -B^p_{qi}(x) y^q$, so this is linear in y. This is the origin of the term "linear connection".

Next, we show that the covariant derivative can be recovered from the horizontal lift map. This will lead to the notion of a general connection, which is based on the horizontal lift.

LEMMA 1.3.1. *Let ∇ be a linear connection on a vector bundle $\mathcal{V} \to M$. Then for any vector field $\xi \in \mathfrak{X}(M)$ and any section $s \in \Gamma(\mathcal{V})$ we have*

$$\nabla_\xi s = Ts \circ \xi - \xi^{hor} \circ s = \tfrac{d}{dt}|_0 \left(\mathrm{Fl}^{\xi^{hor}}_{-t} \circ s \circ \mathrm{Fl}^{\xi}_t\right),$$

where we identify the vertical tangent space to $\mathcal{V}$ in the point $s(x)$ with the fiber $\mathcal{V}_x$.

PROOF. The first equality follows immediately from the coordinate formulae of the covariant derivative and the horizontal lift, and the second equality is a direct computation. □

The *curvature of a linear connection* ∇ on $\mathcal{V}$ is defined by

$$R(\xi, \eta)(s) := \nabla_\xi \nabla_\eta s - \nabla_\eta \nabla_\xi s - \nabla_{[\xi,\eta]} s \tag{1.3}$$

for $\xi, \eta \in \mathfrak{X}(M)$. By construction, this is skew symmetric in ξ and η, and one easily verifies that it is linear over smooth functions in all three entries. Thus, we may view the curvature R as a section of the bundle $\Lambda^2 T^*M \otimes L(\mathcal{V},\mathcal{V})$, i.e. as a two–form with values in endomorphisms of $\mathcal{V}$.

For later use, let us note here that a linear connection ∇ on a vector bundle $\mathcal{V} \to M$ induces operators on $\mathcal{V}$–valued differential forms, called the *covariant exterior derivative*. By definition, the space $\Omega^k(M,\mathcal{V})$ of $\mathcal{V}$–valued k–forms is the space of smooth sections of the bundle $\Lambda^k T^*M \otimes \mathcal{V}$. Alternatively, $\Omega^k(M,\mathcal{V})$ is the space of all k–linear alternating maps $(\mathfrak{X}(M))^k \to \Gamma(\mathcal{V})$ which are linear over $C^\infty(M,\mathbb{R})$ in one (or equivalently any) variable. Now one defines the covariant exterior derivative $d^\nabla : \Omega^k(M,\mathcal{V}) \to \Omega^{k+1}(M,\mathcal{V})$ by taking the formula for the exterior derivative from 1.2.1 and replacing the action of a vector field on a smooth function by a covariant derivative, i.e.

$$\begin{aligned} d^\nabla\omega(\xi_0,\dots,\xi_k) &= \sum_{i=0}^{k}(-1)^i\nabla_{\xi_i}(\omega(\xi_0,\dots,\widehat{\xi_i},\dots,\xi_k)) \\ &+\sum_{i<j}(-1)^{i+j}\omega([\xi_i,\xi_j],\xi_0,\dots,\widehat{\xi_i},\dots,\widehat{\xi_j}\dots,\xi_k). \end{aligned}$$

The verification that this is alternating and linear over $C^\infty(M,\mathbb{R})$ is exactly as in the case of usual differential forms. In contrast to the exterior derivative, the covariant exterior derivative is, however, not a differential, i.e. $d^\nabla \circ d^\nabla \neq 0$ in general.

1.3.2. General connections. A *general connection* on an arbitrary fibered manifold $p : Y \to M$ is a smooth horizontal distribution $\mathcal{H} \subset TY$ which is complementary to the vertical tangent bundle VY. For each $y \in Y$, $\mathcal{H}_y \subset T_yY$ is called the *horizontal subspace* at y of the connection.

There are three further equivalent ways to view this: First, we can consider the induced horizontal lift of vector fields which associates $\xi^{\text{hor}} \in \mathfrak{X}(Y)$ to $\xi \in \mathfrak{X}(M)$. As in the case of linear connections, ξ^{hor} is the unique projectable vector field lying over ξ whose value in each point is horizontal. By construction, the horizontal lift map is linear over smooth functions, and conversely a lift map with this property comes from a smooth horizontal distribution. Second, we have $TY = VY \oplus \mathcal{H}$ and this decomposition can be equivalently described by the smooth *vertical projection* $\Psi : TY \to VY$ with kernel $\mathcal{H}$. Of course, the vertical projection also defines the *horizontal projection* $\chi = \text{id}_{TY} - \Psi$. Finally, as in the case of linear connections, one may view the connection as specifying a unique one–jet of sections $j^1_x s$ with $s(x) = y$ for each $y \in Y$, such that the horizontal lift $T_xM \to T_yY$ is given by T_xs. From that point of view, a connection can equivalently be viewed as a smooth section of the first jet prolongation $J^1Y \to Y$.

The *curvature of a general connection* is defined by $R(\xi,\eta) = -\Psi([\chi(\xi),\chi(\eta)])$ for $\xi,\eta \in \mathfrak{X}(Y)$, so $R(\xi,\eta)$ is minus the vertical projection of the brackets of the horizontal projections of the vector fields. (The sign is used to obtain the usual sign conventions for principal and linear connections.) By definition, R is horizontal, i.e. vanishes upon insertion of one vertical vector field, and has vertical values. Moreover, $\Psi \circ \chi = 0$ immediately implies that R is bilinear over $C^\infty(Y,\mathbb{R})$, so $R \in \Omega^2_{\text{hor}}(Y,VY)$. Moreover, $R(\xi,\eta) = 0$ for all ξ and η if and only if the Lie bracket of any two horizontal vector fields on Y is horizontal, too, i.e. if and only if the horizontal distribution is involutive. By the Frobenius theorem (see 1.2.2), a connection has vanishing curvature if and only if it is locally given by local trivializations $s : \mathbb{R}^m \times \mathbb{R}^n \to Y$. A more sophisticated definition of the curvature of general connections is to define R as the Frölicher–Nijenhuis bracket $[\Psi, \Psi]$ of the vertical projection Ψ; see [**KMS**, section 9].

Since one–dimensional distributions are always involutive, a general connection defines the so–called *parallel transport* along curves on $p : Y \to M$: For each interval $(a, b) \subset \mathbb{R}$ containing zero and each parametrized curve $c : (a, b) \to M$, with $x = c(0)$, and any $y \in Y$ with $p(y) = x$, there is a unique maximal subinterval $(a', b') \subset (a, b)$ containing 0 and a unique curve $\tilde{c}_y : (a', b') \to Y$ such that $\tilde{c}_y(0) = y$, $p \circ \tilde{c}_y = c$, and $T\tilde{c}_y$ has values in the horizontal distribution.

The *absolute derivative* ∇_ξ defined by a general connection is then defined by the formula from Lemma 1.3.1, i.e.

$$\nabla_\xi s = Ts \circ \xi - \xi^{\text{hor}} \circ s = \tfrac{d}{dt}|_0 \left(\text{Fl}^{\xi^{\text{hor}}}_{-t} \circ s \circ \text{Fl}^{\xi}_{t} \right).$$

For a section $s \in \Gamma(Y)$ and a vector field $\xi \in \mathfrak{X}(M)$, we have $\nabla_\xi s \in \Gamma(VY \to M)$. This formula may also be explained in terms of the parallel transport: Take a section s and a direction $\xi_x \in T_x M$. Extend ξ_x into a vector field ξ, consider its flow through x and the corresponding parallel transport. Then the derivative of the pullback of the section by the parallel transport along the flow lines of ξ is exactly given by our formula.

There is also another concept of derivative, the *exterior absolute differential* $d^\nabla : \Omega^k(Y, V) \to \Omega^{k+1}(Y, V)$ for any vector space V, defined by means of the pullback with respect to the horizontal projection, i.e. by the formula $d^\nabla = \chi^* d$. We shall see some explicit relation between the last two concepts under specific geometric assumptions, but they are completely independent in the general setting.

For fiber bundles, there is a nonlinear version of the Christoffel symbols. For a local trivialization $\mathbb{R}^m \times S \to Y$, the horizontal lift of vector fields is described uniquely by the mappings $\gamma_i : TM \to \mathfrak{X}(S)$, so that the horizontal lift of $\xi^i \frac{\partial}{\partial x^i}$ equals $\xi^i \frac{\partial}{\partial x^i} + \xi^i \gamma_i$. Similarly, we obtain the formula for the absolute derivative in a local trivialization.

Finally, in the special case of a vector bundle $Y \to M$, the first jet prolongation $J^1 Y$ carries a vector bundle structure too. It is a simple exercise to verify that a general connection $\sigma : Y \to J^1 Y$ is linear if and only if the section σ is a vector bundle homomorphism.

1.3.3. Principal connections. In the special case of a principal bundle, it is natural to look at connections that are compatible with the principal right action. Let $p : \mathcal{P} \to M$ be a principal fiber bundle. A *principal connection* on $\mathcal{P}$ is a (general) connection whose horizontal distribution is invariant with respect to the principal action of the structure group G, i.e. $\mathcal{H}_{u \cdot g} = Tr^g(\mathcal{H}_u)$ for all $g \in G$. Notice that this immediately implies that the parallel transport along curves is also right invariant. This means that for any curve c, the corresponding parallel transport $\tilde{c}$ satisfies $\tilde{c}_{u \cdot g}(t) = \tilde{c}_u(t) \cdot g$ for all $u \in \mathcal{P}$ and $g \in G$.

There are two further equivalent ways to express the invariance condition:

(1) The horizontal lifts of vector fields are right invariant vector fields on the principal bundle $\mathcal{P}$, i.e. $(r^g)^* \xi^{\text{hor}} = \xi^{\text{hor}}$ for all $g \in G$ and $\xi \in \mathfrak{X}(M)$.
(2) The corresponding section $\sigma : \mathcal{P} \to J^1 \mathcal{P}$ is G–equivariant with respect to the obvious induced G–action on $J^1 \mathcal{P}$.

In the case of a principal bundle, one may reformulate the vertical projection defining a connection in a simple way, since the vertical bundle $V\mathcal{P}$ of a principal G–bundle $p : \mathcal{P} \to M$ is trivialized by the fundamental vector fields. This means that any vertical tangent vector $\xi \in T_u\mathcal{P}$ can be written as the value $\zeta_X(u)$ for a unique element $X \in \mathfrak{g}$, the Lie algebra of the structure group G. Consequently, we may specify the vertical projection of any general connection on $\mathcal{P}$ by a $\mathfrak{g}$–valued one form $\gamma \in \Omega^1(\mathcal{P}, \mathfrak{g})$, such that the vertical projection of $\xi \in T_u\mathcal{P}$ equals $\zeta_{\gamma(\xi)}(u)$. This is a projection if and only if γ reproduces the generators of fundamental vector fields, i.e. $\gamma(\zeta_X) = X$ for all $X \in \mathfrak{g}$. One easily verifies that γ corresponds to a principal connection if and only if $(r^g)^*\gamma = \mathrm{Ad}(g^{-1}) \circ \gamma$. We call the one–form γ the *principal connection form*, or briefly the principal connection, on $\mathcal{P}$.

The trivialization of the vertical bundle by fundamental vector fields also leads to a nice interpretation of the curvature of any connection on a principal fiber bundle. Since the curvature R has vertical values, there is a unique $\mathfrak{g}$–valued two–form $\rho \in \Omega^2(\mathcal{P}, \mathfrak{g})$ such that $R(\xi, \eta)(u) = \zeta_{\rho(\xi,\eta)(u)}(u)$ for any $u \in \mathcal{P}$. If we deal with a principal connection, then the definition of the curvature R by means of the horizontal lifts shows that R, viewed as an element in $\Omega^2(\mathcal{P}, V\mathcal{P})$, is G–equivariant, which in turn immediately implies that $(r^g)^*\rho = \mathrm{Ad}(g^{-1}) \circ \rho$. The form ρ is called the *curvature form* or often simply the curvature of the given principal connection. Since ρ is horizontal and equivariant, it may also be viewed as a two–form on M with values in the associated bundle $\mathcal{P} \times_G \mathfrak{g}$ corresponding to the adjoint action of G on $\mathfrak{g}$; see Corollary 1.2.7.

To compute the curvature ρ from the connection form γ, observe that by definition we have $\rho(\xi, \eta) = -\gamma([\xi - \zeta_{\gamma(\xi)}, \eta - \zeta_{\gamma(\eta)}])$. Since both fields on the right–hand side lie in the kernel of γ, this coincides with $d\gamma(\xi - \zeta_{\gamma(\xi)}, \eta - \zeta_{\gamma(\eta)})$. Note that this by definition means that $\rho = d^\nabla\gamma$, where d^∇ is the exterior absolute derivative from 1.3.2. Now for $X \in \mathfrak{g}$, the flow of the fundamental vector field ζ_X is $r^{\exp(tX)}$, so differentiating the equivariancy property $(r^{\exp(tX)})^*\gamma = \mathrm{Ad}(\exp(-tX)) \circ \gamma$ at $t = 0$, we obtain $\mathcal{L}_{\zeta_X}\gamma = -\operatorname{ad}(X) \circ \gamma$. Moreover, by definition $i_{\zeta_X}\gamma = X$ and thus $di_{\zeta_X}\gamma = 0$, whence we conclude that $d\gamma(\zeta_X, \eta) = -[X, \gamma(\eta)]$. Using this, we immediately conclude from above that

$$\rho(\xi, \eta) = d\gamma(\xi, \eta) + [\gamma(\xi), \gamma(\eta)],$$

which is the usual definition of the curvature of a principal connection.

On the trivial principal bundle $M \times G$ there is the trivial principal connection, whose horizontal subspaces are the kernels of $T\pi_G$, where $\pi_G : M \times G \to G$ is the projection. The connection form of this connection is the pullback of the Maurer–Cartan form by π_G, and for simplicity, we also denote this form by ω_G. Note that then the absolute differential of sections $M \to M \times G$, viewed as functions $M \to G$, is exactly the left logarithmic derivative. Let us further specialize to $M = \mathbb{R}^n$, which, via local trivializations, also describes the local situation for general principal bundles. A general connection form on $\mathbb{R}^n \times G$ can be written as

$$\gamma = \omega_G - \gamma_i dx^i,$$

for $\mathfrak{g}$–valued one forms γ_i. These are called the *Christoffel symbols*, and they are determined by the restriction to the distinguished section $(x, e) \subset \mathbb{R}^n \times G$. In particular, for principal connections on the linear frame bundle $\mathcal{P}^1M$ of a manifold M we obtain the usual Christoffel symbols γ^k_{ji}.

Obviously, the exterior differential of the connection form γ equals

$$d\gamma = d\omega_G - d\gamma_i \wedge dx^i$$

and the evaluation of $d\gamma + \frac{1}{2}[\gamma, \gamma]$ yields the usual coordinate expression for the curvature form. If we write $\gamma_i^p Y_p$ for the expression of $\gamma_i(x, e)$ in the basis Y_p of $\mathfrak{g}$, $\rho = \rho^p Y_p$, and c_{qr}^p for the corresponding structure constants of $\mathfrak{g}$, i.e. $[Y_q, Y_r] = c_{qr}^p Y_p$, we obtain the expression of ρ at the points (x, e) in $M \times G$,

$$\rho^p = \left(\frac{\partial \gamma_i^p}{\partial x^j} + c_{qr}^p \gamma_i^q \gamma_j^r \right) dx^i \wedge dx^j .$$

If the curvature of a principal connection γ vanishes, then locally γ is isomorphic to a trivial principal connection on a trivial principal bundle. There may appear global obstructions to the product structure, however. Next, let us note that the conditions on a general connection on a principal fiber bundle $\mathcal{P}$ to be a principal connection are of affine character. Thus, we may glue together the trivial connections with the help of a cocycle of local trivializations of $\mathcal{P}$ and a subordinated partition of unity, to obtain a globally defined principal connection. In particular, there are always smooth principal connections on any principal fiber bundle $\mathcal{P}$. Let us finally note that the difference of two principal connections γ and $\bar{\gamma}$ is a horizontal equivariant $\mathfrak{g}$–valued one–form. By Corollary 1.2.7 it may equivalently be viewed as a one–form on M with values in the bundle $\mathcal{P} \times_G \mathfrak{g}$.

1.3.4. Induced connections. The main advantage of principal connections is that a single principal connection on a principal bundle gives rise to a connection on any associated bundle, and the connections on various associated bundles are nicely compatible. There are at least two ways to see that this has to work: On the one hand, from 1.2.6 and 1.2.7 we know that the elements of a G–principal bundle $\mathcal{P} \to M$ may be viewed as frames, which give a coordinate–like description of elements of the associated bundles $\mathcal{P} \times_G S$. Given a principal connection on $\mathcal{P}$, we know which local frame fields are constant to first order. We can then define sections of $\mathcal{P} \times_G S$ to be constant to first order, if their coordinates in such frame fields have vanishing derivatives, and this suffices to define a connection on $\mathcal{P} \times_G S$. On the other hand, it is also easy to see that equivariancy of the horizontal distribution of a principal connection implies that it can be pushed down to a horizontal distribution on any associated bundle.

Recall from 1.2.7 that, given a principal G–bundle $p : \mathcal{P} \to M$ and a left action of G on a manifold S, we have the associated bundle $\pi : \mathcal{P} \times_G S \to M$ and a natural projection $q : \mathcal{P} \times S \to \mathcal{P} \times_G S$, which is a G–principal bundle and has the property that $\pi \circ q = p \circ \mathrm{pr}_1$. Now the tangent map of the multiplication makes TG into a Lie group, and the tangent map of the principal right action of G on $\mathcal{P}$ makes $T\mathcal{P}$ into a TG–principal bundle. Then Tq identifies $T(\mathcal{P} \times_G S)$ with $T\mathcal{P} \times_{TG} TS$. Moreover, embedding $\mathcal{P}$ into $T\mathcal{P}$ and G into TG as the zero sections, the restriction of Tq to $\mathcal{P} \times TS$ induces an identification of the vertical bundle $V(\mathcal{P} \times_G S)$ with $\mathcal{P} \times_G TS$.

Now assume that we have given a principal connection on $\mathcal{P}$ with horizontal distribution $\mathcal{H}$. Then one immediately verifies that for any $(u, s) \in \mathcal{P} \times S$ the restriction of $T_{(u,s)}q$ to $\mathcal{H}_u \times \{0_s\}$ is injective, so the image of this subspace defines a candidate for a horizontal subspace in $T_{[\![u,s]\!]}(\mathcal{P} \times_G S)$. Equivariancy of the horizontal distribution easily implies that this subspace is independent of the choice of the representative (u, s), so we get a well–defined horizontal distribution $\mathcal{H}^S$ on the associated bundle $\mathcal{P} \times_G S$. Viewing $T(\mathcal{P} \times_G S)$ as an associated bundle as above, we have

$$\mathcal{H}^S([\![u, s]\!]) = \{[\![\xi(u), 0_s]\!] \in T\mathcal{P} \times_{TG} TS, \xi(u) \in \mathcal{H}(u), u \in \mathcal{P}\}.$$

Thus, any fixed principal connection γ on $\mathcal{P}$ yields on each associated bundle $\mathcal{P}\times_G S$ a general connection γ^S, which is called the *induced connection.*

An induced connection is closely related to the principal connection it comes from. We will prove this only in the case of associated vector bundles, but with straightforward changes in the formulations and proofs, similar statements hold for general induced connections.

PROPOSITION 1.3.4. *Let $p : \mathcal{P} \to M$ be a G–principal bundle, λ a linear representation of G on a vector space V and $\mathcal{V} = \mathcal{P} \times_G V$ the corresponding induced vector bundle. Consider a principal connection on $\mathcal{P}$ with connection form $\gamma \in \Omega^1(\mathcal{P}, \mathfrak{g})$ and curvature $\rho \in \Omega^2(\mathcal{P}, \mathfrak{g})$, and let γ^V be the induced connection on $\mathcal{V}$.*

(1) γ^V is a linear connection on $\mathcal{V}$. Using $T\mathcal{V} = T\mathcal{P} \times_G TV$, its horizontal lift is given by $\xi \mapsto Tq \circ (\xi^{hor}, 0) = [\![\xi^{hor}, 0]\!]$, where $\xi^{hor} \in \mathfrak{X}(\mathcal{P})$ is the horizontal lift of ξ with respect to γ.

(2) Let $s \in \Gamma(\mathcal{V})$ be a section corresponding to the function $f \in C^\infty(\mathcal{P}, V)$, and let $\xi \in \mathfrak{X}(M)$ be a vector field. Then the covariant derivative $\nabla_\xi s \in \Gamma(\mathcal{V})$ corresponds to the function $\xi^{hor} \cdot f : \mathcal{P} \to V$.

(3) In a local trivialization $\mathbb{R}^n \times G \to \mathcal{P}$ with $\gamma = \omega_G - \gamma_i dx^i$ and the corresponding local trivialization $\mathbb{R}^n \times V \to \mathcal{V}$, the absolute derivative is given by

$$\nabla_\xi s = \xi^i \frac{\partial s}{\partial x^i} - \lambda'(\gamma_i \xi^i) \circ s,$$

where $\lambda' : \mathfrak{g} \to \mathfrak{gl}(V)$ is the Lie algebra representation corresponding to λ.

(4) The parallel transport on $\mathcal{V}$ along the curve c on M is given by $\tilde{c}^V_{[\![u,v]\!]}(t) = [\![\tilde{c}_u(t), v]\!]$, where $\tilde{c}_u$ is the parallel transport in $\mathcal{P}$ starting at u.

(5) Let $s \in \Gamma(\mathcal{V})$ be a section corresponding to the function $f : \mathcal{P} \to V$, and let $\xi, \eta \in \mathfrak{X}(M)$ be vector fields. Then the section $\nabla_\xi \nabla_\eta s - \nabla_\eta \nabla_\xi s - \nabla_{[\xi,\eta]} s$ corresponds to the function $\lambda'(\rho(\xi^{hor}, \eta^{hor})) \circ f$.

PROOF. The statement on the horizontal lift in (1) is obvious from the construction. Given a section s corresponding to the function $f : \mathcal{P} \to V$, let us choose a local smooth section σ of $\mathcal{P}$, so that we can locally write s as $q \circ (\sigma, f \circ \sigma)$. By definition of the absolute derivative associated to a general connection, $\nabla_\xi s(x)$ is given by

$$Tq \cdot (T_x\sigma \cdot \xi(x) - \xi^{\mathrm{hor}}(\sigma(x)), Tf \cdot T_x\sigma \cdot \xi(x)).$$

Now, we may fix a frame $u \in \mathcal{P}$ over x and choose σ in such a way that $\xi^{\mathrm{hor}}(u) = T_x\sigma \cdot \xi(x)$ for all ξ, i.e. a section σ which represents the defining jet of γ in x with target u. But then the above expression simplifies to $Tq \cdot (0_u, Tf \cdot \xi^{\mathrm{hor}}(\sigma(x)))$. Identifying the vertical tangent space in $[\![u, f(u)]\!]$ with the vector space V, this shows that the function $\mathcal{P} \to V$ corresponding to $\nabla_\xi s$ has value $\xi^{\mathrm{hor}}(\sigma(x)) \cdot f$ in the point $\sigma(x)$, which proves claim (2). Moreover, since the sum of two sections corresponds to the pointwise sum of the associated functions, this immediately implies that ∇ is actually a covariant derivative on $\mathcal{V}$, and thus defines a linear connection. By construction, this linear connection has associated horizontal distribution $\mathcal{H}^V$, so it coincides with γ^V, which completes the proof of (1).

(3) Let σ be the local smooth section of $\mathcal{P}$ corresponding to the chosen trivialization. Then the associated local trivialization of $\mathcal{V}$ is given by $(x, v) \mapsto [\![\sigma(x), v]\!]$. Consequently, viewing a section s as a V–valued function in this trivialization, one obtains exactly $f \circ \sigma$, where $f : \mathcal{P} \to V$ is the equivariant function corresponding to s. Now the horizontal lift of $\xi^i \frac{\partial}{\partial x^i}$ is given by $\xi^i \frac{\partial}{\partial x^i} + \zeta_{\gamma_i \xi^i}$, and the claim follows from (2) and equivariancy of f.

(4) The curve $[\![\tilde{c}_u(t), v]\!]$ obviously has horizontal derivatives, covers c, and starts at $[\![u, v]\!]$, so the claim follows from uniqueness of the parallel transport.

(5) By (2), the section $\nabla_\xi \nabla_\eta s - \nabla_\eta \nabla_\xi s - \nabla_{[\xi,\eta]} s$ corresponds to the function $([\xi^{\text{hor}}, \eta^{\text{hor}}] - [\xi, \eta]^{\text{hor}}) \cdot f$ and the field in the bracket by definition equals $\zeta_{-\rho(\xi^{\text{hor}}, \eta^{\text{hor}})}$. Equivariancy of f now implies the claim. □

1.3.5. Affine connections on manifolds. The previous construction of induced connections can be often inverted. In particular, any linear connection on a vector bundle $\mathcal{V}$ over M with standard fiber V is induced by a unique principal connection on the frame bundle of $\mathcal{V}$, i.e. the principal bundle of all bases of the fibers of $\mathcal{V}$ with structure group $GL(V)$. (Since a local frame for a vector bundle is made up from local sections, we can simply declare a local frame to be constant to first order in a point, if all the sections that constitute the frame have this property.)

In particular, linear connections on the tangent bundle TM of a smooth manifold M are in one–to–one correspondence with principal connections on the linear frame bundle $\mathcal{P}^1 M$. The latter connections are briefly referred to as *linear connections on M*. Once we fix such a connection γ, then there are the induced connections on all tensor bundles (and more generally any associated bundle to $\mathcal{P}^1 M$). The coordinate formula from part (3) of Proposition 1.3.4 is just the classical formula for the covariant derivative with respect to a linear connection on a manifold.

The next ingredient which is specific to the case of the tangent bundle is the existence of the *canonical form* $\theta \in \Omega^1(\mathcal{P}^1 M, \mathbb{R}^m)$, $m = \dim(M)$. The value of this form in a frame $u \in \mathcal{P}^1 M$ on a tangent vector $\xi \in T_u \mathcal{P}^1 M$ is defined to be the coordinates of the projection $Tp \cdot \xi \in T_{p(u)} M$ in the frame u. Otherwise put, viewing u as linear isomorphism $\mathbb{R}^m \to T_x M$, we have $\theta(u)(\xi) := u^{-1}(T_u p \cdot \xi)$. The names *solder form* and *soldering form* for θ are often used in the literature.

By construction, the canonical form θ is $GL(m, \mathbb{R})$–equivariant with respect to the standard action of $GL(m, \mathbb{R})$ on $\mathbb{R}^m$ and it is strictly horizontal, i.e.

$$(r^g)^* \theta = g^{-1} \circ \theta \text{ for } g \in GL(m, \mathbb{R}), \tag{1.4}$$

$$\theta(\xi) = 0 \text{ if and only if } \xi \text{ is vertical.} \tag{1.5}$$

The second property is obvious from the definition of θ, the first one simply follows from the fact that viewing frames as linear isomorphisms $\mathbb{R}^m \to T_x M$, the principal right action of $GL(m, \mathbb{R})$ is given by composition from the right.

Given a principal connection $\gamma \in \Omega^1(\mathcal{P}^1 M, \mathfrak{gl}(m, \mathbb{R}))$, we may consider the one–form $\omega = \theta + \gamma \in \Omega^1(\mathcal{P}^1 M, \mathbb{R}^m \oplus \mathfrak{gl}(m, \mathbb{R}))$, which is equivariant with respect to the direct sum of the standard action and the adjoint action. Moreover, since the kernel of θ in a point is the vertical subbundle and γ is injective on the vertical bundle, we see that for each $u \in \mathcal{P}^1 M$, the restriction of ω to $T_u \mathcal{P}^1 M$ is injective and hence a linear isomorphism.

In this picture, we can nicely describe the affine m–space A^m as the homogeneous model of such a structure. As a set, $A^m = \mathbb{R}^m$ and the group $A(m, \mathbb{R})$ of affine motions is the group of all maps from $\mathbb{R}^m$ to itself, which are of the form

$x \mapsto Ax + b$ for $A \in GL(m, \mathbb{R})$ and $b \in \mathbb{R}^m$. Viewing A^m as the affine hyperplane $x_1 = 1$ in $\mathbb{R}^{m+1}$ the affine motions are exactly the elements of $GL(m+1, \mathbb{R})$ which map this affine hyperplane to itself, i.e.

$$A(m, \mathbb{R}) = \left\{ \begin{pmatrix} 1 & 0 \\ b & A \end{pmatrix}, \ A \in GL(m, \mathbb{R}), \ b \in \mathbb{R}^m \right\} \subset GL(m+1, \mathbb{R}).$$

The group $A(m, \mathbb{R})$ obviously acts transitively on A^m and the isotropy subgroup of the first unit vector is just the subgroup of all elements with $b = 0$, so $A^m \cong A(m, \mathbb{R})/GL(m, \mathbb{R})$. On the Lie algebra level, we get

$$\mathfrak{a}(m, \mathbb{R}) = \left\{ \begin{pmatrix} 0 & 0 \\ X & B \end{pmatrix}, \ B \in \mathfrak{gl}(m, \mathbb{R}), \ X \in \mathbb{R}^m \right\},$$

so as a vector space this is isomorphic to $\mathbb{R}^m \oplus \mathfrak{gl}(m, \mathbb{R})$. Moreover, since the adjoint action of a matrix group is given by conjugation, one immediately sees that this splitting is invariant under the restriction of the adjoint action to $GL(m, \mathbb{R})$, and the action of this group on $\mathfrak{a}(m, \mathbb{R})$ is the direct sum of the standard representation and the adjoint action.

The natural projection $p : A(m, \mathbb{R}) \to A(m, \mathbb{R})/GL(m, \mathbb{R})$ is a principal bundle with structure group $GL(m, \mathbb{R})$; see 1.2.6. In 1.2.4 we have introduced the Maurer–Cartan from $\omega \in \Omega^1(A(m, \mathbb{R}), \mathfrak{a}(m, \mathbb{R}))$. Now we may split $\omega = \theta + \gamma$ according to the splitting $\mathfrak{a}(m, \mathbb{R}) = \mathbb{R}^m \oplus \mathfrak{gl}(m, \mathbb{R})$, and since this splitting is $GL(m, \mathbb{R})$–invariant, both θ and γ are $GL(m, \mathbb{R})$–equivariant forms. The form θ associates to each element $g \in A(m, \mathbb{R})$ a linear isomorphism $T_{p(g)}A^m \to \mathbb{R}^m$, and thus identifies $A(m, \mathbb{R})$ with the frame bundle $\mathcal{P}^1 A^m$. Hence, the component γ may be viewed as a linear connection on A^m and one immediately sees that this gives the canonical flat connection on A^m.

Returning to a general manifold M, we may now view the data defining a linear connection on M as an analog of the homogeneous space $A(m, \mathbb{R})/GL(m, \mathbb{R}) = A^m$. Indeed, the analog of the principal $GL(m, \mathbb{R})$–bundle $A(m, \mathbb{R}) \to A^m$ is the frame bundle $\mathcal{P}^1 M \to M$. On the other hand, the form $\omega = \theta + \gamma$ as considered above may be viewed as an element of $\Omega^1(\mathcal{P}^1 M, \mathfrak{a}(m, \mathbb{R}))$, which is an analog of the Maurer–Cartan form on $A(m, \mathbb{R})$. In fact, we have already noted above that ω is $GL(m, \mathbb{R})$–equivariant, it reproduces generators of fundamental vector fields, and defines a trivialization of the tangent bundle. More formally, we have

$$(r^g)^* \omega = \operatorname{Ad}(g^{-1}) \circ \omega \text{ for all } g \in GL(m, \mathbb{R}), \tag{1.6}$$

$$\omega(\zeta_X) = X \text{ for all } X \in \mathfrak{gl}(m, \mathbb{R}) \subset \mathfrak{a}(m, \mathbb{R}), \tag{1.7}$$

$$\omega(u) : T_u \mathcal{P}^1 M \to \mathfrak{a}(m, \mathbb{R}) \text{ is a linear isomorphism for all } u \in \mathcal{P}^1 M. \tag{1.8}$$

Here Ad denotes the adjoint action of $A(m, \mathbb{R})$. These are exactly the strongest analogs of the properties of the Maurer–Cartan form that make sense on a general principal $GL(m, \mathbb{R})$–bundle.

To complete this interpretation of linear connections on the tangent bundle, we need one more observation: Suppose that M is an m–dimensional manifold, $p : \mathcal{P} \to M$ is a principal $GL(m, \mathbb{R})$–bundle and $\omega \in \Omega^1(\mathcal{P}, \mathfrak{a}(m, \mathbb{R}))$ is a one–form which satisfies (1.6)–(1.8). Then (1.8) and (1.7) make sure that the kernel of the $\mathbb{R}^m$–component of ω in each point $u \in \mathcal{P}$ is exactly the vertical subspace $V_u \mathcal{P}$. Hence, this component descends to an injective linear map $T_u \mathcal{P}/V_u \mathcal{P} \cong T_{p(u)} M \to \mathbb{R}^m$ which must be an isomorphism by dimensional reasons. We can view this as defining a smooth fiber bundle map $\mathcal{P} \to \mathcal{P}^1M$, and condition (1.6) implies that this must be a homomorphism and thus an isomorphism of principal bundles. By construction, the pullback of the soldering form under this isomorphism is exactly the $\mathbb{R}^m$–component of ω. As before, we can then view the $\mathfrak{gl}(m,\mathbb{R})$–component of ω as a principal connection on $\mathcal{P}$.

This shows that a linear connection on an m–dimensional manifold M can be equivalently described as an *affine structure*, i.e. a principal $GL(m,\mathbb{R})$–bundle $p : \mathcal{P} \to M$, together with a one–form $\omega \in \Omega^1(\mathcal{P}, \mathfrak{a}(m,\mathbb{R}))$ which has the properties (1.6)–(1.8) from above. The basic features of linear connections on manifolds can be nicely phrased in this picture, which motivates many developments for more general Cartan geometries. We continue to write $\omega = \theta + \gamma$ for the splitting into the $\mathbb{R}^m$– and the $\mathfrak{gl}(m,\mathbb{R})$–component.

There is another feature of linear connections on manifolds that we have not discussed yet. For a linear connection γ on TM with covariant derivative ∇, one can define the *torsion* T by $T(\xi,\eta) = \nabla_\xi\eta - \nabla_\eta\xi - [\xi,\eta]$. This expression is visibly skew symmetric and easily seen to be bilinear over $C^\infty(M,\mathbb{R})$, so it defines a tensor field $T \in \Gamma(\Lambda^2T^*M \otimes TM)$. In local coordinates we can describe the linear connection ∇ by the Christoffel symbols γ^i_{jk} (see 1.3.1), and we compute

$$\begin{aligned} T(\xi,\eta) &= \left(\frac{\partial\eta^j}{\partial x^i}\xi^i - \gamma^j_{ik}\eta^i\xi^k - \frac{\partial\xi^j}{\partial x^i}\eta^i + \gamma^j_{ik}\xi^i\eta^k - [\xi,\eta]\right)\tfrac{\partial}{\partial x^j} \\ &= \left((\gamma^j_{ik} - \gamma^j_{ki})\xi^i\eta^k\right)\tfrac{\partial}{\partial x^j}. \end{aligned}$$

Hence, the torsion is given by the antisymmetrization of the Christoffel symbols, $T^j_{ik} = \gamma^j_{ik} - \gamma^j_{ki}$.

To compute the torsion in terms of ω, we observe that by definition of the canonical form θ, for a vector field $\xi \in \mathfrak{X}(M)$ the corresponding function $f : \mathcal{P}^1M \to \mathbb{R}^m$ can be written as $\theta(\tilde{\xi})$, where $\tilde{\xi}$ is any lift of ξ to a vector field on $\mathcal{P}^1M$. In particular, we may use the horizontal lift, and then the definition of the torsion together with part (2) of Proposition 1.3.4 implies that the function $\mathcal{P}^1M \to \mathbb{R}^m$ corresponding to $T(\xi,\eta)$ is given by

$$\xi^{\text{hor}} \cdot \theta(\eta^{\text{hor}}) - \eta^{\text{hor}} \cdot \theta(\xi^{\text{hor}}) - \theta([\xi^{\text{hor}},\eta^{\text{hor}}]) = d\theta(\xi^{\text{hor}},\eta^{\text{hor}}).$$

By definition, this is the absolute exterior derivative $d^\nabla\theta$. As in the case of the curvature dealt with in 1.3.3, one next verifies that by equivariancy of θ, for arbitrary vector fields ξ and η on $\mathcal{P}^1M$, we get

$$d\theta(\xi - \zeta_{\gamma(\xi)}, \eta - \zeta_{\gamma(\eta)}) = d\theta(\xi,\eta) + \gamma(\xi)(\theta(\eta)) - \gamma(\eta)(\theta(\xi)). \tag{1.9}$$

The Lie bracket on $\mathfrak{a}(m,\mathbb{R})$ is given by $[(X,B),(X',B')] = (BX' - B'X, BB' - B'B)$, so (1.9) turns out to be exactly the $\mathbb{R}^m$–component of $d\omega(\xi,\eta) + [\omega(\xi),\omega(\eta)]$. The $\mathfrak{gl}(m,\mathbb{R})$–component of the latter expression is $d\gamma(\xi,\eta) + [\gamma(\xi),\gamma(\eta)]$, and we know from 1.3.4 that this represents the curvature of our connection. Thus, we see that the extent to which ω fails to satisfy the Maurer–Cartan equation is exactly measured by the torsion and the curvature of the corresponding linear connection. In the special case of the homogeneous model A^m the Maurer–Cartan equation expresses the fact that the canonical connection on A^m is torsion free and flat.

Another feature of linear connections on TM that relates nicely to the affine point of view is the concept of geodesics which generalize the straight lines in A^m and the related concept of normal coordinates. Given a linear connection ∇ on M

the geodesics are characterized by the fact that $\nabla_{c'}c' = 0$. (From Lemma 1.3.1 it follows that $\nabla_\xi s$ depends only on the restriction of s to flow lines of ξ, which implies that the above expression makes sense although c' is only defined along c.) In the case of A^m, the geodesics are the (linearly parametrized) straight lines, and it is easy to see that they are exactly the projections of flows of left invariant vector fields on $A(m,\mathbb{R})$. Similarly, one may describe the geodesics for arbitrary affine structures. Indeed, for a point $x \in M$ and a tangent vector $\xi \in T_xM$ choose a point $u \in \mathcal{P}_x$. Then there is a unique element $X \in \mathbb{R}^m$ such that $\xi = T_up\cdot\omega_u^{-1}(X,0)$, and we consider the constant vector field $\tilde{X} := \omega^{-1}(X,0) \in \mathfrak{X}(\mathcal{P})$. Then the curve $\tilde{c}(t) := \mathrm{Fl}_t^{\tilde{X}}(u)$ is defined for $|t|$ sufficiently small, and $c := p \circ \tilde{c}$ satisfies $c(0) = x$, and $c'(0) = \xi$. Moreover, by construction $\tilde{c}$ has horizontal derivatives, so $\tilde{c}'(t)$ is the horizontal lift of $c'(t)$, and $\theta(\tilde{c}'(t)) = X$ for all t, which easily implies that c is a geodesic.

Fixing again $u \in \mathcal{P}$ with $p(u) = x$, we can now consider the mapping $\phi_u(X) := p(\mathrm{Fl}_1^{\tilde{X}}(u))$, which is defined on a neighborhood of $0 \in \mathbb{R}^m$. Moreover, the derivative of ϕ_u at 0 is a linear isomorphism $\mathbb{R}^m \to T_xM$, so possibly restricting to a smaller neighborhood of zero, ϕ_u is a diffeomorphism onto an open neighborhood of x in M. We can view this as defining a local coordinate system around x, which has the property that the straight lines through 0 exactly correspond to the geodesics through x, whence these are exactly the *normal coordinates* associated to the connection γ. Changing from u to $u \cdot g$, the normal coordinates change as $\phi_{u\cdot g} = \phi_u \circ \mathrm{Ad}(g)$, so they are unique up to linear transformations. Alternatively, the last fact may be interpreted in such a way that the normal coordinates define a unique diffeomorphism from an open neighborhood of 0 in T_xM to an open neighborhood of $x \in M$, the affine exponential map.

1.3.6. Connections on G–structures. First order G–structures are among the simplest examples of geometric structures. For any such structure, there is an obvious notion of compatible connections, and these can be interpreted very similarly to affine connections on manifolds. Let $H \subset GL(m,\mathbb{R})$ be a closed subgroup and M an m–dimensional manifold. A *G–structure* with structure group H (or briefly an H–structure) on M is a reduction $i : \mathcal{P} \to \mathcal{P}^1M$ of the frame bundle of M to the structure group H. This means that $\mathcal{P}$ is a principal H–bundle and i is a morphism of principal bundles over the inclusion $H \hookrightarrow GL(m,\mathbb{R})$ which covers the identity on M. Equivalently, one may characterize this as a principal H–bundle $p : \mathcal{P} \to M$ together with a form $\Theta \in \Omega^1(\mathcal{P},\mathbb{R}^m)$ which is strictly horizontal and H–equivariant, i.e. satisfies the analogs of (1.4) and (1.5) from 1.3.5 for elements $g \in H$. Given the reduction $i : \mathcal{P} \to \mathcal{P}^1M$, we obtain the form Θ as the pullback $i^*\theta$ of the canonical form, while in the other direction for $u \in \mathcal{P}$ the isomorphism $T_xM \to \mathbb{R}^m$ corresponding to the frame $i(u)$ is $\Theta(u) : T_u\mathcal{P}/V_u\mathcal{P} \cong T_xM \to \mathbb{R}^m$.

Slightly more generally, we also use the term G–structure with structure group H in the case where H is not a closed subgroup of $GL(m,\mathbb{R})$ but a covering of a virtual subgroup, i.e. if there is a given homomorphism $j : H \to GL(m,\mathbb{R})$ such that the derivative $j' : \mathfrak{h} \to \mathfrak{gl}(m,\mathbb{R})$ is injective. A well–known example of this situation is Riemannian spin structures, corresponding to the universal covering $\mathrm{Spin}(m) \to SO(m) \subset GL(m,\mathbb{R})$. As before, the structure may be either characterized as a reduction of structure group or via an $\mathbb{R}^m$–valued one–form Θ. Since this one form is essentially the same object as the canonical one–form, we will also denote it by

θ and call it the canonical form or the soldering form of the G–structure in the sequel.

G–structures corresponding to subgroups in $GL(m,\mathbb{R})$ allow yet another simple description. From Lemma 1.2.6 we know that reductions of the principal $GL(m,\mathbb{R})$ bundle $\mathcal{P}^1M$ to the structure group $H \subset GL(m,\mathbb{R})$ are in bijective correspondence with smooth sections σ of the bundle $\mathcal{P}^1M/H \to M$. Explicitly, the subbundle $\mathcal{Q} \subset \mathcal{P}^1M$ associated to a section $\sigma \in \Gamma(\mathcal{P}^1M/H)$ is just the preimage of $\sigma(M)$ under the natural projection $\mathcal{P}^1M \to \mathcal{P}^1M/H$. Conversely, the images of local smooth sections of $\mathcal{Q}$ under that projection are immediately seen to piece together a global smooth section of $\mathcal{P}^1M/H$. Finally, the bundle $\mathcal{P}^1M/H$ can be nicely viewed as the associated bundle $\mathcal{P}^1M \times_{GL(m,\mathbb{R})} (GL(m,\mathbb{R})/H)$: Consider the map from $\mathcal{P}^1M$ to this associated bundle that maps u to $[\![u, eH]\!]$. This is clearly smooth and it is surjective since $[\![u, gH]\!] = [\![u \cdot g^{-1}, eH]\!]$. On the other hand, $[\![u, eH]\!] = [\![u', eH]\!]$ if and only if $u' = u \cdot h$ for some $h \in H$, so our map factors to a bijection from $\mathcal{P}^1M/H$ to the associated bundle and one easily verifies that this actually is a diffeomorphism.

Example 1.3.6. There are many well–known G–structures, let us name just a few:

(1) A Riemannian structure is the reduction of $\mathcal{P}^1M$ to orthonormal frames, i.e. to $H = O(m,\mathbb{R}) \subset GL(m,\mathbb{R})$. The group $GL(m,\mathbb{R})$ acts transitively on the space of all inner products on $\mathbb{R}^m$ and the isotropy group of the standard inner product is $O(m)$, so $GL(m,\mathbb{R})/O(m)$ is the space of all inner products on $\mathbb{R}^m$. Hence, the associated bundle $\mathcal{P}^1M/H$ is the bundle of all inner products on the tangent spaces of M, i.e. smooth sections of this bundle are exactly Riemannian metrics on M.

Riemannian spin structures may be similarly interpreted as G–structures corresponding to the homomorphism $j : Spin(m,\mathbb{R}) \to GL(m,\mathbb{R})$. This is an important example in which the structure group is not a subgroup of $GL(m,\mathbb{R})$ but a covering of such a subgroup.

(2) An almost symplectic structure on a smooth manifold M of dimension $2n$ is a reduction of $\mathcal{P}^1M$ to the symplectic group $H = Sp(2n,\mathbb{R})$. As in (1), the homogeneous space $GL(2n,\mathbb{R})/Sp(2n,\mathbb{R})$ is the space of all non–degenerate skew symmetric bilinear maps on $\mathbb{R}^{2n}$. The bundle $\mathcal{P}^1M/H$ is the bundle of non–degenerate 2–forms, and its closed sections are called symplectic forms.

(3) An absolute parallelism is a reduction to the trivial subgroup $\{\mathrm{id}\} \subset GL(m,\mathbb{R})$. In this case, the G–structures are the global trivializations of TM.

A *connection on a G-structure* $\mathcal{P}$ with structure group H is a principal connection on the H–principal bundle $\mathcal{P}$. Notice that any such connection extends to a principal connection on $\mathcal{P}^1M$ by equivariancy. Given $i : \mathcal{P} \to \mathcal{P}^1M$ and $j' : \mathfrak{h} \hookrightarrow \mathfrak{gl}(m,\mathbb{R})$ and a principal connection $\gamma \in \Omega^1(\mathcal{P},\mathfrak{h})$, consider a point $i(u) \in \mathcal{P}^1M$ for $u \in \mathcal{P}$. Since γ reproduces the generators of fundamental vector fields, we get a well–defined linear map $\tilde{\gamma} : T_{i(u)}\mathcal{P}^1M \to \mathfrak{gl}(m,\mathbb{R})$ which coincides with $j' \circ \gamma(u)$ on $T_ui(T_u\mathcal{P})$ and reproduces the generators of fundamental vector fields. Now we can extend $\tilde{\gamma}$ to an element of $\Omega^1(\mathcal{P}^1M, \mathfrak{gl}(m,\mathbb{R}))$ by requiring equivariancy, i.e $(r^g)^*\tilde{\gamma} = \mathrm{Ad}(g^{-1}) \circ \tilde{\gamma}$ for all $g \in GL(m,\mathbb{R})$. This is well–defined by equivariancy of γ.

Conversely, let us assume that γ is a principal connection on $\mathcal{P}^1M$ and we have given a reduction $i : \mathcal{P} \to \mathcal{P}^1M$ corresponding to $j : H \to GL(m,\mathbb{R})$. Then visibly γ is an equivariant extension of a principal connection as constructed above if and only if $\gamma(i(u))(T_u i(T_u\mathcal{P})) \subset j'(\mathfrak{h})$. In the case where H is a closed subgroup of $GL(m,\mathbb{R})$, there is a simple necessary condition. Namely, consider the section $\sigma \in \Gamma(\mathcal{P}^1M/H)$ describing the reduction and the corresponding equivariant function $f : \mathcal{P}^1M \to GL(m,\mathbb{R})/H$. From above we see that by construction this function is constantly equal to eH along the subbundle $\mathcal{P} \subset \mathcal{P}^1M$. For a tangent vector ξ on M and a point $u \in \mathcal{P}$, choose a lift $\tilde{\xi} \in T_u\mathcal{P}$. Then the condition on γ above ensures that $\xi^{\text{hor}}(u) \in T_u\mathcal{P}$, whence we conclude that $\xi^{\text{hor}} \cdot f$ vanishes identically on $\mathcal{P}$ and thus on $\mathcal{P}^1M$ by equivariancy. By the analog of part (2) of Proposition 1.3.4, this means that $\nabla\sigma$ has to vanish identically, i.e. the section defining the reduction must be covariantly constant. In many cases, this condition is also sufficient, e.g. for Riemannian and almost symplectic structures.

We should also remark at this point that there are also general results concerning the question whether a connection on $\mathcal{P}^1M$ comes from some reduction to the structure group H. These results are based on the notion of the holonomy of a connection and are known as the Ambrose–Singer Theorem. A version of this theorem for connections on general fiber bundles can be found in [**KMS**, 9.11].

Connections on G–structures can be treated in a similar style as the affine picture for linear connections on the tangent bundle. Indeed, given $j : H \to GL(m,\mathbb{R})$ such that the infinitesimal homomorphism j' is injective, consider the *affine extension* $B := \mathbb{R}^m \rtimes H$ of H. This means that $B = \mathbb{R}^m \times H$ as a set and the multiplication is given by $(X,g)(Y,h) = (X+j(g)(Y), gh)$. If H is a closed subgroup of $GL(m,\mathbb{R})$, then B is a closed subgroup of $A(m,\mathbb{R})$, in general there is an obvious homomorphism $\tilde{j} : B \to A(m,\mathbb{R})$ such that $\tilde{j}'$ is injective. Now we can equivalently view connections on G–structures with structure group H as principal H–bundles $\mathcal{P} \to M$ endowed with a one–form $\omega \in \Omega^1(\mathcal{P},\mathfrak{b})$ which satisfy the analogs of (1.6)–(1.8) from 1.3.5 with respect to elements $g \in H$. The interpretations of torsion and curvature as well as geodesics and normal coordinates works exactly in the same way as in 1.3.5.

As in the case of the affine space A^m discussed in 1.3.5, one may look at the homogeneous space $B/H \cong \mathbb{R}^m$, and view the Maurer–Cartan form on B as a connection on the G–structure $B \to \mathbb{R}^m$ with structure group H. Connections on G–structures with structure group H can thus be thought of as "curved analogs" of this homogeneous space. In particular, in the case $H = O(m) \subset GL(m,\mathbb{R})$ the affine extension B is exactly the *Euclidean group* $\text{Euc}(n)$ as discussed in 1.1.2, and from example (1) above, we see that this picture leads to viewing m–dimensional Riemannian manifolds as curved analogs of the Euclidean space E^m. This is one of the motivating examples for the concept of Cartan connections.

1.3.7. Partial connections. The starting point for our development of connections was a linear connection on a vector bundle, viewed as an analog of directional derivatives. In this picture, partial connections are operators with similar properties, except that one may only differentiate in directions lying in a fixed distribution $\mathcal{D} \subset TM$. Given a vector bundle $p : \mathcal{V} \to M$ one defines a *partial linear connection* on $\mathcal{V}$ corresponding to the distribution $\mathcal{D}$ as a bilinear operator $\Gamma(\mathcal{D}) \times \Gamma(\mathcal{V}) \to \Gamma(\mathcal{V})$ written as $(\xi, s) \mapsto \nabla_\xi s$ which is linear over $C^\infty(M,\mathbb{R})$ in the first variable and satisfies the Leibniz rule in the second variable.

Most of the developments in this section can be also carried out for partial connections. There are two ways to do this. Either one repeats the development outlined so far, taking into account the obvious changes, or one views a partial connection as an equivalence class of true connections. Here two connections are equivalent if they coincide on $\Gamma(\mathcal{D}) \times \Gamma(\mathcal{V}) \subset \mathfrak{X}(M) \times \Gamma(\mathcal{V})$. In both ways, the development is basically straightforward, so we only state a few main points.

Partial linear connections can be equivalently described by a horizontal lift, which is now only defined on the subbundle $\mathcal{D} \subset TM$. This leads to the notion of a general partial connection on a fibered manifold $p : Y \to M$, which is easiest described as a direct sum decomposition $(Tp)^{-1}(\mathcal{D}) \cong VY \oplus \mathcal{D}^{\text{hor}}$, or equivalently by a vertical projection defined on $(Tp)^{-1}(\mathcal{D})$ only. In the case of a principal bundle $p : \mathcal{P} \to M$ with structure group G, the subbundle $(Tp)^{-1}(\mathcal{D}) \subset T\mathcal{P}$ is G–invariant, so the definition of a partial principal connection is obvious. Such a partial principal connection may be described by a connection form γ which is now a smooth section of the bundle $L((Tp)^{-1}(\mathcal{D}), \mathfrak{g})$. The concept of induced partial connections poses no problems and these behave similarly to true induced connections.

The notion of curvature for partial connection is slightly subtle. The problem is that the bracket of two sections of the distribution $\mathcal{D}$ is not a section of $\mathcal{D}$ in general. Hence, the covariant derivative in the direction of a bracket and the vertical projection of a bracket are not defined in general. The Lie bracket of vector fields induces a bundle map $\Lambda^2\mathcal{D} \to TM/\mathcal{D}$. If we assume that this bundle map has constant rank, then its kernel is a subbundle $\Lambda^2_0\mathcal{D} \subset \Lambda^2\mathcal{D}$. In this case, the curvature of a partial linear connection can be defined by the usual formula (equation (1.3) from 1.3.1) as a section of the bundle $L(\Lambda^2_0\mathcal{D}, L(\mathcal{V}, \mathcal{V}))$. The curvature of partial general connections and partial principal connections can be defined similarly.

A partial linear connection on TM is called a *partial affine connection on M*. In this case, there is at least a well–defined concept of the torsion of the partial connection which is defined by the standard formula $T(\xi, \eta) = \nabla_\xi \eta - \nabla_\eta \xi - [\xi, \eta]$ for all $\xi, \eta \in \Gamma(\mathcal{D})$. This means that the torsion is a bundle map $\Lambda^2\mathcal{D} \to TM$. One can also use this definition of the torsion for partial connections on $\mathcal{D}$.

It is worth mentioning at this point, that a part of the torsion of any (partial) connection which preserves a distribution $\mathcal{D}$ depends only on $\mathcal{D}$ and not on the connection. The Lie bracket of vector fields induces a tensorial map $\Lambda^2\mathcal{D} \to TM/\mathcal{D}$. For any connection preserving $\mathcal{D}$, the composition of the quotient projection $TM \to TM/\mathcal{D}$ with the restriction of the torsion to $\Lambda^2\mathcal{D}$ is evidently given by this map. Notice, that this tensorial map is the obstruction against integrability of $\mathcal{D}$.

1.3.8. Remark. Affine connections on manifolds as treated in 1.3.5 and more generally connections on G–structures as discussed in 1.3.6 are the simplest examples of Cartan connections. In all of these cases, there is a homogeneous space in the background, which is simply $\mathbb{R}^m$ as a homogeneous space of the affine extension $B = \mathbb{R}^m \rtimes H$ of H in the case of G–structures with structure group H and, in particular, affine space A^m in the case of affine connections. The developments in 1.3.5 and 1.3.6 give a first glance on the passage from the homogeneous model to arbitrary Cartan connections, which is fundamental in the theory of Cartan geometries. Any reasonable concept for a Cartan geometry must first of all work for the homogeneous model and many constructions for the homogeneous model carry over to general Cartan connections. In view of this fact, we do not directly move on to general Cartan connections, which would be the next natural item in our chain of various notions of connections, but first study the invariant geometry of homogeneous spaces.

1.4. Geometry of homogeneous spaces

In F. Klein's Erlangen program, homogeneous spaces are the basic setting for classical geometry. Using Cartan connections, one may associate to any homogeneous space a differential geometric structure, called the corresponding Cartan geometry. The given homogeneous space is then called the *homogeneous model* of the Cartan geometry, and it plays a central role in the theory. On the one hand, it provides a distinguished basic object. On the other hand, surprisingly many general geometric properties can be read off directly from the homogeneous model. Otherwise put, many questions about Cartan geometries can be answered by looking at the homogeneous model only. Finally, there is always the subclass of locally flat geometries, in which the relation to the homogeneous model is even closer.

The point of view we take in this chapter is that a homogeneous space G/H carries a geometric structure, whose automorphisms are exactly the actions of elements of G. There are three main directions in our study of these geometric structures in this section. First we study homogeneous bundles and invariant sections of such bundles, which may be viewed as simpler geometric structures underlying the given one. Second, we take some basic steps towards the study of differential operators which are intrinsic to such a structure, which in this setting are G–invariant differential operators. In particular, we discuss the question of existence of invariant connections of various types. Finally, we briefly discuss distinguished curves in homogeneous spaces, which provide generalizations of geodesics of affine connections.

1.4.1. Klein geometries. Let G be a Lie group and let $H \subset G$ be a closed subgroup. Then H is a submanifold and thus a Lie subgroup and the set G/H of all cosets gH is canonically a smooth manifold endowed with a transitive left action of G. Up to the choice of a base point, any transitive action is of this form; see 1.2.5. Here we want to view G/H as carrying a geometric structure whose automorphisms are exactly the left actions ℓ_g for $g \in G$. In this context, the pair (G, H) is referred to as a *Klein geometry*. A careful geometric study of Klein geometries is available in [**Sh97**, Chapter 4].

Given a Klein geometry (G, H) we may first ask whether all of G is "visible" on G/H, i.e. whether the action ℓ of G on G/H is effective. In this case, we call the Klein geometry *effective*. The *kernel* $K \subset G$ of the Klein geometry is defined as the set of all elements $g \in G$ such that $\ell_g = \mathrm{id}_{G/H}$. Since this is the intersection of all isotropy subgroups, it is a closed subgroup of G and since H is the isotropy subgroup of eH, we see that $K \subset H$. Moreover, for $k \in K$ and $g \in G$, we have $\ell_{gkg^{-1}} = \ell_g \circ \ell_k \circ \ell_{g^{-1}}$ and thus $gkg^{-1} \in K$, so the subgroup K is normal in G. On the other hand, suppose that H' is a virtual Lie subgroup of H which is normal in G. Then for $h \in H'$ and $g \in G$ we get $g^{-1}hgH = eH$ and thus $hgH = gH$, so $\ell_h = \mathrm{id}_{G/H}$ and $H' \subset K$. Hence, the kernel K is the maximal normal subgroup of G which is contained in H, and a Klein geometry is effective if and only if there is no non–trivial normal subgroup of G which is contained in H.

Given an arbitrary Klein geometry (G, H) with kernel K, one may pass to the effective quotient $(G/K, H/K)$, i.e. view G/H as a homogeneous space of G/K rather than G. Since the kernel K is a normal subgroup of G its Lie algebra $\mathfrak{k}$ is an ideal in $\mathfrak{g}$ which is contained in $\mathfrak{h}$. If $\mathfrak{h}' \leq \mathfrak{g}$ is an ideal contained in $\mathfrak{h}$, then the corresponding virtual Lie subgroup is normal in G and contained in H, which implies that $\mathfrak{h}' \subset \mathfrak{k}$, so $\mathfrak{k}$ is the maximal ideal in $\mathfrak{g}$ that is contained in $\mathfrak{h}$.

In several situations (for example to treat Spin structures), requiring effectivity of a Klein geometry would be too much. However, one usually wants the geometry to be *infinitesimally effective*, which means that the kernel K has to be discrete. This may be equivalently characterized as the fact that there is no non–zero ideal in the Lie algebra $\mathfrak{g}$ which is contained in the subalgebra $\mathfrak{h}$.

Motivated by the examples from 1.3.5 and 1.3.6, we next define two important subclasses of Klein geometries. Note that the adjoint action Ad of G on its Lie algebra $\mathfrak{g}$ may be restricted to a representation of the subgroup H on $\mathfrak{g}$. Of course, the Lie subalgebra $\mathfrak{h}$ is an H–invariant subspace of $\mathfrak{g}$.

(1) The Klein geometry (G, H) is called *reductive* if there is an H–invariant subspace $\mathfrak{n} \subset \mathfrak{g}$ which is complementary to $\mathfrak{h}$, i.e. such that $\mathfrak{g} = \mathfrak{n} \oplus \mathfrak{h}$ as an H–module.

(2) The Klein geometry (G, H) is called *split* if there is a Lie subalgebra $\mathfrak{g}_- \subset \mathfrak{g}$, which is complementary to $\mathfrak{h}$ as a vector space, i.e. such that $\mathfrak{g} = \mathfrak{g}_- \oplus \mathfrak{h}$ as a vector space.

When dealing with reductive or split Klein geometries, we will usually assume that the complementary space $\mathfrak{n}$, respectively, $\mathfrak{g}_-$ is fixed as part of the geometry.

EXAMPLE 1.4.1. Let $H \subset GL(m, \mathbb{R})$ be a closed subgroup and consider the affine extension $B := \mathbb{R}^m \rtimes H \subset A(m, \mathbb{R})$ as defined in 1.3.6. Representing $A(m, \mathbb{R})$ as a matrix group as in 1.3.5 we see that we may view B as the closed subgroup

$$\left\{ \begin{pmatrix} 1 & 0 \\ X & A \end{pmatrix} : X \in \mathbb{R}^m, A \in H \right\} \subset GL(m+1, \mathbb{R}),$$

with the subgroup H corresponding to elements with $X = 0$. Correspondingly, the Lie algebra $\mathfrak{b}$ is the subalgebra of $\mathfrak{gl}(m+1, \mathbb{R})$ of all elements of the form $\left(\begin{smallmatrix} 0 & 0 \\ Y & C \end{smallmatrix} \right)$ with $Y \in \mathbb{R}^m$ and $C \in \mathfrak{h}$. In particular, there is an obvious decomposition $\mathfrak{b} = \mathbb{R}^m \oplus \mathfrak{h}$ as a vector space, and using that the adjoint action of a matrix group is given by conjugation, one immediately verifies that this decomposition is H–invariant. Thus, any Klein geometry of the form (B, H) is reductive. On the other hand, one immediately verifies that the subspace $\mathbb{R}^m \subset \mathfrak{b}$ is not only a Lie subalgebra but even an abelian ideal in $\mathfrak{b}$, so these geometries are naturally split as well.

1.4.2. Homogeneous bundles. Let G be a Lie group, $H \subset G$ a closed subgroup, $M := G/H$ the corresponding homogeneous space and $p : G \to G/H$ the canonical projection, which is an H–principal bundle; see 1.2.5. By $\ell : G \times M \to M$ we denote the canonical left action $\ell(g, g'H) := gg'H$. The first step towards G–invariant geometric objects on M is to get an appropriate class of fiber bundles over the homogeneous space.

DEFINITION 1.4.2. (1) A *homogeneous bundle* over $M = G/H$ is a locally trivial fiber bundle $\pi : E \to M$ together with a left G–action $\tilde{\ell} : G \times E \to E$, which lifts the action on M, i.e. which satisfies $\pi(\tilde{\ell}(g, y)) = \ell(g, \pi(y))$ for all $g \in G$ and $y \in E$.

(2) A *homogeneous vector bundle* over M is a homogeneous bundle $\pi : E \to M$, which is a vector bundle and such that for each element $g \in G$ the bundle map $\tilde{\ell}_g : E \to E$ is a vector bundle homomorphism, i.e. linear in each fiber.

(3) A *homogeneous principal bundle* is a homogeneous bundle $\pi : E \to M$ which is a principal bundle and such that for each $g \in G$ the bundle map $\tilde{\ell}_g$ is a homomorphism of principal bundles, i.e. equivariant for the principal right action of the structure group.

(4) A *morphism* of homogeneous bundles (respectively homogeneous vector bundles or principal bundles) is a G–equivariant bundle map (respectively G–equivariant homomorphism of vector bundles or principal bundles) which covers the identity on M.

If there is no risk of confusion, we will always denote all actions simply by dots, so the definition of a homogeneous bundle simply reads as $\pi(g\cdot y) = g\cdot\pi(y)$. From the definitions it is also clear that homogeneous fiber bundles (respectively homogeneous vector bundles) form a category $\mathrm{Fib}_G(G/H)$ (respectively $\mathrm{Vect}_G(G/H)$).

EXAMPLE 1.4.2. There are two obvious sources of homogeneous bundles. We shall soon see that the first one is a special case of the second one and that there are no other homogeneous bundles, up to isomorphisms.

(1) Let n be the dimension of the homogeneous space $M = G/H$ and suppose that F is a natural bundle on the category $\mathcal{M}_n$ of n–dimensional manifolds and local diffeomorphisms; see 1.2.8 or [**KMS**, section 14]. Now consider the bundle $p_M : FM \to M$. For any $g \in G$, the left action $\ell_g : M \to M$ of g on M is a diffeomorphism (with inverse $\ell_{g^{-1}}$), so we have the induced bundle map $F(\ell_g) : FM \to FM$ which covers $\ell_g : M \to M$. Since F is a functor, $F(\ell_e) = \mathrm{id}_{FM}$ and $F(\ell_{gg'}) = F(\ell_g \circ \ell_{g'}) = F(\ell_g)\circ F(\ell_{g'})$. Hence, the map $\tilde{\ell} : G\times FM \to FM$, $\tilde{\ell}(g,u) := F(\ell_g)(u)$ defines a left action of G on FM, which is smooth by regularity of F. Thus, any natural bundle over a homogeneous space is a homogeneous bundle.

In particular, we can apply this line of argument to natural vector bundles, for which each of the bundle maps $F(f)$ is a vector bundle homomorphism, to obtain homogeneous vector bundles. In particular, all tensor bundles over G/H are homogeneous vector bundles.

(2) The second basic source of homogeneous bundles is the canonical H–principal bundle $p : G \to G/H$. By definition of the action of G on G/H, this is a homogeneous principal bundle under the action $\tilde{\ell} : G\times G \to G$ which is just given by the multiplication map. Now assume that S is any smooth manifold with a smooth left action $H \times S \to S$. Then we can form the associated (or induced) bundle $E := G\times_H S \to M$. By definition (see 1.2.7) this is the space of orbits in $G\times S$ of the right action $(g,s)\cdot h = (gh, h^{-1}\cdot s)$ of H. We continue to denote the orbit of (g,s) by $[\![g,s]\!]$. Since left and right translations on a Lie group always commute, the map $G\times G\times S \to G\times_H S$ defined by $(g,g',s)\mapsto [\![gg',s]\!]$ descends to a map $G\times(G\times_H S)\to G\times_H S$ which is smooth since $G\times G\times S \to G\times(G\times_H S)$ is a surjective submersion (even an H–principal bundle). From the construction it is clear that this defines a smooth left action of G on E, and since the projection $E\to M$ is simply given by $[\![g,s]\!]\mapsto gH$, this action extends the canonical action of G on M. Thus, any associated bundle to the principal bundle $G\to G/H$ is canonically a homogeneous bundle.

If we start with a representation of H on a vector space V, then the induced bundle $G\times_H V$ is a vector bundle, and the linear structure is determined by

$$[\![g,v]\!] + t[\![g,w]\!] = [\![g, v+tw]\!].$$

Thus, the above construction gives rise to an action of G on $G\times_H V$ by vector bundle homomorphisms, and $G\times_H V \to G/H$ is a homogeneous vector bundle. This construction evidently generalizes to arbitrary homogeneous principal bundles. Any associated bundle (respectively vector bundle) to a homogeneous principal bundle is canonically a homogeneous bundle (respectively homogeneous vector bundle).

1.4.3. Classification of homogeneous bundles. Suppose that $p_E : E \to M$ is a homogeneous fiber bundle over $M = G/H$. Consider the base point $o = eH \in G/H$ and the fiber $S := E_o$ of E over o. For $g \in G$ and $s \in S$ we get $p_E(g \cdot s) = g \cdot o = gH$ by definition of a homogeneous bundle. In particular, $p_E(h \cdot s) = o$ for all $h \in H$, so the left G–action on E restricts to a left action of H on S. Now consider the associated bundle $G \times_H S \to M$, and consider the map $G \times S \to E$ defined by $(g, s) \mapsto g \cdot s$, where we use the action of G on E, and view s as an element of $E_o \subset E$. Since the action of H on S is just the restriction of the action of G on E, this mapping factorizes to a map $\Phi : G \times_H S \to E$ which is smooth since the projection $G \times S \to G \times_H S$ is a surjective submersion. Moreover, by definition $\Phi([\![g, s]\!]) = g \cdot s$, which, on one hand, shows that Φ is a bundle map covering the identity. On the other hand, it implies that Φ is G–equivariant, so $\Phi : G \times_H S \to E$ is a morphism of homogeneous bundles.

If $e \in E$ is any element, then choose an element $g \in G$ such that $g \cdot p_E(e) = o$. Then $p_E(g \cdot e) = o$, so $g \cdot e \in S = E_o$, and we may consider $[\![g^{-1}, g \cdot e]\!] \in G \times_H S$. This is independent of the choice of g, since for another choice $g' \in G$ we must have $(g'g^{-1}) \cdot o = o$, so $g'g^{-1} \in H$, and thus

$$[\![(g')^{-1}, g' \cdot e]\!] = [\![(g')^{-1}g'g^{-1}, g(g')^{-1}g' \cdot e]\!] = [\![g^{-1}, g \cdot e]\!].$$

Thus, we get a well–defined map $\Psi : E \to G \times_H S$. Choosing a local smooth section σ of $G \to G/H$ we can locally write $\Psi(e) = [\![\sigma(p_E(e))^{-1}, \sigma(p_E(e)) \cdot e]\!]$, which shows that Ψ is smooth. One immediately verifies that Ψ and Φ are inverse isomorphisms of homogeneous bundles.

If $f : E \to E'$ is a morphism of homogeneous bundles, then $f(E_o) \subset E'_o$ and the restriction $f|_{E_o} : E_o \to E'_o$ is H–equivariant. Equivariancy of f implies that for $g \in G$ and $s \in S = E_o$, we get $f(g \cdot s) = g \cdot f(s)$ which means that $f(\Phi([\![g, s]\!])) = \Phi'([\![g, f|_{E_o}(s)]\!])$, so we see that identifying E and E' with $G \times_H E_o$, respectively, $G \times_H E'_o$, the map f is induced by $\mathrm{id} \times f|_{E_o}$. Conversely, any H–equivariant map $S \to S'$ clearly induces a morphism $G \times_H S \to G \times_H S'$ of homogeneous bundles.

Finally, note that for a homogeneous vector bundle, the above construction clearly produces an isomorphism $G \times_H E_o \to E$ of homogeneous vector bundles, and homomorphisms of homogeneous vector bundles correspond to linear H–equivariant maps between their standard fibers. Thus, we have proved

PROPOSITION 1.4.3. *The mapping $E \mapsto E_o$ and $f \mapsto f|_{E_o}$ induces equivalences between the category $\mathrm{Fib}_G(G/H)$ and the category of manifolds endowed with left H-actions and H-equivariant smooth maps, as well as between the category $\mathrm{Vect}_G(G/H)$ and the category of finite–dimensional representations of H.*

REMARK 1.4.3. The proposition also implies that these equivalences of categories are compatible with various constructions. For example, the fibered product of homogeneous fiber bundles corresponds to the product of left H–spaces, the Whitney sum of homogeneous vector bundles corresponds to the direct sum of representations and the tensor product of homogeneous vector bundles corresponds to the tensor product of representations. This also works the other way around. If, for example, we start with two homogeneous vector bundles corresponding to indecomposable representations of H, then the decomposition of the tensor product of these two representations into a direct sum of indecomposable representations induces a decomposition of the tensor product of the two homogeneous vector bundles into a Whitney sum.

EXAMPLE 1.4.3. Let us determine the H–representations corresponding to tensor bundles. We start by identifying the tangent bundle, so we have to determine $T_o(G/H)$ as an H–module. The projection map $p : G \to G/H$ is a submersion and thus for each $g \in G$ the tangent map $T_gp : T_gG \to T_{p(g)}(G/H)$ is surjective. In particular, we get a surjection $T_ep : \mathfrak{g} = T_eG \to T_o(G/H)$. Moreover, the representation of H on $T_o(G/H)$ is just obtained as the restriction of the action of G on the homogeneous vector bundle $T(G/H)$, so the action of $h \in H$ on $T_o(G/H)$ is just $T_o\ell_h$. By definition, the action on G/H is induced by left translations in G, so $\ell_h \circ p = p \circ \lambda_h$ and thus $T_o\ell_h \circ T_ep = T_hp \circ T_e\lambda_h$. On the other hand, p commutes with right multiplications by elements from h, and differentiating this, we see that $T_hp = T_ep \circ T_h\rho^{h^{-1}}$. Since $\rho^{h^{-1}} \circ \lambda_h$ is the conjugation by h, whose derivative is the adjoint action, we get $T_o\ell_h \circ T_ep = T_ep \circ \mathrm{Ad}(h)$. Hence, $T_o(G/H)$ is a quotient of $\mathfrak{g}$, with the H–module structure defined by the restriction of the adjoint representation Ad to the subgroup $H \subset G$. On the other hand, the kernel of T_ep is simply the subalgebra $\mathfrak{h}$, which is also an H–submodule by naturality of the adjoint action.

Thus, the homogeneous vector bundle $T(G/H)$ corresponds to the H–representation on $\mathfrak{g}/\mathfrak{h}$ coming from the restriction to H of the adjoint representation of G. By the proof of Proposition 1.4.3, the isomorphism $G \times_H (\mathfrak{g}/\mathfrak{h}) \to T(G/H)$ maps $[\![g, X+\mathfrak{h}]\!]$ to $T_gp \cdot T_e\lambda_g \cdot X$. The opposite isomorphism can be conveniently described using the Maurer–Cartan form $\omega \in \Omega^1(G, \mathfrak{g})$ of the group G. For a tangent vector $\xi_x \in T_x(G/H)$, we have to choose an element $g \in G$ such that $x = g^{-1}H$ and then view $T\ell_g \cdot \xi_x \in T_o(G/H)$ as an element of $\mathfrak{g}/\mathfrak{h}$. Having chosen g, we know that $T_{g^{-1}}p : T_{g^{-1}}G \to T_x(G/H)$ is surjective, so we can choose a tangent vector $\tilde{\xi} \in T_gG$ such that $\xi_x = Tp \cdot \tilde{\xi}$. But then $T\ell_g \cdot \xi_x = Tp \cdot T\lambda_g \cdot \tilde{\xi}$, and identifying T_eG with $\mathfrak{g}$ and $T_o(G/H)$ with $\mathfrak{g}/\mathfrak{h}$, the projection Tp becomes just the canonical projection to the quotient. Since $T\lambda_g \cdot \tilde{\xi}$ by definition equals $\omega(\tilde{\xi})$, we see that the isomorphism $T(G/H) \to G \times_H (\mathfrak{g}/\mathfrak{h})$ maps ξ_x to $[\![g^{-1}, \omega(\tilde{\xi}) + \mathfrak{h}]\!]$.

By the naturality of the correspondence between homogeneous vector bundles and H–representations, this implies that the cotangent bundle $T^*(G/H)$ corresponds to the dual representation $(\mathfrak{g}/\mathfrak{h})^*$. Note that $(\mathfrak{g}/\mathfrak{h})^*$ is just the annihilator of $\mathfrak{h}$ in $\mathfrak{g}^*$, so this is a subrepresentation of the restriction of the coadjoint representation to H. Again, by naturality, the tensor bundle $\otimes^k T(G/H) \otimes \otimes^\ell T^*(G/H)$ corresponds to the representation $\otimes^k(\mathfrak{g}/\mathfrak{h}) \otimes \otimes^\ell(\mathfrak{g}/\mathfrak{h})^*$.

There is an interesting way to rephrase the description of the tangent bundle as an associated bundle, namely as a first order G–structure underlying a Klein geometry. The representation of H on $\mathfrak{g}/\mathfrak{h}$ induced by the restriction of the adjoint action of G may be viewed as a homomorphism $H \to GL(\mathfrak{g}/\mathfrak{h})$. The kernel of this homomorphism is a closed subgroup $K_1 \subset H$ and we define $H_1 := H/K_1$. Then we may consider the homogeneous space G/K_1 which clearly is a principal H_1–bundle over G/H. Viewing the manifold G/H as being modelled on the vector space $\mathfrak{g}/\mathfrak{h}$, we may interpret the frame bundle $\mathcal{P}^1(G/H)$ as the bundle of linear isomorphisms between $\mathfrak{g}/\mathfrak{h}$ and tangent spaces of G/H. Now consider $\pi \circ \omega \in \Omega^1(G, \mathfrak{g}/\mathfrak{h})$, where ω is the Maurer–Cartan form of G and $\pi : \mathfrak{g} \to \mathfrak{g}/\mathfrak{h}$ is the canonical surjection. This one–form is clearly H–equivariant (since the Maurer–Cartan form is even G–equivariant) and its kernel in any point of G is exactly the vertical tangent space in that point. Hence, from 1.3.6 we conclude that there is a unique homomorphism $G \to \mathcal{P}^1(G/H)$ of principal bundles such that $\pi \circ \omega$ is the pullback of the soldering form. This homomorphism factors to a homomorphism $G/K \to \mathcal{P}^1(G/H)$ of principal bundles, which defines a first order H_1–structure on G/H.

There is a natural higher order analog of this construction. We can view $GL(\mathfrak{g}/\mathfrak{h})$ as the group of one–jets at $o = eH \in G/H$ of diffeomorphisms fixing o. The homomorphism $H \to GL(\mathfrak{g}/\mathfrak{h})$ from above is then given by mapping $h \in H$ to the one–jet of ℓ_h in o. Now we can replace $GL(\mathfrak{g}/\mathfrak{h})$ by the group of invertible r–jets in o of diffeomorphisms fixing o. This group is clearly isomorphic to the rth jet group G^r_n introduced in 1.2.8, where $n = \dim(G/H)$. Mapping $h \in H$ to the r–jet of ℓ_h in o defines a homomorphism. Defining K_r to be the kernel of this homomorphism and putting $H_r = H/K_r$, the obvious projection $G/K_r \to G/H$ is a principal bundle with structure group H_r. By construction, this can be viewed as a reduction of the rth order frame bundle $\mathcal{P}^r(G/H)$ (see 1.2.8), i.e. as an rth order G–structure. This point of view has been explored in [**Sl96**].

1.4.4. Sections of homogeneous bundles. Let $p_E : E \to G/H$ be a homogeneous bundle over a homogeneous space G/H, and let $\Gamma(E)$ be the set of all sections of E, i.e. $\Gamma(E) = \{\sigma : G/H \to E : p_E \circ \sigma = \mathrm{id}\}$. Then we get a left action of G on $\Gamma(E)$ defined by $g \cdot \sigma := \tilde{\ell}_g \circ \sigma \circ \ell_{g^{-1}}$, where $\tilde{\ell}$ and ℓ are the actions on E and G/H, respectively. Equivalently, this can be written as $(g \cdot \sigma)(x) = g \cdot (\sigma(g^{-1} \cdot x))$. In particular, if E is a homogeneous vector bundle, then $\Gamma(E)$ is a vector space, and the action of G on $\Gamma(E)$ is clearly linear, and thus we get a representation of G on the space $\Gamma(E)$. In view of Proposition 1.4.3, the homogeneous vector bundle E is completely determined by the representation of H on its standard fiber $V = E_o$. The representation of G on $\Gamma(E)$ is then called the *induced representation* of G corresponding to the representation V of H, and denoted by $\mathrm{Ind}_H^G(V)$.

The correspondence between sections of an associated bundle and functions of the total space of the principal bundle (see 1.2.7) gives rise to another interpretation of the action of G on the set of sections of a homogeneous bundle. This picture is (in the case of homogeneous vector bundles) frequently used in representation theory. Namely, via the isomorphism $E \cong G \times_H E_o$ from Proposition 1.4.3, we can identify $\Gamma(E)$ with $\Gamma(G \times_H E_o)$, which in turn by Proposition 1.2.7 is in bijective correspondence with the set $C^\infty(G, E_o)^H = \{f : G \to E_o : f(gh) = h^{-1} \cdot f(g)\}$. Explicitly, the correspondence between a section σ and a function f is characterized by $\sigma(gH) = [\![g, f(g)]\!] = g \cdot f(g)$ or by $f(g) = g^{-1} \cdot \sigma(gH)$. In the case of a natural vector bundle, this bijection clearly is a linear isomorphism.

To get the action of G on $\Gamma(E)$ in the picture of equivariant functions, we only have to notice that by definition $(g \cdot \sigma)(g'H) = g \cdot \sigma(g^{-1}g'H)$. Consequently, the map $g \cdot f : G \to E_o$ corresponding to $g \cdot \sigma \in \Gamma(E)$ is given by

$$(g \cdot f)(g') = (g')^{-1} \cdot (g \cdot \sigma(g^{-1}g'H)) = (g^{-1}g')^{-1} \cdot \sigma(g^{-1}g'H) = f(g^{-1}g').$$

Our next result is a first version of Frobenius reciprocity, which states that in certain situations one can reduce questions about the (often infinite–dimensional) space $\Gamma(E)$ to questions about the finite–dimensional manifold E_o. In particular, determining G–invariant sections of a homogeneous bundle always reduces to a finite–dimensional problem.

THEOREM 1.4.4 (Geometric version of Frobenius reciprocity). *Let $E \to G/H$ be a homogeneous bundle with standard fiber E_o (viewed as an H–space), and let X be a smooth manifold endowed with a smooth left G–action. Then the evaluation* *at o induces a natural bijection between the set of G–equivariant maps $X \to \Gamma(E)$ and the set of H–equivariant maps $X \to E_o$. In particular, there is a natural bijection between the set $\Gamma(E)^G$ of G–invariant sections of E and the set $(E_o)^H$ of H–invariant elements in the standard fiber.*

If E is the natural vector bundle induced by an H–representation W, and V is a representation of G, then the bijections above are linear and respect the subspaces of linear equivariant maps. Denoting by $\operatorname{Res}^G_H(V)$ the space V viewed as an H–representation, this implies that we get a linear isomorphism

$$\operatorname{Hom}_G(V, \operatorname{Ind}^G_H(W)) \cong \operatorname{Hom}_H(\operatorname{Res}^G_H(V), W),$$

i.e. Res^G_H and Ind^G_H are adjoint functors.

PROOF. Suppose we have given a map $X \to \Gamma(E)$, which we write as $x \mapsto \sigma_x$. This map is G–equivariant if and only if $\sigma_{g\cdot x}(g'H) = g \cdot \sigma_x(g^{-1}g'H)$. In particular, if we consider the map $X \to E_o$ defined by $x \mapsto \sigma_x(o)$, then for $h \in H$, we get $\sigma_{h\cdot x}(o) = h \cdot \sigma_x(h^{-1} \cdot o) = h \cdot \sigma_x(o)$, so this is H–equivariant. Conversely, if $f : X \to E_o$ is H–equivariant, then for $x \in X$ we define $\sigma_x : G/H \to E$ by $\sigma_x(gH) = g \cdot f(g^{-1} \cdot x)$. This is well defined since $f(h^{-1}g^{-1} \cdot x) = h^{-1} \cdot f(g^{-1} \cdot x)$, and thus $gh \cdot f(h^{-1}g^{-1} \cdot x) = g \cdot f(g^{-1} \cdot x)$. Moreover, $\sigma_x(o) = f(x)$ and

$$\sigma_{g\cdot x}(g'H) = g' \cdot f((g')^{-1}g \cdot x) = g \cdot g^{-1}g' \cdot f((g')^{-1}g \cdot x) = g \cdot \sigma_x(g^{-1}g'H),$$

so $x \mapsto \sigma_x$ is G–equivariant. Since for any G–equivariant map $x \mapsto \sigma_x$ we must have $\sigma_x(gH) = g \cdot g^{-1} \cdot \sigma_x(g^{-1}gH) = g \cdot \sigma_{g\cdot x}(o)$, the two constructions are inverse bijections between the set of G–equivariant maps $X \to \Gamma(E)$ and the set of H–equivariant maps $X \to E_o$.

Taking $X = \{pt\}$ a single point space with the trivial G–action, a G–equivariant map $X \to \Gamma(E)$ is the same thing as a section $\sigma_{pt} \in \Gamma(E)$ such that $g \cdot \sigma_{pt} = \sigma_{pt}$, i.e. a G–invariant section, and similarly an H–equivariant map $X \to E_o$ is an H–invariant element in E_o. Thus, we get a bijection $\Gamma(E)^G \to (E_o)^H$.

Finally, if E is a natural vector bundle induced by an H–representation W and X is a G–representation V, then $L(V, \Gamma(E))$ and $L(V, E_o)$ are vector spaces under the pointwise operations, and the evaluation in o induces a linear map $L(V, \Gamma(E)) \to L(V, E_o)$. If we start from a linear map $f : V \to E_o$, the construction above produces $\sigma_v(gH) = g \cdot f(g^{-1} \cdot v)$. Since f is linear and G acts by linear maps, this is linear in v. □

EXAMPLE 1.4.4. The geometric version of Frobenius reciprocity immediately allows us to reduce questions about the existence of invariant Riemannian metrics and other invariant tensor fields to problems of finite–dimensional representation theory: If $M = G/H$ is a homogeneous space, then from 1.4.3 we know that the tensor bundle $\otimes^k TM \otimes \otimes^\ell T^*M$ is the associated bundle to $p : G \to M$ corresponding to the representation $\otimes^k(\mathfrak{g}/\mathfrak{h}) \otimes \otimes^\ell(\mathfrak{g}/\mathfrak{h})^*$. By Frobenius reciprocity, G–invariant sections of this bundle are in bijective correspondence with H–invariant elements in the representation. Since invariant elements in a representation are the same thing as homomorphisms from the trivial representation to the given representation, this can be rephrased in such a way that the dimension of invariant tensor fields of type $\binom{k}{\ell}$ on G/H equals the multiplicity of the trivial representation in $\otimes^k(\mathfrak{g}/\mathfrak{h}) \otimes \otimes^\ell(\mathfrak{g}/\mathfrak{h})^*$. Moreover, the proof of Theorem 1.4.4 gives an explicit construction of the invariant tensor field from the invariant element in the representation. Clearly, the whole construction respects symmetries of any kind, so invariant tensor fields having certain symmetries are in bijective correspondence to invariant elements in the representations having the same symmetries. Finally, if we consider (pointwise) questions of non–degeneracy they clearly reduce to the analogous non–degeneracy properties in the representation. In particular, the set of G–invariant Riemannian metrics on G/H is in bijective correspondence with the set of H–invariant positive definite inner products on the vector space $\mathfrak{g}/\mathfrak{h}$. Hence, we get

COROLLARY 1.4.4. *A homogeneous space G/H admits a G–invariant Riemannian metric if and only if the image $H_1 \subset GL(\mathfrak{g}/\mathfrak{h})$ of H under the map induced by the restriction to H of the adjoint representation of G has compact closure in $GL(\mathfrak{g}/\mathfrak{h})$.*

PROOF. From above we know that existence of an invariant Riemannian metric is equivalent to existence of an H_1–invariant positive definite inner product on $\mathfrak{g}/\mathfrak{h}$. If such an inner product exists, then H_1 is contained in the orthogonal group of this inner product, which is isomorphic to $O(n)$ for some n and thus compact. Hence, H_1 has compact closure.

Conversely, if $K \subset GL(\mathfrak{g}/\mathfrak{h})$ is a compact subgroup containing H_1, then averaging any inner product on $\mathfrak{g}/\mathfrak{h}$ over K gives a K–invariant and thus an H_1–invariant inner product. □

Consider the examples from 1.1 of viewing spheres as homogeneous spaces. In the case of the Riemannian sphere 1.1.1, we have $H = O(n)$ and $\mathfrak{g}/\mathfrak{h} \cong \mathbb{R}^n$ as an H–representation, so the set of $O(n+1)$–invariant Riemannian metrics on S^n consists exactly of all constant positive multiples of the standard metric. The description of invariant Riemannian metrics on $\mathbb{R}^n$, viewed as a homogeneous space of the Euclidean group $\mathrm{Euc}(n)$ is exactly the same. For the projective sphere from 1.1.3, we get $H_1 = GL(\mathfrak{g}/\mathfrak{h})$, so there certainly is no $SL(n+1,\mathbb{R})$–invariant metric on S^n. Similarly, one shows that in the other examples from 1.1.4–1.1.6 there are no invariant Riemannian metrics.

Finally, we remark that invariance often allows us to analyze the situation further in a pretty elementary way. For example, applying natural operators to invariant objects, we get invariant objects. Thus, if ϕ is an invariant differential form on G/H, then the exterior derivative $d\phi$ is invariant as well. Similarly, the curvature of an invariant Riemannian metric has to be an invariant tensor field, and so on.

1.4.5. Homogeneous principal bundles and invariant principal connections. Next we switch to the problem of invariant differential operators on a homogeneous space G/H. A simple special case of this problems is the question whether a given homogeneous bundle admits a connection which is compatible with the action of G. Since connections can be viewed as sections of natural bundles, we could reduce this to the determination of invariant sections. A direct discussion will be easier, however. First, we classify homogeneous principal bundles as defined in 1.4.2.

LEMMA 1.4.5. *Let G and K be Lie groups and let $H \subset G$ be a closed subgroup. Let $\mathcal{P} \to G/H$ be a homogeneous principal bundle with structure group K. Then there is a smooth homomorphism $i : H \to K$ such that $\mathcal{P} \cong G \times_H K$, where the action of H on K is given by $h \cdot k = i(h)k$ for $h \in H$ and $k \in K$.*

The bundles corresponding to two homomorphism $i, \hat{\imath} : H \to K$ are isomorphic (over $\mathrm{id}_{G/H}$) if and only if i and $\hat{\imath}$ are conjugate, i.e. $\hat{\imath}(h) = ki(h)k^{-1}$ for some fixed $k \in K$ and all $h \in H$.

PROOF. Let $\mathcal{P}_o$ be the fiber of $\mathcal{P}$ over the base point $o = eH \in G/H$. As discussed in 1.4.3 the left action of G on $\mathcal{P}$ restricts to a left action of H on $\mathcal{P}_o$, and $\mathcal{P} \cong G \times_H \mathcal{P}_o$ as a homogeneous bundle. Fixing a point $u_0 \in \mathcal{P}_o$, the map $k \mapsto u_0 \cdot k$ is a diffeomorphism $K \to \mathcal{P}_0$, so it remains to describe the H–action in this picture.

For $h \in H$, we have $h \cdot u_0 \in \mathcal{P}_o$, so there is a unique element $i(h) \in K$ such that $h \cdot u_0 = u_0 \cdot i(h)$. By smoothness of the two actions, the map $i : H \to K$ is smooth, and by definition $i(e) = e$. Since H acts by principal bundle maps, we see that $h \cdot (u_0 \cdot k) = (h \cdot u_0) \cdot k = u_0 \cdot i(h)k$. Using this, one immediately concludes that $i(h_1 h_2) = i(h_1)i(h_2)$, so i is a homomorphism.

Suppose that we have given an isomorphism $G \times_i K \to G \times_{\hat{\imath}} K$ of homogeneous principal bundles. Then the restriction to the fibers over o is a diffeomorphism $\phi : K \to K$ which commutes with the principal right action of K and is equivariant for the two left actions of H. By the first property, $\phi(k) = k_0 k$, where $k_0 = \phi(e)$. But then the second property reads as follows:

$$k_0 i(h)k = \phi(i(h)k) = \hat{\imath}(h)\phi(k) = \hat{\imath}(h)k_0 k.$$

In particular, $\hat{\imath}(h) = k_0 i(h) k_0^{-1}$. Conversely, if i and $\hat{\imath}$ are related in this way, right multiplication by k_0 induces a diffeomorphism with the two equivariancy properties. From 1.4.3 we conclude that this gives rise to an isomorphism of homogeneous principal bundles. □

For a homogeneous principal bundle $\mathcal{P} \to G/H$ with structure group K, we can next study invariant principal connections. Recall from 1.3.3 that a principal connection on $\mathcal{P}$ can be described by a one–form $\gamma \in \Omega^1(\mathcal{P}, \mathfrak{k})$, where $\mathfrak{k}$ is the Lie algebra of K. This form has to be K–equivariant and it has to reproduce the generators of fundamental vector fields. Denoting by r^k the principal right action by $k \in K$ and by ζ_A the fundamental vector field generated by $A \in \mathfrak{k}$, these conditions explicitly read as $(r^k)^*\gamma = \mathrm{Ad}(k^{-1}) \circ \gamma$ and $\gamma(\zeta_A) = A$. Denoting by ℓ_g the left action of $g \in G$ on $\mathcal{P}$, we can consider the pullback $(\ell_g)^*\gamma$. Since ℓ_g is a principal bundle automorphism, this is again a principal connection. We call γ an *invariant principal connection* if and only if $(\ell_g)^*\gamma = \gamma$ for all $g \in G$.

THEOREM 1.4.5. *Let $i : H \to K$ be a homomorphism and consider the corresponding homogeneous bundle $\mathcal{P} := G \times_i K \to G/H$. Then invariant principal connections on $\mathcal{P}$ are in bijective correspondence with linear maps $\alpha : \mathfrak{g} \to \mathfrak{k}$ such that:*

(i) $\alpha|_{\mathfrak{h}} = i' : \mathfrak{h} \to \mathfrak{k}$, *the derivative of* i.
(ii) $\alpha \circ \mathrm{Ad}(h) = \mathrm{Ad}(i(h)) \circ \alpha$ *for all* $h \in H$.

Putting $\hat{\imath}(h) = k_0 i(h) k_0^{-1}$, then under the isomorphism $G \times_i K \to G \times_{\hat{\imath}} K$ from the lemma, the homogeneous principal connection on $G \times_i K$ induced by α corresponds to the one on $G \times_{\hat{\imath}} K$ induced by $\hat{\alpha} = \mathrm{Ad}(k_0) \circ \alpha$.

PROOF. Consider the point $u_0 = [\![e, e]\!] \in G \times_i K$. Given an invariant principal connection γ on $\mathcal{P}$, consider its value at u_0, which is a linear map $\gamma(u_0) : T_{u_0}\mathcal{P} \to \mathfrak{k}$. Recall from 1.2.7 that the natural projection $q : G \times K \to G \times_i K$ is an H–principal bundle and, in particular, a surjective submersion. By definition $u_0 = q(e,e)$, so we may consider $\gamma(u_0) \circ T_{(e,e)}q : \mathfrak{g} \times \mathfrak{k} \to \mathfrak{k}$. Now we define $\alpha : \mathfrak{g} \to \mathfrak{k}$ by

$$\alpha(X) := \gamma(u_0) \circ T_{(e,e)}q \cdot (X, 0).$$

First, observe that for $k \in K$ we have $q(e,k) = u_0 \cdot k$. Putting $k = \exp(tA)$ for $A \in \mathfrak{k}$ and differentiating in $t = 0$, we see that $T_{(e,e)}q \cdot (0, A) = \zeta_A(u_0)$. This shows that $\gamma(u_0) \circ T_{(e,e)}q \cdot (X, A) = \alpha(X) + A$, and hence α determines $\gamma(u_0)$. Equivariancy under the principal right action then implies that $\gamma(u_0 \cdot k) = \mathrm{Ad}(k^{-1}) \circ \gamma(u_0) \circ Tr^{k^{-1}}$, so γ is determined along the fiber over o. Further, equivariancy under the left action of G implies

$$\gamma(g \cdot u_0 \cdot k) = \mathrm{Ad}(k^{-1}) \circ \gamma(u_0) \circ Tr^{k^{-1}} \circ T\ell_{g^{-1}}, \tag{1.10}$$

which shows that γ is completely determined by α.

On the other hand, for $h \in H$ we have $q(h,e) = q(e, i(h))$. Putting $h = \exp(tX)$ for $X \in \mathfrak{h}$ and differentiating in $t = 0$ we obtain

$$T_{(e,e)}q \cdot (X, 0) = T_{(e,e)}q \cdot (0, i'(X)),$$

which shows that $\alpha(X) = i'(X)$ for $X \in \mathfrak{h}$.

For $h \in H$ and $g \in G$, we next have

$$q(hgh^{-1}, e) = q(hg, i(h^{-1})) = \ell_h \circ r^{i(h^{-1})} q(g, e).$$

Putting $g = \exp(tX)$ for $X \in \mathfrak{g}$ and differentiating at $t = 0$, we obtain

$$T_{(e,e)}q \cdot (\mathrm{Ad}(h)(X), 0) = T\ell_h \circ Tr^{i(h^{-1})} \cdot T_{(e,e)}q \cdot (X, 0).$$

Applying $\gamma(u_0)$ to the left–hand side, we obtain $\alpha(\mathrm{Ad}(h)(X))$. For the right–hand side, we first note that G–invariance of γ implies that $\gamma(h \cdot u_0) \circ T\ell_h = \gamma(u_0)$. Then K–equivariancy implies that composing with $Tr^{i(h^{-1})}$, one obtains

$$\mathrm{Ad}(i(h)) \circ \gamma(h \cdot u_0 \cdot i(h^{-1})) = \mathrm{Ad}(i(h)) \circ \gamma(u_0).$$

Therefore, the right–hand side evaluates to $\mathrm{Ad}(i(h))(\alpha(X))$ which proves property (ii).

It remains to conversely construct an invariant principal connection γ from a linear map α with properties (i) and (ii). The obvious way is to require

$$\gamma(u_0)(T_{(e,e)}q \cdot (X, A)) := \alpha(X) + A.$$

This is well defined if and only if $(X, A) \mapsto \alpha(X) + A$ vanishes on the kernel of $T_{(e,e)}q$. Since $q^{-1}(u_0) = \{(h, i(h^{-1})) : h \in H\} \subset G \times K$, this kernel evidently consists of all elements of the form $(X, -i'(X))$ for $X \in \mathfrak{h}$. By property (i) we conclude that the above formula uniquely defines $\gamma(u_0) : T_{u_0}\mathcal{P} \to \mathfrak{k}$. We have seen above that $\zeta_A(u_0) = T_{(e,e)}q \cdot (0, A)$, so $\gamma(u_0)(\zeta_A(u_0)) = A$. Now one extends γ to all points using formula (1.10) from above as a definition. Since

$$q(g,k) = g \cdot u_0 \cdot k = \tilde{g} \cdot u_0 \cdot \tilde{k} = q(\tilde{g}, \tilde{k})$$

if and only if $\tilde{g} = gh$ and $\tilde{k} = i(h^{-1})k$ for some $h \in H$, one easily verifies that this is well defined using property (ii). Next, one immediately concludes that the resulting form γ is G–invariant and K–equivariant. Finally, from the definition one verifies that

$$\zeta_A(g \cdot u_0 \cdot k) = T\ell_g \cdot Tr^k \cdot \zeta_{\mathrm{Ad}(k)(A)}(u_0)$$

for all $A \in \mathfrak{k}$. Then $\gamma(\zeta_A) = A$ follows from the definition, whence γ is an invariant principal connection.

For the last statement, observe that the isomorphism $\phi : G \times_{\hat{i}} K \to G \times_i K$ from the proof of the lemma is characterized $\phi(\hat{q}(g,k)) = q(g, k_0^{-1}k)$. In particular,

$$\phi(\hat{q}(g,e)) = q(g, k_0^{-1}) = r^{k_0^{-1}}(q(g,e)).$$

Putting $g = \exp(tX)$ and differentiating at $t = 0$, we obtain $T_{(e,e)}(\phi \circ \hat{q}) \cdot (X, 0) = Tr^{k_0^{-1}} \cdot T_{(e,e)}q \cdot (X, 0)$. Using this, we compute

$$\begin{aligned}\hat{\alpha}(X) &= (\phi^*\gamma)(\hat{u}_0) \circ T_{(e,e)}\hat{q} \cdot (X,0) = \gamma(u_0 \cdot k_0^{-1})(T_{(e,e)}(\phi \circ \hat{q}) \cdot (X,0)) \\ &= \gamma(u_0 \cdot k_0^{-1})(Tr^{k_0^{-1}} \cdot T_{(e,e)}q \cdot (X,0)) = ((r^{k_0^{-1}})^*\gamma)(T_{(e,e)}q \cdot (X,0)) \\ &= \mathrm{Ad}(k_0)(\alpha(X)).\end{aligned}$$

□

Notice that in general, a linear map $\alpha : \mathfrak{g} \to \mathfrak{k}$ with properties (i) and (ii) need not exist. (We shall soon see this in an example.) If there is one such map, however, then for any other map with properties (i) and (ii), the difference vanishes on $\mathfrak{h}$, and hence defines a linear map $\beta : \mathfrak{g}/\mathfrak{h} \to \mathfrak{k}$. By construction, β is equivariant for the actions $h \cdot (X + \mathfrak{h}) := \mathrm{Ad}(h)(X) + \mathfrak{h}$ on the left–hand side and $h \cdot A = \mathrm{Ad}(i(h))(A)$ on the right–hand side. Conversely, given α with properties (i) and (ii) and an equivariant map β, then $\hat{\alpha}(X) = \alpha(X) + \beta(X + \mathfrak{h})$ also has properties (i) and (ii). Hence, we conclude that if one invariant principal connection exists, then the space of all such connections is an affine space modelled on the vector space $\mathrm{Hom}_H(\mathfrak{g}/\mathfrak{h}, \mathfrak{k})$ with actions as above.

1.4.6. The curvature of an invariant principal connection. In 1.3.3 we have seen that the curvature ρ of a principal connection γ on a principal K–bundle $\mathcal{P} \to M$ can be interpreted as a two–form on M with values in the bundle $\mathcal{P} \times_K \mathfrak{k}$. Here K acts on its Lie algebra $\mathfrak{k}$ via the adjoint action. Moreover, we have seen there that ρ is induced by the $\mathfrak{k}$–valued two–form $(\xi, \eta) \mapsto d\gamma(\xi, \eta) + [\gamma(\xi), \gamma(\eta)]$ on $\mathcal{P}$.

Now suppose that $\mathcal{P}$ is a homogeneous principal K–bundle over G/H. From 1.4.5 we know that there is a homomorphism $i : H \to K$ such that $\mathcal{P} = G \times_i K$. Hence, we can identify the bundle $\mathcal{P} \times_K \mathfrak{k}$ with $G \times_H \mathfrak{k}$ where the action of H on $\mathfrak{k}$ is given by $h \cdot A := \mathrm{Ad}(i(h))(A)$. From Example 1.4.3 we know that $T^*(G/H) \cong G \times_H (\mathfrak{g}/\mathfrak{h})^*$, with the action induced by the adjoint action. Hence, the curvature of any principal connection γ on $\mathcal{P}$ has values in the homogeneous vector bundle corresponding to the representation $\Lambda^2(\mathfrak{g}/\mathfrak{h})^* \otimes \mathfrak{k}$, which consists of all skew symmetric bilinear maps $\mathfrak{g}/\mathfrak{h} \times \mathfrak{g}/\mathfrak{h} \to \mathfrak{k}$.

Suppose further that γ is an invariant principal connection on $\mathcal{P}$. Denoting by $\ell_g : \mathcal{P} \to \mathcal{P}$ the action of $g \in G$, by definition we have $(\ell_g)^*\gamma = \gamma$. This immediately implies that the $\mathfrak{k}$–valued two–form on $\mathcal{P}$ which induces the curvature ρ is preserved by each ℓ_g. Hence, $\rho \in \Omega^2(G/H, G \times_H \mathfrak{k})$ is an invariant section. By Theorem 1.4.4, such a section is equivalent to a bilinear map $\mathfrak{g}/\mathfrak{h} \times \mathfrak{g}/\mathfrak{h} \to \mathfrak{k}$, which is H–equivariant. We can determine this bilinear map explicitly:

PROPOSITION 1.4.6. *Consider a homogeneous space G/H and the homogeneous principal K-bundle $\mathcal{P} = G \times_i K \to G/H$ corresponding to a homomorphism $i : H \to K$. Let γ be an invariant principal connection on $\mathcal{P}$ and let $\alpha : \mathfrak{g} \to \mathfrak{k}$ be the corresponding linear map as in Theorem 1.4.5.*

Then the map $\mathfrak{g}\times\mathfrak{g}\to\mathfrak{k}$ defined by $(X,Y)\mapsto[\alpha(X),\alpha(Y)]-\alpha([X,Y])$ descends to an H–equivariant map $(\mathfrak{g}/\mathfrak{h})\times(\mathfrak{g}/\mathfrak{h})\to\mathfrak{k}$. The corresponding invariant section of the homogeneous bundle $G\times_H(\Lambda^2(\mathfrak{g}/\mathfrak{h})^*\otimes\mathfrak{k})$ is exactly the curvature of γ.

PROOF. Property (ii) of α from Theorem 1.4.5 reads as $\alpha\circ\operatorname{Ad}(h)=\operatorname{Ad}(i(h))\circ\alpha$ for all $h\in H$. Using this, we compute

$$\begin{aligned}[\alpha(\operatorname{Ad}(h)(X)),\alpha(\operatorname{Ad}(h)(Y))]&=[\operatorname{Ad}(i(h))(\alpha(X)),\operatorname{Ad}(i(h))(\alpha(Y))]\\&=\operatorname{Ad}(i(h))\left([\alpha(X),\alpha(Y)]\right).\end{aligned}$$

In the same way, one verifies that $(X,Y)\mapsto\alpha([X,Y])$ is equivariant. Further, for $X\in\mathfrak{h}$ we can apply the equation $\alpha\circ\operatorname{Ad}(h)=\operatorname{Ad}(i(h))\circ\alpha$ to $h=\exp(tX)$ and differentiate at $t=0$, to get $\alpha\circ\operatorname{ad}(X)=\operatorname{ad}(i'(X))\circ\alpha$. By property (i) of α from Theorem 1.4.5 we have $\alpha(X)=i'(X)$ for $X\in\mathfrak{h}$, and we get

$$[\alpha(X),\alpha(Y)]-\alpha([X,Y])=\operatorname{ad}(i'(X))(\alpha(Y))-\alpha(\operatorname{ad}(X)(Y))=0.$$

This implies that our bilinear map descends to an H–equivariant mapping as required. It remains to verify that the corresponding invariant section equals the curvature of γ.

To verify this, we have to recall the definition of α. Let $q:G\times K\to G\times_i K$ be the canonical projection. Then we can form the pullback $q^*\gamma\in\Omega^1(G\times K,\mathfrak{k})$, and from the proof of Theorem 1.4.5 we see that $q^*\gamma(e,e)(X,A)=\alpha(X)+A$ for $X\in\mathfrak{g}=T_eG$ and $A\in\mathfrak{k}=T_eK$. Our strategy will be to compute

$$dq^*\gamma(\xi,\eta)+[q^*\gamma(\xi),q^*\gamma(\eta)](e,e).\tag{1.11}$$

Now by definition $q(g,kk')=q(g,k)\cdot k'$. Taking $A\in\mathfrak{k}$, putting $k'=\exp(tA)$ and differentiating at $t=0$, we see that Tq maps $(0,L_A)$ to the (vertical) fundamental vector field ζ_A. Since $d\gamma(\zeta_A,\_)+[A,\gamma(\_)]=0$, we conclude that it suffices to compute (1.11) for $\xi=(L_X,0)$ and $\eta=(L_Y,0)$ with $X,Y\in\mathfrak{g}$.

By definition $q(gg',k)=\ell_g\circ r^k(q(g',e))$. Putting $g'=\exp(tX)$ and differentiating at $t=0$, we see that $T_{(g,k)}q\cdot(L_X,0)=T\ell_g\circ Tr^k\cdot T_{(e,e)}q\cdot(X,0)$. Applying γ and taking into account G–invariance and K–equivariance (see formula (1.10) in 1.4.5), we conclude that $q^*\gamma(g,k)(L_X,0)=\operatorname{Ad}(k^{-1})(\alpha(X))$. In particular, this function is independent of g, so acting on it with a vector field of the form $(L_Y,0)$ we get zero. This shows that

$$\begin{aligned}dq^*\gamma(e,e)((L_X,0),(L_Y,0))&=-q^*\gamma(e,e)([L_X,L_Y],0)\\&=-q^*\gamma(e,e)([X,Y])=-\alpha([X,Y]).\end{aligned}$$

This implies that (1.11) is given by $((X,A),(Y,B))\mapsto-\alpha([X,Y])+[\alpha(X),\alpha(Y)]$. Since the composition $G\times K\to\mathcal{P}\to G/H$ is given by $(g,k)\mapsto gH$, we conclude that (X,A) represents the tangent vector $X+\mathfrak{h}\in\mathfrak{g}/\mathfrak{h}\cong T_oG/H$, and the result follows. □

The most fundamental example of a homogeneous principal bundle is of course the bundle $G\to G/H$ itself. As we shall see below, the existence of an invariant principal connection on this bundle has far reaching consequences.

COROLLARY 1.4.6. *There exists a G–invariant principal connection on the H–principal bundle $p:G\to G/H$, if and only if the Klein geometry (G,H) is reductive. If this is the case and $\mathfrak{n}\subset\mathfrak{g}$ is a fixed H–invariant complement to $\mathfrak{h}$, then the set of all such connections is an affine space modelled on the vector space* $\operatorname{Hom}_H(\mathfrak{n},\mathfrak{h})$.

The curvature of the invariant connection defined by a complement $\mathfrak{n}$ *is induced by the map* $\Lambda^2\mathfrak{n} \to \mathfrak{h}$ *given by* $(X, Y) \mapsto -[X, Y]_{\mathfrak{h}}$. *Here the subscript denotes the* $\mathfrak{h}$*–component with respect to the decomposition* $\mathfrak{g} = \mathfrak{h} \oplus \mathfrak{n}$.

PROOF. The bundle $G \to G/H$ of course corresponds to $i = \mathrm{id}_H : H \to H$. By Theorem 1.4.5 existence of an invariant principal connection is equivalent to existence of an H–equivariant map $\alpha : \mathfrak{g} \to \mathfrak{h}$ such that $\alpha|_{\mathfrak{h}} = i' = \mathrm{id}_{\mathfrak{h}}$. Putting $\mathfrak{n} = \ker(\alpha)$ this condition is equivalent to $\mathfrak{g} = \mathfrak{h} \oplus \mathfrak{n}$ as an H–module. Fixing the choice of α (respectively $\mathfrak{n}$), we have $\mathfrak{n} \cong \mathfrak{g}/\mathfrak{h}$ as an H–module, and the remaining claims follow from Theorem 1.4.5 and Proposition 1.4.6. □

EXAMPLE 1.4.6. (1) As a more specific example, suppose that the subgroup $H \subset G$ is compact. Then averaging any positive definite inner product on $\mathfrak{g}$ over H, we obtain an H–invariant positive definite inner product on $\mathfrak{g}$. Now, the orthogonal projection onto $\mathfrak{h}$ with respect to such an inner product is H–equivariant, thus giving a G–invariant principal connection on $p : G \to G/H$. The complementary subset $\mathfrak{n}$ is then $\mathfrak{h}^{\perp}$.

In particular, consider the case where $H = O(n)$ and $\mathfrak{g}/\mathfrak{h} \cong \mathfrak{n} \cong \mathbb{R}^n$ as an H–module, which occurs both in the case of the Riemannian sphere from 1.1.1 (with $G = O(n+1)$) and the Euclidean space, viewed as a homogeneous space of the group of Euclidean motions from 1.1.2. In both cases, we not only have a G–invariant principal connection on $p : G \to G/H$ but it is also uniquely determined. This is due to the fact that both $\mathfrak{n} \cong \mathbb{R}^n$ and $\mathfrak{h} \cong \mathfrak{o}(n)$ are irreducible $O(n)$–modules and thus the zero map is the only H–homomorphism $\mathfrak{n} \to \mathfrak{h}$. In the case of the Euclidean space 1.1.2, the complement $\mathfrak{n}$ is an abelian subalgebra, so $[\mathfrak{n}, \mathfrak{n}] = \{0\}$ and the principal connection is flat.

On the other hand, in the case of the Riemannian sphere, the decomposition $\mathfrak{g} = \mathfrak{h} \oplus \mathfrak{n}$ derived in 1.1.1 is the unique H–invariant decomposition, and one immediately verifies that $[\mathfrak{n}, \mathfrak{n}] \subset \mathfrak{h}$, but the bracket of two elements of $\mathfrak{n}$ is nonzero in general, so we get nonzero curvature. More precisely, one easily computes that the curvature is induced by the map $\mathbb{R}^n \times \mathbb{R}^n \to \mathfrak{o}(n)$ defined by $(v, w) \mapsto vw^t - wv^t$. Otherwise put, identifying $\mathfrak{o}(n)$ with $\Lambda^2\mathbb{R}^n$, the curvature is simply induced by $(v, w) \mapsto v \wedge w$. Notice that up to scale this is the unique possibility for an invariant tensor field of this type.

(2) Consider the example of the projective sphere from 1.1.3. In this case $G = SL(n+1, \mathbb{R})$ and H is the stabilizer of the line through the first unit vector. The subalgebra $\mathfrak{h} \subset \mathfrak{g}$ is given by

$$\mathfrak{h} = \left\{ \begin{pmatrix} -\operatorname{tr}(A) & Z \\ 0 & A \end{pmatrix} : A \in \mathfrak{gl}(n, \mathbb{R}), Z \in \mathbb{R}^{n*} \right\}.$$

If there were an H–invariant complement $\mathfrak{n}$ to $\mathfrak{h}$ in $\mathfrak{g}$, then, in particular, $[\mathfrak{h}, \mathfrak{n}] \subset \mathfrak{n}$. On the other hand, in order to be a vector space complement, for any $X \in \mathbb{R}^n$, $\mathfrak{n}$ would have to contain a unique element of the form $\begin{pmatrix} * & * \\ X & * \end{pmatrix}$. In particular, the zero matrix would be the only element of $\mathfrak{n}$, whose entry in the left lower block is zero. But one immediately computes that for $Z, W \in \mathbb{R}^{n*}$, $X \in \mathbb{R}^n$ and $A \in \mathfrak{gl}(n, \mathbb{R})$, we get

$$\left[\begin{pmatrix} 0 & Z \\ 0 & 0 \end{pmatrix}, \begin{pmatrix} -\operatorname{tr}(A) & W \\ X & A \end{pmatrix} \right] = \begin{pmatrix} ZX & Z(A + \operatorname{tr}(A)\mathrm{id}) \\ 0 & -XZ \end{pmatrix}.$$

This clearly is incompatible with the existence of an H–invariant complement $\mathfrak{n}$. In particular, for the case of the projective sphere, there is no $SL(n+1,\mathbb{R})$–invariant connection on the H–principal bundle $SL(n+1,\mathbb{R}) \to S^n$. Similarly, one can show that for the examples of the projective contact sphere, the conformal sphere and the CR–sphere from 1.1.4–1.1.6 there never exists a G–invariant principal connection.

1.4.7. Invariant linear connections on homogeneous vector bundles. We have already noted in Example 1.4.2 that any associated (vector) bundle to a homogenous principal bundle is a homogeneous (vector) bundle. It is easy to describe this correspondence in the picture of associate bundles to $G \to G/H$ developed in 1.4.3. We restrict to the case of vector bundles here, general bundles can be dealt with similarly. If $\mathcal{P} \to G/H$ is a homogeneous principal K–bundle, then $\mathcal{P} = G \times_i K$ for some homomorphism $i : H \to K$. If $\rho : K \to GL(V)$ is a representation of K on a vector space V, then of course $\rho \circ i$ is a representation of H on V. From the definitions, one immediately concludes that $\mathcal{P} \times_\rho V \cong G \times_{\rho \circ i} V$.

Conversely, assume that $E \to G/H$ is a homogeneous vector bundle. Then we have a representation $\rho : H \to GL(E_o)$ such that $E \cong G \times_H E_o$. Now we can also use the homomorphism ρ to obtain a homogeneous principal bundle $\mathcal{P} \to G/H$ with structure group $GL(E_o)$. From above, we see that $\mathcal{P} \times_{GL(E_o)} E_o \cong G \times_H E_o \cong E$, so this is exactly the frame bundle of E. Hence, the frame bundle of a homogeneous vector bundle is a homogeneous principal bundle.

The construction of induced connections on associated bundles as described in 1.3.4 is of functorial nature, too. In particular, starting with an invariant principal connection on a homogeneous principal bundle, all induced connections on associated bundles (which we have just seen are homogeneous) are automatically invariant. In particular, we get

OBSERVATION 1.4.7. An invariant principal connection on the H–principal bundle $G \to G/H$ induces an invariant linear connection on *any* homogeneous vector bundle $E \to G/P$.

Conversely, an invariant linear connection on a homogeneous vector bundle induces a unique principal connection on the frame bundle, which is invariant, too. Thus, we see that there is a bijective correspondence between invariant linear connections on a homogeneous vector bundle and invariant principal connections on its frame bundle. Using this, we can now give a complete classification of invariant linear connections.

THEOREM 1.4.7. *Let G be a Lie group, $H \subset G$ a closed subgroup, $\rho : H \to GL(V)$ a representation of H, and consider the corresponding homogeneous vector bundle $E = G \times_H V \to G/H$.*

Then G–invariant linear connections on E are in bijective correspondence with linear maps $\alpha : \mathfrak{g} \to L(V,V)$ such that

(i) *$\alpha|_{\mathfrak{h}} = \rho'$, the derivative of the representation ρ,*
(ii) *$\alpha(\mathrm{Ad}(h)(X)) = \rho(h) \circ \alpha(X) \circ \rho(h^{-1})$ for all $X \in \mathfrak{g}$ and $h \in H$.*

In particular, if there is one such connection, then the space of all of them is an affine space modelled on the vector space $\mathrm{Hom}_H(\mathfrak{g}/\mathfrak{h}, L(V,V))$.

The curvature of the invariant linear connection corresponding to α is the invariant section of $\Omega^2(G/H, L(E,E))$ induced by the map $\Lambda^2(\mathfrak{g}/\mathfrak{h}) \to L(V,V)$ defined by

$$(X+\mathfrak{h}, Y+\mathfrak{h}) \mapsto \alpha(X) \circ \alpha(Y) - \alpha(Y) \circ \alpha(X) - \alpha([X,Y]).$$

PROOF. From above we know that we have to determine all invariant principal connections on the principal $GL(V)$–bundle $G\times_\rho GL(V)$. The Lie algebra of $GL(V)$ is $L(V,V)$ with the commutator of endomorphisms as the Lie bracket. Moreover, the adjoint action of $GL(V)$ on $L(V,V)$ is just given by conjugation. Now all the claims follow from Theorem 1.4.5, Proposition 1.4.6, and the fact that the curvature of an induced connection is induced by the principal curvature. □

Having solved the problem of existence of invariant linear connections, we want to describe them by an explicit formula. We do this in terms of equivariant functions. A linear connection on a vector bundle $E \to G/H$ can be considered as an operator ∇ which maps sections of E to sections of $T^*(G/H) \otimes E$. If E is the homogeneous vector bundle corresponding to a representation V of H, then the target bundle corresponds to $L(\mathfrak{g}/\mathfrak{h}, V)$. Using the correspondence between sections and equivariant functions from Proposition 1.2.7, we can therefore view ∇ as an operator $C^\infty(G,V)^H \to C^\infty(G, L(\mathfrak{g}/\mathfrak{h}, V))^H$. In these terms, there are two nice descriptions:

PROPOSITION 1.4.7. *Let $E \to G/H$ be the homogeneous vector bundle corresponding to a representation $\rho : H \to GL(V)$, and let ∇ be the invariant linear connection on E corresponding to a linear map $\alpha : \mathfrak{g} \to L(V,V)$ as in the theorem. Let $s \in \Gamma(E)$ be a section and let $f : G \to V$ be the corresponding equivariant function.*

(1) The equivariant function $\phi : G \to L(\mathfrak{g}/\mathfrak{h}, V)$ corresponding to ∇s is explicitly given by

$$\phi(g)(X+\mathfrak{h}) = (L_X \cdot f)(g) + \alpha(X)(f(g))$$

for $X \in \mathfrak{g}$ with corresponding left invariant vector field L_X and $g \in G$.

(2) For a vector field $\xi \in \mathfrak{X}(G/H)$ let $\tilde{\xi}$ be a local lift to a vector field on G and let $\omega \in \Omega^1(G,\mathfrak{g})$ be the Maurer–Cartan form of G. Then on the domain of $\tilde{\xi}$ the equivariant function $G \to V$ corresponding to $\nabla_\xi s$ is given by

$$g \mapsto (\tilde{\xi}\cdot f)(g) + \alpha(\omega(\tilde{\xi})(g))(f(g)).$$

PROOF. (1) First note that $\phi(g)$ is well defined, since for $X \in \mathfrak{h}$, equivariancy of f implies that $L_X \cdot f = -\rho'(X)\circ f$ while $\alpha(X) = \rho'(X)$ by assumption on α. Let $\mathcal{P} = G\times_\rho GL(V)$ be the frame bundle of E and let $q : G \times GL(V) \to \mathcal{P}$ be the natural projection. Recall that we can identify E either with $G\times_H V$ or with $\mathcal{P}\times_{GL(V)} V$. The isomorphism between these two bundles is obtained in such a way that the class of $(q(g,A),v)$ in $\mathcal{P}\times_{GL(V)} V$ corresponds to the class of $(g, A(v))$ in $G\times_H V$. This immediately implies that if $F : \mathcal{P} \to V$ is the equivariant function representing a section $s \in \Gamma(E)$, then $f(g) = F(q(g,e))$ is the equivariant function $G \to V$ representing the same section.

In the proof of Proposition 1.4.6, we have verified that for the principal connection γ determined by α, $X \in \mathfrak{g}$ and the left invariant vector field L_X, we have $\gamma(T_{(g,e)}q \cdot (L_X, 0)) = \alpha(X)$. Denoting by $p : G \to G/H$ the projection, this implies that taking the horizontal lift of $T_g p \cdot L_X \in T_{gH}(G/H)$ and using it to differentiate the equivariant function F, we obtain

$$(T_{(g,e)}q \cdot (L_X,0))\cdot F + \alpha(X)(F(q(g,e))) = (L_X\cdot f)(g) + \alpha(X)(f(g)).$$

Using the left action by g to transport $T_g p \cdot L_X$ back to o, we see that this vector corresponds to $X + \mathfrak{h} \in \mathfrak{g}/\mathfrak{p}$. Together with the description of the covariant derivative induced by a principal connection from part (2) of Proposition 1.3.4, this shows that the above formula computes $\phi(g)(X + \mathfrak{h})$.

(2) In Example 1.4.3 we have seen that $\xi(gH) = [\![g, \omega(\tilde{\xi}(g)) + \mathfrak{h}]\!]$. This implies that the equivariant function $G \to \mathfrak{g}/\mathfrak{h}$ corresponding to ξ is given by $g \mapsto \omega(\tilde{\xi}(g)) + \mathfrak{h}$. On the other hand, it says that $\tilde{\xi}(g) = L_{\omega(\tilde{\xi})(g)}(g)$, and this projects onto $\xi(gH)$. Now the result follows from part (1). □

1.4.8. Invariant affine connections. The question of existence of invariant linear connections is particularly important in the case of the tangent bundle $T(G/H)$ of a homogeneous space G/H. From Example 1.4.3 we know that the tangent bundle corresponds to the representation $\underline{\mathrm{Ad}} : H \to GL(\mathfrak{g}/\mathfrak{h})$ induced by the adjoint representation. By Theorem 1.4.7, invariant linear connections on $T(G/H)$ are in bijective correspondence with linear maps $\alpha : \mathfrak{g} \to L(\mathfrak{g}/\mathfrak{h}, \mathfrak{g}/\mathfrak{h})$ such that $\alpha|_{\mathfrak{h}} = \underline{\mathrm{ad}}$ and $\alpha(\mathrm{Ad}(h)(X)) = \underline{\mathrm{Ad}}(h) \circ \alpha(X) \circ \underline{\mathrm{Ad}}(h^{-1})$ for all $X \in \mathfrak{g}$ and $h \in H$.

Theorem 1.4.7 also describes the map $\Lambda^2(\mathfrak{g}/\mathfrak{h}) \to L(\mathfrak{g}/\mathfrak{h}, \mathfrak{g}/\mathfrak{h})$ which induces the curvature of the linear connection determined by a linear map α. For linear connections on the tangent bundle, there is another invariant, the torsion. Recall from 1.3.5 that this is a bundle map $T : \Lambda^2 T(G/H) \to T(G/H)$ characterized by $T(\xi, \eta) = \nabla_\xi \eta - \nabla_\eta \xi - [\xi, \eta]$. For an invariant linear connection ∇, the torsion is an invariant section, so it corresponds to an H–equivariant map $\Lambda^2(\mathfrak{g}/\mathfrak{h}) \to \mathfrak{g}/\mathfrak{h}$. We can easily determine this map explicitly.

PROPOSITION 1.4.8. *Let ∇ be the linear connection on the tangent bundle $T(G/H)$ detemined by a linear map $\alpha : \mathfrak{g} \to L(\mathfrak{g}/\mathfrak{h}, \mathfrak{g}/\mathfrak{h})$. Then the map $\tilde{\tau} : \Lambda^2\mathfrak{g} \to \mathfrak{g}/\mathfrak{h}$ defined by*

$$\tilde{\tau}(X, Y) := \alpha(X)(Y + \mathfrak{h}) - \alpha(Y)(X + \mathfrak{h}) - [X, Y] + \mathfrak{h}$$

factors to an H-invariant map $\tau : \Lambda^2\mathfrak{g}/\mathfrak{h} \to \mathfrak{g}/\mathfrak{h}$. The corresponding invariant section of $\Omega^2(M, TM)$ is the torsion of ∇.

PROOF. From equivariancy of α it follows immediately that $\tilde{\tau}$ is H–equivariant. For $X \in \mathfrak{h}$ we have $\alpha(X) = \underline{\mathrm{ad}}(X)$, which immediately implies that $\tilde{\tau}(X, Y) = 0$ for all $Y \in \mathfrak{g}$. Hence, $\tilde{\tau}$ factors to an H–equivariant map τ as claimed. For $\xi, \eta \in \mathfrak{X}(G/H)$ choose local lifts $\tilde{\xi}$ and $\tilde{\eta}$ on G defined around e. Then locally around e, the function $G \to \mathfrak{g}/\mathfrak{h}$ representing $\nabla_\xi \eta$ is given by $\tilde{\xi} \cdot \omega(\tilde{\eta}) + \mathfrak{h} + \alpha(\omega(\tilde{\xi}))(\omega(\tilde{\eta}) + \mathfrak{h})$. Since $[\tilde{\xi}, \tilde{\eta}]$ is a local lift of $[\xi, \eta]$, the function describing this Lie bracket is, locally around e, given by $\omega([\tilde{\xi}, \tilde{\eta}]) + \mathfrak{h}$. By definition of the exterior derivative, the section $\nabla_\xi \eta - \nabla_\eta \xi - [\xi, \eta]$ corresponds to the function

$$\alpha(\omega(\tilde{\xi}))(\omega(\tilde{\eta}) + \mathfrak{h}) - \alpha(\omega(\tilde{\eta}))(\omega(\tilde{\xi}) + \mathfrak{h}) + d\omega(\tilde{\xi}, \tilde{\eta}) + \mathfrak{h}.$$

Expressing $d\omega$ using the Maurer–Cartan equation, we obtain $\tilde{\tau}(\omega(\tilde{\xi}), \omega(\tilde{\eta}))$. □

Even more special, suppose that the invariant linear connection on $T(G/H)$ is induced by the invariant principal connection on $G \to G/H$ corresponding to an H–equivariant projection $\pi : \mathfrak{g} \to \mathfrak{h}$; see Corollary 1.4.6. Then the corresponding map $\alpha : \mathfrak{g} \to L(\mathfrak{g}/\mathfrak{h}, \mathfrak{g}/\mathfrak{h})$ is simply given by $\alpha(X) = \underline{\mathrm{ad}}(\pi(X))$. Hence, $\alpha(X) = 0$ for $X \in \mathfrak{n} = \ker(\pi)$. Identifying $\mathfrak{g}/\mathfrak{h}$ with $\mathfrak{n}$, the formulae for curvature and torsion simplify considerably. The curvature is induced by the map $\Lambda^2\mathfrak{n} \to \mathfrak{h}$ given by

$(X, Y) \mapsto -\pi([X, Y])$. The torsion is induced by $\tau : \Lambda^2\mathfrak{n} \to \mathfrak{n}$, where $\tau(X, Y) = \pi([X, Y]) - [X, Y] \in \mathfrak{n}$. Hence, the curvature and the torsion are induced by the $\mathfrak{h}$–component, respectively, the $\mathfrak{n}$–component of the negative of the Lie bracket $\Lambda^2\mathfrak{n} \to \mathfrak{g}$.

EXAMPLE 1.4.8. By way of an example, we show that in general there is no invariant linear connection on the tangent bundle of a homogeneous space. Consider the projective sphere from 1.1.3. From there we know that for $\phi \in \mathbb{R}^{n*}$, the element $h := \left(\begin{smallmatrix} 1 & \phi \\ 0 & \mathbb{I} \end{smallmatrix}\right)$ lies in H, where $\mathbb{I}$ denotes the $n \times n$ identity matrix. For $X \in \mathbb{R}^n$ one gets

$$\mathrm{Ad}(h) \cdot \begin{pmatrix} 0 & 0 \\ X & 0 \end{pmatrix} = \begin{pmatrix} \phi(X) & -\phi(X)\phi \\ X & -X \otimes \phi \end{pmatrix}.$$

But in view of the presentation of the Lie algebras in example (2) of 1.4.6, this also implies that $\underline{\mathrm{Ad}}(h) = \mathrm{id}_{\mathfrak{g}/\mathfrak{h}}$. In particular, if $\alpha : \mathfrak{g} \to L(\mathfrak{g}/\mathfrak{h}, \mathfrak{g}/\mathfrak{h})$ is H–equivariant, then we obtain

$$\alpha \begin{pmatrix} 0 & 0 \\ X & 0 \end{pmatrix} = \alpha \begin{pmatrix} \phi(X) & -\phi(X)\phi \\ X & -X \otimes \phi \end{pmatrix}.$$

But if α corresponds to an invariant connection, then the difference of these two elements must be $\underline{\mathrm{ad}} \left(\begin{smallmatrix} \phi(X) & -\phi(X)\phi \\ 0 & -X\otimes\phi \end{smallmatrix}\right)$. If $Y \in \mathbb{R}^n$ is another element, then we get

$$\left[\begin{pmatrix} \phi(X) & -\phi(X)\phi \\ 0 & -X \otimes \phi \end{pmatrix}, \begin{pmatrix} 0 & 0 \\ Y & 0 \end{pmatrix}\right] = \begin{pmatrix} -\phi(X)\phi(Y) & 0 \\ -\phi(Y)X - \phi(X)Y & \phi(X)Y \otimes \phi \end{pmatrix}.$$

This shows that the difference is nonzero, which contradicts the existence of a $SL(n+1, \mathbb{R})$–invariant affine connection on S^n. We shall see in 3.1.4 that this argument extends systematically to the homogeneous models of all parabolic geometries. In particular, also the projective contact, conformal, and CR–spheres from 1.1.4–1.1.6 do not admit invariant affine connections.

1.4.9. Invariant differential operators. From the point of view of geometry, invariant differential operators are the differential operators intrinsic to the given geometric structure, so these provide basic tools for working with such structures. Questions on invariant differential operators will be the main topic of volume two. This and the next subsection, which develop the basic background on homogeneous spaces, should be considered rather as a teaser for volume two, and can be skipped at the first reading.

Consider a homogeneous space G/H and two homogeneous vector bundles E, F over G/H. From 1.4.4 we get G–actions on the spaces $\Gamma(E)$ and $\Gamma(F)$ of sections, and an *invariant differential operator* is a differential operator $D : \Gamma(E) \to \Gamma(F)$ which is equivariant for the G actions, i.e. such that $D(g \cdot s) = g \cdot D(s)$ for all $s \in \Gamma(E)$ and $g \in G$.

The first step towards an algebraic description of invariant differential operators is to pass to jet prolongations. If M is a smooth manifold and $E \to M$ is any vector bundle, then for $k \in \mathbb{N}$ we have the k–jet prolongation J^kE. The fiber of J^kE at $x \in M$ is exactly the vector space of all k–jets at x of sections of E. From 1.2.8 we know that J^kE is a vector bundle over M. If F is another vector bundle over M, then a differential operator $D : \Gamma(E) \to \Gamma(F)$ is of order $\leq k$ if and only if for any two sections $s, t \in \Gamma(E)$ and any point $x \in M$, the equation $j^k_x s = j^k_x t$ implies $D(s)(x) = D(t)(x)$. If D is such an operator, then we get an induced bundle map $\tilde{D} : J^kE \to F$ over the identity on M, defined by $\tilde{D}(j^k_x s) := D(s)(x)$. Conversely, this formula associates to any bundle map $\tilde{D}$ a differential operator D. Clearly, D is linear if and only if $\tilde{D}$ is a homomorphism of vector bundles.

In the special case of a homogeneous vector bundle $E \to G/H$, functoriality of the construction of jets immediately implies that each J^kE is again a homogeneous vector bundle. More precisely, the action is defined by $g \cdot (j^k_x s) := j^k_x(g \cdot s)$. By construction, a differential operator D corresponding to the bundle map $\tilde{D} : J^kE \to F$ is invariant if and only if $\tilde{D}$ is a morphism of homogeneous fiber bundles, i.e. G–equivariant. Hence, we have reduced the determination of linear invariant differential operators to the determination of homomorphisms between homogeneous vector bundles. Since a homomorphism $J^kE \to F$ of homogeneous vector bundles is the same thing as a G–invariant section of the homogeneous vector bundle $L(J^kE, F)$, this reduces to representation theory of the group H by Theorem 1.4.4.

The problem with this is that even if the H–representation V which induces the homogeneous bundle E is very simple, the representations inducing J^kE tend to become very complicated, thus making the problem unmanageable. In any case, there is one possible simple step, namely look at the symbol of the operator. Again, we digress and consider an arbitrary vector bundle $p : E \to M$ over a smooth manifold M with k–jet prolongation $J^kE \to M$. Recall from 1.2.8 that for $\ell < k$ there is an obvious projection $\pi^k_\ell : J^kE \to J^\ell E$, defined by $\pi^k_\ell(j^k_x s) = j^\ell_x s$. Obviously, this is a vector bundle homomorphism and for $\ell = 0$ one simply gets the projection $J^kE \to E$ onto the target of a jet. The symbol of an operator of order $\leq k$ is then the restriction of the corresponding vector bundle homomorphism to the kernel of π^k_{k-1}. It is easy to see in local coordinates that this kernel is isomorphic to $S^kT^*M \otimes E$.

There is a nice description in terms of linear connections. Take a linear connection ∇ on E. For a section $s \in \Gamma(E)$, consider $\nabla s(x) \in T^*_xM \otimes E_x$. This functional depends only on $j^1_x s$ and if, in addition, $s(x) = 0$, then it is independent of ∇. Hence, it identifies $\ker \pi^1_0$ with $T^*M \otimes E$. The identification of $\ker \pi^k_{k-1}$ for $k > 1$ admits a similar description in terms of arbitrary linear connections on E and on the tangent bundle TM. Given these, one can define the k–fold covariant derivative $\nabla^k s$ for each $s \in \Gamma(E)$. For $x \in M$ the value $\nabla^k s(x)$ depends only on $j^k_x s$, and if $j^{k-1}_x s = 0$, then $\nabla^k s(x)$ is totally symmetric and independent of ∇. Hence, it induces the required isomorphism.

If F is another vector bundle over M and $D : \Gamma(E) \to \Gamma(F)$ is a differential operator of order $\leq k$ corresponding to a bundle map $\tilde{D} : J^kE \to F$, then the kth order *symbol* of D is the vector bundle map $\sigma(D) : S^kT^*M \otimes E \to F$ given by the restriction of $\tilde{D}$ to the kernel of π^k_{k-1}.

Returning to the case of a homogeneous vector bundle $E \to G/H$, the projections π^k_ℓ are by construction homomorphisms of homogeneous vector bundles, and a moment of thought shows that also the inclusions $S^k(T^*M) \otimes E \to J^kE$ are homomorphisms of homogeneous vector bundles. In particular, for any invariant linear differential operator D the symbol $\sigma(D)$ is a homomorphism of homogeneous vector bundles. Hence, it corresponds to a G–invariant section of the bundle $L(S^kT^*M \otimes E, F)$, which immediately restricts the possibilities for the existence of invariant differential operators. It is crucial that the representation corresponding to the bundle $L(S^kT^*M \otimes E, F)$ is as manageable as $\mathfrak{g}/\mathfrak{h}$ and the representations V and W inducing E and F. If there are invariant linear connections on E and $T(G/H)$, then things become easy:

PROPOSITION 1.4.9. *Let E and F be homogeneous vector bundles over a homogeneous space G/H, and suppose that there are G–invariant linear connections on E and $T(G/H)$. Fixing such connections, we can construct for each invariant bundle map $\Phi : S^kT^*M \otimes E \to F$ a canonical invariant differential operator $D_\Phi : \Gamma(E) \to \Gamma(F)$ with symbol $\sigma(D_\Phi) = \Phi$. Any G–invariant differential operator $D : \Gamma(E) \to \Gamma(F)$ can be written as a finite sum of operators obtained in that way.*

PROOF. Let us denote the invariant linear connections on E and $T(G/H)$ by ∇. Then for each $k \geq 0$ we get an induced invariant linear connection on the bundle $\otimes^k T^*(G/H) \otimes E$, so we can define iterated covariant derivatives $\nabla^k s \in \Gamma(\otimes^i T^*(G/H) \otimes E)$. We can then symmetrize this section, to obtain a section $\nabla^{(k)} s \in \Gamma(S^kT^*(G/H) \otimes E)$.

By construction, $s \mapsto \nabla^{(k)} s$ is an invariant differential operator. Given a homomorphism $\Phi : S^kT^*(G/H) \otimes E \to F$ of homogeneous vector bundles we put $D_\Phi(s) := \Phi(\nabla^{(k)} s)$. By construction, this is an invariant differential operator with symbol $\sigma(D_\Phi) = \Phi$.

Given a general invariant differential operator D, assume that k is the order of D, i.e. that D has order $\leq k$ and the kth order symbol of D is nonzero. Putting $\Phi = \sigma(D)$ we see that $D - D_\Phi$ has vanishing kth order symbol and thus is of order $\leq k - 1$. Now the result follows by induction. □

In the presence of invariant linear connections on E and $T(G/H)$ we see that each invariant symbol is realized as the symbol of an invariant differential operator. The proposition also reduces the problem of finding invariant differential operators to the description of possible symbols. This is as manageable as the representations $\mathfrak{g}/\mathfrak{h}$, V, and W. Note that the required invariant connections always exist for reductive Klein geometries. More generally, they also exist if there exists an invariant affine connection on G/H and E is a tensor bundle.

1.4.10. Invariant differential operators and homomorphisms of induced modules. Let us now switch to a homogeneous bundle E which does not admit an invariant linear connection. Then the considerations of the last subsection show that the H–representations inducing J^kE will be complicated, and this already occurs for $k = 1$: A linear connection ∇ on E is equivalent to a splitting $J^1E \cong E \oplus (T^*(G/H) \otimes E)$ via $j^1_x s \mapsto (s(x), \nabla s(x))$. The connection ∇ is invariant, if and only if this is an isomorphism of homogenous vector bundles. Passing to the corresponding representations, let V and W be the representations inducing E and J^1E. Then π^1_0 corresponds to an H–equivariant surjection $W \to V$ whose kernel is isomorphic to $(\mathfrak{g}/\mathfrak{h})^* \otimes V$. If this H–invariant subspace would admit an H–invariant complement, this would give an isomorphism $W \cong V \oplus (\mathfrak{g}/\mathfrak{h})^* \otimes V$ of H–modules. From above we know that this would give rise to an invariant linear connection on E. In Example 1.4.8 we have seen that in some cases, there are very simple representations V for which there is no invariant linear connection on the corresponding associated bundle, so in these cases W cannot be completely reducible.

There still is a way to reformulate the classification of linear invariant differential operators as an algebraic problem, which can be solved in some cases. In particular, this is true for the Klein geometries underlying parabolic geometries. In these cases, the Lie group G is semisimple and the subgroup H is a parabolic subgroup. In particular, the representation theory of the Lie algebra $\mathfrak{g}$ is well understood, while the situation with representations of H is much more complicated.

Completely reducible representations of H are manageable since they come from a reductive subgroup, but from above we see that passing to jet prolongation one leaves the realm of completely reducible representations.

The algebraic reformulation that can be successfully used in the parabolic case is in principle available in general. However, since it is technically rather demanding and the parabolic case is the main application, we prefer to only briefly sketch this approach here and give a detailed treatment in the parabolic case in volume two.

The starting point is to look at the universal enveloping algebra $\mathcal{U}(\mathfrak{g})$ of the Lie algebra $\mathfrak{g}$; see 2.1.10. This is a unital associative algebra, endowed with an inclusion $i : \mathfrak{g} \to \mathcal{U}(\mathfrak{g})$ such that $i([X, Y]) = i(X)i(Y) - i(Y)i(X)$. It turns out that representations of $\mathfrak{g}$ are equivalent to representations of $\mathcal{U}(\mathfrak{g})$. The universal enveloping algebra of the subalgebra $\mathfrak{h} \subset \mathfrak{g}$ naturally is a subalgebra in $\mathcal{U}(\mathfrak{g})$. In particular, $\mathcal{U}(\mathfrak{g})$ is naturally a right $\mathcal{U}(\mathfrak{h})$–module.

Given a representation of $\mathfrak{h}$ on a vector space V, this space is a left $\mathcal{U}(\mathfrak{h})$–module, and one can form the *induced module* $\mathcal{U}(\mathfrak{g}) \otimes_{\mathcal{U}(\mathfrak{h})} V$, which is a $\mathcal{U}(\mathfrak{g})$–module under multiplication from the left, and thus a representation of $\mathfrak{g}$ (of infinite dimension). If one wants to take into account the group H, one may consider the induced module as a $(\mathfrak{g}, H)$–module.

The relation to differential operators comes from looking at infinite jets. Given a homogeneous vector bundle $E \to G/H$, one looks at the infinite jet prolongation $J^\infty E$, which is defined as the direct limit of the system $\cdots \to J^k E \to J^{k-1} E \to \dots$. By construction, this is an infinite–dimensional homogeneous vector bundle, so we are naturally led to consider its fiber $J^\infty E_o$ over the point $o = eH$, as an H–module. The advantage of passing to infinite jets is that there is a canonical action of the Lie algebra $\mathfrak{g}$ on this fiber, which comes from differentiation by right invariant vector fields. This makes $J^\infty E_o$ into a $(\mathfrak{g}, H)$–module.

Next, the identification of the Lie algebra $\mathfrak{g}$ as left invariant vector fields on G induces an identification of the universal enveloping algebra $\mathcal{U}(\mathfrak{g})$ with the space of left invariant differential operators on $C^\infty(G, \mathbb{R})$. This gives rise to a pairing between $J^\infty_e(G, \mathbb{R})$ and $\mathcal{U}(\mathfrak{g})$ induced by applying a left invariant differential operator to the representative of an infinite jet and evaluating the result in e. One shows that this induces a linear isomorphism between $\mathcal{U}(\mathfrak{g})$ and the set of those linear maps $J^\infty_e(G, \mathbb{R}) \to \mathbb{R}$ which factor over some $J^k_e(G, \mathbb{R})$.

More generally, looking at functions with values in a finite–dimensional vector space V, one gets an identification of $\mathcal{U}(\mathfrak{g}) \otimes V^*$ and the space of those maps $J^\infty_e(G, V) \to \mathbb{R}$ which factor over some $J^k_e(G, V)$. Analyzing the action on jets of equivariant maps, one finally obtains an identification between the induced module $\mathcal{U}(\mathfrak{g}) \otimes_{\mathcal{U}(\mathfrak{h})} V^*$ and the space of those linear maps $J^\infty_o E \to \mathbb{R}$ which factor over some $J^k_o E$. One verifies directly that this pairing is compatible with the $(\mathfrak{g}, H)$–module structures on both spaces.

Now consider two homogeneous vector bundles E and F corresponding to H–representations V and W. From the discussion in 1.4.9 we conclude that the space of all linear invariant differential operators $D : \Gamma(E) \to \Gamma(F)$ is isomorphic to the space of all H–module maps $J^\infty_o E \to F$ which factor over some $J^k_o E$. Using the above pairing, such maps may be interpreted as H–invariant elements in $(\mathcal{U}(\mathfrak{g}) \otimes_{\mathcal{U}(\mathfrak{h})} V^*) \otimes W$, or equivalently as H–module homomorphisms $W^* \to \mathcal{U}(\mathfrak{g}) \otimes_{\mathcal{U}(\mathfrak{h})} V^*$. This also works the other way around, so one gets an isomorphism between invariant linear differential operators and H–homomorphisms of the above type.

The final ingredient then is an algebraic version of Frobenius reciprocity from 1.4.4, which we will prove in 2.1.10. This leads to a bijective correspondence between H–homomorphisms $W^* \to \mathcal{U}(\mathfrak{g}) \otimes_{\mathcal{U}(\mathfrak{h})} V^*$ and $(\mathfrak{g}, H)$–module homomorphisms $\mathcal{U}(\mathfrak{g}) \otimes_{\mathcal{U}(\mathfrak{h})} W^* \to \mathcal{U}(\mathfrak{g}) \otimes_{\mathcal{U}(\mathfrak{h})} V^*$. Thus, we get the following description of linear invariant differential operators:

THEOREM 1.4.10. *Let G be a Lie group, $H \subset G$ a closed subgroup, and V and W finite–dimensional representations of H. Let E and F denote the homogeneous vector bundles $G \times_H V$ and $G \times_H W$, respectively.*

Then the space of linear invariant differential operators $D : \Gamma(E) \to \Gamma(F)$ is isomorphic to the space $\operatorname{Hom}_{(\mathfrak{g},H)}(\mathcal{U}(\mathfrak{g}) \otimes_{\mathcal{U}(\mathfrak{h})} W^*, \mathcal{U}(\mathfrak{g}) \otimes_{\mathcal{U}(\mathfrak{h})} V^*)$ of homomorphisms of $(\mathfrak{g}, H)$–modules.

In the case of a semisimple Lie algebra $\mathfrak{g}$ and a parabolic subalgebra $\mathfrak{h}$ the induced modules in question are called generalized Verma modules, and they have been intensively studied in representation theory. In particular, a good amount of results on homomorphisms between such modules is available. In the special case where $\mathfrak{h}$ is even a Borel subalgebra, one obtains Verma modules, and there is a complete classification of homomorphisms between such modules. We will discuss these results in more detail in volume two.

1.4.11. Distinguished curves. The final notion that we want to discuss in this section are preferred families of curves, which lead to analogs of geodesics and normal coordinates. Depending on the choice of homogenous space, there may be various interesting families of distinguished curves. We will always require that a family of distinguished curves on G/H is stable under the action of G. This implies that the whole family is determined by the curves through the origin $o = eH$. For this section, we will restrict to a specific concept of distinguished curves which is always available.

For the purpose of motivation, suppose that there is an invariant principal connection on the bundle $p : G \to G/H$. Then we get an induced linear connection on $T(G/H)$ and we can use the geodesics of this linear connection as the distinguished curves. The principal connection on G/H is determined by the choice of an H–invariant complement $\mathfrak{n}$ to $\mathfrak{h}$ in $\mathfrak{g}$; see Corollary 1.4.6. Now for $X \in \mathfrak{n}$ we can consider the curve $t \mapsto p(\exp(tX))$ in G/H. By definition $t \mapsto \exp(tX)$ is a lift of this curve on which the Maurer–Cartan form is constant, so our curve is a geodesic through o.

This suggests a generalization. Assume that we have given a Lie group G, a closed subgroup $H \subset G$, and fix a linear complement $\mathfrak{n}$ of $\mathfrak{h}$ in $\mathfrak{g}$. For any $g \in G$ we then have the notion of horizontal tangent vectors at g, namely $\xi \in T_gG$ is horizontal if and only if $\omega(\xi) = T\lambda_{g^{-1}} \cdot \xi \in \mathfrak{n}$, where ω is the Maurer–Cartan form on G. In particular, for $X \in \mathfrak{n}$ we have the left invariant vector field $L_X \in \mathfrak{X}(G)$ which is horizontal. The projections of the flow lines of these fields are then the basic distinguished curves.

A smooth curve $c : I \to G/H$ defined on an open interval $I \subset \mathbb{R}$ is called distinguished if and only if there are a point $t_0 \in I$ and elements $g \in G$ with $p(g) = c(t_0)$ and $X \in \mathfrak{n}$ such that $c(t) = p(\mathrm{Fl}^{L_X}_{t-t_0}(g)) = p(g \exp((t - t_0)X))$ for all $t \in I$. Note that since left invariant vector fields are always complete, any distinguished curve can be canonically extended to a curve defined on $\mathbb{R}$. Note further, that by the flow property, if $c : I \to G/H$ is a distinguished curve, then for any $t_0 \in I$ there are $g \in G$ and $X \in \mathfrak{n}$ such that $c(t) = p(g \exp((t - t_0)X))$.

PROPOSITION 1.4.11. *Let G be a Lie group, $H \subset G$ a closed subgroup and $\mathfrak{n} \subset \mathfrak{g}$ a linear subspace complementary to $\mathfrak{h}$. Then we have:*

(1) *If $c : I \to M$ is a distinguished curve and $g \in G$ is any element, then $\ell_g \circ c : I \to M$ is a distinguished curve, too.*
(2) *For any point $x \in G/H$ and any tangent vector $\xi \in T_x G/H$ there is at least one distinguished curve $c : \mathbb{R} \to G/H$ such that $c(0) = x$ and $c'(0) = \xi$.*
(3) *If the complement $\mathfrak{n}$ is H–invariant, then the curve c in (2) is uniquely determined. It coincides with the geodesic of the linear connection on $T(G/H)$ induced by $\mathfrak{n}$.*
(4) *For any $g \in G$, the mapping $\mathfrak{n} \to G/H$, $X \mapsto p(g \exp X)$ defines local coordinates around $p(g)$ in which the straight lines through the origin in $\mathfrak{n}$ map to distinguished curves through $p(g)$.*

PROOF. (1) By definition, there are elements $t_0 \in I$, $g_0 \in G$ and $X \in \mathfrak{n}$ such that $c(t) = p(g_0 \exp((t - t_0)X))$. But then by definition of the action ℓ_g, we get $\ell_g(c(t)) = p(gg_0 \exp((t - t_0)X))$, so $\ell_g \circ c$ is distinguished, too.

(2) By (1) it suffices to consider the case $x = o$. Then for $\xi \in T_o G/H$ there is a unique element $X \in \mathfrak{n}$ such that $\xi = T_e p \cdot X$. Thus, $p(\exp(tX))$ is a distinguished curve as required.

(3) Again, we may confine ourselves to the case $x = o$. As above, take $X \in \mathfrak{n}$ such that $T_e p \cdot X = \xi$. Since $p^{-1}(o) = H$, any other distinguished curve through o in direction ξ is of the form $p(h \exp(tY))$ for h in H, with $Y \in \mathfrak{n}$ the unique element such that $T_h p \cdot L_Y(h) = \xi$. But then

$$\xi = T_h p \circ T_e \lambda_h \cdot Y = T_e p \circ T_h \rho^{h^{-1}} \circ T_e \lambda_h \cdot Y = T_e p \cdot \mathrm{Ad}(h)Y.$$

Thus, $X - \mathrm{Ad}(h)Y \in \mathfrak{h}$, but since $\mathfrak{n}$ is H–invariant, we have $\mathrm{Ad}(h)Y \in \mathfrak{n}$ and thus $Y = \mathrm{Ad}(h^{-1})X$. But then $h \exp(t \,\mathrm{Ad}(h^{-1})X) = hh^{-1} \exp(tX)h$, so $p(h \exp(tY)) = p(\exp(tX))$.

(4) is obvious from the definitions. □

REMARK 1.4.11. We may equivalently define the distinguished curves by H–invariant data at the origin $o \in G/H$. The distinguished curves $c(t)$ with $c(0) = o$ form an H–invariant set, and the entire set of the distinguished curves is obtained from them by the left shifts.

More generally, we may fix any H-invariant subset A of curves $\alpha(t)$, $\alpha(0) = o$ and to define the A–distinguished curves as all curves of the form $\ell_g \circ \alpha$ for $g \in G$ and $\alpha \in A$. In particular, each choice of an H–invariant subspace $\mathfrak{a} \subset \mathfrak{n}$ in the complement $\mathfrak{n}$ to $\mathfrak{h}$ with respect to the induced adjoint action leads to a subclass of distinguished curves emanating in directions contained in the distribution $\mathcal{A} \subset T(G/H)$ determined by the subspace $\mathfrak{a}$.

1.5. Cartan connections

Having the necessary background at hand, we can now start to investigate Cartan geometries. Throughout this section we will take Cartan geometries as a given input and develop basic tools for the analysis of such structures. We will look for simpler structures underlying a Cartan geometry, but we will not touch the question to what extent these structures determine the Cartan geometry. This question will be taken up in the next section. It should, however, be kept in mind that in many cases of interest Cartan geometries are equivalent to more conventional geometric structures and hence the tools developed here provide additional approaches to the study of those structures.

Any Cartan geometry is derived from a homogeneous space, called the homogeneous model of the geometry. The interplay between this homogeneous model and general Cartan geometries of the given type is one of the main general features of Cartan geometries and an important topic for this section.

1.5.1. Basic concepts. Let $H \subset G$ be a Lie subgroup in a Lie group G, and let $\mathfrak{g}$ be the Lie algebra of G. A *Cartan geometry* of type (G, H) on a manifold M is a principal fiber bundle $p : \mathcal{P} \to M$ with structure group H, which is endowed with a $\mathfrak{g}$–valued one–form $\omega \in \Omega^1(\mathcal{P}, \mathfrak{g})$, called the *Cartan connection.* We require that ω is H–equivariant, reproduces the generators of fundamental vector fields, and defines an absolute parallelism. More formally, this means that

$$(r^h)^*\omega = \mathrm{Ad}(h^{-1}) \circ \omega \text{ for all } h \in H, \tag{1.12}$$

$$\omega(\zeta_X(u)) = X \text{ for each } X \in \mathfrak{h}, \tag{1.13}$$

$$\omega(u) : T_u\mathcal{P} \to \mathfrak{g} \text{ is a linear isomorphism for all } u \in \mathcal{P}. \tag{1.14}$$

The *homogeneous model* for Cartan geometries of type (G, H) is the canonical bundle $p : G \to G/H$ endowed with the left Maurer–Cartan form $\omega \in \Omega^1(G, \mathfrak{g})$; see 1.2.4. In the terminology of 1.4.1, the homogeneous model for Cartan geometries of type (G, H) is the Klein geometry of that type.

Given a Cartan geometry $(\mathcal{P} \to M, \omega)$, there are the *constant vector fields* $\omega^{-1}(X) \in \mathfrak{X}(\mathcal{P})$ defined for all $X \in \mathfrak{g}$ by $\omega(\omega^{-1}(X)(u)) = X$ for all $u \in \mathcal{P}$. From equivariancy of ω we get

$$\omega^{-1}(X)(u \cdot h) = Tr^h \cdot \omega^{-1}(\mathrm{Ad}(h) \cdot X)(u) \tag{1.15}$$

for all $h \in H$. In the case of the homogeneous model, the constant vector field $\omega^{-1}(X)$ is the left invariant field L_X by definition of the Maurer–Cartan form.

The *curvature form* $K \in \Omega^2(\mathcal{P}, \mathfrak{g})$ of a Cartan geometry $(\mathcal{P} \to M, \omega)$ is defined by the structure equation

$$K(\xi, \eta) := d\omega(\xi, \eta) + [\omega(\xi), \omega(\eta)]. \tag{1.16}$$

Notice that the Maurer–Cartan equation implies that the Maurer–Cartan form on $G \to G/H$ always has zero curvature. Therefore, the homogeneous model is often also referred to as the flat model, but we avoid this terminology since it is sometimes confusing.

Since the Cartan connection ω trivializes $T\mathcal{P}$, any differential form on $\mathcal{P}$ is determined by its values on the constant vector fields $\omega^{-1}(X)$. Thus, the complete information about K is contained in the *curvature function* $\kappa : \mathcal{P} \to \Lambda^2\mathfrak{g}^* \otimes \mathfrak{g}$ defined by $\kappa(u)(X, Y) = K(\omega^{-1}(X)(u), \omega^{-1}(Y)(u))$, and so the standard formula for the exterior derivative d yields

$$\kappa(u)(X, Y) = [X, Y] - \omega\big([\omega^{-1}(X), \omega^{-1}(Y)](u)\big). \tag{1.17}$$

LEMMA 1.5.1. *The curvature form $K \in \Omega^2(\mathcal{P}, \mathfrak{g})$ is horizontal, so the curvature function may be viewed as $\kappa : \mathcal{P} \to \Lambda^2(\mathfrak{g}/\mathfrak{h})^* \otimes \mathfrak{g}$. Moreover, for all $h \in H$, we get*

$$(r^h)^* K = \mathrm{Ad}(h^{-1}) \circ K, \tag{1.18}$$

$$\kappa \circ r^h = \lambda(h^{-1}) \circ \kappa, \tag{1.19}$$

where λ is the tensor product of the actions $\Lambda^2\, \underline{\mathrm{Ad}}^*$ on $\Lambda^2(\mathfrak{g}/\mathfrak{h})^*$ and Ad *on $\mathfrak{g}$.*

PROOF. We just have to imitate the computation from 1.3.3. By definition of a Cartan connection, if $X \in \mathfrak{h}$, then $\omega^{-1}(X) = \zeta_X$, the fundamental vector field. Equivariancy of ω immediately implies that $\mathcal{L}_{\zeta_X}\omega = i_{\zeta_X} d\omega = -\operatorname{ad}(X) \circ \omega$. But this gives

$$d\omega(\omega^{-1}(X), \eta) + [X, \omega(\eta)] = -\operatorname{ad}(X)(\omega(\eta)) + [X, \omega(\eta)] = 0$$

for all X in $\mathfrak{h}$ and all η. Since the fundamental fields span the vertical bundle, we conclude that K is horizontal, and that each $\kappa(u)$ factors to $\Lambda^2(\mathfrak{g}/\mathfrak{h})$.

The equivariancy property (1.18) of K follows directly from the definition and the compatibility of the pullback with the exterior differential d. To prove (1.19), we have to compute $\kappa(u \cdot h)(X, Y)$. By definition, we get

$$\begin{aligned} K(\omega_{u\cdot h}^{-1}(X), \omega_{u\cdot h}^{-1}(Y)) &= K(Tr^h \cdot \omega_u^{-1}(\mathrm{Ad}(h)\cdot X), Tr^h \cdot \omega^{-1}(\mathrm{Ad}(h) \cdot Y)) \\ &= (r^h)^* K(u)(\omega^{-1}(\mathrm{Ad}(h)\cdot X), \omega^{-1}(\mathrm{Ad}(h)\cdot Y)) \\ &= \mathrm{Ad}(h^{-1}) \cdot \kappa(u)(\mathrm{Ad}(h)\cdot X, \mathrm{Ad}(h)\cdot Y). \end{aligned}$$

Passing from $\mathfrak{g}$ to $\mathfrak{g}/\mathfrak{h}$ the two occurrences of $\mathrm{Ad}(h)$ inside of $\kappa(u)$ get replaced by $\underline{\mathrm{Ad}}(h)$, and we obtain the required formula. □

EXAMPLE 1.5.1. (i) Let $A(m, \mathbb{R})$ be the affine group in dimension m. In 1.3.5 we have seen that a Cartan geometry of type $(A(m, \mathbb{R}), GL(m, \mathbb{R}))$ on an m–dimensional manifold M is equivalent to a linear connection on the tangent bundle TM. Moreover, the curvature K as defined above exactly encodes the curvature and torsion of this linear connection.

(ii) For a Lie group H and an infinitesimally injective homomorphism $H \to GL(m, \mathbb{R})$ consider the affine extension $B = \mathbb{R}^m \rtimes H$. In 1.3.6 we have seen that a Cartan geometry of type (B, H) is equivalent to a first order G–structure with structure group H endowed with a connection. The curvature of the Cartan connection again can be interpreted as curvature and torsion of the induced linear connection on the tangent bundle.

(iii) More specifically, let us consider $H = O(m) \subset GL(m, \mathbb{R})$. Then the affine extension $\mathbb{R}^m \rtimes H$ is the Euclidean group $\mathrm{Euc}(m)$ as used in 1.1.2. By Example (1) of 1.3.6 an $O(m)$–structure on an m–dimensional smooth manifold M is equivalent to a Riemannian metric g on M. From (ii) we thus conclude that a Cartan geometry of type $(\mathrm{Euc}(m), O(m))$ is equivalent to a connection on the orthonormal frame bundle for g and hence to a metric linear connection on TM. It is a classical result that there is a unique metric linear connection on TM, which, in addition, is torsion free, namely the Levi–Civita connection. This shows that on each Riemannian manifold of dimension m, we can actually obtain a canonical Cartan geometry of type $(\mathrm{Euc}(m), O(m))$. The curvature in this case coincides with the usual Riemann curvature.

This is a prototypical example for a Cartan geometry which is determined by an underlying structure. We will analyze this case more systematically in 1.6.1.

The interpretation of Riemannian structures as Cartan geometries is one of the motivating examples for the concept. Interesting applications of this point of view can be found in the book [**Sh97**]. We will often use this case as an illustration in this chapter. One has to keep in mind, however, that the geometric structures that we will be ultimately interested in are much more complicated than Riemannian structures.

(iv) By 1.1.1 and 1.1.2, there is no difference between Cartan geometries of the types $(O(m+1), O(m))$, $(\mathrm{Euc}(m), O(m))$, and $(O(m,1), O(m))$. This is because $\mathfrak{o}(m+1)$, $\mathfrak{euc}(m+1)$, and $\mathfrak{o}(m,1)$ are all isomorphic as $O(m)$–modules. However, the notion of curvature is different for these three types of geometries, since the homogeneous models S^m and $\mathcal{H}^m$ of the first and last types have (nonzero) constant curvature for the second type. This is an example of model mutation; see [**Sh97**, 4, §3].

1.5.2. Categories of Cartan geometries. A *morphism* between two Cartan geometries $(\mathcal{P} \to M, \omega)$ and $(\mathcal{P}' \to M', \omega')$ of type (G, H) is a principal bundle morphism $\phi : \mathcal{P} \to \mathcal{P}'$ such that $\phi^*\omega' = \omega$. Notice that compatibility with the Cartan connections implies that any tangent map of ϕ is a linear isomorphism, so ϕ and its base map are local diffeomorphisms. With this definition of morphisms, Cartan geometries of type (G, H) form a category $\mathcal{C}_{(G,H)}$.

LEMMA 1.5.2. *Let $\phi : \mathcal{P} \to \mathcal{P}'$ be a morphism of principal fiber bundles which is a local diffeomorphism. If ω' is a Cartan connection on $\mathcal{P}'$, then $\omega = \phi^*\omega'$ is a Cartan connection on $\mathcal{P}$. If ω' and ω are fixed Cartan connections on $\mathcal{P}$ and $\mathcal{P}'$, then ϕ is a morphism $(\mathcal{P}, \omega) \to (\mathcal{P}', \omega')$ if and only if ϕ preserves the constant vector fields, i.e. $T\phi \circ \omega^{-1}(X) = {\omega'}^{-1}(X) \circ \phi$. In this case the curvature forms K and K' are ϕ–related, and the curvature functions satisfy $\kappa = \kappa' \circ \phi$.*

PROOF. The fundamental vector fields are given by $\zeta_X(u) = \frac{d}{dt}|_0 u \cdot \exp tX$, so equivariancy of ϕ implies that $\phi^*\omega'$ reproduces the generators of fundamental vector fields. Similarly, equivariancy of ϕ and ω' implies equivariancy of $\phi^*\omega'$. Since ϕ is assumed to be a local diffeomorphism, $\phi^*\omega' = \omega' \circ T\phi$ restricts to a linear isomorphism on each tangent space, so we have verified that $\phi^*\omega'$ is a Cartan connection.

The pullback $\phi^*\omega'$ evaluates on a constant field as

$$\phi^*\omega'(\omega^{-1}(X)(u)) = \omega'(\phi(u))(T_u\phi \cdot \omega^{-1}(X)(u))$$

and the right–hand side equals X if and only if $T\phi \circ \omega^{-1}(X)(u) = {\omega'}^{-1}(X)(\phi(u))$. Thus, morphisms are characterized by the fact that they preserve the constant fields. The relatedness of the curvature forms follows immediately from their definition via the structure equation. Finally, the relation between K and K' implies

$$\begin{aligned}\kappa(u)(X,Y) &= K(u)(\omega^{-1}(X), \omega^{-1}(Y)) = K'(\phi(u))({\omega'}^{-1}(X), {\omega'}^{-1}(Y)) \\ &= \kappa'(\phi(u))(X,Y)\end{aligned}$$

for all $u \in \mathcal{P}$, $X, Y \in \mathfrak{g}$. □

Various interesting and useful subcategories in $\mathcal{C}_{(G,H)}$ can be defined by restrictions on curvatures. Such restrictions are usually necessary to characterize Cartan geometries that are equivalent to simpler structures. The simplest way to restrict curvatures is by requiring the curvature function κ to have values in a fixed subspace $\mathfrak{N} \subset \Lambda^2(\mathfrak{g}/\mathfrak{h})^* \otimes \mathfrak{g}$. The simple transformation law $\kappa = \kappa' \circ \phi$ immediately implies that this specifies a full subcategory in $\mathcal{C}_{(G,H)}$. However, as we have seen in 1.5.1, $\kappa(u \cdot g) = g^{-1} \cdot \kappa(u)$, so the values of κ always span an H–invariant subset in $\Lambda^2(\mathfrak{g}/\mathfrak{h})^* \otimes \mathfrak{g}$. Thus, it is natural to require that $\mathfrak{N}$ is an H–submodule. Having chosen a submodule $\mathfrak{N}$, we obtain the full subcategory $\mathcal{C}^{\mathfrak{N}}_{(G,H)}$ of objects whose curvature functions have values in $\mathfrak{N}$.

The appropriate choice of a normalization condition $\mathfrak{N}$ is often a crucial and difficult step in describing geometric structures as Cartan geometries. However, for any pair (G, H) there are two obvious choices available, namely $\mathfrak{N} = 0$ and $\mathfrak{N} = \Lambda^2(\mathfrak{g}/\mathfrak{h})^* \otimes \mathfrak{h}$. In the first case we call the geometries *locally flat* while in the other case we talk about *torsion free* Cartan geometries. The choice of the name "torsion free" should be clear from the examples treated in 1.3.5 and 1.3.6, where it amounts to torsion freeness of the induced linear connection on TM. The name "locally flat" is explained by the first part of the proposition below.

Notice that for any Cartan geometry $(\mathcal{P} \to M, \omega)$ and an open subset $U \subset M$ there is a canonical Cartan geometry $(p^{-1}(U) \to U, \omega|_{p^{-1}(U)})$ on U, so one may restrict Cartan geometries to open subsets.

PROPOSITION 1.5.2. *(1) The curvature of a Cartan geometry $(\mathcal{P} \to M, \omega)$ vanishes identically if and only if any point $x \in M$ has an open neighborhood U such that the restriction $(p^{-1}(U) \to U, \omega)$ is isomorphic to the restriction of the homogeneous model $(G \to G/H, \omega_G)$ to an open neighborhood of o.*

(2) If G/H is connected, then the automorphisms of the Cartan geometry $(G \to G/H, \omega_G)$ are exactly the left multiplications by elements of G.

(3) (Liouville theorem) Suppose that G/H is connected. Then any isomorphism between two restrictions of $(G \to G/H, \omega_G)$ to connected open subsets of G/H uniquely globalizes to an automorphism of the homogeneous model.

PROOF. (1) Assume that the curvature vanishes identically. Then Theorem 1.2.4 implies that for each $u \in \mathcal{P}$, there is a neighborhood V of u in $\mathcal{P}$ and a unique mapping $\phi : V \to G$ such that $\phi(u) = e$ and $\phi^*(\omega_G) = \omega$. In particular, ϕ respects the constant fields restricted to V. This implies that for each $v \in V$, $\phi(v \cdot \exp X) = \phi(v) \cdot \exp X$ on a neighborhood of $0 \in \mathfrak{g}$, and so ϕ can be extended uniquely to a principal bundle morphism over a neighborhood U of $p(u)$. By equivariancy we still have $\phi^*\omega_G = \omega$ on the whole domain of ϕ. The other implication is obvious.

(2) Again, by Theorem 1.2.4, a smooth map $f : G \to G$ satisfies $f^*\omega_G = \omega_G$ if and only if it is left multiplication by an element of G. Conversely, since left and right multiplications commute, any left multiplication is an automorphism of the principal bundle $G \to G/H$.

(3) Let $p : G \to G/H$ be the projection and consider connected open subsets U and V in G/H and a principal bundle automorphism $\phi : p^{-1}(U) \to p^{-1}(V)$ such that $\phi^*\omega_G = \omega_G$. Viewing ϕ as a map $p^{-1}(U) \to G$, the uniqueness part of Theorem 1.2.4 tells us that ϕ differs from the inclusion by a left multiplication with a fixed element of G, which implies the result. □

REMARK 1.5.2. (1) Part (2) of this proposition shows that the Cartan geometry of type (G, H) on the homogeneous space G/H is a geometric structure which has precisely G as its automorphism group, thus justifying the point of view we took in Section 1.4.

(2) While the proof of the Liouville theorem in part (3) of the proposition is very simple, this is a rather impressive general result. It becomes particularly powerful

for Cartan geometries determined by some underlying structure. A simple example is the case of Euclidean space. In this case the Cartan geometry is determined by the Riemannian structure, and we obtain the result that any isometry between open subsets of Euclidean space is the restriction of a unique Euclidean motion. The classical Liouville theorem is the version of this result for the conformal sphere from 1.1.5; see 1.6.9. Of course, to deduce this from the proposition above, one needs the result that conformal structures are equivalent to a Cartan geometry, which we will prove in Section 1.6 below.

(3) Parts (1) and (3) of the proposition can be used to obtain an alternative description of locally flat Cartan geometries of type (G, H). If $(p : \mathcal{P} \to M)$ is such a geometry, we can use part (1) to obtain a covering of M by open subsets U_i and isomorphisms from $p^{-1}(U_i) \to U_i$ onto restrictions of $G \to G/H$. The base maps of these isomorphisms are diffeomorphisms ϕ_i from the U_i onto open subsets of G/H. Viewing $\{(U_i, \phi_i)\}$ as an atlas, the transition functions are the restrictions of left actions of elements of G by part (3) of Proposition 1.5.2.

Conversely, suppose we have given an atlas for a manifold M such that the images of the charts are open subsets in G/H and the transition functions are restrictions of left actions of elements of G. Then we can pull back the appropriate restrictions of $G \to G/H$ to the domains of the charts and glue them via the isomorphism provided by left translations to a principal H–bundle over M. The pullbacks of the Maurer–Cartan form to these pieces can be glued together to a Cartan connection on this H–bundle. The resulting Cartan geometry on M is by construction locally isomorphic to $G \to G/H$ and hence locally flat.

This construction is particularly transparent in the case that the Cartan geometry is actually equivalent to some underlying structure. For example, an atlas on M with images in open subsets of $\mathbb{R}^n$ such that the transition functions are conformal isometries for the flat metric on $\mathbb{R}^n$ evidently gives rise to a locally flat conformal structure on M.

1.5.3. Rigidity of morphisms. By definition, morphisms between Cartan geometries are special principal bundle homomorphisms, and it is natural to ask to what extent they are determined by the underlying maps between the base manifolds. In the case of the homogeneous model we have seen above that automorphisms are exactly given by left multiplications by elements of G and left multiplication by g corresponds to the base map $\ell_g : G/H \to G/H$. Thus, the automorphisms covering the identity map are the actions of the elements of the kernel K of the Klein geometry introduced in 1.4.1. There we saw that K is the maximal normal subgroup of G that is contained in H and its Lie algebra $\mathfrak{k}$ is the maximal ideal of $\mathfrak{g}$ which is contained in $\mathfrak{h}$.

Recall also from 1.4.1 that the Klein geometry of type (G, H) is called effective if its kernel K is trivial and infinitesimally effective if K is discrete. Surprisingly, the kernel K also determines the maximal number of morphisms covering a fixed base map in the case of general Cartan geometries of type (G, H). In particular, if (G, H) is effective, then any morphism is uniquely determined by its base map. We shall prove slightly more than this following [**Sh97**, Chapter 5]:

PROPOSITION 1.5.3. *Let G be a Lie group, $H \subset G$ a closed subgroup such that G/H is connected, and let $K \subset H$ be the kernel of the Klein geometry (G, H). Let ϕ_1 and ϕ_2 be two morphisms between two Cartan geometries $(\mathcal{P} \to M, \omega)$ and $(\mathcal{P}' \to M', \omega')$ of type (G, H) which cover the same base mapping $f : M \to M'$.*

Then there is a smooth map $\psi : \mathcal{P} \to K$ *such that* $\phi_2(u) = \phi_1(u) \cdot \psi(u)$ *for all* $u \in \mathcal{P}$.

In particular, if (G,H) *is effective, then* $\phi_1 = \phi_2$*, and if* (G,H) *is infinitesimally effective, then* ψ *is constant on connected components of* M.

PROOF. Since the statement is local, we may assume that both ϕ_1 and ϕ_2 are diffeomorphisms, and then by assumption $\phi = \phi_2^{-1} \circ \phi_1$ covers the identity mapping on M and $\phi^*\omega = \omega$. Thus, we may, without loss of generality, assume that $\mathcal{P} = \mathcal{P}'$, $\phi := \phi_1$ covers the identity id_M, and $\phi_2 = \mathrm{id}_{\mathcal{P}}$. These assumptions imply that there is a smooth map $\psi : \mathcal{P} \to H$ such that $\phi(u) = u \cdot \psi(u)$, and to prove the result we have to show that $\psi(u) \in K$ for all u.

Let us compute $\phi^*\omega$ in terms of ψ. If c is a smooth curve in $\mathcal{P}$ with $c(0) = u$ and $c'(0) = \xi$, then $\phi^*\omega(\xi)$ is obtained as the evaluation of ω on the vector defined by the curve

$$t \mapsto r(c(t), \psi(c(t))) = r\big(c(t), \psi(u) \cdot (\psi(u)^{-1}\psi(c(t)))\big),$$

where r is the principal right action of H on $\mathcal{P}$. Thus,

$$\phi^*\omega(\xi) = \omega(Tr^{\psi(u)} \cdot \xi) + \omega(\zeta_Z(u \cdot \psi(u))),$$

where $Z = \frac{d}{dt}|_0(\psi(u)^{-1} \cdot \psi(c(t))) \in \mathfrak{h}$. Since the Cartan connection reproduces the generators of fundamental vector fields, this shows that the whole second summand equals $\psi^*\omega_H$, where ω_H is the Maurer–Cartan form on H. Altogether we have proved

$$\phi^*\omega(u) = \mathrm{Ad}_{\psi(u)^{-1}} \circ \omega(u) + \psi^*\omega_H(u).$$

By our assumptions, $\phi^*\omega = \omega$ and so we conclude

$$(\mathrm{Ad}_{\psi(u)^{-1}} - \mathrm{id}_{\mathfrak{g}})(X) = -\psi^*\omega_H(\omega^{-1}(X)(u)) \tag{1.20}$$

for all $X \in \mathfrak{g}$.

For any Lie subalgebra $\mathfrak{q} \subset \mathfrak{h}$ we write

$$K_{\mathfrak{q}} = \{h \in H : \mathrm{Ad}(h^{-1})(X) - X \in \mathfrak{q} \text{ for all } X \in \mathfrak{g}\}.$$

Since

$$\mathrm{Ad}((hg)^{-1})(X) - X = (\mathrm{Ad}(g^{-1})(\mathrm{Ad}(h^{-1})(X)) - \mathrm{Ad}(h^{-1})(X)) + (\mathrm{Ad}(h^{-1})(X) - X)$$

is in $\mathfrak{q}$ whenever both g and h belong to $K_{\mathfrak{q}}$, the subset $K_{\mathfrak{q}}$ is a closed subgroup and thus a Lie subgroup of H. Now we define a series of Lie subgroups of H by $K_0 = H$ and inductively by $K_i = K_{\mathfrak{k}_{i-1}}$, where $\mathfrak{k}_{i-1}$ is the Lie algebra of K_{i-1}, for all $i = 1, 2, \dots$.

Assume that K_i is normal in H for some i. Then $\mathrm{Ad}(h)(\mathfrak{k}_i) \subset \mathfrak{k}_i$ for all $h \in H$ and so

$$\mathrm{Ad}(h^{-1}gh)(X) - X = \mathrm{Ad}(h^{-1})\big(\mathrm{Ad}(g)(\mathrm{Ad}(h)(X)) - \mathrm{Ad}(h)(X)\big) \in \mathrm{Ad}(h^{-1})(\mathfrak{k}_i) \subset \mathfrak{k}_i$$

for all $g \in K_{i+1}$, $h \in H$. Thus, K_{i+1} is normal in H, too. Since $K_0 = H$ is normal, this implies that all K_i, $i = 1, 2, \dots$ are normal Lie subgroups in H.

But now assume that our function ψ has values in a subgroup $Q \subset H$. Then visibly the right–hand side of equation (1.20) lies in the Lie algebra $\mathfrak{q}$ of Q. But for the left–hand side of (1.20), lying in $\mathfrak{q}$ exactly means that $\psi(u) \in K_{\mathfrak{q}}$. Starting from the fact that ψ has values in $K_0 = H$ we conclude that ψ has values in $K_{\mathfrak{h}} = K_1$, thus in K_2, and so on. Therefore, ψ has values in the intersection $K_\infty = \bigcap_{i=0}^\infty K_i$. Moreover, we clearly have $K_i \supset K_{i+1}$ for all i and so the chain of subalgebras

$\mathfrak{h} = \mathfrak{k}_0 \supset \mathfrak{k}_1 \supset \dots$ has to stabilize at some finite i. Hence, K_∞ coincides with some K_i and therefore is a normal Lie subgroup of H. Let us write $\mathfrak{k}_\infty$ for its Lie algebra and notice

$$K_\infty = \{h \in H : \operatorname{Ad}(h)(X) - X \in \mathfrak{k}_\infty \text{ for all } X \in \mathfrak{g}\}.$$

For $Y \in \mathfrak{k}_\infty$ we have $\operatorname{Ad}(\exp tY)(X) - X \in \mathfrak{k}_\infty$ for all $X \in \mathfrak{g}$. Differentiating this at $t = 0$ shows that $[Y, X] \in \mathfrak{k}_\infty$ for all $X \in \mathfrak{g}$, so $\mathfrak{k}_\infty$ is an ideal in $\mathfrak{g}$. This implies that $\operatorname{Ad}(\exp(X))(\mathfrak{k}_\infty) \subset \mathfrak{k}_\infty$ for all $X \in \mathfrak{g}$. As above, one shows that $\exp(X)K_\infty \exp(-X) = K_\infty$. Since G/H is connected, elements of this form together with elements of H generate G. Since we already know that K_∞ is normal in H, we conclude that K_∞ is normal in G, which completes the proof. □

REMARK 1.5.3. It is instructive to look at this result in the case of affine connections on first order G–structures, where a simpler proof is available. Consider a homomorphism $j : H \to GL(m, \mathbb{R})$ such that j' is injective, and let $B = \mathbb{R}^m \rtimes H$ be the corresponding affine extension. The kernel of the Klein geometry (B, H) is simply the kernel of j, so this Klein geometry is always infinitesimally effective and it is effective if H is actually a virtual Lie subgroup of $GL(m, \mathbb{R})$. Given a principal H–bundle $p : \mathcal{P} \to M$ endowed with a Cartan connection $\omega \in \Omega^1(\mathcal{P}, \mathfrak{b})$, we get a homomorphism $\mathcal{P} \to \mathcal{P}^1 M$ to the first order frame bundle of M. This homomorphism is characterized by the fact that the pullback of the soldering form on $\mathcal{P}^1 M$ is the form $\pi \circ \omega \in \Omega^1(\mathcal{P}, \mathbb{R}^m)$, where $\pi : \mathfrak{b} \to \mathfrak{b}/\mathfrak{h} \cong \mathbb{R}^m$ is the canonical surjection.

If $(\mathcal{P} \to M, \omega)$ and $(\mathcal{P}' \to M', \omega')$ are two such structures, and $\Phi : \mathcal{P} \to \mathcal{P}'$ is a morphism, then equivariancy of Φ implies that one gets an induced morphism between the images in $\mathcal{P}^1 M$ and $\mathcal{P}^1 M'$, which can be uniquely extended to a morphism $\tilde{\Phi} : \mathcal{P}^1 M \to \mathcal{P}^1 M'$. From $\Phi^* \omega' = \omega$ one immediately concludes $\tilde{\Phi}$ pulls back the soldering form θ' on $\mathcal{P}^1 M'$ to the soldering form θ on $\mathcal{P}^1 M$. But by definition of the soldering form, this means that $\tilde{\Phi}$ is induced by composition with the tangent map Tf of the base map $f : M \to M'$. This shows that $\tilde{\Phi}$ is uniquely determined by f, and hence Φ is determined up to a smooth function with values in the kernel K.

1.5.4. Local description of Cartan connections. To make contact to the classical literature on Cartan connections, let us describe them in a local picture. Given a Cartan geometry $(p : \mathcal{P} \to M, \omega)$ the point about this approach is to build up and/or study the pullback of the Cartan connection ω along local smooth sections of $\mathcal{P}$. In the classical literature, the bundle $\mathcal{P}$ is often not spelled out explicitly. Rather than that one starts with a certain class of local frames or coframes of the tangent bundle of M or of some auxiliary bundle. The class of frames is usually defined by pointwise conditions, and then admissible frames are parametrized by a Lie group H. This group has to be independent of the point. This exactly means that the admissible frames or coframes form a principal subbundle with structure group H in some frame bundle.

To begin, let us assume that $p : \mathcal{P} \to M$ and $\omega \in \Omega^1(\mathcal{P}, \mathfrak{g})$ are given. For an open subset $U \subset M$ and a local smooth section $\sigma : U \to \mathcal{P}$, consider $\sigma^* \omega \in \Omega^1(U, \mathfrak{g})$. In many classical treatments of Cartan connections, $\mathfrak{g}$ is realized as a Lie algebra of matrices, and $\sigma^* \omega$ is viewed as a matrix of real–valued one–forms rather than a one–form with values in a matrix algebra. Any other section over U is of the form $\hat{\sigma}(x) = \sigma(x) \cdot \psi(x)$ for a smooth function $\psi : U \to H$. As in the proof of Proposition

1.5.3, one shows that for $x \in U$ and $\xi \in T_xM$ one has

$$\hat{\sigma}^*\omega(\xi) = \mathrm{Ad}(\psi(x)^{-1})(\sigma^*\omega(\xi)) + \delta\psi(\xi), \tag{1.21}$$

where $\delta\psi = \psi^*\omega_H \in \Omega^1(U, \mathfrak{h})$ is the left logarithmic derivative of $\psi : U \to H$.

Likewise, we can pull back the curvature form $K \in \Omega^2(\mathcal{P}, \mathfrak{g})$ along σ to obtain $\sigma^*K \in \Omega^2(U, \mathfrak{g})$. By definition, we have

$$\sigma^*K(\xi, \eta) = d\sigma^*\omega(\xi, \eta) + [\sigma^*\omega(\xi), \sigma^*\omega(\eta)],$$

so we obtain the usual structure equation. Changing from σ to $\hat{\sigma}$, the transformation law for the curvature is much easier than the one for the connection. Indeed, since $\delta\psi = \psi^*\omega_H$, it satisfies the Maurer–Cartan equation, and thus this part of the transformation law for the connection does not contribute to the change of curvature. Using that the adjoint action is by Lie algebra homomorphisms, one gets

$$\hat{\sigma}^*K(\xi, \eta) = \mathrm{Ad}(\psi(x)^{-1})(\sigma^*K(\xi, \eta))$$

if $\hat{\sigma}(x) = \sigma(x) \cdot \psi(x)$.

This picture can also be used to construct Cartan connections. One simply tries to associate to an admissible frame (i.e., a local section σ of $\mathcal{P}$) a one–form $\omega_\sigma \in \Omega^1(U, \mathfrak{g})$ (respectively an appropriate matrix of one–forms). Now $U \times H \cong p^{-1}(U)$ via $(x, h) \mapsto \sigma(x) \cdot h$. Using this, it is easy to see that there is a unique one–form $\omega \in \Omega^1(p^{-1}(U), \mathfrak{g})$ which is H–equivariant, reproduces the generators of fundamental vector fields and satisfies $\sigma^*\omega = \omega_\sigma$. Given two local sections with domains overlapping in $V \subset M$, the associated forms induce the same Cartan connection over V, if and only if they transform according to (1.21) for the function ψ relating the two frames.

Associating appropriate matrices of one–forms with the right transformation law to admissible local sections is therefore a way to construct Cartan connections. Having done this, the Cartan curvature can be computed locally as described above.

Finally, one can also analyze morphisms in this picture. One starts with a local diffeomorphism $f : M \to \tilde{M}$ between the bases which pulls back admissible frames to admissible frames. Starting with a local admissible frame $\tilde{\sigma}$ on $\tilde{M}$, one can pull back the corresponding form $\tilde{\omega}_{\tilde{\sigma}}$ along f. Denoting by σ the frame obtained by pulling back $\tilde{\sigma}$ along f, one then has to check whether there is a smooth function ψ for which ω_σ is related to $f^*\tilde{\omega}_{\tilde{\sigma}}$ according to (1.21). If this always works, then f defines a morphism of Cartan geometries.

Doing this in practice often leads to an efficient calculus. Let us sketch this in the case of Riemannian structures. Here one starts with a local orthonormal coframe $\{\sigma^1, \dots, \sigma^n\}$ for an n–dimensional Riemannian manifold M. To construct the Cartan connection with values in $\mathfrak{euc}(n) \cong \mathbb{R}^n \oplus \mathfrak{o}(n)$, one takes the σ^i as the $\mathbb{R}^n$–component. The $\mathfrak{o}(n)$–component then can be written as a family $\omega^i{}_j$ of one–forms such that $\omega^j{}_i = -\omega^i{}_j$.

The Lie bracket in $\mathfrak{euc}(n)$ reads as $[(v, A), (w, B)] = (Aw - Bv, AB - BA)$ for $v, w \in \mathbb{R}^n$ and $A, B \in \mathfrak{o}(n)$. Considering the form $(\sigma^i, \omega^i{}_j)$, the resulting expression for the curvature applied to ξ and η has $\mathbb{R}^n$–component

$$d\sigma^i(\xi, \eta) + \textstyle\sum_j(\omega^i{}_j(\xi)\sigma^j(\eta) - \omega^i{}_j(\eta)\sigma^j(\xi))$$

and $\mathfrak{o}(n)$ component

$$d\omega^i{}_j(\xi, \eta) + \textstyle\sum_k(\omega^i{}_k(\xi)\omega^k{}_j(\eta) - \omega^i{}_k(\eta)\omega^k{}_j(\xi)).$$

Using an analog of the Einstein sum convention, this is usually phrased by saying that the torsion and curvature associated to ${\omega^i}_j$ are represented by the forms $d\sigma^i + {\omega^i}_j \wedge \sigma^j$, respectively, $d{\omega^i}_j + {\omega^i}_k \wedge {\omega^k}_j$.

One then proves that, given a local orthonormal coframe $\{\sigma^1, \dots, \sigma^n\}$, there are unique one–forms ${\omega^i}_j$ such that ${\omega^j}_i = -{\omega^i}_j$ and $d\sigma^i + {\omega^i}_j \wedge \sigma^j = 0$. From this it already follows that the forms ${\omega^i}_j$ transform appropriately: For a second local orthonormal coframe $\{\hat\sigma^1, \dots, \hat\sigma^n\}$, there is an orthogonal matrix ${A^i}_j$ of smooth functions such that $\hat\sigma^i = {A^i}_j\sigma^j$. Denoting by $({B^i}_j)$ the inverse matrix to $({A^i}_j)$ we compute

$$\begin{aligned} d\hat\sigma^i &= d{A^i}_j \wedge \sigma^j + {A^i}_j d\sigma^j \\ &= d{A^i}_j \wedge {B^j}_k \hat\sigma^k - {A^i}_j {\omega^j}_k \wedge \sigma^k \\ &= ({B^j}_k d{A^i}_j - {A^i}_j {\omega^j}_\ell {B^\ell}_k) \wedge \hat\sigma^k. \end{aligned}$$

To interpret this, observe that ${A^i}_j\sigma^j$ is the $\mathbb{R}^n$–component of $\operatorname{Ad}({A^i}_j)(\sigma^i, {\omega^i}_j)$. Consequently, the frame $\hat\sigma$ is obtained from σ by the right action of ${B^i}_j = ({A^i}_j)^{-1}$. Moreover, since ${B^j}_k {A^i}_j = \delta^i_k$ we see that ${B^j}_k d{A^i}_j = -{A^i}_j d{B^j}_k$. Apart from the sign, this is exactly the well–known expression $B^{-1}dB$ for the left logarithmic derivative of B. Then the above equation indeed reads as $\hat\omega = \delta B + \operatorname{Ad}(B^{-1}) \circ \omega$ as required.

Having the forms ${\omega^i}_j$, the Riemannian curvature in the coframe σ can be computed as $d{\omega^i}_j + {\omega^i}_k \wedge {\omega^k}_j \in \Omega^2(M, \mathfrak{o}(TM))$.

1.5.5. Natural bundles. Let us fix a Klein geometry (G, H) and consider the category $\mathcal{C}_{(G,H)}$ of Cartan geometries of type (G, H). The most general definition of a natural bundle in this setting is as a functor which associates to each object $(\mathcal{P} \to M, \omega)$ a fiber bundle $FM \to M$ and to any morphism Φ from $(\mathcal{P} \to M, \omega)$ to $(\mathcal{P}' \to M', \omega')$ covering $f : M \to M'$ a fiber bundle morphism $F\Phi : FM \to FM'$ covering f. This is the concept of *gauge natural bundles* as studied in [**KMS**, Chapter XII].

In the case of Cartan geometries a much simpler concept of natural bundles is sufficient for most purposes. Suppose that F is a natural bundle in the above sense and consider the value $F(G/H)$ on the homogeneous model $(G \to G/H, \omega_G)$. From 1.5.2 we know that the left multiplications by elements $g \in G$ are exactly the automorphisms of this Cartan geometry. Applying F we obtain an action of G on $F(G/H)$ by bundle automorphisms, which lifts the canonical action on G/H. This makes $F(G/H)$ into a homogeneous bundle in the sense of 1.4.2. By Proposition 1.4.3 we get an H–action on the standard fiber $S := F_o(G/H)$ which determines the homogeneous bundle $F(G/H)$ up to isomorphism.

Given the left action of H on S, we can map each Cartan geometry $(\mathcal{P} \to M, \omega)$ of type (G, H) to the associated bundle $\mathcal{P} \times_H S$. This defines a natural bundle on $\mathcal{C}_{(G,H)}$ by the functorial properties of the associated bundle construction; see 1.2.7. Such natural bundles will be called *natural bundles associated to the Cartan bundle.* If not explicitly stated otherwise, in the sequel we will only consider natural bundles of that type.

Formally, natural bundles associated to the Cartan bundle depend only on the principal bundle $\mathcal{P} \to M$ and not the Cartan connection ω. The Cartan connection is, however, necessary to identify natural bundles associated to the Cartan bundle with more traditional geometric objects, for example, with tensor bundles. From

1.4.3 we know that in the case of the homogeneous model $(G \to G/H, \omega_G)$ the tangent bundle is the associated bundle $G \times_H (\mathfrak{g}/\mathfrak{h})$ (and the Maurer–Cartan form is used to identify the two bundles).

The same identification works for a general Cartan geometry $(\mathcal{P} \to M, \omega)$. Consider the mapping $\mathcal{P} \times \mathfrak{g} \to TM$ defined by $(u, X) \mapsto T_u p \cdot \omega_u^{-1}(X)$. For $X \in \mathfrak{h}$, the field $\omega^{-1}(X)$ is the fundamental field ζ_X and thus vertical, so this map factors to $\mathcal{P} \times (\mathfrak{g}/\mathfrak{h})$ and fixing u, one gets a linear isomorphism $\mathfrak{g}/\mathfrak{h} \to T_{p(u)}M$. Equivariancy of ω immediately implies that this factors to a bundle map $\mathcal{P} \times_H (\mathfrak{g}/\mathfrak{h}) \to TM$, where the H–action $\underline{\mathrm{Ad}}$ on $\mathfrak{g}/\mathfrak{h}$ is induced by the restriction of the adjoint action of G. This map induces a linear isomorphism in each fiber and covers the identity on M, and thus is an isomorphism of vector bundles. The cotangent bundle T^*M may then be identified with the natural bundle corresponding to $(\mathfrak{g}/\mathfrak{h})^*$ and similarly for arbitrary tensor bundles.

As in the case of the homogeneous model, there is a way to view this as an underlying G–structure. Consider the action $\underline{\mathrm{Ad}} : H \to GL(\mathfrak{g}/\mathfrak{h})$ from above and let $H^1 \subset H$ be the kernel of this homomorphism, which is a closed normal subgroup of H. In particular, we can form the quotient group $H_0 := H/H^1$ and the space $\mathcal{P}_0 := \mathcal{P}/H^1$ is naturally a principal bundle over M with structure group H_0. There is an evident projection $p_0 : \mathcal{P} \to \mathcal{P}_0$. Now define $\theta \in \Omega^1(\mathcal{P}_0, \mathfrak{g}/\mathfrak{h})$ as follows: for $u_0 \in \mathcal{P}_0$ and $\xi \in T_{u_0}\mathcal{P}_0$ choose a point $u \in \mathcal{P}$ over u_0 and a tangent vector $\tilde{\xi} \in T_u\mathcal{P}$ such that $T_{u_0}p_0 \cdot \tilde{\xi} = \xi$, and put $\theta(\xi) := \omega(\tilde{\xi}) + \mathfrak{h}$.

Two choices for $\tilde{\xi}$ differ by a vector which is vertical for $\mathcal{P} \to M$, so this choice plays no role. Any other choice for the point u is of the form $u \cdot h$ with $h \in H^1$ and we may choose $Tr^h \cdot \tilde{\xi}$ as the lift of ξ. Equivariancy of ω then implies that $\omega(Tr^h \cdot \tilde{\xi}) = \mathrm{Ad}(h^{-1})(\omega(\tilde{\xi}))$ and since $h \in H^1$ we conclude that θ is well defined and strictly horizontal. Similarly, one verifies that θ is H_0–equivariant and using a local smooth section of $\mathcal{P} \to \mathcal{P}_0$ one shows that θ is smooth. Thus, $(\mathcal{P}_0 \to M, \theta)$ is a first order G–structure with structure group H_0; see 1.3.6. Left actions of H_0 are the same thing as left actions of H such that H^1 acts trivially, and the former correspond to natural bundles for G–structures with structure group H_0. Via the Cartan connection ω one can view any such bundle as a natural bundle for Cartan geometries of type (G, H).

The relation to the homogeneous model continues to work in the question of natural sections. Given a natural bundle F on the category $\mathcal{C}_{(G,H)}$ a *natural section* is a family of smooth sections $\sigma_M : M \to FM$ of the values such that for any morphism $\Phi : (\mathcal{P} \to M, \omega) \to (\mathcal{P}' \to M', \omega')$ covering $f : M \to M'$ we have $F\Phi \circ \sigma_M = \sigma_{M'} \circ f$. Looking at automorphisms of the homogeneous model, we see that for any natural section, the section $\sigma_{G/H}$ must be a G–invariant section of the homogeneous bundle $F(G/H)$ as introduced in 1.4.4. By Theorem 1.4.4 such sections are in bijective correspondence with H–invariant elements in the standard fiber S.

But for any H–invariant element $s_0 \in S$ and any Cartan geometry $(p : \mathcal{P} \to M, \omega)$ of type (G, H) we may define $\sigma_M : M \to \mathcal{P} \times_H S$ by $\sigma_M(x) := [\![u, s_0]\!]$ where $u \in \mathcal{P}$ is any point such that $p(u) = x$. Invariance of s_0 implies that this is well defined and using local smooth sections of $\mathcal{P}$ we immediately see that σ_M is smooth. Clearly, the family σ_M defines a natural section. For example, there exists a natural Riemannian metric on Cartan geometries of type (G, H) if and only if there is a

G–invariant Riemannian metric on G/H and any choice of a G–invariant metric on G/H canonically extends to a natural Riemannian metric.

In 1.4.5 we have discussed the existence of G–invariant principal connections on the bundle $G \to G/H$. We have shown that such a connection is equivalent to an H–invariant subspace $\mathfrak{n} \subset \mathfrak{g}$, which is complementary to $\mathfrak{h}$. In particular, such a connection exists if and only if the Klein geometry (G, H) is reductive. Suppose that we have chosen an H–invariant decomposition $\mathfrak{g} = \mathfrak{n} \oplus \mathfrak{h}$. Then any Cartan connection $\omega \in \Omega^1(\mathcal{P}, \mathfrak{g})$ splits as $\omega = \omega_{\mathfrak{n}} + \omega_{\mathfrak{h}}$ and it follows immediately from the definitions that $\omega_{\mathfrak{h}}$ is a principal connection on $\mathcal{P}$. Forming induced connection, we see that there is a natural connection on any natural bundle associated to the Cartan bundle.

1.5.6. Natural connections. We have just seen that an invariant principal connection on $G \to G/H$ gives rise to a principal connection on the principal H–bundle $\mathcal{P}$ for any Cartan geometry $(p : \mathcal{P} \to M, \omega)$ of type (G, H). Via induced connections, one obtains natural (linear) connections on all natural (vector) bundles associated to the Cartan bundle. We next extend this to general invariant principal connections on homogeneous principal bundles as discussed in 1.4.5.

By Lemma 1.4.5 any homogeneous principal K–bundle over G/H is of the form $G \times_i K$ for a homomorphism $i : H \to K$. This is the associated bundle with respect to the action of H on K defined by $h \cdot k := i(h)k$. From 1.5.5 we know that this extends to a natural bundle on the category of Cartan geometries of type (G, H). Given a Cartan geometry $(\mathcal{P} \to M, \omega)$, one simply has to form the associated bundle $\mathcal{P} \times_i K \to M$. This is again a K–principal bundle with principal action induced by multiplication from the right.

In Theorem 1.4.5 we have shown that invariant principal connections on $G \times_i K$ are in bijective correspondence with linear maps $\alpha : \mathfrak{g} \to \mathfrak{k}$ which satisfy two conditions. First, one has to assume that $\alpha|_{\mathfrak{h}} = i'$, the derivative of i. Second, α has to be equivariant, i.e. $\alpha \circ \mathrm{Ad}(h) = \mathrm{Ad}(i(h)) \circ \alpha$ for all $h \in H$. Given such a map, we can define a natural principal connection on the category of Cartan geometries of type (G, H). Observe first that there is an obvious map $j : \mathcal{P} \to \mathcal{P} \times_i K$ induced by mapping $u \in \mathcal{P}$ to the class of (u, e).

THEOREM 1.5.6. *Let G and K be Lie groups, $H \subset G$ a closed subgroup, $i : H \to K$ a homomorphism and $\alpha : \mathfrak{g} \to \mathfrak{k}$ a linear map satisfying conditions (i) and (ii) from Theorem 1.4.5.*

(1) For any Cartan geometry $(p : \mathcal{P} \to M, \omega)$ of type (G, H), there is a unique principal connection γ_α on $\mathcal{P} \times_i K$ such that $j^*\gamma_\alpha = \alpha \circ \omega \in \Omega^1(\mathcal{P}, \mathfrak{k})$.

(2) The assignment from (1) is functorial, i.e. any morphism of Cartan geometries induces a morphism of principal bundles which is compatible with the principal connections.

PROOF. (1) Let us write π for the projection $\mathcal{P} \times_i K \to M$. Since $\pi \circ j = p$ we see that for a point $u \in \mathcal{P}$ the tangent space $T_{j(u)}(\mathcal{P} \times_i K)$ is spanned by vertical vectors and elements of $T_u j(T_u\mathcal{P})$. Hence, there is only one possible definition for $\gamma_\alpha(j(u))$:

$$\gamma_\alpha(j(u))(T_u j \cdot \xi + \zeta_A(j(u))) := \alpha(\omega(u)(\xi)) + A \tag{1.22}$$

for $\xi \in T_u\mathcal{P}$ and $A \in \mathfrak{k}$. We have to show that this is well defined. If $T_u j \cdot \xi$ is vertical, then $T\pi \circ Tj \cdot \xi = Tp \cdot \xi = 0$, so $\xi = \zeta_X(u)$ for some $X \in \mathfrak{h}$. By definition of j, we have $j(u\cdot h)=j(u)\cdot i(h)$. Putting $h=\exp(tX)$ and differentiating at $t=0$ we see that $T_uj\cdot\zeta_X(u)=\zeta_{i'(X)}(j(u))$. Since $\alpha(X)=i'(X)$ for $X\in\mathfrak{h}$ by property (i), we see that (1.22) uniquely defines a linear map $T_{j(u)}(\mathcal{P}\times_i K)\to\mathfrak{k}$.

From the definition, we see that $\gamma_\alpha(j(u))$ reproduces the generators of fundamental vector fields. To ensure equivariancy, we next have to define

$$\gamma_\alpha(j(u)\cdot k)(\eta)=\mathrm{Ad}(k^{-1})(\gamma_\alpha(j(u))(Tr^{k^{-1}}\cdot\eta)). \tag{1.23}$$

To verify that this is well defined, suppose that $j(u)\cdot k=j(\hat{u})\cdot\hat{k}$. Projecting to G/H, we see that $\hat{u}=u\cdot h$ for some $h\in H$. Then $j(\hat{u})\cdot\hat{k}=j(u)\cdot(i(h)\hat{k})$, so $\hat{k}=i(h^{-1})k$. Writing the right–hand side of (1.23) in terms of $\hat{u}$ and $\hat{k}$, we get

$$\mathrm{Ad}(k^{-1})\,\mathrm{Ad}(i(h))\left(\gamma_\alpha(j(u\cdot h))(Tr^{i(h)}\cdot Tr^{k^{-1}}\cdot\eta)\right).$$

To see that γ_α is well defined, we only have to verify that for all η in $T_{j(u)}(\mathcal{P}\times_i K)$ we have

$$\gamma_\alpha(j(u\cdot h))(Tr^{i(h)}\cdot\eta)=\mathrm{Ad}(i(h)^{-1})(\omega(j(u))(\eta)).$$

If $\eta=\zeta_A(j(u))$ for some $A\in\mathfrak{k}$, then this immediately follows from equivariancy of the fundamental vector fields. On the other hand, if $\eta=T_uj\cdot\xi$ for some $\xi\in T_u\mathcal{G}$, then $T_{j(u)}r^{i(h)}\cdot T_uj\cdot\xi=T_{u\cdot h}j\cdot T_ur^h\cdot\xi$, and $\omega(u\cdot h)(T_ur^h\cdot\xi)=\mathrm{Ad}(h^{-1})(\omega(u)(\xi))$, and the result follows from the equivariancy property (ii) of α.

Hence, we have constructed $\gamma_\alpha\in\Omega^1(\mathcal{P}\times_i K,\mathfrak{k})$. By construction, $j^*\gamma_\alpha=\alpha\circ\omega$ and, as a principal connection form, γ_α is uniquely determined by this property. From the definition in (1.23) it follows immediately that $(r^k)^*\gamma_\alpha=\mathrm{Ad}(k^{-1})\circ\gamma_\alpha$. Again by definition, $\gamma_\alpha(j(u))$ reproduces generators of fundamental vector fields, so by equivariancy, this holds on all of $\mathcal{P}\times_i K$.

(2) Let $\Phi:(\mathcal{P}\to M,\omega)\to(\tilde{\mathcal{P}}\to\tilde{M},\tilde{\omega})$ be a morphism of Cartan geometries. Then by definition $\Phi:\mathcal{P}\to\tilde{\mathcal{P}}$ is a principal bundle map. Hence, $\Phi\times\mathrm{id}_K:\mathcal{P}\times K\to\tilde{\mathcal{P}}\times K$ induces a principal bundle map $F(\Phi):\mathcal{P}\times_i K\to\tilde{\mathcal{P}}\times_i K$. Denoting by $\tilde{j}:\tilde{\mathcal{P}}\to\tilde{\mathcal{P}}\times_i K$ the natural map, by definition we get $F(\Phi)\circ j=\tilde{j}\circ\Phi$. Denoting by $\tilde{\gamma}_\alpha$ the connection on $\tilde{\mathcal{P}}\times_i K$ constructed according to (1), we can form the pullback $F(\Phi)^*\tilde{\gamma}_\alpha$. Since $F(\Phi)$ is a principal bundle homomorphism, this is a principal connection on $\tilde{\mathcal{P}}\times_i K$. Now we compute

$$j^*F(\Phi)^*\tilde{\gamma}_\alpha=\Phi^*\tilde{j}^*\tilde{\gamma}_\alpha=\Phi^*(\alpha\circ\tilde{\omega})=\alpha\circ\Phi^*\tilde{\omega}=\alpha\circ\omega.$$

But by part (1), γ_α is the unique principal connection which is pulled back to $\alpha\circ\omega$ along j, so $F(\Phi)^*\tilde{\gamma}_\alpha=\gamma_\alpha$. □

Via associated bundles and induced connections we conclude that any invariant connection on a homogeneous bundle over G/H extends to a natural connection on the corresponding natural bundle on Cartan geometries of type (G,H). Let us describe this for vector bundles and linear connections. Given a homogeneous vector bundle $E\to G/H$, let E_o be the fiber over $o=eH$. Then by 1.4.3 we obtain a representation ρ of H on E_o and $E\cong G\times_H E_o$. In 1.4.7 we have seen that the frame bundle of E is $G\times_\rho GL(E_o)$, and homogeneous linear connections on E are equivalent to homogeneous principal connections on the frame bundle. For a Cartan geometry $(p:\mathcal{P}\to M,\omega)$ of type (G,H), we evidently have $(\mathcal{P}\times_\rho GL(E_o))\times_{GL(E_o)}E_o=\mathcal{P}\times_H E_o$. Starting from an invariant linear connection on E, the theorem gives us a natural principal connection on $\mathcal{P}\times_\rho GL(E_o)$ and hence a linear connection on $\mathcal{P} \times_H E$. Naturality of this connection follows from functoriality of the construction of associated bundles.

1.5.7. Tractor bundles. Starting from an arbitrary Klein geometry (G, H), there is always a class of homogeneous bundles which do admit canonical invariant connections. By 1.5.6 this leads to natural connections on the corresponding bundles on Cartan geometries of type (G, H). We have not studied these bundles in 1.4 since they are canonically trivial on G/H, so the existence of a G–invariant connection is obvious. However, on general Cartan geometries these bundles are nontrivial and the canonical connections on them become a highly interesting and fundamental tool.

The idea here is very simple. In the setting of Theorem 1.4.5, we can simply use the inclusion $i : H \to G$ and $\alpha = \mathrm{id}_{\mathfrak{g}} : \mathfrak{g} \to \mathfrak{g}$. This gives rise to a homogeneous principal connection on the principal bundle $G \times_H G \to G/H$. This extended principal bundle is canonically trivial: The map $G \times G \to (G/H) \times G$ defined by $(g, g') \mapsto (gH, gg')$ factors to an isomorphism $G \times_H G \to (G/H) \times G$ of homogeneous principal bundles. The natural connection on $G \times_H G$ is the pullback along this isomorphism of the canonical flat connection on the product bundle.

Theorem 1.5.6 then implies that for any Cartan geometry $(\mathcal{P} \to M, \omega)$ there is a natural principal connection on the *extended principal bundle* $\tilde{\mathcal{P}} := \mathcal{P} \times_H G \to M$, which is a principal G–bundle. Correspondingly, there are natural connections on all natural bundles associated to the Cartan bundle with respect to an action of H which is the restriction of an action of G. These bundles are not trivial in general, so the existence of natural connections is not evident.

A particularly important special case is natural vector bundles corresponding to the restriction of a representation of G to the subgroup H. These are called *tractor bundles* and from above we know that they carry canonical linear connections, called *tractor connections*

Among all tractor bundles, the *adjoint tractor bundle* is of fundamental importance in the study of Cartan geometries. This is the tractor bundle $\mathcal{A}$ corresponding to the adjoint representation $\mathrm{Ad} : G \to GL(\mathfrak{g})$. For a Cartan geometry $(\mathcal{P} \to M, \omega)$ of type (G, H), we have $\mathcal{A}M = \mathcal{P} \times_H \mathfrak{g}$, where H acts on $\mathfrak{g}$ by the restriction of the adjoint action. The short exact sequence $0 \to \mathfrak{h} \to \mathfrak{g} \to \mathfrak{g}/\mathfrak{h} \to 0$ of H–modules gives rise to the short exact sequence

$$0 \to \mathcal{P} \times_H \mathfrak{h} \to \mathcal{A}M \to TM \to 0.$$

In particular, there is a natural surjective bundle map $\Pi : \mathcal{A}M \to TM$, and we may view the adjoint tractor bundle as an extension of the tangent bundle. Let us explore further important properties of the adjoint tractor bundle.

PROPOSITION 1.5.7. *Let $(\mathcal{P} \to M, \omega)$ be a Cartan geometry of type (G, H), $\mathcal{A}M \to M$ its adjoint tractor bundle and $\Pi : \mathcal{A}M \to TM$ the natural projection. Let $\mathcal{V}M$ be the tractor bundle corresponding to a representation of G on V.*

(1) The Cartan curvature κ of ω can be naturally interpreted as a two–form κ on M with values in $\mathcal{A}M$.

(2) There is a natural bundle map $\{\ ,\ \} : \mathcal{A}M \times \mathcal{A}M \to \mathcal{A}M$, which makes each fiber $\mathcal{A}_xM$ into a Lie algebra isomorphic to $\mathfrak{g}$.

(3) There is an isomorphism between the space $\Gamma(\mathcal{A}M)$ of smooth sections of $\mathcal{A}M$ and the space $\mathfrak{X}(\mathcal{P})^H$ of vector fields on $\mathcal{P}$ which are invariant under the principal right action of H. This induces a Lie bracket $[\ ,\ ]$ on $\Gamma(\mathcal{A}M)$. For

$s_1, s_2 \in \Gamma(\mathcal{A}M)$, one has $\Pi([s_1, s_2]) = [\Pi(s_1), \Pi(s_2)]$, where in the right-hand side we use the Lie bracket of vector fields.

(4) There is a natural bundle map $\bullet : \mathcal{A}M \times \mathcal{V}M \to \mathcal{V}M$. For each point $x \in M$, this makes the fiber $\mathcal{V}_xM$ into a module over the Lie algebra $\mathcal{A}_xM$. In particular, for sections $s_1, s_2 \in \Gamma(\mathcal{A}M)$ and $t \in \Gamma(\mathcal{V}M)$, we get

$$\{s_1, s_2\} \bullet t = s_1 \bullet (s_2 \bullet t) - s_2 \bullet (s_1 \bullet t).$$

(5) The operations introduced in (2) and (4) are parallel for the canonical tractor connections. Denoting them by $\nabla^{\mathcal{A}}$ and $\nabla^{\mathcal{V}}$ we get

$$\nabla^{\mathcal{A}}_\xi \{s_1, s_2\} = \{\nabla^{\mathcal{A}}_\xi s_1, s_2\} + \{s_1, \nabla^{\mathcal{A}}_\xi s_2\},$$
$$\nabla^{\mathcal{V}}_\xi (s \bullet t) = (\nabla^{\mathcal{A}}_\xi s) \bullet t + s \bullet (\nabla^{\mathcal{V}}_\xi t)$$

for sections $s_1, s_2 \in \Gamma(\mathcal{A}M)$ and $t \in \Gamma(\mathcal{V}M)$ and all vector fields $\xi \in \mathfrak{X}(M)$.

PROOF. (1) The curvature function $\kappa : \mathcal{P} \to \Lambda^2(\mathfrak{g}/\mathfrak{h})^* \otimes \mathfrak{g}$ was shown to be H–equivariant in Lemma 1.5.1. Hence, it corresponds to a smooth section of the associated bundle, which by definition is $\Lambda^2 T^*M \otimes \mathcal{A}M$.

(2) $\{\ ,\ \}$ is simply the morphism between associated bundles induced by the H–equivariant map $[\ ,\ ] : \mathfrak{g} \times \mathfrak{g} \to \mathfrak{g}$.

(3) Since $\mathcal{A}M = \mathcal{P} \times_H \mathfrak{g}$, sections of the adjoint tractor bundle are in bijective correspondence with smooth functions $f : \mathcal{P} \to \mathfrak{g}$ such that $f(u \cdot h) = \mathrm{Ad}(h^{-1})(f(u))$. On the other hand, since ω trivializes $T\mathcal{P}$, the map $\xi \mapsto \omega(\xi)$ defines a linear isomorphism between $\mathfrak{X}(\mathcal{P})$ and $C^\infty(\mathcal{P}, \mathfrak{g})$. Now ξ corresponds to an equivariant function if and only if

$$\omega(\xi(u \cdot h)) = \mathrm{Ad}(h^{-1})(\omega(\xi(u))) = \omega(u \cdot h)(Tr^h \cdot \xi(u)),$$

for all $h \in H$, where in the last equality we use equivariancy of ω. Since the values of ω are linear isomorphisms, this is equivalent to $\xi(u \cdot h) = Tr^h \cdot \xi(u)$ and hence $(r^h)^*\xi = \xi$ for all $h \in H$.

Naturality of the Lie bracket implies that $(r^h)^*([\xi, \eta]) = [(r^h)^*\xi, (r^h)^*\eta]$. Therefore, $\mathfrak{X}(\mathcal{P})^H$ is a Lie subalgebra in $\mathfrak{X}(\mathcal{P})$, and we can pull back the Lie bracket via the isomorphism to $\Gamma(\mathcal{A}M)$. Right invariant vector fields on $\mathcal{P}$ are projectable. From the identification $TM \cong \mathcal{P} \times_H (\mathfrak{g}/\mathfrak{h})$ constructed in 1.5.5 we see that $\Pi : \mathcal{A}M \to TM$ corresponds to projecting right invariant vector fields. Thus, $\Pi([s_1, s_2]) = [\Pi(s_1), \Pi(s_2)]$ follows from naturality of the Lie bracket.

(4) Denoting by $\rho : G \to GL(V)$ the representation inducing the tractor bundle, consider its derivative $\rho' : \mathfrak{g} \to L(V, V)$. For $g \in G$ and $X \in \mathfrak{g}$, we have $\exp(t\,\mathrm{Ad}(g)(X)) = g \exp(tX) g^{-1}$. Applying the representation ρ and differentiating at $t = 0$, we see that

$$\rho'(\mathrm{Ad}(g)(X))(\rho(g)(v)) = \rho(g)(\rho'(X)(v)).$$

This means that the bilinear map $\mathfrak{g} \times V \to V$ induced by ρ' is G–equivariant and hence H–equivariant, and thus induces a natural map $\bullet : \mathcal{A}M \times \mathcal{V}M \to \mathcal{V}M$ on associated bundles.

(5) The operations in (2) and (4) are actually induced by G–equivariant maps on the corresponding representations. Hence, we can also view them as being induced on bundles associated to the extended principal bundle $\mathcal{P} \times_H G$. But then the tractor connections are all induced from a fixed principal connection on that bundle. Maps between associated bundles coming from equivariant maps between

the inducing representations are clearly parallel for these connections. Expanding this leads to the claimed formulae. □

The bracket $\{\ ,\ \}$ on $\mathcal{A}M$ from part (2) is called the *algebraic bracket* on adjoint tractors, while the bracket $[\ ,\ ]$ from part (3) is called the *Lie bracket* on adjoint tractors. Note that part (3), in particular, says that $\Pi : \mathcal{A}M \to TM$ makes $(\mathcal{A}M, [\ ,\ ])$ into a Lie algebroid over M.

Composing the curvature $\kappa \in \Omega^2(M, \mathcal{A}M)$ with the projection $\Pi : \mathcal{A}M \to TM$, we obtain a form $T := \Pi \circ \kappa \in \Omega^2(M, TM)$, which is called the *torsion* of the Cartan connection ω. By construction, this torsion vanishes if and only if the Cartan geometry in question is torsion free in the sense introduced in 1.5.1 and then its Cartan curvature is a two–form with values in the bundle $\mathcal{P} \times_H \mathfrak{h}$.

Using the interpretation of κ as a two–form with values in $\mathcal{A}M$, we can give a general description of the curvatures of natural principal connections. For a Cartan geometry $(p : \mathcal{P} \to M, \omega)$ of type (G, H) and a homomorphism $i : H \to K$ consider the bundle $\mathcal{P} \times_i K$. The map $\alpha : \mathfrak{g} \to \mathfrak{k}$ used to determine a natural principal connection on this bundle, in particular, is H–equivariant. Hence, it induces a bundle map

$$\mathcal{A}M = \mathcal{P} \times_H \mathfrak{g} \to \mathcal{P} \times_H \mathfrak{k} \cong (\mathcal{P} \times_i K) \times_K \mathfrak{k},$$

which we also denote by α. On the other hand, by Proposition 1.4.6, the map $(X, Y) \mapsto [\alpha(X), \alpha(Y)] - \alpha([X, Y])$ descends to an H–equivariant map $\Lambda^2(\mathfrak{g}/\mathfrak{h}) \to \mathfrak{k}$. In view of 1.5.5 this gives rise to a natural section of the associated natural bundle, i.e. a natural element $R^0_\alpha \in \Omega^2(M, \mathcal{P} \times_H \mathfrak{k})$.

COROLLARY 1.5.7. *Let $\gamma_\alpha \in \Omega^1(\mathcal{P} \times_i K, \mathfrak{k})$ be the natural principal connection associated to $(\mathcal{P} \to M, \omega)$ via the map $\alpha : \mathfrak{g} \to \mathfrak{k}$ as in Theorem 1.5.6. Then the curvature $R_\gamma \in \Omega^2(M, \mathcal{P} \times_H \mathfrak{k})$ is given by*

$$R_\gamma = \alpha \circ \kappa + R^0_\alpha,$$

where $\kappa \in \Omega^2(M, \mathcal{A}M)$ is the Cartan curvature of ω.

In particular, if $\mathcal{V}M \to M$ is a tractor bundle, then the curvature $R^{\mathcal{V}}$ of the canonical tractor connection $\nabla^{\mathcal{V}}$ is given by

$$R^{\mathcal{V}}(\xi, \eta)(t) = \kappa(\xi, \eta) \bullet t$$

for $\xi, \eta \in \mathfrak{X}(M)$ and $t \in \Gamma(\mathcal{V}M)$

PROOF. Let $j : \mathcal{P} \to \mathcal{P} \times_i K$ be the natural map used in 1.5.6, so the connection γ_α is characterized by $j^*\gamma_\alpha = \alpha \circ \omega$. In particular,

$$[j^*\gamma_\alpha(\xi), j^*\gamma_\alpha(\eta)] = [\alpha(\omega(\xi)), \alpha(\omega(\eta))].$$

On the other hand, we also get $j^*d\gamma_\alpha = \alpha \circ d\omega$, and inserting the definition of the Cartan curvature, we obtain

$$j^*d\gamma_\alpha(\xi, \eta) = \alpha(\kappa(\omega(\xi), \omega(\eta)) - [\omega(\xi), \omega(\eta)]).$$

Summing with the expression above, by definition we obtain the pullback of the curvature of γ_α interpreted as an element of $\Omega^2(\mathcal{P} \times_i K, \mathfrak{k})$; see 1.3.3. The isomorphism $\mathcal{P} \times_H \mathfrak{k} \cong (\mathcal{P} \times_i K) \times_K \mathfrak{k}$ is induced by $[\![u, A]\!] \mapsto [\![j(u), A]\!]$, which shows that this pullback exactly represents R_γ, and the claimed formula follows.

In the special case of a tractor bundle $\alpha = \mathrm{id}_{\mathfrak{g}}$, so $R^0_\alpha = 0$, and we are left with $j^*R_\alpha = \kappa$. To convert to the curvature of the induced connection, we have to interpret the values of κ as acting on $\mathcal{V}M$ via the infinitesimal representation; see 1.3.4. But this was exactly the definition of $\bullet$ in the proposition. □

1.5.8. The fundamental derivative. The next important property of the adjoint tractor bundle is the existence of a basic family of natural differential operators on arbitrary natural bundles. The idea to obtain these operators is simply that differentiating H–equivariant smooth functions with respect to H–invariant vector fields, one obtains again H–equivariant functions. We restrict to natural vector bundles here, the case of general bundles is briefly sketched below.

Let E be a natural vector bundle associated to the Cartan bundle with respect to a representation $\rho : H \to GL(V)$. We define the *fundamental derivative* $D : \Gamma(\mathcal{A}M) \times \Gamma(EM) \to \Gamma(EM)$, which we write as $(s, \sigma) \mapsto D_s\sigma$ as follows: The section $s \in \Gamma(\mathcal{A}M)$ corresponds to an H–invariant vector field $\xi \in \mathfrak{X}(\mathcal{P})^H$, while the section σ corresponds to a smooth equivariant function $\phi : \mathcal{P} \to V$. Now for the function $\xi \cdot \phi : \mathcal{P} \to V$ we compute

$$\xi(u \cdot h) \cdot \phi = (Tr^h \cdot \xi(u)) \cdot \phi = \xi(u) \cdot (\phi \circ r^h) = \rho(h^{-1})(\xi(u) \cdot \phi).$$

Hence, $\xi \cdot \phi$ is H–equivariant, so it corresponds to a smooth section $D_s\sigma$ of EM. By construction, this operator is bilinear, and tensorial and thus linear over $C^\infty(M, \mathbb{R})$ in s.

We next establish some basic properties of the fundamental derivative. Consider the derivative $\rho' : \mathfrak{h} \to L(V, V)$ of the representation ρ inducing E. In the proof of part (4) of Proposition 1.5.7 we have seen that the corresponding bilinear map $\mathfrak{h} \times V \to V$ is H–equivariant. Hence, we obtain a natural bundle map $\bullet : (\mathcal{P} \times_H \mathfrak{h}) \times E \to E$. In the case of a tractor bundle, this is just the restriction of the bundle map from part (4) of Proposition 1.5.7.

PROPOSITION 1.5.8. *(1) For a smooth function $f : M \to \mathbb{R}$ and $s \in \Gamma(\mathcal{A}M)$ we get $D_s f = \Pi(s) \cdot f$.*

(2) If s is a section of the subbundle $\mathcal{P} \times_H \mathfrak{h} \subset \mathcal{A}M$, then $D_s\sigma = -s \bullet \sigma$ for any $\sigma \in \Gamma(E)$.

(3) The fundamental derivative is compatible with all natural bundle maps coming from H–equivariant maps between the inducing representations. In particular, for natural vector bundles E and F, the dual E^* of E, and sections $\sigma \in \Gamma(E)$, $\tau \in \Gamma(F)$ and $\beta \in \Gamma(E^*)$ we get

$$D_s(f\sigma) = (\Pi(s) \cdot f)\sigma + f D_s\sigma,$$
$$D_s(\sigma \otimes \tau) = (D_s\sigma) \otimes \tau + \sigma \otimes D_s\tau,$$
$$\Pi(s) \cdot (\beta(\sigma)) = (D_s\beta)(\sigma) + \beta(D_s\sigma).$$

PROOF. (1) Writing $p : \mathcal{P} \to M$ for the projection, the equivariant function $\mathcal{P} \to \mathbb{R}$ corresponding to f is simply $f \circ p$. But then for $\xi \in \mathfrak{X}(\mathcal{P})$ we get $\xi \cdot (f \circ p) = (Tp \cdot \xi) \cdot f$, and the result follows.

(2) If s is a section of the subbundle $\mathcal{P} \times_H \mathfrak{h}$, then the corresponding vector field ξ has the property that $\omega(\xi)$ has values in $\mathfrak{h}$. Thus, $\xi(u) = \zeta_{\omega(\xi)(u)}(u)$. Let $\phi : \mathcal{P} \to V$ be the equivariant function corresponding to s. Applying $\phi(u \cdot h) = \rho(h)(\phi(u))$ for $h = \exp(tA)$ for $A \in \mathfrak{h}$ and differentiating at $t = 0$, we get $\zeta_A \cdot \phi(u) = -\rho'(A)(\phi(u))$, and the claim follows.

(3) In the picture of equivariant functions, all the operations act only on the values of functions. The natural bundle maps are given by applying linear and

multilinear maps to the values of the functions. Of course, this is compatible in the appropriate sense with differentiation. The three claimed formulae are evident examples of this situation. □

Except for the fact that the tangent bundle has been replaced by the adjoint tractor bundle, the fundamental derivative looks very similar to the family of covariant derivatives by the Levi–Civita connection on a Riemannian manifold. The naturality properties of the fundamental derivative justify the use of the same symbol D to denote all fundamental derivatives.

Of course, we may also leave the algebraic slot of D free, and view $s \mapsto D_s\sigma$ as a section $D\sigma$ of $\mathcal{A}^*M \otimes EM$, and thus the fundamental derivative as a differential operator $\Gamma(EM) \to \Gamma(\mathcal{A}^*M \otimes EM)$. In this version, the fundamental derivative can be iterated, i.e. for $\sigma \in \Gamma(EM)$ and $k \in \mathbb{N}$ we obtain $D^k\sigma \in \Gamma(\otimes^k \mathcal{A}^*M \otimes EM)$.

Now we can use the fundamental derivative to derive a formula for an arbitrary natural linear connection. This generalizes the formula for homogeneous connections from Proposition 1.4.7. As above, let E be the natural vector bundle corresponding to a representation $\rho : H \to GL(V)$. By Theorem 1.4.7, a homogeneous linear connection on $E(G/H) \to G/H$ is induced by a linear map $\alpha : \mathfrak{g} \to L(V, V)$ such that

(i) $\alpha|_{\mathfrak{h}} = \rho'$, the derivative of the representation ρ,
(ii) $\alpha(\mathrm{Ad}(h)(X)) = \rho(h) \circ \alpha(X) \circ \rho(h^{-1})$ for all $X \in \mathfrak{g}$ and $h \in H$.

In 1.5.6 we have noted that, via principal and induced connections, one obtains from α a natural linear connection on E. Now property (ii) says that α, and hence the corresponding bilinear map $\mathfrak{g} \times V \to V$ is H–equivariant. Thus, it induces a natural bundle map $\mathcal{A}M \times EM \to \mathcal{A}M$ which we also denote by α.

THEOREM 1.5.8. *Consider the operation* $\Gamma(\mathcal{A}M) \times \Gamma(EM) \to \Gamma(EM)$ *defined by* $(s, \sigma) \mapsto D_s\sigma + \alpha(s, \sigma)$. *This vanishes identically if* s *is a section of the subbundle* $\mathcal{P} \times_H \mathfrak{h} \subset \mathcal{A}M$. *Hence, it descends to an operator* $\mathfrak{X}(M) \times \Gamma(EM) \to \Gamma(EM)$ *which is exactly the covariant derivative with respect to the natural linear connection induced by* α. *In particular, if* E *is a tractor bundle* $\mathcal{V}$, *then the tractor connection* $\nabla^{\mathcal{V}}$ *is given by*

$$\nabla_{\Pi(s)}t = D_s t + s \bullet t$$

for $s \in \Gamma(\mathcal{A}M)$ *and* $t \in \Gamma(\mathcal{V}M)$.

PROOF. Put $K := GL(V)$, let $\mathcal{P} \times_\rho K$ be the frame bundle of E, and let $j : \mathcal{P} \to \mathcal{P} \times_\rho K$ be the natural map. Then the natural principal connection γ_α on the frame bundle corresponding to the linear map α is characterized by $j^*\gamma_\alpha = \alpha \circ \omega$. Now take a tangent vector $\xi \in T_u\mathcal{P}$. Then the horizontal lift of $T_u p \cdot \xi$ in the point $j(u)$ by definition is $T_u j \cdot \xi - \zeta_{\alpha(\omega(\xi))}(u)$. If $\phi : \mathcal{P} \times_\rho K \to V$ is the equivariant map corresponding to $\sigma \in \Gamma(EM)$, then its derivative with respect to this horizontal lift is

$$(T_u j \cdot \xi) \cdot \phi + \rho'(\alpha(\omega(\xi)))(\phi(j(u))).$$

The first summand can be written as $\xi \cdot (\phi \circ j)$. Viewing EM as $\mathcal{P} \times_H V$, the equivariant map corresponding to σ is $\phi \circ j$. Taking ξ to be the right invariant vector field corresponding to s, the formula follows.

In the case of a tractor bundle, we start with a representation $\rho : G \to GL(V)$, and the map $\alpha : \mathfrak{g} \to L(V, V)$ simply becomes the derivative ρ'. □

We can now use these results to compute the Lie bracket on adjoint tractor fields.

COROLLARY 1.5.8. *For $s_1, s_2 \in \Gamma(\mathcal{A}M)$, the Lie bracket is given by*

$$\begin{aligned}[s_1, s_2] &= D_{s_1}s_2 - D_{s_2}s_1 - \kappa(\Pi(s_1), \Pi(s_2)) + \{s_1, s_2\} \\ &= \nabla^{\mathcal{A}}_{\Pi(s_1)}s_2 - \nabla^{\mathcal{A}}_{\Pi(s_2)}s_1 - \{s_1, s_2\} - \kappa(\Pi(s_1), \Pi(s_2)).\end{aligned}$$

PROOF. For $i = 1, 2$ let $\xi_i \in \mathfrak{X}(\mathcal{P})^H$ be the vector field corresponding to s_i. By definition, the function $\mathcal{P} \to \mathfrak{g}$ corresponding to the Lie bracket $[s_1, s_2]$ is $\omega([\xi_1, \xi_2])$. Inserting the definition of the exterior derivative and of the curvature form, this reads as

$$\xi_1 \cdot \omega(\xi_2) - \xi_2 \cdot \omega(\xi_1) - K(\xi_1, \xi_2) + [\omega(\xi_1), \omega(\xi_2)].$$

Now inserting the definitions of D, κ, and $\{\ ,\ \}$ this is exactly the first claimed formula. To get the second formula, we just have to note that for the adjoint tractor bundle, the formula for the tractor connection from the theorem reads as $\nabla_{\Pi(s_1)}s_2 = D_{s_1}s_2 + \{s_1, s_2\}$ and likewise for the other term. □

REMARK 1.5.8. There also is a nonlinear analog of the fundamental derivative. Since we will use this only rarely, we are brief about it. Let S be a smooth manifold with a left H–action and let F be the corresponding natural bundle. As above, smooth sections of FM may be identified with smooth H–equivariant functions $\mathcal{P} \to S$, but hitting this with a vector field ξ the resulting function $\xi \cdot f$ now has values in TS and is a lift of f. Nevertheless, if $\xi \in \mathfrak{X}(\mathcal{P})^H \cong \Gamma(\mathcal{A}M)$, then the same argument as above shows that $\xi \cdot f$ is equivariant, and defines a smooth section of $\mathcal{P} \times_H TS$ which may be identified with the vertical tangent bundle VFM. Hence, we get a fundamental derivative $D : \Gamma(\mathcal{A}M) \times \Gamma(FM) \to \Gamma(VFM)$, which is linear and tensorial in the first argument. Moreover, for $s \in \Gamma(\mathcal{A}M)$ and $\sigma \in \Gamma(FM)$ the section $D_s\sigma$ of VFM is a lift of σ.

1.5.9. Bianchi and Ricci identities. We next derive the basic differential identities for the curvature and describe iterated fundamental derivatives. In 1.5.8 we saw that for any natural vector bundle E there is a sequence of differential operators $D^k : \Gamma(EM) \to \Gamma(\otimes^k \mathcal{A}^*M \otimes EM)$. Moreover, there is a natural projection $\Pi : \mathcal{A}M \to TM$, so any adjoint tractor field has an underlying vector field. Dually, we get an inclusion $T^*M \to \mathcal{A}^*M$ which means that any differential form canonically extends to taking adjoint tractor fields as an input. Otherwise put, we insert adjoint tractor fields into differential forms by first projecting to the underlying vector fields. We will often suppress the projection Π and simply insert adjoint tractor fields as arguments into differential forms.

PROPOSITION 1.5.9. *Let $(p : \mathcal{P} \to M, \omega)$ be a Cartan geometry of type (G, H) with curvature $\kappa \in \Omega^2(M, \mathcal{A}M)$, let $\nabla^{\mathcal{A}}$ be the adjoint tractor connection and $\{\ ,\ \}$ the algebraic bracket on $\mathcal{A}M$.*

(1) (Bianchi–identity) The curvature κ satisfies

$$\sum_{cycl}\Big(\nabla^{\mathcal{A}}_{\xi_1}(\kappa(\xi_2, \xi_3)) - \kappa([\xi_1, \xi_2], \xi_3)\Big) = 0 \tag{1.24}$$

for all vector fields $\xi_i \in \mathfrak{X}(M)$ or equivalently

$$\sum_{cycl}\Big(\{s_1, \kappa(s_2, s_3)\} - \kappa(\{s_1, s_2\}, s_3) + \kappa(\kappa(s_1, s_2), s_3) + (D_{s_1}\kappa)(s_2, s_3)\Big) = 0 \tag{1.25}$$

for all $s_i \in \Gamma(\mathcal{A}M)$, *where the sums are over all cyclic permutations of the arguments.*

(2) (Ricci–identity) For any natural vector bundle E *and any section* $\sigma \in \Gamma(EM)$ *the alternation of the square of the fundamental derivative is given by*

$$(D^2\sigma)(s_1, s_2) - (D^2\sigma)(s_2, s_1) = -D_{\kappa(s_1,s_2)}\sigma + D_{\{s_1,s_2\}}\sigma.$$

PROOF. (1) Let us first prove the equivalence of (1.24) and (1.25). In view of the fact that $\Pi([s_1, s_2]) = [\Pi(s_1), \Pi(s_2)]$, we may equivalently replace the vector fields ξ_i in (1.24) by adjoint tractor fields s_i. Then the formula for the adjoint tractor connection from Theorem 1.5.8 shows that

$$\nabla^{\mathcal{A}}_{s_1}(\kappa(s_2, s_3)) = D_{s_1}(\kappa(s_2, s_3)) + \{s_1, \kappa(s_2, s_3)\},$$

while the formula for the Lie bracket of adjoint tractors from Corollary 1.5.8 gives

$$-\kappa([s_1, s_2], s_3) = -\kappa(D_{s_1}s_2, s_3) + \kappa(D_{s_2}s_1, s_3) + \kappa(\kappa(s_1, s_2), s_3) - \kappa(\{s_1, s_2\}, s_3).$$

On the other hand, naturality of the fundamental derivative implies

$$(D_{s_1}\kappa)(s_2, s_3) = D_{s_1}(\kappa(s_2, s_3)) - \kappa(D_{s_1}s_2, s_3) - \kappa(s_2, D_{s_1}s_3).$$

Inserting this, we see that, replacing the ξ_i by s_i in (1.24) we obtain the cyclic sum of

$$\begin{aligned}\{s_1, \kappa(s_2, s_3)\} + (D_{s_1}\kappa)(s_2, s_3) + \kappa(\kappa(s_1, s_2), s_3)\\ - \kappa(\{s_1, s_2\}, s_3) + \kappa(s_2, D_{s_1}s_3) + \kappa(D_{s_2}s_1, s_3).\end{aligned}$$

Forming the cyclic sum, the last two terms cancel by skew symmetry of κ, and we obtain (1.25).

Now (1.25) is visibly linear over smooth functions in all arguments s_i, so it can be verified in a point. We may view κ as the curvature function $\mathcal{P} \to L(\Lambda^2\mathfrak{g}, \mathfrak{g})$ (recalling that the result vanishes if one entry is from $\mathfrak{h}$). In these terms, the claimed identity has the form

$$\begin{aligned}0 = \sum_{\text{cycl}}\Big(& [\kappa(X,Y), Z] + \kappa([X,Y], Z)\\ & - \kappa\big(\kappa(X,Y), Z\big) - \omega^{-1}(Z)\cdot\kappa(X,Y)\Big)\end{aligned} \tag{1.26}$$

for all X, Y, $Z \in \mathfrak{g}$. (Observe that since evaluation in (X, Y) is a linear map, there is no difference between $(\omega^{-1}(Z)\cdot\kappa)(X,Y)$ and $\omega^{-1}(Z)\cdot(\kappa(X,Y))$.) Let us evaluate the structure equation on the vector fields $[\tilde{X}, \tilde{Y}]$ and $\tilde{Z}$, where $\tilde{X} = \omega^{-1}(X)$, $\tilde{Y} = \omega^{-1}(Y)$, and $\tilde{Z} = \omega^{-1}(Z)$. Remember, in particular, $\kappa(X,Y) = K(\tilde{X},\tilde{Y})$ and $\omega([\tilde{X},\tilde{Y}]) = -\kappa(X,Y) + [X,Y]$. Thus,

$$\begin{aligned}K([\tilde{X},\tilde{Y}],\tilde{Z}) &= -\tilde{Z}\cdot\omega([\tilde{X},\tilde{Y}]) - \omega\big([[\tilde{X},\tilde{Y}],\tilde{Z}]\big) + \big[\omega([\tilde{X},\tilde{Y}]), Z\big]\\ &= \tilde{Z}\cdot(\kappa(X,Y)) - \omega\big([[\tilde{X},\tilde{Y}],\tilde{Z}]\big) + [[X,Y],Z] - [\kappa(X,Y),Z]\end{aligned}$$

and the left–hand side equals $\kappa([X,Y],Z) - \kappa(\kappa(X,Y),Z)$. Now, let us perform the cyclic permutation over X, Y, and Z. Leaving out the two terms which disappear by virtue of the Jacobi identity for vector fields and the Lie algebra $\mathfrak{g}$, while collecting all remaining terms on one side of the equality, we obtain exactly the required identity.

(2) To prove the Ricci identity, note first that for sections $s_i \in \Gamma(\mathcal{A}M)$ corresponding to $\xi_i \in \mathfrak{X}(\mathcal{P})^H$ the definition of the fundamental derivative immediately

implies that $D_{s_1}(D_{s_2}\sigma) - D_{s_2}(D_{s_1}\sigma) = D_{[s_1,s_2]}\sigma$, where on the right–hand side we have the Lie bracket of adjoint tractor fields. Naturality of the fundamental derivative implies $(D^2\sigma)(s_1, s_2) = D_{s_1}(D_{s_2}\sigma) - D_{D_{s_1}s_2}\sigma$. Alternating this, we see that

$$(D^2\sigma)(s_1, s_2) - (D^2\sigma)(s_2, s_1) = D_{[s_1,s_2]-D_{s_1}s_2+D_{s_2}s_1}\sigma,$$

and the Ricci identity immediately follows from the formula for $[s_1, s_2]$ in Corollary 1.5.8. □

REMARK 1.5.9. (i) The first form of the Bianchi identity matches up nicely with the classical Bianchi identity for a linear connection on a vector bundle which says that the covariant exterior derivative of the curvature vanishes, and in fact this leads to an alternative proof of the identity.

(ii) Since κ does not depend on adjoint tractor fields inserted but only on the underlying vector fields, it follows that the term $\kappa(\kappa(s_1, s_2), s_3)$ depends only on the torsion T, the projection to TM of the values of κ; see 1.5.7. In particular, this term vanishes for torsion–free Cartan geometries.

To get the classical versions of the Bianchi and Ricci identities, let us specialize to Cartan geometries corresponding to a reductive Klein geometry (G, H). Thus, we have to assume that there is a distinguished H–invariant subspace $\mathfrak{n} \subset \mathfrak{g}$ which is complementary to the Lie subalgebra $\mathfrak{h}$. Since we will be mainly interested in the non–reductive case in the sequel, we only give a rough treatment of this case, leaving some details to the reader.

COROLLARY 1.5.9. *Let $(\mathcal{P} \to M, \omega)$ be a Cartan geometry of type (G, H), where (G, H) is a reductive Klein geometry with H–invariant decomposition $\mathfrak{g} = \mathfrak{h} \oplus \mathfrak{n}$. Let R be the curvature of the natural principal connection on $\mathcal{P}$ and let T be the torsion of the induced connection on the tangent bundle TM. Denoting by ∇ the natural covariant derivatives we have:*

(i) *The algebraic Bianchi identity for $\xi_i \in \mathfrak{X}(M)$:*

$$\sum_{cycl}\Big(-R(\xi_1, \xi_2)(\xi_3) + T(T(\xi_1, \xi_2), \xi_3) + (\nabla_{\xi_1}T)(\xi_2, \xi_3)\Big) = 0.$$

(ii) *The differential Bianchi identity*

$$\sum_{cycl}\Big((\nabla_{\xi_1}R)(\xi_2, \xi_3) + R(T(\xi_1, \xi_2), \xi_3)\Big) = 0.$$

(iii) *The Ricci identity for a section σ of an arbitrary natural vector bundle and $\xi, \eta \in \mathfrak{X}(M)$,*

$$(\nabla^2\sigma)(\xi, \eta) - (\nabla^2\sigma)(\eta, \xi) = R(\xi, \eta)(\sigma) - \nabla_{T(\xi,\eta)}\sigma.$$

PROOF. For any Cartan geometry $(\mathcal{P} \to M, \omega)$ of type (G, H), the H–invariant decomposition $\mathfrak{g} = \mathfrak{h} \oplus \mathfrak{n}$ induces a splitting $\mathcal{A}M = (\mathcal{P} \times_H \mathfrak{h}) \oplus (\mathcal{P} \times_H \mathfrak{n})$, and the second summand is isomorphic to TM. According to this, we will write sections of $\mathcal{A}M$ as column vectors with a vector field as lower row and a section of $\mathcal{P} \times_H \mathfrak{h}$ as upper row.

In the setting or Theorem 1.5.8, we have to set $\alpha(s, \sigma) = \pi(s) \bullet \sigma$, where $\pi : \mathfrak{g} \to \mathfrak{h}$ is the projection along $\mathfrak{n}$. Hence, viewing $\xi \in \mathfrak{X}(M)$ as an adjoint tractor, $D_\xi\sigma = \nabla_\xi\sigma$ by Theorem 1.5.8. Via the splitting of $\mathcal{A}M$, also the Bianchi and Ricci identities split into components.

The assumptions on $\mathfrak{h}$ and $\mathfrak{n}$ imply that $[\mathfrak{h},\mathfrak{h}]\subset\mathfrak{h}$ and $[\mathfrak{h},\mathfrak{n}]\subset\mathfrak{n}$, and we may split the remaining bracket $\Lambda^2\mathfrak{n}\to\mathfrak{g}$ according to the decomposition $\mathfrak{g}=\mathfrak{n}\oplus\mathfrak{h}$. From Theorem 1.4.7 we know that the $\mathfrak{n}$–component of the bracket corresponds to $-T_0$, where T_0 is the torsion of the invariant connection on $T(G/H)$ induced by $\mathfrak{n}$, while the $\mathfrak{h}$–component corresponds to $-R_0$, where R_0 is the curvature of the invariant principal connection on $G\to G/H$.

Taking into account that $\alpha:\mathfrak{g}\to\mathfrak{h}$ is the projection along $\mathfrak{n}$, Corollary 1.5.7 shows that the Cartan curvature is given by

$$\kappa(\xi,\eta)=\begin{pmatrix}(R-R_0)(\xi,\eta)\\(T-T_0)(\xi,\eta)\end{pmatrix}$$

for all $\xi,\eta\in\mathfrak{X}(M)$. We can also split the algebraic bracket on $\mathcal{A}M$ according to the decomposition of the bracket on $\mathfrak{g}$. To obtain the two Bianchi identities (i) and (ii), we shall apply formula (1.25) from Proposition 1.5.9 to $\xi_1,\xi_2,\xi_3\in\mathfrak{X}(M)\subset\Gamma(\mathcal{A}M)$. For the first summand $\{\xi_1,\kappa(\xi_2,\xi_3)\}$ we get

$$\left\{\begin{pmatrix}0\\\xi_1\end{pmatrix},\begin{pmatrix}(R-R_0)(\xi_2,\xi_3)\\(T-T_0)(\xi_2,\xi_3)\end{pmatrix}\right\}=\begin{pmatrix}-R_0(\xi_1,(T-T_0)(\xi_2,\xi_3))\\-(R-R_0)(\xi_2,\xi_3)(\xi_1)-T_0(\xi_1,(T-T_0)(\xi_2,\xi_3))\end{pmatrix},$$

and after passing to the cyclic sum, the right–hand side can be replaced by

$$\begin{pmatrix}R_0((T-T_0)(\xi_1,\xi_2),\xi_3)\\-(R-R_0)(\xi_2,\xi_3)(\xi_1)+T_0((T-T_0)(\xi_1,\xi_2),\xi_3)\end{pmatrix}.$$

For the next two summands $-\kappa(\{\xi_1,\xi_2\},\xi_3)+\kappa(\kappa(\xi_1,\xi_2),\xi_3)$, we use that by construction the TM–component of $\kappa(\xi_1,\xi_2)-\{\xi_1,\xi_2\}$ is given by $T(\xi_1,\xi_2)$ to conclude that these two summands contribute

$$\begin{pmatrix}(R-R_0)(T(\xi_1,\xi_2),\xi_3)\\(T-T_0)(T(\xi_1,\xi_2),\xi_3)\end{pmatrix}.$$

For the last summand, we directly get the contribution

$$\begin{pmatrix}(\nabla_{\xi_1}(R-R_0))(\xi_2,\xi_3)\\(\nabla_{\xi_1}(T-T_0))(\xi_2,\xi_3)\end{pmatrix}.$$

Collecting the $\mathcal{P}\times_H\mathfrak{h}$–components, we see that the terms containing R_0 and T cancel, and we conclude that vanishing of this component is equivalent to

$$\sum_{\text{cycl}}\Big((\nabla_{\xi_1}R)(\xi_2,\xi_3)+R(T(\xi_1,\xi_2),\xi_3)\Big)$$
$$=\sum_{\text{cycl}}\Big((\nabla_{\xi_1}R_0)(\xi_2,\xi_3)+R_0(T_0(\xi_1,\xi_2),\xi_3)\Big).$$

Similarly, collecting the TM–components we see that the terms mixing T and T_0 cancel and vanishing of the TM–component is equivalent to

$$\sum_{\text{cycl}}\Big(-R(\xi_1,\xi_2)(\xi_3)+T(T(\xi_1,\xi_2),\xi_3)+(\nabla_{\xi_1}T)(\xi_2,\xi_3)\Big)$$
$$=\sum_{\text{cycl}}\Big(-R_0(\xi_1,\xi_2)(\xi_3)+T_0(T_0(\xi_1,\xi_2),\xi_3)+(\nabla_{\xi_1}T_0)(\xi_2,\xi_3)\Big)$$

To complete the proof, it thus suffices to show that the right–hand sides of these two equations vanish automatically. But this can be easily concluded as follows: Define a new Lie algebra structure on $\mathfrak{h}\oplus\mathfrak{n}$ by keeping the brackets $\mathfrak{h}\times\mathfrak{h}\to\mathfrak{h}$ and

$\mathfrak{h} \times \mathfrak{n} \to \mathfrak{n}$ and declaring the bracket to be zero on $\mathfrak{n} \times \mathfrak{n}$. One immediately verifies that this defines a Lie algebra $\tilde{\mathfrak{g}}$, and $\tilde{\mathfrak{g}} \cong \mathfrak{g}$ as an $\mathfrak{h}$–module. Taking as a Lie group $\tilde{G}$ with this Lie algebra the affine extension of H, we may view $(G \to G/H, \omega_G)$ as a (non–flat) Cartan geometry of type $(\tilde{G}, H)$. Now the curvature and torsion of this geometry is by construction given by R_0 and T_0 from before, while for the new geometry the canonical connection on the homogeneous model is torsion free and flat, so the result follows by using the above formulae in this case.

(iii) Here we just have to observe that for $\xi_1, \xi_2 \in \mathfrak{X}(M) \subset \Gamma(\mathcal{A}M)$, the expression $\kappa(\xi_1, \xi_2) - \{\xi_1, \xi_2\}$ has components $R(\xi_1, \xi_2)$ and $T(\xi_1, \xi_2)$. The claimed identity now immediately follows from part (2) of Proposition 1.5.9 applied to $\xi_i \in \mathfrak{X}(M) \subset \Gamma(\mathcal{A}M)$. □

1.5.10. Fundamental derivative and jet prolongations. Tractor connections and fundamental derivatives as discussed in the last few subsections are examples of invariant differential operators defined for all Cartan geometries. By construction, they are intrinsic to the given geometric structure, and this is the property one tries to capture in the general notion of natural or invariant differential operators for Cartan geometries. It turns out that this leads to very deep and interesting problems and results, which will be the main topic of volume two, so we will discuss the technicalities there and only present an outline here.

For the case of a Klein geometry (G, H), we have discussed the basics on invariant differential operators in 1.4.9 and 1.4.10. We first observe that the situation becomes very simple in the presence of invariant connections. Assume that we have given a homogeneous vector bundle $E \to G/H$ which admits an invariant linear connection and that there also is an invariant linear connection on $T(G/H)$. Then Proposition 1.4.9 gives a complete description of invariant differential operators which map sections of E to sections of an arbitrary homogeneous bundle F. This description is in terms of invariant bundle maps $S^kT^*(G/H) \otimes E \to F$ for $k \geq 0$.

Now all the ingredients generalize to arbitrary Cartan geometries of type (G, H). The bundles E and F give rise to natural bundles on the category $\mathcal{C}_{(G,H)}$, and from 1.5.6 we know that there are natural connections on E and on the tangent bundle. An invariant bundle map $\Phi : S^kT^*(G/H) \otimes E \to F$ is induced by a G–equivaraint map between the inducing representation and hence extends to a natural bundle map on $\mathcal{C}_{(G,H)}$. Applying this bundle map to symmetrized iterated covariant derivatives as in the proof of Proposition 1.4.9 we obtain a natural differential operator, whose symbol is given by the natural bundle map from above. Thus, we see

OBSERVATION 1.5.10. Suppose that (G, H) is a Klein geometry such that there is a G–invariant linear connection on $T(G/H)$. Suppose further that $E \to G/H$ is a homogeneous vector bundle which admits a G–invariant linear connection. Then any invariant linear differential operator mapping $\Gamma(E)$ to sections of some homogeneous vector bundle F canonically extends to a natural differential operator on the category $\mathcal{C}_{(G,H)}$.

In the case of a general homogeneous vector bundle $E \to G/H$, we started by considering a jet prolongation J^rE. This is again homogeneous and hence induced by a representation of H on the standard fiber J^rE_o. For another homogeneous vector bundle F, invariant differential operators of order $\leq r$ are then equivalent to H–module homomorphisms $J^rE_o \to F_o$. Now, again, the representations give rise to natural vector bundles on $\mathcal{C}_{(G,H)}$ and the H–module homomorphism gives rise to a natural bundle map. Hence, it may seem as if invariant differential operators still would automatically extend to natural operators on $\mathcal{C}_{(G,H)}$.

This is not true, however. The problem is that while for the homogeneous model $J^r(G\times_H E_o)$ is isomorphic to $G\times_H J^r E_o$, it is not true for a general Cartan geometry $(\mathcal{P} \to M, \omega)$ of type (G, H) that $\mathcal{P}\times_H J^r E_o$ is naturally isomorphic to $J^r(\mathcal{P}\times_H E_0)$. Already for conformal geometry (which is a rather simple example of a parabolic geometry) there are examples of invariant differential operators on the homogeneous model, which do not extend to the category of all conformal structures; see [**Gr92**] and [**GoHi04**]. These results also prove that in these cases the bundle $\mathcal{P} \times_H J^r E_o$ is really different from $J^r(\mathcal{P} \times_H E_0)$. (The simplest case in which this occurs is $r = 6$ and a one–dimensional representation E_o.) It is this partial breakdown of the correspondence between a Klein geometry and the associated Cartan geometries that makes the theory of invariant differential operators difficult and interesting.

In spite of these difficulties, jet prolongations and the associated representations are an essential ingredient in the theory of natural operators for Cartan geometries. As a first basic result, we show that the fundamental derivative can be used to encode arbitrarily high jets of sections of any bundle associated to the Cartan bundle into a section of a natural bundle. Fix a Klein geometry (G, H) and consider a natural bundle E corresponding to a representation V of H. This means that for any Cartan geometry $(p : \mathcal{G} \to M, \omega)$ of type (G, H) we have $EM = \mathcal{P}\times_H V$. Given a section $\sigma \in \Gamma(EM)$, we can form the fundamental derivative $D\sigma \in \Gamma(\mathcal{A}^*M\otimes EM)$ and iterating we get for any r the operator $D^r\sigma \in \Gamma(\otimes^r\mathcal{A}^*M \otimes EM)$. Formally, we define $D^1\sigma = D\sigma$ and inductively $D^r\sigma = D(D^{r-1}\sigma)$. By construction, D is a first order operator, so we may also view $\sigma \mapsto D^r\sigma$ as a vector bundle map from the rth jet prolongation J^rEM to $\otimes^r\mathcal{A}^*M \otimes EM$. To come closer to the usual jets, the obvious idea is to replace $D^r\sigma$ by its complete symmetrization $\mathrm{Symm}(D^r\sigma)$ defined by

$$\mathrm{Symm}(D^r\sigma)(s_1, \dots, s_r) = \tfrac{1}{r!} \sum_{\tau\in\mathfrak{S}_r} (D^r\sigma)(s_{\tau(1)}, \dots, s_{\tau(r)}),$$

for all $s_j \in \Gamma(\mathcal{A}M)$, where $\mathfrak{S}_r$ denotes the permutation group. This may then be viewed as a section of $S^r\mathcal{A}^*M \otimes EM$.

PROPOSITION 1.5.10. *For any $r \in \mathbb{N}$, the operator*

$$\sigma \mapsto (\sigma, D\sigma, \mathrm{Symm}(D^2\sigma), \dots, \mathrm{Symm}(D^r\sigma))$$

induces an injective bundle map $J^rEM \to \bigoplus_{j=0}^r(S^j\mathcal{A}^*M \otimes EM)$.

PROOF. We first claim that $j^{r-1}_x\sigma = 0$ implies that

$$D^r\sigma(x)(s_1, \dots, s_r) = D_{s_1}D_{s_2}\dots D_{s_r}\sigma(x)$$

for all sections $s_i \in \Gamma(\mathcal{A}M)$. Naturality of the fundamental derivative implies that for any i and arbitrary sections $s_1, \dots, s_r$ of $\mathcal{A}M$ we have

$$D^i\sigma(s_1, \dots, s_i) = D_{s_1}(D^{i-1}\sigma(s_2, \dots, s_i)) - \sum_\ell D^{i-1}\sigma(s_2, \dots, D_{s_1}s_\ell, \dots, s_i).$$

Since D is first order, $j^{r-1}_x\sigma = 0$ implies $j^{r-i}_x(D^{i-1}\sigma) = 0$, and thus

$$j^{r-i}_x(D^i\sigma(s_1, \dots, s_i)) = j^{r-i}_x(D_{s_1}(D^{i-1}\sigma(s_2, \dots, s_i))).$$

Applied to $i = r$, this shows that $D^r\sigma(s_1, \dots, s_r)(x) = D_{s_1}(D^{r-1}\sigma(s_2, \dots, s_r))(x)$. But then we apply the same fact for $i = r - 1$ to conclude that the one–jet in x

of $D^{r-1}\sigma(s_2,\dots,s_r)$ coincides with the one–jet of $D_{s_2}(D^{r-2}(s_3,\dots,s_r))$, and the claim follows by induction.

Still assuming $j^{r-1}_x\sigma = 0$, we next claim that $(D_{s_1}D_{s_2}\dots D_{s_r}\sigma)(x)$ is completely symmetric in the s_j. The Ricci identity from part (2) of Proposition 1.5.9 implies that for any section ϕ of a natural bundle one may compute $(D_{s_1}D_{s_2}\phi)(x)-(D_{s_2}D_{s_1}\phi)(x)$ from $D\phi(x)$. In particular, if $j^\ell_x\phi = 0$, then $j^{\ell-1}_x D\phi = 0$, and thus $j^{\ell-1}_x(D_{s_1}D_{s_2}\phi) = j^{\ell-1}_x(D_{s_2}D_{s_1}\phi)$. Now consider an index ℓ such that $1 \le \ell \le r-1$. Then $j^{r-1}_x\sigma = 0$ implies that $D_{s_{\ell+2}}\dots D_{s_r}\sigma$ has vanishing ℓ–jet in x and thus

$$j^{\ell-1}_x D_{s_\ell}D_{s_{\ell+1}}D_{s_{\ell+2}}\dots D_{s_r}\sigma = j^{\ell-1}_x D_{s_{\ell+1}}D_{s_\ell}D_{s_{\ell+2}}\dots D_{s_r}\sigma.$$

Hence, the value of $D_{s_1}\dots D_{s_r}\sigma(x)$ does not change if one exchanges s_i and s_{i+1}, which implies the claim.

To complete the proof, we only have to verify injectivity, which we do by induction on r. Hence, let us first assume that $\sigma \in \Gamma(EM)$ satisfies $\sigma(x) = 0$ and $D\sigma(x) = 0$ for some $x \in M$. Denoting by $f : \mathcal{P} \to V$ the equivariant function corresponding to σ, let us take any point $u \in \mathcal{P}$ such that $p(u) = x$ and any tangent vector $\xi \in T_u\mathcal{P}$. Then $\xi \cdot f(u) = 0$, since this equals $D_s\sigma(x)$, where $s \in \Gamma(\mathcal{A}M)$ corresponds to any extension of ξ to an H–invariant vector field on $\mathcal{P}$. Thus, $j^1_u f = 0$ and hence $j^1_x\sigma = 0$.

So let us assume that $r > 1$ and $\sigma \in \Gamma(\mathcal{V}M)$ is such that $\sigma(x) = 0$, $D\sigma(x) = 0$ and $\operatorname{Symm}(D^i\sigma)(x) = 0$ for $i \le r$. By induction, this implies $j^{r-1}_x\sigma = 0$ and using the two claims above we conclude that $\operatorname{Symm}(D^r\sigma)(x) = D^r\sigma(x)$, so we conclude that $D^r\sigma$ vanishes in x. By the above considerations, this implies that $D^{r-1}\sigma$ and thus $j^{r-1}\sigma$ has vanishing one–jet in x, whence $j^r_x\sigma = 0$. □

Even in the case $r = 1$ (where there is no issue of symmetrization) the bundle $EM \oplus \mathcal{A}^*M \otimes EM$ is much bigger than the first jet prolongation J^1EM. It is also easy to see directly that the pair $(\sigma, D\sigma)$ for $\sigma \in \Gamma(EM)$ contains redundant information. By part (2) of Proposition 1.5.8, for a section s of the subbundle $\mathcal{P} \times_H \mathfrak{h} \subset \mathcal{A}M$ and any section σ of a natural bundle, $D_s\sigma$ is just the negative of the natural algebraic action of s on σ.

These observations suggest a way to get a better hold on the first jet prolongation. Suppose that V is the representation of H which induces the natural bundle E. Then we define a subspace $J^1V \subset V \oplus L(\mathfrak{g}, V)$ as the space of all (v, ϕ) such that $\phi(A) = -A \cdot v$ for all $A \in \mathfrak{h} \subset \mathfrak{g}$.

THEOREM 1.5.10. *The subspace $J^1V \subset V \oplus L(\mathfrak{g}, V)$ is H–invariant and hence gives rise to a natural subbundle in $EM \oplus \mathcal{A}^*M \otimes EM$. The operator $\sigma \mapsto (\sigma, D\sigma)$ always has values in this subbundle. For any Cartan geometry $(\mathcal{P} \to M, \omega)$, this gives rise to a natural isomorphism $J^1EM \to \mathcal{P} \times_H (J^1V)$.*

In particular, any first order invariant linear differential operator between homogeneous bundles on G/H canonically extends to a natural operator on $\mathcal{C}_{(G,H)}$.

PROOF. The natural H–action on $V \oplus L(\mathfrak{g}, V)$ is given by

$$h \cdot (v, \phi) = (h \cdot v, A \mapsto h \cdot \phi(\operatorname{Ad}(h^{-1})(A))).$$

Assuming that (v, ϕ) lies in the subspace J^1V and that $A \in \mathfrak{h}$, the second component of this is given by

$$A \mapsto -h \cdot (\operatorname{Ad}(h^{-1})(A) \cdot v) = -h \cdot (h^{-1} \cdot A \cdot h \cdot v) = -A \cdot h \cdot v,$$

which implies that $h \cdot (v, \phi) \in J^1V$, so this is an H–submodule. We have noted above that if s is a section of the subbundle $\mathcal{P} \times_H \mathfrak{h} \subset \mathcal{A}M$, then $D_s\sigma$ is given by the negative of the algebraic action on σ, which implies that $(\sigma, D\sigma)$ has values in the subbundle $\mathcal{P} \times_H (J^1V)$. From the proposition we obtain an injective bundle map $J^1EM \to \mathcal{P} \times_H (J^1V)$. Projecting on the first factor induces a surjection $J^1V \to V$, whose kernel by definition is isomorphic to $L(\mathfrak{g}/\mathfrak{h}, V)$. Hence, we see that the dimension of J^1V equals the rank of the bundle J^1EM, so the bundle map $J^1EM \to \mathcal{P} \times_H (J^1V)$ has to be an isomorphism.

Applying this to $E \to G/H$, we see that J^1V is the representation inducing the homogeneous bundle J^1E. Hence, any first order linear invariant operator defined on sections of E is induced by an H–homomorphism from J^1V to some other representation. This homomorphism induces a natural bundle map on J^1EM and hence a natural differential operator on $\mathcal{C}_{(G,H)}$. □

Again, it may seem that this result can be extended to higher orders, but as we have seen already this cannot be true in general. The problem is that the behavior of $\mathrm{Symm}(D^r\sigma)$ under insertion of one section of the subbundle $\mathcal{P} \times_H \mathfrak{h}$ can be computed from $D^i\sigma$ with $i < r$, but it is not sufficent to know $\mathrm{Symm}(D^i\sigma)$ for $i < r$. As we shall see in volume two, there is an analog of the theorem for $r = 2$. This means that the second jet prolongation of any natural bundle associated to the Cartan bundle is again associated to the Cartan bundle. Moreover, any second order invariant operator on G/H canonically extends to a natural operator on $\mathcal{C}_{(G,H)}$.

For higher orders, one can start by looking at *non–holonomic jet prolongations*, i.e. iterated first jet prolongations. For a natural vector bundle bundle EM, we can use the theorem to identify $J^1(J^1EM)$ with the associated bundle corresponding to $J^1(J^1V)$, and similary for higher orders. This can be improved by passing to *semi–holonomic jet prolongations*, which can be constructed from the non–holonomic ones using only functorial properites. For any $r > 0$ this leads to a natural bundle $\bar{J}^rEM$ which is associated to the Cartan bundle with respect to a representation $\bar{J}^rV$. The rth jet prolongation J^rEM naturally includes into $\bar{J}^rEM$. Therefore, H–equivariant maps from $\bar{J}^rV$ to other representations give rise to natural differential operators. However, nonzero maps may lead to the zero operator and not all natural operators are obtained in this way. This circle of ideas will be one of the main topics of volume two.

1.5.11. Automorphisms of Cartan geometries. We next switch to another nice general feature of Cartan geometries. We have seen in 1.5.2 that the automorphisms of the homogeneous model $(G \to G/H, \omega_G)$ are exactly given by left multiplications by elements of G. The aim of this subsection is to show that the group of automorphisms of any Cartan geometry $(\mathcal{P} \to M, \omega)$ is a Lie group (which may have uncountably many connected components) and the dimension of this group at most equals the dimension of G. The first ingredient we need is usually referred to as Lie's second fundamental theorem. This generalizes the fact that Lie algebra homomorphisms integrate to local group homomorphisms (see 1.2.4) to the case of a diffeomorphism group, i.e. it is an analogous result for group actions.

The infinitesimal data for an action of a group G on M is provided by the fundamental vector field mapping $\zeta : \mathfrak{g} \to \mathfrak{X}(M)$, which is a Lie algebra homomorphism for right actions. Now a *local right action* of the group G on M is given by an open neighborhood U of $M \times \{e\}$ in $M \times G$ and a smooth map $r : U \to M$, such that $r(x,e) = x$ and if $x \in M$ and $g,h \in G$ are such that (x,g), (x,gh) and $(r(x,g),h)$ all are in U, then $r(r(x,g),h) = r(x,gh)$. Given a local action $r : U \to M$, then for $x \in M$ and $X \in \mathfrak{g}$ we get a curve $t \mapsto r(x,\exp(tX))$ defined locally around zero, so the derivative $\frac{d}{dt}|_0 r(x,\exp(tX))$ is always well defined.

LEMMA 1.5.11 (Lie's second fundamental theorem). *Let G be a Lie group with Lie algebra $\mathfrak{g}$, let M be a smooth manifold, and let $\phi : \mathfrak{g} \to \mathfrak{X}(M)$ be a homomorphism of Lie algebras. Then there exists a local right action $r : U \to M$ such that $\phi(X)(x) = \frac{d}{dt}|_0 r(x,\exp(tX))$ for all $x \in M$ and $X \in \mathfrak{g}$.*

This is a classical result, which basically goes back to Sophus Lie. A proof of the version above in modern language can be found in [**Pa57**]. The basic idea of the proof is to consider the distribution on $M \times G$ formed by all pairs $(\phi(X)(x), L_X(g))$, where L_X denotes the left invariant vector field corresponding to X. Since ϕ is assumed to be a Lie algebra homomorphism, this distribution is involutive and thus integrable by the Frobenius theorem. By construction the second projection induces a local diffeomorphism from any leaf to G. For $x \in M$ one can then look at the leaf through (x,e) and inverting the second projection, one obtains a smooth map from an open neighborhood of e in G to M. This can be used to define the action on x which visibly produces the correct fundamental vector fields in x. A careful analysis of the equivariancy properties of this foliation shows that this leads to a local action defined on a neighborhood of $M \times \{e\}$ in $M \times G$.

The next step is to prove a sufficient condition for a transformation group to be a Lie group which is due to [**Pa57**].

PROPOSITION 1.5.11. *Let G be a group of diffeomorphisms of a smooth manifold M and let $S \subset \mathfrak{X}(M)$ be the subset of all vector fields ξ whose flows Fl^ξ_t are defined and lie in G for all $t \in \mathbb{R}$. If the Lie subalgebra of $\mathfrak{X}(M)$ generated by S is finite-dimensional, then G is a Lie group of transformations of M and S is the Lie algebra of G.*

PROOF. We present a short proof following [**Ko72**, Theorem 3.1]. Let us write $\mathfrak{g} \subset \mathfrak{X}(M)$ for the Lie algebra of vector fields generated by S and consider the connected and simply connected Lie group $\tilde{G}$ with the Lie algebra $\mathfrak{g}$. By Lie's second fundamental theorem the inclusion $\mathfrak{g} \to \mathfrak{X}(M)$ integrates to a local group action, i.e. there is an open neighborhood U of $M \times \{e\}$ in $M \times \tilde{G}$ and a local action $r : U \to M$ such that $X(x) = \frac{d}{dt}|_0 r(x,\exp(tX))$ for all $X \in \mathfrak{g}$. Of course, this implies that $r(x,\exp tX) = \mathrm{Fl}^X_t(x)$ whenever $(x,\exp tX) \in U$.

First we claim that S generates $\mathfrak{g}$ as a vector space. Denoting by $V \subset \mathfrak{g}$ the vector space generated by S it suffices to prove $[V,V] \subset V$. Consider $X,Y \in S$ and put $Z = \mathrm{Ad}(\exp X)Y \in \mathfrak{g}$. Then $\exp tZ = \exp X \exp tY \exp(-X)$. Now for $x \in M$ and s small enough the action property implies that locally around x one has $r(y,\exp(sX)g\exp(-sX)) = \mathrm{Fl}^X_s(r(\mathrm{Fl}^X_{-s}(y),g))$ and using $\exp(X) = (\exp(\frac{1}{N}X))^N$ for sufficiently large N, we conclude that $r(x,\exp(tZ)) = (\mathrm{Fl}^X_1 \circ \mathrm{Fl}^Y_t \circ \mathrm{Fl}^X_{-1})(x)$ for $|t|$ small enough. But again for $|t|$ small enough we have $\mathrm{Fl}^Z_t = r(x,\exp(tZ))$ and hence

$$\mathrm{Fl}^Z_t = \mathrm{Fl}^X_1 \circ \mathrm{Fl}^Y_t \circ \mathrm{Fl}^X_{-1}.$$

The right–hand side of this equation is defined for all t and is a one–parameter group of diffeomorphisms, so we conclude that Fl^Z_t is defined for all t and thus

$Z \in S$. For $X \in S$ and $t \in \mathbb{R}$ we have $tX \in S$, so our argument shows that $\operatorname{Ad}(\exp(tX))(Y) \in S \subset V$ for all $X, Y \in S$. But differentiating this smooth curve at $t = 0$, the resulting element must also lie in V, so we obtain $[S, S] \subset V$, and since the bracket is bilinear this implies $[V, V] \subset V$.

Now we claim that $S = \mathfrak{g}$. Choose a basis $\{X_1, \dots, X_k\}$ of the vector space $\mathfrak{g}$ consisting of elements of S, and consider the map $\mathfrak{g} \to \tilde{G}$ defined by

$$\sum c^i X_i \mapsto \exp(c^1 X_1) \dots \exp(c^k X_k).$$

This restricts to a diffeomorphism from an open neighborhood of zero in $\mathfrak{g}$ onto an open neighborhood of the unit $e \in \tilde{G}$. For $Y \in \mathfrak{g}$ we thus get smooth functions $c^1, \dots, c^k$, defined for sufficiently small t, such that

$$\exp tY = \exp(c^1(t) X_1) \dots \exp(c^k(t) X_k).$$

Similarly, as above, we next conclude that for each point $x \in M$ we find a neighborhood in M and a bound on $|t|$ up to which we have

$$r(y, \exp tY) = (\mathrm{Fl}^{X_1}_{c^1(t)} \circ \dots \circ \mathrm{Fl}^{X_k}_{c^k(t)})(y)$$

for all y in the neighborhood. Again, the left–hand side coincides with $Fl^Y_t(y)$, while the right–hand side is a one–parameter subgroup defined for all t for which the c^i are defined. Hence, we conclude that the formula for Fl^Y_t is valid for all such t, so for those t the flow Fl^Y_t is defined globally on M. This implies that Fl^Y_t is defined for all $t \in \mathbb{R}$, and hence $Y \in S$.

Now, we know $S = \mathfrak{g}$ so, in particular, S is a Lie algebra and thus $G_0 := \{\mathrm{Fl}^X_t : X \in S, t \in \mathbb{R}\}$ is a subgroup of G. Moreover, $X \mapsto \mathrm{Fl}^X_1$ can be used to define a local chart from an open neighborhood of zero in $\mathfrak{g}$ onto a neighborhood of id in G_0. Transporting this chart around using left multiplications, we obtain an atlas making G_0 into a Lie group. By construction G_0 is connected and since conjugating a one–parameter group of diffeomorphisms by a fixed diffeomorphism gives rise to a one–parameter group, we conclude that G_0 is a normal subgroup of G. Further, it is easy to see that for any $\phi \in G$, conjugation by ϕ defines a continuous homomorphism from G_0 to itself. Thus, we may transport the topology from G_0 to G using either left or right multiplications. This makes G into a topological group which contains G_0 as an open normal subgroup, so G_0 must also be closed and thus the connected component of the identity. Now we can carry over the smooth structure from G_0 to the other connected components, thus making G into a Lie group acting smoothly on M. □

Let us remark, that the topology of the group G from Proposition 1.5.11 is not necessarily second countable since G may have uncountably many connected components. We do not know examples in which this actually occurs.

THEOREM 1.5.11. *Let $(\mathcal{P} \to M, \omega)$ be a Cartan geometry of type (G, H) over a connected manifold M. Then the group* $\operatorname{Aut}(\mathcal{P}, \omega)$ *of all automorphisms of $(\mathcal{P} \to M, \omega)$ is a Lie group of dimension at most* $\dim(G)$.

PROOF. By definition, an automorphism Φ of $(\mathcal{P} \to M, \omega)$ is a principal bundle automorphism of $\mathcal{P}$ such that $\Phi^*\omega = \omega$. Equivalently, Φ is a diffeomorphisms of $\mathcal{P}$ such that $\Phi^*\omega = \omega$ and $\Phi \circ r^h = r^h \circ \Phi$ for all $h \in H$. For a vector field $\xi \in \mathfrak{X}(\mathcal{P})$ such that Fl^ξ_t is defined for all t, the condition that each Fl^ξ_t is an automorphism is equivalent to $\mathcal{L}_\xi \omega = 0$ and $(r^h)^*\xi = \xi$. Let us denote by $\mathfrak{a}$ the space of all *infinitesimal automorphisms*, i.e. the space of all vector fields ξ satisfying these two conditions (without assuming existence of the flow for all times). From the conditions it is obvious that $\mathfrak{a}$ is a Lie subalgebra of $\mathfrak{X}(P)$. For $\xi \in \mathfrak{X}(\mathcal{P})$ and $X \in \mathfrak{g}$ we get $(\mathcal{L}_\xi\omega)(\omega^{-1}(X)) = 0 - \omega([\xi, \omega^{-1}(X)])$. Thus, $\xi \in \mathfrak{a}$ if and only if ξ commutes with each of the fields $\omega^{-1}(X)$ for $X \in \mathfrak{g}$.

For $\xi \in \mathfrak{a}$, the condition that $[\xi, \omega^{-1}(X)] = 0$ implies that the flows of the two fields commute, and thus, in particular, $\xi(\mathrm{Fl}_t^{\omega^{-1}(X)}(u)) = T\,\mathrm{Fl}_t^{\omega^{-1}(X)} \cdot \xi(u)$ whenever the flow is defined. But for $u \in \mathcal{P}$ the map $X \mapsto \mathrm{Fl}_1^{\omega^{-1}(X)}(u)$ defines a diffeomorphism from an open neighborhood of zero in $\mathfrak{g}$ onto an open neighborhood of u in $\mathcal{P}$. This shows that the value of ξ in u determines ξ locally around u, which together with H–invariance of ξ and connectedness of M implies that ξ is uniquely determined by $\xi(u)$ globally. In particular, we have proved that the evaluation mapping $\mathfrak{a} \to T_u\mathcal{P}$ is injective and hence $\dim(\mathfrak{a}) \leq \dim(\mathfrak{g})$.

Next, let $S \subset \mathfrak{a}$ be the subset of those infinitesimal automorphisms whose flow is defined for all $t \in \mathbb{R}$. By construction, the set S and the group $\mathrm{Aut}(\mathcal{P}, \omega)$ of diffeomorphisms of $\mathcal{P}$ satisfy the assumptions of the proposition above and so $\mathrm{Aut}(\mathcal{P}, \omega)$ is a Lie group with Lie algebra S. Of course, we have $\dim(S) \leq \dim(\mathfrak{a}) \leq \dim(\mathfrak{g})$. □

1.5.12. Infinitesimal automorphisms. In the last subsection, we have introduced the Lie algebra $\mathfrak{a}$ of infinitesimal automorphisms of a Cartan geometry of type (G, H) as the Lie algebra of all $\xi \in \mathfrak{X}(\mathcal{P})$ such that $(r^h)^*\xi = \xi$ and $\mathcal{L}_\xi\omega = 0$. Notice, in particular, that $\mathfrak{a} \subset \mathfrak{X}(\mathcal{P})^H \cong \Gamma(\mathcal{A}M)$, so infinitesimal automorphisms can be naturally viewed as sections of the adjoint tractor bundle. In the proof of Theorem 1.5.11, we have seen that the Lie algebra of the automorphism group $\mathrm{Aut}(\mathcal{P}, \omega)$ consists exactly of those $\xi \in \mathfrak{a}$ whose flow is defined for all $t \in \mathbb{R}$. Now we can easily characterize $\mathfrak{a} \subset \Gamma(\mathcal{A}M)$ and at the same time determine the Lie algebra structure on $\mathfrak{a}$ in this picture. By definition, $\mathcal{L}_\xi\omega = 0$ is equivalent to $0 = \xi \cdot (\omega(\eta)) - \omega([\xi, \eta])$ for all $\eta \in \mathfrak{X}(\mathcal{P})$. Of course, it suffices to have this property for $\eta \in \mathfrak{X}(\mathcal{P})^H \cong \Gamma(\mathcal{A}M)$. Hence, we see that $s \in \Gamma(\mathcal{A}M)$ corresponds to an infinitesimal automorphism if and only if $0 = D_s t - [s, t]$ for all $t \in \Gamma(\mathcal{A}M)$. Using the formulae derived in 1.5.8 this equation can be equivalently rewritten as

$$\begin{aligned} 0 &= D_s t - (D_s t - D_t s + \{s, t\} - \kappa(s, t)) \\ &= D_t s + \{t, s\} + \kappa(s, t) \\ &= \nabla^{\mathcal{A}}_{\Pi(t)} s + \kappa(s, t), \end{aligned}$$

where $\nabla^{\mathcal{A}}$ is the adjoint tractor connection. Thus, we have proved

LEMMA 1.5.12. *Let $(\mathcal{P} \to M, \omega)$ be a Cartan geometry of type (G, H). Then for an adjoint tractor field $s \in \Gamma(\mathcal{A}M)$ the following four conditions are equivalent*

1. *The vector field $\xi \in \mathfrak{X}(\mathcal{P})^H$ corresponding to s is an infinitesimal automorphism.*
2. *$D_s t = [s, t]$ for all $t \in \Gamma(\mathcal{A}M)$.*
3. *$D_t s = -\{t, s\} + \kappa(t, s)$ for all $t \in \Gamma(\mathcal{A}M)$.*
4. *$\nabla^{\mathcal{A}} s = -i_{\Pi(s)}\kappa$.*

Viewing formula (4) as a differential equation defining infinitesimal automorphisms, one observes that this actually means that the infinitesimal automorphisms are exactly the parallel sections for the connection $\hat{\nabla}$ on the vector bundle $\mathcal{A}M$ which is defined as $\hat{\nabla}_\xi s = \nabla^{\mathcal{A}}_\xi s - \kappa(\xi, \Pi(s))$.

COROLLARY 1.5.12. *The space of infinitesimal automorphisms of a Cartan geometry $(\mathcal{P} \to M, \omega)$ of type (G, H) is isomorphic to the space of smooth sections of the adjoint tractor bundle $\mathcal{A}M$, which are parallel for the linear connection $\hat{\nabla}$.*

This provides an alternative proof for the fact that any infinitesimal automorphism is determined by its value in a single point, which implies the bound on the dimension of the automorphism group.

The lemma also describes the Lie algebra structure on $\mathfrak{a}$, which is by definition induced by the Lie bracket of vector fields on $\mathcal{P}$ and thus by the Lie bracket of adjoint tractors. Indeed, using (2) and (3) we see that for $s, t \in \mathfrak{a} \subset \Gamma(\mathcal{A}M)$ we have $[s,t] = \kappa(s,t) - \{s,t\}$. Notice that $\Pi(s)(u) = 0$ exactly means that the corresponding vector field ξ is vertical in u, so the underlying point $x \in M$ is a fixed point of the base map of the infinitesimal automorphism. If one of the two infinitesimal automorphisms involved has this property, then the bracket is simply the negative of the algebraic bracket on $\mathcal{A}M$, while in general one gets a curvature correction.

1.5.13. Correspondence spaces. Let G be a Lie group and let $K \subset H \subset G$ be closed subgroups. Then there is an obvious G–equivariant projection $G/K \to G/H$. The fiber of this projection over $o = eH \in G/H$ is simply H/K. Left multiplication by elements of G evidently makes $G/K \to G/H$ into a homogeneous bundle, and from 1.4.3 we conclude that $G/K \cong G \times_H (H/K)$. Let us rephrase this in terms of Klein geometries as introduced in 1.4.1. Starting from the Klein geometry (G, H) and a closed subgroup $K \subset H$, we conclude that the total space of the homogeneous fiber bundle $G \times_H (H/K)$ carries the Klein geometry (G, K).

This simple observation carries over to Cartan geometries, leading to a general construction for natural geometries on the total spaces of certain natural bundles. These are then referred to as *correspondence spaces*, since the concept was first formalized in the context of twistor correspondences. We will take up this concept in the realm of parabolic geometries in Section 4.4, but the basic constructions make sense for arbitrary Cartan geometries.

DEFINITION 1.5.13. Let G be a Lie group and $K \subset H \subset G$ be closed subgroups. Let $(p : \mathcal{P} \to N, \omega)$ be a Cartan geometry of type (G, H). Then we define the *correspondence space* $\mathcal{C}N$ of N for $K \subset H$ to be the quotient space $\mathcal{P}/K$.

On the level of Lie algebras, we of course have $\mathfrak{k} \subset \mathfrak{h} \subset \mathfrak{g}$. In particular, there is an obvious projection $\mathfrak{g}/\mathfrak{k} \to \mathfrak{g}/\mathfrak{h}$.

PROPOSITION 1.5.13. *For a closed subgroup $K \subset H$, let $\mathcal{C}N$ be the correspondence space of a Cartan geometry $(p : \mathcal{P} \to N, \omega)$ of type (G, H). Then we have:*

(1) $\mathcal{C}N$ is the total space of a natural fiber bundle over N with fiber the homogeneous space H/K, and it carries a canonical Cartan geometry $(\pi : \mathcal{P} \to \mathcal{C}N, \omega)$ of type (G, K).

(2) The curvature functions κ^N of $(p : \mathcal{P} \to N, \omega)$ and $\kappa^{\mathcal{C}N}$ of $(\pi : \mathcal{P} \to \mathcal{C}N, \omega)$ are related as follows. For $u \in \mathcal{P}$ and $X, Y \in \mathfrak{g}$ we have

$$\kappa^{\mathcal{C}N}(u)(X + \mathfrak{k}, Y + \mathfrak{k}) = \kappa^N(u)(X + \mathfrak{h}, Y + \mathfrak{h}),$$

so κ^N *and* $\kappa^{\mathcal{C}N}$ *are induced by the same function* $\mathcal{G} \to L(\Lambda^2\mathfrak{g}, \mathfrak{g})$. *In particular,* $(\pi : \mathcal{P} \to \mathcal{C}N, \omega)$ *is locally flat if and only if* $(p : \mathcal{P} \to N, \omega)$ *is locally flat.*

(3) The subspace $\mathfrak{h}/\mathfrak{k} \subset \mathfrak{g}/\mathfrak{k}$ *is a* K*–submodule, which gives rise to a distribution* $V\mathcal{C}N \subset T\mathcal{C}N$. *This is exactly the vertical subbundle of the projection* $\mathcal{C}N \to N$. *The Cartan curvature* $\kappa^{\mathcal{C}N}$ *of* $\mathcal{C}N$ *has the property that* $i_\xi\kappa^{\mathcal{C}N} = 0$ *for any* $\xi \in \Gamma(V\mathcal{C}N) \subset \mathfrak{X}(\mathcal{C}N)$.

(4) The construction of correspondence spaces defines a functor from the category of Cartan geometries of type (G, H) *to the category of Cartan geometries of type* (G, K). *If the homogeneous space* H/K *is connected, then this is an equivalence onto a subcategory, i.e. any morphism between two correspondence spaces comes from a morphism of the original geometries.*

PROOF. (1) Evidently, $\mathcal{C}N = \mathcal{P}/K \cong \mathcal{P} \times_H (H/K)$, so the first claim follows. Since H acts freely on $\mathcal{P}$, the same is true for the subgroup K. Thus, the natural projection $\pi : \mathcal{P} \to \mathcal{C}N$ is a principal fiber bundle with structure group K. By definition the Cartan connection $\omega \in \Omega^1(\mathcal{P}, \mathfrak{g})$ is a trivialization of the tangent bundle $T\mathcal{P}$, it is H–equivariant, and $\omega(\zeta_A) = A$ for $A \in \mathfrak{h}$. But then, of course, ω is K–equivariant and $\omega(\zeta_A) = A$ for $A \in \mathfrak{k}$, and hence defines a Cartan connection on $\pi : \mathcal{P} \to \mathcal{C}N$.

(2) The curvature form $K \in \Omega^2(\mathcal{P}, \mathfrak{g})$ by construction is the same for both geometries. Then the claim about the curvature functions follows from the definition. Since local flatness is equivalent to vanishing of the curvature function, the last statement is evident, too.

(3) Since K is a subgroup of H, we get $\mathrm{Ad}(k)(\mathfrak{h}) \subset \mathfrak{h}$ for all $k \in K$. Hence, $\mathfrak{h}/\mathfrak{k} \subset \mathfrak{g}/\mathfrak{k}$ is K–invariant and $\mathcal{P} \times_K (\mathfrak{h}/\mathfrak{k})$ is a smooth subbundle of $\mathcal{P} \times_K (\mathfrak{g}/\mathfrak{k})$. Since ω defines a Cartan connection on $\pi : \mathcal{P} \to \mathcal{C}N$, we know from 1.5.5 that the latter bundle is isomorphic to $T\mathcal{C}N$. Explicitly, the isomorphism is induced from the map $\mathcal{P} \times \mathfrak{g}/\mathfrak{k} \to T\mathcal{C}N$ defined by $(u, X + \mathfrak{k}) \mapsto T_u\pi \cdot \omega_u^{-1}(X)$. Since the identification of TN with $\mathcal{G} \times_H (\mathfrak{g}/\mathfrak{h})$ is also obtained using ω, we see that the tangent map of the projection $\mathcal{C}N \to N$ corresponds to the canonical projection $\mathfrak{g}/\mathfrak{k} \to \mathfrak{g}/\mathfrak{h}$. Hence, the vertical subbundle of $\mathcal{C}N \to N$ corresponds to the kernel $\mathfrak{h}/\mathfrak{k}$ of $\mathfrak{g}/\mathfrak{k} \to \mathfrak{g}/\mathfrak{h}$. The statement on the curvatures follows immediately from the description of the curvature functions in part (2).

(4) A morphism $(\mathcal{P} \to N, \omega) \to (\tilde{\mathcal{P}} \to \tilde{N}, \tilde{\omega})$ is by definition a principal bundle map $\Phi : \mathcal{P} \to \tilde{\mathcal{P}}$ such that $\Phi^*\tilde{\omega} = \omega$. Since Φ is H–equivariant, it is also K–equivariant. Hence, it induces a smooth map $\mathcal{C}N = \mathcal{P}/K \to \tilde{\mathcal{P}}/K = \mathcal{C}\tilde{N}$ and defines a principal bundle map from $\mathcal{P} \to \mathcal{C}N$ to $\tilde{\mathcal{P}} \to \mathcal{C}\tilde{N}$. Since $\Phi^*\tilde{\omega} = \omega$ by assumption, it is a morphism of Cartan geometries of type (G, K).

Conversely, a morphism $\Phi : (\mathcal{P} \to \mathcal{C}N, \omega) \to (\tilde{\mathcal{P}} \to \mathcal{C}\tilde{N}, \tilde{\omega})$ is a K–equivariant map $\mathcal{P} \to \tilde{\mathcal{P}}$ which is compatible with the Cartan connections. This also defines a morphism $(\mathcal{P} \to N, \omega) \to (\tilde{\mathcal{P}} \to \tilde{N}, \omega)$ if and only if Φ is even H–equivariant. Now observe that since $\tilde{\omega}$ is a Cartan connection on $\tilde{\mathcal{P}} \to \tilde{N}$, the fundamental vector field generated by $A \in \mathfrak{h}$ equals $\tilde{\omega}^{-1}(A)$. Compatibility of Φ with the Cartan connections thus implies that Φ pulls back fundamental vector fields on $\tilde{\mathcal{P}} \to \tilde{N}$ to fundamental vector fields with the same generator on $\mathcal{P} \to N$.

Therefore, Φ commutes with the flows of fundamental vector fields. But the flow of ζ_A is by the principal right action by $\exp(tA)$, so Φ commutes with the principal right actions of elements of the form $\exp(A)$ for $A \in \mathfrak{h}$. Connectedness of H/K implies that K meets each connected component of H, so elements of K together with elements of the form $\exp(A)$ generate the group H. Under this assumption Φ is therefore automatically H–equivariant, which completes the proof. □

EXAMPLE 1.5.13. Consider a Riemannian manifold N of dimension n. By example (iii) of 1.5.1, N carries a canonical Cartan geometry of type $(\mathrm{Euc}(n), O(n))$. Now take the subgroup $O(n-1) \subset O(n)$ and consider the associated correspondence space $\mathcal{C}N$. We can realize $O(n-1)$ as the stabilizer of a unit vector in $\mathbb{R}^n$ and, as in 1.1.1, identify the homogeneous space $O(n)/O(n-1)$ with the unit sphere $S^{n-1} \subset \mathbb{R}^n$. Now the associated bundle $\mathcal{P} \times_{O(n)} \mathbb{R}^n$ is the tangent bundle TN. Hence, $\mathcal{C}N = \mathcal{P} \times_{O(n)} S^{n-1}$ can be identified with the *unit sphere bundle* $SN \subset TN$ of all tangent vectors of length one.

By the proposition, we obtain a natural Cartan geometry $\mathcal{P} \to SN$ of type $(\mathrm{Euc}(n), O(n-1))$ on the unit sphere bundle. Realizing $\mathrm{Euc}(n)$ as a matrix group as in 1.1.2 and $O(n-1) \subset O(n)$ as the stabilizer of the first standard basis vector in $\mathbb{R}^n$, we get

$$\mathrm{Euc}(n) = \left\{ \begin{pmatrix} 1 & 0 \\ v & A \end{pmatrix} : v \in \mathbb{R}^n, A \in O(n) \right\},$$

and the subgroup $O(n-1)$ corresponds to the matrices in which $v = 0$ and A is of the form $\left(\begin{smallmatrix} 1 & 0 \\ 0 & B \end{smallmatrix}\right)$. Looking at the Lie algebras, we see that, as an $O(n-1)$–module, we have $\mathfrak{euc}(n) = \mathbb{R} \oplus \mathbb{R}^{n-1} \oplus \mathbb{R}^{n-1} \oplus \mathfrak{o}(n-1)$, with $\mathfrak{o}(n)$ corresponding to the last two summands. The first three summands provide us with an $O(n-1)$–invariant complement $\mathfrak{n}$ to $\mathfrak{o}(n-1) \subset \mathfrak{euc}(n)$.

Using the standard inner products on these three summands, we obtain an $O(n-1)$–invariant inner product on $\mathfrak{n}$, which induces a canonical Riemannian metric on SN. By construction, TSN decomposes into the orthogonal direct sum of three subbundles. By part (3) of the proposition, the last summand is the vertical subbundle of $SN \to N$. The other two summands constitute the horizontal subbundle for the Levi–Civita connection (with the lifted metric). This decomposes further into the line subbundle formed by multiples of the footpoint and its orthogonal complement.

1.5.14. Characterization of correspondence spaces. We continue working in the setting of a Lie group G with closed subgroups $K \subset H \subset G$. In 1.5.13 we have shown how to associate to a Cartan geometry of type (G, H) a Cartan geometry of type (G, K) on the correspondence space. Now we want to characterize Cartan geometries of type (G, K) which are locally isomorphic to correspondence spaces.

Let $(p : \mathcal{P} \to M, \omega)$ be a Cartan geometry of type (G, K). As in part (3) of Proposition 1.5.13, the K–invariant subspace $\mathfrak{h}/\mathfrak{k} \subset \mathfrak{g}/\mathfrak{k}$ determines a smooth subbundle $VM \subset TM$. In the case of a correspondence space $\mathcal{C}N$, this bundle becomes the vertical subbundle of the projection $\mathcal{C}N \to N$, so, in particular, it must be involutive. If M is locally isomorphic to a correspondence space, then this isomorphism is compatible with the subbundles, so VM must be involutive, too. We can characterize involutivity in terms of the torsion of the Cartan connection ω. Recall from 1.5.7 that the torsion $T \in \Omega^2(M, TM)$ of ω is obtained by applying the projection $\Pi : \mathcal{A}M \to TM$ to the values of the Cartan curvature $\kappa \in \Omega^2(M, \mathcal{A}M)$.

LEMMA 1.5.14. *Let $(p : \mathcal{P} \to M, \omega)$ be a Cartan geometry of type (G, K) with torsion $T \in \Omega^2(M, TM)$, and let $VM \subset TM$ be the subbundle corresponding to $\mathfrak{h}/\mathfrak{k} \subset \mathfrak{g}/\mathfrak{k}$. The VM is integrable if an only if $T(VM, VM) \subset VM$.*

PROOF. Let ξ and η be local sections of $VM \subset TM$, and choose local lifts $\tilde{\xi}, \tilde{\eta} \in \mathfrak{X}(\mathcal{P})$. Then $[\tilde{\xi}, \tilde{\eta}]$ is a local lift of the Lie bracket $[\xi, \eta]$. Thus, we have to check whether $Tp \cdot [\tilde{\xi}, \tilde{\eta}]$ lies in $VM \subset TM$. Since the identification of TM with $\mathcal{P} \times_K (\mathfrak{g}/\mathfrak{k})$ is obtained from $(u, X + \mathfrak{k}) \mapsto T_u p \cdot \omega(u)^{-1}(X)$, this is the case if and only if $\omega([\tilde{\xi}, \tilde{\eta}])$ has values in $\mathfrak{h} \subset \mathfrak{g}$.

The assumptions that ξ and η are sections of VM likewise is equivalent to the fact that $\omega(\tilde{\xi})$ and $\omega(\tilde{\eta})$ have values in $\mathfrak{h} \subset \mathfrak{g}$. In this case, also $\tilde{\xi} \cdot \omega(\tilde{\eta})$ and $\tilde{\eta} \cdot \omega(\tilde{\xi})$ have values in $\mathfrak{h}$. Hence, we see that $[\xi, \eta] \in \Gamma(VM)$ is equivalent to $d\omega(\tilde{\xi}, \tilde{\eta})$ having values in $\mathfrak{h}$. Since $\mathfrak{h}$ is a Lie subalgebra, also $[\omega(\tilde{\xi}), \omega(\tilde{\eta})]$ automatically has values in $\mathfrak{h}$, so we can equivalently replace $d\omega(\tilde{\xi}, \tilde{\eta})$ by $K(\tilde{\xi}, \tilde{\eta})$. But this having values in $\mathfrak{h}$ is equivalent to $T(\xi, \eta)$ having values in the subbundle of TM corresponding to $\mathfrak{h}/\mathfrak{k}$. □

Suppose that the geometry $(p : \mathcal{P} \to M, \omega)$ of type (G, K) satisfies this necessary condition for being locally isomorphic to a correspondence space. Then we can actually construct a candidate for a space N such that M may be locally isomorphic to $\mathcal{C}N$. Namely, in the case of a correspondence space, N is simply the (global) space of leaves of the foliation corresponding to the subbundle $V\mathcal{C}N$. Returning to M, we have to consider spaces which locally parametrize the leaves of the foliation defined by $VM \subset TM$. Due to the origins of this whole circle of ideas in twistor theory, such spaces are called local twistor spaces for M.

DEFINITION 1.5.14. Let $(p : \mathcal{P} \to M, \omega)$ be a Cartan geometry of type (G, K) such that the subbundle $VM \subset TM$ is integrable. Then a (local) *twistor space* for M is a local leaf space for the foliation defined by VM, i.e. a smooth manifold N together with an open subset $U \subset M$ and a surjective submersion $\psi : U \to N$ such that $\ker(T_x\psi) = V_xM$ for all $x \in U$.

Existence of local twistor spaces follows immediately from the local version of the Frobenius theorem (see [**KMS**, Theorem 3.22]) by projecting onto one factor of an adapted chart. Note that for two local twistor spaces $\psi_i : U_i \to N_i$ there is a unique diffeomorphism $\phi : \psi_1(U_1 \cap U_2) \to \psi_2(U_1 \cap U_2)$ such that $\phi \circ \psi_1 = \psi_2$.

We know already that correspondence spaces satisfy a much stronger curvature condition than the one from the lemma, since by part (3) of Proposition 1.5.13 we must have $i_\xi \kappa = 0$ for any section ξ of $VM \subset TM$. Surprisingly, this curvature condition is actually equivalent to local isomorphism to a correspondence space:

THEOREM 1.5.14. *Let $(p : \mathcal{P} \to M, \omega)$ be a Cartan geometry of type (G, K) with curvature κ. Suppose that $i_\xi \kappa = 0$ for all $\xi \in \Gamma(VM)$.*

Then for any sufficiently small local twistor space $\psi : U \to N$ of M, one obtains a Cartan geometry of type (G, H) on N such that $(p^{-1}(U), \omega|_{p^{-1}(U)})$ is isomorphic to an open subspace in the correspondence space $\mathcal{C}N$. If H/K is connected, then this Cartan geometry is uniquely determined.

PROOF. The composition $\psi \circ p : p^{-1}(U) \to N$ is a surjective submersion, so it admits local smooth sections. Choosing U sufficiently small, we may therefore assume that there is a global smooth section $\sigma : N \to p^{-1}(U)$ of $\psi \circ p$. In terms of the curvature form $K \in \Omega^2(\mathcal{P}, \mathfrak{g})$, the condition on κ implies $0 = K(\omega^{-1}(A), \omega^{-1}(B))$ for all $A, B \in \mathfrak{h} \subset \mathfrak{g}$. This can be written as $0 = -\omega([\omega^{-1}(A), \omega^{-1}(B)]) + [A, B]$, which means that $A \mapsto \omega^{-1}(A)$ defines a Lie algebra homomorphism $\mathfrak{h} \to \mathfrak{X}(\mathcal{P})$, i.e. an action of $\mathfrak{h}$ on $\mathcal{P}$. By Lie's second fundamental theorem (Lemma 1.5.11), this

Lie algebra action integrates to a local group action. There is an open neighborhood W of $\mathcal{P} \times \{e\}$ in $\mathcal{P} \times H$ and a smooth map $F: W \to \mathcal{P}$ such that
- $F(u, e) = u$ and $\frac{d}{dt}|_{t=0} F(u, \exp(tA)) = \omega^{-1}(A)(u)$ for all $u \in \mathcal{P}$ and all $A \in \mathfrak{h}$.
- $F(F(u, g), h) = F(u, gh)$ provided that (u, g), (u, gh) and $(F(u, g), h)$ all lie in W.

Possibly shrinking the leaf space further, we find an open neighborhood $\tilde{V}$ of e in H such that $(\sigma(x), g) \in W$ and $(F(\sigma(x), g), e) \in W$ for all $x \in N$ and all $g \in \tilde{V}$. Then we define $\Phi: N \times \tilde{V} \to \mathcal{P}$ by $\Phi(x, g) := F(\sigma(x), g)$. For $x \in N$ the tangent map $T_{(x,e)}\Phi : T_x N \times \mathfrak{h} \to T_{\sigma(x)}\mathcal{G}$ is evidently given by $(\xi, A) \mapsto T_x\sigma \cdot \xi + \omega^{-1}(A)(\sigma(x))$, so it is a linear isomorphism. Possibly shrinking U and $\tilde{V}$, we may assume that Φ is a diffeomorphism onto an open subset $\tilde{U} \subset \mathcal{P}$, and we arrive at the following picture:

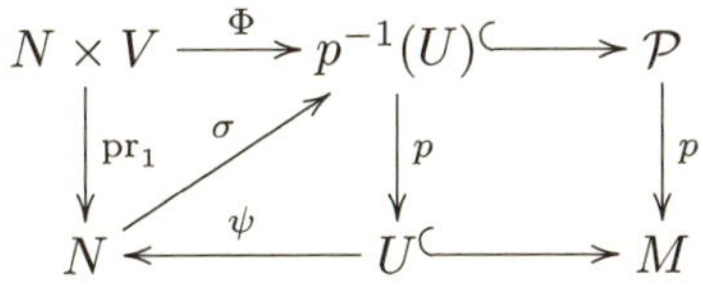

$$\begin{array}{ccccc} N \times V & \xrightarrow{\Phi} & p^{-1}(U) & \hookrightarrow & \mathcal{P} \\ \downarrow{\mathrm{pr}_1} & \sigma \nearrow & \downarrow{p} & & \downarrow{p} \\ N & \xleftarrow{\psi} & \tilde{U} & \hookrightarrow & M \end{array}$$

Possibly shrinking $\tilde{V}$ further, we may assume that it is of special form: Choose a linear subspace $\mathfrak{n} \subset \mathfrak{h}$ which is complementary to $\mathfrak{k}$. Choose an open ball V_1 around 0 in $\mathfrak{n}$ which is so small that $(X, k) \mapsto \exp(X)k$ is a diffeomorphism from $V_1 \times K$ onto an open neighborhood V of K in H. Then assume that there is an open ball V_2 around 0 in $\mathfrak{k}$ such that $(X, B) \mapsto \exp(X)\exp(B)$ is a diffeomorphism $V_1 \times V_2 \to \tilde{V}$.

For a fixed point $x \in N$, any vector tangent to $\{x\} \times \tilde{V}$ can be written as $\zeta_A(x, g) = \frac{d}{dt}|_{t=0}(x, g\exp(tA))$ for some $g \in \tilde{V}$ and $A \in \mathfrak{h}$. For sufficiently small t, by construction we have $\Phi(x, g\exp(tA)) = F(\Phi(x, g), \exp(tA))$, and thus

$$T_{(x,g)}\Phi \cdot \zeta_A(x, g) = \omega^{-1}(A)(\Phi(x, g)).$$

Thus, we see that $T\Phi \circ \zeta_A = \omega^{-1}(A) \circ \Phi$ for all $A \in \mathfrak{h}$. Moreover, $T\Phi \circ \zeta_A$ always lies in $\omega^{-1}(\mathfrak{h}) \subset T\mathcal{P}$, which implies that $\Phi(\{x\} \times \tilde{V})$ is contained in one leaf of the foliation corresponding to the integrable distribution $\omega^{-1}(\mathfrak{h}) \subset T\mathcal{P}$. Hence, the map $\psi \circ \pi \circ \Phi$ is constant on $\{x\} \times \tilde{V}$, and since $\Phi(x, e) = \sigma(x)$, we conclude that $\psi \circ \pi \circ \Phi = \mathrm{pr}_1 : N \times \tilde{V} \to N$.

For $X \in V_1$ and $B \in V_2$ we have $\exp(X)\exp(tB) \in \tilde{V}$ for all $t \in [0, 1]$. Since $\omega^{-1}(B) \in \mathfrak{X}(\mathcal{P})$ is the fundamental vector field generated by $B \in \mathfrak{k}$, the infinitesimal condition $T\Phi \circ \zeta_B = \omega^{-1}(B) \circ \Phi$ implies

$$\Phi(x, \exp(X)\exp(B)) = \Phi(x, \exp(X)) \cdot \exp(B),$$

where in the right-hand side we use the principal right action on $\mathcal{P}$. Since K acts freely both on $N \times V$ and on $\mathcal{P}$ we can uniquely extend Φ to a K-equivariant diffeomorphism from $N \times V$ to the K-invariant open subset

$$\{u \cdot g : u \in \tilde{U}, g \in K\} = p^{-1}(p(\tilde{U})) \subset \mathcal{P}.$$

Since the family of fundamental vector fields on $N \times P$ and the family of the vector fields $\omega^{-1}(A)$ on $\mathcal{P}$ have the same equivariancy property, this extension still satisfies $T\Phi \circ \zeta_A = \omega^{-1}(A) \circ \Phi$ for all $A \in \mathfrak{h}$.

Next, $\Phi^*\omega \in \Omega^1(N \times V, \mathfrak{g})$ restricts to a linear isomorphism on each tangent space. We can consider the restriction of this form to $N \times \{e\}$ and extend it equivariantly to $\tilde{\omega} \in \Omega^1(N \times H, \mathfrak{g})$ by defining

$$\tilde{\omega}(x,h) := \operatorname{Ad}(h^{-1}) \circ (\Phi^*\omega)(x,e) \circ Tr^{h^{-1}},$$

where r denotes the principal right action of H on $N \times H$. By construction, $\tilde{\omega}$ is smooth, it restricts to a linear isomorphism on each tangent space and satisfies $(r^h)^*\tilde{\omega} = \operatorname{Ad}(h^{-1}) \circ \tilde{\omega}$ for all $h \in H$. In points of the form (x,e) we have $\tilde{\omega} = \Phi^*\omega$, which together with $T\Phi \circ \zeta_A = \omega^{-1}(A) \circ \Phi$ implies that $\tilde{\omega}$ reproduces the generators of all fundamental vector fields in such points. By equivariancy, this holds globally, and thus $\tilde{\omega}$ is a Cartan connection on the principal H–bundle $N \times H \to N$. Hence, we have obtained a Cartan geometry of type (G,H) on N.

Since the vector fields ζ_A and $\omega^{-1}(A)$ are Φ–related, their flows are Φ–related, so $\Phi \circ r^{\exp(tA)} = \operatorname{Fl}_t^{\omega^{-1}(A)} \circ \Phi$ whenever defined. In terms of the curvature form K, the condition $i_\xi \kappa = 0$ for all $\xi \in \Gamma(VM)$ reads as $0 = i_{\omega^{-1}(A)}K$ for all $A \in \mathfrak{h}$. Since $i_{\omega^{-1}(A)}\omega$ is constant, this implies $\mathcal{L}_{\omega^{-1}(A)}\omega = -\operatorname{ad}(A) \circ \omega$, and in turn

$$\tfrac{d}{dt}\left(\operatorname{Fl}_t^{\omega^{-1}(A)}\right)^*\omega = -\operatorname{ad}(A) \circ \left(\operatorname{Fl}_t^{\omega^{-1}(A)}\right)^*\omega.$$

Solving this differential equation, we obtain $\left(\operatorname{Fl}_t^{\omega^{-1}(A)}\right)^*\omega = \operatorname{Ad}(\exp(-tA)) \circ \omega$ whenever the flow is defined. Now we compute

$$\begin{aligned}(r^{\exp(tA)})^*\Phi^*\omega =& (\Phi \circ r^{\exp(tA)})^*\omega = (\operatorname{Fl}_t^{\omega^{-1}(A)} \circ \Phi)^*\omega \\ =& \Phi^*(\operatorname{Fl}_t^{\omega^{-1}(A)})^*\omega = \operatorname{Ad}(\exp(-tA)) \circ \Phi^*\omega.\end{aligned}$$

Using for A an element X in the open subset V_1, this makes sense for all $t \in [0,1]$, and we conclude that $\Phi^*\omega$ and $\tilde{\omega}$ coincide on all elements of the form $(x, \exp(X))$ for $x \in V_1$. But then K–equivariancy of Φ shows that $\Phi^*\omega = \tilde{\omega}$ on all of $N \times V$. But this exactly means that Φ defines an isomorphism of Cartan geometries from the open subset $N \times (V/H) \subset \mathcal{C}N = N \times (H/K)$ to the open subset $p(\tilde{U}) \subset M$. Now we can simply replace the leaf space $\psi : U \to N$ by $\psi : p(\tilde{U}) \to N$ to obtain the first part of the claim.

In part (4) of Proposition 1.5.13 we have verified that isomorphism of correspondence space implies isomorphism of the underlying geometries provided that H/K is connected, so the uniqueness statement follows. □

The results on correspondence spaces and their characterizations become particularly interesting if the Cartan geometries are determined by some underlying structures. We will take up these issues in the realm of parabolic geometries in Section 4.4.

1.5.15. Invariant Cartan connections and extension functors. Now we move to a second way to relate Cartan geometries of a different type. This nicely illustrates the interplay between the homogeneous model and the associated Cartan geometries. We consider Lie groups G and L and closed subgroups $H \subset G$ and $K \subset L$. We first describe G–invariant Cartan geometries of type (L,K) on G/H. For this to make sense, we must of course assume that $\dim(G/H) = \dim(L/K)$. Then we show that any such geometry gives rise to a functor mapping Cartan geometries of type (G,H) to Cartan geometries of type (L,K). An exposition of the theory of invariant Cartan geometries and substantial examples can be found in [**Ha06**]; see also [**Ha07**].

Recall first the description of G–homogeneous principal bundles over G/H with structure group K from 1.4.5. Any such bundle is of the form $G \times_i K$, where $i : H \to K$ is a smooth homomorphism, and one forms the associated bundle with respect to the left action $h \cdot k = i(h)k$. Two homomorphisms i and $\hat{i}$ give rise to isomorphic homogeneous principal bundles if an only if there is an element $k \in K$ such that $\hat{i}(h) = ki(h)k^{-1}$ for all $h \in H$.

Multiplication in G induces a left action of G on $G \times_i K$ by principal bundle automorphisms. We write ℓ_g for the action of $g \in G$. Then a Cartan connection ω on $G \times_i K$ is called *invariant* if $(\ell_g)^*\omega = \omega$ for all $g \in G$. This means that G acts by automorphisms on the Cartan geometry $(G \times_i K \to G/H, \omega)$.

PROPOSITION 1.5.15. *Let G and L be Lie groups with Lie algebras $\mathfrak{g}$ and $\mathfrak{l}$, $H \subset G$ and $K \subset L$ closed subgroups such that $\dim(G/H) = \dim(L/K)$, and let $i : H \to K$ be a homomorphism. Then the set of G–invariant Cartan connections on the principal K–bundle $G \times_i K$ is in bijective correspondence with the set of linear maps $\alpha : \mathfrak{g} \to \mathfrak{l}$ such that:*

(i) $\alpha \circ \mathrm{Ad}(h) = \mathrm{Ad}(i(h)) \circ \alpha$ *for all* $h \in H$.
(ii) $\alpha|_{\mathfrak{h}} = i' : \mathfrak{h} \to \mathfrak{k} \subset \mathfrak{l}$, *where i' denotes the derivative of i.*
(iii) *The map $\underline{\alpha} : \mathfrak{g}/\mathfrak{h} \to \mathfrak{l}/\mathfrak{k}$ induced by α is a linear isomorphism.*

For $k_0 \in K$ put $\hat{i}(h) = k_0 i(h) k_0^{-1}$. Then the pullback of the Cartan connection determined by $\alpha : \mathfrak{g} \to \mathfrak{l}$ under the isomorphism $G \times_{\hat{i}} K \to G \times_i K$ corresponds to the map $\hat{\alpha} = \mathrm{Ad}(k_0) \circ \alpha$.

PROOF. This is closely parallel to the description of homogeneous principal connections in Theorem 1.4.5. Let $q : G \times K \to G \times_i K$ be the canonical projection and put $u_0 = q(e,e)$. Given an invariant Cartan connection ω, define $\alpha : \mathfrak{g} \to \mathfrak{l}$ by $\alpha(X) := \omega(u_0)(T_{(e,e)}q \cdot (X,0))$. In the proof of Theorem 1.4.5, we saw that $T_{(e,e)}q \cdot (0,A) = \zeta_A(u_0)$. As in that proof, this implies that $\omega(u_0)$ is determined by α, and by K–equivariance and G–invariance, this in turn determines ω. Still as in the proof of Theorem 1.4.5 we obtain properties (i) and (ii) for α, and conversely having given α with (i) and (ii) we can construct $\omega \in \Omega^1(G \times_i K, \mathfrak{l})$ which is K–equivariant, G–invariant and reproduces the generators of fundamental vector fields.

Now by K–equivariance and G–invariance, the values of ω all are linear isomorphisms if and only if $\omega(u_0)$ is a linear isomorphism. By dimensional reasons, it suffices to show that $\omega(u_0)$ is surjective. Since $\omega(u_0)$ evidently maps onto $\mathfrak{k}$, this is equivalent to the composition with the projection to $\mathfrak{l}/\mathfrak{k}$ being surjective. This in turn is equivalent to surjectivity of $X \mapsto \alpha(X) + \mathfrak{k}$, and hence by condition (ii) and dimensional reasons to (iii). The behavior of α under the change from i to $\hat{i}$ is proved exactly as in Theorem 1.4.5. □

A homogeneous Cartan geometry of type (L,K) on G/H determines a homomorphism $i : H \to K$ and a linear map $\alpha : \mathfrak{g} \to \mathfrak{l}$ which satisfies the conditions (i)–(iii) from the proposition. Suppose now that $(p : \mathcal{P} \to M, \omega)$ is an arbitrary Cartan geometry of type (G,H). As for the homogeneous model, we can of course define a principal K–bundle as $\mathcal{P} \times_i K$. Mapping $u \in \mathcal{P}$ to the class of (u,e) induces a bundle map $j : \mathcal{P} \to \mathcal{P} \times_i K$ which is equivariant over $i : H \to K$.

LEMMA 1.5.15. *In this situation, there is a uniquely determined Cartan connection $\omega_\alpha \in \Omega^1(\mathcal{P} \times_i K, \mathfrak{l})$ such that $j^*\omega_\alpha = \alpha \circ \omega \in \Omega^1(\mathcal{P}, \mathfrak{l})$.*

PROOF. This is closely parallel to the proof of part (1) of Theorem 1.5.6. First for $u \in \mathcal{P}$, the tangent space $T_{j(u)}(\mathcal{P} \times_i K)$ is spanned by vertical vectors and elements of $T_u j(T_u \mathcal{P})$. Then one defines

$$\omega_\alpha(j(u))(T_u j \cdot \xi + \zeta_A(j(u))) := \alpha(\omega(u)(\xi)) + A$$

for $A \in \mathfrak{k}$, and verifies that it is well defined using property (i) of α. By definition, we see that $\omega_\alpha(j(u))$ reproduces the generators of fundamental vector fields. Suppose that $\omega_\alpha(j(u))(T_u j \cdot \xi + \zeta_A(j(u))) = 0$. Projecting to $\mathfrak{l}/\mathfrak{k}$, we see that

$$0 = \alpha(\omega(u)(\xi)) + \mathfrak{k} - \underline{\alpha}(\omega(u)(\xi) + \mathfrak{h}).$$

By condition (iii) on α, this implies that $\omega(u)(\xi) =: X \in \mathfrak{h}$, i.e. $\xi = \zeta_X(u)$. But this means that our original vector was vertical, and by construction $\omega_\alpha(j(u))$ is injective on vertical vectors. Thus, $\omega_\alpha(j(u))$ is a linear isomorphism.

As in the proof of Theorem 1.5.6 we then define

$$\tilde{\omega}(j(u) \cdot k)(\eta) = \mathrm{Ad}(k^{-1})(\tilde{\omega}(j(u))(Tr^{k^{-1}} \cdot \eta))$$

and verify that this is well defined using property (ii) of α. Having defined $\omega_\alpha \in \Omega^1(\mathcal{P} \times_i K, \mathfrak{l})$, we see that by construction the value in each point is a linear isomorphism, $j^* \omega_\alpha = \alpha \circ \omega$ and ω_α is uniquely determined by this property. Equivariancy of ω_α immediately follows from the construction. Finally, the fact that ω_α reproduces generators of fundamental vector fields follows by equivariancy from the fact that $\omega_\alpha(j(u))$ has this property. □

Fixing the data (i, α), which are equivalent to a G–invariant Cartan geometry of type (L, K) on G/H, we can now associate to each Cartan geometry of type (G, H) a Cartan geometry of type (L, K). In fact, this construction is functorial:

THEOREM 1.5.15. *Let G and L be Lie groups with Lie algebras $\mathfrak{g}$ and $\mathfrak{l}$, and let $H \subset G$ and $K \subset L$ be closed subgroups. Fix a homomorphism $i : H \to K$ and a linear map $\alpha : \mathfrak{g} \to \mathfrak{l}$ with properties (i)–(iii) from the proposition.*

Then mapping $(\mathcal{P} \to M, \omega)$ to $(\mathcal{P} \times_i K \to M, \omega_\alpha)$ defines a functor from Cartan geometries of type (G, H) to Cartan geometries of type (L, K). Passing from (i, α) to $(\hat{i}, \hat{\alpha})$ as described in the proposition, one obtains a naturally isomorphic functor.

PROOF. As in the proof of part (2) of Theorem 1.5.6, a morphism $\Phi : (\mathcal{P} \to M, \omega) \to (\tilde{\mathcal{P}} \to \tilde{M}, \tilde{\omega})$ induces a principal bundle map $F(\Phi) : \mathcal{P} \times_i K \to \tilde{\mathcal{P}} \times_i K$. Since Φ is a local diffeomorphism, the same is true for $F(\Phi)$ and hence $F(\Phi)^* \tilde{\omega}_\alpha$ is a Cartan connection on $\mathcal{G} \times_i K$. Then the fact that $F(\Phi)^* \tilde{\omega}_\alpha = \omega_\alpha$ follows as in Theorem 1.5.6, and functoriality is obvious.

Suppose that $\hat{i}(h) = k_0 i(h) k_0^{-1}$ and $\hat{\alpha} = \mathrm{Ad}(k_0) \circ \alpha$. For a fixed geometry $(\mathcal{P} \to M, \omega)$ of type (G, H), the map $\mathcal{P} \times K \to \mathcal{P} \times K$ defined by $(u, k) \mapsto (u, k_0 k)$ induces an isomorphism $\phi_{\mathcal{P}} : \mathcal{P} \times_i K \to \mathcal{P} \times_{\hat{i}} K$. Denoting by j and $\hat{j}$ the two inclusions, this satisfies $r^{k_0} \circ \hat{j} = \phi_{\mathcal{P}} \circ j$. Using this, we compute

$$j^*(\phi_{\mathcal{P}})^* \omega_{\hat{\alpha}} = \hat{j}^*(r^{k_0})^* \omega_{\hat{\alpha}} = \hat{j}^*(\mathrm{Ad}(k_0^{-1}) \omega_{\hat{\alpha}}) = \mathrm{Ad}(k_0^{-1}) \circ \hat{\alpha} \circ \omega = \alpha \circ \omega.$$

By uniqueness in the lemma, this implies that $\phi_{\mathcal{P}}$ defines an isomorphism between the Cartan geometries $(\mathcal{G} \times_i K, \omega_\alpha)$ and $(\mathcal{P} \times_{\hat{i}} K, \omega_{\hat{\alpha}})$. From the constructions it is evident that this defines a natural transformation between the two extension functors. □

Notice that by construction, applying the extension functor determined by (i, α) to the homogeneous model, we exactly recover the invariant Cartan connection on $G \times_i K$ constructed in the proposition.

1.5.16. Extension functors and curvature. Our next task is to compute the curvature of geometries obtained via the functors we have just constructed. Consider a linar map $\alpha : \mathfrak{g} \to \mathfrak{l}$ with the properties required in Proposition 1.5.15. Let us start from $(\mathcal{P} \to M, \omega)$ and consider the adjoint tractor bundles $\mathcal{A}M := \mathcal{P} \times_H \mathfrak{g}$ and $\tilde{\mathcal{A}}M := (\mathcal{G} \times_i K) \times_K \mathfrak{l} \cong \mathcal{P} \times_H \mathfrak{l}$. Here $h \in H$ acts on $\mathfrak{l}$ by $\mathrm{Ad}(i(h))$. Now by assumption α is H–equivariant, and hence induces a natural bundle map $\alpha : \mathcal{A}M \to \tilde{\mathcal{A}}M$. Moreover, from Proposition 1.4.6, we know that conditions (i) and (ii) on α imply that

$$(X + \mathfrak{h}, Y + \mathfrak{h}) \mapsto [\alpha(X), \alpha(Y)] - \alpha([X, Y])$$

defines an H–invariant element of $\Lambda^2(\mathfrak{g}/\mathfrak{h})^* \otimes \mathfrak{l}$. This gives rise to a natural section $\Phi_\alpha \in \Omega^2(M, \tilde{\mathcal{A}}M)$.

PROPOSITION 1.5.16. *In the setting of Proposition 1.5.15, the Cartan curvature $\kappa_\alpha \in \Omega^2(M, \tilde{\mathcal{A}}M)$ of the Cartan geometry $(\mathcal{P} \times_i K, \omega_\alpha)$ is given by*

$$\kappa_\alpha = \alpha \circ \kappa + \Phi_\alpha,$$

where $\kappa \in \Omega^2(M, \mathcal{A}M)$ is the Cartan curvature of $(\mathcal{P} \to M, \omega)$ and $\Phi_\alpha \in \Omega^2(M, \tilde{\mathcal{A}}M)$ is the natural section constructed above.

PROOF. By definition of the extension functor $j^*\omega_\alpha = \alpha \circ \omega$. From this, the result follows exactly as in the proof of Corollary 1.5.7. □

Applying this result to the homogeneous model, we see that Φ_α is the natural section corresponding to the Cartan curvature of the invariant Cartan geometry of type (L, K) on G/H obtained in Proposition 1.5.15. In particular, we see that this invariant Cartan geometry on G/H is locally flat if and only if $\alpha : \mathfrak{g} \to \mathfrak{l}$ is a Lie algebra homomorphism.

EXAMPLE 1.5.16. Here we will only look at a very simple class of examples, a much more substantial one will be discussed in 5.3.13 and 5.3.14. However, in spite of its simple origin, this class of examples leads to quite deep mathematics in specific situations.

The simplest source for the data (i, α) we need is given by homomorphisms $\phi : G \to L$ which have the property that $\phi(H) \subset K$. If this is the case, then ϕ induces a smooth map $\underline{\phi} : G/H \to L/K$. To have a chance for getting an appropriate α, we had to require that G/H and L/K have the same dimension. Therefore, it makes sense to assume that $\underline{\phi}$ is a local diffeomorphism. (Indeed, if $\underline{\phi}$ is a diffeomorphism locally around $o = eH$, then it is a local diffeomorphism everywhere by homogeneity.) If this is the case, then we can put $i := \phi|_H : H \to K$ and $\alpha := \phi' : \mathfrak{g} \to \mathfrak{l}$. Then $\alpha|_\mathfrak{h} = i'$ holds by construction, and $\underline{\alpha}$ can be identified with $T_o\underline{\phi} : \mathfrak{g}/\mathfrak{h} \to \mathfrak{l}/\mathfrak{k}$, which was assumed to be a linear isomorphism. Equivariancy of α then follows from naturality of the adjoint action, i.e. the standard fact that $\mathrm{Ad}(\phi(h))(\phi'(X)) = \phi'(\mathrm{Ad}(h)(X))$.

Even simpler, we can start with the closed subgroup $K \subset L$, and consider another closed subgroup $G \subset L$. Then G naturally acts on L/K and we have to assume that the G–orbit of eK is open. If this is the case, then putting $H = G \cap K$, the inclusion of G into L induces an open embedding of G/H into L/K, so we may use this inclusion as the homomorphism ϕ.

Applying this simplest version in the case that both G and L are semisimple and K is an appropriate parabolic subgroup of L leads, in combination with the construction of correspondence spaces, to a number of deep relations between parabolic geometries of different type; see Section 4.5. The most prominent among these is Fefferman's construction of a natural conformal structure on the total space of a circle bundle over a manifold endowed with a CR–structure; see 4.5.1 and 4.5.2.

1.5.17. Cartan's space $\mathcal{S}$ and development of curves. To conclude this section, we will discuss the generalization of the concept of distinguished curves on homogeneous spaces to general Cartan geometries. A nice conceptual approach to this is based on Cartan's space $\mathcal{S}$. This also provides an interesting example of an associated (nonlinear) fiber bundle corresponding to the restriction of an action of G, which carries a canonical general connection; see 1.5.7. The natural bundle $\mathcal{S}$ is defined as the associated bundle with fiber G/H, with H acting on G/H by the restriction of the natural G–action.

The first remarkable fact about $\mathcal{S}$ is that $x \mapsto O(x) = [\![u, o]\!] \in \mathcal{P} \times_H G/H$ defines a natural section O of $\mathcal{S}M \to M$ for every Cartan geometry $(\mathcal{P} \to M, \omega)$ of type (G, H). Moreover, along the image of O, the vertical subbundle of the projection $\mathcal{S}M \to M$ is the associated bundle $\mathcal{P} \times_H T_o(G/H)$. Since $T_o(G/H)$ is canonically isomorphic with $\mathfrak{g}/\mathfrak{h}$, this can be naturally identified with the tangent bundle TM. Thus, we may view Cartan's space $\mathcal{S}$ as a nonlinear version of the tangent bundle in which the geometry in question is encoded by means of the parallel transport of the induced connection. This point of view goes back to Cartan (cf. [**Car37**]) and it was developed further in an abstract way in the second half of the twentieth century (see e.g. [**Kol71**]).

Let us illustrate the power of this parallelism on the classical concept of the development of curves. Fix a point $x \in M$ and consider a smooth curve $c : I \to M$ defined on some open interval $I \subset \mathbb{R}$ containing zero such that $c(0) = x$. The development dev_c of c around $x = c(0)$ is then a smooth curve into the fiber $\mathcal{S}_x M$, which is defined locally around zero and maps 0 to $O(x)$. The idea of the definition is simple: To obtain $\operatorname{dev}_c(t)$, follow the curve c up to time t. Then consider the unique parallel curve in $\mathcal{S}M$ which lies over c and goes through the point $O(c(t))$. Follow this curve back to $t = 0$, to obtain a unique point $\operatorname{dev}_c(t) \in \mathcal{S}_{c(0)}M = \mathcal{S}_x M$.

More formally, given c, define smooth curves c_t by $c_t(s) := c(t+s)$. As in [**KMS**, 9.8], let us denote by $s \mapsto \operatorname{Pt}_\gamma(u, s)$ the parallel curve lying over $s \mapsto \gamma(s)$ starting at the point $u \in \mathcal{S}_{\gamma(0)}$. Then we have $\operatorname{dev}_c(t) = \operatorname{Pt}_{c_t}(O(c(t)), -t)$. By Theorem 9.8 of [**KMS**], this defines a smooth curve for $t \in \mathbb{R}$ with $|t|$ small enough.

Fixing a point $u \in \mathcal{P}_x$, there is a unique curve $\alpha(t)$ in G/H mapping $0 \in \mathbb{R}$ to $o \in G/H$ such that $\operatorname{dev}_c(t) = [\![u, \alpha(t)]\!]$. Choosing a different frame $u \cdot h \in \mathcal{P}_x$, the curve changes to $\ell_{h^{-1}} \circ \alpha$. Hence, if we fix a family of smooth curves through the origin $o = eH$ in G/H which is H–invariant, then the requirement that a local smooth curve through x develops to a member of this family is unambiguously defined.

To obtain a more explicit description of the development, we consider the extended principal bundle $\tilde{\mathcal{P}} = \mathcal{P} \times_H G$. Then $\mathcal{S}M = \mathcal{P} \times_H (G/H) \cong \tilde{\mathcal{P}} \times_G (G/H)$, and we write $q : \mathcal{P} \times (G/H) \to \mathcal{S}M$ and $\tilde{q} : \tilde{\mathcal{P}} \times (G/H) \to \mathcal{S}M$ for the corresponding projections. They are related by $q = \tilde{q} \circ (j \times \mathrm{id})$, where $j : \mathcal{P} \to \tilde{\mathcal{P}}$ is the natural inclusion. From 1.5.7 and Theorem 1.5.6 we know that the Cartan connection ω on $\mathcal{P}$ induces a principal connection $\tilde{\omega}$ on $\tilde{\mathcal{P}}$ which is characterized by $j^*\tilde{\omega} = \omega$.

Now suppose that $c : I \to M$ is a smooth curve with $c(0) = x$. Choose a point $u \in \mathcal{P}_x$ and consider the horizontal lift c^{hor} of c to $\tilde{\mathcal{P}}$ with initial point $j(u)$, which is defined locally around 0. Choosing an arbitrary lift $\bar{c} : I \to \mathcal{P}$ of c, the curve c^{hor} can be written as $j(\bar{c}(t)) \cdot g(t)$ for some smooth G–valued function g. Any other choice of lift can be written as $\bar{c}(t) \cdot h(t)$ for some smooth H–valued function h, and then the function g gets replaced by $h(t)^{-1}g(t)$. Using this we now formulate:

THEOREM 1.5.17. *Let $c : I \to M$ be a smooth curve with $c(0) = x$, let $u \in \mathcal{P}_x$ be a point, and let g be a smooth G–valued function defined locally around zero such that $g(0) = e$.*

(1) Locally around zero, we have $\operatorname{dev}_c(t) = q(u, g(t)^{-1} \cdot o)$ if and only if there is a lift $\bar{c} : I \to \mathcal{P}$ of c with $\bar{c}(0) = u$ such that the curve $j(\bar{c}(t)) \cdot g(t)$ in $\tilde{\mathcal{P}}$ is horizontal locally around zero.

(2) Mapping c to dev_c defines a bijection between germs of smooth curves through x in M and germs of smooth curves through o in G/H. This bijection is compatible with having contact to some order, i.e. two curves c_1 and c_2 have contact of order r in x if and only if their developments have contact of order r in o.

PROOF. (1) Assume first that we have chosen a lift $\bar{c} : I \to \mathcal{P}$ such that $j(\bar{c}(t)) \cdot g(t)$ is a horizontal curve in $\tilde{\mathcal{P}}$, and we denote by $J \subset \mathbb{R}$ its domain of definition. If we fix some point $y \in G/H$, then by definition of an induced connection, the curve $\alpha(t) := \tilde{q}(j(\bar{c}(t)) \cdot g(t), y)$ is horizontal in $\mathcal{S}M$. Of course, it is a lift of $c : J \to M$. Now for some $t_0 \in J$ put $y = g(t_0)^{-1} \cdot o$. Then

$$\alpha(t_0) = \tilde{q}(j(\bar{c}(t_0)) \cdot g(t_0), g(t_0)^{-1} \cdot o) = \tilde{q}(j(\bar{c}(t_0)), o) = O(c(t_0)).$$

Reading this backwards in time, we see that $\operatorname{Pt}_{c_{t_0}}(O(c(t_0)), s) = \alpha(t_0 - s)$ for all s such that $t_0 - s \in J$. In particular, $\operatorname{dev}_c(t_0) = \alpha(0) = \tilde{q}(j(\bar{c}(0)), g(t_0)^{-1} \cdot o)$. Putting $u = \bar{c}(0)$, we get $\operatorname{dev}_c(t) = q(u, g(t)^{-1} \cdot o)$ as required.

Conversely, assume that $\operatorname{dev}_c(t) = q(u, g(t)^{-1} \cdot o)$ holds locally around zero. Define a smooth function ϕ with values in $\mathfrak{g}$ by $\phi(t) := -\operatorname{Ad}(g(t))(\delta g(t))$, where δ denotes the left logarithmic derivative. Now $\omega(\bar{c}'(t)) = \phi(t)$ is a time dependent first order ODE on $\mathcal{P}$, so locally around 0 we find a unique solution $\bar{c}$ such that $\bar{c}(0) = u$. Now consider $\tilde{\omega}(\frac{d}{dt}(j(\bar{c}(t)) \cdot g(t)))$. As in the proof of Proposition 1.5.3 we see that this is given by

$$\tilde{\omega}\left(Tr^{g(t)} \cdot Tj \cdot \bar{c}'(t) + \zeta_{\delta g(t)}\right) = \operatorname{Ad}(g(t)^{-1})(\omega(\bar{c}'(t))) + \delta g(t),$$

which vanishes by construction. Hence, locally around zero we have constructed a horizontal lift of c as required.

(2) The second part of the proof of (1) actually gives a construction of a curve with given development, at least on the level of germs. Suppose we have given a curve γ through $O(x)$ in $\mathcal{S}_xM$. Fixing $u \in \mathcal{P}$, we can find a smooth G–valued function g such that $\gamma(t) = q(u, g(t)^{-1} \cdot o)$. As in the proof of part (1), we construct a local curve $\bar{c}$ in $\mathcal{P}$, and then we can project this curve down to a curve c through x in M. By construction, $j(\bar{c}(t)) \cdot g(t)$ is the horizontal lift of c starting in u, so $\gamma = \operatorname{dev}_c$ locally around zero.

To prove bijectivity, it therefore remains to show that the curve c is independent of the choices made in its construction. Keeping the point u fixed, any other choice for the function g has the form $h(t)g(t)$ for a smooth H–valued function h. Now the product rule for the left logarithmic derivative reads as

$$\delta(h(t)g(t)) = \mathrm{Ad}(g(t)^{-1})(\delta h(t)) + \delta g(t),$$

compare with the proof of Theorem 1.2.4. Consequently, replacing $g(t)$ by $h(t)g(t)$, the function $\phi(t) = -\mathrm{Ad}(g(t))(\delta g(t))$ gets replaced by

$$-\mathrm{Ad}(h(t)^{-1})(\delta h(t) - \phi(t)) = \mathrm{Ad}((\nu \circ h)(t))(\phi(t)) + \delta(\nu \circ h)(t).$$

Here, ν denotes the inversion on H, and we have used $\delta(\nu\circ h)(t) = -\mathrm{Ad}(h(t))(\delta h(t))$, compare again with 1.2.4. But now suppose that $\bar{c}$ is a local solution of the differential equation $\omega(\bar{c}'(t)) = \phi(t)$. Then as above, we compute that

$$\omega(\tfrac{d}{dt}(\bar{c}(t) \cdot h(t)^{-1}) = \mathrm{Ad}(h(t)^{-1})(\phi(t)) + \delta(\nu \circ h)(t),$$

so we have found the solution for the modified function. But this evidently projects onto the same curve c.

Replacing u by $u{\cdot}h$ for some fixed $h \in H$, we can replace $g(t)$ by $h^{-1}g(t)h$ (which also maps 0 to e). Then the left logarithmic derivative is given by $\delta(h^{-1}g(t)h) = \mathrm{Ad}(h^{-1})(\delta g(t))$, so our function $\phi(t)$ gets replaced by $\mathrm{Ad}(h^{-1})(\phi(t))$. But if $\bar{c}$ is the solution of $\omega(\bar{c}'(t)) = \phi(t)$ with initial value u, then evidently $\bar{c}(t) \cdot h$ is the solution for $\mathrm{Ad}(h^{-1})(\phi(t))$ with initial value $u \cdot h$. Again, this projects to the same curve as $\bar{c}$, so we have established bijectivity of the development.

Finally, suppose that c_1 and c_2 are smooth local curves, which have kth order contact in x. Fixing a point $u \in \mathcal{P}$, the horizontal lifts of c_1 and c_2 to $\tilde{\mathcal{P}}$ with initial value $j(u)$ have kth order contact in $j(u)$. Further, we can choose lifts $\bar{c}_1$ and $\bar{c}_2$ of the curves through u in $\mathcal{P}$, which have kth order contact in u. But then writing the horizontal lifts as $j(\bar{c}_i(t)) \cdot g_i(t)$, we see that the resulting functions g_1 and g_2 have kth order contact in e. Hence, also the developments $q(u, g_1(t)^{-1} \cdot o)$ and $q(u, g_2(t)^{-1} \cdot o)$ have kth order contact in o.

Conversely, if the developments have kth order contact in o, then we can realize them as $q(u, g_i(t)^{-1} \cdot o)$ for G–valued functions g_1 and g_2 which have kth order contact in u. Then the associated $\mathfrak{g}$–valued functions ϕ_i are given in terms of a left logarithmic derivative, so they have contact of order $(k-1)$ in 0. But then the solutions of the associated first order ODEs with initial value u have kth order contact in u, and projecting to M preserves kth order contact. □

1.5.18. Canonical curves. We have discussed canonical curves on a homogeneous space G/H in 1.4.11. Using the development introduced in 1.5.17 above, we can extend any reasonable family of distinguished curves on G/H to a natural family of locally defined curves on each Cartan geometry of type (G, H).

Definition 1.5.18. (1) A family $\mathcal{C}$ of smooth curves through o in G/H is called *admissible* if and only if for each $\gamma \in \mathcal{C}$, each t_0 in the domain of γ and each element $g \in G$ such that $\gamma(t_0) = g^{-1} \cdot o$, the curve $t \mapsto g \cdot \gamma(t + t_0)$ also lies in $\mathcal{C}$.

(2) Let $(\mathcal{P} \to M, \omega)$ be a Cartan geometry of type (G, H), $x \in M$ a point, and let $c : I \to \mathcal{S}_x M$ be a smooth curve defined on an interval containing zero such that $c(0) = O(x)$. Let γ be a smooth curve in G/H. Then we say that *c is represented by γ on I* if and only if γ is defined on I and for some $u \in \mathcal{P}_x$ we have $c(t) = [\![u, \gamma(t)]\!]$ for all $t \in I$.

(3) Fix an admissible family $\mathcal{C}$ of curves and consider a Cartan geometry $(\mathcal{P} \to M, \omega)$ of type (G, H). Then a smooth curve $c : I \to M$ is said to be a *canonical*

curve of type $\mathcal{C}$ if and only if for each $t_0 \in I$, the development $\operatorname{dev}_{c_{t_0}}$ of the curve $c_{t_0}(t) = c(t+t_0)$ is represented by an element of $\mathcal{C}$ on some neighborhood of zero.

The defining property for a canonical curve in (3) can be loosely phrased as "c develops to an element of $\mathcal{C}$ locally around each point". Note that applying the definition of admissibility in part (1) to $t_0 = 0$, we obtain that for $\gamma \in \mathcal{C}$ and $h \in H$ we also have $\ell_h \circ \gamma \in \mathcal{C}$. This property, however, is not sufficient for admissibility. The motivation for the notion of admissibility is the following result, which makes the notion of a canonical curve much more manageable.

PROPOSITION 1.5.18. *Let* $\mathcal{C}$ *be an admissible family of curves in* G/H*, let* $(\mathcal{P} \to M, \omega)$ *be any Cartan geometry of type* (G, H)*, and let* $c : I \to M$ *be a smooth curve.*

Suppose that there is a point $t_0 \in I$ *such that the development of* c_{t_0} *is defined and represented by an element of* $\mathcal{C}$ *on the whole interval* $J = \{t : t + t_0 \in I\}$*. If, in addition,* c *admits a horizontal lift to* $\tilde{\mathcal{P}}$ *which is defined on all of* I *and maps* t_0 *to* $j(\mathcal{P}) \subset \tilde{\mathcal{P}}$*, then* c *is a canonical curve of type* $\mathcal{C}$*.*

In particular, if an arbitrary curve develops to an element of $\mathcal{C}$ *in one point, then its restriction to a sufficiently small interval around that point is a canonical curve.*

PROOF. It suffices to consider $t_0 = 0$ and hence $J = I$. Choose some lift $\bar{c} : I \to \mathcal{P}$ of c. By assumption, we have a horizontal lift mapping 0 to $j(u)$ for some $u \in \mathcal{P}$. Choosing some lift $\bar{c} : I \to \mathcal{P}$ of c such that $\bar{c}(0) = u$, the horizontal lift can be written as $t \mapsto j(\bar{c}(t)) \cdot g(t)$ for some smooth function $g : I \to G$. From the proof of Theorem 1.5.17 we see that $\operatorname{dev}_c(t) = [\![u, g(t)^{-1} \cdot o]\!]$ for all $t \in I$. In particular, the curve $t \mapsto g(t)^{-1} \cdot o$ belongs to $\mathcal{C}$.

Now take $t_1 \in I$, and consider the curve $t \mapsto j(\bar{c}(t+t_1)) \cdot g(t+t_1)g(t_1)^{-1}$ in $\tilde{\mathcal{P}}$. This is obtained by the principal right action of a fixed element of G on a horizontal curve, so it is horizontal, too. Its value in $t = 0$ is $j(\bar{c}(t_1)) \in j(\mathcal{P})$ and it lifts the curve c_{t_1}. By Theorem 1.5.17 the development of c_{t_1} is, locally around zero, represented by $t \mapsto [\![\bar{c}(t_1), g(t_1)g(t+t_1)^{-1} \cdot o]\!]$. But since $g(t)^{-1} \cdot o$ lies in $\mathcal{C}$, by admissibility $g(t_1)g(t+t_1)^{-1} \cdot o$ also lies in $\mathcal{C}$. Since $t_1 \in I$ is arbitrary, the result follows. □

Together with Theorem 1.5.17 we see that the structure of the local canonical curves of type $\mathcal{C}$ through any point x in any Cartan geometry looks exactly as the structure of local curves through o in G/H which are in $\mathcal{C}$. This means that many questions about canonical curves (e.g. how many derivatives in the point x are needed to uniquely specify a canonical curve of type $\mathcal{C}$) can be reduced to looking at $o \in G/H$. Questions of this type in the realm of parabolic geometries will be studied in Section 5.3. Notice further, that the proof of Theorem 1.5.17 also shows how canonical curves of type $\mathcal{C}$ can be constructed as projections of solutions of appropriated ODEs.

As a simple example let us look at the case of exponential curves. In 1.4.11 we have started from a subspace $\mathfrak{n} \subset \mathfrak{g}$ which is complementary to $\mathfrak{h}$ and we have considered the curves $t \mapsto g\exp((t-t_0)X)H$ for $X \in \mathfrak{n}$ and $g \in G$. If this maps 0 to o, then $g\exp(-t_0X) =: h \in H$, and our curve is given by $t \mapsto h\exp(tX) \cdot o = \exp(t\operatorname{Ad}(h)(X)) \cdot o$. More generally, for $X \in \mathfrak{g}$, consider $c^X(t) := \exp(tX) \cdot o$. If $g^{-1} \cdot o = \exp(t_0X) \cdot o$, then as above we see that $g \cdot c^X(t+t_0) = c^{\operatorname{Ad}(h)(X)}(t)$. Hence, if we suppose that $A \subset \mathfrak{g}$ is any subset such that $\mathrm{Ad}(h)(A) \subset A$ for all $h \in H$, then the family $\mathcal{C}_A := \{c^X : X \in A\}$ of curves through o is admissible. Hence, we have the notion of canonical curves of type $\mathcal{C}_A$ on arbitrary Cartan geometries of type (G, H). In this case, we can describe the canonical curves explicitly:

COROLLARY 1.5.18. *Let $A \subset \mathfrak{g}$ be a subset which is invariant under* $\mathrm{Ad}(h)$ *for all $h \in H$. Let $(\mathcal{P} \to M, \omega)$ be any Cartan geometry. Then a curve $c : I \to M$ is canonical of type $\mathcal{C}_A$ if an only if locally it coincides up to a constant shift in parameter with the projection of a flow line of a vector field $\omega^{-1}(X) \in \mathfrak{X}(\mathcal{P})$ for some $X \in A$.*

PROOF. Put $\bar{c}(t) = \mathrm{Fl}_t^{\omega^{-1}(X)}(u)$ for some $u \in \mathcal{P}$ and some $X \in A$. From the definitions and using that $\mathrm{Ad}(\exp(tX))(X) = X$, one immediately verifies that the curve $j(\bar{c}(t)) \cdot \exp(-tX)$ is horizontal in $\tilde{\mathcal{P}}$. By Theorem 1.5.17, the development of the projection to M is represented by $[\![\bar{c}(0), \exp(tX) \cdot o]\!]$, so by the proposition it is canonical of type $\mathcal{C}_A$. Therefore, any curve which locally coincides with such flow lines (up to a constant shift in parameter) is canonical, too.

Conversely, if c is canonical, then around each point, it develops to an element of $\mathcal{C}_A$. Shifting the parameter, the development is given by $[\![u, \exp(tX) \cdot o]\!]$. Now we can locally reconstruct c as in the proof of Theorem 1.5.17. But the resulting data for the ODE is $\phi(t) = -\mathrm{Ad}(\exp(-tX))(-X) = X$, so the solution of the ODE is a flow line of $\omega^{-1}(X)$. □

The simplest possible situation arises if one starts from a reductive Klein geometry (G, H). Then for A one can take a fixed H–invariant complement $\mathfrak{n}$ of $\mathfrak{h}$ in $\mathfrak{g}$. The flow lines of constant vector fields corresponding to elements of $\mathfrak{n}$ are preserved by the principal right action on $\mathcal{P}$. Projecting them down to M, one exactly obtains the geodesics of the natural connection on the tangent bundle induced by $\mathfrak{n}$; compare with 1.5.6 and 1.4.11.

In general, fixing some complementary linear subspace $\mathfrak{n}$ to $\mathfrak{h}$ in $\mathfrak{g}$, one can also construct families of distinguished coordinates. Choosing $u \in \mathcal{P}$, one can take a neighborhood U of $0 \in \mathfrak{n}$ and define a map $U \to M$ by $X \mapsto p(\mathrm{Fl}_1^{\omega^{-1}(X)}(u))$, which is well defined for small U. In the reductive case, this exactly recovers the usual normal coordinates, while in general one obtains larger families of distinguished coordinates. This will be explored in the realm of parabolic geometries in more detail in 5.1 and 5.3.

1.6. Conformal Riemannian structures

In this section, we finally switch to the question of constructing Cartan geometries from underlying structures. Taking pseudo–Riemannian structures as a motivation, we first discuss the case of affine connections on G–structures. While this case is rather easy, since one has already given the bundle, it nicely illustrates the algebraic nature of the problems involved. The main part of the section is devoted to conformal structures, which are the best known example of a parabolic geometry. Many of the general features of parabolic geometries show up in this case in rather simple form, so conformal structures will be a guideline throughout the book. At the same time, this completes the discussion of the example of the conformal sphere from 1.1.5.

We will treat conformal structures in an explicit and classical style, and only try to indicate the Lie algebraic background that will be used systematically in more complicated cases later on. Using the tools from Section 1.5 we could actually shorten the construction of the canonical Cartan geometry associated to a conformal structure considerably; see Remark 1.6.7. We prefer to present the classical presentation here, since it provides more geometric motivation.

1.6.1. The case of affine connections on G–structures. Let H be a Lie group and $j : H \to GL(m, \mathbb{R})$ a smooth homomorphism whose derivative $j' : \mathfrak{h} \to \mathfrak{gl}(m, \mathbb{R})$ is injective. Then a first order G–structure with structure group H is defined as a reduction of the first order frame bundle $\mathcal{P}^1 M \to M$ to the structure group H, i.e. a principal H–bundle $\mathcal{P} \to M$ endowed with a homomorphism $i : \mathcal{P} \to \mathcal{P}^1 M$ over j which covers the identity on M. Equivalently, one may require the existence of an $\mathbb{R}^m$–valued one–form Θ on $\mathcal{P}$ which is strictly horizontal and H–equivariant; see 1.3.6. Forming the affine extension $B = \mathbb{R}^m \rtimes H$, one can consider Cartan geometries of type (B, H). We have seen in 1.3.6 that such a Cartan geometry on M is equivalent to a first order G–structure $\mathcal{P} \to M$ with structure group H endowed with a principal connection $\gamma \in \Omega^1(\mathcal{P}, \mathfrak{h})$. The corresponding Cartan connection is simply given by $\omega = \Theta + \gamma$.

The best known special case of this situation is provided by the orthogonal group $O(p, q)$ of an inner product of signature (p, q) with $p + q = m$ on $\mathbb{R}^m$. For the purpose of motivation, we give a brief direct discussion of this case before switching to the general considerations. A first order G–structure with structure group $O(p, q)$ on an m–dimensional manifold M is equivalent to a pseudo Riemannian metric g of signature (p, q) on M. The principal bundle defining the G–structure is the *orthonormal frame bundle* $\mathcal{O}M$ of M, whose fiber over $x \in M$ may be described as the set of all orthonormal bases of $T_x M$ or equivalently as the set of all orthogonal linear isomorphisms $\mathbb{R}^m \to T_x M$.

There is a bijective correspondence between principal connections on $\mathcal{O}M$ and linear connections on the tangent bundle TM which preserve the metric g, i.e. such that $\nabla g = 0$ for the induced connection on $S^2 T^* M$. But the Levi–Civita connection of the metric g provides a preferred linear connection on TM with this property which is characterized by the additional condition that it is torsion free. Consequently, starting with (M, g) one gets a Cartan geometry $(\mathcal{O}M \to M, \omega)$ of type $(\mathbb{R}^m \rtimes O(p, q), O(p, q))$ by building ω from the canonical form θ and the principal connection corresponding to the Levi–Civita connection; compare with Example (ii) of 1.5.1.

From 1.3.6 we know that torsion freeness of the connection on TM induced by ω is equivalent to torsion freeness of the Cartan connection ω, so we always obtain torsion–free Cartan geometries in this way. Conversely, for any Cartan geometry of the given type, we obtain an underlying pseudo–Riemannian metric. If the Cartan geometry is torsion free, then the induced connection on the tangent bundle must be the Levi–Civita connection. Uniqueness of the Levi–Civita connection also implies that any local isometry is compatible with the Levi–Civita connections, and hence lifts to a morphism of Cartan geometries. Thus, one actually gets an equivalence of categories between pseudo–Riemannian manifolds of signature (p, q) and torsion–free Cartan geometries of type $(\mathbb{R}^m \rtimes O(p, q), O(p, q))$.

Let us analyze the corresponding problem for a general group H with Lie algebra $\mathfrak{h} \subset \mathfrak{gl}(m, \mathbb{R})$, which will also lead to a conceptual proof for the pseudo Riemannian case. As before, we denote the affine extension of H by B and its Lie algebra by $\mathfrak{b}$. Given a first order G–structure $\mathcal{P} \to M$, there is always a principal connection $\gamma \in \Omega^1(\mathcal{P}, \mathfrak{h})$, so using the form $\Theta \in \Omega^1(\mathcal{P}, \mathbb{R}^m)$ from above, we obtain a Cartan connection $\omega \in \Omega^1(\mathcal{P}, \mathfrak{b})$ inducing the given G–structure. Hence, we have to ask how we can prescribe the torsion of ω, and whether fixing the torsion uniquely pins down the Cartan connection.

Both of these questions are rather easy to answer looking at the affine structure on the space of principal connections. Fixing a principal connection γ_0 corresponding to the Cartan connection ω_0 and given any other principal connection γ corresponding to $\omega = \Theta + \gamma$ and a point $u \in \mathcal{P}$, consider the difference $\omega(u) - \omega_0(u) : T_u\mathcal{P} \to \mathfrak{g}$. By construction, this vanishes on the vertical tangent space $V_u\mathcal{P}$ and thus factors to T_xM, where $x = p(u)$, and it has values in $\mathfrak{h}$, since both ω and ω_0 have $\mathbb{R}^m$–component Θ. Hence, we obtain a unique linear map $\psi : \mathbb{R}^m \to \mathfrak{h}$ such that $\omega(u)(\xi) = \omega_0(u)(\xi) + \psi(\Theta(u)(\xi))$. Otherwise put, for $X \in \mathbb{R}^m \subset \mathfrak{b}$ we have $\omega(u)^{-1}(X) = \omega_0(u)^{-1}(X) - \zeta_{\psi(X)}(u)$. Conversely, any linear map $\psi : \mathbb{R}^m \to \mathfrak{h}$ can be used to construct a linear isomorphism $\omega(u)$ from $\omega_0(u)$ as above.

From 1.3.5 and 1.3.6 we know that the torsion part $T \in L(\Lambda^2\mathbb{R}^m, \mathbb{R}^m)$ of the curvature function κ of ω is given by

$$d\Theta(\omega^{-1}(X), \omega^{-1}(Y)) + [X, Y]$$

for $X, Y \in \mathbb{R}^m \subset \mathfrak{b}$. Equivariancy implies that $d\Theta(\zeta_A, \eta) = -A(\Theta(\eta))$ for all $A \in \mathfrak{h}$, so we conclude that $T = T_0 + \partial\psi$, where $\partial\psi : \Lambda^2\mathbb{R}^m \to \mathbb{R}^m$ is the map

$$\partial\psi(X, Y) := \psi(X)(Y) - \psi(Y)(X). \tag{1.27}$$

Using the bracket in $\mathfrak{b}$, we can write $\partial\psi(X, Y) = [X, \psi(Y)] - [Y, \psi(X)]$. This also shows that ∂ is a restriction of the Lie algebra differential in the standard complex computing the cohomology of the abelian Lie algebra $\mathbb{R}^m$ with coefficients in the Lie algebra $\mathfrak{b}$, which will be important for generalizations.

The central object to study, therefore, is the map $\partial : L(\mathbb{R}^m, \mathfrak{h}) \to L(\Lambda^2\mathbb{R}^m, \mathbb{R}^m)$ defined by (1.27). Let us view the source of ∂ as $\mathbb{R}^{m*} \otimes \mathfrak{h}$ and recall that $\mathfrak{h}$ is a subalgebra of $\mathfrak{gl}(m, \mathbb{R}) = \mathbb{R}^{m*} \otimes \mathbb{R}^m$. Then ∂ is the composition

$$\mathbb{R}^{m*} \otimes \mathfrak{h} \to \mathbb{R}^{m*} \otimes \mathbb{R}^{m*} \otimes \mathbb{R}^m \to \Lambda^2\mathbb{R}^{m*} \otimes \mathbb{R}^m,$$

where the first map comes from the inclusion of $\mathfrak{h}$ and the second map comes from the alternation in the first two factors. Suppose that we have given the value $\omega_0(u)$ of one compatible Cartan connection in u and the torsion in this point is $T_0(u)$. Then the possible values of the torsions of compatible Cartan connections are exactly the expressions of the form $T_0(u) + \partial\psi$. On the other hand, the space of all $\omega(u)$ which have torsion $T_0(u)$ in u is parametrized by $\mathfrak{h}^{(1)} := \ker(\partial) \subset \mathbb{R}^{m*} \otimes \mathfrak{h}$. This subspace is called the first *prolongation* of the subalgebra $\mathfrak{h}$ of $\mathfrak{gl}(m, \mathbb{R})$.

To come to global results, we have to take into account the H–module structures on the various spaces involved. Since $\mathbb{R}^m$ and $\mathfrak{h}$ are H–modules, so are $\mathbb{R}^{m*} \otimes \mathfrak{h}$ and $\Lambda^2\mathbb{R}^{m*} \otimes \mathbb{R}^m$. By definition, $(h \cdot \psi)(X) = \mathrm{Ad}(h)(\psi(h^{-1}(X)))$, and thus

$$(h \cdot \psi)(X)(Y) = h(\psi(h^{-1}(X))(h^{-1}(Y))),$$

which immediately implies that ∂ is an H–equivariant map. In particular, $\mathfrak{h}^{(1)} \subset \mathbb{R}^{m*} \otimes \mathfrak{h}$ and $\mathrm{im}(\partial) \subset \Lambda^2\mathbb{R}^{m*} \otimes \mathbb{R}^m$ are H–submodules. Using this, we formulate:

THEOREM 1.6.1. *Let $j : H \to GL(m, \mathbb{R})$ be a smooth homomorphism such that j' is injective, so $\mathfrak{h}$ may be viewed as a subalgebra of $\mathfrak{gl}(m, \mathbb{R})$, and let $B = \mathbb{R}^m \rtimes H$ be the affine extension of H.*

(1) Suppose that $\mathfrak{N} \subset \Lambda^2\mathbb{R}^{m*} \otimes \mathbb{R}^m$ *is an* H*–invariant complement to* $\operatorname{im}(\partial)$*, and the map* $\partial : \mathbb{R}^{m*} \otimes \mathfrak{h} \to \operatorname{im}(\partial)$ *admits an* H*-equivariant split. Then any first order* G*–structure* $(\mathcal{P} \to M, \Theta)$ *with structure group* H *admits an affine connection* $\omega = \Theta + \gamma$*, such that the torsion part of the curvature function* κ *has values in* $\mathfrak{N}$.

(2) If the first prolongation $\mathfrak{h}^{(1)}$ *of* $\mathfrak{h}$ *is trivial, then any affine connection* ω *on a first order* G*–structure with structure group* H *is uniquely determined by its torsion.*

(3) If $\mathfrak{h}^{(1)} = \{0\}$ *and* $\mathfrak{N}$ *is a complement as in (1), then forgetting the* $\mathfrak{h}$*–part of the Cartan connection defines an equivalence between the category of first order* G*–structures with structure group* H *and the category* $\mathcal{C}^{\mathfrak{N}}_{(G,H)}$ *of Cartan geometries of type* (G, H) *such that the torsion part of the curvature function has values in* $\mathfrak{N}$.

PROOF. (1) Given a first order G–structure $(p : \mathcal{P} \to M, \Theta)$, choose a principal connection γ_0 on $\mathcal{P}$, and consider the Cartan connection $\omega_0 = \Theta + \gamma_0$. Let $T_0 : \mathcal{P} \to \Lambda^2\mathbb{R}^{m*} \otimes \mathbb{R}^m$ be the torsion part of the curvature function. From 1.5.1 we know that the full curvature function is H–equivariant, and since $\mathfrak{b} = \mathbb{R}^m \oplus \mathfrak{h}$ as an H–module, we conclude that T_0 is H–equivariant. Since $\mathfrak{N}$ is an H–invariant complement to $\operatorname{im}(\partial)$, we can split T_0 into a $\mathfrak{N}$–component $T_0^{\mathfrak{N}}$ and an $\operatorname{im}(\partial)$ component which are both H–equivariant. Since we assumed existence of an H–equivariant split of ∂, there is a smooth H–equivariant function $\psi : \mathcal{P} \to \mathbb{R}^{m*} \otimes \mathfrak{h}$ such that $T_0(u) = T_0^{\mathfrak{N}}(u) + \partial(\psi(u))$. Now we define $\omega \in \Omega^1(\mathcal{P}, \mathfrak{b})$ by

$$\omega(u)(\xi) := \omega_0(u)(\xi) - \psi(u)(\Theta(u)(\xi)).$$

Equivariancy of ω, ψ and Θ immediately implies that $(r^h)^*\omega = \operatorname{Ad}(h^{-1}) \circ \omega$ for all $h \in H$. On the vertical subbundle, ω coincides with ω_0, so it reproduces the generators of fundamental vector fields. Finally, $\omega(u)(\xi) = 0$ implies that $\omega_0(\xi)$ lies in $\mathfrak{h}$ and thus ξ is vertical, so we see that $\omega(u)$ is a linear isomorphism for all $u \in \mathcal{P}$. Hence, $(\mathcal{P} \to M, \omega)$ is a Cartan geometry of type (B, H). The pointwise computation from above implies that the torsion part T of the curvature function of ω is given by $T(u) = T_0(u) - \partial(\psi(u)) = T_0^{\mathfrak{N}}(u)$ and thus has values in $\mathfrak{N}$.

(2) Having given two affine connections ω and ω_0, their difference in a point $u \in \mathcal{P}$ is described by $\psi : \mathbb{R}^m \to \mathfrak{h}$, and if the torsion parts of the curvature functions agree, we know that $\partial(\psi) = 0$. The condition that $\mathfrak{h}^{(1)} = \{0\}$ exactly says that ∂ is injective, so $\psi = 0$ and hence $\omega(u) = \omega_0(u)$. Since u was arbitrary, the result follows.

(3) If $\mathfrak{h}^{(1)} = \{0\}$, then ∂ is an isomorphism onto its image, so the inverse provides the split required for (1), which together with (2) implies that there is a unique affine connection $\omega = \gamma + \Theta$ on each first order G–structure $(\mathcal{P} \to M, \Theta)$ with the required torsion. Hence, we can associate to any first order G–structure a Cartan geometry. If $\Phi : \mathcal{P} \to \mathcal{P}'$ is a morphism of first order G–structures and ω' is the canonical affine connection on $\mathcal{P}'$, then $\Phi^*\omega'$ is an affine connection on $\mathcal{P}$. The torsion part of the curvature function of $\Phi^*\omega'$ is the composition of the torsion part of the curvature function of ω' with Φ, so it has values in $\mathfrak{N}$, and $\Phi^*\omega' = \omega$ follows. Thus, we actually get a functor from first order G–structures to $\mathcal{C}^{\mathfrak{N}}_{(G,H)}$ which together with the forgetful functor defines an equivalence of categories. □

EXAMPLE 1.6.1. Let us look at the special case $\mathfrak{h} := \mathfrak{o}(p, q)$ of a pseudo–orthogonal subalgebra. Using the standard metric of signature (p, q) to identify $\mathbb{R}^m$ with $\mathbb{R}^{m*}$, the condition that an element $t_{ijk} \in \otimes^3\mathbb{R}^{m*}$ comes from an element in $\mathbb{R}^{m*} \otimes \mathfrak{o}(p, q)$ is simply skew symmetry in the last two indices, i.e. $t_{ijk} = -t_{ikj}$.

Such an element lies in the first prolongation $\mathfrak{o}(p,q)^{(1)}$ if, in addition, it is symmetric in the first two indices, i.e $t_{ijk} = t_{jik}$. But for such an element we now compute

$$t_{ijk} = -t_{ikj} = -t_{kij} = t_{kji} = t_{jki} = -t_{jik} = -t_{ijk},$$

and thus $\mathfrak{o}(p,q)^{(1)} = 0$. Consequently, ∂ is injective, and since $\mathfrak{o}(p,q) \cong \Lambda^2\mathbb{R}^{m*}$, ∂ must be a linear isomorphism for dimensional reasons. Hence, $\{0\}$ is an H–invariant complement to $\operatorname{im}(\partial)$, and part (3) of the theorem gives the equivalence of the category of pseudo–Riemannian manifolds of signature (p,q) and the category of torsion–free Cartan geometries of type $(\mathbb{R}^m \rtimes O(p,q), O(p,q))$.

Notice that for any subalgebra $\mathfrak{h} \leq \mathfrak{o}(p,q)$ we have $\mathfrak{h}^{(1)} = \{0\}$, while for any subalgebra $\mathfrak{h} \subset \mathfrak{gl}(m,\mathbb{R})$ which contains some $\mathfrak{o}(p,q)$ we know that ∂ is surjective.

1.6.2. The Möbius space. Let us fix a signature (p,q) with $p+q=m$. A *conformal structure* of signature (p,q) on an m–dimensional manifold M is defined as an equivalence class of pseudo–Riemannian metrics of that signature on M. Two metrics are considered as equivalent if and only if one is obtained from the other by multiplication by a positive smooth function. The values of these metrics in point $x \in M$ form a ray in $S^2T_x^*M$, i.e. an $\mathbb{R}_+$–subbundle, and conversely any smooth $\mathbb{R}_+$–subbundle whose elements have signature (p,q) in all points defines a conformal structure. Equivalently, conformal structures can be defined as G–structures with structure group $CO(p,q)$, the conformal group of signature (p,q). This G–structure is provided by the conformal frame bundle, whose fiber over a point $x \in M$ can be viewed as the set of orthogonal bases in T_xM consisting of vectors of equal length. Equivalently, it can be viewed as the space of all linear conformal isometries from $\mathbb{R}^m$ with some fixed inner product $\langle\ ,\ \rangle$ of signature (p,q) to T_xM.

There are various ways to see immediately that there is no hope that conformal structures could be in bijective correspondence with a class of Cartan geometries of type $(\mathbb{R}^m \rtimes CO(p,q), CO(p,q))$. Indeed, the homogeneous model of that class of Cartan geometries is $\mathbb{R}^m$ with the action of the affine extension of $CO(p,q)$. But there are well–known examples of local conformal isometries of $\mathbb{R}^m$ which are not of that form, and which do not even globalize to the whole of $\mathbb{R}^m$ (which contradicts being the flat model of a Cartan geometry; see Proposition 1.5.2). For example, consider the open subset $U \subset \mathbb{R}^m$ of all points of nonzero norm and the *inversion* in the unit circle, defined by $x \mapsto \frac{x}{\langle x,x\rangle}$. Clearly, this defines a diffeomorphism ϕ of U, and using

$$T_x\phi \cdot \xi = \frac{\xi}{\langle x,x\rangle} - \frac{2\langle \xi,x\rangle x}{\langle x,x\rangle^2}$$

one immediately verifies that $\langle T_x\phi\cdot\xi, T_x\phi\cdot\eta\rangle = \frac{\langle\xi,\eta\rangle}{\langle x,x\rangle^2}$, so ϕ is a conformal isometry.

The conceptual way to describe the homogeneous model of conformal structures is parallel to the positive definite case discussed in Example 1.1.5. Consider $\mathbb{R}^{m+2}$ endowed with an inner product of signature $(p+1,q+1)$. Let $C \subset \mathbb{R}^{m+2}$ be the cone of nonzero null vectors, and consider the space $S^{(p,q)}$ of lines in C as a subset in the projective space $\mathbb{R}P^{m+1}$. We can consider $S^{(p,q)}$ as $C/\mathbb{R}^*$ and the obvious projection $\pi : C \to S^{(p,q)}$ is a principal bundle with structure group $\mathbb{R}^*$. The space $S^{(p,q)}$ is called the *pseudo–sphere* or the *Möbius space* of signature (p,q). Note that $S^{(p,q)}$ is closed in $\mathbb{R}P^{m+1}$ and thus compact since it consists of all zeros of a homogeneous polynomial of degree two. The orthogonal group $G = O(p+1,q+1)$ of the chosen inner product acts on $\mathbb{R}^{m+2}$, leaving the cone C invariant. Since G acts linearly, the action descends to an action on $S^{(p,q)}$.

PROPOSITION 1.6.2. *The Möbius space* $S^{(p,q)}$ *carries a natural conformal structure of signature* (p,q)*, which is invariant under the action of the group* $G = O(p+1, q+1)$*. This action is transitive, so we can identify* $S^{(p,q)}$ *with the homogeneous space* G/P*, where* $P \subset G$ *is the stabilizer of a null line in* $\mathbb{R}^{m+2}$.

There is a natural two–fold covering $S^p \times S^q \to S^{(p,q)}$*, so as a manifold* $S^{(p,q)} \cong (S^p \times S^q)/\mathbb{Z}_2$*. Under this identification, the conformal structure on* $S^{(p,q)}$ *corresponds to the conformal class of the product of the two round metrics of radius one (with opposite signs).*

PROOF. Let $\pi : C \to S^{(p,q)}$ be the obvious smooth projection. The tangent space of C at $v \in C$ is $v^\perp$, and this contains v since v is null. The tangent map to π induces a linear isomorphism $v^\perp/\mathbb{R}v \to T_{\pi(v)}S^{(p,q)}$. Now the inner product on $\mathbb{R}^{m+2}$ induces an inner product $v^\perp/\mathbb{R}v$, which is nondegenerate of signature (p,q) and can be carried over to $T_{\pi(v)}S^{(p,q)}$. Replacing v by av for some $a \in \mathbb{R}$, this inner product gets multiplied by a^2, so we get a conformal structure on $S^{(p,q)}$. From this description it is evident that G acts by conformal isometries.

Transitivity of the action follows from elementary linear algebra: Let us denote the inner product by $\langle\ ,\ \rangle$. For a null vector v, one can find a vector w such that $\langle v, w\rangle = 1$. By adding an appropriate multiple of v to w, we can further achieve $\langle w, w\rangle = 0$. Then on the subspace generated by v and w, $\langle\ ,\ \rangle$ is non–degenerate of signature $(1,1)$. Hence, we can complete $\{v, w\}$ to a basis by choosing an orthonormal basis of the orthocomplement $\{v, w\}^\perp$. Starting with another null vector $\tilde{v}$ one obtains a basis of the same form. Hence, the linear map which maps the first basis to the second is orthogonal (since the inner product has the same form in both bases). This orthogonal map sends v to $\tilde{v}$, and transitivity follows. From this, it is clear that $S^{(p,q)} \cong G/P$.

To obtain the explicit description of $S^{(p,q)}$, we identify $\mathbb{R}^{m+2}$ with $\mathbb{R}^{p+1} \times \mathbb{R}^{q+1}$ and endow the first vector with the standard inner product and the second factor with the negative of the standard inner product. Then a point $(x,y) \neq (0,0)$ lies in C if and only if $|x| = |y|$. In particular, we get $S^p \times S^q \subset C$. Evidently, any line in C meets $S^p \times S^q$ in exactly two points, namely (x,y) and $(-x,-y)$. This gives the two–fold covering as claimed. Since the inclusion $S^p \times S^q \hookrightarrow C$ is even an isometry, the claim about the conformal classes follows as well. In particular, $S^{(p,0)} = S^p$ with the conformal class of the round metric as we have already seen in 1.1.5. □

1.6.3. The Poincaré conformal group. To work with the homogeneous space $S^{(p,q)}$, it is better not to start with the standard inner product of signature $(p+1, q+1)$ on $\mathbb{R}^{m+2}$, but to use a form of the inner product as it appeared in the proof of Proposition 1.6.2. So we endow $\mathbb{R}^{m+2}$ with the inner product of signature $(p+1, q+1)$ corresponding to the matrix

$$S = \begin{pmatrix} 0 & 0 & 1 \\ 0 & I_{p,q} & 0 \\ 1 & 0 & 0 \end{pmatrix}, \quad I_{p,q} = \begin{pmatrix} I_p & 0 \\ 0 & -I_q \end{pmatrix}, \tag{1.28}$$

where I_n means the identity matrix of rank n. The pseudo–orthogonal group of this inner product consists of all invertible matrices A such that $A^tSA = S$, while its Lie algebra $\mathfrak{g} \cong \mathfrak{o}(p+1, q+1)$ is given by $\{B : B^TS = -SB\}$. Working out the last condition explicitly, we get

$$\mathfrak{g}=\left\{\begin{pmatrix} a & Z & 0\\ X & A & -I_{p,q}Z^t\\ 0 & -X^tI_{p,q} & -a\end{pmatrix} : a\in\mathbb{R}, X\in\mathbb{R}^m, Z\in\mathbb{R}^{m*}, A\in\mathfrak{o}(p,q)\right\}.$$

This block form leads to a vector space decomposition $\mathfrak{g}=\mathfrak{g}_{-1}\oplus\mathfrak{g}_0\oplus\mathfrak{g}_1$ with $\mathfrak{g}_{-1}\cong\mathbb{R}^m$ corresponding to X, $\mathfrak{g}_1\cong\mathbb{R}^{m*}$ corresponding to Z, and $\mathfrak{g}_0\cong\mathfrak{o}(p,q)\oplus\mathbb{R}$ corresponding to a and A. The block form also implies that this defines a grading of $\mathfrak{g}$, i.e. $[\mathfrak{g}_i,\mathfrak{g}_j]\subset\mathfrak{g}_{i+j}$, where we agree that $\mathfrak{g}_\ell=\{0\}$ unless $\ell\in\{-1,0,1\}$.

Denoting elements of $\mathfrak{g}_0$ by (A,a), elements of $\mathfrak{g}_{-1}$ as X, and elements of $\mathfrak{g}_1$ by Z, one immediately verifies that the brackets are given as follows:

$$\begin{array}{ll} \mathfrak{g}_0\times\mathfrak{g}_{-1}\to\mathfrak{g}_{-1} & [(A,a),X]=AX-aX\\ \mathfrak{g}_0\times\mathfrak{g}_0\to\mathfrak{g}_0 & [(A,a),(B,b)]=([A,B],0)\\ \mathfrak{g}_0\times\mathfrak{g}_1\to\mathfrak{g}_1 & [(A,a),Z]=aZ-ZA\\ \mathfrak{g}_{-1}\times\mathfrak{g}_1\to\mathfrak{g}_0 & [X,Z]=(XZ-I_{p,q}(XZ)^tI_{p,q},-ZX). \end{array}\tag{1.29}$$

In particular, we see that acting with elements of $\mathfrak{g}_0$ on $\mathfrak{g}_{-1}$ induces an isomorphism between $\mathfrak{g}_0$ and $\mathfrak{co}(p,q)$, and $\mathfrak{g}_{-1}\cong\mathbb{R}^m$ is the standard representation for this conformal algebra. Further, $\mathfrak{g}_0=\mathfrak{o}(p,q)\oplus\mathbb{R}$ as a Lie algebra. The natural pairing between $\mathfrak{g}_{-1}\cong\mathbb{R}^m$ and $\mathfrak{g}_1\cong\mathbb{R}^{m*}$ occurs as the a–component of the bracket. Finally, the action of $\mathfrak{g}_0$ on $\mathfrak{g}_1$ identifies $\mathfrak{g}_1$ with the dual of $\mathfrak{g}_{-1}$ as a $\mathfrak{g}_0$–module. Indeed, by the Jacobi identity we have $[(A,a),[X,Z]]=[[(A,a),X],Z]+[X,[(A,a),Z]]$ and on the left–hand side the a-component must be zero. The pair $(0,1)$ defines an element $E\in\mathfrak{g}_0$ called the *grading element.* The splitting $\mathfrak{g}=\mathfrak{g}_{-1}\oplus\mathfrak{g}_0\oplus\mathfrak{g}_1$ is exactly the splitting into eigenspaces for $\operatorname{ad}(E)$.

Passing to the group level, it is better to use the group $G:=PO(p+1,q+1)$, the quotient of $O(p+1,q+1)$ by its center $\{\pm\mathrm{id}\}$, since the orthogonal group does not act effectively on $S^{(p,q)}$. The *Poincaré conformal group* P is defined to be the isotropy group of the line through the first basis vector, so $S^{(p,q)}\cong G/P$. Evidently, the Lie algebra $\mathfrak{p}$ of P is $\mathfrak{g}_0\oplus\mathfrak{g}_1\subset\mathfrak{g}$. We can use the commutative subalgebra $\mathfrak{g}_{-1}\subset\mathfrak{g}$ as a complement to $\mathfrak{p}$, thus making (G,P) into a split Klein geometry.

There is an obvious subgroup $G_0\subset P$ consisting of all elements of P which also stabilize the line generated by the last basis vector e_{m+2}. By definition, this is a closed subgroup and thus a Lie subgroup of P (and of G) and its Lie algebra is the subalgebra $\mathfrak{g}_0\subset\mathfrak{p}$.

PROPOSITION 1.6.3. *(1) The elements of the Poincaré conformal group P correspond to matrices of the form*

$$\begin{pmatrix}\lambda & -\lambda w^tI_{p,q}C & -\frac{\lambda}{2}\langle w^t,w^t\rangle\\ 0 & C & w\\ 0 & 0 & \lambda^{-1}\end{pmatrix}$$

with blocks of sizes 1, m, *and* 1, $\lambda\in\mathbb{R}\setminus 0$, $C\in O(p,q)$, *and* $w\in\mathbb{R}^n$. *The subgroup G_0 consists of the matrices for which $w=0$.*

(2) The adjoint action of G_0 preserves the grading on $\mathfrak{g}$, and restricting it to $\mathfrak{g}_{-1}$ induces an isomorphism $G_0\cong CO(\mathfrak{g}_{-1})\cong CO(p,q)$.

(3) The exponential map defines a diffeomorphism from $\mathfrak{g}_1$ onto a closed normal subgroup $P_+\subset P$. The quotient P/P_+ is isomorphic to G_0. Any element $g\in P$ can

be uniquely written as $g_0 \exp(Z)$ for $g_0 \in G_0$ and $Z \in \mathfrak{g}_1$, and P is the semidirect product of the subgroup G_0 and the normal subgroup P_+.

PROOF. (1) By definition any $g \in P$ stabilizes the line generated by the first basis vector e_1, so the first column of the matrix must have the claimed form. By orthogonality, it also stabilizes $e_1^\perp$, which is the subspace generated by $e_1, \dots, e_{m+1}$, so there must be zero in the last row of the second column. Knowing this, one can directly expand the defining equation $A^tSA = S$, to conclude that the matrix must have the claimed form. The statement about G_0 is then obvious.

(2) For $g \in G_0$, the map $\mathrm{Ad}(g)$ is conjugation by the matrix g, so the fact that the grading is preserved follows immediately from the block form. If the entries of the matrix are λ and C, then the adjoint action on $\mathfrak{g}_{-1}$ is given by $X \mapsto \lambda^{-1}CX$. Since $\lambda C = \lambda' C'$ happens if and only if $\lambda = \lambda'$ and $C = C'$ or $\lambda = -\lambda'$ and $C = -C'$, this restriction of the adjoint action induces an injective homomorphism from G_0 to $CO(\mathfrak{g}_{-1})$. On the other hand, any conformal orthogonal map ϕ can be written as $|\det(\phi)|^{\frac{1}{m}}(|\det(\phi)|^{\frac{-1}{m}}\phi)$ and $|\det(\phi)|^{\frac{-1}{m}}\phi$ is orthogonal, so we conclude that $G_0 \cong CO(\mathfrak{g}_{-1})$ via the restriction of the adjoint action.

(3) For $Z \in \mathfrak{g}_1$, $\exp(Z)$ is just the matrix exponential of the corresponding matrix, i.e.

$$\exp(Z) = \begin{pmatrix} 1 & Z & -\frac{1}{2}\langle Z^t, Z^t\rangle \\ 0 & \mathrm{id} & -I_{p,q}Z^t \\ 0 & 0 & 1 \end{pmatrix}.$$

Hence, $Z \mapsto \exp(Z)$ is a diffeomorphism onto the closed subgroup corresponding to the matrices for which $\lambda = 1$ and $C = \mathrm{id}$. Multiplying by the element of G_0 with entries λ and C, we obtain

$$g_0 \exp(Z) = \begin{pmatrix} \lambda & \lambda Z & -\frac{\lambda}{2}\langle Z^t, Z^t\rangle \\ 0 & C & -CI_{p,q}Z^t \\ 0 & 0 & \lambda^{-1} \end{pmatrix}.$$

Evidently, any element of P can be written in this form, and the two factors can be seen from the matrix, so uniqueness of the expression $g = g_0 \exp(Z)$ follows, too. Finally,

$$g_0 \exp(Z) g_0' \exp(Z') = g_0 g_0' \exp(\mathrm{Ad}(g_0')(Z)) \exp(Z') = g_0 g_0' \exp(\mathrm{Ad}(g_0')(Z) + Z'),$$

where in the last step we used that $\mathfrak{g}_1$ is commutative. This immediately implies that P_+ is normal in P, as well as the statement about the semidirect product. □

To continue the study of the Poincaré conformal group, we observe that there is a natural chart on the homogeneous space $S^{(p,q)}$. This comes from the subspace $\mathfrak{g}_{-1} \subset \mathfrak{g}$ which is complementary to the Lie subalgebra $\mathfrak{p}$. Denoting by $\mathcal{C}$ the null cone and by $\pi : \mathcal{C} \to S^{(p,q)}$ the projection, the chart is given by $X \mapsto \pi(\exp(X)e_1)$. As in the proof of the proposition, one immediately computes the matrix exponential to see that this is given by $X \mapsto \pi(1, X, -\frac{1}{2}\langle X, X\rangle) \in S^{(p,q)}$. It defines a diffeomorphism from $\mathfrak{g}_{-1}$ onto the dense open subset of $S^{(p,q)}$ consisting of the images of all points with nonzero first coordinate. From the definition of the inner product in (1.28), we see that $X \mapsto (1, X, -\frac{1}{2}\langle X, X\rangle)$ defines an isometric embedding of $\mathfrak{g}_{-1} = \mathbb{R}^m$ endowed with the standard inner product of signature (p,q). Hence, the standard chart is a conformal isometry onto its image, and we may naturally view $\mathbb{R}^m$ endowed with the flat conformal structure of signature (p,q) as an open dense subset of the Möbius space $S^{(p,q)}$.

Since the Poincaré conformal group P is the isotropy subgroup of $\pi(e_1)$, we can use the standard chart to view the action of any element of P as a conformal isometry between open neighborhoods of $0 \in \mathbb{R}^m$ which maps 0 to itself. As a consequence of existence and uniqueness of a normal Cartan connection for conformal structures, we shall see in 1.6.9 that P is actually the group of all germs at zero of such conformal isometries. Viewing $S^{(p,q)}$ as the homogeneous space G/P, the standard chart $\mathfrak{g}_{-1} \to S^{(p,q)}$ is naturally written as $X \mapsto \exp(X) \cdot o$, where o is the base point. Since $\exp(\mathrm{Ad}(g)X) = g \exp(X) g^{-1}$ and any $g \in G_0$ fixes o, we conclude from part (2) of the proposition that under the inclusion provided by the standard chart, the elements of G_0 exactly act as the linear conformal isometries of $\mathfrak{g}_{-1}$.

Let us finally compute how elements of P_+ act on $\mathfrak{g}_{-1}$ via the standard chart. To do this, one just has to apply the matrix of $\exp(Z)$ from the proof above to $\pi(1, X, -\frac{1}{2}\langle X, X\rangle)$, which gives

$$\pi(1 + ZX + \tfrac{1}{4}\langle X, X\rangle\langle Z^t, Z^t\rangle, X + \tfrac{1}{2}\langle X, X\rangle I_{p,q} Z^t, -\tfrac{1}{2}\langle X, X\rangle).$$

Hence, the corresponding map on $\mathfrak{g}_{-1}$ is only defined on the open subspace consisting of those X, for which the first component is nonzero. There the map is given by

$$X \mapsto \frac{1}{1 + ZX + \frac{1}{4}\langle Z^t, Z^t\rangle\langle X, X\rangle}(X + \tfrac{1}{2}\langle X, X\rangle I_{p,q} Z^t).$$

A straightforward computation shows that the first derivative of this mapping in a general point X in direction Y, is

$$\begin{aligned}-(ZY+\tfrac{1}{2}\langle Y, X\rangle\langle Z^t, Z^t\rangle)(1 + ZX + \tfrac{1}{4}\langle X, X\rangle\langle Z^t, Z^t\rangle)^{-2}(X + \tfrac{1}{2}\langle X, X\rangle I_{p,q} Z^t)\\ +(1 + ZX + \tfrac{1}{4}\langle Z^t, Z^t\rangle\langle X, X\rangle)^{-1}(Y + \langle Y, X\rangle I_{p,q} Z^t).\end{aligned}$$

In particular, the derivative at the origin is the identity, while the second derivative at the origin, evaluated at Y_1 and Y_2, is

$$\langle Y_1, Y_2\rangle I_{p,q} Z^t - (ZY_1)Y_2 - (ZY_2)Y_1.$$

From this, we conclude immediately that this second derivative at the origin determines Z, so we actually obtain an injective homomorphism from P to the second order jet group $G^2_m = J^2_0(\mathbb{R}^m, \mathbb{R}^m)^{\mathrm{inv}}_0$; cf. 1.2.8.

1.6.4. First prolongation and the canonical principal bundle. Let us consider a conformal structure of signature (p, q) on an m–dimensional manifold M, which we view as a first order G–structure with structure group $CO(p, q)$. In other words, we consider the conformal frame bundle $\mathcal{P} \to M$ with structure group $CO(p, q)$ endowed with the (restriction of the) soldering form $\theta \in \Omega^1(\mathcal{P}, \mathbb{R}^m)$; cf. 1.3.6. From 1.6.1 we know that in order to understand affine connections on this bundle, we have to look at the map $\partial : \mathbb{R}^{m*} \otimes \mathfrak{co}(p, q) \to \Lambda^2\mathbb{R}^{m*} \otimes \mathbb{R}^m$ induced by embedding $\mathfrak{co}(p, q)$ into $\mathbb{R}^{m*} \otimes \mathbb{R}^m$ and then alternating the $\mathbb{R}^{m*}$ components. Since $\mathfrak{co}(p, q)$ contains $\mathfrak{o}(p, q)$, we conclude from Example 1.6.1 that ∂ is surjective, and looking at the dimensions we see that the first prolongation $\mathfrak{co}(p, q)^{(1)}$ must be m–dimensional.

To compute this prolongation, consider a tensor $t = t^i_{jk} \in \mathfrak{co}(p, q)^{(1)}$ and the trace mapping

$$t = t^i_{jk} \mapsto \alpha = \alpha_k = t^i_{ik} \in \mathbb{R}^{m*}.$$

If t has trivial trace, then it lies in the first prolongation of $\mathfrak{o}(p, q)$ and thus $t = 0$, so taking the trace induces a linear isomorphism $\mathfrak{co}(p, q)^{(1)} \to \mathbb{R}^{m*}$. To get a concrete description of the prolongation, a bit of experimenting leads to the right guess for a tensor $t = t^i_{jk}$ with the right symmetries and trace α_k:

$$\alpha = \alpha_k \mapsto t = t^i_{jk} = \tfrac{1}{m}(\delta^i_j \alpha_k + \delta^i_k \alpha_j - g_{jk} g^{i\ell} \alpha_\ell),$$

where g_{ij} is the standard pseudo–metric and g^{ij} its inverse. Notice that the choice of the metric in its conformal class does not matter since we always put it together with its inverse. This also provides a $CO(p,q)$–equivariant split of the map ∂. From part (1) of Theorem 1.6.1 we then conclude that there are torsion–free affine connections on $\mathcal{P} \to M$, and any such connection is called a *Weyl connection.*

There is a simpler way to see these facts using the grading $\mathfrak{g} = \mathfrak{g}_{-1} \oplus \mathfrak{g}_0 \oplus \mathfrak{g}_1$ introduced in 1.6.3. We have seen that the adjoint action of elements of $\mathfrak{g}_0$ respects this grading and restricting them to $\mathfrak{g}_{-1}$ gives an isomorphism $\mathfrak{g}_0 \cong \mathfrak{co}(p,q)$. On the other hand, by the grading property, for an element $Z \in \mathfrak{g}_1$ the adjoint action $\operatorname{ad}(Z)$ maps $\mathfrak{g}_{-1}$ to $\mathfrak{g}_0$, so we can consider this as a homomorphism $\mathfrak{g}_1 \to \mathfrak{g}^*_{-1} \otimes \mathfrak{g}_0$. By the formula (1.29) from 1.6.3 for the brackets, this homomorphism is injective. Moreover, by the Jacobi identity, we have

$$0 = [Z,[X,Y]] = [[Z,X],Y] - [[Z,Y],X]$$

for all $X, Y \in \mathfrak{g}_{-1}$, which implies that it has values in the first prolongation $\mathfrak{g}_0^{(1)}$. Since $\mathfrak{g}_1$ is m–dimensional, we see that $\mathfrak{g}_1 \cong \mathfrak{g}_0^{(1)}$, which directly gives an explicit description of the prolongation.

To reformulate what we have done so far in this language, consider the groups $G_0 \subset P \subset G$ from the last subsection. By part (2) of Proposition 1.6.3, the adjoint action on $\mathfrak{g}_{-1}$ identifies G_0 with the group $CO(p,q)$ of linear conformal isometries of $\mathfrak{g}_{-1}$. Thus, we may view the conformal frame bundle as the principal G_0–bundle $p_0 : \mathcal{G}_0 \to M$ of all conformal linear maps from $\mathfrak{g}_{-1}$ to tangent spaces of M. The soldering form θ can be considered as an element of $\Omega^1(\mathcal{G}_0, \mathfrak{g}_{-1})$ and its equivariancy reads as $(r^{g_0})^*\theta = \operatorname{Ad}(g_0)^{-1} \circ \theta$ for all $g_0 \in G_0$. From above we know that there exist torsion–free affine connections on $\mathcal{G}_0$ of the form $\theta + \gamma \in \Omega^1(\mathcal{G}_0, \mathfrak{g}_{-1} \oplus \mathfrak{g}_0)$, and for any point $u \in \mathcal{G}_0$, the map $(\theta + \gamma)(u)$ is a linear isomorphism $T_u\mathcal{G}_0 \to \mathfrak{g}_{-1} \oplus \mathfrak{g}_0$.

For $u \in \mathcal{G}_0$ we now define $\mathcal{G}_u$ to be the space of all linear isomorphisms $\phi = \phi_{-1} + \phi_0 : T_u\mathcal{G}_0 \to \mathfrak{g}_{-1} \oplus \mathfrak{g}_0$ such that $\phi_{-1} = \theta(u)$, $\phi_0(\zeta_{(a,A)})(u) = (a, A)$ for all $(a, A) \in \mathfrak{g}_0$ and such that

$$d\theta(u)(\xi,\eta) + [\phi_0(\xi), \phi_{-1}(\eta)] + [\phi_{-1}(\xi), \phi_0(\eta)] = 0. \tag{1.30}$$

Otherwise put, $\mathcal{G}_u$ consists exactly of the values in u of all Weyl connections on $\mathcal{G}_0$, with the condition (1.30) expressing torsion freeness. We define $\mathcal{G}$ to be the disjoint union of the $\mathcal{G}_u$ for $u \in \mathcal{G}_0$, and by $\pi : \mathcal{G} \to \mathcal{G}_0$ we denote the obvious projection. Finally, we put $p := p_0 \circ \pi : \mathcal{G} \to M$.

LEMMA 1.6.4. *$p : \mathcal{G} \to M$ is a principal bundle with structure group P.*

PROOF. To describe the spaces $\mathcal{G}_u$, we fix a point (ϕ_{-1}, ϕ_0). As in 1.6.1, any other point in $\mathcal{G}_u$ must have the form $(\phi_{-1}, \phi_0 + \psi \circ \phi_{-1})$ for a linear map $\psi : \mathfrak{g}_{-1} \to \mathfrak{g}_0$. This element satisfies (1.30) if and only if $\psi \in \mathfrak{g}_0^{(1)}$, i.e. $\psi = \operatorname{ad}(Z)$ for some $Z \in \mathfrak{g}_1$. Equivalently, if $\phi \in \mathcal{G}_u$ is fixed, then mapping $Z \in \mathfrak{g}_1$ to $(\hat{\phi}_{-1}, \hat{\phi}_0)$, where $\hat{\phi}_{-1} := \phi_{-1}$ and $\hat{\phi}_0(\xi) := \phi_0(\xi) + [Z, \phi_{-1}(\xi)]$, defines a bijection between $\mathfrak{g}_1$ and $\mathcal{G}_u$.

Choosing a Weyl connection on $\mathcal{G}_0$, one obtains a distinguished representative in each $\mathcal{G}_u$ and hence a bijection $\mathcal{G}_0 \times \mathfrak{g}_1 \to \mathcal{G}$. Declaring this to be a diffeomorphism, we obtain a smooth structure on $\mathcal{G}$, which is immediately seen to be independent of the choice of the Weyl connection. This makes $\mathcal{G} \to \mathcal{G}_0$ into a trivial fiber bundle and thus $\mathcal{G} \to M$ into a fiber bundle.

Hence, it remains to construct a right P action on $\mathcal{G}$, which is free and transitive on each fiber. Using the explicit description of P from Proposition 1.6.3, one immediately verifies that $\mathrm{Ad}(g)(\mathfrak{g}_1) \subset \mathfrak{g}_1$ for any $g \in P$. Consequently, restricting the adjoint action of G to the subgroup P induces an action $\underline{\mathrm{Ad}}$ of P on the space $\mathfrak{g}/\mathfrak{g}_1$, which may be identified as $\mathfrak{g}_{-1} \oplus \mathfrak{g}_0$ as a vector space. Otherwise put, $\underline{\mathrm{Ad}}(g)(A)$ is simply obtained by taking the component of $\mathrm{Ad}(g)(A)$ in $\mathfrak{g}_{-1} \oplus \mathfrak{g}_0$. Clearly, the projection onto the $\mathfrak{g}_{-1}$–part is equivariant for the action $\underline{\mathrm{Ad}}$ on $\mathfrak{g}_{-1}$ introduced before, i.e. the action obtained from the identification $\mathfrak{g}_{-1} \cong \mathfrak{g}/\mathfrak{p}$.

By part (3) of Proposition 1.6.3, any element $g \in P$ can be uniquely written as $g_0 \exp(Z)$. Using this, we define for $\phi \in \mathcal{G}_u$ a map $\phi \cdot g : T_{u \cdot g_0}\mathcal{G}_0 \to \mathfrak{g}_{-1} \oplus \mathfrak{g}_0$ by

$$\phi \cdot g = \underline{\mathrm{Ad}}(g^{-1}) \circ \phi \circ Tr^{g_0^{-1}},$$

where $r^{g_0^{-1}}$ denotes the principal right action of g_0^{-1} on $\mathcal{G}_0$. By definition, $\phi \cdot gh = (\phi \cdot g) \cdot h$. To see that $\phi \cdot g \in \mathcal{G}_{u \cdot g_0}$, we proceed as follows. There is a Weyl connection $\theta + \gamma$ on $\mathcal{G}_0$ whose value in u equals ϕ, and equivariancy of $\theta + \gamma$ implies $(\theta + \gamma)(u \cdot g_0) = \mathrm{Ad}(g_0^{-1}) \circ \phi \circ Tr^{g_0^{-1}}$. For $g = g_0 \exp(Z)$ we obtain $\underline{\mathrm{Ad}}(g^{-1}) = e^{\underline{\mathrm{ad}}(-Z)} \circ \underline{\mathrm{Ad}}(g_0^{-1})$, where $\underline{\mathrm{ad}}$ is the infinitesimal action corresponding to $\underline{\mathrm{Ad}}$. Since $g_0 \in G_0$, the adjoint action $\mathrm{Ad}(g_0)$ preserves the decomposition $\mathfrak{g} = \mathfrak{g}_{-1} \oplus \mathfrak{g}_0 \oplus \mathfrak{g}_1$, so $\underline{\mathrm{Ad}}(g_0) = \mathrm{Ad}(g_0)$. On the other hand, we clearly have $\underline{\mathrm{ad}}(-Z)^2 = 0$ (since $\mathrm{ad}(Z)^2$ maps all of $\mathfrak{g}$ to $\mathfrak{g}_1$), whence $e^{\underline{\mathrm{ad}}(-Z)} = \mathrm{id} - \mathrm{ad}(Z)$. This implies that for $\phi \in \mathcal{G}_u$ we have $\phi \cdot g \in \mathcal{G}_{u \cdot g_0}$, so we really have defined a right P–action $\mathcal{G} \times P \to \mathcal{G}$.

If $\phi \cdot g = \phi$, then $u \cdot g_0 = u$ and thus $g_0 = e$; but then we get $\mathrm{ad}(Z) = 0$, so $Z = 0$ and thus $g = e$, so the action is free. On the other hand, if $\phi \in \mathcal{G}_u$ and $\tilde{\phi} \in \mathcal{G}$ lies in the same fiber of $\mathcal{G} \to M$, then $\tilde{\phi} \in \mathcal{G}_{u \cdot g_0}$ for some $g_0 \in G_0$. But then also $\phi \cdot g_0 = \mathrm{Ad}(g_0^{-1}) \circ \phi \circ Tr^{g_0^{-1}} \in \mathcal{G}_{u \cdot g_0}$, so from above we know that there is an element $Z \in \mathfrak{g}_1$ such that $\tilde{\phi} = \phi \cdot g_0 - \mathrm{ad}(Z) \circ (\phi \cdot g_0) = \phi \cdot (g_0 \exp(Z))$. □

Notice that Weyl connections on $\mathcal{G}_0$ are by construction in bijective correspondence with global G_0–equivariant smooth sections of $\mathcal{G} \to \mathcal{G}_0$. For two global smooth sections of this bundle, there is a smooth function $Z : \mathcal{G}_0 \to \mathfrak{g}_1$ such that $\phi'(u) = \phi(u) \cdot \exp(Z(u))$, and if both ϕ and ϕ' are G_0–equivariant, then so is the function Z. Hence, Z corresponds to a smooth section of the bundle $\mathcal{G}_0 \times_{G_0} \mathfrak{g}_1$ which, in view of the duality between $\mathfrak{g}_1$ and $\mathfrak{g}_{-1}$, is canonically isomorphic to T^*M. We conclude that the space of all Weyl connections on $\mathcal{G}_0$ is an affine space modelled on the vector space $\Omega^1(M)$ of all one–forms on M. Looking at the definition of the bracket $\mathfrak{g}_1 \times \mathfrak{g}_{-1} \to \mathfrak{g}_0$, one sees that in terms of the induced connections on the tangent bundle TM, two connections ∇ and $\hat{\nabla}$ whose difference corresponds to $\Upsilon \in \Omega^1(M)$ are explicitly related as

$$\hat{\nabla}_\xi \eta = \nabla_\xi \eta + \Upsilon(\xi)\eta + \Upsilon(\eta)\xi - \gamma(\xi, \eta)\gamma^{-1}(\Upsilon).$$

Here γ is any metric from the conformal class, which we also view as an isomorphism $\gamma : TM \to T^*M$. The expression depends only on the conformal class, since it contains γ and γ^{-1} simultaneously. In abstract index notation, this reads as $\hat{\nabla}_a \eta^b = \nabla_a \eta^b + \Upsilon_a \eta^b + \Upsilon_c \eta^c \delta_a^b - \gamma_{ac} \eta^c \gamma^{bd} \Upsilon_d$, which, as before, depends only on the conformal class of γ.

Now we can complete the first step in the prolongation.

PROPOSITION 1.6.4. *Let $(p_0 : \mathcal{G}_0 \to M, \theta)$ be a principal G_0–bundle describing a conformal structure on M, let $\pi : \mathcal{G} \to \mathcal{G}_0$ be the bundle constructed above, and let $p = p_0 \circ \pi : \mathcal{G} \to M$.*

(1) The principal G_0–bundle $p_0 : \mathcal{G}_0 \to M$ admits a Weyl connection $\theta + \gamma$, i.e. a torsion–free affine connection, and the space of all such connections is in bijective correspondence with the set of all G_0–equivariant sections of $\pi : \mathcal{G} \to \mathcal{G}_0$.

(2) There is a canonical form $\theta_{-1} + \theta_0 \in \Omega^1(\mathcal{G}, \mathfrak{g}_{-1} \oplus \mathfrak{g}_0)$ on the P–principal bundle $p : \mathcal{G} \to M$. This form is P–equivariant and it reproduces the generators of fundamental vector fields, i.e.

$$(r^g)^*(\theta_{-1} + \theta_0) = \underline{\mathrm{Ad}}(g^{-1}) \circ (\theta_{-1} + \theta_0) \text{ for all } g \in P,$$
$$\theta_{-1}(\zeta_B) = 0 \text{ and } \theta_0(\zeta_B) = B + \mathfrak{g}_1 \in \mathfrak{p}/\mathfrak{g}_1 \cong \mathfrak{g}_0 \text{ for all } B \in \mathfrak{p}.$$

The component θ_{-1} is the pullback $\pi^*\theta$ of the soldering form $\theta \in \Omega^1(\mathcal{G}_0, \mathfrak{g}_{-1})$. Finally, the canonical form satisfies the structure equation

$$d\theta_{-1}(\xi, \eta) + [\theta_0(\xi), \theta_{-1}(\eta)] + [\theta_{-1}(\xi), \theta_0(\eta)] = 0$$

for all $\xi, \eta \in T\mathcal{G}$.

PROOF. We have already proved (1) above. The canonical form is defined by

$$(\theta_{-1} + \theta_0)(\phi)(\xi) := (\phi_{-1} + \phi_0)(T\pi \cdot \xi).$$

Clearly, this defines a smooth form and since ϕ is a linear isomorphism, the kernel of $(\theta_{-1} + \theta_0)(\phi)$ is exactly the vertical bundle of $\pi : \mathcal{G} \to \mathcal{G}_0$. Further, since by definition we have $\phi \cdot g = \underline{\mathrm{Ad}}(g^{-1}) \circ \phi \circ Tr^{g_0^{-1}}$, we get

$$(r^g)^*(\theta_{-1} + \theta_0)(\phi)(\xi) = (\theta_{-1} + \theta_0)(\phi \cdot g)(T\pi \cdot Tr^g \cdot \xi) = (\phi \cdot g)(Tr^{g_0} \cdot T\pi \cdot \xi).$$

This immediately implies $(r^g)^*(\theta_{-1} + \theta_0) = \underline{\mathrm{Ad}}(g^{-1}) \circ (\theta_{-1} + \theta_0)$ for all $g \in P$. For $B \in \mathfrak{g}_0$ let $\zeta_B^{\mathcal{G}}$ and $\zeta_B^{\mathcal{G}_0}$ be the corresponding fundamental vector fields. Then G_0–equivariancy of π implies that $\zeta_B^{\mathcal{G}_0}(\pi(\phi)) = T\pi \cdot \zeta_B^{\mathcal{G}}(\phi)$. On the other hand, fundamental fields with generators in $\mathfrak{g}_1$ lie in the kernel of $T\pi$, and the statement on fundamental vector fields follows. The structure equation is just a way to express the torsion–freeness condition in the definition of $\mathcal{G}$. □

1.6.5. The bundle of scales. Although it is not strictly necessary for the construction of the canonical Cartan connection, we take a short detour here to give an alternative description of Weyl connections. This also leads to a different interpretation of the principal bundle $\mathcal{G} \to M$. Since a Weyl connection by definition is a (torsion–free) connection on $\mathcal{G}_0 \to M$, we obtain an induced linear connection on any vector bundle associated to $\mathcal{G}_0$. We have seen in Proposition 1.6.4 above that Weyl connections are in bijective correspondence with equivariant sections of $\pi : \mathcal{G} \to \mathcal{G}_0$. Using this, the space of Weyl connections becomes an affine space modelled on $\Omega^1(M)$. We can then describe the effect of such an affine change both in the picture of the principal connection on $\mathcal{G}_0$ and in the picture of the induced linear connection on an associated bundle in terms of the Lie algebra structure of $\mathfrak{g}$.

COROLLARY 1.6.5. *In terms of G_0–equivariant sections $\sigma : \mathcal{G}_0 \to \mathcal{G}$, the affine structure on the space of Weyl connections is given by $\hat{\sigma}(u) = \sigma(u) \cdot \exp(\Upsilon(u))$ for the equivariant function $\Upsilon : \mathcal{G}_0 \to \mathfrak{g}_1$ corresponding to $\Upsilon \in \Omega^1(M)$.*

The difference of the corresponding principal connection forms on $\mathcal{G}_0$ is given by

$$(\hat{\gamma} - \gamma)(u)(\xi) = [\theta(\xi), \Upsilon(u)]$$

for $u \in \mathcal{G}_0$ and $\xi \in T_u\mathcal{G}_0$.

For a representation V of G_0, the induced covariant derivatives on the associated bundle $\mathcal{G}_0 \times_{G_0} V$ are related as

$$(\hat{\nabla}_X - \nabla_X)s(u) = [X, \Upsilon(u)] \cdot s(u).$$

Here we view a section of $\mathcal{G}_0 \times_{G_0} V$ as an equivariant function $s : \mathcal{G}_0 \to V$ and a vector field on $\mathcal{G}_0$ as an equivariant function $X : \mathcal{G}_0 \to \mathfrak{g}_{-1}$, and the dot denotes the action of $\mathfrak{g}_0$ on V.

Now recall that $G_0 \cong CO(\mathfrak{g}_{-1})$ via the adjoint action, where the conformal class on $\mathfrak{g}_{-1} \cong \mathbb{R}^m$ is the one of the standard inner product of signature (p, q). We have described explicitly the elements of G_0 in part (1) of Proposition 1.6.3. Denoting the entries of the matrix by $\lambda \in \mathbb{R} \setminus 0$ and $C \in O(p, q)$ as introduced there, the action of $\mathfrak{g}_{-1}$ is given by $X \mapsto \lambda CX$. In particular, the elements with $\lambda = \pm 1$ correspond to a closed normal subgroup $G_0' \subset G_0$, which is isomorphic to $O(\mathfrak{g}_{-1}) \cong O(p, q)$ via the adjoint action. In Lie theoretic terms, this is the semisimple part of the reductive group G_0. Recalling that we still have to identify each matrix with its negative, we see that $G_0/G_0' \cong \mathbb{R}_+$. Starting with a conformal structure on M described by $(p_0 : \mathcal{G}_0 \to M, \theta)$, we may therefore form the bundle $\mathcal{L} := \mathcal{G}_0/G_0' \cong \mathcal{G}_0 \times_{G_0} (G_0/G_0')$. This is a principal bundle with structure group $\mathbb{R}_+$ over M. In view of the result of the proposition below, $\mathcal{L}$ is called the *bundle of scales* of the given conformal structure.

PROPOSITION 1.6.5. *Let $\mathcal{L} \to M$ be the bundle of scales associated to a conformal structure $(p_0 : \mathcal{G}_0 \to M, \theta)$.*

(1) There is a natural bundle map $\mathcal{L} \to S^2T^*M$ which restricts to a bijection onto the ray subbundle formed by the metrics in the conformal class. Hence, metrics in the conformal class are in bijective correspondence with global smooth sections of $\mathcal{L}$, so the principal bundle $\mathcal{L}$ is trivial, but there is no natural trivialization.

(2) Mapping a Weyl connection on $\mathcal{G}_0$ to the induced connection on $\mathcal{L}$ gives rise to a bijection between Weyl connections on $\mathcal{G}_0$ and principal connections on $\mathcal{L}$. In this picture, the Levi–Civita connections of metrics in the conformal class correspond to the flat connections induced by global sections.

PROOF. (1) The element of G_0 represented by $(\pm\lambda, \pm\text{id})$ acts on $\mathfrak{g}_{-1}$ by multiplication by λ, so it acts on the inner product on $\mathfrak{g}_{-1}$ by multiplication by λ^2. Hence, we can identify G_0/G_0' with the conformal class of the inner product on $\mathfrak{g}_{-1}$ and passing to associated bundles, the first claim follows. The rest of (1) is then obvious.

(2) We may naturally identify G_0/G_0' with the subgroup of G_0 formed by all elements of the form $\exp(tE)$ where $E \in \mathfrak{g}_0$ is the grading element from 1.6.3 and $t \in \mathbb{R}$. Starting with a Weyl connection $\theta + \gamma$ on $\mathcal{G}_0$, we obtain an induced principal connection on $\mathcal{L}$. The connection form $\gamma_{\mathcal{L}}$ of this connection is given by the E–component of γ. In other words, the central component of γ is G_0'–invariant and thus descends to $\mathcal{L}$.

From the corollary above we know that changing to another Weyl connection $\hat{\gamma}$, we get $\hat{\gamma}(u)(\xi) = \gamma(u)(\xi) + [\theta(\xi), \Upsilon(u)]$ for some smooth equivariant function

$\Upsilon : \mathcal{G}_0 \to \mathfrak{g}_1$. From 1.6.3 we know that taking the central part of the bracket defines a nondegenerate pairing between $\mathfrak{g}_{-1}$ and $\mathfrak{g}_1$. Thus, the mapping which associates to each Weyl connection the induced connection on $\mathcal{L}$ is injective. Since this mapping is given pointwise over M and the corresponding spaces of elements of connections in question have the same dimensions, we have established the claimed bijective correspondence.

Since the injection $\mathcal{L} \hookrightarrow S^2T^*M$ from (1) is induced by a G_0–equivariant map on the standard fibers, it is parallel for connections induced by the same principal connection on $\mathcal{G}_0$. If we start with the Levi–Civita connection of a metric in the conformal class, then this implies that the corresponding global section of $\mathcal{L}$ is parallel for the induced connection. But this characterizes the flat connection determined by a global section. □

This result suggests a conceptual way towards various special classes of Weyl connections. General Weyl connections correspond to G_0–equivariant sections of $\pi : \mathcal{G} \to \mathcal{G}_0$ and to arbitrary principal connections of $\mathcal{L}$. The role of the metrics in the conformal class may be illuminated in another way. The subgroup of G_0 formed by all elements of the form $\exp(tE)$ is the center $Z(G_0)$, and evidently $G_0/Z(G_0) \cong O(p,q)$. Now we can form $\mathcal{G}_0/Z(G_0)$, which is a principal bundle over M with structure group $G_0' \cong O(p,q)$. Sections of $\mathcal{L} \to M$ are in bijective correspondence with G_0'–equivariant sections of $\mathcal{G}_0 \to \mathcal{G}_0/Z(G_0)$, i.e. reductions of $\mathcal{G}_0$ to the structure group G_0'. These reductions correspond to metrics in the conformal class, and by construction, the Weyl–connection associated to a section of $\mathcal{L}$ induces a principal connection on $\mathcal{G}_0/Z(G_0)$. Torsion freeness implies that this must be the Levi–Civita connection of the chosen metric in the conformal class. The situation is illustrated in the following picture:

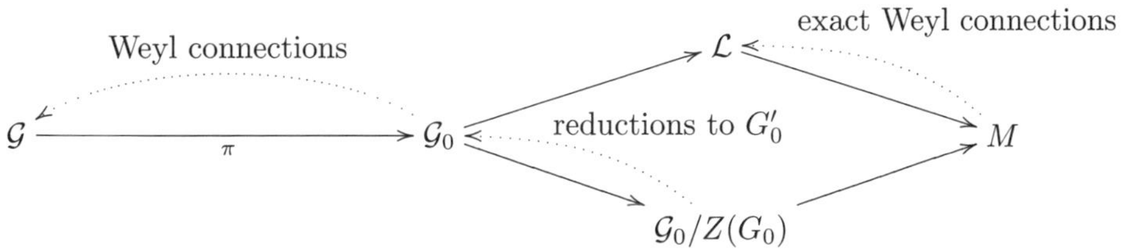

We can also specialize the formula for the change of Weyl connections in the corollary above to the change between two Levi–Civita connections. Starting from a global section σ of $\mathcal{L}$, any other such section has the form $\hat{\sigma} = \sigma h$ for a positive smooth function h on M. From the relation between $\mathcal{L}$ and the bundle of metrics observed in the proof, we see that this corresponds to rescaling from the metric g corresponding to σ to $\hat{g} = h^2 g$. To compute the one form Υ, let us write ∇ and $\hat{\nabla}$ for the induced connections on $\mathcal{L}$. Then for all vectors ξ in TM we get

$$\hat{\nabla}_\xi \sigma = \hat{\nabla}_\xi h^{-1}\hat{\sigma} = -h^{-2}dh(\xi)h\sigma + 0 = -d(\log h)(\xi)\sigma,$$

so this is the change of the central component of the principal connection form of the Weyl connection. But the entire change of γ is given by $X \mapsto [X, \Upsilon]$, and in view of the explicit formula for such a bracket in 1.6.3, this means $\Upsilon = d(\log h) = h^{-1}dh$. Hence, these special connections form an affine space over the space of exact one–forms, whence they are called *exact Weyl connections.* As we have noted, the connections on $\mathcal{L}$ induced by exact Weyl–connections are flat, or equivalently, the central components of the curvatures of the exact Weyl connections vanish. The

Weyl connections corresponding to flat connections on $\mathcal{L}$ are called *closed* Weyl connections. Of course, any closed Weyl connection is locally exact, but this is not true globally in general.

To conclude, we note that the bundle $\mathcal{L}$ of scales can be used to give an alternative description and/or construction of the Cartan bundle $\mathcal{G}$. Note that $\mathcal{L}$ was defined directly from the conformal structure, without using prolongation. Now there is a natural bundle $Q\mathcal{L} \to M$ of all elements of principal connections on $\mathcal{L}$, i.e. principal connections on $\mathcal{L}$ are in bijective correspondence with smooth sections of this connection bundle. The pullback $p_0^*(Q\mathcal{L})$ is a fiber bundle of $\mathcal{G}_0$ whose elements are pairs $(u, \gamma_{\mathcal{L}}(p_0(u)))$ of a frame $u \in \mathcal{G}_0$ and the restriction of the connection form $\gamma_{\mathcal{L}}$ to the fiber $\mathcal{L}_x$ over $x = p_0(u)$. Since the points in $\mathcal{G}_u$ are exactly the values of Weyl connections in u, mapping such a connection to the restriction of the induced connection to $\mathcal{L}_x$ gives us an isomorphism $\mathcal{G} \cong p_0^*(Q\mathcal{L})$. From this point of view it is obvious that there is an isomorphism between global sections of $Q\mathcal{L} \to M$, i.e. principal connections on $\mathcal{L}$ and global G_0–equivariant sections of $\mathcal{G} \cong p_0^*(Q\mathcal{L}) \to \mathcal{G}_0$, i.e. Weyl connections.

To use this as an alternative description for $\mathcal{G}$, one can directly define a right action of P on $p_0^*(Q\mathcal{L})$, which makes that bundle into a principal bundle. Then the first step in the prolongation can be completed by defining the θ_0–component of the canonical form as the connection form of the unique Weyl connection inducing a given principal connection on $\mathcal{L}$. We do not go into the details of this construction here.

1.6.6. Some algebra. The second and last step in the prolongation of conformal structures is the construction of a normal Cartan connection on the principal P–bundle $\mathcal{G} \to M$. Formally, this step is similar to the first one. Due to an algebraic vanishing result, we do not, however, get a bigger bundle but only the next canonical form, which is a Cartan connection. The choice of the appropriate normalization condition, which was trivial in the first step, becomes crucial now. Recall that any P–invariant curvature restriction defines a full subcategory of Cartan geometries (see 1.5.2), and we will establish an equivalence of that subcategory with the category of conformal geometries (as standard G–structures). In particular, this allows us to use invariant constructions with Cartan connections in order to get conformally invariant objects and operations.

Since we already have the canonical form $\theta_{-1}+\theta_0 \in \Omega^1(\mathcal{G}, \mathfrak{g}_{-1}\oplus\mathfrak{g}_0)$, the obvious idea is to look at the possible extensions of this form to a Cartan connection. The missing ingredient is a one–from $\phi_1 \in \Omega^1(\mathcal{G}_0, \mathfrak{g}_1)$ with certain properties. Having made such a choice, we can compute the curvature of the Cartan connection and decompose it according to the decomposition $\mathfrak{g} = \mathfrak{g}_{-1} \oplus \mathfrak{g}_0 \oplus \mathfrak{g}_1$. Since this is a grading of $\mathfrak{g}$, the result evaluated on tangent vectors ξ and η reads as

$$\begin{aligned}
&d\theta_{-1}(\xi,\eta) + [\theta_0(\xi), \theta_{-1}(\eta)] + [\theta_{-1}(\xi), \theta_0(\eta)],\\
&d\theta_0(\xi,\eta) + [\phi_1(\xi), \theta_{-1}(\eta)] + [\theta_0(\xi), \theta_0(\eta)] + [\theta_{-1}(\xi), \phi_1(\eta)],\\
&d\phi_1(\xi,\eta) + [\theta_0(\xi), \phi_1(\eta)] + [\phi_1(\xi), \theta_{-1}(\eta)].
\end{aligned}$$

The $\mathfrak{g}_{-1}$–component is independent of the choice of ϕ_1, and its vanishing was the normalization condition used in the first step of the prolongation. The $\mathfrak{g}_1$–component depends not only on ϕ_1 but also on its exterior derivative, so working with this component, we would have to solve differential equations.

The $\mathfrak{g}_0$–component looks very promising, however. Its value in a point depends only on the value of ϕ_1 in that point, and working in one point, things reduce to linear algebra. Hence, we fix a point $u \in \mathcal{G}$. We have to choose a linear map $\phi_1 : T_u\mathcal{P} \to \mathfrak{g}_1$. To reproduce the generators of fundamental vector fields, we have to require that $\phi_1(\zeta_Z(u)) = Z$ for all $Z \in \mathfrak{g}_1$ while $\phi_1(\zeta_A(u)) = 0$ for $A \in \mathfrak{g}_0$. This also makes sure that $\phi := (\theta_{-1} + \theta_0)(u) + \phi_1$ is a linear isomorphism. Such maps evidently exist. If ϕ and $\hat{\phi}$ are two such maps, then they agree on the vertical subbundle, so there is a unique map $\psi : \mathfrak{g}_{-1} \to \mathfrak{g}_1$ such that

$$\hat{\phi}_1(\xi) = \phi_1(\xi) + \psi(\theta_{-1}(\xi)).$$

Given a choice of ϕ, consider the map $\kappa_\phi : \Lambda^2\mathfrak{g}_{-1} \to \mathfrak{g}_0$, which is given by $\kappa_\phi(X, Y) = d\theta_0(u)(\phi^{-1}(X), \phi^{-1}(Y))$. This represents the $\mathfrak{g}_0$–component of the curvature evaluated on the projections of $\phi^{-1}(X)$ and $\phi^{-1}(Y)$, and by horizontality, it completely determines this $\mathfrak{g}_0$–component. Passing from ϕ to $\hat{\phi}$, the change of κ can computed from equivariancy. Given the map ψ describing the change, we define $\partial\psi : \Lambda^2\mathfrak{g}_{-1} \to \mathfrak{g}_0$ by $\partial\psi(X, Y) = [X, \psi(Y)] + [\psi(X), Y]$. Then the change is explicitly given by $\kappa_{\hat{\phi}}(X, Y) = \kappa_\phi(X, Y) + \partial\psi(X, Y)$.

As before, normalizing amounts to finding a subspace which is complementary to $\operatorname{im}(\partial)$. Having done this, there is a unique possible value for κ in this subspace and the possible choices for isomorphisms ϕ leading to this value are affine over $\ker(\partial)$. Our maps κ_ϕ are elements of $\Lambda^2\mathbb{R}^{m*} \otimes \mathfrak{co}(p, q) \subset \Lambda^2\mathbb{R}^{m*} \otimes \mathbb{R}^{m*} \otimes \mathbb{R}^m$. The simplest invariant subspaces in there are the kernels of the two possible contractions. For the contraction inside of $\mathfrak{co}(p, q)$ the resulting kernel is to big to be a candidate for a complement. The other contraction is similar to the one mapping Riemann curvature to Ricci curvature.

To analyze the latter contraction, we use an abstract index notation, which will directly provide the abstract index versions of the computations for tensors later on. We write a typical element of $\mathfrak{g}_{-1}$ as X^i, elements of $\mathfrak{g}_1$ as Z_i, and so on. In particular, the map κ_ϕ has the form $\kappa_{ij}{}^k{}_\ell$ in this convention, with k and ℓ representing the $\mathfrak{co}(p, q)$–part. The contraction in question reads as $\kappa_{ki}{}^k{}_j$. In accordance to established notation in conformal geometry, we will denote the change caused by ψ by P_{ij} using the convention that $\mathsf{P}(X)_j = X^i\mathsf{P}_{ij}$.

LEMMA 1.6.6. *Let $\kappa = \kappa_{ij}{}^k{}_\ell$ be an arbitrary element of $\Lambda^2\mathbb{R}^{m*} \otimes \mathfrak{co}(p, q)$. Then there is a unique element $\mathsf{P}_{ij} \in \mathbb{R}^{m*} \otimes \mathbb{R}^{m*}$ such that $L := \kappa + \partial(\mathsf{P})$ satisfies $L_{ki}{}^k{}_j = 0$. Explicitly, we have*

$$\mathsf{P}_{ij} = \tfrac{-1}{m-2}\Big(\kappa_{ki}{}^k{}_j + \tfrac{1}{m}(\kappa_{kj}{}^k{}_i - \kappa_{ki}{}^k{}_j) - \tfrac{1}{2(m-1)}g^{ab}\kappa_{ka}{}^k{}_b g_{ij}\Big), \tag{1.31}$$

where g_{ij} and g^{ij} denote the standard inner product of signature (p, q) and its inverse.

PROOF. The first step is to compute $\partial\mathsf{P}$, which needs the bracket $\mathfrak{g}_{-1} \times \mathfrak{g}_1 \to \mathfrak{g}_0$. Using the formulae (1.29) from 1.6.3, one immediately concludes that

$$[X, Z]^i_j = X^iZ_j - g^{ia}Z_ag_{jb}X^b + X^aZ_a\delta^i_j. \tag{1.32}$$

Inserting $Z_j = Y^a\mathsf{P}_{aj}$, alternating in X and Y and renaming indices we conclude

$$(\partial\mathsf{P})_{ij}{}^k{}_\ell = \mathsf{P}_{j\ell}\delta^k_i - \mathsf{P}_{i\ell}\delta^k_j - g^{ka}\mathsf{P}_{ja}g_{\ell i} + g^{ka}\mathsf{P}_{ia}g_{\ell j} + (\mathsf{P}_{ji} - \mathsf{P}_{ij})\delta^k_\ell. \tag{1.33}$$

Applying the contraction, we obtain

$$(\partial \mathsf{P})_{ki}{}^k{}_j = (m-1)\mathsf{P}_{ij} - \mathsf{P}_{ji} + g^{ab}\mathsf{P}_{ab}g_{ij}.$$

In particular, if $\mathsf{P}_{ij} = \lambda g_{ij}$ for some $\lambda \in \mathbb{R}$, then this gives $(2m-2)\lambda g_{ij}$. For P_{ij} symmetric and tracefree, we obtain $(m-2)\mathsf{P}_{ij}$ and for P_{ij} skew symmetric we get $m\mathsf{P}_{ij}$. Since $m \geq 3$, these three factors are all nonzero. Now any tensor $\mathsf{P}_{ij} \in \mathbb{R}^{m*} \otimes \mathbb{R}^{m*}$ may be uniquely decomposed as

$$\mathsf{P}_{ij} = \tfrac{1}{m} g^{ab}\mathsf{P}_{ab}g_{ij} + \tfrac{1}{2}(\mathsf{P}_{ij} + \mathsf{P}_{ji} - \tfrac{2}{m} g^{ab}\mathsf{P}_{ab}g_{ij}) + \tfrac{1}{2}(\mathsf{P}_{ij} - \mathsf{P}_{ji})$$

into trace part, symmetric tracefree part and skew part. Applying the composition of the contraction and ∂, each of these parts is multiplied by a nonzero factor. Hence the map $\mathsf{P}_{ij} \mapsto (\partial \mathsf{P})_{ki}{}^k{}_j$ is bijective. Given κ we therefore find a unique P such that $\kappa_{ki}{}^k{}_j = -(\partial \mathsf{P})_{ki}{}^k{}_j$, which completes the proof of the first part.

Obtaining the explicit formula is now easy. We have to decompose $-\kappa_{ki}{}^k{}_j(u)$ into trace–part, tracefree symmetric part and skew symmetric part. Then we have to multiply these parts by $\frac{1}{2m-2}$, $\frac{1}{m-2}$, and $\frac{1}{m}$, and add them up. This immediately leads to the claimed formula. □

1.6.7. The conformal Cartan connection. Having the necessary algebraic background at hand, we formulate the normalization condition and establish the existence of a canonical Cartan connection.

Definition 1.6.7. A Cartan connection ω of type (G, P) is called *normal* if and only if it is torsion free, and the $\mathfrak{g}_0$–component κ_0 of its curvature function has the property that $(\kappa_0)_{ki}{}^k{}_j = 0$.

Theorem 1.6.7. *Let $(p_0 : \mathcal{G}_0 \to M, \theta)$ be a conformal structure on a smooth manifold M of dimension $m \geq 3$. Let $p : \mathcal{G} \to M$ be the P–principal bundle constructed in 1.6.4. Then the canonical form $\theta_{-1} \oplus \theta_0$ on $\mathcal{G}$ from Proposition 1.6.4 uniquely extends to a normal Cartan connection ω on $\mathcal{G}$.*

This construction induces an equivalence of categories between first order G–structures with structure group $CO(p, q)$ and normal Cartan geometries of type (G, P).

Proof. Using Lemma 1.6.6, we see that for each point $u \in \mathcal{G}$, there is a unique linear map $\omega_1(u) : T_u\mathcal{G} \to \mathfrak{g}_1$ such that $\omega(u) := \theta_{-1}(u) \oplus \theta_0(u) \oplus \omega_1(u) : T_u\mathcal{G} \to \mathfrak{g}$ is a linear isomorphism which reproduces the generators of fundamental vector fields and has the property that the associated map $\kappa_{\omega(u)}$ lies in the kernel of the Ricci type contraction. Looking at the point $u \cdot g$ for $g \in P$, we may consider

$$\phi := \mathrm{Ad}(g^{-1}) \circ \omega(u) \circ Tr^{g^{-1}} : T_{u \cdot g}\mathcal{G} \to \mathfrak{g}.$$

Equivariancy of $\theta_{-1} \oplus \theta_0$ as proved in part (2) of Proposition 1.6.4 shows that the components of this map in $\mathfrak{g}_{-1} \oplus \mathfrak{g}_0$ coincide with $\theta_{-1}(u \cdot g) \oplus \theta_0(u \cdot g)$. Using this and the definition of κ_ϕ, we conclude that writing $g \in P$ as $g_0 \exp(Z)$ according to Proposition 1.6.3, we obtain

$$\kappa_\phi(\mathrm{Ad}(g_0)(X), \mathrm{Ad}(g_0)(Y)) = \mathrm{Ad}(g_0) \circ \kappa_{\omega(u)}(X, Y) \circ \mathrm{Ad}(g_0)^{-1}.$$

But $\mathrm{Ad}(g_0)$ is just the standard action of $CO(p, q)$ on $\mathbb{R}^m$, so κ_ϕ lies in the kernel of the (by construction $CO(p, q)$–equivariant) Ricci type contraction. Uniqueness in the lemma shows that $\phi = \omega(u \cdot g)$.

To prove existence of the canonical normal Cartan connection, it only remains to show that the $\omega(u)$ fit together to define a smooth one–form $\omega \in \Omega^1(\mathcal{G}, \mathfrak{g})$.

Remarkably, this follows from the existence of some Cartan connection extending $\theta_{-1} \oplus \theta_0$. To get such an extension, we use Weyl connections. By part (3) of Proposition 1.6.4, there are Weyl connections on $p_0 : \mathcal{G}_0 \to M$ and they are in bijective correspondence with G_0–equivariant sections $\sigma : \mathcal{G}_0 \to \mathcal{G}$. From the construction of the tautological form it follows immediately that the affine connection corresponding to σ is given by $\sigma^*(\theta_{-1} + \theta_0)$. For a point $u_0 \in \mathcal{G}_0$ and $u = \sigma(u_0)$, the tangent space $T_u\mathcal{G}$ splits as $\operatorname{im}(T_{u_0}\sigma) \oplus \ker(T_u\pi)$. Hence, in this point we can uniquely extend $\theta_{-1} \oplus \theta_0$ to a form ω^σ by requiring that the $\mathfrak{g}_1$–component vanishes on the first summand and reproduces generators of fundamental vector fields. As in the proof of Lemma 1.5.15, one verifies that this extends to a smooth Cartan connection ω^σ on $\mathcal{G}$ which extends $\theta_{-1} \oplus \theta_0$.

Now in each point $u \in \mathcal{G}$, we can write

$$\omega_1(u)(\xi) = \omega_1^\sigma(u)(\xi) + \mathsf{P}(u)(\theta_{-1}(\xi)).$$

By the uniqueness part of the lemma, we can obtain $\mathsf{P}(u)$ by inserting $\kappa^\sigma(u)$ into equation (1.31). Here we have to view the $\mathfrak{g}_0$–component κ^σ of the curvature function of ω^σ as acting on $\mathbb{R}^m$ via θ_{-1}. Anyway, the result evidently depends smoothly on u, so smoothness of ω^σ implies smoothness of ω.

We just sketch the proof of the remaining claims leaving out some straightforward verifications since we will prove a much more general version of this theorem later. Let us assume that $(p : \tilde{\mathcal{G}} \to M, \tilde{\omega})$ is a Cartan geometry of type (G, P). Via the identification $TM = \tilde{\mathcal{G}} \times_P (\mathfrak{g}/\mathfrak{p})$, the P–invariant conformal class of inner products on $\mathfrak{g}/\mathfrak{p} \cong \mathfrak{g}_{-1}$ induces a conformal structure on M. Denoting by $\mathcal{G}_0$ the conformal frame bundle as before, we obtain a homomorphism $\Phi_0 : \tilde{\mathcal{G}} \to \mathcal{G}_0$ over the projection $P \to G_0$ which covers the identity on M and such that $\tilde{\omega}_{-1} = \Phi^*\theta_{-1}$.

Next, for $u \in \tilde{\mathcal{G}}$ the component $(\tilde{\omega}_{-1} + \tilde{\omega}_0)(u)$ induces a linear isomorphism $T_u\tilde{\mathcal{G}}/P_+ \to \mathfrak{g}_{-1} \oplus \mathfrak{g}_0$. On the other hand, $T_u\Phi_0$ induces a linear isomorphism $T_u\tilde{\mathcal{G}}/P_+ \to T_{\Phi_0(u)}\mathcal{G}_0$, so together these two maps induce a linear isomorphism $T_{\Phi_0(u)}\mathcal{G}_0 \to \mathfrak{g}_{-1} \oplus \mathfrak{g}_0$. By construction, the $\mathfrak{g}_{-1}$–component of this isomorphism is $\theta_{-1}(\Phi_0(u))$ and it reproduces the generators of fundamental vector fields. Assuming that $\tilde{\omega}$ is torsion free, this defines an element in $\mathcal{G}$ lying over $\Phi_0(u) \in \mathcal{G}_0$. Hence, we obtain a smooth map $\Phi : \tilde{\mathcal{G}} \to \mathcal{G}$ lifting Φ_0. One easily checks that by construction this map is P–equivariant and thus an isomorphism of principal bundles and from the definition of the tautological form we conclude that $\Phi^*(\theta_{-1} + \theta_0) = \tilde{\omega}_{-1} + \tilde{\omega}_0$. Thus, $(\Phi^{-1})^*\tilde{\omega}$ is an extension of $\theta_{-1} + \theta_0$ to a Cartan connection on $\mathcal{G}$. The curvature function of this Cartan connection is given by $\tilde{\kappa} \circ \Phi^{-1}$. Hence, if we assume that $\tilde{\omega}$ is normal, then also $(\Phi^{-1})^*\tilde{\omega}$ is normal, and hence equals ω.

The first part of this argument shows that any Cartan geometry of type (G, P) has an underlying first order G_0–structure, and clearly this defines a functor from Cartan geometries to G_0–structures. The last part of the argument shows that any morphism of first order G–structures uniquely lifts to a morphism of Cartan geometries, which implies that the functors constructed above establish an equivalence of categories. □

REMARK 1.6.7. The relation to extension functors for Cartan geometries as developed in 1.5.15 that was briefly hinted at in the proof, can be exploited much further. Using this theory, we could have obtained a quicker way towards existence and uniqueness of normal Cartan connections. This would, however, be less transparent from a geometric point of view, so we preferred to use the more complicated traditional approach. The general construction of normal Cartan connections for parabolic geometries in Section 3.1 is closer to the arguments using extension functors, so for comparison we sketch this line of argument here.

Consider the affine extension B of $CO(p,q)$. In view of part (2) of Proposition 1.6.3, we can do this starting from G_0 acting on $\mathfrak{g}_{-1}$ via the adjoint action. Hence, we can view elements of B as pairs (g_0, X) and the multiplication given by $(g_0, X)(g_0', X') := (g_0 g_0', \mathrm{Ad}(g_0')(X) + X')$. Using that $\mathfrak{g}_{-1}$ is abelian, one immediately verifies that $(g_0, X) \mapsto g_0 \exp(X)$ defines a homomorphism $\phi : B \to G$. Following Example 1.5.16, we consider $i = \phi|_{G_0} : G_0 \to P$, which is just the standard inclusion, and $\alpha := \phi' : \mathfrak{b} \to \mathfrak{g}$, which is the inclusion of the subalgebra $\mathfrak{g}_{-1} \oplus \mathfrak{g}_0$ into $\mathfrak{g}$. In particular, α induces a linear isomorphism $\mathfrak{b}/\mathfrak{g}_0 \to \mathfrak{g}/\mathfrak{p}$.

A Cartan geometry of type (B, G_0) on a manifold M is then a conformal structure $(p : \mathcal{G}_0 \to M, \theta_{-1})$ as considered in the last few subsections together with a principal connection γ on $\mathcal{G}_0$. Now we can form the extended bundle $\mathcal{G} = \mathcal{G}_0 \times_i P$, which is a principal P–bundle. By Lemma 1.5.15, there is a unique Cartan connection $\tilde{\omega}$ on $\mathcal{G}$ such that $j^*\tilde{\omega} = \alpha \circ (\theta_{-1} \oplus \gamma)$, where $j : \mathcal{G}_0 \to \mathcal{G}$ is the natural map. Since α is a homomorphism of Lie algebras, Proposition 1.5.16 shows that pulling back the curvature of $\tilde{\omega}$ one obtains the curvature of $\theta_{-1} \oplus \gamma$. In particular, if we use a Weyl connection for γ, then $\tilde{\omega}$ is torsion free.

Next, we can modify $\tilde{\omega}$ by adding a horizontal, P–equivariant one–form P with values in $\mathfrak{g}_1$. Using Lemma 1.6.6 one easily shows that there is a unique choice for P such that $\tilde{\omega} + \mathsf{P}$ is normal. In particular, the result is independent of the Weyl connection we started with. The form P can even be computed explicitly from the curvature of $\tilde{\omega}$ and hence from the curvature of $\theta_{-1} \oplus \gamma$ using formula (1.31). Hence, we have extended a given conformal structure to a normal Cartan geometry of type (G, P).

If $f : M \to \tilde{M}$ is a conformal isometry, then it lifts to a morphism Φ_0 of G_0–structures. Starting with a Weyl connection on $\tilde{\mathcal{G}}_0 \to \tilde{M}$ and considering the pullback of this connection on $\mathcal{G}_0$, the map Φ_0 becomes a morphism of Cartan geometries of type (B, G_0). The functoriality result in Theorem 1.5.15 shows that Φ_0 extends to a morphism of Cartan geometries of type (G, P). Normalizing on one side and pulling back the result to the other side, the pullback is normal and hence has to coincide with the normalized Cartan connection on the other side. Hence, f lifts to a morphism between the associated normal Cartan geometries, and the equivalence of categories follows.

1.6.8. The curvature of the conformal Cartan connection. In the proof of Theorem 1.6.7, we have constructed the conformal Cartan connection ω from the Cartan connections ω^σ associated to a Weyl–structure σ. We want to use this correspondence to determine the curvature of ω. By construction, $\sigma^*\omega^\sigma$ has values in $\mathfrak{g}_{-1} \oplus \mathfrak{g}_0$ and coincides with the (affine) Weyl connection associated to σ. Moreover, the curvature function κ^σ has vanishing $\mathfrak{g}_{-1}$–component, and from this and the definitions of the curvatures, we conclude that $\sigma^*\kappa^\sigma$ is exactly the curvature of the Weyl connection. In particular, the component κ_0^σ, interpreted as a section of the associated bundle $\Lambda^2 T^*M \otimes L(TM, TM)$, is exactly the classical curvature $R_{ij}{}^k{}_\ell$ of the induced linear connection ∇^σ on the tangent bundle TM. Thus, the trace $R_{ki}{}^k{}_j$ is exactly the Ricci curvature R_{ij}, and from above we see that the deformation P_{ij} to the conformal Cartan connection, viewed as a section of $T^*M \otimes T^*M$ is given by

$$\mathsf{P}_{ij} = \tfrac{-1}{m-2}\Big(R_{ij} + \tfrac{1}{m}(R_{ji} - R_{ij}) - \tfrac{1}{2(m-1)} g^{ab} R_{ab} g_{ij}\Big).$$

In particular, taking the Levi–Civita connection of a metric g_{ij} in the conformal class, the Bianchi identity implies symmetry of the Ricci curvature R_{ij}, and $R := g^{ab}R_{ab}$ is exactly the scalar curvature. In this case we obtain

$$\mathsf{P}_{ij} = \tfrac{-1}{m-2}(R_{ij} + \tfrac{1}{2(m-1)} R g_{ij}),$$

which (up to a sign) is the usual *Rho tensor* associated to a metric in the conformal class.

Next, we know that the curvature component κ_0 of the conformal Cartan connection is given by $\kappa^\sigma + \partial(\mathsf{P})$, where P is the Rho tensor as above. Of course, it suffices to compute this for a Levi–Civita connection. In that case, P_{ij} is symmetric, and using formula (1.33) from 1.6.6 for $\partial\mathsf{P}$, we conclude that

$$\kappa_{ij}{}^k{}_\ell = R_{ij}{}^k{}_\ell + \mathsf{P}_{j\ell}\delta_i^k - \mathsf{P}_{i\ell}\delta_j^k - g^{ka}\mathsf{P}_{ja}g_{i\ell} + g^{ka}\mathsf{P}_{ia}g_{j\ell}.$$

By construction, we know that $\kappa_{ki}{}^k{}_j = 0$, so let us look at the other possible trace, $\kappa_{ij}{}^k{}_k$. The corresponding trace for the curvature of the Levi–Civita connection vanishes, since this has values in $\mathfrak{o}(TM)$, while for the remaining terms one obtains $2(\mathsf{P}_{ij} - \mathsf{P}_{ji})$, which vanishes by symmetry of the P–tensor of a Levi–Civita connection. So we see that κ_0 is totally tracefree. On the other hand, κ_0 is obtained from the Riemann curvature by adding trace terms, which shows that we may alternatively characterize κ_0 as the totally tracefree part of the Riemann curvature. Hence, κ_0 is exactly the classical *Weyl curvature*, which is well known to be conformally invariant. Since also in the case of a general Weyl connection we obtain κ_0 from R by adding trace terms, we conclude that the Weyl curvature coincides with the totally tracefree part of the curvature of any Weyl connection.

What remains to be done is interpreting the $\mathfrak{g}_1$–component of the curvature of the conformal Cartan connection. This is a bit more subtle, since this component does not define an equivariant function and thus a geometric object unless the $\mathfrak{g}_0$–component of κ vanishes. A good way to circumvent this problem is to choose a Weyl connection and consider the pullback of this component along the corresponding section $\sigma : \mathcal{G}_0 \to \mathcal{G}$. Equivariancy of κ implies G_0–equivariancy of the individual components κ_0 and κ_1 of the curvature function, so $\kappa_1 \circ \sigma : \mathcal{G}_0 \to \Lambda^2(\mathfrak{g}/\mathfrak{p})^* \otimes \mathfrak{g}_1$ is a G_0–equivariant function and hence corresponds to a T^*M–valued two–form Y^σ on M, called the *Cotton–York tensor* associated to the given Weyl connection. Of course, equivariancy of κ implies that κ_1 is determined by its restriction to the image of σ.

Consider a vector field $\xi \in \mathfrak{X}(M)$ and let $\xi^{\mathrm{hor}} \in \mathfrak{X}(\mathcal{G}_0)$ be its horizontal lift with respect to the chosen Weyl–connection. Then by construction $\sigma^*\omega_1(\xi^{\mathrm{hor}}) = \mathsf{P}(\theta_{-1}(T\sigma \cdot \xi^{\mathrm{hor}}))$, where P is viewed as a function with values in $L(\mathfrak{g}_{-1}, \mathfrak{g}_1)$. But the right–hand side of this may as well be viewed as the G_0–equivariant function $\mathcal{G}_0 \to \mathfrak{g}_1$ representing the section $\mathsf{P}(\xi)$ of T^*M obtained by interpreting P as a T^*M–valued one–form. Since $\omega_0(\xi^{\mathrm{hor}})$ vanishes along the image of σ we conclude that, along this image, the $\mathfrak{g}_1$–component of $[\omega(T\sigma\cdot\xi^{\mathrm{hor}}), \omega(T\sigma\cdot\eta^{\mathrm{hor}})]$ vanishes identically. From the definition of the curvature we thus conclude that $Y^\sigma(\xi, \eta) \in \Gamma(T^*M)$ is represented by the function $\mathcal{G}_0 \to \mathfrak{g}_1$ given by

$$d\sigma^*\omega_1(\xi^{\mathrm{hor}}, \eta^{\mathrm{hor}}) = \xi^{\mathrm{hor}}\cdot\mathsf{P}(\eta) - \eta^{\mathrm{hor}}\cdot\mathsf{P}(\xi) - \mathsf{P}([\xi,\eta]) = \nabla^\sigma_\xi(\mathsf{P}(\eta)) - \nabla^\sigma_\eta(\mathsf{P}(\xi)) - \mathsf{P}([\xi,\eta]).$$

By definition, this is the covariant exterior derivative with respect to ∇^σ of $\mathsf{P} \in \Omega^1(M, T^*M)$. Alternatively, using that any Weyl connection is torsion free, we may express the Lie bracket $[\xi, \eta]$ as $\nabla^\sigma_\xi \eta - \nabla^\sigma_\eta \xi$ (see 1.3.5) and conclude that $Y^\sigma(\xi, \eta) = (\nabla^\sigma \mathsf{P})(\xi, \eta) - (\nabla^\sigma \mathsf{P})(\eta, \xi)$, or in index notation $Y^\sigma_{ijk} = \nabla^\sigma_i \mathsf{P}_{jk} - \nabla^\sigma_j \mathsf{P}_{ik}$.

Collecting the information, we obtain the first two parts of the following result, which completely describes the relation between the curvature of the conformal Cartan connection and the curvature of any Weyl connection.

COROLLARY 1.6.8. *Let $(p_0 : \mathcal{G}_0 \to M, \theta)$ be a conformal structure on a smooth manifold of dimension $m \geq 3$, $p : \mathcal{G} \to M$ the prolongation, $\omega \in \Omega^1(\mathcal{G}, \mathfrak{g})$ the conformal Cartan connection and κ its curvature. Consider a Weyl connection on $\mathcal{G}_0$ and let $\sigma : \mathcal{G}_0 \to \mathcal{G}$ be the corresponding G_0–equivariant section. Let $\mathsf{P} = \mathsf{P}_{ij}$ be the Rho tensor of the Weyl connection, $R = R_{ij}{}^k{}_\ell$ its curvature, and $Y^\sigma = Y^\sigma_{ijk}$ its Cotton–York tensor.*

(1) The totally trace free part $W = W_{ij}{}^k{}_\ell$ of R represents the $\mathfrak{g}_0$–component of the curvature of ω and thus is independent of the choice of the Weyl connection. It is explicitly given by

$$W_{ij}{}^k{}_\ell = R_{ij}{}^k{}_\ell + \mathsf{P}_{j\ell}\delta^k_i - \mathsf{P}_{i\ell}\delta^k_j - g^{ka}\mathsf{P}_{ja}g_{i\ell} + g^{ka}\mathsf{P}_{ia}g_{j\ell} - (\mathsf{P}_{ij} - \mathsf{P}_{ji})\delta^k_\ell,$$

for any metric g_{ij} in the conformal class with inverse g^{ij}.

(2) The Cotton–York tensor Y^σ corresponds to the pullback along σ of the $\mathfrak{g}_1$–component of the curvature of ω. It is the covariant exterior derivative of $\mathsf{P} \in \Omega^1(M, T^*M)$, or equivalently the alternation of $\nabla\mathsf{P}$, i.e.

$$Y^\sigma_{ijk} = \nabla^\sigma_i \mathsf{P}_{jk} - \nabla^\sigma_j \mathsf{P}_{ik}.$$

(3) In dimensions $m > 3$, the conformal structure is flat if and only if W vanishes. For $m = 3$, one always has $W = 0$, the Cotton–York tensor Y^σ is independent of the choice of the Weyl connection, and the geometry is flat if and only if Y vanishes.

PROOF. It remains to prove part (3). First we observe that W vanishes if and only if κ_0 vanishes, and since κ_{-1} is always identically zero, vanishing of κ_0 implies that $\kappa_1 : \mathcal{G} \to \Lambda^2(\mathfrak{g}/\mathfrak{p})^* \otimes \mathfrak{g}_1$ is P–equivariant. But on the right–hand side, the subalgebra $\mathfrak{g}_1 \subset \mathfrak{p}$ acts trivially, so the subgroup $P_+ \subset P$ acts trivially. Thus, κ_1 descends to a G_0–equivariant function on $\mathcal{G}_0$, which clearly coincides with the pullback of κ_1 along any G_0–equivariant section σ. In conclusion, we see that vanishing of the Weyl tensor implies that the Cotton–York tensor is independent of the choice of the Weyl connection and thus a conformal invariant. In this case vanishing of the Cotton–York tensor Y is equivalent to vanishing of κ, which by Proposition 1.5.2 is equivalent to the given Cartan geometry being locally isomorphic to the homogeneous model $(G \to G/P, \omega_G)$. In view of the equivalence of the category of normal Cartan geometries with the category of conformal structures, this is equivalent to M being locally conformally flat.

Let us first consider the case $m = 3$. The curvatures of Levi–Civita connections have values in $\Lambda^2 T^*M \otimes \mathfrak{o}(TM)$, which, for $m = 3$, is a space of dimension nine. The trace map used for the normalization condition has values in $T^*M \otimes T^*M$, which also has dimension nine. For the standard basis $\{e_i\}$ of $\mathbb{R}^3$ with dual basis $\{e^i\}$ one may obtain the element $e^i \otimes e^j \in \mathbb{R}^{3*} \otimes \mathbb{R}^{3*}$ by applying the trace to $e^i \wedge e^k \otimes (e_k \otimes e^j - e_j \otimes e^k) \in \Lambda^2\mathbb{R}^{3*} \otimes \mathfrak{o}(3)$. Thus, the trace is surjective and hence a linear isomorphism by dimensional reasons. In particular, the totally tracefree part of the curvature of any Levi–Civita connection in three dimensions automatically vanishes, so the Weyl curvature is always trivial. From above, we conclude that the Cotton–York tensor Y is a well–defined conformal invariant, which is a complete obstruction to local conformal flatness.

To complete the proof, it remains to show that in dimensions $m > 3$ vanishing of the Weyl tensor implies vanishing of the Cotton–York tensor. This is an application of the Bianchi identity: Let us consider the Bianchi identity in the form (1.26) from the proof of Proposition 1.5.9. We know that $\kappa_{-1} = 0$ and by assumption $W = 0$ and thus $\kappa_0 = 0$. Since $\mathfrak{g}_{-1}$ is abelian and $\kappa_{-1} = 0$, vanishing of the $\mathfrak{g}_0$–component of the Bianchi identity is equivalent to vanishing of $\sum_{\text{cycl}}[\kappa_1(X,Y),Z]$. Let us again pass to an index notation, using the convention that $\kappa_1(X,Y)_k = X^iY^j\kappa_{ijk}$. Formula (1.32) for the bracket from the proof of Lemma 1.6.6 gives

$$[\kappa_1(X,Y),Z]^i_j = -Z^i\kappa_{k\ell j}X^kY^\ell + g^{ia}\kappa_{k\ell a}X^kY^\ell g_{jb}Z^b - Z^a\kappa_{k\ell a}X^kY^\ell\delta^i_j.$$

The first two summands form the tracefree part of this expression, while the last summand is pure trace, and of course the two parts have to vanish individually after forming the cyclic sum over the arguments. Renaming the indices of the entries to a, b, c, we see that the vanishing of the tracefree part of the cyclic sum is equivalent to vanishing of the cyclic sum over a, b, c of $-\delta^i_c\kappa_{abj} + g^{ik}\kappa_{abk}g_{jc}$. Forming this cyclic sum and contracting over i and c, we obtain the equation

$$0 = -m\kappa_{abj} - \kappa_{baj} - \kappa_{baj} + \kappa_{abj} + g^{ik}\kappa_{iak}g_{jb} + g^{ik}\kappa_{bik}g_{ja},$$

and using the κ is skew symmetric in the last two indices, this reduces to

$$0 = (3-m)\kappa_{abj} + g^{ik}(\kappa_{bik}g_{ja} - \kappa_{aik}g_{jb}).$$

Contracting with g^{jb}, we obtain $0 = (4-2m)g^{ik}\kappa_{aik}$. Since $m > 3$, this together with the above equation implies $0 = (3-m)\kappa_{abj}$, and thus vanishing of κ_1. □

REMARK 1.6.8. The approach to the construction of the canonical Cartan connection via extension functors discussed in Remark 1.6.7 above, has another interesting aspect. It may happen, that for some choice of Weyl connection, the extension functor directly produces the normal Cartan connection. This naturally defines a subclass of conformal structures which should be particularly well behaved. By equivariancy, the extended Cartan connection is normal if an only if it is normal along the image of j. Hence, starting with a Weyl connection ∇ with curvature $R_{ij}{}^k{}_\ell$, the extension functor directly produces a normal Cartan connection if and only if $R_{ki}{}^k{}_j = 0$. Hence, the subclass in question is exactly conformal structures admitting a Ricci flat Weyl connection. Using the Bianchi identity, this implies that the curvature has values in $\mathfrak{o}(n) \subset \mathfrak{co}(n)$. In the language introduced in 1.6.5, the given Weyl structure is closed and hence locally exact. Hence, we see that we obtain exactly the subclass of conformal structures, which locally contain Ricci flat representative metrics.

1.6.9. The Liouville theorem. There are various versions of what is called the Liouville theorem. A version for general Cartan geometries can be found in 1.5.2. This says that the automorphisms of the homogeneous model are given by left translations and local automorphisms between connected open subsets of the homogeneous model uniquely globalize. We can apply this to conformal structures using the equivalence to the category of normal Cartan geometries. Hence, the group of conformal isometries of the Möbius space $S^{(p,q)}$ is exactly the group $PO(p+$

$1, q+1)$ which acts on $S^{(p,q)}$ as described in 1.6.2. There we have also used the natural chart $\mathbb{R}^m \cong \mathfrak{g}_{-1} \to S^{(p,q)}$ defined by $X \mapsto \exp(X) \cdot o$. We have noted in 1.6.2 that this chart is a conformal diffeomorphism of $\mathbb{R}^m$ onto a dense subset of $S^{(p,q)}$.

In particular, we conclude that conformal isometry between connected open subsets of $\mathbb{R}^m$ comes, via the natural chart, from the action of a unique element $g \in G = PO(p+1, q+1)$. Using this and some facts we have already proved, we can now easily derive a complete description of the pseudo–group of local conformal isometries for the standard metric of signature (p, q) on $\mathbb{R}^m$. This result is also referred to as the Liouville theorem in the literature. The classical proof is rather involved. One has to completely describe the solutions of the overdetermined system of PDE's which characterizes a conformal isometry.

There are several obvious examples of conformal isometries of $\mathbb{R}^m$, in particular, translations, dilatations $X \mapsto \lambda X$ for $\lambda \in \mathbb{R} \setminus \{0\}$, and orthogonal linear maps $A \in O(p,q)$. Further, we have already noted in 1.6.2 that the inversion ϕ in the unit circle, defined by $\phi(X) = \frac{X}{\langle X,X\rangle}$ is a conformal isometry from the set of all points X such that $\langle X, X\rangle \neq 0$ onto its image. Using this we now formulate

COROLLARY 1.6.9 (Liouville theorem). *For $m = p + q \geq 3$ consider $\mathbb{R}^m$ with the conformal structure of signature (p,q) defined by the standard inner product of that signature. Then the pseudo–group of all local conformal isometries of $\mathbb{R}^m$ is generated by translations, dilatations, orthogonal linear maps and the inversion in the unit circle.*

PROOF. Since any local conformal isometry is induced by an element of $G = PO(p+1, q+1)$ it suffices to show that G is generated by elements which induce the maps listed in the theorem on $\mathbb{R}^m \cong \mathfrak{g}_{-1}$ via the natural chart. Since $\mathfrak{g}_{-1}$ is a commutative subalgebra of $\mathfrak{g}$, we see that for $X, Y \in \mathfrak{g}_{-1}$ we have $\exp(Y)\exp(X) = \exp(X+Y)$ in G. Hence, under the natural chart, translations are given by the action of $\exp(Y)$ for $Y \in \mathfrak{g}_{-1}$. In 1.6.3 we have seen that elements of $G_0 \subset P$ exactly correspond to conformal linear maps on $\mathbb{R}^m$, which are generated by orthogonal maps and dilatations.

It is also not hard to guess the element of G that induces the inversion in the unit circle. Consider the matrix

$$Q := \begin{pmatrix} 0 & 0 & -2 \\ 0 & \mathrm{id} & 0 \\ -\frac{1}{2} & 0 & 0 \end{pmatrix}.$$

It is immediately seen to be orthogonal with respect to the inner product S used in 1.6.2, so we may consider its class in $G = O(p+1, q+1)/\{\pm \mathrm{id}\}$. For $X \in \mathbb{R}^m$, this matrix maps $(1, X, -\frac{1}{2}\langle X, X\rangle)$ to $(\langle X, X\rangle, X, -1/2)$ and if $\langle X, X\rangle \neq 0$, the latter point lies on the line through $(1, \frac{X}{\langle X,X\rangle}, \frac{-1}{2\langle X,X\rangle})$. Thus, the class of Q implements the inversion. Note also that Q has determinant -1, so in case that G has two connected components, Q does not lie in the component of the identity. Further, one immediately verifies that the adjoint action by Q maps $\mathfrak{g}_{-1}$ to $\mathfrak{g}_1$ by a linear isomorphism. Thus, for $Z \in \mathfrak{g}_1$, the element $\exp(Z) \in G$ lies in the subgroup generated by the class of Q and all elements of the form $\exp(Y)$ for $Y \in \mathfrak{g}_{-1}$.

Putting things together, we see that the subgroup in question contains all elements of the form $\exp(Y)g_0\exp(Z)$ for $Y \in \mathfrak{g}_{-1}$, $g_0 \in G_0$ and $Z \in \mathfrak{g}_1$. By part (3) of Proposition 1.6.3, any element of P can be written as $g_0\exp(Z)$, so it contains all elements of the form $\exp(Y)g$ for $Y \in \mathfrak{g}_{-1}$ and $g \in P$. But the images of elements of the form $\exp(Y)$ contain an open neighborhood of o in G/P, so we see that the elements of the form $\exp(Y)g$ as above contain an open neighborhood of e in G. Since these elements also meet each connected component of G, they generate G. □

1.6.10. Invariant operators and invariants of conformal structures. Finding invariants of conformal structures and describing conformally invariant differential operators are two related problems, whose discussion in a more general setting will be among the main topics of volume two. Here we just sketch some basics to give the reader a rough impression of what is going on. We start by describing the problems in classical terms, i.e. without reference to Cartan geometries.

This classical formulation is phrased in Riemannian terms. For simplicity, let us avoid spinors and consider two tensor bundles E and F over a Riemannian manifold (M, g). Then we can use the metric g, its inverse, the Levi–Civita connection and its curvature, and the volume form corresponding to g to write down a differential operator $\Gamma(E) \to \Gamma(F)$. If we can replace all the ingredients determined by g by the corresponding data for a conformally equivalent metric without changing the resulting differential operator, then the operator is called *conformally invariant.* Similarly, the question of (local) invariants consists in using the above ingredients to write down a function which then should be independent of the choice of the metric from the conformal class.

To obtain conformally invariant operators, one has to use density bundles, which rarely are used in Riemannian geometry otherwise. Recall that for a smooth manifold M of dimension m and a number $\alpha \in \mathbb{R}$ the bundle of α–densities is the associated bundle to the full frame bundle $\mathcal{P}^1M$ with respect to the one–dimensional representation $A \mapsto |\det(A)|^{-\alpha}$ of $GL(m, \mathbb{R})$. The sections of the bundle of 1–densities are the right objects to integrate on nonorientable manifolds, while in the oriented case the bundle of 1–densities is canonically isomorphic to $\Lambda^m T^*M$. By construction, any density bundle is trivial, but there is no canonical trivialization. Any Riemannian metric g trivializes the bundle of 1–densities via the volume density, which is locally given by $\sqrt{|\det(g_{ij})|}$. Since this induces a trivialization of all density bundles, they usually are not useful in Riemannian geometry.

Changing from g to a conformally related metric, however, also changes trivialization of density bundles. The usual convention in conformal geometry is to define $\mathcal{E}[w]$ to be the bundle of $(-\frac{w}{m})$–densities, and to add the expression $[w]$ to any bundle in order to indicate the tensor product with $\mathcal{E}[w]$. Such tensor products are referred to as weighted bundles. Any choice of a metric g from the conformal class gives an identification of sections of $\mathcal{E}[w]$ with smooth functions on M but changing from g to $\hat{g} = \Omega^2 g$, these functions transform as $\hat{f} = \Omega^w f$. Otherwise put, the bundle $\mathcal{E}[w]$ may be interpreted as the associated bundle to the bundle $\mathcal{L}$ of scales of the conformal structure with respect to the representation $t \mapsto t^{-w}$ of $\mathbb{R}_+$. From 1.6.5 we know that rescaling $\hat{g} = \Omega^2 g$, the corresponding one–form Υ which describes the change of Weyl connections is given by $\Upsilon = d\log(\Omega)$, i.e. $\Upsilon_a = \Omega^{-1}\nabla_a\Omega$, and we have derived a formula for the change of the principal connection on $\mathcal{L}$ there. This immediately implies that for the covariant derivative of a section f of $\mathcal{E}[w]$, the change under a conformal rescaling as above reads as $\hat{\nabla}_a f = \nabla_a f + w\Upsilon_a f$.

The role of the density bundles is easy to understand in the picture of conformal structures as first order G–structures with structure group $CO(p,q)$. Representations of $CO(p,q)$ can be restricted to the semisimple part $O(p,q)$ and the center, which is isomorphic to $\mathbb{R}_+$. The point is now that the representation of the center may be varied, while the action of $O(p,q)$ remains fixed, and this exactly corresponds to tensoring with different density bundles.

The conformal structure itself may be viewed as a canonical section $\mathbf{g}_{ij}$ of $S^2T^*M[2]$ with inverse $\mathbf{g}^{ij}$, which is a section of $S^2TM[-2]$. Hence, similarly as in the case of Riemannian structures, one may raise and lower indices in conformal geometry, but at the expense of a weight.

Having the possibility to form conformally invariant contractions, we immediately get examples of conformal invariants. Namely, the Weyl tensor $W_{ij}{}^k{}_\ell$ is by construction conformally invariant, so taking any tensor power of W and forming a complete contraction, one obtains a conformally invariant density. Getting more general conformal invariants is a surprisingly hard problem, and only a few other examples are known classically.

A trivial example of a conformally invariant differential operator is given by the exterior derivative of differential forms, which is just the alternation of the covariant derivative. To obtain a slightly less trivial example, let us analyze first order operators on weighted vector fields. The formulae for the change of the Levi–Civita connection under a conformal rescaling from 1.6.4 and the formula for densities above immediately imply that rescaling $\hat{g} = \Omega^2 g$ implies that for a section $\xi = \xi^a$ of $TM[w]$ we obtain

$$\hat{\nabla}_a \xi^b = \nabla_a \xi^b + (w+1)\Upsilon_a \xi^b - \mathbf{g}_{ac}\xi^c \mathbf{g}^{bd}\Upsilon_d + \Upsilon_c \xi^c \delta^b_a$$

as a section of $T^*M \otimes TM[w]$. Tracing over the two free indices, we obtain $\hat{\nabla}_a \xi^a = \nabla_a \xi^a + (m+w)\Upsilon_a \xi^a$, so putting $w = -m$, we obtain a conformally invariant first order differential operator mapping sections of $TM[-m]$ to sections of $\mathcal{E}[-m]$. This is the conformally invariant divergence.

Another possibility is to lower the free upper index in the above equation to obtain

$$\mathbf{g}_{cb}\hat{\nabla}_a \xi^c = \mathbf{g}_{cb}\nabla_a \xi^c + (w+1)\Upsilon_a \mathbf{g}_{bc}\xi^c - \Upsilon_b \mathbf{g}_{ac}\xi^c + \Upsilon_c \xi^c \mathbf{g}_{ab}$$

as a section of $T^*M \otimes T^*M[w+2]$. Obviously, if we put $w = -2$, the deformation term is symmetric, so alternating defines a conformally invariant operator from sections of $TM[-2]$ to $\Lambda^2 T^*M$, which exactly corresponds to the exterior derivative under the identification $TM[-2] \cong T^*M$ induced by $\mathbf{g}_{ab}$. Finally, for $w = 0$, the symmetric tracefree part of the deformation vanishes, so projecting to the symmetric tracefree part gives a conformally invariant differential operator from $\mathfrak{X}(M)$ to $S^2_0 T^*M[2]$. This is exactly the conformal Killing operator, whose solutions are the infinitesimal conformal isometries.

The general theory of conformally invariant operators of first order looks very similar, as long as one restricts to bundles corresponding to irreducible representations of $CO(p,q)$. It is an exercise in representation theory to show that for any irreducible representation V of $O(p,q)$ the tensor product $\mathbb{R}^{m*} \otimes V$ decomposes into a direct sum of pairwise non–isomorphic irreducible representations. Further, one proves that for any of these components there is a unique choice of a weight w such that one obtains a first order conformally invariant operator from sections of $E[w]$ to sections of $F[w]$, where E and F are the bundles corresponding to V and the irreducible representation in question. This result is due to [**Feg79**]. We will prove the vast generalization of this result to all parabolic geometries due to [**SlSo04**] in volume two.

For higher order operators the situation becomes much more complicated, since in many cases one has to include curvature correction terms in order to get conformal invariance. We just mention the simplest and best known example of such an operator, namely the *conformal Laplacian* or *Yamabe operator*. This operator maps sections of $\mathcal{E}[\frac{2-m}{2}]$ to sections of $\mathcal{E}[\frac{-2-m}{2}]$ and is defined in terms of a chosen metric as $\Delta - \frac{2-m}{2}\mathsf{P}$, where $\Delta = \mathbf{g}^{ab}\nabla_a\nabla_b$ and $\mathsf{P} = \mathbf{g}^{ab}\mathsf{P}_{ab}$ is the contraction of the Rho tensor. Proving invariance of this operator directly is a nice exercise that will convince the reader that this direct approach gets out of hand for higher order operators very quickly.

The general calculus for Cartan connections introduced in Section 1.5 offers a completely different approach to this problem. Iterated fundamental derivatives provide an invariant way to capture arbitrarily high jets of sections of any natural bundle as sections of tensor powers of the adjoint tractor bundle and the given natural bundle. Passing to quotients can be used to construct invariant operators with values in simpler bundles. Doing this systematically in a much more general situation leads to the concept of BGG–sequences, which were originally introduced in [**ČSS01**] and will be one of the main objectives of volume two. Similarly, iterated fundamental derivatives of the Cartan curvature provide examples of conformally invariant sections of certain bundles, and constructing natural bundle maps (possibly nonlinear) from these to density bundles leads to conformally invariant densities.

1.6.11. Remarks on further developments. Let us conclude this section with a few remarks on generalizations. We have not explicitly used representation theory in our treatment of conformal structures. The main algebraic ingredients can, however, be proved very efficiently using tools from representation theory. In particular, the analysis of the maps in 1.6.4 and 1.6.6 which were both denoted by ∂ can be obtained as a corollary of the description of some Lie algebra cohomology groups. These groups are described by Kostant's version of the Bott–Borel–Weil theorem. Having switched to this language, one can deal with gradings of the form $\mathfrak{g} = \mathfrak{g}_{-1} \oplus \mathfrak{g}_0 \oplus \mathfrak{g}_1$ of any semisimple Lie algebra $\mathfrak{g}$ in a similar way: Given a group G with Lie algebra $\mathfrak{g}$, one defines subgroups $G_0 \subset P \subset G$ corresponding to the subalgebras $\mathfrak{g}_0 \subset \mathfrak{g}_0 \oplus \mathfrak{g}_1 \subset \mathfrak{g}$. The homomorphism $G_0 \to GL(\mathfrak{g}_{-1})$ defined by the adjoint action is always infinitesimally injective, so one has the notion of a first order G–structure with structure group G_0 on manifolds of dimension $\dim(\mathfrak{g}_{-1})$. Under a cohomological condition which is satisfied in almost all cases, one then obtains an equivalence of categories between such structures and normal Cartan geometries of type (G, P). These structures are called *almost Hermitian symmetric structures* or *AHS structures* in the literature, and a treatment of the canonical Cartan connections for such structures along the lines sketched above can be found in [**ČSS97b**].

Of course, one may try to prolong more general first order G–structures. While in our presentation we used a priori the Lie algebra $\mathfrak{g}$ with the grading, this may be bypassed. Namely, one may start with $\mathfrak{h} \subset \mathfrak{gl}(V)$ and consider $V \oplus \mathfrak{h} \oplus \mathfrak{h}^{(1)}$, where $\mathfrak{h}^{(1)}$ denotes the first prolongation as introduced in 1.6.1. We may view V and $\mathfrak{h}^{(1)}$ as abelian Lie algebras, and since both spaces are $\mathfrak{h}$ modules by construction, we get brackets $\mathfrak{h} \otimes V \to V$ and $\mathfrak{h} \otimes \mathfrak{h}^{(1)} \to \mathfrak{h}^{(1)}$. Finally, by definition, $\mathfrak{h}^{(1)}$ is a subspace of $L(V, \mathfrak{h})$ which can be used as the definition of a bracket $\mathfrak{h}^{(1)} \otimes V \to \mathfrak{h}$. While these brackets do not make $V \oplus \mathfrak{h} \oplus \mathfrak{h}^{(1)}$ into a graded Lie algebra in general, one can build a Lie group H^1 corresponding to the Lie algebra $\mathfrak{h} \oplus \mathfrak{h}^{(1)}$ as a semidirect product. Next, one chooses an appropriate complement to the image of ∂, and it is possible to develop a general version of the prolongation of first order G–structures in that style; see for example [**S64**]. The first prolongation can be viewed as a first order G–structure with structure group H^1 on the total space of the principal bundle one started with. Hence, one may prolong once more, and hope that this stops at some step and gives rise to a Cartan connection.

Unfortunately, there are general results which show that in many interesting cases, this cannot work. A result of S. Kobayashi and T. Nagano (see [**KN64**]) treats the case of a subalgebra $\mathfrak{h} \subset \mathfrak{gl}(V)$ which acts irreducibly on V. This means that there is no nontrivial $\mathfrak{h}$–invariant subspace in V. In this situation, there are exactly three possibilities: Either the first prolongation $\mathfrak{h}^{(1)}$ is trivial, so we are in the situation dealt with in 1.6.1. Secondly, we may have $V = \mathfrak{g}_{-1}$ and $\mathfrak{h} = \mathfrak{g}_0$ for a grading $\mathfrak{g} = \mathfrak{g}_{-1} \oplus \mathfrak{g}_0 \oplus \mathfrak{g}_1$ of a simple Lie algebra $\mathfrak{g}$. Then we are in the situation of AHS–structures discussed above. In all other cases, process of iterated prolongations as outlined above does not stop after finitely many steps, so there is no hope to get a Cartan connection on a finite–dimensional bundle in this way.

In spite of this result, there still is a number of interesting geometric structures that may equivalently be described as normal Cartan geometries. One way to view these structures is as subclasses of first order G–structures. Consider, for example, manifolds of dimension $2n + 1$ endowed with a complex subbundle E of rank n in the tangent bundle TM. Such structures may be equivalently viewed as first order G–structures with structure group a certain subgroup H of $GL(2n + 1, \mathbb{R})$. More precisely, one has to view $\mathbb{R}^{2n+1}$ as $\mathbb{C}^n \oplus \mathbb{R}$, and then define H to be the subgroup of those real linear automorphisms which preserve the subspace $\mathbb{C}^n$ and restrict to complex linear maps on this subspace. It is then obvious how to obtain a reduction of the frame bundle to the structure group H from the complex subbundle $E \subset TM$ and vice versa.

Following the theory of first order G–structures, the first step to understand this structure is to consider the structure function, which in this special case can be described as follows: Consider the quotient bundle TM/E and the canonical projection $q : TM \to TM/E$. For two sections $\xi, \eta \in \Gamma(E)$ consider $q([\xi, \eta]) \in \Gamma(TM/E)$. This is visibly bilinear over smooth functions, and thus defines a tensor $\mathcal{L} : E \times E \to TM/E$, which is equivalent to the structure function. One choice that leads to a Cartan geometry is then to require that for each $x \in M$ the value $\mathcal{L}_x$ is nondegenerate and compatible with the almost complex structure, which forces $\mathcal{L}_x$ to be the imaginary part of a nondegenerate Hermitian form. This leads to the definition of a nondegenerate partially integrable almost CR–structure, for which a canonical Cartan connection turns out to exist. In his pioneering work culminating in [**Tan79**], N. Tanaka showed that such an approach applies in to a large class of examples, arriving at an approach to parabolic geometries; see Appendix A for a survey.

Conceptually, it is preferable to take the different point of view of filtered manifolds. By definition, a filtered manifold is a smooth manifold M endowed with a decreasing filtration $TM = T^{-k}M \supset T^{-k+1}M \supset \cdots \supset T^{-1}M$ of the tangent bundle by smooth subbundles, which is compatible with the Lie bracket in the sense that for $\xi \in \Gamma(T^iM)$ and $\eta \in \Gamma(T^jM)$ we have $[\xi, \eta] \in \Gamma(T^{i+j}M)$. Here we agree that $T^\ell M = TM$ for all $\ell \leq -k$. Then one considers the associated graded bundle $\mathrm{gr}(TM) = \mathrm{gr}_{-k}TM \oplus \cdots \oplus \mathrm{gr}_{-1}(TM)$, where $\mathrm{gr}_i(TM) = T^iM/T^{i+1}M$. Similarly, as above, the Lie bracket of vector fields induces tensorial maps $\mathrm{gr}_i(TM) \times \mathrm{gr}_j(TM) \to \mathrm{gr}_{i+j}(TM)$ (where $\mathrm{gr}_\ell(TM) = 0$ for $\ell < -k$ and $\ell \geq 0$) which, for each $x \in M$, make $\mathrm{gr}(T_xM)$ into a nilpotent graded Lie algebra.

If one requires that all these Lie algebras are isomorphic to a fixed nilpotent graded Lie algebra $\mathfrak{n}$, then one naturally gets a frame bundle for $\mathrm{gr}(TM)$ with structure group the group $\mathrm{Aut}(\mathfrak{n})$ of automorphisms of the Lie algebra $\mathfrak{n}$. Now one can look at reductions of structure group of this frame bundle as a filtered analog of first order G–structures. This is the point of view of [**Mo93**], which constructs canonical normal Cartan connections for a wide variety of structures in this sense. In the example of almost CR–structures above, one would start by looking at manifolds of dimension $2n + 1$ with real rank $2n$ subbundles in the tangent bundle. Requiring that the induced algebraic bracket in each point is nondegenerate means exactly looking at contact structures. The associated graded $\mathrm{gr}(T_xM)$ in each point is then a Heisenberg algebra, and a compatible complex structure on the subbundle can clearly be interpreted as a reduction of structure group of $\mathrm{gr}(TM)$.

In this picture, parabolic geometries can be characterized as those structures in which the nilpotent graded Lie algebra $\mathfrak{n}$ is the nilradical of a parabolic subalgebra of a semisimple Lie algebra $\mathfrak{g}$, while the reduction of structure group is to a group with Lie algebra the Levi part of the parabolic. Equivalently, $\mathfrak{g}$ has to admit a grading of the form $\mathfrak{g} = \mathfrak{g}_{-k} \oplus \cdots \oplus \mathfrak{g}_k$ such that $\mathfrak{n} = \mathfrak{g}_{-k} \oplus \cdots \oplus \mathfrak{g}_{-1}$ and we need a reduction to a structure group G_0 with Lie algebra $\mathfrak{g}_0$. Such a reduction of structure group of the associated graded $\mathrm{gr}(TM)$ is called a regular infinitesimal flag structure.

We will prove in Section 3.1 that in the situations coming from parabolic subalgebras there is a categorical equivalence between regular infinitesimal flag structures and Cartan geometries which satisfy a normalization condition on their curvature. The approach presented there shows that the algebraic properties of the normalization condition are the crucial ingredient for this equivalence result, while the constuction of the Cartan bundle plays only very little role. More traditional prolongation procedures which put more effort into the construction of the Cartan bundle and then obtain the Cartan connection from tautological forms are sketched in Appendix A.

CHAPTER 2

Semisimple Lie algebras and Lie groups

Going through 1.4 one notices that understanding the geometry of homogeneous spaces in most cases boils down to understanding (finite–dimensional) representations of Lie groups and equivariant mappings between such representations. Moreover, from 1.5 we know that this also gives the basis to understanding Cartan geometries. The Lie group / Lie algebra correspondence (see 1.2.3) implies that there is a close relation between representations of a Lie group G and its Lie algebra $\mathfrak{g}$, as well as for equivariant mappings between such representations.

While representations of general Lie algebras may be very complicated, there is a satisfactory representation theory for semisimple Lie algebras, which can also be used to analyze more general situations. Moreover, the analysis of the adjoint representation leads to the classification of semisimple Lie algebras.

We will start this chapter by briefly discussing elementary properties of Lie algebras and the basic structure theory. Next, we will discuss complex simple Lie algebras and their representations. Then we study real forms, arriving at the description of real semisimple Lie algebras via Satake diagrams. We will briefly sketch the classification of simple real Lie algebras and study real representations.

Throughout this chapter, we restrict our attention to real and complex Lie algebras and do not consider other ground fields.

2.1. Basic structure theory of Lie algebras

2.1.1. Abelian, nilpotent and solvable Lie algebras. By definition, a Lie algebra $\mathfrak{g}$ over $\mathbb{K} = \mathbb{R}$ or $\mathbb{C}$ is a vector space together with a $\mathbb{K}$–bilinear mapping $[\ ,\ ] : \mathfrak{g}\times\mathfrak{g} \to \mathfrak{g}$, called the *Lie bracket*, which is skew symmetric, i.e. $[Y, X] = -[X, Y]$ and satisfies the Jacobi identity, i.e. $[X, [Y, Z]] = [[X, Y], Z] + [Y, [X, Z]]$ for all $X, Y, Z \in \mathfrak{g}$. If $\mathfrak{g}$ and $\mathfrak{h}$ are Lie algebras, then a homomorphism $\phi : \mathfrak{g} \to \mathfrak{h}$ of Lie algebras is a $\mathbb{K}$–linear mapping which is compatible with the brackets, i.e. such that $[\phi(X), \phi(Y)] = \phi([X, Y])$ for all $X, Y \in \mathfrak{g}$.

The simplest choice for the bracket is the zero map, and in this way we get an *abelian* Lie algebra, which is just a vector space.

If $(\mathfrak{g}, [\ ,\ ])$ is a Lie algebra and $A, B \subset \mathfrak{g}$ are nonempty subsets, then we denote by $[A, B]$ the linear subspace spanned by all elements of the form $[a, b]$ with $a \in A$ and $b \in B$. Then there is the obvious notion of a *Lie subalgebra* $\mathfrak{h}$ in a Lie algebra $\mathfrak{g}$, namely a linear subspace which is closed under the bracket, i.e. such that $[\mathfrak{h}, \mathfrak{h}] \subset \mathfrak{h}$. We write $\mathfrak{h} \leq \mathfrak{g}$ if $\mathfrak{h}$ is a subalgebra of $\mathfrak{g}$. Clearly, the intersection of an arbitrary family of subalgebras of $\mathfrak{g}$ is again a subalgebra. Thus, for any subset $A \subset \mathfrak{g}$, there is a smallest subalgebra of $\mathfrak{g}$ which contains A, called the *subalgebra generated by* A.

To form quotients of Lie algebras, one needs a strengthening of the notion of a subalgebra. We say that a subalgebra $\mathfrak{h} \leq \mathfrak{g}$ is an *ideal* in $\mathfrak{g}$ and write $\mathfrak{h} \lhd \mathfrak{g}$

if $[\mathfrak{g}, \mathfrak{h}] \subset \mathfrak{h}$. In this case, one easily shows that $[X + \mathfrak{h}, Y + \mathfrak{h}] := [X, Y] + \mathfrak{h}$ is a well–defined Lie bracket on the quotient space $\mathfrak{g}/\mathfrak{h}$ and the canonical mapping $\mathfrak{g} \to \mathfrak{g}/\mathfrak{h}$ is a homomorphism of Lie algebras. As for subalgebras, the intersection of any family of ideals in $\mathfrak{g}$ is again an ideal in $\mathfrak{g}$, so for any subset $A \subset \mathfrak{g}$, there is the *ideal generated* by A. If $\phi : \mathfrak{g} \to \mathfrak{h}$ is a homomorphism of Lie algebras, then obviously the image $\operatorname{im}(\phi)$ is a subalgebra of $\mathfrak{h}$, while the kernel $\ker(\phi)$ is even an ideal in $\mathfrak{g}$.

Next, if $\mathfrak{g}$ is a Lie algebra, then we can consider the subspace $[\mathfrak{g}, \mathfrak{g}] \subset \mathfrak{g}$, which by definition is an ideal in $\mathfrak{g}$ such that the quotient $\mathfrak{g}/[\mathfrak{g}, \mathfrak{g}]$ is abelian. Indeed, this is the largest abelian quotient of $\mathfrak{g}$, since any homomorphism from $\mathfrak{g}$ to an abelian Lie algebra must by definition vanish on $[\mathfrak{g}, \mathfrak{g}]$ and thus factor through this quotient. The ideal $[\mathfrak{g}, \mathfrak{g}] \lhd \mathfrak{g}$ is called the *commutator ideal.* This concept gives rise to two sequences of ideals in $\mathfrak{g}$.

Let us define $\mathfrak{g}^1 = \mathfrak{g}$, $\mathfrak{g}^2 = [\mathfrak{g}, \mathfrak{g}]$ and inductively $\mathfrak{g}^{k+1} = [\mathfrak{g}, \mathfrak{g}^k]$. By the Jacobi identity, if $\mathfrak{h}_1, \mathfrak{h}_2 \subset \mathfrak{g}$ are ideals, then also $[\mathfrak{h}_1, \mathfrak{h}_2] \subset \mathfrak{h}_1 \cap \mathfrak{h}_2$ is an ideal in $\mathfrak{g}$. Inductively, this in particular implies that each $\mathfrak{g}^k$ is an ideal in $\mathfrak{g}$ and $\mathfrak{g}^{k+1} \subset \mathfrak{g}^k$. The sequence $\mathfrak{g} \supset \mathfrak{g}^2 \supset \cdots \supset \mathfrak{g}^k \supset \mathfrak{g}^{k+1} \supset \dots$ is called the *lower central series* of $\mathfrak{g}$. The Lie algebra $\mathfrak{g}$ is called *nilpotent* if $\mathfrak{g}^k = 0$ for some $k \in \mathbb{N}$, i.e. if applying sufficiently many brackets one always ends up with zero.

On the other hand, we define $\mathfrak{g}^{(1)} = \mathfrak{g}$ and inductively $\mathfrak{g}^{(k+1)} := [\mathfrak{g}^{(k)}, \mathfrak{g}^{(k)}]$. As above, we see inductively that each $\mathfrak{g}^{(k)}$ is an ideal in $\mathfrak{g}$, and clearly $\mathfrak{g}^{(k+1)} \subset \mathfrak{g}^{(k)}$. Hence, we get another decreasing sequence $\mathfrak{g} \supset \mathfrak{g}^{(2)} \supset \cdots \supset \mathfrak{g}^{(k)} \supset \mathfrak{g}^{(k+1)} \supset \dots$, which is called the *derived series* of $\mathfrak{g}$. The Lie algebra $\mathfrak{g}$ is called *solvable* if $\mathfrak{g}^{(k)} = 0$ for some $k \in \mathbb{N}$. By construction, we have $\mathfrak{g}^2 = \mathfrak{g}^{(2)} \subset \mathfrak{g}^1 = \mathfrak{g}^{(1)}$, which inductively implies $\mathfrak{g}^{(k)} \subset \mathfrak{g}^k$. In particular, any nilpotent Lie algebra is solvable.

Suppose that $\mathfrak{h} \leq \mathfrak{g}$ is a subalgebra. Then $\mathfrak{h}^k \subset \mathfrak{g}^k$ and $\mathfrak{h}^{(k)} \subset \mathfrak{g}^{(k)}$, so if $\mathfrak{g}$ is nilpotent (respectively solvable), then also $\mathfrak{h}$ is nilpotent (respectively solvable). Similarly, if $\mathfrak{h} \lhd \mathfrak{g}$ is any ideal, then the canonical homomorphism $\mathfrak{g} \to \mathfrak{g}/\mathfrak{h}$ induces surjections $\mathfrak{g}^k \to (\mathfrak{g}/\mathfrak{h})^k$ and $\mathfrak{g}^{(k)} \to (\mathfrak{g}/\mathfrak{h})^{(k)}$. Consequently, quotients of nilpotent (solvable) Lie algebras are nilpotent (solvable).

There is a nice characterization of solvable Lie algebras in terms of extensions which also implies a converse to these results. Suppose that $\mathfrak{g}$ is a Lie algebra and we have a finite sequence $\mathfrak{g} \supset \mathfrak{g}_1 \supset \cdots \supset \mathfrak{g}_{k-1} \supset \mathfrak{g}_k = \{0\}$ such that each $\mathfrak{g}_{j+1}$ is an ideal in $\mathfrak{g}_j$ such that the quotient $\mathfrak{g}_j/\mathfrak{g}_{j+1}$ is abelian. By the universal property of the commutator ideal, since $\mathfrak{g}/\mathfrak{g}_1$ is abelian, we must have $[\mathfrak{g}, \mathfrak{g}] \subset \mathfrak{g}_1$, and inductively it follows that $\mathfrak{g}^{(j)} \subset \mathfrak{g}_j$. In particular, $\mathfrak{g}^{(k)} = 0$, so $\mathfrak{g}$ is solvable. Conversely, the derived series of a solvable Lie algebra is a sequence of this type, so this is a characterization of solvability. From this characterization one easily concludes that if $\mathfrak{g}$ is any Lie algebra which has a solvable ideal $\mathfrak{h} \lhd \mathfrak{g}$ such that the quotient $\mathfrak{g}/\mathfrak{h}$ is solvable, then also $\mathfrak{g}$ itself is solvable.

The basic examples of Lie algebras are given by Lie algebras of matrices. Indeed, if A is any associative algebra, then the *commutator* $[a, b] := ab - ba$ makes A into a Lie algebra. In particular, we may apply this to the algebra $\mathfrak{gl}(V)$ of endomorphisms of a finite–dimensional vector space V. For $V = \mathbb{K}^n$, where $\mathbb{K} = \mathbb{R}$ or $\mathbb{C}$ we obtain the matrix algebra $M_n(\mathbb{K})$, which is commonly denoted by $\mathfrak{gl}(n, \mathbb{K})$. Any subspace of matrices which is closed under the commutator is a Lie subalgebra of $M_n(\mathbb{K})$, for example, the space $\mathfrak{o}(n) \subset M_n(\mathbb{R})$ of skew symmetric matrices or the space

$\mathfrak{u}(n) \subset M_n(\mathbb{C})$ of skew Hermitian matrices. Notice, however, that $\mathfrak{u}(n)$ is not a complex subspace of $M_n(\mathbb{C})$, so it is only a real Lie algebra and not a complex one.

Consider the space $\mathfrak{n} \subset M_n(\mathbb{K})$ of strictly upper triangular matrices, i.e. matrices $A = (a_{ij})$ such that $a_{ij} = 0$ whenever $i \geq j$. These matrices can be equivalently characterized by the fact that they map any element e_j of the standard basis of $\mathbb{K}^n$ into the subspace spanned by $e_1, \dots, e_{j-1}$, which immediately implies that these matrices form a Lie subalgebra $\mathfrak{n}$ of $M_n(\mathbb{K})$ and that $\mathfrak{n}^n = 0$, i.e. this subalgebra is nilpotent. Similarly, the space of upper triangular matrices is a Lie subalgebra $\mathfrak{b}$ of $M_n(\mathbb{K})$. By definition, we have $\mathfrak{n} \leq \mathfrak{b}$ and one immediately sees that $\mathfrak{n}$ is even an ideal and that $\mathfrak{b}/\mathfrak{n}$ is abelian. Thus, $\mathfrak{b}$ is solvable, but for example, the equation $[\left(\begin{smallmatrix} 1 & 0 \\ 0 & -1 \end{smallmatrix}\right), \left(\begin{smallmatrix} 0 & 1 \\ 0 & 0 \end{smallmatrix}\right)] = \left(\begin{smallmatrix} 0 & 2 \\ 0 & 0 \end{smallmatrix}\right)$ shows that $\mathfrak{b}$ cannot be nilpotent. In particular, we see that a Lie algebra may admit a nilpotent ideal such that the quotient is nilpotent without being nilpotent itself.

The following result, whose first part is usually called Engel's theorem, while the second part is known as Lie's theorem, shows that these examples are typical.

THEOREM 2.1.1. *Let V be a finite-dimensional vector space and let $\mathfrak{g} \leq \mathfrak{gl}(V)$ be a Lie subalgebra.*

(1) If each element $X \in \mathfrak{g}$ acts as a nilpotent endomorphism of V, then there is a nonzero vector $v \in V$ such that $X(v) = 0$ for all $X \in \mathfrak{g}$.

(2) If V is a complex vector space and $\mathfrak{g} \leq \mathfrak{gl}_{\mathbb{C}}(V)$ is a solvable (but not necessarily complex) Lie subalgebra, then there is a nonzero vector $v \in V$ which is an eigenvector for all maps $X \in \mathfrak{g}$.

SKETCH OF PROOF. By induction on $n = \dim(\mathfrak{g})$. For $n = 1$, the statements reduce to the fact that any nilpotent linear map has a nontrivial kernel and any complex linear map has a nonzero eigenvector. If $n > 1$, then one shows that $\mathfrak{g}$ contains an ideal $\mathfrak{h}$ of codimension one. For solvable $\mathfrak{g}$, we first observe that $n > 1$ and solvability imply $[\mathfrak{g}, \mathfrak{g}] \neq \mathfrak{g}$. Now the preimage of any codimension one subspace in $\mathfrak{g}/[\mathfrak{g}, \mathfrak{g}]$ is a codimension one ideal in $\mathfrak{g}$.

In the setting of (1), this claim is slightly more subtle to prove: Take a maximal proper subalgebra $\mathfrak{h} \leq \mathfrak{g}$. Applying the induction hypothesis to the set of all endomorphisms of $\mathfrak{g}/\mathfrak{h}$ which are induced by the maps $\mathrm{ad}(X) : \mathfrak{g} \to \mathfrak{g}$ with $X \in \mathfrak{h}$, one finds an element $Y \in \mathfrak{g} \setminus \mathfrak{h}$ such that $[\mathfrak{h}, Y] \subset \mathfrak{h}$. Thus, the subspace of $\mathfrak{g}$ spanned by $\mathfrak{h}$ and Y is a subalgebra which contains $\mathfrak{h}$ as a codimension one ideal, and by maximality of $\mathfrak{h}$ it has to coincide with $\mathfrak{g}$.

Next, one applies the induction hypothesis to the codimension one ideal $\mathfrak{h}$, getting a vector in the common kernel, respectively, a common eigenvector for all elements of $\mathfrak{h}$. Then the result follows from the fact that for any ideal $\mathfrak{h} \lhd \mathfrak{g}$ and any linear functional λ on $\mathfrak{h}$ the eigenspace $W := \{v \in V : X(v) = \lambda(X)v \quad \forall X \in \mathfrak{h}\}$ is $\mathfrak{g}$–invariant; see [**FH91**, Lemma 9.13]. (Note that in (2), W is automatically a complex subspace of V.) This last statement is proved as follows: It is easy to see that the statement is equivalent to showing that $\lambda([X,Y]) = 0$ for $X \in \mathfrak{h}$ and $Y \in \mathfrak{g}$. Taking an element $w \in W$, one now considers the subspace spanned by w, $Y(w)$, $Y^2(w) = Y(Y(w))$ and so on. Using induction, one shows that this subspace is $\mathfrak{h}$–invariant and with respect to an appropriate basis the action of each $X \in \mathfrak{h}$ is by upper triangular matrices with all diagonal entries equal to $\lambda(X)$. Thus, one may compute the value of λ from the trace of the restriction of the action to this subspace. But then the action of $[X,Y]$ is given by a commutator, which is tracefree and hence $\lambda([X,Y]) = 0$. □

An immediate corollary of this result is that for any solvable Lie subalgebra $\mathfrak{g} \subset \mathfrak{gl}_{\mathbb{C}}(V)$, where V is a finite–dimensional complex vector space, there is a basis of V such that all maps in $\mathfrak{g}$ correspond to upper triangular matrices. To see this, one uses induction on the dimension of V. If $v \in V$ is a nonzero vector which is an eigenvector for all $X \in \mathfrak{g}$, then we consider $W = V/\mathbb{C}v$ and the image of $\mathfrak{g}$ in $\mathfrak{gl}(W)$. As a quotient of $\mathfrak{g}$, this image is solvable, so by induction we get an appropriate basis for W. Choosing preimages of the elements of this basis and adding v, we get an appropriate basis for V.

Similarly, given a Lie subalgebra $\mathfrak{g} \leq \mathfrak{gl}(V)$ which satisfies the hypothesis of Engel's theorem, there is a basis of V with respect to which all elements of $\mathfrak{g}$ are represented by strictly upper triangular matrices. Since we have seen above that these matrices form a nilpotent Lie subalgebra, we conclude that $\mathfrak{g}$ is nilpotent, too.

2.1.2. Semisimple, simple, and reductive Lie algebras. A Lie algebra $\mathfrak{g}$ is said to be *semisimple* if it has no nonzero solvable ideal. The Lie algebra $\mathfrak{g}$ is called *simple* if $\mathfrak{g} = [\mathfrak{g}, \mathfrak{g}]$ and the only ideals in $\mathfrak{g}$ are $\{0\}$ and $\mathfrak{g}$. Finally, $\mathfrak{g}$ is called *reductive*, if any solvable ideal of $\mathfrak{g}$ is contained in the *center*

$$\mathfrak{z}(\mathfrak{g}) := \{X \in \mathfrak{g} : [X, Y] = 0 \quad \forall Y \in \mathfrak{g}\}$$

of $\mathfrak{g}$. The condition that $\mathfrak{g} = [\mathfrak{g}, \mathfrak{g}]$ in the definition of a simple Lie algebra is only needed to exclude one–dimensional abelian Lie algebras from being simple. This condition also implies that $\mathfrak{g}$ itself cannot be solvable, since $\mathfrak{g} = [\mathfrak{g}, \mathfrak{g}]$ implies $\mathfrak{g} = \mathfrak{g}^{(k)}$ for all $k \in \mathbb{N}$. In particular, it ensures that any simple Lie algebra is semisimple. On the other hand, it is obvious that the classes of solvable and of semisimple Lie algebras are disjoint.

By definition, the center $\mathfrak{z}(\mathfrak{g})$ of a Lie algebra $\mathfrak{g}$ is an abelian ideal in $\mathfrak{g}$. This immediately implies that any semisimple Lie algebra has trivial center, and that the solvable ideals of a reductive Lie algebra are exactly the subspaces of its center. Finally, if $\mathfrak{h} \lhd \mathfrak{g}$ is an ideal, then by the Jacobi identity, also $[\mathfrak{h}, \mathfrak{h}]$ is an ideal in $\mathfrak{g}$. Inductively, we see that each term $\mathfrak{h}^{(k)}$ in the derived series of $\mathfrak{h}$ is an ideal in $\mathfrak{g}$. In particular, if $\mathfrak{h}$ is solvable, then the last nontrivial term in the derived series is an abelian ideal in $\mathfrak{g}$. Thus, a Lie algebra is semisimple if and only if it does not contain a nontrivial abelian ideal. The fundamental example of a simple Lie algebra is provided by the following result.

PROPOSITION 2.1.2. *Let V be a vector space over $\mathbb{K} = \mathbb{R}$ or $\mathbb{C}$ of dimension at least two. Then the space $\mathfrak{sl}(V)$ of all tracefree endomorphisms of V is a simple Lie algebra.*

PROOF. Since the commutator of two endomorphisms is always tracefree, $\mathfrak{sl}(V)$ is a Lie subalgebra of $\mathfrak{gl}(V)$. To prove simplicity, we may restrict to the case $V = \mathbb{K}^n$ and thus to the Lie algebra $\mathfrak{g} = \mathfrak{sl}(n, \mathbb{K})$ of tracefree $n \times n$–matrices. Let E_{ij} be the elementary matrix, which has a one in the jth column of the ith row and all other entries equal to zero. The multiplication rule for elementary matrices is simply $E_{ij}E_{k\ell} = \delta_{jk}E_{i\ell}$, where δ_{jk} is the Kronecker delta. Clearly, the elements of the forms E_{ij} and $E_{ii} - E_{jj}$ for $i \neq j$ linearly generate $\mathfrak{g}$. Now for $i \neq j$ we have $[E_{ii} - E_{jj}, E_{ij}] = 2E_{ij}$ and $[E_{ij}, E_{ji}] = E_{ii} - E_{jj}$, which shows that $[\mathfrak{g}, \mathfrak{g}] = \mathfrak{g}$.

Now suppose that $\mathfrak{h} \lhd \mathfrak{g}$ is a nonzero ideal in $\mathfrak{g}$. One immediately verifies that for $A \in \mathfrak{g}$ and any i, j the commutator $[E_{ij}, A]$ is obtained by taking the matrix whose ith row is the jth row of A while all other rows are zero, and subtracting from it the matrix whose jth column is the ith column of A while all other columns are zero. Now we first observe that this implies that $\mathfrak{h}$ contains a matrix which has a nonzero off–diagonal entry. Indeed, if $A = (a_{ij}) \in \mathfrak{h}$ is diagonal, then since A is tracefree, we find indices $i \neq j$ such that $a_{ii} - a_{jj} \neq 0$. But then $[E_{ij}, A]$ has in the ith row of the jth column the entry $a_{jj} - a_{ii}$ and thus a nonzero off–diagonal entry. But now suppose that $A = (a_{ij}) \in \mathfrak{h}$ is such that $a_{ji} \neq 0$ for some fixed $i \neq j$. Then using our description one immediately verifies that $[E_{ij}, [E_{ij}, A]] = -2a_{ji}E_{ij}$, which implies $E_{ij} \in \mathfrak{h}$. Thus, also $[E_{ij}, E_{ji}] = E_{ii} - E_{jj} \in \mathfrak{h}$, which in turn implies E_{ji} in $\mathfrak{h}$. If $n = 2$, we are finished here, otherwise we next observe that for $k \neq i, j$ we have $[E_{ij}, E_{jk}] = E_{ik} \in \mathfrak{h}$ and hence $E_{\ell k} = [E_{\ell i}, E_{ik}] \in \mathfrak{h}$ for $\ell \neq i, k$. Since similarly as above we can also get the tracefree diagonal matrices contained in $\mathfrak{h}$, this implies $\mathfrak{h} = \mathfrak{g}$. □

From simplicity of $\mathfrak{sl}(V)$, we can immediately conclude that $\mathfrak{g} := \mathfrak{gl}(V)$ is a reductive Lie algebra. The center of $\mathfrak{g}$ is formed by all multiples of the identity, so $\mathfrak{g} = \mathfrak{z}(\mathfrak{g}) \oplus \mathfrak{sl}(V)$. Since any commutator of matrices is tracefree, we see that $[\mathfrak{g}, \mathfrak{g}] \subset \mathfrak{sl}(V)$, so simplicity of $\mathfrak{sl}(V)$ implies that $[\mathfrak{g}, \mathfrak{g}] = \mathfrak{sl}(V)$. Mapping $A \in \mathfrak{g}$ to its tracefree part $A - \frac{1}{\dim(V)} \operatorname{tr}(A)\mathrm{id}$ defines a projection $\pi : \mathfrak{g} \to \mathfrak{sl}(V)$, which is a homomorphism of Lie algebras and has kernel $\mathfrak{z}(\mathfrak{g})$. If $\mathfrak{h} \subset \mathfrak{g}$ is a solvable ideal, then $\pi(\mathfrak{h})$ is a solvable ideal in $\mathfrak{sl}(V)$, and thus must be zero, which implies $\mathfrak{h} \subset \mathfrak{z}(\mathfrak{g})$ and thus $\mathfrak{g}$ is reductive.

If $\mathfrak{g}$ and $\mathfrak{h}$ are Lie algebras, then we can form the (ideal) *direct sum* $\mathfrak{g} \oplus \mathfrak{h}$. This is the direct sum of the vector spaces $\mathfrak{g}$ and $\mathfrak{h}$ endowed with the componentwise bracket. By definition, $\mathfrak{g}$ and $\mathfrak{h}$ are ideals in $\mathfrak{g} \oplus \mathfrak{h}$, and the obvious projections are Lie algebra homomorphisms. This easily implies that a direct sum of semisimple Lie algebras is semisimple. We can also form direct sums of finitely many factors, and in particular, we see that a (finite) direct sum of simple Lie algebras is semisimple. We shall see below that also the converse is true, i.e. any semisimple Lie algebra is a direct sum of simple ideals.

2.1.3. Representations. A *representation* of a Lie algebra $\mathfrak{g}$ on a vector space V is a homomorphism $\rho : \mathfrak{g} \to \mathfrak{gl}(V)$ of Lie algebras. Equivalently, one may view a representation as a bilinear map $\rho : \mathfrak{g} \times V \to V$ such that

$$\rho([X, Y], v) = \rho(X, \rho(Y, v)) - \rho(Y, \rho(X, v)).$$

If the representation under consideration is clear from the context, then we will often simply write $X \cdot v$ or Xv for $\rho(X)(v)$. A representation is called *trivial* if $X \cdot v = 0$ for all $X \in \mathfrak{g}$ and $v \in V$.

If $\mathfrak{g}$ is real, then by a *complex representation* one simply means a homomorphism from $\mathfrak{g}$ to the Lie algebra of complex linear maps on a complex vector space V. If $\mathfrak{g}$ is complex, then one requires, in addition, that also the map $\rho : \mathfrak{g} \to \mathfrak{gl}(V)$ is complex linear. Therefore, one has to distinguish between complex representations of a complex Lie algebra and complex representations of the underlying real Lie algebra.

If $\rho : \mathfrak{g} \to \mathfrak{gl}(V)$ and $\rho' : \mathfrak{g} \to \mathfrak{gl}(V')$ are representations, then a *morphism* from ρ to ρ' is a linear map $\phi : V \to V'$, which is compatible with the $\mathfrak{g}$ actions. Explicitly, we must have $\phi(\rho(X)(v)) = \rho'(X)(\phi(v))$ or in the simple notation $\phi(X \cdot v) = X \cdot \phi(v)$. Morphisms of representations are also called *intertwining operators* or *equivariant maps*. An *isomorphism* of representations is a bijective morphism between two representations. If $\phi : V \to W$ is an isomorphism, then the inverse map $\phi^{-1} : W \to V$ is automatically a morphism, too.

An important example of a representation of a Lie algebra is the *adjoint representation*, $\mathrm{ad} : \mathfrak{g} \to \mathfrak{gl}(\mathfrak{g})$ defined by $\mathrm{ad}(X)(Y) := [X, Y]$. The Jacobi identity for the bracket exactly states that this is a representation of $\mathfrak{g}$.

A representation $\rho : \mathfrak{g} \to \mathfrak{gl}(V)$ is called *faithful* if the map ρ is injective. If this is not the case, then the kernel of ρ is an ideal in $\mathfrak{g}$. In particular, considering the adjoint representation we see that the kernel by definition is the set of all $X \in \mathfrak{g}$ such that $[X, Y] = 0$ for all $Y \in \mathfrak{g}$, i.e. the center of $\mathfrak{g}$. On the other hand, by definition any nontrivial representation of a simple Lie algebra is automatically faithful.

Suppose that $\rho : \mathfrak{g} \to \mathfrak{gl}(V)$ is a representation. A *subrepresentation* or an *invariant subspace* of V is a linear subspace $W \subset V$ such that $\rho(X)(w) \in W$ for all $X \in \mathfrak{g}$ and all $w \in W$. A representation V is called *irreducible* if $\{0\}$ and V itself are the only invariant subspaces. Given an invariant subspace $W \subset V$ we obtain a representation of $\mathfrak{g}$ on W by restriction, and there is an obvious representation of $\mathfrak{g}$ on the quotient V/W.

Note that an invariant subspace in the adjoint representation of $\mathfrak{g}$ by definition is an ideal $\mathfrak{h} \lhd \mathfrak{g}$. In particular, $\mathfrak{g}$ is simple if and only if $\mathfrak{g} = [\mathfrak{g}, \mathfrak{g}]$ and the adjoint representation is irreducible.

Lie's theorem from 2.1.1 above gives us some information on complex representations of solvable Lie algebras. Indeed, the statement immediately implies that any such representation contains an invariant subspace of dimension one. In particular, irreducible complex representations of a solvable Lie algebra are automatically one–dimensional. More precisely, any such representation is given by a complex–valued linear functional on the abelian Lie algebra $\mathfrak{g}/[\mathfrak{g}, \mathfrak{g}]$.

Note that for a morphism $\phi : V \to W$ between two representations, the kernel $\ker(\phi)$ is a subrepresentation of V and the image $\phi(V)$ is a subrepresentation of W. In particular, this implies that a nonzero morphism with an irreducible source is automatically injective while for an irreducible target it is automatically surjective. Hence, a morphism between irreducible representations is either zero or an isomorphism. A simple consequence of this is Schur's lemma:

LEMMA 2.1.3 (Schur's lemma). *Let V be a complex irreducible representation of a Lie algebra $\mathfrak{g}$. Then any morphism $\phi : V \to V$ is a complex multiple of the identity map.*

PROOF. If $\phi : V \to V$ is a nonzero morphism, then for each $\lambda \in \mathbb{C}$ also $\phi - \lambda\mathrm{id}$ is a morphism. Since ϕ must have an eigenvalue λ_0, we conclude that $\phi - \lambda_0\mathrm{id}$ has nontrivial kernel, so by irreducibility it is identically zero. □

Any functorial construction for finite–dimensional vector spaces can be carried over to a functorial construction for finite–dimensional representations of a given Lie algebra $\mathfrak{g}$. If $\rho : \mathfrak{g} \to \mathfrak{gl}(V)$ is a finite–dimensional representation, then the *dual* or *contragradient* representation ρ^* of $\mathfrak{g}$ on V^* is given by $\rho^*(X) := (\rho(-X))^*$. The direct sum of two representations is simply given by the componentwise action on the direct sum of two vector spaces. For the tensor product, the action is given by $X \cdot (v \otimes w) = (X \cdot v) \otimes w + v \otimes (X \cdot w)$, and similarly for higher tensor powers. If one considers tensor powers of one representation, then one always has the subrepresentations of tensors of various symmetry types. In particular, the k–fold tensor power $\otimes^k V$ contains the subrepresentations $S^k V$ of symmetric and $\Lambda^k V$ of totally antisymmetric elements. Via the natural identification $L(V,W) \cong V^* \otimes W$, we also get canonical representations on spaces of linear and multilinear mappings between representations. For example, the action on $L(V,W)$ is given by $(X \cdot \phi)(v) = X \cdot \phi(v) - \phi(X \cdot v)$, which shows that a linear map $\phi : V \to W$ between two representations is a morphism if and only if it lies in a trivial subrepresentation of $L(V,W)$.

A finite–dimensional representation of a Lie algebra is called *completely reducible* if it can be written as a direct sum of irreducible representations. A completely reducible representation is called *simply reducible* if the irreducible factors showing up in the direct sum are pairwise non–isomorphic. Finally, a representation is called *indecomposable* if it cannot be written as a direct sum of two subrepresentations of positive dimension.

There is a certain subtlety in the relations between indecomposability and irreducibility. More or less by definition, any representation can be written as a direct sum of indecomposable representations. In contrast, complete reducibility is a restrictive condition in general. To see this, consider a complex solvable Lie algebra $\mathfrak{g}$. From above, we know that any irreducible representation comes from a representation of $\mathfrak{g}/[\mathfrak{g},\mathfrak{g}]$, so this continues to hold for completely reducible representations. Hence, on any such representation, the derived algebra $[\mathfrak{g},\mathfrak{g}]$ acts trivially, so they do not see the Lie algebra structure of $\mathfrak{g}$. As we shall see below, the adjoint representation of a Lie algebra $\mathfrak{g}$ is completely reducible if an only if $\mathfrak{g}$ is reductive.

One reason for the popularity of the notion of irreducibility is that large parts of the representation theory of Lie algebras have their origins in the study of *unitary* representations of a Lie group, i.e. representations admitting an invariant positive definite inner product. The corresponding condition on the Lie algebra level is that there is a Hermitian inner product on the representation space V such that any element of $\mathfrak{g}$ acts by a skew Hermitian operator on V, i.e. $\rho(X)^* = -\rho(X)$ for all $X \in \mathfrak{g}$.

Writing the inner product as $\langle\ ,\ \rangle$, this means that $\langle X \cdot v, w\rangle = -\langle v, X \cdot w\rangle$. In particular, for a subrepresentation $W \subset V$ the orthogonal complement $W^\perp$ is an invariant subspace, too. In general, a representation is called *semisimple* if any invariant subspace admits an invariant complement. Thus, unitary representations are automatically semisimple, but the converse is far from being true. On the other hand, on the subcategory of semisimple representations the notions of indecomposability and irreducibility coincide. We shall soon see that any finite–dimensional representation of a semisimple Lie algebra is semisimple.

2.1.4. Complexification. Before we can continue developing the general theory of Lie algebras, we have to discuss the relation between real and complex Lie algebras and, in particular, the technique of complexification in a little more detail. For a real Lie algebra $\mathfrak{g}$ we define the *complexification* $\mathfrak{g}_{\mathbb{C}}$ of $\mathfrak{g}$ as the complex vector space $\mathfrak{g} \otimes_{\mathbb{R}} \mathbb{C}$ (with scalar multiplication given by $z(X \otimes w) = X \otimes zw$) and with the bracket given by the complex bilinear extension of the bracket on $\mathfrak{g}$, i.e. $[X \otimes z, Y \otimes w] := [X,Y] \otimes zw$. The mapping $X \mapsto X \otimes 1$ makes $\mathfrak{g}$ into a real Lie subalgebra of $\mathfrak{g}_{\mathbb{C}}$. Conversely, if $\mathfrak{g}$ is a complex Lie algebra, then a real

Lie subalgebra $\mathfrak{h}$ is called a *real form* of $\mathfrak{g}$, if the mapping $\mathfrak{h} \otimes_{\mathbb{R}} \mathbb{C} \to \mathfrak{g}$ defined by $X \otimes z \mapsto zX$ is a linear isomorphism.

As we shall see below, passing from a real Lie algebra to its complexification loses some information. The simplest way to keep track of this additional information is the conjugation map. On $\mathfrak{g} \otimes_{\mathbb{R}} \mathbb{C}$, we consider $\sigma(X \otimes z) := X \otimes \bar{z}$, which is obviously a conjugate linear involutive automorphism, i.e. σ is linear over the reals, $\sigma(zA) = \bar{z}\sigma(A)$ for all $z \in \mathbb{C}$ and $A \in \mathfrak{g}_{\mathbb{C}}$, $\sigma^2 = \text{id}$ and $\sigma([A, B]) = [\sigma(A), \sigma(B)]$ for all $A, B \in \mathfrak{g}_{\mathbb{C}}$. The real form $\mathfrak{g}$ is then recovered as the fixed point set $\mathfrak{g}_{\mathbb{C}}^{\sigma} = \{A \in \mathfrak{g}_{\mathbb{C}} : \sigma(A) = A\}$. Conversely, let $\mathfrak{g}$ be any complex Lie algebra and $\sigma : \mathfrak{g} \to \mathfrak{g}$ a conjugate linear involutive automorphism. Since $\sigma(iA) = -i\sigma(A)$, we see that $i\mathfrak{g}^{\sigma}$ is exactly the -1–eigenspace of σ and since $\sigma^2 = \text{id}$ we conclude that, as a real vector space, $\mathfrak{g}$ coincides with $\mathfrak{g}^{\sigma} \oplus i\mathfrak{g}^{\sigma}$. Since any element in $\mathfrak{g}^{\sigma} \otimes_{\mathbb{R}} \mathbb{C}$ may be written uniquely as $X \otimes 1 + Y \otimes i$, we conclude that the map $\mathfrak{g}^{\sigma} \otimes_{\mathbb{R}} \mathbb{C} \to \mathfrak{g}$ induced by the complex scalar multiplication on $\mathfrak{g}$ is an isomorphism, so $\mathfrak{g}^{\sigma}$ is a real form of $\mathfrak{g}$.

A complex Lie algebra may have quite different real forms. As a fundamental example, consider the Lie algebra $\mathfrak{sl}(2, \mathbb{C})$. Then $A \mapsto \bar{A}$ and $A \mapsto -A^*$ both are conjugate linear involutive automorphisms, so their fix point spaces $\mathfrak{sl}(2, \mathbb{R})$ and $\mathfrak{su}(2)$ both are real forms of $\mathfrak{sl}(2, \mathbb{C})$, which however are far from being isomorphic. This can be seen for example from the fact that by 1.2.4 isomorphism of these two Lie algebras would imply isomorphism of any two simply connected Lie groups having these Lie algebras. However, $SU(2) \cong S^3$ is compact, while $SL(2, \mathbb{R})$ is connected and noncompact, so its universal covering cannot be compact either. We shall see in Section 2.3 below that any real form of $\mathfrak{sl}(2, \mathbb{C})$ is isomorphic to either $\mathfrak{sl}(2, \mathbb{R})$ or $\mathfrak{su}(2)$.

By definition of the complexification, any real linear map from $\mathfrak{g}$ to a complex vector space V uniquely extends to a complex linear map from $\mathfrak{g}_{\mathbb{C}}$ to V. Moreover, if $\mathfrak{h}$ is a complex Lie algebra, and $\phi : \mathfrak{g} \to \mathfrak{h}$ is a homomorphism of real Lie algebras, then the induced complex linear map $\tilde{\phi} : \mathfrak{g}_{\mathbb{C}} \to \mathfrak{h}$ is automatically a homomorphism of Lie algebras. In particular, any representation of a real Lie algebra on a complex vector space automatically extends to a representation of the complexification. To study real representations of real Lie algebras, given a representation of $\mathfrak{g}$ on V, we will first form the complexification $V_{\mathbb{C}}$ of the representation space. The real representation may as well be viewed as a representation on the complex vector space $V_{\mathbb{C}}$, which then extends to the complexification.

If we form the complexification of a complex Lie algebra $\mathfrak{g}$, then let $J : \mathfrak{g} \to \mathfrak{g}$ denote the linear map given by multiplication with i. Viewing J as a map with values in $\mathfrak{g}_{\mathbb{C}}$ the universal property of the complexification gives us a complex linear map $J_{\mathbb{C}} : \mathfrak{g}_{\mathbb{C}} \to \mathfrak{g}_{\mathbb{C}}$ and since $J^2 = -\text{id}$, uniqueness of the induced map implies $J_{\mathbb{C}}^2 = -\text{id}$. But this immediately implies that $\mathfrak{g}_{\mathbb{C}}$ splits as a direct sum of the $\pm i$–eigenspaces of $J_{\mathbb{C}}$, which we denote by $\mathfrak{g}_{1,0}$ and $\mathfrak{g}_{0,1}$, respectively. Obviously, $\mathfrak{g}_{1,0}$ is the space of all elements of the form $X \otimes 1 - J(X) \otimes i$, while $\mathfrak{g}_{0,1}$ consists of all elements of the form $X \otimes 1 + J(X) \otimes i$. Hence, both spaces are subalgebras conjugate to each other and both isomorphic to $\mathfrak{g}$, so, if $\mathfrak{g}$ is complex, then we get a canonical isomorphism $\mathfrak{g}_{\mathbb{C}} \cong \mathfrak{g} \oplus \mathfrak{g}$. Conversely, if $\mathfrak{g}$ is a real Lie algebra and $\mathfrak{h} \subset \mathfrak{g}_{\mathbb{C}}$ is a complex subalgebra such that $\mathfrak{g}_{\mathbb{C}} = \mathfrak{h} \oplus \sigma(\mathfrak{h})$, then the map $A \mapsto \frac{1}{2}(A + \sigma(A))$ induces a linear isomorphism $\mathfrak{h} \to \mathfrak{g} \subset \mathfrak{g}_{\mathbb{C}}$, which shows that $\mathfrak{g}$ is a complex Lie algebra.

Now let $\mathfrak{g}$ be a real Lie algebra with complexification $\mathfrak{g}_{\mathbb{C}}$. Then we obviously have $[\mathfrak{g}_{\mathbb{C}}, \mathfrak{g}_{\mathbb{C}}] = [\mathfrak{g}, \mathfrak{g}] \otimes_{\mathbb{R}} \mathbb{C}$, and more generally, $\mathfrak{g}^k_{\mathbb{C}} = \mathfrak{g}^k \otimes_{\mathbb{R}} \mathbb{C}$ and $\mathfrak{g}^{(k)}_{\mathbb{C}} = \mathfrak{g}^{(k)} \otimes_{\mathbb{R}} \mathbb{C}$. Hence, a Lie algebra is nilpotent or solvable if and only if its complexification has the same property. In particular, this implies that a real Lie algebra $\mathfrak{g}$ whose complexification $\mathfrak{g}_{\mathbb{C}}$ is semisimple will be semisimple, too, since for any solvable ideal in $\mathfrak{g}$ the complexification would be a solvable ideal in $\mathfrak{g}_{\mathbb{C}}$. We will soon see that the converse result also holds.

2.1.5. The Killing form. Let $\rho : \mathfrak{g} \to \mathfrak{gl}(V)$ be a finite–dimensional representation of a Lie algebra $\mathfrak{g}$. Then we define a bilinear form $B_\rho = B_V : \mathfrak{g} \times \mathfrak{g} \to \mathbb{K}$ by $B_\rho(X, Y) := \operatorname{tr}(\rho(X) \circ \rho(Y))$. Since the trace vanishes on commutators, this form is symmetric. Moreover, we compute

$$\begin{aligned} B_\rho([X,Y], Z) &= \operatorname{tr}((\rho(X) \circ \rho(Y) - \rho(Y) \circ \rho(X)) \circ \rho(Z)) \\ &= \operatorname{tr}(\rho(X) \circ \rho(Y) \circ \rho(Z) - \rho(X) \circ \rho(Z) \circ \rho(Y)) = B_\rho(X, [Y, Z]), \end{aligned}$$

which shows that the form B_ρ is *invariant.* The form B_ρ is called the *trace form* associated to the representation ρ. Applying this construction to the adjoint representation, we obtain the *Killing form* B of the Lie algebra $\mathfrak{g}$, which is one of the main tools for the study of semisimple Lie algebras. The Killing form has stronger invariance properties than the other forms B_ρ. Indeed, let $\phi : \mathfrak{g} \to \mathfrak{g}$ be any automorphism of the Lie algebra $\mathfrak{g}$. Then by definition $\phi \circ \operatorname{ad}(X) = \operatorname{ad}(\phi(X)) \circ \phi$ for all $X \in \mathfrak{g}$. But this implies

$$\operatorname{ad}(\phi(X)) \circ \operatorname{ad}(\phi(Y)) = \phi \circ \operatorname{ad}(X) \circ \operatorname{ad}(Y) \circ \phi^{-1},$$

and thus $B(\phi(X), \phi(Y)) = B(X, Y)$, so the Killing form is invariant under arbitrary automorphisms of $\mathfrak{g}$. For automorphisms of the form $e^{\operatorname{ad}(X)}$ with $X \in \mathfrak{g}$, invariance of any form B_ρ follows from the above infinitesimal invariance property, but in general there are many other automorphisms.

Note that if $\mathfrak{g}$ is a real Lie algebra with complexification $\mathfrak{g}_{\mathbb{C}}$ and V is a complex representation of $\mathfrak{g}$, then we know from 2.1.4 above, that the representation extends to $\mathfrak{g}_{\mathbb{C}}$. Hence, we get an extension of the (complex–valued) trace form B_V to a bilinear form on $\mathfrak{g}_{\mathbb{C}}$. By construction, this is the extension by complex bilinearity. If starting from a real representation of $\mathfrak{g}$ on V, we first complexify V, then $B_{V_{\mathbb{C}}} = B_V$, so this means viewing B_V as a complex form having only real values, and then the extension to $\mathfrak{g}_{\mathbb{C}}$ is again by complex bilinearity. In particular, the Killing form $B_{\mathbb{C}}$ of $\mathfrak{g}_{\mathbb{C}}$ is given by $B_{\mathbb{C}}(X \otimes z, Y \otimes w) = zwB(X, Y)$.

Suppose that $\mathfrak{g}$ is a solvable Lie algebra, V is a finite–dimensional complex vector space, and ρ is a representation of $\mathfrak{g}$ on V. By Lie's Theorem 2.1.1(2), there is a basis of V such that any element of $\mathfrak{g}$ acts by an upper triangular matrix. Since the commutator of two upper triangular matrices is strictly upper triangular, we conclude that any element of $[\mathfrak{g}, \mathfrak{g}]$ acts by a strictly upper triangular matrix. Moreover, the product of an upper triangular matrix with a strictly upper triangular matrix is strictly upper triangular, and hence tracefree, so we see that $B_V(\mathfrak{g}, [\mathfrak{g}, \mathfrak{g}]) = 0$. Passing to a complexification of the representation space, we see that this holds for real representations, too. In particular, we get the result for the Killing form.

Cartan's criterion for solvability states, that this property of the Killing form actually characterizes solvable Lie algebras. This immediately leads also to a nice characterization of semisimple Lie algebras in terms of the Killing form.

We start with a lemma on Jordan decompositions, which we will also need later. Recall that a linear mapping f on a finite–dimensional complex vector space uniquely splits into the sum $f = f_s + f_n$ of two commuting linear maps, such that f_s is diagonalizable and f_n is nilpotent. These are referred to as the semisimple and the nilpotent part of f. Both f_s and f_n can be written as polynomials in f. The map f_s has the same eigenvalues as f and for each of these eigenvalues the corresponding eigenspace of f_s is exactly the generalized eigenspace of f. With respect to a basis, for which the matrix representation of f is in Jordan normal form, f_s is represented by the diagonal part of the matrix and f_n by the off–diagonal part.

LEMMA 2.1.5. *Let V be a finite–dimensional complex vector space and let $X \in \mathfrak{gl}(V)$ be a linear mapping with Jordan decomposition $X = X_s + X_n$. Then the Jordan decomposition of* $\operatorname{ad}(X) : \mathfrak{gl}(V) \to \mathfrak{gl}(V)$ *is given by* $\operatorname{ad}(X) = \operatorname{ad}(X_s) + \operatorname{ad}(X_n)$.

PROOF. The fact that X_s and X_n commute immediately implies that $\operatorname{ad}(X_s)$ and $\operatorname{ad}(X_n)$ commute. Thus, it suffices to show that the map $\operatorname{ad}(X_s)$ is diagonalizable and $\operatorname{ad}(X_n)$ is nilpotent. First, one immediately sees inductively that $\operatorname{ad}(X_n)^k Y$ can be written as a linear combination of terms of the form $X_n^i Y X_n^{k-i}$ for $0 \leq i \leq k$. In particular, if $X_n^N = 0$, then $\operatorname{ad}(X_n)^{2N} = 0$.

To see that $\operatorname{ad}(X_s)$ is diagonalizable, take a basis of V with respect to which X_s is diagonal, and consider the elementary matrices E_{ij} as introduced in 2.1.2. Then one immediately verifies that E_{ij} is an eigenvector for $\operatorname{ad}(X_s)$ with eigenvalue given by the difference of the ith and the jth diagonal entries of X_s. In particular, we obtain a basis of $\mathfrak{gl}(V)$ consisting of eigenvectors for $\operatorname{ad}(X_s)$, so this map is diagonalizable. □

Parts (2) and (3) of the following results are usually referred to as *Cartan's criteria* for solvability and semisimplicity:

THEOREM 2.1.5. *(1) Let V be a vector space and $\mathfrak{g} \subset \mathfrak{gl}(V)$ a Lie subalgebra. If B_V is zero on $\mathfrak{g}$, then $\mathfrak{g}$ is solvable.*

(2) A Lie algebra $\mathfrak{g}$ is solvable if and only if its Killing form has the property that $B(\mathfrak{g}, [\mathfrak{g}, \mathfrak{g}]) = 0$.

(3) A Lie algebra $\mathfrak{g}$ is semisimple if and only if its Killing form is nondegenerate.

PROOF. (1) By complexifying V, we can view $\mathfrak{g}$ as a subalgebra of the Lie algebra of complex endomorphisms of $V_{\mathbb{C}}$, and then the complexification $\mathfrak{g}_{\mathbb{C}}$ of $\mathfrak{g}$ is a complex subalgebra in there. Since solvability of $\mathfrak{g}_{\mathbb{C}}$ implies solvability of $\mathfrak{g}$ (see 2.1.4) we may, without loss of generality, assume that V is a complex vector space and $\mathfrak{g}$ is a complex Lie subalgebra of $\mathfrak{gl}(V)$.

Since $\mathfrak{g}/[\mathfrak{g}, \mathfrak{g}]$ is abelian, it suffices to show that $[\mathfrak{g}, \mathfrak{g}]$ is nilpotent. For an element $X \in [\mathfrak{g}, \mathfrak{g}]$ let X_s be the semisimple part in the Jordan decomposition of $X : V \to V$, and we define $\bar{X}_s$ to be the linear map which has the same eigenspaces as X_s, but complex conjugate eigenvalues. Using a basis of V such that X is in Jordan normal form, we see that $\operatorname{tr}(\bar{X}_s \circ X) = \sum |\lambda_j|^2$, where the λ_j are the eigenvalues of X. If we show that this trace vanishes, then it follows that X is nilpotent, which by Engel's theorem implies that $[\mathfrak{g}, \mathfrak{g}]$ is nilpotent; see 2.1.1. If $\bar{X}_s$ would lie in $\mathfrak{g}$, then this would follow from the vanishing of B_V. Since this is not the case in general, we need an additional argument:

By the lemma, $\operatorname{ad}(X_s)$ is the semisimple part of the endomorphism $\operatorname{ad}(X)$ of $\mathfrak{gl}(V)$, so this can be written as a polynomial in $\operatorname{ad}(X)$, and therefore $\operatorname{ad}(X_s)(\mathfrak{g}) \subset \mathfrak{g}$.

On the other hand, from the proof of the lemma we see that $\mathrm{ad}(\bar{X}_s)$ has the same eigenspaces as $\mathrm{ad}(X_s)$, but complex conjugate eigenvalues. Since the projections onto the eigenspaces of $\mathrm{ad}(X_s)$ can be written as polynomials in $\mathrm{ad}(X_s)$, we see that $\mathrm{ad}(\bar{X}_s)$ is a polynomial in $\mathrm{ad}(X_s)$, so $\mathrm{ad}(\bar{X}_s)(\mathfrak{g}) \subset \mathfrak{g}$. Now since $X \in [\mathfrak{g}, \mathfrak{g}]$, we can write it as a finite sum, $X = \sum [Y_i, Z_i]$. But then

$$\mathrm{tr}(\bar{X}_s \circ X) = \sum \mathrm{tr}(\bar{X}_s \circ [Y_i, Z_i]) = \sum \mathrm{tr}([\bar{X}_s, Y_i] \circ Z_i) = \sum B_V(\mathrm{ad}(\bar{X}_s)(Y_i), Z_i) = 0.$$

(2) We have seen the necessity of the condition already. Conversely, we show that even $B([\mathfrak{g}, \mathfrak{g}], [\mathfrak{g}, \mathfrak{g}]) = 0$ implies solvability of $\mathfrak{g}$. Indeed, by (1) this implies that the image of $[\mathfrak{g}, \mathfrak{g}]$ under the adjoint representation is solvable. Since the kernel of the adjoint representation of $[\mathfrak{g}, \mathfrak{g}]$ is the center of $[\mathfrak{g}, \mathfrak{g}]$, which is an abelian (and hence solvable) ideal, we conclude that $[\mathfrak{g}, \mathfrak{g}]$ is solvable. Since the quotient of $\mathfrak{g}$ by the solvable ideal $[\mathfrak{g}, \mathfrak{g}]$ is abelian, we conclude that $\mathfrak{g}$ is solvable.

(3) For semisimple $\mathfrak{g}$ consider the null space $\mathfrak{h} := \{X \in \mathfrak{g} : B(X, Y) = 0 \quad \forall Y \in \mathfrak{g}\}$ of the Killing form. By invariance of the Killing form, this is an ideal in $\mathfrak{g}$, and by (1) the image $\mathrm{ad}(\mathfrak{h}) \subset \mathfrak{gl}(\mathfrak{g})$ is solvable. Since $\ker(\mathrm{ad}|_{\mathfrak{h}}) = \mathfrak{z}(\mathfrak{g}) \cap \mathfrak{h}$ is abelian, we conclude that $\mathfrak{h}$ is solvable, and thus $\mathfrak{h} = \{0\}$.

Conversely, let us assume that B is nondegenerate and that $\mathfrak{h} \subset \mathfrak{g}$ is an abelian ideal. For $X \in \mathfrak{h}$ and $Y \in \mathfrak{g}$, we see that $\mathrm{ad}(Y) \circ \mathrm{ad}(X)$ maps $\mathfrak{g}$ to $\mathfrak{h}$ and $\mathfrak{h}$ to zero, so this map is nilpotent and thus tracefree. Hence, X lies in the null space of B, so $X = 0$. This shows that $\mathfrak{g}$ has no nontrivial abelian ideal, which by 2.1.2 implies that $\mathfrak{g}$ is semisimple. □

Next we note several important consequences of this result. In particular, as promised in 2.1.4 we show that semisimplicity is well behaved with respect to complexification and we prove that the study of semisimple Lie algebras reduces to the study of simple Lie algebras.

COROLLARY 2.1.5. *(1) If $\mathfrak{g}$ is a semisimple Lie algebra, then there are simple ideals $\mathfrak{g}_1, \dots, \mathfrak{g}_k$ in $\mathfrak{g}$ such that $\mathfrak{g} = \mathfrak{g}_1 \oplus \cdots \oplus \mathfrak{g}_k$ as a Lie algebra. Moreover, $\mathfrak{g} = [\mathfrak{g}, \mathfrak{g}]$ and any ideal in $\mathfrak{g}$ as well as any homomorphic image of $\mathfrak{g}$ is semisimple.*

(2) A real Lie algebra $\mathfrak{g}$ is semisimple if and only if its complexification $\mathfrak{g}_{\mathbb{C}}$ is semisimple.

(3) If $\mathfrak{g}$ is a complex simple Lie algebra and $\Phi : \mathfrak{g} \times \mathfrak{g} \to \mathbb{C}$ is a $\mathfrak{g}$–invariant complex bilinear form, then Φ is a multiple of the Killing form. In particular, if Φ is nonzero, then it is automatically symmetric and nondegenerate.

PROOF. (1) If $\mathfrak{h} \subset \mathfrak{g}$ is an ideal, then the annihilator $\mathfrak{h}^\perp$ with respect to the Killing form is an ideal by invariance of the Killing form. The Killing form restricts to zero on the ideal $\mathfrak{h} \cap \mathfrak{h}^\perp$, so this ideal is solvable, By semisimplicity, $\mathfrak{h} \cap \mathfrak{h}^\perp = \{0\}$, and hence $\mathfrak{g} = \mathfrak{h} \oplus \mathfrak{h}^\perp$ and $[\mathfrak{h}, \mathfrak{h}^\perp] = 0$. The last fact implies that the Killing form of $\mathfrak{h}$ is the restriction of the Killing form of $\mathfrak{g}$ and, in particular, nondegenerate. Thus, $\mathfrak{h}$ is semisimple and cannot be abelian, so the decomposition of $\mathfrak{g}$ into a sum of simple ideals follows by induction. The fact that $\mathfrak{g} = [\mathfrak{g}, \mathfrak{g}]$ then follows immediately, since $\mathfrak{g}_i = [\mathfrak{g}_i, \mathfrak{g}_i]$ for each of the simple ideals.

Finally, let $\mathfrak{a}$ be any Lie algebra and let $\phi : \mathfrak{g} \to \mathfrak{a}$ be a homomorphism. Then $\phi(\mathfrak{g}) \cong \mathfrak{g}/\ker(\phi)$, and from above we see that this is isomorphic to the ideal $\ker(\phi)^\perp \subset \mathfrak{g}$, which is semisimple.

(2) From 2.1.4 we know that $\mathfrak{g}$ is semisimple if $\mathfrak{g}_{\mathbb{C}}$ is semisimple. But the converse now immediately follows from part (3) of the theorem and the fact that the Killing form of $\mathfrak{g}_{\mathbb{C}}$ is the complex bilinear extension of the Killing form of $\mathfrak{g}$.

(3) The adjoint representation of $\mathfrak{g}$ is a complex representation which is irreducible, since a $\mathfrak{g}$–invariant subspace in $\mathfrak{g}$ by definition is an ideal in $\mathfrak{g}$. Now a complex bilinear form $\Phi : \mathfrak{g}\times\mathfrak{g} \to \mathbb{C}$ induces a linear map $\Phi^\vee : \mathfrak{g} \to \mathfrak{g}^* = L(\mathfrak{g}, \mathbb{C})$ defined by $\Phi^\vee(X)(Y) := \Phi(X, Y)$. Invariance of Φ reads as $\Phi([Z, X], Y) = -\Phi(X, [Z, Y])$, which exactly means that $\Phi^\vee$ is a homomorphism of $\mathfrak{g}$–modules. By part (3) of the theorem, the map $B^\vee$ induced by the Killing form is an isomorphism, since B is nondegenerate, and thus $(B^\vee)^{-1} \circ \Phi$ is an endomorphism of the irreducible representation $\mathfrak{g}$ of $\mathfrak{g}$. Now the result follows by Schur's Lemma 2.1.3. □

2.1.6. Complete reducibility. The next essential application of the Killing form is to prove that any finite–dimensional representation of a semisimple Lie algebra is semisimple and thus completely reducible.

THEOREM 2.1.6. *Let $\rho : \mathfrak{g} \to \mathfrak{gl}(V)$ be a finite–dimensional representation of a semisimple Lie algebra $\mathfrak{g}$. Then any invariant subspace $W \subset V$ has an invariant complement.*

PROOF. Let us first assume that V is a complex representation and that $W \subset V$ is a complex subspace. Since by part (1) of Corollary 2.1.5 homomorphic images of semisimple Lie algebras are semisimple, we may assume that $\mathfrak{g} \subset \mathfrak{gl}(V)$. As in the proof of part (1) of Theorem 2.1.5, the null space of the trace form B_V is a solvable ideal in $\mathfrak{g}$, so B_V is nondegenerate. Let $\{X_1, \dots, X_n\}$ be a basis of $\mathfrak{g}$, let $\{Y_1, \dots, Y_n\}$ be the dual basis with respect to B_V, and define the *Casimir operator* $C_V : V \to V$ by $C_V := \sum_{i=1}^n Y_i \circ X_i$. For $X \in \mathfrak{g}$, we compute

$$\begin{aligned} X \circ C_V - C_V \circ X &= \textstyle\sum_i([X, Y_i] \circ X_i + Y_i \circ [X, X_i]) \\ &= \textstyle\sum_{i,j} (B_V([X, Y_i], X_j)Y_j \circ X_i + B_V(Y_j, [X, X_i])Y_i \circ X_j) \\ &= \textstyle\sum_{i,j} (-B_V(Y_i, [X, X_j])Y_j \circ X_i + B_V(Y_j, [X, X_i])Y_i \circ X_j) = 0. \end{aligned}$$

Hence, C_V commutes with the action of any element of $\mathfrak{g}$. Moreover, by construction $\operatorname{tr}(C_V) = \sum_i \operatorname{tr}(Y_i \circ X_i) = \sum_i B_V(Y_i, X_i) = \dim(\mathfrak{g})$, and for any $\mathfrak{g}$–invariant subspace $W \subset V$, we have $C_V(W) \subset W$.

Now we first prove the result if W is irreducible and has codimension one. In this case, the action of $\mathfrak{g}$ on V/W must be trivial since $\mathfrak{g} = [\mathfrak{g}, \mathfrak{g}]$. In particular, C_V acts trivially on V/W. This means that $C_V(V) \subset W$ and hence the subspace $\ker(C_V)$, which is invariant by construction, has to be nontrivial. On the other hand, since W is irreducible, Schur's Lemma 2.1.3 implies that C_V acts by a scalar on W. This scalar must be nonzero, since otherwise we would have $C_V^2 = 0$ and thus $\operatorname{tr}(C_V) = 0$. Hence, the invariant subspace $\ker(C_V)$ is complementary to W.

Next, let us assume that W is not irreducible but still of codimension one. Then we use induction on $\dim(W)$. If $Z \subset W$ is a nontrivial submodule, then by induction we find a complement of W/Z in V/Z, whose preimage in V we denote by Y. Then $Y \subset V$ is a $\mathfrak{g}$–module and $Z \subset Y$ is an invariant codimension one subspace of dimension smaller than the dimension of W. By induction, we find an invariant subspace U such that $Y = U \oplus Z$ and hence $V = U \oplus W$.

Using this second step, we can now prove the result for arbitrary irreducible subspaces $W \subset V$. Namely, consider the set of all linear maps $\phi : V \to W$ whose restriction to W is a scalar multiple of the identity. This is immediately seen to be a

$\mathfrak{g}$–submodule of $L(V, W)$ which contains the codimension one submodule of all maps which restrict to zero on W. From above, we know that there is a complementary submodule, which must be trivial since it is one–dimensional. But this means that we find a $\mathfrak{g}$–invariant element $\psi \in L(V, W)$ (i.e. a $\mathfrak{g}$–homomorphism from V to W) which restricts to the identity on W. Thus, ψ is an equivariant projection onto W and hence $\ker(\psi)$ is an invariant complement to W. For general W one now proceeds inductively as in the second step.

To deal with the case of an invariant subspace W in a real representation V, we pass to the complexification $V_{\mathbb{C}}$. Let us denote the conjugation with respect to V (which clearly is $\mathfrak{g}$–equivariant) simply by $v \mapsto \bar{v}$. By the first part of the proof, there is a $\mathfrak{g}$–invariant complex subspace $U \subset V_{\mathbb{C}}$ which is complementary to $W_{\mathbb{C}}$. Then $U \cap \bar{U} \subset V_{\mathbb{C}}$ is a $\mathfrak{g}$–invariant complex subspace, which is stable under conjugation. Hence, $U \cap \bar{U} = W'_{\mathbb{C}}$ for a $\mathfrak{g}$–invariant subspace $W' \subset V$, which by construction satisfies $W \cap W' = \{0\}$.

Applying the complex result once more, we get a $\mathfrak{g}$–invariant complex subspace $U_1 \subset U$, which is complementary to $U \cap \bar{U}$. By construction, we have $U_1 \cap \bar{U}_1 = \{0\}$. This implies that $W'' = \{u + \bar{u} : u \in U_1\} \subset V$ is a $\mathfrak{g}$–invariant subspace, whose real dimension coincides with the complex dimension of U_1. From the construction, it is clear that any element of V can be written as a sum of elements from W, W', and W''. For dimensional reasons, W'' must be complementary to $W \oplus W'$, so $W' \oplus W''$ is a $\mathfrak{g}$–invariant complement for W. □

This result will be of central importance in the sequel. As an immediate corollary, we can clarify the structure of reductive Lie algebras, and prove a technically important result on derivations. A *derivation* of a Lie algebra $\mathfrak{g}$ is a linear map $D : \mathfrak{g} \to \mathfrak{g}$ such that $D([X, Y]) = [D(X), Y] + [X, D(Y)]$ for all $X, Y \in \mathfrak{g}$. The Jacobi identity says that for any $X \in \mathfrak{g}$ the map $\operatorname{ad}(X)$ is a derivation of $\mathfrak{g}$. Derivations of this form are called *inner derivations.* One immediately verifies that the commutator of two derivations is again a derivation, so the vector space $\mathfrak{Der}(\mathfrak{g})$ of all derivations of $\mathfrak{g}$ is a Lie subalgebra of $\mathfrak{gl}(\mathfrak{g})$. Note that the derivation property can be rewritten as $[D, \operatorname{ad}(X)] = \operatorname{ad}(D(X))$ (where the bracket is in $\mathfrak{gl}(\mathfrak{g})$), which shows that the inner derivations form an ideal $\operatorname{ad}(\mathfrak{g})$ in $\mathfrak{Der}(\mathfrak{g})$.

COROLLARY 2.1.6. *(1) For a Lie algebra $\mathfrak{g}$, the following are equivalent:*

(i) $\mathfrak{g}$ *is reductive,*
(ii) $[\mathfrak{g}, \mathfrak{g}]$ *is semisimple and* $\mathfrak{g} = \mathfrak{z}(\mathfrak{g}) \oplus [\mathfrak{g}, \mathfrak{g}]$ *as a Lie algebra,*
(iii) *the adjoint representation of* $\mathfrak{g}$ *is semisimple.*

(2) For a semisimple Lie algebra $\mathfrak{g}$, the adjoint action is an isomorphism of Lie algebras from $\mathfrak{g}$ onto $\mathfrak{Der}(\mathfrak{g})$. In particular, any derivation of a semisimple Lie algebra is inner.

PROOF. (1) We first show that (i) $\implies$ (iii): Consider the image $\operatorname{ad}(\mathfrak{g})$ of $\mathfrak{g}$ under the adjoint representation. Since the kernel of the adjoint representation is the center $\mathfrak{z}(\mathfrak{g})$, we see that $\operatorname{ad}(\mathfrak{g}) \cong \mathfrak{g}/\mathfrak{z}(\mathfrak{g})$. For a solvable ideal in $\operatorname{ad}(\mathfrak{g})$, the preimage in $\mathfrak{g}$ is a solvable ideal. Since $\mathfrak{g}$ is reductive, this preimage has to be contained $\mathfrak{z}(\mathfrak{g})$. Hence, $\operatorname{ad}(\mathfrak{g})$ is semisimple, and by the theorem the adjoint representation of $\mathfrak{g}$ is semisimple.

(iii) $\implies$ (ii): By assumption, there is an ideal $\mathfrak{h} \subset \mathfrak{g}$ which is complementary to the center $\mathfrak{z}(\mathfrak{g})$. Now for any ideal $\mathfrak{a} \subset \mathfrak{h}$, we have $[\mathfrak{g}, \mathfrak{a}] = [\mathfrak{h}, \mathfrak{a}] \subset \mathfrak{a}$, so $\mathfrak{a}$ is also an ideal in $\mathfrak{g}$. In particular, there is an ideal $\mathfrak{a}' \subset \mathfrak{g}$ such that $\mathfrak{g} = \mathfrak{z}(\mathfrak{g}) \oplus \mathfrak{a} \oplus \mathfrak{a}'$. If $\mathfrak{a}$

is abelian, then this implies $\mathfrak{a} \subset \mathfrak{z}(\mathfrak{g})$. Hence, $\mathfrak{h}$ does not admit a nontrivial abelian ideal and thus is semisimple; see 2.1.2. By construction $[\mathfrak{g},\mathfrak{g}] \subset \mathfrak{h}$ and by Corollary 2.1.5, $\mathfrak{h} = [\mathfrak{h},\mathfrak{h}]$, so $\mathfrak{h} = [\mathfrak{g},\mathfrak{g}]$.

(ii) $\implies$ (i): For a solvable ideal $\mathfrak{a}$ in $\mathfrak{g}$, the image in $\mathfrak{g}/\mathfrak{z}(\mathfrak{g})$ is a solvable ideal, too. But by assumption $\mathfrak{g}/\mathfrak{z}(\mathfrak{g}) \cong [\mathfrak{g},\mathfrak{g}]$ is semisimple, so this image must be trivial, and hence $\mathfrak{a} \subset \mathfrak{z}(\mathfrak{g})$.

(2) Since $\mathfrak{g}$ is semisimple, the adjoint representation is faithful, and $\mathfrak{g} \cong \operatorname{ad}(\mathfrak{g})$. From above we know that $\operatorname{ad}(\mathfrak{g}) \subset \mathfrak{der}(\mathfrak{g}) \subset \mathfrak{gl}(\mathfrak{g})$ are $\mathfrak{g}$–subrepresentations, so by the theorem there exists a $\mathfrak{g}$–invariant complement U to $\operatorname{ad}(\mathfrak{g})$ in $\mathfrak{der}(\mathfrak{g})$. From above, we know that $\operatorname{ad}(\mathfrak{g})$ is an ideal in $\mathfrak{der}(\mathfrak{g})$, so this complementary representation must be trivial. For $D \in U$ this implies $0 = [D, \operatorname{ad}(X)] = \operatorname{ad}(D(X))$ for all $X \in \mathfrak{g}$ and by injectivity of the adjoint action we get $D(X) = 0$ for all $X \in \mathfrak{g}$. □

2.1.7. Examples of reductive and semisimple Lie algebras. We next describe a general result which exhibits certain matrix Lie algebras as reductive. Since verifying triviality of the center of a Lie algebra is usually easy, this result provides an efficient sufficient condition for semisimplicity, too. Let $\mathbb{K}$ be either $\mathbb{R}$ or $\mathbb{C}$ or the skew field $\mathbb{H}$ of quaternions. On $\mathbb{R}$ we define the conjugation to be the identity, while on $\mathbb{C}$ and $\mathbb{H}$ we consider the usual conjugations $\overline{a+ib} = a - ib$ and $\overline{a+ib+jc+kd} = a - ib - jc - kd$. In any case, we have $\overline{xy} = \bar{y}\bar{x}$ and mapping (x, y) to the real part of $x\bar{y}$ defines a positive definite real inner product on $\mathbb{K}$. Next, for $n \in \mathbb{N}$ let us consider the space $M_n(\mathbb{K})$ of $n \times n$–matrices with entries from $\mathbb{K}$. For $A \in M_n(\mathbb{K})$ let A^* be the conjugate transpose of A, i.e. $(a_{ij})^* = (\bar{a}_{ji})$. Then $(AB)^* = B^*A^*$ and $(A, B) \mapsto \operatorname{re}(\operatorname{tr}(AB^*))$ defines a positive definite inner product on $M_n(\mathbb{K})$ which is just the canonical extension to $M_n(\mathbb{K}) = \mathbb{K}^{n^2}$ of the inner product on $\mathbb{K}$ from above.

PROPOSITION 2.1.7. *Let $\mathfrak{g} \subset M_n(\mathbb{K})$ be a Lie subalgebra such that for all $X \in \mathfrak{g}$ also $X^* \in \mathfrak{g}$. Then $\mathfrak{g}$ is reductive.*

PROOF. $\langle X, Y\rangle := \operatorname{re}(\operatorname{tr}(XY^*))$ restricts to a positive definite real inner product on $\mathfrak{g}$. Now suppose that $\mathfrak{h}$ is an ideal in $\mathfrak{g}$, and consider its orthogonal complement $\mathfrak{h}^\perp$. For $X \in \mathfrak{h}^\perp$, $Y \in \mathfrak{g}$ and $Z \in \mathfrak{h}$, we obtain

$$\langle [X,Y], Z\rangle = \operatorname{tr}(XYZ^* - YXZ^*) = \operatorname{tr}(X(YZ^* - Z^*Y)), = \langle X, -[Y^*, Z]\rangle.$$

By assumption, $Y^* \in \mathfrak{g}$, so $[Y^*, Z] \in \mathfrak{h}$, and hence the inner product is zero, which implies that $\mathfrak{h}^\perp$ is an ideal in $\mathfrak{g}$, too. This exactly means that the adjoint representation of $\mathfrak{g}$ is semisimple, so the result follows from part (1) of Corollary 2.1.6. □

By part (1) of Corollary 2.1.6 we conclude that a Lie subalgebra of $M_n(\mathbb{K})$, which is closed under conjugate transpose and has trivial center is semisimple. A useful observation in this direction is that one may often use Schur's lemma from 2.1.3 to conclude triviality of the center. In the case $\mathbb{K} = \mathbb{C}$, one has to show that the obvious representation of $\mathfrak{g} \subset \mathfrak{gl}(n, \mathbb{C})$ on $\mathbb{C}^n$ is irreducible to conclude that the center of $\mathfrak{g}$ coincides with the intersection of $\mathfrak{g}$ with complex multiples of the identity matrix. For $\mathbb{K} = \mathbb{R}$ one can often use the above argument after complexification, while for $\mathbb{K} = \mathbb{H}$, one may use the fact that one may embed $M_n(\mathbb{H})$ into $M_{2n}(\mathbb{C})$.

Let us briefly discuss quaternionic matrices and their relation to complex matrices in a little more detail. Recall that to get the usual conventions for matrix multiplication, one has to view $\mathbb{H}^n$ as a right $\mathbb{H}$–vector space. Writing an element

$v \in \mathbb{H}^n$ as $v = a + bi + cj + dk$ with $a, b, c, d \in \mathbb{R}^n$, we define $z_1, z_2 : \mathbb{H}^n \to \mathbb{C}^n$ by $z_1(v) = a + bi$ and $z_2(v) = c - di$. Then the map $v \mapsto (z_1(v), z_2(v))$ induces an $\mathbb{R}$–linear isomorphism $\mathbb{H}^n \to \mathbb{C}^{2n}$. In this picture, multiplication from the right by the quaternions i and j maps $(z, w) \in \mathbb{C}^{2n}$ to (iz, iw), respectively, $(-\bar{w}, \bar{z})$. Using this, one easily verifies (see [**Kn96**, I.8]) that the above isomorphism induces an isomorphism from $\mathfrak{gl}(n, \mathbb{H})$ to the Lie subalgebra $\mathfrak{u}^*(2n) \subset \mathfrak{gl}(2n, \mathbb{C})$, which is formed by all matrices of the form $\begin{pmatrix} A & B \\ -\bar{B} & \bar{A} \end{pmatrix}$ with $A, B \in \mathfrak{gl}(n, \mathbb{C})$. Now the proposition above implies that $\mathfrak{gl}(n, \mathbb{H})$ is reductive. Moreover, $\mathbb{C}^{2n}$ is easily seen to be an irreducible representation of $\mathfrak{u}^*(2n)$, and thus the center of $\mathfrak{gl}(n, \mathbb{H})$ consists of real multiples of the identity matrix only. In particular, $\mathfrak{sl}(n, \mathbb{H}) := \{X \in \mathfrak{gl}(n, \mathbb{H}) : \mathrm{re}(\mathrm{tr}(X)) = 0\}$ is a semisimple Lie algebra, which under the isomorphism above goes to the Lie algebra $\mathfrak{su}^*(2n)$ of tracefree matrices contained in $\mathfrak{u}^*(2n)$.

Using the proposition and the observation on irreducibility, we can now list a number of examples of reductive and semisimple Lie algebras. The first obvious class of examples is provided by skew Hermitian matrices. These are Lie subalgebras, since from $(AB)^* = B^*A^*$ one immediately concludes that $[A, B]^* = -[A^*, B^*]$. Thus, we see that

$$\begin{aligned} \mathfrak{so}(n) &:= \{X \in M_n(\mathbb{R}) : X^* = -X\}, \\ \mathfrak{u}(n) &:= \{X \in M_n(\mathbb{C}) : X^* = -X\}, \\ \mathfrak{sp}(n) &:= \{X \in M_n(\mathbb{H}) : X^* = -X\} \end{aligned}$$

are reductive Lie algebras. In any case, the defining representation on $\mathbb{K}^n$ is irreducible, so $\mathfrak{so}(n)$ for $n \geq 3$ and $\mathfrak{sp}(n)$ for $n \geq 1$ are semisimple, while the center of $\mathfrak{u}(n)$ consists of all purely imaginary multiples of the identity matrix. The commutator algebra of $\mathfrak{u}(n)$ is $\mathfrak{su}(n) := \{X \in \mathfrak{u}(n) : \mathrm{tr}(X) = 0\}$, which is thus a semisimple Lie algebra.

This easily generalizes to forms with arbitrary signature. More formally, if $\mathbb{I} \in M_n(\mathbb{K})$ is a matrix such that $\mathbb{I}^* = \mathbb{I}$ and $\mathbb{I}^2 = \mathrm{id}$, then one immediately verifies that $\{X \in M_n(\mathbb{K}) : X^*\mathbb{I} = -\mathbb{I}X\}$ is a Lie subalgebra of $\mathfrak{gl}(n, \mathbb{K})$. Multiplying the defining equation from both sides with $\mathbb{I}$ one sees that this Lie subalgebra is stable under conjugate transpose and thus reductive. In particular, putting $\mathbb{I}$ to be the diagonal matrix whose first p diagonal entries are equal to one, while the remaining q diagonal entries are equal to -1, then putting $\mathbb{K} = \mathbb{R}$, $\mathbb{C}$ and $\mathbb{H}$, respectively, we obtain the reductive Lie algebras $\mathfrak{so}(p, q)$, $\mathfrak{u}(p, q)$ and $\mathfrak{sp}(p, q)$, respectively. As above, the first and last of these are semisimple, while $\mathfrak{u}(p, q)$ has one–dimensional center, and the semisimple commutator subalgebra $\mathfrak{su}(p, q)$ is formed by all tracefree elements.

This method also applies to the symplectic Lie algebras over $\mathbb{K} = \mathbb{R}$ and $\mathbb{C}$. Consider $\mathbb{K}^{2n}$ and the matrix $\mathbb{J} = \begin{pmatrix} 0 & 1 \\ -1 & 0 \end{pmatrix} \in M_{2n}(\mathbb{K})$, where 1 denotes the $n \times n$ identity matrix. Then the argument above shows that $\mathfrak{sp}(2n, \mathbb{K}) := \{X \in M_{2n}(\mathbb{K}) : X^*\mathbb{J} = -\mathbb{J}X\}$ is a reductive Lie algebra. Again, one verifies that these Lie algebras are semisimple. Under the mapping from quaternionic matrices to complex matrices from above, one easily verifies that $\mathfrak{sp}(n)$ corresponds to $\mathfrak{sp}(2n, \mathbb{C}) \cap \mathfrak{u}(2n)$.

2.1.8. Radical, nilradical, and the Levi decomposition. Let us return to the study of a general Lie algebra $\mathfrak{g}$. Suppose that $\mathfrak{a}, \mathfrak{b} \subset \mathfrak{g}$ are solvable ideals and consider the sum $\mathfrak{a} + \mathfrak{b}$. Clearly, this is an ideal in $\mathfrak{g}$, and since $\mathfrak{b} \subset \mathfrak{a} + \mathfrak{b}$ is a solvable ideal and $(\mathfrak{a} + \mathfrak{b})/\mathfrak{b} \cong \mathfrak{a}/(\mathfrak{a} \cap \mathfrak{b})$ is solvable, we know from 2.1.1 that $\mathfrak{a} + \mathfrak{b}$

is solvable, too. But this immediately implies that any Lie algebra $\mathfrak{g}$ possesses a largest solvable ideal, which is called the *radical* $\mathfrak{rad}(\mathfrak{g})$ of $\mathfrak{g}$. Indeed, if $\mathfrak{r} \lhd \mathfrak{g}$ is a solvable ideal of maximal dimension and $\mathfrak{a} \subset \mathfrak{g}$ is any solvable ideal, then $\mathfrak{r} + \mathfrak{a}$ is a solvable ideal, which implies $\mathfrak{a} \subset \mathfrak{r}$.

For any solvable ideal in the quotient $\mathfrak{g}/\mathfrak{rad}(\mathfrak{g})$ the preimage is a solvable ideal in $\mathfrak{g}$, which must equal $\mathfrak{rad}(\mathfrak{g})$ by maximality. Consequently, this quotient is semisimple. With a little more effort, one can improve this result, by showing that there is even a Lie subalgebra $\mathfrak{l} \leq \mathfrak{g}$ which is complementary to $\mathfrak{rad}(\mathfrak{g})$. Then $\mathfrak{l} \cong \mathfrak{g}/\mathfrak{rad}(\mathfrak{g})$ is a semisimple Lie algebra. Knowing $\mathfrak{rad}(\mathfrak{g})$ and $\mathfrak{l}$, the bracket on $\mathfrak{g}$ is determined by $[X, Y] \in \mathfrak{rad}(\mathfrak{g})$ for $X \in \mathfrak{l}$ and $Y \in \mathfrak{rad}(\mathfrak{g})$. This bracket defines a representation of $\mathfrak{l}$ on $\mathfrak{rad}(\mathfrak{g})$, and the Jacobi identity implies this action is by derivations. One says that $\mathfrak{g}$ is a semidirect product or semidirect sum of $\mathfrak{l}$ and $\mathfrak{rad}(\mathfrak{g})$. We collect the information in a theorem, which is proved, e.g., in [**FH91**, Appendix E].

THEOREM 2.1.8 (Levi decomposition). *Let $\mathfrak{g}$ be a Lie algebra with radical $\mathfrak{r}$. Then there is a semisimple Lie subalgebra $\mathfrak{l} \leq \mathfrak{g}$ such that $\mathfrak{g} = \mathfrak{r} \oplus \mathfrak{l}$ as a vector space.*

The semisimple subalgebra $\mathfrak{l}$ which shows up in the decomposition is usually called a *Levi factor* of $\mathfrak{g}$. It is not uniquely determined, however, one can show that any two Levi factors for a Lie algebra are conjugate under an inner automorphism.

Note that this also implies some information on representations. In particular, for any one–dimensional representation V of $\mathfrak{g}$, the restriction to $\mathfrak{l}$ is trivial (since semisimplicity implies $\mathfrak{l} = [\mathfrak{l}, \mathfrak{l}]$), so this is induced by a linear map on $\mathfrak{r}/[\mathfrak{r}, \mathfrak{r}]$. On the other hand, suppose that V is a complex irreducible representation of $\mathfrak{g}$. By Lie's theorem, there exists a joint eigenvector for all elements $X \in \mathfrak{r}$, i.e. there is a vector $v \in V$ and a linear functional $\lambda : \mathfrak{r} \to \mathbb{C}$ such that $X \cdot v = \lambda(X)v$ for all $X \in \mathfrak{r}$. From the proof of Lie's theorem we see that the corresponding eigenspace is $\mathfrak{g}$–invariant, so by irreducibility it must coincide with V. Denoting by V_0 the one–dimensional representation of $\mathfrak{g}$ corresponding to the linear functional λ, we see that $V_0^* \otimes V$ is an irreducible representation on which $\mathfrak{r}$ acts trivially, so it may be viewed as an irreducible representation of the quotient $\mathfrak{g}/\mathfrak{r} \cong \mathfrak{l}$. Thus, to understand irreducible representations of arbitrary Lie algebras it is sufficient to understand irreducible representations of semisimple Lie algebras.

Let us also mention the concept of the nilradical of a Lie algebra, which is closely related to the concept of the radical. If $\mathfrak{a}, \mathfrak{b} \subset \mathfrak{g}$ are nilpotent ideals, then we have already noted that $\mathfrak{a} + \mathfrak{b}$ is an ideal. For an ideal $\mathfrak{a} \lhd \mathfrak{g}$, nilpotency is equivalent to the fact there is a natural number k such that any bracket expression $[X_1, [X_2, \dots, [X_{n-1}, X_n] \dots]$ in $\mathfrak{g}$ gives zero if at least k of the entries are from $\mathfrak{a}$. Via this condition it is easy to see that for nilpotent ideals $\mathfrak{a}$ and $\mathfrak{b}$ in $\mathfrak{g}$, also $\mathfrak{a} + \mathfrak{b}$ is a nilpotent ideal in $\mathfrak{g}$.

Using this, we may conclude as in the case of the radical above, that any Lie algebra $\mathfrak{g}$ has a unique maximal nilpotent ideal, called the *nilradical* of $\mathfrak{g}$. Since nilpotent Lie algebras do not have as nice stability properties as solvable ones, the nilradical is less important than the radical. However, in favorable situations (as we shall meet in the case of parabolic subalgebras) there is an analog of the Levi decomposition with the radical replaced by the nilradical. In these cases, one finds a reductive Lie subalgebra $\mathfrak{l} \leq \mathfrak{g}$ (called a *reductive Levi factor*) such that $\mathfrak{g}$ is the direct sum of $\mathfrak{l}$ and the nilradical $\mathfrak{n}$ of $\mathfrak{g}$ as a vector space. As before, this makes $\mathfrak{g}$ into a semidirect product, and the Lie algebra structure of $\mathfrak{g}$ is encoded into the

Lie algebra structures of $\mathfrak{l}$ and $\mathfrak{n}$ and in the representation of $\mathfrak{l}$ on $\mathfrak{n}$ induced by the bracket in $\mathfrak{g}$. This representation is again by derivations.

2.1.9. Homology and cohomology of Lie algebras. Let $\mathfrak{g}$ be a Lie algebra and $\rho : \mathfrak{g} \to \mathfrak{gl}(V)$ a representation on a finite–dimensional vector space V. We are going to define two sequences of vector spaces, the homology groups $H_k(\mathfrak{g}, V)$ and the cohomology groups $H^k(\mathfrak{g}, V)$, both defined for $k \geq 0$. First, for $k \geq 0$ we define the space of *k–chains on $\mathfrak{g}$ with coefficients in V* by $C_k(\mathfrak{g}, V) := \Lambda^k\mathfrak{g} \otimes V$. Next, we define the *boundary operator* $\delta : C_k(\mathfrak{g}, V) \to C_{k-1}(\mathfrak{g}, V)$ by

$$
\begin{aligned}
\delta(X_1 \wedge \dots \wedge X_k \otimes v) :=& \sum_{i=1}^{k}(-1)^i X_1 \wedge \dots \wedge \widehat{X_i} \wedge \dots \wedge X_k \otimes (X_i \cdot v) \\
&+\sum_{i<j}(-1)^{i+j}[X_i, X_j] \wedge X_1 \wedge \dots \wedge \widehat{X_i} \wedge \dots \wedge \widehat{X_j} \wedge \dots \wedge X_k \otimes v,
\end{aligned} \tag{2.1}
$$

where the hats indicate omission. One verifies directly that $\delta^2 = \delta \circ \delta = 0$. Hence, the space $B_k := \delta(C_{k+1}(\mathfrak{g}, V)) \subset C_k(\mathfrak{g}, V)$ of *k–boundaries on $\mathfrak{g}$ with coefficients in V* is contained in the space $Z_k(\mathfrak{g}, V) := \ker(\delta) \subset C_k(\mathfrak{g}, V)$ of *k–cycles*. The *kth homology group* $H_k(\mathfrak{g}, V)$ is then defined as the quotient space $Z_k(\mathfrak{g}, V)/B_k(\mathfrak{g}, V)$.

The space $C_k(\mathfrak{g}, V)$ carries a natural representation of $\mathfrak{g}$ induced from the adjoint representation on $\mathfrak{g}$ and the given representation ρ on V. One verifies directly that δ is equivariant for these actions. Hence, $Z_k(\mathfrak{g}, V)$ and $B_k(\mathfrak{g}, V)$ are $\mathfrak{g}$–submodules of $C_k(\mathfrak{g}, V)$, and we obtain a natural representation of $\mathfrak{g}$ on the homology $H_k(\mathfrak{g}, V)$. In fact, equivariancy of δ is easier to verify for group actions. Denoting by G the unique connected and simply connected Lie group with Lie algebra $\mathfrak{g}$, the representation ρ integrates to a representation of G on V. The resulting representation on $C_k(\mathfrak{g}, V)$ is then characterized by

$$g \cdot (X_1 \wedge \dots \wedge X_k \otimes v) = \mathrm{Ad}(g)(X_1) \wedge \dots \wedge \mathrm{Ad}(g)(X_k) \otimes (g \cdot v).$$

For this action, equivariancy of δ follows easily from the facts that $\mathrm{Ad}(g)$ is a Lie algebra homomorphism and that $\mathrm{Ad}(g)(X) \cdot v = g \cdot X \cdot g^{-1} \cdot v$. Equivariancy for the action of $\mathfrak{g}$ then follows immediately.

Some low–dimensional homology groups have simple direct interpretations. For example, the boundary operator

$$\delta : C_1(\mathfrak{g}, V) = \mathfrak{g} \otimes V \to V = C_0(\mathfrak{g}, V)$$

is given by $\delta(X \otimes v) = X \cdot v$. Hence, $H_0(\mathfrak{g}, V) = V/(\mathfrak{g} \cdot V)$, the quotient of V by the subspace generated by all elements of the form $X \cdot v$. Specializing to $V = \mathfrak{g}$ we see that $H_0(\mathfrak{g}, \mathfrak{g}) = \mathfrak{g}/[\mathfrak{g}, \mathfrak{g}]$, the abelization of $\mathfrak{g}$.

To define cohomology groups, one uses a dual approach. The space $C^k(\mathfrak{g}, V)$ of k–cochains on $\mathfrak{g}$ with values in V is defined as $\Lambda^k\mathfrak{g}^* \otimes V$, so we can view k–cochains as V–valued k–linear alternating maps on $\mathfrak{g}$. In this picture, the *coboundary operator* $\partial : C^k(\mathfrak{g}, V) \to C^{k+1}(\mathfrak{g}, V)$ is defined by

$$
\begin{aligned}
\partial\phi(X_0, \dots, X_k) :=& \sum_{i=0}^{k}(-1)^i X_i \cdot \phi(X_0, \dots, \widehat{X_i}, \dots, X_k) \\
&+\sum_{i<j}(-1)^{i+j}\phi([X_i, X_j], X_0 \dots, \widehat{X_i}, \dots, \widehat{X_j}, \dots, X_k),
\end{aligned} \tag{2.2}
$$

where the hats again indicate omission. It is a well–known fact that this coboundary operator defines a differential, i.e. $\partial^2 = \partial \circ \partial = 0$. Thus, the space of $B^k(\mathfrak{g}, V) := \partial(C^{k-1}(\mathfrak{g}, V))$ of *k–coboundaries* is a subspace of the space $Z^k(\mathfrak{g}, V) = \ker(\partial) \subset C^k(\mathfrak{g}, V)$ of *k–cocycles*. The quotient space $Z^k(\mathfrak{g}, V)/B^k(\mathfrak{g}, V)$ is called the kth *cohomology group* of $\mathfrak{g}$ with coefficients in V and is denoted by $H^k(\mathfrak{g}, V)$.

Some cohomology groups of low degree are easy to interpret: From the definitions one immediately verifies that $H^0(\mathfrak{g}, V) = V^{\mathfrak{g}}$, the subspace of V on which $\mathfrak{g}$ acts trivially. For $k = 1$, a linear map $\phi : \mathfrak{g} \to V$ is a cocycle if and only if $\phi([X, Y]) = X \cdot \phi(Y) - Y \cdot \phi(X)$. A map ϕ which satisfies this condition is called a *derivation* on $\mathfrak{g}$ with values in V. In the case of the adjoint representation one recovers the notion of a derivation introduced in 2.1.6. The definition of a representation can be rephrased as requiring that for any $v \in V$ the map $\rho^v : \mathfrak{g} \to V$, $\rho^v(X) = X \cdot v$ is a derivation. Motivated by the case of the adjoint representation, such derivations are called *inner derivations*. By construction the first cohomology $H^1(\mathfrak{g}, V)$ is exactly the quotient of all V–valued derivations on $\mathfrak{g}$ by all inner derivations. In particular, part (2) of Corollary 2.1.6 can be rephrased as saying that for any semisimple Lie algebra $\mathfrak{g}$, we have $H^1(\mathfrak{g}, \mathfrak{g}) = 0$.

As before, the adjoint representation and the given representation on V make $C^k(\mathfrak{g}, V)$ into a $\mathfrak{g}$–module. Explicitly, the action is given by

$$(X \cdot \phi)(X_1, \dots, X_k) = X \cdot \phi(X_1, \dots, X_k) - \sum_{i=1}^{k} \phi(X_1, \dots, [X, X_i], \dots, X_k).$$

The coboundary map ∂ is equivariant for this action, and as before this is easier to verify in the picture of group actions. Hence, $B^k(\mathfrak{g}, V)$ and $Z^k(\mathfrak{g}, V)$ are subrepresentations of $C^k(\mathfrak{g}, V)$ and there is an induced action of $\mathfrak{g}$ on the cohomology groups.

Using this, we can immediately generalize part (2) of Corollary 2.1.6 to the case of arbitrary (finite–dimensional) representations of semisimple Lie algebras. Indeed, one only has to observe that for a derivation $\phi : \mathfrak{g} \to V$ by definition of the action one gets $X \cdot \phi = \rho^{\phi(X)}$, so the action of $\mathfrak{g}$ maps $Z^1(\mathfrak{g}, V)$ to $B^1(\mathfrak{g}, V)$ and, in particular, the action of $\mathfrak{g}$ on $H^1(\mathfrak{g}, V)$ is trivial. By Theorem 2.1.6 there is an invariant complement U to $B^1(\mathfrak{g}, V)$ in $Z^1(\mathfrak{g}, V)$, and $\mathfrak{g}$ has to act trivially on this complement. Hence, for $\phi \in U$, we must have $\rho^{\phi(X)} = 0$ for all $X \in \mathfrak{g}$, which implies that ϕ must have values in the subspace $Z^0(\mathfrak{g}, V)$ of $\mathfrak{g}$–invariant elements in V. Since ϕ is a derivation, this implies $\phi([X, Y]) = 0$ for all $X, Y \in \mathfrak{g}$, which implies $\phi = 0$, since $[\mathfrak{g}, \mathfrak{g}] = \mathfrak{g}$ by part (1) of Corollary 2.1.5. Thus, $H^1(\mathfrak{g}, V) = 0$ for any semisimple Lie algebra $\mathfrak{g}$ and any finite–dimensional representation V. This result is usually known as the first Whitehead lemma.

We only mention briefly that some of the higher cohomology groups have interesting interpretations for general Lie algebras. In particular, the second cohomology $H^2(\mathfrak{g}, V)$ describes central extensions of $\mathfrak{g}$ with kernel V, i.e. isomorphism classes of Lie algebras $\mathfrak{h}$, such that $V \subset \mathfrak{z}(\mathfrak{h})$ in such a way that $\mathfrak{h}/V \cong \mathfrak{g}$. More general extensions can be also related to Lie algebra cohomology; see [**AMR**]. On the other hand, the second cohomology of $\mathfrak{g}$ with values in the adjoint representation is related to deformations of the Lie algebra $\mathfrak{g}$. Again, there is a general vanishing theorem for the second cohomology of semisimple Lie algebras, which is known as the second Whitehead lemma. In particular, this implies that semisimple Lie algebras are rigid, i.e. cannot be smoothly deformed in a nontrivial way.

2.1.10. Universal enveloping algebra and induced modules. In 2.1.1 we have noted that if A is an associative algebra, then the commutator $[a, b] = ab - ba$ makes A into a Lie algebra. If $\mathfrak{g}$ is a Lie algebra, then one can associate to it a unital associative algebra $\mathcal{U}(\mathfrak{g})$, the *universal enveloping algebra* of $\mathfrak{g}$, together with a homomorphism $i : \mathfrak{g} \to \mathcal{U}(\mathfrak{g})$ of Lie algebras, which has the following universal property: If A is any unital associative algebra and $\phi : \mathfrak{g} \to A$ is a homomorphism of Lie algebras, then there is a unique homomorphism $\tilde{\phi} : \mathcal{U}(\mathfrak{g}) \to A$ of unital associative algebras such that $\tilde{\phi} \circ i = \phi$. In particular, a representation of $\mathfrak{g}$ is a homomorphism to the associative algebra of endomorphisms of a vector space, which is viewed as a Lie algebra under the commutator. Hence, a representation of the Lie algebra $\mathfrak{g}$ is the same thing as a left module over the universal enveloping algebra $\mathcal{U}(\mathfrak{g})$.

Explicitly, $\mathcal{U}(\mathfrak{g})$ can be constructed as the quotient of the tensor algebra $\mathcal{T}(\mathfrak{g})$ by the ideal I generated by all elements of the form $X \otimes Y - Y \otimes X - [X, Y]$ for $X, Y \in \mathfrak{g}$. Then the inclusion $\mathfrak{g} \hookrightarrow \mathcal{T}(\mathfrak{g})$ induces a homomorphism $\mathfrak{g} \to \mathcal{T}(\mathfrak{g})/I$ of Lie algebras and the universal property easily follows from the universal property of the tensor algebra. We will usually suppress the inclusion i and simply view $\mathfrak{g}$ as a subspace of $\mathcal{U}(\mathfrak{g})$.

From this explicit description one sees that the algebra $\mathcal{U}(\mathfrak{g})$ admits a canonical filtration $\mathcal{U}_0(\mathfrak{g}) \subset \cdots \subset \mathcal{U}_i(\mathfrak{g}) \subset \mathcal{U}_{i+1}(\mathfrak{g}) \subset \dots$ by linear subspaces such that it becomes a filtered algebra, i.e. the product of an element of $\mathcal{U}_i(\mathfrak{g})$ and an element of $\mathcal{U}_j(\mathfrak{g})$ lies in $\mathcal{U}_{i+j}(\mathfrak{g})$. To get this filtration, recall that the tensor algebra $\mathcal{T}(\mathfrak{g})$ is canonically graded, $\mathcal{T}(\mathfrak{g}) = \bigoplus_{n \in \mathbb{N}} \mathcal{T}^n(\mathfrak{g})$, where $\mathcal{T}^0(\mathfrak{g}) = \mathbb{K}$, the ground field and $\mathcal{T}^n(\mathfrak{g})$ the tensor product of n copies of $\mathfrak{g}$ for $n \geq 1$. Associated to this grading is an increasing filtration $\mathcal{T}_0(\mathfrak{g}) \subset \mathcal{T}_1(\mathfrak{g}) \subset \dots$, where $\mathcal{T}_n(\mathfrak{g}) = \bigoplus_{i \leq n} \mathcal{T}^i(\mathfrak{g})$. Since the generators of the ideal I from above are linear combinations of two elements of degree two and one element of degree one, I is not homogeneous with respect to the grading of $\mathcal{T}(\mathfrak{g})$, so the quotient $\mathcal{U}(\mathfrak{g}) = \mathcal{T}(\mathfrak{g})/I$ is not a graded algebra, but the filtration passes to the quotient. Otherwise put, an element $x \in \mathcal{U}(\mathfrak{g})$ lies in $\mathcal{U}_n(\mathfrak{g})$ if and only if it can be written as a linear combination of products of the form $X_1 \cdot X_2 \cdots X_\ell$ with $X_i \in \mathfrak{g}$ and $\ell \leq n$.

A basic result on universal enveloping algebras is the Poincaré–Birkhoff–Witt (PBW) theorem (see [**Kn96**, III.2]), which gives a description of the associated graded algebra $\operatorname{gr}\mathcal{U}(\mathfrak{g})$. This associated graded algebra is defined by $\operatorname{gr}\mathcal{U}(\mathfrak{g}) := \bigoplus_{n \geq 0}(\mathcal{U}_n(\mathfrak{g})/\mathcal{U}_{n-1}(\mathfrak{g}))$, where by definition $\mathcal{U}_{-1}(\mathfrak{g}) = \{0\}$. In particular, as a vector space $\operatorname{gr}\mathcal{U}(\mathfrak{g}) \cong \mathcal{U}(\mathfrak{g})$. Note that by definition $\mathcal{U}_0(\mathfrak{g}) = \mathbb{K}$ and $\mathcal{U}_1(\mathfrak{g})/\mathcal{U}_0(\mathfrak{g}) = \mathfrak{g}$, so we have a canonical inclusion $\mathfrak{g} \hookrightarrow \operatorname{gr}\mathcal{U}(\mathfrak{g})$.

THEOREM 2.1.10 (Poincaré–Birkhoff–Witt). *The inclusion $\mathfrak{g} \hookrightarrow \operatorname{gr}\mathcal{U}(\mathfrak{g})$ has the universal property of the symmetric algebra, so $\operatorname{gr}\mathcal{U}(\mathfrak{g})$ is canonically isomorphic to the algebra of polynomials on $\mathfrak{g}^*$. In particular, if $\{X_i : i \in I\}$ is an ordered linear basis for $\mathfrak{g}$, then the set $\{X_{i_1} \cdot X_{i_2} \cdots X_{i_\ell} : \ell \in \mathbb{N}, i_1 \leq i_2 \leq \cdots \leq i_\ell\}$ is a linear basis for $\mathcal{U}(\mathfrak{g})$.*

If $\mathfrak{g}$ is a Lie algebra and $\mathfrak{h} \leq \mathfrak{g}$ is a subalgebra, then the composition $\mathfrak{h} \hookrightarrow \mathfrak{g} \hookrightarrow \mathcal{U}(\mathfrak{g})$ induces a homomorphism $\mathcal{U}(\mathfrak{h}) \to \mathcal{U}(\mathfrak{g})$ of associative algebras. The PBW theorem immediately implies that this homomorphism is injective, so we may view $\mathcal{U}(\mathfrak{h})$ as a subalgebra of $\mathcal{U}(\mathfrak{g})$. This leads to the following concept of induced modules. Via multiplication from the right, we can make $\mathcal{U}(\mathfrak{g})$ into a right $\mathcal{U}(\mathfrak{h})$–module. Given a representation V of $\mathfrak{h}$, we have observed above that

V is canonically a left $\mathcal{U}(\mathfrak{h})$–module. We define the *$\mathfrak{g}$–module induced by V* as $\mathcal{U}(\mathfrak{g}) \otimes_{\mathcal{U}(\mathfrak{h})} V$. By definition, as a vector space this is the quotient of the tensor product $\mathcal{U}(\mathfrak{g}) \otimes V$ by the linear subspace of all elements of the form $ab\otimes v - a\otimes b\cdot v$ for $a \in \mathcal{U}(\mathfrak{g})$, $b \in \mathcal{U}(\mathfrak{h})$ and $v \in V$. This subspace is obviously stable under multiplication by elements of $\mathcal{U}(\mathfrak{g})$ from the left, so $\mathcal{U}(\mathfrak{g}) \otimes_{\mathcal{U}(\mathfrak{h})} V$ is canonically a left $\mathcal{U}(\mathfrak{g})$–module and thus a representation of $\mathfrak{g}$.

If there is a Lie subalgebra $\mathfrak{n} \leq \mathfrak{g}$ which is a linear complement to $\mathfrak{h}$, i.e. such that $\mathfrak{g} = \mathfrak{n} \oplus \mathfrak{h}$ as a vector space, then the PBW theorem easily leads to an explicit description of induced modules. Consider an ordered linear basis $\{X_i, Y_j : i \in I, j \in J\}$ for $\mathfrak{g}$ which is the union of a basis $\{X_i\}$ of $\mathfrak{n}$ and a basis $\{Y_j\}$ of $\mathfrak{h}$. By the PBW theorem, the monomials $X_{i_1} \cdots X_{i_k} \cdot Y_{j_1} \cdots Y_{j_\ell}$ with $i_1 \leq \cdots \leq i_k$ and $j_1 \leq \cdots \leq j_\ell$ form a basis for $\mathcal{U}(\mathfrak{g})$. Taking the tensor product with V and projecting to the induced module, one may move all the Y_j to the other side, so we see that as a vector space $\mathcal{U}(\mathfrak{g})\otimes_{\mathcal{U}(\mathfrak{h})} V$ is isomorphic to $\mathcal{U}(\mathfrak{n})\otimes V$. This is even an isomorphism of $\mathfrak{n}$–modules, which describes part of the action of $\mathfrak{g}$ on the induced module. Hence, it remains to describe the action of the subalgebra $\mathfrak{h} \leq \mathfrak{g}$ in this picture. For $X \in \mathfrak{n}$ and $Y \in \mathfrak{h}$, we have the relation $YX = XY + [Y, X]$ in $\mathcal{U}(\mathfrak{g})$. Splitting the bracket $[Y, X] = [Y, X]_{\mathfrak{n}} + [Y, X]_{\mathfrak{h}}$ according to $\mathfrak{g} = \mathfrak{n} \oplus \mathfrak{h}$, we see that this allows us to step by step commute elements of $\mathfrak{h}$ through elements of $\mathfrak{n}$. Once the elements of $\mathfrak{h}$ are on the right, they can be moved to the V–component.

Induced modules as introduced here are an analog of induced group representations as discussed in 1.4.4, i.e. the natural representation of a Lie group G on smooth sections of the homogeneous vector bundle $G \times_H V \to G/H$ corresponding to a representation V of H. The precise relation between the two concepts is via passing to infinite jets at the base point $o = eH \in G/H$ plus a dualization; see 1.4.10 for a sketch. What we note here is that there is an analog of Frobenius reciprocity from Theorem 1.4.4.

PROPOSITION 2.1.10 (Algebraic version of Frobenius reciprocity). *Let $\mathfrak{g}$ be a Lie algebra, $\mathfrak{h} \leq \mathfrak{g}$ a Lie subalgebra, V a representation of $\mathfrak{h}$, and W a representation of $\mathfrak{g}$. Then restriction to elements of the form $1 \otimes v$ induces a natural isomorphism*

$$\operatorname{Hom}_{\mathfrak{g}}(\mathcal{U}(\mathfrak{g}) \otimes_{\mathcal{U}(\mathfrak{h})} V, W) \to \operatorname{Hom}_{\mathfrak{h}}(V, W).$$

PROOF. Suppose that $\Phi : \mathcal{U}(\mathfrak{g}) \otimes_{\mathcal{U}(\mathfrak{h})} V \to W$ is a $\mathfrak{g}$–module homomorphism, and define $\phi : V \to W$ by $\phi(v) := \Phi(1 \otimes v)$. For $X \in \mathfrak{h}$, we have

$$\phi(X \cdot v) = \Phi(1 \otimes X \cdot v) = \Phi(X \otimes v) = X \cdot \phi(v),$$

so ϕ is a $\mathfrak{h}$–homomorphism. Conversely, assume that $\phi : V \to W$ is a $\mathfrak{h}$–homomorphism. Then consider the bilinear map $\mathcal{U}(\mathfrak{g}) \times V \to W$ defined by $(x, v) \mapsto x \cdot \phi(v)$. For $y \in \mathcal{U}(\mathfrak{h})$, we have $xy \cdot \phi(v) = x \cdot y \cdot \phi(v) = x \cdot \phi(y \cdot v)$ since ϕ is an $\mathfrak{h}$– and thus an $\mathcal{U}(\mathfrak{h})$–homomorphism. Consequently, the induced linear map on the tensor product factors to a linear map $\mathcal{U}(\mathfrak{g}) \otimes_{\mathcal{U}(\mathfrak{h})} V \to W$, which by construction is a $\mathfrak{g}$–homomorphism. Clearly, the two constructions are inverse to each other. □

2.2. Complex semisimple Lie algebras and their representations

In this section we study the structure of complex semisimple Lie algebras and their (complex) representations. We start with a quite detailed presentation of Cartan subalgebras and the corresponding roots, orderings, the resulting concepts of positive and simple roots, and the Dynkin diagram. After a discussion of the classical examples and a detailed study of the Weyl group, we outline the classification of connected Dynkin diagrams and its relation to the classification of complex simple Lie algebras. Next, we switch to finite–dimensional representations. After discussing the concept of weights, we formulate the theorem of the highest weight, prove its uniqueness part, and sketch three different proofs for the existence part (via Verma modules, the Borel–Weil theorem, and via fundamental representations). Then we describe the decomposition of a general representation into isotypical components, the use of the Casimir element for realizing this decomposition, and the infinitesimal character. Formulae for multiplicities of weights, the Weyl character and dimension formulae, and a brief discussion of representations of Lie groups conclude the section.

2.2.1. Jordan decomposition. The first key point in the theory of complex semisimple Lie algebras is the notion of semisimple elements. In 2.1.5 we have met the Jordan–decomposition $f = f_s + f_n$ of an endomorphism of a finite–dimensional complex vector space. For complex semisimple Lie algebras, there is a universal version of the Jordan decomposition, called the *absolute Jordan decomposition*, which gives the Jordan decomposition in any finite–dimensional representation:

LEMMA 2.2.1. *Let $\mathfrak{g}$ be a complex semisimple Lie algebra.*

(1) Suppose that $\mathfrak{g}$ is a complex Lie subalgebra of $\mathfrak{gl}(V)$ for a finite-dimensional complex vector space V. For $X \in \mathfrak{g}$ let $X = S + N$ be the Jordan decomposition of the linear map $X : V \to V$. Then S and N lie in $\mathfrak{g}$.

(2) For any $X \in \mathfrak{g}$ there are unique elements $X_s, X_n \in \mathfrak{g}$ such that $\text{ad}(X) = \text{ad}(X_s) + \text{ad}(X_n)$ *is the Jordan decomposition of* $\text{ad}(X) : \mathfrak{g} \to \mathfrak{g}$.

(3) If $\rho : \mathfrak{g} \to \mathfrak{gl}(V)$ is a finite-dimensional complex representation, then for any element $X \in \mathfrak{g}$, the Jordan decomposition of $\rho(X) : V \to V$ is given by $\rho(X) = \rho(X_s) + \rho(X_n)$.

PROOF. (1) This is done by writing $\mathfrak{g}$ in a smart way as an intersection of Lie subalgebras of $\mathfrak{gl}(V)$. Put $\mathfrak{n} := \{A \in \mathfrak{gl}(V) : \text{ad}(A)(\mathfrak{g}) \subset \mathfrak{g}\}$, the normalizer of $\mathfrak{g}$ in $\mathfrak{gl}(V)$. Of course, this is a subalgebra which contains $\mathfrak{g}$. Moreover, for $A \in \mathfrak{n}$ with Jordan decomposition $A = S + N$, we know from Lemma 2.1.5 that $\text{ad}(A) = \text{ad}(S) + \text{ad}(N)$ is the Jordan decomposition of $\text{ad}(A) : \mathfrak{gl}(V) \to \mathfrak{gl}(V)$. In particular, this implies that $\text{ad}(S)$ and $\text{ad}(N)$ are polynomials in $\text{ad}(A)$, which implies that $S, N \in \mathfrak{n}$. Next, for each $\mathfrak{g}$–invariant linear subspace $W \subset V$ define

$$\mathfrak{s}_W := \{A \in \mathfrak{gl}(V) : A(W) \subset W, \text{tr}(A|_W) = 0\}.$$

This is obviously a Lie subalgebra of $\mathfrak{gl}(V)$. For $X \in \mathfrak{g}$ we have $X(W) \subset W$ by definition, and since $\mathfrak{g} = [\mathfrak{g}, \mathfrak{g}]$ we may write $X|_W$ as a sum of commutators, so it is tracefree. Thus, $\mathfrak{g} \subset \mathfrak{s}_W$ for each $\mathfrak{g}$–invariant subspace W. On the other hand, for $A \in \mathfrak{s}_W$ with Jordan decomposition $A = S+N$, the maps S and N are polynomials in A and thus map W to itself. Moreover, $N|_W$ is nilpotent and thus tracefree, so $S|_W = A|_W - N|_W$ is tracefree, and thus $S, N \in \mathfrak{s}_W$.

Now let $\mathfrak{g}'$ be the intersection of $\mathfrak{n}$ and all $\mathfrak{s}_W$. Then $\mathfrak{g} \subset \mathfrak{g}'$ and we can complete the proof by showing that the two subalgebras are equal. For $X \in \mathfrak{g}$, we have $\text{ad}(X) : \mathfrak{g}' \to \mathfrak{g}'$ so this defines a representation of $\mathfrak{g}$ on $\mathfrak{g}'$, and $\mathfrak{g} \subset \mathfrak{g}'$ is an invariant subspace. By complete reducibility (Theorem 2.1.6), there exists an invariant complement $U \subset \mathfrak{g}'$. Since $\mathfrak{g}' \subset \mathfrak{n}$ we see that $\text{ad}(X)(\mathfrak{g}') \subset \mathfrak{g}$ for all $X \in \mathfrak{g}$, so this complementary representation must be trivial. In particular, taking an element $Y \in U$, we have $[X, Y] = 0$ for all $X \in \mathfrak{g}$. If $W \subset V$ is a $\mathfrak{g}$–irreducible subrepresentation, then $Y \in \mathfrak{s}_W$, so $Y(W) \subset W$ and $\operatorname{tr}(Y|_W) = 0$. On the other hand, the action of Y on W commutes with the action of any $X \in \mathfrak{g}$, so $Y|_W$ must be a multiple of id_W by Schur's lemma, and thus $Y|_W = 0$. By complete reducibility V splits into a direct sum of $\mathfrak{g}$–irreducibles, so we conclude $Y = 0$, and thus $\mathfrak{g} = \mathfrak{g}'$.

(2) Applying (1) to $\operatorname{ad}(\mathfrak{g}) \subset \mathfrak{gl}(\mathfrak{g})$ we see that for $X \in \mathfrak{g}$ the semisimple and nilpotent part of $\operatorname{ad}(X)$ again lie in $\operatorname{ad}(\mathfrak{g})$, and since $\operatorname{ad} : \mathfrak{g} \to \operatorname{ad}(\mathfrak{g})$ is an isomorphism, the result follows.

(3) Observe first, that for $\mathfrak{g} \subset \mathfrak{gl}(V)$ as in (1), the Jordan decomposition $X = S + N$ of the linear map $X : V \to V$ coincides with the absolute Jordan decomposition. Indeed, $\operatorname{ad}(X) = \operatorname{ad}(S) + \operatorname{ad}(N)$ and since S and N commute, also $\operatorname{ad}(S)$ and $\operatorname{ad}(N)$ commute. In view of the uniqueness of the Jordan decomposition of $\operatorname{ad}(X)$ it thus suffices to show that $\operatorname{ad}(S) : \mathfrak{g} \to \mathfrak{g}$ is diagonalizable and $\operatorname{ad}(N) : \mathfrak{g} \to \mathfrak{g}$ is nilpotent. But from Lemma 2.1.5 we know those properties for the adjoint actions on $\mathfrak{gl}(V)$, and they are preserved by passing to an invariant subspace.

In view of this observation, it suffices to show that any surjective homomorphism ρ from $\mathfrak{g}$ to a Lie algebra $\mathfrak{g}'$ (which then is semisimple by Corollary 2.1.5) is compatible with the absolute Jordan decompositions. This is evidently true if ρ is an isomorphism, so it remains to show it for the natural map $\mathfrak{g} \to \mathfrak{g}/\mathfrak{h}$, where $\mathfrak{h} \subset \mathfrak{g}$ is an ideal. (Just observe that any ρ as above is the composition of an isomorphism $\mathfrak{g}/\ker(\rho) \to \mathfrak{g}'$ with the natural map $\mathfrak{g} \to \mathfrak{g}/\ker(\rho)$.) From the proof of Corollary 2.1.5 we know that the orthocomplement $\mathfrak{h}^\perp$ of $\mathfrak{h}$ with respect to the Killing form is an ideal in $\mathfrak{g}$. Further, $\mathfrak{g} = \mathfrak{h} \oplus \mathfrak{h}^\perp$ and hence $\mathfrak{g}/\mathfrak{h} \cong \mathfrak{h}^\perp$.

Now suppose that for $X \in \mathfrak{g}$ we have $X = X^1 + X^2$ with $X^1 \in \mathfrak{h}$ and $X^2 \in \mathfrak{h}^\perp$. Then it is elementary to verify that if $X^i = X^i_s + X^i_n$ is the absolute Jordan decomposition for $i = 1, 2$, then the absolute Jordan decomposition of X is given by $X_s = X^1_s + X^2_s$ and $X_n = X^1_n + X^2_n$. But this implies that the projection $\mathfrak{g} \to \mathfrak{h}^\perp$ along $\mathfrak{h}$ is compatible with the absolute Jordan decomposition. □

2.2.2. Cartan subalgebras. The result on the Jordan decomposition is an essential step towards understanding the representation theory of complex semisimple Lie algebras. An element $X \in \mathfrak{g}$ is called *semisimple* if the linear map $\operatorname{ad}(X) : \mathfrak{g} \to \mathfrak{g}$ is diagonalizable. Observe that nonzero semisimple elements always exist. If not, then by Lemma 2.2.1 we would have $X_s = 0$ for any $X \in \mathfrak{g}$, so $\operatorname{ad}(X) : \mathfrak{g} \to \mathfrak{g}$ would be nilpotent for any $X \in \mathfrak{g}$. But then Engel's Theorem from 2.1.1 implies that $\mathfrak{g}$ is nilpotent, which is a contradiction.

Let $X \in \mathfrak{g}$ be semisimple and $\mathfrak{g} := \bigoplus_\lambda \mathfrak{g}_\lambda(X)$ be the decomposition into eigenspaces for $\operatorname{ad}(X)$. By the Jacobi identity, $[\mathfrak{g}_\lambda(X), \mathfrak{g}_\mu(X)] \subset \mathfrak{g}_{\lambda+\mu}(X)$. From part (3) of Lemma 2.2.1 we then conclude that for any finite–dimensional complex representation $\rho : \mathfrak{g} \to \mathfrak{gl}(V)$ the map $\rho(X) : V \to V$ is diagonalizable. Let $V = \bigoplus V_\mu(X)$ be the resulting decomposition into eigenspaces for the linear map $\rho(X)$. Since $X \cdot Y \cdot v = [X, Y] \cdot v + Y \cdot X \cdot v$, we immediately conclude that for $Y \in \mathfrak{g}_\lambda(X)$ and $v \in V_\mu(X)$ we have $Y \cdot v \in V_{\lambda+\mu}(X)$, so we get information on the possible eigenvalues.

The next step is to refine this decomposition. Recall from linear algebra that two commuting diagonalizable linear maps are simultaneously diagonalizable.

Moreover, for a simultaneously diagonalizable family of linear maps, also the subspace spanned by the family is simultaneously diagonalizable. This suggests the following definition.

DEFINITION 2.2.2. A *Cartan subalgebra* of a complex semisimple Lie algebra is a maximal commutative subalgebra which consists of semisimple elements.

There is a more general notion of Cartan subalgebras in arbitrary Lie algebras (defined as maximal nilpotent subalgebras which coincide with their own normalizer), but we will restrict to the semisimple case here.

To study the questions of existence and uniqueness of Cartan subalgebras, we need a few more concepts. For an element $H \in \mathfrak{g}$, we define the *centralizer of H in* $\mathfrak{g}$ by $\mathfrak{c}(H) := \{X \in \mathfrak{g} : [X, H] = 0\}$. Evidently, $\mathfrak{c}(H) \subset \mathfrak{g}$ is a Lie subalgebra. We will only be interested in the case that H is a semisimple element of $\mathfrak{g}$. The minimal dimension of $\mathfrak{c}(H)$ as H ranges through all semisimple elements of $\mathfrak{g}$ is called the *rank* of $\mathfrak{g}$. Semisimple elements for which this minimal dimension is attained are called *regular.*

On the other hand, we have to make a quick digression on automorphisms of a Lie algebra $\mathfrak{g}$. A linear isomorphism $\phi : \mathfrak{g} \to \mathfrak{g}$ is an automorphism if and only if $\phi([X, Y]) - [\phi(X), \phi(Y)] = 0$ for all $X, Y \in \mathfrak{g}$. Consequently, the group $\operatorname{Aut}(\mathfrak{g})$ of all automorphisms of $\mathfrak{g}$ is a closed subgroup of $GL(\mathfrak{g})$ and thus a Lie subgroup; see 1.2.5. The Lie algebra of this Lie subgroup is given by all maps $A : \mathfrak{g} \to \mathfrak{g}$ which satisfy $A([X, Y]) - [A(X), Y] - [X, A(Y)] = 0$, so this is exactly the Lie algebra $\mathfrak{der}(\mathfrak{g})$ of derivations of $\mathfrak{g}$; see 2.1.6. In the semisimple case, part (2) of Corollary 2.1.6 implies that the Lie algebra of $\operatorname{Aut}(\mathfrak{g})$ is isomorphic to $\mathfrak{g}$. The connected component of the identity in $\operatorname{Aut}(\mathfrak{g})$ is called the group of *inner automorphisms* of $\mathfrak{g}$ and it is often denoted by $\operatorname{Int}(\mathfrak{g})$. It turns out that $\operatorname{Int}(\mathfrak{g})$ is the smallest Lie group with Lie algebra $\mathfrak{g}$, and for any Lie group G with Lie algebra $\mathfrak{g}$ the image of $\operatorname{Ad} : G \to \operatorname{Aut}(\mathfrak{g})$ contains $\operatorname{Int}(\mathfrak{g})$; see 2.2.19.

THEOREM 2.2.2. *Let $\mathfrak{g}$ be a complex semisimple Lie algebra.*

(1) If $H \in \mathfrak{g}$ is a regular semisimple element, then $\mathfrak{c}(H) \leq \mathfrak{g}$ is a Cartan subalgebra.

(2) Any two Cartan subalgebras in $\mathfrak{g}$ are conjugate by an inner automorphism of $\mathfrak{g}$.

PROOF. (1) Assume that H is a regular semisimple element of $\mathfrak{g}$, and let $\mathfrak{g} = \bigoplus \mathfrak{g}_\lambda(H)$ be the decomposition of $\mathfrak{g}$ according to eigenvalues of $\operatorname{ad}(H)$. In particular, $\mathfrak{c}(H) = \mathfrak{g}_0(H)$. We first claim that $\mathfrak{c}(H)$ is a nilpotent Lie subalgebra of $\mathfrak{g}$. If $\mathfrak{c}(H)$ would not be nilpotent, then by Engel's Theorem 2.1.1(1), there is an element $Z \in \mathfrak{c}(H)$ such that the restriction of $\operatorname{ad}(Z)$ to $\mathfrak{c}(H)$ is not nilpotent, i.e. $(\operatorname{ad}(Z)|_{\mathfrak{c}(H)})^{\dim(\mathfrak{c}(H))} \neq 0$. But then the set of all elements of $\mathfrak{c}(H)$ having this property is the complement of the zeros of a polynomial map and hence open and dense. On the other hand, we have observed above, that for any $Z \in \mathfrak{g}_0(H)$, the adjoint action $\operatorname{ad}(Z)$ preserves the decomposition $\mathfrak{g} = \bigoplus \mathfrak{g}_\lambda(H)$, and we can look at those elements, for which the restriction of $\operatorname{ad}(Z)$ to $\bigoplus_{\lambda \neq 0} \mathfrak{g}_\lambda(H)$ is invertible. Again, this set is the complement of the zeros of a polynomial, it is nonempty since it contains H, and thus it is open and dense. Since two dense open subsets of $\mathfrak{c}(H)$ must have a nontrivial intersection, we find an element $Z \in \mathfrak{c}(H)$ whose generalized eigenspace corresponding to the eigenvalue zero is strictly contained in $\mathfrak{c}(H)$. But this generalized eigenspace coincides with $\mathfrak{c}(Z_s)$, where Z_s is the semisimple part of Z. This contradicts regularity of H, so $\mathfrak{c}(H)$ is nilpotent.

The second main ingredient is to observe the behaviour of the Killing form with respect to the decomposition $\mathfrak{g} = \bigoplus \mathfrak{g}_\lambda(H)$. For $X \in \mathfrak{g}_\lambda(H)$ and $Y \in \mathfrak{g}_\mu(H)$ we see that $\operatorname{ad}(X) \circ \operatorname{ad}(Y)$ maps $\mathfrak{g}_\nu(H)$ to $\mathfrak{g}_{\lambda+\mu+\nu}(H)$, and thus is tracefree unless $\lambda + \mu = 0$. In particular, $\mathfrak{c}(H) = \mathfrak{g}_0(H)$ is perpendicular to $\bigoplus_{\lambda \neq 0} \mathfrak{g}_\lambda(H)$, and thus the restriction of the Killing form B to $\mathfrak{c}(H)$ is nondegenerate.

Now $\mathfrak{c}(H)$ is nilpotent and thus solvable, so the image $\operatorname{ad}(\mathfrak{c}(H)) \subset \mathfrak{gl}(\mathfrak{g})$ is solvable, too. By Lie's Theorem 2.1.1(2), there is a basis of $\mathfrak{g}$ with respect to which all $\operatorname{ad}(X)$ for $X \in \mathfrak{c}(H)$ are represented by upper triangular matrices. Hence, for $X, Y, Z \in \mathfrak{c}(H)$ the map $\operatorname{ad}([X,Y]) = [\operatorname{ad}(X), \operatorname{ad}(Y)]$ is represented by a matrix that is strictly upper triangular, and thus $\operatorname{ad}([X,Y]) \circ \operatorname{ad}(Z)$ is tracefree. Thus, $B([X,Y],Z) = 0$ for all $Z \in \mathfrak{c}(H)$, so $[X,Y] = 0$ by nondegeneracy of the restriction of the Killing form. Hence, $\mathfrak{c}(H)$ is an abelian Lie subalgebra of $\mathfrak{g}$. Moreover, $\mathfrak{c}(H)$ is by definition a maximal abelian Lie subalgebra, since any element that commutes with H already lies in $\mathfrak{c}(H)$. It therefore remains to verify that all elements of $\mathfrak{c}(H)$ are semisimple.

Any $Y \in \mathfrak{c}(H)$ commutes with H and thus $\operatorname{ad}(Y)$ preserves the decomposition $\mathfrak{g} = \bigoplus \mathfrak{g}_\lambda(H)$ into $\operatorname{ad}(H)$–eigenspaces. But then any polynomial in $\operatorname{ad}(Y)$ preserves this decomposition and hence commutes with $\operatorname{ad}(H)$. In particular, the nilpotent part and semisimple part in the Jordan decomposition of $\operatorname{ad}(Y)$ have this property, which implies that $Y_n \in \mathfrak{c}(H)$. For any $X \in \mathfrak{c}(H)$ the maps $\operatorname{ad}(X)$ and $\operatorname{ad}(Y_n)$ commute, so since $\operatorname{ad}(Y_n)$ is nilpotent, also $\operatorname{ad}(X) \circ \operatorname{ad}(Y_n)$ is nilpotent and thus tracefree. Hence, $B(Y_n, X) = 0$ for all $X \in \mathfrak{c}(H)$, which implies $Y_n = 0$ by nondegeneracy of the restriction of the Killing form, and hence Y is semisimple.

(2) This is proved using elementary algebraic geometry; see [**FH91**, Appendix D] or [**Kn96**, section II.3]. □

Choosing a Cartan subalgebra $\mathfrak{h} \subset \mathfrak{g}$, we have noted already above that for any finite–dimensional complex representation $\rho : \mathfrak{g} \to \mathfrak{gl}(V)$ the operators $\rho(H) : V \to V$ for $H \in \mathfrak{h}$ are simultaneously diagonalizable. Moreover, on a simultaneous eigenspace, the eigenvalue depends linearly on H, so it is described by a linear functional $\lambda : \mathfrak{h} \to \mathbb{C}$. The functionals corresponding to nontrivial eigenspaces are called the *weights* of the representation V, i.e. $\lambda : \mathfrak{h} \to \mathbb{C}$ is a weight of V if and only if there is a nonzero vector $v \in V$ such that $\rho(H)(v) = \lambda(H)v$ for all $H \in \mathfrak{h}$. If λ is a weight, then the *weight space* V_λ corresponding to λ is defined by $V_\lambda := \{v \in V : \rho(H)(v) = \lambda(H)v \quad \forall H \in \mathfrak{h}\}$. The set of weights of V is a finite subset of $\mathfrak{h}^*$, that will be denoted by $\operatorname{wt}(V)$. Note that the concept of weights also makes sense for infinite–dimensional representations.

In particular, we may consider the adjoint representation. The nonzero weights of the adjoint representation are called the *roots* of the Lie algebra $\mathfrak{g}$, and the weight space corresponding to a root is called the *root space*. We will denote the set of all roots of $\mathfrak{g}$ by Δ. The weight space corresponding to the weight zero is exactly the Cartan subalgebra, so we obtain the *root decomposition* $\mathfrak{g} = \mathfrak{h} \oplus \bigoplus_{\alpha \in \Delta} \mathfrak{g}_\alpha$. The Jacobi identity immediately implies that $[\mathfrak{g}_\alpha, \mathfrak{g}_\beta] \subset \mathfrak{g}_{\alpha+\beta}$ for $\alpha, \beta \in \Delta$, where we agree that $\mathfrak{g}_\gamma = \{0\}$ if γ is not a root. More generally, consider a representation with weight decomposition $V = \bigoplus_{\lambda \in \operatorname{wt}(V)} V_\lambda$. Then for $H \in \mathfrak{h}$, $X \in \mathfrak{g}_\alpha$, and $v \in V_\lambda$ the identitity $H \cdot X \cdot v = X \cdot H \cdot v + [H,X] \cdot v$ shows that $X \cdot v \in V_{\lambda+\alpha}$, and we immediately get some information on the set of all weights.

2.2.3. Digression on $\mathfrak{sl}(2,\mathbb{C})$. Before we can continue the study of the roots of general complex semisimple Lie algebras and the weights of their representations, we have to study the special case $\mathfrak{g} = \mathfrak{sl}(2,\mathbb{C})$ in some detail. On one hand, the representation theory of $\mathfrak{sl}(2,\mathbb{C})$ will be an essential tool in the analysis of general semisimple Lie algebras, and on the other hand, it will be a motivating example for the general theory. In fact, it is possible to also conclude some of the basic general results like complete reducibility of representations from the representation theory of $\mathfrak{sl}(2,\mathbb{C})$, but we preferred to present the linear algebra arguments.

For $\mathfrak{sl}(2,\mathbb{C})$ one can immediately write out the root decomposition. Consider the elements $E := \left(\begin{smallmatrix} 0 & 1 \\ 0 & 0 \end{smallmatrix}\right)$, $F := \left(\begin{smallmatrix} 0 & 0 \\ 1 & 0 \end{smallmatrix}\right)$ and $H := \left(\begin{smallmatrix} 1 & 0 \\ 0 & -1 \end{smallmatrix}\right)$, which obviously form a basis of $\mathfrak{g}$. One immediately verifies the commutation relations $[E, F] = H$, $[H, E] = 2E$ and $[H, F] = -2F$. In particular, this shows that $H \in \mathfrak{g}$ is semisimple and $\mathfrak{h} := \mathbb{C}\cdot H$ is a Cartan subalgebra for $\mathfrak{g}$. Taking $\{H\}$ as a basis for $\mathfrak{h}$, we may view linear functionals on $\mathfrak{h}$ simply as complex numbers. In this picture $\Delta = \{2, -2\}$ and the root decomposition is $\mathfrak{g} = \mathfrak{h} \oplus \mathfrak{g}_{-2} \oplus \mathfrak{g}_2$, where $\mathfrak{g}_2 = \mathbb{C} \cdot E$ and $\mathfrak{g}_{-2} = \mathbb{C} \cdot F$. Now we can completely describe the finite–dimensional irreducible representations of $\mathfrak{sl}(2,\mathbb{C})$.

PROPOSITION 2.2.3. *For any $n \in \mathbb{N}$, there is a unique (up to isomorphism) irreducible representation of $\mathfrak{sl}(2,\mathbb{C})$ of dimension $n+1$. The weights of this representation are $\{n, n-2, \dots, -n+2, -n\}$ and all weight spaces are one–dimensional. More precisely, there is a basis $\{v_0, \dots, v_n\}$ of V such that $H \cdot v_j = (n-2j)v_j$, $F \cdot v_j = v_{j+1}$ for $j < n$ and $F \cdot v_n = 0$, and $E \cdot v_j = j(n-j+1)v_{j-1}$ for all j.*

PROOF. Put $\mathfrak{g} := \mathfrak{sl}(2,\mathbb{C})$ and suppose that $\rho : \mathfrak{g} \to \mathfrak{gl}(V)$ is a finite–dimensional irreducible representation with weight decomposition $V = \bigoplus V_\lambda$. Here the weight λ is a complex number and $V_\lambda = \{v \in V : \rho(H)(v) = \lambda v\}$. Let λ_0 be the weight with maximal real part occurring in V, and let $v_0 \in V_{\lambda_0}$ be a nonzero element. Then $\rho(E)(v_0) = 0$, since if this were nonzero it would be a weight vector of weight $\lambda_0 + 2$. Now consider the subspace of V which is spanned by $\{v_k := \rho(F)^k(v_0) : k \in \mathbb{N}\}$. The nonzero elements in this set are eigenvectors for $\rho(H)$ with different eigenvalues, and thus they are linearly independent. Hence, there must be a minimal index n such that $v_{n+1} = 0$. By construction, $\rho(F)(v_j) = v_{j+1}$, so our subspace is invariant under $\rho(F)$. On the other hand, for $j \geq 1$ we get

$$\rho(E)(v_j) = \rho(E)(\rho(F)(v_{j-1})) = \rho([E, F])(v_{j-1}) + \rho(F)(\rho(E)(v_{j-1})),$$

and the first summand just gives $(\lambda_0 - 2j + 2)v_{j-1}$. Inductively, this easily implies that $\rho(E)(v_j) = j(\lambda_0 - j + 1)v_{j-1}$. Thus, we see that our subspace is invariant, so it has to coincide with V by irreducibility, so $\{v_0, \dots, v_n\}$ is a basis for V.

Next we show that λ_0 is determined by n. Since $\rho(H) = [\rho(E), \rho(F)]$, we know that $\rho(H)$ must be tracefree. On the other hand, $\rho(H)$ is diagonal with respect to the basis $\{v_0, \dots, v_n\}$ and its trace is given by $\sum_{j=0}^{n}(\lambda_0 - 2j) = (n+1)(\lambda_0 - n)$. Hence, $\lambda_0 = n$, and all weights occurring in V are integers. This also shows that an irreducible representation of $\mathfrak{g}$ is determined by its dimension up to isomorphism. Indeed, if we have two irreducible representations V and V' of dimension $n+1$, then we consider the bases $\{v_0, \dots, v_n\}$ and $\{v'_0, \dots, v'_n\}$ as constructed above, and then the mapping defined by $v_j \mapsto v'_j$ is an isomorphism of representations, since the actions of the three generators on each of the basis elements are determined by the dimension.

To complete the proof we have to show that for each $n \in \mathbb{N}$ there exists an irreducible representation of $\mathfrak{g}$ of dimension $n+1$. One way to to this is to define $\rho : \mathfrak{g} \to \mathfrak{gl}(n+1,\mathbb{C})$ by $\rho(F)(e_j) = e_{j+1}$, $\rho(F)(e_n) = 0$, $\rho(H)(e_j) = (n-2j)e_j$ and $\rho(E)(e_j) = j(n-j+1)e_{j-1}$, where $\{e_0,\dots,e_n\}$ is the standard basis of $\mathbb{C}^{n+1}$, and verify directly that this is a representation. More conceptually, we can use the trivial representation for $n=0$ and the standard representation on $\mathbb{C}^2$ for $n=1$. Then we form the kth symmetric power $S^k\mathbb{C}^2$ of the standard representation. This has a basis $e_1^i \vee e_2^j$ for $i+j=k$, so it has dimension $k+1$. Moreover, e_1^k is a weight vector with weight k, which implies that $S^k\mathbb{C}^2$ is the irreducible representation that we were looking for. □

2.2.4. The root system of a semisimple Lie algebra. Let us return to a general semisimple Lie algebra $\mathfrak{g}$ with a Cartan subalgebra $\mathfrak{h}$ and the corresponding root decomposition $\mathfrak{g} = \mathfrak{h} \oplus \bigoplus_{\alpha\in\Delta} \mathfrak{g}_\alpha$. To analyze this decomposition further, we have to study its relation to the Killing form B. We already know that $[\mathfrak{g}_\alpha, \mathfrak{g}_\beta] \subset \mathfrak{g}_{\alpha+\beta}$. In particular, this implies that for $X \in \mathfrak{g}_\alpha$ and $Y \in \mathfrak{g}_\beta$ the map $\operatorname{ad}(X)\circ\operatorname{ad}(Y)$ maps $\mathfrak{g}_\gamma$ to $\mathfrak{g}_{\alpha+\beta+\gamma}$, so this is tracefree unless $\beta = -\alpha$. Hence, $B(\mathfrak{g}_\alpha, \mathfrak{g}_\beta) = 0$ unless $\beta = -\alpha$. Nondegeneracy of the Killing form then implies that for $\alpha \in \Delta$ also $-\alpha \in \Delta$ and the Killing form induces a duality between $\mathfrak{g}_\alpha$ and $\mathfrak{g}_{-\alpha}$. On the other hand, the restriction of the Killing form to $\mathfrak{h}$ is nondegenerate.

In particular, for each linear functional $\lambda \in \mathfrak{h}^*$, there is a unique element $H_\lambda \in \mathfrak{h}$ such that $\lambda(H) = B(H_\lambda, H)$ for all $H \in \mathfrak{h}$. We can also use the Killing form to define a nondegenerate complex bilinear form on $\mathfrak{h}^*$ by $\langle \lambda, \mu\rangle := B(H_\lambda, H_\mu)$. Using this we can now prove the basic properties of the root decomposition.

PROPOSITION 2.2.4. *Let $\mathfrak{g}$ be a complex semisimple Lie algebra, $\mathfrak{h} \leq \mathfrak{g}$ a Cartan subalgebra, $\Delta \subset \mathfrak{h}^*$ the corresponding set of roots, and $\langle\ ,\ \rangle$ the complex bilinear form on $\mathfrak{h}^*$ induced by the Killing form. Then we have:*

(1) *For any $\alpha \in \Delta$, also $-\alpha \in \Delta$ and these are the only complex multiples of α which are roots.*
(2) *For any $\alpha \in \Delta$, the root space $\mathfrak{g}_\alpha$ is one–dimensional and the subspace $\mathfrak{s}_\alpha$ of $\mathfrak{g}$ spanned by $\mathfrak{g}_\alpha$, $\mathfrak{g}_{-\alpha}$, and $[\mathfrak{g}_\alpha, \mathfrak{g}_{-\alpha}] \subset \mathfrak{h}$ is a Lie subalgebra isomorphic to $\mathfrak{sl}(2,\mathbb{C})$.*
(3) *For $\alpha, \beta \in \Delta$ with $\beta \neq -\alpha$ we have $[\mathfrak{g}_\alpha, \mathfrak{g}_\beta] = \mathfrak{g}_{\alpha+\beta}$ if $\alpha+\beta \in \Delta$ and $[\mathfrak{g}_\alpha, \mathfrak{g}_\beta] = \{0\}$ otherwise.*
(4) *For $\alpha, \beta \in \Delta$ with $\beta \neq \pm\alpha$ and $z \in \mathbb{C}$, a functional of the form $\beta + z\alpha$ can only be a root if $z \in \mathbb{Z}$. The roots of this form are an unbroken string*

$$\beta - p\alpha, \beta - (p-1)\alpha, \dots, \beta + (q-1)\alpha, \beta + q\alpha,$$

where $p, q \geq 0$ and $p - q = \frac{2\langle\beta,\alpha\rangle}{\langle\alpha,\alpha\rangle}$.

PROOF. We have already observed that for $\alpha \in \Delta$ also $-\alpha$ is a root. Next, assume that $H \in \mathfrak{h}$ is such that $\alpha(H) = 0$ for all $\alpha \in \Delta$. Then H commutes with any element of any root space, and since $H \in \mathfrak{h}$, it also commutes with any element of $\mathfrak{h}$, so $H \in \mathfrak{z}(\mathfrak{g})$. Since $\mathfrak{g}$ is semisimple, this implies $H = 0$, so the roots generate $\mathfrak{h}^*$.

Let $X \in \mathfrak{g}_\alpha$ be nonzero. By nondegeneracy of the Killing form, there is an element $Y \in \mathfrak{g}_{-\alpha}$ such that $B(X,Y) \neq 0$. For $H \in \mathfrak{h}$ we get $[H,[X,Y]] = 0$ and hence $[X,Y] \in \mathfrak{h}$. Further, we obtain $B(H,[X,Y]) = B([H,X],Y) = \alpha(H)B(X,Y)$

and hence $[X,Y] = B(X,Y)H_\alpha$, where $H_\alpha \in \mathfrak{h}$ is the unique element such that $B(H,H_\alpha) = \alpha(H)$ for all $H \in \mathfrak{h}$. Since $\alpha \neq 0$, this in particular, implies $[X,Y] \neq 0$.

Next, we want to show that $\alpha(H_\alpha) = B(H_\alpha, H_\alpha) \neq 0$. To do this, we introduce an idea which will be used frequently in the sequel. Let $\beta \in \Delta$ be a root and consider the subspace $\mathfrak{k} := \bigoplus_{n\in\mathbb{Z}} \mathfrak{g}_{\beta+n\alpha}$, which is called the *$\alpha$–string through β*. Choosing $X \in \mathfrak{g}_\alpha$ and $Y \in \mathfrak{g}_{-\alpha}$ with $B(X,Y) = 1$, we get $[X,Y] = H_\alpha$. Then $\mathfrak{k}$ is invariant under $\mathrm{ad}(X)$ and $\mathrm{ad}(Y)$, and $\mathrm{ad}(H_\alpha) = [\mathrm{ad}(X), \mathrm{ad}(Y)]$ on $\mathfrak{k}$. In particular, $\mathrm{ad}(H_\alpha)$ is tracefree on $\mathfrak{k}$, which implies that $0 = \sum_{n\in\mathbb{Z}}(\beta(H_\alpha) + n\alpha(H_\alpha)) \dim(g_{\beta+n\alpha})$. Thus, $\alpha(H_\alpha) = 0$ would imply $\beta(H_\alpha) = 0$ for all $\beta \in \Delta$, which would lead to the contradiction $H_\alpha = 0$. Moreover, this equation shows that for any $\beta \in \Delta$ the number $\beta(H_\alpha)$ is a rational multiple of $\alpha(H_\alpha)$.

This is the place where the representation theory of $\mathfrak{sl}(2,\mathbb{C})$ comes into the game. Namely, since $B(H_\alpha, H_\alpha) = \alpha(H_\alpha) \neq 0$, we may choose $X \in \mathfrak{g}_\alpha$ and $Y \in \mathfrak{g}_{-\alpha}$ such that $\alpha([X,Y]) = 2$. But this implies that the elements X, Y and $H := [X,Y]$ span a subalgebra $\mathfrak{s}_\alpha \leq \mathfrak{g}$ which is isomorphic to $\mathfrak{sl}(2,\mathbb{C})$. Now consider the (finite–dimensional) subspace $\mathfrak{h} \oplus \bigoplus_{z\in\mathbb{C}\setminus\{0\}} \mathfrak{g}_{z\alpha}$, which obviously is invariant under the action of $\mathfrak{s}_\alpha$. From complete reducibility and Proposition 2.2.3 we conclude that $\mathfrak{g}_{z\alpha}$ is trivial, unless z is a half integer. Now $\mathfrak{s}_\alpha$ acts trivially on the codimension one subspace $\ker(\alpha) \subset \mathfrak{h}$, and irreducibly on $\mathfrak{s}_\alpha$, which adds a one–dimensional weight space of weight zero spanned by H. Since we have exhausted the zero weight space, there are no other irreducible components of even highest weight. In particular, $\mathfrak{g}_{2\alpha} = \{0\}$, so 2α cannot be a root. But this in turn implies that $\frac{1}{2}\alpha$ cannot be a root either, which shows that the weight space of weight one is trivial. Hence, there cannot be any irreducible components of odd highest weight either, i.e. our space just has the form $\ker(\alpha) \oplus \mathfrak{s}_\alpha$. This completes the proof of (1) and it shows that $\mathfrak{g}_\alpha$ is one–dimensional, which completes the proof of (2).

By construction, the element H in the standard basis of $\mathfrak{s}_\alpha \cong \mathfrak{sl}(2,\mathbb{C})$ constructed above is given by $\frac{2}{\langle\alpha,\alpha\rangle}H_\alpha$. For two roots $\alpha,\beta \in \Delta$ which are not proportional to each other, consider the α–string through β, i.e. the subspace $\bigoplus_{n\in\mathbb{Z}} \mathfrak{g}_{\beta+n\alpha}$. We have already observed that this subspace is invariant under the adjoint action of $\mathfrak{s}_\alpha$, so we may again apply the representation theory of $\mathfrak{sl}(2,\mathbb{C})$. By construction, the eigenvalue for the distinguished element $\frac{2}{\langle\alpha,\alpha\rangle}H_\alpha$ on $\mathfrak{g}_{\beta+n\alpha}$ is given by $\frac{2\langle\beta,\alpha\rangle}{\langle\alpha,\alpha\rangle} + 2n$, so these eigenvalues are all different, and from above we know that each of the space $\mathfrak{g}_{\beta+n\alpha}$ is of dimension at most one.

By complete reducibility, $\bigoplus_{n\in\mathbb{Z}} \mathfrak{g}_{\beta+n\alpha}$ splits into a direct sum of irreducible representations of $\mathfrak{s}_\alpha$. Using the description of irreducible representations in Proposition 2.2.3 and the computation of eigenvalues we conclude that any irreducible component must have the form of an unbroken string $\bigoplus_{n=n_0}^{n_1} \mathfrak{g}_{\beta+n\alpha}$ of 1–dimensional spaces. On the other hand, we know that the eigenvalues of $\frac{2}{\langle\alpha,\alpha\rangle}H_\alpha$ must be symmetric around zero, which implies that there is just one irreducible component. Thus, the α–string through β must have the form $\bigoplus_{n=-p}^{q} \mathfrak{g}_{\beta+n\alpha}$ with $p,q \geq 0$ and the maximal eigenvalue $\frac{2\langle\beta,\alpha\rangle}{\langle\alpha,\alpha\rangle} + 2q$ must equal the dimension minus one, which implies $p - q = \frac{2\langle\beta,\alpha\rangle}{\langle\alpha,\alpha\rangle}$. In particular, $\frac{2\langle\beta,\alpha\rangle}{\langle\alpha,\alpha\rangle} \in \mathbb{Z}$ for all $\alpha,\beta \in \Delta$, and (4) follows. Finally, we see that for $\alpha,\beta \in \Delta$ such that $\alpha+\beta \in \Delta$, we have $[\mathfrak{g}_\alpha, \mathfrak{g}_\beta] = \mathfrak{g}_{\alpha+\beta}$, which implies (3). □

Next, let $V \subset \mathfrak{h}^*$ be the real span of the set Δ of roots and let $\mathfrak{h}_0 \subset \mathfrak{h}$ be the real subspace spanned by the elements H_α for $\alpha \in \Delta$. Of course, we can restrict the Killing form B to $\mathfrak{h}_0$ and the complex bilinear form $\langle\ ,\ \rangle$ to V. We claim that $\mathfrak{h}_0$ is a real form of $\mathfrak{h}$, its (real) dual is exactly V, and the restrictions of the forms are positive definite.

By definition, $\langle \alpha, \alpha \rangle = B(H_\alpha, H_\alpha) = \operatorname{tr}(\operatorname{ad}(H_\alpha) \circ \operatorname{ad}(H_\alpha))$. Now $\operatorname{ad}(H_\alpha)$ acts trivially on $\mathfrak{h}$ and by multiplication by $\beta(H_\alpha)$ on $\mathfrak{g}_\beta$. Therefore, $\langle \alpha, \alpha \rangle = \sum_{\beta \in \Delta} \beta(H_\alpha)^2$. Inserting $\beta(H_\alpha) = \langle \alpha, \beta \rangle$, we obtain

$$\langle \alpha, \alpha \rangle = 2\langle \alpha, \alpha \rangle^2 + \sum_{\beta \in \Delta; \beta \neq \pm\alpha} \langle \beta, \alpha \rangle^2. \tag{2.3}$$

By part (4) of the proposition, each of the numbers $\frac{\langle \beta, \alpha \rangle}{\langle \alpha, \alpha \rangle}$ is rational. Dividing (2.3) by $\langle \alpha, \alpha \rangle^2$, we conclude that $\langle \alpha, \alpha \rangle = \alpha(H_\alpha) \in \mathbb{Q}$. This in turn implies that $\beta(H_\alpha) \in \mathbb{Q}$ for all β. Using this, equation (2.3) shows that $\langle \alpha, \alpha \rangle > 0$ for all α. Hence, the restriction of $\langle\ ,\ \rangle$ to the subspace $\mathfrak{h}_0 \subset \mathfrak{h}$ spanned by the H_α is positive definite. Moreover, we see that the restrictions of all roots to $\mathfrak{h}_0$ are real valued. Hence, the real dimension of $\mathfrak{h}_0$ can be at most the complex dimension of $\mathfrak{h}$, but since the roots span $\mathfrak{h}^*$, it cannot be smaller than that either. Now the claim follows easily from dimensional considerations.

This brings us the final idea in this context, which is to consider the reflections of the real Euclidean space $\mathfrak{h}_0^*$ in the hyperplanes orthogonal to roots. For $\alpha \in \Delta$ the *root reflection* $s_\alpha : \mathfrak{h}_0^* \to \mathfrak{h}_0^*$ on the hyperplane $\alpha^\perp$ is given by $s_\alpha(\phi) = \phi - \frac{2\langle \phi, \alpha \rangle}{\langle \alpha, \alpha \rangle}\alpha$. As a reflection on a hyperplane this is an orthogonal mapping of determinant -1, and $s_\alpha(\alpha) = -\alpha \in \Delta$. For a root $\beta \neq \pm\alpha$ we have $s_\alpha(\beta) = \beta - (p-q)\alpha$, where the α–string through β has the form $\mathfrak{g}_{\beta - p\alpha} \oplus \cdots \oplus \mathfrak{g}_{\beta + q\alpha}$ by part (4) of the proposition. Since $-p \leq q - p \leq q$, we conclude that $s_\alpha(\beta) \in \Delta$.

Collecting our information about the system of roots we get:

THEOREM 2.2.4. *Let $\mathfrak{g}$ be a semisimple Lie algebra of rank r, let $\mathfrak{h} \leq \mathfrak{g}$ be a Cartan subalgebra, $\Delta \subset \mathfrak{h}^*$ the corresponding set of roots, $\mathfrak{h}_0^* \subset \mathfrak{h}^*$ the real span of Δ and let $\langle\ ,\ \rangle$ be the complex bilinear form on $\mathfrak{h}^*$ induced by the Killing form. Then*

1. *$\mathfrak{h}_0^*$ has real dimension r and the restriction of $\langle\ ,\ \rangle$ to $\mathfrak{h}_0^*$ is positive definite.*
2. *The subset $\Delta \subset \mathfrak{h}_0^*$ spans the space $\mathfrak{h}_0^*$, any root reflection s_α for $\alpha \in \Delta$ maps Δ to itself and for $\alpha, \beta \in \Delta$ we have $\frac{2\langle \beta, \alpha \rangle}{\langle \alpha, \alpha \rangle} \in \mathbb{Z}$.*
3. *For $\alpha \in \Delta$ we have $2\alpha \notin \Delta$.*

A finite subset Δ in a Euclidean vector space which satisfies the conditions of part (2) of the theorem is called an *abstract root system.* If in addition, the property of part (3) is satisfied, then this abstract root system is called *reduced.*

There is a final point to make here, which concerns simplicity of $\mathfrak{g}$. Note that there is an obvious notion of direct sum of abstract root systems, by taking the orthogonal direct sum of the surrounding Euclidean spaces and the union of the abstract root systems. An abstract root system is called *irreducible* if and only if it does not decompose as a direct sum in a nontrivial way. Comparing with the proof of Corollary 2.1.5(1) one easily shows that $\mathfrak{g}$ is simple if and only if its root system is irreducible; see [**Kn96**, II, Proposition 2.44].

2.2.5. Positive and simple roots; Dynkin diagram. At this point, we may forget about Lie algebras for a while. The further reductions, which finally lead to the Dynkin diagram and the classification of possible Dynkin diagrams, can be entirely done in the setting of abstract root systems. The main input are elementary considerations from Euclidean geometry. We will only sketch the way from an abstract root system to its Dynkin diagram, referring to [**Kn96**, section II.5] for details.

Since an abstract root system Δ on a Euclidean vector space V is symmetric under root reflections, it follows that $\alpha \in \Delta$ implies $-\alpha \in \Delta$. Thus, a natural idea is to split the set of roots into positive and negative roots. It is better to use a slightly stronger concept, namely to introduce a notion of positivity in V, i.e. selecting a subset $V^+ \subset V \setminus \{0\}$ such that V is the disjoint union of V^+, $\{0\}$ and $-V^+$, and such that V^+ is stable under addition and multiplication by positive scalars. Having chosen such a decomposition, we get a total ordering on V, defined by $v \leq w$ if and only if $v = w$ or $w - v \in V^+$. A simple way to get such an ordering is to choose a basis $\phi_1, \dots, \phi_n$ for the dual space V^* and define $v \in V^+$ if and only if there is an index j such that $\phi_i(v) = 0$ for $i < j$ and $\phi_j(v) > 0$.

Having chosen V^+, we define the set of *positive roots* $\Delta^+ \subset \Delta$ by $\Delta^+ = \Delta \cap V^+$, so Δ is the disjoint union of Δ^+ and $-\Delta^+$. Moreover, sums and positive multiples of positive roots are positive. We will write $\alpha > 0$ to indicate that the root α is positive. Further, we define the subset $\Delta^0 \subset \Delta^+$ of *simple roots* as the set of those positive roots, which cannot be written as the sum of two positive roots. One then shows that the simple roots form a basis of V and that any root can be written as an integral linear combination of simple roots, in which the coefficients are either all positive or all negative. In particular, defining the *root lattice* Λ_R to be the set of all integral linear combinations of roots, we see that $\Lambda_R \cong \mathbb{Z}^{\dim(V)}$ with a basis for Λ_R being formed by the simple roots.

For $\alpha, \beta \in \Delta$, the Cauchy–Schwarz inequality shows that $|\frac{2\langle\beta,\alpha\rangle}{\langle\alpha,\alpha\rangle}\frac{2\langle\beta,\alpha\rangle}{\langle\beta,\beta\rangle}| \leq 4$ and equality is only possible if α and β are proportional. Hence, each of the integers $\frac{2\langle\beta,\alpha\rangle}{\langle\alpha,\alpha\rangle}$ lies between -4 and 4, and for $\alpha \neq \beta$ and reduced abstract root systems, ± 4 cannot occur. Moreover, either $\frac{2\langle\beta,\alpha\rangle}{\langle\alpha,\alpha\rangle}$ or $\frac{2\langle\beta,\alpha\rangle}{\langle\beta,\beta\rangle}$ has to be 0 or ± 1. For $\alpha, \beta \in \Delta^0$ we clearly have $\alpha - \beta \notin \Delta$ and thus the inner product $\langle\alpha, \beta\rangle$ has to be ≤ 0. Finally, one can show also in the case of reduced abstract root systems that for $\alpha, \beta \in \Delta$ with $\beta \neq \pm\alpha$, the elements of Δ of the form $\beta + n\alpha$ form an unbroken string of the form $\beta - p\alpha, \dots, \beta + q\alpha$ with $p - q = \frac{2\langle\beta,\alpha\rangle}{\langle\alpha,\alpha\rangle}$. Replacing β by $\beta - p\alpha$ we conclude from above that such a string has length at most four.

Let us now fix an abstract root system Δ on V and a positive subset $V^+ \subset V$, and let $\Delta^0 = \{\alpha_1, \dots, \alpha_\ell\}$ be the corresponding set of simple roots. Then the *Cartan matrix* of Δ is the $\ell \times \ell$–matrix $A = (a_{ij})$ defined by $a_{ij} = \frac{2\langle\alpha_i,\alpha_j\rangle}{\langle\alpha_i,\alpha_i\rangle}$. Clearly, this matrix depends on the numbering of the simple roots, so it is at best unique up to conjugation with permutation matrices. We already know that all a_{ij} are integers, the diagonal entries are 2, while the off–diagonal ones are ≤ 0, and obviously if $a_{ij} = 0$, then $a_{ji} = 0$ also. Finally, one shows that A is symmetrizable, i.e. there is a diagonal matrix D with positive entries such that DAD^{-1} is symmetric and positive definite. (The entries of D may simply be chosen to be $|\alpha_1|, \dots, |\alpha_\ell|$.) Matrices having these properties are called *abstract*

Cartan matrices. One should, however, be careful, that in the theory of Kac–Moody algebras and their generalizations more general versions of Cartan matrices show up, so there are different meanings of the term "abstract Cartan matrix" used in the literature. The notion used here is the most restrictive one. It is easy to see that the root system Δ is reducible if and only if for an appropriate numbering of the simple roots the Cartan matrix is block diagonal with more than one block.

The last step in the encoding is to associate to a Cartan matrix a type of graph called its *Dynkin diagram.* One simply takes a vertex for each simple root (or each element indexing the Cartan matrix) and then joins the ith and the jth vertex by $a_{ij}a_{ji}$ many lines. Otherwise put, the number of edges joining the vertices corresponding to two simple roots is equal to 4 times the square of the cosine of the angle between the roots. If the two roots are of different length (i.e. if $a_{ij} \neq a_{ji}$), then in addition one orients the edge by an arrow pointing from the longer to the shorter root. From the observations on the inner products made above, it is easy to see that one may recover the Cartan matrix from the Dynkin diagram.

2.2.6. The classical examples. Here we determine the root decomposition, the Cartan matrix and the Dynkin diagram for the classical simple Lie algebras.

(1) Let us start with $\mathfrak{g} = \mathfrak{sl}(n,\mathbb{C})$. As in 2.1.2 we denote by E_{ij} the elementary matrix having an entry one in the jth column of the ith row and all other entries equal to zero. The intersection of $\mathfrak{sl}(n,\mathbb{C})$ with the set of diagonal matrices is a commutative subalgebra of dimension $n-1$. Considering a diagonal matrix H with entries $a_1,\dots,a_n$ (such that $a_1+\dots+a_n=0$) and $i \neq j$, the commutator $[H, E_{ij}]$ is given by $(a_i - a_j)E_{ij}$. This shows that the subset $\mathfrak{h} \subset \mathfrak{g}$ of tracefree diagonal matrices is a maximal commutative subalgebra and that the adjoint action of any $H \in \mathfrak{h}$ is diagonalizable, so $\mathfrak{h} \subset \mathfrak{g}$ is a Cartan subalgebra. In particular, $\mathfrak{sl}(n,\mathbb{C})$ has rank $n-1$. Moreover, denoting by $e_i : \mathfrak{h} \to \mathbb{C}$ the linear functional which extracts the ith entry on the diagonal, we see that the roots are given by $\Delta = \{e_i - e_j : 1 \leq i,j \leq n, i \neq j\} \subset \mathfrak{h}^*$. The real subspace $\mathfrak{h}_0$ is exactly the subspace of real tracefree diagonal $n\times n$–matrices. The root decomposition has the form $\mathfrak{g} = \mathfrak{h} \oplus \bigoplus_{i\neq j} \mathfrak{g}_{e_i - e_j}$ and $\mathfrak{g}_{e_i-e_j}$ is the one–dimensional subspace spanned by E_{ij}.

Using $\{E_{11} - E_{nn}, \dots, E_{n-1,n-1} - E_{nn}\}$ as the ordered basis for $\mathfrak{h}_0$ to construct a positive subset in $\mathfrak{h}_0^*$ as described in 2.2.5, we obtain $\Delta^+ = \{e_i - e_j : i < j\}$. The corresponding set of simple roots is then $\{\alpha_1,\dots,\alpha_{n-1}\}$ with $\alpha_i = e_i - e_{i+1}$. In terms of this set of simple roots, the positive roots are exactly all combinations of the form $\alpha_i + \alpha_{i+1} + \dots + \alpha_j$ for $1 \leq i \leq j \leq n-1$.

Next, we can compute the restriction of the Killing form to $\mathfrak{h}$ from the definition. Given diagonal matrices H_1 with entries $a_1,\dots,a_n$ and H_2 with entries $b_1,\dots,b_n$ we get from the definition

$$B(H_1,H_2) = \sum_{i\neq j}(a_i - a_j)(b_i - b_j) = \sum_i \sum_{j\neq i}(a_ib_i + a_jb_j - a_ib_j - a_jb_i).$$

Using $\sum_{j\neq i} a_j = -a_i$ and the analogous equation for the b's this adds up to $B(H_1,H_2) = 2n\sum_i a_ib_i$. Up to the factor $2n$, this is the usual inner product. In particular, $H_{e_i-e_j} = \frac{1}{2n}(E_{ii} - E_{jj})$, and hence $\langle \alpha_i, \alpha_i\rangle = \frac{1}{n}$ for all i, $\langle \alpha_i, \alpha_{i+1}\rangle = \frac{-1}{2n}$ for $i = 1,\dots,n-2$ and $\langle \alpha_i,\alpha_j\rangle = 0$ if $|i-j| > 1$. Thus, the Cartan matrix of $\mathfrak{sl}(n,\mathbb{C})$ is the $(n-1)\times(n-1)$–matrix with $a_{ii} = 2$ for all i, $a_{i,i+1} = a_{i+1,i} = -1$

for $i = 1, \dots, n-2$ and all other entries are zero. Finally, the Dynkin diagram has the simple form $\circ\!\!-\!\!\!-\!\!\circ\!\!-\cdots-\!\!\circ\!\!-\!\!\!-\!\!\circ$ with $n-1$ vertices, which is usually called A_{n-1}.

(2) Let us next consider the Lie algebras $\mathfrak{so}(n,\mathbb{C})$ of linear mappings which are skew symmetric with respect to a nondegenerate complex bilinear form on $\mathbb{C}^n$. This is the complexification of the real Lie algebra $\mathfrak{so}(n)$ from 2.1.7. There we have shown that $\mathfrak{so}(n)$ is semisimple, so semisimplicity of $\mathfrak{so}(n,\mathbb{C})$ follows from part (3) of Corollary 2.1.5. Recall from linear algebra that up to the obvious notion of equivalence there is only one nondegenerate symmetric complex bilinear form in each dimension. In the analysis of the orthogonal Lie algebras, one has to distinguish the cases of even and odd dimensions, and we start with the even–dimensional case. The best presentation of a nondegenerate symmetric complex bilinear form on $\mathbb{C}^{2n}$ for our purposes is to view $\mathbb{C}^{2n}$ as $\mathbb{C}^n \oplus \mathbb{C}^{n*}$ with the bilinear form $((z,\phi),(w,\psi)) = \phi(w) + \psi(z)$, which is obviously nondegenerate. Choosing dual bases of $\mathbb{C}^n$ and $\mathbb{C}^{n*}$ we arrive at a presentation of $\mathfrak{so}(2n,\mathbb{C})$ as block matrices with blocks of size n, and skew symmetry is immediately seen to be equivalent to a block form $\left(\begin{smallmatrix} A & B \\ C & -A^t \end{smallmatrix}\right)$, where $A, B, C \in \mathfrak{gl}(n,\mathbb{C})$, $B^t = -B$, and $C^t = -C$.

For block diagonal matrices (i.e. matrices with trivial B and C–part), the adjoint action preserves this block decomposition, and explicitly we have

$$\left[\begin{pmatrix} A & 0 \\ 0 & -A^t \end{pmatrix}, \begin{pmatrix} A' & B' \\ C' & -(A')^t \end{pmatrix}\right] = \begin{pmatrix} [A,A'] & AB' + B'A^t \\ -A^tC' - C'A & -[A,A']^t \end{pmatrix}.$$

An obvious choice for a commutative subalgebra $\mathfrak{h} \subset \mathfrak{g}$ is again the intersection with the space of diagonal matrices. The diagonal entries of such a matrix are $a_1, \dots, a_n, -a_1, \dots, -a_n$ and we denote by e_j the linear functional which extracts a_j. For $i \neq j$ we can consider the matrix $E_{ij} - E_{n+j,n+i}$ which is an eigenvector corresponding to the eigenvalue $a_i - a_j$. Together with $\mathfrak{h}$, these matrices span the block diagonal part. The block above the main diagonal is spanned by the matrices $E_{i,n+j} - E_{j,n+i}$ for $i < j$, which are eigenvectors with eigenvalue $a_i + a_j$. The block below the main diagonal is spanned by the matrices $E_{n+i,j} - E_{n+j,i}$ for $i < j$, which are eigenvectors with eigenvalues $-a_i - a_j$.

In particular, we conclude that $\mathfrak{h} \subset \mathfrak{g}$ is a Cartan subalgebra, so $\mathfrak{so}(2n,\mathbb{C})$ has rank n, and the set of roots is $\Delta = \{\pm e_i \pm e_j : 1 \leq i < j \leq n\}$. As in (1), $\mathfrak{h}_0$ is the subspace of real diagonal matrices and defining positivity using the basis $\{E_{11} - E_{n+1,n+1}, \dots, E_{nn} - E_{2n,2n}\}$, we get $\Delta^+ := \{e_i \pm e_j : i < j\}$. To determine the simple roots, we first note that the elements $\alpha_1, \dots, \alpha_{n-1}$, defined by $\alpha_j := e_j - e_{j+1}$ clearly are simple. A moment of thought shows that among the roots $e_i + e_j$ only $\alpha_n := e_{n-1} + e_n$ can be simple, so we have found the set $\Delta^0 = \{\alpha_1, \dots, \alpha_n\}$ of simple roots.

To express all positive roots as linear combinations of simple roots, one first may use all expressions of the form $\alpha_i + \alpha_{i+1} + \cdots + \alpha_j$ for $i < j < n$, which give the roots $e_i - e_{j+1}$. Next, the roots $e_i + e_n$ for $i < n-1$ are obtained as $\alpha_i + \cdots + \alpha_{n-2} + \alpha_n$. The remaining positive roots $e_i + e_j$ for $i < j < n$ are obtained as $\alpha_i + \cdots + \alpha_{j-1} + 2\alpha_j + \cdots + 2\alpha_{n-2} + \alpha_{n-1} + \alpha_n$.

Next, we have to compute the Killing form. From the definition and our computation of the roots we conclude that for matrices H_1 corresponding to $a_1, \dots, a_n$ and H_2 corresponding to $b_1, \dots, b_n$ the Killing form is given by

$$B(H_1, H_2) = 2\sum_{i<j} \big((a_i - a_j)(b_i - b_j) + (a_i + a_j)(b_i + b_j)\big).$$

This equals $4\sum_{i<j}(a_ib_i + a_jb_j) = (4n-4)\sum_j a_jb_j$, so up to the factor $4(n-1)$ we again get the standard inner product. In particular, we see that $\langle \alpha_j, \alpha_j\rangle = \frac{1}{2(n-1)}$ for all j, so all simple roots have the same length. As for the other inner products, we see that for $j = 1, \dots, n-2$ we have $\langle \alpha_j, \alpha_{j+1}\rangle = \frac{-1}{4(n-1)}$, but $\langle \alpha_{n-1}, \alpha_n\rangle = 0$ and $\langle \alpha_{n-2}, \alpha_n\rangle = \frac{-1}{4(n-1)}$, while all other inner products are zero. For the Cartan matrix, this means that the upper part looks like the Cartan matrix of A_n, while the bottom right–hand 3×3–block has the form $\begin{pmatrix} 2 & -1 & -1 \\ -1 & 2 & 0 \\ -1 & 0 & 2 \end{pmatrix}$. For the Dynkin diagram we get

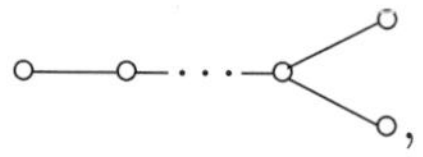

which is usually called D_n. Our convention for the Dynkin diagram will be that the upper root is α_{n-1} and the lower is α_n. For $n \geq 3$, this Dynkin diagram is connected (the case $n = 2$ will be discussed immediately) which implies that $\mathfrak{so}(2n, \mathbb{C})$ is simple for $n \geq 3$.

For small n, there is some overlap with what we have done before. For $n = 1$, the Lie algebra $\mathfrak{so}(2, \mathbb{C})$ is isomorphic to $\mathbb{C}$ and thus not semisimple. If $n = 2$, then the roots are just $\pm e_1 \pm e_2$, so this is simply an orthogonal direct sum of two copies of A_1. (In terms of the Dynkin diagram only the upper and the lower root, which are not connected, are there.) Indeed, this reflects an isomorphism $\mathfrak{so}(4, \mathbb{C}) \cong \mathfrak{sl}(2, \mathbb{C}) \times \mathfrak{sl}(2, \mathbb{C})$. This isomorphism is easiest to see on the group level in the real picture. Viewing $SU(2)$ as the group of quaternions of unit length, mapping (p, q) to the map $r \mapsto prq$ identifies $SU(2) \times SU(2)$ with a two–fold covering of $SO(\mathbb{H}) = SO(4)$. In particular, $\mathfrak{su}(2) \oplus \mathfrak{su}(2) \cong \mathfrak{so}(4)$ and passing to complexifications, we obtain the required isomorphism.

For $n = 3$, one immediately sees that the Dynkin diagram D_3 coincides with A_3, and with a little more work, one also verifies that the root system of $\mathfrak{so}(6, \mathbb{C})$ coincides with the root system of $\mathfrak{sl}(4, \mathbb{C})$. Again, this expresses an isomorphism of the two Lie algebras. This isomorphism can be understood either by showing that the image of the Lie algebra $\mathfrak{so}(6, \mathbb{C})$ under the spin representation (which is four–dimensional in this case) is the full Lie algebra $\mathfrak{sl}(4, \mathbb{C})$. The other way to get it is to study the action of $\mathfrak{sl}(4, \mathbb{C})$ on the second exterior power $\Lambda^2\mathbb{C}^4$, which is of dimension six and admits a nondegenerate symmetric bilinear form defined by the wedge product $\Lambda^2\mathbb{C}^4 \times \Lambda^2\mathbb{C}^4 \to \Lambda^4\mathbb{C}^4 \cong \mathbb{C}$.

(3) For the odd orthogonal algebras $\mathfrak{so}(2n+1, \mathbb{C})$ the situation is very similar to the even orthogonal case. As the basic vector space we take an orthogonal direct sum of $\mathbb{C}^{2n}$ with the bilinear form used in (2) and $\mathbb{C}$ with the bilinear form given by multiplication. For this choice of bilinear form we get

$$\mathfrak{so}(2n+1, \mathbb{C}) = \left\{ \begin{pmatrix} A & v \\ -v^t & 0 \end{pmatrix} : A \in \mathfrak{so}(2n, \mathbb{C}), v \in \mathbb{C}^{2n} \right\}.$$

In particular, we have $\mathfrak{so}(2n, \mathbb{C})$ in the form presented in (2) included as a subalgebra. Denoting by $\mathfrak{h}$ the intersection of $\mathfrak{so}(2n+1, \mathbb{C})$ with the space of diagonal matrices we get the same subalgebra as in (2). Elements of $\mathfrak{h}$ have diagonal entries of the form $a_1, \dots, a_n, -a_1, \dots, -a_n, 0$, and we denote by e_j the functional which extracts a_j. From the subalgebra $\mathfrak{so}(2n, \mathbb{C})$ we get eigenspaces for the adjoint action of an element of $\mathfrak{h}$ with eigenvalues $\pm a_i \pm a_j$ with $i < j$, and the additional row and column are spanned by eigenvectors with eigenvalues $\pm e_j$ for $j = 1, \dots, n$. Hence, we see that $\mathfrak{h}$ is a Cartan subalgebra, so $\mathfrak{so}(2n+1,\mathbb{C})$ has rank n, and the set of roots is given by $\Delta = \{\pm e_i \pm e_j : 1 \leq i < j \leq n\} \cup \{\pm e_j : 1 \leq j \leq n\}$. Defining positivity as in (2), we get $\Delta^+ = \{e_i \pm e_j : i < j\} \cup \{e_j\}$, and the simple subsystem $\Delta^0 = \{\alpha_1, \dots, \alpha_n\}$ is given by $\alpha_j = e_j - e_{j+1}$ for $j = 1, \dots, n-1$ and $\alpha_n = e_n$.

To express arbitrary positive roots as linear combinations of simple roots, we first need the expressions $\alpha_i + \alpha_{i+1} + \cdots + \alpha_j$, which give the roots $e_i - e_{j+1}$ for $j < n$ and e_i for $j = n$. The remaining positive roots $e_i + e_j$ for $i < j$ are given by $\alpha_i + \cdots + \alpha_{j-1} + 2\alpha_j + \cdots + 2\alpha_n$. The computation of the Killing form is as in (2) above. For H_1, H_2 corresponding to $a_1, \dots, a_n$ and $b_1, \dots, b_n$ the Killing form is given by $B(H_1, H_2) = (4n-2)\sum_j a_j b_j$, so we again obtain a multiple of the standard inner product. In particular, for $j < n$ we get $\langle \alpha_j, \alpha_j \rangle = \frac{1}{2n-1}$ while $\langle \alpha_n, \alpha_n \rangle = \frac{1}{4n-2}$, so the last simple root is shorter than the others. On the other hand, for all $j < n$ we have $\langle \alpha_j, \alpha_{j+1} \rangle = \frac{-1}{4n-2}$, and all other inner products are zero. Thus, the Cartan matrix looks like the one for A_n, with the only exception that the bottom right–hand 2×2–block has the form $\left(\begin{smallmatrix} 2 & -1 \\ -2 & 2 \end{smallmatrix}\right)$ instead of $\left(\begin{smallmatrix} 2 & -1 \\ -1 & 2 \end{smallmatrix}\right)$. The Dynkin diagram of $\mathfrak{so}(2n+1,\mathbb{C})$ is thus given by  and this diagram is usually called B_n. Connectedness of this diagram shows that $\mathfrak{so}(2n+1,\mathbb{C})$ is simple for $n \geq 1$.

Again, something special happens in low dimensions, namely for $n = 1$, i.e. for $\mathfrak{so}(3,\mathbb{C})$. The root system in this case consists of $\pm e_1$ only, so it is isomorphic to A_1. This isomorphism comes from an isomorphism $\mathfrak{so}(3,\mathbb{C}) \cong \mathfrak{sl}(2,\mathbb{C})$ of Lie algebras: The adjoint representation of $\mathfrak{sl}(2,\mathbb{C})$ is three–dimensional and the Killing form is an invariant complex bilinear form, so it maps $\mathfrak{sl}(2,\mathbb{C})$ to $\mathfrak{so}(3,\mathbb{C})$. Since the adjoint representation is faithful and both algebras are three–dimensional, this is an isomorphism.

(4) The final classical example of complex simple Lie algebras is provided by the complex symplectic Lie algebras $\mathfrak{sp}(2n,\mathbb{C})$. We know from 2.1.7 that this Lie algebra is semisimple. By definition, $\mathfrak{sp}(2n,\mathbb{C})$ consists of all endomorphisms of $\mathbb{C}^{2n}$ which are skew symmetric with respect to a nondegenerate skew symmetric bilinear form. It is well known that such a form is uniquely determined up to equivalence (in the obvious sense). Similarly, as in the case of the even orthogonal algebras, we view $\mathbb{C}^{2n}$ as $\mathbb{C}^n \oplus \mathbb{C}^{n*}$ and consider the form $\omega((z,\lambda),(w,\mu)) := \lambda(w) - \mu(z)$, which obviously is skew symmetric and nondegenerate. Choosing dual bases, we see that $\mathfrak{sp}(2n,\mathbb{C})$ consists of all matrices of the form $\left(\begin{smallmatrix} A & B \\ C & -A^t \end{smallmatrix}\right)$, where $A, B, C \in \mathfrak{gl}(n,\mathbb{C})$, $B^t = B$ and $C^t = C$. As before, we denote by $\mathfrak{h}$ the space of diagonal matrices contained in $\mathfrak{sp}(2n,\mathbb{C})$, so these have diagonal entries $(a_1, \dots, a_n, -a_1, \dots, -a_n)$, and we denote by e_j the linear functional that extracts a_j. The formula for the adjoint action of such an element from (2) remains valid. Using this one computes the eigenvalues as before.

This shows that $\mathfrak{h}$ is a Cartan subalgebra, so $\mathfrak{sp}(2n,\mathbb{C})$ has rank n, and that the set of roots is given by $\Delta = \{\pm e_i \pm e_j : 1 \leq i < j \leq n\} \cup \{\pm 2e_j : 1 \leq j \leq n\}$. Defining positivity as in (2), we get $\Delta^+ = \{e_i \pm e_j : i < j\} \cup \{2e_j\}$. A moment of thought shows that the set $\Delta^0 = \{\alpha_1, \dots, \alpha_n\}$ of simple roots is given by $\alpha_j = e_j - e_{j+1}$ for $j < n$ and $\alpha_n = 2e_n$. To get all positive roots from simple roots, one first has to consider the expressions $\alpha_i + \alpha_{i+1} + \cdots + \alpha_j$, which equals $e_i - e_{j+1}$ for $j < n$ and $e_i + e_n$ for $j = n$. The roots $e_i + e_j$ for $i < j < n$ are then given by $\alpha_i + \cdots + \alpha_{j-1} + 2\alpha_j + \cdots + 2\alpha_{n-1} + \alpha_n$, while for $j < n$ the root $2e_j$ equals $2\alpha_i + 2\alpha_{i+1} + \cdots + 2\alpha_{n-1} + \alpha_n$.

Computing the Killing form for $H_1, H_2 \in \mathfrak{h}$ corresponding to $a_1, \ldots, a_n$ and $b_1, \ldots, b_n$ as before, one obtains $B(H_1, H_2) = 4n \sum_j a_j b_j$. Again, this is a multiple of the standard inner product. This implies that for $j < n$ we have $\langle \alpha_j, \alpha_j \rangle = \frac{1}{2n}$, while $\langle \alpha_n, \alpha_n \rangle = \frac{1}{n}$, so this time one simple root is longer than the others. The remaining nontrivial inner products are $\langle \alpha_j, \alpha_{j+1} \rangle = \frac{-1}{4n}$ for $j < n-1$ and $\langle \alpha_{n-1}, \alpha_n \rangle = \frac{-1}{2n}$. Hence, the Cartan matrix is just the transpose of the Cartan matrix of $\mathfrak{so}(2n+1, \mathbb{C})$, and the Dynkin diagram is ○—○⋯○⇐○, which is usually called C_n. Connectedness of the Dynkin diagram shows simplicity of $\mathfrak{sp}(2n, \mathbb{C})$ for all $n \geq 1$.

Concerning low–dimensional special cases, $n = 1$ i.e. $\mathfrak{sp}(2, \mathbb{C})$ by our description coincides with $\mathfrak{sl}(2, \mathbb{C})$. A more interesting case is $n = 2$. Here the roots are $\pm e_1 \pm e_2$, $\pm 2e_1$ and $\pm 2e_2$. Putting $u = e_1 + e_2$ and $v = e_1 - e_2$, these are $\pm u$, $\pm v$ and $\pm u \pm v$, which looks like the root system B_2 of $\mathfrak{so}(5, \mathbb{C})$. Indeed, there is an isomorphism $\mathfrak{sp}(4, \mathbb{C}) \cong \mathfrak{so}(5, \mathbb{C})$ underlying this, which can be obtained as follows: Consider the dual of the standard representation, which is a representation of $\mathfrak{sp}(4, \mathbb{C})$ on $\mathbb{C}^{4*}$. The induced representation on the second exterior power $\Lambda^2 \mathbb{C}^{4*}$ leaves the line generated by the symplectic form ω invariant, so by complete reducibility there is a complementary five–dimensional representation of $\mathfrak{sp}(4, \mathbb{C})$. This admits an invariant symmetric bilinear form coming from the wedge product, so we obtain a map to $\mathfrak{so}(5, \mathbb{C})$, which is an isomorphism by simplicity of the algebras in question.

2.2.7. The Weyl group. The idea of the classification of finite–dimensional simple Lie algebras is to reduce the problem to the classification of Dynkin diagrams. Up to now, we have seen how to associate to a semisimple Lie algebra a Dynkin diagram making two choices. First, we have chosen a Cartan subalgebra $\mathfrak{h} \leq \mathfrak{g}$, and second we have the subset $\Delta^+ \subset \Delta$ of positive roots. That the Dynkin diagram does not depend on the choice of the Cartan subalgebra follows from the fact that any two Cartan subalgebras are conjugate under an inner automorphism of $\mathfrak{g}$; see 2.2.2. To see that the Dynkin diagram does not depend on the choice of the order, we introduce the Weyl group W of an abstract root system. Having given a complex semisimple Lie algebra $\mathfrak{g}$ and a Cartan subalgebra $\mathfrak{h}$, the Weyl group of the corresponding abstract root system is called the Weyl group $W = W(\mathfrak{g}, \mathfrak{h})$ of $\mathfrak{g}$ (with respect to $\mathfrak{h}$).

If $\Delta \subset V$ is an abstract root system, then for any root $\alpha \in \Delta$, we have the root reflection $s_\alpha : V \to V$, and we know from 2.2.5 that $s_\alpha(\Delta) \subset \Delta$. Now we define the *Weyl group* $W = W(\Delta)$ of Δ to be the subgroup of the orthogonal group $O(V)$ generated by all the reflections s_α. Then any element $w \in W$ maps the root system Δ to itself. If $w \in W$ fixes all elements of Δ, then w must be the identity map, since Δ spans V. Hence, we may also view W as a subgroup of the permutation group of Δ, so W is finite. For later use, we define the *sign* of an element $w \in W$ as the determinant of w, viewed as a linear automorphism of V. Since w is an orthogonal map, the sign really is either 1 or -1, depending on whether w is a product of an even or odd number of reflections.

Alternatively to what we have done so far, one may also define simple roots directly and then use these to define positive roots. A *simple subsystem* $\Delta^0 = \{\alpha_1, \ldots, \alpha_n\} \subset \Delta$ is defined to be a basis of V consisting of reduced roots (i.e. $2\alpha_j$

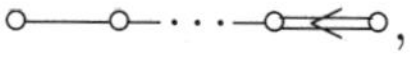

is not a root for all j) such that any element of Δ can be written as a linear combination of the α_j with all coefficients of the same sign. Having given such a simple system, one defines the corresponding positive roots $\Delta^+ \subset \Delta$ as consisting of those roots for which all coefficients are positive. For a given simple subsystem Δ^0, one defines the *height* of a root by $\mathrm{ht}(a_1\alpha_1 + \dots + a_n\alpha_n) = \sum_j a_j$.

PROPOSITION 2.2.7. *Let Δ be an abstract root system and let $\Delta^0 = \{\alpha_1, \dots, \alpha_n\}$ be a simple subsystem.*

(1) The Weyl group W of Δ is generated by the reflections s_{α_i} corresponding to simple roots.

(2) Mapping w to $w(\Delta^0)$ induces a bijection between W and the family of all simple subsystems of Δ.

PROOF. (1) Let $W' \subset W$ be the subgroup generated by the reflections s_{α_j} for $j = 1, \dots, n$. Suppose that $\alpha = \sum a_j\alpha_j \in \Delta^+$ is reduced but not simple, i.e. $a_j \geq 0$ for all j and $a_j > 0$ for at least two j. Since $0 < \langle \alpha, \alpha \rangle = \sum a_j \langle \alpha, \alpha_j \rangle$, there is a simple root α_i such that $\langle \alpha, \alpha_i \rangle > 0$. Now $s_{\alpha_i}(\alpha)$ is a root, and expanding this as $\sum b_j\alpha_j$ we have $b_j = a_j$ for $j \neq i$ and $b_i = a_i - \frac{2\langle \alpha, \alpha_i \rangle}{\langle \alpha_i, \alpha_i \rangle}$. By construction, $\frac{2\langle \alpha, \alpha_i \rangle}{\langle \alpha_i, \alpha_i \rangle}$ is a positive integer. Since for at least one j we have $b_j > 0$, we conclude that $s_{\alpha_i}(\alpha) \in \Delta^+$. We can continue this process until we reach a simple root.

On the one hand, this shows that all the coefficients a_j are integers. On the other hand, we conclude that any positive reduced root α may be written as $w(\alpha_j)$ for some $w \in W'$. Since $-\alpha_j = s_{\alpha_j}(\alpha_j)$ we conclude that any reduced root may be written in this form. Finally, if $\alpha = w(\alpha_j)$, then $s_\alpha = w \circ s_{\alpha_j} \circ w^{-1}$ and since $s_{2\alpha} = s_\alpha$ this implies $W' = W$.

(2) It is almost obvious that for $w \in W$, also $w(\Delta^0) = \{w(\alpha_1), \dots, w(\alpha_n)\}$ is a simple subsystem. (To obtain a representation of α in terms of the $w(\alpha_j)$ use a representation of $w^{-1}(\alpha) \in \Delta$ in terms of the α_j.) Hence, we obtain a well-defined map from W to the family of simple subsystems. Suppose that $w \in W$ has the property that $w(\alpha_i) \in \Delta^0$ for all i. Writing temporarily s_i for s_{α_i} we know from above that we may write $w = s_{i_\ell} \circ \dots \circ s_{i_1}$ for some choice of indices i_j. Now by assumption $w(\alpha_{i_1}) \in \Delta^0 \subset \Delta^+$ while $s_{i_1}(\alpha_{i_1}) = -\alpha_{i_1} \in -\Delta^+$. Thus, there is a minimal $r \geq 2$ such that for $w' = s_{i_{r-1}} \circ \dots \circ s_{i_1}$ we have $w'(\alpha_{i_1}) \in -\Delta^+$, but $s_{i_r}(w'(\alpha_{i_1})) \in \Delta^+$. We have seen above that for $\alpha \in \Delta^+$ with $\alpha \neq \alpha_{i_r}$ we have $s_{i_r}(\alpha) \in \Delta^+$, so we must have $w'(\alpha_{i_1}) = -\alpha_{i_r}$. Since $s_{i_r} = s_{-\alpha_{i_r}}$, we obtain $s_{i_r} = s_{w'(\alpha_{i_1})} = w' \circ s_{i_1} \circ (w')^{-1}$. Hence, $s_{i_r} \circ w' = w' \circ s_{i_1} = s_{i_{r-1}} \circ \dots \circ s_{i_2}$, which implies that w can be written as the product of $r - 2$ simple reflections. Iterating this, we see that $w = \mathrm{id}$ if r is even and w is a simple reflection if r is odd. But the latter case cannot occur, since $s_i(\alpha_i) = -\alpha_i \notin \Delta^0$. Hence, we conclude that the mapping $w \mapsto w(\Delta^0)$ is injective.

On the other hand, suppose that $A \subset \Delta$ is any simple subsystem and $D^+ \subset \Delta$ is the corresponding system of positive roots. Then $A \subset \Delta^+$ implies (and thus is equivalent to) $D^+ \subset \Delta^+$ and hence to $D^+ = \Delta^+$. If this is the case, then $\Delta^0 \subset D^+$, so any element of Δ^0 may be written as a linear combination of elements of A with nonnegative integral coefficients and vice versa, which easily implies $A = \Delta^0$. Hence, if $A \neq \Delta^0$ we can find a root $\alpha \in A \cap -\Delta^+$. Since $\alpha \in A$, we see that $s_\alpha(D^+)$ is obtained from D^+ by replacing α by $-\alpha$, so $s_\alpha(D^+) \cap \Delta^+$ is strictly larger than $D^+ \cap \Delta^+$. Clearly, $s_\alpha(D^+)$ is the positive system associated to the simple subsystem

$s_\alpha(A)$. Inductively, this implies that we find an element $w \in W$ such that the positive system associated to $w(A)$ equals Δ^+ and thus $w(A) = \Delta^0$. □

This immediately implies that the Cartan matrix and hence the Dynkin diagram is independent of the choice of the order (or equivalently of the simple subsystem). We have just seen that any two simple subsystems are related by an orthogonal isomorphism and since the Cartan matrix depends on inner products only, it does not change. We can also conclude that a reduced root system is completely determined by any simple subsystem, since the roots are exactly given by the images of the simple roots under the group generated by the simple root reflections.

Finally, let us discuss the concept of dominant elements and Weyl chambers which will be important in the sequel. Let $\Delta \subset V$ be an abstract root system with a distinguished simple subsystem Δ^0 (or equivalently a distinguished positive subsystem Δ^+). Then an element $v \in V$ is called *dominant* (or Δ^0–dominant) if and only if $\langle v, \alpha \rangle \geq 0$ for all $\alpha \in \Delta^0$ (or equivalently for all $\alpha \in \Delta^+$). The closed *dominant Weyl chamber* is defined to be the set of all dominant elements in V.

In general, one defines an open Weyl chamber to be one of the connected components of the complement of all hyperplanes orthogonal to the roots. Hence, an element of $v \in V$ lies in some open Weyl chamber if and only if $\langle v, \alpha \rangle \neq 0$ for all $\alpha \in \Delta$. Two such elements v and v' lie in the same open Weyl chamber if $\langle v, \alpha \rangle$ and $\langle v', \alpha \rangle$ have the same sign for all α. Obviously, the closed dominant Weyl chamber is the closure of the open Weyl chamber for which all the inner products are positive.

By construction, any root reflection permutes the open Weyl chambers, so the same is true for any element of the Weyl group. On the other hand, choosing an open Weyl chamber $\mathcal{C}$, one obtains a positive subsystem Δ^+ as those roots whose inner product with all elements in the chamber $\mathcal{C}$ are positive. This positive subsystem in turn determines a simple subsystem Δ^0. From the construction it is clear that for $w \in W$ the simple system associated to $w(\mathcal{C})$ is $w(\Delta^0)$, so from above we conclude that the Weyl group acts simply transitively on the set of all open Weyl chambers. In particular, given any element $v \in V$ there is an element $w \in W$ such that $w(v)$ is dominant.

We shall analyze the Weyl group in much more detail in Section 3.2. Here we just note what the Weyl groups and dominant Weyl chambers look like in the classical examples from 2.2.6.

For the root system A_{n-1} of $\mathfrak{sl}(n,\mathbb{C})$ from 2.2.6(1), the dual space $\mathfrak{h}_0^*$ is the quotient of the space of all $\sum a_j e_j$ by the line generated by $e_1 + \cdots + e_n$, so we may view it as the space of all $\sum a_j e_j$ such that $\sum a_j = 0$. One easily verifies that the root reflection $s_{e_i - e_j} : \mathfrak{h}_0^* \to \mathfrak{h}_0^*$ is induced by the map which exchanges e_i and e_j and leaves the e_k for $k \neq i, j$ untouched. Hence, the Weyl group of A_{n-1} is the permutation group $\mathfrak{S}_n$ of n elements. Elements of the dominant Weyl chamber by definition are represented by expressions $\sum a_j e_j$ such that $a_1 \geq a_2 \geq \cdots \geq a_n$.

For the root system B_n of $\mathfrak{so}(2n+1,\mathbb{C})$ from 2.2.6(3), the reflections in $e_i - e_j$ again exchanges e_i and e_j, while the reflection in e_j changes the sign of e_j and leaves the other e_k untouched. Thus, we may view the Weyl group W as the subgroup of all permutations σ of the $2n$ elements $\pm e_j$ such that $\sigma(-e_j) = -\sigma(e_j)$ for all $j = 1, \ldots, n$. Otherwise put, W is a semidirect product of $\mathfrak{S}_n$ and $(\mathbb{Z}_2)^n$, so in particular, W has $n!2^n$ elements. The dominant Weyl chamber by construction consists of all element $\sum a_j e_j$ such that $a_1 \geq \cdots \geq a_n \geq 0$.

Since the reflection corresponding to $2e_j$ coincides with the reflection for e_j, for the root system C_n from 2.2.6(4) we get the same Weyl group and the same positive Weyl chamber as for B_n.

Finally, for the even orthogonal root system D_n from 2.2.6(2), the reflections in the roots $e_i - e_j$ again generate permutations of the e_j, while the reflection in $e_i + e_j$ maps e_i to $-e_j$ and e_j to $-e_i$ while all other e_k remain untouched. Consequently, W can be viewed as the subgroup of those permutations π of the elements $\pm e_j$ which satisfy $\pi(-e_j) = -\pi(e_j)$ and have the property that the number of j such that $\pi(e_j) = -e_k$ for some k is even. In particular, the number of elements in W equals $n!2^{n-1}$. The positive Weyl chamber consists of all $\sum a_j e_j$ such that $a_1 \geq a_2 \cdots \geq a_n$ and $a_{n-1} \geq -a_n$.

2.2.8. The classification of Dynkin diagrams. To classify reduced abstract root systems, it suffices by our observations in 2.2.7 to classify abstract Cartan matrices or equivalently Dynkin diagrams. Clearly, we may restrict ourselves to irreducible reduced root systems and thus to connected Dynkin diagrams. Apart from the classical examples A_n, B_n, C_n and D_n that we have met in 2.2.6, there are five exceptional irreducible root systems which are called G_2, F_4, E_6, E_7 and E_8. The index refers to the dimension of the Euclidean space in which these systems sit. The Dynkin diagram of G_2 is , for F_4 one obtains . For the E series, the Dynkin diagram of E_8 is 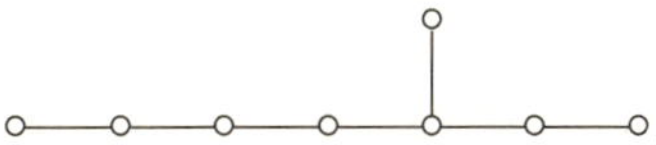, and the diagrams for E_7 and E_6 are obtained from this by removing the leftmost, respectively, the two leftmost vertices and edges. Now the classification of abstract root systems reads as:

THEOREM 2.2.8. *Any irreducible reduced root system is isomorphic to exactly one of the systems A_n $(n \geq 1)$, B_n $(n \geq 2)$, C_n $(n \geq 3)$, D_n $(n \geq 4)$, E_6, E_7, E_8, F_4, or G_2. Any non-reduced irreducible root system is isomorphic to $(BC)_n := B_n \cup C_n$ for some $n \geq 2$.*

SKETCH OF PROOF. For the reduced case, there is a very elegant proof based on extended Dynkin diagrams which we have taken from [**CSM95**]: We have seen in 2.2.5 that for the Cartan matrix A of an abstract root system there is a diagonal matrix D with positive entries, such that the matrix $S := DAD^{-1}$ is symmetric and positive definite. One immediately sees that the entries s_{ij} of S are given by $s_{ii} = 2$ and $s_{ij} = -\sqrt{n_{ij}}$ for $i \neq j$, where n_{ij} is the number of edges in the Dynkin diagram joining the ith and the jth vertex. The proof is done by showing that the Dynkin diagrams listed in the theorem are the only connected diagrams such that the associated matrix S is positive definite. Since S is obviously independent of the arrows in the diagram, we may forget about them for the purpose of the classification.

It is elementary to see that the quadratic form associated to S remains positive definite if some of the n_{ij} are decreased. On the other hand, if we take a connected subset of a diagram corresponding to a positive definite S, then the associated matrix just describes the restriction of the inner product to some subspace, and hence is positive definite, too. For the rest of this proof, we use the term "subdiagram" for any diagram which can be obtained from a given one by these two operations.

The upshot of this is that if we find a diagram for which the corresponding matrix S has zero determinant, then this cannot occur as a subdiagram in the Dynkin diagram of any irreducible abstract root system. A smart way to construct such diagrams is to add to the Dynkin diagram of an irreducible abstract root system one more vertex corresponding to the largest root of the system (in the given ordering) and to add the edges corresponding to the inner products with the simple roots. This leads to the following diagrams, which are called the *extended Dynkin diagrams.* (Any diagram with index k in this list has $k+1$ vertices.)

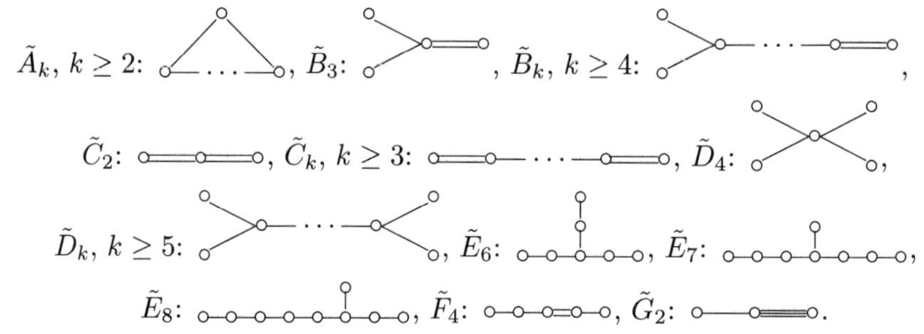

The matrix associated to any of these diagrams by construction has linearly dependent lines and thus zero determinant. This may also be verified directly for each of the extended Dynkin diagrams.

With the extended Dynkin diagrams at hand, the classification proceeds very quickly: The diagram $\tilde{A}_k$ shows that there are no cycles in the Dynkin diagram of an abstract root system. We already know that there may be at most triple edges. The diagrams $\tilde{B}_3$, $\tilde{C}_2$, $\tilde{D}_4$ and $\tilde{G}_2$ show that there are at most three edges connected to one point. In particular, if there is a triple edge, then G_2 is the only possibility. Let us call a vertex in which three single edges meet a branch point. Suppose that a Dynkin diagram contains a double edge. Then in view of $\tilde{B}_k$ and $\tilde{C}_k$ there may only be one double edge and no branch points. The diagram $\tilde{F}_4$ shows that F_4 is the only possible Dynkin diagram in which further edges are attached to both vertices joined by a double edge. The remaining possibilities in case of a double edge are only B_n and C_n.

Thus, we are left with the case that the Dynkin diagram contains only single edges. If there is no branch point, the absence of loops implies that the diagram is of type A_n. Moreover, in view of $\tilde{D}_k$, there may be at most one branch point in each Dynkin diagram. From the diagram $\tilde{E}_6$ we see that at least one of the three chains meeting at a branch point must consist of a single vertex, and $\tilde{E}_7$ shows that one of the two remaining chains consists of at most two vertices. But then $\tilde{E}_8$ shows that D_n, E_6, E_7 and E_8 are the only possible diagrams.

In the non–reduced case, one shows that the subset of all roots α such that 2α is not a root forms an irreducible reduced abstract root system. Using this, the result easily follows from the reduced case; see [**Kn96**, II.8]. □

2.2.9. The classification of complex simple Lie algebras. We have seen how to pass from a Lie algebra to a root system and further to a Dynkin diagram. We have also noted in 2.2.7 that this Dynkin diagram does not depend on the choices of a Cartan subalgebra and a set of positive roots. In particular, this implies that isomorphic Lie algebras lead to the same Dynkin diagram. Hence, there are two remaining questions. On the one hand, we do not know whether there exist complex simple Lie algebras corresponding to the exceptional root systems of type E, F and G from 2.2.8. On the other hand, we do not know whether two Lie algebras having the same Dynkin diagram must already be isomorphic. Both questions can be simultaneously answered (positively) by giving a universal construction for a simple Lie algebra with a given Dynkin diagram, using the so–called Serre relations. These will also be important in the study of real semisimple Lie algebras later on.

Let us start from a complex simple Lie algebra $\mathfrak{g}$ with a chosen Cartan subalgebra $\mathfrak{h}$, the corresponding set Δ of roots and a chosen simple subsystem $\Delta^0 = \{\alpha_1, \dots, \alpha_n\}$. For any $j = 1, \dots, n$ choose elements E_j and F_j in the root spaces $\mathfrak{g}_{\alpha_j}$, respectively, $\mathfrak{g}_{-\alpha_j}$ such that $B(E_j, F_j) = \frac{2}{\langle\alpha_j,\alpha_j\rangle}$. Recall from 2.2.4 that this means that $H_j := [E_j, F_j]$ satisfies $\alpha_j(H_j) = 2$, so $\{E_j, F_j, H_j\}$ is a standard basis for the subalgebra $\mathfrak{s}_{\alpha_j} \cong \mathfrak{sl}(2, \mathbb{C})$. Moreover, the elements H_j for $j = 1, \dots, n$ span the Cartan subalgebra $\mathfrak{h}$. By part (3) of Proposition 2.2.4 we have $\mathfrak{g}_{\alpha+\beta} = [\mathfrak{g}_\alpha, \mathfrak{g}_\beta]$ for all $\alpha, \beta \in \Delta$. Together with the above, this easily implies that $\{E_j, F_j, H_j : 1 \le j \le n\}$ is a set of generators for the Lie algebra $\mathfrak{g}$. Such a set of generators is called a set of *standard generators* for $\mathfrak{g}$.

Next, there are some obvious relations. Since all H_j lie in $\mathfrak{h}$, we have $[H_i, H_j] = 0$ for all i, j. By definition and the fact that the difference of two positive roots is not a root, we further have $[E_i, F_j] = \delta_{ij} H_i$. Next, by definition of the Cartan matrix $A = (a_{ij})$ of $\mathfrak{g}$, we have $[H_i, E_j] = a_{ij} E_j$ and $[H_i, F_j] = -a_{ij} F_j$. Finally, the formula for the length of the α_i–string through α_j from 2.2.4 implies that $\operatorname{ad}(E_i)^{-a_{ij}+1}(E_j) = 0$ and $\operatorname{ad}(F_i)^{-a_{ij}+1}(F_j) = 0$ for $i \neq j$. These six families of relations are called the *Serre relations* for $\mathfrak{g}$. The essential point for our questions now is that this is a complete set of relations. To be more precise one proves

THEOREM 2.2.9. *Let $A = (a_{ij})$ be an abstract $n \times n$ Cartan matrix. Let $\mathfrak{F}$ be the free complex Lie algebra generated by $3n$ elements E_j, F_j and H_j for $j = 1, \dots, n$, and let $\mathfrak{R}$ be the ideal generated by the Serre relations. Then $\mathfrak{g} := \mathfrak{F}/\mathfrak{R}$ is a finite-dimensional simple Lie algebra. The elements H_j span a Cartan subalgebra of $\mathfrak{g}$, and the functionals $\alpha_j \in \mathfrak{h}^*$ defined by $\alpha_j(H_i) = a_{ij}$ form a simple subsystem of the corresponding root system. In particular the Cartan matrix of $\mathfrak{g}$ is exactly A.*

PROOF. The proof is rather involved; see [**Kn96**, II.9–II.11]. □

COROLLARY 2.2.9. *(1) Any reduced irreducible abstract root system is isomorphic to the root system of some finite–dimensional complex simple Lie algebra.*

(2) Two complex simple Lie algebras are isomorphic if and only if their root systems are isomorphic, i.e. if and only if they have the same Dynkin diagram.

PROOF. (1) is obvious from the theorem in view of the bijective correspondence between Cartan matrices and reduced irreducible abstract root systems described in 2.2.7.

(2) Let $\mathfrak{g}$ be any complex simple Lie algebra and let A be its Cartan matrix. Let $\mathfrak{F}/\mathfrak{R}$ be the Lie algebra constructed from A in the theorem. By the universal property of a free Lie algebra, choosing a set of standard generators for $\mathfrak{g}$ gives a surjective homomorphism $\mathfrak{F} \to \mathfrak{g}$, which factors to $\mathfrak{F}/\mathfrak{R}$ since the Serre relations hold in $\mathfrak{g}$. But from the theorem we know that $\mathfrak{F}/\mathfrak{R}$ is simple, which implies that this homomorphism must be injective, and thus an isomorphism. □

REMARK 2.2.9. (1) The description of a complex simple Lie algebra by generators and relations corresponding to a Cartan matrix is the basis for various stages of generalizations of these Lie algebras in the direction of Kac–Moody algebras. Weakening the conditions on an abstract Cartan matrix from 2.2.5 slightly and then defining a Lie algebra by generators and relations as in the theorem, one obtains infinite–dimensional Lie algebras, which, however, behave similarly to complex simple Lie algebras in several respects; see for example [**Kac90**].

(2) While the theorem asserts the existence of the exceptional complex simple Lie algebras, i.e. Lie algebras corresponding to the exceptional root systems, it does not offer a good description of these. There are various ways to give more explicit descriptions of the exceptional Lie algebras. One way is to describe them in the form $W \oplus \mathfrak{g}_0 \oplus W^*$, where $\mathfrak{g}_0$ is a certain complex semisimple Lie algebra, W is a representation of $\mathfrak{g}_0$ with dual representation W^*. The bracket is described by the bracket of $\mathfrak{g}_0$, the action of $\mathfrak{g}_0$ on W and W^*, and certain pairings on the representation spaces. This description of the exceptional algebras is outlined in [**FH91**, §22.4].

Other conceptual descriptions of the exceptional algebras are related to the octonions (or Cayley numbers) $\mathbb{O}$, the unique 8–dimensional (non–associative) normed real division algebra, and the exceptional Jordan algebra $\mathbb{J}$ of dimension 27 formed by Hermitian 3×3–matrices with entries from $\mathbb{O}$. The simplest case is the Lie algebra G_2, which is the complexification of the Lie algebra of derivations of $\mathbb{O}$. Similarly, the complexification of the Lie algebra of derivations of $\mathbb{J}$ is a Lie algebra of type F_4. For a discussion of the relation between octonions and exceptional Lie algebras see [**Ba02**].

Finally, one should also note that the Lie algebra of type G_2 can also be viewed as the algebra of endomorphisms of a seven–dimensional vector space V, which preserves a generic element of the third exterior power $\Lambda^3 V$. This point of view is important for the study of G_2–structures on seven–dimensional manifolds.

2.2.10. Finite–dimensional representations. Let us fix a complex simple Lie algebra $\mathfrak{g}$, a Cartan subalgebra $\mathfrak{h} \leq \mathfrak{g}$ and let Δ be the corresponding set of roots. Fix an order on $\mathfrak{h}^*$, and let Δ^+ and Δ^0 be the corresponding sets of positive respectively simple roots. By $\mathfrak{h}_0^* \subset \mathfrak{h}^*$ we denote the real span of the roots. Let $\langle\ ,\ \rangle$ be the complex bilinear form on $\mathfrak{h}^*$ induced by the Killing form. By part (1) of Theorem 2.2.4, the restriction of $\langle\ ,\ \rangle$ to $\mathfrak{h}_0^*$ is positive definite. From 2.2.2 we know that any finite–dimensional representation of $\mathfrak{g}$ admits a weight decomposition into a direct sum of joint eigenspaces for the actions of the elements of $\mathfrak{h}$. These are the weight spaces and the eigenvalues, called weights, are linear functionals on $\mathfrak{h}$. We will also use the term weights for general linear functionals on $\mathfrak{h}$.

To describe properties of the weights of finite–dimensional representations, we have to introduce a few notions. A weight $\lambda \in \mathfrak{h}^*$ is called *real* if it lies in the subspace $\mathfrak{h}_0^*$, or equivalently, if $\langle\lambda, \alpha\rangle \in \mathbb{R}$ for all roots $\alpha \in \Delta$. A (real) weight λ is called *algebraically integral* if $\frac{2\langle\lambda,\alpha\rangle}{\langle\alpha,\alpha\rangle} \in \mathbb{Z}$ for all $\alpha \in \Delta$.

Recall from 2.2.7 that the simple roots form a complex basis for $\mathfrak{h}^*$ and a real basis for $\mathfrak{h}_0^*$. Writing $\Delta^0 = \{\alpha_1, \dots, \alpha_n\}$, we define elements $\omega_1, \dots, \omega_n \in \mathfrak{h}_0^*$ by requiring that $\frac{2\langle\omega_i,\alpha_j\rangle}{\langle\alpha_j,\alpha_j\rangle} = \delta_{ij}$. These are called the *fundamental weights* corresponding to the simple system Δ^0. Recall further that an element $\lambda \in \mathfrak{h}_0^*$ is called *dominant* if $\langle\lambda, \alpha_j\rangle \geq 0$ for all j.

LEMMA 2.2.10. *(1) A weight $\lambda \in \mathfrak{h}^*$ is algebraically integral if and only if $\frac{2\langle\lambda,\alpha\rangle}{\langle\alpha,\alpha\rangle} \in \mathbb{Z}$ holds for all simple roots $\alpha \in \Delta^0$.*

(2) The fundamental weights $\omega_1, \dots, \omega_n$ form a real basis for $\mathfrak{h}_0^*$ and a complex basis for $\mathfrak{h}^*$. For $\lambda \in \mathfrak{h}^*$ let $\lambda = \sum_i \lambda_i\omega_i$ be the expansion in this basis. Then λ is real, respectively, algebraically integral, respectively, dominant if and only if all $\lambda_i \in \mathbb{R}$, respectively, $\lambda_i \in \mathbb{Z}$, respectively, $\lambda_i \geq 0$ for all i.

PROOF. (1) Suppose that $\frac{2\langle\lambda,\alpha_j\rangle}{\langle\alpha_j,\alpha_j\rangle} \in \mathbb{Z}$ for all $\alpha_j \in \Delta_0$. Consider the reflection s_{α_k} corresponding to the simple root α_k. Then by definition

$$\frac{2\langle s_{\alpha_k}(\lambda), \alpha_j\rangle}{\langle\alpha_j,\alpha_j\rangle} = \frac{2\langle\lambda,\alpha_j\rangle}{\langle\alpha_j,\alpha_j\rangle} - \frac{2\langle\lambda,\alpha_k\rangle}{\langle\alpha_k,\alpha_k\rangle}\frac{2\langle\alpha_k,\alpha_j\rangle}{\langle\alpha_j,\alpha_j\rangle} \in \mathbb{Z}.$$

By part (1) of Proposition 2.2.7, the reflections corresponding to simple roots generate the Weyl group W of $\mathfrak{g}$. Hence, we conclude that $\frac{2\langle w(\lambda),\alpha_j\rangle}{\langle\alpha_j,\alpha_j\rangle} \in \mathbb{Z}$ for all $w \in W$ and all $\alpha_j \in \Delta^0$. For an arbitrary root $\alpha \in \Delta$ we know from 2.2.7 that there exists a simple root α_j and an element $w \in W$ such that $\alpha = w(\alpha_j)$. Since w is an orthogonal map, $\frac{2\langle w^{-1}(\lambda),\alpha_j\rangle}{\langle\alpha_j,\alpha_j\rangle} \in \mathbb{Z}$ immediately implies $\frac{2\langle\lambda,\alpha\rangle}{\langle\alpha,\alpha\rangle} \in \mathbb{Z}$.

(2) The elements $\omega_i \in \mathfrak{h}_0^*$ are linearly independent by construction. From 2.2.5 we know that the simple roots form a basis of $\mathfrak{h}_0^*$, so the fundamental weights form a basis, too. For $\lambda = \sum_i \lambda_i\omega_i$ we by definition have $\frac{2\langle\lambda,\alpha_i\rangle}{\langle\alpha_i,\alpha_i\rangle} = \lambda_i$, which implies the rest of (2). □

From part (2) we see that the set Λ_W of all algebraically integral elements of $\mathfrak{h}_0^*$ forms a lattice in $\mathfrak{h}_0^*$. This is called the *weight lattice* of $\mathfrak{g}$. By Proposition 2.2.4, any root is an algebraically integral weight, so the root lattice Λ_R (see 2.2.5) is contained in Λ_W. It turns out that the quotient Λ_W/Λ_R is a finite group, whose order equals the determinant of the Cartan matrix of $\mathfrak{g}$; see [**FH91**, Lemma 23.15].

Part (2) of the proposition also suggests a notation for weights. We can expand each weight as $\lambda = \sum_i \lambda_i\omega_i$ in terms of fundamental weights. We will denote the weight λ by taking the Dynkin diagram of $\mathfrak{g}$ and writing the coefficient λ_i over the node corresponding to the simple root dual to ω_i. For example, consider the Lie algebra $\mathfrak{sl}(4,\mathbb{C})$. By 2.2.6(1) the Dynkin diagram is $A_3 = \circ\!\!-\!\!\!-\!\!\circ\!\!-\!\!\!-\!\!\circ$. Now, for example, the weight $\lambda = \overset{1}{\circ}\!\!-\!\!\!-\!\!\overset{-2}{\circ}\!\!-\!\!\!-\!\!\overset{1}{\circ}$ can be easily computed to be given (in the notation of 2.2.6(1)) by $e_1 - 2(e_1+e_2) + (e_1+e_2+e_3) = -e_2+e_3 = -\alpha_2$, the negative of the second simple root. In general, the weight given by a simple root α_j is given by the entries in the jth column of the Cartan matrix. In particular, the entry over the jth node is 2 and the number over a node is nonzero if and only if the node is connected to the jth node. If it is connected by a single edge, the number is -1, if there are multiple edges, then the number depends on the direction of the arrow.

Now let $\rho : \mathfrak{g} \to \mathfrak{gl}(V)$ be a complex representation of $\mathfrak{g}$ on a finite–dimensional vector space V. Let $\mathrm{wt}(V) \subset \mathfrak{h}^*$ be the sets of weights and let $V = \bigoplus_{\lambda\in\mathrm{wt}(V)} V_\lambda$ be the weight decomposition. The dimension of the weight space V_λ is called the *multiplicity of the weight λ in V*. By definition of a weight space, for $v \in V_\lambda$ and $X \in \mathfrak{g}_\alpha$, we have $X \cdot v \in V_{\lambda+\alpha}$ (so, in particular, $X \cdot v = 0$ if $\lambda + \alpha$ is not a weight of V).

PROPOSITION 2.2.10. *Let $\mathfrak{g}$ be a finite–dimensional complex semisimple Lie algebra, $\mathfrak{h} \leq \mathfrak{g}$ a Cartan subalgebra, and Δ the corresponding set of roots. Fix a choice of positive roots Δ^+ with simple roots Δ^0, and let W be the Weyl group. Let V be a finite–dimensional complex representation of $\mathfrak{g}$ with weight decomposition $V = \bigoplus_{\lambda \in \mathrm{wt}(V)} V_\lambda$. Then we have:*

1. *Any weight λ of V is algebraically integral and at least one weight of V is dominant.*
2. *For any weight $\lambda \in \mathrm{wt}(V)$ and any $w \in W$, also $w(\lambda) \in \mathrm{wt}(V)$ and the two weights have the same multiplicity in V.*
3. *Suppose that for $\lambda \in \mathrm{wt}(V)$ and $\alpha \in \Delta$ the integer $k := \frac{2\langle\lambda,\alpha\rangle}{\langle\alpha,\alpha\rangle}$ is positive. Then for each $\ell \in \{1, \dots, k\}$ we have $\lambda - \ell\alpha \in \mathrm{wt}(V)$ with at least the same multiplicity as λ.*

PROOF. Consider a fixed positive root $\alpha \in \Delta^+$, let $\mathfrak{s}_\alpha = \mathfrak{g}_{-\alpha} \oplus [\mathfrak{g}_\alpha, \mathfrak{g}_{-\alpha}] \oplus \mathfrak{g}_{-\alpha} \cong \mathfrak{sl}(2, \mathbb{C})$ be the corresponding subalgebra; see Proposition 2.2.4. For $\lambda \in \mathrm{wt}(V)$ consider $V' := \bigoplus_{n \in \mathbb{Z}} V_{\lambda + n\alpha} \subset V$. This subspace is invariant under the action of $\mathfrak{s}_\alpha$, so we can apply Proposition 2.2.3. In 2.2.4 we have constructed a standard basis of $\mathfrak{s}_\alpha$ which contained the element $\frac{2}{\langle\alpha,\alpha\rangle} H_\alpha \in [\mathfrak{g}_\alpha, \mathfrak{g}_{-\alpha}]$. By Proposition 2.2.3, this element has integral eigenvalues on any finite–dimensional representation, which shows that $\frac{2\langle\lambda,\alpha\rangle}{\langle\alpha,\alpha\rangle} \in \mathbb{Z}$. Further, the eigenvalues of $\frac{2}{\langle\alpha,\alpha\rangle} H_\alpha$ form an unbroken string of the form $a, a-2, a-4, \dots, -a$. Now,

$$\frac{2\langle\lambda - \ell\alpha, \alpha\rangle}{\langle\alpha,\alpha\rangle} = \frac{2\langle\lambda,\alpha\rangle}{\langle\alpha,\alpha\rangle} - 2\ell,$$

so we see that, starting from λ, we must also have the weights $\lambda - \ell\alpha$ for $\ell = 1, \dots, \frac{2\langle\lambda,\alpha\rangle}{\langle\alpha,\alpha\rangle}$, which proves (3). Moreover, for the maximal value of ℓ, by definition, we obtain the weight $s_\alpha(\lambda)$. We also see that V_λ and $V_{s_\alpha(\lambda)}$ must have the same dimension. Since the Weyl group is generated by the reflections s_α, this implies (2). Since any weight can be mapped to the dominant Weyl chamber by an element of W, we obtain the last claim in (1). □

2.2.11. Highest weight vectors. For the next step, we will briefly leave the realm of finite–dimensional representations, and also allow infinite–dimensional ones. However, we stay in a purely algebraic context, so we do not need topologies on the representation spaces. We continue to fix $\mathfrak{g}$ and $\mathfrak{h}$ and use the notation of the last subsection. Assume that V is a representation of $\mathfrak{g}$ such that the action of any element of the Cartan subalgebra $\mathfrak{h}$ is diagonalizable. This means that V admits a (possibly infinite) decomposition into weight spaces. A *highest weight vector* in V is a weight vector $v \in V$ such that $X \cdot v = 0$ for any element X lying in a root space $\mathfrak{g}_\alpha$ with $\alpha \in \Delta^+$.

Let us fix a set $\{E_i, F_i, H_i : i = 1, \dots, n\}$ of standard generators for $\mathfrak{g}$; see 2.2.9. Then a weight vector $v \in V$ evidently is a highest weight vector if and only if $E_i \cdot v = 0$ for all $i = 1, \dots, n$.

THEOREM 2.2.11. *Let $\mathfrak{g}$ be a finite–dimensional complex semisimple Lie algebra. Let $\mathfrak{h} \leq \mathfrak{g}$ be a fixed Cartan subalgebra and let Δ be the corresponding set of roots. Fix an order on the real span $\mathfrak{h}_0^*$ of Δ, let Δ^+ and Δ^0 be the sets of positive and simple roots with respect to this order, and let $\{E_i, F_i, H_i : i = 1, \dots, n\}$ be a standard set of generators. Let V be a representation of $\mathfrak{g}$ such that the elements of $\mathfrak{h}$ act simultaneously diagonalizable.*

(1) For any highest weight vector $v \in V$ the elements of the form $F_{i_1} \cdots F_{i_\ell} \cdot v$ span an indecomposable subrepresentation V' of V. Denoting by λ the weight of v, all weights occurring in V' have the form $\lambda - \sum n_i \alpha_i$ for $\alpha_i \in \Delta^0$ and nonnegative integers n_i, and the weight space V'_λ is one–dimensional.

(2) Suppose further that V is finite-dimensional. Then the weight of any highest weight vector is dominant and algebraically integral, and there exists at least one highest weight vector. In this case, the submodule $V' \subset V$ from (1) is irreducible.

PROOF. (1) The subspace V' spanned by the elements of the claimed form is by construction invariant under the action of all F_j. Now for $H \in \mathfrak{h}$ and $w \in V$ we compute

$$H \cdot F_i \cdot w = [H, F_i] \cdot w + F_i \cdot H \cdot w = -\alpha_i(H) F_i \cdot w + F_i \cdot H \cdot w.$$

Inductively, this shows that V' is $\mathfrak{h}$–invariant and that $F_{i_1} \cdots F_{i_\ell} \cdot v$ is a weight vector of weight $\lambda - \alpha_{i_1} - \cdots - \alpha_{i_\ell}$. Likewise, $E_j \cdot (F_{i_1} \cdots F_{i_\ell} \cdot v)$ is a linear combination of $F_{i_1} \cdot E_j \cdot F_{i_2} \cdots F_{i_\ell} \cdot v$ and $H_j \cdot F_{i_2} \cdots F_{i_\ell} \cdot v$. Again, by induction, this implies that V' is invariant under the action of each E_j. Since the elements E_j, F_j, and H_j generate $\mathfrak{g}$ as a Lie algebra, we see that $V' \subset V$ is a $\mathfrak{g}$–invariant subspace. Moreover, the weight spaces of V' have the claimed form and $\dim(V'_\lambda) = 1$.

To prove that $V' \subset V$ is indecomposable, suppose that $V' = V'_1 \oplus V'_2$ as a representation of $\mathfrak{g}$. By assumption, V is a direct sum of weight spaces, so any element in V can be written as a finite sum of weight vectors of different weights. Exchanging the two summands if necessary, we may assume that V'_1 contains an element of the form $v + v_1 + \cdots + v_N$, where the v_j are weight vectors of pairwise different weights, which are all different from λ. Fix an element $H \in \mathfrak{h}$. Then each v_j is an eigenvector for the action of H and only finitely many eigenvalues $a_1, \dots, a_r$ occur. Now by $\mathfrak{g}$–invariance, the element

$$(\rho(H) - a_1) \circ \ldots \circ (\rho(H) - a_r)(v + v_1 + \cdots + v_N)$$

lies in V'_1. By construction, this is a nonzero multiple of $v + v_{i_1} + \cdots + v_{i_s}$, where the v_{i_s} are those v_j on which the action of H has the same eigenvalue as on v. Doing this construction step by step for the elements of a basis of $\mathfrak{h}$, we obtain a nonzero multiple of v plus the sum of those v_j on which all elements of $\mathfrak{h}$ have the same eigenvalue as on v. But since the weight space V'_λ is one–dimensional, this means that $v \in V'_1$ and hence $V'_1 = V'$.

(2) By Proposition 2.2.10, all weights of V lie in $\mathfrak{h}_0^*$ and the choice of positive roots leads to an ordering on this space. Since V has only finitely many weights, there is a weight $\lambda_0 \in \mathrm{wt}(V)$ such that $\lambda \leq \lambda_0$ for all $\lambda \in \mathrm{wt}(V)$. If v is any nonzero element of V_{λ_0}, then v must be a highest weight vector, since $E_i \cdot v$ has weight $\lambda_0 + \alpha_i$, which is strictly larger than λ_0. If $\langle \lambda_0, \alpha_j \rangle < 0$ for some simple root α_j, then $s_{\alpha_j}(\lambda_0) = \lambda_0 - \frac{2\langle \lambda_0, \alpha_j \rangle}{\langle \alpha_j, \alpha_j \rangle} \alpha_j$ is strictly larger than λ_0. By Proposition 2.2.10, this is a weight of V which again contradicts maximality of λ_0. Finally, by Theorem 2.1.6, finite–dimensional indecomposable representations are irreducible. □

For a finite–dimensional representation V, the maximal weight used in the proof of part (2) is often called the *highest weight* of V, in particular if V is irreducible.

2.2.12. Existence of finite–dimensional irreducible representations. Let V be a finite–dimensional irreducible representation of a semisimple Lie algebra $\mathfrak{g}$. Then by Theorem 2.2.11 there is exactly one weight space V_λ which contains a highest weight vector, and the weight λ is dominant and algebraically integral. Moreover, if V and V' are irreducible representations with the same highest weight λ, then we choose highest weight vectors $v \in V$ and $v' \in V'$. Then (v, v') is a highest weight vector in the finite–dimensional representation $V \oplus V'$ of $\mathfrak{g}$, so it generates an irreducible subrepresentation $\tilde{V} \subset V \oplus V'$. The restrictions of the two projections to $\tilde{V}$ define homomorphisms $\tilde{V} \to V$ and $\tilde{V} \to V'$. The homomorphism $\tilde{V} \to V$ is nonzero, since v lies in the image. By irreducibility it must be an isomorphism. Similarly, the other homomorphism $\tilde{V} \to V'$ is an isomorphism, so $V \cong V'$. Conversely, isomorphic irreducible representations obviously have the same highest weight.

Thus, to get a complete hand on the finite–dimensional irreducible representations (and thus by complete reducibility 2.1.6 on all finite–dimensional representations), the remaining question is for which dominant algebraically integral weights λ there exists a finite–dimensional irreducible representation with highest weight λ.

THEOREM 2.2.12 (Theorem of the highest weight). *If $\mathfrak{g}$ is a finite–dimensional complex semisimple Lie algebra, then for any dominant algebraically integral weight $\lambda \in \mathfrak{h}_0^*$ there is a (up to isomorphism) unique finite–dimensional irreducible representation with highest weight λ.*

We will next sketch two different general proofs for the existence part of this theorem, namely via Verma modules and using the Borel–Weil theorem. A more pedestrian approach to the proof on a case by case basis will be discussed in 2.2.13 below.

Verma modules. The first approach uses the universal enveloping algebra $\mathcal{U}(\mathfrak{g})$ of the Lie algebra $\mathfrak{g}$ and induced modules as discussed in 2.1.10. Fix a Cartan subalgebra $\mathfrak{h} \leq \mathfrak{g}$ with roots Δ and an ordering on $\mathfrak{h}_0^*$ with corresponding sets Δ^+ and $\Delta^0 = \{\alpha_1, \dots, \alpha_n\}$ of positive and simple roots. Let $\mathfrak{n}_\pm \subset \mathfrak{g}$ be the sum of all positive (respectively negative) root spaces. Clearly, these are nilpotent subalgebras of $\mathfrak{g}$. Now the *standard Borel subalgebra* $\mathfrak{b} \leq \mathfrak{g}$ corresponding to these choices is defined as $\mathfrak{b} = \mathfrak{h} \oplus \mathfrak{n}_+$. Clearly, this is a subalgebra of $\mathfrak{g}$ and the commutator algebra $[\mathfrak{b}, \mathfrak{b}]$ is the nilpotent subalgebra $\mathfrak{n}_+$, so $\mathfrak{b}$ is solvable. It is easy to see that, in fact, $\mathfrak{b}$ is a maximal solvable subalgebra of $\mathfrak{g}$. The fact that any two Cartan subalgebras as well as the choice of the order can be absorbed in an inner automorphism of $\mathfrak{g}$ (see 2.2.2 and 2.2.7) can be rephrased as the fact that any two maximal solvable subalgebras of $\mathfrak{g}$ are conjugate under an inner automorphism.

It is easy to describe all irreducible representations of $\mathfrak{b}$. In 2.1.3 we have observed that such representations are given by linear functionals on $\mathfrak{b}/[\mathfrak{b}, \mathfrak{b}] \cong \mathfrak{h}$. Choosing $\lambda \in \mathfrak{h}^*$ we thus obtain an irreducible representation $\mathbb{C}_\lambda$ of $\mathfrak{b}$. Since $\mathfrak{b} \subset \mathfrak{g}$ we may view $\mathcal{U}(\mathfrak{b})$ as a subalgebra of $\mathcal{U}(\mathfrak{g})$ and we define the *Verma module* $M_\mathfrak{b}(\lambda)$ with highest weight λ as the $\mathcal{U}(\mathfrak{g})$–module induced by $\mathbb{C}_\lambda$, i.e. $M_\mathfrak{b}(\lambda) = \mathcal{U}(\mathfrak{g}) \otimes_{\mathcal{U}(\mathfrak{b})} \mathbb{C}_\lambda$; see 2.1.10. The action of $\mathcal{U}(\mathfrak{g})$ (and thus also the action of $\mathfrak{g}$) comes from multiplication from the left in $\mathcal{U}(\mathfrak{g})$.

Since the subalgebra $\mathfrak{n}_- \subset \mathfrak{g}$ is complementary to $\mathfrak{b}$, we see from 2.1.10 that as an $\mathfrak{n}_-$–module the Verma module $M_\mathfrak{b}(\lambda)$ is isomorphic to $\mathcal{U}(\mathfrak{n}_-) \otimes \mathbb{C}_\lambda$. Let us number the positive roots as $\Delta^+ = \{\beta_1, \dots, \beta_\ell\}$ and choose an element $F_{\beta_i} \in \mathfrak{g}_{-\beta_i}$ for each $i = 1, \dots, \ell$. Then from 2.1.10 we know that the elements $F_{\beta_1}^{i_1} \cdots F_{\beta_\ell}^{i_\ell} \otimes 1$ form a linear basis of $M_\mathfrak{b}(\lambda)$. To obtain the action of an element $H \in \mathfrak{h}$ in this picture, one has to use the commutation relation $HF_{\beta_i} - F_{\beta_i}H = [H, F_{\beta_i}] = -\beta_i(H)F_{\beta_i}$ to commute H through the F's, and finally bring H across the tensor product, which adds a factor $\lambda(H)$. In particular, this implies that $F_{\beta_1}^{i_1} \cdots F_{\beta_\ell}^{i_\ell} \otimes 1$ is a weight vector of weight $\lambda - (i_1\beta_1 + \cdots + i_\ell\beta_\ell)$. Hence, $M_{\mathfrak{b}}(\lambda)$ is a direct sum of weight spaces and each weight space is obviously finite–dimensional. By construction, $1 \otimes 1$ is a highest weight vector in $M_{\mathfrak{b}}(\lambda)$ which generates the whole module.

In particular, $M_{\mathfrak{b}}(\lambda)$ is indecomposable by part (1) of Theorem 2.2.11. Note also that Frobenius reciprocity becomes particularly simple for Verma modules. By Proposition 2.1.10, restriction to $1 \otimes \mathbb{C}_\lambda$ induces an isomorphism $\mathrm{Hom}_{\mathfrak{g}}(M_{\mathfrak{b}}(\lambda), V) \cong \mathrm{Hom}_{\mathfrak{b}}(\mathbb{C}_\lambda, V)$ for any $\mathfrak{g}$–module V. Of course, a linear map $\mathbb{C}_\lambda \to V$ is determined by the image v of $1 \in \mathbb{C}_\lambda$, and v gives rise to a homomorphism if and only if $\mathfrak{n}_+$ acts trivially on v, while each $H \in \mathfrak{h}$ acts by multiplication by $\lambda(H)$. Hence, we conclude that $\mathfrak{g}$–homomorphisms $M_{\mathfrak{b}}(\lambda) \to V$ are in bijective correspondence with highest weight vectors of weight λ in V.

From the proof of Theorem 2.2.11 we see that for a $\mathfrak{g}$–submodule $N \subset M_{\mathfrak{b}}(\lambda)$ an element $x \in N$ can be written as a finite sum of weight vectors of different weights and each of the components of this sum again lies in N. This immediately implies that for a proper submodule N there may never be a nonzero component of weight λ. This in turn shows that the subspace spanned by an arbitrary family of proper submodules of $M_{\mathfrak{b}}(\lambda)$ is again a proper submodule, and hence $M_{\mathfrak{b}}(\lambda)$ contains a unique maximal proper submodule. Let $L(\lambda)$ be the quotient of $M_{\mathfrak{b}}(\lambda)$ by this maximal proper submodule. Then by construction $L(\lambda)$ is irreducible and has highest weight λ. Hence, we see that for any $\lambda \in \mathfrak{h}^*$ there is an irreducible representation of highest weight λ, so to prove the theorem of the highest weight it remains to show that $L(\lambda)$ is finite–dimensional if λ is dominant and algebraically integral. To do this, the main step (see [**Kn96**, Theorem 5.16]) is to show that under this assumption the set of weights of $L(\lambda)$ (including multiplicities) is stable under the action of the Weyl group. Having shown this, it is easy to see from the explicit description of $M_{\mathfrak{b}}(\lambda)$ above that there are only finitely many dominant weights, which implies that $L(\lambda)$ is finite–dimensional.

The description above shows that any irreducible finite–dimensional representation of $\mathfrak{g}$ is a quotient of a Verma module. In fact, a much more precise description is possible. It turns out that any finite–dimensional irreducible representation of $\mathfrak{g}$ admits a finite resolution by Verma modules, the so–called Bernstein–Gelfand–Gelfand resolution, and the Verma modules showing up in this resolution can be described precisely in terms of the Weyl group. We will study this resolution and give a geometric construction in volume two.

The Borel–Weil theorem. The Borel–Weil theorem offers a geometric construction of finite–dimensional irreducible representations. We only briefly outline the statement here, since we will discuss the extension of this theorem due to Bott in more detail in Section 3.3. Starting with a complex semisimple Lie algebra $\mathfrak{g}$, let G be a connected and simply connected complex Lie group with Lie algebra $\mathfrak{g}$. Let $\mathfrak{b} \leq \mathfrak{g}$ be a Borel subalgebra (which gives a choice of a Cartan subalgebra and a notion of positivity), and let $B \subset G$ be the normalizer of $\mathfrak{b}$, i.e. $B = \{g \in G : \mathrm{Ad}(g)(\mathfrak{b}) \subset \mathfrak{b}\}$. This is obviously a closed subgroup and thus a Lie subgroup of G, and one easily shows that the Lie algebra of this subgroup is $\mathfrak{b} \subset \mathfrak{g}$. The homogeneous space G/B turns out to be a compact Kähler manifold (see 3.2.6 and 3.2.8), which is called the *full flag variety* of G. Next, one shows that for a dominant integral weight $\lambda \in \mathfrak{h}_0^*$ the one–dimensional representation $\mathbb{C}_{-\lambda}$ of $\mathfrak{b}$ integrates to a representation of B, so we can form the associated line bundle $G \times_B \mathbb{C}_{-\lambda} \to G/B$. From 1.4.4 we know that the space of smooth sections of this bundle naturally is a G–representation. Since G acts holomorphically on G/B, the (global) holomorphic sections form a subrepresentation. The Borel–Weil theorem then states that this representation is irreducible and the corresponding irreducible representation of $\mathfrak{g}$ has highest weight λ.

2.2.13. Fundamental representations – examples. The case–by–case approach to proving existence of irreducible finite–dimensional representations is based on the following observation: Suppose that V and W are finite–dimensional irreducible representations of $\mathfrak{g}$ with highest weights λ and μ, respectively. If $v \in V$ and $w \in W$ are highest weight vectors, then $v \otimes w$ is a highest weight vector in the tensor product representation $V \otimes W$ with weight $\lambda + \mu$. By Theorem 2.2.10, this vector generates an irreducible subrepresentation of $V \otimes W$ with highest weight $\lambda + \mu$. Moreover, if $V = \bigoplus V_{\lambda'}$ and $W = \bigoplus W_{\mu'}$ are the weight decompositions, then the weight spaces in $V \otimes W$ have the form $\bigoplus_{\lambda'+\mu'=\nu} V_{\lambda'} \otimes W_{\mu'}$. In particular, any weight ν of $V \otimes W$ is $\leq \lambda + \mu$. The subrepresentation generated by the highest weight vector above is usually called the *Cartan product* of V and W and is denoted by $V \circledcirc W$.

Recall from 2.2.10 that the dominant algebraically integral weights of $\mathfrak{g}$ are exactly the linear combinations of the fundamental weights $\omega_1, \dots, \omega_n$, with non–negative integral coefficients. The irreducible finite–dimensional representation V_i with highest weight ω_i is called the ith *fundamental representation.* Suppose that we have constructed the fundamental representations $V_1, \dots, V_n$. Given a dominant integral weight $\lambda = a_1\omega_1 + \dots + a_n\omega_n$, consider the representation $V_1^{\otimes a_1} \otimes \dots \otimes V_n^{\otimes a_n}$. From above we see that this contains a unique (up to scale) highest weight vector of weight λ.

We can also use the symmetric powers $S^{a_i}V_i$ of the fundamental representations rather than the tensor powers. In any case, we obtain a unique irreducible subrepresentation of highest weight λ. In simple cases we can even obtain an explicit description of this highest weight representation, since we have to find all homomorphisms to irreducible representations of lower highest weight and consider the intersection of their kernels. Hence, the essential step that remains is to construct the fundamental representations, which, at least for the classical algebras, is fairly simple. Let us discuss this for the classical examples using the notation of 2.2.6.

Representations of $\mathfrak{sl}(n, \mathbb{C})$. As in 2.2.6(1), the Cartan subalgebra $\mathfrak{h}$ consists of all tracefree diagonal matrices, and the functionals e_i for $i = 1, \dots, n$ are given by extracting the ith diagonal entry. The simple roots $\alpha_1, \dots, \alpha_{n-1}$ are given by $\alpha_i = e_i - e_{i+1}$, and the Killing form satisfies $\langle \alpha_i, e_j \rangle = 2n(\delta_{i,j} - \delta_{i+1,j})$. By definition of a fundamental weight, we must have $\frac{2\langle \omega_i, \alpha_j \rangle}{\langle \alpha_j, \alpha_j \rangle} = \delta_{ij}$, from which one easily concludes that $\omega_i = e_1 + \dots + e_i$ for $i = 1, \dots, n-1$.

Let $V = \mathbb{C}^n$ be the standard representation of $\mathfrak{sl}(n, \mathbb{C})$. If $\{v_1, \dots, v_n\}$ is the standard basis, then obviously each v_i is a weight vector of weight e_i. Consequently, the highest weight of V is e_1, so V contains the fundamental representation V_1 as an irreducible subrepresentation. Next, consider the exterior powers $\Lambda^j V$ of the standard representation for $j = 2, \dots, n-1$. The elements $v_{i_1} \wedge \dots \wedge v_{i_j}$ for

$1 \leq i_1 < \cdots < i_j \leq n$ form a basis for $\Lambda^j V$, which implies that the weights of $\Lambda^j V$ are given by all expressions of the form $e_{i_1} + \cdots + e_{i_j}$ for $1 \leq i_1 < \cdots < i_j \leq n$. In particular, $\omega_j = e_1 + \cdots + e_j$ is a weight (and in fact the highest weight) of $\Lambda^j V$, so this contains the fundamental representation V_j as an irreducible subrepresentation.

By Proposition 2.2.10, the set of weights of a finite–dimensional representation must be invariant under the action of the Weyl group. From 2.2.7 we know that the Weyl group of $\mathfrak{sl}(n, \mathbb{C})$ is the permutation group $\mathfrak{S}_n$ which permutes the e_i. But this shows that for each of the representations $\Lambda^j V$ all weights are obtained from the highest weight by the action of the Weyl group. In particular, the fundamental representation cannot be strictly smaller, so $\Lambda^j V = V_j$ for $j = 1, \dots, n-1$, and we have found all fundamental representations.

Note that $\Lambda^n V$ is the trivial representation, since the action of $\mathfrak{gl}(n, \mathbb{C})$ on $\Lambda^n \mathbb{C}$ is given by multiplication by the trace. This implies that the wedge product induces a duality between the $\mathfrak{sl}(n, \mathbb{C})$ representations $\Lambda^j V$ and $\Lambda^{n-j} V$, which shows that $\Lambda^{n-j} V \cong \Lambda^j V^*$.

From the description of the fundamental representations we see that any finite–dimensional irreducible representation shows up as a subrepresentation of a sufficiently high tensor power of the standard representation $V = \mathbb{C}^n$. There is a general method how to obtain these subrepresentations, which is based on the description of dominant algebraically integral weights in terms of partitions of integers. These in turn may be described by Young diagrams, which define Young symmetrizers in an appropriate permutation group. This symmetrizer then acts on a tensor power of V by permutations of the factors, and the image is exactly the irreducible representation corresponding to the partition. This construction goes under the name of *Schur functors*, an account can be found in [**FH91**, §15.3].

Representations of $\mathfrak{so}(2n, \mathbb{C})$. Following 2.2.6(2), the simple roots $\alpha_1, \dots, \alpha_n$ are given by $\alpha_j = e_j - e_{j+1}$ for $j < n$ and $\alpha_n = e_{n-1} + e_n$. Using that the Killing form is a multiple of the standard form, one easily verifies that $\omega_j = e_1 + \cdots + e_j$ for $j < n-1$, $\omega_{n-1} = \frac{1}{2}(e_1 + \cdots + e_{n-1} - e_n)$ and $\omega_n = \frac{1}{2}(e_1 + \cdots + e_n)$. For the standard representation $V = \mathbb{C}^{2n}$, one immediately sees that the weights are $\pm e_i$ for $i = 1, \dots, n$. Since by 2.2.7 the Weyl group acts by permuting the e_i and changing the sign of an even number of e's, we immediately see that the orbit of the highest weight e_1 under the Weyl group is exactly the set of all $\pm e_j$, so we see that $\mathbb{C}^{2n}$ is the first fundamental representation V_1. As in the case of $\mathfrak{sl}(n, \mathbb{C})$ above, we next conclude that for $j = 2, \dots, n-2$ the exterior power $\Lambda^j V$ has to contain the fundamental representation V_j. It turns out that $\Lambda^j V$ is irreducible for $j = 1, \dots, n-1$ and splits into two irreducible components for $j = n$. This can be either verified directly (see [**FH91**, Theorem 19.2]) or deduced from the Weyl–dimension formula; see 2.2.18 below. In particular, for $j = 1, \dots, n-2$ the representation $\Lambda^j V$ is the jth fundamental representation.

The remaining two fundamental representations are the two spin representations. They cannot show up in any tensor power of the standard representation, since a tensor power contains only weights which are integral linear combinations of the e_j, while half integers cannot occur. A construction of the spin representations is described in detail in [**FH91**, Lecture 20]. For irreducible representations whose highest weight is an integral linear combination of the e_j, there is again a version of Schur functors; see [**FH91**, §19.5].

Representations of $\mathfrak{so}(2n+1,\mathbb{C})$. From 2.2.6(3) we know that the simple roots $\alpha_1,\dots,\alpha_n$ are given by $\alpha_j = e_j - e_{j+1}$ for $j < n$ and $\alpha_j = e_n$, and the Killing form is a multiple of the standard form. Using this, one easily verifies that $\omega_j = e_1+\dots+e_j$ for $j = 1,\dots,n-1$, while $\omega_n = \frac{1}{2}(e_1+\dots+e_n)$. So here only the last fundamental representation is of different nature than the others. Similarly, as for the even orthogonal algebras, one shows that for the standard representation $V = \mathbb{C}^{2n+1}$ the exterior power $\Lambda^j V$ is the jth fundamental representation for $j = 1,\dots,n-1$. In fact, even $\Lambda^n V$ is irreducible, but not a fundamental representation. It should be noted, however, that for the odd orthogonal algebras the weights of the standard representation are not in one orbit of the Weyl group any more, since apart from $\pm e_j$ also 0 is a weight. The last fundamental representation is the spin representation; see [**FH91**, Lecture 20]. A version of Schur functors is available for representations of $\mathfrak{so}(2n+1,\mathbb{C})$ whose highest weight is an integral linear combination of $e_1,\dots,e_n$.

Representations of $\mathfrak{sp}(2n,\mathbb{C})$. Following 2.2.6(4), the simple roots $\alpha_1,\dots,\alpha_n$ are given by $\alpha_j = e_j - e_{j+1}$ for $j < n$ and $\alpha_n = 2e_n$. Since the Killing form is again a multiple of the standard form, one concludes that $\omega_j = e_1+\dots+e_j$ for all $j = 1,\dots,n$, so the situation here is simpler than for the orthogonal algebras. The weights of the standard representation $V = \mathbb{C}^{2n}$ are given by $\pm e_j$ for $j = 1,\dots,n$, so they lie on one orbit of the Weyl group, which acts by permutations and sign changes of the e_j; see 2.2.7. Thus, V is irreducible and hence coincides with the fundamental representation V_1. Obviously, the fundamental weight ω_j shows up as the weight of a highest weight vector in $\Lambda^j V$, but in contrast to the earlier cases, the exterior powers are not irreducible any more. The point here is that the symplectic form on V is an invariant element of $\Lambda^2 V^*$. Contracting with this form defines a $\mathfrak{sp}(2n,\mathbb{C})$–homomorphism $\Lambda^j V \to \Lambda^{j-2} V$. Since this map is clearly nonzero, its kernel is a nontrivial subrepresentation V_j, which turns out to be the jth fundamental representation. This can be proved directly (see [**FH91**, Theorem 17.5]) or using the Weyl dimension formula. Again, the formalism of Schur functors can be adapted to the symplectic case.

2.2.14. The isotypical decomposition. We next discuss some tools which can be used to decompose representations into components. From Theorem 2.1.6 we know that any finite dimensional representation of a complex semisimple Lie algebra $\mathfrak{g}$ splits into a direct sum of irreducible representations, but the proof of this result does not tell us how to construct such a splitting. For most purposes, it is better to stick to a slightly coarser but more natural splitting. The point here is that for a direct sum of copies of a single representation, there is no canonical choice of the summands. To avoid this problem, it is better to only split into components corresponding to different irreducible representations, which leads to the isotypical decomposition.

From part (2) of Theorem 2.2.11 we know that any finite–dimensional representation V of $\mathfrak{g}$ contains at least one highest weight vector. For a given weight λ, any linear combination of highest weight vectors of weight λ again is a highest weight vector of that weight, so these form a subspace $V^0_\lambda \subset V$. The dimension of this subspace is called the *multiplicity* $m_\lambda(V)$ of the irreducible representation Γ_λ with highest weight λ in V. By Theorem 2.2.11, any element $v \in V^0_\lambda$ generates an irreducible subrepresentation of V which is isomorphic to Γ_λ, so we see that fixing a highest weight vector in Γ_λ leads to an isomorphism $\operatorname{Hom}_{\mathfrak{g}}(\Gamma_\lambda, V) \cong V_\lambda^0$. Thus, the multiplicity of Γ_λ in V can also be interpreted as the dimension of the space of all homomorphisms $\Gamma_\lambda \to V$.

Next, we define the *isotypical component* $V^\lambda \subset V$ of highest weight λ to be the $\mathfrak{g}$–subrepresentation of V generated by V_λ^0. Choosing a basis of V_λ^0 gives rise to an isomorphism $V^\lambda \cong (\Gamma_\lambda)^{m_\lambda(V)}$. Alternatively, one may also view the isotypical component V^λ as the image of the evaluation homomorphism $\Gamma_\lambda \otimes \operatorname{Hom}_{\mathfrak{g}}(\Gamma_\lambda, V) \to V$ defined by $x \otimes \phi \mapsto \phi(x)$.

THEOREM 2.2.14. *Let V and W be finite–dimensional representations of a complex semisimple Lie algebra $\mathfrak{g}$. Then we have:*

(1) $V = \bigoplus V^\lambda \cong \bigoplus (\Gamma_\lambda)^{m_\lambda(V)}$, *where the sum goes over all weights λ such that $m_\lambda(V) > 0$.*
(2) *Restriction to the subspaces V_λ^0 induces an isomorphism*

$$\operatorname{Hom}_{\mathfrak{g}}(V, W) \cong \bigoplus_{\lambda : m_\lambda(V) > 0} L(V_\lambda^0, W_\lambda^0).$$

PROOF. (1) The sum of all isotypical components clearly is an invariant subspace of V. By complete reducibility (Theorem 2.1.6) this admits an invariant complement, which by construction cannot contain any nonzero highest weight vector. Thus, it must be zero by part (2) of Theorem 2.2.11, so the isotypical components span V.

On the other hand, the highest weight vectors in V^λ are exactly the elements of V_λ^0. Starting with the maximal weight λ for which $m_\lambda(V) > 0$, we see from Theorem 2.2.11 that all weights in the sum of the other isotypical components are strictly smaller than λ. Hence, the intersection of V^λ with this sum cannot contain any highest weight vectors and must be zero. This shows that V^λ splits off as a direct summand and inductively we conclude that V is the direct sum of the isotypical components.

(2) If $\phi : V \to W$ is a $\mathfrak{g}$–homomorphism and $v \in V_\lambda^0$, then $\phi(v)$ must be a highest weight vector of weight λ, and thus $\phi(V_\lambda^0) \subset W_\lambda^0$. Moreover, the restriction of ϕ to V^λ is determined by the restriction to V_λ^0, so by part (1) we obtain a well–defined injective linear map $\operatorname{Hom}_{\mathfrak{g}}(V, W) \to \oplus_\lambda L(V_\lambda^0, W_\lambda^0)$.

Conversely, assume that for each λ such that $m_\lambda(V) > 0$, we have given a linear map $\phi_\lambda : V_\lambda^0 \to W_\lambda^0$. Then for each such λ and each $v \in V_\lambda^0$, the element $(v, \phi_\lambda(v))$ is a highest weight vector of weight λ in $V \oplus W$. Let $\tilde{V} \subset V \oplus W$ be the $\mathfrak{g}$–submodule generated by all of these elements. Then one immediately verifies that the restriction of the first projection to $\tilde{V}$ is an isomorphism $\tilde{V} \to V$. The composition of the restriction of the second projection with the inverse of this isomorphism clearly defines a $\mathfrak{g}$–homomorphism $\phi : V \to W$, which restricts to ϕ_λ on each V_λ^0. □

Note that the first part of this theorem implies that V is determined up to isomorphism by the multiplicities $m_\lambda(V)$. On the other hand, the second part evidently is a generalization of Schur's lemma from 2.1.3.

2.2.15. The Casimir element. Recall from 2.1.10 that representations of a Lie algebra $\mathfrak{g}$ are the same thing as representations of the universal enveloping algebra $\mathcal{U}(\mathfrak{g})$. Now if u is any element in the center of $\mathcal{U}(\mathfrak{g})$, then the action of u on any representation commutes with the action of any element of $\mathfrak{g}$. In particular, by

Schur's lemma, u has to act by a scalar multiple of the identity on any irreducible representation of $\mathfrak{g}$. It even turns out (see [**Kn96**, Proposition 5.19]) that there is an analog of Schur's lemma for arbitrary irreducible $\mathcal{U}(\mathfrak{g})$–modules, so also on these modules elements of the center of $\mathcal{U}(\mathfrak{g})$ have to act by a scalar.

At first sight it is not clear how large the center of $\mathcal{U}(\mathfrak{g})$ is, but we can easily construct one nontrivial element. Observe first that there is a canonical action ad of $\mathfrak{g}$ on $\mathcal{U}(\mathfrak{g})$, defined by $\mathrm{ad}(X)(u) := Xu - uX$. Since $XY - YX = [X,Y]$ holds in $\mathcal{U}(\mathfrak{g})$, this indeed defines an action. Since $\mathcal{U}(\mathfrak{g})$ is generated by $\mathfrak{g}$, it follows that u lies in the center $\mathcal{Z}$ if and only if $\mathrm{ad}(X)(u) = 0$ for all $X \in \mathfrak{g}$. Recall that $\mathcal{U}(\mathfrak{g})$ can be realized as the quotient of the tensor algebra $T(\mathfrak{g})$ by the ideal generated by all elements of the form $X \otimes Y - Y \otimes X - [X,Y]$ for $X, Y \in \mathfrak{g}$. Now consider an element $Y_1 \otimes \cdots \otimes Y_n \in \otimes^n \mathfrak{g} \subset T(\mathfrak{g})$. The natural action of $\mathfrak{g}$ on this space is given by

$$X \cdot (Y_1 \otimes \cdots \otimes Y_n) = \sum_i Y_1 \otimes \cdots \otimes [X, Y_i] \otimes \cdots \otimes Y_n.$$

Under the projection to $\mathcal{U}(\mathfrak{g})$, this element goes to $\sum_i Y_1 \cdots (XY_i - Y_iX) \cdots Y_n$, and this telescopic sum reduces to $XY_1 \cdots Y_n - Y_1 \cdots Y_n X$. Hence, the action ad on $\mathcal{U}(\mathfrak{g})$ really comes from natural action on $T(\mathfrak{g})$.

In particular, we may look at $\mathfrak{g} \otimes \mathfrak{g}$, which via the Killing form is isomorphic to $\mathfrak{g}^* \otimes \mathfrak{g} = L(\mathfrak{g}, \mathfrak{g})$ as a $\mathfrak{g}$–representation . The identity map is a $\mathfrak{g}$–invariant element in $L(\mathfrak{g}, \mathfrak{g})$, which gives an invariant element of $\mathfrak{g} \otimes \mathfrak{g}$. Projecting to $\mathcal{U}(\mathfrak{g})$, we obtain a $\mathfrak{g}$–invariant element $\Omega \in \mathcal{U}(\mathfrak{g})$, called the *Casimir element* of $\mathfrak{g}$.

To describe the properties of the Casimir element, we need one more ingredient: The *lowest form* δ of a complex semisimple Lie algebra $\mathfrak{g}$ is defined to be half the sum of all positive roots. This lowest form has another important description: In 2.2.7 we have seen that for a simple root α_i the reflection s_{α_i} maps α_i to $-\alpha_i$ and permutes the other positive roots. Using this, we compute

$$s_{\alpha_i}(\delta) = \tfrac{1}{2} \sum_{\alpha \in \Delta^+} s_{\alpha_i}(\alpha) = -\alpha_i + \tfrac{1}{2} \sum_{\alpha \in \Delta^+} \alpha = \delta - \alpha_i.$$

By definition of s_{α_i}, this implies that $\frac{2\langle \delta, \alpha_i \rangle}{\langle \alpha_i, \alpha_i \rangle} = 1$ for each simple root α_i, so we see that δ is the sum of all fundamental weights. In particular, δ is the smallest algebraically integral element lying in the interior of the dominant Weyl chamber and if λ is any dominant weight, then $\lambda + \delta$ lies in the interior of the dominant Weyl chamber.

PROPOSITION 2.2.15. *Let $\mathfrak{g}$ be a complex semisimple Lie algebra with Casimir element $\Omega \in \mathcal{U}(\mathfrak{g})$, and let V be a finite–dimensional representation of $\mathfrak{g}$. Then Ω acts on the isotypical component V^λ of highest weight λ by multiplication with $|\lambda|^2 + 2\langle \lambda, \delta \rangle$, where δ is the lowest form and the inner product is induced by the Killing form B.*

PROOF. Let $\{X_i\}$ be any basis of $\mathfrak{g}$ and let $\{\tilde{X}_i\}$ be the dual basis with respect to the Killing form, i.e. $B(\tilde{X}_i, X_j) = \delta_{ij}$. Then we may write the element of $\mathfrak{g} \otimes \mathfrak{g}$ corresponding to the identity map as $\sum_j \tilde{X}_j \otimes X_j$, so $\Omega = \sum_j \tilde{X}_j X_j$. Specializing the basis, let $\{H_1, \dots, H_n\}$ be an orthonormal basis for $\mathfrak{h}$ (recall that the restriction of B to $\mathfrak{h}_0$ is positive definite). Further, for any root $\alpha \in \Delta$ choose vectors $E_\alpha \in \mathfrak{g}_\alpha$ and $E_{-\alpha} \in \mathfrak{g}_{-\alpha}$ such that $B(E_\alpha, E_{-\alpha}) = 1$. Since $\mathfrak{h}$ is perpendicular to all root spaces, we get $\tilde{H}_j = H_j$ for all j, and since $\mathfrak{g}_\alpha$ and $\mathfrak{g}_\beta$ are perpendicular unless

$\beta = -\alpha$, we see that $\tilde{E}_\alpha = E_{-\alpha}$. Thus, we obtain

$$\Omega = \sum_j H_j^2 + \sum_{\alpha\in\Delta} E_\alpha E_{-\alpha} = \sum_j H_j^2 + \sum_{\alpha\in\Delta^+} (E_\alpha E_{-\alpha} + E_{-\alpha}E_\alpha)$$
$$= \sum_j H_j^2 + \sum_{\alpha\in\Delta^+} ([E_\alpha, E_{-\alpha}] + 2E_{-\alpha}E_\alpha).$$

From 2.2.4 we know that $[E_\alpha, E_{-\alpha}] = H_\alpha$, the element characterized by $\alpha(H) = B(H_\alpha, H)$. Summing over all $\alpha \in \Delta^+$, these elements add up to $2H_\delta$ by definition of the lowest form δ. Hence, we finally arrive at the expression

$$\Omega = \sum_j H_j^2 + 2H_\delta + 2\sum_{\alpha\in\Delta^+} E_{-\alpha}E_\alpha.$$

Now suppose that $v \in V$ is a highest weight vector of weight λ. Then v is killed by all E_α with $\alpha \in \Delta^+$, so the last sum acts trivially. The first sum acts on v by multiplication by $\sum_j \lambda(H_j)^2$, and since the H_j form an orthonormal basis, this equals $|\lambda|^2 = \langle\lambda, \lambda\rangle$. The middle term simply gives $2\langle\lambda, \delta\rangle$. Thus, we conclude that Ω acts on V_λ^0 by multiplication with $|\lambda|^2 + 2\langle\lambda, \delta\rangle$. Since the action of Ω by construction is a $\mathfrak{g}$–homomorphism, it acts in the same way on the whole isotypical component V^λ. □

Suppose that for a given representation V of $\mathfrak{g}$ we know the weights λ_i for which $m_{\lambda_i}(V) > 0$. Then we can compute the eigenvalue a_i of the Casimir element on the isotypical component V^{λ_i}. If these are all different, then we may write the projection on the a_i–eigenspace of Ω as $\prod_{j\neq i} \frac{1}{(a_i - a_j)}(\Omega - a_j \mathrm{id})$, and this gives us the projection onto the isotypical component V^{λ_i}. Hence, in this case we get an explicit realization of the isotypical decomposition. Let us add another useful observation: For a dominant weight λ we have $\langle\lambda, \delta\rangle \geq 0$, which implies that the eigenvalue of Ω is positive unless $\lambda = 0$. Hence, we see that the kernel of Ω always equals the trivial isotypical component in V with highest weight 0.

2.2.16. Tensor products of irreducible representations. A problem that one often meets in applications of representation theory is to decompose the tensor product of two representations. As before, let us denote by Γ_λ the irreducible representation of highest weight λ. For any finite–dimensional representation V we have the multiplicities $m_\lambda(V)$ of Γ_λ in V as introduced in 2.2.14 and $V \cong \bigoplus_{\lambda: m_\lambda(V)>0} \Gamma_\lambda^{m_\lambda(V)}$. For another representation W we therefore get

$$m_\nu(V \otimes W) = \sum_{\lambda,\mu} m_\lambda(V) m_\mu(W) m_\nu(\Gamma_\lambda \otimes \Gamma_\mu).$$

Hence, it suffices to deal with the case of the tensor product of two irreducible representations. Recall from 2.2.10 that the multiplicity $n_\lambda(V)$ of the weight λ in the representation V is the dimension of the weight space V_λ.

PROPOSITION 2.2.16. *Let $\mathfrak{g}$ be a complex semisimple Lie algebra, and for each dominant algebraically integral weight λ let Γ_λ be the irreducible representation with highest weight λ. Then we have:*

If $m_\nu(\Gamma_\lambda \otimes \Gamma_\mu) > 0$, then $\nu = \lambda + \mu'$ for some weight μ' of Γ_μ. If this is the case, then, in addition, $m_\nu(\Gamma_\lambda \otimes \Gamma_\mu) \leq n_{\mu'}(\Gamma_\mu)$.

PROOF. Any weight vector in $\Gamma_\lambda \otimes \Gamma_\mu$ of weight ν can be written in the form $v_1 \otimes w_1 + \cdots + v_k \otimes w_k$, where the w_i are linearly independent weight vectors in Γ_μ and the v_i are weight vectors in Γ_λ. Denoting by λ_i the weight of v_i we order the sum in such a way that $\lambda_1 \geq \cdots \geq \lambda_k$. Now assume that $\sum v_i \otimes w_i$ is a highest weight vector and let E_α be an element in a positive root space $\mathfrak{g}_\alpha$. Then by definition E_α annihilates $v_1 \otimes w_1 + \cdots + v_k \otimes w_k$, whence we may write $(E_\alpha \cdot v_1) \otimes w_1$ as a linear combination of elements of the form $v_j \otimes (E_\alpha \cdot w_j)$ for $j \geq 1$ and $(E_\alpha \cdot v_j) \otimes w_j$ for $j \geq 2$. Applying the tensor product of the identity on Γ_λ with a linear functional, which is one on w_1 and vanishes on all the other w_j, we conclude that we may write $E_\alpha \cdot v_1$ as a linear combination of the v_j for $j \geq 1$. But if $E_\alpha \cdot v_1$ is nonzero, then it would be a weight vector of weight $\lambda_1 + \alpha$, which is strictly larger than all λ_j, so we get a contradiction. This implies that v_1 is a highest weight vector, and thus ν is the sum of λ and the weight of w_1.

Hence, we see that any highest weight vector of weight ν contains a nonzero component of the form $v_0 \otimes w$ for a fixed highest weight vector $v_0 \in \Gamma_\lambda$ and $w \in (\Gamma_\mu)_{\mu'}$. If we could find more than $n_{\mu'}(\Gamma_\mu)$ linearly independent highest weight vectors, then we could evidently obtain a nontrivial linear combination in which the terms of the form $v_0 \otimes w$ add up to zero. But this linear combination would be a highest weight vector of weight ν, which is a contradiction. □

Let us discuss the application of this result in a simple special case. Take $\mathfrak{g} = \mathfrak{sl}(n, \mathbb{C})$, let λ be an arbitrary dominant integral weight and put $\mu = \omega_1 = e_1$. Then Γ_μ is the standard representation $\mathbb{C}^n$ of $\mathfrak{g}$. The weights of $\mathbb{C}^n$ are $e_1, \dots, e_n$ and each of these has multiplicity one. We conclude that if $m_\nu(\Gamma_\lambda \otimes \mathbb{C}^n) > 0$, then $\nu = \lambda + e_i$ for some $i = 1, \dots, n$ and $m_\nu(\Gamma_\lambda \otimes \mathbb{C}^n) \leq 1$. Thus, $\Gamma_\lambda \otimes \mathbb{C}^n$ is simply reducible, i.e. it splits into a direct sum of pairwise non–isomorphic irreducibles. Moreover, we can see immediately that this splitting into irreducibles can be realized using the Casimir element Ω from 2.2.15. There we have seen that on the isotypical component with highest weight $\lambda + e_j$, the Casimir element acts by the scalar

$$|\lambda + e_j|^2 + 2\langle \lambda + e_j, \delta \rangle = |\lambda|^2 + 2\langle \lambda, \delta \rangle + |e_j|^2 + 2\langle \lambda, e_j \rangle + 2\langle e_j, \delta \rangle.$$

Since all e_j have the same length, these scalars being the same for $i < j$ would imply $\langle e_i - e_j, \lambda + \delta \rangle = 0$. This is impossible, since $e_i - e_j$ is a positive root and $\lambda + \delta$ lies in the interior of the dominant Weyl chamber. Hence, we see that, whatever irreducible components really show up, the Casimir element has different eigenvalues on different components.

2.2.17. Infinitesimal character. Let us return to the general question of decomposing a representation into isotypical components. We have indicated in 2.2.15 that elements of the center $\mathcal{Z}$ of the universal enveloping algebra $\mathcal{U}(\mathfrak{g})$ act by scalars on irreducible representations of $\mathcal{U}(\mathfrak{g})$. Rather than considering a single element of $\mathcal{Z}$, we will next look at the action of the whole center. Denoting by $\chi(a)$ the number by which an element $a \in \mathcal{Z}$ acts, one obtains a map $\chi : \mathcal{Z} \to \mathbb{C}$, which clearly is a homomorphism of unital algebras. This homomorphism is called the *infinitesimal character* or the *central character* of the representation V. As we shall see immediately, the center is fairly large, so the infinitesimal character contains a good amount of information, even for infinite–dimensional representations.

If V is a general representation of $\mathcal{U}(\mathfrak{g})$ (or equivalently of $\mathfrak{g}$), then we say that *V has an infinitesimal character* if all elements of $\mathcal{Z}$ act by scalars on V. Then one gets a homomorphism $\chi : \mathcal{Z} \to \mathbb{C}$ as above, which is the infinitesimal character of

V. One should be aware of the fact that in spite of the similar name, the properties of the infinitesimal character are quite different from the properties of the character to be discussed below.

Now there is a result of Harish–Chandra, which at the same time gives a complete description of the center $\mathcal{Z}$ of $\mathcal{U}(\mathfrak{g})$ and a description of all unital homomorphisms $\mathcal{Z} \to \mathbb{C}$, so one knows all possible infinitesimal characters in advance. We only outline this and refer to [**Kn96**, V.5] for details. To formulate the result, note first that the inclusion of a Cartan subalgebra $\mathfrak{h} \leq \mathfrak{g}$ gives an inclusion of the universal enveloping algebra $\mathcal{U}(\mathfrak{h})$ into $\mathcal{U}(\mathfrak{g})$. Since $\mathfrak{h}$ is abelian, $\mathcal{U}(\mathfrak{h}) = S(\mathfrak{h})$, the symmetric algebra. Noting that $S(\mathfrak{h})$ can be viewed as the algebra of polynomials on $\mathfrak{h}^*$, so the action of the Weyl group W of $\mathfrak{g}$ on $\mathfrak{h}^*$ induces an action on $S(\mathfrak{h})$ by $(w \cdot p)(\phi) = p(w^{-1} \cdot \phi)$.

Consider the subalgebras $\mathfrak{n}_\pm$ formed by all positive, respectively, negative root spaces. The Poincaré–Birkhoff–Witt theorem immediately implies that as a vector space $\mathcal{U}(\mathfrak{g})$ is isomorphic to the direct sum of $S(\mathfrak{h})$ and $\mathcal{U}(\mathfrak{g})\mathfrak{n}_+ + \mathfrak{n}_-\mathcal{U}(\mathfrak{g})$. For $u \in \mathcal{U}(\mathfrak{g})$ we denote by u_0 the component in $S(\mathfrak{h})$ according to this decomposition. Next, consider the linear map $\mathfrak{h} \to S(\mathfrak{h})$ defined by $H \mapsto H - \delta(H)1$, where δ is the lowest form of $\mathfrak{g}$. This induces an algebra homomorphism $\tau : S(\mathfrak{h}) \to S(\mathfrak{h})$, and we define the *Harish–Chandra map* $\gamma : \mathcal{U}(\mathfrak{g}) \to S(\mathfrak{h})$ by $\gamma(u) = \tau(u_0)$. Now Harish–Chandra's theorem reads as follows.

THEOREM 2.2.17 (Harish–Chandra). *(1) The Harish–Chandra map $\gamma : \mathcal{U}(\mathfrak{g}) \to S(\mathfrak{h})$ restricts to an isomorphism of algebras from the center $\mathcal{Z}$ of $\mathcal{U}(\mathfrak{g})$ to the subalgebra $S(\mathfrak{h})^W$ of polynomials which are invariant under the action of the Weyl group.*

(2) For a linear functional $\lambda : \mathfrak{h} \to \mathbb{C}$, the composition of the induced algebra homomorphism $S(\mathfrak{h}) \to \mathbb{C}$ with the Harisch–Chandra map restricts to a homomorphism $\chi_\lambda : \mathcal{Z} \to \mathbb{C}$ of unital algebras, and any such homomorphism is of this form. Finally, for $\lambda, \mu \in \mathfrak{h}^*$ we have $\chi_\lambda = \chi_\mu$ if and only if there is an element w in the Weyl group such that $w(\lambda) = \mu$.

Our main use of the infinitesimal character is that it provides obstructions to the existence of homomorphisms between Verma modules. In view of Theorem 1.4.10, this leads to obstructions against the existence of invariant differential operators. From the definition of the Harish–Chandra map it follows easily that a module generated by a single highest weight vector of highest weight λ has infinitesimal character $\chi_{\lambda+\delta}$, so, in particular, this holds for the Verma module $M_{\mathfrak{b}}(\lambda)$. Clearly, a $\mathcal{U}(\mathfrak{g})$–homomorphism (or equivalently a $\mathfrak{g}$–homomorphism) $M_{\mathfrak{b}}(\lambda) \to M_{\mathfrak{b}}(\mu)$ can exist only if both modules have the same infinitesimal character. By Harish–Chandra's theorem, this is the case if and only if there exists a $w \in W$ such that $w(\lambda + \delta) = \mu + \delta$, or $\mu = w(\lambda + \delta) - \delta =: w \cdot \lambda$. This action of the Weyl group on weights is called the *affine action.*

2.2.18. Formulae for multiplicities, characters and dimensions. We now switch to the question of getting additional information on the irreducible representations of a complex semisimple Lie algebra $\mathfrak{g}$. This is also very helpful for dealing with general representations. A particularly important question is to determine the multiplicities of weights in irreducible representations. In view of Proposition 2.2.16 this has consequences for the decomposition of tensor products. It can also be helpful for deciding irreducibility, compare with 2.2.13.

A basic example of a formula for the multiplicity of a weight in an irreducible representation is the *Freudenthal multiplicity formula.* It expresses the multiplicity $n_\mu(\Gamma_\lambda)$ in terms of the multiplicities of weights which are higher than μ. By symmetry of the weights under the Weyl group, it suffices to determine the multiplicities of dominant weights, so one may use the formula to recursively compute all multiplicities. Explicitly, the Freudenthal multiplicity formula reads as

$$(2\langle\lambda-\mu,\mu+\delta\rangle+\|\lambda-\mu\|^2)n_\mu(\Gamma_\lambda)=2\sum_{\alpha\in\Delta^+}\sum_{k\geq 1}\langle\mu+k\alpha,\alpha\rangle n_{\mu+k\alpha}(\Gamma_\lambda), \tag{2.4}$$

where δ is the lowest form of the Lie algebra $\mathfrak{g}$ from 2.2.15. If $\mu\neq\lambda$ is a dominant weight of Γ_λ, then $\lambda-\mu$ is a linear combination of simple roots with nonnegative integral coefficients. This immediately implies that $\langle\lambda-\mu,\mu+\delta\rangle>0$, so we see that the numerical factor in the left–hand side of (2.4) is positive.

The Freudenthal multiplicity formula may be proved by analyzing the action of the Casimir element Ω on Γ_λ. Since this is a $\mathfrak{g}$–homomorphism, it maps the weight space $(\Gamma_\lambda)_\mu$ to itself, and one may look at the trace of the action of Ω on this weight space. On the one hand, one knows that Ω acts by a scalar, which expresses this trace as a multiple of the dimension $n_\mu(\Gamma_\lambda)$ of the weight space. On the other hand, for a positive root α one may look at the corresponding subalgebra $\mathfrak{s}_\alpha\cong\mathfrak{sl}(2,\mathbb{C})$. This subalgebra naturally acts on the sum of weight spaces for weights of the form $\mu+n\alpha$, and one may compute the trace analyzing this representation. Details of the proof can be found in [**FH91**, §25.1].

It is possible to capture the multiplicities of all weights of Γ_λ into a closed expression called the Weyl character formula, however, in a rather involved way. Fix a choice of Cartan subalgebra $\mathfrak{h}\subset\mathfrak{g}$. Consider a representation V of $\mathfrak{g}$ on which the elements of $\mathfrak{h}$ act simultaneously diagonalizable. Then V splits into the direct sum of weight spaces V_λ. If each of these weight spaces is finite–dimensional, then one says that V *has a character.* If this is the case, then one defines the *character* $\mathrm{char}(V):\mathfrak{h}^*\to\mathbb{Z}$ by $\mathrm{char}(V)(\lambda)=\dim(V_\lambda)$. It follows easily from the definition that if V has a character and $W\subset V$ is a subrepresentation, then W and V/W have a character and $\mathrm{char}(V)=\mathrm{char}(W)+\mathrm{char}(V/W)$. More generally, in a finite exact sequence of representations having characters, the alternating sum of the characters is zero.

If V is finite–dimensional, then of course V has a character and since $\mathrm{char}(V)$ has finite support, one may view it as an element of the group ring $\mathbb{Z}[\mathfrak{h}^*]$ of the abelian group $\mathfrak{h}^*$. The multiplication on $\mathbb{Z}[\mathfrak{h}^*]$ is given by the convolution of functions, i.e. $fg(\phi)=\sum_{\psi+\psi'=\phi}f(\psi)g(\psi')$. Denoting by $e^\phi\in\mathbb{Z}[\mathfrak{h}^*]$ the function which is one on ϕ and zero on all other elements of $\mathfrak{h}^*$, one gets $e^\phi e^\psi=e^{\phi+\psi}$. By construction, the elements e^ϕ form a linear basis of $\mathbb{Z}[\mathfrak{h}^*]$, so we may view elements of $\mathbb{Z}[\mathfrak{h}^*]$ as expressions $f=\sum_{\phi\in\mathfrak{h}^*}a_\phi e^\phi$ with $a_\phi\in\mathbb{Z}$ and only finitely many a_ϕ nonzero. In particular, the element e^0 is a unit element in $\mathbb{Z}[\mathfrak{h}^*]$. From the definition of the convolution it is obvious that for finite–dimensional representations V and W one obtains $\mathrm{char}(V\otimes W)=\mathrm{char}(V)\,\mathrm{char}(W)$.

To formulate the first version of the Weyl character formula, we define for each weight $\lambda\in\mathfrak{h}_0^*$ an element $A_\lambda\in\mathbb{Z}[\mathfrak{h}^*]$ by $A_\lambda:=\sum_{w\in W}\mathrm{sgn}(w)e^{w(\lambda)}$, where W denotes the Weyl group of $\mathfrak{g}$. Denoting by δ the lowest form of $\mathfrak{g}$, the *Weyl character formula* states that

$$A_\delta\,\mathrm{char}(\Gamma_\lambda)=A_{\lambda+\delta}. \tag{2.5}$$

Weyl's original proof of this character formula involves what is nowadays called Weyl's unitary trick, i.e. that any complex semisimple Lie algebra has a compact real form. The representation theory of this compact real form is equivalent to the representation theory of $\mathfrak{g}$ (see Remark 2.3.2) and passing to the group level, the character formula is proved using the Peter–Weyl theorem and integration; see [**FH91**, §26.2]. A purely algebraic proof can be based on the Freudenthal multiplicity formula. Expressing $\text{char}(\Gamma_\lambda)$ using the Freudenthal formula, one shows that $A_\delta \,\text{char}(\Gamma_\lambda)$ is nonzero only on elements in the Weyl orbit of $\lambda+\delta$. Evidently, $A_\delta\,\text{char}(\Gamma_\lambda)$ is alternating under the action of the Weyl group, and its value on λ equals 1, which then implies the character formula. Details of this proof can be found in [**FH91**, §25.2]. We will prove the character formula as a consequence of Kostant's version of the Bott–Borel–Weil theorem in 3.3.9.

It is not obvious that (2.5) determines $\text{char}(\Gamma_\lambda)$. To see this, one has to extend the group ring $\mathbb{Z}[\mathfrak{h}^*]$. Let Q^+ be the set of all linear combinations of positive roots with nonnegative integral coefficients. Denote by $\mathbb{Z}\langle\mathfrak{h}^*\rangle \subset \mathbb{Z}^{\mathfrak{h}^*}$ the set of those functions, whose support is contained in a finite union of sets of the form $\lambda_0 - Q^+ = \{\lambda_0 - \phi : \phi \in Q^+\}$. Then it is easy to see that for $f, g \in \mathbb{Z}\langle\mathfrak{h}^*\rangle$ and $\phi \in \mathfrak{h}^*$, there are only finitely many pairs $\psi, \psi' \in \mathfrak{h}^*$ such that $\phi = \psi + \psi'$, $f(\psi) \neq 0$, and $g(\psi') \neq 0$. Hence, the convolution of two elements of $\mathbb{Z}\langle\mathfrak{h}^*\rangle$ makes sense and one shows that the result again lies in $\mathbb{Z}\langle\mathfrak{h}^*\rangle$. Thus, the convolution makes $\mathbb{Z}\langle\mathfrak{h}^*\rangle$ into an associative commutative ring with unit e^0. We continue to use the notation $f = \sum_{\phi\in\mathfrak{h}^*} a_\phi e^\phi$ for elements of this larger ring. Note that if V and W have a character which lies in $\mathbb{Z}\langle\mathfrak{h}^*\rangle$, then $\text{char}(V\otimes W) = \text{char}(V)\,\text{char}(W)$ continues to hold.

Consider a Verma module $M_\mathfrak{b}(\lambda)$ for $\lambda \in \mathfrak{h}^*$. Putting $\Delta^+ = \{\alpha_1, \dots, \alpha_\ell\}$, we know from 2.2.12 that the monomials $F_{\alpha_1}^{i_1}\dots F_{\alpha_\ell}^{i\ell} \otimes 1$ form a linear basis of $M_\mathfrak{b}(\lambda)$ and such an element is a weight vector of weight $\lambda - \sum_j i_j\alpha_j$. Hence, we see that $M_\mathfrak{b}(\lambda)$ has a character, which lies in $\mathbb{Z}\langle\mathfrak{h}^*\rangle$. To write down this character explicitly, define the *Kostant partition function* $\mathcal{P} : \mathfrak{h}^* \to \mathbb{N}$ by putting $\mathcal{P}(\phi)$ the number of all tuples $(a_\alpha)_{\alpha\in\Delta^+}$ with each $a_\alpha \in \mathbb{N}$ such that $\phi = \sum_{\alpha\in\Delta^+} a_\alpha\alpha$. Thus, $\mathcal{P}(\phi)$ is the number of different representations of ϕ as a linear combination of positive roots with nonnegative integral coefficients. By convention, one puts $\mathcal{P}(0) = 1$, and of course we have $\mathcal{P}(\phi) = 0$ unless $\phi \in Q^+$. Using this it is clear that

$$\text{char}(M_\mathfrak{b}(\lambda)) = \sum_{\mu\in\mathfrak{h}^*} \mathcal{P}(\lambda-\mu)e^\mu = \sum_{\phi\in Q^+} \mathcal{P}(\phi)e^{\lambda-\phi} = e^\lambda \sum_{\phi\in Q^+} \mathcal{P}(\phi)e^{-\phi}.$$

Now it turns out that the *Weyl denominator* A_δ can be written as $A_\delta = e^\delta \prod_{\alpha\in\Delta^+}(1-e^{-\alpha})$. Using this, one verifies that in $\mathbb{Z}\langle\mathfrak{h}^*\rangle$ one has $Ke^{-\delta}A_\delta = 1$, where $K = \sum_{\phi\in Q^+}\mathcal{P}(\phi)e^{-\phi}$. Hence, A_δ is invertible in $\mathbb{Z}\langle\mathfrak{h}^*\rangle$, and the inverse is exactly the character of the Verma module $M_\mathfrak{b}(-\delta)$. Hence, we see that from (2.5) we can explicitly express $\text{char}(\Gamma_\lambda)$ as

$$\text{char}(\Gamma_\lambda) = e^{-\delta}\Big(\sum_{\phi\in Q^+}\mathcal{P}(\phi)e^{-\phi}\Big)\Big(\sum_{w\in W}\text{sgn}(w)e^{w(\lambda+\delta)}\Big). \tag{2.6}$$

Extracting information on $\text{char}(\Gamma_\lambda)$ from this formula is not a straightforward task. The character is expressed as the product of an infinite sum with a sum having as many elements as the Weyl group, so there is a lot of cancellation in the formula. Thus, it is highly desirable to get simpler results giving at least partial information. One of these results is the *Weyl dimension formula* which computes the dimension of Γ_λ. To get this dimension, one has to apply the homomorphism $\mathbb{Z}[\mathfrak{h}^*] \to \mathbb{Z}$ which sends any e^ϕ to 1 to the character formula. Doing this naively, one obtains 0/0, but with a bit of work (see [**FH91**, Corollary 24.6] or [**Kn96**, Theorem 5.84]) on obtains

$$\dim(\Gamma_\lambda) = \prod_{\alpha\in\Delta^+} \frac{\langle \lambda+\delta, \alpha\rangle}{\langle \delta, \alpha\rangle}. \tag{2.7}$$

As an example for using this formula, let us compute the dimensions of the irreducible representation of $\mathfrak{so}(2n, \mathbb{C})$ with highest weight $\lambda = e_1 + \dots + e_k$ for $k < n$, thus verifying the claims of irreducibility of the exterior powers of the standard representation from 2.2.13. From the list of fundamental weights in 2.2.13 one immediately sees that (using the notation from 2.2.6 and 2.2.13) we have $\delta = (n-1)e_1 + (n-2)e_2 + \dots + 2e_{n-2} + e_{n-1}$. Thus,

$$\lambda + \delta = ne_1 + \dots + (n-k+1)e_k + (n-k-1)e_{k+1} + \dots + e_{n-1}.$$

By 2.2.6 the positive roots are $e_i \pm e_j$ for $1 \le i < j \le n$ and since we may clearly replace the Killing form by a multiple, we can use the standard form in terms of the e_j. Thus, we see that for the roots $e_i - e_j$ we get a nontrivial contribution to the dimension formula only if $i \le k$ and $j > k$, so these roots contribute

$$\prod_{i=1}^{k}\prod_{j=k+1}^{n} \frac{j-i+1}{j-i} = \prod_{i=1}^{k} \frac{n-i+1}{k-i+1} = \frac{n!}{(n-k)!k!}.$$

For the roots $e_i + e_j$, we only get a nontrivial contribution if $i \le k$, but we have to distinguish the cases $j \le k$ and $j > k$. From the roots with $i < j \le k$ we get the contribution

$$\begin{aligned}\prod_{i=1}^{k-1}\prod_{j=i+1}^{k} \frac{2n+2-i-j}{2n-i-j} &= \prod_{i=1}^{k-1} \frac{(2n+1-2i)(2n-2i)}{(2n-k-i)(2n+1-k-i)}\\ &= \frac{(2n-1)!(2n-2k)!}{(2n-k-1)!(2n-k)!},\end{aligned}$$

and the ones with $i \le k < j$ give us

$$\prod_{i=1}^{k}\prod_{j=k+1}^{n} \frac{2n+1-i-j}{2n-i-j} = \prod_{i=1}^{k} \frac{2n-k-i}{n-i} = \frac{(2n-k-1)!(n-k-1)!}{(2n-2k-1)!(n-1)!}.$$

Multiplying these three expressions and cancelling, we obtain $\frac{(2n)!}{k!(2n-k)!}$, the dimension of $\Lambda^k\mathbb{C}^{2n}$.

The Weyl character formula also leads to a formula for the multiplicity of a weight in an irreducible representation, called the *Kostant multiplicity formula.* This formula has the advantage that it is not recursive, but the price one has to pay is a summation over the entire Weyl group:

$$n_\mu(\Gamma_\lambda) = \sum_{w\in W} \operatorname{sgn}(w)\mathcal{P}(w(\lambda+\delta) - (\mu+\delta)).$$

A proof for this multiplicity formula which shows that it is equivalent to the character formula can be found in [**FH91**, Proposition 25.1]. To successfully apply this formula, the first step is usually to verify that only a few Weyl group elements lead to a nontrivial contribution in the sum.

The Weyl character formula can also be used to derive formulae for the multiplicities of irreducible components in representations obtained by various constructions. As an example we mention a particularly useful formula for the multiplicity of irreducible components in a tensor product which is usually called *Klimyk's formula* or *Racah's formula*:

$$m_\mu(\Gamma_\lambda \otimes V) = \sum_{w \in W} \operatorname{sgn}(w) n_{\mu+\delta-w(\lambda+\delta)}(V).$$

Of course, it suffices to verify this in the case $V = \Gamma_\nu$, which is done in [**FH91**, Exercise 25.31]. Further formulae in similar directions can be found in [**FH91**, §25.3].

2.2.19. Complex semisimple Lie groups and their representations. Having developed the representation theory for Lie algebras, it is rather easy to discuss the group case. Let us fix a complex semisimple Lie algebra $\mathfrak{g}$. We start by describing all connected Lie groups with Lie algebra $\mathfrak{g}$. By Lie's third theorem (see 1.2.5) there exists a connected complex Lie group (i.e. a complex manifold such that multiplication and inversion are holomorphic) with Lie algebra $\mathfrak{g}$. Passing to the the universal covering, we obtain a simply connected Lie group $\tilde{G}$ with Lie algebra $\mathfrak{g}$. Given any other connected Lie group G with Lie algebra $\mathfrak{g}$, there is a homomorphism $\pi : \tilde{G} \to G$ such that π' is the identity on $\mathfrak{g}$. Hence, π is a local diffeomorphism and since G is connected, π is surjective. The kernel N of π is a discrete normal subgroup of $\tilde{G}$ and hence π is a covering map. For each $n \in N$, the set $\{gng^{-1} : g \in \tilde{G}\}$ is a connected subspace of N, and hence consists of n only. Thus, N is contained in the center of $\tilde{G}$.

On the other hand, recall from 2.2.2 that there is another natural choice of a Lie group with Lie algebra $\mathfrak{g}$, namely the group $\operatorname{Int}(\mathfrak{g})$ of inner automorphisms of $\mathfrak{g}$. By definition this is the connected component of the identity in the automorphism group $\operatorname{Aut}(\mathfrak{g})$. For any connected Lie group G with Lie algebra $\mathfrak{g}$, the adjoint action defines a homomorphism onto $\operatorname{Int}(\mathfrak{g})$. The kernel of Ad is the center $Z(G)$. Thus, we see that any connected Lie group G with Lie algebra $\mathfrak{g}$ lies between the simply connected group $\tilde{G}$ and the adjoint group $\operatorname{Int}(\mathfrak{g}) = \tilde{G}/Z(\tilde{G})$.

Now let us fix a connected Lie group G with Lie algebra $\mathfrak{g}$. First, observe that for a complex representation of $\mathfrak{g}$ which integrates to G the induced group homomorphism $G \to GL(V)$ has complex linear derivative and hence is automatically holomorphic. Conversely, the derivative of a holomorphic group representation defines a complex representation of the Lie algebra. Now consider the universal covering $\pi : \tilde{G} \to G$. Any finite–dimensional representation $\rho : \mathfrak{g} \to \mathfrak{gl}(V)$ then integrates to a representation $\tilde{\rho} : \tilde{G} \to GL(V)$; see 1.2.4. There is a representation $G \to GL(V)$ with differential ρ if and only if $\ker(\pi) \subset \ker(\tilde{\rho})$. Thus, it remains to understand the kernel of π, which turns out to be rather easy. We have seen above that this kernel is a subgroup of the center $Z(\tilde{G})$, and it turns out that this center is contained in $\exp(\mathfrak{h})$, where $\mathfrak{h} \leq \mathfrak{g}$ is a Cartan subalgebra. Hence, we have to determine the $\mathcal{N}_G = \{H \in \mathfrak{h} : \exp(H) = e \in G\}$.

For an irreducible representation with highest weight $\lambda \in \mathfrak{h}^*$ and $H \in \mathfrak{h}$ the action on a highest weight vector v is given by $\exp(H) \cdot v = e^{\operatorname{ad}(H)}(v) = e^{\lambda(H)}v$. If the representation ρ integrates to G, we must have $\lambda(H) \in 2i\pi\mathbb{Z}$ for all $H \in \mathcal{N}_G$.

Weights which have this property are called *analytically integral* for the group G. Since the adjoint representation does integrate to G, we see that all roots $\alpha \in \Delta$ are analytically integral for G. Since all other weights of the representation have the form $\lambda - \sum a_i\alpha_i$ for simple roots α_i and $a_i \in \mathbb{Z}$ with $a_i \geq 0$, we easily conclude inductively that any representation with analytically integral highest weight integrates to G.

One may determine explicitly $\mathcal{N}_{\tilde{G}}$ and $\mathcal{N}_{\mathrm{Int}(\mathfrak{g})}$ for any semisimple Lie algebra $\mathfrak{g}$. It turns out, that for the simply connected group $\tilde{G}$ one gets $\mathcal{N}_{\tilde{G}} = \{H \in \mathfrak{h} : \lambda(H) = 2i\pi\mathbb{Z} \quad \forall\lambda \in \Lambda_W\}$. For the adjoint group $\mathrm{Int}(\mathfrak{g})$ we get the analogous description with the weight lattice Λ_W replaced by the root lattice Λ_R. In particular, the analytically integral weights for $\tilde{G}$ are exactly the algebraically integral weights. The center of $\tilde{G}$, which by construction is isomorphic to $\mathcal{N}_{\tilde{G}}/\mathcal{N}_{\mathrm{Int}(\mathfrak{g})}$, is therefore isomorphic to Λ_W/Λ_R. We have noted in 2.2.10 that Λ_W/Λ_R is a finite group whose order equals the determinant of the Cartan matrix. Hence, we see the order of the center of the universal covering group from the Cartan matrix. In particular, this order turns out to be one in the case of G_2, F_4, and E_8, two for E_7, and three for E_6. Hence, for the first three types the adjoint group is simply connected and there is (up to isomorphism) only one connected Lie group with the given Lie algebra. For E_6 and E_7, any connected Lie group is either simply connected or isomorphic to the adjoint group.

For the classical series of Lie algebras, it is easier to compute the fundamental groups and centers of the classical Lie groups. To compute the fundamental groups, one proceeds by induction using the long exact homotopy sequence of a fibration. More precisely, one only needs the fact that in a fibration with connected base whose first two homotopy groups vanish, the fundamental group of the total space is isomorphic to the fundamental group of the fiber. For $SL(n,\mathbb{C})$ one uses the transitive action on $\mathbb{C}^n \setminus \{0\}$ coming from the standard representation. The stabilizer of the first unit vector is isomorphic to $SL(n-1,\mathbb{C}) \times \mathbb{C}^{n-1}$ and thus homotopy equivalent to $SL(n-1,\mathbb{C})$. For $n = 2$, the stabilizer is $\mathbb{C}$, so we get a fibration $SL(2,\mathbb{C}) \to \mathbb{C}^2 \setminus \{0\}$ with contractible fiber. Since $\mathbb{C}^2 \setminus \{0\}$ is homotopy equivalent to the unit sphere S^3, we see that also $SL(2,\mathbb{C})$ has this property and ,in particular, is simply connected. From the fibration above, we thus conclude that $SL(n,\mathbb{C})$ is simply connected for all $n \geq 2$.

On the other hand, Schur's lemma implies that the center of $SL(n,\mathbb{C})$ consists exactly of those multiples of the identity which are contained in $SL(n,\mathbb{C})$, so the possible factors are the nth roots of unity. Hence, the center is isomorphic to $\mathbb{Z}_n$, and for any integer m dividing n, we get a connected Lie group G with Lie algebra $\mathfrak{sl}(n,\mathbb{C})$. This is the quotient of $SL(n,\mathbb{C})$ by the subgroup $\mathbb{Z}_m$ of the center $\mathbb{Z}_n$, and up to isomorphism, these exhaust all connected groups with Lie algebra $\mathfrak{sl}(n,\mathbb{C})$. From 2.2.7 we know that the highest weight of any finite–dimensional representation of $\mathfrak{sl}(n,\mathbb{C})$ can be uniquely written as $a_1e_1 + \cdots + a_{n-1}e_{n-1}$ with $a_1 \geq a_2 \geq \cdots \geq a_{n-1} \geq 0$ (normalizing the coefficient of e_n to be zero). In this picture, one immediately concludes from the explicit description of the Cartan subalgebra in 2.2.6(1) that the representation with this highest weight integrates to the group G corresponding to $\mathbb{Z}_m \subset \mathbb{Z}_n$ if and only if $a_1 + \cdots + a_{n-1}$ is a multiple of m.

With the other classical groups one deals similarly (see [**FH91**, §23.1]). For $n \geq 1$, the group $Sp(2n,\mathbb{C})$ is simply connected and irreducibility of the standard representation immediately implies that its center consists of plus and minus the identity only, so there are only two connected Lie groups with Lie algebra $\mathfrak{sp}(2n, \mathbb{C})$. In the orthogonal case, it turns out that $SO(3, \mathbb{C})$ admits a two sheeted covering by $SL(2, \mathbb{C})$ (which comes from the adjoint representation of $SL(2, \mathbb{C})$, see 2.2.6), and thus has fundamental group $\mathbb{Z}_2$. A similar argument as for the SL groups above shows that $\pi_1(SO(n, \mathbb{C})) = \mathbb{Z}_2$ for all $n \geq 1$. The universal covering of $SO(n, \mathbb{C})$ is the spin group $\operatorname{Spin}(n, \mathbb{C})$, which is described in detail in [**FH91**, Lecture 20].

For the center, we conclude from irreducibility of the standard representation that $SO(2n + 1, \mathbb{C})$ has trivial center for $n \geq 1$, while for $SO(2n, \mathbb{C})$ the center is $\mathbb{Z}_2 = \{\pm \mathrm{id}\}$. For the center of the universal covering one obtains $\mathbb{Z}_2 \times \mathbb{Z}_2$ for $\operatorname{Spin}(4n, \mathbb{C})$ ($n \geq 1$) and $\mathbb{Z}_4$ for $\operatorname{Spin}(4n + 2, \mathbb{C})$ ($n \geq 1$). Hence, in dimension $4n + 2$ we get three groups, since $\mathbb{Z}_4$ has only one nontrivial subgroup, which is isomorphic to $\mathbb{Z}_2$, and these are $\operatorname{Spin}(4n + 2, \mathbb{C})$, $SO(4n + 2, \mathbb{C})$, and the quotient $PSO(4n + 2, \mathbb{C})$ of SO by its center. In the other case, besides $SO(4n, \mathbb{C})$ there are two other Lie groups which both are double covered by $\operatorname{Spin}(4n, \mathbb{C})$ and both double cover $PSO(4n, \mathbb{C})$. The characterization of representations that integrate is straightforward in each case.

Finally, if we consider a group G with Lie algebra $\mathfrak{g}$ that is not connected, then we denote by G_0 the connected component of the identity, and we have the (discrete) component group G/G_0. Using the methods from above, we can decide whether a given irreducible representation of $\mathfrak{g}$ integrates to G_0. Now if A is any connected component of G, then left multiplication with any element of A defines a diffeomorphism $G_0 \to A$, so in particular, any representation of G is determined by its restriction to G_0 and its value on one element in each of the other components. A case of interest is the automorphism group $\operatorname{Aut}(\mathfrak{g})$ of a simple Lie algebra $\mathfrak{g}$. Its connected component of the identity is the adjoint group $\operatorname{Int}(\mathfrak{g})$. The quotient $\operatorname{Aut}(\mathfrak{g})/\operatorname{Int}(\mathfrak{g})$ turns out to be isomorphic to the group of automorphisms of the Dynkin diagram, see [**FH91**, Proposition D.40].

The diagrams B_n, C_n, G_2, F_4, E_7 and E_8 do not admit nontrivial automorphisms. For A_n with $n \geq 2$, D_n with $n \geq 5$ and E_6 there is one evident nontrivial automorphism. A special case is the Dynkin diagram D_4 of $\mathfrak{so}(8, \mathbb{C})$, which has automorphism group isomorphic to the group $\mathfrak{S}_3$ of permutations of three elements (permuting the three "outer" nodes). This is the basis of the concept of triality; see [**FH91**, §20.3]. It is related to the fact that apart from the standard representation the two spin representations also have dimension eight, and any permutation between these three representations can be realized by an outer automorphism of $\mathfrak{so}(8, \mathbb{C})$.

2.3. Real semisimple Lie algebras and their representations

We already know that the complexification of a real semisimple Lie algebra is semsimple, too. Hence, real semisimple Lie algebras can be always viewed as real forms of complex semisimple Lie algebras. Since we understand complex semisimple Lie algebras, we have to see how to distinguish different real forms. The first step is to show that any complex semisimple Lie algebra has a compact real form, which is unique up to isomorphism. The main aim of this section is to describe how non–compact real forms of a given complex semisimple Lie algebra can be encoded into data related to the system of simple roots. This leads to the notion of the Satake diagram of a real semisimple Lie algebra. We also briefly outline the classification of real semisimple Lie algebras, the details of which are not of much interest for our purposes. Finally, we study complex and real representations of real semisimple Lie algebras.

2.3.1. Split form and compact real form. In 2.1.4 we have introduced the complexification $\mathfrak{g}_{\mathbb{C}} = \mathfrak{g} \otimes \mathbb{C}$ of a real Lie algebra $\mathfrak{g}$, and the corresponding involutive automorphism σ of $\mathfrak{g}_{\mathbb{C}}$ such that $\mathfrak{g}_{\mathbb{C}}^{\sigma} = \{A \in \mathfrak{g}_{\mathbb{C}} : \sigma(A) = A\} = \mathfrak{g}$. Thus, $\mathfrak{g}$ is a *real form* of $\mathfrak{g}_{\mathbb{C}}$ and any real form can be described as $\mathfrak{g}_{\mathbb{C}}^{\sigma}$ for an appropriate conjugate linear involutive automorphism σ. The Killing form of $\mathfrak{g}_{\mathbb{C}}$ is the complex bilinear extension of the Killing form of $\mathfrak{g}$, so by Cartan's criterion (part (3) of Theorem 2.1.5) $\mathfrak{g}_{\mathbb{C}}$ is semisimple if and only if $\mathfrak{g}$ is semisimple.

We have also noted in 2.1.4 that the complex simple Lie algebra $\mathfrak{sl}(2, \mathbb{C})$ has $\mathfrak{su}(2)$ and $\mathfrak{sl}(2, \mathbb{R})$ as real forms, and these two real forms are certainly non–isomorphic. In fact, these two real forms represent opposite extremes, which exist for any complex simple Lie algebra, namely the compact real form and the split real form.

Let us start with a brief digression on compact Lie algebras. Suppose that G is a compact Lie group with Lie algebra $\mathfrak{g}$. Then there is an inner product on $\mathfrak{g}$ such that for each $g \in G$ the map $\operatorname{Ad}(g) : \mathfrak{g} \to \mathfrak{g}$ is orthogonal. (Such an inner product can be obtained for example by averaging an arbitrary inner product over G.) Likewise, for any element $A \in \mathfrak{g}$, the map $\operatorname{ad}(A) : \mathfrak{g} \to \mathfrak{g}$ is skew symmetric. As a map on the complexification, $\operatorname{ad}(A)$ is therefore skew Hermitian, and hence diagonalizable with purely imaginary eigenvalues. Consequently, $\operatorname{ad}(A)^2$ is diagonalizable over the reals with nonpositive real eigenvalues, which shows that the Killing form B of $\mathfrak{g}$ is negative semidefinite. Since the maps $\operatorname{ad}(A)$ are all skew symmetric, the orthogonal complement of any ideal in $\mathfrak{g}$ is an ideal, too. Corollary 2.1.6 then shows that $\mathfrak{g}$ is reductive, $[\mathfrak{g}, \mathfrak{g}]$ is semisimple, and $\mathfrak{g} = \mathfrak{z}(\mathfrak{g}) \oplus [\mathfrak{g}, \mathfrak{g}]$ as a Lie algebra. Cartan's criterion shows that for a compact Lie group with semisimple Lie algebra, the Killing form of $\mathfrak{g}$ must be negative definite.

Conversely, if $\mathfrak{g}$ is a Lie algebra with negative definite Killing form, then by Cartan's criterion $\mathfrak{g}$ is semisimple. Then the adjoint group $\operatorname{Int}(\mathfrak{g})$ has Lie algebra $\mathfrak{g}$ (see 2.2.19). For $X, Y \in \mathfrak{g}$, we have $\operatorname{Ad}(\exp(X))(Y) = e^{\operatorname{ad}(X)}(Y)$. Since $\operatorname{ad}(X)$ is skew symmetric with respect to the Killing form, we conclude that $\operatorname{Ad}(\exp(X)) : \mathfrak{g} \to \mathfrak{g}$ is an orthogonal map. These elements generate $\operatorname{Int}(\mathfrak{g})$, so $\operatorname{Int}(\mathfrak{g})$ is a subgroup of the orthogonal group of the Killing form and hence compact. By a well–known theorem of H. Weyl (see [**Kn96**, IV.8]), the universal covering group of a compact semisimple Lie group G is compact, too (so, in particular, G has finite fundamental group). This shows that any connected Lie group with Lie algebra $\mathfrak{g}$ is compact, and this property is equivalent to the fact that the Killing form is negative definite. A semisimple Lie algebra which satisfies these equivalent conditions is called *compact.*

There is an opposite extremal class of real forms, which in a certain sense is maximally noncompact. Such a real form is called a split real form and it is characterized by the following property. For a real Lie algebra $\mathfrak{g}$ with complexification $\mathfrak{g}_{\mathbb{C}}$ we define a *Cartan subalgebra* in $\mathfrak{g}$ as a subalgebra $\mathfrak{h} \leq \mathfrak{g}$ whose complexification $\mathfrak{h}_{\mathbb{C}}$ is a Cartan subalgebra in $\mathfrak{g}_{\mathbb{C}}$. Given a Cartan subalgebra $\mathfrak{h} \leq \mathfrak{g}$ one can then look at the roots of $\mathfrak{g}_{\mathbb{C}}$ with respect to $\mathfrak{h}_{\mathbb{C}}$ and restrict these linear functionals to $\mathfrak{h}$. We call $\mathfrak{g}$ a *split real form* of $\mathfrak{g}_{\mathbb{C}}$ if there exists a Cartan subalgebra $\mathfrak{h} \leq \mathfrak{g}$ for which all these restrictions are real valued.

PROPOSITION 2.3.1. *Any complex semisimple Lie algebra $\mathfrak{g}$ admits at least one split real form and at least one compact real form.*

PROOF. Choose a Cartan subalgebra $\mathfrak{h} \leq \mathfrak{g}$ and consider the corresponding root decomposition $\mathfrak{g} = \mathfrak{h} \oplus \bigoplus_{\alpha\in\Delta} \mathfrak{g}_\alpha$. Let $\mathfrak{h}_0 \subset \mathfrak{h}$ be the subspace on which all roots have real values, which by 2.2.4 is a real form of $\mathfrak{h}$. Choosing nonzero elements $E_\alpha \in \mathfrak{g}_\alpha$ for all $\alpha \in \Delta$, we know from 2.2.4 that if α, β and $\alpha+\beta$ are roots, then $[E_\alpha, E_\beta] = a_{\alpha,\beta}E_{\alpha+\beta}$ for some nonzero number $a_{\alpha,\beta} \in \mathbb{C}$.

If we further normalize in such a way that $B(E_\alpha, E_{-\alpha}) = 1$, then we know from 2.2.4 that $[E_\alpha, E_{-\alpha}] = H_\alpha \in \mathfrak{h}_0$, the unique element of $\mathfrak{h}$ such that $\alpha(H) = B(H_\alpha, H)$ for all $H \in \mathfrak{h}$. Now one shows (see [**Kn96**, VI, Theorem 6.6]) that the generators E_α can be normalized in such that way that all $a_{\alpha,\beta}$ are real. Hence, there are elements $X_\alpha \in \mathfrak{g}_\alpha$ for each root α, such that $[X_\alpha, X_{-\alpha}] = H_\alpha$, and for $\alpha, \beta \in \Delta$ with $\beta \neq -\alpha$ we have $[X_\alpha, X_\beta] = 0$ if $\alpha+\beta$ is not a root and $[X_\alpha, X_\beta] = N_{\alpha,\beta}X_{\alpha+\beta}$ for some $N_{\alpha,\beta} \in \mathbb{R}$ with $N_{\alpha,\beta} = -N_{-\alpha,-\beta}$. For any such choice one has $N^2_{\alpha,\beta} = \frac{1}{2}q(p+1)|\alpha|^2$, where the α–string through β has the form $\beta - p\alpha, \dots, \beta + q\alpha$. The main point for us here is that $\mathfrak{g}_0 := \mathfrak{h}_0 \oplus \bigoplus_{\alpha\in\Delta} \mathbb{R}X_\alpha$ obviously is a real form of $\mathfrak{g}$. Since $\mathfrak{h}_0 \leq \mathfrak{g}_0$ is evidently a Cartan subalgebra, this is a split real form.

Now we can construct a compact real form from $\mathfrak{g}_0 = \mathfrak{h}_0 \oplus \bigoplus_{\alpha\in\Delta} \mathbb{R}X_\alpha$ by defining

$$\mathfrak{u} := i\mathfrak{h}_0 \oplus \bigoplus_{\alpha\in\Delta^+} \left(\mathbb{R}(X_\alpha - X_{-\alpha}) + i\mathbb{R}(X_\alpha + X_{-\alpha})\right).$$

Using the fact that $N_{\alpha,\beta} = -N_{-\alpha,-\beta}$ from above, one verifies that this is a real Lie subalgebra of $\mathfrak{g}$. By construction $\mathfrak{u} \cap i\mathfrak{u} = \{0\}$, so $\mathfrak{u}$ is a real form of $\mathfrak{g}$. Using a real basis of $\mathfrak{u}$ as a complex basis for $\mathfrak{g}$, we see that the Killing form of $\mathfrak{u}$ is the restriction of the real part of the Killing form of $\mathfrak{g}$. Since the Killing form is positive definite on $\mathfrak{h}_0$, it is negative definite on $i\mathfrak{h}_0$, and this is orthogonal to the rest of $\mathfrak{u}$. On the other hand, by construction, $B(X_\alpha, X_{-\alpha}) = 1$, which implies that $B(X_\alpha - X_{-\alpha}, X_\alpha - X_{-\alpha}) = -2$ and $B(i(X_\alpha + X_{-\alpha}), i(X_\alpha + X_{-\alpha})) = -2$. Hence, the Killing form is negative definite on $\mathfrak{u}$, and we have constructed a compact real form of $\mathfrak{g}$. □

For the classical complex simple Lie algebras, the real forms constructed in the proof can be immediately seen from the presentations introduced in 2.2.6. In all cases discussed there, the Lie algebra $\mathfrak{g}$ consists of complex block matrices in which the blocks, in addition, may be required to be tracefree, symmetric, or skew symmetric. This immediately implies that the subalgebra $\mathfrak{g}_0$ of real matrices contained in $\mathfrak{g}$ is a real form of $\mathfrak{g}$. Moreover, in each of the examples in 2.2.6, the subspace $\mathfrak{h}_0$ of the Cartan subalgebra $\mathfrak{h}$ on which all roots are real, was the subspace of real matrices contained in $\mathfrak{h}$. This shows that the real matrices contained in $\mathfrak{g}$ always are a split real form $\mathfrak{g}_0$. For the algebras $\mathfrak{sl}(n,\mathbb{C})$ and $\mathfrak{sp}(2n,\mathbb{C})$ this just leads to the real analogs $\mathfrak{sl}(n,\mathbb{R})$, respectively, $\mathfrak{sp}(2n,\mathbb{R})$. In the case of $\mathfrak{so}(2n,\mathbb{C})$ we get all linear maps on $\mathbb{R}^{2n} = \mathbb{R}^n \oplus \mathbb{R}^{n*}$, which preserve the symmetric form $\langle(x,\phi),(y,\psi)\rangle = \phi(y) + \psi(x)$. This form has signature (n,n), so $\mathfrak{so}(n,n)$ is a split real form of $\mathfrak{so}(2n,\mathbb{C})$. Similarly, $\mathfrak{so}(n+1,n)$ is a split real form of $\mathfrak{so}(2n+1,\mathbb{C})$.

To describe the compact real forms, we again start with $\mathfrak{sl}(n,\mathbb{C})$. The subspace $i\mathfrak{h}_0$ consists of all tracefree purely imaginary diagonal matrices. The other generators are either real and skew symmetric or purely imaginary and symmetric. Hence, the compact real form is the Lie algebra $\mathfrak{su}(n)$ of tracefree skew Hermitian matrices. For the orthogonal algebras $\mathfrak{so}(n,\mathbb{C})$, it is better two switch from the presentation of 2.2.6 to the presentation using the standard bilinear form $\langle z,w\rangle = \sum z_j w_j$. In this picture, $\mathfrak{so}(n,\mathbb{C})$ is the space of skew symmetric complex $n\times n$–matrices, and evidently the subalgebra $\mathfrak{so}(n)$ of all real skew symmetric matrices is a compact real form. It is easy to verify that actually this real form is obtained by the above construction starting from the Cartan subalgebra of $\mathfrak{so}(n,\mathbb{C})$ given (in the picture with the standard form) by block diagonal matrices with 2×2–blocks of the form $\begin{pmatrix} 0 & a_j \\ -a_j & 0 \end{pmatrix}$ along the main diagonal.

In the case of $\mathfrak{sp}(2n,\mathbb{C})$, we use the picture as block matrices $\begin{pmatrix} A & B \\ C & -A^t \end{pmatrix}$ with $B^t = B$ and $C^t = C$ from 2.2.6(4). From the description of the root spaces there, one concludes that our construction for a compact real form from above produces the subalgebra of those matrices for which $A^* = -A$ and $C = -B^*$. We obtain the compact real form $\mathfrak{sp}(2n,\mathbb{C})\cap\mathfrak{u}(2n)$, and we have already noted in 2.1.7 that this subalgebra is isomorphic to $\mathfrak{sp}(n)$.

2.3.2. Cartan involutions. Having the two extremal real forms of a complex semisimple Lie algebra, we start to study general real forms. The first step is to construct a vector space decomposition of a real semisimple Lie algebra into a maximal compact subalgebra and a complementary subspace.

Let $\mathfrak{g}$ be a real semisimple Lie algebra. A *Cartan involution* for $\mathfrak{g}$ is an involutive automorphism $\theta:\mathfrak{g}\to\mathfrak{g}$ such that the bilinear form $B_\theta(X,Y) := -B(X,\theta Y)$ is positive definite. A *Cartan decomposition* of $\mathfrak{g}$ is a vector space decomposition $\mathfrak{g}=\mathfrak{k}\oplus\mathfrak{p}$ such that $\mathfrak{k}$ is a subalgebra, $[\mathfrak{k},\mathfrak{p}]\subset\mathfrak{p}$ and $[\mathfrak{p},\mathfrak{p}]\subset\mathfrak{k}$, and such that the restriction of the Killing form to $\mathfrak{k}$ is negative definite and the restriction to $\mathfrak{p}$ is positive definite. Defining $\mathfrak{k}$ and $\mathfrak{p}$ to be the eigenspaces of θ with eigenvalue $+1$ and -1 is easily seen to give a bijection between Cartan involutions and Cartan decompositions of $\mathfrak{g}$.

Consider the underlying real Lie algebra $\mathfrak{g}_\mathbb{R}$ of a complex semisimple Lie algebra $\mathfrak{g}$. Let $\mathfrak{u}\subset\mathfrak{g}_\mathbb{R}$ be a compact real form of $\mathfrak{g}$ and let τ be the corresponding conjugation. The Killing form of $\mathfrak{g}_\mathbb{R}$ is the real part of the Killing form of $\mathfrak{g}$, whose restriction to $\mathfrak{u}$ is negative definite. Now each element of $\mathfrak{g}_\mathbb{R}$ can be written as $X+iY$ for $X,Y\in\mathfrak{u}$, and we compute

$$B_\tau(X+iY,X+iY) = -\operatorname{re}(B(X+iY,X-iY)) = -B(X,X)-B(Y,Y).$$

This shows that τ is a Cartan involution on $\mathfrak{g}_\mathbb{R}$ corresponding to $\mathfrak{k}=\mathfrak{u}$ and $\mathfrak{p}=i\mathfrak{u}$.

Next, suppose that $\mathfrak{g}$ is a real semisimple Lie algebra with complexification $\mathfrak{g}_\mathbb{C}$ and that $\mathfrak{g}=\mathfrak{k}\oplus\mathfrak{p}$ is a Cartan decomposition. Then $\mathfrak{k}$ is a maximal subspace of $\mathfrak{g}$ on which the Killing form is negative definite, and thus a maximal compact subalgebra. On the other hand, viewing $\mathfrak{g}$ as a subspace of $\mathfrak{g}_\mathbb{C}$, we can also consider $\mathfrak{u}:=\mathfrak{k}\oplus i\mathfrak{p}\subset\mathfrak{g}_\mathbb{C}$. Since the Killing form of $\mathfrak{g}_\mathbb{C}$ is the complex bilinear extension of the Killing form of $\mathfrak{g}$, we conclude that the Killing form is negative definite on $\mathfrak{u}$, so this is a compact real form of $\mathfrak{g}_\mathbb{C}$. Let σ and τ be the conjugations of $\mathfrak{g}_\mathbb{C}$ corresponding to the real forms $\mathfrak{g}$ and $\mathfrak{u}$, respectively. Then both σ and τ are either id or $-$id on each of the subspaces $\mathfrak{k}$, $i\mathfrak{k}$, $\mathfrak{p}$, and $i\mathfrak{p}$. This immediately implies that $\sigma\circ\tau=\tau\circ\sigma$, so in particular, $\tau(\mathfrak{g})\subset\mathfrak{g}$, and the restriction of τ to $\mathfrak{g}$ is exactly the Cartan involution θ corresponding to the Cartan decomposition $\mathfrak{g}=\mathfrak{k}\oplus\mathfrak{p}$.

Conversely, let us assume that $\mathfrak{u} \subset \mathfrak{g}_{\mathbb{C}}$ is a compact real form of the complexification of $\mathfrak{g}$ such that the involutions σ and τ corresponding to $\mathfrak{g}$ and $\mathfrak{u}$ commute. Then $\tau(\mathfrak{g}) \subset \mathfrak{g}$, and we claim that τ restricts to a Cartan involution θ for $\mathfrak{g}$. Indeed, since τ is an involutive automorphism of the real Lie algebra $\mathfrak{g}_{\mathbb{C}}$, θ is an involutive automorphism of $\mathfrak{g}$, so we only have to check the condition on the Killing form. Since $\mathfrak{u}$ is a real form of $\mathfrak{g}_{\mathbb{C}}$, any element of $\mathfrak{g}$ can be uniquely written as $X + iY$ with $X, Y \in \mathfrak{u}$. The involution θ maps this element to $X - iY$, so we get $B_\theta(X + iY, X + iY) = -B(X + iY, X - iY) = -B(X, X) - B(Y, Y)$, which is > 0, since $X, Y \in \mathfrak{u}$. This reduces the construction of a Cartan involution to the construction of a compact real form of the complexification whose involution commutes with the involution corresponding to $\mathfrak{g}$.

LEMMA 2.3.2. *Let $\mathfrak{g}$ be a real Lie algebra, θ a Cartan involution of $\mathfrak{g}$, and σ any involutive automorphism of $\mathfrak{g}$. Then there is an inner automorphism $\psi \in \mathrm{Int}(\mathfrak{g})$, such that $\psi\theta\psi^{-1}$ commutes with σ.*

PROOF. Since θ is a Cartan involution, B_θ is a positive definite inner product on $\mathfrak{g}$, and by assumption $\sigma \circ \theta$ is an automorphism of $\mathfrak{g}$. Now $B_\theta(\sigma\theta X, Y) = -B(\sigma\theta X, \theta Y)$. Since σ is an involutive automorphism of $\mathfrak{g}$, invariance of the Killing form implies that this equals $-B(\theta X, \sigma\theta Y) = B_\theta(X, \sigma\theta Y)$.

Hence, $\sigma\theta : \mathfrak{g} \to \mathfrak{g}$ is a symmetric linear map, so the square $\phi := \sigma\theta\sigma\theta$ is positive and thus diagonalizable with positive eigenvalues. Then we can form ϕ^r for all $r \in \mathbb{R}$ just by taking rth powers of the eigenvalues, and these maps form a one–parameter group, i.e. $\phi^{r+s} = \phi^r \circ \phi^s$ for all $r, s \in \mathbb{R}$. The fact that ϕ is an automorphism is equivalent to the fact that the Lie bracket on $\mathfrak{g}$ maps the product of the λ–eigenspace and the μ–eigenspace of ϕ to the $\lambda\mu$–eigenspace. This condition then holds for ϕ^r for all r, so each of these maps is an automorphism of $\mathfrak{g}$, too. Since $r \mapsto \psi^r$ is a smooth curve through the identity, it has values in $\mathrm{Int}(\mathfrak{g})$.

By construction $\phi^{-1} = \theta\sigma\theta\sigma$, and thus $\phi\theta = \theta\phi^{-1}$. This can be equivalently expressed as the fact that θ maps the λ–eigenspace of ϕ to the λ^{-1}–eigenspace, and hence $\phi^r\theta = \theta\phi^{-r}$ for all $r \in \mathbb{R}$. On the other hand, by construction ϕ commutes with $\sigma\theta$, which implies that $\sigma\theta$ preserves the eigenspaces of ϕ, so $\phi^r\sigma\theta = \sigma\theta\phi^r$ for all $r \in \mathbb{R}$. Now we put $\psi := \phi^{1/4}$, and compute

$$(\psi\theta\psi^{-1})\sigma = \psi^2\theta\sigma = \psi^{-2}\phi\theta\sigma = \psi^{-2}\sigma\theta = \sigma\theta\psi^{-2} = \sigma(\psi\theta\psi^{-1}),$$

which completes the proof. □

THEOREM 2.3.2. *Let $\mathfrak{g}$ be a real semisimple Lie algebra.*

(1) There exists a Cartan involution θ on $\mathfrak{g}$.

(2) If θ and θ' are Cartan involutions on $\mathfrak{g}$, then there is a $\phi \in \mathrm{Int}(\mathfrak{g})$ such that $\theta' = \phi\theta\phi^{-1}$.

PROOF. (1) Let $\mathfrak{g}_{\mathbb{C}}$ be the complexification of $\mathfrak{g}$, and let $\mathfrak{u}_0 \subset \mathfrak{g}_{\mathbb{C}}$ be a compact real form. Let σ and τ be the involutions of $\mathfrak{g}_{\mathbb{C}}$ corresponding to the real forms $\mathfrak{g}$ and $\mathfrak{u}_0$, respectively. Then we know from above that τ is a Cartan involution for the real Lie algebra $\mathfrak{g}_{\mathbb{C}}$. By the lemma, there is an automorphism $\phi \in \mathrm{Int}(\mathfrak{g}_{\mathbb{C}})$ such that $\phi\tau\phi^{-1}$ commutes with σ. The fixed point set of $\phi\tau\phi^{-1}$ is just $\phi(\mathfrak{u}_0)$, so this is a compact real form of $\mathfrak{g}_{\mathbb{C}}$. We have already observed above, that, together with the fact that $\phi\tau\phi^{-1}$ commutes with σ, this implies that the restriction of $\phi\tau\phi^{-1}$ to $\mathfrak{g}$ is a Cartan involution.

(2) By the lemma, we find an inner automorphism ϕ, such that $\phi\theta\phi^{-1}$ commutes with θ'. One immediately verifies that $\phi\theta\phi^{-1}$ is a Cartan involution of $\mathfrak{g}$, so to complete the proof it suffices to show that if θ and θ' are commuting Cartan involutions of $\mathfrak{g}$, then $\theta = \theta'$. If θ and θ' commute, then their eigenspace decompositions are compatible, so any element $X \in \mathfrak{g}$ stabilized by θ splits into the sum $Y + Z$ of two elements stabilized by θ such that $\theta'(Y) = Y$ and $\theta'(Z) = -Z$. But then $0 \leq B_\theta(Z, Z) = -B(Z, Z)$ and $0 \leq B_{\theta'}(Z, Z) = -B(Z, -Z)$, which is possible only for $Z = 0$. Thus, we see that the $+1$–eigenspace of θ is contained in the $+1$–eigenspace of θ', so by symmetry the $+1$–eigenspaces coincide. Similarly, we conclude that the -1 eigenspaces coincide, which concludes the proof. □

COROLLARY 2.3.2. *If $\mathfrak{g}$ is a complex semisimple Lie algebra and $\mathfrak{u}, \mathfrak{u}' \subset \mathfrak{g}$ are compact real forms, then there is an inner automorphism $\phi \in \operatorname{Int}(\mathfrak{g})$, such that $\phi(\mathfrak{u}) = \mathfrak{u}'$. In particular, all compact real forms of $\mathfrak{g}$ are isomorphic, so the classification of compact semisimple Lie algebras is equivalent to the classification of complex semisimple Lie algebras, and thus given by Dynkin diagrams.*

PROOF. Let τ and τ' be the conjugations with respect to the two compact real forms. Then these are Cartan involutions on the underlying real Lie algebra $\mathfrak{g}_\mathbb{R}$ of $\mathfrak{g}$. By the theorem, there is an element $\phi \in \operatorname{Int}(\mathfrak{g}_\mathbb{R})$ such that $\tau' = \phi\tau\phi^{-1}$. Since the group $\operatorname{Int}(\mathfrak{g}_\mathbb{R})$ is generated by the maps $e^{\operatorname{ad}(A)}$ for $A \in \mathfrak{g} = \mathfrak{g}_\mathbb{R}$, we see that $\operatorname{Int}(\mathfrak{g}_\mathbb{R}) = \operatorname{Int}(\mathfrak{g})$. Hence, the fix point set $\mathfrak{u}'$ of $\phi\tau\phi^{-1}$ is the image under ϕ of the fix point set $\mathfrak{u}$ of τ. □

REMARK 2.3.2. (1) Let $\mathfrak{g}$ be a real semisimple Lie algebra, θ a Cartan involution on $\mathfrak{g}$, and consider the inner product B_θ on $\mathfrak{g}$. For $X, Y, Z \in \mathfrak{g}$, we get

$$B_\theta(\operatorname{ad}(\theta X)Y, Z) = -B([\theta X, Y], \theta Z) = B(Y, [\theta X, \theta Z]).$$

Since θ is an automorphism of $\mathfrak{g}$, the last term coincides with $-B_\theta(Y, [X, Z])$, which shows that $\operatorname{ad}(\theta X)$ is the negative of the adjoint map of $\operatorname{ad}(X)$. In particular, choosing any basis of $\mathfrak{g}$, the image of $\mathfrak{g}$ under the adjoint representation is a Lie subalgebra of $\mathfrak{gl}(\mathfrak{g})$, which is closed under transposition, and in this picture the Cartan involution θ is given by the negative transpose. Hence, Proposition 2.1.7 implies that any subalgebra of $\mathfrak{g}$ which is stable under a Cartan involution is reductive.

Conversely, suppose that $\mathfrak{g} \subset \mathfrak{gl}(n, \mathbb{R})$ is a Lie subalgebra, such that $X \in \mathfrak{g}$ implies $X^t \in \mathfrak{g}$. Then $\theta(X) = -X^t$ defines an involutive automorphism of $\mathfrak{g}$. In the associated decomposition $\mathfrak{g} = \mathfrak{k} \oplus \mathfrak{p}$, the factor $\mathfrak{k}$ consists of all skew symmetric elements of $\mathfrak{g}$, while $\mathfrak{p}$ consists of all symmetric elements of $\mathfrak{g}$. For $X \in \mathfrak{k}$ and $Y \in \mathfrak{p}$, the map $\operatorname{ad}(X) \circ \operatorname{ad}(Y)$ clearly maps $\mathfrak{k}$ to $\mathfrak{p}$ and $\mathfrak{p}$ to $\mathfrak{k}$, so $\mathfrak{g} = \mathfrak{k} \oplus \mathfrak{p}$ is an orthogonal decomposition with respect to the Killing form. Finally, one easily verifies that B is negative definite on $\mathfrak{k}$ and positive definite on $\mathfrak{p}$, so this is a Cartan decomposition and hence θ is a Cartan involution. In particular, this gives examples of Cartan involutions on many of the classical real Lie algebras.

(2) The fact that a complex semisimple Lie algebra has a (up to isomorphism) unique compact real form is the basis for an approach to representation theory of complex semisimple Lie algebras known as *Weyl's unitary trick.* For a real semisimple Lie algebra $\mathfrak{g}$ with complexification $\mathfrak{g}_\mathbb{C}$, there is a bijective correspondence between complex representations of $\mathfrak{g}$ and $\mathfrak{g}_\mathbb{C}$, induced by restriction, respectively, complex linear extension. Applying this correspondence once more to a compact

real form $\mathfrak{u}$ of $\mathfrak{g}_{\mathbb{C}}$, we end up with a bijection between complex representations of $\mathfrak{g}$ and $\mathfrak{u}$.

Passing to the simply connected groups, we get a relation between smooth representations of a real semisimple Lie group, holomorphic representations of a complex semisimple Lie group and smooth representations of a compact Lie group. Many problems are much easier to treat in the compact setting, using for example the existence of invariant inner products. Also, the original proof of the Weyl character formula from 2.2.18 is based on the unitary trick and the use of integration on compact groups. See [**FH91**, §26.2] for a discussion of the unitary trick and the resulting proof of the Weyl character formula.

2.3.3. Cartan decomposition on the group level. There is an analog of the Cartan involution and the Cartan decomposition on the group level. Let $\mathfrak{g}$ be a real semisimple Lie algebra, and let G be a connected Lie group with Lie algebra $\mathfrak{g}$. Let θ be a Cartan involution on $\mathfrak{g}$ and let $\mathfrak{g} = \mathfrak{k} \oplus \mathfrak{p}$ be the corresponding Cartan decomposition. The main point about the Cartan decomposition on the group level for our purposes is that there is a nice subgroup $K \subset G$ corresponding to the Lie subalgebra $\mathfrak{k} \subset \mathfrak{g}$, which is a maximal compact subgroup in most situations of interest.

THEOREM 2.3.3. *Let $\mathfrak{g}$ be a real semisimple Lie algebra, θ a Cartan involution on $\mathfrak{g}$, $\mathfrak{g} = \mathfrak{k} \oplus \mathfrak{p}$ the corresponding Cartan decomposition, and let G be a connected Lie group with Lie algebra $\mathfrak{g}$. Then there is a unique automorphism $\Theta : G \to G$ with differential θ. The fixed point group $K := \{g \in G : \Theta(g) = g\} \subset G$ is a connected closed subgroup of G, which contains the center $Z(G)$ and has Lie algebra $\mathfrak{k}$. The subgroup K is compact if and only if $Z(G)$ is finite and in that case K is a maximal compact subgroup of G. Moreover, the mapping $(g, X) \mapsto g\exp(X)$ defines a diffeomorphism $K \times \mathfrak{p} \to G$.*

SKETCH OF PROOF. (See [**Kn96**, VI, Theorem 6.31] for details.) One starts by considering the case $G = \operatorname{Int}(\mathfrak{g})$ of the adjoint group. In that case, we may work in the group $GL(\mathfrak{g})$, respectively, in the Lie algebra $\mathfrak{gl}(\mathfrak{g})$, using the positive definite inner product B_θ on $\mathfrak{g}$ to define adjoints. As we have noticed in 2.3.2 above, in this picture $\theta(X) = -X^*$, so it is natural to try to define $\Theta(g) = (g^*)^{-1}$. Using invariance of the Killing form, one easily verifies that for $g \in \operatorname{Aut}(\mathfrak{g})$ and $X, Y, Z \in \mathfrak{g}$ one gets $B_\theta([g^*(X), g^*(Y)], Z) = B_\theta(g^*([X, Y]), Z)$, which shows that $\operatorname{Aut}(\mathfrak{g})$ is closed under adjoints. Applying this to $g \in \operatorname{Int}(\mathfrak{g})$, we conclude that g^*g is a positive definite map lying in $\operatorname{Aut}(\mathfrak{g})$. As in the proof of Lemma 2.3.2, this implies that $(g^*g)^r$ for $r \in \mathbb{R}$ is a one–parameter group of automorphisms of $\mathfrak{g}$. In particular, this one–parameter group is contained in $G = \operatorname{Int}(\mathfrak{g})$, so G is closed under adjoints, $\Theta(g) = (g^*)^{-1}$ is a well–defined involutive automorphism of G, and clearly the derivative of Θ is θ.

On the other hand, the one–parameter group $(g^*g)^r$ must be of the form $\exp(rX)$ for some $X \in \mathfrak{g}$. By definition $\Theta(g^*g) = (g^*g)^{-1}$, which implies $\theta(X) = -X$, and thus $X \in \mathfrak{p}$. The fixed point group K of Θ is a closed subgroup of G and clearly it has Lie algebra $\mathfrak{k}$. By construction, $g \in K$ if and only if $g^* = g^{-1}$, so K is the intersection of G with the orthogonal group of the inner product B_θ, which implies that K is compact.

Next, $\phi(k, X) := k\exp(X)$ defines a smooth map $K \times \mathfrak{p} \to G$. If $g \in G$ is arbitrary, as above, we find an element $X \in \mathfrak{p}$ such that $\exp rX = (g^*g)^r$ for all

$r \in \mathbb{R}$ and we put $p := \exp(\frac{1}{2}X)$. Then $p \in G$ satisfies $p^* = p$ and putting $k = gp^{-1}$, we see that $k^*k = p^{-1}g^*gp^{-1} = \mathrm{id}$. Thus, we get $g = kp = k\exp\frac{1}{2}X$ and $k \in K$, whence ϕ is surjective. On the other hand, if $g = k\exp(X)$, then $g^*g = \exp 2X$, so we see that X is uniquely determined as half the generator of the one–parameter subgroup $(g^*g)^r$. Then $k = g\exp(-X)$ is uniquely determined, too, so ϕ is bijective. The exponential map restricts to a diffeomorphism from $\mathfrak{p}$ onto $\exp(\mathfrak{p})$, and this easily implies that ϕ is a diffeomorphism.

Connectedness of G together with the diffeomorphism $K \times \mathfrak{p} \to G$ implies that K is connected. Since G has trivial center, we only have to prove that K is maximally compact to complete the proof for the adjoint group. If $K' \supset K$ is a subgroup of G, which properly contains K, then it must contain at least one element of the form $\exp X$, for a nonzero element $X \in \mathfrak{p}$. But then it must contain the one–parameter subgroup $\exp rX = \exp(X)^r$. From above we know that this is a group of positive operators, so it has unbounded eigenvalues, which shows that K' cannot be compact, so the proof for the adjoint group is complete.

If G is a general connected Lie group with Lie algebra $\mathfrak{g}$, then put $\underline{G} = \mathrm{Int}(\mathfrak{g})$, $\underline{\Theta}$ the involutive automorphism of $\underline{G}$ constructed above and $\underline{K}$ its fixed point group. Then we have a covering $\pi : G \to \underline{G}$, and we put $K := \pi^{-1}(\underline{K})$. Via π, the group G acts transitively on $\underline{G}/\underline{K}$ with isotropy group K, so we get a diffeomorphism $G/K \cong \underline{G}/\underline{K}$. From $\underline{G} \cong \underline{K} \times \mathfrak{p}$ we immediately conclude that $\underline{G}/\underline{K}$ is simply connected, and since G/K is simply connected it follows that K is connected, and by construction the Lie algebra of K is $\mathfrak{k}$. Again by construction, K is a closed subgroup of G and since $Z(G)$ is exactly the fiber of π over the unit element, we see that K contains $Z(G)$ and is compact if and only if $Z(G)$ is finite.

Now $\pi \times \mathrm{id} : K \times \mathfrak{p} \to \underline{K} \times \mathfrak{p}$ is a covering, and the map $\phi_G(k, X) = k\exp(X)$ is a covering of the map $\phi_{\underline{G}}(k, X) = k\exp(X)$. We know that $\phi_{\underline{G}}$ is a diffeomorphism, and one easily verifies that ϕ_G is bijective, so we conclude that ϕ_G is a diffeomorphism, too. Next, let $\tilde{G}$ be the universal covering of G and $\tilde{K} \subset \tilde{G}$ the corresponding subgroup, which we already know to be the analytic subgroup with Lie algebra $\mathfrak{k}$. From 1.2.4 we know that there is a unique involutive automorphism $\tilde{\Theta} : \tilde{G} \to \tilde{G}$ with differential θ. By construction, $\tilde{K}$ is contained in the fixed point group of $\tilde{\Theta}$, which immediately implies that $\tilde{\Theta}$ descends to an involutive automorphism Θ of G, which contains K in its fixed point group. On the other hand, if $g\exp X$ lies in the fixed point group of Θ, then one immediately verifies $\exp 2X = 0$, whence $X = 0$, so K is exactly the fixed point group of Θ. Finally, if $Z(G)$ is finite, then the fact that K is maximal compact in G immediately follows from the fact that $\underline{K}$ is maximal compact in $\underline{G}$, so the proof is complete. □

REMARK 2.3.3. The involution $\Theta : G \to G$ is called the *global Cartan involution*, and the diffeomorphism $K \times \mathfrak{p} \to G$ is called the *global Cartan decomposition.* In particular, this shows that G is homotopy equivalent to K, so (in the case of finite center) all the topology of a semisimple Lie group is already visible in the maximal compact subgroup. For example, $SL(n, \mathbb{C})$ is homotopy equivalent to $SU(n)$, and $SO(n, \mathbb{C})$ and $SL(n, \mathbb{R})$ are homotopy equivalent to $SO(n)$. The disadvantage of this decomposition is that it is not satisfactory from the point of view of the group structure of G, which is encoded in K and $\mathfrak{p}$ in a nontrivial way. This drawback is removed by the Iwasawa decomposition to be discussed below.

2.3.4. Restricted roots. We next move to analogs of the root decomposition for noncompact real Lie algebras. Let $\mathfrak{g}$ be a real semisimple Lie algebra with a Cartan decomposition $\mathfrak{g} = \mathfrak{k} \oplus \mathfrak{p}$ such that $\mathfrak{p} \neq \{0\}$ and corresponding Cartan involution θ. From Remark 2.3.2 (1), we know that with respect to the inner product B_θ we have $\operatorname{ad}(\theta X) = -\operatorname{ad}(X)^t$. Hence, $\operatorname{ad}(X)$ is skew symmetric for $X \in \mathfrak{k}$ and symmetric for $X \in \mathfrak{p}$. In particular, $\operatorname{ad}(X)$ is never diagonalizable (over the reals) for $X \in \mathfrak{k}$ and always diagonalizable for $X \in \mathfrak{p}$.

Let $\mathfrak{a} \subset \mathfrak{p}$ be a maximal abelian subalgebra (which has to exist since $\mathfrak{p}$ is finite–dimensional). Then the maps $\operatorname{ad}(A)$ for $A \in \mathfrak{a}$ form a family of commuting symmetric linear maps which is thus simultaneously diagonalizable. Moreover, the eigenvalue on a joint eigenspace depends linearly on A, and by symmetry different eigenspaces are automatically orthogonal to each other. Now for a linear functional $\lambda : \mathfrak{a} \to \mathbb{R}$, we denote by $\mathfrak{g}_\lambda$ the set of all $X \in \mathfrak{g}$ such that $[A, X] = \lambda(A)X$ for all $A \in \mathfrak{a}$. The nonzero functionals λ such that $\mathfrak{g}_\lambda \neq \{0\}$ are called the *restricted roots* of $\mathfrak{g}$ with respect to $\mathfrak{a}$. We denote the set of restricted roots by Δ_r. By construction, we obtain a decomposition of $\mathfrak{g}$ into an orthogonal (with respect to B_θ) direct sum $\mathfrak{g} = \mathfrak{g}_0 \oplus \bigoplus_{\lambda \in \Delta_r} \mathfrak{g}_\lambda$. If λ is a restricted root, and $X \in \mathfrak{g}_\lambda$, then for $A \in \mathfrak{a}$ we get

$$[A, \theta X] = \theta[\theta A, X] = -\theta[A, X] = -\lambda(A)\theta X,$$

so we see that θ restricts to a linear isomorphism between $\mathfrak{g}_\lambda$ and $\mathfrak{g}_{-\lambda}$. In particular, $\lambda \in \Delta_r$ implies $-\lambda \in \Delta_r$. On the other hand, we obviously have $[\mathfrak{g}_\lambda, \mathfrak{g}_\mu] \subset \mathfrak{g}_{\lambda+\mu}$. Finally, we can also easily describe the subalgebra $\mathfrak{g}_0$. By definition $\mathfrak{a} \subset \mathfrak{g}_0$, and from above we know that $\theta(\mathfrak{g}_0) \subset \mathfrak{g}_0$. Thus, $\mathfrak{g}_0 = (\mathfrak{g}_0 \cap \mathfrak{k}) \oplus (\mathfrak{g}_0 \cap \mathfrak{p})$, and clearly this decomposition is orthogonal with respect to B_θ. By definition, $\mathfrak{a}$ is maximal abelian in $\mathfrak{p}$, whence $\mathfrak{g}_0 \cap \mathfrak{p} = \mathfrak{a}$. On the other hand, by definition $\mathfrak{m} := \mathfrak{g}_0 \cap \mathfrak{k}$ consists of all elements X of the subalgebra $\mathfrak{k}$ such that $[X, A] = 0$ for all $A \in \mathfrak{a}$, so this is exactly the centralizer $Z_\mathfrak{k}(\mathfrak{a})$ of the subspace $\mathfrak{a}$ in the $\mathfrak{k}$–module $\mathfrak{p}$. (Recall that by definition of a Cartan decomposition we have $[\mathfrak{k}, \mathfrak{p}] \subset \mathfrak{p}$, so $\mathfrak{p}$ is a $\mathfrak{k}$–module under the restriction of the adjoint action.)

Examples. (1) Let $\mathfrak{g}$ be the underlying real Lie algebra of a complex semisimple Lie algebra. Then we know that the conjugation with respect to a compact real form $\mathfrak{u} \subset \mathfrak{g}$ is a Cartan involution. Choose a Cartan subalgebra $\mathfrak{h} \subset \mathfrak{g}$ and consider the compact real form $\mathfrak{u} = i\mathfrak{h}_0 \oplus \bigoplus_{\alpha \in \Delta^+} (\mathbb{R}(X_\alpha - X_{-\alpha}) \oplus i\mathbb{R}(X_\alpha + X_{-\alpha}))$ as constructed in 2.3.1. The Cartan decomposition we obtain has the form $\mathfrak{g} = \mathfrak{k} \oplus \mathfrak{p}$ with $\mathfrak{k} = \mathfrak{u}$ and $\mathfrak{p} = i\mathfrak{u}$. In particular, we have $\mathfrak{h}_0 \subset \mathfrak{p}$, and the decomposition of $\mathfrak{g}$ into eigenspaces for the adjoint action of $\mathfrak{h}_0$ coincides with the root decomposition of the complex simple Lie algebra $\mathfrak{g}$. From this, one readily concludes that $\mathfrak{h}_0$ is a maximal abelian subalgebra of $\mathfrak{p}$. In particular, the restricted roots are exactly the restrictions of the roots to the real subspace $\mathfrak{h}_0$, and any restricted root space has real dimension two in this case.

(2) Next, we consider various real forms of $\mathfrak{sl}(n, \mathbb{C})$. For the split real form $\mathfrak{sl}(n, \mathbb{R})$ we can simply use the subspace $\mathfrak{h}_0$ of tracefree real diagonal matrices as the subspace $\mathfrak{a}$, and clearly the restricted roots in this case are exactly the restrictions of the roots of $\mathfrak{sl}(n, \mathbb{C})$ to the subspace $\mathfrak{a}$. In particular, the restricted roots form an abstract root system of type A_{n-1} for the split real form. Here, all restricted root spaces have real dimension one.

For the compact real form $\mathfrak{su}(n)$ there are no restricted roots since $\mathfrak{p} = \{0\}$. The next obvious real forms are the algebras $\mathfrak{su}(p, q)$ with $p + q = n$ of complex

$n \times n$–matrices which are skew Hermitian with respect to a Hermitian form of signature (p,q). Since $\mathfrak{su}(q,p)$ is isomorphic to $\mathfrak{su}(p,q)$, we may restrict to the case $p \geq q$. Since the case of $\mathfrak{su}(p,p)$ is slightly special, we start by assuming $p > q$. To get a nice description of the restricted roots for these algebras, the simplest presentation is to view $\mathbb{C}^n$ as $\mathbb{C}^q \oplus \mathbb{C}^q \oplus \mathbb{C}^{p-q}$ with the Hermitian form defined by

$$\langle (x,y,z),(x',y',z')\rangle := \sum_{j=1}^{q}(x_j\overline{y'_j} + y_j\overline{x'_j}) + \sum_{j=1}^{p-q} z_j\overline{z'_j}.$$

Clearly, the first part defines a Hermitian form of signature (q,q) on $\mathbb{C}^{2q}$, while the second part defines a positive definite Hermitian form on $\mathbb{C}^{p-q}$, so the whole form has signature (p,q). The special unitary algebra of this Hermitian form consists of all tracefree matrices M, such that $M^*\mathbb{J} = -\mathbb{J}M$, where $\mathbb{J}$ is the block matrix $\begin{pmatrix} 0 & \mathbb{I}_q & 0 \\ \mathbb{I}_q & 0 & 0 \\ 0 & 0 & \mathbb{I}_{p-q}\end{pmatrix}$, with $\mathbb{I}_k$ denoting the $k \times k$–identity matrix. A short computation shows that we obtain all tracefree block matrices with blocks of size q, q and $p-q$ of the form $\begin{pmatrix} A & B & C \\ D & -A^* & E \\ -E^* & -C^* & F\end{pmatrix}$, with $A \in \mathfrak{gl}(q,\mathbb{C})$, C and E arbitrary complex $q\times(p-q)$–matrices, B and D in $\mathfrak{u}(q)$ and $F \in \mathfrak{u}(p-q)$. The advantage of this presentation is that the Lie algebra is closed under conjugate transpose, which implies that putting $\mathfrak{k}$ and $\mathfrak{p}$ the subspace of skew–Hermitian, respectively, Hermitian matrices in $\mathfrak{g}$, we obtain a Cartan decomposition. In particular, the $\mathfrak{p}$–component consists of all matrices of the form $\begin{pmatrix} A & B & C \\ -B & -A & -C \\ C^* & -C^* & 0\end{pmatrix}$, where $A^* = A$ and $B^* = -B$. Now consider the q–dimensional subspace $\mathfrak{a}$ of all real diagonal matrices contained in $\mathfrak{p}$, and for $j = 1,\dots,q$ let $e_j : \mathfrak{a} \to \mathbb{R}$ be the functional which extracts the jth diagonal entry. For a matrix A' such that $(A')^* = A'$ we get

$$\left[\begin{pmatrix} A' & 0 & 0 \\ 0 & -A' & 0 \\ 0 & 0 & 0\end{pmatrix}, \begin{pmatrix} A & B & C \\ D & -A^* & E \\ -E^* & -C^* & F\end{pmatrix}\right] = \begin{pmatrix} [A',A] & A'B+BA' & A'C \\ -A'D-DA' & -[A',A]^* & -A'E \\ (A'E)^* & -(A'C)^* & 0\end{pmatrix}.$$

From this, one concludes that $\mathfrak{a} \subset \mathfrak{p}$ is a maximal abelian subspace. Moreover, the A–block gives us restricted roots $e_i - e_j$ for $i \neq j$, the B–block leads to $e_i + e_j$ for all i,j, from the C–block we get all e_j, while the D–block gives $-e_i - e_j$ and the E–block contains restricted root spaces for $-e_j$. Thus, the restricted roots are exactly $\pm e_i \pm e_j$ for $i \neq j$ as well as $\pm e_j$ and $\pm 2e_j$ for $j = 1,\dots,q$, so they form the non–reduced abstract root system $(BC)_q$; see 2.2.8. Moreover, the real dimensions of the restricted root spaces all are equal to two, except for the restricted roots $\pm 2e_j$, where the real dimension is one.

The case of $\mathfrak{g} := \mathfrak{su}(p,p)$ can be dealt with very similarly. Here the last block from above simply does not occur, so we may view $\mathfrak{g}$ as the algebra of block matrices of the form $\begin{pmatrix} A & B \\ D & -A^*\end{pmatrix}$, with $A \in \mathfrak{gl}(p,\mathbb{C})$, B and D in $\mathfrak{u}(p)$, but to assure that the matrix is tracefree, we have now to assume that the trace of A is real. This is closed under conjugate transpose, so we may again use the decomposition into skew–Hermitian and Hermitian part as the Cartan decomposition. The p–dimensional subspace $\mathfrak{a}$ of real diagonal matrices contained in $\mathfrak{g}$ lies in $\mathfrak{p}$ and is maximal abelian in there. We get the functionals e_i for $i = 1,\dots,q$, as before and we may use the above computation for the adjoint action. We get the restricted roots $e_i - e_j$ for $i \neq j$ (from the A–block), $e_i + e_j$ for all i,j (from the B–block), and $-e_i - e_j$ for all i,j (from the D–block). Hence, the restricted roots form an abstract root system of type C_p, compare with 2.2.6(4). From the description one immediately concludes that for $\pm e_i \pm e_j$ with $i \neq j$, the restricted root space has real dimension 2, while the restricted root spaces for the restricted roots $\pm 2e_i$ are of real dimension one.

(3) Replacing Hermitian forms by real bilinear forms, we can treat the orthogonal Lie algebra $\mathfrak{so}(p,q)$ similarly. Again, we start by assuming $p > q$, and consider $\mathbb{R}^{p+q} = \mathbb{R}^q \oplus \mathbb{R}^q \oplus \mathbb{R}^{p-q}$, with the obvious real bilinear analog of the Hermitian form from (2), which has signature (p,q). The special orthogonal algebra of this inner product consists of all block matrices with blocks of sizes q, q, and $p-q$ of the form $\begin{pmatrix} A & B & C \\ D & -A^t & E \\ -E^t & -C^t & F \end{pmatrix}$, with $A \in \mathfrak{gl}(n,\mathbb{R})$, $B, D \in \mathfrak{so}(q)$, $F \in \mathfrak{so}(p-q)$ and C, E arbitrary real $q \times (p-q)$–matrices. This algebra is closed under forming transposes, and thus a Cartan decomposition is given by putting $\mathfrak{k}$ and $\mathfrak{p}$ the subspaces of skew symmetric, respectively, symmetric matrices contained in $\mathfrak{g}$.

As in (2) we get a q–dimensional commutative subalgebra $\mathfrak{a}$ of real diagonal matrices contained in $\mathfrak{p}$, and we define the functionals $e_1, \dots, e_q : \mathfrak{a} \to \mathbb{R}$ as before. For the commutator we get the analogous formula as in (2), but with adjoints replaced by transposes. This shows that $\mathfrak{a}$ is a maximal abelian subspace of $\mathfrak{p}$ and the restricted roots are $e_i - e_j$ for $i \neq j$ from the A–block, and $\pm e_j$ for $j = 1, \dots, q$ from the C– and E–blocks. However, since the B– and D–blocks now consist of skew symmetric matrices (which have zeros on the main diagonal) these only lead to the restricted roots $\pm(e_i + e_j)$ for $i \neq j$. Consequently, in this case the restricted roots form an abstract root system of type B_q, and all restricted root spaces have real dimension one.

The case of $\mathfrak{so}(p,p)$ is closely parallel, but one only has the four blocks in the upper left part. The maximal abelian subspace $\mathfrak{a}$ is again formed by real diagonal matrices, and the functionals e_i for $i = 1, \dots, p$ are defined as above. The discussion above shows that the restricted roots now are $\pm e_i \pm e_j$ for $i \neq j$, so they form an abstract root system of type D_p and all restricted root spaces are of real dimension one.

2.3.5. The Iwasawa decomposition. If $\mathfrak{g} = \mathfrak{k} \oplus \mathfrak{p}$ is a Cartan decomposition of a real semisimple Lie algebra $\mathfrak{g}$, then $\mathfrak{k}$ is a subalgebra of $\mathfrak{g}$. If we further choose a maximal abelian subspace $\mathfrak{a} \subset \mathfrak{p}$, then of course $\mathfrak{a}$ also is a subalgebra of $\mathfrak{g}$ and $\mathfrak{k} \cap \mathfrak{a} = \{0\}$. Choose a notion of positivity in $\mathfrak{a}^* = L(\mathfrak{a}, \mathbb{R})$ (see 2.2.5), and let Δ_r^+ be the corresponding subset of positive restricted roots. Define $\mathfrak{n} \subset \mathfrak{g}$ to be the direct sum of all positive restricted root spaces. By construction, $\mathfrak{n} \subset \mathfrak{g}$ is a nilpotent Lie subalgebra and $\mathfrak{a} \oplus \mathfrak{n} \subset \mathfrak{g}$ is a subalgebra with $[\mathfrak{a} \oplus \mathfrak{n}, \mathfrak{a} \oplus \mathfrak{n}] = \mathfrak{n}$. In particular, $\mathfrak{a} \oplus \mathfrak{n} \subset \mathfrak{g}$ is solvable.

PROPOSITION 2.3.5 (Iwasawa decomposition of a real semisimple Lie algebra). *In the notation above, we have* $\mathfrak{g} = \mathfrak{k} \oplus \mathfrak{a} \oplus \mathfrak{n}$. *This decomposition is called the Iwasawa decomposition of* $\mathfrak{g}$.

PROOF. It only remains to show that $\mathfrak{k}$ and $\mathfrak{a} \oplus \mathfrak{n}$ are complementary subspaces of $\mathfrak{g}$. If $X \in \mathfrak{k} \cap (\mathfrak{a} \oplus \mathfrak{n})$, then $X = \theta X$, and since $\mathfrak{a} \subset \mathfrak{p}$ is stable under θ, we must have $\theta X \in \mathfrak{a} \oplus \theta\mathfrak{n}$. But we have observed in 2.3.4 that θ maps the restricted root space $\mathfrak{g}_\lambda$ to $\mathfrak{g}_{-\lambda}$, whence $(\mathfrak{a} \oplus \mathfrak{n}) \cap (\mathfrak{a} \oplus \theta\mathfrak{n}) = \mathfrak{a}$. Hence, we conclude that $X \in \mathfrak{k} \cap \mathfrak{a} = \{0\}$.

On the other hand, the restricted root decomposition reads as $\mathfrak{g} = \mathfrak{g}_0 \oplus \mathfrak{n} \oplus \theta\mathfrak{n}$, and again from 2.3.4 we know that $\mathfrak{g}_0 = \mathfrak{a} \oplus \mathfrak{m}$, where $\mathfrak{m} = Z_{\mathfrak{k}}(\mathfrak{a})$. Hence, given an arbitrary element $X \in \mathfrak{g}$, we find elements $H \in \mathfrak{a}$, $X_0 \in \mathfrak{m}$ and $X_\lambda \in \mathfrak{g}_\lambda$ for all $\lambda \in \Delta_r$, such that $X = H + X_0 + \sum_{\lambda\in\Delta_r} X_\lambda$. But this sum can be rewritten as

$$X = \left(X_0 + \sum_{\lambda\in\Delta_r^+} (X_{-\lambda} + \theta X_{-\lambda}) \right) + H + \left(\sum_{\lambda\in\Delta_r^+} (X_\lambda - \theta X_{-\lambda}) \right),$$

and by construction the right–hand side lies in $\mathfrak{k} + \mathfrak{a} + \mathfrak{n}$. □

In the examples treated in 2.3.4, we can immediately see the form of the Iwasawa decomposition. In particular, for $\mathfrak{sl}(n,\mathbb{C})$ (respectively $\mathfrak{sl}(n,\mathbb{R})$), we obtain $\mathfrak{k} = \mathfrak{u}(n)$ (respectively $\mathfrak{o}(n)$), $\mathfrak{a}$ is the subalgebra of real diagonal matrices, and $\mathfrak{n}$ is the subalgebra of complex (respectively real) strictly upper triangular matrices.

As in the case of the Cartan decomposition, the Iwasawa decomposition admits an analog on the group level. Suppose that G is a connected real semisimple Lie group with Lie algebra $\mathfrak{g}$, and let $\mathfrak{g} = \mathfrak{k} \oplus \mathfrak{a} \oplus \mathfrak{n}$ be an Iwasawa decomposition. Then we denote by A and N the analytic subgroups of G corresponding to the Lie subalgebra $\mathfrak{a}$ and $\mathfrak{n}$, and we use the subgroup $K \leq G$ from Theorem 2.3.3. Then we have:

THEOREM 2.3.5 (Global Iwasawa decomposition). *With G, K, A, and N as above, the multiplication $(k, a, n) \mapsto kan$ is a diffeomorphism $K \times A \times N \to G$. Moreover, the subgroups A and N of G are contractible.*

SKETCH OF PROOF. As in the case of the Cartan decomposition, we start with the case $G = \mathrm{Int}(\mathfrak{g}) \leq GL(\mathfrak{g})$ of the adjoint group, and we consider adjoints with respect to the inner product B_θ on $\mathfrak{g}$. From Theorem 2.3.3 we know that in this case K is compact and consists of all maps in G which are orthogonal for B_θ. Using an orthonormal basis of $\mathfrak{g}$ which is compatible with the restricted root decomposition $\mathfrak{g} = \mathfrak{g}_0 \oplus \bigoplus_{\lambda\in\Delta_r} \mathfrak{g}_\lambda$, we see that all elements of $\mathfrak{a}$ act diagonally. Let us order the basis in such a way that the first elements correspond to the largest restricted roots (with respect to the fixed chosen ordering), elements from $\mathfrak{g}_0$ are in the middle and elements corresponding to negative restricted roots are last. Then elements of $\mathfrak{n}$ act by strictly upper triangular matrices.

Consequently, A and N are analytic subgroups of the group of positive diagonal matrices, respectively, the group of upper triangular matrices with diagonal entries equal to one. Since both of these matrix groups are nilpotent and simply connected, their exponential maps are global diffeomorphisms. From this, one easily concludes that A and N are closed in $GL(\mathfrak{g})$ and thus in G, and simply connected. Moreover, this explicit description implies that $(a, n) \mapsto an$ is a bijection from $A \times N$ onto a closed subgroup AN of G. This closed subgroup has Lie algebra $\mathfrak{a} \oplus \mathfrak{n}$, which implies that the map $A \times N \to AN$ is a diffeomorphism.

The image of $(k, a, n) \mapsto kan$ is the product of the compact subset K with the closed subset AN of G, so it is closed in G. Moreover, since $\mathfrak{g} = \mathfrak{k} \oplus \mathfrak{a} \oplus \mathfrak{n}$, the map $K \times A \times N \to G$ has invertible differential in any point, so the image is open, too. Since G is connected, we conclude that $K \times A \times N \to G$ is a surjective local diffeomorphism. Finally, from the explicit description of the subgroups K and AN above, we immediately conclude that $K \cap AN = \{\mathrm{id}\}$, which implies the theorem for $G = \mathrm{Int}(\mathfrak{g})$.

If G is a general connected Lie group with Lie algebra $\mathfrak{g}$, then we put $\underline{G} = \mathrm{Int}(\mathfrak{g})$ and we denote by $\underline{K}$, $\underline{A}$ and $\underline{N}$ the subgroups of $\underline{G}$ constructed above. As in the case of the adjoint group, the map $K \times A \times N \to G$ has invertible differential in each point. Moreover, restricting the covering $\pi : G \to \underline{G}$ to the subgroups A and N we get coverings, which have to be diffeomorphisms since $\underline{A}$ and $\underline{N}$ are simply connected. Thus, A and N are simply connected closed subgroups of G. Using the covering π and the results for $\underline{G}$ one easily verifies that $K \times A \times N \to G$ is bijective, which completes the proof. □

2.3.6. Uniqueness of the Iwasawa decomposition. Next we have to show that the Iwasawa decomposition is independent (up to isomorphism) of the choices made in its construction. First, we need some additional information on the system of restricted roots.

PROPOSITION 2.3.6. *Let $\mathfrak{g}$ be a real semisimple Lie algebra endowed with a Cartan decomposition $\mathfrak{g} = \mathfrak{k} \oplus \mathfrak{p}$ and a fixed maximal abelian subspace $\mathfrak{a} \subset \mathfrak{p}$.*

Then the system $\Delta_r \subset \mathfrak{a}^*$ of restricted roots is an abstract root system. The action of any element in the Weyl group of this root system can be realized by an inner automorphism of $\mathfrak{g}$ which fixes $\mathfrak{k}$ and $\mathfrak{a}$.

PROOF. Since $\mathfrak{a} \subset \mathfrak{p}$, the restriction of the Killing form to $\mathfrak{a}$ is positive definite (and coincides with the restriction of the inner product B_θ, where θ is the Cartan involution). In particular, for any restricted root $\lambda \in \Delta_r$, there is a unique element $H_\lambda \in \mathfrak{a}$ such that $\lambda(H) = B(H, H_\lambda)$ for all $H \in \mathfrak{a}$. Choose a nonzero element E_λ in the restricted root space $\mathfrak{g}_\lambda$, and consider its image θE_λ under the Cartan involution. Then the bracket $[E_\lambda, \theta E_\lambda]$ is contained in $\mathfrak{g}_0$. Evidently, we have $\theta[E_\lambda, \theta E_\lambda] = -[E_\lambda, \theta E_\lambda]$, so this element lies in $\mathfrak{g}_0 \cap \mathfrak{p} = \mathfrak{a}$. For $H \in \mathfrak{a}$ we get $B([E_\lambda, \theta E_\lambda], H) = -B(\theta E_\lambda, [E_\lambda, H]) = \lambda(H) B(E_\lambda, \theta E_\lambda)$, so $[E_\lambda, \theta E_\lambda] = B(E_\lambda, \theta E_\lambda) H_\lambda$, and $B(E_\lambda, \theta E_\lambda) = -B_\theta(E_\lambda, E_\lambda) < 0$. One immediately concludes that the elements E_λ, θE_λ and H_λ span a three–dimensional Lie subalgebra of $\mathfrak{g}$ which is isomorphic to $\mathfrak{sl}(2, \mathbb{R})$.

Normalize the element E_λ in such a way that $B(E_\lambda, \theta E_\lambda) = \frac{-2}{|\lambda|^2}$ (with the norm on $\mathfrak{a}^*$ induced by the Killing form). Consider the group $\mathrm{Int}(\mathfrak{g})$ of inner automorphisms and let $K \subset \mathrm{Int}(\mathfrak{g})$ be the (compact) subgroup corresponding to $\mathfrak{k}$. Since $E_\lambda + \theta E_\lambda \in \mathfrak{k}$, we can form $k_\lambda = \exp(\frac{\pi}{2}(E_\lambda + \theta E_\lambda)) \in K$. For an element $H \in \mathfrak{a}$ such that $\lambda(H) = 0$, we have $[E_\lambda, H] = 0$ and $[\theta E_\lambda, H] = 0$. Therefore, we get $\mathrm{ad}(\frac{\pi}{2}(E_\lambda + \theta E_\lambda))(H) = 0$, and thus $\mathrm{Ad}(k_\lambda)(H) = e^{\mathrm{ad}(\frac{\pi}{2}(E_\lambda + \theta E_\lambda))}(H) = H$. On the other hand, to get the action of $\mathrm{Ad}(k_\lambda)$ on the element H_λ, observe that in $\mathfrak{sl}(2, \mathbb{R})$ we have

$$\left[\begin{pmatrix} 0 & \pi/2 \\ \pi/2 & 0 \end{pmatrix}, \begin{pmatrix} 1 & 0 \\ 0 & -1 \end{pmatrix}\right] = \begin{pmatrix} 0 & -\pi \\ \pi & 0 \end{pmatrix},$$

and applying the adjoint action once more, we obtain $\begin{pmatrix} \pi^2 & 0 \\ 0 & -\pi^2 \end{pmatrix}$. One easily concludes that $\mathrm{Ad}(k_\lambda)(H_\lambda) = \cos(\pi) H_\lambda = -H_\lambda$. Hence, $\mathrm{Ad}(k_\lambda)(\mathfrak{a}) \subset \mathfrak{a}$, and the adjoint action of k_λ on $\mathfrak{a}$ is the reflection in the hyperplane orthogonal to H_λ. Transferring this to the dual, we see that the coadjoint action of k_λ on $\mathfrak{a}^*$ is given by the reflection s_λ in the hyperplane orthogonal to the restricted root λ.

If $\lambda(H) = 0$ for some $H \in \mathfrak{a}$ and all $\lambda \in \Delta_r$, then $[H, \mathfrak{g}_\lambda] = 0$ for all λ, so H lies in the center of $\mathfrak{g}$. Since $\mathfrak{g}$ is semisimple, this implies $H = 0$, and hence Δ_r spans $\mathfrak{a}^*$. For a restricted root $\lambda \in \Delta_r$, consider an element $E_\lambda \in \mathfrak{g}_\lambda$ and the corresponding subalgebra of $\mathfrak{g}$ as constructed above. This subalgebra is isomorphic to $\mathfrak{sl}(2, \mathbb{R})$ and acts on $\mathfrak{g}$, so we obtain an action of $\mathfrak{sl}(2, \mathbb{C})$ on the complexification of $\mathfrak{g}$. In this action, the element $\frac{-2}{|\lambda|^2}H_\lambda$ acts diagonalizably and all eigenvalues are integers. But by construction, E_μ is an eigenvector for this action with eigenvalue $2\frac{\langle\lambda,\mu\rangle}{|\lambda|^2}$, whence this is an integer. Finally, for restricted roots λ and μ, a nonzero vector $E_\mu \in \mathfrak{g}_\mu$, and an element $H \in \mathfrak{a}$, we get

$$[\mathrm{Ad}(k_\lambda)E_\mu, H] = \mathrm{Ad}(k_\lambda)([E_\mu, \mathrm{Ad}(k_\lambda^{-1})(H)]) = \mu(\mathrm{Ad}(k_\lambda^{-1})(H))\,\mathrm{Ad}(k_\lambda)E_\mu,$$

which shows that $\mathrm{Ad}^*(k_\lambda)(\mu) = s_\lambda(\mu)$ is a restricted root. This shows that Δ_r is an abstract root system and since the Weyl group is generated by the root reflections the second statement follows, too. □

REMARK 2.3.6. (1) The abstract root system Δ_r is not reduced in general; see Example (2) of 2.3.4.

(2) The proof shows that any element of $W(\Delta_r)$ can be realized by the action of an element of $N_K(\mathfrak{a}) = \{k \in K : \mathrm{Ad}(k)(\mathfrak{a}) \subset \mathfrak{a}\}$, the normalizer of $\mathfrak{a}$ in K. Defining the centralizer of $\mathfrak{a}$ in K as $Z_K(\mathfrak{a}) = \{k \in K : \mathrm{Ad}(k)|_{\mathfrak{a}} = \mathrm{id}_{\mathfrak{a}}\}$, one easily shows that $Z_K(\mathfrak{a}) \subset N_K(\mathfrak{a})$ is a normal subgroup. It turns out that the qoutient $N_K(\mathfrak{a})/Z_K(\mathfrak{a})$ (with its obvious action on $\mathfrak{a}$) is isomorphic to $W(\Delta_r)$. A proof of this fact can be found in [**Kn96**, VI,6.57].

THEOREM 2.3.6. *Let $\mathfrak{g}$ be a real semisimple Lie algebra and let $\mathfrak{g} = \mathfrak{k}\oplus\mathfrak{a}\oplus\mathfrak{n}$ and $\mathfrak{g} = \tilde{\mathfrak{k}}\oplus\tilde{\mathfrak{a}}\oplus\tilde{\mathfrak{n}}$ be two Iwasawa decompositions. Then there is an inner automorphism of $\mathfrak{g}$ which intertwines between the two decompositions.*

PROOF. By part (2) of Theorem 2.3.2, any two Cartan involutions are conjugate by an inner automorphism, so we may assume that $\tilde{\mathfrak{k}} = \mathfrak{k}$ and that $\mathfrak{p}$ remains fixed, too. Consider the restricted root decomposition $\mathfrak{g} = \mathfrak{g}_0 \oplus \bigoplus_\lambda \mathfrak{g}_\lambda$ with respect to the maximal abelian subspace $\mathfrak{a} \subset \mathfrak{p}$. Since $\mathfrak{g}_0$ is stable under the Cartan involution, we may further decompose it as $\mathfrak{g}_0 = \mathfrak{a} \oplus \mathfrak{m}$, where $\mathfrak{m} = Z_{\mathfrak{k}}(\mathfrak{a})$. Since Δ_r is finite, we can choose an element $H \in \mathfrak{a}$ such that $\lambda(H) \neq 0$ for all $\lambda \in \Delta_r$. Now we claim that $\mathfrak{a} = Z_{\mathfrak{g}}(H) \cap \mathfrak{p}$.

By the Iwasawa decomposition any element $X \in \mathfrak{g}$ can be written as $X_0 + H_0 + \sum X_\lambda$ with $X_0 \in \mathfrak{k}$, $H_0 \in \mathfrak{a}$ and each X_λ in a positive restricted root space. Then $[H, X] = [H, X_0] + \sum_\lambda \lambda(H)X_\lambda$, the first summand lies in $\mathfrak{k}$, and the rest lies in $\mathfrak{n}$. Hence, we see that $[H, X] = 0$ implies $X = X_0 + H_0$ and if this, in addition, lies in $\mathfrak{p}$, then $X = H_0 \in \mathfrak{a}$, which proves the claim.

Now choose an element $\tilde{H} \in \tilde{\mathfrak{a}}$ on which all restricted roots with respect to $\tilde{\mathfrak{a}}$ are nonzero. By Theorem 2.3.3, the subgroup $K \subset \mathrm{Int}(\mathfrak{g})$ corresponding to $\mathfrak{k}$ is compact. Hence, we can choose an element $k_0 \in K$ such that the function $k \mapsto B(\mathrm{Ad}(k)\tilde{H}, H)$ assumes a local extremum at k_0. Then for any $X \in \mathfrak{k}$, the map $t \mapsto B(\mathrm{Ad}(\exp(tX)k_0)\tilde{H}, H)$ assumes a local extremum at $t = 0$. Writing $\mathrm{Ad}(\exp(tX)k_0) = \mathrm{Ad}(\exp tX)\,\mathrm{Ad}(k_0)$, and differentiating with respect to t at $t = 0$, we see that $0 = B([X, \mathrm{Ad}(k_0)\tilde{H}], H) = B(X, [\mathrm{Ad}(k_0)\tilde{H}, H])$ holds for all $X \in \mathfrak{k}$. Since $[\mathrm{Ad}(k_0)\tilde{H}, H] \in \mathfrak{k}$ and the Killing form is negative definite on $\mathfrak{k}$, we conclude that $[\mathrm{Ad}(k_0)\tilde{H}, H] = 0$, whence $\mathrm{Ad}(k_0)\tilde{H} \in Z_{\mathfrak{g}}(H) \cap \mathfrak{p} = \mathfrak{a}$. Since $\mathfrak{a}$ is abelian, we conclude that $\mathfrak{a} \subset Z_{\mathfrak{g}}(\mathrm{Ad}(k_0)\tilde{H}) \cap \mathfrak{p} = \mathrm{Ad}(k_0)(\tilde{\mathfrak{a}})$. Since both $\mathfrak{a}$ and $\tilde{\mathfrak{a}}$ are maximally abelian, we conclude that $\mathfrak{a} = \mathrm{Ad}(k_0)(\tilde{\mathfrak{a}})$.

Since $\mathrm{Ad}(k_0)$ preserves the subalgebra $\mathfrak{k}$, we may thus assume that $\tilde{\mathfrak{a}} = \mathfrak{a}$, and we are left with understanding the effect of the choice of positive restricted roots. By the lemma, we can apply the theory of abstract root systems as developed in

Section 2.2.7 to Δ_r. In particular, the choice of a positive subsystem is equivalent to the choice of a simple subsystem. Starting from one simple subsystems one obtains all simple subsystems by the action of the Weyl group $W = W(\Delta_r)$. Again, by the lemma, two such choices are related by an inner automorphism of $\mathfrak{g}$ that fixes $\mathfrak{k}$ and $\mathfrak{a}$. □

2.3.7. Cartan subalgebras. If $\mathfrak{g}$ is a real semisimple Lie algebra, then a *Cartan subalgebra* of $\mathfrak{g}$, is an abelian subalgebra $\mathfrak{h}$ such that the complexification $\mathfrak{h}_{\mathbb{C}}$ is a Cartan subalgebra of the complex semisimple Lie algebra $\mathfrak{g}_{\mathbb{C}}$. If, moreover, θ is a Cartan involution on $\mathfrak{g}$, then a Cartan subalgebra $\mathfrak{h}$ is called *θ–stable* if $\theta(\mathfrak{h}) \subset \mathfrak{h}$. If this is the case, then $\mathfrak{h} = (\mathfrak{h} \cap \mathfrak{k}) \oplus (\mathfrak{h} \cap \mathfrak{p})$, where $\mathfrak{g} = \mathfrak{k} \oplus \mathfrak{p}$ is the Cartan decomposition corresponding to θ. For a θ–stable Cartan subalgebra $\mathfrak{h}$, the dimension of $\mathfrak{h} \cap \mathfrak{k}$ is called the *compact dimension* of $\mathfrak{h}$, and the dimension of $\mathfrak{h} \cap \mathfrak{p}$ is called the *noncompact dimension* of $\mathfrak{h}$. A θ–stable Cartan subalgebra $\mathfrak{h} \leq \mathfrak{g}$ is called *maximally compact* (respectively *maximally noncompact*) if and only if its compact (respectively noncompact) dimension is maximal possible.

In contrast to the complex case, Cartan subalgebras in real semisimple Lie algebras are not unique up to conjugation. Indeed, since the compact and noncompact dimensions can be read off the signature of the restriction of the Killing form, θ–stable Cartan subalgebras of different compact dimensions cannot be conjugate. This already shows up for the simplest possible example $\mathfrak{g} = \mathfrak{sl}(2, \mathbb{R})$. Since the complexification $\mathfrak{g}_{\mathbb{C}} = \mathfrak{sl}(2, \mathbb{C})$ is of rank one, any Cartan subalgebra $\mathfrak{h} \leq \mathfrak{g}$ must be of real dimension one. Now consider the subspaces $\mathfrak{h}_1$ of tracefree real diagonal matrices and $\mathfrak{h}_2$ of all matrices of the form $\left(\begin{smallmatrix} 0 & t \\ -t & 0 \end{smallmatrix}\right)$ with $t \in \mathbb{R}$. The complexification of $\mathfrak{h}_1$ is just the space of tracefree complex diagonal matrices, i.e. the standard Cartan subalgebra of $\mathfrak{sl}(2, \mathbb{C})$. Obviously, $\mathfrak{h}_1$ is θ–stable for the standard Cartan involution $\theta X = -X^t$ and $\mathfrak{h}_1 \subset \mathfrak{p}$, so it is maximally noncompact. On the other hand, one easily verifies that the elements of $\mathfrak{h}_2$ act diagonalizably under the adjoint action, so this is a Cartan subalgebra, too. This Cartan subalgebra is obviously θ–stable and contained in $\mathfrak{k}$ and thus maximally compact, whence $\mathfrak{g}$ has two Cartan subalgebras which are not conjugate.

THEOREM 2.3.7. *Let $\mathfrak{g}$ be a real semisimple Lie algebra endowed with a Cartan involution θ, $\mathfrak{g} = \mathfrak{k} \oplus \mathfrak{p}$ the corresponding Cartan decomposition, G a connected Lie group with Lie algebra $\mathfrak{g}$, and $K \subset G$ the Lie subgroup with Lie algebra $\mathfrak{k}$ from Theorem 2.3.3.*

(1) $\mathfrak{g}$ has a Cartan subalgebra and any Cartan subalgebra of $\mathfrak{g}$ is conjugate to a θ–stable Cartan subalgebra by an element of $\operatorname{Int}(\mathfrak{g})$.

(2) Any two maximally compact (maximally noncompact) θ–stable Cartan subalgebras of $\mathfrak{g}$ are conjugate by an element of K.

(3) Up to conjugation by elements of $\operatorname{Int}(\mathfrak{g})$, *there are only finitely many Cartan subalgebras of $\mathfrak{g}$.*

SKETCH OF PROOF. (1) To prove existence of a Cartan subalgebra, let $\mathfrak{a} \subset \mathfrak{p}$ be a maximal abelian subspace, consider $\mathfrak{m} = Z_{\mathfrak{k}}(\mathfrak{a})$, let $\mathfrak{t}$ be a maximal abelian subalgebra of $\mathfrak{m}$, and put $\mathfrak{h} = \mathfrak{t} \oplus \mathfrak{a}$. We have to show that $\mathfrak{h}_{\mathbb{C}}$ is maximally abelian and consists of semisimple elements. From 2.3.5 we know that in an appropriate basis of $\mathfrak{g}$ the adjoint action of elements of $\mathfrak{a}$ is diagonal, while elements of $\mathfrak{t}$ act by skew symmetric matrices. All of these actions are diagonalizable in the complexification, so we see that for any element of $\mathfrak{h}_{\mathbb{C}}$ the adjoint action is diagonalizable.

To see that $\mathfrak{h}_{\mathbb{C}}$ is maximally abelian in $\mathfrak{g}_{\mathbb{C}}$ it suffices to show that $\mathfrak{h}$ is maximally abelian in $\mathfrak{g}$, which is easy.

If $\mathfrak{h} \subset \mathfrak{g}$ is any Cartan subalgebra, let $\mathfrak{u} \subset \mathfrak{g}_{\mathbb{C}}$ be the compact real form constructed from the Cartan subalgebra $\mathfrak{h}_{\mathbb{C}} \leq \mathfrak{g}_{\mathbb{C}}$ as in 2.3.1. Let σ and τ be the conjugations of $\mathfrak{g}_{\mathbb{C}}$ corresponding to the real forms $\mathfrak{g}$ and $\mathfrak{u}$. By construction, both of these involutions map $\mathfrak{h}_{\mathbb{C}}$ to itself, and τ is a Cartan involution on $\mathfrak{g}_{\mathbb{C}}$. By Lemma 2.3.2, there is an element $\phi \in \operatorname{Int}(\mathfrak{g}_{\mathbb{C}})$ such that the Cartan involution $\tilde{\tau} = \phi\tau\phi^{-1}$ commutes with σ. From the proof of this lemma we further know that $\phi = (\sigma\tau\sigma\tau)^{1/4}$, which implies that $\phi(\mathfrak{h}_{\mathbb{C}}) \subset \mathfrak{h}_{\mathbb{C}}$. Hence, $\tilde{\tau}(\mathfrak{h}_{\mathbb{C}}) \subset \mathfrak{h}_{\mathbb{C}}$ and since $\tilde{\tau}$ commutes with σ, we have $\tilde{\tau}(\mathfrak{g}) \subset \mathfrak{g}$. From the fact that $\tilde{\tau}$ is a Cartan involution of $\mathfrak{g}_{\mathbb{C}}$ one easily concludes that $\tilde{\tau}$ restricts to a Cartan involution on $\mathfrak{g}$. This Cartan involution is conjugate to θ by an inner automorphism ψ and since $\mathfrak{h}$ is by construction $\tilde{\tau}$–stable, the Cartan subalgebra $\psi(\mathfrak{h})$ is θ–stable.

(2) Suppose that $\mathfrak{h}, \mathfrak{h}'$ are maximally noncompact θ–stable Cartan subalgebras in $\mathfrak{g}$. From the first part of the proof of (1) above, we conclude that $\mathfrak{h} \cap \mathfrak{p}$ and $\mathfrak{h}' \cap \mathfrak{p}$ both have to be maximally abelian subspaces of $\mathfrak{p}$. From 2.3.6 we know that any two such subspaces are conjugate by an element of K, so we may, without loss of generality, assume that $\mathfrak{h} \cap \mathfrak{p} = \mathfrak{h}' \cap \mathfrak{p}$. Writing $\mathfrak{a}$ for the latter space, we see that $\mathfrak{h} = \mathfrak{a} \oplus \mathfrak{t}$ and $\mathfrak{h}' = \mathfrak{a} \oplus \mathfrak{t}'$, where $\mathfrak{t}, \mathfrak{t}' \subset Z_{\mathfrak{k}}(\mathfrak{a})$ are maximally abelian. Thus, $\mathfrak{t}$ and $\mathfrak{t}'$ are maximally abelian subspaces in the Lie algebra of the compact group $Z_K(\mathfrak{a})$, whose conjugacy is a classical result (see [**Kn96**, IV, Theorem 4.34]). The maximally compact case is less relevant for our purposes and can be dealt with similarly.

(3) A similar argument as in the proof of (2) shows that a θ–stable Cartan subalgebra $\mathfrak{h} \leq \mathfrak{g}$ is determined up to conjugacy by the subspace $\mathfrak{h} \cap \mathfrak{p}$ (see [**Kn96**, VI, Lemma 6.62]). As above, we may assume that $\mathfrak{h} \cap \mathfrak{p} \subset \mathfrak{a}$ for a fixed maximal abelian subspace $\mathfrak{a} \subset \mathfrak{p}$. Finally, one proves that $\mathfrak{h} \cap \mathfrak{p}$ must be the intersection of the kernels of some restricted roots (see [**Kn96**, VI, Lemma 6.63]), which leaves only finitely many possibilities. □

2.3.8. Satake diagrams. Let $\mathfrak{g}$ be a noncompact real semisimple Lie algebra endowed with a Cartan involution θ and let $\mathfrak{g} = \mathfrak{k} \oplus \mathfrak{p}$ be the corresponding Cartan decomposition. From Theorem 2.3.7 we know that there is a θ–stable maximally noncompact Cartan subalgebra $\mathfrak{h} \leq \mathfrak{g}$, and that $\mathfrak{a} = \mathfrak{h} \cap \mathfrak{p}$ is a maximal abelian subspace of $\mathfrak{p}$. By definition, $\mathfrak{h}_{\mathbb{C}} \leq \mathfrak{g}_{\mathbb{C}}$ is a Cartan subalgebra, so we can consider the corresponding set $\Delta = \Delta(\mathfrak{g}_{\mathbb{C}}, \mathfrak{h}_{\mathbb{C}})$ of roots. On the other hand, we have the system of restricted roots Δ_r encoding the action of the elements of $\mathfrak{a} \subset \mathfrak{h} \subset \mathfrak{h}_{\mathbb{C}}$. Since both the roots and the restricted roots describe eigenvalues of the adjoint action, we see that the restricted roots are exactly the nonzero restrictions of roots to $\mathfrak{a} \subset \mathfrak{h}_{\mathbb{C}}$. Likewise, the restricted root spaces are given as

$$\mathfrak{g}_\lambda = \mathfrak{g} \cap \bigoplus_{\alpha : \alpha|_{\mathfrak{a}} = \lambda} (\mathfrak{g}_{\mathbb{C}})_\alpha .$$

Since elements of $\mathfrak{a}$, respectively, $\mathfrak{t}$ act by selfadjoint, respectively, skew symmetric maps, all roots are real on $i\mathfrak{t} \oplus \mathfrak{a}$. By definition $\mathfrak{g}$ is a split real form of $\mathfrak{g}_{\mathbb{C}}$ if and only if it contains a Cartan subalgebra on which all roots are real; see 2.3.1. Using Theorem 2.3.7 we see that this is the case if and only if $\mathfrak{m} := Z_{\mathfrak{k}}(\mathfrak{a}) = \{0\}$.

Let σ be the conjugation of $\mathfrak{g}_{\mathbb{C}}$ with respect to the real form $\mathfrak{g}$. For $\alpha \in \Delta$, we define $\sigma^*\alpha$ by $\sigma^*\alpha(H) := \overline{\alpha(\sigma H)}$ for all $H \in \mathfrak{h}_{\mathbb{C}}$. Note that since σ is conjugate linear, $\sigma^*\alpha$ is again complex linear, and identifying $L_{\mathbb{C}}(\mathfrak{h}_{\mathbb{C}}, \mathbb{C})$ with $L_{\mathbb{R}}(\mathfrak{h}, \mathbb{C})$, the map σ^* coincides with complex conjugation. Since σ is an automorphism of the real Lie algebra $\mathfrak{g}_{\mathbb{C}}$, one immediately concludes that for a root vector $E_\alpha \in (\mathfrak{g}_{\mathbb{C}})_\alpha$, the element σE_α is an eigenvector for the adjoint action of $\mathfrak{h}_{\mathbb{C}}$ with eigenvalue $\sigma^*\alpha$, so σ^* restricts to an involutive automorphism of the root system Δ. Now a root $\alpha \in \Delta$ is called *compact* if $\sigma^*\alpha = -\alpha$. The set of compact roots is denoted by $\Delta_c \subset \Delta$.

PROPOSITION 2.3.8. *The set of compact roots is given by* $\Delta_c = \{\alpha \in \Delta : \alpha|_{\mathfrak{a}} = 0\}$. *It is an abstract root system on the Euclidean subspace of* $i\mathfrak{t} \oplus \mathfrak{a}$ *spanned by its elements. For* $\alpha \in \Delta_c$, *the root space* $(\mathfrak{g}_{\mathbb{C}})_\alpha$ *is contained in* $\mathfrak{k}_{\mathbb{C}} \subset \mathfrak{g}_{\mathbb{C}}$.

PROOF. On $\mathfrak{h} \subset \mathfrak{h}_{\mathbb{C}}$, the map σ^* is given by conjugation, so we see that $\alpha \in \Delta_c$ if and only if $\alpha|_{\mathfrak{h}}$ has purely imaginary values. Since we know that α is real on $\mathfrak{a}$ and purely imaginary on $\mathfrak{t}$, we see that this is equivalent to $\alpha|_{\mathfrak{a}} = 0$. Consequently, if $\alpha, \beta \in \Delta_c$ such that $\alpha + \beta \in \Delta$ then $\alpha + \beta \in \Delta_c$. This easily implies that Δ_c is an abstract root system as claimed.

The Cartan involution θ on $\mathfrak{g}$ is the restriction of the conjugation τ on $\mathfrak{g}_{\mathbb{C}}$ with respect to the compact real form $\mathfrak{k} \oplus i\mathfrak{p}$. The intersection of $\mathfrak{h}_{\mathbb{C}}$ with $\mathfrak{k} \oplus i\mathfrak{p}$ is $\mathfrak{t} \oplus i\mathfrak{a}$, so all roots are purely imaginary on this intersection. Considering the involution $\tau^* : \Delta \to \Delta$ as for σ above, we obtain $\tau^*\alpha = -\alpha$ for all $\alpha \in \Delta$. Moreover, the involutions τ and σ commute, and $\tau\sigma = \sigma\tau$ is exactly the complex linear extension of θ. For a compact root $\alpha \in \Delta_c$, we now have $(\sigma\tau)^*\alpha = \tau^*(-\alpha) = \alpha$. Thus, the (complex one–dimensional) root space $(\mathfrak{g}_{\mathbb{C}})_\alpha$ is stable under the involution $\sigma\tau$, and thus either contained in $\mathfrak{k}_{\mathbb{C}}$ or in $\mathfrak{p}_{\mathbb{C}}$. Correspondingly, $\sigma\tau$ acts as id or as $-$id on this root space. For $0 \neq X \in (g_{\mathbb{C}})_\alpha$ we of course have $Y := X + \sigma X \in \mathfrak{g}$ and we must have $\theta(Y) = \pm Y$, so Y must lie either in $\mathfrak{k}$ or in $\mathfrak{p}$. But since $\alpha|_{\mathfrak{a}} = 0$, Y commutes with any element from $\mathfrak{a}$. If $Y \in \mathfrak{p}$, then $Y \in \mathfrak{a}$ since $\mathfrak{a}$ is maximally abelian. Since this would lead to the contradiction $X \in \mathfrak{a}_{\mathbb{C}} \subset \mathfrak{h}_{\mathbb{C}}$, we conclude that $Y \in \mathfrak{k}$. □

Next, we need a choice of positive roots adapted to our situation. The condition to require is that for $\alpha \in \Delta^+ \setminus \Delta_c$, we have $\sigma^*\alpha \in \Delta^+$. Such an ordering is actually easy to find: Consider the subspace $i\mathfrak{t} \oplus \mathfrak{a} \subset \mathfrak{h}_{\mathbb{C}}$ on which all roots are real. Choose a basis $\{H_1, \dots, H_p\}$ of $\mathfrak{a}$, extend it to a basis $\{H_1, \dots, H_\ell\}$ of $i\mathfrak{t} \oplus \mathfrak{a}$, and consider the lexicographic ordering with respect to this basis as in 2.2.5. By definition, $\alpha \in \Delta^+$ if and only if there is an index j such that $\alpha(H_i) = 0$ for all $i < j$ and $\alpha(H_j) > 0$. By definition $\alpha \in \Delta^+ \setminus \Delta_c$ implies that $\alpha(H_i) \neq 0$ for some $i \leq p$. Moreover, since all $\alpha(H_j)$ are real, $\sigma H_i = H_i$ for $i \leq p$ and $\sigma(H_i) = -H_i$ for $i > p$, we conclude that the required condition is satisfied.

Having chosen such an admissible ordering, consider the system Δ^0 of simple roots and the intersection $\Delta^0_c := \Delta^0 \cap \Delta_c$. Obviously, an integral linear combination of simple roots with all coefficients having the same sign lies in Δ_c if and only if all the simple roots lie in Δ_c, which implies that Δ^0_c is a simple system for Δ_c. Then we order our simple system $\Delta^0 = \{\alpha_1, \dots, \alpha_\ell\}$ in such a way, that the elements of Δ^0_c come last. To proceed further, we need the following lemma due to I. Satake; see [**Sa60**].

LEMMA 2.3.8. *(1) For any* $\alpha \in \Delta$, *the element* $\sigma^*\alpha - \alpha$ *is not a root.*

(2) For $\alpha \in \Delta^0 \setminus \Delta^0_c$, *there is a unique element* $\alpha' \in \Delta^0 \setminus \Delta^0_c$ *such that* $\sigma^*\alpha - \alpha'$ *is a linear combination of compact roots.*

PROOF. (1) For $\alpha \in \Delta_c$, we have $\sigma^*\alpha - \alpha = -2\alpha$, which is not a root since Δ is a reduced root system. For $\alpha \in \Delta \setminus \Delta_c$, let us assume that $\sigma^*\alpha - \alpha \in \Delta$, and thus $\sigma^*\alpha - \alpha \in \Delta_c$. Then from above we know that the root space $(\mathfrak{g}_{\mathbb{C}})_{\sigma^*\alpha-\alpha}$ is contained in $\mathfrak{k}_{\mathbb{C}}$, so the involution $\sigma\tau$ acts as the identity on that root space. On the other hand, if X is a nonzero element of $(\mathfrak{g}_{\mathbb{C}})_\alpha$, then τX and σX are nonzero elements in $(\mathfrak{g}_{\mathbb{C}})_{-\alpha}$ and $(\mathfrak{g}_{\mathbb{C}})_{\sigma^*\alpha}$, respectively. By 2.2.4 the element $[\tau X, \sigma X] \in (\mathfrak{g}_{\mathbb{C}})_{\sigma^*\alpha-\alpha}$ is nonzero. But since σ and τ commute, we get

$$\sigma\tau[\tau X, \sigma X] = \sigma[X, \tau\sigma X] = [\sigma X, \tau X] = -[\tau X, \sigma X],$$

a contradiction to $\sigma\tau$ acting as the identity on $(\mathfrak{g}_{\mathbb{C}})_{\sigma^*\alpha-\alpha}$.

(2) Suppose that we have n simple roots, ℓ of which are not compact. Order them as $\alpha_1, \dots, \alpha_n$ such that the compact simple roots come last. Then for $i, j = 1, \dots, \ell$ we find unique nonnegative integers a_{ij} such that $\sigma^*\alpha_i - \sum_j a_{ij}\alpha_j$ is a linear combination of compact roots for all i. Since $\sigma^*(\Delta_c) \subset \Delta_c$, we can apply σ^* to this expression to conclude that $\alpha_i - \sum_j a_{ij}\sigma^*\alpha_j$ is a linear combination of compact roots. Uniqueness of the coefficients implies that the $\ell \times \ell$–matrix (a_{ij}) must coincide with its inverse. Hence, $\sum_j a_{ij}a_{ji} = 1$ for all i and since the coefficients are nonnegative integers there is a single j such that $a_{ij} = a_{ji} = 1$, while all other a_{ik} are zero. Hence, the result follows. □

DEFINITION 2.3.8. The *Satake diagram* of the real Lie algebra $\mathfrak{g}$ is defined as follows: In the Dynkin diagram associated to the simple system Δ^0, represent compact simple roots by a black dot $\bullet$ and roots in $\Delta^0 \setminus \Delta^0_c$ by a white dot $\circ$. Moreover, for any $\alpha \in \Delta^0 \setminus \Delta^0_c$ such that $\sigma^*\alpha \neq \alpha$, connect α by an arrow to the unique simple root $\alpha' \in \Delta^0 \setminus \Delta^0_c$ such that $\sigma^*\alpha - \alpha'$ is a linear combination of compact roots. A consistent extension of this notation is to define the Satake diagram of a compact semisimple Lie algebra to be the Dynkin diagram of the complexification $\mathfrak{g}_{\mathbb{C}}$ with all dots black.

2.3.9. Examples of Satake diagrams. (1) Let $\mathfrak{g}$ be the underlying real Lie algebra of a complex semisimple Lie algebra, $\mathfrak{h} \leq \mathfrak{g}$ a Cartan subalgebra, Δ the corresponding set of roots, Δ^+ a choice of positive roots, and $\Delta^0 = \{\alpha_1, \dots, \alpha_n\}$ the corresponding set of simple roots. Constructing a compact real form $\mathfrak{u}$ of $\mathfrak{g}$ from $\mathfrak{h}$ as in 2.3.1, the conjugation of $\mathfrak{g}$ with respect to $\mathfrak{u}$ is a Cartan involution, and by construction $\theta(\mathfrak{h}) = \mathfrak{h}$. Since any two Cartan subalgebras of $\mathfrak{g}$ are conjugate, the Cartan subalgebra $\mathfrak{h}$ is automatically maximally noncompact, with the subspace $\mathfrak{a}$ given as the subspace $\mathfrak{h}_0$ on which all roots are real. From 2.1.4 we know that the complexification $\mathfrak{g}_{\mathbb{C}}$ is isomorphic to $\mathfrak{g} \oplus \mathfrak{g}$ with the conjugation given by swapping the two factors. Now $\mathfrak{h}_{\mathbb{C}}$ is $\mathfrak{h} \oplus \mathfrak{h}$ and consequently the set $\Delta(\mathfrak{g}_{\mathbb{C}}, \mathfrak{h}_{\mathbb{C}})$ of roots for $\mathfrak{g}_{\mathbb{C}}$ with respect to $\mathfrak{h}_{\mathbb{C}}$ is simply given by $\{(\alpha, 0), (0, \alpha) : \alpha \in \Delta\}$. For the obvious choice of positive roots, the simple roots are $(\alpha_i, 0)$ and $(0, \alpha_i)$ for $\alpha_i \in \Delta^0$. Since no root vanishes on $\mathfrak{h}_0$, there are no compact roots in this case. Moreover, we obviously have $\sigma^*(\alpha_i, 0) = (0, \alpha_i)$. Hence, the Satake diagram for $\mathfrak{g}$ is given by taking two copies of the Dynkin diagram of $\mathfrak{g}$ and connecting each node in one copy to its counterpart in the other copy by an arrow.

(2) Suppose that $\mathfrak{g}$ is a split real form. By definition, this means that there is a θ–stable Cartan subalgebra $\mathfrak{h} \subset \mathfrak{g}$, such that all roots of the complexification are real on $\mathfrak{h}$. In particular, this implies that in the Satake diagram there are no black dots. Moreover, on a real root, the involution σ^* clearly acts as the identity, so we

see that the Satake diagram of a split real form simply coincides with the Dynkin diagram of the complexification.

(3) Consider the algebra $\mathfrak{g} := \mathfrak{su}(p,q)$ for $p > q$, using the presentation and the Cartan involution of this Lie algebra from Example (2) of 2.3.4. Thus, we view $\mathfrak{g}$ as all tracefree block matrices with blocks of size q, q and $p-q$ of the form

$$\begin{pmatrix} A & B & C \\ D & -A^* & E \\ -E^* & -C^* & F \end{pmatrix},$$

with $A \in \mathfrak{gl}(q,\mathbb{C})$, C and E arbitrary complex $q \times (p-q)$–matrices, B and D in $\mathfrak{u}(q)$ and $F \in \mathfrak{u}(p-q)$. This describes $\mathfrak{g}$ as a subalgebra of its complexification $\mathfrak{sl}(p+q,\mathbb{C})$, and the Cartan decomposition $\mathfrak{g} = \mathfrak{k} \oplus \mathfrak{p}$ is given by the splitting into skew–Hermitian and Hermitian part. We have also noticed already that the subset $\mathfrak{a}$ of real diagonal matrices lying in $\mathfrak{g}$ is a maximal abelian subspace of $\mathfrak{p}$. To complete this to a maximally noncompact Cartan subalgebra, we have to choose a maximal abelian subspace $\mathfrak{t}$ of $Z_{\mathfrak{k}}(\mathfrak{a})$. The subspace of purely imaginary diagonal matrices lying in $\mathfrak{g}$ is a commutative subspace contained in this centralizer. Since this has dimension $q + (p-q-1) = p-1$, we see that it is a good choice for $\mathfrak{t}$. Thus, $\mathfrak{h} = \mathfrak{t} \oplus \mathfrak{a}$ is the subspace of all diagonal matrices contained in $\mathfrak{g}$, with the splitting given by imaginary and real part. The complexification $\mathfrak{h}_{\mathbb{C}} \subset \mathfrak{sl}(p+q,\mathbb{C})$ is the standard Cartan subalgebra of all tracefree diagonal matrices. Hence, we know from 2.2.6(1) that the roots are all expressions of the form $e_i - e_j$ for $i \neq j$, where e_i is the functional extracting the ith entry of a diagonal matrix. Obviously, the compact roots are exactly the roots $e_i - e_j$ with $i, j > 2q$. Unfortunately, the standard ordering of roots is not appropriate for our purposes, so we have to consider a different ordering instead.

To get such an ordering, we have to choose a basis of the space of real diagonal matrices, which starts with a basis of $\mathfrak{a}$. Denoting by $E_{i,j}$ the elementary matrix, we define $H_i := E_{i,i} - E_{q+i,q+i}$ for $i = 1,\dots,q$ and $H_i := E_{i,i} - E_{p+q,p+q}$ for $i = 2q+1,\dots,p+q-1$. With respect to the resulting ordering, the positive roots are $e_i - e_j$ for $i \leq q$ and $i < j$ and for $2q < i < j$ and $-e_i + e_j$ for $q < i \leq 2q$ and $i < j$. A moment of thought shows that the resulting simple roots (ordered in such a way to get the usual Dynkin diagram for $\mathfrak{sl}(p+q,\mathbb{C})$) are $e_1 - e_2$, $\dots$, $e_{q-1} - e_q$, $e_q - e_{2q+1}$, $e_{2q+1} - e_{2q+2}$, $\dots$, $e_{p+q-1} - e_{p+q}$, $-e_{2q} + e_{p+q}$, $-e_{2q-1} + e_{2q}$, $\dots$, $-e_{q+1} + e_{q+2}$. Hence, in the Satake diagram we have q white dots, followed by $p-q-1$ black dots and after that again q white dots, and it remains to determine the permutation of the white dots induced by σ^*. Keeping in mind that for $i = 1,\dots,q$ the $(q+i)$th entry of a diagonal matrix lying in $\mathfrak{h}$ is minus the conjugate of the ith entry and that $\sigma^*\alpha|_{\mathfrak{h}} = \overline{\alpha}|_{\mathfrak{h}}$, we see that $\sigma^*(e_i - e_{i+1}) = -e_{q+i} + e_{q+i+1}$ for $i < q$. Similarly, $\sigma^*(e_q - e_{2q+1}) = -e_{2q} + e_{2q+1}$. The difference of this and $-e_{2q} + e_{p+q}$ is the compact root $e_{2q+1} - e_{p+q}$, so we see that our permutation exchanges the ith and the $(p+q-i)$th node for all $i = 1,\dots,q$. To draw the Satake diagram it is better to fold it, and the result is

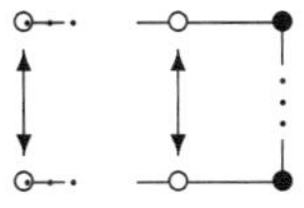

with q white roots in each of the two rows and $p-q-1$ black roots in the middle. Note that in the case of $\mathfrak{su}(p+1,p)$ we get a Satake diagram with only (an even number of) white dots but connected by arrows describing the unique nontrivial automorphism of the underlying Dynkin diagram. Hence, the resulting Satake diagram differs (by these arrows) from the Satake diagram of the split real from $\mathfrak{sl}(2p+1,\mathbb{R})$.

(4) The case $\mathfrak{g} = \mathfrak{su}(p,p)$ for $p \geq 2$ can be dealt with similarly. We use the presentation, the Cartan decomposition and the maximal abelian subspace $\mathfrak{a} \subset \mathfrak{p}$ as in example (2) of 2.3.4. So we view $\mathfrak{g}$ as the algebra of all block matrices of the form $\begin{pmatrix} A & B \\ D & -A^* \end{pmatrix}$, with $A \in \mathfrak{gl}(p,\mathbb{C})$ having real trace, and $B, D \in \mathfrak{u}(p)$. The Cartan decomposition is the decomposition into skew–Hermitian and Hermitian part, while the maximally abelian subspace $\mathfrak{a}$ is given by the space of real diagonal matrices contained in $\mathfrak{g}$. As a Cartan subalgebra $\mathfrak{h} \leq \mathfrak{g}$ we may use the space of all diagonal matrices contained in $\mathfrak{g}$ (which has the "correct" dimension $2p-1$ since the imaginary part of the trace of A has to vanish). As above, this leads to the usual Cartan subalgebra $\mathfrak{h}_{\mathbb{C}}$ in the complexification $\mathfrak{g}_{\mathbb{C}} = \mathfrak{sl}(2p,\mathbb{C})$, so we get the roots $e_i - e_j$ for $i \neq j$. Denoting the first p diagonal entries of an element of $\mathfrak{a}$ by $a_1, \dots, a_p$, any of the roots produces an expression of one of the forms $a_i - a_j$ for $i \neq j$, or $\pm(a_i + a_j)$ for arbitrary i, j, which implies that there are no compact roots in this case.

To get an adapted ordering, we proceed as in (3) above, which leads to the positive roots $e_i - e_j$ for $i \leq p$ and $i < j$, and $-e_i + e_j$ for $p < i < j$. The resulting simple roots (ordered in such a way that one gets the usual Dynkin diagram for the complexification) are $e_1 - e_2, \dots, e_{p-1} - e_p, e_p - e_{2p}, e_{2p} - e_{2p-1}, \dots, e_{p+2} - e_{p+1}$. To analyze the action of σ^* on the simple roots, we have to recall that σ^* amounts to conjugation on $\mathfrak{h} \subset \mathfrak{h}^{\mathbb{C}}$. Taking into account the form of $\mathfrak{h}$, one immediately verifies that for $i < p$ we get $\sigma^*(e_i - e_{i+1}) = e_{p+i+1} - e_{p+i}$, while $e_p - e_{2p}$ is fixed by σ^*. Thus, we see that σ^* is the unique nontrivial automorphism of the Dynkin diagram A_{2p-1}. To draw the Satake diagram, it is again better to fold it, and we obtain

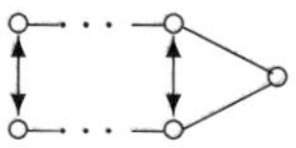

(5) As we shall see from the classification of real semisimple Lie algebras in 2.3.11 below, the only real form of a complex Lie algebra of type A_ℓ that we have not yet considered is $\mathfrak{g} = \mathfrak{sl}(n,\mathbb{H})$ for $n \geq 2$. From 2.1.7 we know that we may identify $\mathfrak{g}$ with the Lie algebra $\mathfrak{u}^*(2n)$ of complex $2n \times 2n$–matrices of the block form $\begin{pmatrix} A & B \\ -\bar{B} & \bar{A} \end{pmatrix}$ such that $A, B \in \mathfrak{gl}(n,\mathbb{C})$ and the real part of the trace of A vanishes. This subalgebra of $\mathfrak{sl}(2n,\mathbb{C})$ is closed under conjugate transpose, so we may use the splitting into skew–Hermitian and Hermitian part as the Cartan decomposition. In particular, the $\mathfrak{p}$–component is given by those matrices, for which A is Hermitian and B is skew symmetric. Denoting elements of $\mathfrak{g}$ as pairs (A, B) one easily verifies that for an element of the form $(A', 0)$ with A' real (and thus tracefree), one gets $[(A',0),(A,B)] = ([A',A],[A',B])$. Since $\mathfrak{p}$ has skew symmetric matrices in the B–block, this immediately implies that the space $\mathfrak{a}$ of real diagonal matrices contained in $\mathfrak{g}$ is a maximal abelian subspace of $\mathfrak{p}$ (of dimension $n-1$). From this description, one easily reads off that the restricted root system is of type A_{n-1} with each restricted root space of real dimension four.

We also conclude immediately that the space $\mathfrak{h}$ of all diagonal matrices contained in $\mathfrak{g}$ is a maximally noncompact Cartan subalgebra, whose complexification $\mathfrak{h}_{\mathbb{C}}$ is the standard Cartan subalgebra of $\mathfrak{g}_{\mathbb{C}} = \mathfrak{sl}(2n,\mathbb{C})$. Thus, we again obtain the roots $e_i - e_j$ for $i \neq j$. To get an appropriate subset of positive roots, we need a basis of $\mathfrak{h}$ starting with a basis of $\mathfrak{a}$. In terms of elementary matrices, we define a basis of $\mathfrak{a}$ by $H_i := E_{i,i} - E_{n,n} + E_{n+i,n+i} - E_{2n,2n}$ for $i = 1, \ldots, n-1$. Independent of the extension of this to a basis of $\mathfrak{h}$ we see that the roots $e_i - e_j$ are positive if $1 \leq i < j \leq n$, or $n + 1 \leq i < j \leq 2n$, or $1 \leq i \leq n$ and $n + i < j \leq 2n$. Also, the roots $-e_i + e_j$ are positive for $1 \leq i \leq n$ and $n + 1 \leq j < n + i$ for any extension. Thus, the only roots for which we have to decide about positivity are the roots $\pm(e_i - e_{n+i})$ for $i = 1, \ldots, n$. It turns out that the more convenient choice is to complete the basis by $H_i := -E_{i-n+1, i-n+1} + E_{i+1, i+1}$ for $i = n, \ldots, 2n-1$. This forces $-e_i + e_{n+i}$ for $i = 1, \ldots, n$ to be the remaining positive roots. From the construction it is clear that these last n roots are exactly the compact positive roots.

Now $-e_1 + e_{n+1}$ is the only positive root involving $-e_1$, so it has to be simple. Apart from $-e_2 + e_{n+2}$ the only other positive root involving $-e_2$ is $e_1 - e_2$, so one of these two roots must be simple. But $e_1 - e_2 = e_1 - e_{n+2} + (-e_2 + e_{n+2})$, so we see that $-e_2 + e_{n+2}$ must be simple. Similarly, one sees that $-e_i + e_{n+i}$ is simple for all $i = 1, \ldots, n$, so we have n compact simple roots. Next, $e_1 - e_{n+2}$ and $e_{n+1} - e_{n+2}$ are the only positive roots involving $-e_{n+2}$, and one easily sees that $e_1 - e_{n+2}$ is simple. Similarly, $e_i - e_{n+i+1}$ is simple for $i = 2, \ldots, p - 2$, which means that we have found all simple roots. Ordered in such a way that one gets the usual Dynkin diagram for the complexification, they are $-e_1 + e_{n+1}, e_1 - e_{n+2}, -e_2 + e_{n+2}, e_2 - e_{n+3}, \ldots, e_{n-1} - e_{2n}, -e_n + e_{2n}$, so the Satake diagram has black and white dots alternating, starting and ending with a black dot. Finally, one immediately verifies that

$$\sigma^*(e_i - e_{n+i+1}) = -e_{i+1} + e_{n+i} = e_i - e_{n+i+1} + (-e_i + e_{n+i}) + (-e_{i+1} + e_{n+i+1}).$$

Since the last two roots are compact, the involutive permutation induced by σ^* is the identity and we obtain the Satake diagram ●—○—●—⋯—●—○—●.

2.3.10. σ–systems of roots. Since we will not need details in that direction, we only outline the classification of real simple Lie algebras. We follow the paper [**Ar62**], which describes a classification directly related to Satake diagrams. The classification consists essentially of two parts: On one hand, one has to show that two real semisimple Lie algebras are isomorphic if and only if their Satake diagrams are isomorphic (in the obvious sense). The other part is to describe the possible Satake diagrams. As in the complex case, one also needs the abstract version of the appropriate notion of a root system.

This abstract concept is a so-called σ-system of roots. By definition, this is given by an abstract root system Δ on a Euclidean vector space V (see 2.2.4 and 2.2.5) together with an involutive isometry σ of V such that $\sigma(\Delta) \subset \Delta$, i.e. σ restricts to an involution of Δ. If Δ and $\tilde{\Delta}$ are σ-systems on V, respectively, $\tilde{V}$ with involutions σ respectively $\tilde{\sigma}$, then an isomorphism of the σ-systems is a linear isometry $\phi : V \to \tilde{V}$, which restricts to a bijection $\Delta \to \tilde{\Delta}$ and has the property that $\tilde{\sigma} \circ \phi = \phi \circ \sigma$. A σ-system of roots is called *normal* if for any root $\alpha \in \Delta$, the element $\sigma(\alpha) - \alpha$ is not a root.

Now let $\mathfrak{g}$ be a real semisimple Lie algebra with complexification $\mathfrak{g}_\mathbb{C}$ endowed with a Cartan involution θ and a maximally noncompact θ-stable Cartan subalgebra $\mathfrak{h} \subset \mathfrak{g}$. Then we have the root system $\Delta = \Delta(\mathfrak{g}_\mathbb{C}, \mathfrak{h}_\mathbb{C})$, which by Theorem 2.2.4 is an abstract root system on the real span V of the elements of Δ. From

2.3.8 we know that the conjugation on $\mathfrak{g}_{\mathbb{C}}$ with respect to the real form $\mathfrak{g}$ preserves Δ, which implies that it restricts to an involution of the space $(\mathfrak{h}_{\mathbb{C}})^0$ on which all roots are real. This involution is an isometry since the conjugation is an automorphism of the real Lie algebra $\mathfrak{g}_{\mathbb{C}}$, so it dualizes to an isometry σ of V which preserves Δ. (In the notation of 2.3.8 this isometry was denoted by σ^* rather than σ.) Thus, a real semisimple Lie algebra gives rise to a σ–system of roots and by Lemma 2.3.8(1) these σ–systems are always normal. To obtain this σ–system, we have made two choices, namely the Cartan involution θ and the maximally noncompact θ–stable Cartan subalgebra $\mathfrak{h}$. But by Theorems 2.3.2 and 2.3.7 Cartan involutions and maximally noncompact Cartan subalgebras are unique up to conjugation, which immediately implies that the σ–system of roots associated to $\mathfrak{g}$ is uniquely determined up to isomorphism.

The next step is the analog of the passage from a root system to a simple subsystem. To do this on the level of (abstract) normal σ–systems of roots, one first has to prove some properties of normal σ–systems that we have already verified in the case of σ–systems coming from real semisimple Lie algebras. Since σ is an involutive isometry it is selfadjoint, so the space V splits into a direct sum $V = V_+ \oplus V_-$ of ± 1–eigenspaces with respect to σ. Now one defines $\Delta_c := \Delta \cap V_- \subset \Delta$ and observes that this is an abstract root system on the subspace of V_- spanned by Δ_c. On the other hand, we get a projection $V \to V_+$ and we denote by $\Delta_r \subset V_+$ the set of nonzero elements in the image of Δ under this projection. (This is just the analog of passing from roots to restricted roots.) Now one can prove that if the original σ–system is normal, then Δ_r is an abstract root system. Moreover, one can analyze the relation between Δ and Δ_r quite precisely on the abstract level, which is an important ingredient in the description of all possible Satake diagrams.

Knowing these facts, we can now proceed as in 2.3.8 to call a subset $\Delta^+ \subset \Delta$ of positive roots admissible if and only if for any $\alpha \in \Delta^+ \setminus \Delta_c$ we have $\sigma(\alpha) \in \Delta^+$. Such orderings can, for example, be obtained as the lexicographic ordering with respect to a basis of V^*, whose first elements form a basis of V_+^*. Having chosen such an order, we can next pass to the corresponding simple subsystem and use part (2) of Lemma 2.3.8, whose proof works in the abstract setting without changes. We obtain an involutive permutation of the noncompact simple roots, and thus associate to any normal σ–system of roots a Satake diagram. To see that a normal σ–system of roots is completely determined by it's Satake diagram, one has to analyze the effect of the choice of the admissible positive system, which (as in the complex case) is done using the Weyl group:

In the Weyl group W of Δ one has two obvious subgroups, namely the Weyl group W_c of Δ_c, and the subgroup W_σ of elements which commute with σ. By construction, W_c is a subgroup of W_σ and it turns out that it is automatically a normal subgroup. On the other hand, since any element of W_σ preserves the decomposition $V = V_+ \oplus V_-$, we get a homomorphism from W_σ to the Weyl group W_r of the restricted root system Δ_r. It turns out (see [**Sa60**, Appendix, Lemmas 1 and 2]) that this homomorphism is surjective with kernel W_c, so $W_r \cong W_\sigma / W_c$. Moreover, by [**Sa60**, Appendix, Proposition A] the group W_σ acts transitively on the set of simple systems obtained from admissible positive systems. From this, one immediately concludes that two normal σ–systems are isomorphic if and only if their Satake diagrams are isomorphic.

Assuming that two real semisimple Lie algebras $\mathfrak{g}$ and $\tilde{\mathfrak{g}}$ have isomorphic Satake diagrams, the isomorphism of the underlying Dynkin diagrams is induced by an isomorphism of the complexifications $\mathfrak{g}_{\mathbb{C}}$ and $\tilde{\mathfrak{g}}_{\mathbb{C}}$, respecting the Cartan subalgebras. Hence, we may assume that $\mathfrak{g}$ and $\tilde{\mathfrak{g}}$ are real forms contained in the same complex Lie algebra $\mathfrak{g}_{\mathbb{C}}$ such that the Cartan subalgebras correspond to the same complex Cartan subalgebra and such that the Cartan involutions come from the same compact real form of $\mathfrak{g}_{\mathbb{C}}$. In this situation the analysis of the restricted root systems allows one to show that $\mathfrak{g}$ and $\tilde{\mathfrak{g}}$ are conjugate by an inner automorphism of $\mathfrak{g}_{\mathbb{C}}$; see [**Ar62**, Theorem 2.14].

2.3.11. The classification of real simple Lie algebras. Knowing that two real semisimple Lie algebras are isomorphic if and only if their Satake diagrams are isomorphic, it remains to describe which Satake diagrams arise from real simple Lie algebras. Recall that by Cartan's criterion $\mathfrak{g}_{\mathbb{C}}$ is semisimple if and only if $\mathfrak{g}$ is semisimple. Regarding simplicity, we have

LEMMA 2.3.11. *(1) The underlying real Lie algebra $\mathfrak{g}_{\mathbb{R}}$ of a complex simple Lie algebra $\mathfrak{g}$ is simple.*

(2) Let $\mathfrak{g}$ be a real simple Lie algebra with complexification $\mathfrak{g}_{\mathbb{C}}$. Then $\mathfrak{g}_{\mathbb{C}}$ is simple unless $\mathfrak{g}$ is the underlying real Lie algebra of a complex simple Lie algebra, in which case $\mathfrak{g}_{\mathbb{C}} = \mathfrak{g} \oplus \mathfrak{g}$.

PROOF. (1) Suppose that $\mathfrak{a} \subset \mathfrak{g}_{\mathbb{R}}$ is a real ideal. Then $\mathfrak{a}$ is semisimple, so $\mathfrak{a} = [\mathfrak{a}, \mathfrak{a}]$ and thus $\mathfrak{a} = [\mathfrak{a}, \mathfrak{g}_{\mathbb{R}}]$. Hence, $X \in \mathfrak{a}$ may be written as $\sum [X_j, Y_j]$ for certain elements $X_j \in \mathfrak{a}$ and $Y_j \in \mathfrak{g}_{\mathbb{R}}$. Complex bilinearity of the bracket then implies that $\sum [X_j, iY_j] = iX$, but this element lies in $[\mathfrak{a}, \mathfrak{g}_{\mathbb{R}}] = \mathfrak{a}$. Thus, $\mathfrak{a}$ is a complex ideal, so $\mathfrak{a} = \{0\}$ or $\mathfrak{a} = \mathfrak{g}$ by simplicity of $\mathfrak{g}$.

(2) We have seen in 2.1.4 that the complexification of a complex Lie algebra $\mathfrak{g}$ is isomorphic to $\mathfrak{g} \oplus \bar{\mathfrak{g}}$ which implies that it is never simple. On the other hand, a Cartan involution on $\mathfrak{g}$ (i.e. the conjugation with respect to a compact real form) may be viewed as an isomorphism $\bar{\mathfrak{g}} \to \mathfrak{g}$, which completes the discussion of the complex case.

Conversely, assume that $\mathfrak{g}$ is real and $\mathfrak{a}$ is a nontrivial (complex) ideal in $\mathfrak{g}_{\mathbb{C}}$. Denoting by σ the conjugation of $\mathfrak{g}_{\mathbb{C}}$ with respect to the real form $\mathfrak{g}$, the intersection $\mathfrak{a} \cap \sigma(\mathfrak{a})$ and the sum $\mathfrak{a} + \sigma(\mathfrak{a})$ are ideals in $\mathfrak{g}_{\mathbb{C}}$, which are both stable under σ and thus complexifications of ideals in $\mathfrak{g}$. Since $\mathfrak{a}$ is nontrivial, the only possibility is that $\mathfrak{a} \cap \sigma(\mathfrak{a}) = \{0\}$ and $\mathfrak{a} + \sigma(\mathfrak{a}) = \mathfrak{g}_{\mathbb{C}}$ in the second case. But this implies that $\mathfrak{g}_{\mathbb{C}} = \mathfrak{a} \oplus \sigma(\mathfrak{a})$. By construction $\mathfrak{a}$ is a complex Lie algebra and $X \mapsto X + \sigma(X)$ defines an isomorphism $\mathfrak{a} \to \mathfrak{g}$. □

From this we see that apart from the compact real forms of complex simple Lie algebras we have another obvious class of real simple Lie algebras namely the underlying real Lie algebras of complex simple Lie algebras. Eliminating these two trivial classes, it remains to discuss noncompact real forms of a given complex simple Lie algebra. Assume that $\mathfrak{g}$ is a complex simple Lie algebra endowed with a Cartan subalgebra $\mathfrak{h} \subset \mathfrak{g}$ and a compact real form $\mathfrak{u} \subset \mathfrak{g}$ such that the involution τ corresponding to $\mathfrak{u}$ leaves $\mathfrak{h}$ invariant. For example, one may take the compact real form obtained from $\mathfrak{h}$ as in 2.3.1. Since by part (2) of Theorem 2.2.2 and Corollary 2.3.2 any two Cartan subalgebras and any two compact real forms are conjugate, we conclude that any real form of $\mathfrak{g}$ is conjugate to a real form $\mathfrak{g}^{\sigma} \subset \mathfrak{g}$ such that the corresponding involution σ commutes with τ and preserves $\mathfrak{h}$.

Now considering the root system $\Delta = \Delta(\mathfrak{g}, \mathfrak{h})$ and the fixed involution τ, we have to study isometric involutions σ^* of the real span V of Δ, which make (Δ, σ^*) into a σ–system of roots, and are such that the dual map $\sigma : \mathfrak{h}_0 \to \mathfrak{h}_0$ (the subspace on which all roots are real) commutes with the restriction of τ. Then the σ–system (Δ, σ^*) of roots is called *normally extendible* if and only if σ extends to a conjugate linear involutive automorphism of the real Lie algebra $\mathfrak{g}$, such that the Cartan subalgebra $\mathfrak{h} \cap \mathfrak{g}^\sigma$ is maximally noncompact. Having given such an extension, $\mathfrak{g}^\sigma$ is a real form of $\mathfrak{g}$, such that the associated σ–system of roots is (Δ, σ^*). Hence, it suffices to classify normally extendible σ–systems (Δ, σ^*) starting from a reduced root system Δ with connected Dynkin diagram.

The essential step in [**Ar62**] is that the problem of classifying normally extendible σ–systems can be reduced to the case of restricted rank 1, where the restricted rank of a σ–system Δ is defined as the rank of the associated restricted root system Δ_r. In terms of Lie algebras, this means that one may reduce to the situation where the maximal abelian subspace $\mathfrak{a} \subset \mathfrak{p}$ has dimension one. Essentially, this is done as follows: Consider the projection $\Delta \to \Delta_r$ onto the restricted root system. For a fixed restricted root $\lambda \in \Delta_r$, one denotes by $\Delta_\lambda \subset \Delta$ the preimage of the set of all multiples of λ under this projection. This is easily seen to be a σ–system of roots on its real span, and it turns out that the Dynkin diagram of Δ_λ is connected. The theorem that allows the reduction to restricted rank one then states that a σ–system Δ is normally extendible if and only if the corresponding systems Δ_λ are normally extendible for all restricted roots λ.

As we have noted in 2.3.10 above, one may obtain quite detailed information on the relation between Δ and Δ_r in a general setting. Using this information, one may next determine explicitly the possible Satake diagrams of normal σ–systems of roots of restricted rank one. Among these possible Satake diagrams, one may single out the normally extendible ones by a direct analysis. Knowing the Satake diagrams of normally extendible σ–systems of restricted rank one, it is then rather easy to determine (case by case) the Satake diagrams corresponding to normally extendible σ–systems for any connected Dynkin diagram. The complete result in the language of Satake diagrams is presented in Table B.4 in Appendix B. For any given real simple Lie algebra, one may determine the Satake diagram similarly as in Example (3) of 2.3.9, thus obtaining a complete list of non–isomorphic real simple Lie algebras. The result may be phrased as follows (see [**Kn96**, VI, Theorem 6.105]).

THEOREM 2.3.11 (Classification of real simple Lie algebras). *The following is a complete list of real simple Lie algebras and any two entries in this list are pairwise non–isomorphic:*

(1) *The underlying real Lie algebras of complex simple Lie algebras of type A_n for $n \geq 1$, B_n for $n \geq 2$, C_n for $n \geq 3$, D_n for $n \geq 4$, E_6, E_7, E_8, F_4, and G_2.*
(2) *The compact real forms of the complex simple Lie algebras from (1).*
(3) *The classical matrix Lie algebras*

$\mathfrak{sl}(n, \mathbb{R})$	$n \geq 2$
$\mathfrak{sl}(n, \mathbb{H})$	$n \geq 2$
$\mathfrak{su}(p, q)$	$p \geq q > 0$, $p + q \geq 3$
$\mathfrak{so}(p, q)$	$p > q > 0$, $p + q$ *odd*, $p + q \geq 3$

$\mathfrak{sp}(2n, \mathbb{R})$	$n \geq 3$
$\mathfrak{sp}(p, q)$	$p \geq q > 0,\ p+q \geq 3$
$\mathfrak{so}(p, q)$	$p \geq q > 0,\ p+q$ *even*, $p+q \geq 8$
$\mathfrak{so}^*(2n)$	$n \geq 5$

(4) *12 exceptional noncomplex, noncompact simple real Lie algebras; see Table B.4 in Appendix B.*

Except for the last entry $\mathfrak{so}^*(2n)$, we have met all the classical Lie algebras which occur in this classification in 2.1.7, 2.2.6, and 2.3.1. To describe $\mathfrak{so}^*(2n)$, recall that for a usual complex or quaternionic Hermitian form the real part is symmetric and the imaginary part is skew symmetric. Now one may also consider Hermitian forms with skew symmetric real part and symmetric imaginary part. In the complex case, these are just purely imaginary multiples of usual Hermitian forms, so nothing new is obtained. This is not true over the quaternions, however. Indeed, it turns out that for each n there is an (up to isomorphism) unique such form on $\mathbb{H}^n$. The Lie algebra $\mathfrak{so}^*(2n)$ is then defined as the space of those matrices in $\mathfrak{sl}(n, \mathbb{H})$ (see 2.1.7) which are skew–Hermitian with respect to this form. In particular, we can view $\mathfrak{so}^*(2n)$ as a Lie subalgebra of $\mathfrak{sl}(2n, \mathbb{C})$, and it is easy to see that an explicit realization is given by

$$\left\{\begin{pmatrix} A & B \\ -B^t & \bar{A} \end{pmatrix} : A^t = -A, B^* = B\right\},$$

Evidently, this consists of skew symmetric matrices, so we can view it as a Lie subalgebra of $\mathfrak{so}(2n, \mathbb{C})$, which turns out to be the complexification of $\mathfrak{so}^*(2n)$.

2.3.12. Other methods of classification – Vogan diagrams. Following the original approach of Cartan (see [**Car14**]), there are several ways to obtain the classification of real simple Lie algebras based on a maximally compact Cartan subalgebra rather than a maximally noncompact one. We will briefly outline the approach presented in the book [**Kn96**], in which the author also introduces a new diagrammatic representation of real semisimple Lie algebras by so–called Vogan diagrams. From the point of view of the classification, Vogan diagrams are probably simpler than Satake diagrams, since the possible Vogan diagrams are easier to describe (see below). On the other hand, several aspects of the structure of a real semisimple Lie algebra are encoded in the Satake diagram in a much more transparent way than in the Vogan diagram, which is why we prefer to use Satake diagrams.

Let us start with a real semisimple Lie algebra $\mathfrak{g}$ endowed with a Cartan involution θ and a maximally compact θ–stable Cartan subalgebra $\mathfrak{h} \leq \mathfrak{g}$. Let $\mathfrak{g}_{\mathbb{C}}$ be the complexification of $\mathfrak{g}$ with Cartan subalgebra $\mathfrak{h}_{\mathbb{C}}$, and consider the associated set Δ of roots. Denoting by $\mathfrak{h} = \mathfrak{t} \oplus \mathfrak{a}$ the splitting of $\mathfrak{h}$ according to θ, as before all roots are real on $i\mathfrak{t} \oplus \mathfrak{a}$. The fact that $\mathfrak{h}$ is maximally compact implies that there are no real roots, i.e. no root vanishes identically on $\mathfrak{t}$. Next, one observes that the Cartan involution θ acts on the set of roots. Since θ acts by 1 on $\mathfrak{t}$ and by -1 on $\mathfrak{a}$, we conclude that it fixes the *imaginary roots*, i.e. those which vanish on $\mathfrak{a}$, and it defines an involutive permutation of the other roots (called "complex roots").

The appropriate positive subsystems $\Delta^+ \subset \Delta$ now are those induced by a basis of $(\mathfrak{h}_{\mathbb{C}})_0 = i\mathfrak{t} \oplus \mathfrak{a}$, which starts with a basis of $i\mathfrak{t}$. This choice and the fact that no root vanishes on $\mathfrak{t}$ ensures that $\theta(\Delta^+) \subset \Delta^+$ holds for the corresponding positive system, which implies that θ defines an involutive permutation of the simple roots.

Finally, for an imaginary root (i.e. one that vanishes on $\mathfrak{a}$) one sees as in 2.3.8 that the corresponding root space lies either in $\mathfrak{k}_\mathbb{C}$ or in $\mathfrak{p}_\mathbb{C}$ (and this time both situations are really possible). Now the *Vogan diagram* associated to $(\mathfrak{g}, \mathfrak{h}, \Delta^+)$ is defined as the Dynkin diagram of $\mathfrak{g}_\mathbb{C}$ with the two element orbits of θ on the simple roots indicated by arrows and an imaginary root painted if its root space lies in $\mathfrak{p}_\mathbb{C}$. Note that the action of θ must be the restriction of an automorphism of the Dynkin diagram. Thus, Vogan diagrams are made up of the same ingredients as Satake diagrams. One shows that if two triples $(\mathfrak{g}, \mathfrak{h}, \Delta^+)$ and $(\tilde{\mathfrak{g}}, \tilde{\mathfrak{h}}, \tilde{\Delta}^+)$ lead to the same Vogan diagram, then the Lie algebras $\mathfrak{g}$ and $\tilde{\mathfrak{g}}$ must be isomorphic.

Next, one introduces abstract Vogan diagrams as Dynkin diagrams endowed with an automorphism and some fixed points of the automorphism painted, and proves that any such abstract Vogan diagram comes from some triple $(\mathfrak{g}, \mathfrak{h}, \Delta^+)$ as above. In this respect Vogan diagrams behave much nicer than Satake diagrams (for which the determination of the possible abstract diagrams is the main difficulty in the classification). However, various choices of Δ^+ (for fixed $\mathfrak{g}$ and $\mathfrak{h}$) may lead to different Vogan diagrams, and one is left with the problem to describe when two Vogan diagrams come from isomorphic real Lie algebras. This is solved by a theorem of Borel and de Siebenthal. One may choose the positive system in such a way that there is at most one painted vertex, and in the case of the trivial automorphism, one obtains further restrictions on the possible painted vertices (which are only relevant for the exceptional algebras). Given these restrictions, one can then push through the classification by a case–by–case analysis.

2.3.13. Relation to symmetric spaces. Much of the interest in the classification of real semisimple Lie algebras among differential geometers comes from the relation to symmetric spaces. Since this ties in nicely with the geometry of homogeneous space that we have studied in Section 1.4, we briefly outline the relation here. Recall that a Riemannian symmetric space is a connected Riemannian manifold (M, g) such that for each point $x \in M$ there exists an isometry σ_x which has x as an isolated fixed point and satisfies $\sigma_x^2 = \text{id}$. Since an isometry of a connected Riemannian manifold is determined by its value and its tangent map in one point (see 1.1.1), we see that we must have $T_x\sigma_x = -\text{id}_{T_xM}$ and together with $\sigma_x(x) = x$, this uniquely determines σ_x. In particular, if γ is a geodesic in M starting at x, then $\sigma_x(\gamma(t)) = \gamma(-t)$, whence σ_x is called the geodesic reflection in x.

The first observation now is that the group $\text{Isom}(M)$ of isometries of M acts transitively on M. To see this, take two points x and y in M. Since M is connected, x and y may be joined by a broken sequence of geodesics, i.e. there are points $x = x_1, x_2, \dots, x_n = y$, elements $t_1, \dots, t_{n-1} \in \mathbb{R}$ and geodesics $\gamma_1, \dots, \gamma_{n-1}$ in M such that $\gamma_i(0) = x_i$ and $\gamma_i(t_i) = x_{i+1}$. Defining σ_i to be $\sigma_{\gamma_i(t_i/2)}$, one immediately concludes that $\sigma_i(x_i) = x_{i+1}$. Thus, $\sigma_{n-1} \circ \dots \circ \sigma_1$ is an isometry that maps x to y. In particular, choosing a point $x_0 \in M$, we may identify M with G/G_{x_0}, where G is the group of isometries of M (which is a Lie group by 1.5.11) and G_{x_0} is the isotropy subgroup of the point x_0. Since the isometry group acts transitively and M is connected, one easily concludes that the connected component of the identity in the isometry group still acts transitively on M.

So let us assume that $M = G/H$ is a connected symmetric space with G the connected component of the identity in $\text{Isom}(M)$. Then H is the isotropy subgroup of a base point $o \in M$, so, in particular, H is compact; compare with 1.4.4. Now consider the geodesic reflection σ_o in the base point. Denoting by ℓ the left action of

G on M, consider the map $\sigma_o \circ \ell_g \circ \sigma_o$ for some element $g \in G$. This is an isometry of M and since ℓ_g lies in the connected component of the identity and $\sigma_o^2 = \text{id}$, there must be a unique element $\Theta(g)$ such that $\ell_{\Theta(g)} = \sigma_o \circ \ell_g \circ \sigma_o$. From this defining equation, one immediately verifies that Θ is an involutive automorphism of the Lie group G. Moreover, for $h \in H$, we see that $\ell_{\Theta(h)}$ stabilizes the point o, so we have $\Theta(h) \in H$. Taking the tangent map in o, and using that $T_o\sigma_o = -\text{id}$, we conclude that $T_o\ell_{\Theta(h)} = T_o\ell_h$, whence $\Theta(h) = h$ for all $h \in H$, so H is contained in the fixed point group of Θ. The derivative $\theta : \mathfrak{g} \to \mathfrak{g}$ of the automorphism Θ is an involutive automorphism of the Lie algebra $\mathfrak{g}$ and since Θ restricts to the identity on the subgroup H, θ has to restrict to the identity on the Lie algebra $\mathfrak{h}$ of H. On the other hand, the defining equation for Θ immediately implies that $\sigma_o(\ell_g(o)) = \ell_{\Theta(g)}(o)$, so we see that $\sigma_o : G/H \to G/H$ is induced by the map $\Theta : G \to G$. But this implies that the mapping $\mathfrak{g}/\mathfrak{h} \to \mathfrak{g}/\mathfrak{h}$ induced by θ is just $T_o\sigma_o = -\text{id}$. On the other hand, since H is compact, there is an H–invariant complement $\mathfrak{m}$ to the Lie subalgebra $\mathfrak{h} \leq \mathfrak{g}$, and clearly θ must act as minus the identity on $\mathfrak{m}$. In particular, this implies that the fixed point group of Θ has Lie algebra $\mathfrak{h}$, so we conclude that H must lie between the fixed point group of Θ and its connected component of the identity.

Conversely, let us assume that G is a Lie group, $\Theta : G \to G$ is an involutive automorphism, $\mathfrak{h}$ is the Lie algebra of the fixed point group of Θ, and $H \leq G$ is a subgroup lying between the fixed point group of Θ and its connected component of the identity. Then the derivative θ of Θ must be diagonalizable, and one easily verifies that the corresponding decomposition $\mathfrak{g} = \mathfrak{h} \oplus \mathfrak{m}$ into ± 1–eigenspaces for θ is H–invariant. If there is an H–invariant inner product on $\mathfrak{m}$, then it extends to a G–invariant Riemannian metric on $M = G/H$. The involutive automorphism Θ descends to an involutive diffeomorphism $\sigma_o : M \to M$, and by construction $T_o\sigma_o = -\text{id}$. Since Θ is an automorphism, we get $\sigma_o \circ \ell_g = \ell_{\Theta(g)} \circ \sigma_o$. Thus, for a point $x = gH \in M$, we get $T_x\sigma_o = T_o\ell_{\Theta(g)} \circ T_o\sigma_o \circ T_x\ell_{g^{-1}}$. Since the right–hand side of this is the composition of three mappings preserving inner products, we conclude that σ_o is an isometry, so it defines a geodesic reflection in the point $o \in M$. For $x = gH$ as above, $\sigma_x = \ell_g \circ \sigma_o \circ \ell_{g^{-1}}$ defines a geodesic reflection in x, so $M = G/H$ is a symmetric space. Hence, we have arrived at a Lie theoretic description of Riemannian symmetric spaces.

The theory of real Lie groups that we have developed immediately gives rise to an important class of examples: Suppose that G is a noncompact connected semisimple Lie group, $\Theta : G \to G$ is a global Cartan involution and $K \leq G$ is the fixed point group of Θ; see 2.3.3. (Recall that K is automatically connected in this case.) Then the derivative θ of Θ is a Cartan involution, and the decomposition of $\mathfrak{g}$ into eigenspace is the Cartan decomposition $\mathfrak{g} = \mathfrak{k} \oplus \mathfrak{p}$. The Killing form is positive definite on $\mathfrak{p}$ and K–invariant (since it is even G–invariant), so by 1.3.3 we have found all ingredients necessary to make G/K into a (noncompact) symmetric space.

Similarly, one can construct compact symmetric spaces starting in the situation above with the compact real form $\mathfrak{u} = \mathfrak{k} + i\mathfrak{p}$ of the complexification $\mathfrak{g}_{\mathbb{C}}$ of $\mathfrak{g}$. Further, one shows that all examples of Riemannian symmetric spaces such that the connected component of the isometry group is semisimple can be obtained in that way, which shows that the classification of symmetric spaces of that type is essentially equivalent to the classification of real semisimple Lie algebras.

2.3.14. Basics on representations of real Lie algebras. We conclude this chapter by discussing representations of real semisimple Lie algebras. We start with some generalities on the relation between representations of a real Lie algebra and of its complexification. Recall from 2.1.3 that for a real Lie algebra $\mathfrak{g}$ a complex representation is given by a homomorphism from $\mathfrak{g}$ to the Lie algebra $\mathfrak{gl}_{\mathbb{C}}(V)$ of complex linear endomorphisms of a complex vector space V. Since this is a complex Lie algebra, we know from 2.1.4 that such a homomorphism uniquely extends to a complex linear homomorphism from the complexification $\mathfrak{g}_{\mathbb{C}}$ to $\mathfrak{gl}_{\mathbb{C}}(V)$. This homomorphism defines a complex representation of the complex Lie algebra $\mathfrak{g}_{\mathbb{C}}$. Conversely, given a complex representation of $\mathfrak{g}_{\mathbb{C}}$ on V, we can of course restrict the corresponding homomorphism $\rho : \mathfrak{g}_{\mathbb{C}} \to \mathfrak{gl}_{\mathbb{C}}(V)$ to the Lie subalgebra $\mathfrak{g} \subset \mathfrak{g}_{\mathbb{C}}$, thus obtaining a complex representation of $\mathfrak{g}$. Since ρ by definition is complex linear, these two constructions are inverse to each other and thus define a bijection between complex representations of the real Lie algebra $\mathfrak{g}$ and of the complex Lie algebra $\mathfrak{g}_{\mathbb{C}}$.

Suppose that $\rho : \mathfrak{g}_{\mathbb{C}} \to \mathfrak{gl}_{\mathbb{C}}(V)$ is a complex representation, and let $W \subset V$ be a complex subspace, which is invariant under the action of the Lie subalgebra $\mathfrak{g}$. Since any $Z \in \mathfrak{g}_{\mathbb{C}}$ can be written as $Z = X + iY$ for $X, Y \in \mathfrak{g}$ we get $\rho(Z) = \rho(X) + i\rho(Y)$, so W is invariant under the action of $\mathfrak{g}_{\mathbb{C}}$. In particular, V is irreducible as a complex representation of $\mathfrak{g}$ if and only if it is irreducible as a complex representation of $\mathfrak{g}_{\mathbb{C}}$. Hence, the complex representation theory of $\mathfrak{g}$ and $\mathfrak{g}_{\mathbb{C}}$ are completely equivalent.

Since for a real Lie algebra $\mathfrak{g}$ the term "complex representation" only means that any element acts by a complex linear map, one may view a complex representation of $\mathfrak{g}$ as a real representation endowed with an invariant complex structure. For this, denote by $J : V \to V$ the linear map defined by multiplication by $i = \sqrt{-1}$. Then $J^2 = J \circ J = -\mathrm{id}$, and the fact that elements of $\mathfrak{g}$ act by complex linear maps is equivalent to J being a $\mathfrak{g}$–homomorphism. Conversely, a real representation endowed with an invariant complex structure can be viewed as a complex representation.

For a complex representation V of $\mathfrak{g}$ let $J : V \to V$ be the complex structure. Then $-J : V \to V$ defines a $\mathfrak{g}$–invariant complex structure, too. Let us denote by $\bar{V}$ the real vector space underlying V endowed with the complex structure $-J$. Then we have a natural complex representation of $\mathfrak{g}$ on $\bar{V}$, which is called the *conjugate representation* of V. Notice that the identity of V is an isomorphism between the real representations underlying V and $\bar{V}$, but as we shall see below V and $\bar{V}$ are not isomorphic as complex representations in general. Note also, that if we extend V and $\bar{V}$ to representations of $\mathfrak{g}_{\mathbb{C}}$, then the identity map is not compatible with the actions of all elements of $\mathfrak{g}_{\mathbb{C}}$, since their actions are defined by complex linear extension. Fixing the real form $\mathfrak{g}$, forming the conjugate representation gives rise to an operation on complex representations of $\mathfrak{g}_{\mathbb{C}}$: One first restricts a representation to $\mathfrak{g}$, then forms the conjugate, and extends back to $\mathfrak{g}_{\mathbb{C}}$.

Suppose that $\rho : \mathfrak{g} \to \mathfrak{gl}(V)$ is a real representation of $\mathfrak{g}$. Then we can form the complexification $V_{\mathbb{C}} = V \otimes_{\mathbb{R}} \mathbb{C}$. Since any linear endomorphism of V uniquely extends to a complex linear endomorphism of $V_{\mathbb{C}}$, we obtain a complex representation $\rho_{\mathbb{C}} : \mathfrak{g} \to \mathfrak{gl}_{\mathbb{C}}(V_{\mathbb{C}})$ (which then of course extends to the complexification $\mathfrak{g}_{\mathbb{C}}$ as above). Now there is a natural *real structure* $R : V_{\mathbb{C}} \to V_{\mathbb{C}}$ on $V_{\mathbb{C}}$ defined by $R(v \otimes z) := v \otimes \bar{z}$ for $v \in V$ and $z \in \mathbb{C}$ or equivalently $R(v + iw) = v - iw$ for

$v, w \in V \subset V_{\mathbb{C}}$. By definition $R^2 = R \circ R = \text{id}$, R is conjugate linear and compatible with the action of $\mathfrak{g}$.

If $W \subset V$ is a $\mathfrak{g}$–invariant subspace, then $W_{\mathbb{C}} \subset V_{\mathbb{C}}$ is a complex subspace which is invariant under the actions of $\mathfrak{g}$ and $\mathfrak{g}_{\mathbb{C}}$. In particular, if $V_{\mathbb{C}}$ is an irreducible complex representation, then the real representation V is irreducible, too. The converse assertion is not true in general:

PROPOSITION 2.3.14. *Let $\mathfrak{g}$ be a finite–dimensional real Lie algebra.*

(1) For a complex representation W of $\mathfrak{g}$, there is a real representation V of $\mathfrak{g}$ such that $W \cong V_{\mathbb{C}}$ if and only if there is a $\mathfrak{g}$–invariant real structure R on W. If this is the case, then R defines an isomorphism $W \cong \bar{W}$.

(2) Let V be an irreducible real representation of $\mathfrak{g}$. Then either $V_{\mathbb{C}}$ is an irreducible complex representation or there exists an invariant complex structure J on V and $V_{\mathbb{C}} \cong V \oplus \bar{V}$ is the decomposition into irreducible components.

PROOF. (1) We have seen above that for a real representation V, the complexification $V_{\mathbb{C}}$ admits an invariant real structure R. Since R is conjugate linear, we may view it as an isomorphism $V_{\mathbb{C}} \to \overline{V_{\mathbb{C}}}$ of complex representations.

Conversely, assume that W is a complex representation and $R : W \to W$ is a $\mathfrak{g}$–equivariant conjugate linear isomorphism such that $R^2 = \text{id}$. Since $R^2 = \text{id}$ the real vector space W splits into the direct sum of the ± 1–eigenspaces of R and we denote by $V \subset W$ the $+1$–eigenspace. Since $v \in V \subset W$ if and only if $R(v) = v$ and R is $\mathfrak{g}$–equivariant, the real subspace $V \subset W$ is $\mathfrak{g}$–invariant. Since R is conjugate linear, multiplication by i maps V to the -1–eigenspace and vice versa. Hence, the two eigenspaces have the same dimension, and $W = V \oplus iV$ is naturally isomorphic to $V_{\mathbb{C}}$ as a $\mathfrak{g}$–representation.

(2) Suppose that $W \subset V_{\mathbb{C}}$ is a proper complex subspace, which is $\mathfrak{g}$–invariant. Let R be the natural real structure on $V_{\mathbb{C}}$. Then $R(W)$ is a $\mathfrak{g}$–invariant complex subspace, too, so also $W \cap R(W)$ and $W + R(W)$ are such subspaces. Since the latter two subspaces evidently are invariant under R, we conclude from (1) that they are the complexifications of their intersections with $V \subset V_{\mathbb{C}}$. Since by construction these intersections are $\mathfrak{g}$–invariant and V is irreducible, this is only possible if $W \cap R(W) = \{0\}$ and $W + R(W) = V_{\mathbb{C}}$. Hence, we get $V_{\mathbb{C}} = W \oplus R(W)$ as a $\mathfrak{g}$–representation and both summands have trivial intersection with $V \subset V_{\mathbb{C}}$. Restricting the projection onto W to V, we obtain a $\mathfrak{g}$–equivariant linear isomorphism $V \to W$, which shows that V admits an invariant complex structure and W is irreducible. As above, $R(W) \cong \bar{W}$, and the last claim follows. □

From part (1) of this proposition we see that general complex representations do not arise as complexifications of real representations. Indeed, any representation arising as a complexification must by isomorphic to its conjugate representation. This condition is not sufficient, however, and we have to discuss this next.

Let us assume that W is a complex irreducible representation of $\mathfrak{g}$ such that $W \cong \bar{W}$ as a $\mathfrak{g}$–representation. An isomorphism of the two representations may also be viewed as a $\mathfrak{g}$–equivariant, conjugate linear isomorphism $\Phi : W \to W$. Then Φ^2 is complex linear and thus an automorphism of W, so $\Phi^2 = \lambda\text{id}$ for some $\lambda \in \mathbb{C} \setminus \{0\}$ by Schur's lemma from 2.1.3. Rescaling Φ by a complex number α, the map Φ^2 is rescaled by $\alpha\bar{\alpha}$, so we may, without loss of generality, assume that $|\lambda| = 1$. Extending Φ to a complex linear map $\tilde{\Phi} : W_{\mathbb{C}} \to W_{\mathbb{C}}$, we obtain $\tilde{\Phi}^2 = \lambda\text{id}$, so $\tilde{\Phi}$ is diagonalizable with eigenvalues the two possible square roots of λ. But on the other hand, nonreal eigenvalues of $\tilde{\Phi}$ must show up in pairs of conjugate complex numbers, so these two square roots must be either both real or complex conjugate. Hence, the only remaining possibilities are $\lambda = \pm 1$. For $\lambda = 1$, the map Φ is a real structure as defined above.

A $\mathfrak{g}$–equivariant conjugate linear isomorphism $J : W \to W$ such that $J^2 = -\text{id}$ is called a *quaternionic structure* on the complex representation W. If $J : W \to W$ is a quaternionic structure, then denote by $I : W \to W$ the given complex structure, and define $K := I \circ J = -J \circ I$. Then these maps can be used to define the multiplication by the standard imaginary quaternions, hence making W into a quaternionic vector space such that each element of $\mathfrak{g}$ acts by a quaternionic linear map on W.

Finally, notice that an irreducible complex representation W cannot admit a real structure and a quaternionic structure at the same time. Assuming that R and J are such structures, $J \circ R : W \to W$ is an isomorphism of complex representations and thus $J \circ R = \lambda\text{id}$ for some nonzero $\lambda \in \mathbb{C}$ by Schur's lemma. But now there is a nonzero element $w \in W$ such that $R(w) = w$ and hence $J(w) = \lambda w$. But this implies $-w = J^2(w) = |\lambda|^2 w$, which is a contradiction. Consequently, for a complex irreducible representation W of $\mathfrak{g}$ such that $W \cong \bar{W}$ we can define the *index* $\epsilon(\mathfrak{g}, W)$ of W with respect to $\mathfrak{g}$ to be $+1$ if W admits a real structure and -1 if W admits a quaternionic structure.

2.3.15. Representations of real semisimple Lie algebras. It remains to adapt our results to the semisimple case, which essentially means to formulate them in terms of highest weights. We will mainly be interested in deducing the decomposition of a real representation into irreducible components from the decomposition of its complexification. Thus, we will not need a very specific description of all real representations, which can be found in [**On04**] and [**Ti67**]. Also in most examples it is possible to get the necessary information directly rather then by applying general results. Therefore, we put more emphasis on the tools needed to deal with examples.

Let $\mathfrak{g}$ be a real semisimple Lie algebra with complexification $\mathfrak{g}_{\mathbb{C}}$. Then by 2.3.14 complex representations for $\mathfrak{g}$ and $\mathfrak{g}_{\mathbb{C}}$ are the same, so we can apply the complex representation theory developed in 2.2. Complex irreducible representations of $\mathfrak{g}_{\mathbb{C}}$ are classified by their highest weights. For an irreducible complex representation W we can form the conjugate representation $\bar{W}$, which is irreducible, too. If λ is the highest weight of W, then we denote by $\bar{\lambda}$ the highest weight of $\bar{W}$, so we can view passing to the conjugate representation as an operation on dominant integral weights. The first step to understand real representations of $\mathfrak{g}$ is to describe the weights for which $\lambda = \bar{\lambda}$.

Suppose that W_1 and W_2 are two complex representations of $\mathfrak{g}$. Then clearly $\overline{W_1 \otimes W_2} = \overline{W_1} \otimes \overline{W_2}$. Applying this to irreducible representations, one easily deduces that $\overline{\lambda + \mu} = \bar{\lambda} + \bar{\mu}$ holds for all dominant integral weights λ and μ. In particular, this shows that the conjugate of a fundamental weight (see 2.2.13) will again be a fundamental weight. On the other hand, since any dominant integral weight can be written as a linear combination of fundamental weights, it suffices to determine the conjugate representations of the fundamental representations.

The second step is to understand the indices of those irreducible representations which are isomorphic to their conjugate. We write $\epsilon(\mathfrak{g}, \lambda)$ for the index $\epsilon(\mathfrak{g}, \Gamma_\lambda)$ of the irreducible representation with highest weight λ. Now there is one more simple observation to make: Suppose that W_1 and W_2 are (not necessarily irreducible) complex representations of $\mathfrak{g}$, which both admit invariant real structures R_1 and R_2. Then $R_1 \otimes R_2$ defines a real structure on $W_1 \otimes W_2$. Similarly, if J_1 and J_2 are quaternionic structures on W_1 and W_2, then $J_1 \otimes J_2$ is a real structure on $W_1 \otimes W_2$. Finally, the tensor product of one real structure and one quaternionic structure is a quaternionic structure. If there is a complex irreducible representation V which occurs with multiplicity one in $W_1 \otimes W_2$, then any real or quaternionic structure on this tensor product must map V to itself, and thus defines a real or quaternionic structure on V. In particular, $\epsilon(\mathfrak{g}, \lambda + \mu) = \epsilon(\mathfrak{g}, \lambda)\epsilon(\mathfrak{g}, \mu)$ for all self–conjugate dominant integral weights λ and μ. Similarly, for a real or quaternionic structure on W, the induced structure on $\otimes^k W$ preserves the subrepresentations $S^k W$ and $\Lambda^k W$. Hence, if W admits an invariant real structure, then so do $S^k W$ and $\Lambda^k W$ for each k. If W admits an invariant quaternionic structure, then $S^k W$ and $\Lambda^k W$ admit an invariant real structure if k is even and an invariant quaternionic structure if k is odd.

As a last observation, we remark that for any complex representation V with conjugate $\overline{V}$, the tensor product $V \otimes \overline{V}$ has an invariant real structure defined by swapping the two factors in the tensor product. In particular, this implies that for any dominant weight λ of $\mathfrak{g}$, the self–conjugate weight $\lambda + \bar{\lambda}$ has index $+1$.

The necessary information for determining conjugate and dual representations as well as indices is collected in Table B.4 in Appendix B. Given a dominant integral weight λ, let us expand λ as $\sum_i \Lambda_i \omega_i$ in terms of fundamental weights. Then it turns out that both the dual weight λ^* and the conjugate weight $\bar{\lambda}$ can be obtained by applying an automorphism of the Satake diagram to the coefficients Λ_i. In Table B.4, the automorphism inducing duality is denoted by ν, while the automorphism giving rise to the conjugate weight is denoted by s. Now for the diagrams of types B, C, E_7, E_8, F_4, and G_2, there are no nontrivial automorphisms, so in these cases any irreducible representation is self–dual and self–conjugate. In all other cases except the split real form D_4 (for which $s = \nu = \mathrm{id}$), there is at most one nontrivial automorphism of the Dynkin diagram. Hence, in Table B.4 it is only indicated whether the given automorphism is nontrivial (indicated by "$\neq e$") or trivial (indicated by "e"). For self–conjugate representations (those with $s = e$), the last column of Table B.4 shows the index $\epsilon(\mathfrak{g}, \sum \Lambda_i \omega_i)$.

EXAMPLE 2.3.15. Let us verify the entries in Table B.4 for the real forms of $\mathfrak{sl}(n, \mathbb{C})$.

(1) For the split real form $\mathfrak{g} := \mathfrak{sl}(n, \mathbb{R})$ the usual conjugation defines an invariant real structure on the standard representation on $\mathbb{C}^n$. Since the other fundamental representations are exterior powers of $\mathbb{C}^n$, they admit an invariant real structure, too. Consequently, any complex irreducible representation of $\mathfrak{g}$ admits an invariant real structure and thus is the complexification of a real irreducible representation. This also happens for the split real forms of the other classical complex simple Lie algebras.

(2) Next we consider the real forms $\mathfrak{g} := \mathfrak{su}(p, q)$ with $p + q = n$. Here the main point is that the standard representation $V := \mathbb{C}^n$ admits a nondegenerate Hermitian form. This Hermitian form induces an isomorphism between $\bar{V}$ and the dual V^*, which in turn induces an isomorphism $\overline{\Lambda^k V} = \Lambda^k \overline{V} \cong \Lambda^k V^* \cong (\Lambda^k V)^*$. Hence, for each of the fundamental representations the conjugate coincides with the dual. Otherwise put, denoting by $\omega_1, \dots, \omega_{n-1}$ the fundamental weights, we

get $\overline{\omega}_i = \omega_{n-i}$, so both duality and conjugacy are induced by the unique nontrivial automorphism of the Satke diagram. For $\lambda = a_1\omega_1 + \cdots + a_{n-1}\omega_{n-1}$ we get $\lambda = \bar{\lambda}$ if and only if $a_i = a_{n-i}$ for all $i = 1, \dots, n-1$.

If $p + q = n$ is odd, then none of the fundamental weights ω_i is self–conjugate, and we can write any self–conjugate weight as $\sum a_i(\omega_i + \omega_{n-i})$ where $i = 1, \dots, \frac{n-1}{2}$ and the a_i are nonnegative integers. Since $\omega_{n-i} = \bar{\omega}_i$, we know from above that $\epsilon(\mathfrak{g}, \omega_i + \omega_{n-i}) = 1$, so $\epsilon(\mathfrak{g}, \lambda) = 1$ for any self–conjugate weight λ.

If $p + q = n$ is even, then the fundamental weight $\omega_{n/2}$ is self–conjugate, and each self–conjugate weight λ can be written as a linear combination of $\omega_{n/2}$ and the weights $\omega_i + \omega_{n-i}$ for $i = 1, \dots, \frac{n}{2} - 1$. As above, the latter weights have index one, so the index of λ depends only on the coefficient of $\omega_{n/2}$. The fundamental representation $V_{n/2}$ with highest weight $\omega_{n/2}$ is $\Lambda^{n/2}\mathbb{C}^n$. Now the wedge product $\Lambda^{n/2}\mathbb{C}^n \otimes \Lambda^{n/2}\mathbb{C}^n \to \Lambda^n\mathbb{C}^n$ defines a nondegenerate $\mathfrak{g}$–invariant bilinear form on $V_{n/2}$, which is symmetric if $n = 4k$ and skew symmetric if $n = 4k + 2$. The resulting isomorphism $V_{n/2} \to V_{n/2}^*$ can be composed with the isomorphism $V_{n/2}^* \to \bar{V}_{n/2}$ coming from the induced Hermitian inner product. It is easy to see that the result is an invariant real structure on $V_{n/2}$ for $n = 4k$ and an invariant quaternionic structure for $n = 4k + 2$. Consequently, any self–conjugate weight has index $+1$ for $n = 4k$, while for $n = 4k + 2$ the index is $+1$ if the coefficient on $\omega_{n/2}$ is even and -1 if this coefficient is odd.

(3) We conclude the discussion with the real form $\mathfrak{g} := \mathfrak{sl}(n, \mathbb{H})$ of $\mathfrak{sl}(2n, \mathbb{C})$. Here the situation is simple, since the standard representation $\mathbb{C}^{2n} \cong \mathbb{H}^n$ has an invariant quaternionic structure. Looking at the exterior powers we see that $\overline{\omega}_i = \omega_i$ for all $i = 1, \dots, 2n-1$ and $\epsilon(\mathfrak{g}, \omega_i) = (-1)^i$. Thus, any dominant integral weight is self–conjugate, and expanding such a weight λ as $\lambda = \sum a_i\omega_i$, we see that $\epsilon(\mathfrak{g}, \lambda) = (-1)^{a_1 + a_3 + \cdots + a_{2n-1}}$.

Part 2

General theory

CHAPTER 3

Parabolic geometries

This chapter is devoted to the definition and study of the fundamental properties of parabolic geometries. They are defined as Cartan geometries of type (G, P) for semisimple Lie groups G and parabolic subgroups P. The usual definition of parabolic subgroups is (via the notion of parabolic subalgebras) based on the structure theory of semisimple Lie algebras. Alternatively, parabolic subalgebras may also be described in terms of so–called $|k|$–gradings of semisimple Lie algebras. Starting from the latter description, the basic theory of parabolic geometries can be developed using almost exclusively the elementary theory of semisimple Lie algebras as presented in Section 2.1. This is the point of view taken in Section 3.1.

Analyzing $|k|$–gradings and the corresponding subgroups, one is lead to a sequence of structures underlying a parabolic geometry, the weakest of which is called an infinitesimal flag structure. The basic question addressed in Section 3.1 is to what extent a parabolic geometry is determined by the underlying infinitesimal flag structure. To obtain results in this direction, we have to impose the technical condition of regularity, which avoids particularly bad types of torsion. On the other hand, to single out a unique (up to isomorphism) parabolic geometry with a fixed underlying infinitesimal flag structure one has to assume that the Cartan connection satisfies a normalization condition. It is one of the key features of parabolic geometries that there is a conceptual choice for a normalization condition, and the normalization conditions for all parabolic geometries can be described in a uniform way.

Under a cohomological restriction on the $|k|$–graded Lie algebra it is then shown that passing to the underlying infinitesimal flag structure induces an equivalence of categories between regular normal parabolic geometries and regular infinitesimal flag structures. This is the central result of Section 3.1. If the cohomological condition is not satisfied, we describe how a more complicated underlying structure is equivalent to a regular normal parabolic geometry. A brief discussion of complex parabolic geometries (the holomorphic version of the theory) and of an alternative approach to parabolic geometries via abstract tractor bundles concludes Section 3.1.

To understand the possible $|k|$–gradings of a given semisimple Lie algebra $\mathfrak{g}$ and to deal with examples efficiently, one has to invoke the structure theory of semisimple Lie algebras as developed in Sections 2.2 and 2.3. Section 3.2 starts by showing that both in the real and complex case the subalgebras obtained from $|k|$–gradings are exactly the parabolic subalgebras in the sense of representation theory. Since the possible parabolic subalgebras (up to conjugation) can be read off the Dynkin diagram, respectively, the Satake diagram of $\mathfrak{g}$, we get a complete overview over possible gradings.

Next, we study the homogeneous models of parabolic geometries, the generalized flag manifolds. Using representation theory, we obtain various realizations of these homogeneous spaces.

The structure theory also leads to an efficient way of dealing with irreducible representations of a parabolic subalgebra and their relation to irreducible representations of $\mathfrak{g}$. The last ingredient coming from the structure theory is related to the Weyl group W of $\mathfrak{g}$. Any parabolic subalgebra in $\mathfrak{g}$ gives rise to a subset of W, which can be naturally viewed as an oriented graph. This is called the Hasse graph or Hasse diagram of the parabolic and it is a central ingredient in the theory of parabolic geometries. In the end of Section 3.2 we describe algorithms to determine the Hasse diagram explicitly. We show how the integral homology and cohomology of generalized flag varieties can be described in terms of the Hasse diagram.

Section 3.3 contains a complete proof of Kostant's version of the Bott–Borel–Weil theorem. For any parabolic subalgebra in $\mathfrak{g}$, this theorem describes the Lie algebra cohomology of the nilradical of the parabolic with coefficients in the restriction of an arbitrary finite–dimensional irreducible representation of $\mathfrak{g}$. The description is in terms of the Hasse diagram, so we get an algorithm for computing the cohomologies explicitly. In particular, we get complete information on the cohomological conditions used in Section 3.1. We show how to deduce the classical Bott–Borel–Weil theorem, which describes the cohomology of the sheaf of holomorphic sections of an irreducible homogeneous vector bundle on a generalized flag variety from Kostant's version. Finally, we also obtain a short proof of the Weyl character formula.

3.1. Underlying structures and normalization

We start this section by discussing $|k|$–gradings of a semisimple Lie algebra $\mathfrak{g}$. The nonnegative parts of such gradings form distinguished Lie subalgebras of $\mathfrak{g}$ and hence give rise to distinguished Lie subgroups in any Lie group with Lie algebra $\mathfrak{g}$. As we shall see later, these are exactly the parabolic subalgebras, respectively, subgroups in the sense of representation theory.

Given a semisimple Lie group G and a parabolic subgroup $P \subset G$, parabolic geometries of type (G, P) are then defined as Cartan geometries of the given type. Following the ideas from Section 1.5, we find several geometric structures underlying a parabolic geometry, the weakest of which is called an infinitesimal flag structure. One ingredient of an infinitesimal flag structure is a filtration of the tangent bundle of the manifold carrying the parabolic geometry. Requiring this filtration to be compatible with Lie bracket of vector fields leads to the notion of regularity for infinitesimal flag structures and parabolic geometries.

It is easy to describe the set of all parabolic geometries having a fixed underlying regular infinitesimal flag structure. This description suggests necessary properties for a normalization condition, which singles out one of these parabolic geometries. We then construct a normalization condition with the required properties using a bit of Lie theory. Assuming normality, one can project the Cartan curvature to a certain quotient, thus obtaining the harmonic curvature of the geometry. Using the Bianchi identity, we prove that the harmonic curvature still is a complete obstruction against local flatness.

Understanding the algebraic properties of the normalization condition leads to the central results of this section, which concern existence and uniqueness of a regular normal parabolic geometry inducing a given regular infinitesimal flag structure. Under a cohomological condition, this leads to an equivalence of categories between regular normal parabolic geometries and regular infinitesimal flag structures. Without assuming this condition, we still get an equivalence to a slightly more complicated underlying structure.

Next, we discuss complex parabolic geometries, which are the holomorphic version of the theory. We characterize complex parabolic geometries among real parabolic geometries via their curvature. This is used to prove an equivalence between parabolic geometries and underlying structures in the holomorphic category. The last part of the section is devoted to the alternative description of parabolic geometries in terms of abstract tractor bundles and tractor connections.

It should be mentioned here that there is an alternative description of the geometric structures underlying a regular parabolic geometry. In his pioneering work, N. Tanaka called them "$G_0^\#$–structures of type $\mathfrak{m}$". Together with further alternative constructions for the canonical Cartan connections, this approach is outlined in Appendix A.

3.1.1. Filtrations. The basic data needed for the definition of a parabolic geometry are a semisimple Lie algebra endowed with a certain type of grading and a Lie group with that Lie algebra. However, the grading on the Lie algebra is rather an auxiliary object, while the main structure is the filtration associated to this grading. Relations between filtered objects and the associated graded objects will play a central role in the theory, so for the convenience of the reader we collect here some basic facts about filtered vector spaces, filtered Lie algebras and filtered vector bundles.

A *filtered vector space* is a vector space V together with a sequence $\{V^i : i \in \mathbb{Z}\}$ of subspaces $V^i \subset V$, such that $V^i \supset V^{i+1}$ for all $i \in \mathbb{Z}$, $\bigcup_{i\in\mathbb{Z}} V^i = V$ and $\bigcap_{i\in\mathbb{Z}} V^i = \{0\}$. We will usually deal with the case of finite–dimensional spaces and finite filtrations, which we will write as $V = V^j \supset V^{j+1} \supset \dots \supset V^{k-1} \supset V^k$. Using this notation, it is always assumed that $V^\ell = V$ for all $\ell < j$ and $V^\ell = \{0\}$ for all $\ell > k$. The *trivial filtration* on a vector space V is given by $V^i = V$ for $i \le 0$ and $V^i = \{0\}$ for $i > 0$.

From a filtration $\{V^i\}$ of a vector space V, one can canonically construct a graded vector space $\mathrm{gr}(V) = \bigoplus_{i\in\mathbb{Z}} \mathrm{gr}_i(V)$ by putting $\mathrm{gr}_i(V) := V^i/V^{i+1}$ for all $i \in \mathbb{Z}$. The graded vector space $\mathrm{gr}(V)$ is called the *associated graded* to the filtered vector space V. In the case of a finite filtration $V = V^j \supset \dots \supset V^k$, we obtain a finite grading $\mathrm{gr}(V) = \mathrm{gr}_j(V) \oplus \dots \oplus \mathrm{gr}_k(V)$ by omitting the zero summands $\mathrm{gr}_i(V)$ for $i < j$ and $i > k$.

An important point to note here is that although the name "associated graded" might suggest this, for a filtered vector space $(V, \{V^i\})$ with associated graded $\mathrm{gr}(V)$ there is neither a natural linear map from V to $\mathrm{gr}(V)$ nor a natural linear map in the opposite direction. The only natural linear maps available are the canonical projections $V^i \to \mathrm{gr}_i(V) = V^i/V^{i+1}$. One may, however, construct a linear isomorphism between V and $\mathrm{gr}(V)$ by making choices. Indeed, choosing for each $i \in \mathbb{Z}$ a subspace $V_i \subset V^i$ which is complementary to the subspace $V^{i+1} \subset V^i$, the restriction of the canonical projection induces a linear isomorphism $V_i \to \mathrm{gr}_i(V)$. The inverses of these linear isomorphisms give rise to a linear map $\mathrm{gr}(V) = \bigoplus_i \mathrm{gr}_i(V) \to V$, which is easily seen to be a linear isomorphism. Choosing such a linear isomorphism corresponds exactly to choosing a grading $V = \bigoplus_{i\in\mathbb{Z}} V_i$ which induces the given filtration, i.e. which has the property that $V^i = \bigoplus_{j\geq i} V_j$.

While at this stage the distinction between a filtered vector space and the associated graded vector space may seem artificial, it immediately becomes important if there is some additional structure on V. Suppose, for example, that we have given a representation of a Lie group G or a Lie algebra $\mathfrak{g}$ on V which preserves the filtration, i.e., which is such that the action of any element preserves each of the subspaces $V^i \subset V$. Then each V^i is a subrepresentation of V. In particular, we get an induced representation on $\operatorname{gr}_i(V) = V^i/V^{i+1}$, and hence a representation on $\operatorname{gr}(V)$. While the vector spaces V and $\operatorname{gr}(V)$ are linearly isomorphic, the representations on V and $\operatorname{gr}(V)$ are far from being equivalent in general. For example, by Theorem 2.1.1 any complex representation V of a solvable Lie algebra $\mathfrak{g}$ admits a $\mathfrak{g}$–invariant filtration such that each of the components $\operatorname{gr}_i(V)$ of the associated graded is one–dimensional. In particular, $[\mathfrak{g},\mathfrak{g}]$ then acts trivially on $\operatorname{gr}(V)$.

Many natural constructions with vector spaces can be carried out in the filtered setting. If $(V, \{V^i\})$ is a filtered vector space and $W \subset V$ is a linear subspace, then we get a filtration on W by putting $W^i := W \cap V^i$. By definition, the inclusion $W \hookrightarrow V$ is a filtration preserving linear map. Since W^i is a subspace in V^i, its image in $\operatorname{gr}_i(V) = V^i/V^{i+1}$ is a linear subspace isomorphic to $W^i/(V^{i+1}\cap W^i) = gr_i(W)$, so we can naturally view $\operatorname{gr}(W)$ as a subspace of $\operatorname{gr}(V)$. Moreover, we clearly get a filtration on the quotient V/W by defining $(V/W)^i$ to be the image of V^i in V/W. This simply means that $(V/W)^i = V^i/(W \cap V^i) = V^i/W^i$. Now the natural projection $V^i \to (V/W)^i$ maps V^{i+1} to $(V/W)^{i+1}$ and thus induces a surjective linear map $\operatorname{gr}_i(V) \to \operatorname{gr}_i(V/W)$. The kernel of this map consists exactly of those elements which have a representative in $W^i \subset V^i$, and thus is given by $\operatorname{gr}_i(W)$. In conclusion, we get a natural isomorphism $\operatorname{gr}(V/W) = \operatorname{gr}(V)/\operatorname{gr}(W)$ and $\operatorname{gr}_i(V/W) = \operatorname{gr}_i(V)/\operatorname{gr}_i(W)$.

Consider two filtered vector spaces $(V, \{V^i\})$ and $(W, \{W^j\})$ and a linear map $f : V \to W$. We say that f *has homogeneity* $\geq j$ for some $j \in \mathbb{Z}$, if $f(V^i) \subset W^{i+j}$ holds for all i. In particular, f has homogeneity ≥ 0 if and only if it is filtration preserving. The maps of homogeneity $\geq j$ form a linear subspace $L(V,W)^j \subset L(V,W)$ and we obtain a filtration of $L(V,W)$.

For $\phi \in L(V,W)^j$, we have $\phi(V^i) \subset W^{i+j}$ and $\phi(V^{i+1}) \subset W^{i+j+1}$, so for each i we get an induced map $\operatorname{gr}_i(V) \to \operatorname{gr}_{i+j}(W)$. These fit together to define a linear map $\operatorname{gr}_j(\phi) : \operatorname{gr}(V) \to \operatorname{gr}(W)$, which is homogeneous of degree j. By definition, $\operatorname{gr}_j(\phi) = 0$ if and only if $\phi \in L(V,W)^{j+1}$, so we get an injective linear map $\operatorname{gr}(L(V,W)) \to L(\operatorname{gr}(V), \operatorname{gr}(W))$, which is compatible with the gradings. Choosing linear isomorphisms $V \to \operatorname{gr}(V)$ and $\operatorname{gr}(W) \to W$ as above, we see that this map is also surjective, so $\operatorname{gr}(L(V,W)) \cong L(\operatorname{gr}(V), \operatorname{gr}(W))$ with the grading by homogeneous degree. All of this easily extends to multilinear maps and also to the subspaces of symmetric and antisymmetric k–linear maps $V^k \to W$. Notice that passing from $\phi \in L(V,W)^j$ to $\operatorname{gr}_j(\phi)$ has functorial properties. More precisely, if Z is another filtered vector space and $\psi \in L(W,Z)^k$, then obviously $\psi\circ\phi \in L(V,Z)^{j+k}$ and $\operatorname{gr}_{j+k}(\psi \circ \phi) = \operatorname{gr}_k(\psi) \circ \operatorname{gr}_j(\phi)$.

Specializing to the case $W = \mathbb{K}$ with the trivial filtration, we get a natural filtration on the dual V^* of a filtered vector space. This means that $(V^*)^i$ is the annihilator of V^{-i+1}. From above, we know that $\operatorname{gr}(V^*) \cong L(\operatorname{gr}(V), \mathbb{K}) = (\operatorname{gr}(V))^*$ with the grading by homogeneous degrees of maps, i.e. $\operatorname{gr}_i(V^*) = (\operatorname{gr}_{-i}(V))^*$.

Finally, if $(V, \{V^i\})$ and $(W, \{W^j\})$ are filtered vector spaces, then for all i, j we have the subspace $V^i \otimes W^j$ in the tensor product $V \otimes W$. We define $(V \otimes W)^k$ to be the sum of all components $V^i \otimes W^j$ such that $i + j = k$. One can show in general, that this makes $V \otimes W$ into a filtered vector space and that

$$\operatorname{gr}_k(V \otimes W) = \bigoplus_{i+j=k} \operatorname{gr}_i(V) \otimes \operatorname{gr}_j(W).$$

We will only need this for finite–dimensional vector spaces, for which it follows easily from the above considerations using that $V \otimes W$ can be identified with the dual of the space of bilinear maps $V \times W \to \mathbb{K}$. This easily extends to tensor products of more than two factors. In the case of tensor powers of a single filtered vector space we get induced filtrations on the symmetric and alternating powers. In particular, we get a natural filtration on the kth exterior power $\Lambda^k V$ of a filtered space V, and the associated graded is naturally isomorphic to $\Lambda^k \operatorname{gr}(V)$, with the grading characterized by the fact that for v_i of degree j_i, the element $v_1 \wedge \cdots \wedge v_k$ has degree $i_1 + \cdots + i_k$.

Filtered vector bundles. Everything we have said about filtered vector spaces extends without essential changes to (finite–dimensional) filtered vector bundles over a smooth manifold. A *filtered vector bundle* over a smooth manifold M is a smooth vector bundle $p : E \to M$ together with a sequence $\{E^i : i \in \mathbb{Z}\}$ of smooth subbundles such that $E^i \supset E^{i+1}$ and such that there are $i_0 < j_0 \in \mathbb{Z}$ such that $E^i = E$ for $i \leq i_0$ and $E^i = M$ (the zero vector bundle) for $i > j_0$. Given such a filtered bundle, we may form the quotient bundles $\operatorname{gr}_i(E) := E^i/E^{i+1}$ and the *associated graded vector bundle* $\operatorname{gr}(E) = \bigoplus_{i \in \mathbb{Z}} \operatorname{gr}_i(E)$. As before, we ignore the zero summands, so $\operatorname{gr}(E) = \operatorname{gr}_{i_0}(E) \oplus \cdots \oplus \operatorname{gr}_{j_0}(E)$. Choosing a covering of M by open subsets over which all the subbundles E^i are simultaneously trivial, we see that we may actually view E as a bundle modelled on a filtered vector space $(V, \{V^i\})$, i.e. there are vector bundle charts $\psi : p^{-1}(U) \to U \times V$, for E such that $p^{-1}(U) \cap E^i = \psi^{-1}(U \times V^i)$, so we have charts compatible with the filtration. Then it is natural to view the bundle $\operatorname{gr}(E)$ as being modelled on $\operatorname{gr}(V)$ and in charts as above the natural projections $E^i \to \operatorname{gr}_i(E)$ simply correspond to the natural projections $V^i \to \operatorname{gr}_i(V)$. Notice, however, that for a vector bundle E modelled on a vector space V, choosing a filtration on V usually does not lead to a filtration of E. This works only, if the transition functions of a given vector bundle atlas have values in filtration preserving endomorphisms of V.

Of course, there is neither a natural bundle map from E to $\operatorname{gr}(E)$ nor a natural bundle map from $\operatorname{gr}(E)$ to E. However, as in the case of vector spaces, there is a distinguished class of bundle isomorphisms between E and $\operatorname{gr}(E)$. Since any exact sequence of vector bundles splits, one can always find a smooth subbundle E_i in the filtration component $E^i \subset E$ such that $E^i = E_i \oplus E^{i+1}$. The natural projection $E^i \to \operatorname{gr}_i(E)$ of course induces an isomorphism $E_i \cong \operatorname{gr}_i(E)$, and putting these together we get an isomorphism $E \cong \operatorname{gr}(E)$. Also, the natural constructions work exactly as in the case of vector spaces, so we have canonical filtrations on subbundles, quotient bundles and the dual bundle of a filtered vector bundle, as well as on tensor products of filtered vector bundles, and bundles of linear and multilinear bundle maps between filtered vector bundles. The descriptions of the associated graded bundles is analogous to the case of vector spaces.

Filtered Lie algebras. Finally, we want to consider algebraic structures on filtered vector spaces. We only discuss Lie algebras, but other types of algebra structures can be treated similarly. A *filtered Lie algebra* is a Lie algebra $(\mathfrak{g}, [\ ,\ ])$ together with a filtration $\{\mathfrak{g}^i : i \in \mathbb{Z}\}$ of the vector space $\mathfrak{g}$ such that for all $i, j \in \mathbb{Z}$ we have $[\mathfrak{g}^i, \mathfrak{g}^j] \subset \mathfrak{g}^{i+j}$.

In particular, for each $i \geq 0$, the subspace $\mathfrak{g}^i \subset \mathfrak{g}$ is a Lie subalgebra, and for each $i > 0$ the subspace $\mathfrak{g}^i$ is an ideal in the Lie algebra $\mathfrak{g}^0$. Moreover, if the filtration is finite (in the positive direction), i.e. if there is a $j \in \mathbb{Z}$ such that $\mathfrak{g}^j = \{0\}$, then the subalgebra $\mathfrak{g}^1 \subset \mathfrak{g}$ is nilpotent.

Next, consider the associated graded vector space $\mathrm{gr}(\mathfrak{g})$. The Lie bracket may be viewed as a filtration preserving map $\mathfrak{g} \otimes \mathfrak{g} \to \mathfrak{g}$, so we get an induced mapping on the associated graded, i.e. a map $\mathrm{gr}(\mathfrak{g}) \otimes \mathrm{gr}(\mathfrak{g}) \to \mathrm{gr}(\mathfrak{g})$. By construction, this is skew symmetric, and using the functorial properties of the maps induced on the associated graded vector space, one easily verifies that it satisfies the Jacobi identity, thus making $\mathrm{gr}(\mathfrak{g})$ into a graded Lie algebra.

We will mostly deal with the case where the filtration on $\mathfrak{g}$ actually comes from a grading, i.e. we start with a graded Lie algebra $\mathfrak{g} = \mathfrak{g}_k \oplus \cdots \oplus \mathfrak{g}_\ell$ for some $k < \ell \in \mathbb{Z}$ such that $[\mathfrak{g}_i, \mathfrak{g}_j] \subset \mathfrak{g}_{i+j}$ (where $\mathfrak{g}_n = \{0\}$ for $n < k$ or $n > \ell$) and define $\mathfrak{g}^i = \bigoplus_{j \geq i} \mathfrak{g}_j$. Clearly, $[\mathfrak{g}^i, \mathfrak{g}^j] \subset \mathfrak{g}^{i+j}$, so this makes $\mathfrak{g}$ into a filtered Lie algebra, which is isomorphic to the associated graded $\mathrm{gr}(\mathfrak{g})$ as a Lie algebra. However, even in this simple situation it will be important to carefully distinguish between the filtration and the associated grading.

One way to obtain a filtration on a Lie algebra is to choose a filtration of a representation space. Indeed, suppose that $\phi : \mathfrak{g} \to L(V, V)$ is a representation of a Lie algebra $\mathfrak{g}$, and that $\{V^i : i \in \mathbb{Z}\}$ is a filtration on the vector space V. Then we have the induced filtration on $L(V, V)$ and thus on the linear subspace $\phi(\mathfrak{g})$. Explicitly, $A \in \mathfrak{g}^i$ if and only if $A(V^j) \subset V^{i+j}$ for all $j \in \mathbb{Z}$. Then the filtration $\{\mathfrak{g}^i : i \in \mathbb{Z}\}$ automatically makes $\mathfrak{g}$ into a filtered Lie algebra: For $A \in \mathfrak{g}^i$, $B \in \mathfrak{g}^j$ and $v \in V^k$, we have $[A, B]\cdot v = A\cdot B\cdot v - B\cdot A\cdot v$, and by definition both summands lie in V^{i+j+k}, and thus $[A, B] \in \mathfrak{g}^{i+j}$.

Similarly, a grading $V = \bigoplus V_i$ on a representation space for $\mathfrak{g}$ induces a grading on the Lie algebra $\mathfrak{g}$, by defining $\mathfrak{g}_j$ as the space of those elements $A \in \mathfrak{g}$ such that for each $i \in \mathbb{Z}$ and $v \in V_i$ we have $A \cdot v \in V_{i+j}$. The same computation as above shows that for $A \in \mathfrak{g}_i$ and $B \in \mathfrak{g}_j$ we get $[A, B] \in \mathfrak{g}_{i+j}$, whence this makes $\mathfrak{g}$ into a graded Lie algebra.

These constructions are compatible with the passage from a grading to a filtration. Indeed, if a grading on $\mathfrak{g}$ is induced by a grading of the representation space V, then the associated filtration on $\mathfrak{g}$ comes from the associated filtration on V.

3.1.2. $|k|$–graded semisimple Lie algebras.

DEFINITION 3.1.2. Let $\mathfrak{g}$ be a semisimple Lie algebra and let $k > 0$ be an integer. A *$|k|$–grading* on $\mathfrak{g}$ is a decomposition $\mathfrak{g} = \mathfrak{g}_{-k} \oplus \cdots \oplus \mathfrak{g}_k$ of $\mathfrak{g}$ into a direct sum of subspaces such that

- $[\mathfrak{g}_i, \mathfrak{g}_j] \subset \mathfrak{g}_{i+j}$, where we agree that $\mathfrak{g}_i = \{0\}$ for $|i| > k$,
- the subalgebra $\mathfrak{g}_- := \mathfrak{g}_{-k} \oplus \cdots \oplus \mathfrak{g}_{-1}$ is generated (as a Lie algebra) by $\mathfrak{g}_{-1}$,
- $\mathfrak{g}_{-k} \neq \{0\}$ and $\mathfrak{g}_k \neq \{0\}$.

By definition, if $\mathfrak{g} = \mathfrak{g}_{-k} \oplus \cdots \oplus \mathfrak{g}_k$ is a $|k|$–grading, then $\mathfrak{p} := \mathfrak{g}_0 \oplus \cdots \oplus \mathfrak{g}_k$ is a subalgebra of $\mathfrak{g}$, and $\mathfrak{p}_+ := \mathfrak{g}_1 \oplus \cdots \oplus \mathfrak{g}_k$ is a nilpotent ideal in $\mathfrak{p}$. Similarly, the subalgebra $\mathfrak{g}_- = \mathfrak{g}_{-k} \oplus \cdots \oplus \mathfrak{g}_{-1}$ is nilpotent by the grading property.

By the grading property, $\mathfrak{g}_0 \subset \mathfrak{g}$ is a subalgebra, and the adjoint action makes each $\mathfrak{g}_i$ into a $\mathfrak{g}_0$–module, such that the bracket $[\ ,\ ] : \mathfrak{g}_i \otimes \mathfrak{g}_j \to \mathfrak{g}_{i+j}$ is a $\mathfrak{g}_0$–homomorphism. The central object for the further study will be the pair $(\mathfrak{g}, \mathfrak{p})$, while $\mathfrak{g}_0$ is an auxiliary object, which is usually easier to deal with. We will always have to distinguish carefully between $\mathfrak{g}_0$–invariant data and $\mathfrak{p}$–invariant data in the sequel. What makes life more complicated is that $\mathfrak{g}_0 = \mathfrak{p}/\mathfrak{p}_+$, so $\mathfrak{g}_0$ naturally is a quotient of $\mathfrak{p}$. Hence, we can view any $\mathfrak{g}_0$ module at the same time as a $\mathfrak{p}$–module with trivial action of $\mathfrak{p}_+$.

Since the object of main interest is the subalgebra $\mathfrak{p}$, it is clear that the grading of $\mathfrak{g}$ (which of course is not $\mathfrak{p}$–invariant) will be of minor importance, while the main object is the associated filtration $\mathfrak{g} = \mathfrak{g}^{-k} \supset \mathfrak{g}^{-k+1} \supset \cdots \supset \mathfrak{g}^k$ defined by $\mathfrak{g}^i := \bigoplus_{j \geq i} \mathfrak{g}_j$. Then by the grading property $(\mathfrak{g}, \{\mathfrak{g}^i\})$ is a filtered Lie algebra and by definition $\mathfrak{p} = \mathfrak{g}^0$ and $\mathfrak{p}_+ = \mathfrak{g}^1$. In particular, any filtration component $\mathfrak{g}^i \subset \mathfrak{g}$ is a $\mathfrak{p}$ submodule. Hence, the quotient $\operatorname{gr}_i(\mathfrak{g}) = \mathfrak{g}^i/\mathfrak{g}^{i+1}$ naturally is a $\mathfrak{p}$–module, and by the filtration property $\mathfrak{p}_+$ acts trivially on this quotient. Hence, $\operatorname{gr}_i(\mathfrak{g})$ is simply $\mathfrak{g}_i$ with the $\mathfrak{g}_0$–action trivially extended to $\mathfrak{p}$. Thus, $\operatorname{gr}(\mathfrak{g})$ becomes a $\mathfrak{p}$–module with trivial action of $\mathfrak{p}_+$.

Let us collect some basic properties of $\mathfrak{g}$ in the following:

PROPOSITION 3.1.2. *Let $\mathfrak{g} = \mathfrak{g}_{-k} \oplus \cdots \oplus \mathfrak{g}_k$ be a $|k|$–graded semisimple Lie algebra over $\mathbb{K} = \mathbb{R}$ or $\mathbb{C}$ and let $B : \mathfrak{g} \times \mathfrak{g} \to \mathbb{K}$ be a nondegenerate invariant bilinear form; see 2.1.5. Then we have:*

(1) *There is a unique element $E \in \mathfrak{g}$, called the grading element, such that $[E, X] = jX$ for all $X \in \mathfrak{g}_j$, $j = -k, \dots, k$. The element E lies in the center of the subalgebra $\mathfrak{g}_0 \leq \mathfrak{g}$.*
(2) *The $|k|$–grading on $\mathfrak{g}$ induces a $|k_i|$–grading for some $k_i \leq k$ on each ideal $\mathfrak{s} \subset \mathfrak{g}$. In particular, $\mathfrak{g}$ is direct sum of $|k_i|$–graded simple Lie algebras, where $k_i \leq k$ for all i and $k_i = k$ for at least one i.*
(3) *The isomorphism $\mathfrak{g} \to \mathfrak{g}^*$ provided by B is compatible with the filtration and the grading of $\mathfrak{g}$. In particular, B induces dualities of $\mathfrak{g}_0$–modules between $\mathfrak{g}_i$ and $\mathfrak{g}_{-i}$, and the filtration component $\mathfrak{g}^i$ is exactly the annihilator (with respect to B) of $\mathfrak{g}^{-i+1}$. Hence, B induces a duality of $\mathfrak{p}$–modules between $\mathfrak{g}/\mathfrak{g}^{-i+1}$ and $\mathfrak{g}^i$, and in particular between $\mathfrak{g}/\mathfrak{p}$ and $\mathfrak{p}_+$.*
(4) *For $i < 0$ we have $[\mathfrak{g}_{i+1}, \mathfrak{g}_{-1}] = \mathfrak{g}_i$. If no simple ideal of $\mathfrak{g}$ is contained in $\mathfrak{g}_0$, then this also holds for $i = 0$.*
(5) *Let $A \in \mathfrak{g}_i$ with $i > 0$ be an element such that $[A, X] = 0$ for all $X \in \mathfrak{g}_{-1}$. Then $A = 0$. If no simple ideal of $\mathfrak{g}$ is contained in $\mathfrak{g}_0$, then this also holds for $i = 0$.*

PROOF. (1) Consider the map $D : \mathfrak{g} \to \mathfrak{g}$ which is defined by $D(X) = jX$ for $X \in \mathfrak{g}_j$, $j = -k, \dots, k$. Since $[\mathfrak{g}_i, \mathfrak{g}_j] \subset \mathfrak{g}_{i+j}$, we immediately conclude that $D([X, Y]) = [D(X), Y] + [X, D(Y)]$ for all $X, Y \in \mathfrak{g}$, i.e. D is a derivation. By part (2) of Corollary 2.1.6, semisimplicity of $\mathfrak{g}$ implies that there is a unique element $E \in \mathfrak{g}$ such that $D(X) = [E, X]$. Decomposing $E = E_{-k} + \cdots + E_k$ with $E_i \in \mathfrak{g}_i$, we get $0 = [E, E] = \sum_{j=-k}^{k} [E, E_j] = \sum_{j=-k}^{k} jE_j$, whence $E = E_0 \in \mathfrak{g}_0$. By definition $[E, A] = 0$ for all $A \in \mathfrak{g}_0$, so E lies in the center of $\mathfrak{g}_0$.

(2) Let $\mathfrak{s} \subset \mathfrak{g}$ be an ideal. By (1), the grading components $\mathfrak{g}_i$ are the eigenspaces for $\operatorname{ad}(E)$, where E denotes the grading element. The projections onto these eigenspaces can be written as polynomials in $\operatorname{ad}(E)$. Any such polynomial maps the ideal $\mathfrak{s}$ to itself. Hence, if $X \in \mathfrak{s}$ decomposes as $X = X_{-k} + \cdots + X_k$ according to the grading, then each X_j lies in $\mathfrak{s}$. This implies that we get an induced grading on $\mathfrak{s}$. The second statement follows since by Corollary 2.1.5, $\mathfrak{g}$ can be written as a direct sum of simple ideals.

(3) Invariance of B implies that $B([E,X],Y) = -B(X,[E,Y])$ for all $X, Y \in \mathfrak{g}$. For $X \in \mathfrak{g}_i$ and $Y \in \mathfrak{g}_j$, we get $0 = (i+j)B(X,Y)$, whence $B(X,Y) = 0$ unless $i + j = 0$. Nondegeneracy of B now immediately implies that its restrictions to $\mathfrak{g}_0 \times \mathfrak{g}_0$ and $\mathfrak{g}_j \times \mathfrak{g}_{-j}$ for $j = 1, \dots, k$ are nondegenerate. Hence, the isomorphism $\mathfrak{g} \to \mathfrak{g}^*$ induced by B is compatible with the grading. The resulting dualities are $\mathfrak{g}_0$–equivariant by invariance of B.

The compatibility with the grading also implies that the restriction of B to $\mathfrak{g}^{-i+1} \times \mathfrak{g}^i$ vanishes, so B induces a bilinear form $\mathfrak{g}/\mathfrak{g}^{-i+1} \times \mathfrak{g}^i \to \mathbb{K}$. Now $\mathfrak{g}/\mathfrak{g}^{-i+1}$ is linearly isomorphic to $\mathfrak{g}_{-k} \oplus \cdots \oplus \mathfrak{g}_{-i}$ and hence has the same dimension as $\mathfrak{g}^i = \mathfrak{g}_i \oplus \cdots \oplus \mathfrak{g}_k$. Nondegeneracy of B hence implies that we get a duality between the two spaces, which is compatible with the $\mathfrak{p}$–actions by invariance of B. The final statement is just the special case $i = 1$.

(4) The condition that $\mathfrak{g}_-$ is generated by $\mathfrak{g}_{-1}$ immediately implies the statement for $i \leq -2$. Since $[E, X] = -X$ for $X \in \mathfrak{g}_{-1}$, we get the statement for $i = -1$. Finally, using the fact that $\mathfrak{g}_-$ is generated by $\mathfrak{g}_{-1}$, one immediately verifies that $[\mathfrak{g}_1, \mathfrak{g}_{-1}] \oplus \bigoplus_{i \neq 0} \mathfrak{g}_i$ is an ideal in $\mathfrak{g}$. By part (2), there is a complementary ideal which has to be contained in $\mathfrak{g}_0$, so the last part follows.

(5) Let B be the Killing form of $\mathfrak{g}$, which is nondegenerate and invariant; see 2.1.5. For any $Z \in \mathfrak{g}_{-i+1}$ and $X \in \mathfrak{g}_{-1}$ we get $0 = B([A,X],Z) = B(A,[X,Z])$, so A is orthogonal with respect to the Killing form to $[\mathfrak{g}_{-1}, \mathfrak{g}_{-i+1}]$, so the result follows from (4) and (3). □

REMARK 3.1.2. The proofs of parts (1) and (3) show that any grading on a semisimple Lie algebra $\mathfrak{g}$ must be symmetric around zero, i.e. it must have the form $\mathfrak{g}_{-k} \oplus \cdots \oplus \mathfrak{g}_k$ for some k.

EXAMPLE 3.1.2. Using the structure theory of semisimple Lie algebras, one gets a nice description of all possible $|k|$–gradings. We will discuss this in detail in Section 3.2, so we just describe some examples that we have met already.

(1) Put $\mathfrak{g} = \mathfrak{so}(p+1, q+1)$. In our study of conformal structures in 1.6.3 we have met the decomposition $\mathfrak{g} = \mathfrak{g}_{-1} \oplus \mathfrak{g}_0 \oplus \mathfrak{g}_1$ with $\mathfrak{g}_{-1} \cong \mathbb{R}^{p+q}$, $\mathfrak{g}_1 \cong \mathbb{R}^{(p+q)*}$ and $\mathfrak{g}_0 = \mathfrak{co}(p,q)$, the conformal algebra of signature (p,q), so this gives an example of a $|1|$–grading. Notice that the adjoint action of $\mathfrak{g}_0$ on $\mathfrak{g}_{-1}$ is exactly the standard action of $\mathfrak{co}(p,q)$ on $\mathbb{R}^{p+q}$ which is the basis for the relation of this grading to conformal geometry.

This example can be nicely understood in the setting of filtrations and gradings on a Lie algebra induced by filtrations and gradings on a representation space. The representation involved here is the standard representation $V = \mathbb{R}^{p+q+2}$. The filtration on V has the form $V \supset V^0 \supset V^1$, and it is given by choosing a null–line V^1 (which is spanned by the first basis vector in the presentation of 1.6.3) and putting $V^0 := (V^1)^\perp$. In particular, we see immediately that $\mathfrak{p}$ is exactly the stabilizer of the null line V^1 in this case. To come to the grading, one simply has to choose a null line V_{-1} such that the inner product is nondegenerate on $V_{-1}\oplus V^1$. In the presentation of 1.6.3, the null line V_{-1} is spanned by the last basis vector. Having chosen V_{-1}, one defines $V_0:=(V_{-1})^\perp\cap V^0$ and $V_1=V^1$ to get the grading $V=V_{-1}\oplus V_0\oplus V_1$. Naively, one would expect that this leads to a grading of the form $\mathfrak{g}=\mathfrak{g}_{-2}\oplus\cdots\oplus\mathfrak{g}_2$; but it turns out that $\mathfrak{g}_{\pm 2}=\{0\}$, so we obtain the grading $\mathfrak{g}=\mathfrak{g}_{-1}\oplus\mathfrak{g}_0\oplus\mathfrak{g}_1$.

(2) Consider a $\mathbb{K}$–vector space V of dimension $p+q$ and the Lie algebra $\mathfrak{g}=\mathfrak{sl}(V)$. Choosing a subspace $V^1\subset V$ of dimension p, we get a filtration of the form $V=V^0\supset V^1$, which induces a filtration $\mathfrak{g}=\mathfrak{g}^{-1}\supset\mathfrak{g}^0\supset\mathfrak{g}^1$ on $\mathfrak{g}$ as described in 3.1.1. Choosing a complementary space V_0 such that $V=V_0\oplus V_1$ with $V_1=V^1$ (whence $\dim(V_0)=q$) the procedure from 3.1.1 induces a $|1|$–grading on $\mathfrak{g}$. Putting $V=\mathbb{K}^{p+q}$, V_1 the span of the first p–basis vectors and V_0 the span of the last q basis vectors, we get a nice block presentation of this $|1|$–grading. Writing elements of $\mathfrak{sl}(p+q,\mathbb{K})$ as block matrices $\left(\begin{smallmatrix} A & B\\ C & D\end{smallmatrix}\right)$ with blocks of size p and q, the block corresponding to C has degree -1, the block corresponding to A and D has degree zero and the block corresponding to B has degree one. In particular, $\mathfrak{p}$ is the stabilizer of the subspace $\mathbb{K}^p\subset\mathbb{K}^{p+q}$. The special case $p=1$ is exactly the situation which occurred in the discussion of the projective sphere in 1.1.3.

This example can be easily generalized using more complicated filtrations, respectively, block decompositions. The extremal case is to consider the filtration $\mathbb{K}^n\supset\mathbb{K}^{n-1}\supset\cdots\supset\mathbb{K}^2\supset\mathbb{K}$, respectively, the grading $\mathbb{K}^n=\mathbb{K}\oplus\cdots\oplus\mathbb{K}$. In matrix terms, this means that in $\mathfrak{g}:=\mathfrak{sl}(n,\mathbb{K})$ one assigns to the entry in the ith row and the jth column degree $j-i$. One obtains an $|n-1|$–grading, for which $\mathfrak{p}$ is the subalgebra of tracefree upper triangular matrices and $\mathfrak{p}_+$ is the subalgebra of strictly upper triangular matrices.

(3) A complex analog of the construction in example (1) above leads to a $|2|$–grading on the Lie algebra $\mathfrak{g}=\mathfrak{su}(n+1,1)$ which underlies CR–geometry; compare to 1.1.6. Take a complex vector space V of dimension $n+2$ endowed with a Hermitian form $\langle\ ,\ \rangle$ of signature $(n+1,1)$ and choose a null line $V^1\subset V$. Putting $V^0:=(V^1)^\perp$, we get a filtration $V\supset V^0\supset V^1$ of V by complex subspaces. This comes from the grading $V=V_{-1}\oplus V_0\oplus V_1$ given by choosing a null line V_{-1} such that the Hermitian form is nondegenerate on $V_{-1}\oplus V_1$ and putting $V_0:=V^0\cap(V_{-1})^\perp$ and $V_1=V^1$. Via the construction from 3.1.1 this gives rise to a $|2|$–grading on $\mathfrak{g}$, and the associated filtration has the property that $\mathfrak{p}=\mathfrak{g}^0$ is exactly the stabilizer of the null line V^1.

Again, this may also be nicely presented by a block decomposition of matrices: Put $V=\mathbb{C}^{n+2}$ and the Hermitian form $\langle z,w\rangle:=z_0\bar w_{n+1}+z_{n+1}\bar w_0+\sum_{j=1}^n z_j\bar w_j$, which is of signature $(n+1,1)$. This form is chosen in such a way that we may take V^1 to be spanned by the basis vector e_0 and V_{-1} to be spanned by e_{n+1}. Then the Lie algebra $\mathfrak{g}$ is the space of all tracefree matrices M such that $M^*\mathbb{J}=-\mathbb{J}M$, where $\mathbb{J}=\left(\begin{smallmatrix} 0 & 0 & 1\\ 0 & \mathbb{I}_n & 0\\ 1 & 0 & 0\end{smallmatrix}\right)$ with $\mathbb{I}_n$ the $n\times n$ unit matrix. A short computation shows that M has to be of the form $\begin{pmatrix} a & Z & iz\\ X & A-\frac{2i}{n}\operatorname{im}(a)\mathbb{I}_n & -Z^*\\ ix & -X^* & -\bar a\end{pmatrix}$ with $A\in\mathfrak{su}(n)$, $a\in\mathbb{C}$, $X\in\mathbb{C}^n$, $Z\in\mathbb{C}^{n*}$, and $x,z\in\mathbb{R}$. By construction, the entry ix has degree -2, X has degree -1, a and A have degree zero, Z has degree one, and iz has degree two.

3.1.3. The group level. Consider a $|k|$–graded semisimple Lie algebra $\mathfrak{g} = \mathfrak{g}_{-k} \oplus \cdots \oplus \mathfrak{g}_k$ and a Lie group G with Lie algebra $\mathfrak{g}$. Our next task is to study subgroups of G corresponding to the Lie subalgebras $\mathfrak{g}_0 \subset \mathfrak{p} \subset \mathfrak{g}$. Recall that for a subset $A \subset \mathfrak{g}$, the normalizer $N_G(A)$ of A in G is defined as $\{g \in G : \mathrm{Ad}(g)(A) \subset A\}$. As before, we denote by $\{\mathfrak{g}^i\}$ the filtration induced by the $|k|$–grading.

LEMMA 3.1.3. *(1) The Lie subalgebras $\mathfrak{g}_0 \subset \mathfrak{p} \subset \mathfrak{g}$ can be characterized as*

$$\mathfrak{g}_0 = \{X \in \mathfrak{g} : \mathrm{ad}(X)(\mathfrak{g}_i) \subset \mathfrak{g}_i \textit{ for all } i = -k, \dots, k\},$$
$$\mathfrak{p} = \{X \in \mathfrak{g} : \mathrm{ad}(X)(\mathfrak{g}^i) \subset \mathfrak{g}^i \textit{ for all } i = -k, \dots, k\}.$$

(2) Let G be a (not necessarily connected) Lie group with Lie algebra $\mathfrak{g}$. Then $P := \bigcap_{i=-k}^{k} N_G(\mathfrak{g}^i) \subset G$ is a closed subgroup with Lie algebra $\mathfrak{p}$.

PROOF. (1) In both statements, the inclusion $\subset$ is clear. Conversely, take $X \in \mathfrak{g}$ and decompose it as $X_{-k}+\cdots+X_k$ according to the grading. For the grading element $E \in \mathfrak{g}_0 \subset \mathfrak{p}$ we obtain $\mathrm{ad}(X)(E) = kX_{-k} + \cdots + X_{-1} - X_1 - \cdots - kX_k$. Consequently, if $\mathrm{ad}(X)(\mathfrak{g}_0) \subset \mathfrak{g}_0$, then $X_i = 0$ for $i \neq 0$ and hence $X \in \mathfrak{g}_0$. Likewise, if $\mathrm{ad}(X)(\mathfrak{p}) \subset \mathfrak{p}$ we must have $X_i = 0$ for all $i < 0$, and hence $X \in \mathfrak{p}$.

(2) For each i, the subset $\mathfrak{g}^i \subset \mathfrak{g}$ is closed, and hence the normalizer $N_G(\mathfrak{g}^i)$ is a closed subgroup of G, so $P \subset G$ is a closed subgroup. Since $\mathrm{Ad}(\exp(X)) = e^{\mathrm{ad}(X)}$, we see that the Lie algebra of P is formed by all elements $X \in \mathfrak{g}$ such that $\mathrm{ad}(X)(\mathfrak{g}^i) \subset \mathfrak{g}^i$ for all i. Now the result follows from (1). □

DEFINITION 3.1.3. Let $\mathfrak{g} = \mathfrak{g}_{-k}\oplus\cdots\oplus\mathfrak{g}_k$ be a k–graded semisimple Lie algebra and let G be a Lie group with Lie algebra $\mathfrak{g}$.

(1) A *parabolic subgroup* of G corresponding to the given $|k|$–grading is a subgroup $P \subset G$ which lies between $\bigcap_{i=-k}^{k} N_G(\mathfrak{g}^i)$ and its connected component of the identity.

(2) Given a parabolic subgroup $P \subset G$ we define the *Levi subgroup* $G_0 \subset P$ by

$$G_0 := \{g \in P : \mathrm{Ad}(g)(\mathfrak{g}_i) \subset \mathfrak{g}_i \text{ for all } i = -k, \dots, k\}.$$

Note that by definition, any parabolic subgroup $P \subset G$ is closed and has Lie algebra $\mathfrak{p}$. Further, $G_0 \subset P$ is an intersection of normalizers of closed subsets, and hence a closed subgroup of P. From the lemma and its proof we see that $G_0 \subset P$ corresponds to the Lie subalgebra $\mathfrak{g}_0 \subset \mathfrak{p}$. The names "parabolic subgroup" and "Levi subgroup" are motivated by the relation to the structure theory of semisimple Lie algebras which will be discussed in Section 3.2.

THEOREM 3.1.3. *Let $\mathfrak{g} = \mathfrak{g}_{-k} \oplus \cdots \oplus \mathfrak{g}_k$ be a k–graded semisimple Lie algebra, and let G be a Lie group with Lie algebra $\mathfrak{g}$. Let $P \subset G$ be a parabolic subgroup for the given grading and let $G_0 \subset P$ be the Levi subgroup.*

Then $(g_0, Z) \mapsto g_0 \exp(Z)$ defines a diffeomorphism $G_0 \times \mathfrak{p}_+ \to P$, and

$$(g_0, Z_1, \dots, Z_k) \mapsto g_0 \exp(Z_1) \cdots \exp(Z_k)$$

is a diffeomorphism $G_0 \times \mathfrak{g}_1 \times \dots \times \mathfrak{g}_k \to P$.

PROOF. Both maps are obviously smooth, and their tangent maps in $(e, 0)$ (respectively $(e, 0, \dots, 0)$) are just the identity $\mathfrak{g}_0 \times \mathfrak{p}_+ \to \mathfrak{p}$, respectively, $\mathfrak{g}_0 \times \dots \times \mathfrak{g}_k \to \mathfrak{p}$. More generally, since for all elements from $\mathfrak{p}_+$ the adjoint action is a nilpotent endomorphism of $\mathfrak{p}$, and thus has only zero eigenvalues, the tangent map of the exponential mapping in any point of $\mathfrak{p}_+$ is injective; see [**KMS**, Corollary

4.28]. This implies that both maps are diffeomorphisms locally around any point of the form (e, Z) (respectively $(e, Z_1, \dots, Z_k)$. Since both maps are equivariant for the left multiplication of G_0, we conclude that they both are diffeomorphisms locally around each point in their domain of definition. Hence, we only have to show that the maps are bijective.

We first show that the second map is surjective. Given $g \in P$, consider the adjoint action $\mathrm{Ad}(g) : \mathfrak{g} \to \mathfrak{g}$. This is an automorphism of the filtered Lie algebra $\mathfrak{g}$, so it induces a linear map $\phi_0 = \mathrm{gr}_0(\mathrm{Ad}(g))$ on the associated graded $\mathrm{gr}(\mathfrak{g})$, which is homogeneous of degree zero; see 3.1.1. Explicitly, for $X \in \mathfrak{g}_j$ the element $\phi_0(X)$ is the $\mathfrak{g}_j$–component of $\mathrm{Ad}(g)(X) \in \mathfrak{g}^j$. One immediately verifies that ϕ_0 is a Lie algebra homomorphism, and an inverse to ϕ_0 can be constructed in the same way starting from $\mathrm{Ad}(g^{-1})$. Hence, ϕ_0 is an automorphism of the *graded* Lie algebra $\mathfrak{g}$. By construction, for an element $Y \in \mathfrak{g}_j$ we have $\mathrm{Ad}(g)(Y) - \phi_0(Y) \in \mathfrak{g}^{j+1}$.

Put $\phi_1 := \phi_0^{-1} \circ \mathrm{Ad}(g)$ and consider $Y \in \mathfrak{g}_j$. By construction, $\phi_1(Y) - Y \in \mathfrak{g}^{j+1}$. In particular, for the grading element E, we have $E - \phi_1(E) \in \mathfrak{g}^1$, and we denote by Z_1 the $\mathfrak{g}_1$–component of this element. Then $\phi_1(E)$ is congruent to $E - Z_1$ modulo $\mathfrak{g}^2$. Moreover, since $Z_1 \in \mathfrak{g}_1$, we get $\mathrm{ad}(-Z_1)(E - Z_1) = Z_1$, so $\mathrm{ad}(-Z_1)^2((E - Z_1)) = 0$, and thus

$$\mathrm{Ad}(\exp(-Z_1))(E - Z_1) = e^{\mathrm{ad}(-Z_1)}(E - Z_1) = E.$$

Putting $\phi_2 = \mathrm{Ad}(\exp(-Z_1)) \circ \phi_1$ we see that $\phi_2(E)$ is congruent to E modulo $\mathfrak{g}^2$. Further, for each $Y \in \mathfrak{g}_j$ the element $\phi_2(Y)$ is congruent to Y modulo $\mathfrak{g}^{j+1}$. Inductively, we find elements $Z_i \in \mathfrak{g}_i$ and automorphisms ϕ_i of $\mathfrak{g}$ of the form $\phi_i = \mathrm{Ad}(\exp(-Z_{i-1})) \circ \phi_{i-1}$, such that $\phi_i(E)$ is congruent to E modulo $\mathfrak{g}^i$, and $\phi_i(Y)$ is congruent to Y modulo $\mathfrak{g}^{j+1}$ for each $Y \in \mathfrak{g}_j$.

The automorphism ϕ_{k+1} then by construction satisfies $\phi_{k+1}(E) = E$. Hence, for $Y \in \mathfrak{g}_j$ we see that $[E, \phi_{k+1}(Y)] = \phi_{k+1}([E, Y]) = j\phi_{k+1}(Y)$, so $\phi_{k+1}(Y) \in \mathfrak{g}_j$. But by construction, $\phi_{k+1}(Y)$ is congruent to Y modulo $\mathfrak{g}^{j+1}$, so $\phi_{k+1}(Y) = Y$. Thus, we can write the identity map as

$$\mathrm{Ad}(\exp(-Z_k)) \circ \cdots \circ \mathrm{Ad}(\exp(-Z_1)) \circ \phi_0^{-1} \circ \mathrm{Ad}(g).$$

This shows that ϕ_0 is the adjoint action of $g_0 := g \exp(-Z_k) \cdots \exp(-Z_1)$. By definition, this implies $g_0 \in G_0$ and $g = g_0 \exp(Z_1) \cdots \exp(Z_k)$ as required. By the Baker–Campbell–Hausdorff formula, we may rewrite $\exp(Z_1) \dots \exp(Z_k)$ as $\exp(Z)$ for some element $Z \in \mathfrak{p}_+$, so we have proved surjectivity for both maps.

To prove injectivity, note first that for $Z \in \mathfrak{p}_+$ and $Y \in \mathfrak{g}_j$, the element $\mathrm{Ad}(\exp(Z))(Y)$ is congruent to Y modulo $\mathfrak{g}^{j+1}$. Hence, for $Y \in \mathfrak{g}_j$ we can recover $\mathrm{Ad}(g_0)(Y)$ as the $\mathfrak{g}_j$–component of $\mathrm{Ad}(g_0 \exp(Z))(Y)$. Knowing $\mathrm{Ad}(g_0)$, we also get $\mathrm{Ad}(\exp(Z))$. Next, we can determine the lowest nonzero homogeneous component of Z as the lowest nonzero homogeneous component of $\mathrm{Ad}(\exp(Z))(E) - E \in \mathfrak{p}_+$. Step by step we may then determine the higher homogeneous components of Z in the same way, and once we have determined Z we can also recover g_0. Hence, we have proved that the first map is injective and thus a diffeomorphism.

For the second map, one may recover $\mathrm{Ad}(g_0)$ from $\mathrm{Ad}(g_0 \exp(Z_1) \cdots \exp(Z_k))$ as above. Splitting off $\mathrm{Ad}(g_0)$ we can recover Z_1 as the negative of the $\mathfrak{g}_1$–component of $\mathrm{Ad}(\exp(Z_1) \cdots \exp(Z_k))(E)$. Knowing this, we step by step compute all the Z_j, and thus also g_0. □

The second part of this theorem immediately suggests how to define further subgroups of P. Namely, we define $P_+ \subset P$ to be the image of $\mathfrak{p}_+$ under the exponential mapping. By part (2) of the theorem, $\exp : \mathfrak{p}_+ \to P_+$ is a global diffeomorphism and $P_+ \subset P$ is a closed nilpotent subgroup. For $Z \in \mathfrak{p}_+$ and $g \in G$ we have $g\exp(Z)g^{-1} = \exp(\mathrm{Ad}(g)(Z))$, and for $g \in P$ we have $\mathrm{Ad}(g)(Z) \in \mathfrak{p}_+$, which shows that $P_+ \subset P$ is a normal subgroup. Finally, again by part (2) of the above theorem $P/P_+ \cong G_0$. In particular, this shows that P is the semidirect product of the subgroup G_0 and the nilpotent normal vector subgroup P_+.

This can be easily generalized by replacing $\mathfrak{p}_+ = \mathfrak{g}^1$ by $\mathfrak{g}^i$ for $i \geq 2$. As we shall see later on, $\mathfrak{g}^2 = [\mathfrak{p}_+, \mathfrak{p}_+]$, and more generally $\mathfrak{g}^i$ is the ith power of $\mathfrak{p}_+$ for all $i = 1, \dots, k$, so we denote the exponential image of $\mathfrak{g}^i$ by P^i_+. Then $\exp : \mathfrak{g}^i \to P^i_+$ is a global diffeomorphism, $P^i_+ \subset P$ is a closed normal nilpotent subgroup. In particular, for any $i = 2, \dots, k$ we may consider the quotient group P/P^i_+, which is the semidirect product of the subgroup G_0 and the nilpotent normal vector subgroup $P_+/P^i_+ \subset P/P^i_+$.

3.1.4. Definition and basic properties of parabolic geometries.

DEFINITION 3.1.4. A *parabolic geometry* is a Cartan geometry of type (G, P), where G is a semisimple Lie group and $P \subset G$ is a parabolic subgroup corresponding to some $|k|$–grading of the Lie algebra $\mathfrak{g}$ of G. We will use the terminology "parabolic geometry of type (G, P)" in this situation.

We can easily deduce some basic properties of parabolic geometries by applying the theory developed in Section 1.5. These properties are determined by properties of the corresponding Klein geometry (see 1.4.1), so we have to look at the homogeneous models G/P.

PROPOSITION 3.1.4. *Let $\mathfrak{g}$ be a $|k|$–graded semisimple Lie algebra, G a Lie group with Lie algebra $\mathfrak{g}$, and $P \subset G$ a parabolic subgroup corresponding to the $|k|$–grading. Then we have:*

(1) The Klein geometry (G, P) is infinitesimally effective if and only if no simple ideal of $\mathfrak{g}$ is contained in $\mathfrak{g}_0$. Assuming this condition, the kernel K of this Klein geometry is a discrete normal subgroup of G_0 and any morphism between parabolic geometries of type (G, P) is determined by its base map up to multiplication by a locally constant function with values in K.

(2) The Klein geometry (G, P) is naturally split using the subalgebra $\mathfrak{g}_- \subset \mathfrak{g}$ as a complement to $\mathfrak{p}$. However, it is very far from being reductive: For any subspace $\mathfrak{n} \subset \mathfrak{g}$ which is complementary to $\mathfrak{p}$, the $\mathfrak{p}$–module generated by $\mathfrak{n}$ is the direct sum of all simple ideals of $\mathfrak{g}$ which are not contained in $\mathfrak{g}_0$.

(3) Parabolic geometries of type (G, P) do not admit natural linear connections on the tangent bundle. Hence, they do not admit natural pseudo–Riemannian metrics either.

PROOF. (1) Recall from 1.4.1 that the kernel K of the Klein geometry (G, P) is the subgroup of G consisting of all elements which act as the identity on G/P. It is the largest normal subgroup of G which is contained in P. Infinitesimal effectivity means that K is discrete, or equivalently that there is no ideal in $\mathfrak{g}$ that is contained in $\mathfrak{p}$; see 1.4.1. Now if $\mathfrak{s} \subset \mathfrak{g}$ is an ideal, then by Corollary 2.1.5, $\mathfrak{s}$ is itself semisimple, and it inherits a $|k_i|$–grading by part (2) of Proposition 3.1.2. But then $\mathfrak{s} \subset \mathfrak{p}$ is only possible if $\mathfrak{s} \subset \mathfrak{g}_0$, and the first statement follows.

If no simple ideal is contained in $\mathfrak{g}_0$, then the normal subgroup $K \subset G$ is discrete. In particular, for $X \in \mathfrak{g}$ and $k \in K$ we must have $\exp(-tX)k\exp(tX) = k$, since the left–hand side is a smooth curve in K. Differentiating at $t = 0$, we get $\mathrm{Ad}(k)X = X$, so K lies in the kernel of the adjoint action and hence in G_0. The statement on morphisms now follows directly from Proposition 1.5.3.

(2) By definition, the subalgebra $\mathfrak{g}_-$ is complementary to $\mathfrak{p}$. Let us next determine the $\mathfrak{p}$–submodule generated by $\mathfrak{g}_-$. For $X, Y \in \mathfrak{g}_-$ and $Z \in \mathfrak{p}$, the Jacobi identity gives $[X,[Z,Y]] = [[X,Z],Y] + [Z,[X,Y]]$. Splitting $[X,Z]$ into a $\mathfrak{g}_-$– and a $\mathfrak{p}$–component, we see that the first summand on the right–hand side splits into the sum of an element of $\mathfrak{g}_-$ and the bracket of an element of $\mathfrak{p}$ with an element of $\mathfrak{g}_-$, so we conclude that $[X,[Z,Y]]$ is contained in the $\mathfrak{p}$–module generated by $\mathfrak{g}_-$. Inductively, this implies that this module is stable under the adjoint action of $\mathfrak{g}_-$, and thus it is an ideal in $\mathfrak{g}$. As above we conclude that this implies that the $\mathfrak{p}$–submodule of $\mathfrak{g}$ generated by $\mathfrak{g}_-$ is the sum of all simple ideals of $\mathfrak{g}$, which are not contained in $\mathfrak{g}_0$.

Now let us assume that $\mathfrak{n} \subset \mathfrak{g}$ is a linear subspace complementary to the subalgebra $\mathfrak{p}$. Since the projection onto an eigenspace of an operator can be written as a polynomial in the operator and the grading element E is contained in $\mathfrak{p}$, we see that the $\mathfrak{g}_-$–component of any element of $\mathfrak{n}$ is contained in the $\mathfrak{p}$–module generated by $\mathfrak{n}$. Since $\mathfrak{n}$ is complementary to $\mathfrak{p}$, this implies that $\mathfrak{g}_-$ is contained in the $\mathfrak{p}$–module generated by $\mathfrak{n}$, so from above we conclude that this module contains the sum of all simple ideals of $\mathfrak{g}$ which are not contained in $\mathfrak{g}_0$. This completes the proof of (2).

(3) It suffices to show that there is no G–invariant linear connection on the tangent bundle $T(G/P)$. Let us denote by $\underline{\mathrm{Ad}}$ and $\underline{\mathrm{ad}}$ the actions of P and $\mathfrak{p}$ on $\mathfrak{g}/\mathfrak{p}$ induced by the adjoint action. According to Theorem 1.4.7 we have to prove that there is no P–equivariant map $\Phi : \mathfrak{g} \to L(\mathfrak{g}/\mathfrak{p}, \mathfrak{g}/\mathfrak{p})$ (which we write as $X \mapsto \Phi_X$) such that $\Phi_A = \underline{\mathrm{ad}}(A)$ for all $A \in \mathfrak{p} \subset \mathfrak{g}$. Equivariancy of Φ can be written explicitly as $\Phi_{\mathrm{Ad}(g)(X)} = \underline{\mathrm{Ad}}(g) \circ \Phi_X \circ \underline{\mathrm{Ad}}(g^{-1})$ for all $X \in \mathfrak{g}$ and $g \in P$. Assume that such a map Φ does exist, and choose a nonzero element $Z \in \mathfrak{g}_k$. Then $\mathrm{ad}(Z)(\mathfrak{g}) \subset \mathfrak{p}$, which implies that $\underline{\mathrm{Ad}}(\exp(Z)) = \mathrm{id}_{\mathfrak{g}/\mathfrak{h}}$.

We claim that there is a nonzero element $A \in \mathfrak{g}_0$ with $\underline{\mathrm{ad}}(A) = 0$. By part (5) of Proposition 3.1.2, there is an element $X \in \mathfrak{g}_{-1}$ such that $0 \neq [Z,X] \in \mathfrak{g}_{k-1}$. On the other hand, equivariancy of Φ implies that $\Phi_{\mathrm{Ad}(\exp(Z))(X)} = \Phi_X$. For $k = 1$, we get $\mathrm{Ad}(\exp(Z))(X) = X + [Z,X] + \frac{1}{2}[Z,[Z,X]]$, and $\underline{\mathrm{ad}}([Z,[Z,X]]) = 0$ since $[Z,[Z,X]] \in \mathfrak{g}_1$. Thus, $\Phi_{\mathrm{Ad}(\exp(Z))(X)} = \Phi_X + \underline{\mathrm{ad}}([Z,X])$, and since $[Z,X] \in \mathfrak{g}_0$, this is an element as required. For $k > 1$, we get $\mathrm{Ad}(\exp(Z))(X) = X + [Z,X]$, and thus again $\underline{\mathrm{ad}}([Z,X]) = 0$. (We shall see later that this is already a contradiction, but this needs more detailed structure theory for $|k|$–graded Lie algebras.) Since $\underline{\mathrm{ad}}([Z,X]) = 0$ implies $\underline{\mathrm{Ad}}(\exp([Z,X])) = \mathrm{id}$, we can repeat the argument with Z replaced by $[Z,X]$, finding an element $X_1 \in \mathfrak{g}_{-1}$ such that $[[Z,X],X_1] \neq 0$, but since $\underline{\mathrm{ad}}([[Z,X],X_1])$ is the lowest homogeneous component of $\Phi_{\underline{\mathrm{Ad}}(\exp([Z,X]))(X_1)} - \Phi_{X_1}$ it must be the zero map. Iterating this argument we obtain a nonzero element $A \in \mathfrak{g}_0$ with $\underline{\mathrm{ad}}(A) = 0$ as claimed.

But for any element $Y \in \mathfrak{g}_{-1}$ we then get $\underline{\mathrm{ad}}(A)(Y + \mathfrak{p}) = [A,Y] + \mathfrak{p}$, since $A \in \mathfrak{g}_0$. This implies that $[A,Y] = 0$ for all $Y \in \mathfrak{g}_{-1}$, which contradicts part (5) of Proposition 3.1.2 applied to the direct sum of all simple ideals of $\mathfrak{g}$ which are not contained in $\mathfrak{g}_0$. □

We will usually assume that none of the simple ideals of $\mathfrak{g}$ are contained in $\mathfrak{g}_0$, thus ensuring infinitesimal effectivity. Indeed, any such ideal can be simply left out without essential changes. Note, however, that the Klein geometry (G, P) will not be effective in general. In fact, it is very desirable to have non–effective geometries to deal with spin–structures and similar structures.

3.1.5. The underlying infinitesimal flag structure. As described in Sections 1.5 and 1.6, Cartan geometries (satisfying some normalization condition) are sometimes determined by weaker underlying structures, and these are the cases of most interest. The prototypical example of this situation are conformal structures. As described in 1.6 the conformal frame bundle together with its soldering form can be recovered from an appropriate Cartan geometry and if this geometry is normal, then it can in turn be uniquely constructed from the conformal frame bundle. Therefore, it is natural to look for structures underlying parabolic geometries. There is a whole family of such structures which can be constructed step by step from one another. However, in most situations the weakest and simplest underlying structure is already equivalent to a normal parabolic geometry, and we discuss only this weakest structure at this point. The more complicated underlying structures will be described in 3.1.15 below.

Let us fix a $|k|$–graded semisimple Lie algebra $\mathfrak{g} = \mathfrak{g}_{-k} \oplus \cdots \oplus \mathfrak{g}_k$, a Lie group G with Lie algebra $\mathfrak{g}$, and a parabolic subgroup $P \subset G$ corresponding to the grading. We continue to use the notation of 3.1.2 and 3.1.3. Consider a parabolic geometry $(p : \mathcal{G} \to M, \omega)$ of type (G, P). From 3.1.3 we know that we have the reductive subgroup G_0 and the nilpotent normal subgroup P_+ of P, which decompose P as a semidirect product. Since P acts freely on $\mathcal{G}$, the same is true for P_+, so we can form the orbit space $\mathcal{G}_0 := \mathcal{G}/P_+$. By construction, the projection p factors to a smooth map $p_0 : \mathcal{G}_0 \to M$. If $U \subset M$ is open such that there is a principal bundle chart $\psi : p^{-1}(U) \to U \times P$, then ψ is equivariant for the principal right action, so it factors to a diffeomorphism $p_0^{-1}(U) = p^{-1}(U)/P_+ \to U \times (P/P_+)$. This is obviously equivariant for the right action of G_0, so we conclude that $p_0 : \mathcal{G}_0 \to M$ is a smooth principal bundle with structure group $P/P_+ \cong G_0$. On the other hand, the inclusion of G_0 into P leads to local smooth sections of the projection $\mathcal{G} \to \mathcal{G}_0$, so this is a principal bundle with structure group P_+.

The second step is to descend parts of the Cartan connection ω to the bundle $\mathcal{G}_0$. To formulate this, we need some more observations. Let us return to the principal bundle $p : \mathcal{G} \to M$ and consider the filtration $\mathfrak{g} = \mathfrak{g}^{-k} \supset \mathfrak{g}^{-k+1} \supset \cdots \supset \mathfrak{g}^k \supset \{0\}$ from 3.1.2. Since the Cartan connection ω induces an isomorphism $T\mathcal{G} \cong \mathcal{G} \times \mathfrak{g}$, we see that for each $i = -k, \dots, k$ we get a smooth subbundle $T^i\mathcal{G} := \omega^{-1}(\mathfrak{g}^i)$ of $T\mathcal{G}$. This defines a filtration $T\mathcal{G} = T^{-k}\mathcal{G} \supset \cdots \supset T^k\mathcal{G}$ of the tangent bundle $T\mathcal{G}$. Moreover, since the filtration $\{\mathfrak{g}^i\}$ is P–invariant, equivariancy of ω implies that each of the subbundles $T^i\mathcal{G}$ is stable under the principal right action, i.e. $Tr^g(T^i\mathcal{G}) \subset T^i\mathcal{G}$ for all $g \in P$ and all $i = -k, \dots, k$. Since ω reproduces the generators of fundamental vector fields, we further conclude that for $i \geq 0$ the subbundle $T^i\mathcal{G}$ is spanned by the fundamental vector fields with generators in $\mathfrak{g}^i \subset \mathfrak{g}$. In particular, $T^0\mathcal{G}$ is the vertical bundle of $p : \mathcal{G} \to M$, while $T^1\mathcal{G}$ is the vertical bundle of $\mathcal{G} \to \mathcal{G}_0$.

Since the filtration of $T\mathcal{G}$ is stable under the principal right action it can be pushed down to $\mathcal{G}_0$ and to M, so we obtain filtrations $T\mathcal{G}_0 = T^{-k}\mathcal{G}_0 \supset \cdots \supset T^0\mathcal{G}_0$ and $TM = T^{-k}M \supset \cdots \supset T^{-1}M$ by smooth subbundles. By construction, the tangent maps to all the bundle projections are filtration preserving and $T^0\mathcal{G}_0$ is exactly the vertical bundle of $p_0 : \mathcal{G}_0 \to M$.

Finally, observe that once one has a filtration of the tangent bundle of a manifold, it makes sense to consider partially defined differential forms, i.e. sections of a bundle of the form $L(T^iM,V)$, where V is some finite–dimensional vector space or a vector bundle over M. For $i=-k$ we get $T^{-k}M=TM$ and sections of $L(T^{-k}M,V)$ are V–valued differential forms on M. Using this we formulate:

PROPOSITION 3.1.5. *Let $(p:\mathcal{G}\to M,\omega)$ be a parabolic geometry of type (G,P) corresponding to the $|k|$–grading $\mathfrak{g}=\mathfrak{g}_{-k}\oplus\dots\oplus\mathfrak{g}_k$ of the Lie algebra $\mathfrak{g}$ of G. Let $(p_0:\mathcal{G}_0\to M)$ be the underlying G_0–principal bundle.*

Then for each $i=-k,\dots,-1$, the Cartan connection ω descends to a smooth section ω_i^0 of the bundle $L(T^i\mathcal{G}_0,\mathfrak{g}_i)$. For each $u\in\mathcal{G}_0$ and $i=-k,\dots,-1$ the kernel of $\omega_i^0(u):T_u^i\mathcal{G}_0\to\mathfrak{g}_i$ is $T_u^{i+1}\mathcal{G}_0$, and each ω_i^0 is equivariant in the sense that for $g\in G_0$ we have $(r^g)^*\omega_i^0=\mathrm{Ad}(g^{-1})\circ\omega_i^0$.

PROOF. Let us denote by $\pi:\mathcal{G}\to\mathcal{G}_0$ the natural projection. For a point $u_0\in\mathcal{G}_0$, some $i=-k,\dots,-1$ and a tangent vector $\xi\in T^i_{u_0}\mathcal{G}_0$ choose a point $u\in\mathcal{G}$ with $\pi(u)=u_0$ and a tangent vector $\tilde{\xi}\in T_u\mathcal{G}$ such that $T\pi\cdot\tilde{\xi}=\xi$. By construction of the filtrations, $\tilde{\xi}\in T_u^i\mathcal{G}$ and thus $\omega(\tilde{\xi})\in\mathfrak{g}^i=\mathfrak{g}_i\oplus\dots\oplus\mathfrak{g}_k$. Define $\omega_i^0(\xi)$ to be the $\mathfrak{g}_i$–component of $\omega(\tilde{\xi})$. Fixing the choice of u, two possible choices for $\tilde{\xi}$ differ by an element in the kernel of $T_u\pi$, and we have observed that this kernel equals $T^1\mathcal{G}$. Hence, the difference of the values of ω lies in $\mathfrak{g}^1$, and therefore does not influence the $\mathfrak{g}_i$–component.

On the other hand, any other possible choice for the point in $\mathcal{G}$ is of the form $u\cdot g$ for some $g\in P_+$, and from 3.1.3 we know that $g=\exp(Z)$ for some $Z\in\mathfrak{p}_+$. Given a lift $\tilde{\xi}\in T_u\mathcal{G}$ of ξ, also $T_ur^g\cdot\tilde{\xi}\in T_{u\cdot g}\mathcal{G}$ is a lift of ξ. But then equivariancy of ω implies that $\omega(u\cdot g)(T_ur^g\cdot\tilde{\xi})=e^{\mathrm{ad}(-Z)}(\omega(u)(\xi))$. Since $Z\in\mathfrak{p}_+$ we know that $\mathrm{ad}(Z)(\mathfrak{g}^i)\subset\mathfrak{g}^{i+1}$, so again the $\mathfrak{g}_i$–component remains unchanged, and we get a well–defined map $\omega_i^0:T^i\mathcal{G}_0\to\mathfrak{g}_i$.

Once we know that ω_i^0 is well defined, we can locally write it as follows: Choose a local smooth section σ of $\pi:\mathcal{G}\to\mathcal{G}_0$, and consider the $\mathfrak{g}_i$–component of $(\sigma^*\omega)|_{T^i\mathcal{G}}$. By definition, this maps $\xi\in T_{u_0}\mathcal{G}_0$ to the $\mathfrak{g}_i$–component of $\omega(\sigma(u_0))(T_{u_0}\sigma\cdot\xi)$, so it coincides with $\omega_i^0(\xi)$. Hence, we see that each ω_i^0 is smooth.

By construction, $\omega_i^0(\xi)$ vanishes if and only if ξ admits a lift $\tilde{\xi}$ such that $\omega(\tilde{\xi})\in\mathfrak{g}^{i+1}$, which is equivalent to $\tilde{\xi}\in T^{i+1}\mathcal{G}$ and thus to $\xi\in T^{i+1}\mathcal{G}_0$. Finally, note that since $P_+\subset P$ is a normal subgroup, it follows immediately that the projection $\pi:\mathcal{G}\to\mathcal{G}_0$ is G_0–equivariant. Hence, given a tangent vector $\xi\in T^i_{u_0}\mathcal{G}_0$, a lift $\tilde{\xi}\in T^i_u\mathcal{G}$ and $g\in G_0$, we see that $T(r^g)\cdot\tilde{\xi}$ is a lift of $T(r^g)\cdot\xi$ (where we denote the principal right action on both bundles by r). Thus, equivariancy of ω implies equivariancy of each ω_i^0. □

3.1.6. Infinitesimal flag structures. There is an abstract version of the structures found as underlying a parabolic geometry in the last subsection. Fix a semisimple Lie group G, a $|k|$–grading of the Lie algebra $\mathfrak{g}$ of G, and a parabolic subgroup $P\subset G$ as before. Then we have the Levi subgroup $G_0\subset P$, and the normal subgroup $P_+\subset P$. Consider a smooth manifold M with a filtration $TM=T^{-k}M\supset T^{-k+1}M\supset\dots\supset T^{-1}M$ such that the rank of T^iM equals the dimension of $\mathfrak{g}^i/\mathfrak{p}$ for all $i=-k,\dots,-1$. Let $p:E\to M$ be a principal fiber bundle with structure group G_0. Then one gets a filtration of the tangent bundle TE of the form $TE=T^{-k}E\supset\dots\supset T^0E$ by letting T^0E be the vertical bundle and $T^iE:=$

$(Tp)^{-1}(T^iM)$ for $i<0$. By construction, the map $Tp:TE\to TM$ is filtration preserving and each of the subbundles T^iE is stable under the principal right action.

DEFINITION 3.1.6. (1) An *infinitesimal flag structure* of type (G,P) on a smooth manifold M is given by:

(i) A filtration $TM=T^{-k}M\supset\cdots\supset T^{-1}M$ of the tangent bundle of M such that the rank of T^iM equals the dimension of $\mathfrak{g}^i/\mathfrak{p}$ for all $i=-k,\dots,-1$.
(ii) A principal G_0–bundle $p:E\to M$.
(iii) A collection $\theta=(\theta_{-k},\dots,\theta_{-1})$ of smooth sections $\theta_i\in\Gamma(L(T^iE,\mathfrak{g}_i))$ which are G_0–equivariant in the sense that $(r^g)^*\theta_i=\operatorname{Ad}(g^{-1})\circ\theta_i$ for all $g\in G_0$, and such that for each $u\in E$ and $i=-k,\dots,-1$ the kernel of $\theta_i(u):T^i_uE\to\mathfrak{g}_i$ is $T^{i+1}_uE\subset T^i_uE$.

(2) Let M and $\tilde M$ be smooth manifolds endowed with infinitesimal flag structures $(\{T^iM\},p:E\to M,\theta)$ and $(\{T^i\tilde M\},\tilde p:\tilde E\to\tilde M,\tilde\theta)$ of type (G,P). Then a *morphism of infinitesimal flag structures* is a principal bundle homomorphism $\Phi:E\to\tilde E$ which covers a local diffeomorphism $f:M\to\tilde M$ such that Tf is filtration preserving and $\Phi^*\tilde\theta_i=\theta_i$ for all $i=-k,\dots,-1$.

Notice that in part (2) the condition that $Tf:TM\to T\tilde M$ is filtration preserving implies that $T\Phi:TE\to T\tilde E$ is filtration preserving, so the section $\tilde\theta^i$ of $L(T^i\tilde E,\mathfrak{g}_i)$ pulls back to a section $\Phi^*\tilde\theta_i$ of $L(T^iE,\mathfrak{g}_i)$.

In this language, the results of 3.1.5 say that any parabolic geometry $(p:\mathcal{G}\to M,\omega)$ of type (G,P) gives rise to an underlying infinitesimal flag structure $(\{T^iM\},p_0:\mathcal{G}_0\to M,\omega^0)$ of type (G,P). From the construction it follows immediately that any morphism of parabolic geometries descends to a morphism of infinitesimal flag structures. Hence, we obtain a functor from the category of parabolic geometries to the category of infinitesimal flag structures of the same type. One of the main aims of this section will be to show that under weak assumptions this functor restricts to an equivalence between appropriate subcategories.

We can clarify the geometric meaning of an infinitesimal flag structure. Fix a filtration $TM=T^{-k}M\supset\cdots\supset T^{-1}M$ such that the rank of T^iM equals the dimension of $\mathfrak{g}^i/\mathfrak{p}$ for all $i=-k,\dots,-1$. Then the rank of $\operatorname{gr}_i(TM)=T^iM/T^{i+1}M$ equals the dimension of $\mathfrak{g}_i$, so we can use $\mathfrak{g}_-=\mathfrak{g}_{-k}\oplus\cdots\oplus\mathfrak{g}_{-1}$ as the modelling vector space for the graded vector bundle $\operatorname{gr}(TM)$. On the other hand, via the adjoint action the group G_0 acts on $\mathfrak{g}_-$ and the action preserves the grading, so we get a homomorphism $G_0\to GL_{\mathrm{gr}}(\mathfrak{g}_-)$. Hence, it makes sense to talk about a reduction to the structure group G_0 of the bundle $\operatorname{gr}(TM)$.

PROPOSITION 3.1.6. *Let $\mathfrak{g}=\mathfrak{g}_{-k}\oplus\cdots\oplus\mathfrak{g}_k$ be a $|k|$–graded semisimple Lie algebra, let G be a Lie group with Lie algebra $\mathfrak{g}$, $P\subset G$ a parabolic subgroup corresponding to the grading, and $G_0\subset P$ the Levi subgroup; see 3.1.3.*

(1) An infinitesimal flag structure of type (G,P) on a smooth manifold M is equivalent to a filtration $TM=T^{-k}M\supset\cdots\supset T^{-1}M$ of the tangent bundle of M such that for each i the rank of T^iM equals the dimension of $\mathfrak{g}^i/\mathfrak{p}$ and a reduction of the structure group of the associated graded bundle $\operatorname{gr}(TM)$ to the structure group G_0 with respect to the homomorphism $\operatorname{Ad}:G_0\to GL_{\mathrm{gr}}(\mathfrak{g}_-)$.

(2) Suppose that no simple ideal of $\mathfrak{g}$ is contained in $\mathfrak{g}_0$. Consider two infinitesimal flag structures $(\{T^iM\},p:E\to M,\theta)$ and $(\{T^i\tilde M\},\tilde p:\tilde E\to\tilde M,\tilde\theta)$ of type (G,P). Let $\Phi_1,\Phi_2:E\to\tilde E$ be two morphisms of infinitesimal flag structures

covering the same base map $f : M \to \tilde{M}$. *Then there is a locally constant map* $\phi : M \to K$ *such that* $\Phi_2(u) = \Phi_1(u) \cdot \phi(p(u))$. *Here,* $K \subset G_0$ *is the subgroup of all elements* $g \in G$ *such that* $\mathrm{Ad}(g) = \mathrm{id}_{\mathfrak{g}}$.

PROOF. (1) Let us fix a filtration $\{T^iM\}$ of TM such that the rank of T^iM equals the dimension of $\mathfrak{g}^i/\mathfrak{p}$. Then a reduction of $\mathrm{gr}(TM)$ to the structure group G_0 is given by a principal G_0–bundle $p : E \to M$ together with a homomorphism Φ from E to the graded linear frame bundle $\mathcal{P} := GL_{\mathrm{gr}}(\mathfrak{g}_-, \mathrm{gr}(TM))$ of $\mathrm{gr}(TM)$, which induces the identity on M and is equivariant for the homomorphism $\mathrm{Ad} : G_0 \to GL_{\mathrm{gr}}(\mathfrak{g}_-)$. To prove (1), we have to show that such a homomorphism is equivalent to a collection $\theta = (\theta_i)$ of sections as required in the definition of an infinitesimal flag structure. This is analogous to the case of G–structures discussed in 1.3.6.

Given an infinitesimal flag structure $(\{T^iM\}, p : E \to M, \theta)$, take a point $u \in E$. For each i we have the linear map $\theta_i(u) : T^i_uE \to \mathfrak{g}_i$. By definition, the kernel of this map is T^{i+1}_uE, so for dimensional reasons it has to descend to a linear isomorphism $T^i_uE/T^{i+1}_uE \to \mathfrak{g}_i$. The tangent map T_up induces a linear isomorphism $T^i_uE/T^{i+1}_u \cong T^i_xM/T^{i+1}_xM$, where $x = p(u)$. Hence, we may interpret $\theta(u) = (\theta_{-k}(u), \dots, \theta_{-1}(u))$ as a linear isomorphism $\mathrm{gr}(T_xM) \to \mathfrak{g}_-$, whose inverse defines an element of $\mathcal{P}_x$. Hence, we obtain a smooth fiber bundle homomorphism $\Phi : E \to \mathcal{P}$, which covers the identity on M. Equivariancy of the form θ reads as $\theta(u \cdot g) = \mathrm{Ad}(g^{-1}) \circ \theta(u)$, which is exactly the condition required for a principal bundle homomorphism.

For the other direction, we observe that the bundle $\mathcal{P} = GL_{\mathrm{gr}}(\mathfrak{g}_-, \mathrm{gr}(TM))$ carries a natural analog of the soldering form described in 1.3.5. Since $\pi : \mathcal{P} \to M$ is a principal bundle, we can lift the filtration of TM to a filtration $T\mathcal{P} = T^{-k}\mathcal{P} \supset \dots \supset T^0\mathcal{P}$ as above. Having this filtration at hand, it is obvious how to obtain an analog $\Theta = (\Theta_{-k}, \dots, \Theta_{-1})$ of the soldering form, where each Θ_i is a smooth section of $L(T^i\mathcal{P}, \mathfrak{g}_i)$: For $x \in M$, a point in $\mathcal{P}_x$ is a linear isomorphism $\phi : \mathfrak{g}_- \to \mathrm{gr}(T_xM)$ which is homogeneous of degree zero. Given a tangent vector $\xi \in T^i_\phi\mathcal{P}$, we have $T_\phi\pi \cdot \xi \in T^i_xM$. Denoting by $[T_u\pi \cdot \xi] \in \mathrm{gr}_i(T_xM)$ its equivalence class, we define $\Theta_i(\xi) := \phi^{-1}([T_u\pi \cdot \xi]) \in \mathfrak{g}_i$. By construction, this is smooth, the kernel of $\Theta_i(u)$ is $T^{i+1}_u\mathcal{P}$, and we get equivariancy in the sense that $(r^\psi)^*\Theta_i = \psi^{-1} \circ \Theta_i$ for each $\psi \in GL_{\mathrm{gr}}(\mathfrak{g}_-)$.

Given a reduction of structure group $\Phi : E \to \mathcal{P}$ we see that by construction $T\Phi : TE \to T\mathcal{P}$ is filtration preserving, so we can form the pullback $\Phi^*\Theta = (\Phi^*\Theta_{-k}, \dots, \Phi^*\Theta_{-1})$ of the generalized soldering form. One immediately checks that $(\{T^iM\}, E \to M, \Phi^*\Theta)$ is an infinitesimal flag structure, and this construction is inverse to the one described above.

(2) By definition, f is a local diffeomorphism, and for an open subset $U \subset M$ such that $f : U \to f(U)$ is a diffeomorphism, Φ_1 and Φ_2 are principal bundle isomorphisms $p^{-1}(U) \to \tilde{p}^{-1}(f(U))$. Since the locally constant maps ϕ can be pieced together, we may assume that f is a diffeomorphism and thus Φ_1 and Φ_2 are isomorphisms of principal bundles, and replacing Φ_2 by $\Phi_2^{-1} \circ \Phi_1$, we may assume that we deal with an automorphism Φ of the infinitesimal flag structure on M that covers the identity.

By definition, this means that there is a smooth map $\phi : E \to G_0$ such that $\Phi(u) = u \cdot \phi(u)$. For $u \in E$ and $\xi \in T_uE$, we get $\Phi^*\theta_i(u)(\xi) = \theta_i(u \cdot \phi(u))(T_u\Phi \cdot \xi)$. As in the proof of Proposition 1.5.3 one shows that $T_u\Phi \cdot \xi$ equals the sum of

$T_u r^{\phi(u)} \cdot \xi$ and some vertical vector field. Since each θ_i vanishes on vertical fields, we conclude that $\Phi^*\theta_i(u)(\xi) = (r^{\phi(u)})^*\theta_i(u)(\xi) = \mathrm{Ad}(\phi(u)^{-1})(\theta_i(u)(\xi))$. For Φ to be a morphism of infinitesimal flag structures we must have $\Phi^*\theta_i = \theta_i$ for all $i = -k, \dots, -1$, which implies that the adjoint action of $\phi(u)$ on $\mathfrak{g}_i$ is the identity for all $i = -k, \dots, -1$. Since for $j > 0$, the G_0–module $\mathfrak{g}_j$ is dual to $\mathfrak{g}_{-j}$, we conclude that $\mathrm{Ad}(\phi(u))$ also is the identity on all $\mathfrak{g}_j$ with $j > 0$. Thus, $\mathrm{Ad}(\phi(u))$ also acts as the identity on the submodule $[\mathfrak{g}_1, \mathfrak{g}_{-1}] \subset \mathfrak{g}_0$, which coincides with $\mathfrak{g}_0$ by part (4) of Proposition 3.1.2 since no simple ideal of $\mathfrak{g}$ is contained in $\mathfrak{g}_0$. Consequently, the function ϕ has values in K, which has trivial Lie algebra and hence is a discrete subgroup of G_0, so ϕ must be locally constant. □

Note that part (1), in particular, implies that for an infinitesimal flag structure $(\{T^iM\}, p : E \to M, \theta)$ we have a natural isomorphism $\mathrm{gr}_i(TM) \cong E \times_{G_0} \mathfrak{g}_i$. Explicitly, this identification is induced by the map $E \times \mathfrak{g}_i \to \mathrm{gr}_i(TM)$ defined by $(u, X) \mapsto [T_u p \cdot \xi]$, where $\xi \in T^i_u E$ is any tangent vector such that $\theta_i(\xi) = X$ and $[\ ]$ denotes the class in $\mathrm{gr}_i(TM) = T^iM/T^{i+1}M$.

Example 3.1.6. Since in all other cases we need restrictions on infinitesimal flag structures to characterize the relevant examples, here we stick to the simplest case of a $|1|$–grading. Let $\mathfrak{g} = \mathfrak{g}_{-1} \oplus \mathfrak{g}_0 \oplus \mathfrak{g}_1$ be a $|1|$–graded semisimple Lie algebra, G a Lie group with Lie algebra $\mathfrak{g}$, $P \subset G$ a parabolic subgroup for the given grading and $G_0, P_+ \subset P$ the usual subgroups. In this case, the situation becomes very simple, since the filtration degenerates to $TM = T^{-1}M$. Hence, from above we conclude that infinitesimal flag structures of type (G, P) are simply reductions of structure group of TM to the group G_0. Here, we view M as being modeled on the vector space $\mathfrak{g}_{-1}$ and the reduction is with respect to the homomorphism $\mathrm{Ad} : G_0 \to GL(\mathfrak{g}_{-1})$. Also, morphisms of infinitesimal flag structures simply are morphisms of the reductions. Let us look at two special cases in a little more detail:

(1) Consider the $|1|$–grading on the Lie algebra $\mathfrak{g} = \mathfrak{so}(p+1, q+1)$ discussed in 1.6.3 and in Example 3.1.2 (1). Then $\mathfrak{g}_{-1} \cong \mathbb{R}^{p+q}$, $\mathfrak{g}_0 \cong \mathfrak{co}(p, q)$ and $\mathfrak{g}^1 \cong \mathbb{R}^{(p+q)*}$, and the adjoint action of $\mathfrak{g}_0$ on $\mathfrak{g}_{-1}$ is the standard action. Put $G = PO(p + 1, q + 1)$ and let $P \subset G$ the stabilizer of an isotropic line. Then by part (2) of Proposition 1.6.3, $G_0 = CO(\mathfrak{g}_{-1}) \cong CO(p, q)$ via the adjoint action. Consequently, an infinitesimal flag structure of type (G, P) in this case is a first order $CO(p, q)$–structure, and hence is equivalent to the choice of a conformal class of pseudo–Riemannian metrics of signature (p, q) on M. We will discuss other possible choices of groups in 4.1.2.

The prolongation procedure presented in 1.6.4 and 1.6.7 shows that an infinitesimal flag structure of type (G, P) uniquely determines a Cartan bundle endowed with a normalized Cartan connection. Hence, in this case, there is an equivalence of categories between normalized Cartan geometries and infinitesimal flag structures.

(2) We shall soon see that surprisingly a similar correspondence between normal Cartan connections and certain infinitesimal flag structures exists for almost all parabolic geometries. There are, however, two series of exceptions, one of which corresponds to a $|1|$–grading: Consider $\mathfrak{g} = \mathfrak{sl}(1+n, \mathbb{R})$ with the $|1|$–grading obtained from the block decomposition $\begin{pmatrix} -\mathrm{tr}(A) & \phi \\ v & A \end{pmatrix}$, with $A \in \mathfrak{gl}(n, \mathbb{R})$, $\phi \in \mathbb{R}^{n*}$, and $v \in \mathbb{R}^n$, as in Example 3.1.2 (2), i.e v corresponds to $\mathfrak{g}_{-1}$, A to $\mathfrak{g}_0$ and ϕ to $\mathfrak{g}_1$. The adjoint action of $A \in \mathfrak{g}_0$ on $v \in \mathfrak{g}_{-1}$ is immediately seen to be given by $(A + \mathrm{tr}(A))v$. It is easy to see (and we will do this explicitly in 4.1.5) that choosing $G := PSL(n+1, \mathbb{R})$

and P the stabilizer of the line spanned by the first basis vector, we get $G_0 = GL(\mathfrak{g}_{-1}) \cong GL(n, \mathbb{R})$ via the adjoint action. Hence, in this case an infinitesimal flag structure contains no information, so there is no hope to get a Cartan connection from it. We will discuss the right notion of underlying structures for this parabolic geometry in 3.1.15 below.

3.1.7. Regularity. Regularity is a restriction on infinitesimal flag structures (and hence on parabolic geometries), which is of crucial importance in the theory. It expresses a tight relation between the two ingredients of an infinitesimal flag structure, the filtration of the tangent bundle and the reduction to the structure group G_0. Let $(\{T^iM\}, p : E \to M, \theta)$ be an infinitesimal flag structure of some fixed type (G, P). We have seen in 3.1.6 that $\mathrm{gr}_i(TM) \cong E \times_{G_0} \mathfrak{g}_i$ and thus $\mathrm{gr}(TM) \cong E \times_{G_0} \mathfrak{g}_-$. Via this identification, the Lie bracket on $\mathfrak{g}_-$ (which is preserved by the adjoint action) induces a bilinear bundle map

$$\{\ ,\ \} : \mathrm{gr}(TM) \times \mathrm{gr}(TM) \to \mathrm{gr}(TM),$$

which is compatible with the grading, i.e. $\{\mathrm{gr}_i(TM), \mathrm{gr}_j(TM)\} \subset \mathrm{gr}_{i+j}(TM)$. This makes $\mathrm{gr}(TM)$ into a bundle of nilpotent graded Lie algebras modelled on $\mathfrak{g}_-$.

Under additional assumptions a similar structure is already intrinsic to the filtration $\{T^iM\}$. Assume that this filtration is compatible with the Lie bracket in the sense that for $\xi \in \Gamma(T^iM)$ and $\eta \in \Gamma(T^jM)$ the Lie bracket $[\xi, \eta]$ is a section of $T^{i+j}M$. (We follow the usual convention that $T^\ell M = TM$ for all $\ell \leq -k$.) Notice that this condition is automatically satisfied if the filtration is of length at most 2.

Then for each $i = -k, \dots, -1$ let us denote by $q_i : T^iM \to \mathrm{gr}_i(TM)$ the natural quotient map, and consider the operator $\Gamma(T^iM) \times \Gamma(T^jM) \to \Gamma(\mathrm{gr}_{i+j}(TM))$ defined by $(\xi, \eta) \mapsto q_{i+j}([\xi, \eta])$. For a smooth function $f \in C^\infty(M, \mathbb{R})$ we have $[\xi, f\eta] = (\xi \cdot f)\eta + f[\xi, \eta]$. Since $i \leq -1$, we see that $T^jM \subset T^{i+j+1}M$, so the first term lies in the kernel of q_{i+j}. Hence, we conclude that the mapping defined above is bilinear over smooth functions, so it is induced by a bilinear bundle map $T^iM \times T^jM \to \mathrm{gr}_{i+j}(TM)$. Moreover, if $\xi \in T^{i+1}M$ or $\eta \in T^{j+1}M$ then $[\xi, \eta] \in T^{i+j+1}M$, so again this lies in the kernel of q_{i+j}. Thus, our map further descends to a bundle map $\mathrm{gr}_i(TM) \times \mathrm{gr}_j(TM) \to \mathrm{gr}_{i+j}(TM)$. Taking these maps together, we obtain a bundle map $\mathcal{L} : \mathrm{gr}(TM) \times \mathrm{gr}(TM) \to \mathrm{gr}(TM)$, which is compatible with the gradings. Since $\mathcal{L}$ is induced by the Lie bracket of vector fields, it follows immediately that it makes each fiber $\mathrm{gr}(T_xM)$ into a nilpotent graded Lie algebra.

DEFINITION 3.1.7. (1) A *filtered manifold* is a smooth manifold M together with a filtration $TM = T^{-k}M \supset \cdots \supset T^{-1}M$ of its tangent bundle by smooth subbundles, which is compatible with the Lie bracket in the sense that $[\xi, \eta] \in \Gamma(T^{i+j}M)$ for any $\xi \in \Gamma(T^iM)$ and $\eta \in \Gamma(T^jM)$.

(2) For a filtered manifold $(M, \{T^iM\})$ the tensorial map

$$\mathcal{L} : \mathrm{gr}(TM) \times \mathrm{gr}(TM) \to \mathrm{gr}(TM)$$

induced by the Lie bracket of vector fields as described above is called the *(generalized) Levi bracket*. For $x \in M$, the nilpotent graded Lie algebra $(\mathrm{gr}(T_xM), \mathcal{L}_x)$ is called the *symbol algebra* of the filtered manifold at x. The bundle $(\mathrm{gr}(TM), \mathcal{L})$ of nilpotent graded Lie algebras obtained in this way is called the *bundle of symbol algebras*.

(3) An infinitesimal flag structure $(\{T^iM\}, E \to M, \theta)$ is called *regular* if $(M, \{T^iM\})$ is a filtered manifold and the algebraic bracket $\{\ ,\ \} : \operatorname{gr}(TM) \times \operatorname{gr}(TM) \to \operatorname{gr}(TM)$ coincides with the Levi bracket $\mathcal{L}$.

(4) A parabolic geometry is called *regular* if the underlying infinitesimal flag structure constructed in Proposition 3.1.5 is regular.

Suppose that $(M, \{T^iM\})$ and $(\tilde{M}, \{T^i\tilde{M}\})$ are filtered manifolds and that $f : M \to \tilde{M}$ is a local diffeomorphism such that $Tf : TM \to T\tilde{M}$ is filtration preserving. Then for each $x \in M$, the tangent map T_xf induces a linear map $\operatorname{gr}(T_xf) : \operatorname{gr}(T_xM) \to \operatorname{gr}(T_{f(x)}\tilde{M})$. From the fact that the pullback of vector fields is compatible with the Lie bracket, one immediately concludes that this map pulls back $\mathcal{L}_{f(x)}$ to $\mathcal{L}_x$, so it is an isomorphism between the symbol algebras. Hence, the symbol algebras are the basic invariants of a filtered manifold, which replace the tangent space for ordinary manifolds.

In general, the isomorphism type of the symbol algebra may change from point to point, so $(\operatorname{gr}(TM), \mathcal{L})$ is not necessarily locally trivial as a bundle of Lie algebras. If we suppose that this is the case, however, then we get a smaller canonical frame bundle for $\operatorname{gr}(TM)$. Assume that the bundle of symbol algebras is locally trivial with typical fiber a nilpotent graded Lie algebra $\mathfrak{a}$. Then there is a natural frame bundle for $\operatorname{gr}(TM)$ with structure group the group $\operatorname{Aut}_{\operatorname{gr}}(\mathfrak{a})$ of all automorphisms of the graded Lie algebra $\mathfrak{a}$. The fiber of this frame bundle in $x \in M$ is the set of all Lie algebra isomorphisms from $\mathfrak{a}$ to the symbol algebra at x.

If the filtration $\{T^iM\}$ is part of a regular infinitesimal flag structure of type (G, P), then the bundle of symbol algebras is isomorphic to $(\operatorname{gr}(TM), \{\ ,\ \})$ as a bundle of Lie algebras, so it is locally trivial with typical fiber $\mathfrak{g}_-$. Moreover, the adjoint action actually defines a homomorphism $G_0 \to \operatorname{Aut}_{\operatorname{gr}}(\mathfrak{g}_-)$, which is infinitesimally injective if no simple ideal of $\mathfrak{g}$ is contained in $\mathfrak{g}_0$. The definition of regularity means that the construction of Proposition 3.1.6 gives a reduction of structure group of the natural frame bundle of $\operatorname{gr}(TM)$ corresponding to $\operatorname{Ad} : G_0 \to \operatorname{Aut}_{\operatorname{gr}}(\mathfrak{g}_-)$. Thus, we make the following:

OBSERVATION 3.1.7. A regular infinitesimal flag structure of type (G, P) on a smooth manifold M is equivalent to:

- A filtration $\{T^iM\}$ of the tangent bundle which makes M into a filtered manifold such that the bundle of symbol algebras is locally trivial and modelled on the nilpotent graded Lie algebra $\mathfrak{g}_-$.
- A reduction of structure group of the natural frame bundle of $\operatorname{gr}(TM)$ with respect to $\operatorname{Ad} : G_0 \to \operatorname{Aut}_{\operatorname{gr}}(\mathfrak{g}_-)$.

A similar characterization holds for morphisms, i.e. they are equivalent to filtration preserving local diffeomorphisms which are compatible with the G_0–structure on the associated graded bundles to the tangent bundles.

Thus, regular infinitesimal flag structures provide examples of the natural analog of first order G–structures in the setting of filtered manifolds. This makes their geometric interpretation easy. It should be mentioned at this point that in many cases $\operatorname{Ad} : G_0 \to \operatorname{Aut}_{\operatorname{gr}}(\mathfrak{g}_-)$ is an isomorphism; see 4.3.1. In these cases, a regular infinitesimal flag structure is only given by an appropriate filtration of the tangent bundle. This leads to highly interesting examples of parabolic geometries.

For later use, we characterize regularity of an infinitesimal flag structure in terms of the frame form θ.

PROPOSITION 3.1.7. *Let $(\{T^iM\}, p : E \to M, \theta)$ be an infinitesimal flag structure such that $(M, \{T^iM\})$ is a filtered manifold. Then the structure is regular if and only if for all $i, j < 0$ such that $i + j \geq -k$ and all sections $\xi \in \Gamma(T^iE)$ and $\eta \in \Gamma(T^jE)$ we have*

$$\theta_{i+j}([\xi, \eta]) = [\theta_i(\xi), \theta_j(\eta)].$$

PROOF. If we choose ξ and η to be projectable, then $[\xi, \eta]$ lifts the bracket of the projections. Since $(M, \{T^iM\})$ is a filtered manifold, this implies that $[\xi, \eta] \in \Gamma(T^{i+j}E)$. For general sections ξ and η and a point $x \in E$, the class of $[\xi, \eta](x)$ in $T_xE/T_x^{i+j}E$ depends only on $\xi(x)$ and $\eta(x)$. Hence, we conclude that we always have $[\xi, \eta] \in \Gamma(T^{i+j}E)$, so $\theta_{i+j}([\xi, \eta])$ makes sense. Since T^iE and T^jE are contained in $T^{i+j+1}E$, we see that $\theta_{i+j}([\xi, \eta](x))$ depends only on $\xi(x)$ and $\eta(x)$. Since the same is evidently true for the other side of the equation we see that $\theta_{i+j}([\xi, \eta]) = [\theta_i(\xi), \theta_j(\eta)]$ holds for arbitrary vector fields ξ and η if and only if it holds for projectable fields.

But if ξ and η project to sections $\underline{\xi} \in \Gamma(T^iM)$ and $\underline{\eta} \in \Gamma(T^jM)$, then the functions $\theta_i(\xi)$ represents the section $q_i(\underline{\xi}) \in \Gamma(\operatorname{gr}_i(TM))$ and likewise for $\theta_j(\eta)$. Hence, the right–hand side of the equation represents $\{q_i(\underline{\xi}), q_j(\underline{\eta})\}$. On the other hand, since $[\xi, \eta]$ projects to $[\underline{\xi}, \underline{\eta}]$, we conclude that the function $\theta_{i+j}([\xi, \eta])$ represents $q_{i+j}([\underline{\xi}, \underline{\eta}]) = \mathcal{L}(q_i(\underline{\xi}), q_j(\underline{\eta}))$. □

REMARK 3.1.7. (1) Since the components θ_i of a frame form are only partially defined, their exterior derivative is not well defined in general. However, if $(M, \{T^iM\})$ is a filtered manifold, then from the proof of the proposition we see that also $(E, \{T^iE\})$ is a filtered manifold. In particular, for $i, j < 0$ such that $i + j \geq -k$ we can use the standard formula for the exterior derivative to define $d\theta_{i+j}$ on $T^iE \times T^jE$. Alternatively, we can extend θ_{i+j} arbitrarily to a $\mathfrak{g}_{i+j}$–valued one form on E and observe that the restriction of the exterior derivative of this form is independent of the choice of extension. In this language, the equation in the proposition can be written as $d\theta_{i+j}(\xi, \eta) + [\theta_i(\xi), \theta_j(\eta)] = 0$.

EXAMPLE 3.1.7. Let us briefly describe the geometric interpretation of infinitesimal flag structures and the geometric relevance of regularity in an important example. Details of this example will be discussed in 4.2.4.

Consider the Lie algebra $\mathfrak{g} = \mathfrak{su}(n+1, 1)$ endowed with the $|2|$–grading from Example (3) of 3.1.2, so we have the following description of $\mathfrak{g}$:

$$\left\{ \begin{pmatrix} a & Z & iz \\ X & A - \frac{2i}{n}\operatorname{im}(a)\mathbb{I}_n & -Z^* \\ ix & -X^* & -\bar{a} \end{pmatrix} : A \in \mathfrak{su}(n), a \in \mathbb{C}, X \in \mathbb{C}^n, Z \in \mathbb{C}^{n*}, x, z \in \mathbb{R} \right\}.$$

From this block form, one immediately sees that the bracket $\mathfrak{g}_{-1} \times \mathfrak{g}_{-1} \to \mathfrak{g}_{-2}$ (which completely describes $\mathfrak{g}_-$) is given by $[X, Y] = -X^*Y + Y^*X$. This is twice the imaginary part of the standard Hermitian inner product on $\mathbb{C}^n$. In particular, we have $[iX, iY] = [X, Y]$, and the bracket is nondegenerate, i.e. $[X, Y] = 0$ for all Y implies $X = 0$. On the other hand, $\mathfrak{g}_0 \cong \mathfrak{su}(n) \oplus \mathbb{C}$ and the action of the semisimple part $\mathfrak{su}(n)$ on $\mathfrak{g}_{-1} \cong \mathbb{C}^n$ is the standard representation.

Put $G = PSU(n+1, 1)$, the quotient of $SU(n+1, 1)$ by its center, and let P be the stabilizer of the isotropic line spanned by the first basis vector. Then it turns out that the adjoint action identifies the subgroup $G_0 \subset G$ with the group of all pairs (ϕ_1, ϕ_2) of linear isomorphisms $\phi_1 : \mathfrak{g}_{-1} \to \mathfrak{g}_{-1}$ and $\phi_2 : \mathfrak{g}_{-2} \to \mathfrak{g}_{-2}$ such that

ϕ_1 is complex linear and $[\phi_1(X), \phi_1(Y)] = \phi_2([X,Y])$ for all $X, Y \in \mathfrak{g}_{-1}$. (This implies that ϕ_2 is uniquely determined by ϕ_1, but this is not important here.)

To obtain an infinitesimal flag structure of type (G, P), we have to start with a smooth manifold M of dimension $2n+1$, and the filtration of the tangent bundle TM we need in this case is simply given by a subbundle $T^{-1}M \subset TM$ of real rank $2n$. The associated graded of the tangent bundle then has the form $\mathrm{gr}(TM) = \mathrm{gr}_{-2}(TM) \oplus \mathrm{gr}_{-1}(TM)$ and $\mathrm{gr}_{-2}(TM) = TM/T^{-1}M$ is a real line bundle, while $\mathrm{gr}_{-1}(TM) = T^{-1}M$. From the description of G_0 given above it is easy to see that a reduction to the structure group G_0 of $\mathrm{gr}(TM)$ is equivalent to an almost complex structure J on $T^{-1}M$ and a skew symmetric bundle map $\{\ ,\ \} : T^{-1}M \times T^{-1}M \to \mathrm{gr}_{-2}(TM)$, which is (in appropriate trivializations) the imaginary part of a definite Hermitian form.

Therefore, an infinitesimal flag structure in this case is given by a complex subbundle of rank n plus the choice of the tensorial map $\{\ ,\ \}$. In the language of CR–structures, a rank n complex subbundle in a manifold of dimension $2n+1$ is called a hypersurface type *almost CR–structure* of CR–dimension n. However, in this situation the tensorial map $\{\ ,\ \}$ is an additional ingredient, which is not deeply related to the almost CR–structure.

The situation changes completely, if one imposes the regularity condition. If the filtration $TM \supset T^{-1}M$ is part of a regular infinitesimal flag structure, then the Levi bracket $\mathcal{L} : T^{-1}M \times T^{-1}M \to TM/T^{-1}M$ has to be equal to the bracket $\{\ ,\ \}$. In particular, $\mathcal{L}$ has to be nondegenerate, which is equivalent to the fact that $T^{-1}M \subset TM$ is a contact structure. In terms of CR–geometry, this means that the almost CR–structure is nondegenerate. Since $\{\ ,\ \}$ comes from the imaginary part of a Hermitian form, we further conclude that $\mathcal{L}(J\xi, J\eta) = \mathcal{L}(\xi, \eta)$ for all ξ, η. In CR terms, this means that the CR–structure is partially integrable. Under this assumption, it turns out that $\mathcal{L}$ is the imaginary part of the classical Levi form, and the Levi form being definite means that the almost CR–structure is strictly pseudoconvex. Summarizing, we see that a regular infinitesimal flag structure of type (G, P) is equivalent to a strictly pseudoconvex partially integrable almost CR–structure.

This is also easily expressed in terms of filtered manifolds. A filtration such that any symbol algebra is isomorphic to $\mathfrak{g}_-$ (as a real Lie algebra) is just a contact structure. The additional reduction to $G_0 \subset \mathrm{Aut}_{\mathrm{gr}}(\mathfrak{g}_-)$ is equivalent to a complex structure on $T^{-1}M$, which makes M into a partially integrable almost CR–manifold. The morphisms are equivalent to filtration preserving local diffeomorphisms such that the restriction to $T^{-1}M$ of each tangent map is complex linear, i.e. to local CR diffeomorphisms.

3.1.8. A characterization of regular parabolic geometries. Our next goal is to characterize regularity of parabolic geometries in terms of their curvature. Recall from 1.5.1 that the curvature is the basic invariant for any Cartan geometry. For a parabolic geometry $(p : \mathcal{G} \to M, \omega)$ it can be either encoded in the curvature form $K \in \Omega^2(\mathcal{G}, \mathfrak{g})$ or in the curvature function $\kappa : \mathcal{G} \to L(\Lambda^2\mathfrak{g}_-, \mathfrak{g})$. By definition, $K(\xi, \eta) = d\omega(\xi, \eta) + [\omega(\xi), \omega(\eta)]$, while $\kappa(u)(X, Y) = K(\omega_u^{-1}(X), \omega_u^{-1}(Y))$. Recall further that K is horizontal and both K and κ are equivariant in an appropriate sense; see 1.5.1.

We will obtain the characterization of regularity from a more general result, which will be very useful for the geometric interpretation of the torsion of parabolic geometries in the sequel. The tools used in this proof are closely related to normal Weyl structures, which will be discussed in 5.1.12.

PROPOSITION 3.1.8. *Let $\mathfrak{g} = \mathfrak{g}_{-k} \oplus \cdots \oplus \mathfrak{g}_k$ be a $|k|$–graded semisimple Lie algebra, G a Lie group with Lie algebra $\mathfrak{g}$, $P \subset G$ a parabolic subgroup corresponding to the grading, and $G_0 \subset P$ the Levi subgroup. Let $(p : \mathcal{G} \to M, \omega)$ be a parabolic geometry of type (G, P) with curvature function $\kappa : \mathcal{G} \to L(\Lambda^2\mathfrak{g}_-, \mathfrak{g})$, $u \in \mathcal{G}$ any point, and put $x = p(u) \in M$.*

Then there is an open neighborhood U of $x \in M$ and a linear extension operator $T_xM \to \mathfrak{X}(U)$, written as $\xi \mapsto \tilde{\xi}$, which is compatible with all structures on TM obtained from the Cartan connection ω and has the following property: For $\xi, \eta \in T_xM$ let $X, Y \in \mathfrak{g}_-$ be the unique elements such that $T_up \cdot \omega_u^{-1}(X) = \xi$ and $T_up \cdot \omega_u^{-1}(Y) = \eta$. Then

$$[\tilde{\xi}, \tilde{\eta}](x) = T_up \cdot \omega_u^{-1}\big([X, Y] - \kappa(u)(X, Y)\big).$$

PROOF. Note first that the elements X and Y are well defined since two lifts of a tangent vector on M differ by some vertical vector and $\mathfrak{g}_-$ is complementary to $\mathfrak{p}$. For any element $Z \in \mathfrak{g}_-$ consider the vector field $\omega^{-1}(Z)$ and its flow $\mathrm{Fl}_t^{\omega^{-1}(Z)}(u)$ which is defined for sufficiently small t. Then there is a neighborhood V of 0 in $\mathfrak{g}_-$ on which the map $\phi(Z) := \mathrm{Fl}_1^{\omega^{-1}(Z)}(u)$ is well defined. The tangent map of $p \circ \phi$ in $0 \in V$ is given by $Z \mapsto T_up \cdot \omega_u^{-1}(Z)$, so this is a linear isomorphism. Possibly shrinking V we may therefore assume that $p \circ \phi$ is a diffeomorphism from V onto an open neighborhood U of $x \in M$. The composition of ϕ with the inverse of this diffeomorphism defines a local smooth section $\sigma : U \to \mathcal{G}$ of the principal bundle $\mathcal{G}$. Consequently, $(y, g) \mapsto \sigma(y) \cdot g$ defines a diffeomorphism $U \times P \to p^{-1}(U)$.

Given a tangent vector $\xi \in T_xM$ and the corresponding element $X \in \mathfrak{g}_-$ we now define a vector field $\hat{\xi}$ on $p^{-1}(U)$ by $\hat{\xi}(\sigma(y) \cdot g) := Tr^g \cdot \omega_{\sigma(y)}^{-1}(X)$. By construction, $\hat{\xi}$ is projectable, and we denote by $\tilde{\xi} \in \mathfrak{X}(U)$ the projection. From the construction it is also clear that the operator $\xi \mapsto \tilde{\xi}$ is linear and compatible with all structures on TM induced from the Cartan connection, so it remains to verify the condition on the Lie bracket.

Since $\hat{\xi}$ is a lift of $\tilde{\xi}$ we conclude that for two tangent vectors $\xi, \eta \in T_xM$ the vector field $[\hat{\xi}, \hat{\eta}]$ on $p^{-1}(U)$ is a lift of the Lie bracket $[\tilde{\xi}, \tilde{\eta}]$. In particular, this implies that $[\tilde{\xi}, \tilde{\eta}](x) = T_up \cdot [\hat{\xi}, \hat{\eta}](u)$, so it remains to compute $\omega([\hat{\xi}, \hat{\eta}](u))$. By definition of the exterior derivative, this may be computed as

$$\hat{\xi}(u) \cdot \omega(\hat{\eta}) - \hat{\eta}(u) \cdot \omega(\hat{\xi}) - d\omega(u)(\hat{\xi}, \hat{\eta}).$$

By construction, along the image of the section σ, the vector fields $\hat{\xi}$ and $\hat{\eta}$ coincide with $\omega^{-1}(X)$ and $\omega^{-1}(Y)$, respectively, and the flow lines of these fields through u stay within the image of σ. But this implies that $\omega(\hat{\eta})$ is constant along the flow line of $\hat{\xi}$ through u, and thus $\hat{\xi}(u) \cdot \omega(\hat{\eta}) = 0$. Similarly, $\hat{\eta}(u) \cdot \omega(\hat{\xi}) = 0$, and therefore $\omega([\hat{\xi}, \hat{\eta}](u)) = -d\omega(u)(\hat{\xi}, \hat{\eta})$. By definition of the curvature function we have $-d\omega(u)(\hat{\xi}, \hat{\eta}) = [X, Y] - \kappa(u)(X, Y)$, which completes the proof. □

While the statement of this proposition is fairly technical, its use is easy to describe. Suppose that in some way we construct a tensorial mapping out of Lie brackets. Then given tangent vectors at a point we can use the extensions provided by the proposition to compute the value of that map and immediately get a relation to the curvature of the parabolic geometry. As a first application, we can characterize regularity of a parabolic geometry:

COROLLARY 3.1.8. *Let $(p : \mathcal{G} \to M, \omega)$ be a parabolic geometry of type (G, P) with curvature form $K \in \Omega^2(\mathcal{G}, \mathfrak{g})$ and curvature function $\kappa : \mathcal{G} \to L(\Lambda^2\mathfrak{g}_-, \mathfrak{g})$, and let $TM = T^{-k}M \supset \cdots \supset T^{-1}M$ be the induced filtration of the tangent bundle TM. Then we have:*

(1) *$(M, \{T^iM\})$ is a filtered manifold if and only if $K(T^i\mathcal{G}, T^j\mathcal{G}) \subset \mathfrak{g}^{i+j}$, or equivalently $\kappa(\mathfrak{g}_i, \mathfrak{g}_j) \subset \mathfrak{g}^{i+j}$, for all $i, j < 0$.*

(2) *The geometry $(p : \mathcal{G} \to M, \omega)$ is regular if and only if $K(T^i\mathcal{G}, T^j\mathcal{G}) \subset \mathfrak{g}^{i+j+1}$, or equivalently $\kappa(\mathfrak{g}_i, \mathfrak{g}_j) \subset \mathfrak{g}^{i+j+1}$, for all $i, j < 0$.*

PROOF. For $i, j \leq -1$ both T^iM and T^jM are contained in $T^{i+j+1}M$. As in the construction of the Levi bracket in 3.1.7, this implies that the map

$$\Gamma(T^iM) \times \Gamma(T^jM) \to \Gamma(TM/T^{i+j+1}M),$$

which takes (ξ, η) to the class of $[\xi, \eta]$ is bilinear over smooth functions. Hence, we get an induced bundle map $\Phi : T^iM \times T^jM \to TM/T^{i+j+1}M$. Fixing a point $x \in M$ and a point $u \in \mathcal{G}$ which lies over x and taking tangent vectors $\xi \in T^i_xM$ and $\eta \in T^j_xM$, we can compute $\Phi(\xi, \eta)$ using the extensions $\tilde{\xi}$ and $\tilde{\eta}$ provided by the proposition. This implies that $\Phi(\xi, \eta)$ is given by the class in T_xM/T^{i+j+1}_xM of $T_up \cdot \omega_u^{-1}([X, Y] - \kappa(u)(X, Y))$, where $X, Y \in \mathfrak{g}_-$ are the elements corresponding to ξ and η.

Compatibility of the filtration of TM with the Lie bracket is equivalent to the fact that Φ has values in $T^{i+j}_xM/T^{i+j+1}_xM \subset T_xM/T^{i+j+1}_xM$. By definition of the filtration, this means that $[X, Y] - \kappa(u)(X, Y) \in \mathfrak{g}^{i+j}$. By construction, $X \in \mathfrak{g}^i$ and $Y \in \mathfrak{g}^j$ and therefore $[X, Y] \in \mathfrak{g}^{i+j}$, so the condition reduces to $\kappa(u)(X, Y) \in \mathfrak{g}^{i+j}$. The equivalence to the condition on K in (1) follows immediately from horizontality, which completes the proof of (1).

(2) If the condition of (1) is satisfied, then Φ has values in $\mathrm{gr}_{i+j}(TM)$. Denoting by q the projections to the associated graded, we by construction have $\Phi(\xi, \eta) = \mathcal{L}(q_i(\xi), q_j(\eta))$. Thus, we conclude that

$$\mathcal{L}(q_i(\xi), q_j(\eta)) = q_{i+j}(T_up \cdot \omega_u^{-1}([X, Y] - \kappa(u)(X, Y))).$$

Of course, $q_{i+j}(T_up \cdot \omega_u^{-1}([X, Y]))$ depends only on the $\mathfrak{g}_{i+j}$–component of $[X, Y]$ which equals the bracket of the $\mathfrak{g}_i$–component of X and the $\mathfrak{g}_j$–component of Y. But this implies by construction that $q_{i+j}(T_up \cdot \omega_u^{-1}([X, Y])) = \{q_i(\xi), q_j(\eta)\}$, so we obtain

$$\mathcal{L}(q_i(\xi), q_j(\eta)) - \{q_i(\xi), q_j(\eta)\} = -q_{i+j}(T_up \cdot \omega_u^{-1}(\kappa(u)(X, Y))),$$

and the characterizations in (2) follow as above. □

3.1.9. The adjoint tractor bundle. From the examples in 3.1.6 and 3.1.7, we see that regular infinitesimal flag structures are interesting geometric structures. (Note that in the case of a $|1|$–grading the regularity condition is automatically satisfied, since the underlying filtration is trivial.) Moreover, in the case of conformal structures we know from Theorem 1.6.7 that any regular infinitesimal flag structure underlies a parabolic geometry, which is uniquely determined if we assume that its curvature satisfies a normalization condition. Hence, we are naturally led to the question of whether analogous results are available in a more general setting. One necessary step in that direction is to find an appropriate normalization condition for the curvature. In the conformal case, we have simply chosen a normalization condition and proved that this condition is of the right nature. However, it will be very important for some of the further developments to get a uniform construction of a normalization condition for all parabolic geometries, since this leads to relations between geometries of different type.

The first step towards this is to interpret the Cartan curvature of a parabolic geometry as an object on the base manifold. Let $(p : \mathcal{G} \to M, \omega)$ be a parabolic geometry of type (G, P). Then the fact that the Cartan curvature $K \in \Omega^2(\mathcal{G}, \mathfrak{g})$ is horizontal and equivariant implies that it defines an element $\kappa \in \Omega^2(M, \mathcal{A}M)$, where $\mathcal{A}M = \mathcal{G} \times_P \mathfrak{g}$ is the adjoint tractor bundle; see Proposition 1.5.7. When viewed as having values in $\Lambda^2(\mathfrak{g}/\mathfrak{p})^* \otimes \mathfrak{g}$, the curvature function is the equivariant function corresponding to this section. The P–invariant filtration $\mathfrak{g} = \mathfrak{g}^{-k} \supset \cdots \supset \mathfrak{g}^k$ gives rise to a filtration of the associated bundle $\mathcal{A}M$ of the form $\mathcal{A}M = \mathcal{A}^{-k}M \supset \cdots \supset \mathcal{A}^kM$ by smooth subbundles. It is evident how to reformulate the conditions of Corollary 3.1.8 in terms of this filtration. The filtration of TM is compatible with the Lie bracket of vector fields if and only if $\kappa(T^iM, T^jM) \subset \mathcal{A}^{i+j}M$ and the parabolic geometry is regular if and only if $\kappa(T^iM, T^jM) \subset \mathcal{A}^{i+j+1}M$ for all $i, j < 0$. Moreover, since the bundle $\mathcal{A}^0M$ corresponds to the subalgebra $\mathfrak{p} \leq \mathfrak{g}$, we see that the parabolic geometry is torsion free if and only if $\kappa(\xi, \eta) \in \mathcal{A}^0M$ for all $\xi, \eta \in TM$.

The adjoint tractor bundle is a rather complicated object which depends on the full Cartan geometry, but there are some relations to simpler bundles. By definition, $\mathcal{A}M/\mathcal{A}^0M$ is the associated bundle $\mathcal{G} \times_P (\mathfrak{g}/\mathfrak{p})$, so $\mathcal{A}M/\mathcal{A}^0M \cong TM$. In particular, there is a natural bundle map $\Pi : \mathcal{A}M \to TM$ with kernel $\mathcal{A}^0M$; see 1.5.7. On the other hand, from part (3) of Proposition 3.1.2 we know that the choice of a nondegenerate $\mathfrak{g}$–invariant bilinear form B on $\mathfrak{g}$ leads to an isomorphism $\mathfrak{p}_+ \cong (\mathfrak{g}/\mathfrak{p})^*$ of $\mathfrak{p}$–modules. If we assume that B is $\mathrm{Ad}(P)$–invariant (which is satisfied for example for the Killing form), then this is a duality of P–modules. Hence, we get an identification $\mathcal{A}^1M = \mathcal{G} \times_P \mathfrak{p}_+ \cong T^*M$, so $\mathcal{A}M$ contains the cotangent bundle as a subbundle. The filtration $\mathcal{A}^1M \supset \cdots \supset \mathcal{A}^kM$ may be viewed as a filtration of T^*M by smooth subbundles, which can be immediately characterized in terms of the filtration of TM. Indeed, the proof of part (3) of Proposition 3.1.2 shows that for $i > 0$, the filtration component $\mathfrak{g}^i$ is exactly the annihilator (with respect to B) of $\mathfrak{g}^{-i+1}$. Hence, we may characterize the subbundle $\mathcal{A}^iM$ as the annihilator of $T^{-i+1}M$, i.e. $\mathcal{A}^i_xM = \{\phi \in T^*_xM : \phi(\xi) = 0 \quad \forall \xi \in T^{-i+1}_xM\}$ for $i > 0$.

We have noted in 1.5.7 that the Lie bracket on $\mathfrak{g}$ induces a bundle map $\{\ ,\ \} : \Lambda^2\mathcal{A}M \to \mathcal{A}M$, called the algebraic bracket, which makes each fiber $\mathcal{A}_xM$ into a Lie algebra isomorphic to $\mathfrak{g}$. By construction, $\{\ ,\ \}$ is compatible with the filtration of $\mathcal{A}M$, i.e. $\{\mathcal{A}^iM, \mathcal{A}^jM\} \subset \mathcal{A}^{i+j}M$. In particular, for each $x \in M$, the subspace $\mathcal{A}^1_xM \cong T^*_xM$ is a Lie subalgebra, so we get a tensorial Lie bracket on T^*M. This makes T^*M into a bundle of filtered Lie algebras modelled on $\mathfrak{p}_+$. As before, we can pass to the associated graded bundle to obtain a simpler version of this bracket. By construction, the quotient $\mathrm{gr}_i(\mathcal{A}M) = \mathcal{A}^iM/\mathcal{A}^{i+1}M$ is the associated bundle $\mathcal{G} \times_P (\mathfrak{g}^i/\mathfrak{g}^{i+1})$, which may as well be viewed as $\mathcal{G}_0 \times_{G_0} \mathfrak{g}_i$. Hence, it suffices to know the underlying G_0–bundle to construct these vector bundles.

The bundles $\text{gr}_i(\mathcal{A}M)$ can be easily described explicitly. For $i < 0$, we have $\text{gr}_i(\mathcal{A}M) = \text{gr}_i(TM)$, while for $i > 0$ we have

$$\text{gr}_i(\mathcal{A}M) = (T^{-i}M/T^{-i+1}M)^* = \text{gr}_{-i}(TM)^* = \text{gr}_i(T^*M).$$

Finally, the bundle $\text{gr}_0(\mathcal{A}M) = \mathcal{G}_0 \times_{G_0} \mathfrak{g}_0$ can be interpreted as follows: The adjoint action gives us a map $\mathfrak{g}_0 \to \mathfrak{gl}(\mathfrak{g}_-)$, which by Proposition 3.1.2 is injective, provided that no simple ideal of $\mathfrak{g}$ is contained in $\mathfrak{g}_0$. Thus, we may view $\text{gr}_0(\mathcal{A}M)$ as a subbundle of the bundle $\text{End}(\text{gr}(TM))$ of linear endomorphisms of $\text{gr}(TM)$, so we also write $\text{End}_0(\text{gr}(TM)) := \text{gr}_0(\mathcal{A}M)$. Describing this bundle explicitly amounts exactly to characterizing $\mathfrak{g}_0$ as a subspace of $\mathfrak{gl}(\mathfrak{g}_-)$, which is usually easy. Hence, we see that

$$\text{gr}(\mathcal{A}M) = \text{gr}(TM) \oplus \text{End}_0(\text{gr}(TM)) \oplus \text{gr}(T^*M).$$

Note that since $\mathfrak{g}_-$ is generated by $\mathfrak{g}_{-1}$ and $\mathfrak{g}_0$ acts on $\mathfrak{g}_-$ by derivations, the adjoint action of an element of $\mathfrak{g}_0$ on $\mathfrak{g}_-$ is determined by its restriction to $\mathfrak{g}_{-1}$. This means that we can also view $\text{gr}_0(\mathcal{A}M)$ as a subbundle of $\text{End}(T^{-1}M)$.

Viewing $\text{gr}(\mathcal{A}M)$ as $\mathcal{G}_0 \times_{G_0} \mathfrak{g}$, the Lie bracket on $\mathfrak{g}$ induces a tensorial Lie bracket on $\text{gr}(\mathcal{A}M)$, which we also denote by $\{\ ,\ \}$, since it extends the algebraic bracket on $\text{gr}(TM)$ introduced in 3.1.7. By construction this can alternatively be obtained from the algebraic bracket on $\mathcal{A}M$ by passing to the associated graded bundle. In particular, we see that this bracket on $\text{gr}(\mathcal{A}M)$ can be entirely described in terms of the underlying infinitesimal flag structure. This description can be easily made explicit as follows:

We know that the bracket on $\text{gr}(TM)$ is given by the Levi brackets, and the identification of $\text{End}_0(\text{gr}(TM))$ as a subbundle of the linear endomorphisms of $\text{gr}(TM)$ implies that the bracket $\text{End}_0(\text{gr}(TM)) \times \text{gr}(TM) \to \text{gr}(TM)$ is given by $\{\Phi, \xi\} = \Phi(\xi)$. Invariance of B then immediately implies that the bracket $\text{End}_0(\text{gr}(TM)) \times \text{gr}(T^*M) \to \text{gr}(T^*M)$ is given by $\{\Phi, \phi\} = -\Phi^*(\phi)$, where Φ^* is the dual map to Φ. By the Jacobi identity, the bracket on $\text{End}_0(\text{gr}(TM))$ is given by the commutator of endomorphisms. For $i > 1$, we get the bracket on $T^{-1}M \times \text{gr}_i(T^*M)$ from invariance of B, which implies that for $\xi \in T^{-1}M$, $\phi \in \text{gr}_i(T^*M)$ and $\eta \in \text{gr}_{-i+1}(TM)$ (note that by assumption $-i+1 < 0$), we get $\{\xi, \phi\}(\eta) = -\phi(\{\xi, \eta\})$. To get the bracket $T^{-1}M \times \text{gr}_1(T^*M)$, one has to compute explicitly the trilinear map $\mathfrak{g}_{-1} \times \mathfrak{g}_{-1} \times \mathfrak{g}_1 \to \mathfrak{g}_{-1}$ defined by $(X, Y, Z) \mapsto [[X, Z], Y]$, since this is exactly the endomorphism of $\mathfrak{g}_{-1}$ induced by $[X, Z] \in \mathfrak{g}_0$. The fact that $\mathfrak{g}_-$ is generated by $\mathfrak{g}_{-1}$ implies that as a Lie algebra $\text{gr}(TM)$ is generated by $T^{-1}M$. Using the Jacobi identity, we can then compute all brackets on $\text{gr}(TM) \times \text{gr}(T^*M)$. Hence, we are only missing the brackets defined on $\text{gr}(T^*M) \times \text{gr}(T^*M)$, which now follow from invariance of B. For $i, j > 0$, $\phi \in \text{gr}_i(T^*M)$, $\psi \in \text{gr}_j(T^*M)$, and $\xi \in \text{gr}_{-i-j}(TM)$, we have $\{\phi, \psi\}(\xi) = \phi(\{\psi, \xi\})$.

3.1.10. Normalization. With the information about the adjoint tractor bundle at hand, we may now return to the task of finding a normalization condition which leads to a unique parabolic geometry with fixed underlying regular infinitesimal flag structure. So we start with a regular parabolic geometry $(p : \mathcal{G} \to M, \omega)$ of type (G, P), and consider the underlying infinitesimal flag structure $(p_0 : \mathcal{G}_0 \to M, \omega^0)$ introduced in 3.1.5. We want to study the set of Cartan connections $\tilde{\omega} \in \Omega^1(\mathcal{G}, \mathfrak{g})$ which induce the same infinitesimal flag structure. To do this, we start by analyzing the affine structure on the space of Cartan connections on a given principal bundle.

By definition, the difference $\tilde{\omega} - \omega$ of two Cartan connections is an element of $\Omega^1(\mathcal{G}, \mathfrak{g})$. Since both Cartan connections reproduce generators of fundamental vector fields, this form has to vanish upon insertion of a vertical tangent vector, so it is horizontal. Further, since both $\tilde{\omega}$ and ω are P–equivariant, so is their difference. By Corollary 1.2.7 we can interpret $\Phi := \tilde{\omega} - \omega$ as a one–form on M with values in $\mathcal{A}M$. Conversely, a one–form $\Phi \in \Omega^1(M, \mathcal{A}M)$ can be interpreted as a horizontal equivariant $\mathfrak{g}$–valued one–form on $\mathcal{G}$. As such, we can add it to a given Cartan connection ω to obtain a $\mathfrak{g}$–valued one–form $\tilde{\omega}$ on $\mathcal{G}$ which is equivariant and reproduces the generators of fundamental vector fields. Hence, $\tilde{\omega}$ is a Cartan connection, provided that it restricts to a linear isomorphism on each tangent space.

The set of all $f \in L(\mathfrak{g}/\mathfrak{p}, \mathfrak{g})$, such that $\mathrm{id} + f \circ \pi$ is a linear isomorphism of $\mathfrak{g}$, is an open neighborhood of zero, and one immediately verifies that it is P–invariant. Passing to associated bundles, this invariant open subset gives rise to an open subbundle of $L(TM, \mathcal{A}M)$, and $\tilde{\omega}$ is a Cartan connection if and only if Φ is a section of this open subbundle. Thus, the space of all Cartan connections on the bundle $\mathcal{G}$ is an affine space modelled on the space of smooth sections of that open subbundle of $L(TM, \mathcal{A}M)$. We will simply write $\tilde{\omega} - \omega = \Phi$, respectively, $\tilde{\omega} = \omega + \Phi$ to indicate this affine structure over a space of one–forms with values in the adjoint tractor bundle. Notice that if we have a linear map $f : \mathfrak{g}/\mathfrak{p} \to \mathfrak{g}$ which is homogeneous of degree ≥ 1, i.e., which has the property that $f(\mathfrak{g}^i/\mathfrak{p}) \subset \mathfrak{g}^{i+1}$ for all $i < 0$, then $\mathrm{id} + f \circ \pi$ is homogeneous of degree ≥ 0, and the induced linear map on $\mathrm{gr}(\mathfrak{g})$ is the identity. In particular, $\mathrm{id} + f \circ \pi$ is a linear isomorphism. Hence, if a one–form Φ with values in $\mathcal{A}M$ is actually a section of the filtration component $L(TM, \mathcal{A}M)^1$, then for any Cartan connection ω the construction above again leads to a Cartan connection.

We need a final piece of information. From 3.1.1 we know that the filtrations on TM and $\mathcal{A}M$ induce a filtration of the vector bundle $L(\Lambda^\ell TM, \mathcal{A}M)$ for any $\ell = 0, \dots, \dim(M)$. The sections of this bundle are the ℓ–forms with values in the adjoint tractor bundle. The filtration on this bundle is given by homogeneous degree, i.e. $\tau \in L(\Lambda^\ell TM, \mathcal{A}M)^m$ if and only if $\tau(T^{i_1}M, \dots, T^{i_\ell}M) \subset \mathcal{A}^{i_1+\dots+i_\ell+m}M$ for all $i_1, \dots, i_\ell < 0$. We also know from 3.1.1 that the associated graded bundle $\mathrm{gr}(L(\Lambda^\ell TM, \mathcal{A}M))$ is given as $L(\Lambda^\ell \mathrm{gr}(TM), \mathrm{gr}(\mathcal{A}M))$ with the grading induced by homogeneous degrees. Hence, $\mathrm{gr}(L(\Lambda^\ell TM, \mathcal{A}M))$ may be viewed as the associated bundle to $\mathcal{G}_0$ corresponding to the G_0–module $L(\Lambda^\ell \mathfrak{g}_-, \mathfrak{g})$, so this depends only on the underlying infinitesimal flag structure. On the other hand, G_0–equivariant maps between these modules give rise to well–defined bundle maps.

The spaces $L(\Lambda^2 \mathfrak{g}_-, \mathfrak{g})$ are the chain spaces in the standard complex for computing the Lie algebra cohomology of $\mathfrak{g}_-$ with coefficients in $\mathfrak{g}$ (viewed as a $\mathfrak{g}_-$–module via the adjoint action); see 2.1.9 for the definition of Lie algebra cohomology. Moreover, since the adjoint action of G_0 on $\mathfrak{g}_-$ and on $\mathfrak{g}$ is by Lie algebra homomorphisms, the Lie algebra differential $\partial : L(\Lambda^\ell \mathfrak{g}_-, \mathfrak{g}) \to L(\Lambda^{\ell+1} \mathfrak{g}_-, \mathfrak{g})$ as defined in formula (2.2) in 2.1.9 is a G_0–homomorphism. Thus, we get induced bundle maps $\partial : \mathrm{gr}(L(\Lambda^\ell TM, \mathcal{A}M)) \to \mathrm{gr}(L(\Lambda^{\ell+1} TM, \mathcal{A}M))$ for all $\ell \geq 0$. By construction, they are given by the usual formula for the Lie algebra coboundary, but with all Lie brackets replaced by the algebraic bracket, i.e.

$$\begin{aligned}\partial\phi(\xi_0,\dots,\xi_\ell) :=& \sum_{i=0}^{\ell}(-1)^i\{\xi_i,\phi(\xi_0,\dots,\widehat{\xi_i},\dots,\xi_\ell)\}\\ &+\sum_{i<j}(-1)^{i+j}\phi(\{\xi_i,\xi_j\},\xi_0,\dots,\widehat{\xi_i},\dots,\widehat{\xi_j},\dots,\xi_\ell),\end{aligned}$$

where the hats denote omission. From this formula, it is obvious that ∂ preserves homogeneities, i.e. if ϕ is homogeneous, then also $\partial(\phi)$ is homogeneous of the same degree. Using this, we can now formulate:

PROPOSITION 3.1.10. *Let $(p : \mathcal{G} \to M, \omega)$ be a regular parabolic geometry of type (G, P), let $\tilde{\omega} \in \Omega^1(\mathcal{G}, \mathfrak{g})$ be another Cartan connection and put $\Phi := \tilde{\omega} - \omega \in \Omega^1(M, \mathcal{A}M)$.*

(1) The Cartan connections ω and $\tilde{\omega}$ induce the same filtration of TM if and only if $\Phi \in \Omega^1(M, \mathcal{A}M)^0$ and they induce the same underlying infinitesimal flag structure if and only if $\Phi \in \Omega^1(M, \mathcal{A}M)^1$.

(2) Suppose that $\Phi(T^iM) \subset \mathcal{A}^{i+\ell}M$ for some fixed $\ell \geq 1$. Then the difference $\tilde{\kappa} - \kappa$ of the curvatures of the two Cartan connections maps $T^iM \times T^jM$ to $\mathcal{A}^{i+j+\ell}M$. The induced section $\mathrm{gr}_\ell(\tilde{\kappa} - \kappa)$ of $\mathrm{gr}_\ell(L(\Lambda^2TM, \mathcal{A}M))$ is given by $\partial(\mathrm{gr}_\ell(\Phi))$, where $\mathrm{gr}_\ell(\Phi) \in \Gamma(\mathrm{gr}_\ell(L(TM, \mathcal{A}M)))$ is the section induced by Φ.

PROOF. (1) Since for $i < 0$ we may characterize T^iM as the image under Tp of $T^i\mathcal{G} = \omega^{-1}(\mathfrak{g}^i)$ and both ω and $\tilde{\omega}$ restrict to linear isomorphisms on each tangent space, we see that the fact that $\tilde{\omega}$ induces the same filtration as ω is equivalent to the fact that $\tilde{\omega}(T^i\mathcal{G}) \subset \mathfrak{g}^i$ for all $i < 0$. Since $\tilde{\omega} - \omega$ by definition vanishes on $T^0\mathcal{G}$, we see that ω and $\tilde{\omega}$ induce the same filtration if and only if their difference maps each $T^i\mathcal{G}$ to $\mathfrak{g}^i$. From the construction in 3.1.5 it is clear that ω and $\tilde{\omega}$ induce the same infinitesimal flag structure if and only if their difference maps $T^i\mathcal{G}$ to $\mathfrak{g}^{i+1}$ for all $i < 0$. From the definition of Φ, one immediately concludes that this is equivalent to $\Phi(T^iM) \subset \mathcal{A}^{i+1}M$.

(2) We start by working in the picture of forms on $\mathfrak{g}$. In terms of the function $\phi : \mathcal{G} \to L(\mathfrak{g}/\mathfrak{p}, \mathfrak{g})$ representing Φ, we have $\tilde{\omega} = \omega + \phi \circ \omega$, or more precisely,

$$\tilde{\omega}(u)(\xi) = \omega(u)(\xi) + \phi(u)(\omega(u)(\xi)).$$

From the definition of the exterior derivative, one concludes that this implies $d\tilde{\omega}(\xi,\eta) = d\omega(\xi,\eta) + d\phi(\xi)(\omega(\eta)) - d\phi(\eta)(\omega(\xi)) + \phi(d\omega(\xi,\eta))$. On the other hand, $[\tilde{\omega}(\xi),\tilde{\omega}(\eta)] = [\omega(\xi),\omega(\eta)] + [\phi(\omega(\xi)),\omega(\eta)] + [\omega(\xi),\phi(\omega(\eta))] + [\phi(\omega(\xi)),\phi(\omega(\eta))]$. Thus, we obtain

$$\begin{aligned}\tilde{K}(\xi,\eta) - K(\xi,\eta) =& d\phi(\xi)(\omega(\eta)) - d\phi(\eta)(\omega(\xi)) + \phi(d\omega(\xi,\eta))\\ &+ [\phi(\omega(\xi)),\omega(\eta)] + [\omega(\xi),\phi(\omega(\eta))] + [\phi(\omega(\xi)),\phi(\omega(\eta))].\end{aligned} \tag{3.1}$$

The relation between Φ and ϕ reads as $\Phi(Tp \cdot \xi) = \omega^{-1}(\phi(\omega(\xi)))$. Thus, the condition on homogeneity of Φ exactly means that for $\xi \in T^i\mathcal{G}$ and $\eta \in T^j\mathcal{G}$ with $i, j < 0$, we have $\phi(\omega(\xi)) \in \mathfrak{g}^{i+\ell}$ and $\phi(\omega(\eta)) \in \mathfrak{g}^{j+\ell}$. On the other hand, since $\phi(T^n\mathcal{G}) \subset \mathfrak{g}^{n+\ell}$, the same is true for $d\phi(\xi)$ and $d\phi(\eta)$. In particular, the first two terms in the right–hand side of (3.1) have values in $\mathfrak{g}^{j+\ell}$ and $\mathfrak{g}^{i+\ell}$, respectively. By regularity, $d\omega(\xi,\eta) \in \mathfrak{g}^{i+j}$, which together with the above implies that the next three terms in (3.1) have values in $\mathfrak{g}^{i+j+\ell}$, while the last term has values in $\mathfrak{g}^{i+j+2\ell}$. Since $i, j < 0$ and $\ell > 0$ by assumption, this implies that the difference

$\tilde{K}(\xi,\eta)-K(\xi,\eta)$ has values in $\mathfrak{g}^{i+j+\ell}$, and its class in $\mathrm{gr}_{i+j+\ell}(\mathfrak{g})$ coincides with the class of

$$\phi(d\omega(\xi,\eta))+[\phi(\omega(\xi)),\omega(\eta)]+[\omega(\xi),\phi(\omega(\eta))].$$

By regularity, $d\omega(\xi,\eta)$ is congruent to $-[\omega(\xi),\omega(\eta)]$ modulo $\mathfrak{g}^{i+j+1}$, and hence the first summand is congruent to $-\phi([\omega(\xi),\omega(\eta)])$ modulo $\mathfrak{g}^{i+j+\ell+1}$. On the level of the associated graded to the tangent bundle, the bracket $[\ ,\ ]$ on $\mathfrak{g}$ corresponds to the algebraic bracket on $\mathrm{gr}(\mathcal{A}M)$ and $\mathrm{gr}(TM)$. Now by construction, the function $\mathcal{G}\to\mathrm{gr}_\ell(L(\Lambda^2(\mathfrak{g}_-),\mathfrak{g}))$ corresponding to $\mathrm{gr}_\ell(\tilde{\kappa}-\kappa)$ is characterized by $(\tilde{\kappa}-\kappa)(\xi,\eta)=\omega^{-1}(\tilde{K}(\tilde{\xi},\tilde{\eta})-K(\tilde{\xi},\tilde{\eta}))$, where $\tilde{\xi}$ projects onto a representative of the class ξ in TM, and similarly for $\tilde{\eta}$. But then the class of $-\phi([\omega(\xi),\omega(\eta)])$ precisely corresponds to $-\mathrm{gr}_\ell(\Phi)(\{\xi,\eta\})$, while the other two terms correspond to $-\{\eta,\mathrm{gr}_\ell(\Phi)(\xi)\}$ and $\{\xi,\mathrm{gr}_\ell(\Phi)(\eta)\}$, respectively. Thus, we get the required formula $\mathrm{gr}_\ell(\tilde{\kappa}-\kappa)(\xi,\eta)=\partial(\mathrm{gr}_\ell(\Phi))(\xi,\eta)$. □

We can now summarize what properties we need for a normalization condition for regular parabolic geometries, i.e. a condition on the curvature, which together with the underlying infinitesimal flag structure pins down the Cartan connection as much as possible. To be of geometric nature, the normalization condition should require that the curvature has values in a natural subbundle of $\Lambda^2T^*M\otimes\mathcal{A}M$. Such a subbundle corresponds to a P–submodule in $\Lambda^2(\mathfrak{g}/\mathfrak{p})^*\otimes\mathfrak{g}$. The above proposition tells us that passing to the associated graded, identifying it with $\Lambda^2\mathfrak{g}_-^*\otimes\mathfrak{g}$, and restricting to a fixed homogeneous degree, our submodule should be complementary to the G_0–submodule $\mathrm{im}(\partial)$, where ∂ denotes the Lie algebra differential. An obvious choice for a submodule complementary to the image of a linear map is the kernel of the adjoint map with respect to some inner product. Hence, the best way to get a normalization condition would be to construct a P–module homomorphism $\Lambda^2(\mathfrak{g}/\mathfrak{p})^*\otimes\mathfrak{g}\to(\mathfrak{g}/\mathfrak{p})^*\otimes\mathfrak{g}$ which is compatible with homogeneous degrees and has the property that the induced map on the associated graded is adjoint (with respect to some inner product) to the Lie algebra differential ∂. Such an adjoint fortunately is provided by Kostant's harmonic theory on the standard complex $C^*(\mathfrak{g}_-,\mathfrak{g})$. Since this harmonic theory will be an essential ingredient in a more general situation later, we just give a short description of the result in the special case we need, and postpone the proofs to 3.3.1 below.

3.1.11. The Kostant codifferential. Let $\mathfrak{g}=\mathfrak{g}_{-k}\oplus\cdots\oplus\mathfrak{g}_k$ be a $|k|$–graded semisimple Lie algebra, G a group with Lie algebra $\mathfrak{g}$, and let B be the Killing form of $\mathfrak{g}$. Then B is a G–invariant nondegenerate bilinear form on $\mathfrak{g}$. Hence, B induces an isomorphism $\mathfrak{g}\cong\mathfrak{g}^*$ of G–modules (and thus of P–modules), and from Proposition 3.1.2 we know that B induces an isomorphism $(\mathfrak{g}/\mathfrak{p})^*\cong\mathfrak{p}_+$ of P–modules. Consequently, we can identify $\Lambda^j(\mathfrak{g}/\mathfrak{p})^*\otimes\mathfrak{g}$ with the dual P–module of $\Lambda^j\mathfrak{p}_+^*\otimes\mathfrak{g}=C^j(\mathfrak{p}_+,\mathfrak{g})$. By definition, $\partial_\mathfrak{p}:C^j(\mathfrak{p}_+,\mathfrak{g})\to C^{j+1}(\mathfrak{p}_+,\mathfrak{g})$ is a P–homomorphism, and from 2.1.9 we know that $\partial_\mathfrak{p}^2=\partial_\mathfrak{p}\circ\partial_\mathfrak{p}=0$. Dualizing this map, we obtain a P–homomorphism $\partial^*:\Lambda^j(\mathfrak{g}/\mathfrak{p})^*\otimes\mathfrak{g}\to\Lambda^{j-1}(\mathfrak{g}/\mathfrak{p})^*\otimes\mathfrak{g}$ which satisfies $\partial^*\circ\partial^*=0$, and is called the *Kostant codifferential.*

We can easily obtain a formula for ∂^* on decomposable elements. Taking into account that $(\mathfrak{g}/\mathfrak{p})^*=\mathfrak{p}_+$, we can write a decomposable element of $\Lambda^{n+1}(\mathfrak{g}/\mathfrak{p})^*\otimes\mathfrak{g}$ as $Z_0\wedge\cdots\wedge Z_n\otimes A$ with $Z_i\in\mathfrak{p}_+$ and $A\in\mathfrak{g}$. The pairing of $\psi\in C^{n+1}(\mathfrak{p}_+,\mathfrak{g})$ with that element is given by $B(\psi(Z_0,\dots,Z_n),A)$. Hence, for $\phi\in C^n(\mathfrak{p}_+,\mathfrak{g})$, the pairing of $\partial\phi$ with our element is given by

$$\sum_{i=0}^{n}(-1)^i B([Z_i,\phi(Z_0,\dots,\widehat{Z_i},\dots,Z_n)],A)$$
$$+\sum_{i<j}(-1)^{i+j}B(\phi([Z_i,Z_j],Z_0,\dots,\widehat{Z_i},\dots,\widehat{Z_j},\dots,Z_n),A).$$

Using invariance of B, we may rewrite each of the summands in the first sum as $(-1)^{i+1}B(\phi(Z_0,\dots,\widehat{Z_i},\dots,Z_n),[Z_i,A])$, whence we immediately see that

$$\partial^*(Z_0\wedge\dots\wedge Z_n\otimes A)=\sum_{i=0}^{n}(-1)^{i+1}Z_0\wedge\dots\wedge\widehat{Z_i}\wedge\dots\wedge Z_n\otimes[Z_i,A]$$
$$+\sum_{i<j}(-1)^{i+j}[Z_i,Z_j]\wedge Z_0\wedge\dots\wedge\widehat{Z_i}\wedge\dots\wedge\widehat{Z_j}\wedge\dots\wedge Z_n\otimes A.$$

This is the Lie algebra homology differential as defined in formula (2.1) in 2.1.9, with the different signs due to the different conventions for numbering the elements.

The formula for ∂^* shows that it is filtration preserving and it provides another proof for the fact that ∂^* is P–equivariant, which will be crucial in the sequel. From part (3) of Proposition 3.1.2, we know that the dualities provided by B are compatible with filtrations and gradings. In particular, there is an induced map $\mathrm{gr}_0(\partial^*)$ between the associated graded spaces, which we may identify with the spaces $L(\Lambda^n\mathfrak{g}_-,\mathfrak{g})$, on which the Lie algebra differential is naturally defined. But the above formula shows that ∂^* even is homogeneous of degree zero with respect to the gradings, so $\mathrm{gr}_0(\partial^*)$ is given by the same formula. Therefore, when working on the Lie algebra, we will simply identify ∂^* and $\mathrm{gr}_0(\partial^*)$.

For later applications, the case $n=2$ will be of particular importance. It will be helpful to have a formula for the action of ∂^* on general maps in this case.

LEMMA 3.1.11. *Consider an element $\phi\in L(\Lambda^2(\mathfrak{g}/\mathfrak{p}),\mathfrak{g})$. Interpreting ϕ as a bilinear map on $\mathfrak{g}$, which vanishes if one of its entries lies in $\mathfrak{p}$, we can compute $\partial^*\phi:\mathfrak{g}/\mathfrak{p}\to\mathfrak{g}$ as follows. Choose elements $X_1,\dots,X_\ell\in\mathfrak{g}$ such that $\{X_1+\mathfrak{p},\dots,X_\ell+\mathfrak{p}\}$ is a basis of $\mathfrak{g}/\mathfrak{p}$, and let $Z_1,\dots,Z_\ell$ be the dual basis of $\mathfrak{p}_+$. Then for each $X\in\mathfrak{g}$ we get*

$$\partial^*\phi(X+\mathfrak{p})=2\sum_i[Z_i,\phi(X,X_i)]-\sum_i\phi([Z_i,X],X_i).$$

PROOF. It suffices to prove this in the case that ϕ is decomposable. Now denoting by B the Killing form, a decomposable element $\phi:=U\wedge V\otimes A\in\Lambda^2\mathfrak{p}_+\otimes\mathfrak{g}$ acts as a bilinear map as

$$\phi(X,Y)=\tfrac{1}{2}\left(B(U,X)B(V,Y)-B(V,X)B(U,Y)\right)A.$$

Thus,

$$2[Z_i,\phi(X,X_i)]=\left(B(U,X)B(V,X_i)-B(V,X)B(U,X_i)\right)[Z_i,A],$$

and summing over all i and using that $\sum_i B(V,X_i)Z_i=V$ and likewise for U, we obtain $B(U,X)[V,A]-B(V,X)[U,A]$.

Using that $B([Z_i,X],U)=-B(X,[Z_i,U])$, and similarly for V, we get

$$\sum_i\phi([Z_i,X],X_i)=\tfrac{1}{2}(-B(X,[V,U])+B(X,[U,V]))A=B(X,[U,V])A.$$

Hence, the right–hand side of the claimed formula gives

$$-B(V,X)[U,A]-B(U,X)[V,A]-B([U,V],X)A,$$

which is exactly the action of $-V\otimes[U,A]+U\otimes[V,A]-[U,V]\otimes A$ on X. □

Next, one defines the associated Laplacian $\square:=\partial\circ\partial^*+\partial^*\circ\partial$, which is called the *Kostant Laplacian.* By construction, this is a G_0–homomorphism mapping each of the spaces $C^n(\mathfrak{g}_-,\mathfrak{g})$ to itself. In 3.3.1 we will construct (in a more general setting) an inner product with respect to which the operators ∂ and ∂^* are adjoint and use this to prove the following Hodge decomposition

PROPOSITION 3.1.11. *For each $n\geq 0$, the chain space $C^n(\mathfrak{g}_-,\mathfrak{g})$ naturally splits into a direct sum of G_0–submodules as*

$$C^n(\mathfrak{g}_-,\mathfrak{g})=\operatorname{im}(\partial^*)\oplus\ker(\square)\oplus\operatorname{im}(\partial),$$

with the first two summands adding up to $\ker(\partial^*)$ *and the last two summands adding up to* $\ker(\partial)$.

This result has several important consequences:

COROLLARY 3.1.11. *(1) For each $n\geq 0$, we may naturally identify the G_0–module $H^n(\mathfrak{g}_-,\mathfrak{g})$ with* $\ker(\square)\subset C^n(\mathfrak{g}_-,\mathfrak{g})$.

(2) We may also view $H^n(\mathfrak{g}_-,\mathfrak{g})$ as $\ker(\partial^*)/\operatorname{im}(\partial^*)$, *which naturally makes the cohomology groups into P–modules. The subgroup $P_+\subset P$ acts trivially on the cohomology groups, so their P–module structure is obtained by trivially extending the natural G_0–module structure.*

PROOF. Part (1) and the fact that $H^n(\mathfrak{g}_-,\mathfrak{g})\cong\ker(\partial^*)/\operatorname{im}(\partial^*)$ are evident from the proposition. Since ∂^* is P–equivariant, the cohomology groups inherit a P–module structure from this description. To prove the last statement, we observe that for $Z_j\in\mathfrak{p}_+$ and $A\in\mathfrak{g}$ we get

$$\begin{aligned}&\partial^*(Z_0\wedge\dots\wedge Z_n\otimes A)+Z_0\wedge\partial^*(Z_1\wedge\dots\wedge Z_n\otimes A)\\&=-Z_1\wedge\dots\wedge Z_n\otimes[Z_0,A]+\sum_{j=1}^{n}(-1)^j[Z_0,Z_j]\wedge Z_1\wedge\dots\wedge\widehat{Z_j}\wedge\dots\wedge Z_n\otimes A.\end{aligned}$$

In the last sum we can move the bracket behind Z_{j-1} (which leads to a sign $(-1)^{j-1}$) to conclude that the right–hand side gives the negative of the action of Z_0 on $Z_1\wedge\dots\wedge Z_n\otimes A$. Hence, for any $\phi\in L(\Lambda^n(\mathfrak{g}/\mathfrak{p}),\mathfrak{g})$ and any $Z\in\mathfrak{p}_+$ we get $\partial^*(Z\wedge\phi)=-Z\wedge\partial^*(\phi)-Z\cdot\phi$. In particular, the action of an element of $\mathfrak{p}_+$ maps $\ker(\partial^*)$ to $\operatorname{im}(\partial^*)$, whence $\mathfrak{p}_+$ acts trivially on the quotient. Since $P_+=\exp(\mathfrak{p}_+)$, the result follows. □

REMARK 3.1.11. Part (1) of the corollary is the first step towards the computation of these (and of more general) cohomologies in the proof of Kostant's version of the Bott–Borel–Weil theorem.

Part (2) also shows that the Lie algebra cohomology group $H^n(\mathfrak{g}_-,\mathfrak{g})$ is just the dual space of the Lie algebra homology group $H_n(\mathfrak{p}_+,\mathfrak{g})$.

3.1.12. Normal parabolic geometries and harmonic curvature. We can directly carry over the constructions of the last subsections to any manifold endowed with a parabolic geometry. Suppose that $(p : \mathcal{G} \to M, \omega)$ is a parabolic geometry of type (G, P), and for $n = 0, \dots, \dim(M)$ consider the bundle $\Lambda^n T^*M \otimes \mathcal{A}M$, whose sections are the n–forms with values in the adjoint tractor bundle. By definition, this bundle is associated to the P–module $L(\Lambda^n(\mathfrak{g}/\mathfrak{p}), \mathfrak{g})$, so the Kostant codifferential induces bundle maps $\partial^* : \Lambda^{n+1}T^*M \otimes \mathcal{A}M \to \Lambda^n T^*M \otimes \mathcal{A}M$ for any n. We denote by the same symbol the induced tensorial operator $\Omega^{n+1}(M, \mathcal{A}M) \to \Omega^n(M, \mathcal{A}M)$. The explicit formula for ∂^* on decomposable elements from 3.1.11 above can be directly translated into geometric terms. Denoting by $\{\ ,\ \}$ the algebraic bracket on $\mathcal{A}M$ and also the induced algebraic bracket on $T^*M \cong \mathcal{A}^1 M$, we get for $\phi_0, \dots, \phi_n \in \Omega^1(M)$ and $s \in \Gamma(\mathcal{A}M)$ the formula

$$\begin{aligned}\partial^*(\phi_0 \wedge \dots \wedge \phi_n \otimes s) &= \sum_{i=0}^{n} (-1)^{i+1} \phi_0 \wedge \dots \wedge \widehat{\phi_i} \wedge \dots \wedge \phi_n \otimes \{\phi_i, s\} \\ &+ \sum_{i<j} (-1)^{i+j} \{\phi_i, \phi_j\} \wedge \phi_0 \wedge \dots \wedge \widehat{\phi_i} \wedge \dots \wedge \widehat{\phi_j} \wedge \dots \wedge \phi_n \otimes s.\end{aligned}$$

From 3.1.11 we know that the P–module homomorphism ∂^* is compatible with the filtrations, so the induced bundle map is compatible with the natural filtrations of the bundles $\Lambda^n T^*M \otimes \mathcal{A}M$. Thus, we may pass to the associated graded bundles and get an induced bundle map

$$\operatorname{gr}_0(\partial^*) : L(\Lambda^{n+1} \operatorname{gr}(T^*M), \operatorname{gr}(\mathcal{A}M)) \to L(\Lambda^n \operatorname{gr}(T^*M), \operatorname{gr}(\mathcal{A}M)),$$

which is homogeneous of degree zero. We already know that on the associated graded bundle $L(\Lambda^n \operatorname{gr}(T^*M), \operatorname{gr}(\mathcal{A}M))$ the subgroup P_+ acts trivially, so we may view this as $\mathcal{G}_0 \times_{G_0} C^n(\mathfrak{g}_-, \mathfrak{g})$. In particular, it depends only on the underlying infinitesimal flag structure. We also have the bundle map ∂ induced by the Lie algebra differential, and the Kostant Laplacian induces bundle maps $\square$ mapping each $L(\Lambda^n \operatorname{gr}(T^*M), \operatorname{gr}(\mathcal{A}M))$ to itself. Note that the bundle maps ∂ and $\square$ are homogeneous of degree zero. Moreover, the Hodge decomposition from Proposition 3.1.11 induces a Hodge decomposition

$$L(\Lambda^n \operatorname{gr}(T^*M), \operatorname{gr}(\mathcal{A}M)) = \operatorname{im}(\operatorname{gr}_0(\partial^*)) \oplus \ker(\square) \oplus \operatorname{im}(\partial) \tag{3.2}$$

into a direct sum of smooth subbundles. The subbundle $\ker(\square)$ is by construction isomorphic to the associated bundle $\mathcal{G}_0 \times_{G_0} H^n(\mathfrak{g}_-, \mathfrak{g})$, with respect to the canonical action of G_0 on the cohomology space. Moreover, the first two summands add up to the subbundle $\ker(\operatorname{gr}_0(\partial^*))$, while the last two summands add up to the subbundle $\ker(\partial)$. The asymmetry in this version of the Hodge decomposition (namely that we have $\operatorname{gr}_0(\partial^*)$ and ∂ involved) is due to the fact that the codifferential on the associated graded bundles is induced by a bundle map on the filtered bundles, while the differential ∂ and the Laplacian $\square$ are only defined on the associated graded bundles.

The subspaces $\operatorname{im}(\partial^*) \subset \ker(\partial^*) \subset L(\Lambda^n(\mathfrak{g}, \mathfrak{p}), \mathfrak{g})$, are P–submodules, so they give rise to smooth subbundles $\operatorname{im}(\partial^*) \subset \ker(\partial^*) \subset \Lambda^n T^*M \otimes \mathcal{A}M$, which are exactly the image and the kernel of the bundle maps ∂^*. Thus, we may form the quotient bundles $\ker(\partial^*)/\operatorname{im}(\partial^*)$ and from Corollary 3.1.11 we know that they correspond to a representation with trivial P_+–action. Hence, they are naturally associated to $\mathcal{G}_0$, and again from Corollary 3.1.11 we see that

$$\ker(\partial^*)/\operatorname{im}(\partial^*) \cong \mathcal{G}_0 \times_{G_0} H^n(\mathfrak{g}_-, \mathfrak{g}).$$

So as before we end up with bundles which only depend on the underlying infinitesimal flag structure only. In particular, any smooth form $\tau \in \Omega^n(M, \mathcal{A}M)$ such that $\partial^*(\tau) = 0$ is a section of the subbundle $\ker(\partial^*)$, so it may be projected to a section τ_H of the bundle $\mathcal{G}_0 \times_{G_0} H^n(\mathfrak{g}_-, \mathfrak{g})$.

DEFINITION 3.1.12. (1) A parabolic geometry $(p : \mathcal{G} \to M, \omega)$ is called *normal* if its curvature $\kappa \in \Omega^2(M, \mathcal{A}M)$ satisfies $\partial^*(\kappa) = 0$.

(2) Let $(p : \mathcal{G} \to M, \omega)$ be a normal parabolic geometry with curvature $\kappa \in \Omega^2(M, \mathcal{A}M)$. Then the *harmonic curvature* κ_H is defined to be the image of κ in the space of sections of the bundle $\mathcal{G}_0 \times_{G_0} H^2(\mathfrak{g}_-, \mathfrak{g})$.

The harmonic curvature of a normal parabolic geometry is a much simpler object than the curvature, since it is a section of a bundle depending only on the underlying infinitesimal flag structure. It turns out, however, that for a regular normal parabolic geometry, the harmonic curvature contains the complete information about the curvature. Proving this requires a considerable amount of technology and will be done in volume two. Here we can prove a first step in this direction, which, in particular, implies that the harmonic curvature is a complete obstruction to local flatness of a regular normal parabolic geometry. This is based on the Bianchi identity, which was proved for arbitrary Cartan geometries in 1.5.9.

From Corollary 3.1.8 and from 3.1.9 we know that the curvature κ of a regular parabolic geometry $(p : \mathcal{G} \to M, \omega)$ lies in $\Omega^2(M, \mathcal{A}M)^1$, i.e. $\kappa(T^iM, T^jM) \subset \mathcal{A}^{i+j+1}M$. If we assume that $\kappa \in \Omega^2(M, \mathcal{A}M)^\ell$ for some $\ell \geq 1$, then we may consider the associated section $\operatorname{gr}_\ell(\kappa)$ of $\operatorname{gr}(\Lambda^2T^*M \otimes \mathcal{A}M) = L(\Lambda^2 \operatorname{gr}(TM), \operatorname{gr}(\mathcal{A}M))$.

THEOREM 3.1.12. *Let $(p : \mathcal{G} \to M, \omega)$ be a regular parabolic geometry of type (G, P) such that for some $\ell > 0$, the curvature κ lies in $\Omega^2(M, \mathcal{A}M)^\ell$, i.e. such that $\kappa(T^iM, T^jM) \subset \mathcal{A}^{i+j+\ell}M$ for all $i, j < 0$. Then the induced section $\operatorname{gr}_\ell(\kappa)$ has the property that $\partial(\operatorname{gr}_\ell(\kappa)) = 0$.*

If the parabolic geometry is normal, then $\operatorname{gr}_\ell(\kappa)$ is a section of the subbundle $\ker(\square) \subset L(\Lambda^2 \operatorname{gr}(TM), \operatorname{gr}(\mathcal{A}M))$, and under the natural identification of this bundle with $\ker(\partial^*)/\operatorname{im}(\partial^*)$ the section $\operatorname{gr}_\ell(\kappa)$ coincides with the homogeneous component of degree ℓ of the harmonic curvature κ_H. In particular, for normal parabolic geometries vanishing of κ_H implies vanishing of κ.

PROOF. Consider sections $\xi_j \in \Gamma(\operatorname{gr}_{i_j}(TM))$ for $j = 1, 2, 3$ and $i_1, i_2, i_3 < 0$. Since $\operatorname{gr}_{i_j}(TM) = \operatorname{gr}_{i_j}(\mathcal{A}M)$, we may choose smooth sections $s_j \in \Gamma(\mathcal{A}^{i_j}M)$ representing the ξ_j. Denoting by $\{\ ,\ \}$ the algebraic brackets and by D the fundamental derivative on $\mathcal{A}M$ (see 1.5.8), the form (1.25) of the Bianchi identity from 1.5.9 reads as

$$0 = \sum_{cycl} \Big(\{\kappa(s_1, s_2), s_3\} + \kappa(\{s_1, s_2\}, s_3) - \kappa(\kappa(s_1, s_2), s_3) - D_{s_3}(\kappa(s_1, s_2)) \Big).$$

The right–hand side of this expression is a trilinear map $\mathcal{A}M \times \mathcal{A}M \times \mathcal{A}M \to \mathcal{A}M$. By our assumptions on κ, the first two summands are visibly homogeneous of degree $\geq \ell$, while the third summand is homogeneous of degree $\geq 2\ell > \ell$. Finally, for the last summand, $\kappa(s_1, s_2)$ is by assumption a section of $\mathcal{A}^{i_1+i_2+\ell}M$, so also $D_{s_3}(\kappa(s_1, s_2))$ is a section of that subbundle, and all terms coming from these summands are homogeneous of degree $> \ell$. Hence, computing gr_ℓ of this expression, we only have to consider the first two terms in each summand of the cyclic sum. Since the algebraic bracket on $\mathcal{A}M$ induces the algebraic bracket on $\mathrm{gr}(\mathcal{A}M)$ (and thus also on $\mathrm{gr}(TM)$), passing to gr_ℓ we simply get

$$0 = \sum_{cycl} (\{\mathrm{gr}_\ell(\kappa)(\xi_1, \xi_2), \xi_3\} + \mathrm{gr}_\ell(\kappa)(\{\xi_1, \xi_2\}, \xi_3)).$$

Using skew symmetry of $\mathrm{gr}_\ell(\kappa)$ and the algebraic bracket, one immediately sees that the right–hand side equals $-\partial(\mathrm{gr}_\ell(\kappa))(\xi_1, \xi_2, \xi_3)$.

If we now assume that the parabolic geometry in question is normal (and regular), then by definition $\partial^*(\kappa) = 0$. Since ∂^* is compatible with homogeneities, we get $0 = \mathrm{gr}_\ell(\partial^*(\kappa)) = \mathrm{gr}_0(\partial^*)(\mathrm{gr}_\ell(\kappa))$, so we see that $\mathrm{gr}_\ell(\kappa)$ is a section of the subbundle $\ker(\mathrm{gr}_0(\partial^*)) \cap \ker(\partial) = \ker(\Box)$. On the other hand, the homogeneous component of degree ℓ of κ_H by construction equals the image of $\mathrm{gr}_\ell(\kappa) \in \Gamma(\ker(\mathrm{gr}_0(\partial^*)))$ in the quotient modulo the image of ∂^*, which implies that this component coincides with $\mathrm{gr}_\ell(\kappa)$. Inductively, starting from $\ell = 1$, this shows that $\kappa_H = 0$ implies $\kappa = 0$. □

This result is remarkable in various respects. First, it reduces the question of local flatness from considering the curvature κ to considering the harmonic curvature κ_H, which is much simpler to understand. Second, it immediately leads to an overview on what the essential curvature quantities for any parabolic geometry are. As we shall see in Section 3.3, there is a simple algorithm to compute the cohomology $H^2(\mathfrak{g}_-, \mathfrak{g})$ explicitly as a $\mathfrak{g}_0$–representation. Moreover, one even gets complete information on how this cohomology (viewed as $\ker(\Box)$) sits within the space $L(\Lambda^2\mathfrak{g}_-, \mathfrak{g})$. In most cases, $H^2(\mathfrak{g}_-, \mathfrak{g})$ consists of few components only, so one gets quite effective information on possible obstructions to local flatness.

3.1.13. Existence of normal Cartan connections. We are ready to take the first step towards proving existence of normal parabolic geometries with a given underlying regular infinitesimal flag structure. By normalizing a given Cartan connection, we prove that any regular infinitesimal flag structure that comes from some parabolic geometry also comes from a normal parabolic geometry. This uses only the algebraic properties of the normalization condition as described at the end of 3.1.10.

PROPOSITION 3.1.13. *Let $(p : \mathcal{G} \to M, \omega)$ be a regular parabolic geometry with curvature $\kappa \in \Omega^2(M, \mathcal{A}M)$ and suppose that $\partial^*\kappa \in \Omega^1(M, \mathcal{A}M)^\ell$ for some $\ell \geq 1$. Then there is a normal Cartan connection $\tilde{\omega} \in \Omega^1(\mathcal{G}, \mathfrak{g})$ such that $\tilde{\omega} - \omega \in \Omega^1(M, \mathcal{A}M)^\ell$. In particular, there is always a normal Cartan connection $\tilde{\omega}$ which induces the same underlying infinitesimal flag structure as ω.*

PROOF. We show that we can find a Cartan connection $\tilde{\omega}$ such that $\tilde{\omega} - \omega \in \Omega^1(M, \mathcal{A}M)^\ell$ and such that the curvature $\tilde{\kappa}$ of $\tilde{\omega}$ satisfies $\partial^*\tilde{\kappa} \in \Omega^1(M, \mathcal{A}M)^{\ell+1}$. Inductively, this implies that we can find a normal $\tilde{\omega}$ such that $\tilde{\omega} - \omega \in \Omega^1(M, \mathcal{A}M)^\ell$.

So let us assume that $\partial^*\kappa \in \Omega^1(M, \mathcal{A}M)^\ell$ for some fixed $\ell \geq 1$. Then we have an induced section $\mathrm{gr}_\ell(\partial^*\kappa)$ of the bundle $L(\mathrm{gr}(TM), \mathrm{gr}(\mathcal{A}M))$, whose value in each point is homogeneous of degree ℓ. The bundle map

$$\partial^* : \Lambda^2 T^*M \otimes \mathcal{A}M \to T^*M \otimes \mathcal{A}M$$

is compatible with homogeneities, so it induces a bundle map between the filtration components in degree ℓ. In particular, this implies that we find an element $\psi \in$

$\Omega^2(M, \mathcal{A}M)^\ell$ such that $\partial^*\psi = \partial^*\kappa$. Hence, we obtain

$$\mathrm{gr}_\ell(\partial^*\kappa) = \mathrm{gr}_\ell(\partial^*\psi) = \mathrm{gr}_0(\partial^*)(\mathrm{gr}_\ell(\psi)),$$

so $\mathrm{gr}_\ell(\partial^*\kappa)$ is a section of the subbundle $\mathrm{im}(\mathrm{gr}_0(\partial^*)) \subset L(\mathrm{gr}(TM), \mathrm{gr}(\mathcal{A}M))$. The Hodge decomposition (3.2) in 3.1.12 for $n = 2$ shows that the subbundle $\mathrm{im}(\partial) \subset L(\Lambda^2 \mathrm{gr}(TM), \mathrm{gr}(\mathcal{A}M))$ is complementary to the subbundle $\ker(\mathrm{gr}_0(\partial^*))$. Hence, $\mathrm{gr}_0(\partial^*)$ restricts to a bundle isomorphism $\mathrm{im}(\partial) \to \mathrm{im}(\mathrm{gr}_0(\partial^*))$. Thus, we can find a smooth section ϕ of $L(\mathrm{gr}(TM), \mathrm{gr}(\mathcal{A}M))$ whose value in each point is homogeneous of degree ℓ such that

$$\mathrm{gr}_0(\partial^*)(\partial\phi) = \mathrm{gr}_\ell(\partial^*\kappa).$$

Next, we can find a one–form $\Phi \in \Omega^1(M, \mathcal{A}M)^\ell$ such that $\mathrm{gr}_\ell(\Phi) = \phi$. As we have observed in 3.1.10, interpreting Φ as a horizontal equivariant form on $\mathcal{G}$, we can form a Cartan connection $\tilde{\omega} = \omega - \Phi$. By part (1) of Proposition 3.1.10, $\tilde{\omega}$ and ω induce the same underlying infinitesimal flag structure. On the other hand, from part (2) of the same proposition, we see that the curvature $\tilde{\kappa}$ of $\tilde{\omega}$ has the property that $\tilde{\kappa} - \kappa \in \Omega^2(M, \mathcal{A}M)^\ell$ and the induced section $\mathrm{gr}_\ell(\tilde{\kappa} - \kappa)$ of $L(\Lambda^2 \mathrm{gr}(TM), \mathrm{gr}(\mathcal{A}M))$ is given by $-\partial(\mathrm{gr}_\ell(\Phi)) = -\partial\phi$. Now

$$\partial^*\tilde{\kappa} = \partial^*\kappa + \partial^*(\tilde{\kappa} - \kappa) \in \Omega^1(M, \mathcal{A}M)^\ell,$$

and we compute

$$\mathrm{gr}_\ell(\partial^*\tilde{\kappa}) = \mathrm{gr}_\ell(\partial^*\kappa) + \mathrm{gr}_0(\partial^*)(\mathrm{gr}_\ell(\tilde{\kappa} - \kappa)) = \mathrm{gr}_\ell(\partial^*\kappa) - \mathrm{gr}_0(\partial^*)(\partial\phi) = 0.$$

This shows that $\partial^*\tilde{\kappa} \in \Omega^1(M, \mathcal{A}M)^{\ell+1}$, and the result follows.

The last statement follows since regularity of ω implies that $\kappa \in \Omega^2(M, \mathcal{A}M)^1$ and thus $\partial^*\kappa \in \Omega^1(M, \mathcal{A}M)^1$. Hence, there is a normal Cartan connection $\tilde{\omega}$ such that $\tilde{\omega} - \omega \in \Omega^1(M, \mathcal{A}M)^1$, which implies that $\tilde{\omega}$ and ω induce the same underlying infinitesimal flag structure. □

We have already mentioned an analogy between infinitesimal flag structures and first order G–structures in 3.1.6 and 3.1.7. To prove existence of normal parabolic geometries with a given underlying regular infinitesimal flag structure, we have to make this analogy more precise and consider the analog of connections on first order G–structures as discussed in 1.3.6. There we started with a Lie group H and an infinitesimally injective homomorphism $H \to GL(m, \mathbb{R})$. We can view this homomorphism as a representation of H on $\mathbb{R}^m$ and form the semidirect product $B = \mathbb{R}^m \rtimes H$. This is a Lie group which contains H as a closed subgroup, $B/H \cong \mathbb{R}^m$ and the resulting action of B on $\mathbb{R}^m$ consists of all transformations generated by translations and the elements of H. A first order G–structure with structure group H is then a reduction of the linear frame bundle of an m–dimensional manifold M to this structure group. We have seen in 1.3.6 and Example 1.5.1 (ii) that principal connections on such G–structures are equivalent to Cartan geometries of type (B, H).

In our current situation, we have to replace the homomorphism $H \to GL(m, \mathbb{R})$ by the homomorphism $G_0 \to \mathrm{Aut}_{\mathrm{gr}}(\mathfrak{g}_-)$ induced by the restriction of the adjoint action. To take into account the nontrivial Lie algebra structure on $\mathfrak{g}_-$, we have to replace $\mathbb{R}^m$ by the (uniquely determined) simply connected Lie group G_- with Lie algebra $\mathfrak{g}_-$. Via the left trivialization $TG_- \cong G_- \times \mathfrak{g}_-$ of the tangent bundle of G_-, the grading of $\mathfrak{g}_-$ induces a decomposition of TG_- into a direct sum of smooth subbundles. The group G_- endowed with these subbundles is sometimes called the *Carnot group* associated to the graded nilpotent Lie algebra $\mathfrak{g}_-$. Since G_- is simply connected, any automorphism of $\mathfrak{g}_-$ integrates to a group automorphism of G_-. Hence, we can view $\mathrm{Aut}_{\mathrm{gr}}(\mathfrak{g}_-)$ as a group of automorphisms of G_- and use this to define a semidirect product of G_- and G_0.

However, in the parabolic case, a simpler direct description is available. Consider a $|k|$–graded semisimple Lie algebra $\mathfrak{g} = \mathfrak{g}_{-k} \oplus \cdots \oplus \mathfrak{g}_k$, a Lie group G with Lie algebra $\mathfrak{g}_-$, and a parabolic subgroup $P \subset G$ corresponding to the grading. Let $G_0 \subset P$ be the Levi subgroup as defined in 3.1.3. Similarly, as in 3.1.3, we can defined a subgroup respecting the filtration which is opposite to $\{\mathfrak{g}^i\}$. More formally, consider

$$L := \{g \in G : \mathrm{Ad}(g)(\mathfrak{g}_i) \subset \mathfrak{g}_{-k} \oplus \cdots \oplus \mathfrak{g}_i \text{ for all } i = -k, \dots, k\}.$$

This is a closed subgroup of G and as in 3.1.3 one shows that it has Lie algebra $\mathfrak{p}^{op} := \mathfrak{g}_- \oplus \mathfrak{g}_0$. By construction G_0 is a subgroup of L corresponding to the subalgebra $\mathfrak{g}_0$ and we define P^{op} to be the union of all those connected components of L which meet G_0. This is called the *opposite parabolic subgroup* to P.

As in the proof of Theorem 3.1.3, one shows that any element $g \in P^{op}$ can be uniquely written as $g = \exp(X)g_0$ for $X \in \mathfrak{g}_-$ and $g_0 \in G_0$. (Getting a representation with g_0 on the right rather than on the left can be realized by passing to inverses.) Moreover, the exponential map is a diffeomorphism from $\mathfrak{g}_-$ onto its image, which is a closed subgroup in P^{op} isomorphic to G_-. This identifies P^{op} with a semidirect product of G_- and G_0 as required. In this setting, the concept of a Cartan geometry of type (P^{op}, G_0) makes sense. Note that the decomposition $\mathfrak{p}^{op} = \mathfrak{g}_- \oplus \mathfrak{g}_0$ is G_0–invariant, so Cartan connections of this type decompose into a soldering form and a principal connection.

THEOREM 3.1.13. *Let $\mathfrak{g} = \mathfrak{g}_{-k} \oplus \cdots \oplus \mathfrak{g}_k$ be $|k|$–graded semisimple Lie algebra, G a Lie group with Lie algebra $\mathfrak{g}$, $P \subset G$ a parabolic subgroup corresponding to the grading and $G_0 \subset P$ the Levi subgroup. Then any regular infinitesimal flag structure of type (G, P) on a smooth manifold M is induced by a normal parabolic geometry of type (G, P).*

PROOF. Let $(\{T^iM\}, p_0 : E \to M, \theta)$ be a regular infinitesimal flag structure. In view of Proposition 3.1.13, it suffices to construct some parabolic geometry $(p : \mathcal{G} \to M, \omega)$ which induces this structure. Making several choices, this can be done directly.

First, choose any principal connection $\gamma \in \Omega^1(E, \mathfrak{g}_0)$ on the principal G_0–bundle $p_0 : E \to M$. This splits TE as the direct sum of the vertical bundle VE and the horizontal subbundle $HE = \ker(\gamma)$. For each $u \in E$ the map $T_up_0 : H_uE \to T_xM$ is a linear isomorphism, where $x = p_0(u)$. Consider the filtration $TE = T^{-k}E \supset \cdots \supset T^0E = VE$ from 3.1.6 and put $H^iE := T^iE \cap HE$ for $i < 0$. By construction, each H^iE is a smooth subbundle and $T_up_0 : H_uE \to T_xM$ is an isomorphism of filtered vector spaces for each $u \in E$.

Next, for each $i = -k, \dots, -1$ we have the smooth subbundle $T^iM \subset TM$, and we may choose a smooth homomorphism $\underline{\pi}_i : TM \to T^iM$ of vector bundles which is a projection onto this subbundle. As we have noted above, for each $u \in E$ with $p_0(u) = x$ and each $i < 0$, the mapping $T_up_0 : H_uE \to T_xM$ is an isomorphism of filtered vector spaces. Using this, we can lift each $\underline{\pi}_i$ to a smooth bundle map $\pi_i : TE \to H^iE$, which is characterized by $Tp_0 \circ \pi_i = \underline{\pi}_i \circ Tp_0$. In particular, π_i is a projection onto the subbundle $H^iE \subset TE$. Moreover, denoting by r^{g_0} the principal right action of $g_0 \in G_0$ we obtain $T(r^{g_0}) \circ \pi_i = \pi_i \circ T(r^{g_0})$, so this projection is G_0–equivariant. Having made these choices, we can now define $\tilde{\theta} = (\tilde{\theta}_{-k}, \dots, \tilde{\theta}_{-1}, \tilde{\theta}_0) \in \Omega^1(E, \mathfrak{g}_- \oplus \mathfrak{g}_0)$ by $\tilde{\theta}_0 = \gamma$ and $\tilde{\theta}_i(\xi) := \theta_i(\pi_i(\xi))$ for each $i < 0$. We claim that $\tilde{\theta}$ is a Cartan connection of type (P^{op}, G_0) on E.

If $\tilde{\theta}(\xi) = 0$, then $\gamma(\xi) = 0$ and hence $\xi \in HE$. On this subbundle, π_{-k} is the identity, and hence $\theta_{-k}(\xi) = 0$, which means that $\xi \in T^{-k+1}E$. This implies $\pi_{-k+1}(\xi) = \xi$, so $\theta_{-k+1}(\xi) = 0$ and hence $\xi \in T^{-k+2}E$. Iterating this, we get $\xi \in T^0E = VE$, and since we have already seen that $\xi \in HE$, we get $\xi = 0$. Hence, $\tilde{\theta}$ is injective and thus a linear isomorphism on each tangent space. For $A \in \mathfrak{g}_0$, the fundamental vector field ζ_A lies in T^0E and thus $\pi_i(\zeta_A) = 0$ for all $i < 0$ and $\gamma(\zeta_A) = A$, so $\tilde{\theta}$ reproduces the generators of fundamental vector fields. Finally, the adjoint action of G_0 preserves the decomposition $\mathfrak{g}_{-k} \oplus \dots \oplus \mathfrak{g}_0$, so we can verify G_0–equivariancy componentwise. For the component $\tilde{\theta}_0 = \gamma$ this holds by definition, while for the other components it follows immediately from invariance of π_i and equivariancy of θ_i. Hence, we have proved our claim.

Consider the inclusion $P^{op} \hookrightarrow G$. By construction, $P^{op} \cap P = G_0$ and infinitesimally, we obtain the inclusion $\mathfrak{p}^{op} \hookrightarrow \mathfrak{g}$ which induces a linear isomorphism $\mathfrak{p}^{op}/\mathfrak{g}_0 \to \mathfrak{g}/\mathfrak{p}$. Following Example 1.5.16, we can use the inclusions $i : G_0 \to P$ and $\alpha : \mathfrak{p}^{op} \to \mathfrak{g}$ to obtain the data for an extension functor as defined in 1.5.15, which maps Cartan geometries of type (P^{op}, G_0) to Cartan geometries of type (G, P). Applying this functor to $(E, \tilde{\theta})$, we obtain the principal bundle $\mathcal{G} := E \times_{G_0} P$ and a Cartan connection $\omega \in \Omega^1(\mathcal{G}, \mathfrak{g})$ which is characterized by the fact that $j^*\omega = \tilde{\theta}$, where $j : E \to \mathcal{G}$ is the inclusion. In particular, the Cartan connection ω induces the given filtration on TM.

Composing the projection $\mathcal{G} \to \mathcal{G}/P_+$ with j induces an isomorphism $E \to \mathcal{G}/P_+$ of principal G_0–bundles. For $u \in E$ and $\xi \in T_uE$ we can use $j(u) \in \mathcal{G}$ and $T_uj \cdot \xi \in T_{j(u)}\mathcal{G}$ as representatives. But then by construction $\omega(T_uj \cdot \xi) = \tilde{\theta}(\xi)$, and if $\xi \in T^iE$, then the $\mathfrak{g}_i$–component of this is just $\theta_i(\xi)$. But this exactly shows that the underlying infinitesimal flag structure of $(\mathcal{G} \to M, \omega)$ is $(E \to M, \theta)$. □

3.1.14. Uniqueness of normal Cartan connections. The remaining point to understand the relation between regular infinitesimal flag structures and normal regular parabolic geometries is the question of uniqueness of normal Cartan connections. One can expect uniqueness at best up to isomorphism. Namely, suppose that $(p : \mathcal{G} \to M, \omega)$ is a normal parabolic geometry and that $\Psi : \mathcal{G} \to \mathcal{G}$ is an automorphism of the principal bundle $\mathcal{G}$ covering the identity on M. Then $\Psi^*\omega$ is a Cartan connection, and Ψ is a homomorphism from the parabolic geometry $(\mathcal{G}, \Psi^*\omega)$ to $(\mathcal{G}, \omega)$. From the definition of the curvature one immediately concludes that the curvature of $\Psi^*\omega$ is just $\Psi^*\kappa$, where κ is the curvature of ω. Naturality of ∂^* then implies that $\partial^*(\Psi^*\kappa) = \Psi^*(\partial^*\kappa) = 0$, so also $\Psi^*\omega$ is a normal Cartan connection. If Ψ has the property that it induces the identity on the underlying infinitesimal flag structure (and we shall see in the proof of the following proposition that there are many automorphisms having this property), then we have two normal Cartan connections inducing the same underlying infinitesimal flag structure.

We will soon show, however, that under a cohomological condition this is the only freedom left. To formulate this, we have to observe that the Lie algebra differential $\partial : C^n(\mathfrak{g}_-, \mathfrak{g}) \to C^{n+1}(\mathfrak{g}_-, \mathfrak{g})$ is homogeneous of degree zero for the natural gradings on the chain spaces. Consequently, each cohomology space $H^n(\mathfrak{g}_-, \mathfrak{g})$ is naturally graded, i.e. it decomposes as $H^n(\mathfrak{g}_-, \mathfrak{g}) = \bigoplus_\ell H^n(\mathfrak{g}_-, \mathfrak{g})_\ell$ according to homogeneous degrees of representative cocycles. Of course, we also have the associated filtration given by the subspaces $H^n(\mathfrak{g}_-,\mathfrak{g})^\ell := \bigoplus_{m\geq\ell} H^n(\mathfrak{g}_-,\mathfrak{g})_m$.

PROPOSITION 3.1.14. *Let $\mathfrak{g}=\mathfrak{g}_{-k}\oplus\dots\oplus\mathfrak{g}_k$ be a $|k|$–graded Lie algebra such that $H^1(\mathfrak{g}_-,\mathfrak{g})^\ell=0$ for some $\ell\geq 1$. Let G be a Lie group with Lie algebra $\mathfrak{g}$, $P\subset G$ a parabolic subgroup for the given grading, $G_0\subset P$ the Levi subgroup, and let $(p:\mathcal{G}\to M,\omega)$ be a normal regular parabolic geometry of type (G,P). Suppose further that $\tilde\omega\in\Omega^1(\mathcal{G},\mathfrak{g})$ is another normal Cartan connection such that for each $i=-k,\dots,-1$ the difference $\tilde\omega-\omega$ maps $T^i\mathcal{G}$ to $\mathfrak{g}^{i+\ell}$.*

Then there is an automorphism Ψ of the principal bundle $\mathcal{G}$, which induces the identity on the underlying infinitesimal flag structure such that $\Psi^*\tilde\omega=\omega$.

PROOF. Putting $\Phi:=\tilde\omega-\omega$, the assumption on $\tilde\omega$ can be phrased as $\Phi\in\Omega^1(M,\mathcal{A}M)^\ell$. By part (2) of Proposition 3.1.10, the difference $\tilde\kappa-\kappa$ of the curvatures of $\tilde\omega$ and ω lies in $\Omega^2(M,\mathcal{A}M)^\ell$, and $\operatorname{gr}_\ell(\tilde\kappa-\kappa)=\partial(\operatorname{gr}_\ell(\Phi))$. Moreover, since both ω and $\tilde\omega$ are normal, $\partial^*(\tilde\kappa-\kappa)=0$, and passing to the associated graded, this implies $0=\operatorname{gr}_\ell(\partial^*(\tilde\kappa-\kappa))=\operatorname{gr}_0(\partial^*)(\operatorname{gr}_\ell(\tilde\kappa-\kappa))$. Formula (3.2) from 3.1.12 shows that the subspaces $\ker(\operatorname{gr}_0(\partial^*))$ and $\operatorname{im}(\partial)$ of $L(\Lambda^2\operatorname{gr}(TM),\operatorname{gr}(\mathcal{A}M))$ have zero intersection, so we see that $\operatorname{gr}_\ell(\tilde\kappa-\kappa)=\partial(\operatorname{gr}_\ell(\Phi))=0$. Since by assumption $H^1(\mathfrak{g}_-,\mathfrak{g})_\ell=0$, this implies that $\operatorname{gr}_\ell(\Phi)$ must lie in the image of ∂.

Let us first assume that $\ell>k$. Then $\mathcal{A}^\ell M=0$ and hence $\operatorname{gr}_\ell(\Phi)=0$, so $\Phi\in\Omega^1(M,\mathcal{A}M)^{\ell+1}$. Repeating the argument finitely many times and using that $\Omega^1(M,\mathcal{A}M)^i=0$ for $i>2k$, we obtain $\Phi=0$ and thus $\tilde\omega=\omega$.

It remains to discuss the case $\ell\leq k$. Since $\operatorname{gr}_\ell(\Phi)$ lies in the image of ∂, we can choose a section ψ of $\mathcal{A}^\ell M$ such that $\partial(\operatorname{gr}_\ell(\psi))=-\operatorname{gr}_\ell(\Phi)$. The section ψ corresponds to a smooth function $\mathcal{G}\to\mathfrak{g}^\ell$ that we denote by the same symbol, which satisfies $\psi(u\cdot g)=\operatorname{Ad}(g^{-1})(\psi(u))$ for all $u\in\mathcal{G}$ and $g\in P$. Now we define a smooth map $\Psi:\mathcal{G}\to\mathcal{G}$ by $\Psi(u):=u\cdot\exp(\psi(u))$. Equivariancy of ψ implies that $\exp(\psi(u\cdot g))=g^{-1}\exp(\psi(u))g$, which in turn implies $\Psi(u\cdot g)=\Psi(u)\cdot g$, so Ψ is an automorphism of the principal bundle $\mathcal{G}$ which covers the identity on M. Thus, $\Psi^*\tilde\omega\in\Omega^1(\mathcal{G},\mathfrak{g})$ is a Cartan connection, which is normal by construction. For a point $u\in\mathcal{G}$ and a tangent vector $\xi\in T^i_u\mathcal{G}$, consider

$$\Psi^*\tilde\omega(u)(\xi)=\tilde\omega(u\cdot\exp(\psi(u)))(T_u\Psi\cdot\xi).$$

We can compute the right–hand side as in the proof of Proposition 1.5.3 as

$$\operatorname{Ad}(\exp(-\psi(u)))(\tilde\omega(u)(\xi))+\delta(\exp\circ\psi)(u)(\xi),$$

where δ denotes the left logarithmic derivative; see 1.2.4. By assumption, ψ has values in $\mathfrak{g}^\ell$, which is a Lie subalgebra of $\mathfrak{g}$ since $\ell\geq 1$. Thus, $\exp\circ\psi$ has values in the corresponding subgroup of P, and $\delta(\exp\circ\psi)$ has values in $\mathfrak{g}^\ell$. Hence, we conclude that $\Psi^*\tilde\omega(u)(\xi)$ is congruent to $\operatorname{Ad}(\exp(-\psi(u))(\tilde\omega(u)(\xi))$ modulo $\mathfrak{g}^\ell\subset\mathfrak{g}^{i+\ell+1}$. Using $\operatorname{Ad}(\exp(-\psi(u)))=e^{-\operatorname{ad}(\psi(u))}$, we see that $\Psi^*\tilde\omega(u)(\xi)$ is congruent to $\tilde\omega(u)(\xi)-[\psi(u),\tilde\omega(u)(\xi)]$ modulo $\mathfrak{g}^{i+\ell+1}$. But $\tilde\omega(u)(\xi)=\omega(u)(\xi)+\Phi(u)(\xi)$, while in the second term, we may replace $\tilde\omega$ by ω without changing the class modulo $\mathfrak{g}^{i+\ell+1}$. Thus, we see that

$$(\Psi^*\tilde\omega-\omega)(u)(\xi)\equiv\Phi(u)(\xi)-[\psi(u),\omega(u)(\xi)] \text{ modulo } \mathfrak{g}^{i+\ell+1}.$$

Hence, $\Psi^*\tilde\omega-\omega\in\Omega^1(M,\mathcal{A}M)^\ell$ and

$$\operatorname{gr}_{i+\ell}((\Psi^*\tilde\omega-\omega)(u)(\xi))=\{\operatorname{gr}_\ell(\psi),\operatorname{gr}_i(\xi)\}=-\partial(\operatorname{gr}_\ell(\psi))(\operatorname{gr}_i(\xi)).$$

Hence, $\mathrm{gr}_\ell(\Psi^*\tilde{\omega} - \omega) = \mathrm{gr}_\ell(\Phi) - \partial(\mathrm{gr}_\ell(\psi)) = 0$ and $\Psi^*\tilde{\omega} - \omega \in \Omega^1(M, \mathcal{A}M)^{\ell+1}$. Iterating this argument, we can make the difference to be in $\Omega^1(M, \mathcal{A}M)^{k+1}$ and we have already dealt with this case before. □

From this uniqueness result, we directly get the general theorem establishing the relation between parabolic geometries and infinitesimal flag structures in most cases, namely under the assumption that $H^1(\mathfrak{g}_-, \mathfrak{g})^1 = 0$. We will clarify the meaning of this condition completely in 3.3. There we will show that if $\mathfrak{g}$ does not contain a simple summand isomorphic to $\mathfrak{sl}(2)$, then we always have $H^1(\mathfrak{g}_-, \mathfrak{g})^2 = 0$. Moreover, $H^1(\mathfrak{g}_-, \mathfrak{g})^1 \neq 0$ happens only if $\mathfrak{g}$ contains a simple factor that belongs to one of two specific series of simple graded Lie algebras. Geometrically, these two series correspond to classical projective structures (see 1.1.3, 4.1.5, and Example 3.1.6 (2) for the corresponding $|1|$–grading), respectively, contact projective structures (see 1.1.4 and 4.2.6).

THEOREM 3.1.14. *Let $\mathfrak{g} = \mathfrak{g}_{-k} \oplus \cdots \oplus \mathfrak{g}_k$ be a $|k|$–graded semisimple Lie algebra such that none of the simple ideals of $\mathfrak{g}$ is contained in $\mathfrak{g}_0$, and such that $H^1(\mathfrak{g}_-, \mathfrak{g})^1 = 0$. Suppose that G is a Lie group with Lie algebra $\mathfrak{g}$, and $P \subset G$ is a parabolic subgroup corresponding to the grading with Levi subgroup $G_0 \subset P$.*

Then associating to a parabolic geometry the underlying infinitesimal flag structure and to any morphism of parabolic geometries the induced morphism of the underlying infinitesimal flag structures defines an equivalence between the category of normal regular parabolic geometries of type (G, P) and the category of regular infinitesimal flag structures of type (G, P).

PROOF. We have noted in 3.1.6 that passing to the underlying infinitesimal flag structure of a parabolic geometry defines a functor, and by definition this functor preserves regularity. Thus, we get a well–defined functor from normal regular parabolic geometries to regular infinitesimal flag structures. From Theorem 3.1.13 we see that conversely given a regular infinitesimal flag structure, we can construct a normal regular parabolic geometry inducing it. Now we claim that for two normal regular parabolic geometries any morphism between the underlying infinitesimal flag structures uniquely lifts to a morphism of parabolic geometries. Having proved this, any construction like the one from Theorem 3.1.13 extends to a functor from regular infinitesimal flag structures to normal parabolic geometries, which evidently establishes the claimed equivalence of categories.

So let us assume that $(p : \mathcal{G} \to M, \omega)$ and $(\tilde{p} : \tilde{\mathcal{G}} \to \tilde{M}, \tilde{\omega})$ are two normal regular parabolic geometries. Let $(\{T^iM\}, p_0 : \mathcal{G}_0 \to M, \omega^0)$ be the underlying infinitesimal flag structure of $(\mathcal{G} \to M, \omega)$ and likewise for the other geometry. Assume that $\Phi_0 : \mathcal{G}_0 \to \tilde{\mathcal{G}}_0$ is a morphism of infinitesimal flag structures covering $f : M \to \tilde{M}$, i.e. a homomorphism of G_0–principal bundles such that $T\Phi_0$ is filtration preserving and $\Phi_0^*\tilde{\omega}_i^0 = \omega_i^0$ for all $i = -k, \dots, -1$. Choose open coverings $\{U_i : i \in I\}$ of M and $\{\tilde{U}_j : j \in J\}$ of $\tilde{M}$ which trivialize the bundles $p : \mathcal{G} \to M$ and $\tilde{p} : \tilde{\mathcal{G}} \to \tilde{M}$, respectively. Over each U_i we get isomorphisms $p^{-1}(U_i) \cong U_i \times P$ and $p_0^{-1}(U_i) \cong U_i \times G_0$, and the natural projection $\pi : \mathcal{G} \to \mathcal{G}_0$ in this picture is just given by the natural projection $P \to G_0 \cong P/P_+$ in the second factor. Consequently, there is a G_0–equivariant section $\sigma_i : p_0^{-1}(U_i) \to p^{-1}(U_i)$ of the projection π, and similarly we get $\tilde{\sigma}_j : \tilde{p}_0^{-1}(\tilde{U}_j) \to \tilde{p}^{-1}(\tilde{U}_j)$.

For indices $i \in I$ and $j \in J$ such that $U_{ij} := U_i \cap f^{-1}(\tilde{U}_j) \neq \emptyset$, we can define a map $\Phi_{ij} : p^{-1}(U_{ij}) \to \tilde{p}^{-1}(f(U_{ij}))$ by $\Phi_{ij}(\sigma(u) \cdot g) = \tilde{\sigma}(\Phi_0(u)) \cdot g$ for

$u \in p_0^{-1}(U_{ij})$ and $g \in P_+$. Since σ and $\tilde{\sigma}$ are G_0–equivariant, this is a principal bundle homomorphism and by construction $\tilde{\pi} \circ \Phi_{ij} = \Phi_0 \circ \pi$. Then $\Phi_{ij}^*\tilde{\omega}$ is a normal Cartan connection on the principal P–bundle $p^{-1}(U_{ij}) \to U_{ij}$ and since Φ_{ij} induces the homomorphism $\Phi_0|_{p_0^{-1}(U_{ij})}$ of infinitesimal flag structures, we conclude that the infinitesimal flag structure induced by $(p^{-1}(U_{ij}) \to U_{ij}, \Phi_{ij}^*\tilde{\omega})$ equals the restriction of $(\{T^iM\}, \mathcal{G}_0 \to M, \omega^0)$. Thus, we may invoke Proposition 3.1.14 to get a principal bundle isomorphism $\Psi_{ij} : p^{-1}(U_{ij}) \to p^{-1}(U_{ij})$, which induces the identity on the underlying infinitesimal flag structure, such that $\Psi_{ij}^*(\Phi_{ij}^*\tilde{\omega}) = \omega$.

Now for each i, j, the homomorphism $\Phi_{ij} \circ \Psi_{ij} : p^{-1}(U_{ij}) \to \tilde{p}^{-1}(f(U_{ij}))$ of principal P–bundles induces the restriction of Φ_0 on the level of the underlying infinitesimal flag structures and pulls back $\tilde{\omega}$ to ω. Now we may invoke the general result of Proposition 1.5.3 for Cartan geometries, which tells us that by the action on the Cartan connections, these principal bundle maps are determined up to multiplication by a smooth map with values in the kernel K of the Cartan geometry, i.e. the maximal normal subgroup of G, which is contained in P. From 3.1.4 we know that K contained in G_0, so any such multiplication changes the induced morphism of infinitesimal flag structures. Thus, the bundle maps are actually uniquely determined by these two conditions, and in particular, they fit together to form a unique bundle map $\Phi : \mathcal{G} \to \tilde{\mathcal{G}}$ which induces Φ_0 on the underlying infinitesimal flag structures and pulls back $\tilde{\omega}$ to ω. □

3.1.15. Finer underlying structures. We will next discuss the question of underlying structures which are equivalent to regular normal parabolic geometries in the case that $H^1(\mathfrak{g}_-, \mathfrak{g})_1 \neq 0$. The main line of the theory will be taken up in 3.1.17 below, and for the first reading it may be preferable to continue there.

It $H^1(\mathfrak{g}_-, \mathfrak{g})_1 \neq 0$, then the underlying infinitesimal flag structure is too weak to uniquely determine a normal regular parabolic geometry. Indeed, we have seen in Example 3.1.6 (2) that there are cases in which the underlying infinitesimal flag structure contains no information at all. For these cases, one has to keep track of larger parts of the Cartan connection by using finer underlying structures. Since these exist in all cases, we describe them in general.

Fix a $|k|$–graded semisimple Lie algebra $\mathfrak{g} = \mathfrak{g}_{-k} \oplus \cdots \oplus \mathfrak{g}_k$, a Lie group G with Lie algebra $\mathfrak{g}$, and a parabolic subgroup $P \subset G$ corresponding to the grading, and let $G_0 \subset P$ be the Levi subgroup. From 3.1.3 we know that apart from the normal subgroup $P_+ \subset P$ that was used in the construction of the underlying infinitesimal flag structure, we also have the smaller normal subgroups $P_+^\ell = \exp(\mathfrak{g}^\ell)$ for $\ell = 2, \dots, k$. Now we may proceed similarly as in 3.1.5 using one of these groups rather than P_+. Given a parabolic geometry $(p : \mathcal{G} \to M, \omega)$ we observe that $P_+^{\ell+1}$ acts freely on $\mathcal{G}$, so we can form $\mathcal{G}_\ell := \mathcal{G}/P_+^{\ell+1}$ for $\ell = 0, \dots, k-1$. The projection $p : \mathcal{G} \to M$ descends to a projection $p_\ell : \mathcal{G}_\ell \to M$, which is a principal fiber bundle with group $P/P_+^{\ell+1}$. Moreover, the diffeomorphism $G_0 \times \mathfrak{g}_1 \times \cdots \times \mathfrak{g}_k \to P$ constructed in part (2) of Theorem 3.1.3 factors to a diffeomorphism $G_0 \times \mathfrak{g}_1 \times \cdots \times \mathfrak{g}_\ell \to P/P_+^{\ell+1}$. Thus, the inclusion of $G_0 \times \mathfrak{g}_1 \times \cdots \times \mathfrak{g}_i$ into $G_0 \times \mathfrak{g}_1 \times \cdots \times \mathfrak{g}_k$ induces a smooth section $P/P_+^{\ell+1} \to P$ of the canonical projection. While this is not a group homomorphism, it shows that the natural projection $\pi_\ell : \mathcal{G} \to \mathcal{G}_\ell$ admits local smooth sections, so it is a principal bundle with structure group $P_+^{\ell+1}$. Of course, for $\ell = 0$ we recover the bundle of the underlying infinitesimal flag structure.

Next, the filtration $T\mathcal{G} = T^{-k}\mathcal{G} \supset \cdots \supset T^k\mathcal{G}$ descends to a filtration $T\mathcal{G}_\ell = T^{-k}\mathcal{G}_\ell \supset \cdots \supset T^\ell\mathcal{G}_\ell$ of the tangent bundle $T\mathcal{G}_\ell$. By construction, for $i < 0$ we have $T^i\mathcal{G}_\ell = (Tp_\ell)^{-1}(T^iM)$, and $T^0\mathcal{G}_\ell$ is the vertical bundle of $\mathcal{G}_\ell \to M$. For $i > 0$ we may characterize $T^i\mathcal{G}_\ell$ as the subbundle spanned by the fundamental vector fields with generators in $\mathfrak{g}^i/\mathfrak{g}^{\ell+1} \subset \mathfrak{p}/\mathfrak{g}^{\ell+1}$. Now we can construct partially defined differential forms on $\mathcal{G}_\ell$ induced by the Cartan connection ω similarly as in 3.1.5. Notice that for each i, the Lie subalgebra $\mathfrak{g}^{\ell+1} \subset \mathfrak{p}$ acts trivially on the quotient $\mathfrak{g}^i/\mathfrak{g}^{i+\ell+1}$. Thus, also the subgroup $P_+^{\ell+1} \subset P$ acts trivially on that quotient, so the P action is actually induced by an action of $P/P_+^{\ell+1}$, which we will also denote by Ad.

PROPOSITION 3.1.15. *Let $(p : \mathcal{G} \to M, \omega)$ be a parabolic geometry of type (G, P) corresponding to the $|k|$-grading $\mathfrak{g} = \mathfrak{g}_{-k} \oplus \cdots \oplus \mathfrak{g}_k$ of the Lie algebra $\mathfrak{g}$ of G. Fix an integer ℓ such that $0 \leq \ell \leq k-1$ and consider the underlying bundle $p_\ell : \mathcal{G}_\ell \to M$.*

(1) For each $i = -k, \dots, 0$, the Cartan connection ω descends to a well-defined smooth section ω_i^ℓ of the bundle $L(T^i\mathcal{G}_\ell, \mathfrak{g}^i/\mathfrak{g}^{i+\ell+1})$. For any point $u \in \mathcal{G}_\ell$ the kernel of $\omega_i^\ell(u)$ is exactly the filtration component $T_u^{i+\ell+1}\mathcal{G}_\ell$. Each of the forms ω_i^ℓ is $P/P_+^{\ell+1}$-equivariant in the sense that $(r^g)^*\omega_i^\ell = \mathrm{Ad}(g^{-1}) \circ \omega_i^\ell$. The forms are mutually compatible, i.e. for $\xi \in T^{i+1}\mathcal{G}$ the value $\omega_i^\ell(\xi)$ coincides with the class of $\omega_{i+1}^\ell(\xi)$ in $\mathfrak{g}^{i+1}/\mathfrak{g}^{i+\ell+1} \subset \mathfrak{g}^i/\mathfrak{g}^{i+\ell+1}$. Finally, the form ω_0^ℓ is the inverse of the fundamental vector field map.

(2) For $\ell = k-1$, the form $\omega_{-k}^{k-1} \in \Omega^1(\mathcal{G}_{k-1}, \mathfrak{g}/\mathfrak{p})$ defines a first order G-structure with structure group P/P_+^k on M.

PROOF. (1) Let $u \in \mathcal{G}_\ell$ be a point and $\xi \in T_u^i\mathcal{G}_\ell$ be a tangent vector. Choose a point $\tilde{u} \in \mathcal{G}$ over u and a tangent vector $\tilde{\xi} \in T_{\tilde{u}}\mathcal{G}$ such that $\xi = T_{\tilde{u}}\pi_\ell \cdot \tilde{\xi}$. Since $\pi_\ell : \mathcal{G} \to \mathcal{G}_\ell$ is filtration preserving, we have $\tilde{\xi} \in T_{\tilde{u}}^i\mathcal{G}$. Hence, $\omega(\tilde{\xi}) \in \mathfrak{g}^i$, and we define $\omega_i^\ell(\xi)$ to be the class of $\omega(\tilde{\xi})$ in $\mathfrak{g}^i/\mathfrak{g}^{i+\ell+1}$. Fixing $\tilde{u}$, the tangent vector $\tilde{\xi}$ is unique up to elements in the kernel of $T_{\tilde{u}}\pi_\ell$, which equals $T_{\tilde{u}}^{\ell+1}\mathcal{G}$. So this can only add an element of $\mathfrak{g}^{\ell+1}$, and since $i \leq 0$ the class in $\mathfrak{g}^i/\mathfrak{g}^{i+\ell+1}$ remains unchanged.

On the other hand, any other choice for the point $\tilde{u}$ is of the form $\tilde{u} \cdot g$ for some $g \in P_+^{\ell+1}$, and a tangent vector at that point can be chosen as $T_{\tilde{u}}r^g \cdot \tilde{\xi}$, where r^g denotes the principal right action by g. Equivariancy of ω reads as

$$\omega(\tilde{u} \cdot g)(Tr^g \cdot \tilde{\xi}) = \mathrm{Ad}(g^{-1})(\omega(\tilde{u})(\tilde{\xi})),$$

and since the adjoint action of $\mathfrak{p}_+^{\ell+1}$ maps $\mathfrak{g}^i$ to $\mathfrak{g}^{i+\ell+1}$, we conclude that $\omega_i^\ell(\xi)$ is well defined. As in the proof of Proposition 3.1.5, this implies that one may locally write ω_i^ℓ in terms of the pullback of ω along a local smooth section of π_ℓ, so each ω_i^ℓ is smooth.

From this description, the compatibility of ω_i^ℓ and ω_{i+1}^ℓ is obvious, since both can be constructed from the same pullback of ω. Since $\omega(\tilde{u})(\tilde{\xi}) \in \mathfrak{g}^{i+\ell+1}$ by definition holds if and only if $\tilde{\xi} \in T^{i+\ell+1}\mathcal{G}$ and $T\pi_\ell$ is filtration preserving, we see that the kernel of $\omega_i^\ell(u)$ is exactly $T_u^{i+\ell+1}\mathcal{G}_\ell$.

Next, the projection π is equivariant in the sense that $\pi(\tilde{u} \cdot g) = \pi(\tilde{u}) \cdot [g]$, where for $g \in P$ we denote by $[g]$ the class in $P/P_+^{\ell+1}$. Hence, for $u \in \mathcal{G}_\ell$ and $\xi \in T_u^i\mathcal{G}_\ell$ and choices $\tilde{u}$ and $\tilde{\xi}$ as above, we see that $\pi(\tilde{u} \cdot g) = u \cdot [g]$ and $T\pi \cdot Tr^g \cdot \tilde{\xi} = Tr^{[g]} \cdot \xi$. Equivariancy of ω implies $\omega(Tr^g \cdot \tilde{\xi}) = \mathrm{Ad}(g^{-1})(\omega(\tilde{\xi}))$, and since $\omega(\tilde{\xi}) \in \mathfrak{g}^i$, the class of $\mathrm{Ad}(g^{-1})(\omega(\tilde{\xi}))$ modulo $\mathfrak{g}^{i+\ell+1}$ depends only on $[g]^{-1}$. Thus, we obtain the required equivariancy $(r^{[g]})^*\omega_i^\ell = \mathrm{Ad}([g]^{-1})\omega_i^\ell$.

Finally, equivariancy of π immediately implies that for the fundamental vector fields we get $T\pi \cdot \zeta_X(\tilde{u}) = \zeta_{[X]}(\pi(\tilde{u}))$, where $[X] \in \mathfrak{p}/\mathfrak{g}^{\ell+1}$ denotes the class of $X \in \mathfrak{p}$. By construction, this immediately implies that an appropriate lift for $\zeta_{[X]}(u) \in T_u^0\mathcal{G}_\ell$ is $\zeta_X(\tilde{u})$ for any point $\tilde{u}$ over u and any representative X of the class. This implies that ω_0^ℓ is the inverse of the fundamental vector field map.

(2) By part (1), ω_{-k}^{k-1} is a smooth section of the bundle $L(T^{-k}\mathcal{G}_{k-1}, \mathfrak{g}^{-k}/\mathfrak{g}^0)$ and thus an element of $\Omega^1(\mathcal{G}_{k-1}, \mathfrak{g}/\mathfrak{p})$. Moreover, the kernel of $\omega_{-k}^{k-1}(u)$ is the subbundle $T_u^0\mathcal{G}_{k-1}$, i.e. the vertical bundle of the principal P/P_+^k–bundle $p_{k-1} : \mathcal{G}_{k-1} \to M$. Hence, for any point $u \in \mathcal{G}_{k-1}$ with $x = p_{k-1}(u)$, the value $\omega_{-k}^{k-1}(u)$ may be viewed as a linear isomorphism from $T_u\mathcal{G}_{k-1}/T_u^0\mathcal{G}_{k-1} \cong T_xM$ to $\mathfrak{g}/\mathfrak{p}$. Since $\dim(M) = \dim(\mathfrak{g}/\mathfrak{p})$, we may view the manifold M as being modelled on the vector space $\mathfrak{g}/\mathfrak{p}$. Associating to u the inverse of the above linear isomorphism gives a bundle map Φ from $\mathcal{G}_{k-1}$ to the linear frame bundle of TM, i.e. the frame bundle of M. By construction, this covers the identity on M and is equivariant for the homomorphism $\mathrm{Ad} : P/P_+^k \to GL(\mathfrak{g}/\mathfrak{p})$. Hence, this is a reduction of structure group and ω_{-k}^{k-1} is the pullback along Φ of the soldering form. □

As in the case of infinitesimal flag structures discussed in 3.1.6 there is an obvious abstract notion for this type of structures, and we just need one more observation to formulate this: Suppose that $TM = T^{-k}M \supset \cdots \supset T^{-1}M$ is a filtration of the tangent bundle of a smooth manifold M such that the rank of T^iM equals the dimension of $\mathfrak{g}^i/\mathfrak{p}$, and let $p : E \to M$ be a principal bundle with structure group $P/P_+^{\ell+1}$. Then we get a canonical filtration $TE = T^{-k}E \supset \cdots \supset T^\ell E$ as follows. For $i < 0$ we put $T^iE := (Tp_\ell)^{-1}(T^iM)$, we define T^0E to be the vertical bundle of $p_\ell : E \to M$, and for $i > 0$ we let T^iE be the subbundle spanned by all fundamental vector fields with generators in $\mathfrak{g}^i/\mathfrak{g}^{\ell+1} \subset \mathfrak{p}/\mathfrak{g}^{\ell+1}$.

DEFINITION 3.1.15. (1) A *P–frame bundle* of degree $\ell \geq 0$ of type (G, P) on a smooth manifold M is given by:

- (i) A filtration $TM = T^{-k}M \supset \cdots \supset T^{-1}M$ of the tangent bundle of M such that the rank of T^iM equals the dimension of $\mathfrak{g}^i/\mathfrak{p}$ for all $i = -k, \dots, -1$.
- (ii) A principal bundle $p : E \to M$ with structure group $P/P_+^{\ell+1}$.
- (iii) A family $\theta = (\theta_{-k}, \dots, \theta_0)$ of smooth sections $\theta_i \in \Gamma(L(T^iE, \mathfrak{g}^i/\mathfrak{g}^{i+\ell+1}))$, called the *frame form* of E, which have the following properties:
  - $(r^g)^*\theta_i = \mathrm{Ad}(g^{-1}) \circ \theta_i$ for all $g \in P/P_+^{\ell+1}$ and all $i = -k, \dots, -1$.
  - For each $u \in E$ and $i = -k, \dots, -1$ the kernel of $\theta_i(u)$ is exactly $T_u^{i+\ell+1}E \subset T_u^iE$.
  - The forms are mutually compatible in the sense that for each $i = -k, \dots, -1$, and each $\xi \in T_u^{i+1}E$ we have $\theta_i(\xi) \in \mathfrak{g}^{i+1}/\mathfrak{g}^{i+\ell+1}$ and this coincides with the class of $\theta_{i+1}(\xi) \in \mathfrak{g}^{i+1}/\mathfrak{g}^{i+\ell+2}$.
  - $\theta_0 \in \Gamma(L(T^0E, \mathfrak{p}/\mathfrak{g}^{i+\ell+1}))$ is the inverse of the fundamental vector field map.

(2) A *morphism of P–frame bundles* of type (G, P) is a morphism of principal bundles whose base map is a local diffeomorphism with filtration preserving tangent maps, and which is compatible with the frame forms in the obvious sense.

In these terms, the proposition says that for each $\ell = 0, \dots, k-1$, a parabolic geometry of type (G, P) determines an underlying P–frame bundle of degree ℓ. From the construction it is also clear that morphisms of parabolic geometries descend to morphisms of the underlying P–frame bundles, so we again get a functor.

Let us look at the special case $\ell = 0$. Then the condition on compatibility of the components of the frame form becomes vacuous and the component θ_0 which is not present in the definition of an infinitesimal flag structure is simply determined as the inverse of the fundamental vector field map. Thus, we recover infinitesimal flag structures as P–frame bundles of degree zero. The component θ_0 is included in the general notion of a P–frame bundle since via the compatibility condition it determines parts of the other components.

Finally, we observe that there also is a functor from P–frame bundles of degree ℓ to infinitesimal flag structures. Given a P–frame bundle $(\{T^iM\}, p : E \to M, \theta)$ one takes the normal subgroup $P_+/P_+^{\ell+1} \subset P/P_+^{\ell+1}$ and defines E_0 as the orbit space $E/(P_+/P_+^{\ell+1})$. As in 3.1.5 one shows that the obvious projection $p_0 : E_0 \to M$ is a principal bundle with structure group $P/P_+ \cong G_0$ and the frame form θ descends to $\theta^0 = (\theta^0_{-k}, \dots, \theta^0_{-1})$, which makes $(\{T^iM\}, p_0 : E_0 \to M, \theta^0)$ into an infinitesimal flag structure. In particular, there is an obvious notion of *regular P–frame bundles* of type (G, P). It is also clear from the construction that passing first from a parabolic geometry to an underlying P–frame bundle and then to the underlying infinitesimal flag structure one obtains the same structure as going down directly as in 3.1.5.

3.1.16. The equivalence of categories in the case $H^1(\mathfrak{g}_-, \mathfrak{g})_1 \neq 0$. The need for the cohomological condition in the proof of the equivalence of categories in 3.1.14 actually came from Proposition 3.1.14. In the case $H^1(\mathfrak{g}_-, \mathfrak{g})^2 = 0$, but $H^1(\mathfrak{g}_-, \mathfrak{g})_1 \neq 0$ (which is the only remaining case of interest, see 3.1.14) a normal Cartan connection ω is unique up to isomorphism only in the class of those Cartan connections whose difference to ω maps $T^i\mathcal{G}$ to $\mathfrak{g}^{i+2}$ for all $i = -k, \dots, -1$. From the proof of Proposition 3.1.15 it is obvious that these are exactly those Cartan connections which induce the same underlying P–frame bundle of degree 1 as ω, compare also with Proposition 3.1.10. Thus, the obvious idea is to take the P–frame bundle of degree 1 as the basic underlying structure in this case.

To get an equivalence of categories we will again have to restrict to normal regular parabolic geometries, and thus to P–frame bundles of degree 1 which may underlie such geometries. As we have noticed in 3.1.15, regularity is no problem. There is, however, a new ingredient here, since in contrast to the underlying infinitesimal flag structure a part of the normality condition is visible on the level of the underlying P–frame bundle of degree 1.

Suppose that $(\{T^iM\}, p : E \to M, \theta)$ is a P–frame bundle of degree 1 of type (G, P). Then for each $i \leq -2$, the frame form θ gives us an identification $T^iM/T^{i+2}M \cong E \times_{P/P_+^2} \mathfrak{g}^i/\mathfrak{g}^{i+2}$. As before, this is induced by mapping $(u, X + \mathfrak{g}^{i+2})$ to the class of $T_up \cdot \xi$, where $\xi \in T^i_uE$ is any tangent vector such that $\theta_i(\xi) = X + \mathfrak{g}^{i+2}$. This is well defined since the kernel of $\theta_i(u)$ is T^{i+2}_uE and equivariancy of θ_i implies that it descends to an isomorphism as claimed. An immediate consequence is that for a section $\xi \in \Gamma(T^iM)$ and a chosen (local) lift $\tilde{\xi} \in \Gamma(T^iE)$, the function $\theta_i(\tilde{\xi}) : E \to \mathfrak{g}^i/\mathfrak{g}^{i+2}$ corresponds to the section of $T^iM/T^{i+2}M$ induced by ξ. Likewise, we get isomorphisms $\operatorname{gr}_i(TM) = T^iM/T^{i+1}M \cong E \times_{P/P^2_+} \mathfrak{g}^i/\mathfrak{g}^{i+1}$ using the forms $\underline{\theta}_i$, defined as the composition of the projection to $\mathfrak{g}^i/\mathfrak{g}^{i+1}$ with θ_i.

Similarly, as in the proof of Theorem 3.1.13, we can extend each of the sections $\theta_i \in \Gamma(L(T^iE, \mathfrak{g}^i/\mathfrak{g}^{i+2}))$ for $i < 0$ to a P/P^2_+–equivariant form $\tilde{\theta}_i \in \Omega^1(E, \mathfrak{g}^i/\mathfrak{g}^{i+2})$. To do this, we choose a principal connection γ on E, thus equivariantly splitting $TE = HE \oplus VE$. For each i, choose a projection $\underline{\pi}_i : TM \to T^iM$, and define π_i on TE to be the identity on VE and the lift of $\underline{\pi}_i$ on HE. Then π_i is a projection onto the subbundle T^iE, which by construction is (P/P^2_+)–equivariant. Putting $\tilde{\theta}_i = \theta_i \circ \pi_i$, we obtain forms as required. Moreover,

$$\tilde{\theta}_i(Tr^g \cdot \xi) = \theta_i(\pi_i(Tr^g \cdot \xi)) = \theta_i(Tr^g \cdot \pi_i(\xi))$$

together with equivariancy of θ_i immediately implies that $(r^g)^*\tilde{\theta}_i = \operatorname{Ad}(g^{-1}) \circ \tilde{\theta}_i$ for all $i < 0$.

Note that by compatibility of the Lie bracket on $\mathfrak{g}$ with the filtration $\{\mathfrak{g}^i\}$, we obtain for each i and j an induced map $\mathfrak{g}^i/\mathfrak{g}^{i+2} \times \mathfrak{g}^j/\mathfrak{g}^{j+2} \to \mathfrak{g}^{i+j}/\mathfrak{g}^{i+j+2}$, which we again denote by $[\ ,\ ]$. Using this, we formulate

LEMMA 3.1.16. *Let $(\{T^iM\}, p : E \to M, \theta)$ be a P–frame bundle of degree 1, and choose equivariant extensions $\tilde{\theta}_i \in \Omega^1(E, \mathfrak{g}^i/\mathfrak{g}^{i+2})$ of the components of the frame form as above.*

(1) If the filtration $\{T^iM\}$ is compatible with the Lie bracket, then for each i and j the expression

$$(\xi, \eta) \mapsto d\tilde{\theta}_{i+j}(\xi, \eta) + [\theta_i(\xi), \theta_j(\eta)] \tag{3.3}$$

defines a horizontal map $T^iE \times T^jE \to \mathfrak{g}^{i+j}/\mathfrak{g}^{i+j+2}$, which depends only on θ and not on the extension $\tilde{\theta}$.

(2) If the P–frame bundle is regular, then the expression (3.3) *always has values in $\mathfrak{g}^{i+j+1}/\mathfrak{g}^{i+j+2}$. It induces a well-defined bundle map $t_\theta : \Lambda^2 \operatorname{gr}_i(TM) \to \operatorname{gr}(TM)$, which is homogeneous of degree one.*

PROOF. (1) By assumption, for $\xi \in \Gamma(T^iM)$ and $\eta \in \Gamma(T^jM)$, we have $[\xi, \eta] \in \Gamma(T^{i+j}M)$. From the definition of the exterior derivative we then conclude that $d\tilde{\theta}_{i+j}(\xi, \eta)$ depends only on the restriction of $\tilde{\theta}_{i+j}$ to $T^{i+j}M$, which coincides with θ_{i+j}.

Horizontality is evident unless $i = -1$ or $j = -1$ and by skew symmetry it suffices to look at $i = -1$. For $A \in \mathfrak{p}/\mathfrak{g}^2$ the value of $\theta_{-1}(\zeta_A)$ by definition equals the class of A in $\mathfrak{p}/\mathfrak{g}^1$. Thus, $[\theta_{-1}(\zeta_A), \theta_j(\eta)] \in \mathfrak{g}^{-1+j}/\mathfrak{g}^{j+1}$ equals the class of $[A, \theta_j(\eta)]$, which equals $\operatorname{ad}(A)(\theta_{-1+j}(\eta))$. On the other hand, the flow of ζ_A is given by the principal right action of $\exp(tA)$. The infinitesimal version of equivariancy of $\tilde{\theta}$ reads as $\mathcal{L}_{\zeta_A}\tilde{\theta}_i = -\operatorname{ad}(A) \circ \tilde{\theta}_i$, where $\mathcal{L}_{\zeta_A}$ denotes the Lie derivative. Expanding the left–hand side as $d(i_{\zeta_A}\tilde{\theta}_i) + i_{\zeta_A} d\tilde{\theta}_i$, the first term vanishes since $i_{\zeta_A}\tilde{\theta}_i$ is constant. Thus, $d\tilde{\theta}_{-1+j}(\zeta_A, \eta) = -\operatorname{ad}(A)(\theta_{-1+j}(\eta))$, and horizontality follows.

(2) The class of $d\tilde{\theta}_{i+j}(\xi, \eta) + [\theta_i(\xi), \theta_j(\eta)]$ in $\mathfrak{g}^{i+j}/\mathfrak{g}^{i+j+1}$ depends only on the classes of $\theta_i(\xi)$ in $\mathfrak{g}^i/\mathfrak{g}^{i+1}$, of $\theta_j(\eta)$ in $\mathfrak{g}^j/\mathfrak{g}^{j+1}$ and of $\theta_{i+j}([\xi, \eta])$ in $\mathfrak{g}^{i+j}/\mathfrak{g}^{i+j+1}$. All of these classes are determined by the underlying infinitesimal flag structure. Since we already know that (3.3) is horizontal, we may always use lifts of vector fields on the underlying G_0–bundle. For such lifts vanishing of the expression follows directly from 3.1.7. This proves the claim about the image.

Compatibility of the components of the frame form then implies that the expression vanishes if $\xi \in T^{i+1}E$ or $\eta \in T^{j+1}E$, whence it factors to a map $T^iE/T^{i+1}E \times T^jE/T^{j+1}E \to \mathfrak{g}^{i+j+1}/\mathfrak{g}^{i+j+2}$. For $u \in E$, the quotient T^i_uE/T^{i+1}_uE is isomorphic to $\mathrm{gr}_i(TM)$. Likewise, fixing u, we can identify $\mathfrak{g}^{i+j+1}/\mathfrak{g}^{i+j+2}$ with

$$T^{i+j+1}_uE/T^{i+j+2}_uE \cong \mathrm{gr}_{i+j+1}(TM)$$

using $\theta_{i+1}(u)$. Fixing u and putting $x = p(u)$, we obtain a map

$$\mathrm{gr}_i(T_xM) \times \mathrm{gr}_j(T_xM) \to \mathrm{gr}_{i+j+1}(T_xM).$$

Equivariancy of the forms θ_ℓ easily implies that this map depends only on x and not on u. Collecting the components together, we obtain a bundle map t_θ as claimed. □

DEFINITION 3.1.16. A regular P–frame bundle $(\{T^iM\}, p : E \to M, \theta)$ of degree one is called *normal* if $\mathrm{gr}_0(\partial^*)(t_\theta) = 0$.

Suppose that we start with a regular normal parabolic geometry $(p : \mathcal{G} \to M, \omega)$ and consider the underlying P–frame bundle $(\{T^iM\}, p_1 : \mathcal{G}_1 \to M, \omega^1)$ of degree one. Then from Corollary 3.1.8 we know that the curvature κ lies in $\Omega^2(M, \mathcal{A}M)^1$. Hence, we can form $\mathrm{gr}_1(\kappa) \in \Gamma(\mathrm{gr}(\Lambda^2T^*M \otimes \mathcal{A}M))$, which is homogeneous of degree 1. From homogeneity it follows that $\mathrm{gr}_1(\kappa)$ must actually have values in the associated graded of $\Lambda^2T^*M \otimes TM$, and hence is a bundle map $\Lambda^2\,\mathrm{gr}(TM) \to \mathrm{gr}(TM)$. From the construction above it follows easily that this bundle map coincides with t_{ω^1}. In particular, this implies that $\mathrm{gr}_0(\partial^*)(t_{\omega^1}) = \mathrm{gr}_1(\partial^*\kappa) = 0$, so P–frame bundles underlying regular normal parabolic geometries are always normal. Having the notion of normality at hand, we may now proceed as in 3.1.13 and 3.1.14:

THEOREM 3.1.16. *Let $\mathfrak{g} = \mathfrak{g}_{-k} \oplus \cdots \oplus \mathfrak{g}_k$ be a $|k|$–graded semisimple Lie algebra such that none of the simple ideals of $\mathfrak{g}$ is contained in $\mathfrak{g}_0$, and such that $H^1(\mathfrak{g}_-, \mathfrak{g})^2 = 0$. Suppose that G is any Lie group with Lie algebra $\mathfrak{g}$ and that $P \subset G$ is a parabolic subgroup for the given grading with Levi-subgroup $G_0 \subset P$.*

Then associating to a parabolic geometry $(p : \mathcal{G} \to M, \omega)$ the underlying P–frame bundle of degree one defines an equivalence between the category of normal regular parabolic geometries of type (G, P) and the category of regular normal P–frame bundles of degree one of that type.

PROOF. We have already observed that forming the underlying P–frame bundle of degree one defines a functor from normal regular parabolic geometries to normal regular P–frame bundles. As in 3.1.13, we next show that any normal regular P–frame bundle of degree one lies in the image of this functor.

Let $(\{T^iM\}, p_1 : E \to M, \theta)$ be such a P–frame bundle and consider the underlying G_0–principal bundle $p_0 : E_0 \to M$, where $E_0 := E/(P_+/P_+^2)$. We first show that there is a global G_0–equivariant section $\sigma : E_0 \to E$ of the natural projection $E \to E_0$. Taking an open subset $U \subset M$ over which E is trivial, we get isomorphisms $p_1^{-1}(U) \cong U \times (P/P_+^2)$ and $p_0^{-1}(U) \cong U \times G_0$, so the inclusion $G_0 \to P \to P/P_+^2$ induces a G_0–equivariant section $p_0^{-1}(U) \to p_1^{-1}(U)$. Since any principal bundle admits a finite atlas, we can take a finite covering $\{U_1, \dots, U_N\}$ of M by such subsets and the corresponding sections σ_i. Moreover, there are open subsets $V_i \subset U_i$ which still cover M and are such that $\bar{V}_i \subset U_i$. Over the intersection $U_1 \cap U_2$, the sections σ_1 and σ_2 are related by the principal right action of an element of P_+/P_+^2. In view of Theorem 3.1.3 there is a smooth function $Z : U_1 \cap U_2 \to$

$\mathfrak{g}^1/\mathfrak{g}^2$ such that $\sigma_2(u) = \sigma_1(u)\exp(Z(u))$. The fact that both σ_1 and σ_2 are G_0–equivariant immediately implies that $Z(u \cdot g_0) = \mathrm{Ad}(g_0^{-1})(Z(u))$ for all $u \in U_1 \cap U_2$ and $g_0 \in G_0$. Choose a smooth function f with support in U_2 which is identically 1 on V_2 and define $\sigma : p_0^{-1}(U_1 \cup V_2) \to E$ by $\sigma(u) = \sigma_1(u)\exp(f(p_0(u))Z(u))$ for $u \in p_0^{-1}(U_1)$ and $\sigma(u) = \sigma_2(u)$ for $u \in p_0^{-1}(V_2)$. On the intersection the two definitions obviously coincide, so this defines a smooth section σ. Equivariancy of Z implies that σ is G_0–equivariant. Similarly, one extends σ to $U_1 \cup V_2 \cup V_3$, and so on. By induction one reaches a global G_0–equivariant section σ.

Fix an equivariant section $\sigma : E_0 \to E$, and consider $\sigma^*\theta_i \in \Gamma(L(T^iE_0, \mathfrak{g}^i/\mathfrak{g}^{i+2}))$. By construction, this section is G_0–equivariant, and as a G_0–module, we can naturally identify $\mathfrak{g}^i/\mathfrak{g}^{i+2}$ with $\mathfrak{g}_i \oplus \mathfrak{g}_{i+1}$. By construction, the $\mathfrak{g}_i$–component of $\sigma^*\theta_i$ is exactly the canonical form on the underlying infinitesimal flag structure. Consider the component $\sigma^*\theta_{-1} \in L(T^{-1}E_0, \mathfrak{g}_{-1} \oplus \mathfrak{g}_0)$. For $A \in \mathfrak{g}_0$ we conclude from equivariancy of σ that $T\sigma \cdot \zeta_A(u)$ coincides with the fundamental vector field on E generated by A, which implies that $\sigma^*\theta_{-1}(\zeta_A) = A$. Moreover, the kernel of $\sigma^*\theta_{-1}$ is $T^1E_0 = \{0\}$, so we see that in each point u, the map $\sigma^*\theta_{-1}(u) : T^{-1}E_0 \to \mathfrak{g}_{-1} \oplus \mathfrak{g}_0$ is a linear isomorphism. In particular, we can use it to define a projection $T^{-1}E_0 \to T^0E_0$. As in the proof of Theorem 3.1.13 we can choose any G_0–invariant projection $TE_0 \to T^{-1}E_0$, and we let $\pi_0 : TE_0 \to T^0E_0$ be the composition. For $\xi \in TE_0$ we now define $\tilde{\theta}_0(\xi) := \sigma^*\theta_{-1}(\pi_0(\xi)) \in \mathfrak{g}_0$. By construction, this is a principal connection on E_0 and $\tilde{\theta}_0(\xi)$ coincides with the $\mathfrak{g}_0$–component of $\sigma^*\theta_{-1}(\xi)$ for each $\xi \in T^{-1}E_0$.

Next, $\sigma^*\theta_{-2} \in \Gamma(L(T^{-2}E_0, \mathfrak{g}_{-2} \oplus \mathfrak{g}_{-1}))$ induces in each point u a linear isomorphism $T_u^{-2}E_0/T_u^0E_0 \cong \mathfrak{g}_{-2} \oplus \mathfrak{g}_{-1}$. We can identify $T^{-2}E_0/T^0E_0$ with the kernel of $\pi_0|_{T^{-2}E_0}$ and thus use $\sigma^*\theta_{-2}$ to construct a projection $T^{-2}E_0 \to T^{-1}E_0$. Choosing any G_0–equivariant projection $TE_0 \to T^{-2}E_0$, we define $\pi_{-1} : TE_0 \to T^{-1}E_0$ as the composition as before. Then we put $\tilde{\theta}_{-1}(\xi) := \sigma^*\theta_{-2}(\pi_{-1}(\xi)) \in \mathfrak{g}_{-1}$. By construction, $\tilde{\theta}_{-1}(\xi)$ coincides with the $\mathfrak{g}_{-1}$–component of $\sigma^*\theta_{-2}(\xi)$ for $\xi \in T^{-2}E_0$. On the other hand, for $\xi \in T^{-1}E_0$ compatibility of the components of θ implies that $\sigma^*\theta_{-1}(\xi) = (\tilde{\theta}_{-1}(\xi), \tilde{\theta}_0(\xi))$.

Now we continue in the same way step by step, using $\sigma^*\theta_i$ to construct a projection $T^iE_0 \to T^{i+1}E_0$ and composing with any G_0–equivariant projection $TE_0 \to T^iE_0$ to obtain $\pi_{i+1} : TE_0 \to T^{i+1}E_0$. Then we define $\tilde{\theta}_{i+1}(\xi) := \sigma^*\theta_i(\pi_{i+1}(\xi)) \in \mathfrak{g}_i$. Putting $\tilde{\theta} = (\tilde{\theta}_{-k}, \dots, \tilde{\theta}_0)$ we see that $\sigma^*\theta_i(\xi) = (\tilde{\theta}_i(\xi), \tilde{\theta}_{i+1}(\xi))$ for all $\xi \in T^iE_0$. Of course, $\tilde{\theta}$ restricts to a linear isomorphism on each tangent space, it is G_0–equivariant, and it reproduces the generators of fundamental vector fields.

Hence, we have constructed a Cartan connection of type (P^{op}, G_0), compare with 3.1.13, and we can use the extension functor considered there to obtain a Cartan connection ω of type (G, P) on the bundle $\mathcal{G} := E_0 \times_{G_0} P$. For the inclusion $j : E_0 \to \mathcal{G}$ we have $j^*\omega = \tilde{\theta}$, and this characterizes ω. Now define $\pi : \mathcal{G} \to E$ by $\pi(u, g) := \sigma(u) \cdot \underline{g}$, where $\underline{g}$ denotes the class of g in P_+/P_+^2. This induces an isomorphism $\mathcal{G}_1 = \mathcal{G}/P_+^2 \to E$ of principal P/P_+^2–bundles. Again by construction, we have $\pi \circ j = \sigma$, so $j^*\pi^*\theta = \sigma^*\theta$. By construction, we have $\sigma^*\theta_i = (\tilde{\theta}_i, \tilde{\theta}_{i+1})$ on T^iE_0. This easily implies that on the subbundle $T^i\mathcal{G}$, the difference $\omega - \pi^*\theta_i$ has values in $\mathfrak{g}^{i+2}$. Hence, our map actually identifies (E, θ) with the underlying P–frame bundle of degree one of $(\mathcal{G}, \omega)$.

Denoting by κ the curvature of ω, we know by regularity that $\kappa \in \Omega^2(M, \mathcal{A}M)^1$, and from above we know that $\mathrm{gr}_1(\kappa) = t_{\omega^1} = t_\theta$. Since $(\{T^iM\}, E \to M, \theta)$ is assumed to be normal, we see that $\mathrm{gr}_0(\partial^*)(\mathrm{gr}_1(\kappa)) = 0$, and thus $\partial^*(\kappa) \in \Omega^1(M, \mathcal{A}M)^2$. Hence, by Proposition 3.1.13 there is a normal Cartan connection $\tilde{\omega}$ on $\mathcal{G}$ such that $\tilde{\omega} - \omega \in \Omega^1(M, \mathcal{A}M)^2$, so in particular, $\tilde{\omega}$ has the same underlying P–frame bundle of degree one as ω. Thus, any normal regular P–frame bundle of degree one is induced by some regular normal parabolic geometry, and we can complete the proof by showing the unique lifting of morphisms as in Theorem 3.1.14.

Consider two normal regular parabolic geometries, say $(p : \mathcal{G} \to M, \omega)$ and $(\tilde{p} : \tilde{\mathcal{G}} \to \tilde{M}, \tilde{\omega})$, and the underlying P–frame bundles $(\{T^iM\}, p_1 : \mathcal{G}_1 \to M, \omega^1)$ and $(\{T^i\tilde{M}\}, \tilde{p}_1 : \tilde{\mathcal{G}}_1 \to \tilde{M}, \tilde{\omega}^1)$ of degree one. Suppose that Φ_1 is a morphism between the two P–frame bundles, i.e. a homomorphism of principal bundles which preserves the filtrations on the tangent bundles and has the property that $\Phi_1^*\tilde{\omega}_i^1 = \omega_i^1$ for all $i = -k, \dots, -1$. Then we claim that locally over M, Φ_1 lifts to a principal bundle homomorphism $\mathcal{G} \to \tilde{\mathcal{G}}$.

To see this, consider the underlying infinitesimal flag structures $p_0 : \mathcal{G}_0 \to M$ and $\tilde{p}_0 : \tilde{\mathcal{G}}_0 \to \tilde{M}$. Clearly, Φ_1 induces a morphism $\Phi_0 : \mathcal{G}_0 \to \tilde{\mathcal{G}}_0$ of infinitesimal flag structures which covers a local diffeomorphism $f : M \to \tilde{M}$. From the proof of Theorem 3.1.14, we know that we can find open coverings $\{U_i : i \in I\}$ of M and $\{\tilde{U}_j : j \in J\}$ of $\tilde{M}$ such that for $U_{ij} := U_i \cap f^{-1}(\tilde{U}_j)$ we get a principal bundle homomorphisms $\Psi_{ij} : p^{-1}(U_{ij}) \to \tilde{p}^{-1}(f(U_{ij}))$ lifting the restriction of Φ_0. Denoting by π and $\tilde{\pi}$ the natural projections $\mathcal{G} \to \mathcal{G}_1$ and $\tilde{\mathcal{G}} \to \tilde{\mathcal{G}}_1$, the fact that Ψ_{ij} covers Φ_0 immediately implies that for any point $u \in p^{-1}(U_{ij})$ there is an element $Z(u) \in \mathfrak{g}^1/\mathfrak{g}^2$ such that $\Phi_1(\pi(u)) = \tilde{\pi}(\Psi_{ij}(u)) \exp_{P/P_+^2}(Z(u))$. This defines a smooth map $Z : p^{-1}(U_{ij}) \to \mathfrak{g}^1/\mathfrak{g}^2$. The equivariancy properties of the maps involved imply that for an element $g \in P$ with image $\underline{g} \in P/P_+^2$, we must have $\underline{g} \exp_{P/P_+^2}(Z(u \cdot g)) = \exp(Z(u))\underline{g}$, and thus $Z(u \cdot g) = \underline{\mathrm{Ad}}(g^{-1})(Z(u))$. Hence, the smooth map Z corresponds to a smooth section over U_{ij} of the associated bundle $\mathcal{G} \times_P (\mathfrak{g}^1/\mathfrak{g}^2) = \mathrm{gr}_1(\mathcal{A}M)$. Choose a section W of $\mathcal{A}^1M$ over U_{ij} such that $\mathrm{gr}_1(W) = Z$, consider the corresponding equivariant function $W : p^{-1}(U_{ij}) \to \mathfrak{g}^1$ and define $\Phi_{ij} : p^{-1}(U_{ij}) \to \tilde{p}^{-1}(f(U_{ij}))$ by $\Phi_{ij}(u) := \Psi_{ij}(u) \cdot \exp_P(W(u))$. Equivariancy of W immediately implies that Φ_{ij} is a homomorphism of principal bundles, and by construction we get $\Phi_1(\pi(u)) = \tilde{\pi}(\Phi_{ij}(u))$.

Now we may conclude the proof exactly as the one of Theorem 3.1.14: $\Phi_{ij}^*\tilde{\omega}$ is a normal Cartan connection on $p^{-1}(U_{ij})$, which by construction has the same underlying frame form of length two as the restriction of ω. Then we may invoke Proposition 3.1.14 to modify Φ_{ij} to a morphism of parabolic geometries, and as in the proof of Theorem 3.1.14 these morphisms are shown to be uniquely determined. Hence, we get a unique lift of morphisms and as in the proof of Theorem 3.1.14 this establishes the required equivalence of categories. □

3.1.17. Complex parabolic geometries. Up to now we have worked in a real, smooth setting and this will be the main subject of the book. However, in some cases it is important to consider the holomorphic version of the theory, which we refer to as complex parabolic geometries. Some background on complex manifolds can be found in 1.2.10.

Suppose that $\mathfrak{g} = \mathfrak{g}_{-k} \oplus \cdots \oplus \mathfrak{g}_k$ is a complex $|k|$–graded Lie algebra (i.e. the bracket of $\mathfrak{g}$ is complex bilinear and each $\mathfrak{g}_i$ is a complex subspace), and let G be a complex Lie group with Lie algebra $\mathfrak{g}$. Then all the subgroups P, P_+^i and G_0 are complex, too, and in particular, the homogeneous space G/P is a complex manifold. A *complex parabolic geometry* of type (G, P) on a complex manifold M is then given by a holomorphic principal P–bundle $p : \mathcal{G} \to M$ endowed with a holomorphic Cartan connection $\omega \in \Omega^{1,0}(\mathcal{G}, \mathfrak{g})$. Morphisms of complex parabolic geometries are holomorphic principal bundle maps that are compatible with the Cartan connections.

Any complex parabolic geometry of type (G, P) can also be viewed as a real parabolic geometry of the same type. The main tool for our study of complex parabolic geometry will be a characterization of these geometries among real ones and the results we have already available in the real case.

Suppose that $(p : \mathcal{G} \to M, \omega)$ is a real parabolic geometry of type (G, P). Then ω induces an isomorphism $TM \cong \mathcal{G} \times_P (\mathfrak{g}/\mathfrak{p})$. By assumption, $\mathfrak{g}/\mathfrak{p}$ is a complex vector space, and the adjoint action of P on $\mathfrak{g}$ is by complex linear maps. Hence, the above identification makes TM into a complex vector bundle, and thus defines an almost complex structure J on M; see 1.2.10. Similarly, ω induces an isomorphism $T\mathcal{G} \cong \mathcal{G} \times \mathfrak{g}$, and since $\mathfrak{g}$ is a complex vector space, this gives us an almost complex structure $J^{\mathcal{G}}$ on $\mathcal{G}$. Notice that this means that we define ω to be a $(1,0)$–form. Since the isomorphism $TM \cong \mathcal{G} \times_P (\mathfrak{g}/\mathfrak{p})$ is induced by mapping $(u, A) \in \mathcal{G} \times \mathfrak{g}$ to $Tp \cdot \omega(u)^{-1}(A)$, we conclude that $Tp \circ J^{\mathcal{G}} = J \circ Tp$.

Now $(p : \mathcal{G} \to M, \omega)$ is actually a complex parabolic geometry if and only if the almost complex structures J and $J^{\mathcal{G}}$ are integrable, the bundle is holomorphic and the Cartan connection is holomorphic, too. We can characterize when this is satisfied in terms of the curvature κ of ω. These results should be considered as preliminary. Using the machinery of Bernstein–Gelfand–Gelfand sequences, we will show in volume two that there is an analogous characterization in terms of the harmonic curvature, which is much easier to verify. The latter characterization can be found in [**Čap05**]. To formulate the preliminary characterization we just need one more observation: The curvature κ of ω is a (real) two–form on M with values in the adjoint tractor bundle $\mathcal{A}M$, which by construction is a complex vector bundle. Consequently, $\mathcal{A}M$–valued forms split into (p, q)–types; see 1.2.10. In particular, κ splits as $\kappa = \kappa_{0,2} + \kappa_{1,1} + \kappa_{2,0}$, with $\kappa_{2,0}$ being complex bilinear, $\kappa_{0,2}$ being conjugate linear in both arguments and $\kappa_{1,1}$ has the property that $\kappa_{1,1}(J\xi, J\eta) = \kappa_{1,1}(\xi, \eta)$. Explicitly, the components are given by

$$\kappa_{0,2}(\xi, \eta) = \tfrac{1}{4}\big(\kappa(\xi, \eta) - \kappa(J\xi, J\eta) + i\kappa(J\xi, \eta) + i\kappa(\xi, J\eta)\big),$$
$$\kappa_{1,1}(\xi, \eta) = \tfrac{1}{2}\big(\kappa(\xi, \eta) + \kappa(J\xi, J\eta)\big),$$
$$\kappa_{2,0}(\xi, \eta) = \tfrac{1}{4}\big(\kappa(\xi, \eta) - \kappa(J\xi, J\eta) - i\kappa(J\xi, \eta) - i\kappa(\xi, J\eta)\big).$$

PROPOSITION 3.1.17. *Let $\mathfrak{g} = \mathfrak{g}_{-k} \oplus \cdots \oplus \mathfrak{g}_k$ be a complex $|k|$–graded semisimple Lie algebra, G a complex Lie group with Lie algebra $\mathfrak{g}$, and $P \subset G$ a parabolic subgroup for the given grading with Levi subgroup $G_0 \subset P$. Let $(p : \mathcal{G} \to M, \omega)$ be a real parabolic geometry of type (G, P) with curvature $\kappa = \kappa_{0,2} + \kappa_{1,1} + \kappa_{2,0} \in \Omega^2(M, \mathcal{A}M)$. Then we have:*

(1) *The almost complex structure J on M induced by ω is integrable if and only if $\kappa_{0,2}(\xi, \eta) \in \mathcal{A}^0M$ for all $\xi, \eta \in TM$.*

(2) *The almost complex structure* $J^{\mathcal{G}}$ *on* $\mathcal{G}$ *induced by* ω *is integrable if and only if* $\kappa_{0,2} = 0$. *If this is the case, then* $p : \mathcal{G} \to M$ *is a holomorphic principal* P*–bundle over the complex manifold* M.
(3) $(p : \mathcal{G} \to M, \omega)$ *is a complex parabolic geometry of type* (G, P) *if and only if* $\kappa_{0,2} = 0$ *and* $\kappa_{1,1} = 0$, *i.e. if and only if the curvature* κ *is of type* $(2, 0)$.

PROOF. (1) and (2): By the Newlander–Nirenberg theorem (see 1.2.10) integrability of an almost complex structure is equivalent to vanishing of the Nijenhuis tensor. By definition, the Nijenhuis tensor of an almost complex manifold is given by $N(\xi, \eta) = [\xi, \eta] - [J\xi, J\eta] + J[J\xi, \eta] + J[\xi, J\eta]$. Hence, N is just the $(0, 2)$ component of the Lie bracket of vector fields, which is tensorial. Now let $N^{\mathcal{G}}$ be the Nijenhuis tensor of $J^{\mathcal{G}}$ and let us compute $\omega(N^{\mathcal{G}}(\omega^{-1}(A), \omega^{-1}(B)))$ for $A, B \in \mathfrak{g}$. Since ω is constant on the vector fields $\omega^{-1}(A)$ and $\omega^{-1}(B)$, the definition of the exterior derivative implies that

$$\omega([\omega^{-1}(A), \omega^{-1}(B)]) = -d\omega(\omega^{-1}(A), \omega^{-1}(B)) = -\kappa(A + \mathfrak{p}, B + \mathfrak{p}) + [A, B],$$

where we also denote by κ the curvature function; see 3.1.8. Again by definition $J^{\mathcal{G}}(\omega^{-1}(A)) = \omega^{-1}(iA)$, and since the bracket on $\mathfrak{g}$ is complex bilinear, we get $[A, B] - [iA, iB] + i[iA, B] + i[A, iB] = 0$, so

$$\omega(N^{\mathcal{G}}(\omega^{-1}(A), \omega^{-1}(B))) = -4\kappa_{0,2}(A + \mathfrak{p}, B + \mathfrak{p}).$$

Hence, vanishing of $N^{\mathcal{G}}$ is equivalent to vanishing of $\kappa_{0,2}$. On the other hand, starting with vector fields $\xi, \eta \in \mathfrak{X}(M)$ and choosing local lifts $\tilde{\xi}, \tilde{\eta}$, we have already observed that $J^{\mathcal{G}}\tilde{\xi}$ is a local lift of $J\xi$ and similarly for η, and brackets of the lifts are lifts of the brackets. Thus, we conclude that for $x \in M$ and $u \in \mathcal{G}$ with $p(u) = x$, we have $N(\xi, \eta)(x) = T_u p \cdot N^{\mathcal{G}}(\tilde{\xi}, \tilde{\eta})(u)$. Hence, integrability of J is equivalent $\omega(N^{\mathcal{G}})$ having values in $\mathfrak{p}$, whenever lifts of vector fields on M are inserted. Taking into account that κ vanishes upon insertion of a vertical vector field, this together with the above implies (1).

If $\kappa_{0,2} = 0$, then we know that J and $J^{\mathcal{G}}$ are integrable, so $p : \mathcal{G} \to M$ is a holomorphic map which, in addition, is a surjective submersion. By the implicit function theorem, p admits local holomorphic sections. On the other hand, consider the principal right action $r : \mathcal{G} \times P \to \mathcal{G}$. The derivative of r in a point (u, g) is given by $T_{(u,g)}r \cdot (\xi, \eta) = T_u r^g \cdot \xi + T_g r_u \cdot \eta$, where $r^g : \mathcal{G} \to \mathcal{G}$ is the right action by g and $r_u : P \to \mathcal{G}$ is the right action on u. Equivariancy of ω reads as $\omega(u \cdot g)(T_u r^g \cdot \xi) = \mathrm{Ad}(g)^{-1}(\omega(u)(\xi))$, and since $\mathrm{Ad}(g^{-1}) : \mathfrak{g} \to \mathfrak{g}$ is complex linear, we see that Tr^g is complex linear for the complex structure $J^{\mathcal{G}}$. Now $r_u = r_{u \cdot g} \circ \lambda_{g^{-1}}$ and thus $T_g r_u = T_e r_{u \cdot g} \circ T_g \lambda_{g^{-1}}$. Since ω reproduces the generators of fundamental vector fields, we see that $\omega(u \cdot g) \circ T_e r_{u \cdot g}$ is just the inclusion $\mathfrak{p} \to \mathfrak{g}$, which is complex linear, and $T_g \lambda_{g^{-1}}$, is complex linear, too. Thus, the map r is holomorphic, and hence $p : \mathcal{G} \to M$ is a holomorphic principal bundle and the proof of (2) is complete.

(3): By (2) we know that $\mathcal{G}$ is a complex manifold and $p : \mathcal{G} \to M$ is a holomorphic principal bundle if and only if $\kappa_{0,2} = 0$. By construction $\omega(u)$ is complex linear for any $u \in \mathcal{G}$, so ω is a smooth $\mathfrak{g}$–valued $(1, 0)$–form on $\mathcal{G}$. But then ω is holomorphic, if and only if $\bar{\partial}\omega = 0$ i.e. if and only if the $(1, 1)$–part of $d\omega$ vanishes; see 1.2.10. But since the bracket on $\mathfrak{g}$ is complex bilinear, this $(1, 1)$–part is described by the $(1, 1)$–part of the curvature function κ, and the result follows. □

3.1.18. The equivalence to underlying structures in the complex case. We continue to work in the setting of 3.1.17 and consider a complex parabolic geometry $(p : \mathcal{G} \to M, \omega)$ of type (G, P). Since ω is a holomorphic $(1,0)$–form, we may identify the associated bundle $\mathcal{G} \times_P (\mathfrak{g}/\mathfrak{p})$ with the holomorphic tangent bundle $T^{1,0}M$ of M, and a complex parabolic geometry gives rise to a filtration of the holomorphic tangent bundle by holomorphic subbundles. The constructions of underlying structures done in 3.1.5 and 3.1.15 carries over to the holomorphic setting without changes. There are also obvious holomorphic versions of (abstract) infinitesimal flag structures and P–frame bundles of degree one, and underlying structures of complex parabolic geometries are by construction of that type. Regularity can be defined and characterized in the holomorphic setting completely parallel to the real case.

By definition the Lie algebra differential ∂ maps complex multilinear maps to complex multilinear maps. For the codifferential ∂^*, this property follows either by dualization or from the explicit formula in 3.1.11. The Hodge decomposition on the Lie algebra level from 3.1.11 works for the complex multilinear maps case without any changes. Hence, the concept of normality of parabolic geometries works in the holomorphic category without problems.

To get the equivalence of categories between complex normal regular parabolic geometries and underlying structures, one has to use slightly different arguments than in the real case. The problem is that in the proofs of this equivalence in the real case, we frequently used that for a vector bundle homomorphism $\Phi : E \to F$ and a smooth section $s \in \Gamma(F)$ which has values in the image of Φ, one can find a smooth section $\tilde{s} \in \Gamma(E)$ such that $s = \Phi(\tilde{s})$. Otherwise put, we used that any exact sequence of smooth vector bundle homomorphisms splits, and this is not true in the holomorphic category. Thus, we will use a different approach, which is based on Proposition 3.1.17:

THEOREM 3.1.18. *Let $\mathfrak{g} = \mathfrak{g}_{-k} \oplus \cdots \oplus \mathfrak{g}_k$ be a complex $|k|$–graded semisimple Lie algebra such that none of the simple ideals of $\mathfrak{g}$ is contained in $\mathfrak{g}_0$, let G be a complex Lie group with Lie algebra $\mathfrak{g}$, $P \subset G$ a parabolic subgroup for the given grading and $G_0 \subset P$ the Levi subgroup.*

Then the equivalences of categories from Theorems 3.1.14 and 3.1.16 restrict to equivalences from the category of complex regular normal parabolic geometries to the category of holomorphic regular infinitesimal flag structures, respectively, the category of holomorphic regular normal P–frame bundles of degree one.

PROOF. We have already observed that the structures underlying complex parabolic geometries are automatically holomorphic, so we obtain restrictions of functors as claimed. Moreover, any smooth morphism between complex parabolic geometries is automatically holomorphic, since compatibility of a principal bundle map with the Cartan connections implies complex linearity of its tangent map. Now start with a holomorphic regular infinitesimal flag structure, respectively, regular normal P–frame bundle of degree one over a complex manifold M. The procedures from 3.1.14–3.1.16 lead to a regular normal real parabolic geometry $(p : \mathcal{G} \to M, \omega)$. If we can show that this parabolic geometry is actually complex, then this defines a functor in the opposite direction, and the fact that we have obtained an equivalence of categories follows as in the proof of Theorem 3.1.14. In view of Proposition 3.1.17 this can be done by showing that the curvature κ of ω has complex bilinear values.

Now this is a local property and locally we can get analogs of 3.1.14–3.1.16 in the holomorphic category: Let us first assume that $(\{T^iM\}, p_0 : E \to M, \theta)$ is a holomorphic regular infinitesimal flag structure of type (G, P). Let $U \subset M$ be an open subset such that all the filtration components T^iM as well as the bundle E are holomorphically trivial over U. From a simultaneous holomorphic trivialization of the bundles T^iU we obtain holomorphic projections $\pi_i : TU \to T^iU$, while a holomorphic trivialization of $E|_U$ provides us with a holomorphic principal connection γ on $p_0^{-1}(U)$. As in the proof of Theorem 3.1.13, this gives us an extension of θ to a holomorphic Cartan connection $\tilde{\theta} \in \Omega^{1,0}(p_0^{-1}(U), \mathfrak{g}_- \oplus \mathfrak{g}_0)$ on the holomorphic principal bundle $p_0^{-1}(U) \to U$. The mechanism of extension functors from Cartan geometries of type (P^{op}, G_0) to Cartan geometries of type (G, P) works without changes in the holomorphic setting, since it just uses equivariant extension. Hence, we see that the restriction to U of the regular infinitesimal flag structure can be obtained from a complex parabolic geometry on U.

If $(\{T^iM\}, p_1 : E \to M, \theta)$ is a holomorphic regular normal P–frame bundle of degree one, then the underlying regular infinitesimal flag structure $(\{T^iM\}, p_0 : E_0 \to M, \underline{\theta})$ is holomorphic, too. Restricting to an open subset $U \subset M$ over which all the bundles T^iM, E and E_0 are trivial, we can similarly imitate the first part of the proof of Theorem 3.1.16 locally in the holomorphic category. This leads to a complex parabolic geometry on U, such that the underlying P–frame bundle of degree one is isomorphic to the restriction of $(\{T^iM\}, p_1 : E \to M, \theta)$ to U.

Now assume that we have given complex parabolic geometry $(\{T^iU\}, U \times P, \tilde{\omega})$ with holomorphically trivial Cartan bundle, which is what we have obtained in the above constructions. Then we get a holomorphic trivialization of all filtration components of the adjoint tractor bundle $\mathcal{A}U$ and thus of all filtration components of the tangent bundle. In particular, we can identify holomorphic differential forms with values in $\mathcal{A}U$ with holomorphic functions with values in appropriate vector spaces, and the bundle maps like ∂ and ∂^* are just given by acting on the values of these maps. Taking this into account, we see that the proof of Proposition 3.1.13 works in the holomorphic category in the case of a trivial Cartan bundle. Thus, we conclude that we always obtain a normal complex parabolic geometry $(U \times P, \omega_U)$ of type (G, P) over U, which induces the restriction of the regular infinitesimal flag structure, respectively, the regular normal P–frame bundle of degree one that we have started with.

From the last part of the proofs of Theorems 3.1.14 respectively 3.1.16 we conclude that there is a (unique) isomorphism $\Phi_U : \mathcal{G}|_U \to U \times P$, which induces the identity on the underlying infinitesimal flag structure, respectively, the underlying P–frame bundle of degree one, such that $\omega = \Phi_U^* \omega_U$. This of course implies that the restriction of the curvature function κ of ω to $p^{-1}(U)$ is given by $\kappa = \kappa_U \circ \Phi_U$, where κ_U is the curvature function of ω_U. Since ω_U is holomorphic, this has complex bilinear values, and hence κ has complex bilinear values, which completes the proof. □

3.1.19. Abstract adjoint tractor bundles. To finish this section, we discuss an alternative approach to parabolic geometries in which the principal bundle and the Cartan connection is replaced by an induced vector bundle and a linear connection on that bundle. The passage from the Cartan connection to an induced connection was described for general Cartan geometries in 1.5.7, where general bundles associated to actions of the "big" group G were considered. Here we specialize to associated vector bundles. The advantage of the tractor approach is that knowing the normal tractor connection on a tractor bundle one immediately gets access to the fundamental derivative on that tractor bundle and all its subquotients; see 1.5.8. The starting point for the tractor description of parabolic geometries is to introduce an abstract version of the adjoint tractor bundle and its induced linear connection, which is rather straightforward:

Let $\mathfrak{g} = \mathfrak{g}_{-k} \oplus \cdots \oplus \mathfrak{g}_k$ be a $|k|$–graded semisimple Lie algebra. Let M be a smooth manifold of the same dimension as $\mathfrak{g}/\mathfrak{p}$, and consider a bundle of filtered Lie algebras $p : \mathcal{A} \to M$ modelled on $\mathfrak{g} = \mathfrak{g}^{-k} \supset \mathfrak{g}^{-k+1} \supset \cdots \supset \mathfrak{g}^k$. By definition, this means that $\mathcal{A} \to M$ is a vector bundle, whose rank equals the dimension of $\mathfrak{g}$, which is endowed with a filtration $\mathcal{A} = \mathcal{A}^{-k} \supset \mathcal{A}^{-k+1} \supset \cdots \supset \mathcal{A}^k$ by smooth subbundles and with a tensorial bracket $\{\ ,\ \} : \Lambda^2\mathcal{A} \to \mathcal{A}$, which makes each fiber into a Lie algebra, such that there are local charts with values in $\mathfrak{g}$ that are compatible with the filtration and the bracket. For sections $s_1, s_2 \in \Gamma(\mathcal{A})$ one can use the pointwise bracket to obtain a smooth section $\{s_1, s_2\}$ of $\mathcal{A}$, and this makes the space $\Gamma(\mathcal{A})$ into a filtered Lie algebra.

DEFINITION 3.1.19. Let $p : \mathcal{A} \to M$ be a bundle of filtered Lie algebras modelled on $(\mathfrak{g}, \{\mathfrak{g}^i\})$ as above.

(1) A linear connection ∇ on $\mathcal{A}$ is called *nondegenerate* if for any point $x \in M$ and any tangent vector $\xi \in T_xM$ there is an index i with $|i| \leq k$ and a smooth (local) section $s \in \Gamma(\mathcal{A}^i)$ such that $\nabla_\xi s(x) \notin \mathcal{A}^i_x$.

(2) A nondegenerate linear connection ∇ on $\mathcal{A}$ is called a *tractor connection* if it is compatible with the bracket on $\mathcal{A}$, i.e. if for each vector field $\xi \in \mathfrak{X}(M)$ and sections $s_1, s_2 \in \Gamma(\mathcal{A})$ one has $\nabla_\xi(\{s_1, s_2\}) = \{\nabla_\xi s_1, s_2\} + \{s_1, \nabla_\xi s_2\}$.

(3) The bundle $\mathcal{A}$ is called an *(abstract) adjoint tractor bundle* on M if it admits a tractor connection.

Let G be a Lie group with Lie algebra $\mathfrak{g}$, $P \subset G$ a parabolic subgroup for the given grading, $G_0 \subset P$ the Levi subgroup, and $(p : \mathcal{G} \to M, \omega)$ a parabolic geometry of type (G, P). Since the adjoint action of P on $\mathfrak{g}$ is by filtration preserving Lie algebra automorphisms, the associated bundle $\mathcal{A}M = \mathcal{G} \times_P \mathfrak{g}$ is a bundle of filtered Lie algebras modelled on $\mathfrak{g}$. From 1.5.7 we know that the Cartan connection ω induces a linear connection ∇ on the bundle $\mathcal{A}M \to M$. Compatibility of ∇ with the bracket $\{\ ,\ \}$ was proved in Proposition 1.5.7.

To prove nondegeneracy of ∇, we use the relation to the fundamental derivative D. Given sections $s_1, s_2 \in \Gamma(\mathcal{A}M)$ and denoting by $\Pi : \mathcal{A}M \to TM$ the natural projection, we have $\nabla_{\Pi(s_1)} s_2 = D_{s_1} s_2 + \{s_1, s_2\}$ by Theorem 1.5.8. If $s_2 \in \Gamma(\mathcal{A}^0 M)$, then by naturality of the fundamental derivative also $D_{s_1} s_2 \in \Gamma(\mathcal{A}^0 M)$. Now suppose that for a point $x \in M$ and all $s_2 \in \Gamma(\mathcal{A}^0 M)$ we have $\nabla_{\Pi(s_1)} s_2(x) \in \mathcal{A}^0_x M$. Then $\{s_1(x), s_2(x)\} \in \mathcal{A}^0_x M$ for all $s_2(x) \in \mathcal{A}^0_x M$. Since the normalizer of $\mathfrak{p}$ in $\mathfrak{g}$ is $\mathfrak{p}$, this implies $s_1(x) \in \mathcal{A}^0_x M$ and hence $\Pi(s_1)(x) = 0$, and nondegeneracy follows. Hence, ∇ is a tractor connection on $\mathcal{A}M$ in the sense of the above definition, which therefore provides an abstract version of the adjoint tractor bundle and its canonical linear connection.

3.1.20. Adapted frame bundles. In this setting, there is a natural choice of a group G with Lie algebra $\mathfrak{g}$. We have noted in 2.2.2 that the automorphism group $G = \mathrm{Aut}(\mathfrak{g})$ has Lie algebra $\mathfrak{g}$. Since $\mathrm{Aut}(\mathfrak{g})$ is a subgroup of $GL(\mathfrak{g})$, the adjoint action is given by conjugation, so for $\phi \in \mathrm{Aut}(\mathfrak{g})$ and $A \in \mathfrak{g}$, we get

$\mathrm{Ad}(\phi)(\mathrm{ad}(A)) = \phi \circ \mathrm{ad}(A) \circ \phi^{-1} = \mathrm{ad}(\phi(A))$. Hence, the adjoint action of G on $\mathfrak{g}$ is given by applying automorphisms. Choosing P to be the maximal parabolic subgroup $\bigcap_{i=-k}^{k} N_G(\mathfrak{g}^i)$ (see 3.1.3) we see that $P = \mathrm{Aut}_f(\mathfrak{g})$, the group of automorphisms of the filtered algebra $(\mathfrak{g}, \{\mathfrak{g}^i\})$. For the Levi subgroup, we obtain $G_0 = \mathrm{Aut}_{\mathrm{gr}}(\mathfrak{g})$, the automorphism group of the graded Lie algebra $\mathfrak{g} = \mathfrak{g}_{-k} \oplus \cdots \oplus \mathfrak{g}_k$. For this choice of G, we can naturally associate a principal P–bundle to any abstract adjoint tractor bundle $\mathcal{A}$.

PROPOSITION 3.1.20. *Let $\mathfrak{g}$ be a $|k|$–graded Lie algebra, put $G = \mathrm{Aut}(\mathfrak{g})$, $P = \mathrm{Aut}_f(\mathfrak{g})$ and $G_0 = \mathrm{Aut}_{\mathrm{gr}}(\mathfrak{g})$. Let $\pi : \mathcal{A} \to M$ be a bundle of filtered Lie algebras modelled on $\mathfrak{g}$. Then there is a natural principal P–bundle $p : \mathcal{G} \to M$ such that $\mathcal{A} = \mathcal{G} \times_P \mathfrak{g}$.*

PROOF. For $x \in M$ define $\mathcal{G}_x$ to be the set of all isomorphisms $\psi : \mathfrak{g} \to \mathcal{A}_x$ of filtered Lie algebras, where $\mathcal{A}_x$ denotes the fiber of $\mathcal{A}$ at x, and define $\mathcal{G}$ to be the disjoint union of all $\mathcal{G}_x$ as x ranges over M. Then we can naturally view $\mathcal{G}$ as a subset of the linear frame bundle $GL(\mathfrak{g}, \mathcal{A})$ of the vector bundle $\mathcal{A}$, and the restriction of the projection of the frame bundle defines a map $p : \mathcal{G} \to M$. By assumption, the bundle $\mathcal{A}$ admits local charts with values in $\mathfrak{g}$ which are compatible with the filtration and the bracket, which exactly means that there are local smooth sections of $GL(\mathfrak{g}, \mathcal{A}) \to M$ that have values in the subset $\mathcal{G}$. On the other hand, $GL(\mathfrak{g}, \mathcal{A})$ is by definition a principal bundle with structure group $GL(\mathfrak{g})$, so the subgroup $P = \mathrm{Aut}_f(\mathfrak{g})$ acts smoothly and freely on $GL(\mathfrak{g}, \mathcal{A})$ by composition from the right. By construction, this action preserves the subset $\mathcal{G}$, and for $\psi_1, \psi_2 \in \mathcal{G}_x$, the composition $\phi := \psi_1^{-1} \circ \psi_2$ is an automorphism of the filtered Lie algebra $\mathfrak{g}$, and hence $\phi \in P$. But then $\psi_2 = \psi_1 \circ \phi$, so the action of P is transitive on each of the fibers $\mathcal{G}_x$ of $\mathcal{G}$. Together with the existence of local smooth $\mathcal{G}$–valued sections, this implies that $\mathcal{G}$ is a submanifold of the bundle $GL(\mathfrak{g}, \mathcal{A})$ and the projection $p : \mathcal{G} \to \mathcal{A}$ is a principal P–bundle.

Finally, we get a well–defined smooth map $\mathcal{G} \times \mathfrak{g} \to \mathcal{A}$ given by $(\psi, A) \mapsto \psi(A)$, and since $\mathrm{Ad}(\phi)(A) = \phi(A)$ for $\phi \in P$ and $A \in \mathfrak{g}$, we see that (ψ, A) and $(\psi \circ \phi, \mathrm{Ad}(\phi^{-1})(A))$ have the same image under this map. Thus, our mapping factors to a bundle map $\mathcal{G} \times_P \mathfrak{g} \to \mathcal{A}$ covering the identity on M, which by construction is bijective on each fiber and thus an isomorphism of vector bundles. □

To get similar results for other choices of a group G with Lie algebra $\mathfrak{g}$, one proceeds as follows: For any Lie group G with Lie algebra $\mathfrak{g}$, the adjoint action defines a covering map from G to a subgroup of $GL(\mathfrak{g})$ which lies between $\mathrm{Aut}(\mathfrak{g})$ and $\mathrm{Int}(\mathfrak{g})$, the connected component of the identity in $\mathrm{Aut}(\mathfrak{g})$. The elements of this subgroup can usually be characterized as those automorphisms of $\mathfrak{g}$, which preserve an additional structure, for example, an orientation. An analog of the above proposition for this subgroup can be obtained by looking at abstract adjoint tractor bundles endowed with that additional structure, and isomorphism compatible with it. To pass further to a nontrivial covering G of that subgroup, one usually needs, in addition, a structure on M which is similar to a spin structure, and the analog of the proposition is obtained via a fibered product construction. Since such versions of the proposition cannot be derived in a uniform way, we require the existence of an appropriate principal bundle as an additional ingredient:

DEFINITION 3.1.20. Let $\mathfrak{g} = \mathfrak{g}_{-k} \oplus \cdots \oplus \mathfrak{g}_k$ be a $|k|$–graded semisimple Lie algebra, M a smooth manifold of the same dimension as $\mathfrak{g}/\mathfrak{p}$ and $p : \mathcal{A} \to M$

a bundle of filtered Lie algebras modelled on $\mathfrak{g}$. Let G be a Lie group with Lie algebra $\mathfrak{g}$, and let $P \subset G$ be a parabolic subgroup for the given $|k|$–grading with Levi subgroup $G_0 \subset P$.

An *adapted frame bundle of type* (G, P) for $\mathcal{A}$ corresponding to the group G is a principal P–bundle $\mathcal{G} \to M$ such that $\mathcal{A} = \mathcal{G} \times_P \mathfrak{g}$, the associated bundle with respect to the adjoint action of P on $\mathfrak{g}$.

3.1.21. Abstract tractor bundles. We could now directly derive an equivalent description of parabolic geometries via abstract adjoint tractor bundles and normal tractor connections, but in many situations it is easier to pass to simpler tractor bundles. Typically, these are associated to the standard representation of a classical Lie algebra. Hence, we will work in the setting of general abstract tractor bundles and tractor connections, and the most natural way to do this is via $(\mathfrak{g}, P)$–modules (rather than G–modules). Recall that a $(\mathfrak{g}, P)$–module is a vector space V endowed with representations of the Lie algebra $\mathfrak{g}$ and the Lie group P which are compatible in the sense that the restriction of the $\mathfrak{g}$–representation to $\mathfrak{p}$ is the derivative of the P–representation and for $A \in \mathfrak{g}$, $g \in P$, and $v \in V$ we have $(\mathrm{Ad}(g) \cdot A) \cdot v = g^{-1} \cdot A \cdot g \cdot v$. For any representation of the group G on V, the restriction to P together with the derivative make V into a $(\mathfrak{g}, P)$–module.

Now assume that $\mathfrak{g}$ is a $|k|$–graded Lie algebra, G is a Lie group with Lie algebra $\mathfrak{g}$, and $P \subset G$ is a parabolic subgroup for the given grading with Levi subgroup $G_0 \subset P$. Further, let M be a smooth manifold of the same dimension as $\mathfrak{g}/\mathfrak{p}$, $\pi : \mathcal{A} \to M$ a bundle of filtered Lie algebras modelled on $\mathfrak{g}$ and $p : \mathcal{G} \to M$ an adapted frame bundle for $\mathcal{A}$. For a $(\mathfrak{g}, P)$–module V we can then consider the associated bundle $\mathcal{T} := \mathcal{G} \times_P V$. Since $\mathcal{A} = \mathcal{G} \times_P \mathfrak{g}$, the representation of $\mathfrak{g}$ on V induces a bundle map $\bullet : \mathcal{A} \otimes \mathcal{T} \to \mathcal{T}$, which defines an action of the bundle $\mathcal{A}$ of Lie algebras, i.e. $\{s_1, s_2\} \bullet t = s_1 \bullet (s_2 \bullet t) - s_2 \bullet (s_1 \bullet t)$ for $s_1, s_2 \in \mathcal{A}_x$ and $t \in \mathcal{T}_x$, and similarly on the level of smooth sections; compare with 1.5.7.

Next, we want to define an analog of tractor connections on such bundles. Since $\mathcal{T}$ is associated to $\mathcal{G}$, any point $u \in \mathcal{G}$ with $p(u) = x \in M$ gives rise to a linear isomorphism $\underline{u} : V \to \mathcal{T}_x$, which is characterized by $\underline{u}(v) = [\![u, v]\!]$. In these terms, the correspondence between smooth sections $t \in \Gamma(\mathcal{T})$ and equivariant functions $\tilde{t} \in C^\infty(\mathcal{G}, V)$ reads as $t(p(u)) = \underline{u}(\tilde{t}(u))$. Now suppose that $\nabla^{\mathcal{T}}$ is a linear connection on $\mathcal{T}$, $u \in \mathcal{G}$ is a point with $p(u) = x$ and $\xi \in T_u\mathcal{G}$ is a tangent vector at the point u. For a smooth section $t \in \Gamma(\mathcal{T})$ corresponding to the equivariant map $\tilde{t} : \mathcal{G} \to V$, consider $\underline{u}^{-1}(\nabla^{\mathcal{T}}_{Tp\cdot\xi} t(x)) - (\xi \cdot \tilde{t})(u) \in V$. If $f : M \to \mathbb{R}$ is a smooth function, then $\widetilde{ft} = (f \circ p)\tilde{t}$, which immediately implies that changing t to ft, the above element of V changes only by multiplication by $f(x)$. Thus, we conclude that this element depends only on $t(x)$ or equivalently on $\tilde{t}(u)$, whence we get a well–defined linear map $\Phi(\xi) : V \to V$ which is characterized by

$$\underline{u}^{-1}(\nabla^{\mathcal{T}}_{Tp\cdot\xi} t(x)) - (\xi \cdot \tilde{t})(u) = \Phi(\xi)(\tilde{t}(u));$$

compare with 1.5.8.

DEFINITION 3.1.21. (1) The linear connection $\nabla^{\mathcal{T}}$ is called a $\mathfrak{g}$*–connection* if and only if for any $u \in \mathcal{G}$ and any $\xi \in T_u\mathcal{G}$, there is an element $A \in \mathfrak{g}$ such that $\Phi(\xi)(v) = A \cdot v$ for all $v \in V$.

Of course, any structure on V that is invariant under the action of P leads to a corresponding structure on the bundle $\mathcal{T}$, and if it is, in addition, $\mathfrak{g}$–invariant, this structure is preserved by any $\mathfrak{g}$–connection on $\mathcal{T}$. For concrete choices the Lie algebra $\mathfrak{g}$, the group G and the representation V it is usually easy to characterize $\mathfrak{g}$–connections as those connections which preserve such induced structures. Note further, that in the case of the adjoint tractor bundle itself, this condition is equivalent to compatibility of the connection with the bracket: For $s_1, s_2 \in \Gamma(\mathcal{A})$ with corresponding functions $\tilde{s}_1, \tilde{s}_2 : \mathcal{G} \to \mathfrak{g}$, the bracket $\{s_1, s_2\}$ corresponds to the function $u \mapsto [\tilde{s}_1(u), \tilde{s}_2(u)]$. Hitting this function with a tangent vector ξ, bilinearity of the bracket immediately implies $\xi \cdot \widetilde{\{s_1, s_2\}} = [\xi \cdot \tilde{s}_1, \tilde{s}_2] + [s_1, \xi \cdot \tilde{s}_2]$. Thus, from the defining equation for $\Phi(\xi)$ above, one immediately concludes that compatibility of a linear connection $\nabla^{\mathcal{A}}$ on $\mathcal{A}$ with the bracket is equivalent to the fact that $\Phi(\xi)([\tilde{s}_1(u), \tilde{s}_2(u)]) = [\Phi(\xi)(\tilde{s}_1(u)), \tilde{s}_2(u)] + [\tilde{s}_1(u), \Phi(\xi)(\tilde{s}_2(u))]$. So this is equivalent to $\Phi(\xi)$ being a derivation of $\mathfrak{g}$ for all $\xi \in T\mathcal{G}$ and since any derivation of the semisimple Lie algebra $\mathfrak{g}$ is inner, this is equivalent to $\nabla^{\mathcal{A}}$ being a $\mathfrak{g}$–connection.

Let us now assume that the $\mathfrak{g}$–action on V is effective, which just means that none of the simple ideals of $\mathfrak{g}$ acts trivially on V. Then the representation identifies $\mathfrak{g}$ with a Lie subalgebra of $\mathfrak{gl}(V)$ and passing to the associated bundles this means that $\mathcal{A}$ is naturally a subbundle of the vector bundle $L(\mathcal{T}, \mathcal{T})$ of linear endomorphisms of $\mathcal{T}$. Now any linear connection $\nabla^{\mathcal{T}}$ on $\mathcal{T}$ induces a linear connection on $L(\mathcal{T}, \mathcal{T})$ characterized by $(\nabla_\xi \Psi)(t) = \nabla^{\mathcal{T}}_\xi(\Psi(t)) - \Psi(\nabla^{\mathcal{T}}_\xi t)$. If we assume that $\nabla^{\mathcal{T}}$ is a $\mathfrak{g}$–connection, then this induced connection preserves the subbundle $\mathcal{A} \subset L(\mathcal{T}, \mathcal{T})$: By construction, the linear connection $\nabla^{\mathcal{T}}$ is given by

$$\nabla^{\mathcal{T}}_{Tp\cdot\xi} t(x) = \underline{u}(\xi \cdot \tilde{t}(u) - \Phi(\xi)(\tilde{t}(u))).$$

For $s \in \Gamma(\mathcal{A})$, we have $\widetilde{s \bullet t}(u) = \tilde{s}(u) \cdot \tilde{t}(u)$, and hitting this with a tangent vector ξ, bilinearity of the action implies $\xi \cdot (\widetilde{s \bullet t})(u) = (\xi \cdot \tilde{s}(u)) \cdot \tilde{t}(u) + \tilde{s}(u) \cdot (\xi \cdot \tilde{t}(u))$. Using this, we immediately compute that

$$\nabla^{\mathcal{T}}_{Tp\cdot\xi}(s \bullet t)(x) - s \bullet \nabla^{\mathcal{T}}_{Tp\cdot\xi} t(x) = \underline{u}((\xi \cdot \tilde{s}(u) - [\Phi(\xi), \tilde{s}(u)]) \cdot \tilde{t}(u)),$$

and since $\Phi(\xi)$ is given by the action of an element of $\mathfrak{g}$, the right–hand side is given by the action of a section of $\mathcal{A}$ on t. Thus, we get an induced connection $\nabla^{\mathcal{A}}$ on $\mathcal{A}$, which is compatible with the algebraic bracket since the induced connection on $L(\mathcal{T}, \mathcal{T})$ is compatible with the commutator of endomorphisms.

DEFINITION 3.1.21. (2) A $\mathfrak{g}$–connection $\nabla^{\mathcal{T}}$ on $\mathcal{T}$ is called a *tractor connection* if and only if the induced connection $\nabla^{\mathcal{A}}$ on $\mathcal{A}$ is nondegenerate and thus a tractor connection.

(3) The bundle $\mathcal{T}$ is called the *V-tractor bundle* corresponding to $\mathcal{A}$ (and $\mathcal{G}$), if it admits a tractor connection (and thus $\mathcal{A}$ is an abstract adjoint tractor bundle).

Let us remark that for concrete choices of $\mathfrak{g}$, G and V it is usually easy to directly characterize nondegeneracy of $\mathfrak{g}$–connections on $\mathcal{T}$ in terms of natural filtrations available on $\mathcal{T}$.

3.1.22. Tractor description of parabolic geometries. At this stage it is already visible how to obtain a relation between tractor connections on tractor bundles and Cartan connections on the corresponding adapted frame bundles. If $\mathcal{T} = \mathcal{G} \times_P V$ for a $(\mathfrak{g}, P)$–module V with effective $\mathfrak{g}$–action, and $\nabla^{\mathcal{T}}$ is a $\mathfrak{g}$–connection on $\mathcal{T}$, then for any tangent vector $\xi \in T\mathcal{G}$, the action of $\Phi(\xi)$ on V is completely determined by $\nabla^{\mathcal{T}}$, so by effectivity of the $\mathfrak{g}$–action, there is a unique element

$\omega(\xi) \in \mathfrak{g}$ such that $\Phi(\xi)(v) = \omega(\xi) \cdot v$. Clearly, this defines a smooth one–form $\omega \in \Omega^1(\mathcal{G}, \mathfrak{g})$.

THEOREM 3.1.22. *Let $\mathfrak{g}$ be a $|k|$–graded Lie algebra, G a group with Lie algebra $\mathfrak{g}$, $P \subset G$ a parabolic subgroup for the given grading with Levi–subgroup $G_0 \subset P$, M a manifold of the same dimension as $\mathfrak{g}/\mathfrak{p}$, $\pi : \mathcal{A} \to M$ a bundle of filtered Lie algebras modelled on $\mathfrak{g}$ and $p : \mathcal{G} \to M$ an adapted frame bundle for $\mathcal{A}$ of type (G, P). Let V be a $(\mathfrak{g}, P)$–module which is effective as a $\mathfrak{g}$–module and $\mathcal{T} = \mathcal{G} \times_P V$ the corresponding induced bundle. Then we have:*

1. *For any $\mathfrak{g}$–connection $\nabla^{\mathcal{T}}$ on $\mathcal{T}$, the corresponding one–form $\omega \in \Omega^1(\mathcal{G}, \mathfrak{g})$ is P–equivariant and it reproduces the generators of fundamental vector fields. Moreover, ω is a Cartan connection if and only if $\nabla^{\mathcal{T}}$ is a tractor connection.*
2. *Conversely, any Cartan connection $\omega \in \Omega^1(\mathcal{G}, \mathfrak{g})$ induces a tractor connection $\nabla^{\mathcal{T}}$ on $\mathcal{T}$.*
3. *If $\nabla^{\mathcal{T}}$ is a tractor connection, then its curvature R is given by $R(\xi, \eta)(t) = \kappa(\xi, \eta) \bullet t$, where $\kappa \in \Omega^2(M, \mathcal{A})$ is the curvature of the corresponding Cartan connection $\omega \in \Omega^1(\mathcal{G}, \mathfrak{g})$.*

PROOF. (1) From above we know that the one–form ω is characterized by

$$\nabla^{\mathcal{T}}_{Tp\cdot\xi} t(x) = \underline{u}\big((\xi \cdot \tilde{t})(u) - \omega(\xi) \cdot \tilde{t}(u)\big). \tag{3.4}$$

If $\xi = \zeta_A(u)$ for some $A \in \mathfrak{p}$, then the left–hand side vanishes, while equivariancy of $\tilde{t}$ implies that $\zeta_A \cdot \tilde{t}(u) = (\mathcal{L}_{\zeta_A}\tilde{t})(u) = -A \cdot \tilde{t}(u)$, and since $\underline{u}$ is injective, this implies $\omega(\zeta_A) = A$. On the other hand, replacing u by $u \cdot g$ and ξ by $Tr^g \cdot \xi$ in the above equation, the left–hand side remains unchanged. In the right–hand side, we have $\underline{u \cdot g}(v) = \underline{u}(g \cdot v)$, while equivariancy of $\tilde{t}$ reads as $\tilde{t}(u \cdot g) = g^{-1} \cdot \tilde{t}(u)$ and thus implies $Tr^g \cdot \xi \cdot \tilde{t} = g^{-1} \cdot (\xi \cdot \tilde{t})$, so this term also produces the same value as before. Thus, by injectivity of $\underline{u}$ we are left with $\omega(\xi) \cdot \tilde{t}(u) = g \cdot \omega(Tr^g \cdot \xi) \cdot g^{-1} \cdot \tilde{t}(u)$, and since V is a $(\mathfrak{g}, P)$–module, the right–hand side coincides with $(\mathrm{Ad}(g) \cdot \omega(Tr^g \cdot \xi)) \cdot \tilde{t}(u)$. Effectivity of the $\mathfrak{g}$–action then implies $\omega(Tr^g \cdot \xi) = \mathrm{Ad}(g^{-1})(\omega(\xi))$, i.e. equivariancy of ω.

From above, we also know that the induced connection $\nabla^{\mathcal{A}}$ on $\mathcal{A}$ corresponds to the same one–form ω, so to prove the characterization of ω being a Cartan connection, we may assume $\mathcal{T} = \mathcal{A}$. Let us first assume that $\nabla^{\mathcal{A}}$ is a tractor connection, and that $\omega(\xi) = 0$. If $Tp \cdot \xi \neq 0$, nondegeneracy implies that we can find an index i and a smooth section $s \in \Gamma(\mathcal{A}^i)$ such that $\nabla^{\mathcal{A}}_{Tp\cdot\xi} s(x) \notin \mathcal{A}^i$ (and thus, in particular, is nonzero). But since $s \in \Gamma(\mathcal{A}^i)$, the corresponding function $\tilde{s} : \mathcal{G} \to \mathfrak{g}$ has values in $\mathfrak{g}^i$, and thus also $\xi \cdot \tilde{s}$ has values in $\mathfrak{g}^i$. Hence, that class of $\nabla^{\mathcal{A}}_{\xi} s(x)$ in $\mathcal{A}/\mathcal{A}^i$ coincides with the class of $-\underline{u}^{-1}(\omega(\xi) \cdot \tilde{s}(u))$, and this being nonzero contradicts $\omega(\xi) = 0$. Hence, $Tp \cdot \xi = 0$, so ξ is vertical and thus of the form ζ_A, whence $\omega(\xi) = A$ so $\omega(\xi) = 0$ implies $\xi = 0$. Thus, the restriction of ω to any tangent space is injective and hence a linear isomorphism for dimensional reasons. Conversely, if ω is a Cartan connection, then $\nabla_{Tp\cdot\xi} s(x) \in \mathcal{A}^0$ for all $s \in \mathcal{A}^0$ implies $[\omega(\xi), \mathfrak{p}] \subset \mathfrak{p}$, thus $\omega(\xi) \in \mathfrak{p}$ and $Tp \cdot \xi = 0$, so $\nabla^{\mathcal{A}}$ is nondegenerate.

(2) If $\omega \in \Omega^1(M)$ is a Cartan connection, then we use formula (3.4) from above to define a linear connection $\nabla^{\mathcal{T}}$ on $\mathcal{T}$. Equivariancy of ω immediately implies that the right–hand side of (3.4) is independent of the choice of u with $p(u) = x$ and since ω reproduces the generators of fundamental vector fields, the right–hand

side vanishes if ξ is vertical (compare with the computations made in the proof of (1) and with 1.5.8). Thus, the left–hand side really depends only on x and $Tp \cdot \xi$. Moreover, starting with a vector field $\underline{\xi}$ downstairs and choosing a lift $\xi \in \mathfrak{X}(\mathcal{G})$, (3.4) clearly defines a smooth section $\nabla_{\underline{\xi}} t \in \Gamma(\mathcal{T})$, so we get a well–defined operator $\mathfrak{X}(M) \times \Gamma(\mathcal{T}) \to \Gamma(\mathcal{T})$. Replacing $\underline{\xi}$ by $f\underline{\xi}$ for $f \in C^\infty(M, \mathbb{R})$, we may replace ξ by $(f \circ p)\xi$, and (3.4) then immediately implies that $\nabla^{\mathcal{T}}_{f\underline{\xi}} t = f \nabla^{\mathcal{T}}_{\underline{\xi}} t$. On the other hand, replacing t by ft, we get $\widetilde{ft} = (f \circ p)\tilde{t}$ and since $\xi \cdot (f \circ p)\tilde{t} = (Tp \cdot \xi \cdot f)\tilde{t} + (f \circ p)\xi \cdot \tilde{t}$ equation (3.4) implies $\nabla^{\mathcal{T}}_{\underline{\xi}} ft = (\underline{\xi} \cdot f)t + f\nabla^{\mathcal{T}}_{\underline{\xi}} t$, so $\nabla^{\mathcal{T}}$ is a linear connection. By construction, $\nabla^{\mathcal{T}}$ is a $\mathfrak{g}$–connection corresponding to the one–form ω, so from (1) we know that $\nabla^{\mathcal{T}}$ is a tractor connection since ω is a Cartan connection.

(3) Let $\nabla^{\mathcal{T}}$ be the tractor connection corresponding to the Cartan connection $\omega \in \Omega^1(\mathcal{G}, \mathfrak{g})$, take vector fields $\underline{\xi}$ and $\underline{\eta}$ on M and choose lifts $\xi, \eta \in \mathfrak{X}(\mathcal{G})$. For a section $t \in \Gamma(\mathcal{T})$ with corresponding function $\tilde{t} : \mathcal{G} \to V$, the function corresponding to $\nabla^{\mathcal{T}}_{\underline{\eta}} t$ is given by $u \mapsto (\eta \cdot \tilde{t})(u) - \omega(\eta) \cdot \tilde{t}(u)$. Since $[\xi, \eta]$ is a lift of $[\underline{\xi}, \underline{\eta}]$, the section $\nabla^{\mathcal{T}}_{[\underline{\xi}, \underline{\eta}]} t$ corresponds to the function $[\xi, \eta] \cdot \tilde{t} - \omega([\xi, \eta]) \cdot \tilde{t}$. On the other hand, $\nabla^{\mathcal{T}}_{\underline{\xi}} \nabla^{\mathcal{T}}_{\underline{\eta}} t$ corresponds to the function

$$\xi \cdot (\eta \cdot \tilde{t}) - \omega(\xi) \cdot (\eta \cdot \tilde{t}) - \xi \cdot (\omega(\eta) \cdot \tilde{t}) + \omega(\xi) \cdot \omega(\eta) \cdot \tilde{t}. \tag{3.5}$$

Bilinearity of the action implies that the third term in (3.5) may be rewritten as $(\xi \cdot \omega(\eta)) \cdot \tilde{t} - \omega(\eta) \cdot (\xi \cdot \tilde{t})$. Hence, subtracting from (3.5) the same term with ξ and η exchanged and the function corresponding to $\nabla^{\mathcal{T}}_{[\underline{\xi}, \underline{\eta}]} t$ from above, the terms in which vector fields differentiate $\tilde{t}$ all cancel out and we are left with $R(\underline{\xi}, \underline{\eta})t$ corresponding to the function

$$\big(\xi \cdot \omega(\eta) - \eta \cdot \omega(\xi) - \omega([\xi, \eta])\big) \cdot \tilde{t} + \omega(\xi) \cdot \omega(\eta) \cdot t - \omega(\eta) \cdot \omega(\xi) \cdot \tilde{t}.$$

By definition of the exterior derivative, the first terms give $d\omega(\xi, \eta) \cdot \tilde{t}$, while the last ones add up to $[\omega(\xi), \omega(\eta)] \cdot t$, so by definition of the curvature we get $R(\underline{\xi}, \underline{\eta})t = \kappa(\underline{\xi}, \underline{\eta}) \bullet t$ as required. □

Having given an abstract adjoint tractor bundle $\pi : \mathcal{A} \to M$, an adapted frame bundle $p : \mathcal{G} \to M$ of type (G, P) and a tractor connection $\nabla^{\mathcal{T}}$ on the V–tractor bundle $\mathcal{T} = \mathcal{G} \times_P V$, we get a tractor connection $\nabla^{\mathcal{A}}$ on $\mathcal{A}$ and a Cartan connection $\omega \in \Omega^1(\mathcal{G}, \mathfrak{g})$. For the parabolic geometry $(p : \mathcal{G} \to M, \omega)$ of type (G, P) on M, we have $\mathcal{A} = \mathcal{A}M$. In particular, we get the filtration $TM = T^{-k}M \supset \dots \supset T^{-1}M$ coming from the fact that the Cartan connection ω gives the identification $TM = \mathcal{G} \times_P (\mathfrak{g}/\mathfrak{p}) \cong \mathcal{A}/\mathcal{A}^0$. This filtration can be nicely seen directly from the tractor connection $\nabla^{\mathcal{A}}$. By definition, a tangent vector $\xi \in T_xM$ lies in T^i_xM if and only if for any lift $\tilde{\xi} \in T_u\mathcal{G}$, we have $\omega(\tilde{\xi}) \in \mathfrak{g}^i$. By the formula for $\nabla^{\mathcal{A}}$, this implies that $\nabla^{\mathcal{A}}_\xi s \in \Gamma(\mathcal{A}^{i+j})$ for all $s \in \Gamma(\mathcal{A}^j)$.

Conversely, if $\nabla^{\mathcal{A}}_\xi s \in \Gamma(\mathcal{A}^i)$ for all $s \in \Gamma(\mathcal{A}^0)$, this formula implies that $[\omega(\tilde{\xi}), \mathfrak{p}] \subset \mathfrak{g}^i$ for any lift $\tilde{\xi}$ of ξ. Inserting the grading element E, we see that this implies $\omega(\tilde{\xi}) \in \mathfrak{g}^i$, so we see that this gives a characterization of T^iM. It remains to characterize regularity and normality of the parabolic geometry corresponding to a tractor connection $\nabla^{\mathcal{T}}$. Note in particular, that having verified these conditions, the corresponding tractor bundle and its normal tractor connection are uniquely determined by the appropriate underlying geometric structure by Theorems 3.1.14 and 3.1.16. For the characterizations, we need one more observation. In the definition of the codifferential, we used the Killing form B, which is a nondegenerate invariant bilinear form B on $\mathfrak{g}$. Of course, this form gives us an identification $\mathcal{A} \cong \mathcal{A}^*$ and thus a trace map $\mathcal{A} \otimes \mathcal{A} \to M \times \mathbb{R}$. Using this, we formulate.

PROPOSITION 3.1.22. *Let $\nabla^{\mathcal{T}}$ be a tractor connection on the V–tractor bundle $\mathcal{T} = \mathcal{G} \times_P V$, where V is a $(\mathfrak{g}, P)$–module which is effective as a $\mathfrak{g}$–module, and let $\kappa \in \Omega^2(M, \mathcal{A})$ be the two form such that $R(\xi, \eta)t = \kappa(\xi, \eta) \bullet t$ for all $\xi, \eta \in \mathfrak{X}(M)$ and $t \in \Gamma(\mathcal{T})$. Then the parabolic geometry $(p : \mathcal{G} \to M, \omega)$ induced by $\nabla^{\mathcal{T}}$ is regular if and only if $\kappa(T^iM, T^jM) \subset \mathcal{A}^{i+j+1}M$ and it is normal if and only if the trace over the first and third entry of the trilinear map $\mathcal{A} \otimes \mathcal{A} \otimes \mathcal{A} \to \mathcal{A}$ given by $(s_1, s_2, s_3) \mapsto \{s_1, \kappa(\Pi(s_2), \Pi(s_3))\} - \frac{1}{2}\kappa(\Pi(\{s_1, s_2\}), s_3)$ vanishes. Here $\Pi : \mathcal{A} \to TM$ denotes the canonical projection.*

PROOF. Since κ is the curvature of ω, the characterization of regularity follows immediately from Corollary 3.1.8. For the normality condition, let us analyze the Kostant codifferential on the Lie algebra level. From 3.1.11 we know that for a decomposable element $Z \wedge W \otimes A$ in $\Lambda^2\mathfrak{p}_+ \otimes \mathfrak{g}$ and $X \in \mathfrak{g}$ we have

$$\partial^*(Z \wedge W \otimes A)(X) = -B(W, X)[Z, A] + B(Z, X)[W, A] - B([Z, W], X)A,$$

and this depends only on the class of X in $\mathfrak{g}/\mathfrak{p}$. Taking dual bases (with respect to B) $\{Z^\ell\}$ of $\mathfrak{p}_+$ and $\{X_\ell\}$ of $\mathfrak{g}/\mathfrak{p}$ we may rewrite $[Z, A]$ as $\sum_\ell B(Z, X_\ell)[Z^\ell, A]$ and similarly for $[W, A]$. But then the first two terms simply add up to $2[Z^\ell, (Z \wedge W \otimes A)(X, X_\ell)]$. Moreover, interpreting $Z \wedge W \otimes A$ as a bilinear map $\Lambda^2\mathfrak{g} \to \mathfrak{g}$ which vanishes if one of its entries lies in $\mathfrak{p}$, we may extend the bases to dual bases of $\mathfrak{g}$ and sum over all elements of these bases without changing the result.

This expression then makes sense in the same form for arbitrary bilinear maps of that type and the corresponding bundle map applied to κ is given by taking the trace over the first and last entry of $(s_1, s_2, s_3) \mapsto 2\{s_1, \kappa(\Pi(s_2), \Pi(s_3))\}$. On the other hand, using invariance of B we may rewrite $-B([Z, W], X)$ as $-\frac{1}{2}(B(Z, [W, X]) - B(W, [Z, X]))$. Rewriting brackets as above, we get $-\frac{1}{2}(B(Z, [Z^\ell, X])B(W, X_\ell) - B(W, [Z^\ell, X])B(Z, X_\ell))$ so the last summand from above corresponds to $-\sum_\ell(Z \wedge W \otimes A)([Z^\ell, X], X_\ell)$. Extending X_ℓ by elements of $\mathfrak{p}$ to a basis of $\mathfrak{g}$ and correspondingly extending Z^ℓ to the dual basis, we again may sum over all elements without changing the result. This then makes sense for all bilinear maps of the above type and on the bundle level applied to κ gives the trace over the first and last entry of $-\kappa(\Pi(\{s_1, s_2\}), s_3)$. Thus, we see that the claimed expression is exactly the extension of $2\partial^*\kappa$ to a section of $\mathcal{A}^* \otimes \mathcal{A}$ which vanishes on $(\mathcal{A}^0)^* \otimes \mathcal{A}$, and the result follows. □

3.2. Structure theory and classification

In this section, we use the structure theory of semisimple Lie algebras developed in Chapter 2 to give a complete description of the available $|k|$–gradings on a semisimple Lie algebra in terms of the Dynkin diagram, respectively, the Satake diagram. We also describe realizations of the homogeneous space G/P using representation theory. Next, we study representations of the subalgebras $\mathfrak{p} = \mathfrak{g}^0$ coming from such gradings. Finally, we describe the Hasse diagram associated to the pair $(\mathfrak{g}, \mathfrak{p})$, which is a major tool in the study of parabolic geometries. As a first application, we describe the integral homology and cohomology of generalized flag varieties in terms of the Hasse diagram.

3.2.1. Complex parabolic subalgebras and complex $|k|$–gradings. The basic structural result on complex semisimple Lie algebras is that they are determined up to isomorphism by the associated root system, which in turn is completely determined by the configuration of a subsystem of simple roots. To get these data one has to make two choices, which, however, do not influence the result. Starting from a complex semisimple Lie algebra $\mathfrak{g}$, one first has to choose a Cartan subalgebra $\mathfrak{h} \subset \mathfrak{g}$, i.e. a maximal Abelian subalgebra such that the adjoint action $\mathrm{ad}(H) : \mathfrak{g} \to \mathfrak{g}$ is diagonalizable for any element $H \in \mathfrak{h}$; see 2.2.2. Cartan subalgebras exist and any two Cartan subalgebras are conjugate by an inner automorphism of $\mathfrak{g}$; see Theorem 2.2.2. Having chosen $\mathfrak{h}$, one gets a set of roots, i.e. the finite set Δ of linear functionals $\alpha \in \mathfrak{h}^*$, such that the *root space* $\mathfrak{g}_\alpha = \{A \in \mathfrak{g} : [H, A] = \alpha(H)A \quad \forall H \in \mathfrak{h}\}$ is nonzero. The Lie algebra $\mathfrak{g}$ decomposes as $\mathfrak{g} = \mathfrak{h} \oplus \bigoplus_{\alpha\in\Delta} \mathfrak{g}_\alpha$; see 2.2.2. The basic facts about this root decomposition of $\mathfrak{g}$ are, on one hand, that all of the root spaces $\mathfrak{g}_\alpha$ are one–dimensional and, on the other hand, that for $\alpha, \beta \in \Delta$ we have $[\mathfrak{g}_\alpha, \mathfrak{g}_\beta] = \mathfrak{g}_{\alpha+\beta}$ if $\alpha + \beta \in \Delta$ and $[\mathfrak{g}_\alpha, \mathfrak{g}_\beta] = 0$ if $\alpha + \beta \notin \Delta$; see 2.2.4. From the fact that the projections onto eigenspaces of an operator are polynomials in the operator, one concludes that any subalgebra of $\mathfrak{g}$ which contains the Cartan subalgebra $\mathfrak{h}$ is (as a vector space) automatically the direct sum of $\mathfrak{h}$ and some root spaces.

The subspace $\mathfrak{h}_0 \subset \mathfrak{h}$ on which all roots are real is a real form of $\mathfrak{h}$. Choosing an ordered basis $\{H_1, \dots, H_r\}$ of $\mathfrak{h}_0$ one defines a real linear functional $\phi : \mathfrak{h}_0 \to \mathbb{R}$ to be positive, if for some $i = 1, \dots, r$ one has $\phi(H_j) = 0$ for all $j < i$ and $\phi(H_i) > 0$. This induces a total ordering on $L(\mathfrak{h}_0, \mathbb{R})$ by defining $\phi < \psi$ if and only if $\psi - \phi$ is positive. In particular, one then gets the set Δ^+ of positive roots and Δ is the disjoint union of Δ^+ and $\{-\alpha : \alpha \in \Delta^+\}$; see 2.2.5 for details.

These two choices can be equivalently encoded as the choice of a *Borel subalgebra*, i.e. a maximal solvable subalgebra, $\mathfrak{b} \leq \mathfrak{g}$. In terms of the Cartan subalgebra $\mathfrak{h}$ and the positive system Δ^+, the associated Borel subalgebra $\mathfrak{b}$ is given as $\mathfrak{b} = \mathfrak{h} \oplus \mathfrak{n}_+$, where $\mathfrak{n}_+ := \bigoplus_{\alpha\in\Delta^+} \mathfrak{g}_\alpha$ is the sum of all positive root spaces. Obviously, $\mathfrak{n}_+$ is a nilpotent subalgebra of $\mathfrak{g}$ and $[\mathfrak{b}, \mathfrak{b}] \subset \mathfrak{n}_+$, so $\mathfrak{b}$ is solvable. On the other hand, any subalgebra of $\mathfrak{g}$ which strictly contains $\mathfrak{b}$ has to contain at least one negative root space, say $\mathfrak{g}_{-\alpha}$ for some $\alpha \in \Delta_+$. But then $\mathfrak{g}_{-\alpha} \oplus [\mathfrak{g}_{-\alpha}, \mathfrak{g}_\alpha] \oplus \mathfrak{g}_\alpha$ is a subalgebra isomorphic to the simple Lie algebra $\mathfrak{sl}(2, \mathbb{C})$; see 2.2.4. Since any subalgebra of a solvable Lie algebra is solvable, we conclude that $\mathfrak{b}$ really is a maximal solvable Lie subalgebra of $\mathfrak{g}$. The Borel subalgebra $\mathfrak{b}$ is called the *standard Borel subalgebra* associated to $\mathfrak{h}$ and $\Delta^+ \subset \Delta$.

The fact that Cartan subalgebras as well as the choice of positive roots are unique up to conjugation can be nicely rephrased as the fact that any two Borel subalgebras of a complex semisimple Lie algebra are conjugate by an inner automorphism of $\mathfrak{g}$; see 2.2.2.

DEFINITION 3.2.1. Let $\mathfrak{g}$ be a complex semisimple Lie algebra. A *parabolic subalgebra* $\mathfrak{p}$ of $\mathfrak{g}$ is a Lie subalgebra that contains a Borel subalgebra.

Fixing a choice of a Cartan subalgebra $\mathfrak{h}$ for $\mathfrak{g}$ and a system of positive roots, one obtains the corresponding standard Borel subalgebra $\mathfrak{b}$ as above. Subalgebras of $\mathfrak{g}$ containing this Borel subalgebra are called *standard parabolic subalgebras*. Since any Borel subalgebra is conjugate to $\mathfrak{b}$, any parabolic subalgebra of $\mathfrak{g}$ is conjugate to a standard one. To understand parabolic subalgebras it therefore suffices to deal with the standard parabolic subalgebras.

To give a complete description of all standard parabolic subalgebras of a complex semisimple Lie algebra $\mathfrak{g}$, we need one more ingredient from the structure theory: A positive root $\alpha \in \Delta^+$ is called a *simple root* if it cannot be written as the sum of two positive roots. The set of simple roots is denoted by Δ^0. Denoting the simple roots by $\alpha_1, \dots, \alpha_r$, one may write any root $\alpha \in \Delta$ uniquely as a linear combination $\alpha = a_1\alpha_1 + \dots + a_r\alpha_r$ with integral coefficients $a_1, \dots, a_r \in \mathbb{Z}$, which are either all ≥ 0 or all ≤ 0. Moreover, if $\alpha \in \Delta^+$ is not simple, then there is a simple root α_i such that $\alpha - \alpha_i \in \Delta$. Using this we can now state

PROPOSITION 3.2.1. *Let $\mathfrak{g}$ be a complex semisimple Lie algebra, $\mathfrak{h} \leq \mathfrak{g}$ a Cartan subalgebra, Δ the corresponding set of roots and Δ^0 the set of simple roots for some choice of a positive subsystem. Then standard parabolic subalgebras $\mathfrak{p} \leq \mathfrak{g}$ are in bijective correspondence with subsets $\Sigma \subset \Delta^0$.*

Explicitly, we associate to $\mathfrak{p}$ the subset $\Sigma_{\mathfrak{p}} = \{\alpha \in \Delta^0 : \mathfrak{g}_{-\alpha} \not\subset \mathfrak{p}\}$. Conversely, the standard parabolic subalgebra $\mathfrak{p}_\Sigma$ corresponding to a subset Σ is the sum of the standard Borel subalgebra $\mathfrak{b}$ and all negative root spaces corresponding to roots which can be written as a linear combination of elements of $\Delta^0 \setminus \Sigma$.

PROOF. Let $\Sigma \subset \Delta^0$ be an arbitrary subset and consider $\mathfrak{p}_\Sigma \subset \mathfrak{g}$. Using that for $\alpha, \beta \in \Delta$ such that $\alpha + \beta \in \Delta$ we have $[\mathfrak{g}_\alpha, \mathfrak{g}_\beta] = \mathfrak{g}_{\alpha+\beta}$, one easily verifies that $\mathfrak{p}_\Sigma$ is a subalgebra of $\mathfrak{g}$. But then $\mathfrak{p}_\Sigma$ is evidently a standard parabolic subalgebra.

Conversely, let $\mathfrak{p} \subset \mathfrak{g}$ be a standard parabolic subalgebra. By definition $\mathfrak{b} \subset \mathfrak{p}$ and we have noted above that since $\mathfrak{p}$ is a subalgebra containing $\mathfrak{h}$, it must be the direct sum of $\mathfrak{b}$ and some negative root spaces. Let $\Phi \subset \Delta^+$ be the set of all α such that $\mathfrak{g}_{-\alpha} \subset \mathfrak{p}$. For $\alpha, \beta \in \Phi$ such that $\alpha + \beta \in \Delta$, we have $\mathfrak{g}_{-\alpha-\beta} = [\mathfrak{g}_{-\alpha}, \mathfrak{g}_{-\beta}] \subset \mathfrak{p}$ and hence $\alpha + \beta \in \Phi$. Conversely, for $\alpha \in \Phi$ and $\beta \in \Delta^+$ such that $\alpha - \beta \in \Delta^+$, we have $\mathfrak{g}_{-(\alpha-\beta)} = [\mathfrak{g}_{-\alpha}, \mathfrak{g}_\beta]$, and hence $\alpha - \beta \in \Phi$.

Let us write $\Delta^0 = \{\alpha_1, \dots, \alpha_r\}$ and suppose that $\alpha \in \Phi$ is not simple and decomposes as $\alpha = a_1\alpha_1 + \dots + a_r\alpha_r$. Then there is a simple root α_i such that $\alpha - \alpha_i \in \Delta$ and hence in Δ^+, so from above we see that both α_i and $\alpha - \alpha_i$ lie in Φ. Inductively, this shows that for each i such that $a_i \neq 0$, we must have $\alpha_i \in \Phi$. Hence, we see that Φ is completely determined by $\Sigma := \Delta^0 \setminus (\Phi \cap \Delta^0)$, and that $\mathfrak{p} = \mathfrak{p}_\Sigma$. □

Note that the two obvious choices $\Sigma = \emptyset$ and $\Sigma = \Delta^0$ lead to the subalgebra $\mathfrak{g}$ and the standard Borel subalgebra, respectively. Note also that if $\Sigma \subset \Sigma' \subset \Delta^0$ are two subsets, then $\mathfrak{p}_{\Sigma'} \subset \mathfrak{p}_\Sigma$.

The description of standard parabolic subalgebras $\mathfrak{p} \subset \mathfrak{g}$ immediately suggest a relation to $|k|$–gradings as introduced in 3.1.2. Namely, having given the subset $\Sigma \subset \Delta^0 = \{\alpha_1, \dots, \alpha_r\}$ of simple roots and a root $\alpha \in \Delta$, we define the *Σ–height* $\mathrm{ht}_\Sigma(\alpha)$ of α by

$$\mathrm{ht}_\Sigma\left(\sum_i a_i\alpha_i\right) := \sum_{i:\alpha_i \in \Sigma} a_i.$$

For $0 \neq i \in \mathbb{Z}$ define $\mathfrak{g}_i := \bigoplus_{\alpha:\mathrm{ht}_\Sigma(\alpha)=i} \mathfrak{g}_\alpha$ and put $\mathfrak{g}_0 := \mathfrak{h} \oplus \bigoplus_{\alpha:\mathrm{ht}_\Sigma(\alpha)=0} \mathfrak{g}_\alpha$. Recall from above that we have a total ordering on the set of roots. In particular, there is a maximal root in this ordering, and we define k to be the Σ–height of this root. Of course, we then have $\mathfrak{g}_i = \{0\}$ for $|i| > k$, so the grading has the form $\mathfrak{g} = \mathfrak{g}_{-k} \oplus \dots \oplus \mathfrak{g}_k$.

THEOREM 3.2.1. *Let $\mathfrak{g}$ be a complex semisimple Lie algebra, $\mathfrak{h} \subset \mathfrak{g}$ a Cartan algebra with corresponding roots Δ, Δ^+ a set of positive roots and $\Delta^0 \subset \Delta^+$ the set of simple roots.*

(1) For any standard parabolic subalgebra $\mathfrak{p} \leq \mathfrak{g}$ corresponding to the subset $\Sigma \subset \Delta^0$, the decomposition $\mathfrak{g} = \mathfrak{g}_{-k} \oplus \cdots \oplus \mathfrak{g}_k$ according to Σ–height makes $\mathfrak{g}$ into a $|k|$–graded Lie algebra such that $\mathfrak{p} = \mathfrak{g}^0 = \mathfrak{g}_0 \oplus \cdots \oplus \mathfrak{g}_k$. Moreover, the subalgebra $\mathfrak{g}_0 \subset \mathfrak{g}$ is reductive and the dimension of its center $\mathfrak{z}(\mathfrak{g}_0)$ coincides with the number of elements of Σ.

(2) Conversely, for any $|k|$–grading $\mathfrak{g} = \mathfrak{g}_{-k} \oplus \cdots \oplus \mathfrak{g}_k$, the subalgebra $\mathfrak{g}^0$ is parabolic, and choosing a Cartan subalgebra and positive roots in such a way that $\mathfrak{g}^0$ is a standard parabolic subalgebra $\mathfrak{p}_\Sigma$, the grading is given by the Σ–height.

PROOF. (1) By the properties of the root decomposition, we have $[\mathfrak{h}, \mathfrak{g}_\alpha] = \mathfrak{g}_\alpha$, $[\mathfrak{g}_\alpha, \mathfrak{g}_{-\alpha}] \subset \mathfrak{h}$, and $[\mathfrak{g}_\alpha, \mathfrak{g}_\beta] = \mathfrak{g}_{\alpha+\beta}$ for roots α and β such that $\alpha + \beta \in \Delta$. Since obviously $\mathrm{ht}_\Sigma(\alpha + \beta) = \mathrm{ht}_\Sigma(\alpha) + \mathrm{ht}_\Sigma(\beta)$ in the latter case, it follows immediately that $[\mathfrak{g}_i, \mathfrak{g}_j] \subset \mathfrak{g}_{i+j}$, so we have defined a grading on $\mathfrak{g}$. Next, the subalgebra $\mathfrak{g}^0 = \mathfrak{g}_0 \oplus \cdots \oplus \mathfrak{g}_k$ by definition consists of $\mathfrak{h}$, all positive root spaces, and all negative root spaces corresponding to roots with zero Σ–height, so $\mathfrak{g}^0 = \mathfrak{p}_\Sigma$. Thus, it remains to verify that the Lie subalgebra $\mathfrak{g}_- = \mathfrak{g}_{-k} \oplus \cdots \oplus \mathfrak{g}_{-1}$ is generated by $\mathfrak{g}_{-1}$.

Let $\mathfrak{a} \subset \mathfrak{g}_-$ be the subalgebra generated by $\mathfrak{g}_{-1}$. Then $[\mathfrak{g}_{-1}, \mathfrak{a}] \subset \mathfrak{a}$ and $[\mathfrak{g}_0, \mathfrak{a}] \subset \mathfrak{a}$. If $\mathfrak{a} \neq \mathfrak{g}_-$, then there has to be a negative root α such that $\mathfrak{g}_\alpha \not\subset \mathfrak{a}$, and we choose a root α of maximal height (not Σ–height) with this property. Then there is a simple root α_i such that $\alpha + \alpha_i$ is a root, and by construction this root has a larger height than α, so $\mathfrak{g}_{\alpha+\alpha_i} \subset \mathfrak{a}$. But then $\mathfrak{g}_\alpha = [\mathfrak{g}_{-\alpha_i}, \mathfrak{g}_{\alpha+\alpha_i}]$, which is a contradiction since for $\alpha_i \in \Sigma$, we have $\mathfrak{g}_{-\alpha_i} \subset \mathfrak{g}_{-1}$ while for $\alpha_i \notin \Sigma$ we have $\mathfrak{g}_{-\alpha_i} \subset \mathfrak{g}_0$.

By construction, the subalgebra $\mathfrak{g}_0$ decomposes (as a vector space) into the direct sum of $\mathfrak{h}$ and the root spaces $\mathfrak{g}_\alpha$ such that $\mathrm{ht}_\Sigma(\alpha) = 0$. Now consider the subspace

$$\mathfrak{h}' := \{H \in \mathfrak{h} : \alpha_i(H) = 0 \quad \forall \alpha_i \in \Delta^0 \setminus \Sigma\}.$$

For $H \in \mathfrak{h}'$ and all α with $\mathrm{ht}_\Sigma(\alpha) = 0$, we have $\alpha(H) = 0$, and thus $\mathfrak{h}' \subset \mathfrak{z}(\mathfrak{g}_0)$. Since the simple roots form a basis of $\mathfrak{h}^*$, the dimension of $\mathfrak{h}'$ coincides with the number of elements of Σ.

On the other hand, for $\alpha_i \in \Delta^0$ we have the canonical element $H_{\alpha_i} \in \mathfrak{h}$, which generates $[\mathfrak{g}_{\alpha_i}, \mathfrak{g}_{-\alpha_i}]$ (see 2.2.4), and these elements form a basis of $\mathfrak{h}$. Defining $\mathfrak{h}''$ to be the span of the H_{α_i} for $\alpha_i \in \Delta^0 \setminus \Sigma$, the fact that $\alpha_i(H_{\alpha_i}) = 2$ implies $\mathfrak{h}' \cap \mathfrak{h}'' = \{0\}$, and thus $\mathfrak{h} = \mathfrak{h}' \oplus \mathfrak{h}''$ by dimensional reasons.

To prove that $\mathfrak{g}_0$ is reductive, we have to show that any solvable ideal of $\mathfrak{g}_0$ is contained in $\mathfrak{z}(\mathfrak{g}_0)$; see 2.1.2. Now suppose that I is any ideal in $\mathfrak{g}_0$ that is not contained in $\mathfrak{h}'$. Since the decomposition $\mathfrak{g}_0 = \mathfrak{h} \oplus \bigoplus \mathfrak{g}_\alpha$ is the decomposition into eigenspaces for the adjoint action of $\mathfrak{h}$ on $\mathfrak{g}_0$ and the projection onto an eigenspace of an operator is a polynomial in the operator, we conclude that I is the direct sum of a subspace of $\mathfrak{h}$ and some root spaces. If I is not contained in $\mathfrak{h}'$, then it has to contain at least one of the root spaces $\mathfrak{g}_\alpha$, since if it contains an element of $\mathfrak{h} \setminus \mathfrak{h}'$, then there is a root space on which this element acts nontrivially and that root space then is contained in I. But if I contains $\mathfrak{g}_\alpha$, then it also contains $[\mathfrak{g}_{-\alpha}, \mathfrak{g}_\alpha]$ and thus also $\mathfrak{g}_{-\alpha} = [[\mathfrak{g}_{-\alpha}, \mathfrak{g}_\alpha], \mathfrak{g}_{-\alpha}]$. Hence, I contains the subalgebra $\mathfrak{g}_{-\alpha} \oplus [\mathfrak{g}_{-\alpha}, \mathfrak{g}_\alpha] \oplus \mathfrak{g}_\alpha \cong \mathfrak{sl}(2, \mathbb{C})$ and cannot be solvable. Thus, we see that any solvable ideal I of $\mathfrak{g}_0$ is contained in $\mathfrak{h}' \subset \mathfrak{z}(\mathfrak{g}_0)$, so $\mathfrak{g}_0$ is reductive. Moreover, since

$\mathfrak{z}(\mathfrak{g}_0)$ is a solvable ideal in $\mathfrak{g}_0$, we also get $\mathfrak{z}(\mathfrak{g}_0) = \mathfrak{h}'$, which proves the claim about the dimension of the center. Finally, note that by construction the semisimple part $\mathfrak{g}_0^{ss} = [\mathfrak{g}_0, \mathfrak{g}_0]$ of $\mathfrak{g}_0$ is given by $\mathfrak{h}'' \oplus \bigoplus \mathfrak{g}_\alpha$.

(2) Assume that $\mathfrak{g} = \mathfrak{g}_{-k} \oplus \cdots \oplus \mathfrak{g}_k$ is a $|k|$–grading. By part (1) of Proposition 3.1.2, there is the grading element $E \in \mathfrak{z}(\mathfrak{g}_0)$ whose adjoint action restricts to multiplication by j on the grading component $\mathfrak{g}_j$. In particular, $\mathrm{ad}(E)$ is diagonalizable, so we may extend $\mathbb{C}E$ to a maximal abelian subalgebra of $\mathfrak{g}$ consisting of semisimple elements, thus obtaining a Cartan subalgebra that contains E. Since the eigenvalues of $\mathrm{ad}(E)$ are $\{-k, \dots, k\}$ we, in particular, conclude that E lies in the subspace of this Cartan subalgebra on which all roots are real. Taking a basis for this subspace which starts with E, the resulting positive roots have nonnegative values on E.

Conjugating by an appropriate inner automorphism, we may therefore assume that $E \in \mathfrak{h}$ and $\alpha(E) \geq 0$ for all $\alpha \in \Delta^+$. Then $[E, \mathfrak{h}] = 0$ which shows that $\mathfrak{h} \subset \mathfrak{g}_0$ and all positive root spaces lie in $\mathfrak{g}^0$, so $\mathfrak{g}^0 \subset \mathfrak{g}$ is a standard parabolic subalgebra. Since the element E lies in $\mathfrak{h}$, it acts by a scalar on each root space, so any root space is contained in some grading component. For $\alpha \in \Delta^0$ consider the root space $\mathfrak{g}_{-\alpha_j}$. By construction $\mathfrak{g}_{-\alpha_j} \in \mathfrak{g}_i$ for some $i \leq 0$. In fact, we must have either $i = 0$ or $i = -1$ since otherwise the fact that $\mathfrak{g}_-$ is generated by $\mathfrak{g}_{-1}$ contradicts the fact that α_i cannot be written as a nontrivial sum of positive roots. Consequently, all simple root spaces are contained either in $\mathfrak{g}_0$ or in $\mathfrak{g}_1$, and defining $\Sigma \subset \Delta^0$ to be the set of those simple roots whose root spaces lie in $\mathfrak{g}_1$, we see that $\mathfrak{g}^0$ is the standard parabolic $\mathfrak{p}_\Sigma$ and the grading is given by the Σ–height. □

This result immediately gives us additional information on the structure of $|k|$–graded Lie algebras. In particular, we see that the situation between the subalgebras $\mathfrak{g}_- = \mathfrak{g}_{-k} \oplus \cdots \oplus \mathfrak{g}_{-1}$ and $\mathfrak{p}_+ = \mathfrak{g}_1 \oplus \cdots \oplus \mathfrak{g}_k$ is completely symmetric, since changing the sign of the grading just amounts to using the opposite order for the roots, so $\mathfrak{g}_-$ and $\mathfrak{p}_+$ are isomorphic as Lie algebras. Moreover, we may conclude that the filtration of $\mathfrak{g}$ is completely determined by the parabolic subalgebra $\mathfrak{p} = \mathfrak{g}^0$. In particular, this implies that given a Lie group G with Lie algebra $\mathfrak{g}$, the parabolic subgroups corresponding to the grading as defined in 3.1.3 coincide with the parabolic subgroups used in representation theory.

COROLLARY 3.2.1. *Let $\mathfrak{g} = \mathfrak{g}_{-k} \oplus \cdots \oplus \mathfrak{g}_k$ be a $|k|$–graded semisimple Lie algebra over $\mathbb{K} = \mathbb{R}$ or $\mathbb{C}$. Then we have:*

(1) *For $i > 0$ we have $[\mathfrak{g}_{i-1}, \mathfrak{g}_1] = \mathfrak{g}_i$. In particular, the filtration component $\mathfrak{g}^i$ is the ith power of $\mathfrak{p}_+ = \mathfrak{g}^1$ and $\mathfrak{p}_+ \supset \mathfrak{g}^2 \supset \cdots \supset \mathfrak{g}^k$ is the lower central series of $\mathfrak{p}_+$.*

(2) *If for some $i < 0$, an element $X \in \mathfrak{g}_i$ satisfies $[X, Z] = 0$ for all $Z \in \mathfrak{g}_1$, then $X = 0$. If no simple ideal of $\mathfrak{g}$ is contained in $\mathfrak{g}_0$, this also holds for $i = 0$.*

(3) *The filtration component $\mathfrak{g}^1 = \mathfrak{p}_+$ is the nilradical of $\mathfrak{p} = \mathfrak{g}^0$.*

(4) *For any Lie group G with Lie algebra $\mathfrak{g}$, the parabolic subgroups defined in 3.1.3 are exactly the subgroups which lie between the normalizer $N_G(\mathfrak{p})$ of $\mathfrak{p}$ in G and its connected component of the identity.*

PROOF. As we have noted above, flipping the sign of the grading leads to an isomorphic $|k|$–graded Lie algebra in the complex case. Since the complexification of a real $|k|$–graded Lie algebra is a complex semisimple $|k|$–graded Lie algebra this also holds in the real case. From this, (1) and (2) follow immediately, since these are just parts (4) and (5) of Proposition 3.1.2 for the flipped grading.

(3) By the grading property, $\mathfrak{g}^1$ is a nilpotent ideal in $\mathfrak{p}$, and thus contained in the nilradical $\mathfrak{n}$ of $\mathfrak{p}$. In the complex case, since the Cartan subalgebra $\mathfrak{h}$ is contained in $\mathfrak{p}$, we conclude that any ideal in $\mathfrak{p}$ is the direct sum of a subspace of $\mathfrak{h}$ and some root spaces. Assume that $\mathfrak{n}$ contains a root space $\mathfrak{g}_\alpha \subset \mathfrak{g}_0$. Then $\mathfrak{n}$ also contains $[\mathfrak{g}_\alpha, \mathfrak{g}_{-\alpha}]$ and $[[\mathfrak{g}_\alpha, \mathfrak{g}_{-\alpha}], \mathfrak{g}_{-\alpha}] = \mathfrak{g}_{-\alpha}$. Hence, $\mathfrak{n}$ contains a subalgebra isomorphic to $\mathfrak{sl}(2, \mathbb{C})$, which is a contradiction.

Hence, $\mathfrak{n} \subset \mathfrak{h} \oplus \mathfrak{g}^1$. But if an element $H \in \mathfrak{h}$ lies in $\mathfrak{n}$, then it must have trivial bracket with all root spaces contained in $\mathfrak{g}_0$, since otherwise one of these root spaces would be contained in $\mathfrak{n}$. But, on the other hand, H must also have trivial bracket with any of the root spaces contained in $\mathfrak{g}^1$, since otherwise we would get nonzero brackets of arbitrary length, which again contradicts $\mathfrak{n}$ being nilpotent. Hence, $H = 0$, which completes the proof in the complex case. For the real case we just have to note that the complexification of the nilradical is a nilpotent ideal in the complexification and thus has to be contained in the nilradical of the complexification.

(4) Consider the subgroup $\tilde{P} := \bigcap_{i=-k}^{k} N_G(\mathfrak{g}^i) \subset G$, which by definition is contained in $N_G(\mathfrak{p})$. To complete the proof, it suffices to show that $\tilde{P} = N_G(\mathfrak{p})$. If $g \in N_G(\mathfrak{p})$, then $\mathrm{Ad}(g)(\mathfrak{g}^1) \subset \mathrm{Ad}(g)(\mathfrak{p}) = \mathfrak{p}$ is a nilpotent ideal, and hence contained in the nilradical $\mathfrak{g}^1$. For $i \geq 2$, the filtration component $\mathfrak{g}^i$ is just the ith power of $\mathfrak{g}^1$ by part (1), so $\mathrm{Ad}(g)(\mathfrak{g}^i) \subset \mathfrak{g}^i$ for all $i \geq 0$. For $i < 0$, we know from part (3) of Proposition 3.1.2 that we may characterize $\mathfrak{g}^i$ as the annihilator with respect to the Killing form of $\mathfrak{g}^{-i+1}$. Invariance of the Killing form then implies that $\mathrm{Ad}(g)(\mathfrak{g}^i) \subset \mathfrak{g}^i$ for $i < 0$, and thus $g \in \tilde{P}$. □

3.2.2. Notation. By Theorem 3.2.1, the classification of complex $|k|$–graded Lie algebras (up to conjugation by an inner automorphism of $\mathfrak{g}$) reduces to the classification of standard parabolic subalgebras of $\mathfrak{g}$, which in turn are determined by subsets $\Sigma \subset \Delta^0$ of simple roots. This immediately suggests the following notation.

DEFINITION 3.2.2 (Notation for complex parabolic subalgebras and for complex $|k|$–gradings). Let $\mathfrak{g}$ be a complex semisimple Lie algebra endowed with a Cartan subalgebra $\mathfrak{h} \subset \mathfrak{g}$ and a set Δ^+ of positive roots. Then we denote the standard parabolic subalgebra $\mathfrak{p}_\Sigma \subset \mathfrak{g}$ corresponding to $\Sigma \subset \Delta^0$ as well as the corresponding $|k|$–grading by Σ–height by representing in the Dynkin diagram of $\mathfrak{g}$ the nodes corresponding to elements of Σ by a cross instead of a dot.

From Theorem 3.2.1 we know that for any given $|k|$–grading the subalgebra $\mathfrak{g}_0 \subset \mathfrak{g}$ is reductive, so it is the direct sum of a semisimple Lie algebra $\mathfrak{g}_0^{ss}$ and the center $\mathfrak{z}(\mathfrak{g}_0)$. From the Dynkin diagram representing the $|k|$–grading, we immediately get a complete description of the structure of $\mathfrak{g}_0$.

PROPOSITION 3.2.2. *Let $\mathfrak{g} = \mathfrak{g}_{-k} \oplus \cdots \oplus \mathfrak{g}_k$ be a complex $|k|$–graded Lie algebra. Then the dimension of the center of $\mathfrak{g}_0$ coincides with the number of crosses in the diagram describing the $|k|$–grading, and the Dynkin diagram of the semisimple part $\mathfrak{g}_0^{ss}$ is obtained by removing all crossed nodes and all edges connected to crossed nodes.*

PROOF. By Theorem 3.2.1, the dimension of $\mathfrak{z}(\mathfrak{g}_0)$ equals the number of elements of Σ and hence the number of crosses in the Dynkin diagram representing the $|k|$–grading. In the proof of Theorem 3.2.1, we have seen that $\mathfrak{g}_0^{ss}$ is the direct sum of the subspace $\mathfrak{h}'' \subset \mathfrak{h}$ spanned by the elements H_{α_i} for $\alpha_i \in \Delta^0 \setminus \Sigma$ and the root spaces $\mathfrak{g}_\alpha$ corresponding to roots α of Σ–height zero. A root of Σ–height zero is by definition an integral linear combination of simple roots contained in $\Delta^0 \setminus \Sigma$, and by definition any such root vanishes on $\mathfrak{h}' \subset \mathfrak{h}$. This immediately implies that $\mathfrak{h}'' \subset \mathfrak{g}_0^{ss}$ is a Cartan subalgebra, and $\mathfrak{g}_0^{ss} = \mathfrak{h}'' \oplus \bigoplus_{\mathrm{ht}_\Sigma(\alpha)=0} \mathfrak{g}_\alpha$ is the corresponding root decomposition of $\mathfrak{g}_0^{ss}$.

Using the induced order on $\mathfrak{h}''$, the positive roots of $\mathfrak{g}_0^{ss}$ are exactly the positive roots of $\mathfrak{g}$ of Σ–height zero, and the corresponding simple roots are exactly the elements of $\Delta^0 \setminus \Sigma$. To determine the Dynkin diagram of $\mathfrak{g}_0^{ss}$, we only have to compute the entries $a_{ij} = \frac{2\langle \alpha_i, \alpha_j \rangle}{\langle \alpha_i, \alpha_i \rangle}$ of the Cartan matrix; see 2.2.5. The inner products in this expression are induced by the Killing form on $\mathfrak{g}_0^{ss}$. Replacing this by the Killing form of $\mathfrak{g}$, we obtain the Cartan integers of $\mathfrak{g}$, so we have to compare the two Killing forms. By part (1) of Corollary 2.1.5, $\mathfrak{g}_0^{ss}$ decomposes into a sum of simple ideals, and this decomposition is compatible with the root decomposition.

Suppose first that $\alpha_i, \alpha_j \in \Delta^0 \setminus \Sigma$ are such that $\mathfrak{g}_{\alpha_i}$ and $\mathfrak{g}_{\alpha_j}$ lie in different simple ideals of $\mathfrak{g}_0^{ss}$. Then the two roots are orthogonal with respect to the Killing form of $\mathfrak{g}_0^{ss}$. On the other hand, $[\mathfrak{g}_{\alpha_i}, \mathfrak{g}_{\alpha_j}] = 0$, so $\alpha_i + \alpha_j$ cannot be a root of $\mathfrak{g}$. Since $\alpha_i - \alpha_j$ cannot be a root of $\mathfrak{g}$ either, the two roots must also be orthogonal with respect to the Killing form of $\mathfrak{g}$; see part (4) of Proposition 2.2.4.

On the other hand, on a simple Lie algebra the Killing form is uniquely determined up to scale by invariance; see part (3) of Corollary 2.1.5. Since the restriction of the Killing form of $\mathfrak{g}$, is invariant, too, the numbers a_{ij} from above coincide with the Cartan integers of $\mathfrak{g}$. □

3.2.3. Complex $|1|$–gradings. Let us start by considering the simplest case of $|1|$–gradings. These $|1|$–gradings are also of interest because of their relation to Hermitian symmetric spaces, which will be discussed in 3.2.7. First, we clarify how to reduce from the semisimple case to the simple case.

LEMMA 3.2.3. *Let $\mathfrak{g} = \mathfrak{g}_{-1} \oplus \mathfrak{g}_0 \oplus \mathfrak{g}_1$ be a $|1|$–graded semisimple Lie algebra such that no simple ideal is contained in $\mathfrak{g}_0$, then:*

(i) *$\mathfrak{g}$ is the sum of $|1|$–graded simple Lie algebras $\mathfrak{g}^{(j)}$.*
(ii) *The dimension of the center $\mathfrak{z}(\mathfrak{g})$ coincides with the number of simple factors.*
(iii) *The decomposition of the $\mathfrak{g}_0$–module $\mathfrak{g}_{-1}$ into irreducible components is given by $\mathfrak{g}_{-1} = \bigoplus_j \mathfrak{g}_{-1}^{(j)}$.*

PROOF. By part (2) of Proposition 3.1.2, any simple ideal of $\mathfrak{g}$ inherits a $|1|$–grading or has to be contained in $\mathfrak{g}_0$ and the second case is ruled out by assumption, so (i) follows. Writing the decomposition into simple ideals as $\mathfrak{g} = \bigoplus_j \mathfrak{g}^{(j)}$, we get $\mathfrak{g}_i = \bigoplus_j \mathfrak{g}_i^{(j)}$ for $i = -1, 0, 1$ and also $\mathfrak{z}(\mathfrak{g}_0) = \bigoplus_j \mathfrak{z}(\mathfrak{g}_0^{(j)})$. To complete the proof, it therefore suffices to show that for a simple $|1|$–graded Lie algebra $\mathfrak{g}$, the center $\mathfrak{z}(\mathfrak{g}_0)$ has dimension 1 and $\mathfrak{g}_{-1}$ is an irreducible $\mathfrak{g}_0$–module.

Representing the highest root as a linear combination of simple roots, any simple root has a nonzero coefficient. Hence, a $|1|$–grading on a simple Lie algebra can be only obtained if a single simple root is crossed out. Then $\dim(\mathfrak{z}(\mathfrak{g}_0)) = 1$ by Theorem 3.2.1. Since we deal with a $|1|$–grading, the subalgebra $\mathfrak{p}_+ = \mathfrak{g}_1$ acts

$\mathfrak{g}$	root	Dynkin diagram	$\mathfrak{g}_0^{ss}$
A_ℓ, $\ell \geq 1$	α_1		$A_{\ell-1}$
A_ℓ, $\ell \geq 3$	α_i, $1 < i \leq (\ell+1)/2$		$A_{i-1} \times A_{\ell-i}$
B_ℓ, $\ell \geq 2$	α_1		$B_{\ell-1}$
C_ℓ, $\ell \geq 3$	α_ℓ		$A_{\ell-1}$
D_ℓ, $\ell \geq 4$	α_1	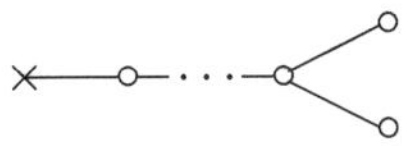	$D_{\ell-1}$
D_ℓ, $\ell \geq 5$	α_n	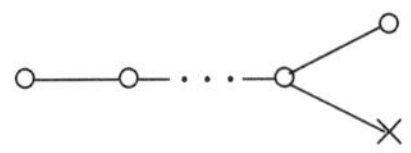	$A_{\ell-1}$
E_6	α_1	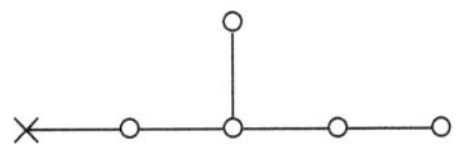	D_5
E_7	α_1	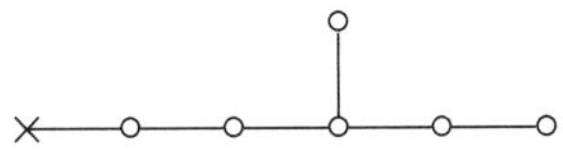	E_6

Table of complex $|1|$–gradings in Dynkin diagram notation.

trivially on $\mathfrak{g}/\mathfrak{p}$. Hence, the $\mathfrak{p}$–representation $\mathfrak{g}/\mathfrak{p}$ is obtained by extending the $\mathfrak{g}_0$–representation $\mathfrak{g}_{-1}$ trivially to $\mathfrak{p}$. In particular, $\mathfrak{g}/\mathfrak{p}$ is completely reducible as a $\mathfrak{p}$–module. But from the construction it is clear that in the simple case $\mathfrak{g}_{-1}$ is generated as a $\mathfrak{g}_0$–module by the root space corresponding to the unique crossed root, so $\mathfrak{g}/\mathfrak{p}$ is an irreducible $\mathfrak{p}$–module. □

PROPOSITION 3.2.3. *The complete list of non-isomorphic $|1|$–gradings of complex simple Lie algebras is given in the table above.*

PROOF. We have seen in the proof of the claim above that a $|1|$–grading in the simple case can be only obtained by crossing out a single node in the Dynkin diagram. Moreover, crossing out one node, we obtain a $|1|$–grading if and only if the coefficient of the corresponding root in the representation of the highest root is one, so $|1|$–gradings can be classified by looking at the expressions for the highest roots.

In the A_ℓ case, i.e. $\mathfrak{g} = \mathfrak{sl}(\ell+1, \mathbb{C})$, with simple roots $\alpha_1, \dots, \alpha_\ell$, the highest root is $\alpha_1 + \dots + \alpha_\ell$ by example (1) of 2.2.6. Thus, any choice of a single crossed node leads to a $|1|$–grading. Notice that choosing the ith node and the $(\ell - i)$th node is related by an automorphism of the Dynkin diagram and thus by an outer automorphism of $\mathfrak{g}$, so one obtains isomorphic $|1|$–graded Lie algebras in the two cases. Hence, there are $\ell/2$ different possibilities for even ℓ and $(\ell+1)/2$ different possibilities for odd ℓ.

In the case of B_ℓ, i.e. $\mathfrak{g} = \mathfrak{so}(2\ell+1, \mathbb{C})$ with simple roots $\alpha_1, \dots, \alpha_\ell$ numbered according to example (3) of 2.2.6, we know from this example that the highest root is $\alpha_1 + 2\alpha_2 + \dots + 2\alpha_\ell$. Thus, there is only one possibility for a $|1|$–grading here, namely crossing the first node in the Dynkin diagram.

For type C_ℓ, i.e. $\mathfrak{g} = \mathfrak{sp}(2\ell, \mathbb{C})$, the relevant data are in example (4) of 2.2.6. For the numbering $\alpha_1, \dots, \alpha_\ell$ of the simple roots used there, the highest root is

$2\alpha_1 + \cdots + 2\alpha_{\ell-1} + \alpha_\ell$. As for B_ℓ, there is only one $|1|$–grading available here, namely the one corresponding to crossing the last root.

Type D_ℓ, i.e. $\mathfrak{g} = \mathfrak{so}(2\ell, \mathbb{C})$, has been treated in example (2) of 2.2.6. For the numbering $\alpha_1, \dots, \alpha_\ell$ of simple roots introduced there, the highest root is given by $\alpha_1 + 2\alpha_2 + \cdots + 2\alpha_{\ell-2} + \alpha_{\ell-1} + \alpha_\ell$, so there are three possible $|1|$–gradings in this case. For $\ell = 4$, all these three possibilities are conjugate by automorphisms of the Dynkin diagram and thus by outer automorphisms of $\mathfrak{g}$, so up to isomorphism there is only one $|1|$–graded simple Lie algebra of type D_4. For $\ell > 4$, crossing α_{n-1} and α_n differs only by an automorphism of the Dynkin diagram, so there are up to isomorphism exactly two $|1|$–graded simple Lie algebras of type D_ℓ with $\ell > 4$. For the exceptional Lie algebras, the expressions for the highest roots can be found in Table B.2 in Appendix B. □

3.2.4. Complex contact gradings. These are a special class of complex $|2|$–gradings. From the point of view of parabolic geometries, they are interesting since the corresponding geometries have an underlying complex contact structure. They are also of independent interest due to their relation to quaternionic symmetric spaces, which will be discussed in 3.2.7.

A *contact grading* is a $|2|$–grading $\mathfrak{g} = \mathfrak{g}_{-2} \oplus \cdots \oplus \mathfrak{g}_2$ such that the dimension of $\mathfrak{g}_{\pm 2}$ is equal to one and such that the bracket $[\ ,\ ] : \mathfrak{g}_{-1} \times \mathfrak{g}_{-1} \to \mathfrak{g}_{-2}$ is nondegenerate. Explicitly, this means that if $X \in \mathfrak{g}_{-1}$ is such that $[X, Y] = 0$ for all $Y \in \mathfrak{g}_{-1}$, then $X = 0$.

PROPOSITION 3.2.4. *Contact gradings can only exist on simple (and not on general semisimple) Lie algebras. On each complex simple Lie algebra of rank larger than one, there is a unique (up to inner automorphism) contact grading. In Dynkin diagram notation, the complete list of complex contact gradings is given by*

$\mathfrak{g}$	Σ	*Dynkin diagram*	$\mathfrak{g}_0$
A_ℓ, $\ell \geq 2$	$\{\alpha_1, \alpha_\ell\}$		$\mathbb{C}^2 \oplus A_{\ell-2}$
B_ℓ, $\ell \geq 3$	$\{\alpha_2\}$		$\mathbb{C} \oplus A_1 \oplus B_{\ell-2}$
C_ℓ, $\ell \geq 2$	$\{\alpha_1\}$		$\mathbb{C} \oplus C_{\ell-1}$
D_4	$\{\alpha_2\}$	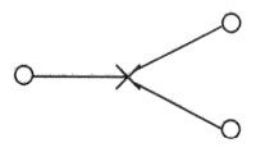	$\mathbb{C} \oplus A_1 \oplus A_1 \oplus A_1$
D_ℓ, $\ell \geq 5$	$\{\alpha_2\}$	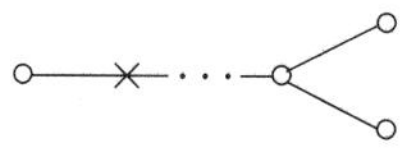	$\mathbb{C} \oplus A_1 \oplus D_{\ell-2}$
E_6	$\{\alpha_6\}$	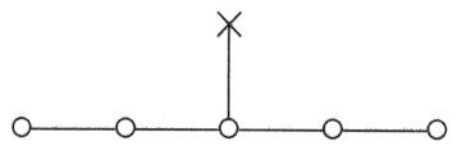	$\mathbb{C} \oplus A_5$
E_7	$\{\alpha_6\}$	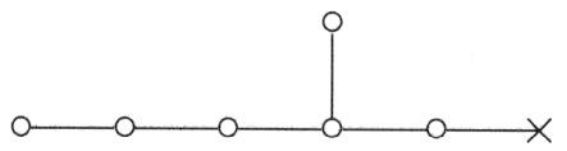	$\mathbb{C} \oplus D_6$
E_8	$\{\alpha_1\}$	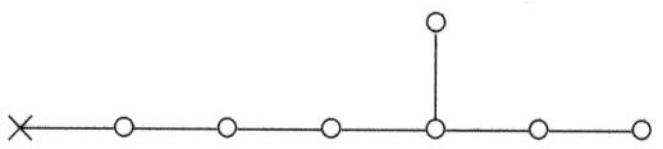	$\mathbb{C} \oplus E_7$
F_4	$\{\alpha_4\}$		$\mathbb{C} \oplus C_3$
G_2	$\{\alpha_2\}$		$\mathbb{C} \oplus A_1$

PROOF. If $\mathfrak{g}$ is a $|2|$–graded semisimple Lie algebra such that no simple ideal of $\mathfrak{g}$ is contained in $\mathfrak{g}_0$, then by part (2) of Proposition 3.1.2, $\mathfrak{g}$ splits into a direct sum of $|1|$–graded and $|2|$–graded simple ideals. If $\mathfrak{g}_{-2}$ is one–dimensional, then only one of the simple ideals may be $|2|$–graded. But then the (-1)–components of the other simple ideals lie in the null space of the bracket $\mathfrak{g}_{-1} \times \mathfrak{g}_{-1} \to \mathfrak{g}_2$, so contact gradings exist only on simple Lie algebras.

If $\mathfrak{g} = \mathfrak{g}_{-2} \oplus \cdots \oplus \mathfrak{g}_2$ is a contact grading on a simple Lie algebra, then $\mathfrak{g}_2$ consists only of the root space of the highest root. Denoting this highest root by α_0, we have the corresponding element $H_{\alpha_0} \in \mathfrak{h}$, which is dual to α_0 with respect to the Killing form. The element $H := \frac{2}{\langle \alpha_0, \alpha_0 \rangle} H_{\alpha_0}$ has the property that $[H, X] = 2X$ for all $X \in \mathfrak{g}_{\alpha_0}$ and $[H, Y] = -2Y$ for all $Y \in \mathfrak{g}_{-\alpha_0}$; see 2.2.4. We claim that H is the grading element for the given grading.

For any root $\alpha \in \Delta$, we obtain $\alpha(H) = \frac{2\langle \alpha, \alpha_0 \rangle}{\langle \alpha_0, \alpha_0 \rangle}$, which is an integer, see again 2.2.4. We further know from there that $\alpha(H) = p - q$, where the α_0–string through α, i.e. the set of all roots of the form $\alpha + m\alpha_0$ for $m \in \mathbb{Z}$, has the form $\alpha - p\alpha_0, \dots, \alpha + q\alpha_0$. For $\alpha \in \Delta^+$ we must have $q = 0$ and thus $\alpha(H) \geq 0$ since α_0 is the highest root. Also, $-\alpha_0$ is the lowest root, so $p \leq 2$ and $p = 2$ is only possible for $\alpha = \alpha_0$. Hence, we see that the adjoint action of H on any positive root space $\mathfrak{g}_\alpha$ with $\alpha \neq \alpha_0$ is by multiplication by zero or one. Since we deal with a contact grading, the bracket $[\ ,\ ] : \mathfrak{g}_1 \times \mathfrak{g}_1 \to \mathfrak{g}_2$ is nondegenerate as well, so for $0 \neq Z \in \mathfrak{g}_1$ we can find an element $W \in \mathfrak{g}_1$ such that $[Z, W] \neq 0$. But then $2[Z, W] = [H, [Z, W]] = [[H, Z], W] + [Z, [H, W]]$ is possible only for $[H, Z] = Z$ and $[H, W] = W$, so $\mathrm{ad}(H)$ is the identity on $\mathfrak{g}_1$. Finally, for $A \in \mathfrak{g}_0$ and $Z \in \mathfrak{g}_1$ we get $[A, Z] \in \mathfrak{g}_1$ and thus $[H, [A, Z]] = [A, Z]$ and since $[H, Z] = Z$, this implies $[[H, A], Z] = 0$. Since this holds for all $Z \in \mathfrak{g}_1$ we conclude $[H, A] = 0$ from part (2) of Corollary 3.2.1, and the claim follows.

Conversely, let $\mathfrak{g}$ be an arbitrary complex simple Lie algebra different from $\mathfrak{sl}(2, \mathbb{C})$. Let α_0 be the highest root of $\mathfrak{g}$ and put $H = \frac{2}{\langle \alpha_0, \alpha_0 \rangle} H_{\alpha_0}$. Since $H \in \mathfrak{h}$, the adjoint action of H on $\mathfrak{g}$ is diagonalizable. The above arguments show that the eigenvalues are integers between -2 and 2, on each positive root space the eigenvalue is positive, and the eigenspaces for the eigenvalues ± 2 are $\mathfrak{g}_{\pm\alpha_0}$. Thus, decomposing $\mathfrak{g}$ into eigenspaces for $\mathrm{ad}(H)$ defines a $|2|$–grading with $\dim(\mathfrak{g}_{\pm 2}) = 1$. Since $\mathfrak{g} \neq \mathfrak{sl}(2, \mathbb{C})$, the root α_0 is not simple. Hence, $\mathfrak{g}_{\pm 1} \neq 0$, and $\mathfrak{g}_-$ is generated by $\mathfrak{g}_{-1}$. If $\alpha \in \Delta$ is a root such that $\mathfrak{g}_\alpha \subset \mathfrak{g}_1$, then $\alpha(H_{\alpha_0}) = 1$ implies that $\alpha_0 - \alpha$ is a root. Since $(\alpha_0 - \alpha)(H_{\alpha_0}) = 1$, we see that $\mathfrak{g}_{\alpha_0 - \alpha} \subset \mathfrak{g}_1$. But then $[\mathfrak{g}_\alpha, \mathfrak{g}_{\alpha_0 - \alpha}] = \mathfrak{g}_{\alpha_0}$ implies nondegeneracy of the bracket $\mathfrak{g}_1 \times \mathfrak{g}_1 \to \mathfrak{g}_2$, so we have constructed a contact grading.

For $\mathfrak{sl}(2, \mathbb{C})$ this construction gives a decomposition $\mathfrak{g} = \mathfrak{g}_{-2} \oplus \mathfrak{g}_0 \oplus \mathfrak{g}_2$, which is not useful for our purposes. On any other complex simple Lie algebra, we obtain a unique contact grading. To find the Dynkin diagrams for these gradings, one has to compute the inner products of the highest root with the simple roots. The crossed nodes are exactly those nodes which are connected to the node representing the highest root in the extended Dynkin diagrams; see 2.2.8. □

3.2.5. Complex parabolics as stabilizers of lines and flags. The theorem of the highest weight in 2.2.12 describes all finite–dimensional irreducible representations of a complex semisimple Lie algebra. Given a choice of a Cartan subalgebra and of positive roots, any such representation contains a distinguished complex line consisting of highest weight vectors. As we shall see, the stabilizer subalgebra of this line is always a standard parabolic subalgebra and all standard parabolics are obtained in this way. For the classical simple Lie algebras, this can be used to obtain a fairly uniform description of all standard parabolic subalgebras as stabilizers flags in the standard representation. This will also be very useful for the description of the homogeneous models of parabolic geometries.

Let $\mathfrak{g}$ be a complex semisimple Lie algebra, $\mathfrak{h} \leq \mathfrak{g}$ a Cartan subalgebra, Δ the corresponding set of roots. Assume we have chosen an ordering on $\mathfrak{h}^*$, and let Δ^+ be the corresponding set of positive roots. If V is a finite–dimensional representation of $\mathfrak{g}$, and $\lambda : \mathfrak{h} \to \mathbb{C}$ is maximal among all the weights of V, then there is a nonzero vector $v_0 \in V$, which is a highest weight vector of weight λ. This means, that for $X \in \mathfrak{g}_\alpha$ with $\alpha \in \Delta^+$ we have $X \cdot v_0 = 0$, while for $H \in \mathfrak{h}$ we have $H \cdot v_0 = \lambda(H)v_0$. If V is irreducible, then (up to scale) v_0 is the unique highest weight vector in V. Observe that for any nonzero vector $v \in V$, the space of all elements $A \in \mathfrak{g}$ such that $A \cdot v$ is a multiple of v is a subalgebra of $\mathfrak{g}$. We will refer to this subalgebra as the stabilizer of the line through v.

PROPOSITION 3.2.5. *Let $\mathfrak{g}$ be a complex semisimple Lie algebra, V a complex irreducible representation of $\mathfrak{g}$ of highest weight λ, and $v_0 \in V$ a highest weight vector. Then the stabilizer of the line through v_0 is the standard parabolic subalgebra $\mathfrak{p}_\Sigma \subset \mathfrak{g}$ corresponding to the set $\Sigma = \{\alpha \in \Delta^0 : \langle \lambda, \alpha \rangle \neq 0\}$ of simple roots.*

PROOF. Let $\mathfrak{p}$ be the stabilizer of the line through v_0. By definition of a highest weight vector, $\mathfrak{p}$ contains $\mathfrak{h}$ and all positive root spaces, so it is a standard parabolic subalgebra. By 3.2.1 the corresponding set of simple roots is $\{\alpha \in \Delta^0 : \mathfrak{g}_{-\alpha} \not\subset \mathfrak{p}\}$.

To complete the proof we thus have to show that for a simple root α and a nonzero element $X \in \mathfrak{g}_{-\alpha}$ we have $X \cdot v_0 = 0$ if and only if $\langle \lambda, \alpha \rangle = 0$. Note first that we can choose $Y \in \mathfrak{g}_\alpha$ such that $[Y, X] = H_\alpha \in \mathfrak{h}$, the element dual to α. By definition $Y \cdot v_0 = 0$, so $Y \cdot X \cdot v_0 = [Y, X] \cdot v_0 = \lambda(H_\alpha)v_0$. Again by definition, $\lambda(H_\alpha) = \langle \lambda, \alpha \rangle$, so we see that $X \cdot v_0 \neq 0$ unless the inner product vanishes.

On the other hand, if $X \cdot v_0 \neq 0$, then it is a weight vector of weight $\lambda - \alpha$. From 2.2.10 we know that the set of weights of V is invariant under the Weyl group. In particular, denoting by s_α the reflection corresponding to α we conclude that $s_\alpha(\lambda - \alpha) = s_\alpha(\lambda) + \alpha$ is a weight of V. But if $\langle \lambda, \alpha \rangle = 0$, then $s_\alpha(\lambda) = \lambda$, so $X \cdot v_0 \neq 0$ would lead to a contradiction to λ being the highest weight of V. □

There is a useful simple way to rephrase the description of the subset Σ. Recall from 2.2.10 that the dominant integral weights for $\mathfrak{g}$ are exactly the linear combinations of the fundamental weights with nonnegative integral coefficients. The fundamental weights are the dual elements to the simple roots, so for $\alpha \in \Delta^0$ the coefficient of the corresponding fundamental weight in the expansion of λ is $\frac{2\langle \lambda, \alpha \rangle}{\langle \alpha, \alpha \rangle}$. In particular, the set Σ consists of those simple roots for which the corresponding fundamental weights have a nonzero coefficient in the expansion of λ. In the Dynkin diagram notation for representations this means that the "right" parabolic subalgebra is obtained by crossing those nodes of the Dynkin diagram over which there is a nonzero number.

Hence, the simplest way to realize the standard parabolic subalgebra $\mathfrak{p}$ corresponding to $\Sigma \subset \Delta^0$ as the stabilizer of a highest weight line is to use for the highest weight the sum of the fundamental weights corresponding to the elements of Σ. In particular, maximal parabolic subalgebras (corresponding to a one element subset Σ) can be realized using fundamental representations.

There is an alternative description of arbitrary parabolics in terms of maximal parabolics, which is based on the following simple observation: Let $\mathfrak{g}$ be a complex semisimple Lie algebra, let $\Sigma', \Sigma'' \subset \Delta^0$ be some sets of simple roots, put $\Sigma = \Sigma' \cup \Sigma''$, and let $\mathfrak{p}'$, $\mathfrak{p}''$, and $\mathfrak{p}$ be the associated standard parabolic subalgebras of $\mathfrak{g}$. Then each of the parabolics is a direct sum of the Cartan subalgebra $\mathfrak{h}$ and some root spaces. A root space is contained in $\mathfrak{p}$ if and only if the Σ–height of the corresponding root is nonnegative, which clearly is the case if and only if both it's Σ'–height and its Σ''–height are nonnegative. This immediately implies that $\mathfrak{p} = \mathfrak{p}' \cap \mathfrak{p}''$. Similarly, the Levi component is formed by $\mathfrak{h}$ and all root spaces corresponding to roots of Σ–height zero, which immediately implies that $\mathfrak{g}_0 = \mathfrak{g}_0' \cap \mathfrak{g}_0''$. On the other hand, the subalgebra $\mathfrak{p}_+$ consists of all root spaces of positive Σ–height, which implies $\mathfrak{p}_+ = \mathfrak{p}_+' + \mathfrak{p}_+''$. Using this, we can now describe all standard parabolic subalgebras of the classical simple Lie algebras in terms of stabilizers of flags.

Let us start with the case $\mathfrak{g} = A_{n-1} = \mathfrak{sl}(n, \mathbb{C})$, using the Cartan subalgebra and positive roots from example (1) of 2.2.6. From 2.2.13 we know that the kth fundamental representation of $\mathfrak{g}$ is the kth exterior power $\Lambda^k V$ of the standard representation $V = \mathbb{C}^n$ of $\mathfrak{g}$. Moreover, denoting by $\{v_1, \dots, v_n\}$ the standard basis of V, the element $v_1 \wedge \dots \wedge v_k$ is a highest weight vector in $\Lambda^k V$. Elementary linear algebra shows that a matrix $A \in \mathfrak{g}$ maps the element $v_1 \wedge \dots \wedge v_k$ to a multiple of itself if and only if it maps any of the vectors v_i for $i = 1, \dots, k$ to a linear combination of $v_1, \dots, v_k$. Otherwise put, the stabilizer of the line through $v_1 \wedge \dots \wedge v_k$ coincides with the stabilizer of the subspace $\mathbb{C}^k \subset \mathbb{C}^n$ generated by $\{v_1, \dots, v_k\}$. Thus, the standard parabolic corresponding to $\Sigma = \{\alpha_k\}$ is the stabilizer of $\mathbb{C}^k \subset \mathbb{C}^n$.

In matrix terms this means that the maximal standard parabolic corresponding to α_k is the subalgebra of all tracefree block upper triangular matrices with block sizes k and $n-k$. The subalgebras $\mathfrak{g}_0$ and $\mathfrak{p}_+$ of $\mathfrak{p}$ are the subalgebras of all block diagonal matrices, respectively, strictly block upper triangular matrices with the same block sizes.

Therefore, the standard parabolic corresponding to $\Sigma = \{\alpha_{i_1}, \dots, \alpha_{i_\ell}\}$ with $1 \le i_1 < i_2 < \dots < i_\ell < n$ is the intersection of the stabilizers the subspaces $\mathbb{C}^{i_j} \subset \mathbb{C}^n$, i.e. the stabilizer of the standard flag $\mathbb{C}^{i_1} \subset \mathbb{C}^{i_2} \subset \dots \subset \mathbb{C}^{i_\ell} \subset \mathbb{C}^n$. In matrix terms, these are block upper triangular matrices with block sizes i_1, $i_2 - i_1$, $\dots i_\ell - i_{\ell-1}$, $n - i_\ell$, and the subalgebras $\mathfrak{g}_0$ and $\mathfrak{p}_+$ consist of the block diagonal matrices respectively the strictly block upper triangular matrices with these block sizes.

For the symplectic algebra $C_n = \mathfrak{sp}(2n, \mathbb{C})$ the situation is completely parallel. As in example (4) of 2.2.6 we consider the standard representation $V = \mathbb{C}^{2n}$ with standard basis $\{v_1, \dots, v_{2n}\}$ and the nondegenerate skew symmetric bilinear form ω such that $\omega(v_i, v_j) = \delta_{j,n+i}$ for $1 \le i \le n$. The description of the fundamental weights in 2.2.13 implies that $v_1 \wedge \dots \wedge v_k \in \Lambda^k V$ is a highest weight vector whose weight is the kth fundamental weight for $k = 1, \dots n$. So while the exterior powers are no more irreducible in this case, these highest weight vectors generate irreducible subrepresentations. Hence, we can still characterize the standard parabolics as the stabilizers of the corresponding lines. As in the A_n case, a matrix preserves the vector $v_1 \wedge \dots \wedge v_k$ up to scale if and only if it preserves the subspace $\mathbb{C}^k \subset V$

generated by $\{v_1, \dots, v_k\}$. Notice that this subspace now is not arbitrary but isotropic, i.e. the restriction of the bilinear form to the subspace vanishes identically. The passage to general parabolics is exactly as before, so we see that the standard parabolics in $\mathfrak{sp}(2n, \mathbb{C})$ can be characterized as the stabilizers of isotropic flags in the standard representation.

For the odd orthogonal algebra $\mathfrak{g} = B_n = \mathfrak{so}(2n+1, \mathbb{C})$ we follow example (3) of 2.2.6. Denote by $V = \mathbb{C}^{2n+1}$ the standard representation with standard basis $\{v_1, \dots, v_{2n+1}\}$. The inner product is such that the subspaces generated by $\{v_1, \dots, v_n\}$ and $\{v_{n+1}, \dots, v_{2n}\}$ are isotropic and $\langle v_i, v_{n+j} \rangle = \delta_{i,j}$ for $i, j = 1, \dots, n$, while $\langle v_{2n+1}, v_j \rangle = \delta_{j,2n+1}$. From 2.2.13 we see that the vector $v_1 \wedge \cdots \wedge v_k \in \Lambda^k V$ is a highest weight vector whose weight coincides with the kth fundamental weight for $k = 1, \dots, n-1$. On the other hand, $v_1 \wedge \cdots \wedge v_n \in \Lambda^n V$ is a highest weight vector whose weight is twice the nth fundamental weight. As before we conclude that the maximal standard parabolic corresponding to α_k can be realized as the stabilizer of the isotropic subspace of V which is generated by $\{v_1, \dots, v_k\}$, and all standard parabolics can be realized as the stabilizers of flags of nested isotropic subspaces of the standard representation.

The last case of the even orthogonal algebra $\mathfrak{g} = D_n = \mathfrak{so}(2n, \mathbb{C})$ is slightly more complicated. We follow example (2) of 2.2.6 and consider the standard representation $V = \mathbb{C}^{2n}$ with standard basis $\{v_1, \dots, v_{2n}\}$. The inner product is such that the subspaces generated by $\{v_1, \dots, v_n\}$ and $\{v_{n+1}, \dots, v_{2n}\}$ are isotropic and such that $\langle v_i, v_{n+j} \rangle = \delta_{i,j}$ for $i, j = 1, \dots, n$. From the list of fundamental weights in 2.2.13 we see that $v_1 \wedge \cdots \wedge v_k \in \Lambda^k V$ are highest weight vectors whose weight is the kth fundamental weight if $k < n-1$. On the other hand, $v_1 \wedge \cdots \wedge v_n$ and $v_1 \wedge \cdots \wedge v_{n-1} \wedge v_{2n}$ are highest weight vectors in $\Lambda^n V$ whose weight is twice the $(n-1)$st, respectively, the nth fundamental weight. These two highest weight vectors correspond to the splitting of $\Lambda^n V$ into two irreducible components mentioned in 2.2.13. As above we conclude that any maximal parabolic subalgebra in $\mathfrak{so}(2n, \mathbb{C})$ can be realized as the stabilizer of an isotropic subspace in the standard representation V.

Let us finally mention that $v_1 \wedge \cdots \wedge v_{n-1} \in \Lambda^{n-1} V$ is a highest weight vector whose weight is the sum of the last two fundamental weights. Thus, we see that the standard parabolic corresponding to $\Sigma = \{\alpha_{n-1}, \alpha_n\}$ is exactly the stabilizer of the isotropic subspace $\mathbb{C}^{n-1} \subset \mathbb{C}^{2n}$. It is a nice exercise to verify directly that a matrix in $\mathfrak{g}$ stabilizes the subspace generated by $\{v_1, \dots, v_{n-1}\}$ if and only if it stabilizes both the subspace generated by $\{v_1, \dots, v_n\}$ and the one generated by $\{v_1, \dots, v_{n-1}, v_{2n}\}$.

3.2.6. Complex generalized flag varieties. The description of standard parabolic subalgebras as stabilizers of lines in representations can be extended to the group level. Let us fix a complex $|k|$–graded semisimple Lie algebra $\mathfrak{g} = \mathfrak{g}_{-k} \oplus \cdots \oplus \mathfrak{g}_k$. For any Lie group G with Lie algebra $\mathfrak{g}$ and any parabolic subgroup $P \subset G$ associated to the grading, we have the homogeneous space G/P. These spaces are usually referred to as (complex) *generalized flag varieties.*

We start by observing that, fixing $\mathfrak{g}$, the choice of the group G and the parabolic subgroup P has only little influence on the structure of the homogeneous space G/P. If G is not connected, then let G^0 be the connected component of the identity, which also is a Lie group with Lie algebra $\mathfrak{g}$. For any choice of P associated to the given grading, we may restrict the canonical left action of G on G/P to the subgroup

G^0. The stabilizer of the base point $o = eP \in G/P$ in G^0 is $P \cap G^0$, which clearly is a parabolic subgroup of G^0 for the given grading. For any point $x \in G/P$, the map $G^0 \to G/P$ defined by $g \mapsto g \cdot x$ is the restriction of the corresponding map for G and thus it is a submersion. Hence, any G^0–orbit is open in G/P, and since the complement of an orbit is the union of all other orbits, each orbit is closed, too. Hence, the orbit $G^0 \cdot o \subset G/P$ is exactly the connected component of o. Since different connected components of a homogeneous spaces are diffeomorphic, we see that connectedness of G is not a serious restriction.

Assuming that G is connected, the adjoint action defines a covering homomorphism π from G onto the group $\mathrm{Int}(\mathfrak{g})$ of inner automorphisms of $\mathfrak{g}$, the connected component of the identity in the automorphism group $\mathrm{Aut}(\mathfrak{g})$; see 2.2.19. The kernel of this covering homomorphism is the center $Z(G)$ of G. This implies that the normalizer $N_G(\mathfrak{p})$ is the preimage $\pi^{-1}(N_{\mathrm{Int}(\mathfrak{g})}(\mathfrak{p}))$. In particular, π induces a diffeomorphism $G/N_G(\mathfrak{p}) \to \mathrm{Int}(\mathfrak{g})/N_{\mathrm{Int}(\mathfrak{g})}(\mathfrak{p})$. Choosing the maximal parabolic subgroups available for a given grading, all connected Lie groups with Lie algebra $\mathfrak{g}$ give rise to the same generalized flag manifold.

Finally, in most cases the normalizer $N_G(\mathfrak{p})$ is connected, so there actually is no freedom in the choice of the parabolic subgroup. Passing from $N_G(\mathfrak{p})$ to a subgroup containing the connected component of the identity only replaces the homogeneous space by a covering. In particular, the connected component $N_G(\mathfrak{p})^0$ of the identity leads to the universal covering. As we shall see below, this universal covering is compact, so in any case, we only obtain finite coverings of $G/N_G(\mathfrak{p})$.

The first step to understand generalized flag varieties is to prove that they are compact. The proof given here will also work in the real case, and it actually needs a bit of real Lie theory. Let θ be a Cartan involution on $\mathfrak{g}$; see 2.3.2. Since $\mathfrak{g}$ is complex, we can use the conjugation with respect to any compact real form. Given a connected Lie group G with Lie algebra $\mathfrak{g}$, Theorem 2.3.3 shows that there is an involutive group automorphism $\Theta : G \to G$ with derivative θ. The fixed point group K of Θ is a closed subgroup of G, which is compact provided that $Z(G)$ is finite. In the latter case, it is a maximal compact subgroup of G.

PROPOSITION 3.2.6. *Let $\mathfrak{g} = \mathfrak{g}_{-k} \oplus \cdots \oplus \mathfrak{g}_k$ be a complex $|k|$–graded semisimple Lie algebra such that $\mathfrak{p} = \mathfrak{g}^0$ is a standard parabolic subalgebra with respect to a Cartan subalgebra $\mathfrak{h}$, which is stable under a Cartan involution θ of $\mathfrak{g}$. Let G be a connected Lie group with Lie algebra $\mathfrak{g}$ and let $P \subset G$ be a parabolic subgroup for the given grading. Let $\Theta : G \to G$ be the unique automorphism of G with differential θ and K the fixed point group of Θ.*

Then the subgroup K acts transitively on G/P. In particular, G/P is compact and canonically diffeomorphic to $K/(K \cap P)$.

PROOF. We need the Iwasawa decomposition on the group level as described in 2.3.5. Since $\mathfrak{g}$ is complex, the Cartan decomposition simply reads as $\mathfrak{g} = \mathfrak{k} \oplus i\mathfrak{k}$ for a compact real form $\mathfrak{k}$ of $\mathfrak{g}$. As a maximal abelian subspace $\mathfrak{a} \subset i\mathfrak{k}$ we can use the subspace $\mathfrak{h}_0 \subset \mathfrak{h}$ on which all roots are real, and then the restricted root decomposition coincides with the root decomposition; see example (1) of 2.3.4. In particular, the subalgebra $\mathfrak{n} \subset \mathfrak{g}$ used in the Iwasawa decomposition is given by $\mathfrak{n} = \bigoplus_{\alpha \in \Delta^+} \mathfrak{g}_\alpha$. In particular, $\mathfrak{a} \oplus \mathfrak{n}$ is contained in the standard Borel subalgebra and hence in $\mathfrak{p}$.

Let A and N be the analytic subgroups of G corresponding to $\mathfrak{a}$ and $\mathfrak{n}$. Then by Theorem 2.3.5 the multiplication $(k, a, n) \mapsto kan$ defines a diffeomorphism $K \times$

$A \times N \to G$. Since the subgroups A and N are by definition connected, they are contained in P, which implies that $kanP = kP \in G/P$. Hence, we see that K acts transitively on G/P and thus $G/P \cong K/(K \cap P)$. Applying this in the case $G = \operatorname{Int}(\mathfrak{g})$, the corresponding subgroup K is compact since $\operatorname{Int}(\mathfrak{g})$ has trivial center. Since the homogeneous spaces for various choices of a connected group with Lie algebra $\mathfrak{g}$ are diffeomorphic, it follows that G/P is compact for any connected group G. □

This result gives rise to a first realization of a class of generalized flag varieties. The group $P = N_G(\mathfrak{p})$ by definition is the stabilizer of the subspace $\mathfrak{p} \subset \mathfrak{g}$ under the adjoint action. Hence, we can identify G/P with the orbit of $\mathfrak{p}$ under the adjoint action of G. This orbit is the variety of all parabolic subalgebras of $\mathfrak{g}$, which are conjugate to $\mathfrak{p}$ via an automorphism from $\operatorname{Ad}(G)$.

3.2.7. Remark: Relation of $|k|$–gradings to special symmetric spaces. We now have the necessary background to describe the relation of $|1|$–gradings to Hermitian symmetric spaces and of contact gradings to quaternionic symmetric spaces. Since this is outside of the main line of this book, we only give a brief discussion. We have already discussed the basics about symmetric spaces in 2.3.13. In particular, we have seen that any symmetric space is homogeneous, and if K/L is symmetric, then there is an involutive automorphism σ of the Lie algebra $\mathfrak{k}$ of K, which has the Lie algebra $\mathfrak{l}$ of L as its eigenspace with eigenvalue 1.

The first topic to discuss here are compact Hermitian symmetric spaces. Here, one looks for a compact Riemannian symmetric space K/L which admits a K–invariant orthogonal complex structure. In terms of the geometry of homogeneous spaces discussed in Section 1.4, it is easy to describe the algebraic data needed for this. In addition to $\mathfrak{k}$ and an involutive automorphism σ with fixed point set $\mathfrak{l}$ and appropriate groups K and $L \subset K$, we need a complex structure and a Hermitian inner product on $\mathfrak{k}/\mathfrak{l}$, which both are L–invariant.

Now suppose that $\mathfrak{g} = \mathfrak{g}_{-1} \oplus \mathfrak{g}_0 \oplus \mathfrak{g}_1$ is a complex semisimple $|1|$–graded Lie algebra, G is a simply connected Lie group with Lie algebra $\mathfrak{g}$, and $P := N_G(\mathfrak{p})^0 \subset G$ is the smallest parabolic subgroup for the given grading. Then $X := G/P$ is simply connected, and from Proposition 3.2.6 we know that denoting by $K \subset G$ the maximal compact subgroup, the inclusion $K \hookrightarrow G$ induces a diffeomorphism $K/L \to G/P$, where $L := K \cap P$. Infinitesimally, this means that $\mathfrak{k}/\mathfrak{l} \cong \mathfrak{g}/\mathfrak{p}$, so this carries a L–invariant (and even P–invariant) complex structure. Choosing any Hermitian inner product for this complex structure and averaging over the compact group L, we obtain an L–invariant Hermitian metric on $\mathfrak{k}/\mathfrak{l}$.

So it remains to realize $\mathfrak{l}$ as the $(+1)$–eigenspace of an involutive automorphism of $\mathfrak{k}$. Evidently, the map $\sigma : \mathfrak{g} \to \mathfrak{g}$, which is the identity on $\mathfrak{g}_0$ and minus the identity on $\mathfrak{g}_{-1} \oplus \mathfrak{g}_1$ defines an involutive automorphism of $\mathfrak{g}$ and it is easy to see that $\sigma(\mathfrak{k}) \subset \mathfrak{k}$. By definition, $\mathfrak{l} = \mathfrak{k} \cap \mathfrak{p}$. From the relation of a compact real form with the root decomposition described in 2.3.1, one easily concludes that $\mathfrak{k} \cap \mathfrak{p} = \mathfrak{k} \cap \mathfrak{g}_0$, which is exactly the required result.

Hence, we conclude that, viewed as a homogeneous space of K, the generalized flag manifold G/P constructed from a $|1|$–grading as described above naturally is a simply connected compact Hermitian symmetric space. The main point in the classification of such spaces is to show that this leads to all irreducible simply connected compact Hermitian symmetric spaces; see [**Wo64**].

Second, we discuss the relation between complex contact gradings and compact quaternionic symmetric spaces, which are also called Wolf spaces. These are symmetric spaces which admit a compatible quaternionic structure; see 4.1.8 for details on almost quaternionic and quaternionic structures. Suppose that K is a Lie group with Lie algebra $\mathfrak{k}$, σ is an involutive automorphism of $\mathfrak{k}$ with fixed point subalgebra $\mathfrak{l} \subset \mathfrak{k}$ and $L \subset K$ is a subgroup corresponding to $\mathfrak{l}$. Using the theory developed in Section 1.4, a K–invariant almost quaternionic structure on K/L can be described as follows. One needs a three–dimensional, L–invariant subspace $Q \subset L(\mathfrak{k}/\mathfrak{l}, \mathfrak{k}/\mathfrak{l})$ which admits a basis of the form $\{I, J, I \circ J\}$ for linear maps $I, J : \mathfrak{k}/\mathfrak{l} \to \mathfrak{k}/\mathfrak{l}$ such that $I^2 = J^2 = -\mathrm{id}$ and $J \circ I = -I \circ J$. The elements I and J are not supposed to be L–invariant, however.

Now suppose that $\mathfrak{g}$ is a complex simple Lie algebra endowed with a complex contact grading $\mathfrak{g} = \mathfrak{g}_{-2} \oplus \cdots \oplus \mathfrak{g}_2$ as described in 3.2.4. The identity on $\mathfrak{g}_{-2} \oplus \mathfrak{g}_0 \oplus \mathfrak{g}_2$ together with minus the identity on $\mathfrak{g}_{-1} \oplus \mathfrak{g}_1$ defines an involutive automorphism σ of $\mathfrak{g}$. Similarly, as above, there is a compact real form $\mathfrak{k} \subset \mathfrak{g}$ which is invariant under σ. For the fixed point subalgebra we obtain $\mathfrak{l} = \mathfrak{k} \cap (\mathfrak{g}_{-2} \oplus \mathfrak{g}_0 \oplus \mathfrak{g}_2)$ and we can naturally identify $\mathfrak{k}/\mathfrak{l}$ with the $\mathfrak{k}$–invariant subspace $\mathfrak{k} \cap (\mathfrak{g}_{-1} \oplus \mathfrak{g}_1)$.

Recall from 3.2.4 that $[\mathfrak{g}_{-2}, \mathfrak{g}_2] = \mathbb{C} \cdot E$, where $E \in \mathfrak{g}_0$ is the grading element for our contact grading. Now $\mathfrak{g}_{-2} \oplus \mathbb{C} \cdot E \oplus \mathfrak{g}_2$ evidently is an ideal in $\mathfrak{g}_{-2} \oplus \mathfrak{g}_0 \oplus \mathfrak{g}_2$, so its intersection with $\mathfrak{l}$ is a three–dimensional ideal $\mathfrak{q} \subset \mathfrak{l}$, which by compactness has to be isomorphic to $\mathfrak{su}(2) \cong \mathfrak{sp}(1)$. The adjoint action embeds $\mathfrak{q}$ into the set of endomorphisms of $\mathfrak{k} \cap (\mathfrak{g}_{-1} \oplus \mathfrak{g}_1)$ and one easily concludes that this subspace has the required properties.

As before, we can then take as a group K the maximal compact subgroup of the simply connected Lie group with Lie algebra $\mathfrak{g}$, and one shows that $\mathfrak{l} \subset \mathfrak{k}$ corresponds to a closed subgroup $L \subset K$, which we may take to be connected. Now we can take an arbitrary inner product on $\mathfrak{k} \cap (\mathfrak{g}_{-1} \oplus \mathfrak{g}_1)$, which is totally real with respect to any of the complex structures in $\mathfrak{q}$ and average it over the compact group L. Then this gives rise to a quaternion Kähler metric on K/L whose underlying almost quaternionic structure is given by $\mathfrak{q}$. In particular, this almost quaternionic structure has to be quaternionic, so we have found a quaternionic symmetric space. As above, the main step in the classification of such spaces is to show that they are all obtained from these examples; see [**Wo65**].

3.2.8. Projective realization of complex generalized flag varieties. Next, we discuss the realization of the homogeneous spaces G/P as G–orbits in projectivized representations. In the complex case, one not only obtains one such realization, but the projectivization of any finite–dimensional irreducible representation contains a G–orbit which is isomorphic to a generalized flag manifold. If G is a complex Lie group with Lie algebra $\mathfrak{g}$ and $P \subset G$ is a parabolic subgroup, then G/P is a (by 3.2.6 compact) complex manifold on which G acts holomorphically.

Now assume that G is connected and V is a complex irreducible representation of G. Let us fix a Cartan subalgebra $\mathfrak{h} \leq \mathfrak{g}$ with corresponding roots Δ and a subset Δ^+ of positive roots. By Theorem 2.2.11, there is a highest weight vector $v_0 \in V$, which is unique up to scale. Thus, we obtain a well–defined point $[v_0]$ in the complex projectivization $\mathcal{P}V$, i.e. the space of complex lines in V. Let us denote the corresponding weight, which is also called the highest weight of V, by $\lambda : \mathfrak{h} \to \mathbb{C}$.

THEOREM 3.2.8. *Let $\mathfrak{g}$ be a complex semisimple Lie algebra, G a connected Lie group with Lie algebra $\mathfrak{g}$, V a finite–dimensional holomorphic irreducible representation of G with highest weight $\lambda : \mathfrak{h} \to \mathbb{C}$, and $v_0 \in V$ a highest weight vector. Let $\mathfrak{p} \leq \mathfrak{g}$ be the standard parabolic subalgebra corresponding to the set $\Sigma = \{\alpha \in \Delta^0 : \langle \lambda, \alpha \rangle \neq 0\}$ of simple roots of $\mathfrak{g}$, and put $P := N_G(\mathfrak{p}) \subset G$.*

Then P coincides with the stabilizer of the point $[v_0]$ in the complex projectivization $\mathcal{P}V$ and we get a biholomorphism between G/P and the orbit $G \cdot [v_0] \subset \mathcal{P}V$.

PROOF. By Proposition 3.2.5, $\mathfrak{p}$ is the stabilizer subalgebra of $[v_0]$. For $g \in P$ and $X \in \mathfrak{p}$ and $v \in V$ we obtain $\mathrm{Ad}(g)(X) \cdot v = g \cdot X \cdot g^{-1} \cdot v$, and hence

$$X \cdot g \cdot v_0 = g \cdot \mathrm{Ad}(g)(X) \cdot v_0 = \mu(X) g \cdot v_0,$$

for some linear functional $\mu : \mathfrak{p} \to \mathbb{C}$, where we have used that $\mathrm{Ad}(g)(X) \in \mathfrak{p}$ stabilizes the line spanned by v_0. Applying this to $X \in \mathfrak{h}$ we see that $g \cdot v_0$ is a weight vector. For X in a positive root space, also $X \cdot g \cdot v_0$ is a weight vector of different weight, whence $\mu(X) = 0$. This shows that $g \cdot v_0$ is a highest weight vector and thus a multiple of v_0. Thus, P is contained in the isotropy subgroup of $[v_0]$. On the other hand, any element in an isotropy subgroup must normalize the corresponding isotropy subalgebra, so we conclude that P coincides with the isotropy subgroup of $[v_0]$.

Now the mapping $g \mapsto g \cdot [v_0] = [g \cdot v_0]$ is a holomorphic submersion from G onto the orbit $G \cdot [v_0]$. Since P is the isotropy subgroup of $[v_0]$, this factors to a holomorphic bijection $G/P \to G \cdot [v_0]$, which is a diffeomorphism and thus a biholomorphism by compactness of G/P. □

COROLLARY 3.2.8. *Let $\mathfrak{g}$ be a complex semisimple Lie algebra, $\mathfrak{p} \leq \mathfrak{g}$ a complex parabolic subalgebra, G a connected Lie group with Lie algebra $\mathfrak{g}$, and $P = N_G(\mathfrak{p})$. Then the generalized flag manifold G/P is a compact Kähler manifold and a projective algebraic variety.*

PROOF. We may assume that $\mathfrak{p} \leq \mathfrak{g}$ is a standard parabolic, and we denote by Σ the corresponding subset of simple roots. Let $\delta^{\mathfrak{p}}$ be the sum of all fundamental weights corresponding to elements of Σ, compare with 3.2.16. Since this is a dominant integral weight, there is a finite–dimensional complex irreducible representation V of $\mathfrak{g}$ with highest weight $\delta^{\mathfrak{p}}$. In view of 3.2.6 we may assume that G is simply connected, so any representation of $\mathfrak{g}$ integrates to a holomorphic representation of G. From the theorem, we get a biholomorphism of G/P with the (closed) orbit of a highest weight vector in the projectivization $\mathcal{P}V$, so G/P is biholomorphic to a closed submanifold of a complex projective space. Since complex projective spaces are Kähler, we see that G/P is Kähler. Further, G/P is an analytic subvariety of projective space, which is algebraic by Chow's theorem (see [**GrHa78**, 1.3]). □

EXAMPLE 3.2.8. Let us start by discussing the A_n–case, i.e. $G = SL(n+1, \mathbb{C})$ and the maximal parabolic $P_k = N_G(\mathfrak{p}_\Sigma)$ for $\Sigma = \{\alpha_k\}$ for $1 \leq k \leq n$. Denoting by $V = \mathbb{C}^{n+1}$ the standard representation and by $\{v_1, \dots, v_{n+1}\}$ the standard basis of V, we have seen in 3.2.5 that the kth fundamental representation is $\Lambda^k V$ and a highest weight vector is given by $v_1 \wedge \cdots \wedge v_k$. By the theorem, G/P_k is biholomorphic to the G–orbit of the point $[v_1 \wedge \cdots \wedge v_k] \in \mathcal{P}(\Lambda^k V) \cong \mathbb{C}P^{\binom{n+1}{k}-1}$. For arbitrary linearly independent vectors $w_1, \dots, w_k \in V$ there is an element $g \in G$ such that

$gv_i = w_i$ for all $i = 1, \dots, k$. Thus, we see that the orbit of $[v_1 \wedge \cdots \wedge v_k]$ is exactly the projectivization of the cone of all decomposable elements in $\Lambda^k V$.

On the other hand, elementary linear algebra shows that P_k also is the stabilizer in G of the subspace $\mathbb{C}^k \subset \mathbb{C}^n$ generated by $v_1, \dots, v_k$. Since G acts transitively on the set of all k–dimensional subspaces of $\mathbb{C}^n$, we may identify G/P_k with the Grassmannian $Gr_k(\mathbb{C}^n)$ of all k–dimensional complex subspaces of $\mathbb{C}^n$. These two realizations of G/P_k give rise to a holomorphic embedding of $Gr_k(\mathbb{C}^n)$ into $\mathcal{P}(\Lambda^k V) \cong \mathbb{C}P^{\binom{n+1}{k}-1}$, whose image is exactly the projectivization of the cone of decomposable elements. This is the well–known *Plücker embedding.* Notice that Proposition 3.2.6 in this case states the well–known fact that the group $SU(n+1)$ acts transitively on $Gr_k(\mathbb{C}^{n+1})$, which leads to the identification of the Grassmannian with the homogeneous space $SU(n+1)/S(U(k) \times U(n+1-k))$.

Similarly, denoting by $P = N_G(\mathfrak{p}_\Sigma)$ with $\Sigma = \{\alpha_{i_1}, \dots, \alpha_{i_k}\}$ and $1 \leq i_1 < i_2 < \cdots < i_k \leq n$, we see that G/P can be realized as an orbit in an appropriate projective space or as the space of flags of nested complex subspaces of dimensions $i_1, \dots, i_k$. Thus, we obtain a holomorphic embedding of this partial flag manifold into a projective space, which generalizes the Plücker embedding. These examples are the reason why the homogeneous spaces of semisimple groups by parabolic subgroups are called generalized flag manifolds.

For the other classical simple groups everything can be done in a very similar way following the treatment of the Lie algebra case in 3.2.5. In any case, one may realize the generalized flag manifolds corresponding to maximal parabolics as the space of decomposable null elements in the projectivization of some exterior power of the standard representation. Alternatively, they may be realized as the space of all isotropic subspaces of some fixed dimension. This leads to a realization of an isotropic Grassmannian or flag manifold generalizing the Plücker embedding. The only exception occurs in the case of D_n, i.e. $G = SO(2n, \mathbb{C})$, where there are two orbits of isotropic n–dimensional subspaces and correspondingly two orbits of isotropic decomposable elements in $\Lambda^n \mathbb{C}^{2n}$.

REMARK 3.2.8. The results of the theorem can be pushed considerably further by using the fact that any complex semisimple Lie group is an algebraic group defined over $\mathbb{C}$ and any finite–dimensional irreducible representation is algebraic. This implies that the resulting actions on projective spaces are algebraic, too. One then shows that the orbits of such an action are affine subvarieties, and the closure of an orbit is obtained by adding orbits of lower dimensions. In particular, this implies that all orbits of lowest possible dimension have to be closed. It turns out, however, that the orbit of the line through the highest weight vector is the unique closed orbit in the projectivization of an irreducible representation, and thus the unique orbit of lowest possible dimension.

This is proved using the *Borel fixed point theorem*, which states that if V is a representation of a connected solvable algebraic group B and X is a B–invariant projective subvariety of $\mathcal{P}V$, then X contains a fix point of the B–action. For complex groups this is rather easy to prove: Using Lie's theorem (part (2) of Theorem 2.1.1) one shows that for each $1 \leq k \leq \dim(V)$ one finds a k–dimensional B–invariant subspace V_k of V such that $V_k \supset V_{k-1}$ for all k. Elementary algebraic geometry shows that if one looks at the minimal i such that $X \cap \mathcal{P}(V_i) \neq \emptyset$, then this intersection is a finite set of points. Since both V_i and X are B–invariant, so is the intersection, and since B is connected, all points in the intersection must be fixed by the B–action.

This applies to our situation since it turns out that the Borel subgroup of a complex semisimple group is always connected; see 3.2.19. If we have a closed orbit in the projectivization of a finite–dimensional G–representation, then as in the proof of Corollary 3.2.8 one concludes that it is a projective subvariety. Applying the Borel fixed point theorem we obtain a fixed point of the action of the Borel subgroup, which by definition is the line through a highest weight vector.

3.2.9. Parabolic subalgebras in real semisimple Lie algebras. The description of real $|k|$–gradings proceeds via complexification. We start be reviewing the main ingredients of the real structure theory. Given a real semisimple Lie algebra $\mathfrak{g}$, one first chooses a Cartan involution θ on $\mathfrak{g}$; see 2.3.2. This is an involutive automorphism such that the bilinear form $B_\theta(X, Y) := -B(X, \theta Y)$ is positive definite, where B denotes the Killing form of $\mathfrak{g}$. The decomposition of $\mathfrak{g}$ into eigenspaces for θ is called the Cartan decomposition. To avoid confusion with parabolics, we denote the -1–eigenspace by $\mathfrak{q}$, so the Cartan decomposition reads as $\mathfrak{g} = \mathfrak{k} \oplus \mathfrak{q}$. Now one looks at θ–stable Cartan subalgebras $\mathfrak{h} \leq \mathfrak{g}$, i.e. abelian subalgebras such that $\theta(\mathfrak{h}) = \mathfrak{h}$ and such that the complexification $\mathfrak{h}_{\mathbb{C}}$ is a Cartan subalgebra of $\mathfrak{g}_{\mathbb{C}}$. Then $\mathfrak{h} = (\mathfrak{h} \cap \mathfrak{k}) \oplus (\mathfrak{h} \cap \mathfrak{q})$, and $\mathfrak{h}$ is called maximally noncompact if the dimension of $\mathfrak{a} := \mathfrak{h} \cap \mathfrak{q}$ is maximal among all θ–stable Cartan subalgebras. Such Cartan subalgebras can be found by first choosing a maximal abelian subspace $\mathfrak{a} \subset \mathfrak{q}$, then looking at the centralizer $\mathfrak{m} := \mathfrak{z}_{\mathfrak{k}}(\mathfrak{a})$ of $\mathfrak{a}$ in $\mathfrak{k}$, choosing a maximal abelian subspace $\mathfrak{t} \subset \mathfrak{m}$ and putting $\mathfrak{h} := \mathfrak{t} \oplus \mathfrak{a}$, see 2.3.7.

Having chosen θ and $\mathfrak{h}$, one can then look at the root system Δ associated to the Cartan subalgebra $\mathfrak{h}_{\mathbb{C}} \leq \mathfrak{g}_{\mathbb{C}}$. Let σ be the conjugation of $\mathfrak{g}_{\mathbb{C}}$ with respect to the real form $\mathfrak{g}$. Then σ induces an involutive automorphism $\sigma^* : \Delta \to \Delta$; see 2.3.8. A positive subsystem $\Delta^+ \subset \Delta$ is called admissible if for $\alpha \in \Delta^+$ we either have $\sigma^*\alpha = -\alpha$ or $\sigma^*\alpha \in \Delta^+$.

DEFINITION 3.2.9. Let $\mathfrak{g}$ be a real semisimple Lie algebra with complexification $\mathfrak{g}_{\mathbb{C}}$, θ a Cartan involution, $\mathfrak{h} \leq \mathfrak{g}$ a θ–stable maximally noncompact Cartan subalgebra, Δ the set of roots for the Cartan subalgebra $\mathfrak{h}_{\mathbb{C}} \leq \mathfrak{g}_{\mathbb{C}}$, and $\Delta^+ \subset \Delta$ an admissible positive subsystem.

A Lie subalgebra $\mathfrak{p} \leq \mathfrak{g}$ is called a *standard parabolic subalgebra* with respect to the choices of $\mathfrak{h}$ and Δ^+ if and only if the complexification $\mathfrak{p}_{\mathbb{C}}$ is a standard parabolic subalgebra of $\mathfrak{g}_{\mathbb{C}}$ with respect to $\mathfrak{h}_{\mathbb{C}}$ and Δ^+.

To formulate the results on real standard parabolics we need a bit more structure theory. First, we need the restricted root decomposition of $\mathfrak{g}$. Consider the abelian subspace $\mathfrak{a} = \mathfrak{h} \cap \mathfrak{q}$. For $A \in \mathfrak{a}$ we by definition have $\theta(A) = -A$, which easily implies that $\operatorname{ad}(A) : \mathfrak{g} \to \mathfrak{g}$ is symmetric for the inner product B_θ. Thus, the family $\{\operatorname{ad}(A) : A \in \mathfrak{a}\}$ is simultaneously diagonalizable over $\mathbb{R}$. The corresponding eigenvalues are given by linear functionals $\lambda : \mathfrak{a} \to \mathbb{R}$, and the nonzero eigenvalues are called the *restricted roots.* The set of all restricted roots is denoted by Δ_r. The eigenspaces are called *restricted root spaces* and they define the *restricted root decomposition* of $\mathfrak{g}$. In 2.3.6 we have seen that the set $\Delta_r \subset \mathfrak{a}^*$ is an abstract root system, but it is not reduced in general. Still the notions of positive and simple subsystem pose no problems for Δ_r. In contrast to the root decomposition in the complex case, there is no general result on the dimension of restricted root spaces.

Second, we briefly describe how to obtain the Satake diagram of $\mathfrak{g}$. Having given θ, $\mathfrak{h} \leq \mathfrak{g}$, Δ and the conjugation σ as above, we define $\Delta_c := \{\alpha : \sigma^*\alpha = -\alpha\} \subset \Delta$. By 2.3.8, all roots are real on $i\mathfrak{t} \oplus \mathfrak{a} \subset \mathfrak{h}_{\mathbb{C}}$. Restricting the roots to $\mathfrak{h} \subset \mathfrak{h}^{\mathbb{C}}$, the map σ^* becomes complex conjugation, so $\Delta_c = \{\alpha \in \Delta : \alpha|_{\mathfrak{a}} = 0\}$. The condition of admissibility of a positive subsystem $\Delta^+ \subset \Delta$ then reads as $\sigma^*\alpha \in \Delta^+$ for all $\alpha \in \Delta^+ \setminus \Delta_c$. Passing to the associated simple system Δ^0, it turns out that $\Delta^0_c := \Delta^0 \cap \Delta_c$ is a simple system for Δ_c, and for any $\alpha \in \Delta^0 \setminus \Delta^0_c$ there is a unique $\alpha' \in \Delta^0 \setminus \Delta^0_c$ such that $\sigma^*\alpha - \alpha'$ is a linear combination of compact roots. Mapping α to α' defines an involutive automorphism of $\Delta^0 \setminus \Delta^0_c$. The Satake diagram of $\mathfrak{g}$ is then obtained by taking the Dynkin diagram of Δ^0 with elements of Δ^0_c indicated by black dots $\bullet$ and elements of $\Delta^0 \setminus \Delta^0_c$ by white dots $\circ$. Moreover, for any element $\alpha \in \Delta^0 \setminus \Delta^0_c$ such that $\alpha' \neq \alpha$, one connects α and α' by an arrow; see 2.3.8.

Finally, let us describe the relation between the Satake diagram and restricted roots. Since all roots are real on $i\mathfrak{t} \oplus \mathfrak{a}$, the restricted roots are exactly the nonzero restrictions of roots to $\mathfrak{a} \subset \mathfrak{h}$. Thus, we obtain a surjective restriction map $\Delta \setminus \Delta_c \to \Delta_r$. Since for $\alpha \in \Delta$, the restrictions to $\mathfrak{h}$ of α and $\sigma^*\alpha$ are conjugate, we see that $\sigma^*\alpha|_{\mathfrak{a}} = \alpha|_{\mathfrak{a}}$. For an admissible choice of $\Delta^+ \subset \Delta$, the image in Δ_r of Δ^+ under the restriction map is a positive subsystem. This easily implies that the corresponding simple system Δ^0_r for Δ_r is the quotient of $\Delta^0 \setminus \Delta^0_c$ obtained by identifying each simple root α with α'.

Having these ingredients at hand, we can now state the classification of real standard parabolics.

THEOREM 3.2.9. *Let $\mathfrak{g}$ be a real semisimple Lie algebra, θ a Cartan involution with associated Cartan decomposition $\mathfrak{g} = \mathfrak{k} \oplus \mathfrak{q}$, $\mathfrak{h} = \mathfrak{t} \oplus \mathfrak{a} \subset \mathfrak{g}$ a maximally noncompact θ–stable Cartan subalgebra. Put $\Delta = \Delta(\mathfrak{g}_{\mathbb{C}}, \mathfrak{h}_{\mathbb{C}})$, let σ^* be the involutive automorphism of Δ induced by the conjugation with respect to $\mathfrak{g} \subset \mathfrak{g}_{\mathbb{C}}$ and let $\Delta^+ \subset \Delta$ be a positive subsystem such that for $\alpha \in \Delta^+$ we either have $\sigma^*\alpha \in \Delta^+$ or $\sigma^*\alpha = -\alpha$. Then we have:*

(1) Put $\mathfrak{m} = \mathfrak{z}_{\mathfrak{k}}(\mathfrak{a})$ and let $\mathfrak{n} \subset \mathfrak{g}$ be the direct sum of all positive restricted root spaces. Then $\mathfrak{p}_0 := \mathfrak{m} \oplus \mathfrak{a} \oplus \mathfrak{n}$ is a subalgebra of $\mathfrak{g}$, and the standard parabolic subalgebras of $\mathfrak{g}$ are exactly the subalgebras containing $\mathfrak{p}_0$.

(2) Let Δ^0 be the set of simple roots and let $\Delta^0_r \subset \Delta_r$ be the corresponding set of simple restricted roots. Then subsets of Δ^0_r are in bijective correspondence with subsets of Δ^0 that are disjoint to Δ^0_c and stable under the involution induced by σ^*. On the other hand, the set of all subsets of Δ^0 with these two properties is in bijective correspondence with the set of all standard parabolic subalgebras of $\mathfrak{g}$.

Explicitly, the parabolic subalgebra corresponding to $\Sigma \subset \Delta^0$ is the sum of $\mathfrak{p}_0$ and the restricted root spaces for those negative restricted roots which can be written as linear combinations of the simple restricted roots which are outside of the image of Σ in Δ^0_r.

PROOF. (1) By definition $\mathfrak{m}$ and $\mathfrak{a} \oplus \mathfrak{n}$ are subalgebras of $\mathfrak{g}$ and $[\mathfrak{m}, \mathfrak{a}] = 0$. On the other hand, $[\mathfrak{m}, \mathfrak{a}] = 0$ immediately implies that $\operatorname{ad}(\mathfrak{m})$ maps any restricted root space to itself, so in particular, $[\mathfrak{m}, \mathfrak{n}] \subset \mathfrak{n}$. Thus, $\mathfrak{p}_0 = \mathfrak{m} \oplus \mathfrak{a} \oplus \mathfrak{n}$ is a subalgebra of $\mathfrak{g}$. From the proof of Proposition 2.3.5 we know that $\mathfrak{m} \oplus \mathfrak{a}$ is the centralizer of $\mathfrak{a}$ in $\mathfrak{g}$, so $\mathfrak{p}_0$ is the direct sum of all nonnegative restricted root spaces.

Now consider the complexification $(\mathfrak{p}_0)_{\mathbb{C}}$. By construction, $\mathfrak{h} \subset \mathfrak{m} \oplus \mathfrak{a} \subset \mathfrak{p}_0$, so $\mathfrak{h}_{\mathbb{C}} \subset (\mathfrak{p}_0)_{\mathbb{C}}$. We have noted above that $\sigma^*\alpha|_{\mathfrak{a}} = \alpha|_{\mathfrak{a}}$ for all $\alpha \in \Delta$. If $\alpha \in \Delta^+$, then this restriction is either zero (so $\alpha \in \Delta_c$) or a positive restricted root. If $\alpha \in \Delta_c$, then $(\mathfrak{g}_{\mathbb{C}})_\alpha \subset \mathfrak{k}_{\mathbb{C}}$ (see 2.3.8), so $\mathfrak{l} := \mathfrak{t}_{\mathbb{C}} \oplus \bigoplus_{\alpha\in\Delta_c}(\mathfrak{g}_{\mathbb{C}})_\alpha \subset \mathfrak{k}_{\mathbb{C}}$. Since σ maps $(\mathfrak{g}_{\mathbb{C}})_\alpha$ to $(\mathfrak{g}_{\mathbb{C}})_{\sigma^*\alpha}$ (see 2.3.8), we get $\sigma(\mathfrak{l}) \subset \mathfrak{l}$ and hence $\mathfrak{l}$ is the complexification of $\mathfrak{l}\cap\mathfrak{g}$. By construction, any element of $\mathfrak{l}$ commutes with any $A \in \mathfrak{a}$, so $\mathfrak{l}\cap\mathfrak{g} \subset \mathfrak{m}$ and hence $\mathfrak{l} \subset \mathfrak{m}_{\mathbb{C}}$. One easily verifies that actually $\mathfrak{l} = \mathfrak{m}_{\mathbb{C}}$.

If $\alpha \in \Delta^+$ is such that $\sigma^*\alpha \in \Delta^+$, then the σ–stable subspace $(\mathfrak{g}_{\mathbb{C}})_\alpha \oplus (\mathfrak{g}_{\mathbb{C}})_{\sigma^*\alpha}$ intersects $\mathfrak{g}$ in a subspace of a sum of positive restricted root spaces. Conversely, for a positive restricted root λ, the complexification of the corresponding restricted root space has to be contained in $\bigoplus_{\alpha:\alpha|_{\mathfrak{a}}=\lambda}(\mathfrak{g}_{\mathbb{C}})_\alpha$. Putting the information together, we conclude that

$$(\mathfrak{p}_0)_{\mathbb{C}} = \mathfrak{h}_{\mathbb{C}} \oplus \bigoplus_{\alpha\in\Delta^+}((\mathfrak{g}_{\mathbb{C}})_\alpha + (\mathfrak{g}_{\mathbb{C}})_{\sigma^*\alpha}).$$

Thus, $(\mathfrak{p}_0)_{\mathbb{C}}$ is exactly the sum of the standard Borel subalgebra and its conjugate, which implies the claim.

(2) The standard parabolic subalgebras of $\mathfrak{g}$ are exactly the intersections of $\mathfrak{g}$ with σ–stable standard parabolic subalgebras $\tilde{\mathfrak{p}} \subset \mathfrak{g}_{\mathbb{C}}$. As we saw in 3.2.1, parabolic subalgebras of $\mathfrak{g}_{\mathbb{C}}$ are in bijective correspondence with subsets of Δ^0, so we only have to describe which subsets correspond to σ–stable parabolics.

If $\tilde{\mathfrak{p}}$ is σ–stable, then for $\alpha \in \Delta$ such that $(\mathfrak{g}_{\mathbb{C}})_\alpha \subset \tilde{\mathfrak{p}}$, we also have $(\mathfrak{g}_{\mathbb{C}})_{\sigma^*\alpha} \subset \tilde{\mathfrak{p}}$. For $\alpha \in \Delta^0_c$ we have $\sigma^*\alpha = -\alpha$, which implies that the subset $\Sigma \subset \Delta^0$ corresponding to $\tilde{\mathfrak{p}}$ is disjoint from Δ^0_c. On the other hand, the involutive permutation $\alpha \mapsto \alpha'$ on Δ^0 induced by σ^*, is characterized by the fact that $\sigma^*\alpha - \alpha'$ a linear combination of compact roots. Since Σ is disjoint from Δ^0_c, we see that $\mathrm{ht}_\Sigma(\sigma^*\alpha) = \mathrm{ht}_\Sigma(\alpha')$, so $\alpha \in \Sigma$ implies $\alpha' \in \Sigma$. Hence, Σ is stable under the involutive permutation induced by σ^*.

Conversely, if Σ is disjoint from Δ^0_c and stable under the involutive permutation induced by σ^*, then $\mathrm{ht}_\Sigma(\alpha) = \mathrm{ht}_\Sigma(\sigma^*\alpha)$ holds for all $\alpha \in \Delta^0$. But this immediately implies the same statement for all $\alpha \in \Delta$, and hence the corresponding parabolic is stable under σ. The explicit description of the parabolic subalgebra of $\mathfrak{g}$ associated to Σ follows as in (1) above.

Finally, the bijection between these subsets of Δ^0 and all subsets of Δ^0_r follows immediately from the description of Δ^0_r as a quotient of $\Delta^0 \setminus \Delta^0_c$. □

REMARK 3.2.9. In the complex case, the number of standard parabolic subalgebras depends only on the number of simple roots. This equals the (complex) dimension of a Cartan subalgebra and thus the rank of the Lie algebra. For a real semisimple Lie algebra $\mathfrak{g}$, the essential quantity is the number of elements of Δ^0_r, which equals the (real) dimension of $\mathfrak{a}$. This dimension is usually referred to as the *real rank* of $\mathfrak{g}$. Note that the real rank is different for different real forms of a given complex semisimple Lie algebra. For a compact real Lie algebra, the real rank is zero, so there are no nontrivial standard parabolics. On the other hand, for a split real form $\mathfrak{g}$, there is by definition a maximally noncompact Cartan subalgebra $\mathfrak{h}$ on which all roots are real, so $\mathfrak{h} = \mathfrak{a}$. Hence, the real rank equals the rank for a split real form, and there are as many standard parabolics as in the complex case. As a final example, we see from 2.3.9 that for $p \geq q$ the real rank of $\mathfrak{su}(p,q)$ equals q.

The relation between $|k|$–gradings and standard parabolics is similar to the complex case:

PROPOSITION 3.2.9. *Let $\mathfrak{g}$ be a real semisimple Lie algebra endowed with a Cartan involution θ, a θ–stable maximally noncompact Cartan subalgebra $\mathfrak{h}$ and an admissible positive subsystem Δ^+. Then we have:*

(1) *For a standard parabolic subalgebra $\mathfrak{p} \leq \mathfrak{g}$ corresponding to a subset $\Sigma \subset \Delta_r^0$, the Σ–height determines a $|k|$–grading of $\mathfrak{g}$, such that $\mathfrak{p} = \mathfrak{g}^0$. Here, k is the Σ–height of the maximal restricted root.*
(2) *Given a $|k|$–grading of $\mathfrak{g}$, there is an automorphism $\phi \in \mathrm{Int}(\mathfrak{g})$ such that $\phi(\mathfrak{g}^0)$ is a standard parabolic subalgebra of $\mathfrak{g}$. Denoting by $\Sigma \subset \Delta_r^0$ the corresponding subset, the given grading on $\mathfrak{g}$ corresponds to the grading by Σ–height under ϕ.*

PROOF. (1) is clear.

(2) A $|k|$–grading of $\mathfrak{g}$ gives rise to a $|k|$–grading of the complexification $\mathfrak{g}_{\mathbb{C}}$. The grading element $E \in \mathfrak{g}$ also is the grading element for $\mathfrak{g}_{\mathbb{C}}$. Since the map $\mathrm{ad}(E)$ is diagonalizable on $\mathfrak{g}_{\mathbb{C}}$, we can find a Cartan subalgebra $\tilde{\mathfrak{h}} \subset \mathfrak{g}_{\mathbb{C}}$ containing E. Construct a compact real form of $\mathfrak{g}_{\mathbb{C}}$ from $\tilde{\mathfrak{h}}$ as in 2.3.1 and let $\tilde{\tau}$ be the corresponding conjugation of $\mathfrak{g}_{\mathbb{C}}$. Since all eigenvalues of $\mathrm{ad}(E)$ are real, we get $\tilde{\tau}(E) = -E$. Denoting by σ the conjugation of $\mathfrak{g}_{\mathbb{C}}$ with respect to $\mathfrak{g}$, we obtain $\sigma\tilde{\tau}\sigma\tilde{\tau}(E) = E$. Hence, $\psi := (\sigma\tilde{\tau}\sigma\tilde{\tau})^{1/4}$ (see 2.3.2) also satisfies $\psi(E) = E$. By the proofs of Lemma 2.3.2 and Theorem 2.3.2, $\tilde{\theta} := \psi\tilde{\tau}\psi^{-1}$ restricts to a Cartan involution on $\mathfrak{g}$, and by construction, $\tilde{\theta}(E) = -E$. By part (2) of Theorem 2.3.2, any two Cartan involutions are conjugate by an inner automorphism. Hence, we find $\phi_1 \in \mathrm{Int}(\mathfrak{g})$ such that $\phi_1(E)$ lies in the (-1)–eigenspace $\mathfrak{q}$ of θ. Choosing a maximal abelian subspace $\tilde{\mathfrak{a}} \subset \mathfrak{q}$ containing $\phi_1(E)$, we know from Theorem 2.3.6 that there is $\phi_2 \in \mathrm{Int}(\mathfrak{g})$ such that $\phi_2(E) \in \mathfrak{a} = \mathfrak{h} \cap \mathfrak{q}$. By Proposition 2.3.6 any two choices of positive restricted roots are conjugate, so we finally get $\phi \in \mathrm{Int}(\mathfrak{g})$ such that $\phi(E) \in \mathfrak{a}$ and $\lambda(\phi(E)) \geq 0$ for all positive restricted roots λ.

Since $\mathrm{ad}(E)$ is diagonalizable on $\mathfrak{g}$ with real eigenvalues, the same holds for $\mathrm{ad}(\phi(E)) = \phi \circ \mathrm{ad}(E) \circ \phi^{-1}$ and obviously ϕ maps $\mathfrak{g}^0$ to the sum of all eigenspaces of $\mathrm{ad}(\phi(E))$ corresponding to nonnegative eigenvalues. But this exactly means that $\phi(\mathfrak{g}^0)$ contains the minimal standard parabolic $\mathfrak{p}_0$ and hence is a standard parabolic subalgebra. The relation between the original grading and the Σ–height is then evident. □

REMARK 3.2.9. As in the complex case, one may directly obtain information about the subalgebra $\mathfrak{g}_0$ directly from the Satake diagram describing the parabolic. The (real) dimension of the center of $\mathfrak{g}_0$ again equals the number of crossed nodes in the Satake diagram. Parallel to what we have done in the complex case in Proposition 3.2.2, one may also show that in the real case the Satake diagram of the semisimple part $\mathfrak{g}_0^{ss}$ of $\mathfrak{g}_0$ is obtained by erasing all crossed nodes as well as all edges and arrows connecting to these nodes from the Satake diagram describing the parabolic subalgebra. Details about this can be found in [**Kane93**].

3.2.10. Notation for real parabolics and examples. We will use the obvious analog of the notation for complex parabolics in the real case. To describe a standard parabolic subalgebra $\mathfrak{p} \leq \mathfrak{g}$, we consider the Satake diagram of $\mathfrak{g}$ and denote all the simple roots corresponding to elements of the subset $\Sigma \subset \Delta^0$ by crosses. Since we know that Σ is disjoint from Δ_c, we can recover the original Satake diagram by replacing all crosses by white dots. In these terms, the classification of real parabolics can be rephrased in terms of replacing white dots in a

Satake diagram by crosses. The only rule one has to take into account is that two roots joined by an arrow either have to be both crossed or both uncrossed.

EXAMPLE 3.2.10. We now go through the most important examples following the discussion of the complex case in 3.2.2–3.2.5.

Real $|1|$–gradings. To get a complete list of real $|1|$–gradings, we only have to look at the list of complex $|1|$–gradings in 3.2.3 and check which of the $|1|$–gradings are present for the various real forms. Since a complex $|1|$–grading has only one crossed root, the Satake diagram of a real form has to have a white dot, which is not connected to any other dot by an arrow, at the right place in order to admit a $|1|$–grading. The quickest way to see the result is to compare the table of complex $|1|$–gradings in 3.2.3 with the Satake diagrams in Table B.4 in Appendix B.

In the A_n–case, crossing any simple root leads to a $|1|$–grading, so for a given real form, the available $|1|$–gradings correspond exactly to the white roots that are not connected to another root by an arrow. For the split real form $\mathfrak{sl}(n,\mathbb{R})$, the real $|1|$–gradings are in bijective correspondence with complex $|1|$–gradings of $\mathfrak{sl}(n,\mathbb{C})$. For $\mathfrak{su}(p,q)$ with $p>q$, there are no appropriate roots by example (3) of 2.3.9, so this real simple Lie algebra does not admit any $|1|$–grading. Similarly, example (4) of 2.3.9 implies that there is a unique $|1|$–grading on the real simple Lie algebra $\mathfrak{su}(p,p)$ for $p\geq 2$. Example (5) of 2.3.9 shows that there are $n-1$ $|1|$–gradings on the real form $\mathfrak{sl}(n,\mathbb{H})$ of $\mathfrak{sl}(2n,\mathbb{C})$, and by Theorem 2.3.11 the above exhaust the list of noncompact real forms of A_n.

For B_n, the unique complex $|1|$–grading corresponds to the first simple root, and going through the tables of noncompact real forms, one sees that this root is white and not connected to another root by an arrow for any of the noncompact real forms. Thus any of the algebras $\mathfrak{so}(p,q)$ with $p,q\neq 0$ and $p+q$ odd admits a unique real $|1|$–grading.

In the C_n–case, there is also only one complex $|1|$–grading which corresponds to the last root in the Dynkin diagram. This time one sees from the tables of noncompact real forms that $|1|$–gradings are only available on the split real form $\mathfrak{sp}(2n,\mathbb{R})$ and on the real form $\mathfrak{sp}(n,n)$ of $\mathfrak{sp}(4n,\mathbb{C})$. In both cases, this grading is unique up to isomorphism.

Finally, for the D_n case, there are three complex $|1|$–gradings corresponding to the first and the last two roots. Here, the tables imply that the real $|1|$–grading corresponding to the first root is available for all the noncompact real forms except $\mathfrak{so}^*(2n)$, while the gradings for the last two roots are available on the split real form $\mathfrak{so}(p,p)$ and one of them is available on the real form $\mathfrak{so}^*(4n)$ of $\mathfrak{so}(4n,\mathbb{C})$.

For exceptional algebras, a unique $|1|$–grading is available on the split real forms of E_6 and E_7 as well as on the noncompact real forms EIV of E_6 and $EVII$ of E_7.

Real contact gradings. Similarly, as for $|1|$–gradings above, we can analyze the contact gradings on real simple Lie algebras, which are by definition $|2|$–gradings such that $\dim(\mathfrak{g}_{\pm 2})=1$ and the bracket $\mathfrak{g}_{-1}\times\mathfrak{g}_{-1}\to\mathfrak{g}_{-2}$ is nondegenerate. As above, we only have to go through the list of complex contact gradings in 3.2.4 and check to which real form they descend using the Satake diagrams in Table B.4 of Appendix B. The result of this is that except $\mathfrak{sl}(n,\mathbb{H})$, $\mathfrak{so}(n-1,1)$, $\mathfrak{sp}(p,q)$, and the real forms EIV of E_6 and FII of F_4, any noncompact noncomplex real simple Lie algebra admits a unique real contact grading.

Real parabolics as stabilizers of lines and flags. The descriptions of complex parabolics as stabilizers of lines and flags from 3.2.5 have analogs in the real case. Suppose that the inclusion $\mathfrak{g} \to \mathfrak{g}_{\mathbb{C}}$ is chosen in such a way that a θ–stable maximally noncompact Cartan subalgebra of $\mathfrak{g}$ complexifies to the standard Cartan subalgebra of $\mathfrak{g}_{\mathbb{C}}$ and the usual positive system is admissible. Then the complexification of a standard parabolic in $\mathfrak{g}$ is the complex standard parabolic in $\mathfrak{g}_{\mathbb{C}}$ corresponding to the same set of simple roots, so one may directly carry over the descriptions from 3.2.5 to the real case. In some examples, we have used a different positive system, but the changes caused by this are easy to analyze. We outline this for the real forms of $\mathfrak{sl}(n, \mathbb{C})$.

For the split real form $\mathfrak{g} = \mathfrak{sl}(n, \mathbb{R})$ of $\mathfrak{sl}(n, \mathbb{C})$ the real diagonal matrices form a maximally noncompact θ–stable Cartan subalgebra, which complexifies to the standard Cartan subalgebra of $\mathfrak{g}_{\mathbb{C}}$. Moreover, the condition of admissibility of a positive subsystem is vacuous in the case of a split form, so we may use the standard positive and simple roots. Hence, we may conclude directly from 3.2.5 that the standard parabolic subalgebra of $\mathfrak{g}$ corresponding to the set $\{\alpha_{i_1}, \dots, \alpha_{i_k}\}$ of simple roots with $1 \leq i_1 < \dots < i_k < n$ is the intersection of $\mathfrak{g}$ with the stabilizer of the flag $\mathbb{C}^{i_1} \subset \dots \subset \mathbb{C}^{i_k} \subset \mathbb{C}^n$. Since $\mathfrak{g}$ itself is the stabilizer in $\mathfrak{g}_{\mathbb{C}}$ of the real subspace $\mathbb{R}^n \subset \mathbb{C}^n$, we conclude that this parabolic is the stabilizer in $\mathfrak{g}$ of the flag $\mathbb{R}^{i_1} \subset \dots \subset \mathbb{R}^{i_k} \subset \mathbb{R}^n$.

In example (2) of 2.3.4 and example (3) of 2.3.9, we have realized the Lie algebra $\mathfrak{g} := \mathfrak{su}(p, q)$ for $p > q$ as the subalgebra of $\mathfrak{g}_{\mathbb{C}} = \mathfrak{sl}(p+q, \mathbb{C})$ consisting of all matrices of the block form $\begin{pmatrix} A & B & C \\ D & -A^* & E \\ -E^* & -C^* & F \end{pmatrix}$, with $A \in \mathfrak{gl}(q, \mathbb{C})$, C and E arbitrary complex $q \times (p-q)$–matrices, B and D in $\mathfrak{u}(q)$ and $F \in \mathfrak{u}(p-q)$. There we have also verified that the tracefree diagonal matrices contained in $\mathfrak{g}$ form a θ–stable maximally noncompact Cartan subalgebra whose complexification is the standard Cartan subalgebra of $\mathfrak{g}_{\mathbb{C}}$. However, one has to use a nonstandard set of positive roots, for which the simple roots are given by $e_1 - e_2, \dots, e_{q-1} - e_q$, $e_q - e_{2q+1}$, $e_{2q+1} - e_{2q+2}, \dots, e_{n-1} - e_n$, $-e_{2q} + e_n$, $-e_{2q-1} + e_{2q}, \dots, -e_{q+1} + e_{q+2}$. Looking at the Satake diagram, we see that there is a basic set of $|2|$–gradings on $\mathfrak{g}$, which correspond to crossing the ith and the $(p+q-i)$th nodes in the Satake diagram (which are connected by an arrow). These gradings exist for $i = 1, \dots, q$, and for $i = 1$ we obtain the real contact grading from the example above. As in 3.2.5, one easily verifies that the standard parabolic subalgebra of $\mathfrak{g}_{\mathbb{C}}$ corresponding to the ith simple root is the stabilizer of the subspace generated by the elements $v_1, \dots, v_i$ of the standard basis, while for the $(p+q-i)$th simple root one obtains the stabilizer of the subspace generated by $v_1, \dots, v_q, v_{q+i+1}, \dots, v_n$.

Thus, we see that the ith basic standard parabolic in $\mathfrak{g}$ is the stabilizer in $\mathfrak{g}$ of the flag formed by these two subspaces. The explicit description of the Hermitian form used for this realization of $\mathfrak{g}$ can be found in 2.3.4. In particular, the first subspace in the flag is isotropic and the second one is the orthocomplement of the first. Since a skew–Hermitian matrix that stabilizes a subspace automatically stabilizes the orthogonal complement, we conclude that the ith basic parabolic in $\mathfrak{g}$ is the stabilizer in $\mathfrak{g}$ of the isotropic subspace $\mathbb{C}^i \subset \mathbb{C}^n$. Thus, the standard parabolic subalgebras in $\mathfrak{su}(p, q)$ for $p > q$ are exactly the stabilizers of isotropic flags in $\mathbb{C}^{p+q}$, and the existence of parabolics (depending on the signature (p, q)) corresponds exactly to the existence of isotropic subspaces of the appropriate dimensions.

For the real form $\mathfrak{g} := \mathfrak{su}(p,p)$ of $\mathfrak{sl}(2p,\mathbb{C})$ the situation is completely parallel. The parabolics in $\mathfrak{g}$ are exactly the stabilizers of isotropic flags in $\mathbb{C}^{2p}$. The contact grading corresponds to the stabilizer of a null line, while the unique $|1|$–grading corresponds to the stabilizer of the maximal isotropic subspace $\mathbb{C}^p \subset \mathbb{C}^{2p}$.

Finally, let us consider the real form $\mathfrak{g} := \mathfrak{sl}(n,\mathbb{H})$ sitting inside $\mathfrak{g}_{\mathbb{C}} = \mathfrak{sl}(2n,\mathbb{C})$ as in example (5) of 2.3.9. Thus, we view $\mathfrak{g}$ as the subalgebra of matrices of the form $\left(\begin{smallmatrix} A & B \\ -\bar{B} & \bar{A} \end{smallmatrix}\right)$ such that $A, B \in \mathfrak{gl}(n,\mathbb{C})$ and the real part of the trace of A vanishes. Again, the diagonal matrices contained in $\mathfrak{g}$ form a θ–stable maximally noncompact Cartan subalgebra whose complexification is the standard Cartan subalgebra of $\mathfrak{g}_{\mathbb{C}}$. As before, one has to use an ordering of the roots that is different from the usual one, and which leads to the simple roots $-e_1+e_{n+1}$, e_1-e_{n+2}, $-e_2+e_{n+2}$, $e_2-e_{n+3}, \dots$, $e_{n-1}-e_{2n}$, $-e_n+e_{2n}$. These roots are ordered in such a way that one obtains the usual Dynkin diagram.

Looking at the Satake diagram, we see only the nodes corresponding to the roots $e_i - e_{n+i+1}$ may be crossed, and we obtain $n-1$ basic parabolics. Each of them gives rise to a $|1|$–grading on $\mathfrak{g}$. As in 3.2.5, one easily verifies that the standard parabolic of $\mathfrak{g}_{\mathbb{C}}$ corresponding to the simple root $e_i - e_{n+i+1}$ is the stabilizer of the subspace generated by the elements $v_1, \dots, v_i, v_{n+1}, \dots, v_{n+i}$ of the standard basis of $\mathbb{C}^{2n}$. Thus, the ith basic parabolic in $\mathfrak{g}$ is the stabilizer of this subspace in $\mathfrak{g}$. Looking at the description of the isomorphism $\mathbb{H}^n \to \mathbb{C}^{2n}$ used to obtain the embedding of $\mathfrak{sl}(n,\mathbb{H})$ into $\mathfrak{sl}(2n,\mathbb{C})$, we see that this is the quaternionic subspace $\mathbb{H}^i \subset \mathbb{H}^n$. Hence, the standard parabolic subalgebras in $\mathfrak{sl}(n,\mathbb{H})$ are exactly the stabilizers of the standard quaternionic flags $\mathbb{H}^{i_1} \subset \cdots \subset \mathbb{H}^{i_\ell} \subset \mathbb{H}^n$ for $1 \le i_1 < i_2 < \cdots < i_\ell < n$.

REMARK 3.2.10. Let $\mathfrak{g}$ be a real semisimple Lie algebra and V a finite–dimensional complex irreducible representation of $\mathfrak{g}$. Then the representation extends to the complexification $\mathfrak{g}_{\mathbb{C}}$ of $\mathfrak{g}$, and of course the stabilizers in $\mathfrak{g}$ of the line through a highest weight vector coincides with the intersection of $\mathfrak{g}$ with the stabilizer in $\mathfrak{g}_{\mathbb{C}}$ of that line. According to 3.2.5 we obtain the intersection of $\mathfrak{g}$ with a parabolic subalgebra of $\mathfrak{g}_{\mathbb{C}}$. This intersection is, however, not a parabolic subalgebra of $\mathfrak{g}$ in general.

3.2.11. Real generalized flag varieties. We switch to the discussion of the homogeneous spaces G/P in the real case, which are again called *generalized flag varieties*. The considerations on the dependence of the homogeneous space G/P on the choice of the groups G and P (corresponding to a fixed choice of a $|k|$–graded semisimple Lie algebra $\mathfrak{g}$) carry over to the real case without big changes. The main difference is that even for connected G, the normalizer $N_G(\mathfrak{p})$ is not connected in many cases of interest. As in the complex case, this will only replace $G/N_G(\mathfrak{p})$ by a (usually finite) covering, so it causes only a minor change.

To prove compactness of G/P, we can proceed very similarly as in the complex case. Fixing a Cartan involution θ on $\mathfrak{g}$ and a connected subgroup G with Lie algebra $\mathfrak{g}$, there is a unique involutive automorphism Θ with derivative θ. The fixed point group K of Θ is a closed subgroup of G which is compact if $Z(G)$ is finite. In that case, K is a maximal compact subgroup of G. As in the proof of Proposition 3.2.6, we can next use the global Iwasawa decomposition $G \cong K \times A \times N$. Part (1) of Theorem 3.2.9 shows that the Lie subalgebra $\mathfrak{a} \oplus \mathfrak{n}$ is contained in $\mathfrak{p}$, so $A \times N$ is contained in any parabolic subgroup corresponding to the grading. Hence, K

acts transitively on G/P, and, provided that $Z(G)$ is finite, compactness follows as in Proposition 3.2.6. This also leads to an interpretation of G/P as the variety of parabolic subalgebras of $\mathfrak{g}$ which are conjugate to $\mathfrak{p}$ via an element of $\mathrm{Ad}(G)$.

Now we switch to the interpretation of generalized flag varieties as orbits in projectivized representations. In 3.2.10 we have seen that real parabolic subalgebras can be realized as stabilizers of lines and flags, but the stabilizer of a highest weight space in a general complex irreducible representation of a real semisimple Lie algebra is not a parabolic subalgebra in general. Consequently, we may still realize the corresponding homogeneous space G/P as orbits in certain projectivized representations, but this does not give as much information about orbits in general irreducible representations as in the complex case.

Given a real semisimple Lie algebra $\mathfrak{g}$ and a standard parabolic subalgebra $\mathfrak{p} \leq \mathfrak{g}$, we can consider the complexification $\mathfrak{p}_{\mathbb{C}}$, which is a standard parabolic subalgebra in the complex semisimple Lie algebra $\mathfrak{g}_{\mathbb{C}}$. Let us denote by Σ the set of simple roots of $\mathfrak{g}_{\mathbb{C}}$ corresponding to this standard parabolic, and consider a dominant integral weight λ for $\mathfrak{g}_{\mathbb{C}}$, such that for a simple root α we have $\langle \lambda, \alpha \rangle \neq 0$ if and only if $\alpha \in \Sigma$. For example, we may take the sum of all fundamental weights corresponding to elements of Σ. Consider a finite–dimensional representation V of $\mathfrak{g}_{\mathbb{C}}$ with highest weight λ and a highest weight vector $v_0 \in V$. From 3.2.5 we know that the stabilizer in $\mathfrak{g}_{\mathbb{C}}$ of the line through v_0 is exactly $\mathfrak{p}_{\mathbb{C}}$, which immediately implies that the stabilizer of this line in $\mathfrak{g}$ is $\mathfrak{p}$. As in 3.2.8 above, we see that the isotropy subgroup of the line through this point in any connected group G to which the representation integrates coincides with $P = N_G(\mathfrak{p})$. Consequently, the G–orbit of the point $[v_0] \in \mathcal{P}(V)$ is diffeomorphic to G/P, and we get a realization as a compact submanifold of a complex projective space. Of course, if the representation V is of real type, i.e. if it is the complexification of a real representation W, then we can choose the highest weight vector in W. Then the G–orbit of $[v_0]$ is contained in the real projectivization of W, so we obtain a realization as a compact submanifold in some real projective space in this case.

Let us discuss this for the A_n–series, where we have looked at the Lie algebraic counterpart in 3.2.10. For the split form $\mathfrak{sl}(n+1, \mathbb{R})$ the parabolics look exactly the same as for $\mathfrak{sl}(n+1, \mathbb{C})$. As in 3.2.6 we obtain real projective space, the real Grassmannians, and all partial real flag manifolds in this way, each of them coming with an embedding into some real projective space.

For the real forms $\mathfrak{su}(p, q)$ with $p + q = n + 1$, $p \geq q > 0$ we know from 3.2.8 that the available parabolic subalgebras corresponds to subsets of the first q simple roots, and with the ith root, always the $(n + 1 - i)$th root has to be crossed. Looking at the basic parabolics (with two crossed nodes) we see that the procedure above leads to a realization of G/P as the orbit of a highest weight line in the projectivization of the highest weight component in $\Lambda^i\mathbb{C}^{n+1} \otimes \Lambda^{n+1-i}\mathbb{C}^{n+1}$. In terms of flags, this means that P is the stabilizers of a flag consisting of an i–dimensional isotropic subspace sitting inside its orthogonal complement. Thus, the orthogonal complement is superfluous, and P is the stabilizer of the i–dimensional isotropic subspace spanned by the first i vectors in the standard basis. Otherwise put, P is the stabilizer of the highest weight line in the projectivization of $\Lambda^i\mathbb{C}^{n+1}$, and G/P can be realized as the orbit of this point.

Similar realizations can be constructed for more general parabolics. Let P be the parabolic subgroup corresponding to $\Sigma = \{\alpha_{i_1}, \dots, \alpha_{i_\ell}, \alpha_{n+1-i_1}, \dots, \alpha_{n+1-i_\ell}\}$

with $1 \leq i_1 < \cdots < i_\ell \leq q$. Elementary linear algebra shows that the group $G = SU(p,q)$ acts transitively on all space of isotropic flags in $\mathbb{C}^{p+q}$. Using this one shows as above that G/P is the manifold of all flags $V_1 \subset \cdots \subset V_\ell \subset \mathbb{C}^{p+q}$ where each V_j is isotropic and has dimension i_j. A particularly important example of this situation is the case of the maximal parabolic corresponding to $\Sigma = \{\alpha_1, \alpha_n\}$. In this case, G/P is the space of all isotropic lines in $\mathbb{C}^{p+q}$, i.e. a quadric in $\mathbb{C}P^{p+q-1}$. Hence, G/P is a real hypersurface in the complex manifold $\mathbb{C}P^{p+q-1}$, which indicates the relation of this generalized flag manifold to CR–geometry.

Finally, in the case of the real forms $\mathfrak{sl}(m,\mathbb{H})$ of $\mathfrak{sl}(2m,\mathbb{C})$ we can use the group $G = SL(m,\mathbb{H})$. As above, one shows that the parabolic subgroups P of G can all be realized as the stabilizers of the standard quaternionic flags $\mathbb{H}^{i_1} \subset \cdots \subset \mathbb{H}^{i_\ell} \subset \mathbb{H}^m$ for $1 \leq i_1 < \cdots < i_\ell < m$. Hence, the homogeneous spaces G/P are the quaternionic flag manifolds. As before, we also get embeddings of flag manifolds into the projectivizations of appropriate complex representations of $SL(m,\mathbb{H})$.

3.2.12. Representations of $\mathfrak{p}$. From 1.5.5 we know that any representation of P gives rise to a natural bundle on the category of parabolic geometries of type (G,P), so representations of the parabolic subalgebra $\mathfrak{p}$ are of central interest in the theory. The Lie algebra $\mathfrak{p}$ is rather complicated, so there is little hope to get a complete picture for general representations of $\mathfrak{p}$. Some basic properties are, however, easy to prove. Moreover, there is a complete description of completely reducible representations.

There is a standing assumption that we will have to make on representations of $\mathfrak{p}$, which comes from the representation theory of $\mathfrak{g}_0$. Since this subalgebra is only reductive and not semisimple, finite–dimensional representations of $\mathfrak{g}_0$ are *not* automatically completely reducible. Indeed, a finite–dimensional representation W of $\mathfrak{g}_0$ is completely reducible if and only if the center $\mathfrak{z}(\mathfrak{g}_0)$ acts diagonalizably on W: Since $\mathfrak{z}(\mathfrak{g}_0)$ acts by a character on any irreducible representation of $\mathfrak{g}_0$ the necessity of this condition is evident. Conversely, if $\mathfrak{z}(\mathfrak{g}_0)$ acts diagonalizably, then any eigenspace for this action is invariant under $\mathfrak{g}_0^{ss}$. Thus, by Theorem 2.1.6 each of these eigenspaces splits into a direct sum of irreducible representations of $\mathfrak{g}_0^{ss}$ and this gives a splitting of W into a direct sum of irreducible $\mathfrak{g}_0$–modules. Studying representations of $\mathfrak{p}$, we will therefore always assume that $\mathfrak{z}(\mathfrak{g}_0)$ acts diagonalizably.

PROPOSITION 3.2.12. *Let $\mathfrak{g} = \mathfrak{g}_{-k} \oplus \cdots \oplus \mathfrak{g}_k$ be a $|k|$–graded semisimple Lie algebra, $\mathfrak{p} = \mathfrak{g}^0$ the corresponding parabolic subalgebra and $E \in \mathfrak{z}(\mathfrak{g}_0)$ the grading element.*

(1) Any finite–dimensional completely reducible representation W of $\mathfrak{p}$ is obtained by trivially extending a completely reducible representation of $\mathfrak{g}_0$ to $\mathfrak{p}$. Moreover, E acts by a scalar on each irreducible component of W.

(2) Let V be a finite–dimensional representation of $\mathfrak{p}$ such that $\mathfrak{z}(\mathfrak{g}_0)$ acts diagonalizably. Then V admits a $\mathfrak{p}$–invariant filtration $V = V^0 \supset V^1 \supset \cdots \supset V^N \supset V^{N+1} = \{0\}$ such that each of the quotients V^i/V^{i+1} is completely reducible.

PROOF. (1) It suffices to consider the case that W is irreducible, and we first assume that W is a complex representation. Then the action of E on W must have at least one eigenvalue. Let s_0 be an eigenvalue with maximal real part and let W_{s_0} be the corresponding eigenspace. Since $E \in \mathfrak{z}(\mathfrak{g}_0)$, the subspace $W_{s_0} \subset W$ is $\mathfrak{g}_0$–invariant. For $w \in W_{s_0}$ and $X \in \mathfrak{g}_j \subset \mathfrak{p}$ the equation $E \cdot X \cdot w = [E,X] \cdot w + X \cdot E \cdot w$ implies that $X \cdot w$ is an eigenvector corresponding to the eigenvalue $s_0 + j$, so

$X \cdot w = 0$ for $j > 0$ and $X \cdot w \in W_{s_0}$ for $j = 0$. Thus, $W_{s_0} \subset W$ is $\mathfrak{p}$–invariant and by irreducibility $W_{s_0} = W$. This completes the proof in the complex case.

For a real irreducible representation W of $\mathfrak{p}$ which is not complex, the complexification $W_{\mathbb{C}}$ is irreducible, which immediately implies that E acts by a (real) scalar on W and the subalgebra $\mathfrak{p}_+ = \mathfrak{g}_1 \oplus \cdots \oplus \mathfrak{g}_k$ acts trivially.

(2) Define $V^N := \{v \in V : Z \cdot v = 0 \quad \forall Z \in \mathfrak{p}_+\}$, where as usual $\mathfrak{p}_+ = \mathfrak{g}_1 \oplus \cdots \oplus \mathfrak{g}_k$. For $A \in \mathfrak{p}$ and $Z \in \mathfrak{p}_+$ we have $[Z, A] \in \mathfrak{p}_+$, and since $Z \cdot A \cdot v = A \cdot Z \cdot v + [Z, A] \cdot v$ we see that V^N is a $\mathfrak{p}$–invariant subspace of V. Since $\mathfrak{z}(\mathfrak{g}_0)$ acts diagonalizably on V, the same is true on the invariant subspace V^N. In particular, V^N is completely reducible as a $\mathfrak{g}_0$–representation. Since $\mathfrak{p}_+$ acts trivially on V^N the decomposition into $\mathfrak{g}_0$–irreducibles is a decomposition into $\mathfrak{p}$–irreducibles, so V^N is a completely reducible $\mathfrak{p}$–module. Next for $i = N, N-1, \ldots$ we inductively define

$$V^{i-1} := \{v \in V : Z \cdot v \in V^i \quad \forall Z \in \mathfrak{p}_+\}.$$

As before, we conclude inductively that each V^i is a $\mathfrak{p}$–invariant subspace of V. By construction, $\mathfrak{p}_+$ acts trivially on the quotient V^i/V^{i+1}. Moreover, $\mathfrak{z}(\mathfrak{g}_0)$ acts diagonalizably on V^i and V^{i+1} so it also acts diagonalizably on the quotient. As before, this implies that V^i/V^{i+1} is a completely reducible $\mathfrak{p}$–module.

Hence, it remains to show that $V^i = V$ for sufficiently small i. Since $\mathfrak{z}(\mathfrak{g}_0)$ acts diagonalizably on V it suffices to show that any eigenvector for the action of the grading element E lies in some V^i. But if $E \cdot v = sv$ and $Z \in \mathfrak{g}_j$, then $E \cdot Z \cdot v = (s + j)Z \cdot v$. If $\ell \in \mathbb{N}$ is such that $s + \ell$ is the maximal eigenvalue of E on V of the form $s + n$ with $n \in \mathbb{N}$, then by construction $v \in V^{N-\ell}$. Thus, we conclude that $V^i = V$ for an appropriate i and we can change N in such a way that the largest i with this property is 0. □

We next want to describe complex irreducible representations of $\mathfrak{p}$ in terms of highest weights. Since such representations extend to the complexification, we may restrict to the case of complex $|k|$–graded Lie algebras. Let us first briefly recall the theory for the complex semisimple Lie algebra $\mathfrak{g}$. Let $\mathfrak{h} \leq \mathfrak{g}$ be a Cartan subalgebra, Δ the corresponding set of roots, and Δ^+ a choice of positive subsystem. We will later assume that these choices have been made in such a way that $\mathfrak{p}$ is a standard parabolic subalgebra, so $\mathfrak{h}$ and all positive root spaces are contained in $\mathfrak{p}$.

On any finite–dimensional complex representation V of $\mathfrak{g}$, the Cartan subalgebra $\mathfrak{h}$ acts diagonalizably. The corresponding eigenvalues $\lambda \in \mathfrak{h}^*$ are called the *weights* of V and the eigenspaces are called *weight spaces.* A *highest weight vector* in V is an element of a weight space which is annihilated by the action of all elements of positive root spaces. Such vectors exist in any finite dimensional complex representation and in irreducible representations they are unique up to scale; see 2.2.11. The *highest weight* of a finite–dimensional irreducible representation V is the weight of its highest weight vectors. These highest weights are always *dominant* and *algebraically integral,* which means that for the inner product induced by the Killing form the expression $\frac{2\langle\lambda,\alpha\rangle}{\langle\alpha,\alpha\rangle}$ is a nonnegative integer for each simple root α. The theorem of the highest weight (Theorem 2.2.12) says that there is a bijective correspondence between dominant integral weights and isomorphism classes of finite–dimensional complex irreducible representations of $\mathfrak{g}$.

There is a useful reformulation of the condition of being dominant and integral. Namely, if $\alpha_1, \ldots, \alpha_n$ are the simple roots of $\mathfrak{g}$, one defines the *fundamental weights*

$\lambda_1, \dots, \lambda_n$ of $\mathfrak{g}$ by $\frac{2\langle\lambda_i,\alpha_j\rangle}{\langle\alpha_j,\alpha_j\rangle} = \delta_{ij}$. Then any weight can be written as a linear combination of the fundamental weights. Dominant integral weights are exactly those, for which all coefficients in this expansion are nonnegative integers. The Dynkin diagram notation for weights and representations is then obtained by writing the coefficient of λ_i in this expansion over the node of the Dynkin diagram of $\mathfrak{g}$ that corresponds to the simple root α_i.

Complex irreducible representations of $\mathfrak{p}$ can be dealt with in a very similar way. By part (1) of the proposition, these coincide with complex irreducible representations of $\mathfrak{g}_0$, which in turn are given by irreducible representations of the semisimple part $\mathfrak{g}_0^{ss}$ and linear functionals on the center $\mathfrak{z}(\mathfrak{g}_0)$. Assuming that $\mathfrak{p}$ is a standard parabolic, we obtain the corresponding subset $\Sigma = \{\alpha \in \Delta^0 : \mathfrak{g}_\alpha \subset \mathfrak{g}_1\}$ of simple roots. From 3.2.2 we know we can naturally split the Cartan algebra $\mathfrak{h} \subset \mathfrak{g}$ as $\mathfrak{h} = \mathfrak{h}' \oplus \mathfrak{h}''$, with $\mathfrak{h}' := \{H \in \mathfrak{h} : \alpha(H) = 0 \quad \forall\alpha \in \Delta^0 \setminus \Sigma\}$ and $\mathfrak{h}''$ the span of the elements H_α with $\alpha \in \Delta^0 \setminus \Sigma$. Then $\mathfrak{h}' = \mathfrak{z}(\mathfrak{g}_0)$, while $\mathfrak{h}''$ is a Cartan subalgebra for $\mathfrak{g}_0^{ss}$. Hence, complex irreducible representations of $\mathfrak{g}_0$ are again in bijective correspondence with a set of functionals on $\mathfrak{h}$, but the dominance and integrality conditions refer only to the restriction to $\mathfrak{h}''$. In analogy to the usual notions we define a weight $\lambda : \mathfrak{h} \to \mathbb{C}$ to be $\mathfrak{p}$*–dominant*, respectively, $\mathfrak{p}$*–algebraically integral* if $\frac{2\langle\lambda,\alpha\rangle}{\langle\alpha,\alpha\rangle}$ is real and nonnegative, respectively, an integer for all $\alpha \in \Delta^0 \setminus \Sigma$. Thus, we obtain

COROLLARY 3.2.12. *Let $\mathfrak{p} \leq \mathfrak{g}$ be a standard parabolic subalgebra in a complex semisimple Lie algebra. Then isomorphism classes of finite–dimensional complex irreducible representations of $\mathfrak{p}$ are in bijective correspondence with weights $\lambda : \mathfrak{h} \to \mathbb{C}$ which are $\mathfrak{p}$–dominant and $\mathfrak{p}$–algebraically integral.*

The condition of $\mathfrak{p}$–dominance and integrality can again be rephrased in terms of fundamental weights as the requirement that the coefficients of all fundamental weights corresponding to simple roots not contained in Σ must be nonnegative integers. In the Dynkin diagram notation this means that the coefficients over all uncrossed nodes are nonnegative integers.

As we have noted in the propostion, the grading element acts by a scalar on each irreducible representation. Since this is often needed in applications, let us remark at this point that there is a simple way to compute this scalar. Since the simple roots form a basis for $\mathfrak{h}^*$, we can expand any weight in terms of the simple roots. Having determined this expansion, the action of E on a highest weight vector (and hence on all of W) is the sum of the coefficients of the crossed roots. To convert from the expansion in terms of fundamental weights to the one in terms of simple roots, we proceed as follows. Suppose that the fundamental weight λ_i can be written as $\sum_k b_{ki}\alpha_k$. Then the defining property of λ_i reads as

$$\delta_{ij} = \frac{2\langle\lambda_i, \alpha_j\rangle}{\langle\alpha_j, \alpha_j\rangle} = \sum_k b_{ki} \frac{2\langle\alpha_k,\alpha_j\rangle}{\langle\alpha_j,\alpha_j\rangle}.$$

But the last term is just the coefficient a_{jk} of the Cartan matrix of $\mathfrak{g}$, so the numbers b_{ki} form the ith column of the inverse of the Cartan matrix of $\mathfrak{g}$. Hence, the action of E on W can be computed by adding those rows of the inverse of the Cartan matrix which correspond to crossed roots, and pairing the result with the vector of coefficients with respect to the fundamental weights. The inverse Cartan matrices for all complex simple Lie algebras can be found in Table B.4 in Appendix B.

In the realm of Lie algebra representations, there is no restriction on the coefficients over the crossed nodes. However, if one wants the representation to integrate to at least one parabolic subgroup, then the coefficients over the crossed nodes have to be integers. To see this, note that for a simple root α_j, we have the corresponding subalgebra $\mathfrak{s}_{\alpha_j} := \mathfrak{g}_{-\alpha_j} \oplus [\mathfrak{g}_{-\alpha_j}, \mathfrak{g}_{\alpha_j}] \oplus \mathfrak{g}_{\alpha_j}$, which is isomorphic to $\mathfrak{sl}(2,\mathbb{C})$; see 2.2.4. Under this identification, the element H_{α_i} corresponds to the matrix $H = \left(\begin{smallmatrix} 1 & 0 \\ 0 & -1 \end{smallmatrix}\right)$. Since $SL(2,\mathbb{C})$ is simply connected, the inclusion $\mathfrak{s}_{\alpha_j} \to \mathfrak{g}$ induces a Lie group homomorphism $SL(2,\mathbb{C}) \to G$ for any group G with Lie algebra $\mathfrak{g}$. Now obviously $\exp(2\pi i H) = \mathrm{id}$ in $SL(2,\mathbb{C})$, whence $\exp(2\pi i H_{\alpha_j}) = e$ in G and thus also in the parabolic subgroup P. If the $\mathfrak{p}$–representation with highest weight λ integrates to P, then the action of $\exp(2\pi i H_{\alpha_j})$ on a highest weight vector is given by multiplication by $e^{2\pi i \lambda(H_{\alpha_j})}$, which shows that $\lambda(H_{\alpha_j}) \in \mathbb{Z}$ and $\lambda(H_{\alpha_j})$ is exactly the coefficient of λ_j. Hence, a representation that integrates to some parabolic subgroup has to have integer coefficients over all nodes. Note that in the case of real parabolics there are usually less (or even no) integrality conditions since in this case it may happen that $\mathfrak{z}(\mathfrak{g}_0)$ integrates to a subgroup isomorphic to $\mathbb{R}^\ell$.

3.2.13. The relation between representations of $\mathfrak{g}$ and $\mathfrak{p}$. Let $\mathfrak{g}$ be a complex simple Lie algebra and let $\mathfrak{p} \subset \mathfrak{g}$ be a standard parabolic. For a finite–dimensional complex representation V of $\mathfrak{g}$, the center $\mathfrak{z}(\mathfrak{g}_0)$ acts diagonalizably on V since $\mathfrak{z}(\mathfrak{g}_0) \subset \mathfrak{h}$. Hence, by part (2) of Proposition 3.2.12, we get a $\mathfrak{p}$–invariant filtration $V = V^0 \supset V^1 \supset \cdots \supset V^N$ in which V^N is the space $V^{\mathfrak{p}_+}$ of elements $v \in V$, which are $\mathfrak{p}_+$–invariant i.e. which satisfy $Z \cdot v = 0$ for all $Z \in \mathfrak{p}_+$. Notice that $V^{\mathfrak{p}_+}$ may also be interpreted cohomologically as $H^0(\mathfrak{p}_+, V)$; see 2.1.9.

PROPOSITION 3.2.13. *Let V be a finite-dimensional complex representation of $\mathfrak{g}$ and let $V^{\mathfrak{p}_+} \subset V$ be the subspace of $\mathfrak{p}_+$–invariant elements. Then there is a bijective correspondence between $\mathfrak{p}$–invariant subspaces of $V^{\mathfrak{p}_+}$ and $\mathfrak{g}$–invariant subspaces of V. In particular, if V is the irreducible representation of $\mathfrak{g}$ with highest weight λ, then $V^{\mathfrak{p}_+}$ is the irreducible $\mathfrak{p}$–representation with the same highest weight.*

PROOF. If $\tilde{V} \subset V$ is $\mathfrak{g}$–invariant, then evidently $\tilde{V}^{\mathfrak{p}_+} = \tilde{V} \cap V^{\mathfrak{p}_+}$ is $\mathfrak{p}$–invariant. Conversely, for a subspace $W \subset V^{\mathfrak{p}_+}$ we can consider the $\mathfrak{g}$–submodule of V generated by W, which is a $\mathfrak{g}$–invariant subspace by definition. It suffices to show that these two constructions are inverse to each other.

If $v \in V$ is a highest weight vector for $\mathfrak{g}$, then $v \in V^{\mathfrak{p}_+}$ by definition. On $V^{\mathfrak{p}_+}$ we have to understand the representation of $\mathfrak{g}_0^{ss}$, since we know that $\mathfrak{z}(\mathfrak{g}_0)$ acts diagonalizably. This is determined by the highest weight vectors for $\mathfrak{g}_0^{ss}$. But if such a highest weight vector is contained in $V^{\mathfrak{p}_+}$, then it is by definition a highest weight vector for $\mathfrak{g}$.

From 2.2.11 we know that any finite–dimensional $\mathfrak{g}$–representation is generated by its highest weight vectors. In particular, any $\mathfrak{g}$–invariant subspace $\tilde{V} \subset V$ has to be generated by $\tilde{V}^{\mathfrak{p}_+}$. If $W \subset V^{\mathfrak{p}_+}$ is $\mathfrak{p}$–invariant, then W is generated as a $\mathfrak{p}$–module by the $\mathfrak{g}_0$–highest weight vectors it contains. Any such vector is a highest weight vector for $\mathfrak{g}$, and hence by Theorem 2.2.11 generates an irreducible $\mathfrak{g}$–submodule in V. In this submodule, the highest weight vector is unique up to scale. This implies that if $\hat{W}$ is the $\mathfrak{g}$–submodule generated by W, then $\hat{W}^{\mathfrak{p}_+} = W$. Thus, we have established the correspondence and the last statement follows immediately. □

We can nicely rephrase these facts in terms of the universal enveloping algebra. Let $\mathcal{U}(\mathfrak{g})$ be the universal enveloping algebra of $\mathfrak{g}$; see 2.1.10. Choosing a basis of $\mathfrak{g}$ that is the union of a basis $\{X_j\}$ of $\mathfrak{g}_-$ and a basis $\{A_k\}$ of $\mathfrak{p}$, the Poincaré–Birkhoff–Witt theorem (see 2.1.10) implies that the monomials of the form $X_1^{i_1}\dots X_n^{i_n}A_1^{j_1}\dots A_m^{j_m}$ form a linear basis of $\mathcal{U}(\mathfrak{g})$. In particular, this implies that the multiplication in $\mathcal{U}(\mathfrak{g})$ defines an isomorphism $\mathcal{U}(\mathfrak{g})\cong\mathcal{U}(\mathfrak{g}_-)\otimes\mathcal{U}(\mathfrak{p})$ of vector spaces. Since V is generated by $V^{\mathfrak{p}+}$ as a $\mathfrak{g}$–module, we get $V=\mathcal{U}(\mathfrak{g})\cdot V^{\mathfrak{p}+}$. Since $V^{\mathfrak{p}+}\subset V$ is a $\mathfrak{p}$–submodule, we have $\mathcal{U}(\mathfrak{p})\cdot V^{\mathfrak{p}+}\subset V^{\mathfrak{p}+}$ and hence $V=\mathcal{U}(\mathfrak{g}_-)\cdot V^{\mathfrak{p}+}$.

These ideas naturally lead to the concept of generalized Verma modules. Parallel to the case of ordinary Verma modules, let W be an irreducible representation of the parabolic subalgebra $\mathfrak{p}$ with highest weight λ. Then we define the *generalized Verma module* $M_{\mathfrak{p}}(\lambda)$ as $\mathcal{U}(\mathfrak{g})\otimes_{\mathcal{U}(\mathfrak{p})}W$, compare with 2.2.12. By construction, this is a $\mathcal{U}(\mathfrak{g})$–module under left multiplication. The vector space decomposition $\mathcal{U}(\mathfrak{g})\cong\mathcal{U}(\mathfrak{g}_-)\otimes\mathcal{U}(\mathfrak{p})$ from above immediately implies that $M_{\mathfrak{p}}(\lambda)\cong\mathcal{U}(\mathfrak{g}_-)\otimes W$ as a vector space and a $\mathfrak{g}_-$–module.

The generalized Verma module has a universal property. Namely, mapping $w\in W$ to $1\otimes w$ defines an inclusion $i:W\to M_{\mathfrak{p}}(\lambda)$ of $\mathfrak{p}$–modules. Suppose that V is any representation of $\mathfrak{g}$ and $\phi:W\to V$ is a homomorphism of $\mathfrak{p}$–modules. Then consider the map $\mathcal{U}(\mathfrak{g})\otimes W\to V$ defined by $A\otimes w\mapsto A\cdot\phi(w)$. Obviously, this is a $\mathfrak{g}$–homomorphism and since ϕ is a $\mathcal{U}(\mathfrak{p})$–homomorphism, it factors to a $\mathfrak{g}$–homomorphism $\tilde{\phi}:M_{\mathfrak{p}}(\lambda)\to V$ such that $\tilde{\phi}\circ i=\phi$, and since $M_{\mathfrak{p}}(\lambda)$ is visibly generated by $i(W)$ as a $\mathcal{U}(\mathfrak{g})$–module, $\tilde{\phi}$ is uniquely determined by this property. Taking V to be the irreducible $\mathfrak{g}$–representation of highest weight λ and $\phi:W\to V$ to be the inclusion of $V^{\mathfrak{p}+}$, the homomorphism $\tilde{\phi}$ must be surjective by irreducibility, so V can be naturally realized as a quotient of $M_{\mathfrak{p}}(\lambda)$ for any parabolic subalgebra $\mathfrak{p}$ of $\mathfrak{g}$.

For later use, we also want to clarify the relation between ordinary and generalized Verma modules. Of course, mapping $1\otimes 1\in M_{\mathfrak{b}}(\lambda)$ to $1\otimes w_0\in M_{\mathfrak{p}}(\lambda)$, where $w_0\in W$ is a highest weight vector induces a homomorphism $M_{\mathfrak{b}}(\lambda)\to M_{\mathfrak{p}}(\lambda)$ of $\mathcal{U}(\mathfrak{g})$–modules. Moreover, from 2.2.12 we know that, as a vector space, $M_{\mathfrak{b}}(\lambda)$ is isomorphic to $\mathcal{U}(\mathfrak{n}_-)$, where $\mathfrak{n}_-$ is the sum of all negative root spaces. Choosing an appropriate basis for $\mathfrak{n}_-$ we see that as a vector space $\mathcal{U}(\mathfrak{n}_-)\cong\mathcal{U}(\mathfrak{g}_-)\otimes\mathcal{U}(\mathfrak{p}_-)$. Here, $\mathfrak{p}_-$ is the direct sum of all negative root spaces contained in $\mathfrak{g}_0$. Since W is $\mathfrak{p}$–irreducible, we have $W=\mathcal{U}(\mathfrak{p}_-)\cdot w_0$, and this immediately implies that the homomorphism $M_{\mathfrak{b}}(\lambda)\to M_{\mathfrak{p}}(\lambda)$ is surjective, so any generalized Verma module is a quotient of the Verma module with the same highest weight.

3.2.14. On the Weyl group. Our next aim is to define and analyze the Hasse diagram associated to a parabolic subalgebra $\mathfrak{p}$ in a complex semisimple Lie algebra $\mathfrak{g}$, which encodes an amazing amount of information about a parabolic geometry and its homogeneous model. It is based on the Weyl group of $\mathfrak{g}$ and we first have to prove some facts about the Weyl group.

As before, we consider a complex semisimple Lie algebra $\mathfrak{g}$ endowed with a Cartan subalgebra $\mathfrak{h}\leq\mathfrak{g}$ and an ordering on $\mathfrak{h}^*$, and we denote by Δ, Δ^+, and Δ^0 the corresponding sets of roots, positive roots, and simple roots, respectively. Recall from 2.2.4 that the (real) subspace $\mathfrak{h}_0\subset\mathfrak{h}$ on which all roots are real defines a real form of $\mathfrak{h}$, and the Killing form restricts to a positive definite inner product on $\mathfrak{h}_0$. The set Δ of roots is then a finite subset of the real dual space $\mathfrak{h}_0^*$, and via the duality, we can carry over the Killing form to a positive definite inner product

$\langle\ ,\ \rangle$ on $\mathfrak{h}_0^*$. For any $\alpha \in \Delta$ we can consider the *root reflection* $s_\alpha : \mathfrak{h}_0^* \to \mathfrak{h}_0^*$, defined by $s_\alpha(\phi) = \phi - 2\frac{\langle \phi,\alpha\rangle}{\langle \alpha,\alpha\rangle}\alpha$, which maps Δ to itself.

The Weyl group $W = W_\mathfrak{g}$ of $\mathfrak{g}$ is then by definition the subgroup of the orthogonal group $O(\mathfrak{h}_0^*)$ generated by these root reflections. In 2.2.7 we have noted that we may view W as a subgroup of the group of bijections of Δ, whence W is finite. Further we noted that W is actually generated by the reflections s_{α_i} corresponding to simple roots α_i. An expression of $w \in W$ as a composition of simple root reflections is called *reduced* if it has the least possible number of factors. This number is called the *length* $\ell(w)$ of the element w. On the other hand, the *sign* $\operatorname{sgn}(w)$ of $w \in W$ is defined as the determinant of the linear map $w : \mathfrak{h}_0^* \to \mathfrak{h}_0^*$. By definition, we have $\operatorname{sgn}(w) = (-1)^{\ell(w)}$.

For later use, we have to view the Weyl group W not only as a group but also as a directed graph. The vertices of this graph are elements $w \in W$, and we have a directed edge $w \xrightarrow{\alpha} w'$ labeled with $\alpha \in \Delta^+$ if and only if $\ell(w') = \ell(w) + 1$ and $w' = s_\alpha w$. To analyze this graph structure, we need one more ingredient, which is closely related to the fact that the Weyl group acts transitively on the set of all simple systems for Δ, respectively, the set of all Weyl–chambers. Namely, for $w \in W$ define $\Phi_w := \{\alpha \in \Delta^+ : w^{-1}(\alpha) \in -\Delta^+\}$, or equivalently $\Phi_w = w(-\Delta^+) \cap \Delta^+$. Suppose that $\alpha, \beta \in \Phi_w$ are such that $\alpha + \beta \in \Delta$ (and hence $\alpha + \beta \in \Delta^+$). Then $w^{-1}(\alpha + \beta) = w^{-1}(\alpha) + w^{-1}(\beta)$ lies in Δ, and since it is the sum of two negative roots, it must be negative, too. Consequently, the set Φ_w is *saturated*, i.e. if $\alpha, \beta \in \Phi_w$ and $\alpha + \beta \in \Delta$, then $\alpha + \beta \in \Phi_w$. Similarly, if $\alpha, \beta \in \Delta^+ \setminus \Phi_w$ are such that $\alpha + \beta \in \Delta$, then $w^{-1}(\alpha + \beta)$ must be a positive root. Hence, $\alpha + \beta \notin \Phi_w$, so the complement $\Delta^+ \setminus \Phi_w$ is saturated, too. Further, we define $\langle \Phi_w \rangle := \sum_{\alpha \in \Phi_w} \alpha$. Finally, recall from 2.2.18 that the *lowest form* $\delta = \delta_\mathfrak{g}$ of the Lie algebra $\mathfrak{g}$ is defined as $\delta = \frac{1}{2}\sum_{\alpha \in \Delta^+} \alpha$, and can also be expressed as the sum of all fundamental weights. Hence, δ lies in the interior of the dominant Weyl chamber, and since the Weyl group W acts simply transitively on the set of all Weyl chambers, we conclude that mapping $w \in W$ to $w(\delta)$ is an injection from W to $\mathfrak{h}_0^*$. Using this, we can now state

PROPOSITION 3.2.14. *Let $\mathfrak{g}$ be a complex semisimple Lie algebra with a chosen simple system Δ^0, Weyl group W and lowest form $\delta \in \mathfrak{h}_0^*$. Then we have:*

(1) *$w(\delta) = \delta - \langle \Phi_w \rangle$ for all $w \in W$.*
(2) *The map $w \mapsto \Phi_w$ defines a bijection between W and the set of all subsets $\Phi \subset \Delta^+$ such that both Φ and $\Delta^+ \setminus \Phi$ are saturated.*
(3) *$|\Phi_w| = \ell(w)$ for all $w \in W$.*
(4) *For $w \in W$ and $\alpha \in \Delta^+$, the element α is contained in exactly one of the sets Φ_w and $\Phi_{s_\alpha w}$, and $\ell(s_\alpha w) > \ell(w)$ if and only if $\alpha \notin \Phi_w$. In particular, if $\alpha \in \Delta^0$, then $\ell(s_\alpha w) = \ell(w) + 1$ if $\alpha \notin \Phi_w$ and $\ell(s_\alpha w) = \ell(w) - 1$ if $\alpha \in \Phi_w$.*
(5) *Let $w, w' \in W$ be two elements such that $|\Phi_{w'}| = |\Phi_w| + 1$. Then for $\alpha \in \Delta^+$ we have $w \xrightarrow{\alpha} w'$ if and only if $\langle \Phi_{w'} \rangle = \langle \Phi_w \rangle + k\alpha$ for some $k \in \mathbb{Z}$.*

PROOF. For $\alpha \in \Delta$ let us during this proof put $\operatorname{sgn}(\alpha)$ to be 1 if $\alpha \in \Delta^+$ and -1 if $\alpha \in -\Delta^+$.

(1) Since any element $w \in W$ permutes the roots, we get

$$\delta = \tfrac{1}{4}\textstyle\sum_{\alpha \in \Delta} \operatorname{sgn}(\alpha)\alpha = \tfrac{1}{4}\sum_{\alpha \in \Delta} \operatorname{sgn}(w(\alpha))w(\alpha).$$

On the other hand, applying w to the first expression for δ, we obtain $w(\delta) = \frac{1}{4}\sum_{\alpha\in\Delta}\operatorname{sgn}(\alpha)w(\alpha)$, and consequently

$$\delta - w(\delta) = \frac{1}{4}\sum_{\alpha\in\Delta}(\operatorname{sgn}(w(\alpha)) - \operatorname{sgn}(\alpha))w(\alpha).$$

Splitting the sum into the sum over negative roots and the sum over positive roots, we see that for $\alpha \in -\Delta^+$ the expression $\operatorname{sgn}(w(\alpha)) - \operatorname{sgn}(\alpha)$ equals 2 for $w(\alpha) \in \Phi_w$ and zero otherwise, whence this first sum adds up to $2\langle\Phi_w\rangle$. On the other hand, if $\alpha \in \Delta^+$, then $\operatorname{sgn}(w(\alpha)) - \operatorname{sgn}(\alpha)$ equals -2 if $w(\alpha) \in -\Delta^+$ and zero otherwise. But $w(\alpha) \in -\Delta^+$ if and only if $w(\alpha) \in -\Phi_w$, so the sum adds up to $-2(-\langle\Phi_w\rangle)$ and the claim follows.

(2) Since we have observed above that both Φ_w and $\Delta^+ \setminus \Phi_w$ are saturated, we get a well–defined map as claimed. If $\Phi_w = \Phi_{w'}$ for some elements $w, w' \in W$, then (1) implies $w(\delta) = w'(\delta)$, and we have observed above that this implies $w = w'$, so our mapping is injective.

For a subset $\Phi \subset \Delta^+$ such that Φ and $\Delta^+ \setminus \Phi$ are saturated define $\Psi := -\Phi \cup (\Delta^+ \setminus \Phi) \subset \Delta$. By construction, Δ is the disjoint union of Ψ and $-\Psi$. We claim that if $\alpha, \beta \in \Psi$ are such that $\alpha + \beta \in \Delta$, then $\alpha + \beta \in \Psi$. If both roots lie either in $-\Phi$ or in $(\Delta^+ \setminus \Phi)$, then this follows since the two sets are saturated. Hence, suppose that $\alpha \in -\Phi$ and $\beta \in (\Delta^+ \setminus \Phi)$. If $\alpha + \beta \in \Phi$, then $-\alpha \in \Phi$ implies $\beta \in \Phi$, a contradiction. Likewise, $\alpha+\beta \in -(\Delta^+\setminus\Phi)$ together with $-\beta \in -(\Delta^+\setminus\Phi)$ would imply the contradiction $\alpha \in -(\Delta^+ \setminus \Phi)$, so the claim follows.

Hence, we can use Ψ as a positive subsystem of Δ. By part (2) of Proposition 2.2.7, we can map Δ^0 to the the corresponding simple subset by an element $w \in W$. This element then satisfies $w(-\Delta^+) = -\Psi = \Phi \cup -(\Delta^+ \setminus \Phi)$, and hence $\Phi_w = \Phi$ by definition.

(3) and (4): Fix a positive root $\alpha \in \Delta^+$ and for $\beta \in \Delta^+$ define $\phi_\alpha(\beta) := \operatorname{sgn}(s_\alpha(\beta))s_\alpha(\beta)$. By construction $\phi_\alpha(\Delta^+) \subset \Delta^+$ and $\phi_\alpha^2 = \mathrm{id}$, so this is a bijection. We first claim that for any element $w \in W$ the map ϕ_α restricts to bijections

$$(\Phi_w \setminus \Phi_{s_\alpha}) \cong (\Phi_{s_\alpha w} \setminus \Phi_{s_\alpha}) \quad\text{and}\quad (\Phi_{s_\alpha} \setminus \Phi_w) \cong (\Phi_{s_\alpha w} \cap \Phi_{s_\alpha}). \tag{3.6}$$

If $\beta \in \Phi_{s_\alpha}$, then $\phi_\alpha(\beta) = -s_\alpha(\beta)$, which under s_α is mapped to $-\beta \in -\Delta^+$, so ϕ_α preserves Φ_{s_α} and thus also $\Delta^+ \setminus \Phi_{s_\alpha}$. Next, suppose that $\beta \in \Phi_w \setminus \Phi_{s_\alpha}$. Then $\phi_\alpha(\beta) = s_\alpha(\beta)$, and $(s_\alpha w)^{-1}(\phi_\alpha(\beta)) = w^{-1}(\beta)$, so $\phi_\alpha(\beta) \in \Phi_{s_\alpha w}$. Hence, ϕ_α maps $\Phi_w \setminus \Phi_{s_\alpha}$ to $\Phi_{s_\alpha w} \setminus \Phi_{s_\alpha}$. Replacing w by $s_\alpha w$, the first bijection in (3.6) follows. On the other hand, for $\beta \in \Phi_{s_\alpha}$ we have $\phi_\alpha(\beta) = -s_\alpha(\beta)$, and thus $w^{-1}s_\alpha(\phi_\alpha(\beta)) = -w^{-1}(\beta)$, so $\phi_\alpha(\beta) \in \Phi_{s_\alpha w}$ if and only if $\beta \notin \Phi_w$. Replacing w by $s_\alpha w$, the second bijection in (3.6) follows, which completes the proof of the claim.

The bijections in (3.6) immediately imply that

$$|\Phi_{s_\alpha w}| = |\Phi_w \setminus \Phi_{s_\alpha}| + |\Phi_{s_\alpha} \setminus \Phi_w| = |\Phi_w| + |\Phi_{s_\alpha}| - 2|\Phi_w \cap \Phi_{s_\alpha}|. \tag{3.7}$$

Let us, in addition, assume that $\alpha \in \Phi_w$, and consider $\beta \in \Phi_{s_\alpha} \setminus \Phi_w$. By definition, $\phi_\alpha(\beta) = -s_\alpha(\beta) = \frac{2\langle\beta,\alpha\rangle}{\langle\alpha,\alpha\rangle}\alpha - \beta$, and the coefficient of α must be positive, since $\phi_\alpha(\beta) \in \Delta^+$. But then $w^{-1}(\phi_\alpha(\beta))$ is a root which equals $\frac{2\langle\beta,\alpha\rangle}{\langle\alpha,\alpha\rangle}w^{-1}(\alpha) - w^{-1}(\beta)$. Since $\alpha \in \Phi_w$, we have $w^{-1}(\alpha) \in -\Delta^+$, and $\beta \notin \Phi_w$ implies $-w^{-1}(\beta) \in -\Delta^+$. Since $\frac{2\langle\beta,\alpha\rangle}{\langle\alpha,\alpha\rangle} > 0$, we get $w^{-1}(\phi_\alpha(\beta)) \in -\Delta^+$, and hence $\phi_\alpha(\beta) \in \Phi_w \cap \Phi_{s_\alpha}$. Together with the above, this implies that

$$\Phi_{s_\alpha w} \cap \Phi_{s_\alpha} \subset \Phi_w \cap \Phi_{s_\alpha} \quad \text{if } \alpha \in \Phi_w, \tag{3.8}$$

and since $\alpha \notin \Phi_{s_\alpha w}$ we conclude that $|\Phi_{s_\alpha w}| < |\Phi_w|$ if $\alpha \in \Phi_w$.

Let us, in particular, consider the case that $\alpha \in \Delta^0$ is a simple root. Then we have observed in 2.2.7 that the reflection s_α maps α to $-\alpha$ and permutes the other positive roots. Hence, $\Phi_{s_\alpha} = \{\alpha\}$ and (3) holds for simple reflections. For simple α and $w \in W$ assume next that $\alpha \notin \Phi_w$. On the one hand, this implies $w^{-1}(\alpha) \in \Delta^+$ and hence $(s_\alpha w)^{-1}(\alpha) = w^{-1}(-\alpha) \in -\Delta^+$. On the other hand, for $\beta \in \Phi_w$ we get $s_\alpha(\beta) \in \Delta^+$ and $(s_\alpha w)^{-1}(s_\alpha(\beta)) = w^{-1}(\beta) \in -\Delta^+$. Hence, $\{\alpha\} \cup s_\alpha(\Phi_w) \subset \Phi_{s_\alpha w}$ and (3.7) shows that the two sets must be equal. This proves that α is contained in exactly one of the two sets Φ_w and $\Phi_{s_\alpha w}$. For simple α, (3.7) shows that $\alpha \in \Phi_w$ implies $|\Phi_{s_\alpha w}| = |\Phi_w| - 1$. Replacing w by $s_\alpha w$, we see that $\alpha \notin \Phi_w$ implies $|\Phi_{s_\alpha w}| = |\Phi_w| + 1$.

Next, we observe that any nonempty set Φ_w contains at least one simple root. If this is not the case, then take a root $\beta \in \Phi_w$ of minimal height, and a simple root α such that $\beta - \alpha$ is a root. By minimality of the height of β, we have $\beta - \alpha \in \Delta^+ \setminus \Phi_w$, and then $\beta = (\beta - \alpha) + \alpha$ contradicts the fact that $\Delta^+ \setminus \Phi_w$ is saturated. Consequently, we can always find a simple root α such that $|\Phi_{s_\alpha w}| = |\Phi_w| - 1$, and inductively we can find simple roots $\alpha_{i_1}, \dots, \alpha_{i_k}$ with $k = |\Phi_w|$ such that $\Phi_{s_{\alpha_{i_k}} \cdots s_{\alpha_{i_1}} w} = \emptyset$. This implies that $s_{\alpha_{i_k}} \cdots s_{\alpha_{i_1}} w = \mathrm{id}$, which shows that $\ell(w) \leq |\Phi_w|$.

Let us assume inductively that we have proved $\ell(w') = |\Phi_{w'}|$ if $\ell(w') < k$. Assume that $\ell(w) = k$ and that $w = s_{\alpha_{i_1}} \cdots s_{\alpha_{i_k}}$ for simple roots α_{i_j}. If $\alpha_{i_1} \in \Phi_{s_{\alpha_{i_2}} \cdots s_{\alpha_{i_k}}}$, then we obtain $|\Phi_w| = |\Phi_{s_{\alpha_{i_2}} \cdots s_{\alpha_{i_k}}}| - 1 = k - 2$ and thus $\ell(w) \leq k - 2$, a contradiction. Hence, $\alpha_{i_1} \notin \Phi_{s_{\alpha_{i_2}} \cdots s_{\alpha_{i_k}}}$ and $|\Phi_w| = k = \ell(w)$ by induction, which completes the proof of (3). Having proved (3), part (4) follows from the observations on the cardinalities of the sets Φ_w and $\Phi_{s_\alpha w}$ made above.

(5) By part (3) we already know that $\ell(w') = \ell(w) + 1$, so it remains to show that the condition in (5) is equivalent to $w' = s_\alpha w$. Let us first assume that $w' = s_\alpha w$. Since $\mathrm{sgn}(s_\alpha) = -1$, $\ell(s_\alpha)$ must be odd and we put $\ell(s_\alpha) = 2k + 1$. Using equation (3.7) we see that $|\Phi_{s_\alpha w}| = |\Phi_w| + 1$ implies $|\Phi_w \cap \Phi_{s_\alpha}| = k$. From (3.6) we see that $|\Phi_{s_\alpha w} \cap \Phi_{s_\alpha}| = |\Phi_{s_\alpha} \setminus \Phi_w| = k+1$. From (4) we know that $\alpha \notin \Phi_w$, so $\alpha \in \Phi_{s_\alpha w}$, while (3.8) (with w replaced by $s_\alpha w$) implies $\Phi_w \cap \Phi_{s_\alpha} \subset \Phi_{s_\alpha w} \cap \Phi_{s_\alpha}$, so we obtain $\Phi_{s_\alpha w} \cap \Phi_{s_\alpha} = \{\alpha\} \cup (\Phi_w \cap \Phi_{s_\alpha})$. Finally, from (3.6) we know that $\Phi_{s_\alpha w} \setminus \Phi_{s_\alpha} = \phi_\alpha(\Phi_w \setminus \Phi_{s_\alpha})$, and on this subset ϕ_α coincides with s_α. Altogether, we see that

$$(3.9) \qquad |\Phi_{s_\alpha w}| = |\Phi_w| + 1 \implies \Phi_{s_\alpha w} = \{\alpha\} \cup (\Phi_w \cap \Phi_{s_\alpha}) \cup s_\alpha(\Phi_w \setminus \Phi_{s_\alpha}).$$

Since $s_\alpha(\beta)$ is the sum of β and an integer multiple of α, this immediately implies $\langle \Phi_{w'} \rangle = \langle \Phi_w \rangle + k\alpha$ for some integer k.

Conversely, if $\langle \Phi_{w'} \rangle = \langle \Phi_w \rangle + k\alpha$, then part (1) shows that $w(\delta)$ and $w'(\delta)$ differ by a multiple of α. Since the maps w and w' both are orthogonal, $w(\delta)$ and $w'(\delta)$ have the same length. Elementary Euclidean geometry then implies that $w'(\delta) = s_\alpha(w(\delta))$. Since different elements of W act differently on δ, this implies that $w' = s_\alpha w$. □

This result has some important immediate consequences: From parts (2) and (3) we conclude that there is a unique element $w_0 \in W$ of maximal length $\ell(w_0) = |\Delta^+|$, which corresponds to $\Phi_{w_0} = \Delta^+$. Moreover, for any representation of $w \in W$ as a composition of simple reflections, the composition in the opposite order equals w^{-1}, which implies that $\ell(w) = \ell(w^{-1})$ for all $w \in W$. Since w_0 is the unique element of maximal length, we conclude that $w_0 = (w_0)^{-1}$. Moreover, since $w_0(\Delta^+) = -\Delta^+$ it must map the simple system Δ^0 to a simple system for $-\Delta^+$, whence $w_0(\Delta^0) = -\Delta^0$. (This, however, does not imply that $w_0(\alpha) = -\alpha$ for $\alpha \in \Delta^0$.) Since w_0 exchanges positive and negative roots, we see that for $w \in W$ and $\alpha \in \Delta^+$ we have $w_0(w^{-1}(\alpha)) \in \Delta^+$ if and only if $w^{-1}(\alpha) \in -\Delta^+$, i.e. if and only if $\alpha \in \Phi_w$. Consequently, $\Phi_{ww_0} = \Delta^+ \setminus \Phi_w$, so complementation of the set Φ_w corresponds to multiplication from the right by w_0.

There is another important point about the element $w_0 \in W$, which is connected to the behaviour of weights under the actions of elements of the Weyl group. To formulate this, let us first note that the directed graph structure on W gives rise to a partial order on the set W, which is called the *Bruhat order*. For $w, w' \in W$ we put $w \leq w'$ if either $w = w'$ or there is a chain

$$w \overset{\alpha_1}{\to} w_1 \overset{\alpha_2}{\to} \cdots \overset{\alpha_{n-1}}{\to} w_{n-1} \overset{\alpha_n}{\to} w'$$

for some (not necessarily different) roots $\alpha_i \in \Delta^+$. Of course, in this ordering the identity is the smallest element and w_0 is the largest element. Moreover, $w \leq w'$ implies $\ell(w) \leq \ell(w')$ but not conversely.

By definition, any element $w \in W$ acts as an orthogonal transformation on $\mathfrak{h}_0^*$, so, in particular, for any weight λ, we may consider $w(\lambda)$. Since we have chosen a total ordering on $\mathfrak{h}_0^*$ when choosing the positive subsystem, we may compare any two weights, and we claim that $w \leq w'$ implies $w(\lambda) \geq w'(\lambda)$ for any *dominant* weight λ. To verify this, it suffices to deal with the case that $w \overset{\alpha}{\to} w'$ for some $\alpha \in \Delta^+$. If this is the case, then $w' = s_\alpha w$, so $w'(\lambda) = s_\alpha(w(\lambda)) = w(\lambda) - \frac{2\langle w(\lambda), \alpha\rangle}{\langle \alpha, \alpha\rangle}\alpha$. Now since w is orthogonal, we have $\langle w(\lambda), \alpha\rangle = \langle \lambda, w^{-1}(\alpha)\rangle$, and part (4) of the proposition implies $\alpha \notin \Phi_w$, whence $w^{-1}(\alpha) \in \Delta^+$. But the condition that λ is dominant by definition means that λ has nonnegative inner product with all simple roots and hence with all positive roots. Consequently, $w'(\lambda)$ differs from $w(\lambda)$ by a nonpositive multiple of a positive root, whence $w'(\lambda) \leq w(\lambda)$ and our claim follows. Using this we now get

COROLLARY 3.2.14. *Let $\mathfrak{g}$ be a complex semisimple Lie algebra with Weyl group W, $w_0 \in W$ the longest element, λ a dominant integral weight and V the irreducible complex finite–dimensional representation of highest weight λ. Then the highest weight of the dual representation V^* is $-w_0(\lambda)$.*

PROOF. From Theorem 2.2.10 we know that for $w \in W$ and any weight μ of V also $w(\mu)$ is a weight of V. Moreover, the weights of V^* are exactly the negatives of the weights of V. In particular, we see that $-w_0(\lambda)$ is a weight of V^*. Denoting by μ_0 the highest weight of V^*, we know again from Theorem 2.2.10 that there are nonnegative integers n_i such that $-w_0(\lambda) = \mu_0 - \sum_{\alpha_i \in \Delta^0} n_i\alpha_i$. Applying $-w_0$, we see that $-w_0(\mu_0) = \lambda - \sum n_i w_0(\alpha_i)$. But from above we know that for each i, the element $w_0(\alpha_i)$ is the negative of some simple root, which means $-w_0(\mu_0) = \lambda + \sum m_i\alpha_i$ with $m_i \geq 0$. Since λ is the highest weight of V, this is only possible if all m_j are zero, and hence $\mu_0 = -w_0(\lambda)$. □

3.2.15. The Hasse diagram associated to a complex parabolic. Let $\mathfrak{g}$ be a complex semisimple Lie algebra and let $\mathfrak{p} \subset \mathfrak{g}$ be a standard parabolic subalgebra corresponding to a Cartan subalgebra $\mathfrak{h} \leq \mathfrak{g}$ and a positive subsystem Δ^+. Let Δ^0 be the set of simple roots and $\Sigma \subset \Delta^0$ the subset determined by $\mathfrak{p}$.

Then we obtain the $|k|$–grading of $\mathfrak{g}$ by Σ–height, (where k is the Σ–height of the highest root), and in particular, the subalgebras $\mathfrak{g}_0$ and $\mathfrak{p}_+ = \mathfrak{g}^1$ of $\mathfrak{p}$. The reductive subalgebra $\mathfrak{g}_0$ splits as a direct sum of its center $\mathfrak{Z}(\mathfrak{g}_0)$ and its semisimple part $\mathfrak{g}_0^{ss}$. The basis for the definition of the Hasse diagram is that the Weyl group of $\mathfrak{g}_0^{ss}$ can be naturally viewed as a subgroup of the Weyl group $W = W_\mathfrak{g}$ of $\mathfrak{g}$. The Hasse diagram is then a distinguished set of representatives for the corresponding coset space.

The Cartan subalgebra $\mathfrak{h}$ splits as $\mathfrak{h}'\oplus\mathfrak{h}''$, where $\mathfrak{h}' = \mathfrak{z}(\mathfrak{g}_0)$ is the common kernel of all elements of $\Delta^0 \setminus \Sigma$, while $\mathfrak{h}''$ is spanned by the elements H_α for $\alpha \in \Delta^0 \setminus \Sigma$. From 3.2.1 we know that $\mathfrak{h}''$ is a Cartan subalgebra for the semisimple part $\mathfrak{g}_0^{ss}$. Let $\mathfrak{h}_0 \subset \mathfrak{h}$ be the subspace on which all roots are real. By construction, the element H_α for $\alpha \in \Delta^0 \setminus \Sigma$ lies in $\mathfrak{h}'' \cap \mathfrak{h}_0$, so we may identify this space with $\mathfrak{h}_0''$. On the other hand, $\mathfrak{h}_0' := \mathfrak{h}' \cap \mathfrak{h}_0$ is a real form of $\mathfrak{h}'$, so we get $\mathfrak{h}_0 = \mathfrak{h}_0' \oplus \mathfrak{h}_0''$. By construction, the inner product on $\mathfrak{h}_0$ induced by the Killing form satisfies $\langle H_\alpha, H\rangle = \alpha(H)$, so $\mathfrak{h}_0'$ and $\mathfrak{h}_0''$ are orthogonal. Passing to the duals we get an orthogonal decomposition of $\mathfrak{h}_0^*$, and in particular, any simple reflection s_{α_j} with $\alpha_j \in \Delta^0 \setminus \Sigma$ acts as the identity on $(\mathfrak{h}_0')^*$. Defining $W_\mathfrak{p}$ to be the Weyl group of $\mathfrak{g}_0^{ss}$, we see that we may naturally view this as the subgroup of $W_\mathfrak{g}$ generated by the simple reflections s_{α_j} for $\alpha_j \in \Delta^0 \setminus \Sigma$.

The Hasse diagram is a set of distinguished representatives for the set $W_\mathfrak{p}\backslash W_\mathfrak{g}$ of right cosets. To see how such representatives can be obtained, let us decompose $\Delta^+ = \Delta^+(\mathfrak{g}_0) \sqcup \Delta^+(\mathfrak{p}_+)$ according to the subalgebra containing the corresponding root space. Otherwise put, for $\alpha \in \Delta^+$ we have $\alpha \in \Delta^+(\mathfrak{g}_0)$ if and only if $\mathrm{ht}_\Sigma(\alpha) = 0$. Since the Σ–height is additive, both $\Delta^+(\mathfrak{g}_0)$ and $\Delta^+(\mathfrak{p}_+)$ are saturated in the sense introduced in 3.2.14. Now assume that $\alpha \in \Delta^+(\mathfrak{g}_0)$ and $\beta \in \Delta^+(\mathfrak{p}_+)$. Then $s_\alpha(\beta)$ differs from β by a multiple of α, and thus $\mathrm{ht}_\Sigma(s_\alpha(\beta)) = \mathrm{ht}_\Sigma(\beta)$. Thus, s_α maps $\Delta^+(\mathfrak{p}_+)$ to itself, so the same holds for any element $w \in W_\mathfrak{p}$. In particular, $\Phi_w \subset \Delta^+(\mathfrak{g}_0)$ for any $w \in W_\mathfrak{p}$. Conversely, if $w \in W_\mathfrak{g}$ is such that $\Phi_w \subset \Delta^+(\mathfrak{g}_0)$, then Φ_w is saturated. Since $\Delta^+ \setminus \Phi_w$ and $\Delta^+(\mathfrak{g}_0)$ both are saturated, also $\Delta^+(\mathfrak{g}_0) \setminus \Phi_w$ is saturated. Applying part (2) of Proposition 3.2.14 to $\mathfrak{g}_0^{ss}$ we find an element in $W_\mathfrak{p}$ corresponding to this subset, and again by part (2) of Proposition 3.2.14 we conclude that this element coincides with w, whence $w \in W_\mathfrak{p}$. Having characterized $W_\mathfrak{p}$ as those elements $w \in W$ for which $\Phi_w \subset \Delta^+(\mathfrak{g}_0)$ the following definition is natural.

DEFINITION 3.2.15. The Hasse diagram $W^\mathfrak{p}$ of the standard parabolic subalgebra $\mathfrak{p} \leq \mathfrak{g}$ is the subset of $W_\mathfrak{g}$ consisting of all elements w such that $\Phi_w \subset \Delta^+(\mathfrak{p}_+)$. We endow $W^\mathfrak{p}$ with the structure of a directed graph induced from the structure on $W_\mathfrak{g}$ constructed in 3.2.14.

There is a nice alternative characterization of $W^\mathfrak{p}$: Recall that a weight $\lambda \in \mathfrak{h}_0^*$ is $\mathfrak{g}$–dominant if $\langle\lambda, \alpha\rangle \geq 0$ for all $\alpha \in \Delta^0$ and $\mathfrak{p}$–dominant if the same holds for all $\alpha \in \Delta^0 \setminus \Sigma = \Delta^0 \cap \Delta^+(\mathfrak{g}_0)$. Equivalently, one may require these conditions for all elements of Δ^+, respectively, $\Delta^+(\mathfrak{g}_0)$, since they can be written as linear combinations of the corresponding simple elements with nonnegative coefficients. Since any element $w \in W$ acts as an orthogonal transformation on $\mathfrak{h}_0^*$, we get $\langle w(\lambda), \alpha\rangle = \langle\lambda, w^{-1}(\alpha)\rangle$. But this shows that $w(\lambda)$ is $\mathfrak{p}$–dominant for any $\mathfrak{g}$–dominant weight λ if and only if $w^{-1}(\alpha) \in \Delta^+$ for any $\alpha \in \Delta^+(\mathfrak{g}_0)$ i.e. if and only if $w \in W^\mathfrak{p}$. Hence, $w \in W^\mathfrak{p}$ if and only if $w(\lambda)$ is $\mathfrak{p}$–dominant for any $\mathfrak{g}$–dominant weight λ.

PROPOSITION 3.2.15. *Let $w \in W$ be any element. Then there are unique elements $w_{\mathfrak{p}} \in W_{\mathfrak{p}}$ and $w^{\mathfrak{p}} \in W^{\mathfrak{p}}$ such that $w = w_{\mathfrak{p}} w^{\mathfrak{p}}$. Moreover, $\ell(w) = \ell(w_{\mathfrak{p}}) + \ell(w^{\mathfrak{p}})$.*

PROOF. Given $w \in W$ define $\Phi := \Phi_w \cap \Delta^+(\mathfrak{g}_0)$. Then Φ is saturated since it is the intersection of two saturated subsets of Δ^+. Assume that $\alpha, \beta \in \Delta^+ \setminus \Phi$ are such that $\alpha + \beta \in \Delta$. If one of the elements lies in $\Delta^+(\mathfrak{p}_+)$, then so does the sum, so in particular, $\alpha + \beta \notin \Phi$. On the other hand, if both α and β lie in $\Delta^+(\mathfrak{g}_0)$, then they both do not lie in Φ_w, so $\alpha + \beta \notin \Phi$ follows since $\Delta^+ \setminus \Phi_w$ is saturated. Consequently, also the complement of Φ in Δ^+ is saturated, so there is a unique element $w_{\mathfrak{p}} \in W$ such that $\Phi = \Phi_{w_{\mathfrak{p}}}$. Since $\Phi \subset \Delta^+(\mathfrak{g}_0)$, we conclude that $w_{\mathfrak{p}} \in W_{\mathfrak{p}}$.

Now take any element $\alpha \in \Delta^+(\mathfrak{g}_0)$. If $\alpha \in \Phi_{w_{\mathfrak{p}}^{-1}}$, then

$$-w_{\mathfrak{p}}(\alpha) \in \Delta^+(\mathfrak{g}_0) \cap \Phi_{w_{\mathfrak{p}}} \subset \Phi_w.$$

Thus, $w^{-1}(-w_{\mathfrak{p}}(\alpha)) \in -\Delta^+$, and thus $w^{-1}(w_{\mathfrak{p}}(\alpha)) \in \Delta^+$. If $\alpha \notin \Phi_{w_{\mathfrak{p}}^{-1}}$, then $w_{\mathfrak{p}}(\alpha) \in \Delta^+(\mathfrak{g}_0) \setminus \Phi_{w_{\mathfrak{p}}}$. Hence, by construction $w^{-1}(w_{\mathfrak{p}}(\alpha)) \in \Delta^+$. Thus, we see that the element $w^{\mathfrak{p}} := w_{\mathfrak{p}}^{-1} w$ has the property that its inverse maps any element of $\Delta^+(\mathfrak{g}_0)$ to a positive root. By definition $w^{\mathfrak{p}} \in W^{\mathfrak{p}}$ and we obtain a decomposition of the required form.

To prove uniqueness, assume that $w = w_1 w_2$ with $w_1 \in W_{\mathfrak{p}}$ and $w_2 \in W^{\mathfrak{p}}$, and assume that $\alpha \in \Delta^+(\mathfrak{g}_0)$. Then $w_1^{-1}(\alpha)$ is a root of Σ–height zero. Since $w_2 \in W^{\mathfrak{p}}$, we conclude that $\alpha \in \Phi_w$, i.e. $w_2^{-1}(w_1^{-1}(\alpha)) \in -\Delta^+$ if and only if $w_1^{-1}(\alpha) \in -\Delta^+$, i.e. $\alpha \in \Phi_{w_1}$. Consequently, $\Delta^+(\mathfrak{g}_0) \cap \Phi_w = \Delta^+(\mathfrak{g}_0) \cap \Phi_{w_1}$, and since $w_1 \in W_{\mathfrak{p}}$ the latter set coincides with Φ_{w_1}. By part (2) of Proposition 3.2.14 this implies that w_1 coincides with the element $w_{\mathfrak{p}}$ from above, and thus we also get $w_2 = w^{\mathfrak{p}}$.

To prove the statement on the length, one just has to note that by construction $\Phi_{w_{\mathfrak{p}}} = \Delta^+(\mathfrak{g}_0) \cap \Phi_w$. On the other hand, $\Phi_{w^{\mathfrak{p}}} \subset \Delta^+(\mathfrak{p}_+)$ and thus also $w_{\mathfrak{p}}(\Phi_{w^{\mathfrak{p}}}) \subset \Delta^+(\mathfrak{p}_+)$. But for $\alpha \in w_{\mathfrak{p}}(\Phi_{w^{\mathfrak{p}}})$ we have $w_{\mathfrak{p}}^{-1}(\alpha) \in \Phi_{w^{\mathfrak{p}}}$ and thus $\alpha \in \Phi_w$. This shows that $|\Phi_w| \geq |\Phi_{w_{\mathfrak{p}}}| + |w_{\mathfrak{p}}(\Phi_{w^{\mathfrak{p}}})|$. Using part (3) of Proposition 3.2.14, we obtain $\ell(w) \geq \ell(w_{\mathfrak{p}}) + \ell(w^{\mathfrak{p}})$, and the opposite inequality is obvious. □

The existence and uniqueness of the decomposition $w = w_{\mathfrak{p}} w^{\mathfrak{p}}$ tells us that $w^{\mathfrak{p}}$ is the unique element in the right coset $W_{\mathfrak{p}} w$ that lies in $W^{\mathfrak{p}}$. Thus, $W^{\mathfrak{p}}$ is a set of distinguished representatives for the right coset space $W_{\mathfrak{p}} \backslash W_{\mathfrak{g}}$. The statement about the length then tells us that these representatives are the unique elements of minimal length in each coset.

Let us note two simple facts about the Hasse diagram. If $w, w' \in W^{\mathfrak{p}}$ and $w \xrightarrow{\alpha} w'$, then $\alpha \in \Phi_{w'}$ whence $\alpha \in \Delta^+(\mathfrak{p}_+)$. On the other hand, since both the sets $\Delta^+(\mathfrak{p}_+)$ and $\Delta^+(\mathfrak{g}_0)$ are saturated, there is a unique longest element $w_0^{\mathfrak{p}} \in W^{\mathfrak{p}}$ with $\Phi_{w_0^{\mathfrak{p}}} = \Delta^+(\mathfrak{p}_+)$. Since $W_{\mathfrak{p}}$ is the Weyl group of a semisimple Lie algebra, it contains a unique longest element $w_{\mathfrak{p}}^0$, and $w_{\mathfrak{p}}^0 w_0^{\mathfrak{p}} = w_0$, the longest element in the Weyl group $W_{\mathfrak{g}}$.

3.2.16. Determining the Hasse diagram. We will describe a two step procedure to determine the Hasse diagram. In the first step, we determine the points of $W^{\mathfrak{p}}$ and some of the arrows in the Hasse diagram, while in the second step we determine the sets Φ_w, the remaining arrows, and the labels over all arrows. In many applications of the Hasse diagram, one is mainly interested in (a part of) the orbit of a given weight under the action of the elements of $W^{\mathfrak{p}}$. For these applications, only the first step of the procedure is needed.

First, we observe that it is easy to determine the action of a simple reflection on a weight in the Dynkin diagram notation. By definition, for a simple root $\alpha \in \Delta^0$ and a weight λ, we have $s_\alpha(\lambda) = \lambda - \frac{2\langle\lambda,\alpha\rangle}{\langle\alpha,\alpha\rangle}\alpha$. Writing λ as a linear combination of the fundamental weights, the number $\frac{2\langle\lambda,\alpha\rangle}{\langle\alpha,\alpha\rangle}$ is the coefficient of the fundamental weight corresponding to α. In the Dynkin diagram notation, this is the number over the node representing α. Hence, to determine $s_\alpha(\lambda)$ as a linear combination of the fundamental weights, we only have to write a simple root $\alpha \in \Delta^0$ as a linear combination of fundamental weights. Writing the simple roots as $\alpha_1, \ldots, \alpha_n$ and the corresponding fundamental weights as $\omega_1, \ldots, \omega_n$, this linear combination is clearly given by $\alpha = \sum_{i=1}^n \frac{2\langle\alpha,\alpha_i\rangle}{\langle\alpha_i,\alpha_i\rangle}\omega_i$. These coefficients form exactly the column corresponding to α in the Cartan matrix of the Lie algebra $\mathfrak{g}$. Of course, the coefficient of the fundamental weight corresponding to α itself is 2, from which we conclude that applying s_α to λ, the coefficient over the node corresponding to α changes sign. Moreover, the coefficient over a node may change only if the inner product of α with the corresponding simple root is nontrivial, i.e. if there is an edge between the two nodes in question. Looking at the description of the passage from the Cartan matrix to the Dynkin diagram in 2.2.5, one sees that $\frac{2\langle\alpha,\beta\rangle}{\langle\beta,\beta\rangle} = -1$ if between the nodes corresponding to α or β there is a single edge or a multiple edge with the arrow pointing towards α. On the other hand, if there is a multiple edge with the arrow pointing towards β, then $\frac{2\langle\alpha,\beta\rangle}{\langle\beta,\beta\rangle}$ equals minus the number of edges. Thus, we obtain the following examples for the action of the simple reflection corresponding to the node with coefficient b (with all coefficients not explicitly indicated remaining unchanged by the reflection):

$$s_{\alpha_i}\left(\cdots\overset{a}{\circ}\!\!-\!\!\overset{b}{\circ}\!\!-\!\!\overset{c}{\circ}\cdots\right) = \cdots\overset{a+b}{\circ}\!\!-\!\!\overset{-b}{\circ}\!\!-\!\!\overset{b+c}{\circ}\cdots$$

$$s_{\alpha_i}\left(\cdots\overset{a}{\circ}\!\!-\!\!\overset{b}{\circ}\!\!\Leftarrow\!\!\overset{c}{\circ}\cdots\right) = \cdots\overset{a+b}{\circ}\!\!-\!\!\overset{-b}{\circ}\!\!\Leftarrow\!\!\overset{b+c}{\circ}\cdots$$

$$s_{\alpha_i}\left(\cdots\overset{a}{\circ}\!\!-\!\!\overset{b}{\circ}\!\!\Rightarrow\!\!\overset{c}{\circ}\cdots\right) = \cdots\overset{a+b}{\circ}\!\!-\!\!\overset{-b}{\circ}\!\!\Rightarrow\!\!\overset{2b+c}{\circ}\cdots$$

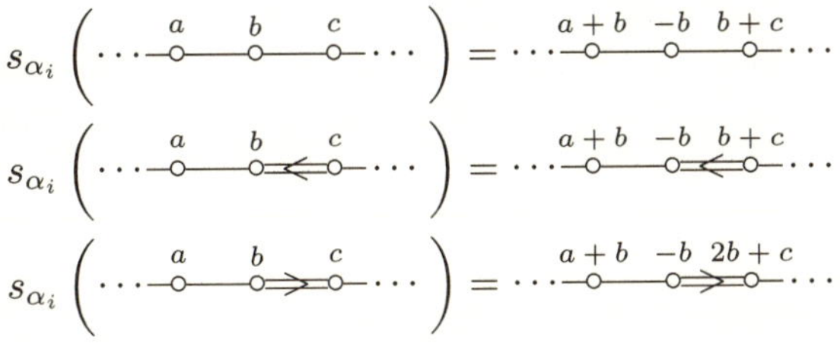

Since the Weyl group $W_\mathfrak{g}$ is generated by the simple reflections, we may determine the orbit of any weight under $W_\mathfrak{g}$ by applying step by step all simple reflections that do not lead "backwards" i.e. we must not apply the same simple reflection twice without another simple reflection in between. If we choose a weight lying in the interior of a Weyl chamber, than the orbit will be in bijective correspondence with $W_\mathfrak{g}$, since we know that the action is simply transitive on the set of Weyl chambers. In particular, one may use the lowest form δ to determine the Weyl group in this way. Choosing a weight adapted to a standard parabolic, one may use the same procedure to determine the points of the subset $W^\mathfrak{p} \subset W_\mathfrak{g}$:

PROPOSITION 3.2.16. *Let $\mathfrak{g}$ be a complex semisimple Lie algebra and let $\mathfrak{p} \leq \mathfrak{g}$ be the standard parabolic subalgebra corresponding to a set Σ of simple roots. Let $\delta^\mathfrak{p}$ be the sum of all fundamental weights corresponding to elements of Σ. Then we have:*

(1) *The map* $w \mapsto w^{-1}(\delta^{\mathfrak{p}})$ *restricts to a bijection between* $W^{\mathfrak{p}}$ *and the orbit of* $\delta^{\mathfrak{p}}$ *under* $W_{\mathfrak{g}}$.
(2) *Suppose that* $w \in W^{\mathfrak{p}}$ *and* $\alpha \in \Delta^0$ *is a simple root such that* $\alpha \notin \Phi_{w^{-1}}$ *and* $s_\alpha(w^{-1}(\delta^{\mathfrak{p}})) \neq w^{-1}(\delta^{\mathfrak{p}})$. *Then* $ws_\alpha \in W^{\mathfrak{p}}$, $w \overset{w(\alpha)}{\to} ws_\alpha$, *and* $\Phi_{ws_\alpha} = \Phi_w \cup \{w(\alpha)\}$.

PROOF. (1) For $\alpha \in \Delta^0 \setminus \Sigma$ by definition we have $\langle \delta^{\mathfrak{p}}, \alpha\rangle = 0$ and thus $s_\alpha(\delta^{\mathfrak{p}}) = \delta^{\mathfrak{p}}$. Since these simple reflections generate $W_{\mathfrak{p}}$, it follows that $w(\delta^{\mathfrak{p}}) = \delta^{\mathfrak{p}}$ for all $w \in W_{\mathfrak{p}}$. If $w \in W$ is arbitrary, then by Proposition 3.2.15 there are unique elements $w_{\mathfrak{p}} \in W_{\mathfrak{p}}$ and $w^{\mathfrak{p}} \in W^{\mathfrak{p}}$ such that $w = w_{\mathfrak{p}}w^{\mathfrak{p}}$. Since $w_{\mathfrak{p}}^{-1} \in W_{\mathfrak{p}}$, we conclude from above that $w^{-1}(\delta^{\mathfrak{p}}) = (w^{\mathfrak{p}})^{-1}(\delta^{\mathfrak{p}})$, so we get a surjection from $W^{\mathfrak{p}}$ onto the orbit of $\delta^{\mathfrak{p}}$.

On the other hand, assume that $\mathrm{id} \neq w \in W^{\mathfrak{p}}$. Then $\Phi_w \neq \emptyset$, so there is an element $\alpha \in -\Delta^+$ such that $w(\alpha) \in \Phi_w \subset \Delta^+(\mathfrak{p}_+)$. But this implies $\langle \delta^{\mathfrak{p}}, \alpha\rangle \leq 0$ and $\langle \delta^{\mathfrak{p}}, w(\alpha)\rangle > 0$, and since the last expression equals $\langle w^{-1}(\delta^{\mathfrak{p}}), \alpha\rangle$, we see that $w^{-1}(\delta^{\mathfrak{p}}) \neq \delta^{\mathfrak{p}}$. Using Proposition 3.2.15, we see that $w^{-1}(\delta^{\mathfrak{p}}) = \delta^{\mathfrak{p}}$ implies $w \in W_{\mathfrak{p}}$. Now, if $w_1, w_2 \in W^{\mathfrak{p}}$ are such that $w_1(\delta^{\mathfrak{p}}) = w_2(\delta^{\mathfrak{p}})$, we conclude $w_2^{-1}w_1 \in W_{\mathfrak{p}}$, which implies $w_1 = w_2$ by the uniqueness in Proposition 3.2.15, so we get the claimed bijection.

(2) By part (4) of Proposition 3.2.14, our assumptions imply $\ell(s_\alpha w^{-1}) = \ell(w^{-1}) + 1$, and thus $\ell(ws_\alpha) = \ell(w) + 1$. Moreover, by assumption $w(\alpha) \in \Delta^+$ and it is elementary to verify that $s_{w(\alpha)} = ws_\alpha w^{-1}$, which implies $w \overset{w(\alpha)}{\to} ws_\alpha$. Next, for $\beta \in \Delta^+$ the root $s_\alpha(w^{-1}(\beta))$ is negative if and only if either $w^{-1}(\beta)$ is negative and thus $\beta \in \Phi_w$ or $\beta = w(\alpha)$. Hence, we see that $\Phi_{ws_\alpha} = \Phi_w \cup \{w(\alpha)\}$, so to complete the proof we only have to show that $w(\alpha) \in \Delta^+(\mathfrak{p}_+)$. By assumption $s_\alpha(w^{-1}(\delta^{\mathfrak{p}})) \neq w^{-1}(\delta^{\mathfrak{p}})$, so in particular, $\langle w^{-1}(\delta^{\mathfrak{p}}), \alpha\rangle \neq 0$. Moreover, since $w^{-1} \leq s_\alpha w^{-1}$ and $\delta^{\mathfrak{p}}$ is dominant, we know from 3.2.14 that $s_\alpha(w^{-1}(\delta^{\mathfrak{p}})) \leq \delta^{\mathfrak{p}}$, which implies that $\langle w^{-1}(\delta^{\mathfrak{p}}), \alpha\rangle > 0$. But this inner product equals $\langle \delta^{\mathfrak{p}}, w(\alpha)\rangle$, so $w(\alpha) \in \Delta^+(\mathfrak{p}_+)$. □

This result immediately allows us to determine the elements of $W^{\mathfrak{p}}$ and some of the arrows in the Hasse diagram. Take the weight $\delta^{\mathfrak{p}}$, which in Dynkin diagram notation has 1 over each crossed node and 0 over each uncrossed node. Now we apply successive simple reflections to this weight. To really change the weight and avoid going "backwards" exactly means to apply only reflections corresponding to nodes which have a positive coefficient. At some stage this procedure stops, since there are no longer any positive coefficients, and we have found the full orbit of the weight $\delta^{\mathfrak{p}}$ under $W_{\mathfrak{g}}$. The elements of $W^{\mathfrak{p}}$ are then obtained by composing the simple reflections used to get to some place in the opposite order, and we know that any of the simple reflections corresponds to an arrow in the Hasse diagram. In particular, this implies that the Hasse diagram is always a connected subgraph of $W_{\mathfrak{g}}$.

Let us carry this out in a nontrivial example: Consider $\mathfrak{g} = \mathfrak{sl}(4, \mathbb{C})$ and the parabolic ×—×—∘ corresponding to the first two simple roots. Let us write $\alpha_1, \alpha_2, \alpha_3$ for the three simple roots, and in our diagram we put the number of the simple root used for reflecting over the arrow. Then we obtain the pattern:

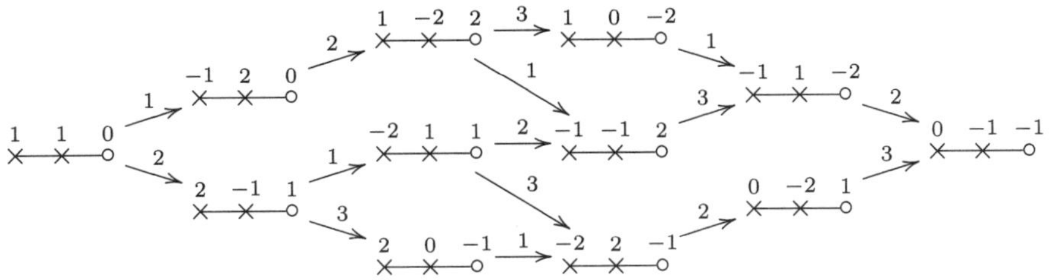

We can read off any required element of $W^{\mathfrak{p}}$ directly from this diagram. For example, the longest element of $W^{\mathfrak{p}}$ can be written as $s_{\alpha_1}s_{\alpha_2}s_{\alpha_3}s_{\alpha_1}s_{\alpha_2}$, or as $s_{\alpha_2}s_{\alpha_1}s_{\alpha_2}s_{\alpha_3}s_{\alpha_2}$, and so on.

Having this at hand, we can now proceed to determine the sets Φ_w as well as the labels over the arrows in the Hasse diagram that we have already found. The starting point in the pattern is the identity corresponding to the empty set. Having determined Φ_w for some element w in the pattern, we know from the proposition that $\Phi_{ws_\alpha} = \Phi_w \cup \{w(\alpha)\}$. Hence, it suffices to determine the image of α under the reflections along a path leading backwards to the identity, and label the arrow as $w \overset{w(\alpha)}{\to} ws_\alpha$. Having determined the set Φ_{ws_α} we may read off the labels on all arrows ending in this point by looking at what root was added to the set along the given arrow.

To carry this out in the example of the parabolic ×—×—○, we number the additional positive roots by $\alpha_4 = \alpha_1 + \alpha_2$, $\alpha_5 = \alpha_2 + \alpha_3$ and $\alpha_6 = \alpha_1 + \alpha_2 + \alpha_3$. In our example $\Delta^+(\mathfrak{p}_+)$ consists of all positive roots except α_3. Moreover, one immediately verifies that s_{α_1} exchanges α_2 and α_4 as well as α_5 and α_6, and leaves α_3 fixed. Similarly, s_{α_2} exchanges α_1 and α_4 as well as α_3 and α_5 and leaves α_6 fixed. Finally, s_{α_3} exchanges α_2 and α_5 as well as α_4 and α_6 and leaves α_1 fixed. Using this, we obtain the following pattern for the sets Φ_w and arrows in the Hasse diagram, where in the sets we only note the indices of the roots

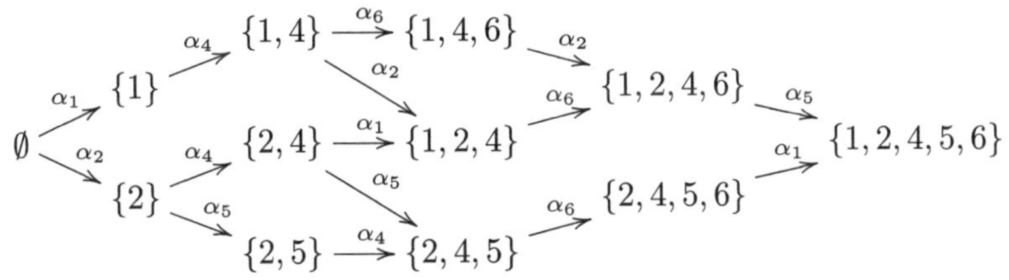

The last thing to do is to find the remaining arrows and their labels, which can be done using part (5) of Proposition 3.2.14. We simply have to determine the sums of the elements in each of the subsets. If for two sets in adjacent columns these sums differ by a multiple of one of the roots $\alpha_1, \dots, \alpha_6$, then we have to add an arrow between the two sets labeled by that root. For example, in the third column, these sums are, from top to bottom, given by $2\alpha_1 + \alpha_2$, $\alpha_1 + 2\alpha_2$, and $2\alpha_2 + \alpha_3$. This shows that we have to add the arrows $\{1\} \overset{\alpha_2}{\to} \{2,4\}$ and $\{2\} \overset{\alpha_1}{\to} \{1,4\}$. In the fourth column, the sums are, from top to bottom, $3\alpha_1 + 2\alpha_2 + \alpha_3$, $2\alpha_1 + 2\alpha_2$, and $\alpha_1 + 3\alpha_2 + \alpha_3$. Hence, we only have to add the arrow $\{2,5\} \overset{\alpha_1}{\to} \{1,4,6\}$. Similarly, one deals with the arrows between the fourth and fifth columns to see that the complete Hasse diagram for the parabolic ×—×—○ has the form

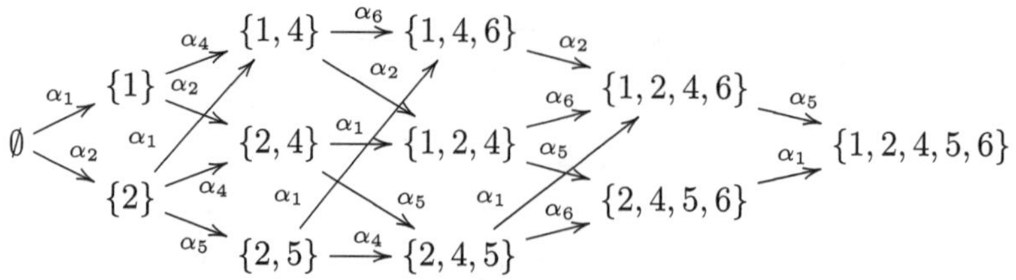

3.2.17. Simplifications in special cases. The procedure for determining the Hasse diagram outlined in 3.2.16 can be simplified considerably in special cases. The point is that one may skip the first step in the procedure, i.e. determining the orbit of the weight $\delta^{\mathfrak{p}}$ if the structure of saturated subsets of Δ^+ is understood well enough.

From part (2) of Proposition 3.2.14 and the definition of $W^{\mathfrak{p}}$, we know that the sets Φ_w for $w \in W^{\mathfrak{p}}$ are exactly those saturated subsets of $\Delta^+(\mathfrak{p}_+)$ whose complement in Δ^+ is saturated, too. Moreover, all the arrows obtained in the first step of the algorithm from 3.2.16 have the property that they correspond to an inclusion of Φ_w into $\Phi_{w'}$, and there is just one additional positive root in $\Phi_{w'}$. Conversely, if $\Phi_{w'} = \Phi_w \cup \{\alpha\}$ for some $\alpha \in \Delta^+$, then $w \xrightarrow{\alpha} w'$ by part (5) of Proposition 3.2.14.

Consequently, we see that once we are able to determine the saturated subsets of $\Delta^+(\mathfrak{p}_+)$ whose complement in Δ^+ is saturated, too, then we can draw the diagram of all these sets with arrows corresponding to inclusions of a k–element set into a $(k+1)$–element set and labeled by the additional root. This gives all points and some of the arrows in the Hasse diagram and the remaining arrows can then be determined as described in 3.2.16 above.

This procedure is particularly effective in the case of parabolics corresponding to $|1|$–gradings, in which there are some further simplifications:

PROPOSITION 3.2.17. *Let $\mathfrak{g}$ be a complex semisimple Lie algebra and $\mathfrak{p} \leq \mathfrak{g}$ a standard parabolic subalgebra corresponding to a $|1|$–grading. Then we have:*

(1) *Any subset of $\Delta^+(\mathfrak{p}_+)$ is saturated.*
(2) *If $\Phi \subset \Delta^+(\mathfrak{p}_+)$ has the property that $\Delta^+ \setminus \Phi$ is saturated, and $\alpha \in \Phi$ and $\beta \in \Delta^+(\mathfrak{p}_+)$ are such that $\alpha - \beta \in \Delta^+$, then $\beta \in \Phi$.*
(3) *If $w \xrightarrow{\alpha} w'$ is an arrow in the Hasse diagram of $\mathfrak{p}$, then $\Phi_{w'} = \Phi_w \cup \{\alpha\}$.*

PROOF. The assumptions on $\mathfrak{p}$ imply that the corresponding set Σ of simple roots consists of only one element, the only possible Σ–heights of roots are -1, 0, and 1, and $\Delta^+(\mathfrak{p}_+)$ consists exactly of the roots of Σ–height 1. Consequently, the sum of two elements of $\Delta^+(\mathfrak{p}_+)$ can never be a root, since otherwise it would have Σ–height 2, and (1) follows.

(2) Since both α and β have Σ–height one by assumption, we see that $\alpha - \beta$ has Σ–height zero, and hence lies in $\Delta^+ \setminus \Phi$. But then $\beta \notin \Phi$ would imply the contradiction $\alpha \in \Delta^+ \setminus \Phi$ since this subset is saturated by assumption.

(3) For $\alpha, \beta \in \Delta^+(\mathfrak{p}_+)$ the root $s_\alpha(\beta)$ is given by the sum of β and an integral multiple of α. Since both α and β have Σ–height one, $s_\alpha(\beta) \in \Delta^+(\mathfrak{p}_+)$ is only possible if $s_\alpha(\beta) = \beta$. But then equation (3.9) from the proof of Proposition 3.2.14 shows that (3) holds. □

Parts (1) and (3) show that in the case of a $|1|$–grading, the Hasse diagram is given by the pattern of all subsets of $\Delta^+(\mathfrak{p}_+)$ whose complement in Δ^+ is saturated, and the arrows correspond to the inclusions of sets into sets with one more element. The label of each arrow is given by the additional element.

EXAMPLE 3.2.17. Let us analyze the case $\mathfrak{g} = \mathfrak{sl}(n+1, \mathbb{C})$, in which there are many $|1|$–gradings available. From 3.2.3 we know that for each simple root α_i the standard parabolic $\mathfrak{p}$ corresponding to $\Sigma = \{\alpha_i\}$ gives rise to a $|1|$–grading. Denoting the simple roots by $\alpha_1, \dots, \alpha_n$, the positive roots in the A_n–case are given by the unbroken strings of the form $\alpha_j + \alpha_{j+1} + \dots + \alpha_{k-1} + \alpha_k$ for $j \leq k$ and such a root lies in $\Delta^+(\mathfrak{p}_+)$ if and only if $j \leq i \leq k$. From the proof of Proposition 3.2.14 we know that any set Φ_w for $w \in W_{\mathfrak{g}}$ contains at least one simple root, so we conclude that $\alpha_i \in \Phi_w$ for all $w \in W^{\mathfrak{p}}$.

Now we can determine when the complement in Δ^+ of a subset $\Phi \subset \Delta^+(\mathfrak{p}_+)$ is saturated. Suppose that $\alpha_j + \dots + \alpha_k \in \Phi$ with $j \leq i \leq k$. If $j < i$, we can apply part (2) of the proposition to $\alpha = \alpha_j + \dots + \alpha_k$ and $\beta = \alpha_{j+1} + \dots + \alpha_k$ to see that the latter element lies in Φ. Likewise, if $k > i$, then $\alpha_j + \dots + \alpha_{k-1} \in \Phi$. Visualizing the roots as being arranged in such a way as the corresponding root spaces sit as entries in a matrix, this means that for any root contained in Φ, any root below the given one or left of the given one which still lies in $\Delta^+(\mathfrak{p}_+)$ has to be contained in Φ. Otherwise put, this condition can be expressed as $\alpha \in \Phi$ and $\beta \in \Delta^+(\mathfrak{p}_+)$ such that $\mathrm{ht}_{\{\alpha_1,\dots,\alpha_{i-1}\}}(\beta) < \mathrm{ht}_{\{\alpha_1,\dots,\alpha_{i-1}\}}(\alpha)$ or $\mathrm{ht}_{\{\alpha_{i+1},\dots,\alpha_n\}}(\beta) < \mathrm{ht}_{\{\alpha_{i+1},\dots,\alpha_n\}}(\alpha)$ implies $\beta \in \Phi$. But this condition visibly implies that $\Delta^+ \setminus \Phi$ is saturated, so we have found the description that we need.

Returning to the picture of matrices, we see that a subset has saturated complement if and only if it is a Young diagram turned upside down. Thus, we see that the Hasse diagram of the parabolic corresponding to α_i may be identified with the pattern of all Young diagrams with at most i rows of length $\leq n - i + 1$ and the arrows correspond to diagrams obtained from each other by adding one box.

To have some concrete examples, consider first the simplest case of the parabolic ×—○—⋯—○—○ corresponding to the first simple root. The nonempty subsets Φ_w in this case are exactly the sets of the form $\{\alpha_1, \alpha_1 + \alpha_2, \dots, \alpha_1 + \dots + \alpha_i\}$, so from the proposition above we see that we get the Hasse diagram

$$\emptyset \xrightarrow{\alpha_1} \{\alpha_1\} \xrightarrow{\alpha_1+\alpha_2} \{\alpha_1, \alpha_1+\alpha_2\} \xrightarrow{\alpha_1+\alpha_2+\alpha_3} \dots \xrightarrow{\alpha_1+\dots+\alpha_n} \{\alpha_1, \dots, \alpha_1 + \dots + \alpha_n\}.$$

Choosing other simple roots, the resulting Hasse diagrams become significantly more complicated for general n, so we restrict to an example with small n here. Consider the standard parabolic ○—×—○ in $\mathfrak{sl}(4, \mathbb{C})$, and let us number the positive roots as in 3.2.16 above as $\alpha_4 = \alpha_1 + \alpha_2$, $\alpha_5 = \alpha_2 + \alpha_3$ and $\alpha_6 = \alpha_1 + \alpha_2 + \alpha_3$. As before, we use the notation that we just indicate the indices of positive roots in the subsets. From the description above, we see that the subsets of $\Delta^+(\mathfrak{p}_+)$ whose complement in Δ^+ is saturated are $\emptyset$, $\{2\}$, $\{2,4\}$, $\{2,5\}$, $\{2,4,5\}$, and $\{2,4,5,6\}$, so we obtain the Hasse diagram

$$\begin{array}{ccccccc} & & & \{2,4\} & & & \\ & & \overset{\alpha_4}{\nearrow} & & \overset{\alpha_5}{\searrow} & & \\ \emptyset \xrightarrow{\alpha_2} \{2\} & & & & & & \{2,4,5\} \xrightarrow{\alpha_6} \{2,4,5,6\} \\ & & \underset{\alpha_5}{\searrow} & & \underset{\alpha_4}{\nearrow} & & \\ & & & \{2,5\} & & & \end{array}$$

Comparing with 3.2.16 we see that, as expected, the Hasse diagrams of both ×—○—○ and ○—×—○ sit as subdiagrams in the Hasse diagram of ×—×—○.

To present a more substantial example, we consider the standard parabolic in $\mathfrak{sl}(6,\mathbb{C})$ corresponding to the third (middle) root. The vertices in its Hasse diagram are all Young diagrams with at most three rows and boxes and the arrows are given by inclusion, which leads to the pattern

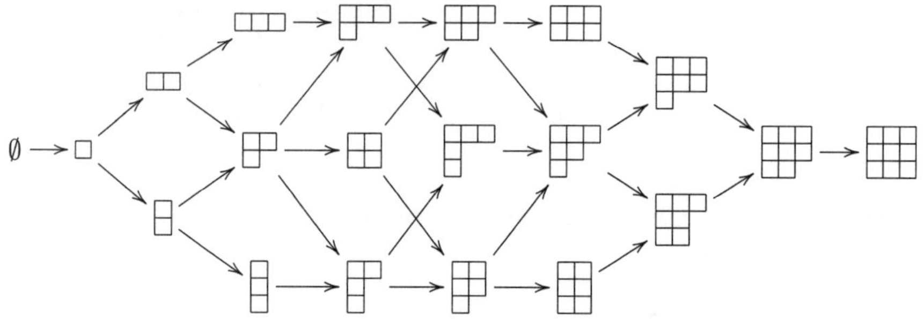

Reinterpreting this in terms of the sets Φ_w and obtaining the labels of the arrows is straightforward.

3.2.18. Recipes for determining the Hasse diagram. For the convenience of the reader we collect the steps needed to determine the Hasse diagram of a parabolic subalgebra in a complex simple Lie algebra in the form of two recipes. The first of these works for arbitrary parabolics while the second one works only for parabolics corresponding to $|1|$–gradings and is quite different in nature.

Recipe for general parabolics. A model example for the application of this recipe can be found in 3.2.16.

(A) Determine the Dynkin diagram of the parabolic, i.e. the Dynkin diagram of $\mathfrak{g}$ with those simple roots crossed whose root spaces are contained in $\mathfrak{g}_1$.

(B) Determine the elements of $W^{\mathfrak{p}}$.

Take the weight $\delta^{\mathfrak{p}}$, i.e. the weight which has coefficient 1 over the crossed nodes and zero over the uncrossed nodes. Apply simple reflections to this weight according to the following rules which describe the action of the reflection corresponding to the simple root with coefficient b (all coefficients not shown in the picture remain unchanged under this reflection):

$$\cdots \underset{a}{\circ}\!\!-\!\!\underset{b}{\circ}\!\!-\!\!\underset{c}{\circ}\cdots \;\mapsto\; \cdots \underset{a+b}{\circ}\!\!-\!\!\underset{-b}{\circ}\!\!-\!\!\underset{b+c}{\circ}\cdots$$

$$\cdots \underset{a}{\circ}\!\!-\!\!\underset{b}{\circ}\!\!\Leftarrow\!\!\underset{c}{\circ}\cdots \;\mapsto\; \cdots \underset{a+b}{\circ}\!\!-\!\!\underset{-b}{\circ}\!\!\Leftarrow\!\!\underset{b+c}{\circ}\cdots$$

$$\cdots \underset{a}{\circ}\!\!-\!\!\underset{b}{\circ}\!\!\Rightarrow\!\!\underset{c}{\circ}\cdots \;\mapsto\; \cdots \underset{a+b}{\circ}\!\!-\!\!\underset{-b}{\circ}\!\!\Rightarrow\!\!\underset{2b+c}{\circ}\cdots$$

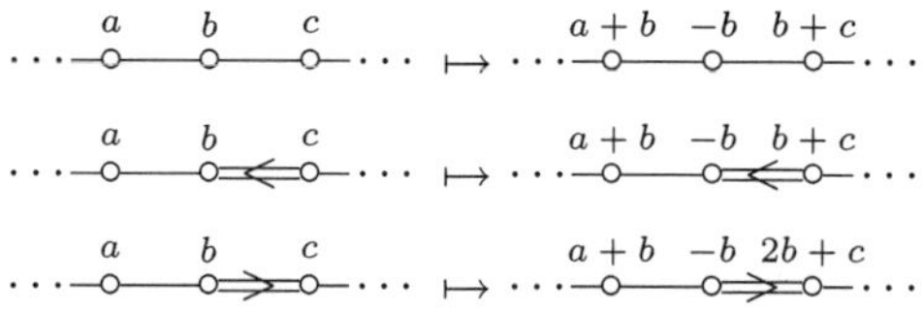

(In the case of a triple edge, the second rule remains unchanged, while in the last rule one obtains $3b + c$ rather than $2b + c$.) One only has to apply reflections corresponding to nodes with positive coefficients. Record the reflection by putting the number of the simple root over the arrow.

The resulting pattern gives all elements of the Hasse diagram and some of the arrows, so if one is only interested in the elements and not in the arrows, one may stop here. The element in the Hasse diagram corresponding to the weight obtained by applying first $s_{\alpha_{i_1}}$, then $s_{\alpha_{i_2}}$ on so on up to $s_{\alpha_{i_\ell}}$ to $\delta^{\mathfrak{p}}$ is given by $s_{\alpha_{i_1}} \ldots s_{\alpha_{i_\ell}}$,

so one has to revert the order of the compositions. Moreover, the length of this element is ℓ, so if one is only interested in elements of some fixed length, one has to apply exactly that many simple reflections to $\delta^{\mathfrak{p}}$.

If one needs the complete set of arrows in the Hasse diagram, then two more things have to be done:

(C) For each element w in the pattern, determine the corresponding set Φ_w of roots, as well as the labels of the arrows determined so far.

Start with the empty set for the point corresponding to $\delta^{\mathfrak{p}}$. Having determined the sets for all elements of length $< \ell$ and the labels of the arrows leading to these sets, consider a point corresponding to an element of length ℓ in the original diagram determined in step (B). Choose a sequence of arrows leading from $\delta^{\mathfrak{p}}$ to the given point, take the simple root indicated on the last arrow in that path, and apply the simple reflections corresponding to the other arrows in the path going back to $\delta^{\mathfrak{p}}$. The resulting root has to be contained in $\Delta^+(\mathfrak{p}_+)$ and the set corresponding to the chosen point is given by adding this root to the set corresponding to the source of the last arrow in the chosen sequence. Now for any of the arrows determined so far which ends in the given element, the set corresponding to the source of the arrow has to be obtained by deleting one element from the set corresponding to the target of the arrow, and this element is the right label for the arrow.

(D) Determine the remaining arrows.

For each of the sets determined in step (C), compute the sum of all roots contained in the set. For two sets in adjacent columns which are not yet joined by an arrow, check whether the difference between the two corresponding expressions is a multiple of a root. If yes, then add an arrow between the two sets labelled by that root.

Recipe for parabolics corresponding to $|1|$–gradings. Two typical examples for the application of this recipe can be found in 3.2.17.

Determine the set Δ^+ of all positive roots, and the subspace $\Delta^+(\mathfrak{p}_+)$ of all positive roots whose root spaces are contained in $\mathfrak{g}_1$. Determine those subsets of $\Delta^+(\mathfrak{p}_+)$, whose complement in Δ^+ is closed under sums. Then the Hasse diagram coincides with the pattern of these subsets, with the arrows given by inclusions into sets with one more element, and the label on each arrow is this additional element. The Weyl group element corresponding to a set in this pattern is determined as follows: Take a path of arrows from the empty set to the given set. Go along this path from the empty set to Φ and for each arrow in the path multiply by the reflection corresponding to the label of the arrow from the left.

3.2.19. Bruhat decomposition and Schubert cells. We next show that the integral homology of complex generalized flag manifolds can be completely described in terms of the Hasse diagram. The first step towards this is a group theoretic decomposition of a complex semisimple Lie group, which leads to a cell–decomposition of its generalized flag varieties. This and the next two subsections are a bit outside of the main line of development. At a first reading, one may skip them and proceed directly to Section 3.3.

The way to obtain the integral homology is via a cell decomposition into so–called Schubert cells. Let $\mathfrak{g}$ be a complex semisimple Lie algebra, $\mathfrak{h} \leq \mathfrak{g}$ a Cartan subalgebra, Δ the corresponding set of roots, Δ^+ a choice of positive roots and Δ^0

the corresponding set of simple roots. To start, let $\mathfrak{b}$ be the standard Borel subalgebra, i.e. the sum of $\mathfrak{h}$ and all positive root spaces. Further, let G be the connected and simply connected Lie group with Lie algebra $\mathfrak{g}$ and put $B = N_G(\mathfrak{b})$. The basis for the decomposition of G/B into Schubert cells is the *Bruhat decomposition* of the group G, which describes the set $B\backslash G/B$ of double cosets of B in G:

THEOREM 3.2.19. *Let G be a connected and simply connected complex semisimple Lie group with Lie algebra $\mathfrak{g}$, $\mathfrak{h} \subset \mathfrak{g}$ a Cartan subalgebra, $\mathfrak{b} \subset \mathfrak{g}$ the standard Borel subalgebra and $B = N_G(\mathfrak{b})$ the corresponding Borel subgroup. Then we have:*

(1) Let $N_G(\mathfrak{h})$ be the normalizer of $\mathfrak{h}$ in G and

$$Z_G(\mathfrak{h}) = \{g \in G : \mathrm{Ad}(g)(H) = H \quad \forall H \in \mathfrak{h}\}$$

its centralizer. Then via the restriction of the coadjoint action Ad^* to $\mathfrak{h}^*$, the quotient $N_G(\mathfrak{h})/Z_G(\mathfrak{h})$ is naturally isomorphic to the Weyl group W of $\mathfrak{g}$. Moreover, $N_G(\mathfrak{h}) \cap N_G(\mathfrak{b}) = Z_G(\mathfrak{h})$ and this intersection is connected.

(2) (Bruhat–decomposition) For any $w \in W$, the double coset $B\tilde{w}B$ of a representative $\tilde{w} \in N_G(\mathfrak{h})$ of w depends only on w. Denoting this double coset by BwB, we get $G = \bigsqcup_{w\in W} BwB$ (disjoint union).

SKETCH OF PROOF. (1) For a positive root $\alpha \in \Delta^+$, consider the root reflection s_α. Choose elements $E_\alpha \in \mathfrak{g}_\alpha$ and $F_\alpha \in \mathfrak{g}_{-\alpha}$ such that $[E_\alpha, F_\alpha] = H_\alpha$, i.e. such that $\alpha([E_\alpha, F_\alpha]) = 2$. Then consider the element $n_\alpha := \exp(E_\alpha)\exp(-F_\alpha)\exp(E_\alpha) \in G$. The adjoint action $\mathrm{Ad}(n_\alpha)$ is given by $e^{\mathrm{ad}(E_\alpha)} \circ e^{\mathrm{ad}(-F_\alpha)} \circ e^{\mathrm{ad}(E_\alpha)}$. Using this, one immediately verifies that for $H \in \mathfrak{h}$ such that $\alpha(H) = 0$ one has $\mathrm{Ad}(n_\alpha)(H) = H$, while $\mathrm{Ad}(n_\alpha)(H_\alpha) = -H_\alpha$. Hence, $n_\alpha \in N_G(\mathfrak{h})$ and $\mathrm{Ad}^*(n_\alpha) = s_\alpha$. Since W is generated by root reflections, this shows that the action of any $w \in W$ on $\mathfrak{h}$ can be realized as the coadjoint action of an element $\tilde{w} \in N_G(\mathfrak{h})$.

Now let $g \in N_G(\mathfrak{h})$ be arbitrary. For $\alpha \in \Delta$ and $H \in \mathfrak{h}$ we have by definition $(\mathrm{Ad}^*(g)(\alpha))(\mathrm{Ad}(g)(H)) = \alpha(H)$. For an element $X \in \mathfrak{g}_\alpha$, we then obtain

$$\alpha(H)\,\mathrm{Ad}(g)(X) = \mathrm{Ad}(g)([H, X]) = [\mathrm{Ad}(g)(H), \mathrm{Ad}(g)(X)].$$

Hence, $\mathrm{Ad}(g)(X)$ is an eigenvector for the adjoint action of $\mathfrak{h}$ with eigenvalue $\mathrm{Ad}^*(g)(\alpha)$, so $\mathrm{Ad}^*(g)$ permutes the roots as well as the root spaces $\mathfrak{g}_\alpha$. In particular, if $g \in Z_G(\mathfrak{h})$, then $\mathrm{Ad}(g)$ preserves each root space and hence $Z_G(\mathfrak{h}) \subset N_G(\mathfrak{b}) = B$.

Still fixing $g \in N_G(\mathfrak{h})$, let $\alpha_1, \dots, \alpha_n$ be the simple roots of $\mathfrak{g}$. Since $\mathrm{Ad}^*(g)$ permutes the roots, it is easy to see that $\{\mathrm{Ad}^*(g)(\alpha_i) : i = 1, \dots, n\} \subset \Delta$ is a simple subsystem. By part (2) of Proposition 2.2.7, the Weyl group acts simply transitively on the set of all simple subsystems, so there is a unique element $w \in W$ which maps $\{\mathrm{Ad}^*(g)(\alpha_i) : i = 1, \dots, n\}$ back to the original simple system $\{\alpha_i\}$. Choosing a representative $\tilde{w} \in N_G(\mathfrak{h})$, we conclude that $\tilde{w}g \in N_G(\mathfrak{h})$ permutes the simple roots α_i. In particular, this element preserves the set Δ^+ of positive roots and the corresponding root spaces, so it also lies in $N_G(\mathfrak{b}) = B$.

Hence, we can complete the proof of part (1) by showing that $N_G(\mathfrak{h}) \cap B$ is contained in the connected component of the identity of $Z_G(\mathfrak{h})$. From 2.3.1 we know that there is a compact real form $\mathfrak{k} \subset \mathfrak{g}$ which contains the subspace $i\mathfrak{h}_0$ on which all roots are purely imaginary. The conjugation θ with respect to $\mathfrak{k}$ is a Cartan involution on the real Lie algebra underlying $\mathfrak{g}$ such that $\theta(\mathfrak{h}) = \mathfrak{h}$; see 2.3.2. Let $\Theta : G \to G$ be the global Cartan involution corresponding to θ and $K \subset G$ its fixed point group. From 2.2.19, we know that G has finite center (whose order equals the determinant of the Cartan matrix of $\mathfrak{g}$), so K is a maximal compact subgroup of G.

Theorem 2.3.3 shows that any element $g \in G$ can be uniquely written as $k\exp(X)$ with $k \in K$ and $X \in i\mathfrak{k}$.

Now assume that $g = k\exp(X) \in N_G(\mathfrak{h}) \cap B$. Since Θ is an involutive automorphism, we get $\Theta(g)x\Theta(g)^{-1} = \Theta(g\Theta(x)g^{-1})$ for all $x \in G$. This shows that $\mathrm{Ad}(\Theta(g)) = \theta \circ \mathrm{Ad}(g) \circ \theta$, so this preserves $\mathfrak{h}$, too. Then also $\Theta(g)^{-1}g$ preserves $\mathfrak{h}$, and since $\Theta(g) = k\exp(-X)$, we see that $\mathrm{Ad}(\exp(2X))(\mathfrak{h}) \subset \mathfrak{h}$. Now $\mathrm{Ad}(\exp(2X)) = \mathrm{Ad}(\exp(X))^2$ is symmetric with respect to B_θ and hence diagonalizable with positive real eigenvalues. Since $\theta(2X) = -2X$, also $\mathrm{ad}(2X)$ is symmetric with respect to B_θ and hence diagonalizable with real eigenvalues. Since $\mathrm{Ad}(\exp(2X)) = e^{\mathrm{ad}(2X)}$, we see that $\mathrm{ad}(2X)$ has the same eigenspaces as $\mathrm{Ad}(\exp(2X))$ so, in particular, $\mathrm{ad}(2X)(\mathfrak{h}) \subset \mathfrak{h}$. Since $\mathfrak{h}$ is a Cartan subalgebra, this implies $X \in \mathfrak{h}$, and hence $\exp(X)$ lies in the connected component of the identity of $Z_G(\mathfrak{h})$.

Hence, we see that $k \in N_G(\mathfrak{h}) \cap B$, so $\mathrm{Ad}(k)$ preserves $\mathfrak{h} \cap \mathfrak{k} = i\mathfrak{h}_0$ as well as the positive roots. But then it is a well–known fact from the theory of compact groups (see e.g. [**Kn96**, Corollary 4.52]) that k actually lies in the maximal torus corresponding to $i\mathfrak{h}_0$ and hence in the connected component of the identity of $Z_G(\mathfrak{h})$.

(2) In (1) we have seen that $Z_G(\mathfrak{h}) \subset B$, so $B\tilde{w}B$ is independent of the choice of the element $\tilde{w} \in N_G(\mathfrak{h})$ representing $w \in W$. The disjointness of the double cosets is also easy to see: Let $\mathfrak{g} = \mathfrak{g}_{-k} \oplus \cdots \oplus \mathfrak{g}_k$ be the $|k|$–grading defined by $\mathfrak{b}$. By Theorem 3.1.3, any element $g \in B$ can be uniquely written as $g = g_0\exp(Z)$ for $g_0 \in G_0$ and $Z \in \mathfrak{b}_+$, where G_0 is the reductive Levi component of B and $\mathfrak{b}_+$ is the nilradical of $\mathfrak{b}$. Since $\mathfrak{g}_0 = \mathfrak{h}$, we see from (1) that $G_0 = B \cap N_G(\mathfrak{h}) = Z_G(\mathfrak{h})$ and that G_0 is connected. In particular, for any $H \in \mathfrak{h}$, the $\mathfrak{h}$–component of $\mathrm{Ad}(g)(H) \in \mathfrak{b}$ equals H.

Now assume that $B\tilde{w}_1B \cap B\tilde{w}_2B \neq \emptyset$. Then we find $g, h \in B$ such that $g\tilde{w}_1 = \tilde{w}_2h$. From above, we see that for $H \in \mathfrak{h}$ the $\mathfrak{h}$–component of $\mathrm{Ad}(g\tilde{w}_1)(H)$ equals $\mathrm{Ad}(\tilde{w}_1)(H)$ and similarly the $\mathfrak{h}$–component of $\mathrm{Ad}(\tilde{w}_2h)(H)$ equals $\mathrm{Ad}(\tilde{w}_2)(H)$. Hence, we see that $\mathrm{Ad}(\tilde{w}_1)$ and $\mathrm{Ad}(\tilde{w}_2)$ act in the same way on $\mathfrak{h}$, whence they represent the same element of W.

The proof that the double cosets cover G is a bit more involved. Let $B_+ = \exp(\mathfrak{b}_+)$ be the subgroup corresponding to the nilradical. For $g \in B_+$ and $H \in \mathfrak{h}$, $\mathrm{Ad}(g)(H)$ by definition is congruent to H modulo $\mathfrak{b}_+$. For fixed H, we can then define a map $B_+ \to \mathfrak{b}_+$ by $g \mapsto \mathrm{Ad}(g)(H) - H$. First, one proves that if H is regular (i.e. if $\alpha(H) \neq 0$ for all $\alpha \in \Delta$), then this mapping is surjective.

Second, for an arbitrary element $g \in G$, one considers $(\mathrm{Ad}(g)(\mathfrak{b}) \cap \mathfrak{b}) + \mathfrak{b}_+ \subset \mathfrak{b}$ and verifies that this subspace has the same dimension as $\mathfrak{b}$ and hence coincides with $\mathfrak{b}$. In particular, the grading element E for the grading induced by $\mathfrak{b}$ can be written as $E = X + Y$ for $X \in \mathfrak{b}_+$ and $Y \in \mathrm{Ad}(g)(\mathfrak{b}) \cap \mathfrak{b}$. Since $\mathfrak{b}_+$ consists of all positive root spaces, we see that $\alpha(E) \neq 0$ for all $\alpha \in \Delta$. Hence, there is element $n_1 \in B_+$ such that $\mathrm{Ad}(n_1)(E) - E = -X$, and thus $\mathrm{Ad}(n_1)(E) = E - X = Y$. By construction, $\mathrm{Ad}(g^{-1})(Y) \in \mathfrak{b}$, whence $\mathrm{Ad}(g^{-1}n_1)(E) =: Z \in \mathfrak{b}$. Let us write $Z = H' + X'$ for $H' \in \mathfrak{h}$ and $X' \in \mathfrak{b}_+$. Taking a basis of $\mathfrak{g}$ consisting of root vectors with smaller roots coming before larger roots, we see that the matrix of $\mathrm{ad}(Z)$ is upper triangular, and its diagonal part is the matrix of $\mathrm{ad}(H')$, so both maps have the same eigenvalues. But by construction, $\mathrm{ad}(Z)$ is conjugate to $\mathrm{ad}(E)$, so its zero eigenspace is conjugate to $\mathfrak{h}$. This implies that the kernel of $\mathrm{ad}(H')$ has dimension at most $\dim(\mathfrak{h})$ and thus $\alpha(H') \neq 0$ for all $\alpha \in \Delta$. Hence, we find $n_2 \in B_+$ such that $\mathrm{Ad}(n_2^{-1})(H') - H' = X'$, and $H' = \mathrm{Ad}(n_2)(Z) = \mathrm{Ad}(n_2 g^{-1} n_1)(E)$. Putting $u := n_1^{-1} g n_2^{-1} \in BgB$, we obtain $H' = \mathrm{Ad}(u^{-1})(E)$. For an arbitrary element $H \in \mathfrak{h}$ we then get

$$[\mathrm{Ad}(u)(H), E] = \mathrm{Ad}(u)([H, \mathrm{Ad}(u^{-1})(E)]) = \mathrm{Ad}(u)([H, H']) = 0,$$

so $\mathrm{Ad}(u)(H) \in \mathfrak{h}$ and thus $u \in N_G(\mathfrak{h})$ and $u = \tilde{w}$ for some $w \in W$. □

Note that the beginning of the proof of part (2), in particular, shows that B is diffeomorphic to $N_G(\mathfrak{b}) \cong G_0 \times \mathfrak{b}_+$ and hence connected.

Passing from the Bruhat decomposition of G to the decomposition of G/P into Schubert cells is rather easy.

COROLLARY 3.2.19. *Let G be a simply connected complex semisimple Lie group with Lie algebra $\mathfrak{g}$, $\mathfrak{h} \subset \mathfrak{b} \subset \mathfrak{g}$ a Cartan and a Borel subalgebra, and let $\mathfrak{p} \subset \mathfrak{g}$ be a standard parabolic subalgebra. Put $B := N_G(\mathfrak{b})$ and $P := N_G(\mathfrak{p}) \subset G$. Let W be the Weyl group of $\mathfrak{g}$ and $W^{\mathfrak{p}} \subset W$ the Hasse diagram of $\mathfrak{p}$. For each $w \in W$ choose a representative $\tilde{w} \in N_G(\mathfrak{h})$. Let $o = eP \in G/P$ be the base point of the generalized flag variety determined by P.*

Then the point $\tilde{w} \cdot o \in G/P$ depends only on w, and

$$G/P = \bigsqcup_{w \in W^{\mathfrak{p}}} B \cdot \tilde{w}^{-1} \cdot o.$$

Moreover, the B–orbit $B \cdot \tilde{w} \cdot o$ is diffeomorphic to $\mathbb{C}^{\ell(w)}$, where $\ell(w)$ denotes the length of w.

SKETCH OF PROOF. Let us start with the case $\mathfrak{p} = \mathfrak{b}$ and hence $W^{\mathfrak{p}} = W$. Since $G = \bigsqcup_{w \in W} B\tilde{w}B$ we get $G/B = \bigsqcup_{w \in W} B\tilde{w}^{-1} \cdot o$. Now we use the projective realization of G/B from 3.2.8. Let V be the irreducible representation of G with highest weight δ and $v_0 \in V$ a nonzero highest weight vector. For $H \in \mathfrak{h}$ we then get $H \cdot \tilde{w}^{-1} \cdot v_0 = \tilde{w}^{-1} \cdot \mathrm{Ad}(\tilde{w})(H) \cdot v_0$, which shows that $\tilde{w}^{-1} \cdot v_0$ is a weight vector of weight $w^{-1}(\delta)$. The weight space of this weight is one–dimensional, so we see that the points $\tilde{w} \cdot o \in G/B$ correspond to the points in the projectivization $\mathcal{P}(V)$ which are given by the weight spaces of the weights in the Weyl orbit of δ.

Let $\alpha \in \Delta^+$ be a positive root and $E_\alpha \in \mathfrak{g}_\alpha$ an element in the corresponding root space. As above, $E_\alpha \cdot \tilde{w}^{-1} \cdot v_0 = \tilde{w}^{-1} \cdot \mathrm{Ad}(\tilde{w})(E_\alpha) \cdot v_0$. This vanishes if and only if $\mathrm{Ad}(\tilde{w})(E_\alpha) \in \mathfrak{b}$. But from the proof of the theorem above, we know that $\mathrm{Ad}(\tilde{w})(E_\alpha)$ lies in the root space $\mathfrak{g}_{w^{-1}(\alpha)}$, whence $E_\alpha \cdot \tilde{w}^{-1} \cdot v_0 = 0$ is equivalent to $w^{-1}(\alpha) \in \Delta^+$. In the notation introduced in 3.2.14, this exactly means that $\alpha \notin \Phi_w$.

Thus, the stabilizer in $\mathfrak{b}$ of the line through $\tilde{w}^{-1} \cdot v_0$ is given by $\mathfrak{h} \oplus \bigoplus_{\alpha \notin \Phi_w} \mathfrak{g}_\alpha$. This suggests looking at the complementary space $\bigoplus_{\alpha \in \Phi_w} \mathfrak{g}_\alpha$, and one shows that $Z \mapsto \exp(Z) \cdot [\tilde{w}^{-1} \cdot v_0]$ is a diffeomorphism from this vector space onto the orbit of the line $[\tilde{w}^{-1} \cdot v_0]$. Hence, this orbit is diffeomorphic to $\mathbb{C}^{|\Phi_w|} = \mathbb{C}^{\ell(w)}$, where $\ell(w)$ denotes the length of the Weyl group element w; see part (3) of Proposition 3.2.14. Note that $\ell(w) = \ell(w^{-1})$, so there are no problems with the inversion.

The analogous result for a general parabolic is an easy consequence: Of course, we have $G/P = \bigcup_{w \in W} B\tilde{w}^{-1} \cdot o$, and two such orbits are either equal or disjoint. Again, this can be expressed nicely in the projective realization, starting with a representation V of highest weight $\delta^{\mathfrak{p}}$. For a highest weight vector v_0 we then get $\tilde{w} \cdot [v_0] = [v_0]$ if and only if $w \in W_{\mathfrak{p}}$. Using Proposition 3.2.15 we conclude that

$G/P = \bigsqcup_{w \in W^\mathfrak{p}} B \cdot \tilde{w}^{-1} \cdot [v_0]$ in the projective realization. As above, these are exactly the lines associated to the weight spaces with weights in the Weyl orbit of $\delta^\mathfrak{p}$. The proof that any such orbit is diffeomorphic to $\mathbb{C}^{\ell(w)}$ is exactly as above. □

DEFINITION 3.2.19. For an element $w \in W^\mathfrak{p}$ the B–orbit $B \cdot \tilde{w}^{-1} \cdot o \cong \mathbb{C}^{\ell(w)} \subset G/P$ is called the *Schubert cell* corresponding to w.

3.2.20. Integral homology of complex generalized flag varieties. If the decomposition $G/P = \bigsqcup_{w \in W^\mathfrak{p}} B \cdot \tilde{w}^{-1} \cdot o$ established in Corollary 3.2.19 above were a CW decomposition, then we could immediately see that the integral homology $H_*(G/P)$ is isomorphic to $\mathbb{Z}^{W^\mathfrak{p}}$, with each element $w \in W^\mathfrak{p}$ giving rise to a generator in degree $2\ell(w)$. While this will turn out to be the correct result, its needs quite a few technicalities to establish it. Since it is not at all clear at this point whether one has attaching maps for the Schubert cells, we have to take two more nontrivial steps.

First, we have to describe the closures of the Schubert cells. Let us again start in the case of G/B, and look at the projective realization. For a highest weight vector v_0, the B–orbit of the line $[\tilde{w}^{-1} \cdot v_0]$ coincides with the B_+ orbit. Moreover, for $Z \in \mathfrak{b}_+$ we see that $\exp(Z) \cdot \tilde{w}^{-1} \cdot v_0$ is a linear combination of $\tilde{w}^{-1} \cdot v_0$ and weight vectors of higher weight. Choose a basis of V^* dual to a basis of V consisting of weight vectors and denote by ψ_w the element dual to the weight vector $\tilde{w}^{-1} \cdot v_0$. Then one easily concludes that the Schubert cell $X_w := B \cdot \tilde{w}^{-1} \cdot [v_0]$ is the intersection of $G \cdot [v_0]$ with the set $\{[v] : v \in \mathcal{U}(\mathfrak{b}_+) \cdot \tilde{w}^{-1} \cdot v_0, \psi_w(v) \neq 0\}$. This immediately implies that the closure $\bar{X}_w$ (which is called the *Schubert variety* corresponding to w) is the intersection of $G \cdot [v_0]$ with $\{[v] : v \in \mathcal{U}(\mathfrak{b}_+) \cdot \tilde{w}^{-1} \cdot v_0\}$.

Since $\bar{X}_w$ is a B–invariant subset of G/B it must be the union of some Schubert cells. Of course, $X_{w'} \subset \bar{X}_w$ if and only if $(\tilde{w}')^{-1} \cdot v_0 \in \mathcal{U}(\mathfrak{b}_+) \cdot \tilde{w}^{-1} \cdot v_0$, and in [**BGG73**] it is shown that this is equivalent to $(w')^{-1} \leq w^{-1}$ in the Bruhat ordering. This in turn is easily seen to be equivalent to $w' \leq w$. Hence, we conclude that $\bar{X}_w = \bigcup_{w' \leq w} X_{w'}$, and in particular, it is the union of X_w and cells of strictly lower dimension. The analogous result for G/P can be easily proved using the fact that there is a natural fibration $G/B \to G/P$, which by construction maps Schubert cells to Schubert cells.

The second ingredient is the use of Borel–Moore homology. The application of this homology theory to related problems was first introduced in [**BH61**], an exposition can be found in an appendix of the book [**Fu97**]. This homology is defined only for spaces X which admit a closed embedding into some $\mathbb{R}^N$, and one defines the ith *Borel–Moore homology group* of X as $\bar{H}_i(X) := H^{N-i}(\mathbb{R}^N, \mathbb{R}^N \setminus X)$. This is independent of the choice of the embedding into $\mathbb{R}^N$ and more generally, for any embedding into an oriented m–dimensional manifold M, one has $\bar{H}_i(X) \cong H^{m-i}(M, M \setminus X)$. In particular, if X itself is a compact smooth manifold, then Borel–Moore homology coincides with singular homology by Poincaré duality.

Further, one shows that this homology is functorial for proper mappings and thus, in particular, for closed embeddings. The main advantage of passing to Borel–Moore homology is the existence of the following exact sequence: Let X be a space admitting a closed embedding into some $\mathbb{R}^N$, $Y \subset X$ a closed subset and $U := X \setminus Y$. Then Borel–Moore homology is defined for X, Y, and U, since embedding X as a closed subset into some m–manifold M, we obtain closed embeddings of Y into M and of U into $M \setminus Y$. The long exact cohomology sequence of the triple

$M \setminus X \subset M \setminus Y \subset M$ then gives a long exact sequence

$$\cdots \to \bar{H}_{i+1}(U) \to \bar{H}_i(Y) \to \bar{H}_i(X) \to \bar{H}_i(U) \to \cdots$$

of Borel–Moore homology groups. Finally, the Mayer–Vietoris sequence in cohomology immediately shows that Borel–Moore homology is additive for disjoint unions of open subsets. Now we consider G/P and define X_j to be the union of all Schubert cells X_w for $\ell(w) \leq j$. Then X_0 is a single point, each X_j is a closed subset of G/P, $X_j \subset X_{j+1}$ and $X_{j+1} \setminus X_j$ is the disjoint union of the (open) Schubert cells X_w for $\ell(w) = j+1$. For such a cell we have $\bar{H}_k(X_w) = \bar{H}_k(\mathbb{C}^{\ell(w)}) \cong H^{2\ell(w)-k}(\mathbb{C}^{\ell(w)})$, so this is $\mathbb{Z}$ for $k = 2\ell(w)$ and zero for all other k. From the above exact sequence one now concludes inductively that $H_*(X_j)$ has one free generator for each element $w \in W$ with $\ell(w) \leq j$ and this generator sits in degree $2\ell(w)$. For j equal to the length of the longest element in $W^{\mathfrak{p}}$ we get $X_j = G/P$ and since G/P is a compact orientable manifold, $\bar{H}_*(G/P) \cong H_*(G/P)$, and we conclude:

THEOREM 3.2.20. *Let G be a connected complex semisimple Lie group with Lie algebra $\mathfrak{g}$, $\mathfrak{p} \leq \mathfrak{g}$ a parabolic subalgebra, and $P = N_G(\mathfrak{p})$ the corresponding parabolic subgroup. Then the integral homology $H_*(G/P)$ has one free generator for each element w in the Hasse diagram $W^{\mathfrak{p}}$ of $\mathfrak{p}$, and this generator sits in $H_{2\ell(w)}(G/P)$, where $\ell(w)$ is the length of w.*

Let us remark that there is an analog of the construction of Schubert cells for real generalized flag manifolds, which leads to a quite detailed description of their integral homology; see [**CaSt99**]. This result, however, uses rather heavy machinery.

3.2.21. Integral cohomology of complex generalized flag varieties. The results on integral homology from Theorem 3.2.20 show that, as an abelian group, $H^*(G/P, \mathbb{Z}) \cong \mathbb{Z}^{W^{\mathfrak{p}}}$, with the generator corresponding to $w \in W^{\mathfrak{p}}$ sitting in degree $2\ell(w)$. This also implies that the integral cohomology groups $H^*(G/P, \mathbb{Z})$ inject into the de Rham cohomology groups $H^*(G/P, \mathbb{R})$ and we can use this injection to understand the multiplicative structure on integral cohomology. From the description of Schubert cells in 3.2.19 we conclude that for a Borel subgroup $B \subset G$ and a parabolic $P \subset G$ which contains B, the natural projection $p : G/B \to G/P$ induces a surjection in integral homology. Dually we obtain an injection $p^* : H^*(G/P, \mathbb{R}) \to H^*(G/B, \mathbb{R})$, so it suffices to consider the case of the Borel subgroup.

By Proposition 3.2.6, the maximal compact subgroup $K \subset G$ acts transitively on any generalized flag manifold G/P and we may identify G/P with $K/(K \cap P)$. Therefore, one may invoke the general machinery to compute the de Rham cohomology of homogeneous spaces of compact Lie groups, which is presented, for example, in the books [**On93**] and [**On94**]. The basic idea here is that for a homogeneous space $M = K/L$ of a compact Lie group K, one may look at the subalgebra $\Omega^*(M)^K$ of K–invariant differential forms on M, which is immediately seen to be closed under the exterior derivative. An averaging argument shows that the subcomplex $(\Omega^*(M)^K, d)$ of the de Rham complex still computes the de Rham cohomology of M. By Theorem 1.4.4, K–invariant k–forms on M are in bijective correspondence with L–invariant elements in $\Lambda^k(\mathfrak{k}/\mathfrak{l})^*$ and it is easy to express the exterior derivative in this picture. Thus the computation of the cohomology is reduced to a finite–dimensional algebraic problem.

This algebraic problem can be studied for general homogeneous spaces of compact Lie groups using the Weyl algebra and the Cartan algebra associated to the pair $(\mathfrak{k}, \mathfrak{l})$; see chapter 3 of [**On94**]. In particular, one may use transgression arguments to pass from skew symmetric invariants to symmetric invariants.

In the case of the generalized flag manifold G/B, all these developments can be neatly phrased in terms of Chern classes of homogeneous line bundles. We may assume that G is simply connected and $B \subset G$ is a standard Borel subgroup. By $\mathfrak{h} \subset \mathfrak{g}$ we denote the corresponding Cartan subalgebra and by $\mathfrak{h}_0^*$ the real subspace of the dual $\mathfrak{h}^*$ spanned by the roots. Then the symmetric algebra $R = S(\mathfrak{h}_0^*)$ may be viewed as the algebra of polynomial functions on the space $\mathfrak{h}_0 \subset \mathfrak{h}$, and the natural action of the Weyl group W of $\mathfrak{g}$ on $\mathfrak{h}_0^*$ extends to an action on R. Let $I \subset R$ be the subring of W–invariant polynomials, $I_+ \subset I$ the subspace of W–invariant polynomials without constant term and $J = I_+R$ the right ideal generated by I_+. Observe that the ideal J is invariant under the natural action of W, so W naturally acts on R/J.

Now one constructs an algebra homomorphism $R \to H^*(G/B, \mathbb{R})$ as follows: Any linear functional $\chi : \mathfrak{h} \to \mathbb{C}$ can be viewed as a Lie algebra homomorphism $\mathfrak{b} \to \mathbb{C}$. One proves that χ integrates to a group homomorphism $B \to \mathbb{C}^*$ if and only if χ is algebraically integral as introduced in 2.2.10. In particular, this implies that $\chi : \mathfrak{h}_0 \to \mathbb{R}$. Passing to the associated line bundle we obtain a homogeneous complex line bundle $E_\chi \to G/B$, and one defines $\alpha(\chi)$ to be the first Chern class $c_1(E_\chi) \in H^2(G/B, \mathbb{Z})$ of this line bundle. This uniquely extends to an algebra homomorphism $\alpha : R \to H^*(G/B, \mathbb{R})$. The result on the real cohomology then reads as follows (see [**Bor53, AtHi61**] for part (1) and [**BGG73**] for part(2)):

PROPOSITION 3.2.21. *(1) The homomorphism $\alpha : R \to H^*(G/B, \mathbb{R})$ descends to an isomorphism $R/J \cong H^*(G/B, \mathbb{R})$ of rings.*

(2) Let $P \subset G$ be a standard parabolic and $W_\mathfrak{p}$ the corresponding subgroup of the Weyl group W. Then the isomorphism $R/J \to H^*(G/B, \mathbb{R})$ from (1) restricts to an isomorphism between the subring of $W_\mathfrak{p}$–invariant elements in R/J and the image of $H^*(G/P, \mathbb{R})$ in $H^*(G/B, \mathbb{R})$.

REMARK 3.2.21. (1) Since $H_*(G/P, \mathbb{Z})$ is free abelian, the real cohomology $H^*(G/P, \mathbb{R})$ is isomorphic to the space $\mathrm{Hom}_\mathbb{Z}(H_*(G/P, \mathbb{Z}), \mathbb{R})$. In particular, there is a basis of $H^*(G/P, \mathbb{R})$, which is dual to the basis of the homology formed by the Schubert cells. The question of describing this dual basis explicitly has been studied intensively; see [**Kos63**] and [**BGG73**] for different descriptions.

(2) The algebraic methods for dealing with invariant differential forms mentioned above do not only lead to a description of real cohomology but also provide information on the real homotopy type of a homogeneous space of a compact Lie group; see [**On94**] for several results in that direction.

3.3. Kostant's version of the Bott–Borel–Weil theorem

In this section, we give a complete proof of Kostant's version of the Bott–Borel–Weil theorem and study some of its consequences. Given a complex semisimple Lie algebra $\mathfrak{g}$ and a standard parabolic subalgebra $\mathfrak{p} \leq \mathfrak{g}$, one has the nilradical $\mathfrak{p}_+$ and the reductive part $\mathfrak{g}_0$ of $\mathfrak{p}$, which both are subalgebras of $\mathfrak{g}$. Any finite–dimensional representation V of $\mathfrak{g}$ may be viewed as a representation of $\mathfrak{p}_+$ and of $\mathfrak{g}_0$ by restriction, so one may consider the cohomology $H^*(\mathfrak{p}_+, V)$ of $\mathfrak{p}_+$ with coefficients in V. It is easy to see that all spaces in the standard complex that computes this cohomology are $\mathfrak{g}_0$–modules and the differentials are $\mathfrak{g}_0$–equivariant, so any of the cohomology spaces $H^k(\mathfrak{p}_+, V)$ is naturally a $\mathfrak{g}_0$–module. Kostant's version of the Bott–Borel–Weil theorem gives a complete description of the $\mathfrak{g}_0$–module structure of these cohomology spaces in the case of a complex irreducible representation V of $\mathfrak{g}$. This description is in terms of the Hasse diagram $W^{\mathfrak{p}}$, so there is an algorithm to compute the cohomologies.

The cohomology spaces $H^k(\mathfrak{p}_+, V)$ are naturally dual to the cohomology spaces $H^k(\mathfrak{g}_-, V^*)$ of the subalgebra $\mathfrak{g}_-$ with coefficients in the dual $\mathfrak{g}$–module V^*, so we also get a description of the latter cohomologies. In the case $V = \mathfrak{g}$ of the adjoint representation, this immediately relates to the developments of Section 3.1. On the one hand, it allows us to analyze the conditions on $H^1(\mathfrak{g}_-, \mathfrak{g})$ needed in Theorems 3.1.14 and 3.1.16 to obtain an equivalence between regular normal parabolic geometries and underlying structures. On the other hand, we know from 3.1.12 that $H^2(\mathfrak{g}_-, \mathfrak{g})$ determines the possible values for components of the harmonic curvature, which are the essential curvature and torsion quantities for a regular normal parabolic geometry. The cohomology spaces for general representations V will also play an important role in volume two, since they determine the bundles showing up in Bernstein–Gelfand–Gelfand sequences.

In the end of the section, we show how to deduce the Weyl character formula and the Bott–Borel–Weil theorem from Kostant's theorem.

3.3.1. Codifferential and Hodge theory. Let $\mathfrak{g} = \mathfrak{g}_{-k} \oplus \cdots \oplus \mathfrak{g}_k$ be a $|k|$–graded semisimple Lie algebra, $\mathfrak{p} = \mathfrak{g}^0 = \mathfrak{g}_0 \oplus \cdots \oplus \mathfrak{g}_k$ the corresponding parabolic subalgebra, and $\mathfrak{p}_+ = \mathfrak{g}^1$ its nilradical. From part (1) of Proposition 3.1.2 we know that there is a unique element $E \in \mathfrak{z}(\mathfrak{g}_0)$ such that $[E, A] = jA$ for $A \in \mathfrak{g}_j$, the grading element. Assume that V is a finite–dimensional irreducible representation of $\mathfrak{g}$. Then from 3.2.12 we know that we get a decomposition $V = V_0 \oplus \cdots \oplus V_\ell$, where V_0 is the eigenspace for the E–action with the lowest possible eigenvalue s_0, and each V_j is the eigenspace for the eigenvalue $s_0 + j$. This direct sum decomposition is by construction $\mathfrak{g}_0$–invariant and the corresponding filtration $V = V^0 \supset V^1 \supset \cdots \supset V^\ell$, where $V^j = V_j \oplus V_{j+1} \oplus \cdots \oplus V_\ell$, is $\mathfrak{p}$–invariant.

Now consider the standard complex $(C^*(\mathfrak{g}_-, V), \partial)$ for computing the Lie algebra cohomology $H^*(\mathfrak{g}_-, V)$ as introduced in 2.1.9. By definition, $C^n(\mathfrak{g}_-, V)$ is the space of n–linear, alternating maps $(\mathfrak{g}_-)^n \to V$, and the differential ∂ is given by

$$\begin{aligned} \partial\phi(X_0, \dots, X_n) := &\sum_{i=0}^{n} (-1)^i X_i \cdot \phi(X_0, \dots, \widehat{X_i}, \dots, X_n) \\ &+ \sum_{i<j} (-1)^{i+j} \phi([X_i, X_j], X_0, \dots, \widehat{X_i}, \dots \widehat{X_j}, \dots, X_n), \end{aligned} \tag{3.10}$$

for $X_0, \dots, X_n \in \mathfrak{g}_-$, with the hat over an argument denoting omission. By definition, any of the spaces $C^n(\mathfrak{g}_-, V)$ is a $\mathfrak{g}_0$–module and we claim that ∂ is $\mathfrak{g}_0$–equivariant. This is easier to see for the corresponding group action. So let G be the connected and simply connected Lie group with Lie algebra $\mathfrak{g}$ and define $G_0 := \bigcap_{i=-k}^{k} N_G(\mathfrak{g}_i) \subset G$; see 3.1.3. Then G_0 acts on $\mathfrak{g}_-$ via the adjoint action and hence by Lie algebra homomorphisms. Since G is simply connected, the $\mathfrak{g}$–action on V integrates to G, and we can restrict this action to G_0, thus obtaining an induced action on $C^n(\mathfrak{g}_-, V)$. Explicitly, this action is given by

$(g \cdot \phi)(X_1, \dots, X_n) = g \cdot (\phi(\mathrm{Ad}(g^{-1})(X_1), \dots, \mathrm{Ad}(g^{-1})(X_n)))$. From the above formula it is evident that ∂ is equivariant for this action, and thus it is also a homomorphism of $\mathfrak{g}_0$–modules. In particular, each of the cohomology spaces $H^n(\mathfrak{g}_-, V)$ is naturally a $\mathfrak{g}_0$–module.

The gradings on $\mathfrak{g}_-$ and V give rise to a $\mathfrak{g}_0$–invariant grading on each of the spaces $C^n(\mathfrak{g}_-, V)$, which is given by the homogeneous degree of maps. This means that $\phi \in C^n(\mathfrak{g}_-, V)_m$ if and only if for $X_j \in \mathfrak{g}_{i_j}$ for $j = 1, \dots, n$ we have $\phi(X_1, \dots, X_n) \in V_{i_1+\dots+i_n+m}$. We may also make the spaces $C^n(\mathfrak{g}_-, V)$ into $\mathfrak{p}$–modules by identifying $\mathfrak{g}_-$ with $\mathfrak{g}/\mathfrak{p}$, but ∂ is *not* a homomorphism of $\mathfrak{p}$–modules. Also, the grading on $C^n(\mathfrak{g}_-, V)$ is not $\mathfrak{p}$–invariant, but the associated filtration has this property.

Now let $B : \mathfrak{g} \times \mathfrak{g} \to \mathbb{K}$ be the Killing form of $\mathfrak{g}$. By part (3) of Proposition 3.1.2, B induces dualities of $\mathfrak{g}_0$–modules between $\mathfrak{g}_-$ and $\mathfrak{p}_+$ and of $\mathfrak{p}$–modules between $\mathfrak{g}/\mathfrak{p}$ and $\mathfrak{p}_+$. Hence, we may naturally view the spaces $C^n(\mathfrak{g}_-, V)$ and $C^n(\mathfrak{p}_+, V^*)$ as dual $\mathfrak{g}_0$ modules and as dual $\mathfrak{p}$–modules. We have to keep in mind that the $\mathfrak{p}$–action on $\mathfrak{g}_-$ comes from the identification with $\mathfrak{g}/\mathfrak{p}$. Moreover, the dualities are compatible with the invariant gradings respectively filtrations available on the two spaces.

Consider any Lie group G with Lie algebra $\mathfrak{g}$, and let $P \subset G$ be a parabolic subgroup for the given grading; see 3.1.3. Then P acts on $\mathfrak{p}_+$ by the restriction of the adjoint action. If the $\mathfrak{g}$–action on V integrates to G, then V^* naturally is a P–module. As above we conclude that $\partial_{\mathfrak{p}_+} : C^n(\mathfrak{p}_+, V^*) \to C^{n+1}(\mathfrak{p}_+, V^*)$ is P–equivariant and thus a $\mathfrak{p}$–homomorphism. Using the duality introduced above, $\partial_{\mathfrak{p}_+}$ dualizes to a P–homomorphism $\partial^* : C^{n+1}(\mathfrak{g}_-, V) \to C^n(\mathfrak{g}_-, V)$. To obtain an explicit formula for ∂^* on decomposable elements, we proceed exactly as in the special case dealt with in 3.1.11. Using that $C^{n+1}(\mathfrak{g}_-, V) \cong \Lambda^{n+1}\mathfrak{p}_+ \otimes V$, we may write a decomposable element in this space as $Z_0 \wedge \dots \wedge Z_n \otimes v$ for $Z_1, \dots, Z_n \in \mathfrak{p}_+$ and $v \in V$. The pairing of that element with $\phi \in C^{n+1}(\mathfrak{p}_+, V^*)$ is then given by the paring of $\phi(Z_0, \dots, Z_n) \in V^*$ with v. Now putting $\phi = \partial_{\mathfrak{p}_+}\psi$ for $\psi \in C^n(\mathfrak{p}_+, V^*)$, and using that

$$(Z_i \cdot \psi(Z_0, \dots, \widehat{Z_i}, \dots, Z_n))(v) = -\psi(Z_0, \dots, \widehat{Z_i}, \dots, Z_n)(Z_i \cdot v),$$

we immediately obtain the explicit formula

$$\begin{aligned} \partial^*(Z_0 \wedge \dots \wedge Z_n \otimes v) &= \sum_{i=0}^{n} (-1)^{i+1} Z_0 \wedge \dots \wedge \widehat{Z_i} \wedge \dots \wedge Z_n \otimes Z_i \cdot v \\ &+ \sum_{i<j} (-1)^{i+j} [Z_i, Z_j] \wedge Z_0 \wedge \dots \wedge \widehat{Z_i} \wedge \dots \wedge \widehat{Z_j} \wedge \dots \wedge Z_n \otimes v. \end{aligned} \tag{3.11}$$

Renumbering the indices we see that ∂^* coincides with the differential δ in the standard complex for the Lie algebra homology $H_*(\mathfrak{p}_+, V)$; see formula (2.1) in 2.1.9. In accordance with most of the literature on parabolic geometry, we keep the name "codifferential" and the notation ∂^*.

The first essential step towards understanding the cohomologies $H^*(\mathfrak{g}_-, V)$ and $H^*(\mathfrak{p}_+, V)$ is to show that the operators ∂ and ∂^* are adjoint with respect to a positive definite inner product on $C^*(\mathfrak{g}_-, V)$.

PROPOSITION 3.3.1. *Let $\mathfrak{g} = \mathfrak{g}_{-k} \oplus \dots \oplus \mathfrak{g}_k$ be a real or complex $|k|$–graded semisimple Lie algebra. Then there is a natural positive definite inner product (Hermitian in the complex case) on the cochain spaces $C^n(\mathfrak{g}_-, V)$ for all n, with respect* *to which the operators* $\partial : C^n(\mathfrak{g}_-, V) \to C^{n+1}(\mathfrak{g}_-, V)$ *and* $\partial^* : C^{n+1}(\mathfrak{g}_-, V) \to C^n(\mathfrak{g}_-, V)$ *are adjoint.*

PROOF. Recall from Section 2.3 that for a Cartan involution θ on $\mathfrak{g}$, $B_\theta(X, Y) := -B(X, \theta(Y))$ defines a positive definite inner product on $\mathfrak{g}$. In the complex case, θ is conjugate linear, so the inner product becomes Hermitian. We first claim that θ can be chosen in such a way that it restricts to a linear isomorphism $\mathfrak{g}_- \to \mathfrak{p}_+$, which then is exactly the isomorphism between $\mathfrak{p}_+$ and $\mathfrak{p}_+^* \cong \mathfrak{g}_-$ induced by B_θ.

If $\mathfrak{g}$ is complex, then choose a Cartan subalgebra $\mathfrak{h} \leq \mathfrak{g}$ which contains the grading element E (compare with the proof of Theorem 3.2.1). Let $\mathfrak{h}_0 \subset \mathfrak{h}$ be the subspace on which all roots are real, which is a real form of $\mathfrak{h}$. Then $E \in \mathfrak{h}_0$ (since all roots even have integral values on E). From 2.3.1 we know that there is a compact real form $\mathfrak{u}$ of $\mathfrak{g}$ such that $i\mathfrak{h}_0 \subset \mathfrak{u}$, and from 2.3.2 we know that the conjugation θ of $\mathfrak{g}$ with respect to the real form $\mathfrak{u}$ is a Cartan involution. Since $E \in \mathfrak{h}_0$ and $i\mathfrak{h}_0 \subset \mathfrak{u}$ we obtain $\theta(E) = -E$, and since θ is a homomorphism of the real Lie algebra underlying $\mathfrak{g}$, this implies $\theta(\mathfrak{g}_i) = \mathfrak{g}_{-i}$. This proves the claim in the complex case.

If $\mathfrak{g}$ is real, then by the proof of part (2) of Proposition 3.2.9, there is a Cartan involution θ on $\mathfrak{g}$ and a θ–stable maximally noncompact Cartan subalgebra $\mathfrak{h} \leq \mathfrak{g}$ which contains the grading element E. Decomposing $\mathfrak{h} = \mathfrak{t} \oplus \mathfrak{a}$ into ± 1–eigenspaces of the restriction of θ, we know from 2.3.8 that all roots are real on $\mathfrak{a}$ and purely imaginary on $\mathfrak{t}$, so $E \in \mathfrak{a}$, and $\theta(E) = -E$. Hence, as above we get $\theta(\mathfrak{g}_i) = \mathfrak{g}_{-i}$, which proves the claim in the real case.

Second, we claim that the representation V admits a positive definite inner product $\langle\ ,\ \rangle_V$ (Hermitian in the complex case) such that the induced isomorphism $\Phi : V \to V^*$ (which is conjugate linear in the complex case) has the property that $\Phi(\theta(X) \cdot v) = X \cdot \Phi(v)$ for all $X \in \mathfrak{g}$ and $v \in V$.

Suppose first that $\mathfrak{g}$ is a complex Lie algebra and V is a complex representation of $\mathfrak{g}$. Then the Cartan involution θ is the conjugation with respect to a compact real form $\mathfrak{u}$ of $\mathfrak{g}$; see 2.3.2. There is a positive definite Hermitian inner product $\langle\ ,\ \rangle_V$ on V for which the elements of $\mathfrak{u}$ act as skew Hermitian operators. (This is the case for any Hermitian inner product which is invariant under the action of the integrated representation of the simply connected Lie group with Lie algebra $\mathfrak{u}$.) Any element $A \in \mathfrak{g}$ can be written as $X + iY$ for $X, Y \in \mathfrak{u}$, and by construction of the inner product we get $\langle (X + iY) \cdot v, w \rangle_V = -\langle v, (X - iY) \cdot w \rangle_V$, which is exactly the claim.

In the general case, we pass to the complexifications and then $V_{\mathbb{C}}$ is a complex representation of $\mathfrak{g}_{\mathbb{C}}$. The Cartan involution θ is then the restriction to $\mathfrak{g}$ of a conjugation of $\mathfrak{g}_{\mathbb{C}}$ with respect to an appropriate compact real form $\mathfrak{u} \subset \mathfrak{g}_{\mathbb{C}}$; see again 2.3.2. Now we can proceed as above and then restrict the Hermitian inner product to the $\mathfrak{g}$–invariant real subspace $V \subset V_{\mathbb{C}}$, which proves the claim in general.

Together with the above, we now obtain a positive definite inner product (Hermitian in the complex case) $\langle\ ,\ \rangle$ on each of the spaces $C^j(\mathfrak{g}_-, V) \cong \Lambda^j \mathfrak{p}_+ \otimes V$. The isomorphism $\mathcal{F}$ between $C^j(\mathfrak{g}_-, V)$ and its dual $C^j(\mathfrak{p}_+, V^*)$ induced by this inner product is simply given by $(\mathcal{F}(\phi)(Z_1, \dots, Z_j))(v) = \langle v, \phi(\theta(Z_1), \dots, \theta(Z_j)) \rangle_V$, for $Z_\ell \in \mathfrak{p}_+$ and $v \in V$. Using the fact that $\theta([Z_1, Z_2]) = [\theta(Z_1), \theta(Z_2)]$ and the compatibility of $\langle\ ,\ \rangle_V$ with the action of $\mathfrak{g}$ from above, one immediately verifies that $\mathcal{F}(\partial_{\mathfrak{g}_-} \phi) = \partial_{\mathfrak{p}_+} \mathcal{F}(\phi)$. This immediately implies the result, since for

$\phi_1 \in C^{n+1}(\mathfrak{g}_-, V)$ and $\phi_2 \in C^n(\mathfrak{g}_-, V)$ we get

$$\langle \phi_1, \partial(\phi_2) \rangle = \mathcal{F}(\partial_{\mathfrak{g}_-}(\phi_2))(\phi_1) = \partial_{\mathfrak{p}_+}(\mathcal{F}(\phi_2))(\phi_1) = \mathcal{F}(\phi_2)(\partial^*(\phi_1)) = \langle \partial^*(\phi_1), \phi_2 \rangle.$$

□

From this adjointness result, we may now easily deduce the Hodge decomposition. For each n, we define the *algebraic Laplacian* or *Kostant Laplacian* $\square = \square_n : C^n(\mathfrak{g}_-, V) \to C^n(\mathfrak{g}_-, V)$ by $\square = \partial\partial^* + \partial^*\partial$. By construction, $\square$ is a homomorphism of $\mathfrak{g}_0$–modules.

THEOREM 3.3.1. *For a representation V of $\mathfrak{g}$ and each $n = 0, \dots, \dim(\mathfrak{g}_-)$, we obtain a decomposition*

$$C^n(\mathfrak{g}_-, V) = \operatorname{im}(\partial^*) \oplus \ker(\square) \oplus \operatorname{im}(\partial).$$

Moreover, $\ker(\partial^*) = \operatorname{im}(\partial^*) \oplus \ker(\square)$ *and* $\ker(\partial) = \ker(\square) \oplus \operatorname{im}(\partial)$.

PROOF. For $\phi \in C^*(\mathfrak{g}_-, V)$, we have $\langle \partial^*\partial\phi, \phi \rangle = \langle \partial\phi, \partial\phi \rangle$, which shows that $\partial^*\partial\phi = 0$ implies $\partial\phi = 0$. This property is called *disjointness* by Kostant. Equivalently, it can be phrased as $\ker(\partial^*) \cap \operatorname{im}(\partial) = \{0\}$. In the same way, $\ker(\partial) \cap \operatorname{im}(\partial^*) = \{0\}$. Since $\partial^2 = 0$ and $(\partial^*)^2 = 0$, this in particular implies that $\operatorname{im}(\partial) \cap \operatorname{im}(\partial^*) = \{0\}$. From the definition of $\square$ we therefore conclude that $\square\phi = 0$ implies $\partial\partial^*\phi = 0$ and $\partial^*\partial\phi = 0$, so we obtain $\ker(\square) = \ker(\partial) \cap \ker(\partial^*)$. Obviously, the restriction of $\square$ to $\operatorname{im}(\partial)$ has values in $\operatorname{im}(\partial)$, and $\ker(\square) \cap \operatorname{im}(\partial) = \{0\}$, so $\square$ restricts to an automorphism on the image of ∂. Similarly, $\square$ restricts to an automorphism of $\operatorname{im}(\partial^*)$, so we find $\mathfrak{g}_0$–endomorphisms of $\operatorname{im}(\partial)$ and $\operatorname{im}(\partial^*)$, which we both denote by Q, that are inverse to the restriction of $\square$ to the respective subspace. But then we may write any element $\phi \in C^n(\mathfrak{g}_-, V)$ as

$$Q\partial^*\partial\phi + (\phi - Q\partial^*\partial\phi - Q\partial\partial^*\phi) + Q\partial\partial^*\phi.$$

By construction, the first summand lies in $\operatorname{im}(\partial^*)$, while the last one lies in $\operatorname{im}(\partial)$, and the middle summand is immediately seen to be contained in $\ker(\square)$. Since we already know from above that $\operatorname{im}(\partial) \cap \operatorname{im}(\partial^*) = \{0\}$ and $\ker(\square) \cap (\operatorname{im}(\partial) \oplus \operatorname{im}(\partial^*)) = \{0\}$, the Hodge decomposition follows.

Finally, we have noticed already that $\ker(\partial) \cap \operatorname{im}(\partial^*) = \{0\}$, while obviously $\ker(\square) \oplus \operatorname{im}(\partial) \subset \ker(\partial)$. Together with the analogous statement for ∂^*, this implies the last part of the theorem. □

COROLLARY 3.3.1. *For any representation V of $\mathfrak{g}$, the cohomology groups $H^k(\mathfrak{g}_-, V)$ can be naturally identified with* $\ker(\square_k)$ *as a $\mathfrak{g}_0$–module. This in turn can be naturally identified with the quotient* $\ker(\partial^*)/\operatorname{im}(\partial^*)$, *which endows the cohomology groups with the structure of a $\mathfrak{p}$–module. Finally, we get natural identifications of $\mathfrak{g}_0$–modules*

$$H^k(\mathfrak{g}_-, V) \cong H_k(\mathfrak{p}_+, V) \cong (H^k(\mathfrak{p}_+, V^*))^*.$$

PROOF. The first two statements are evident from the Hodge decomposition. By construction, ∂^* was obtained as the dual map of the Lie algebra cohomology differential $\partial_{\mathfrak{p}_+}$, which shows that its homology is dual to $H^*(\mathfrak{p}_+, V^*)$. On the other hand, we have noted that ∂^* coincides with the differential in the standard complex computing the Lie algebra homology $H_*(\mathfrak{p}_+, V)$. □

3.3.2. The next step is to find a formula for the algebraic Laplacian $\Box$ that can be nicely interpreted using representation theory. To do this, we will work in the space $\Lambda^*\mathfrak{g}\otimes V$. As we have noted in 3.3.1 above, we may identify $C^k(\mathfrak{g}_-,V)$ with $\Lambda^k\mathfrak{p}_+\otimes V$. Hence, there is a natural inclusion $j : C^k(\mathfrak{g}_-,V)\hookrightarrow\Lambda^k\mathfrak{g}\otimes V$. Moreover, the explicit formula for ∂^* from 3.3.1 actually defines a linear operator $\partial^*_{\mathfrak{g}}:\Lambda^k\mathfrak{g}\otimes V\to\Lambda^{k-1}\mathfrak{g}\otimes V$, and as in 3.3.1 one verifies that $\partial^*_{\mathfrak{g}}$ is a $\mathfrak{g}$–homomorphism. By construction we have $\partial^*_{\mathfrak{g}}\circ j=j\circ\partial^*_{\mathfrak{g}_-}$, so if there is no risk of confusion we will denote both maps by ∂^* in the sequel.

On the other hand, using the Killing form B, we may identify $\Lambda^k\mathfrak{g}\otimes V$ with the space $C^k(\mathfrak{g},V)$ of V–valued k–cochains on $\mathfrak{g}$. Explicitly, viewed as a multilinear map, a decomposable element $X_1\wedge\dots\wedge X_k\otimes v\in\Lambda^k\mathfrak{g}\otimes V$ acts as

$$(Y_1,\dots,Y_k)\mapsto\det(B(X_i,Y_j))v.$$

Let us interpret the inclusion $C^k(\mathfrak{g}_-,V)\hookrightarrow\Lambda^k\mathfrak{g}\otimes V$ in this picture. Of course, any element $Y\in\mathfrak{g}$ may be written as $Y=Y^-+Y^{\mathfrak{p}}$ according to the decomposition $\mathfrak{g}=\mathfrak{g}_-\oplus\mathfrak{p}$, and for $Z\in\mathfrak{p}_+$ we get $B(Z,Y)=B(Z,Y^-)$; see part (3) of Proposition 3.1.2. On the other hand, the identification of $C^k(\mathfrak{g}_-,V)$ with $\Lambda^k\mathfrak{p}_+\otimes V$ was obtained using the linear isomorphism $\mathfrak{g}_-\cong\mathfrak{g}/\mathfrak{p}$ corresponding to the decomposition $\mathfrak{g}=\mathfrak{g}_-\oplus\mathfrak{p}$. Consequently, for $\phi\in C^k(\mathfrak{g}_-,V)$ the multilinear map $\mathfrak{g}^k\to V$ corresponding to $j(\phi)\in\Lambda^k\mathfrak{g}\otimes V$ is given by $(Y_1,\dots,Y_k)\mapsto\phi(Y_1^-,\dots,Y_k^-)$.

Via the identification of $\Lambda^k\mathfrak{g}\otimes V$ with $C^k(\mathfrak{g},V)$, we obtain for each k the Lie algebra differential $\partial_{\mathfrak{g}}:\Lambda^k\mathfrak{g}\otimes V\to\Lambda^{k+1}\mathfrak{g}\otimes V$. In contrast to the case of the codifferential discussed above, the relation between the Lie algebra differentials on $\mathfrak{g}$ and on $\mathfrak{g}_-$ is more subtle, and clarifying this relation actually goes a long way towards the formula for $\Box$. To do this, we first have to introduce two more operators and discuss their properties, which are closely parallel to the familiar formulae relating the various operators on differential forms.

First, for any $X\in\mathfrak{g}$, we denote by $\epsilon_X:\Lambda^k\mathfrak{g}\otimes V\to\Lambda^{k+1}\mathfrak{g}\otimes V$ the exterior product with X, i.e. $\epsilon_X(X_1\wedge\dots\wedge X_k\otimes v):=X\wedge X_1\wedge\dots\wedge X_n\otimes v$. To interpret this operator in terms of multilinear functions, take elements $Y_0,\dots,Y_k\in\mathfrak{g}$. By definition, $\epsilon_X(X_1\wedge\dots\wedge X_k\otimes v)$ maps $(Y_0,\dots,Y_k)$ to

$$\det\begin{pmatrix}B(X,Y_0)&\dots&B(X,Y_k)\\B(X_1,Y_0)&\dots&B(X_1,Y_k)\\\vdots&&\vdots\\B(X_k,Y_0)&\dots&B(X_k,Y_k)\end{pmatrix}v.$$

Computing this determinant by expanding along the first row, we obtain a formula that clearly extends to linear combinations of decomposable elements and gives us

$$(\epsilon_X\phi)(Y_0,\dots,Y_k)=\sum_{i=0}^{k}(-1)^iB(X,Y_i)\phi(Y_0,\dots,\widehat{Y_i},\dots,Y_k).\tag{3.12}$$

Dually, for $Y\in\mathfrak{g}$ we define the insertion operator $i_Y:C^{k+1}(\mathfrak{g},V)\to C^k(\mathfrak{g},V)$ by $(i_Y\phi)(Y_1,\dots,Y_k):=\phi(Y,Y_1,\dots,Y_k)$. To compute the resulting insertion operator $i_Y:\Lambda^{k+1}\mathfrak{g}\otimes V\to\Lambda^k\mathfrak{g}\otimes V$, consider the matrix

$$\begin{pmatrix}B(X_0,Y)&B(X_0,Y_1)&\dots&B(X_0,Y_k)\\\vdots&\vdots&&\vdots\\B(X_k,Y)&B(X_k,Y_1)&\dots&B(X_k,Y_k)\end{pmatrix}.$$

Computing its determinant by expansion along the first column, we immediately conclude that

$$i_Y(X_0 \wedge \dots \wedge X_k \otimes v) = \sum_{i=0}^{k}(-1)^i B(X_i, Y) X_0 \wedge \dots \wedge \widehat{X_i} \wedge \dots \wedge X_k \otimes v. \tag{3.13}$$

The equation $i_Y \epsilon_X = B(X,Y)\mathrm{id} - \epsilon_X i_Y$ is then easily verified directly for decomposable elements.

On the other hand, any element $X \in \mathfrak{g}$ canonically acts both on $\Lambda^*\mathfrak{g}$ (via the exterior powers of the adjoint action) as well as on V, and thus also on each of the tensor products $\Lambda^k\mathfrak{g} \otimes V$. To emphasize the similarity with the case of differential forms, we will denote these actions by $\mathcal{L}_X$. Since we will need both the action on the tensor product and the actions on the individual factors, we will write $\mathcal{L}_X$ for the tensor product action, $\mathcal{L}_X^{\mathfrak{g}}$ for the action on $\Lambda^k\mathfrak{g}$ tensored with the identity on V and $\mathcal{L}_X^V$ for the identity of $\Lambda^k\mathfrak{g}$ tensored with the action of X on V. In this notation, we clearly have $\mathcal{L}_X = \mathcal{L}_X^{\mathfrak{g}} + \mathcal{L}_X^V$. Concerning the relations of these operations to the codifferential we get

LEMMA 3.3.2. *Let $\mathfrak{g}$ be a semisimple Lie algebra, $\{\xi_\ell\}$ any basis for $\mathfrak{g}$ and $\{\eta_\ell\}$ the dual basis with respect to B. Then for any $X \in \mathfrak{g}$ we have:*

(1) $\partial^* \circ \epsilon_X + \epsilon_X \circ \partial^* = -\mathcal{L}_X$.
(2) $\partial^* \circ \mathcal{L}_X^V - \mathcal{L}_X^V \circ \partial^* = \sum_\ell i_{[\eta_\ell, X]} \circ \mathcal{L}_{\xi_\ell}^V$.
(3) $\sum_\ell \epsilon_{\eta_\ell} i_{[X,\xi_\ell]} = -\mathcal{L}_X^{\mathfrak{g}}$.

PROOF. (1) We compute $\partial^*\epsilon_X(X_0 \wedge \dots \wedge X_k \otimes v)$. Applying formula (3.11) from 3.3.1 for ∂^* to $X \wedge X_0 \wedge \dots \wedge X_k \otimes v$ we obtain

$$\begin{aligned}&-X_0 \wedge \dots \wedge X_k \otimes X \cdot v + \textstyle\sum_{i=0}^k (-1)^i X \wedge X_0 \wedge \dots \wedge \widehat{X_i} \wedge \dots \wedge X_k \otimes X_i \cdot v\\ &+ \textstyle\sum_{j=0}^k (-1)^{j+1}[X, X_j] \wedge X_0 \wedge \dots \wedge \widehat{X_j} \wedge \dots \wedge X_k \otimes v\\ &+ \textstyle\sum_{i<j} (-1)^{i+j}[X_i, X_j] \wedge X \wedge X_0 \wedge \dots \wedge \widehat{X_i} \wedge \dots \wedge \widehat{X_j} \wedge \dots \wedge X_k \otimes v.\end{aligned}$$

In the second line, we may move $[X, X_j]$ behind X_{j-1}, thus picking up a sign $(-1)^j$, so these terms add up to $-\mathcal{L}_X^{\mathfrak{g}}(X_0 \wedge \dots \wedge X_k \otimes v)$. Together with the very first term, we get $-\mathcal{L}_X(X_0 \wedge \dots \wedge X_k \otimes v)$. Exchanging $[X_i, X_j]$ and X in the last sum (and thus picking up a sign -1), the remaining terms visibly add up to $-\epsilon_X(\partial^*(X_0 \wedge \dots \wedge X_k \otimes v))$, and the result follows.

(2) We compute $\partial^*\mathcal{L}_X^V(X_0 \wedge \dots \wedge X_k \otimes v)$. Applying formula (3.11) from 3.3.1 for ∂^* to $X_0 \wedge \dots \wedge X_k \otimes X \cdot v$, we obtain

$$\begin{aligned}&\textstyle\sum_{i=0}^k (-1)^{i+1} X_0 \wedge \dots \wedge \widehat{X_i} \wedge \dots \wedge X_k \otimes X_i \cdot X \cdot v\\ &+ \textstyle\sum_{i<j} (-1)^{i+j}[X_i, X_j] \wedge X_0 \wedge \dots \wedge \widehat{X_i} \wedge \dots \wedge \widehat{X_j} \wedge \dots \wedge X_k \otimes X \cdot v.\end{aligned}$$

Now we may write $X_i \cdot X \cdot v = X \cdot X_i \cdot v + [X_i, X] \cdot v$. Inserting this in the first line, the terms involving $X \cdot X_i \cdot v$ together with the second line add up to $\mathcal{L}_X^V(\partial^*(X_0 \wedge \dots \wedge X_k \otimes v))$.

On the other hand, $i_{[\eta_\ell, X]}$ maps $X_0 \wedge \dots \wedge X_k \otimes \xi_\ell \cdot v$ to

$$\textstyle\sum_{i=0}^k (-1)^i B([\eta_\ell, X], X_i) X_0 \wedge \dots \wedge \widehat{X_i} \wedge \dots \wedge X_k \otimes \xi_\ell \cdot v.$$

Invariance of B implies that $B([\eta_\ell, X], X_i) = B(\eta_\ell, [X, X_i])$. Since $\{\xi_\ell\}$ and $\{\eta_\ell\}$ are dual bases we conclude that $\sum_\ell B(\eta_\ell, [X, X_i])\xi_\ell = [X, X_i]$, which implies the result.

(3) Formula (3.12) from above shows that $\epsilon_{\eta_\ell} i_{[X,\xi_\ell]}\phi$ maps $(X_0,\dots,X_k)$ to

$$\sum_{i=0}^k (-1)^i B(\eta_\ell, X_i)(i_{[X,\xi_\ell]}\phi)(X_0,\dots\widehat{X_i},\dots,X_k) \\ = \sum_{i=0}^k B(\eta_\ell, X_i)\phi(X_0,\dots,[X,\xi_\ell],\dots,X_k),$$

and the claim follows immediately since $\sum_\ell B(\eta_\ell, X_i)\xi_\ell = X_i$. □

3.3.3. Formulae for ∂ and $\square$. Our aim is to write the operator $j\circ\partial_{\mathfrak{g}_-}$ as the composition of some operator on $\Lambda^*\mathfrak{g}\otimes V$ with the inclusion $j : C^*(\mathfrak{g}_-, V) \hookrightarrow \Lambda^*\mathfrak{g}\otimes V$ introduced in 3.3.2 above. To do this, it is convenient to split the Lie algebra differential into two parts. For $\phi\in C^k(\mathfrak{g}_-, V)$ we define $\partial^{(1)}_{\mathfrak{g}_-}\phi \in C^{k+1}(\mathfrak{g}_-, V)$ by

$$\partial^{(1)}_{\mathfrak{g}_-}\phi(X_0,\dots,X_k) := \sum_{i=0}^k (-1)^i X_i\cdot\phi(X_0,\dots,\widehat{X_i},\dots,X_k),$$

and we put $\partial^{(2)}_{\mathfrak{g}_-} = \partial_{\mathfrak{g}_-} - \partial^{(1)}_{\mathfrak{g}_-}$. Similarly, we split the Lie algebra differential on $C^*(\mathfrak{g}, V)\cong\Lambda^*\mathfrak{g}\otimes V$ into two parts denoted by $\partial^{(1)}_{\mathfrak{g}}$ and $\partial^{(2)}_{\mathfrak{g}}$.

By definition, for $\phi\in C^k(\mathfrak{g}, V)$, we may write $(\partial^{(1)}_{\mathfrak{g}}\phi)(X_0,\dots,X_k)$ as

$$\sum_{i=0}^k (-1)^i \mathcal{L}^V_{X_i}(\phi)(X_0,\dots,\widehat{X_i},\dots,X_k).$$

Taking a basis $\{\xi_\ell\}$ of $\mathfrak{g}$ with dual basis $\{\eta_\ell\}$, any element $X\in\mathfrak{g}$ may be written as $X = \sum_\ell B(\eta_\ell, X)\xi_\ell$. Together with the formula (3.12) for ϵ_X from 3.3.2 this implies that $\partial^{(1)}_{\mathfrak{g}} = \sum_\ell \epsilon_{\eta_\ell}\circ\mathcal{L}^V_{\xi_\ell}$.

To obtain an interpretation of $j\circ\partial^{(1)}_{\mathfrak{g}_-}$, we have to refine the choice of dual bases. Namely, let us choose our basis $\{\xi_\ell\}$ to be the union of a basis of $\mathfrak{g}_-$ and a basis of $\mathfrak{p}$. For an element $X\in\mathfrak{g}$ then the decomposition $X = X^- + X^{\mathfrak{p}}$ according to the decomposition $\mathfrak{g}=\mathfrak{g}_-\oplus\mathfrak{p}$ can be written as $X^- = \sum_{\ell:\xi_\ell\in\mathfrak{g}_-} B(\eta_\ell, X)\xi_\ell$ and $X^{\mathfrak{p}} = \sum_{\ell:\xi_\ell\in\mathfrak{p}} B(\eta_\ell, X)\xi_\ell$. From 3.3.2 above, we know that applying j to a multilinear map corresponds to replacing all entries by their component in $\mathfrak{g}_-$, from which we immediately conclude that

$$j\circ\partial^{(1)}_{\mathfrak{g}_-} = \sum_{\ell:\xi_\ell\in\mathfrak{g}_-} \epsilon_{\eta_\ell}\circ\mathcal{L}^V_{\xi_\ell}\circ j.$$

For future purposes, it will be better to write this in a more sophisticated way. Observe that the sum over all ℓ plus the sum over all ℓ such that $\xi_\ell\in\mathfrak{g}_-$ minus the sum over all ℓ such that $\xi_\ell\in\mathfrak{p}$ gives twice the sum over all ℓ such that $\xi_\ell\in\mathfrak{g}_-$, which leads to

$$j\circ\partial^{(1)}_{\mathfrak{g}_-} = \frac{1}{2}\left(\partial^{(1)}_{\mathfrak{g}} + \sum_{\ell:\xi_\ell\in\mathfrak{g}_-}\epsilon_{\eta_\ell}\circ\mathcal{L}^V_{\xi_\ell} - \sum_{\ell:\xi_\ell\in\mathfrak{p}}\epsilon_{\eta_\ell}\circ\mathcal{L}^V_{\xi_\ell}\right)\circ j. \tag{3.14}$$

With the other summand of the Lie algebra differential, the situation is a bit more subtle. Let us again start with $\partial^{(2)}_{\mathfrak{g}}$ to see what is going on. For $X\in\mathfrak{g}$ and $\phi\in C^k(\mathfrak{g}, V)$, the operator $\mathcal{L}^{\mathfrak{g}}_X$ acts only on the entries of ϕ, and using skew symmetry of ϕ we get

$$(\mathcal{L}^{\mathfrak{g}}_X\phi)(X_0,\dots,X_{k-1}) = -\sum_{i=0}^{k-1}(-1)^i\phi([X,X_i],X_0,\dots,\widehat{X_i},\dots,X_{k-1}).$$

Hence, for $X_0, \dots, X_k \in \mathfrak{g}$ we get

$$\sum_{i=0}^{k}(-1)^i(\mathcal{L}^{\mathfrak{g}}_{X_i}\phi)(X_0,\dots,\widehat{X_i},\dots,X_k)$$

$$= -\sum_{i=0}^{k}\sum_{j<i}(-1)^{i+j}\phi([X_i,X_j],X_0,\dots,\widehat{X_j},\dots,\widehat{X_i},\dots,X_k)$$

$$-\sum_{i=0}^{k}\sum_{i<j}(-1)^{i+j-1}\phi([X_i,X_j],X_0,\dots,\widehat{X_i},\dots,\widehat{X_j},\dots,X_k)$$

$$=\sum_{i<j}(-1)^{i+j}\phi([X_i,X_j]-[X_j,X_i],X_0,\dots,\widehat{X_i},\dots,\widehat{X_j},\dots,X_k),$$

so this adds up to $2\partial^{(2)}_{\mathfrak{g}}\phi(X_0,\dots,X_k)$. As before, consider a basis $\{\xi_\ell\}$ of $\mathfrak{g}$ and let $\{\eta_\ell\}$ be the dual basis with respect to B. Using formula (3.12) from 3.3.2, we see that $\epsilon_{\eta_\ell}\mathcal{L}^{\mathfrak{g}}_{\xi_\ell}\phi$ maps $X_0,\dots,X_k$ to

$$\sum_{i=0}^{k}(-1)^iB(\eta_\ell,X_i)\mathcal{L}^{\mathfrak{g}}_{\xi_\ell}\phi(X_0,\dots,\widehat{X_i},\dots,X_k).$$

Together with the above, this shows that

$$\partial^{(2)}_{\mathfrak{g}} = \tfrac{1}{2}\sum_\ell \epsilon_{\eta_\ell}\circ\mathcal{L}^{\mathfrak{g}}_{\xi_\ell}.$$

What we actually need is an operator that, when acting on $j\phi$ for some $\phi \in C^k(\mathfrak{g}_-,V)$, produces $j\partial^{(2)}_{\mathfrak{g}_-}\phi$. This means that instead of the Lie bracket $[X_i,X_j]$ we need an expression, whose $\mathfrak{g}_-$–component coincides with $[X_i^-,X_j^-]$. Fortunately, there is a nice trick to obtain such an expression. Since $\mathfrak{p}$ is a subalgebra of $\mathfrak{g}$, the $\mathfrak{g}_-$–component of $[X_i^-,X_j^-]$ coincides with the $\mathfrak{g}_-$–component of $[X_i^-,X_j^-]-[X_i^{\mathfrak{p}},X_j^{\mathfrak{p}}]$. But one immediately verifies directly that

$$[X_i^-,X_j^-]-[X_i^{\mathfrak{p}},X_j^{\mathfrak{p}}] = \tfrac{1}{2}\left([X_i^- - X_i^{\mathfrak{p}},X_j]-[X_j^- - X_j^{\mathfrak{p}},X_i]\right).$$

If we assume as above that our basis $\{\xi_\ell\}$ is the union of a basis of $\mathfrak{g}_-$ and a basis of $\mathfrak{p}$ and use the formulae for the projections onto $\mathfrak{g}_-$ and $\mathfrak{p}$ from before, we conclude that

$$j\circ\partial^{(2)}_{\mathfrak{g}_-} = \frac{1}{2}\left(\sum_{\ell:\xi_\ell\in\mathfrak{g}_-}\epsilon_{\eta_\ell}\circ\mathcal{L}^{\mathfrak{g}}_{\xi_\ell} - \sum_{\ell:\xi_\ell\in\mathfrak{p}}\epsilon_{\eta_\ell}\circ\mathcal{L}^{\mathfrak{g}}_{\xi_\ell}\right)\circ j. \tag{3.15}$$

Putting this together with the interpretation of the other summand, we obtain

$$j\circ\partial_{\mathfrak{g}_-} = \frac{1}{2}\left(\partial^{(1)}_{\mathfrak{g}} + \sum_{\ell:\xi_\ell\in\mathfrak{g}_-}\epsilon_{\eta_\ell}\circ\mathcal{L}_{\xi_\ell} - \sum_{\ell:\xi_\ell\in\mathfrak{p}}\epsilon_{\eta_\ell}\circ\mathcal{L}_{\xi_\ell}\right)\circ j. \tag{3.16}$$

From this, we can directly derive an explicit formula for the Kostant Laplacian $\square$. The fact that the codifferential $\partial^* = \partial^*_{\mathfrak{g}}$ is a $\mathfrak{g}$–homomorphism simply means $\partial^*\circ\mathcal{L}_X = \mathcal{L}_X\circ\partial^*$. Consequently, for a fixed index ℓ, we get

$$\partial^*\circ\epsilon_{\eta_\ell}\circ\mathcal{L}_{\xi_\ell} + \epsilon_{\eta_\ell}\circ\mathcal{L}_{\xi_\ell}\circ\partial^* = (\partial^*\circ\epsilon_{\eta_\ell}+\epsilon_{\eta_\ell}\circ\partial^*)\circ\mathcal{L}_{\xi_\ell},$$

and by part (1) of Lemma 3.3.2, this gives $-\mathcal{L}_{\eta_\ell}\circ\mathcal{L}_{\xi_\ell}$.

To obtain the formula for $\square$, it thus suffices to compute $\partial^* \circ \partial^{(1)}_{\mathfrak{g}} + \partial^{(1)}_{\mathfrak{g}} \circ \partial^*$. From above, we know that $\partial^{(1)}_{\mathfrak{g}}$ can be written as $\sum_\ell \epsilon_{\eta_\ell} \circ \mathcal{L}^V_{\xi_\ell}$. Using Lemma 3.3.2, we compute

$$\begin{aligned}\textstyle\sum_\ell \left(\partial^* \circ \epsilon_{\eta_\ell} \circ \mathcal{L}^V_{\xi_\ell} + \epsilon_{\eta_\ell} \circ \mathcal{L}^V_{\xi_\ell} \circ \partial^*\right) &= \textstyle -\sum_\ell \mathcal{L}_{\eta_\ell} \circ \mathcal{L}^V_{\xi_\ell} - \sum_{\ell,m} \epsilon_{\eta_\ell} \circ i_{[\eta_m,\xi_\ell]} \circ \mathcal{L}^V_{\xi_m} \\ &= \textstyle -\sum_\ell \mathcal{L}_{\eta_\ell} \circ \mathcal{L}^V_{\xi_\ell} + \sum_m \mathcal{L}^{\mathfrak{g}}_{\eta_m} \circ \mathcal{L}^V_{\xi_m} = -\sum_\ell \mathcal{L}^V_{\eta_\ell} \circ \mathcal{L}^V_{\xi_\ell}.\end{aligned}$$

Collecting our results, we obtain:

PROPOSITION 3.3.3. *Let $\mathfrak{g} = \mathfrak{g}_- \oplus \mathfrak{p}$ be a $|k|$–graded semisimple Lie algebra, let $\{\xi_\ell\}$ be a basis for $\mathfrak{g}$ which is the union of a basis of $\mathfrak{g}_-$ and a basis of $\mathfrak{p}$. Let $j : C^*(\mathfrak{g}_-, V) \cong \Lambda^*\mathfrak{p}_+ \otimes V \hookrightarrow \Lambda^*\mathfrak{g} \otimes V$ be the inclusion. Then*

$$j \circ \square = \frac{1}{2}\left(-\sum_\ell \mathcal{L}^V_{\eta_\ell}\mathcal{L}^V_{\xi_\ell} - \sum_{\ell:\xi_\ell\in\mathfrak{g}_-} \mathcal{L}_{\eta_\ell}\mathcal{L}_{\xi_\ell} + \sum_{\ell:\xi_\ell\in\mathfrak{p}} \mathcal{L}_{\eta_\ell}\mathcal{L}_{\xi_\ell}\right) \circ j.$$

3.3.4. The action of the Laplacian on an isotypical component. The formula for the algebraic Laplacian $\square$ can now be interpreted in terms of representation theory. To do this, we have to restrict to the case of a complex parabolic subalgebra $\mathfrak{p}$ in a complex semisimple Lie algebra $\mathfrak{g}$ and the case where the representation V of $\mathfrak{g}$ is irreducible. Moreover, one gets a nicer description for the cohomology groups $H^*(\mathfrak{p}_+, V)$ rather than $H^*(\mathfrak{g}_-, V)$, so we will switch to considering the Laplacian on $C^*(\mathfrak{p}_+, V) \cong \Lambda^*\mathfrak{g}_- \otimes V$. Of course, the formula in Proposition 3.3.3 above has to be modified only by replacing the condition $\xi_\ell \in \mathfrak{g}_-$ by $\xi_\ell \in \mathfrak{p}_+$ and the condition $\xi_\ell \in \mathfrak{p}$ by $\xi_\ell \in \mathfrak{g}_{\leq 0}$. Here, $\mathfrak{g}_{\leq 0}$ is the subalgebra $\mathfrak{g}_{-k} \oplus \cdots \oplus \mathfrak{g}_0$ of $\mathfrak{g}$, i.e. the opposite parabolic to $\mathfrak{p}$.

Recall from 2.2.14 the decomposition of representations into isotypical components. (To apply this to the reductive Lie algebra $\mathfrak{g}_0$, one first has to use the decomposition for the semisimple part, and then decompose further according to the action of the center.) For a weight $\nu : \mathfrak{h} \to \mathbb{C}$ let $W^\nu_0 \subset C^*(\mathfrak{p}_+, V)$ be the subspace of all highest weight vectors of weight ν. By definition, a weight vector of weight ν lies in W^ν_0 if and only if it is killed by all elements in positive root spaces of $\mathfrak{g}_0$. The *isotypical component* $W^\nu \subset C^*(\mathfrak{p}_+, V)$ of highest weight ν is defined as the $\mathfrak{g}_0$–submodule generated by W^ν_0. Evidently, there are only finitely many weights for which the isotypical component is nonzero.

From Theorem 2.2.14 we know that $C^*(\mathfrak{p}_+, V)$ is the direct sum of its nonzero isotypical components and the isotypical component W^ν is the direct sum of copies of the irreducible representation Γ_ν with highest weight ν. The number of these copies is called the *multiplicity* of Γ_ν in the representation $C^*(\mathfrak{p}_+, V)$. Further, if $\phi : C^*(\mathfrak{p}_+, V) \to C^*(\mathfrak{p}_+, V)$ is $\mathfrak{g}_0$–equivariant, then $\phi(W^\nu) \subset W^\nu$ and $\phi(W^\nu_0) \subset W^\nu_0$, and ϕ is uniquely determined by its restrictions to the subspaces W^ν_0.

PROPOSITION 3.3.4. *Let $\mathfrak{g}$ be a complex semisimple Lie algebra, $\mathfrak{p} = \mathfrak{g}_0 \oplus \mathfrak{p}_+$ a standard parabolic subalgebra of $\mathfrak{g}$, and V a complex irreducible representation of $\mathfrak{g}$ with highest weight λ. Let W^ν be the $\mathfrak{g}_0$–isotypical component of highest weight ν for the natural $\mathfrak{g}_0$–action on $C^*(\mathfrak{p}_+, V)$. Then the Kostant Laplacian $\square$ acts on W^ν by multiplication by the scalar $-\frac{1}{2}(\|\lambda+\delta\|^2 - \|\nu+\delta\|^2)$, where δ is the lowest form and the norm is induced by the Killing form.*

PROOF. Since $\square$ is a $\mathfrak{g}_0$–homomorphism, it suffices to show that it acts on a highest weight vector of weight ν by multiplication by $-\frac{1}{2}(\|\lambda+\delta\|^2 - \|\nu+\delta\|^2)$. To show this, we apply the formula from Proposition 3.3.3 with appropriately chosen dual bases. Let $\mathfrak{h} \subset \mathfrak{g}$ be the Cartan subalgebra and let $\mathfrak{h}_0 \subset \mathfrak{h}$ be the subspace on which all roots are real. From 2.2.4 we know that the restriction of the Killing form B to $\mathfrak{h}_0$ is positive definite, and we choose an orthonormal basis $\{H_1, \dots, H_r\}$ for $\mathfrak{h}_0$. For any root $\alpha \in \Delta$, choose elements $E_\alpha \in \mathfrak{g}_\alpha$ and $F_\alpha \in \mathfrak{g}_{-\alpha}$ such that $B(E_\alpha, F_\alpha) = 1$. Then $\{E_\alpha, H_j, F_\beta\}$ is a basis for $\mathfrak{g}$ (a so–called Chevalley–basis) with dual basis $\{F_\alpha, H_j, E_\beta\}$. For $H \in \mathfrak{h}$ we get

$$B(H, [E_\alpha, F_\alpha]) = B([H, E_\alpha], F_\alpha) = \alpha(H)B(E_\alpha, F_\alpha) = \alpha(H)$$

by invariance of B. Otherwise put, $[E_\alpha, F_\alpha] \in \mathfrak{h}$ is dual to α with respect to B. This means that if we denote by $\langle\ ,\ \rangle$ the (positive definite) inner product on $\mathfrak{h}_0^*$ induced by B, then for any weight $\mu \in \mathfrak{h}_0^*$ we get $\mu([E_\alpha, F_\alpha]) = \langle \mu, \alpha \rangle$.

The action of the first sum in the formula for $j \circ \square$ in Proposition 3.3.3 actually represents the Casimir element of $\mathfrak{g}$ acting on the irreducible representation V, so this just contributes an overall factor, which we have already computed in 2.2.15. However, since the idea for analyzing the other summands is very closely related to the ideas used for the Casimir element, we redo this computation here. For our choice of dual bases, the first sum is

$$\textstyle\sum_{j=1}^r \mathcal{L}^V_{H_j} \circ \mathcal{L}^V_{H_j} + \sum_{\alpha \in \Delta^+} (\mathcal{L}^V_{F_\alpha} \circ \mathcal{L}^V_{E_\alpha} + \mathcal{L}^V_{E_\alpha} \circ \mathcal{L}^V_{F_\alpha}).$$

Acting on a highest weight vector for $\mathfrak{g}$ of weight λ, the first sum simply gives a factor $\sum_j \lambda(H_j)^2$, and since the H_j form an orthonormal basis, this may be rewritten as $\langle \lambda, \lambda \rangle = \|\lambda\|^2$. For the second sum, one simply rewrites the individual summands as

$$2\mathcal{L}^V_{F_\alpha} \circ \mathcal{L}^V_{E_\alpha} + (\mathcal{L}^V_{E_\alpha} \circ \mathcal{L}^V_{F_\alpha} - \mathcal{L}^V_{F_\alpha} \circ \mathcal{L}^V_{E_\alpha}) = 2\mathcal{L}^V_{F_\alpha} \circ \mathcal{L}^V_{E_\alpha} + \mathcal{L}^V_{[E_\alpha, F_\alpha]}.$$

On a highest weight vector for $\mathfrak{g}$ of weight λ, any E_α acts trivially by definition, while from above we know that $[E_\alpha, F_\alpha]$ acts by multiplication by $\langle \lambda, \alpha \rangle$. Now recall from 3.2.14 and 2.2.18 that the lowest form δ of $\mathfrak{g}$ is defined to be half the sum of all positive roots. Using this, we see that the action of the first summand in the formula for $j \circ \square$ is given by multiplication by $-\frac{1}{2}\langle \lambda, \lambda \rangle - \langle \lambda, \delta \rangle$ on the whole of $C^k(\mathfrak{p}_+, V)$, where λ is the highest weight of the irreducible representation V of $\mathfrak{g}$. Notice that we may also write this factor as $-\frac{1}{2}(\|\lambda + \delta\|^2 - \|\delta\|^2)$.

The other parts of the formula from Proposition 3.3.3 are interpreted similarly. The next part in this formula reads as

$$-\tfrac{1}{2} \sum_{\ell : \xi_\ell \in \mathfrak{p}_+} \mathcal{L}_{\eta_\ell} \circ \mathcal{L}_{\xi_\ell} = -\tfrac{1}{2} \sum_{\alpha \in \Delta^+(\mathfrak{p}_+)} \mathcal{L}_{F_\alpha} \circ \mathcal{L}_{E_\alpha}. \tag{3.17}$$

The last part, $\frac{1}{2}\sum_{\ell : \xi_\ell \in \mathfrak{p}} \mathcal{L}_{\eta_\ell} \mathcal{L}_{\xi_\ell}$ in the formula from Proposition 3.3.3 can be expanded as

$$\frac{1}{2}\left(\sum_{\alpha \in \Delta^+(\mathfrak{p}_+)} \mathcal{L}_{E_\alpha} \circ \mathcal{L}_{F_\alpha} + \sum_j \mathcal{L}_{H_j} \circ \mathcal{L}_{H_j} + \sum_{\alpha \in \Delta^+(\mathfrak{g}_0)} (\mathcal{L}_{F_\alpha} \circ \mathcal{L}_{E_\alpha} + \mathcal{L}_{E_\alpha} \circ \mathcal{L}_{F_\alpha}) \right) \tag{3.18}$$

where the first sum corresponds to root spaces in $\mathfrak{g}_-$, the second to $\mathfrak{h} \subset \mathfrak{g}_0$, and the last one to root spaces contained in $\mathfrak{g}_0$. The first sum in (3.18) adds up with the right–hand side of (3.17) to $\frac{1}{2}\sum_{\alpha \in \Delta^+(\mathfrak{p}_+)} \mathcal{L}_{[E_\alpha, F_\alpha]}$. As for the Casimir element, the last sum in (3.18) can be rewritten as a term acting trivially on any $\mathfrak{g}_0$–highest weight vector and $\frac{1}{2}\sum_{\alpha\in\Delta^+(\mathfrak{g}_0)}\mathcal{L}_{[E_\alpha,F_\alpha]}$. Thus, we see that the actions of (3.17) and (3.18) on a $\mathfrak{g}_0$–highest weight vector of weight ν are given by multiplication by

$$\tfrac{1}{2}\langle\nu,\nu\rangle+\langle\nu,\delta\rangle=\tfrac{1}{2}(\|\nu+\delta\|^2-\|\delta\|^2).$$

□

Using the Hodge decomposition from Theorem 3.3.1, we conclude that the cohomology $H^*(\mathfrak{p}_+,V)$ is the direct sum of all $\mathfrak{g}_0$–isotypical components $W^\nu\subset C^*(\mathfrak{p}_+,V)$ such that the highest weight ν has the property that $\|\nu+\delta\|=\|\lambda+\delta\|$, where λ is the highest weight of V. This, for example, implies that the fact whether an irreducible component of $C^*(\mathfrak{p}_+,V)$ is contained in the cohomology or not depends only on the highest weight.

3.3.5. Analysis of the weight condition. By Proposition 3.3.4, determining the cohomology $H^*(\mathfrak{p}_+,V)$ is equivalent to finding those irreducible components of $C^*(\mathfrak{p}_+,V)$, whose highest weights ν satisfy $\|\nu+\delta\|=\|\lambda+\delta\|$. To attack this problem, we will start by looking at arbitrary weights of the $\mathfrak{g}_0$–representation $C^*(\mathfrak{p}_+,V)$. The point here is that we get a good enough hand on all these weights to find the ones satisfying the norm condition. In a second step, we will show that all the occurring weights are actually highest weights of irreducible components.

Using the Killing form B, we may identify $C^*(\mathfrak{p}_+,V)$ with $\Lambda^*\mathfrak{g}_-\otimes V$. Now $\mathfrak{g}_-$ is the direct sum of all root spaces $\mathfrak{g}_{-\alpha}$ for $\alpha\in\Delta^+(\mathfrak{p}_+)$, so the weights of $\mathfrak{g}_-$ are exactly the negatives of elements of $\Delta^+(\mathfrak{p}_+)$, each occurring with multiplicity one. Choosing a basis of $\mathfrak{g}_-$ consisting of weight vectors $F_\alpha\in\mathfrak{g}_{-\alpha}$, the natural basis for $\Lambda^*\mathfrak{g}_-$ is indexed by all subsets $\Psi\subset\Delta^+(\mathfrak{p}_+)$, and we denote by F_Ψ the corresponding basis vector (which is determined up to a complex multiple). Denoting by $\langle\Psi\rangle$ as in 3.2.14 the sum of all elements of Ψ, the weight of F_Ψ is $-\langle\Psi\rangle$. Consequently, we see that any weight of $C^*(\mathfrak{p}_+,V)$ is of the form $\mu-\langle\Psi\rangle$ for a weight μ of V and a subset Ψ of $\Delta^+(\mathfrak{p}_+)$. Recall from 3.2.14 that any element w of the Weyl group W of $\mathfrak{g}$ determines the subset $\Phi_w:=\{\alpha\in\Delta^+:w^{-1}(\alpha)\in-\Delta^+\}\subset\Delta^+$.

LEMMA 3.3.5. *For any weight ν of $C^*(\mathfrak{p}_+,V)$ we have $\|\nu+\delta\|^2\leq\|\lambda+\delta\|^2$, where λ is the highest weight of V. Moreover, equality occurs if an only if there is an element w in the Weyl group of $\mathfrak{g}$, such that $\nu=\nu_w:=w(\lambda)-\langle\Phi_w\rangle$. If the weight ν_w occurs in $C^*(\mathfrak{p}_+,V)$, then it occurs with multiplicity one, and for $w\neq w'$ we have $\nu_w\neq\nu_{w'}$.*

PROOF. Consider an arbitrary weight $\nu=\mu-\langle\Psi\rangle$ of $C^*(\mathfrak{p}_+,V)$. Since the Weyl group W of $\mathfrak{g}$ acts transitively on the set of Weyl chambers, we can find an element $w\in W$ such that the weight $w(\nu+\delta)$ is dominant; see 2.2.7. Since any Weyl group element acts by an orthogonal mapping, we have

$$\|\nu+\delta\|=\|w(\nu+\delta)\|=\|w(\mu)+w(\delta-\langle\Psi\rangle)\|.$$

Since the set of weights of any finite–dimensional representation is invariant under the Weyl group, $w(\mu)$ is a weight of V. By part (2) of Theorem 2.2.11, the weight $w(\mu)$ may be written as $\lambda-\sum_{i=1}^r n_i'\alpha_i$, where $\{\alpha_1,\dots,\alpha_r\}$ are the simple roots of $\mathfrak{g}$ and the n_i' are nonnegative integers.

To study $w(\delta-\langle\Psi\rangle)$, we associate to any subset $\Psi\subset\Delta^+$ the subset $\hat\Psi:=\Psi\cup-(\Delta^+\setminus\Psi)\subset\Delta$. By construction, $\Delta=\hat\Psi\sqcup-\hat\Psi$, so $\hat\Psi$ can be used as a positive subsystem. Conversely, if $A \subset \Delta$ satisfies $\Delta = A \sqcup -A$, then $A = \widehat{A \cap \Delta^+}$. For a subset $\Psi \subset \Delta^+$ consider $\delta - \langle\Psi\rangle$. Inserting the definition of δ, we get

$$\delta - \langle\Psi\rangle = -\tfrac{1}{2}\langle\Psi\rangle + \tfrac{1}{2}\langle\Delta^+ \setminus \Psi\rangle = -\tfrac{1}{2}\langle\hat{\Psi}\rangle. \tag{3.19}$$

Now $w(-\frac{1}{2}\langle\hat{\Psi}\rangle) = -\frac{1}{2}\langle w(\hat{\Psi})\rangle$. Evidently, we have $\Delta = w(\hat{\Psi}) \sqcup -w(\hat{\Psi})$, so we can apply (3.19) to finally conclude that

$$w(\delta - \langle\Psi\rangle) = \delta - \langle w(\hat{\Psi}) \cap \Delta^+\rangle.$$

In particular, there are nonnegative integers n_i'' such that $w(\delta - \langle\Psi\rangle) = \delta - \sum_{i=1}^r n_i''\alpha_i$. Putting $n_i = n_i' + n_i''$, we get $w(\nu + \delta) = \lambda + \delta - \sum_i n_i\alpha_i$, which implies that

$$\|\lambda + \delta\|^2 = \|\nu + \delta\|^2 + 2\sum_{i=1}^r n_i\langle w(\nu + \delta), \alpha_i\rangle + \|\sum_i n_i\alpha_i\|^2.$$

By assumption, $n_i \geq 0$ for all i and $w(\nu + \delta)$ is dominant, so $\langle w(\nu + \delta), \alpha_i\rangle \geq 0$ for all i. This shows that $\|\lambda+\delta\|^2 \geq \|\nu+\delta\|^2$ for all weights ν of $C^*(\mathfrak{p}_+, V)$. Moreover, we get equality if and only if all n_i are zero and hence all n_i' and n_i'' are zero. Vanishing of the n_i' is equivalent to $\lambda = w(\mu)$, i.e. $\mu = w^{-1}(\lambda)$. Vanishing of the n_i'' is equivalent to $w(\hat{\Psi}) \cap \Delta^+ = \emptyset$. This in turn is equivalent to the fact that for a positive root $\alpha \in \Delta^+$, we have $w(\alpha) \in \Delta^-$ if and only if $\alpha \in \Psi$. In the notation of 3.2.14 this means $\Psi = \Phi_{w^{-1}}$, and hence $\nu = \nu_{w^{-1}}$.

Conversely, if $\nu = w(\lambda) - \langle\Phi_w\rangle$, then by part (1) of Proposition 3.2.14 we have $\delta - \langle\Phi_w\rangle = w(\delta)$, which implies $\nu+\delta = w(\lambda+\delta)$ and hence the two weights have the same norm. Since λ is dominant, the sum $\lambda + \delta$ lies in the interior of the dominant Weyl chamber. Hence, $\nu_w = \nu_{w'}$ which explicitly means $w(\lambda + \delta) = w'(\lambda + \delta)$ implies $w = w'$; see 2.2.7.

It remains to show that the weight ν_w occurs with multiplicity at most one in $C^*(\mathfrak{p}_+, V)$. Since V is irreducible, the weight $w(\lambda)$ occurs with multiplicity one in V. Further, if $\Psi \subset \Delta^+$ satisfies $\langle\Psi\rangle = \langle\Phi_w\rangle$, then $w(\delta) = \delta - \langle\Psi\rangle = -\frac{1}{2}\langle\hat{\Psi}\rangle$. Applying w^{-1}, we obtain $-\frac{1}{2}\langle w^{-1}(\hat{\Psi})\rangle = \delta$ and hence $w^{-1}(\hat{\Psi}) \cap \Delta^+ = \emptyset$. This means that $w^{-1}(\hat{\Psi}) = -\Delta^+$, so $\Psi = \Phi_w$. □

From this, we can deduce the full statement of Kostant's version of the Bott–Borel–Weil theorem:

THEOREM 3.3.5. *Let $\mathfrak{g}$ be a complex semisimple Lie algebra, $\mathfrak{p} = \mathfrak{g}_0 \oplus \mathfrak{p}_+$ a standard parabolic subalgebra, W the Weyl group of $\mathfrak{g}$, $W^\mathfrak{p} \subset W$ the Hasse diagram of the parabolic $\mathfrak{p}$, and δ the lowest form of $\mathfrak{g}$. For a finite-dimensional complex irreducible representation V of $\mathfrak{g}$ with highest weight λ consider the Lie algebra cohomology $H^*(\mathfrak{p}_+, V)$, and for a $\mathfrak{g}_0$-dominant weight ν let $H^*(\mathfrak{p}_+, V)^\nu$ be the isotypical component of highest weight ν for the natural $\mathfrak{g}_0$-representation on the cohomology. Then we have:*

(1) $H^*(\mathfrak{p}_+, V)^\nu \neq \{0\}$ if and only if there is an element $w \in W^\mathfrak{p}$ such that $\nu = \nu_w := w(\lambda + \delta) - \delta$.

(2) For any $w \in W^\mathfrak{p}$ the isotypical component $H^*(\mathfrak{p}_+, V)^{\nu_w}$ is irreducible, and even the multiplicity of ν_w as a weight of $C^*(\mathfrak{p}_+, V)$ is one. In particular, the set of irreducible components of $H^*(\mathfrak{p}_+, V)$ is in bijective correspondence with $W^\mathfrak{p}$.

(3) For $w \in W^\mathfrak{p}$, the isotypical component $H^*(\mathfrak{p}_+, V)^{\nu_w}$ is contained in the space $H^{\ell(w)}(\mathfrak{p}_+, V)$, where $\ell(w)$ is the length of w; see 3.2.14. A highest weight

vector in the isotypical component is given by the cohomology class containing the harmonic representative $F_{\Phi_w} \otimes v$, where F_{Φ_w} is the wedge of one nonzero element from each of the root spaces $\mathfrak{g}_{-\alpha}$ with $\alpha \in \Phi_w$ and $v \in V$ is a nonzero weight vector of weight $w(\lambda)$.

PROOF. By the lemma, the possible highest weights of irreducible components of $C^*(\mathfrak{p}_+, V)$, which are contained in $\ker(\Box) \cong H^*(\mathfrak{p}_+, V)$, are of the form $\nu_w = w(\lambda + \delta) - \delta = w(\lambda) - \langle \Phi_w \rangle$. If such a weight occurs, then this is the only way to write that weight as the sum of a weight of V and $-\langle \Psi \rangle$ for some subset $\Psi \subset \Delta^+$. Since $-\langle \Psi \rangle$ is a weight of $\Lambda^* \mathfrak{g}_-$ only if $\Psi \subset \Delta^+(\mathfrak{p}_+)$, the weight ν_w may only show up in $C^*(\mathfrak{p}_+, V)$ if $\Phi_w \subset \Delta^+(\mathfrak{p}_+)$, i.e. if $w \in W^{\mathfrak{p}}$. On the other hand, if $w \in W^{\mathfrak{p}}$, then $\Phi_w \subset \Delta^+(\mathfrak{p}_+)$, so we have the corresponding element $F_{\Phi_w} \in \Lambda^* \mathfrak{g}_-$. By construction, this is a weight vector of weight $-\langle \Phi_w \rangle$ and it lies in $\Lambda^\ell \mathfrak{g}_-$, where ℓ is the cardinality of Φ_w, which equals the length of w by part (3) of Proposition 3.2.14. Since $w(\lambda)$ is a weight of V, so we can find a nonzero weight vector v of that weight, and hence $F_{\Phi_w} \otimes v \in C^{\ell(w)}(\mathfrak{p}_+, V)$ is a nonzero weight vector of weight ν_w.

Next, we claim that for any $w \in W^{\mathfrak{p}}$, the weight ν_w is actually a highest weight of $C^*(\mathfrak{p}_+, V)$. To see this, it suffices to prove that for any root $\alpha \in \Delta^+(\mathfrak{g}_0)$, the weight $\nu_w + \alpha$ does not occur in $C^*(\mathfrak{p}_+, V)$, since this implies that any element of a positive root space of $\mathfrak{g}_0$ has to act trivially on any weight vector of weight ν_w. To see this, we compute

$$\|\nu_w + \alpha + \delta\|^2 = \|w(\lambda + \delta) + \alpha\|^2 = \|\lambda + \delta\|^2 + 2\langle w(\lambda + \delta), \alpha \rangle + \|\alpha\|^2.$$

But now $\langle w(\lambda + \delta), \alpha \rangle = \langle \lambda + \delta, w^{-1}(\alpha) \rangle$, and since $w \in W^{\mathfrak{p}}$ and $\alpha \in \Delta^+(\mathfrak{g}_0)$, we have $w^{-1}(\alpha) \in \Delta^+$. Since λ is dominant, the inner product $\langle \lambda + \delta, w(\alpha) \rangle$ is positive, which implies $\|\nu_w + \alpha + \delta\| > \|\lambda + \delta\|$, so $\nu_w + \alpha$ is not a weight of $C^*(\mathfrak{p}_+, V)$ by the lemma.

Using this, (1) follows immediately. Since $\lambda + \delta$ lies in the interior of the dominant Weyl chamber, different elements of $W^{\mathfrak{p}}$ lead to different weights. Together with the multiplicity one result in the lemma, this implies (2), while (3) is now clear from the fact that we have constructed a weight vector above. □

3.3.6. Some simple consequences. As it stands, Kostant's version of the Bott–Borel–Weil theorem is not strong enough for our purposes, since it only deals with the case of complex Lie algebras and complex irreducible representations. Using some basic tricks of the trade in cohomology theory, it is, however, easy to deduce information on the case of real Lie algebras and real representations.

First of all, the irreducibility assumption is not a real restriction. For a complex semisimple $|k|$–graded Lie algebra $\mathfrak{g}$, any finite–dimensional complex representation is completely reducible, i.e. a direct sum of irreducible representations; see 2.1.6. From the definition of the chain groups it follows immediately that $C^n(\mathfrak{g}_-, \bigoplus_i V_i) \cong \bigoplus_i C^n(\mathfrak{g}_-, V_i)$ and this is an isomorphism of $\mathfrak{g}_0$–modules. The definition of the Lie algebra differential immediately implies compatibility with this decomposition, which shows that $H^*(\mathfrak{g}_-, \oplus_i V_i) \cong \bigoplus_i H^*(\mathfrak{g}_-, V_i)$ as a $\mathfrak{g}_0$–module. Thus, Theorem 3.3.5 can be used to compute the cohomology of $\mathfrak{g}_-$ with coefficients in arbitrary finite dimensional complex representations V.

Recall that for a real $|k|$–graded semisimple Lie algebra $\mathfrak{g} = \mathfrak{g}_{-k} \oplus \cdots \oplus \mathfrak{g}_k$ the complexification $\mathfrak{g}^{\mathbb{C}} = \mathfrak{g} \otimes_{\mathbb{R}} \mathbb{C}$ is naturally $|k|$–graded. The negative part $\mathfrak{g}_-^{\mathbb{C}}$ is the complexification of $\mathfrak{g}_-$ and $\mathfrak{g}_0^{\mathbb{C}}$ is the complexification of $\mathfrak{g}_0$. Recall further that any complex representation of a real Lie algebra naturally extends to the complexification. If W is such a complex representation, then there is a natural conjugate representation $\overline{W}$; see 2.3.14.

PROPOSITION 3.3.6. *Let $\mathfrak{g}$ be a real $|k|$–graded semisimple Lie algebra with complexification $\mathfrak{g}^{\mathbb{C}}$.*

(1) Let V be a complex representation of $\mathfrak{g}$. Then the real cohomology spaces $H^*_{\mathbb{R}}(\mathfrak{g}_-, V)$ are naturally complex vector spaces and $H^*_{\mathbb{R}}(\mathfrak{g}_-, V) \cong H^*_{\mathbb{C}}(\mathfrak{g}^{\mathbb{C}}_-, V)$ as a module over $\mathfrak{g}_0 \subset \mathfrak{g}^{\mathbb{C}}_0$.

(2) If V is a real representation of $\mathfrak{g}$, then for the complexification $V \otimes \mathbb{C}$ we have

$$H^*_{\mathbb{C}}(\mathfrak{g}^{\mathbb{C}}_-, V \otimes \mathbb{C}) \cong H^*_{\mathbb{R}}(\mathfrak{g}_-, V) \otimes \mathbb{C}.$$

(3) In the situation of (2) assume that V is irreducible and does not admit a $\mathfrak{g}$–invariant complex structure, and let $W \subset H^*_{\mathbb{C}}(\mathfrak{g}^{\mathbb{C}}_-, V \otimes \mathbb{C})$ be a $\mathfrak{g}^{\mathbb{C}}_0$–irreducible component.

If $W \cong \overline{W}$, then W is the complexification of a real irreducible component in $H^*_{\mathbb{R}}(\mathfrak{g}_-, V)$. If W is not isomorphic to $\overline{W}$, then $H^*_{\mathbb{C}}(\mathfrak{g}^{\mathbb{C}}_-, V \otimes \mathbb{C})$ contains $\overline{W}$ as an irreducible component and $W \oplus \overline{W}$ is the complexification of one complex irreducible component in $H^*_{\mathbb{R}}(\mathfrak{g}_-, V)$.

PROOF. (1) Since V is complex, any real n–linear map $\phi : \mathfrak{g}_- \times \cdots \times \mathfrak{g}_- \to V$ uniquely extends to a complex n–linear map $\tilde{\phi} : \mathfrak{g}^{\mathbb{C}}_- \times \cdots \times \mathfrak{g}^{\mathbb{C}}_- \to V$. Thus, the chain group $C^n_{\mathbb{R}}(\mathfrak{g}_-, V)$ (which is a complex vector space under pointwise operations) is isomorphic to $C^n_{\mathbb{C}}(\mathfrak{g}^{\mathbb{C}}_-, V)$, and by construction this identification is compatible with the Lie algebra differentials. From this the result follows immediately.

(2) From the definition of cohomology, one immediately concludes that the complex representation $H^*_{\mathbb{R}}(\mathfrak{g}_-, V^{\mathbb{C}})$ is the complexification of $H^*_{\mathbb{R}}(\mathfrak{g}_-, V)$, so the claim follows from (1).

(3) By Proposition 2.3.14, our assumptions on V imply that $V^{\mathbb{C}}$ is an irreducible representation of $\mathfrak{g}^{\mathbb{C}}$. Hence, Kostant's Version of the Bott–Borel–Weil Theorem tells us that any $\mathfrak{g}^{\mathbb{C}}_0$–irreducible component in $H^*_{\mathbb{C}}(\mathfrak{g}^{\mathbb{C}}_-, V \otimes \mathbb{C})$ occurs with multiplicity one. Now if we have an irreducible component $U \subset H^*_{\mathbb{R}}(\mathfrak{g}_-, V)$, which admits an invariant complex structure, then its complexification is isomorphic to $U \oplus \overline{U}$. By multiplicity one, this is only possible if U and $\overline{U}$ are non–isomorphic. On the other hand, if U does not admit an invariant complex structure, then $U \otimes \mathbb{C}$ is isomorphic to its conjugate; see 2.3.14. From this the description follows. □

REMARK 3.3.6. (1) Part (3) of the proposition implies that the real representation $H^*_{\mathbb{R}}(\mathfrak{g}_-, V)$ does not contain irreducible components which are quaternionic in the sense introduced in 2.3.14.

(2) While Theorem 3.3.5 is valid for semisimple Lie algebras, it is often convenient to reduce to the case of simple Lie algebras. Assume that $\mathfrak{g} = \mathfrak{g}' \oplus \mathfrak{g}''$ is a direct sum of a $|k'|$–graded semisimple Lie algebra $\mathfrak{g}'$ and a $|k''|$–graded semisimple Lie algebra $\mathfrak{g}''$, which both are ideals in $\mathfrak{g}$. Then any irreducible representation V of $\mathfrak{g}$ is isomorphic to an exterior tensor product $V \cong V' \boxtimes V''$ for irreducible representations V' of $\mathfrak{g}'$ and V'' of $\mathfrak{g}''$. The underlying vector space of $V' \boxtimes V''$ is the tensor product $V' \otimes V''$ and the action is given by

$$(A', A'') \cdot (v' \otimes v'') = (A' \cdot v') \otimes v'' + v' \otimes (A'' \cdot v'').$$

This follows immediately from the fact that for Cartan subalgebras $\mathfrak{h}' \leq \mathfrak{g}'$ and $\mathfrak{h}'' \leq \mathfrak{g}''$, also $\mathfrak{h}' \oplus \mathfrak{h}'' \leq \mathfrak{g}$ is a Cartan subalgebra and the theorem of the highest weight. Indeed, if $\lambda : \mathfrak{h}' \oplus \mathfrak{h}'' \to \mathbb{C}$ is a dominant integral weight we define λ' and λ'' to be the appropriate restrictions of λ, then the exterior tensor product of irreducible representations V' and V'' with highest weights λ' and λ'' is the irreducible representation with highest weight λ. Moreover, it follows immediately from the definitions that the Weyl group W of $\mathfrak{g}$ is the product $W' \times W''$ of the Weyl groups of the two factors, and the length of (w', w'') is the sum of the lengths of the two elements. Hence, from Theorem 3.3.5 we immediately conclude that

$$H^n(\mathfrak{g}_-, V' \boxtimes V'') \cong \bigoplus_{i+j=n} \Big(H^i(\mathfrak{g}'_-, V') \boxtimes H^j(\mathfrak{g}''_-, V'') \Big),$$

as a module over $\mathfrak{g}_0 \cong \mathfrak{g}'_0 \oplus \mathfrak{g}''_0$. In particular, the adjoint representation $V = \mathfrak{g}$ is not irreducible in this case, but we have $\mathfrak{g} = \mathfrak{g}' \boxtimes \mathbb{K} \oplus \mathbb{K} \boxtimes \mathfrak{g}''$. Consequently, we obtain

$$H^n(\mathfrak{g}_-, \mathfrak{g}) \cong \bigoplus_{i+j=n} \Big(H^i(\mathfrak{g}'_-, \mathfrak{g}') \boxtimes H^j(\mathfrak{g}''_-, \mathbb{K}) \oplus H^i(\mathfrak{g}'_-, \mathbb{K}) \boxtimes H^j(\mathfrak{g}''_-, \mathfrak{g}'') \Big). \tag{3.20}$$

These results can be used to compute the real cohomologies of complex simple Lie algebras, since for a complex simple Lie algebra $\mathfrak{g}$, we have $\mathfrak{g}^{\mathbb{C}} = \mathfrak{g} \oplus \mathfrak{g}$.

3.3.7. Zeroth and first cohomologies. The cohomology groups $H^*(\mathfrak{g}_-, V)$ will play an important role in many parts of the subsequent developments. As a first application of Theorem 3.3.5 we study the first cohomology groups with coefficients in the adjoint representation. These cohomology groups determine whether a normal parabolic geometry is uniquely determined by the underlying infinitesimal flag structure; see 3.1.14 and 3.1.16. On the way to these results, we will compute $H^0(\mathfrak{g}_-, V)$ for any V and $H^1(\mathfrak{g}_-, \mathbb{K})$.

The zeroth cohomology $H^0(\mathfrak{g}_-, V)$ can be determined directly for arbitrary V. By definition, this cohomology group is simply the kernel of $\partial : V \to L(\mathfrak{g}_-, V)$ and thus the $\mathfrak{g}_0$–module $V^{\mathfrak{g}_-}$ of $\mathfrak{g}_-$–invariant elements in V. Alternatively, we may describe $H^0(\mathfrak{g}_-, V)$ as $V/\operatorname{im}(\partial^*)$, and $\partial^*(Z \otimes v) = -Z \cdot v$ for $Z \in \mathfrak{p}_+ = \mathfrak{g}_-^*$ and $v \in V$. Thus, $\operatorname{im}(\partial^*) = \mathfrak{p}_+ \cdot V$, and $H^0(\mathfrak{g}_-, V) \cong V/(\mathfrak{p}_+ \cdot V)$, which is a more natural description from the point of view of the $\mathfrak{p}$–module structure. In particular, for the trivial representation $\mathbb{K}$, we obtain $H^0(\mathfrak{g}_-, \mathbb{K}) = \mathbb{K}$. In the case of the adjoint representation we conclude from 3.3.6 that $H^0(\mathfrak{g}_-, \mathfrak{g})$ is the direct sum of the zeroth cohomologies of the simple ideals of $\mathfrak{g}$, so let us assume that $\mathfrak{g} = \mathfrak{g}_{-k} \oplus \cdots \oplus \mathfrak{g}_k$ is simple. Of course, $[\mathfrak{p}_+, \mathfrak{g}] \subset \mathfrak{g}_{-k+1} \oplus \cdots \oplus \mathfrak{g}_k = \mathfrak{g}^{-k+1}$. Moreover, Theorem 3.3.5 together with 3.3.6 tells us that $H^0(\mathfrak{g}_-, \mathfrak{g}) = \mathfrak{g}/([\mathfrak{p}_+, \mathfrak{g}])$ is irreducible, which implies that there is no $\mathfrak{p}$–submodule of $\mathfrak{g}$ which lies strictly between $\mathfrak{g}$ and $[\mathfrak{p}_+, \mathfrak{g}]$. But this immediately implies that $[\mathfrak{p}_+, \mathfrak{g}] = \mathfrak{g}^{-k+1}$, and thus $H^0(\mathfrak{g}_-, \mathfrak{g}) \cong \mathfrak{g}_{-k} \cong \mathfrak{g}/\mathfrak{g}^{-k+1}$ for simple $\mathfrak{g}$.

The explicit description of the zeroth cohomology in Theorem 3.3.5 tells us that for an irreducible $\mathfrak{g}$–representation V with highest weight λ the cohomology $H^0(\mathfrak{p}_+, V) \cong V^{\mathfrak{p}_+}$ is the irreducible $\mathfrak{p}$–representation of highest weight λ, which we have noticed in 3.2.13. To obtain the corresponding statement for $H^0(\mathfrak{g}_-, V)$, recall that $H^0(\mathfrak{g}_-, V) = (H^0(\mathfrak{p}_+, V^*))^*$ by Corollary 3.3.1. Hence, if V is dual to the irreducible $\mathfrak{g}$–representation with highest weight μ (i.e. V has lowest weight $-\mu$), then $V^{\mathfrak{g}_-}$ is dual to the irreducible $\mathfrak{p}$–representation with highest weight μ.

Let us next consider the cohomology group $H^1(\mathfrak{g}_-, \mathbb{K})$ with trivial coefficients. By definition, $\partial : \mathbb{K} \to L(\mathfrak{g}_-, \mathbb{K})$ is the zero map, so $H^1(\mathfrak{g}_-, \mathbb{K})$ coincides with the kernel of $\partial : L(\mathfrak{g}_-, \mathbb{K}) \to L(\Lambda^2\mathfrak{g}_-, \mathbb{K})$. Since the action is trivial, we simply get $\partial\phi(X,Y) = -\phi([X,Y])$, which immediately implies $H^1(\mathfrak{g}_-, \mathbb{K}) = L(\mathfrak{g}_-/[\mathfrak{g}_-, \mathfrak{g}_-], \mathbb{K})$. Clearly, $[\mathfrak{g}_-, \mathfrak{g}_-] \subset \mathfrak{g}_{-k} \oplus \cdots \oplus \mathfrak{g}_{-2}$, and these two spaces coincide since $\mathfrak{g}_-$ is generated by $\mathfrak{g}_{-1}$. Thus, we conclude that $H^1(\mathfrak{g}_-, \mathbb{K}) \cong (\mathfrak{g}_{-1})^* \cong \mathfrak{g}_1$.

Now we are ready to prove the main result on the cohomological obstructions showing up in Theorems 3.1.14 and 3.1.16. Recall from 3.1.14 that the cohomology groups $H^n(\mathfrak{g}_-, \mathfrak{g})$ are naturally graded as $H^n(\mathfrak{g}_-, \mathfrak{g}) = \bigoplus_\ell H^n(\mathfrak{g}_-, \mathfrak{g})_\ell$ by the homogeneous degree of maps. In particular, for the first cohomology this grading is simply induced by the natural grading on $\mathfrak{p}_+ \otimes \mathfrak{g}$.

PROPOSITION 3.3.7. *Let $\mathfrak{g} = \mathfrak{g}_{-k} \oplus \cdots \oplus \mathfrak{g}_k$ be a $|k|$-graded semisimple Lie algebra over $\mathbb{K} = \mathbb{R}$ or $\mathbb{C}$ such that none of the simple ideals of $\mathfrak{g}$ is contained in $\mathfrak{g}_0$. Then we have:*

(1) $H^1(\mathfrak{g}_-, \mathfrak{g})_\ell = \{0\}$ for all $\ell > 1$ unless $\mathfrak{g}$ contains a simple ideal isomorphic to $\mathfrak{sl}(2,\mathbb{R})$ or $\mathfrak{sl}(2,\mathbb{C})$ with the (unique) grading induced by the Borel subalgebra.

(2) $H^1(\mathfrak{g}_-, \mathfrak{g})_\ell = \{0\}$ for all $\ell > 0$ unless $\mathfrak{g}$ contains a simple ideal $\mathfrak{g}'$ such that either $\mathfrak{g}'$ or $\mathfrak{g}'_{\mathbb{C}}$ is isomorphic to  *or* .

PROOF. We first reduce the problem to the complex simple case. From 3.3.6 we know that for a sum $\mathfrak{g} = \mathfrak{g}' \oplus \mathfrak{g}''$ of ideals, the cohomology $H^1(\mathfrak{g}_-, \mathfrak{g})$ splits into the direct sum of

$$H^1(\mathfrak{g}'_-, \mathfrak{g}') \boxtimes H^0(\mathfrak{g}''_-, \mathbb{K}) \oplus H^0(\mathfrak{g}'_-, \mathfrak{g}') \boxtimes H^1(\mathfrak{g}''_-, \mathbb{K})$$

and the corresponding terms with the roles of $\mathfrak{g}'$ and $\mathfrak{g}''$ exchanged. From above we know that $H^0(\mathfrak{g}'_-, \mathfrak{g}')$ is concentrated in negative degrees, and $H^1(\mathfrak{g}''_-, \mathbb{K})$ is concentrated in homogeneous degree one, so their tensor product sits in nonpositive homogeneity. This also holds for the symmetric term and using that $H^0(\mathfrak{g}'_-, \mathbb{K}) = \mathbb{K}$, we conclude that $H^1(\mathfrak{g}_-, \mathfrak{g})_\ell \cong H^1(\mathfrak{g}'_-, \mathfrak{g}')_\ell \oplus H^1(\mathfrak{g}''_-, \mathfrak{g}'')_\ell$ for all $\ell > 0$.

In particular, the first cohomology with coefficients in the adjoint representation of a $|k|$-graded semisimple Lie algebra $\mathfrak{g}$ has a nonzero component in some positive homogeneous degree if and only if the same is true for one of the simple ideals of $\mathfrak{g}$. On the other hand, the splitting into homogeneous degrees is obviously compatible with complexifications, so it suffices to determine the complex simple $|k|$-graded Lie algebras $\mathfrak{g}$ such that $H^1(\mathfrak{g}_-, \mathfrak{g})_\ell \neq \{0\}$ for some $\ell > 0$. Now the isomorphism $H^1(\mathfrak{g}_-, \mathfrak{g}) \cong H^1(\mathfrak{p}_+, \mathfrak{g})^*$ is compatible with homogeneities, i.e. we may equivalently determine the cases in which $H^1(\mathfrak{p}_+, \mathfrak{g})_\ell$ is nontrivial for some $\ell < 0$. We assume that $\mathfrak{p}$ is the standard parabolic corresponding to a set Σ of simple roots.

Now according to Theorem 3.3.5, the irreducible components of $H^1(\mathfrak{p}_+, \mathfrak{g})$ are in bijective correspondence with the elements of length one in $W^{\mathfrak{p}}$, i.e. with the reflections σ_α corresponding to simple roots $\alpha \in \Sigma$. Moreover, denoting by λ the highest weight of the adjoint representation, the highest weight of the irreducible component corresponding to σ_α is given by $\sigma_\alpha(\lambda+\delta)-\delta = \sigma_\alpha(\lambda)-\alpha$. By construction, the root space $\mathfrak{g}_\alpha$ is contained in $\mathfrak{g}_1$, from which we immediately conclude that the homogeneous degree of a weight vector of the above weight is given by $\mathrm{ht}_\Sigma(\sigma_\alpha(\lambda)) - 1$. Now λ simply is the highest root, so $\sigma_\alpha(\lambda)$ is a root, too, and by definition $\sigma_\alpha(\lambda)$ differs from λ by a multiple of α. If there is more than one simple root in $\mathfrak{g}$, then $\lambda \neq \alpha$, and thus $\sigma_\alpha(\lambda)$ is a positive root and thus $\mathrm{ht}_\Sigma(\sigma_\alpha(\lambda)) \geq 0$ and (1) follows.

To prove (2), it remains to determine those $\mathfrak{g}$ and Σ such that there is an element $\alpha \in \Sigma$ with $\mathrm{ht}_\Sigma(\sigma_\alpha(\lambda)) = 0$. We have noted above that $\sigma_\alpha(\lambda)$ is a positive root and expanding it as a linear combination of simple roots, the coefficients of all simple roots $\neq \alpha$ are the same as for λ. In particular, this implies that $\mathrm{ht}_\Sigma(\sigma_\alpha(\lambda)) > 0$ unless $\Sigma = \{\alpha\}$. Given that $\Sigma = \{\alpha\}$, the condition that $\mathrm{ht}_\Sigma(\sigma_\alpha(\lambda)) = 0$ by definition of the simple reflection is equivalent to the fact that $\frac{2\langle\lambda,\alpha\rangle}{\langle\alpha,\alpha\rangle}$ coincides with the coefficient of α in the expansion of λ as a linear combination of simple roots.

Now we have to go through the list of complex simple Lie algebras. In the case of A_n for $n \geq 2$, we have simple roots $\alpha_1, \dots, \alpha_n$ and the highest root λ is given by $\alpha_1 + \dots + \alpha_n$; see Example (1) of 2.2.6. Consequently, $\frac{2\langle\lambda,\alpha_i\rangle}{\langle\alpha_i,\alpha_i\rangle}$ equals 1 for $i = 1$ and $i = n$ and 0 otherwise. Since the $|1|$–gradings corresponding to $\{\alpha_1\}$ and $\{\alpha_n\}$ are isomorphic, this is compatible with the statement of (2).

Let us next consider the case of C_n with $n \geq 2$. As in example (4) of 2.2.6 we have simple roots $\alpha_1, \dots, \alpha_n$ and $\lambda = 2\alpha_1 + \dots + 2\alpha_{n-1} + \alpha_n$. From this description and the Dynkin diagram, one immediately concludes that $\frac{2\langle\lambda,\alpha_i\rangle}{\langle\alpha_i,\alpha_i\rangle}$ equals 2 for $i = 1$ and 0 for $i \neq 1$, so we get the second class of exceptions in (2).

Since $B_2 = C_2$ and this has been dealt with above, we may next consider B_n for $n \geq 3$. According to example (3) of 2.2.6, the usual Dynkin diagram with simple roots $\alpha_1, \dots, \alpha_n$ leads to $\lambda = \alpha_1 + 2\alpha_2 + \dots + 2\alpha_n$. Assuming $n \geq 3$, this implies that $\frac{2\langle\lambda,\alpha_i\rangle}{\langle\alpha_i,\alpha_i\rangle}$ equals 1 for $i = 2$ and 0 for $i \neq 2$, so we get no exception in the case of B_n with $n \geq 3$.

For D_n, we may restrict to the case $n \geq 4$. Using the simple roots $\alpha_1, \dots, \alpha_n$ as in example (2) of 2.2.6, the highest root λ is given as $\alpha_1 + 2\alpha_2 + \dots + 2\alpha_{n-2} + \alpha_{n-1} + \alpha_n$. Using this, one immediately verifies that $\frac{2\langle\lambda,\alpha_i\rangle}{\langle\alpha_i,\alpha_i\rangle}$ equals 1 for $i = 2$ and 0 for $i \neq 2$, so we get no exception in the case of D_n with $n \geq 4$.

For the exceptional Lie algebras one may read off from Table B.2 in Appendix B that there always is a unique simple root α such that $\frac{2\langle\lambda,\alpha\rangle}{\langle\alpha,\alpha\rangle} \neq 0$, but the resulting number never coincides with the coefficient of α in the expression of λ as a linear combination of simple roots. □

REMARK 3.3.7. Following the lines of the proof of the above proposition, one may also give a complete list of all simple $|k|$–graded Lie algebras such that $H^1(\mathfrak{g}_-, \mathfrak{g})_\ell = \{0\}$ for all $\ell \geq 0$. This is of considerable interest from the point of view of parabolic geometries, since it means that the corresponding geometry is essentially determined by the filtration of the tangent bundle only. We will prove this result and study the corresponding geometries in 4.3.1.

There are also classification results concerning the second cohomology groups with values in the adjoint representation available. These are important from the point of view of parabolic geometries, since $H^2(\mathfrak{g}_-, \mathfrak{g})$ determines the possible components of the harmonic curvature, which is a complete obstruction to local flatness; see 3.1.12. The article [**Ya93**] contains a complete list of the gradings for which there is an irreducible component of $H^2(\mathfrak{g}_-, \mathfrak{g})$, which is contained in positive homogeneity. If there is no such component, then regular normal parabolic geometries of that type are automatically locally flat (since by Theorem 3.1.12 the lowest nonvanishing component of the Cartan curvature has to be harmonic). Still the results on equivalence to underlying structures can be interesting in these cases; see 4.3.5.

3.3.8. The Bott–Borel–Weil theorem. We next discuss how Kostant's theorem is related to the original Bott–Borel–Weil theorem, mainly following Kostant's original article [**Kos61**]. This theorem computes the sheaf cohomology groups of a complex generalized flag variety G/P with coefficients in the sheaves of local holomorphic sections of certain homogeneous holomorphic vector bundles. As in the smooth case treated in 1.4.3, holomorphic homogeneous vector bundles over G/P are in bijective correspondence with holomorphic representations of the parabolic subgroup P, and in particular, we may look at irreducible representations. From 3.2.12 we know that finite–dimensional complex representations of the Lie algebra $\mathfrak{p}$ are in bijective correspondence with $\mathfrak{p}$–dominant and $\mathfrak{p}$–algebraically integral weights λ, which may be viewed as weights for $\mathfrak{g}$. In the formulation of the theorem it will be more convenient to work with lowest weights rather than highest weights, but symmetry of the weights of a $\mathfrak{p}$–representation under the Weyl group $W_{\mathfrak{p}}$ implies that these are just the negatives of the $\mathfrak{p}$–dominant algebraically integral weights.

The G–action on a homogeneous holomorphic vector bundle is holomorphic, so it induces an action on the cohomology groups of G/P with values in the sheaf of local holomorphic sections. (This is evident by viewing sheaf cohomology groups as Dolbeault cohomology groups.) Hence, we may analyze the cohomology groups as representations of G.

THEOREM 3.3.8 (Bott–Borel–Weil). *Let G be a connected complex semisimple Lie group with Lie algebra $\mathfrak{g}$, $P \subset G$ a standard parabolic subgroup with Lie algebra $\mathfrak{p} \subset \mathfrak{g}$. Consider an irreducible representation of P, let $-\lambda$ be its lowest weight, and let $\mathcal{O}(\lambda)$ be the sheaf of local holomorphic sections of the corresponding homogeneous holomorphic vector bundle over G/P. Denoting by δ the sum of all fundamental weights of $\mathfrak{g}$, we have:*

(1) If $\lambda + \delta$ lies in a wall of some Weyl chamber, then the sheaf cohomology $H^*(G/P, \mathcal{O}(\lambda))$ is trivial.

(2) If $\lambda + \delta$ lies in the interior of some Weyl chamber, then let $w \in W$ be the unique element such that $w \cdot \lambda := w(\lambda + \delta) - \delta$ is dominant, and let $\ell(w)$ be the length of w. Then $H^k(G/P, \mathcal{O}(\lambda)) = 0$ for $k \neq \ell(w)$ and $H^{\ell(w)}(G/P, \mathcal{O}(\lambda))$ is an irreducible representation of G with lowest weight $-w \cdot \lambda$ (and hence dual to the representation with highest weight $w \cdot \lambda$).

SKETCH OF PROOF. By Proposition 3.2.6, the maximal compact subgroup $K \subset G$ acts transitively on G/P and $G/P \cong K/L$ where $L := K \cap P$. Since G is complex, we know from 2.3.2 that the Lie algebra $\mathfrak{k}$ of K is a compact real form of $\mathfrak{g}$. Let $\mathfrak{h} \subset \mathfrak{g}$ be the Cartan subalgebra and let $\mathfrak{h}_0 \subset \mathfrak{h}$ be the subspace on which all roots are real. Then from 2.3.1 we know that for any positive root $\alpha \in \Delta^+$ we can choose generators $X_{\pm\alpha} \in \mathfrak{g}_{\pm\alpha}$ such that

$$\mathfrak{k} = i\mathfrak{h}_0 \oplus \bigoplus_{\alpha \in \Delta^+} (\mathbb{R}(X_\alpha - X_{-\alpha}) + i\mathbb{R}(X_\alpha + X_{-\alpha})).$$

In particular, $\mathfrak{l} = \mathfrak{k} \cap \mathfrak{p} \subset \mathfrak{g}_0$, and $\mathfrak{l}$ is a real form of $\mathfrak{g}_0$.

Hence, $\mathfrak{g}/\mathfrak{g}_0$ is the complexification of $\mathfrak{k}/\mathfrak{l}$, and we may view $T_{\mathbb{C}}(K/L)$ as $K \times_L \mathfrak{g}/\mathfrak{g}_0$. Since $L \subset G_0$, we get $\mathfrak{g}/\mathfrak{g}_0 \cong \mathfrak{g}_- \oplus \mathfrak{p}_+$ as an L–module. The complex structure on the tangent bundle $T(K/L)$ comes from the linear isomorphism $\mathfrak{k}/\mathfrak{l} \cong \mathfrak{g}/\mathfrak{p}$. One verifies that this induces

$$J(a(X_\alpha - X_{-\alpha}) + ib(X_\alpha + X_{-\alpha})) = -b(X_\alpha - X_{-\alpha}) - ia(X_\alpha + X_{-\alpha}).$$

This implies that $\mathfrak{g}/\mathfrak{g}_0 \cong \mathfrak{g}_- \oplus \mathfrak{p}_+$ is the splitting describing the given complex structure on K/L, with $\mathfrak{g}_-$ corresponding to the holomorphic part and $\mathfrak{p}_+$ corresponding to the anti–holomorphic part.

Now let V be an irreducible holomorphic representation of P. By restriction, we may view this as a representation of L, and hence we can view the holomorphic vector bundle $G \times_P V$ also as $K \times_L V$. In view of the above discussion and the standard correspondence between sections of associated bundles and equivariant functions on the group, we may view the space $\Omega^{0,q}(G/P, G\times_P V)$ of bundle–valued $(0,q)$–forms as the space $C^\infty(K, \Lambda^q \mathfrak{p}_+^* \otimes V)^L$ of L–equivariant smooth functions. Via inserting fixed elements of $\mathfrak{p}_+$ into such functions, we get the alternative picture of this space as $\mathrm{Hom}_L(\Lambda^k \mathfrak{p}_+, C^\infty(K, V))$. From above we know that $T_{\mathbb{C}}K \cong K \times \mathfrak{g}$, so we may view elements of $\mathfrak{p}_+$ as sections of the complexified tangent bundle of K, and thus obtain a natural action of $\mathfrak{p}_+$ on $C^\infty(K, V)$. We may further identify $C^\infty(K, V)$ with $C^\infty(K) \otimes V$ and since V is an irreducible P–module, $\mathfrak{p}_+$ acts trivially on V. From the definition, one easily sees that the Dolbeault differential corresponds to

$$\partial \otimes \mathrm{id}_V : \Lambda^q \mathfrak{p}_+^* \otimes C^\infty(K) \otimes V \to \Lambda^{q+1} \mathfrak{p}_+^* \otimes C^\infty(K) \otimes V,$$

where ∂ denotes the Lie algebra differential for the representation $C^\infty(K)$ of $\mathfrak{p}_+$. By construction, the natural K–action on sections is given by acting only on the $C^\infty(K)$–factor by $(g \cdot f)(g') = f(g^{-1}g')$. Putting this together, we see that we may identify the Dolbeault cohomology group $H^q(G/P, (G \times_P V))$ with the space $\left(H^q(\mathfrak{p}_+, C^\infty(K)) \otimes V\right)^L$ of L–invariant elements. Here, the action of L on the cohomology groups is obtained from the natural (tensor product) action on $\Lambda^q \mathfrak{p}_+^* \otimes C^\infty(K)$ and L acts on $C^\infty(K)$ via $(g \cdot f)(g') = f(g'g)$.

Since G/P is compact, the sheaf cohomology is known to be finite–dimensional, so we may replace $C^\infty(K)$ by the subspace of K–finite vectors, i.e. the subspace of all those functions which are contained in a finite–dimensional K–invariant subspace of $C^\infty(K)$. This subspace is described explicitly by the Peter–Weyl theorem as follows.

Consider the space $C(K)$ of complex–valued continuous functions on K. This carries a representation of $K \times K$ defined by $((g,h) \cdot f)(g') := f(g^{-1}g'h)$. Now suppose that W is a finite–dimensional complex representation of K. Since K is compact, there is a K–invariant inner product on W, and for $\xi, \eta \in W$, we define the matrix coefficient $f_{\xi,\eta} : K \to \mathbb{C}$ by $f_{\xi,\eta}(g) = \langle \xi, g \cdot \eta \rangle$. Of course, this is a smooth function, and invariance of the inner product immediately implies that $(g,h) \cdot f_{\xi,\eta} = f_{g^{-1}\cdot\xi, h\cdot\eta}$. In particular, if W is irreducible, then $f_{\xi,\eta} = 0$ implies that either ξ or η has to vanish, and we obtain an injection of the $K \times K$–representation $W^* \boxtimes W$ into $C^\infty(K)$ and $C(K)$. Consider the set of matrix coefficients of all finite–dimensional representations of K. One immediately checks that a linear combination of matrix coefficients of two representations can be obtained as a matrix coefficient of the direct sum, while the product can be obtained as a matrix coefficient of the tensor product of representations. Hence, this is a subalgebra of $C^\infty(K)$ and of $C(K)$. Provided that K admits a faithful finite–dimensional representation (which can be proved for general compact Lie groups and is obvious for the groups we are concerned with here), this subalgebra separates points and hence is dense in $C(K)$ by the Stone–Weierstrass theorem.

The Peter–Weyl theorem now states that the subspace of all matrix coefficients is exactly the subspace of K–finite vectors in $C(K)$ (and thus also in $C^\infty(K)$), and it can be explicitly described as a $K \times K$–representation as $\bigoplus_W (W^* \boxtimes W)$, where W runs through all irreducible representations of K (which are automatically finite–dimensional). See [**CSM95**] for a nice exposition of the Peter–Weyl theorem.

Returning to the description of Dolbeault cohomology, we see that it is given as $\bigoplus_W \big(H^q(\mathfrak{p}_+, W^* \boxtimes W) \otimes V\big)^L$. By construction, L acts only on the W component and the same is true for $\mathfrak{p}_+$, since the $\mathfrak{p}_+$–action comes from differentiation along left invariant vector fields. On the other hand, the natural K–action only hits the W^*–component, and

$$H^q(G/P, G \times_P V) \cong \bigoplus_W \left(W^* \otimes (H^q(\mathfrak{p}_+, W) \otimes V)^L\right),$$

with K (and hence G) acting only via the W^*–factor. Since V is irreducible and $H^q(\mathfrak{p}_+, W)$ is completely reducible, we conclude that $(H^q(\mathfrak{p}_+, W)\otimes V)^L$ is either 0 or $\mathbb{C}$, and the second possibility occurs if and only if $H^q(\mathfrak{p}_+, W)$ contains an irreducible component isomorphic to V^*. The G_0–irreducible components (which coincide with the L–irreducible components) of $H^q(\mathfrak{p}_+, W)$ are described by Kostant's version of the Bott–Borel–Weil theorem in 3.3.5. In particular, their highest weights always lie in the affine $W^{\mathfrak{p}}$–orbit of the highest weight of W and thus in the interior of some Weyl chamber, and the result follows. □

REMARK 3.3.8. (1) Let us consider the special case of the Borel subgroup B and a dominant integral weight λ. Then $\lambda + \delta$ lies in the interior of the dominant Weyl chamber, so part (2) of the theorem applies with $w = \mathrm{id}$. This says that the space of global holomorphic sections of the homogeneous line bundle corresponding to the representation defined by $-\lambda$ is dual to the representation of highest weight λ. This is commonly referred to as the Borel–Weil theorem, and it gives a uniform construction of all finite–dimensional irreducible representations of complex semisimple Lie algebras.

(2) The general version of the Bott–Borel–Weil theorem was proved in [**Bott57**]. There are several alternative proofs, using sheaf cohomology techniques and the fact that the case of $SL(2, \mathbb{C})$ is easy to deal with; see [**Dem76**] and [**BE89**]. The Bott–Borel–Weil theorem is a fundamental ingredient in the study of Penrose transforms.

3.3.9. The Weyl character formula. To finish this section, we show how Theorem 3.3.5 can be used to derive the Weyl character formula (see 2.2.18). To do this, one only needs Theorem 3.3.5 for a Borel subalgebra $\mathfrak{b} \leq \mathfrak{g}$ and some basic observations on characters. Any irreducible representation of $\mathfrak{b}$ is one–dimensional, so the character of the irreducible $\mathfrak{b}$–representation of highest weight $\lambda : \mathfrak{h} \to \mathbb{C}$ is simply $e^\lambda \in \mathbb{Z}[\mathfrak{h}^*]$, see 2.2.18 for the notation. Consequently, Theorem 3.3.5 implies that if V is the irreducible $\mathfrak{g}$–representation with highest weight λ, then the character of the $\mathfrak{h}$–module $H^n(\mathfrak{b}_+, V)$ is given by $\sum_{w\in W:\ell(w)=n} e^{w(\lambda+\delta)-\delta}$ or, equivalently,

$$e^\delta \operatorname{char}(H^n(\mathfrak{b}_+, V)) = \sum_{w\in W:\ell(w)=n} e^{w(\lambda+\delta)}.$$

We have observed in 2.2.18 that for a finite exact sequence of modules and equivariant maps, the alternating sum of the characters vanishes. For a slight generalization, assume that $\cdots \xrightarrow{\partial_{i-1}} V_i \xrightarrow{\partial_i} V_{i+1} \to \cdots$ is a finite complex of $\mathfrak{h}$–modules with finite–dimensional weight spaces and $\mathfrak{h}$–equivariant differential and denote the cohomology modules by $H_i := \ker(\partial_i)/\operatorname{im}(\partial_{i-1})$. Then we claim that

$$\textstyle\sum_i (-1)^i \operatorname{char}(V_i) = \sum_i (-1)^i \operatorname{char}(H_i).$$

Indeed, $V_i/\ker(\partial_i) \cong \operatorname{im}(\partial_i)$ and thus $\operatorname{char}(V_i) = \operatorname{char}(\ker(\partial_i)) + \operatorname{char}(\operatorname{im}(\partial_i))$. On the other hand, by definition of H_i we get $\operatorname{char}(H_i) = \operatorname{char}(\ker(\partial_i)) - \operatorname{char}(\operatorname{im}(\partial_{i-1}))$, and the claim follows immediately by forming alternating sums. Using this we get

$$\sum_n (-1)^n \operatorname{char}(H^n(\mathfrak{b}_+, V)) = \sum_n (-1)^n \operatorname{char}(\Lambda^n \mathfrak{b}_+^* \otimes V)$$
$$= \operatorname{char}(V) \sum_n (-1)^n \operatorname{char}(\Lambda^n \mathfrak{b}_+^*) = \operatorname{char}(V) \sum_n (-1)^n \operatorname{char}(H^n(\mathfrak{b}_+, \mathbb{C})).$$

Now for $w \in W$ we have $(-1)^{\ell(w)} = \operatorname{sgn}(w)$, so we may rewrite the above equation as

$$\operatorname{char}(V) \sum_{w \in W} \operatorname{sgn}(w) e^{w(\delta)} = \sum_{w \in W} \operatorname{sgn}(w) e^{w(\lambda+\delta)},$$

which is an equivalent version of the Weyl–character formula; see [**Kn96**, V, Theorem 5.77]. This version follows from the one in 2.2.18 by using the trivial representation to compute the Weyl–denominator d.

Historical remarks and references for Chapter 3

The idea to associate normal Cartan geometries to $|k|$–graded semisimple Lie algebras goes back to the pioneering work of N. Tanaka in the 1960s and 1970s. Starting from special cases, in particular, partially integrable almost CR–structures of hypersurface type (see [**Tan62**]), he developed the general theory for the adjoint group of a $|k|$–graded simple Lie algebra $\mathfrak{g} = \mathfrak{g}_{-k} \oplus \cdots \oplus \mathfrak{g}_k$, which is the algebraic prolongation of $\mathfrak{g}_{-k} \oplus \cdots \oplus \mathfrak{g}_0$. The latter condition is equivalent to the fact that $H^1(\mathfrak{g}_-, \mathfrak{g})$ is concentrated in nonpositive homogeneous degrees. Tanaka worked in the setting of differential systems and described the underlying geometric structures as special first order G–structures as discussed in Appendix A. The most general version of Tanaka's results can be found in [**Tan79**]. It was also Tanaka's idea to use the Kostant codifferential (which he derived independently) as a normalization condition for the Cartan connection.

Unfortunately, Tanaka's papers were not well received in the mathematical community, probably due to their technical and general nature. His results were applied to geometric problems in some works of Japanese authors, for example, [**SaYa88, SaYa93, Tak94, Mi91a, Mi91b**], but otherwise went rather unnoticed. In fact, several special cases of Tanaka's results have been reproved independently by other authors, usually using a priori descriptions of the normalization condition for the Cartan connection rather than Tanaka's conceptual conditions. The most prominent example is certainly the construction of the normal Cartan connection for hypersurface type CR–structures in [**ChMo76**].

The systematic use of filtrations of the tangent bundle and the associated graded of that bundle in the description of the geometric structures underlying parabolic geometries is, to our knowledge, due to T. Morimoto. He is developing a general theory of analysis and geometry of filtered manifolds. In particular, he obtained a construction for normal Cartan connections on filtered manifolds in [**Mo93**], which contains parabolic geometries as a special case. We will sketch Morimoto's construction in Appendix A. The construction shows that the main thing needed to obtain a Cartan connection is a normalization condition with certain (purely algebraic) properties (called "condition (C)" by Morimoto). While our presentation in Section 3.1 is tailored to the parabolic case, we believe that the procedure used in 3.1.13 and 3.1.14 can be used locally to obtain an alternative proof of Morimoto's general result.

Much of the renewed interest in parabolic geometries during the last decades came from the conformally invariant nature of the Penrose transform. The Penrose transform on Minkowski space is induced by the double fibration of complex generalized flag manifolds obtained by fibering the flag manifold of lines in planes in $\mathbb{C}^4$ over $\mathbb{C}P^3$ on one side and over the Grassmannian of two–planes in $\mathbb{C}^4$ on the other side; see [**BE89**, Chapter 1] for an outline. This naturally leads to the study of complex conformal manifolds. On the other hand, in the book [**BE89**] it is shown that the Penrose transform naturally generalizes to double fibrations between complex generalized flag manifolds so looking for curved analogs, one is naturally lead to complex parabolic geometries. The characterization of complex parabolic geometries among real ones described in 3.1.18 was first observed (in a stronger form based on harmonic curvature that we will discuss in volume two) in [**Čap05**]. It should be remarked that apart from the local theory of complex parabolic geometries, which is similar to the real theory, there are also interesting global aspects, which do not have counterparts in a smooth setting. There are many results providing strong restrictions on the existence of holomorphic Cartan geometries on compact complex manifolds, respectively, of complete holomorphic geometries on more general complex manifolds; see e.g. [**McK06**].

Another important motivation for studying parabolic geometries came from Ch. Fefferman's work on the Bergman kernel, which showed that invariant theory of parabolic subgroups plays an important role in complex analysis; see [**Fe79**]. Together with the work related to Penrose transforms, this led to the desire to have a description of the relevant geometric structures in terms of canonical linear connections on certain vector bundles rather than Cartan connections on principal bundles. The first modern presentation of such a description for projective and conformal structures was [**BEG94**]. After completing their results, the authors found out that this theory had already been developed by T. Thomas in the 1920s and 30s as an alternative to Cartan's approach to these structures; see the papers [**Tho26, Tho31**] and the book [**Tho34**].

To honor the influence of Tracy Thomas and since "traction" is the next step after "tension" the vector bundles in question were named tractor bundles. These bundles and the resulting differential calculus were subsequently studied for special examples of parabolic geometries. The general version and the characterization of parabolic geometries via tractor bundles as described in 3.1.19–3.1.21 were introduced in [**ČGo02**].

Much of the material in Section 3.2 is classical and the origins are hard to trace back. The material on projective realizations of generalized flag manifolds in 3.2.6 and 3.2.8 as well as the Bruhat decomposition and Schubert cells presented in 3.2.19 and 3.2.20 is of very classical origin in special cases. It goes back to the Plücker embedding of Grassmannians, the classical work of Schubert, and Ehresmann's computation of the cohomology of Grassmannians. While in the case of Grassmannians one can show that the decomposition into Schubert cells is a CW decomposition, and hence immediately gives the integral homology, in the general case the detour using Borel–Moore homology, which goes back to [**BH61**], seems to be necessary. An outline of the necessary facts on Borel–Moore homology can be found in [**Fu97**, Appendix B]. The computation of the multiplicative structure of the integral cohomology of complex generalized flag manifolds in 3.2.21 goes back to [**Bor53**]. Further results on the algebraic topology of homogeneous spaces of compact groups based on the study of invariant differential forms can be found in [**On94**].

Most of the results on the Weyl group presented in 3.2.14 and 3.2.15 can be found either in [**Kos61**] or in [**BGG75**]. Parts of the algorithm used for determining the Hasse diagram are taken from the book [**BE89**]. Part (5) of Theorem 3.2.14 and the resulting algorithm for determining all arrows in the Hasse diagram was derived independently here, but it may well be available in the literature. This also applies to the material in 3.2.17.

The proof of Kostant's version of the Bott–Borel–Weil theorem in Section 3.3 follows Kostant's original article [**Kos61**] with a few simplifications, which are partly based on Cartier's remarks [**Cart61**] on Kostant's article. (In particular, Lemma 3.3.5 is proved in [**Cart61**].) A version for real Lie algebras was first obtained in [**Ši04**]. Surprisingly, although Tanaka explicitly points out the relation to the work of Kostant in [**Tan79**], it seems that he did not use Kostant's theorem to compute the cohomologies needed for his theory. This was done systematically by K. Yamaguchi in [**Ya93**], which contains the results of 3.3.7 and several more general classification results. It was observed by Bott that knowledge of the cohomology of the nilradical of the Borel subalgebra $\mathfrak{b} \subset \mathfrak{g}$ with coefficients in an irreducible representation of $\mathfrak{g}$ leads to a proof of the Weyl character formula as described in 3.3.9.

The realization of irreducible representations of a complex simple Lie group as global holomorphic sections of a homogeneous holomorphic line bundle on the full flag variety seemingly was a folklore result known as Borel–Weil theorem without a published proof. (See, for example, M. Atiyah's MR review of [**Bott57**].) The general version of the Bott–Borel–Weil theorem as stated in 3.3.8 was proved in [**Bott57**]; our proof follows [**Kos61**].

CHAPTER 4

A panorama of examples

In this chapter we describe a number of examples of parabolic geometries. Some of the examples are studied in detail, while for others we only sketch the basic features. On the one hand, this illustrates the large variety of (apparently unrelated) geometric structures that fall into this class. On the other hand, it illustrates that the methods developed so far quickly lead to quite detailed information on these geometries.

There are several reasons for the choice of the examples we present. Of course, we present those examples which have already been studied intensively in the literature. Second, we have chosen structures which illustrate general features of parabolic geometries in simple cases. The example of conformal structures in 4.1.2 represents links of the general theory to our elementary exposition in Section 1.6, but it is also used to illustrate the effect of different choices of Lie groups with a given $|k|$–graded semisimple Lie algebra. The relation between various real forms of the same complex $|k|$–grading can be seen from the examples of almost Grassmannian structures in 4.1.3 and of almost quaternionic structures in 4.1.8. An example of real parabolic geometries corresponding to a complex $|k|$–graded semisimple Lie algebra is provided by the elliptic partially integrable almost CR–structures of CR–dimension and codimension two in 4.3.10. The hyperbolic structures of the same type, which we discuss briefly in 4.3.9, are an example of structures associated to a non–simple $|k|$–graded Lie algebra. We also discuss the two basic examples of parabolic geometries which are not determined by the underlying infinitesimal flag structure, namely classical projective structures in 4.1.5 and contact projective structures in 4.2.6. Finally, in Section 4.3 we discuss several examples of the large class of parabolic geometries which are completely determined by the filtration of the tangent bundle. This shows that most generic types of distributions in low dimensions are equivalent to parabolic geometries.

In Sections 4.4 and 4.5, we specialize the two constructions relating Cartan geometries of different types from 1.5 to the case of parabolic geometries. The construction of correspondence spaces associates to a parabolic geometry several more complicated structures on the total spaces of certain natural bundles. This also leads to a general approach to a twistorial description of these more complicated structures. On the other hand, we discuss analogs of the Fefferman construction, which are of a more subtle nature.

4.1. Structures corresponding to $|1|$–gradings

The structures associated to $|1|$–gradings are the simplest examples of parabolic geometries, and several examples of such structures have been studied intensively. Motivated mainly by conformal geometry, a uniform geometric study of these structures was initiated independently by A. Goncharov in [**Gon87**] under the name "generalized conformal structures" and by R. Baston in [**Ba91**] under the name "almost Hermitian symmetric structures" (or AHS–structures). These structures have also been named "irreducible parabolic geometries" and "abelian parabolic geometries" in the literature. In this special case, we can obtain a very detailed description with the tools already available. Theorem 4.1.1 explicitly describes both the canonical Cartan connection and the harmonic curvature.

4.1.1. General principles; distinguished connections. Compared to general parabolic geometries, there are several obvious simplifications in the case of structures associated to $|1|$–gradings. Let $\mathfrak{g} = \mathfrak{g}_{-1} \oplus \mathfrak{g}_0 \oplus \mathfrak{g}_1$ be a $|1|$–graded semisimple Lie algebra, let G be a Lie group with Lie algebra $\mathfrak{g}$, and let $P \subset G$ be a parabolic subgroup for the given grading with Levi sugroup $G_0 \subset P$; see 3.1.3. Then parabolic geometries of type (G, P) can exist on manifolds M whose dimension coincides with the dimension of $\mathfrak{g}_-$.

Let us consider infinitesimal flag structures $(\{T^iM\}, E \to M, \theta)$ of type (G, P) as defined in 3.1.6. By definition, the filtration has the form $TM = T^{-1}M$, so this contains no information. Moreover, the regularity condition becomes vacuous, so any infinitesimal flag structure is automatically regular. Finally, the frame form θ simply is a differential form $\theta = \theta_{-1} \in \Omega^1(E, \mathfrak{g}_{-1})$ whose kernel in each point is the vertical subbundle and which satisfies $(r^g)^*\theta = \mathrm{Ad}(g^{-1}) \circ \theta$ for all $g \in P$. We have already observed in Example 3.1.6 that this implies that $(E \to M, \theta)$ is a first order G–structure on M with structure group G_0 acting via the homomorphism $G_0 \to GL(\mathfrak{g}_{-1})$ defined by the adjoint action. If no simple ideal of $\mathfrak{g}$ is contained in $\mathfrak{g}_0$ (which is the only case of interest), then this homomorphism is infinitesimally injective by part (5) of Proposition 3.1.2. Thus, we are in the setting of first order G–structures as introduced in 1.3.6.

We first need a few observations on associated bundles. Of course, $E \times_{G_0} \mathfrak{g}_{-1}$ is just the tangent bundle TM. Fixing a $\mathfrak{g}$–invariant bilinear form on $\mathfrak{g}$, we obtain an identification $\mathfrak{g}_1 \cong \mathfrak{g}_{-1}^*$ of G_0–modules (see 3.1.2) and thus an identification of $E \times_{G_0} \mathfrak{g}_1$ with the cotangent bundle T^*M. Since the adjoint action induces an injection $\mathfrak{g}_0 \to L(\mathfrak{g}_{-1}, \mathfrak{g}_{-1})$ of G_0–modules we see that we can view $E \times_{G_0} \mathfrak{g}_0$ as a subbundle $\mathrm{End}_0(TM)$ of the bundle $L(TM, TM)$. The Lie bracket on $\mathfrak{g}$ induces tensorial maps $\{\ ,\ \}$ on these bundles; see 3.1.9. By construction, the bracket $\mathrm{End}_0(TM) \times TM \to TM$ is simply given by the application of endomorphisms, while the bracket $\mathrm{End}_0(TM) \times T^*M \to T^*M$ is given by applying the negative of the dual of an endomorphism. Hence, the main interesting point is the bracket $TM \times T^*M \to \mathrm{End}_0(TM)$, whose explict form depends on the choice of $\mathfrak{g}$. Again from 3.1.9 we know that $E \times_{G_0} \mathfrak{g} = TM \oplus \mathrm{End}_0(TM) \oplus T^*M$ may be viewed as the associated graded $\mathrm{gr}(\mathcal{A}M)$ of the adjoint tractor bundle $\mathcal{A}M$. The operators ∂, $\mathrm{gr}_0(\partial^*)$, and $\square$, which were used to define the normalization condition on parabolic geometries can then be immediately expressed in terms of the bracket $\{\ ,\ \}$ using the obvious analogs of the algebraic formulae for these operators; see 3.1.11 and 3.1.12. Note that the possible homogeneities of components of a linear map $\Lambda^2\mathfrak{g}_{-1} \to \mathfrak{g}$ are 1, 2, or 3, according to the values lying in $\mathfrak{g}_{-1}$, $\mathfrak{g}_0$, and $\mathfrak{g}_1$, respectively.

LEMMA 4.1.1. *Suppose that (G, P) corresponds to a $|1|$–grading of $\mathfrak{g}$, and let $(E \to M, \theta)$ be an infintesimal flag structure of type (G, P). Suppose further that no simple ideal of $\mathfrak{g}$ is contained in $\mathfrak{g}_0$ and that $H^1(\mathfrak{g}_{-1}, \mathfrak{g})$ is contained in homogeneous degrees ≤ 1. Then we have:*

(1) *There is a principal connection* γ *on* E *such that the torsion of the induced linear connection on* TM *has values in* $\ker(\Box) \subset L(\Lambda^2 TM, TM)$.
(2) *Let* γ *be an arbitrary principal connection on* E *and let* $R \in \Gamma(\Lambda^2 T^*M \otimes \mathrm{End}_0(TM))$ *be its curvature. Then there is a unique element* $\mathsf{P} \in \Gamma(T^*M \otimes T^*M)$ *such that* $\mathrm{gr}_0(\partial^*)(R + \partial(\mathsf{P})) = 0$.

PROOF. (1) The map $\partial : L(\mathfrak{g}_{-1}, \mathfrak{g}_0) \to L(\Lambda^2\mathfrak{g}_{-1}, \mathfrak{g}_{-1})$ introduced in 1.6.1 coincides with the restriction of the Lie algebra differential from 2.1.9 and 3.1.11. Hence, if we consider $\partial^* : L(\Lambda^2\mathfrak{g}_{-1}, \mathfrak{g}_{-1}) \to L(\mathfrak{g}_{-1}, \mathfrak{g}_0)$ as introduced in 3.1.12, then from 3.3.1 we see that $L(\mathfrak{g}_{-1}, \mathfrak{g}_0) = \ker(\partial^*) \oplus \mathrm{im}(\partial)$ and that $L(\Lambda^2\mathfrak{g}_{-1}, \mathfrak{g}_{-1}) = \mathrm{im}(\partial^*) \oplus \ker(\partial)$. In particular, $\partial^* \circ \partial$ restricts to a G_0–equivariant automorphism of $\mathrm{im}(\partial^*)$ and composing the inverse of this automorphism with $\partial^*|_{\mathrm{im}(\partial)}$ we obtain a G_0–equivariant split of ∂. Part (1) of Theorem 1.6.1 now shows that there exists a principal connection γ on E whose torsion function has values in $\ker(\partial^*) \subset L(\Lambda^2\mathfrak{g}_{-1}, \mathfrak{g}_{-1})$. By homogeneity $L(\Lambda^2\mathfrak{g}_{-1}, \mathfrak{g}_{-1})$ is automatically contained in $\ker(\partial)$ so $\ker(\partial^*) = \ker(\Box) \subset L(\Lambda^2\mathfrak{g}_{-1}, \mathfrak{g}_{-1})$. Now the result follows by passing to associated bundles.

(2) In the language of associated bundles we have

$$\partial : L(TM, T^*M) \to L(\Lambda^2 TM, \mathrm{End}_0(TM))$$

and $\mathrm{gr}_0(\partial^*)$ going in the opposite direction. As before, the kernel of $\mathrm{gr}_0(\partial^*)$ is complementary to the image of ∂, so in a point $x \in M$ we can find an element $\mathsf{P}(x)$ with the required properties. This element is unique up to elements of $\ker(\partial) \subset L(T_xM, T_x^*M)$. Now $L(TM, T^*M)$ sits in homogeneity two, so by assumption there is no cohomology in this homogeneity, whence $\mathsf{P}(x)$ is unique up to elements in $\mathrm{im}(\partial)$. But for a $|1|$–grading, this image of ∂ sits in homogeneities ≤ 1, so $\mathsf{P}(x)$ is uniquely determined. This also implies that P is smooth. $\Box$

Notice that for the section P there is even an explicit formula as $\mathsf{P} = \Box^{-1}\partial^* R$, but in practice it is usually easier to determine P by solving the equation from part (2) of the lemma.

DEFINITION 4.1.1. Suppose that (G, P) corresponds to a $|1|$–grading of $\mathfrak{g}$, that no simple ideal of $\mathfrak{g}$ is contained in $\mathfrak{g}_0$, and that $H^1(\mathfrak{g}_{-1}, \mathfrak{g})$ is concentrated in homogeneous degrees ≤ 1. Let $(E \to M, \theta)$ be an infinitesimal flag structure of type (G, P). Let γ be a principal connection on E with curvature R.

(1) The unique element $\mathsf{P} \in \Omega^1(M, T^*M)$ such that $\mathrm{gr}_0(\partial^*)(R + \partial(\mathsf{P}))=0$ is called the *Rho tensor* of γ.

(2) The section $R + \partial(\mathsf{P})$ of $\ker(\partial^*) \subset \Lambda^2 T^*M \otimes \mathrm{End}_0(TM)$ is called the *Weyl–curvature* of γ.

(3) The *Cotton–York tensor* of γ is $d^\gamma \mathsf{P} \in \Omega^2(M, T^*M)$, where d^γ denotes the covariant exterior derivative.

(4) If $H^1(\mathfrak{g}_{-1}, \mathfrak{g})$ is concentrated in homogeneous degrees ≤ 0, then the principal connection γ is called a *distinguished connection* if and only if its torsion is a section of $\ker(\Box) \subset \Lambda^2 T^*M \otimes TM$.

In view of the general equivalence result in Theorem 3.1.14, the cohomological condition in part (4) exactly means that normal parabolic geometries of type (G, P) are equivalent to regular infinitesimal flag structures of type (G, P). We will mainly focus on the case that this condition is satisfied in the rest of the subsection, but the results will be useful in general.

Our aim is to describe the harmonic curvature, which we know to be a complete obstruction to local isomorphism with the homogeneous model; see 3.1.12. In fact, we will also obtain a complete description of the normal Cartan connection. Recall that if $(\mathcal{G} \to M, \omega)$ is a normal Cartan geometry of type (G, P), then the harmonic curvature has values in the associated bundle for the representation $H^2(\mathfrak{g}_{-1}, \mathfrak{g})$. Since this is a completely reducible representation, this bundle can also be obtained as an associated bundle to the underlying inifinitesimal flag structure (E, θ). Since $H^2(\mathfrak{g}_{-1}, \mathfrak{g})$ is algorithmically computable as a representation of $\mathfrak{g}_0$, we can determine this bundle in advance. It splits into a direct sum according to the splitting of $H^2(\mathfrak{g}_{-1}, \mathfrak{g})$ into irreducible components.

THEOREM 4.1.1. *Suppose that (G, P) corresponds to a $|1|$–grading of $\mathfrak{g}$, that no simple ideal of $\mathfrak{g}$ is contained in $\mathfrak{g}_0$, and that $H^1(\mathfrak{g}_{-1}, \mathfrak{g})$ is concentrated in homogeneous degrees ≤ 0. Let $(E \to M, \theta)$ be an infinitesimal flag structure of type (G, P) and let γ be a distinguished connection on (E, θ) with Rho tensor P.*

(1) There is a normal Cartan connection ω on the principal P–bundle $E \times P_+ \to M$ such that denoting by $i : E \to E \times P_+$ the inclusion we get $i^*\omega = (\theta, \gamma, \mathsf{P})$. Here we denote by P also the $\mathfrak{g}_1$–valued form representing the Rho tensor.

(2) For $j = 1, 2, 3$ let $(\kappa_H)_j$ be the component in homogeneity j of the harmonic curvature κ_H of a normal parabolic geometry with underlying infinitesimal flag structure (E, θ). Then

- (i) *$(\kappa_H)_1$ equals the torsion of γ. This coincides with the component in $\ker(\Box)$ of the torsion of an arbitrary principal connection on E. In particular, vanishing of $(\kappa_H)_1$ is equivalent to existence of a torsion–free connection on E.*
- (ii) *$(\kappa_H)_2$ is the component in $\ker(\Box) \subset L(\Lambda^2 TM, \mathrm{End}_0(TM))$ of the Weyl curvature of γ. If $(\kappa_H)_1 = 0$, this coincides with the full Weyl curvature.*
- (iii) *$(\kappa_H)_3$ is the component in $\ker(\Box) \subset L(TM, T^*M)$ of the Cotton–York tensor of γ. If $(\kappa_H)_1 = (\kappa_H)_2 = 0$, then this coincides with the full Cotton–York tensor.*

PROOF. This is mainly an interpretation of the procedure used to construct a normal parabolic geometry from an infinitesimal flag structure which we have described in Section 3.1.

As in the proof of Theorem 3.1.13 we can then view $\theta \oplus \gamma \in \Omega^1(E, \mathfrak{g}_{-1} \oplus \mathfrak{g}_0)$ as a Cartan connection on E and then obtain a Cartan connection $\tilde{\omega}$ on $\mathcal{G} = E \times P_+$ via an extension functor. Then we can use the normalization procedure from Proposition 3.1.13 to modify $\tilde{\omega}$ to a normal Cartan connection ω. Let $j : E \hookrightarrow \mathcal{G}$ be the inclusion. Then by construction $j^*\tilde{\omega} = \theta \oplus \gamma$. Since the inclusion of $\mathfrak{g}_{-1} \oplus \mathfrak{g}_0$ into $\mathfrak{g}$ is a Lie algebra homomorphims, Proposition 1.5.16 shows that pulling back the curvature $\tilde{\kappa}$ of $\tilde{\omega}$ along j, we obtain the curvature of the Cartan connection $\theta \oplus \gamma$. By 1.3.6 the latter equals the torsion and curvature of the principal connection γ. Since γ is a distinguished connection, its torsion lies in $\ker(\mathrm{gr}_0(\partial^*))$. This is a P–invariant subspace, so we see that the degree one component of $\tilde{\kappa}$ must have values in this subspace. Now the degree one component of $\partial^*(\tilde{\kappa})$ is obtained by applying $\mathrm{gr}_0(\partial^*)$ to the degree one component of $\tilde{\kappa}$, so $\partial^*(\tilde{\kappa})$ is homogeneous of degree ≥ 2. For later use we observe that this exactly means that $(E, \theta \oplus \gamma)$ is a normal P–frame bundle of degree one, and this is all we need for the rest of this proof.

This homogeneity observation means that in the procedure of Proposition 3.1.13 the next (and actually last) step is to construct an appropriate element $\Phi \in \Omega^1(M, \mathcal{A})^2 = \Omega^1(M, T^*M)$. Since this is the highest homogeneity, Φ coincides with its lowest homogeneous component, and the condition in Proposition 3.1.13 simply boils down to $\partial^*\partial\Phi$ being congruent to $\partial^*\kappa$ modulo elements of homogenity ≥ 3. By equivariancy it again suffices to satisfy this condition along $j(E) \subset \mathcal{G}$, and there $-\mathsf{P}$ by definition does the job. Putting $\omega = \tilde{\omega} - \Phi = \tilde{\omega} + \mathsf{P}$, the procedure of Proposition 3.1.13 stops, and part (1) evidently follows.

(2) Let $\kappa : \mathcal{G} \to L(\Lambda^2\mathfrak{g}_{-1}, \mathfrak{g})$ be the curvature function of ω. By normality, κ has values in $\ker(\partial^*)$ and by definition the harmonic curvature κ_H is represented by the function $\pi_H \circ \kappa : \mathcal{G} \to H^2(\mathfrak{g}_{-1}, \mathfrak{g})$, where $\pi_H : \ker(\partial^*) \to \ker(\partial^*)/\operatorname{im}(\partial^*) \cong H^2(\mathfrak{g}_{-1}, \mathfrak{g})$ is the natural map; see 3.1.12. Since P_+ acts trivially on this quotient, $\pi_H \circ \kappa$ is the P_+–invariant extension of $(\pi_H \circ \kappa)|_E$. But this means that viewing $\mathcal{G} \times_P H^2(\mathfrak{g}_{-1}, \mathfrak{g})$ as $E \times_{G_0} H^2(\mathfrak{g}_{-1}, \mathfrak{g})$, the harmonic curvature corresponds to the function $(\pi_H \circ \kappa)|_E : E \to H^2(\mathfrak{g}_{-1}, \mathfrak{g})$.

Now we can view $\kappa|_E$ as a G_0–equivariant function from E to $L(\Lambda^2\mathfrak{g}_{-1}, \mathfrak{g})$, and, as a G_0–module, the latter space splits as $\operatorname{im}(\partial^*) \oplus \ker(\Box) \oplus \operatorname{im}(\partial)$ by Proposition 3.1.11. Since $\kappa|_E$ has values in $\ker(\partial^*) = \operatorname{im}(\partial^*) \oplus \ker(\Box)$, we may replace it by the component in $\ker(\Box)$ without changing $(\pi_H \circ \kappa)|_E$. Hence, we conclude that the function $E \to H^2(\mathfrak{g}_{-1}, \mathfrak{g})$ representing the harmonic curvature κ_H can be taken to be the component in $\ker(\Box) \cong H^2(\mathfrak{g}_{-1}, \mathfrak{g})$ of the restriction $\kappa|_E$.

Now part (i) immediately follows from the fact that κ differs from $\tilde{\kappa}$ by elements of homogeneity ≥ 2 and the observations about $\tilde{\kappa}$ above. For the description of $(\kappa_H)_2$, we just have to note that by Proposition 3.1.10 the degree 2 component of $\kappa - \tilde{\kappa}$ equals $-\partial(\mathsf{P})$. To get $(\kappa_H)_3$ we observe that denoting by i the inclusion $E \hookrightarrow \mathcal{G}$, we have by part (1)

$$(i^*\omega)(\xi) = (\theta(\xi), \gamma(\xi), \mathsf{P}(\theta(\xi))),$$

where we also denote by P the function to $L(\mathfrak{g}_{-1}, \mathfrak{g}_1)$ representing the Rho tensor. For the component of $di^*\omega = i^*d\omega$ in $\mathfrak{g}_1$ we obtain by definition of the exterior derivative

$$(\xi, \eta) \mapsto \xi \cdot \mathsf{P}(\theta(\eta)) - \eta \cdot \mathsf{P}(\theta(\xi)) - \mathsf{P}(\theta([\xi, \eta])).$$

Choosing ξ and η in the kernel of γ, i.e. as horizontal lifts of vector fields $\underline{\xi}$ and $\underline{\eta}$ on M, this expression represents

$$\nabla_{\underline{\xi}}(\mathsf{P}(\underline{\eta})) - \nabla_{\underline{\eta}}(\mathsf{P}(\underline{\xi})) - \mathsf{P}([\underline{\xi}, \underline{\eta}]) = d^\gamma\mathsf{P}(\underline{\xi}, \underline{\eta}).$$

Since the bracket of two elements with trivial $\mathfrak{g}_0$–component has vanishing $\mathfrak{g}_1$–component, we see that this already represents the $\mathfrak{g}_1$–component of κ, and the description of $(\kappa_H)_3$ follows.

The last statements in (ii) and (iii) follow immediately from Theorem 3.1.12, which states that the lowest nontrivial homogeneous component of κ has to lie in $\ker(\Box)$. □

Having this description of the harmonic curvature at hand, we can now describe the procedure to analyze normal parabolic geometries corresponding to |1|–gradings (assuming that $H^2(\mathfrak{g}_{-1}, \mathfrak{g})$ is concentrated in homogeneous degrees ≤ 0):

(A) Choose a |1|–graded semisimple Lie algebra $\mathfrak{g} = \mathfrak{g}_{-1} \oplus \mathfrak{g}_0 \oplus \mathfrak{g}_1$, a Lie group G with Lie algebra $\mathfrak{g}$ and a parabolic subgroup $P \subset G$ for the given grading. Then determine the Levi subgroup $G_0 \subset P$ and its adjoint action on $\mathfrak{g}_{-1}$. Optionally, describe the homogeneous model G/P.

(B) Identify the basic irreducible representations of G_0, based on the fundamental representations of $\mathfrak{g}_0$.

(C) Via the description of $\mathrm{Ad} : G_0 \to GL(\mathfrak{g}_{-1})$ from step (A) describe first order structures with structure group G_0 geometrically. Use step (B) to describe the basic irreducible natural vector bundles on manifolds endowed with such structures.

(D) Compute $H^2(\mathfrak{g}_{-1}, \mathfrak{g})$ using Theorem 3.3.5 and determine how the irreducible components are realized within $L(\Lambda^2\mathfrak{g}_{-1}, \mathfrak{g})$. The homogeneities of the components can be determined from their highest weights using the action of the grading element. Alternatively, one may describe the components directly from the explicit representative described in Theorem 3.3.5.

(E) Use the theorem above to determine the geometric interpretation of the harmonic curvature, thus obtaining a complete set of obstructions to local isomorphism with the homogeneous model G/P.

(F) Optionally, describe the basic tractor bundles starting from fundamental representations of $\mathfrak{g}$, and their associated graded bundles.

4.1.2. Conformal pseudo–Riemannian structures. Rather than following the classification of Lie algebras, we will start the presentation of examples with conformal pseudo–Riemannian structures, which are the guiding example throughout this book. We have discussed these structures in detail in Section 1.6. At that time, we had to work out the necessary constructions and computations in an elementary way. Now we can illustrate our theoretical achievements by recovering the results from Section 1.6 using Theorem 4.1.1. At the same time, we shall illustrate how the freedom in the choice of groups and parabolic subgroups (for a fixed grading on a Lie algebra) can be used to encode minor variations of geometric structures.

Step (A): Put $\mathfrak{g} = \mathfrak{so}(p+1, q+1)$, with $p+q = m \geq 3$, and $p \geq q \geq 0$, endowed with the $|1|$–grading defined in 1.6.3, which has the form

$$\mathfrak{g} = \mathfrak{g}_{-1} \oplus \mathfrak{g}_0 \oplus \mathfrak{g}_1 = \mathbb{R}^m \oplus \mathfrak{cso}(p, q) \oplus \mathbb{R}^{m*}.$$

For odd m, say $m = 2n+1$, we saw in Example 3.2.10 that these are the only $|1|$–gradings on real forms of the complex simple Lie algebras of type B_n. For even $m = 2n$, they are the $|1|$–gradings corresponding to the first simple root on real forms of complex simple Lie algebras of type D_n. (Here, we have to identify D_3 with A_3 and the first simple root of D_3 is the second simple root of A_3).

As discussed in 1.6.3, the natural choice for a group with Lie algebra $\mathfrak{g}$ is to put $G = PO(p+1, q+1)$ and for the parabolic subgroup $P \subset G$ we take the stabilizer of an isotropic line. In the proof of Proposition 1.6.3 we have seen that the Levi subgroup $G_0 \subset P$ is formed by the classes of block diagonal matrices of the form $(\lambda, C, \lambda^{-1})$ with $\lambda \in \mathbb{R} \setminus 0$ and $C \in O(p, q)$ and we have to identify each such matrix with its negative. From this, we easily concluded that $\mathrm{Ad} : G_0 \to GL(\mathfrak{g}_{-1})$ induces an isomorphism $G_0 \cong CO(\mathfrak{g}_{-1})$. The homogeneous model G/P is the space of null lines in $\mathbb{R}^{(p+1,q+1)}$, i.e. the Möbius space as discussed in 1.6.2.

While for this choice of (G, P) nothing else has to be done in this step, we want to briefly discuss other possible choices of groups. The obvious choices for G are the full orthogonal group $O(p+1, q+1)$, the subgroup $SO(p+1, q+1)$, the connected component of the identity $SO_0(p+1, q+1)$, and the spin group $Spin(p+1,$

$q+1$). For the orthogonal group and its subgroups, the maximal parabolic subgroup corresponding to the grading is formed by the block upper triangular matrices as in part (1) of Proposition 1.6.3 which lie in the given subgroup. In general, this subgroup does not have to be connected, but there may be two connected components according to the sign of the matrix entry λ. In this case, one may choose between using the full parabolic subgroup P or its connected component P_0 of the identity. In each case, the Levi subgroup then consists of the block diagonal matrices with the allowed entries for λ and C.

Rather than discussing all the possiblities, let us restrict to a few interesting ones. Choosing $G = SO(p+1, q+1)$, the only restriction on the matrices in P is that $C \in SO(p,q)$. Taking the connected component P_0 as our parabolic subgroup, we arrive at a Levi subgroup G_0 which consists of block diagonal matrices with $\lambda > 0$ and $C \in SO(p,q)$. Thus, the adjoint action maps G_0 injectively to $GL(\mathfrak{g}_{-1})$ and the image is generated by $SO(\mathfrak{g}_{-1})$ and positive multiples of the identity, so this is exactly the group $CSO(\mathfrak{g}_{-1})$ of orientation preserving confomal linear maps. Likewise, taking $G = SO_0(p+1, q+1)$ and the connected component of the identity for the parabolic subgroup, the adjoint action is injective with image the connected component $CSO_0(\mathfrak{g}_{-1})$ of the group of conformal isometries. Finally, $Spin(p+1, q+1)$ is a two–fold covering of $SO_0(p+1, q+1)$, so if we argue as before and take the preimage of P_0 as our parabolic subgroup, then the adjoint action identifies G_0 with a two–fold covering of $CSO_0(\mathfrak{g}_{-1})$, and this is exactly the conformal spin group.

Step (B): To describe the irreducible representations of $\mathfrak{p}$ in terms of highest weights, one way is to start with representations of $\mathfrak{g}$. For complex representations we obtain an extension to the complexifications, and then by Proposition 3.2.13 the subspace of $(\mathfrak{p}_+)_{\mathbb{C}}$–invariant elements in an irreducible representation of $\mathfrak{g}_{\mathbb{C}}$ with highest weight λ is an irreducible representation of $\mathfrak{p}_{\mathbb{C}}$ with the same highest weight.

We have discussed the fundamental representations of the complex orthogonal algebras in 2.2.13. These fundamental representations can be realized as exterior powers of the standard representation except for the last fundamental representation of the odd orthogonal algebras and the last two fundamental representations of the even orthgonal algebras. The latter are the spin representation. We will assume that $m = p + q > 4$, the cases $m = 3$ and $m = 4$ need simple adaptations which are left to the reader.

A good way to start is with the adjoint representation of $\mathfrak{so}(m+2, \mathbb{C})$. This is the second fundamental representation, so its highest weight is the second fundamental weight λ_2. Now the grading of $\mathfrak{g}$ induces a grading of $\mathfrak{g}_{\mathbb{C}}$ and the elements in there which are invariant under the adjoint action of $(\mathfrak{p}_+)_{\mathbb{C}} = (\mathfrak{g}_1)_{\mathbb{C}}$ are exactly the elements of $(\mathfrak{p}_+)_{\mathbb{C}}$ themselves. Hence, we conclude that $\mathfrak{g}_1 \cong \mathbb{R}^{(p,q)*}$ is a real representation of $\mathfrak{g}_0$ whose complexification has highest weight λ_2. For $k \in \mathbb{N}$ the highest weight $k\lambda_2$ then corresponds to $S^k_0\mathbb{R}^{(p,q)*}$, the tracefree (with respect to the inner products in the conformal class) part in the kth symmetric power of $\mathfrak{g}_1$.

For the representation $\mathfrak{g}_{-1} \cong \mathbb{R}^{(p,q)}$ of $\mathfrak{g}_0$ (which we can view as the irreducible representation $\mathfrak{g}/\mathfrak{p}$ of $\mathfrak{p}$) it is evident that the highest weight of the complexification is the negative of the first simple root, and hence given by $-2\lambda_1 + \lambda_2$. Next, the line spanned by the standard inner product defines a one–dimensional $\mathfrak{g}_0$–invariant subspace in $S^2\mathfrak{g}^*_{-1}$. There is an evident isomorphism from the tensor product of this line with $\mathfrak{g}_{-1}$ to $\mathfrak{g}_{-1}^* \cong \mathfrak{g}_1$, so this one–dimensional representation has to correspond to the highest weight $2\lambda_1$. Representing G_0 by block diagonal matrices with entries λ, C, and λ^{-1}, one immediately computes that the natural action on inner products on $\mathfrak{g}_{-1}$ is given by multiplication by λ^2. The usual convention in conformal geometry is to denote this representation by $\mathbb{R}[-2]$. Hence, for $w \in \mathbb{R}$ we can define a representation $\mathbb{R}[w]$ of G_0 by acting with $|\lambda|^{-w}$, and the complexification of this has highest weight $-w\lambda_1$. (Since λ_1 corresponds to a crossed root, the weight $w\lambda_1$ is $\mathfrak{p}$–integral.) In this language, $\mathfrak{g}_1 \cong \mathfrak{g}_{-1} \otimes \mathbb{R}[-2]$, and we will often write $V[w]$ for $V \otimes \mathbb{R}[w]$.

Using this, we can identify the $\mathfrak{g}_0$–irreducible components in the standard representation $\mathbb{R}^{(p+1,q+1)}$ of $\mathfrak{g}$. On the top part (the $\mathfrak{p}$–invariant line) the action is by multiplication by λ, so this is a square root of $\mathbb{R}[-2]$. For the connected component of the parabolic, this root is isomorphic to $\mathbb{R}[-1]$, but in general it is a nontrivial root. For the middle part, the action is given by $X \mapsto CX = \lambda\lambda^{-1}CX$, so it is isomorphic to the tensor product of the top part with $\mathbb{R}^{(p,q)}$. For the connected part of the parabolic, it thus is $\mathbb{R}^{(p,q)}[-1] \cong \mathbb{R}^{(p,q)*}[1]$. The last component is evidently dual to the top one, and hence $\mathbb{R}[1]$ for the connected component of the parabolic.

Now for $k \leq \frac{m}{2} - 1$ (even m), respectively, $k \leq \frac{m-1}{2}$ (odd m) the kth fundamental representation of $\mathfrak{so}(m+2, \mathbb{C})$ is $\Lambda^k \mathbb{C}^{m+2}$. Now $\Lambda^k \mathbb{R}^{(p+1,q+1)}$ is a real representation of $\mathfrak{g}$ having this as its complexification. From our description of the $\mathfrak{g}_0$–irreducible components, it follows immediately, that the $\mathfrak{p}_+$–invariant part in there has the form

$$\Lambda^{k-1}(\mathbb{R}^{(p,q)}[-1]) \otimes \mathbb{R}[-1] \cong \Lambda^{k-1}\mathbb{R}^{(p,q)}[-k].$$

Hence, under the given restrictions on k, the latter representation has highest weight λ_k, which shows that $\Lambda^{k-1}\mathbb{R}^{(p,q)}$ has highest weight $-k\lambda_1 + \lambda_k$, while $\Lambda^{k-1}\mathbb{R}^{(p,q)*}$ has highest weight $(k-2)\lambda_1 + \lambda_k$.

Now suppose that $m = p + q$ is odd. Then for $k = \frac{m+1}{2}$, the representation $\Lambda^k \mathbb{C}^{m+2}$ of $\mathfrak{so}(m+2, \mathbb{C})$ is irreducible with highest weight twice the last fundamental weight. So as before, we conclude that $\Lambda^{k-1}\mathbb{R}^{(p,q)}[-k]$ is a real irreducible representation of $\mathfrak{p}$, which corresponds to the highest weight $2\lambda_k$. This implies that $\Lambda^{(m-1)/2}\mathbb{R}^{(p,q)}$ has highest weight $-\frac{m+1}{2}\lambda_1 + \lambda_{(m-1)/2}$, while $\Lambda^{(m-1)/2}\mathbb{R}^{(p,q)*}$ has highest weight $\frac{m-3}{2}\lambda_1 + \lambda_{(m-1)/2}$. Forming irreducible components in tensor products, we get representations of all highest weights for which the coefficient of the last fundamental weight is even. The remaining representations can be realized via tensor products with one spin representation, but we will not go into detail here.

If $p + q = m$ is even, the situation is slightly more complicated, since the last two fundamental weights correspond to spin representations. In this case, the representation $\Lambda^{m/2}\mathbb{C}^{m+2}$ is irreducible with highest weight the sum of the last two fundamental weights. Hence, $\Lambda^{\frac{m}{2}-1}\mathbb{R}^{(p,q)}[m/2]$ is a real irreducible representation corresponding to that highest weight. As we have remarked in 2.2.13, for $k = \frac{m+2}{2}$, the exterior power $\Lambda^k \mathbb{C}^{m+2}$ is not irreducible but splits as $\Lambda^k_+ \mathbb{C}^{m+2} \oplus \Lambda^k_- \mathbb{C}^{m+2}$. The components are the eigenspaces of the Hodge star operator, which is characterized by $\phi \wedge *\psi = \langle \phi, \psi \rangle \mathrm{vol}$, where the inner product comes by extension from the one on $\mathbb{C}^{m+2}$ and vol is the volume form determined by a positively oriented orthonormal basis.

The spaces $\Lambda_\pm^{\frac{m+2}{2}} \mathbb{C}^{m+2}$ are called the spaces of *self–dual*, respectively, *anti–self–dual* forms in middle degree. They are irreducible with highest weight twice the mth, respectively, the $(m+1)$st fundamental weight. In the same way, one defines real forms of these representations for $\mathfrak{so}(p+1,q+1)$ (at least for appropriate parities of p and q). The $\mathfrak{p}_+$–invariant part in these representations then is $\Lambda^{m/2}_{\pm}\mathbb{R}^{(p,q)}[-\frac{m+2}{2}]$, so these correspond to twice the mth, respectively, twice the $(m+1)$st fundamental weight. Hence, we again can realize representations as real tensor representations provided that the sum of the coefficients of the last two fundamental weights is even. A bit of care is needed with the question of integrating to groups here. In particular, the representations $\Lambda^{m/2}_{\pm}\mathbb{R}^{(p,q)}$ integrate to $SO(p,q)$ but not to $O(p,q)$, since their definition involves the volume form. The remaining representations can be realized in a tensor product of a tensor representation with one of the two basic spin representations. Whether the spinor representations exist as real or only as complex representations depends on the parities of p and q; see [**Bau81**]. We will not go into details here.

Step (C): As we have seen in step (A) above, $\mathrm{Ad} : G_0 \to GL(\mathfrak{g}_{-1})$ identifies G_0 (depending on the choice of G and P) with $CO(p,q)$, $CSO(p,q)$, $CSO_0(p,q)$, respectively, $CSpin(p,q)$. Now a first order structure with structure group $CO(p,q)$ clearly amounts to fixing an inner product of signature (p,q) on each tangent space up to positive multiples, in a way that depends smoothly on the point. This smooth dependence means that there are global smooth sections, so the structure amounts to an equivalence class of (pseudo–)Riemannian metrics. Of course, two metrics in the class are related by multiplication by a positive smooth function, so we obtain the usual definition of a conformal structure.

For the subgroups of $CO(p,q)$ we get a conformal structure plus some additional data. Namely, for $CSO(p,q)$ we have to fix an orientation, for $CSO_0(p,q)$ a space–orientation and a time–orientation, and for $CSpin(p,q)$ these two orientations plus a compatible spin structure.

Concerning the basic natural bundles, the representation $\mathfrak{g}_{-1} \cong \mathbb{R}^{(p,q)}$ of course corresponds to the tangent bundle TM. Likewise, $\mathfrak{g}_1 \cong \mathbb{R}^{(p,q)*}$ gives rise to the cotangent bundle T^*M. Let us denote by $\mathcal{E}[w]$ the natural line bundle corresponding to the representation $\mathbb{R}[w]$. This is usually called the bundle of functions of *conformal weight* w. In step (B) we have realized $\mathbb{R}[-2]$ as the line spanned by the inner product. Hence, we can view the conformal structure as an inclusion of $\mathcal{E}[-2]$ into S^2T^*M. Taking into account that $\mathcal{E}[-2]^* = \mathcal{E}[2]$, we can equivalently view this as a section of $S^2T^*M[2]$, which is called the *conformal metric.* Contracting with the conformal metric induces an isomorphism $TM \cong T^*M[2]$ as expected. The conformal metric is invertible, so its pointwise inverse defines a canonical section of $S^2TM[-2]$. Hence, in conformal geometry we can raise and lower indices at the expense of a change of the conformal weight.

The bundles $\mathcal{E}[w]$ admit an interpretation as classical density bundles. Recall that 1–densities are the objects that can be canonically integrated also on nonoriented manifolds. They can be defined as associated bundles to the full linear frame bundle. Then one defines α–densities for $\alpha \in \mathbb{R}$ by taking appropriate powers. The relation to $\mathcal{E}[w]$ comes from the fact that a Riemannian metric on a manifold M gives rise to a canonical volume density, which in local coordinates has the form $\sqrt{\det(g_{ij})}$, where g_{ij} is the metric. Evidently, the conformal groups acts on this by multiplication by $|\lambda|^m$, which shows that $\mathcal{E}[w]$ can be viewed as the bundle of $-\frac{w}{m}$–densities, and therefore makes sense without the choice of a conformal structure. Therefore, the bundles $\mathcal{E}[w]$ are often referred to as density bundles.

Steps (D) and (E): The detailed description of the curvature of the normal Cartan conformal connection was computed explicitly in 1.6.8, while the computation of the corresponding cohomologies was hidden in the elementary approach to the normalizations during the classical prolongation procedure for the chosen G_0-structure in 1.6.4 and 1.6.7. In fact, we proceeded straight with our step (E) there, and we made all the necessary normalizations and computations on the way. Let us now compare how this relates to our current general procedure. As we saw in Theorem 3.3.5, the first and second cohomologies correspond to elements w of length one and two in the subset $W^{\mathfrak{p}}$ in the Weyl group W while the highest weight of the corresponding G_0-representation is the image of the highest weight of the adjoint representation under the affine action by w. Thus, the first cohomologies are given by the first reflection s_1, and the affine action on the highest root of the algebra itself provides

$$\begin{array}{c}\overset{0}{\times}\!\!-\!\!\overset{1}{\circ}\!\cdots\!\overset{0}{\circ}\!\!\Rightarrow\!\!\overset{0}{\circ}\end{array} \;\mapsto\; \begin{array}{c}\overset{-2}{\times}\!\!-\!\!\overset{2}{\circ}\!\cdots\!\overset{0}{\circ}\!\!\Rightarrow\!\!\overset{0}{\circ}\end{array}$$

$$\begin{array}{c}\overset{0}{\times}\!\!-\!\!\overset{1}{\circ}\!\cdots\!\overset{0}{\circ}\!\!<\!\!\overset{0}{\underset{0}{}}\end{array} \;\mapsto\; \begin{array}{c}\overset{-2}{\times}\!\!-\!\!\overset{2}{\circ}\!\cdots\!\overset{0}{\circ}\!\!<\!\!\overset{0}{\underset{0}{}}\end{array}$$

and their homogeneities are zero. The diagrams and coefficients in dimensions three and four are different, but the result is again of homogeneity zero. (Of course, we know all this from the general result in Proposition 3.3.7.) This computation reveals that the prolongation computed in 1.6 has to lead to a canonical Cartan connection; cf. the uniqueness and existence results discussed in 3.1.14. We have seen above, how the choice of Lie groups G and P with the given algebras influences the geometry in questions, but the construction, existence, and uniqueness of the normal Cartan connection is unaffected.

The curvature is described by the second cohomologies and, except for dimension four, the only element of length two in the subset $W^{\mathfrak{p}}$ is the composition $s_1 \circ s_2$ of the first two reflections; cf. the recipes in 3.2.18. Thus, we easily compute the highest weights with coefficients $(-4, 0, 2)$ at the first three fundamental weights for both odd and even dimensions greater than 6 while $(-4, 0, 4)$ are the coeffients in the B_3 case. In step (B) we have seen that $\Lambda^2 \mathbb{R}^{(p,q)}$ corresponds to the highest weight $(-3, 0, 1)$, respectively, $(-3, 0, 2)$. The adjoint representation $\mathfrak{g}_0$ can be obtained from that by converting one of the two factors $\mathbb{R}^{(p,q)}$ into its dual, so the highest weight of $\mathfrak{g}_0$ is $(-1, 0, 1)$, respectively, $(-1, 0, 2)$. Hence, we see that the weights correspond to the highest weight component in $\Lambda^2 \mathbb{R}^{(p,q)} \otimes \mathfrak{so}(p, q)$, so they are of homogeneity -2. These weights correspond to the duals of the bundles where the curvatures of the normal Cartan connection live. Thus, the harmonic curvature has one irreducible component in all dimensions bigger than 4 and its homogeneity is two. Therefore, the distinguished connections constructed by the general procedure in Lemma 4.1.1 and Theorem 4.1.1 and explicitly described in 1.6.4 have vanishing torsion and the invariant part of their curvature is the Weyl curvature, the tracefree part as computed in 1.6.8. We have also seen there that vanishing of the Weyl curvature is equivalent to the local flatness for all dimensions at least four, and we computed the normal Cartan connections out of any of the distinguished connections by solving the equation $\partial^*(R + \partial(P)) = 0$ from Lemma 4.1.1; see 1.6.7.

In dimension four the harmonic curvature is also concentrated in homogeneity two, but split into two irreducible components. This corresponds to the fact that there are two different elements in $W^{\mathfrak{p}}$ of length two, namely $s_1 \circ s_2$ and $s_1 \circ s_3$. In fact, from the algebraic point of view, this geometry belongs to the A_3 algebra and we shall come back to this later in 4.1.4, 4.1.9, and 4.1.10.

In dimension three, there is again a single component of the curvature corresponding to the reflection $s_1 \circ s_2$, but the coefficients of the highest weight computed by Kostant's theorem (the dual to the curvature component) expressed via the fundamental representations are $(-5, 4)$. This yields homogeneity -3 and so there is no Weyl curvature in this dimension and the harmonic curvature coincides with the Cotton–York tensor.

Step (F): The standard tractor bundle $\mathcal{T}$ is defined by the standard representation of the group G on $\mathbb{R}^{p+q+2}$. Thus, the P–invariant standard scalar product of signature $(p+1, q+1)$ on $\mathbb{R}^{p+q+2}$ yields a natural metric of the same signature on $\mathcal{T}$. Furthermore, the bundle comes equipped by the natural filtration coming from the P–invariant filtration $\mathbb{R}^{p+q+2} = V \supset V^0 \supset V^1$, where V^1 is the line stabilized by P and V^0 is its orthocomplement. The associated graded vector bundle $\operatorname{gr}\mathcal{T} = \mathcal{T}/\mathcal{T}^0 \oplus \mathcal{T}^0/\mathcal{T}^1 \oplus \mathcal{T}^1$ has components of homogeneities -1, 0, and 1, respectively (as computed easily by actions of the grading element $E \in \mathfrak{g}$ on the corresponding parts of the standard representation). From the description of the constituents in step (A), we see that this has the form

$$\operatorname{gr}\mathcal{T} = \mathcal{E}[1] \oplus TM[-1] \oplus \mathcal{E}[-1].$$

4.1.3. Almost Grassmannian structures. After this introductory example, we switch to a systematic presentation of geometries corresponding to $|1|$–gradings. We have determined the complete list of complex simple $|1|$–graded Lie algebras in Subsection 3.2.3, and from 3.2.10 we know which of these gradings exist for which real form. We start our discussion with the A_n case. From 3.2.3 we know that the choice of any of the simple roots leads to a $|1|$–grading in this case. Choosing the first or last simple root, we know from Proposition 3.3.7 that one gets a nontrivial component of $H^1(\mathfrak{g}_{-1}, \mathfrak{g})$ in homogeneity one. This exceptional case will be discussed in 4.1.5 below, and we start with the cases of one of the other simple roots. From Theorem 2.3.11 we know that in the A_n case we have the real forms $\mathfrak{sl}(n,\mathbb{R})$ of $\mathfrak{sl}(n,\mathbb{C})$, $\mathfrak{sl}(n,\mathbb{H})$ of $\mathfrak{sl}(2n,\mathbb{C})$ and $\mathfrak{su}(p,q)$ of $\mathfrak{sl}(p+q,\mathbb{C})$. We first discuss the split real form $\mathfrak{sl}(n,\mathbb{R})$.

Step (A): Fix numbers $p, q \geq 2$ and consider $\mathfrak{g} := \mathfrak{sl}(p+q,\mathbb{R})$. For each such choice, we obtain a $|1|$–grading on $\mathfrak{g}$ as described in example 3.1.2 (2). Consider $V = \mathbb{R}^{p+q}$ as $V_1 \oplus V_0$, where V_1 is spanned by the first p vectors and V_0 is spanned by the last q vectors in the standard basis of V. For $i = -1, 0, 1$ define $\mathfrak{g}_i \subset \mathfrak{g}$ to consist of those maps which map V_j to V_{j+i} for $j = 0, 1$, where we agree that $V_\ell = \{0\}$ for $\ell \neq 0, 1$. In matrix form, this means that we view elements of $\mathfrak{g}$ as block matrices $\left(\begin{smallmatrix} A & Z \\ X & B \end{smallmatrix}\right)$ with block sizes p and q, respectively. Then $\mathfrak{g}_{-1}$ consists of those matrices for which only the X–block is nonzero, $\mathfrak{g}_1$ of those for which only the Z–block is nonzero, while $\mathfrak{g}_0$ consists of the block diagonal matrices. Hence, we see that $\mathfrak{g}_{-1} = L(\mathbb{R}^p, \mathbb{R}^q)$, $\mathfrak{g}_0 = \mathfrak{s}(\mathfrak{gl}(p,\mathbb{R}) \oplus \mathfrak{gl}(q,\mathbb{R}))$ and $\mathfrak{g}_1 = L(\mathbb{R}^q, \mathbb{R}^p)$. Of course, the parabolic subalgebra $\mathfrak{p} = \mathfrak{g}_0 \oplus \mathfrak{g}_1$ is the stabilizer of $\mathbb{R}^p$ in $\mathfrak{g}$. This immediately implies that $\mathfrak{p}$ is the standard parabolic corresponding to the pth simple root. Since the Satake diagram of the split form $\mathfrak{g}$ coincides with the Dynkin diagram of $\mathfrak{sl}(p+q,\mathbb{C})$, the diagram describing $\mathfrak{p}$ is A_{p+q-1} with the pth node crossed. Hence, there are $p-1$ white nodes left of the crossed node and $q-1$ white nodes right of the crossed node.

As a Lie group G with Lie algebra $\mathfrak{g}$ we take $SL(p+q,\mathbb{R})$. Since the parabolic subalgebra $\mathfrak{p} \subset \mathfrak{g}$ is the stabilizer of $\mathbb{R}^p$ in $\mathbb{R}^{p+q}$ the stabilizer P of this subspace in G is a parabolic subgroup for the given grading. It is easy to verify directly that P is the normalizer of $\mathfrak{p}$ in G, so it is the maximal parabolic subgroup for this grading. In terms of matrices, P is the subgroup of block upper triangular matrices with block sizes p and q. The resulting Levi subgroup $G_0 \subset P$ is then the group of block diagonal matrices with these block sizes. Hence, $G_0 = S(GL(p,\mathbb{R}) \times GL(q,\mathbb{R})) \subset SL(p+q,\mathbb{R})$. From the definition of P we conclude that the homogeneous model G/P of our geometry is the Grassmannian $\mathrm{Gr}_p(\mathbb{R}^{p+q})$ of p–dimensional subspaces of $\mathbb{R}^{p+q}$.

Step (B): Any complex irreducible representation of $SL(p+q,\mathbb{R})$ is the complexification of a real irreducible representation. As we shall see immediately this implies that the same is true for G_0, so we will only describe the real irreducible representations here. Of course, the two basic real representations of G_0 are provided by $\mathbb{R}^p$ and $\mathbb{R}^q$ sitting inside $\mathbb{R}^{p+q}$ which correspond to the standard representations of the two simple factors. The fact that we deal with $S(GL(p,\mathbb{R}) \times GL(q,\mathbb{R}))$ rather than with the full product can be expressed as triviality of the representation $\Lambda^p\mathbb{R}^p \otimes \Lambda^q\mathbb{R}^q$ or equivalently as having a fixed G_0–invariant isomorphism $\Lambda^p\mathbb{R}^p \cong \Lambda^q\mathbb{R}^{q*}$. To describe the basic representations of G_0 in terms of highest weights, let $\lambda_1,\dots,\lambda_{p+q-1}$ be the fundamental weights of $\mathfrak{g}$. From 3.2.13 we know how to describe the irreducible representation of $\mathfrak{p}$ (and thus of $\mathfrak{g}_0$) with highest weight λ for a $\mathfrak{g}$–dominant weight λ. This is given as the subspace in the $\mathfrak{g}$–irreducible representation V with highest weight λ on which $\mathfrak{g}_1$ acts trivially.

For each i, the irreducible representations of $\mathfrak{g}$ with highest weight λ_i is the ith exterior power $\Lambda^i\mathbb{R}^{p+q}$ of the standard representation. In particular, for λ_1, we have the standard representation, and the subspace on which $\mathfrak{g}_1$ acts trivially is of course $\mathbb{R}^p$, so this is the irreducible representation of G_0 with highest weight λ_1.

The decomposition $\mathbb{R}^{p+q} = \mathbb{R}^p \oplus \mathbb{R}^q$ leads to a decomposition $\Lambda^k\mathbb{R}^{p+q} = \bigoplus_{i+j=k} \Lambda^i\mathbb{R}^p \otimes \Lambda^j\mathbb{R}^q$. The subspace on which $\mathfrak{g}_1$ acts trivially is evidently given by the part in which i is as large as possible. In particular, for $k = 2,\dots,p-1$, the irreducible representation of G_0 with highest weight λ_k is simply $\Lambda^k\mathbb{R}^p$. For $k = p$, we obtain $\Lambda^p\mathbb{R}^p \cong \Lambda^q\mathbb{R}^{q*}$, which is one–dimensional in accordance with the fact that this is the fundamental weight corresponding to a crossed root, so only the center of $\mathfrak{g}_0$ acts nontrivially. Consequently, for $\ell \in \mathbb{Z}$ we can construct the irreducible G_0–representation with highest weight $\ell\lambda_p$ as the ℓth tensor power of $\Lambda^p\mathbb{R}^p$ for $\ell > 0$ and as the $-\ell$th tensor power of $\Lambda^p\mathbb{R}^{p*}$ for $\ell < 0$. For $k > p$, the $\mathfrak{g}_1$–invariant part in $\Lambda^k\mathbb{R}^{p+q}$ is $\Lambda^p\mathbb{R}^p \otimes \Lambda^{k-p}\mathbb{R}^q$. Hence, we conclude that the basic representation $\mathbb{R}^q$ has highest weight $\lambda_{p+1} - \lambda_p$ and more generally $\Lambda^k\mathbb{R}^q$ has highest weight $\lambda_{p+k} - \lambda_p$ for $k < q$.

Concerning the duals of these basic representations and their exterior powers, we observe that the wedge product gives a G_0–invariant nondegenerate map $\Lambda^k\mathbb{R}^p \otimes \Lambda^{p-k}\mathbb{R}^p \to \Lambda^p\mathbb{R}^p$. This immediately leads to an isomorphism $\Lambda^k\mathbb{R}^{p*} \cong \Lambda^{p-k}\mathbb{R}^p \otimes (\Lambda^p\mathbb{R}^p)^*$. Hence, the representation $\Lambda^k\mathbb{R}^{p*}$ is the irreducible G_0–representation with highest weight $\lambda_{p-k} - \lambda_p$ for all $k < p$. Similarly, for $1 \leq k < q$ we get that $\Lambda^k\mathbb{R}^{q*}$ is irreducible with highest weight λ_{p+q-k}.

Let us next look at the parts of the adjoint representation. For $\mathfrak{g}_{-1}$ one may either use the isomorphism $\mathfrak{g}_{-1} \cong L(\mathbb{R}^p, \mathbb{R}^q) = \mathbb{R}^{p*} \otimes \mathbb{R}^q$ or the fact that the highest weight in $\mathfrak{g}_{-1}$ is obviously the negative of the pth simple root to conclude that this is the irreducible G_0–representation of highest weight $\lambda_{p-1} - 2\lambda_p + \lambda_{p+1}$. Similarly, either by viewing $\mathfrak{g}_1$ as the tensor product $\mathbb{R}^{q*} \otimes \mathbb{R}^p$ or by observing that it is the $\mathfrak{g}_1$–invariant part in the adjoint representation of $\mathfrak{g}$, we see that $\mathfrak{g}_1$ is irreducible with highest weight $\lambda_1 + \lambda_{p+q-1}$. Finally, let us consider the adjoint representation of G_0 on $\mathfrak{g}_0$. The G_0–action on the one–dimensional center of $\mathfrak{g}_0$ is trivial, while the semisimple part splits as $\mathfrak{sl}(p, \mathbb{R}) \oplus \mathfrak{sl}(q, \mathbb{R})$. This may be identified with the highest weight components in $\mathbb{R}^{p*} \otimes \mathbb{R}^p$, respectively, $\mathbb{R}^{q^*} \otimes \mathbb{R}^q$, or one may observe that the highest weights are the sums of the first $p-1$, respectively, the last $q-1$ simple roots. Either of these two approaches shows that these two components are irreducible with highest weights $\lambda_1 + \lambda_{p-1} - \lambda_p$ for $p > 2$ and $2\lambda_1 - \lambda_2$ for $p = 2$, respectively, $-\lambda_p + \lambda_{p+1} + \lambda_{p+q-1}$ for $q > 2$ and $-\lambda_p + 2\lambda_{p+1}$ for $q = 2$.

Step (C): From the description above, we see that

$$G_0 = \{(C_1, C_2) \in GL(p, \mathbb{R}) \times GL(q, \mathbb{R}) : \det(C_1)\det(C_2) = 1\}.$$

Viewing elements $X \in \mathfrak{g}_{-1}$ as linear maps $\mathbb{R}^p \to \mathbb{R}^q$, the adjoint action is immediately seen to be given by $\mathrm{Ad}(C_1, C_2)(X) = C_2 X C_1^{-1}$. A moment of thought shows that $\mathrm{Ad} : G_0 \to GL(\mathfrak{g}_{-1})$ is injective if $p+q$ is odd, while for $p+q$ even its kernel is given by $\{(\mathrm{id}, \mathrm{id}), (-\mathrm{id}, -\mathrm{id})\}$.

A first order structure with structure group G_0 can be defined on manifolds whose dimension equals $\dim(\mathfrak{g}_{-1}) = pq$. Given such a structure $(p_0 : \mathcal{G}_0 \to M, \theta)$, we can of course form the associated bundles corresponding to the basic representations $\mathbb{R}^p$ and $\mathbb{R}^q$ of G_0. Hence, we obtain a rank p vector bundle $E \to M$ and a rank q vector bundle $F \to M$. Fixing a G_0–invariant isomorphism $\Lambda^p\mathbb{R}^p \otimes \Lambda^q\mathbb{R}^q \to \mathbb{R}$, we obtain a preferred trivialization of $\Lambda^p E \otimes \Lambda^q F$. The fact that the tangent bundle TM can be identified (using θ) with $\mathcal{G}_0 \times_{G_0} \mathfrak{g}_{-1}$ implies that we get an isomorphism $\Phi : E^* \otimes F \to TM$.

Let us conversely assume that on a manifold M of dimension pq we have given vector bundles E and F of rank p, respectively, q, a trivialization $\phi : \Lambda^p E \otimes \Lambda^q F \to M \times \mathbb{R}$ and an isomorphism $\Phi : E^* \otimes F \to TM$. Then we consider the fibered product $GL(\mathbb{R}^p, E) \times_M GL(\mathbb{R}^q, F)$ of the linear frame bundles of E and F. The fiber of this bundle over $x \in M$ consists of pairs (ψ_1, ψ_2) of linear isomorphisms $\psi_1 : \mathbb{R}^p \to E_x$ and $\psi_2 : \mathbb{R}^q \to F_x$. We define a subspace $\mathcal{G}_0$ in this bundle as the set of those pairs for which the second component of $\phi \circ (\Lambda^p \psi_1 \otimes \Lambda^q \psi_2)$, which is a linear isomorphism $\Lambda^p\mathbb{R}^p \otimes \Lambda^q\mathbb{R}^q \to \mathbb{R}$ coincides with the fixed G_0–invariant isomorphism. The group G_0 acts on $\mathcal{G}_0$ by composition from the right, and visibly the action is free and transitive on the fibers of the natural projection $p_0 : \mathcal{G}_0 \to M$. Hence, this becomes a smooth principal G_0–bundle. Next, we define $\theta \in \Omega_1(\mathcal{G}_0, \mathfrak{g}_{-1})$ as follows: For a tangent vector $\xi \in T_{(\psi_1,\psi_2)}\mathcal{G}_0$ we have $Tp_0 \cdot \xi \in T_x M$, and thus $\Phi^{-1}(Tp_0 \cdot \xi) \in L(E_x, F_x)$. Thus, we may define

$$\theta(\xi) := \psi_2^{-1} \circ \Phi^{-1}(Tp_0 \cdot \xi) \circ \psi_1 \in L(\mathbb{R}^p, \mathbb{R}^q) = \mathfrak{g}_{-1}.$$

This is evidently smooth, its kernel in each point is the vertical subbundle, and by construction it is equivariant in the sense that $(r^g)^*\theta = \mathrm{Ad}(g^{-1}) \circ \theta$ for each $g \in G_0$. Hence, $(p_0 : \mathcal{G}_0 \to M, \theta)$ is a first order G_0–structure, and we conclude that such a structure is equivalent to the choice of E, F, ϕ and Φ. Such a structure is usually referred to as an *almost Grassmannian structure* of type (p,q) in view of the homogeneous model being the Grassmannian $\mathrm{Gr}_p(\mathbb{R}^{p+q})$.

Of course, for the homogeneous model, the auxiliary bundles E and F are simply the two tautological bundles over the Grassmannian. The bundle E is the subbundle in $\mathrm{Gr}_p(\mathbb{R}^{p+q})\times\mathbb{R}^{p+q}$ whose fiber over a p–dimensional subspace is given by that subspace, while F is simply the quotient of the trivial $\mathbb{R}^{p+q}$–bundle by the subbundle E. It is a well–known fact that the tangent bundle of the Grassmannian can be canonically identified with the bundle $L(E,F)$ of linear maps.

Step (D): According to Theorem 3.3.5, the irreducible components of $H^2(\mathfrak{g}_1,\mathfrak{g})$ are in bijective correspondence with elements of length two in the Hasse diagram $W^{\mathfrak{p}}$ of the parabolic $\mathfrak{p}$ (or more precisely the complexification $\mathfrak{p}^{\mathbb{C}}\subset\mathfrak{sl}(p+q,\mathbb{C})$). To determine these elements we may use either of the two recipes from 3.2.18, but the recipe for $|1|$–gradings is simpler. We have to look for two–element subsets in $\Delta^+(\mathfrak{g}_1)$ whose complement in Δ^+ is saturated, and we have determined all such subsets in Example 3.2.17. Since our grading corresponds to the simple root α_p, this simple root has to be contained in any of the subsets. In the picture of matrices, for each root in the subset, any root in $\Delta^+(\mathfrak{g}_1)$ whose root space lies below or left of the given one, has to be in the subset, too. Hence, the only possible two element subsets are $\{\alpha_p,\alpha_{p-1}+\alpha_p\}$ and $\{\alpha_p,\alpha_p+\alpha_{p+1}\}$. These correspond to the Weyl group elements $w_1:=s_{\alpha_{p-1}+\alpha_p}\circ s_{\alpha_p}$ and $w_2:=s_{\alpha_p+\alpha_{p+1}}\circ s_{\alpha_p}$, and the sets are recovered as the sets Φ_{w_i} associated to the elements w_i as in 3.2.14.

From the explicit representatives provided by Theorem 3.3.5 we see that the highest weight of the irreducible component corresponding to $w\in W^{\mathfrak{p}}$ is given by the negative of the sum of the roots contained in Φ_w plus the image of the highest weight of the adjoint representation under w. From the sets Φ_{w_i} we therefore get the contributions $-\alpha_{p-1}-2\alpha_p$, respectively, $-2\alpha_p-\alpha_{p+1}$ to the highest weight, and these are easy to analyze: We have noticed above the $-\alpha_p$ is the sum of the highest weights of $\mathbb{R}^{p*}$ and $\mathbb{R}^q$. Similarly, $-\alpha_{p-1}-\alpha_p$ is the sum of the second highest weight of $\mathbb{R}^{p*}$ with the highest weight of $\mathbb{R}^q$, while $-\alpha_p-\alpha_{p+1}$ is the sum of the highest weight of $\mathbb{R}^{p*}$ with the second highest weight of $\mathbb{R}^q$. Thus, we conclude that $-\alpha_{p-1}-2\alpha_p$ is the sum of the highest and the second highest weight of $\mathbb{R}^{p*}$ with twice the highest weight of $\mathbb{R}^q$, so it is the highest weight of $\Lambda^2\mathbb{R}^{p*}\otimes S^2\mathbb{R}^q$. Likewise, $-2\alpha_p-\alpha_{p+1}$ is the highest weight of $S^2\mathbb{R}^{p*}\otimes\Lambda^2\mathbb{R}^q$. Notice that $\mathfrak{g}_{-1}\cong\mathbb{R}^{p*}\otimes\mathbb{R}^q$ implies that the second exterior power $\Lambda^2\mathfrak{g}_{-1}$ decomposes as $(\Lambda^2\mathbb{R}^{p*}\otimes S^2\mathbb{R}^q)\oplus(S^2\mathbb{R}^{p*}\otimes\Lambda^2\mathbb{R}^q)$, so the two representations naturally sit inside this exterior power.

On the other hand, the highest weight λ of the adjoint representation of $\mathfrak{sl}(p+q,\mathbb{C})$ is $\lambda_1+\lambda_{p+q-1}=\alpha_1+\dots+\alpha_{p+q-1}$. This immediately implies that for $p>2$ we have $\langle\lambda,\alpha_i\rangle=0$ for $i=p-1,p$ and thus $w_1(\lambda)=\lambda$. Similarly, for $q>2$ we get $w_2(\lambda)=\lambda$, and λ is the highest weight of $\mathfrak{g}_1=\mathbb{R}^p\otimes\mathbb{R}^{q*}$. In particular, we conclude that for $p>2$, the irreducible component corresponding to w_1 is the highest weight component in $(\Lambda^2\mathbb{R}^{p*}\otimes S^2\mathbb{R}^q)\otimes\mathfrak{g}_1$. (Recall that once we have found an irreducible component in $\Lambda^*\mathfrak{g}_{-1}\otimes\mathfrak{g}_1$ with the right highest weight, then it must be the cohomology component by the multiplicity one result in Theorem 3.3.5.) From 3.3.1 we know that $H^2(\mathfrak{g}_{-1},\mathfrak{g})$ is the dual representation to $H^2(\mathfrak{g}_1,\mathfrak{g}^*)\cong H^2(\mathfrak{g}_1,\mathfrak{g})$. Hence, the irreducible component in $H^2(\mathfrak{g}_{-1},\mathfrak{g})$ corresponding to w_1 (still for $p>2$) is the highest weight component in

$$(\Lambda^2\mathbb{R}^p\otimes S^2\mathbb{R}^{q*})\otimes\mathfrak{g}_{-1}\subset\Lambda^2(\mathfrak{g}_{-1})^*\otimes\mathfrak{g}_{-1}.$$

This highest weight component can be easily described explicitly as follows: According to $\mathfrak{g}_{-1} = \mathbb{R}^{p*} \otimes \mathbb{R}^q$, we get

$$(\Lambda^2\mathbb{R}^p \otimes S^2\mathbb{R}^{q*}) \otimes \mathfrak{g}_{-1} \cong (\Lambda^2\mathbb{R}^p \otimes \mathbb{R}^{p*}) \otimes (S^2\mathbb{R}^{q*} \otimes \mathbb{R}^q).$$

Visibly, the first component admits a unique contraction with values in $\mathbb{R}^p$, while for the second component there is a unique contraction with values in $\mathbb{R}^{q*}$. Hence, from the tensor product we obtain two contractions, one with values in $\mathbb{R}^p \otimes (S^2\mathbb{R}^{q*} \otimes \mathbb{R}^q)$ and the other one with values in $(\Lambda^2\mathbb{R}^p \otimes \mathbb{R}^{p*}) \otimes \mathbb{R}^{q*}$, and the highest weight component is the intersection of the kernels of these two contractions. Similarly, for $q > 2$ the irreducible component corresponding to w_2 is the highest weight component in $(S^2\mathbb{R}^p \otimes \Lambda^2\mathbb{R}^{q*}) \otimes \mathfrak{g}_{-1}$, and the description of this highest weight component is completely parallel to the other one.

Let us next consider the case $p = 2$. Then $w_1 = s_{\alpha_1+\alpha_2} \circ s_{\alpha_2}$, and for the highest weight $\lambda = \lambda_1 + \lambda_{q+1}$ of the adjoint representation we get $s_{\alpha_2}(\lambda) = \lambda$ and $w_1(\lambda) = \lambda - \alpha_1 - \alpha_2 = \alpha_3 + \dots + \alpha_{q+1}$. This evidently is the highest weight of the simple component $\mathfrak{sl}(q,\mathbb{R}) \subset \mathfrak{g}_0$. Since this is selfdual we conclude that the irreducible component in $H^2(\mathfrak{g}_{-1},\mathfrak{g})$ corresponding to w_1 in the case $p = 2$ is the highest weight component in

$$(\Lambda^2\mathbb{R}^2 \otimes S^2\mathbb{R}^{q*}) \otimes \mathfrak{sl}(q,\mathbb{R}) \subset \Lambda^2(\mathfrak{g}_{-1})^* \otimes \mathfrak{g}_0.$$

As before, this highest weight component can be easily described explicitly. The space $\mathfrak{sl}(q,\mathbb{R})$ sits as the tracefree part in $\mathbb{R}^{q*} \otimes \mathbb{R}^q$. The highest weight component then must be contained in

$$\Lambda^2\mathbb{R}^2 \otimes S^3\mathbb{R}^{q*} \otimes \mathbb{R}^q \subset \Lambda^2\mathbb{R}^2 \otimes S^2\mathbb{R}^{q*} \otimes \mathbb{R}^{q*} \otimes \mathbb{R}^q.$$

On that space there is a unique contraction with values in $\Lambda^2\mathbb{R}^2 \otimes S^2\mathbb{R}^{q*}$, and the highest weight component is exactly the kernel of this contraction.

For $q = 2$ we obtain the parallel description of the irreducible component in $H^2(\mathfrak{g}_{-1},\mathfrak{g})$ corresponding to w_2 as the highest weight component in

$$(S^2\mathbb{R}^p \otimes \Lambda^2\mathbb{R}^{2*}) \otimes \mathfrak{sl}(p,\mathbb{R}).$$

Step (E): Using the correspondence between representations and bundles established in Step (C), we can directly translate the information on $H^2(\mathfrak{g}_{-1},\mathfrak{g})$ obtained in Step (D) above into a description of the harmonic curvature components. This leads to the following table:

$p = 2,\ q = 2$	two curvatures	$\rho_1 \in \Gamma(\Lambda^2 E \otimes S^2 F^* \otimes \mathfrak{sl}(F))$ $\rho_2 \in \Gamma(S^2 E \otimes \mathfrak{sl}(E) \otimes \Lambda^2 F^*)$
$p = 2,\ q > 2$	one torsion, one curvature	$\tau_2 \in \Gamma(S^2 E \otimes E^* \otimes \Lambda^2 F^* \otimes F)$ $\rho_1 \in \Gamma(\Lambda^2 E \otimes S^2 F^* \otimes \mathfrak{sl}(F))$
$p > 2,\ q = 2$	one torsion, one curvature	$\tau_1 \in \Gamma(\Lambda^2 E \otimes E^* \otimes S^2 F^* \otimes F)$ $\rho_2 \in \Gamma(S^2 E \otimes \mathfrak{sl}(E) \otimes \Lambda^2 F^*)$
$p > 2,\ q > 2$	two torsions	$\tau_1 \in \Gamma(\Lambda^2 E \otimes E^* \otimes S^2 F^* \otimes F)$ $\tau_2 \in \Gamma(S^2 E \otimes E^* \otimes \Lambda^2 F^* \otimes F)$

To interpret the harmonic curvature geometrically using Theorem 4.1.1, we have to clarify the meaning of a principal connection on a G_0–structure $(p_0 : \mathcal{G}_0 \to M, \theta)$. As we have seen in Step (C), we may interpret the bundle $\mathcal{G}_0$ as $S(GL(\mathbb{R}^p, E) \times_M GL(\mathbb{R}^q, F))$. From this description it follows immediately that a principal connection γ on this bundle is equivalent to a pair (∇^E, ∇^F) of linear connections on the bundles E and F, such that the constant global sections of the bundle $\Lambda^p E \otimes \Lambda^q F$ determined by the trivialization ϕ are parallel for the induced connection.

Having chosen such a pair of connections, the induced connection ∇ on the tangent bundle comes from the isomorphism $\Phi : E^* \otimes F \to TM$. More explicitly, this means that for a vector field ξ, and a section $X \in \Gamma(L(E,F))$ we have $\nabla_\xi \Phi(X) = \Phi(\tilde{X})$, where for $e \in \Gamma(E)$ we have $\tilde{X}(e) = \nabla^F_\xi(X(e)) - X(\nabla^E_\xi e)$. The torsion T of the connection ∇ on TM is by definition a section of $\Lambda^2 T^*M \otimes TM$, and via Φ we may identify this bundle with $\Lambda^2(E \otimes F^*) \otimes (E^* \otimes F)$. Now viewing $(E \otimes F^*) \otimes (E \otimes F^*)$ as $(E \otimes E) \otimes (F^* \otimes F^*)$, we can split $E \otimes E = S^2 E \oplus \Lambda^2 E$ and likewise for $F^* \otimes F^*$. A tensor product will be skew symmetric if one of the components is skew and the other one is symmetric, so

$$\Lambda^2(E \otimes F^*) = (\Lambda^2 E \otimes S^2 F^*) \oplus (S^2 E \otimes \Lambda^2 F^*).$$

Thus, we may decompose the torsion T into two components, which have values in $(\Lambda^2 E \otimes E^*) \otimes (S^2 F^* \otimes F)$, respectively, $(S^2 E \otimes E^*) \otimes (\Lambda^2 F^* \otimes F)$. If $p > 2$, then by Theorem 4.1.1 and our description of $\ker(\Box)$, the torsion τ_1 is the highest weight part of the component of T in $(\Lambda^2 E \otimes E^*) \otimes (S^2 F^* \otimes F)$. From Step (D) we know that this is the component in the intersection of the kernels of the two traces on that space which have values in $E \otimes (S^2 F^* \otimes F)$, respectively, in $(\Lambda^2 E \otimes E^*) \otimes F^*$.

To describe the totally tracefree part more explicitly, we use an abstract index notation, which is similar to the spinorial abstract index notation used in 4–dimensional conformal geometry. We put $E := \mathcal{E}^A$ and $F = \mathcal{E}^{A'}$ and follow the usual convention that dualizing makes upper indices into lower indices and vice versa, and concatenation of indices corresponds to tensor products of bundles. Via the isomorphism Φ we can view TM as $E^* \otimes F$, so $TM = \mathcal{E}_A^{A'}$. Then the torsion of the connection ∇ is a section T of the bundle $\Lambda^2 T^*M \otimes TM$, so it has the form $T^{A}_{A'}{}^{B}_{B'}{}^{C'}_{C}$. Indicating alternations by square brackets and symmetrizations by round brackets, the component of the torsion in $(\Lambda^2 E \otimes E^*) \otimes (S^2 F^* \otimes F)$ is given by

$$S^{A}_{A'}{}^{B}_{B'}{}^{C'}_{C} = T^{[A}_{(A'}{}^{B]}_{B')}{}^{C'}_{C} = \tfrac{1}{2}\left(T^{A}_{A'}{}^{B}_{B'}{}^{C'}_{C} - T^{B}_{A'}{}^{A}_{B'}{}^{C'}_{C}\right). \tag{4.1}$$

By construction, $S^{A}_{A'}{}^{B}_{B'}{}^{C'}_{C}$ is skew symmetric in the two upper unprimed indices, which easily implies that

$$S^{A}_{A'}{}^{B}_{B'}{}^{C'}_{C} - \tfrac{1}{p-1}\left(S^{A}_{A'}{}^{D}_{B'}{}^{C'}_{D}\delta^B_C - S^{B}_{A'}{}^{D}_{B'}{}^{C'}_{D}\delta^A_C\right) \tag{4.2}$$

lies in the kernel of the unique trace over the unprimed indices. Moreover, this expression is skew symmetric in the two unprimed upper indices and symmetric in the two primed lower indices. Similarly, one verifies that for each $U^{A}_{A'}{}^{B}_{B'}{}^{C'}_{C}$ with these symmetry properties the expression

$$U^{A}_{A'}{}^{B}_{B'}{}^{C'}_{C} - \tfrac{1}{q+1}\left(U^{A}_{A'}{}^{B}_{D'}{}^{D'}_{C}\delta^{C'}_{B'} + U^{A}_{B'}{}^{B}_{D'}{}^{D'}_{C}\delta^{C'}_{A'}\right)$$

has the same symmetry properties and lies in the kernel of the unique trace over the primed indices. Moreover, if U already has trivial trace over the unprimed indices, then the same is visibly true for this expression. Hence, we see that we obtain a formula for the totally tracefree part by applying this operation to (4.2). To formulate the result efficiently, we introduce

$$S^{A}_{A'} := S^{A}_{A'}{}^{B}_{B'}{}^{B'}_{B} = \tfrac{1}{2}(T^{A}_{A'}{}^{B}_{B'}{}^{B'}_{B} - T^{B}_{A'}{}^{A}_{B'}{}^{B'}_{B}). \tag{4.3}$$

Using this, a short computation shows that $(\tau_1)^{A\ B\ C'}_{A'B'C}$ is given by

$$S^{A\ B\ C'}_{A'B'C} + \tfrac{1}{2(p-1)}\delta^{[A}_C S^{B]\,D\ C'}_{A'B'D} - \tfrac{1}{2(q+1)}\delta^{C'}_{(A'}S^{A\ \ B\ D'}_{B')D'C} - \tfrac{1}{4(p-1)(q+1)}\delta^{[A}_C S^{B]}_{(A'}\delta^{C'}_{B')},$$

where the tensors S are defined in (4.1) and (4.3) above.

The second torsion can be dealt with in a similar way. We define

$$\tilde{S}^{A\ B\ C'}_{A'B'C} = T^{[A\ B]\ C'}_{(A'B')C} = \tfrac{1}{2}\left(T^{A\ B\ C'}_{A'B'C} + T^{B\ A\ C'}_{A'B'C}\right),$$
$$\tilde{S}^A_{A'} := \tilde{S}^{A\ B\ B'}_{A'B'B} = \tfrac{1}{2}(T^{A\ B\ B'}_{A'B'B} + T^{B\ A\ B'}_{A'B'B}),$$

and in terms of these, $(\tau_2)^{A\ B\ C'}_{A'B'C}$ is given by

$$\tilde{S}^{A\ B\ C'}_{A'B'C} - \tfrac{1}{2(p+1)}\delta^{(A}_C \tilde{S}^{B)\,D\ C'}_{A'\,B'D} + \tfrac{1}{2(q-1)}\delta^{C'}_{[A'}\tilde{S}^{A\ \ B\ D'}_{B']D'C} + \tfrac{1}{4(p+1)(q-1)}\delta^{(A}_C \tilde{S}^{B)}_{[A'}\delta^{C'}_{B']}.$$

If $p > 2$ and $q > 2$, then the two torsions τ_1 and τ_2 are the complete obstructions against local isomorphism to the homogeneous model $\mathrm{Gr}(p, \mathbb{R}^{p+q})$, so from our current point of view we are done with that case.

To continue the discussion of the remaining cases, we have to construct a principal connection γ on $\mathcal{G}_0$ such that the torsion of the induced connection ∇ on TM equals the degree one part of the harmonic curvature. The existence of such a connection is proved in Lemma 4.1.1, in general, and it will be useful later to have such a connection at hand in the general case. Moreover, the construction is essentially independent of p and q, so we do it in general and specialize for the further discussion of the harmonic curvature below.

From above we know that a principal connection γ on $\mathcal{G}_0$ is equivalent to a certain pair (∇^E, ∇^F) of linear connections on E and F, and we continue to work in that picture. The difference $\hat{\nabla}^E - \nabla^E$ of two linear connections on E is described by a section Ψ of the bundle $T^*M \otimes E^* \otimes E$ characterized by $\hat{\nabla}^E_\xi e = \nabla^E_\xi e + \Psi(\xi, e)$. In abstract index notation, this has the form $\Psi^{A\ C}_{A'B}$, and we use the convention that $\Psi(\xi, e)^C = \Psi^{A\ C}_{A'B}\xi^{A'}_A e^B$. Similarly, the difference of two linear connections on F is described by a section of $T^*M \otimes F^* \otimes F$, which we also denote by Ψ but with indices $\Psi^{A\ C'}_{A'B'}$, so it follows from the indices which part of the change we consider. The condition that $(\hat{\nabla}^E, \hat{\nabla}^F)$ corresponds to a principal connection $\hat{\gamma}$ on $\mathcal{G}_0$, i.e. that the natural section of $\Lambda^p E \otimes \Lambda^q F$ remains parallel, of course, reads as $\Psi^{A\ B}_{A'B} + \Psi^{A\ B'}_{A'B'} = 0$. Notice that the two sections Ψ can be naturally viewed as defining an element of $\Omega^1(M, \mathrm{End}_0(TM))$. The change of the torsion of the induced connection on TM can be either computed by applying ∂ to this element, or by plugging into the definition $(\hat{\nabla}_\xi \eta)(e) = \hat{\nabla}^F_\xi(\eta(e)) - \eta(\hat{\nabla}^E_\xi e)$, where we identify TM with $L(E, F)$ via Φ. In any case, one obtains

$$\hat{T}^{A\ B\ C'}_{A'B'C} - T^{A\ B\ C'}_{A'B'C} = \Psi^{A\ C'}_{A'B'}\delta^B_C - \Psi^{B\ C'}_{B'A'}\delta^A_C - \Psi^{A\ B}_{A'C}\delta^{C'}_{B'} + \Psi^{B\ A}_{B'C}\delta^{C'}_{A'}.$$

Our task is to choose Ψ in such a way that $\hat{T}^{A\ B\ C'}_{A'B'C}$ lies in the kernel of all traces, since this implies that it lies in $\ker(\Box)$ and thus has to coincide with $\tau_1 + \tau_2$. What makes life a bit difficult at this stage is that this condition will not pin down Ψ uniquely. Indeed, we know that the change of the torsion is given by $\partial(\Psi)$, so it is at best unique up to addition of elements in $\ker(\partial)$. Since $H^1(\mathfrak{g}_{-1}, \mathfrak{g})$ is concentrated in homogeneous degree 0, we know that $\ker(\partial) = \mathrm{im}(\partial)$, so the non-uniqueness is given by adding elements of the form $\partial(\alpha)$ for $\alpha \in \Omega^1(M)$. From the presentation of the Lie algebra $\mathfrak{g}$ one immediately sees that, viewing $\mathrm{End}_0(TM)$ as $\mathfrak{s}(\mathfrak{gl}(E) \times \mathfrak{gl}(F))$, the algebraic bracket $\{\ ,\ \} : T^*M \times TM \to \mathrm{End}_0(TM)$ is given by $\{\alpha,\xi\}(e,f)=(\alpha(\xi(e)),-\xi(\alpha(f)))$, where $\xi:E\to F$ and $\alpha:F\to E$. From this we immediately conclude that the change of connections corresponding to $\partial(\alpha)$ is given by $\Psi^{A}_{A'}{}^{C}_{B}=\alpha^{C}_{A'}\delta^{A}_{B}$ and $\Psi^{A}_{A'}{}^{C'}_{B'}=-\alpha^{A}_{B'}\delta^{C'}_{A'}$. One immediately verifies that such a change of connections leaves the torsion of the induced connection on TM unchanged. Forming the contractions over B and C, respectively, B' and C' in these expressions, we obtain $\alpha^{A}_{A'}$, respectively, $-\alpha^{A}_{A'}$.

The upshot of this is that there is a unique Ψ such that all traces of $\hat{T}$ vanish if we assume, in addition, that $\Psi^{A}_{A'}{}^{B}_{B}=\Psi^{A}_{A'}{}^{B'}_{B'}=0$. Under these assumptions, the solution can indeed be easily determined explicitly: Vanishing of $\hat{T}^{A}_{A'}{}^{D}_{D'}{}^{D'}_{D}$, respectively, $\hat{T}^{D}_{A'}{}^{A}_{D'}{}^{D'}_{D}$ is equivalent to

$$T^{A}_{A'}{}^{D}_{D'}{}^{D'}_{D}=\Psi^{A}_{D'}{}^{D'}_{A'}-\Psi^{D}_{A'}{}^{A}_{D},\qquad T^{D}_{A'}{}^{A}_{D'}{}^{D'}_{D}=p\Psi^{A}_{D'}{}^{D'}_{A'}+q\Psi^{D}_{A'}{}^{A}_{D}.$$

This can be solved as

$$\Psi^{D}_{A'}{}^{A}_{D}=-\tfrac{p}{p+q}T^{A}_{A'}{}^{D}_{D'}{}^{D'}_{D}+\tfrac{1}{p+q}T^{D}_{A'}{}^{A}_{D'}{}^{D'}_{D},\qquad \Psi^{A}_{D'}{}^{D'}_{A'}=\tfrac{q}{p+q}T^{A}_{A'}{}^{D}_{D'}{}^{D'}_{D}+\tfrac{1}{p+q}T^{D}_{A'}{}^{A}_{D'}{}^{D'}_{D}.$$

Next, vanishing of $\hat{T}^{A}_{A'}{}^{B}_{D'}{}^{D'}_{C}$ and of $\hat{T}^{A}_{D'}{}^{B}_{A'}{}^{D'}_{C}$ is equivalent to

$$T^{A}_{A'}{}^{B}_{D'}{}^{D'}_{C}=\Psi^{B}_{D'}{}^{D'}_{A'}\delta^{A}_{C}+q\Psi^{A}_{A'}{}^{B}_{C}-\Psi^{B}_{A'}{}^{A}_{C},$$
$$T^{A}_{D'}{}^{B}_{A'}{}^{D'}_{C}=-\Psi^{A}_{D'}{}^{D'}_{A'}\delta^{B}_{C}+\Psi^{A}_{A'}{}^{B}_{C}-q\Psi^{B}_{A'}{}^{A}_{C}.$$

Solving for $\Psi^{A}_{A'}{}^{B}_{C}$ and inserting for the trace terms from above, we obtain

$$\Psi^{A}_{A'}{}^{B}_{C}=\tfrac{1}{q^2-1}\Big(qT^{A}_{A'}{}^{B}_{D'}{}^{D'}_{C}-T^{A}_{D'}{}^{B}_{A'}{}^{D'}_{C}-\tfrac{q^2}{p+q}T^{B}_{A'}{}^{D}_{D'}{}^{D'}_{D}\delta^{A}_{C}$$
$$-\tfrac{q}{p+q}T^{D}_{A'}{}^{B}_{D'}{}^{D'}_{D}\delta^{A}_{C}-\tfrac{q}{p+q}T^{A}_{A'}{}^{D}_{D'}{}^{D'}_{D}\delta^{B}_{C}-\tfrac{1}{p+q}T^{D}_{A'}{}^{A}_{D'}{}^{D'}_{D}\delta^{B}_{C}\Big).$$

From vanishing of the two traces over the unprimed indices, one similarly deduces

$$\Psi^{A}_{A'}{}^{C'}_{B'}=\tfrac{-1}{p^2-1}\Big(pT^{A}_{A'}{}^{D}_{B'}{}^{C'}_{D}-T^{D}_{A'}{}^{A}_{B'}{}^{C'}_{D}-\tfrac{p^2}{p+q}T^{A}_{B'}{}^{D}_{D'}{}^{D'}_{D}\delta^{C'}_{A'}$$
$$+\tfrac{p}{p+q}T^{D}_{B'}{}^{A}_{D'}{}^{D'}_{D}\delta^{C'}_{A'}-\tfrac{p}{p+q}T^{A}_{A'}{}^{D}_{D'}{}^{D'}_{D}\delta^{C'}_{B'}+\tfrac{1}{p+q}T^{D}_{A'}{}^{A}_{D'}{}^{D'}_{D}\delta^{C'}_{B'}\Big).$$

Changing an arbitrary intial connection by Ψ as computed in the last two formulae, we obtain a principal connection γ, respectively, a pair (∇^E,∇^F) of linear connections, which has torsion $\tau_1+\tau_2$.

For the further analysis of curvature, the cases $p=2$ and $q>2$, respectively, $p>2$ and $q=2$ are completely parallel so we only discuss the first of these. In that case, we must have $\tau_1=0$ and τ_2 is the basic torsion. Structures for which τ_2 vanishes identically are called *Grassmannian structures* of type $(2,q)$ (as opposed to almost Grassmannian structures). If $\tau_2=0$, then the above procedure produces a torsion–free connection on $\mathcal{G}_0$. Conversely, by Theorem 4.1.1 existence of a torsion–free connection on $\mathcal{G}_0$ implies vanishing of τ_2. Thus, Grassmannian structures of type $(2,q)$ are exactly those almost Grassmannian structures which are integrable in the sense of G–structures.

The second basic obstruction to local flatness for type $(2,q)$ with $q>2$ is the curvature $\rho_1\in\Gamma(\Lambda^2E\otimes S^2F^*\otimes\mathfrak{sl}(F))$. From part (2) of Theorem 4.1.1 we know that we can compute ρ_1 as the appropriate component of the curvature of γ. Since the bundle F corresponds to the natural representation of G_0 on $\mathbb{R}^q$, and ∇^F is the linear connection induced by γ, the curvature of ∇^F corresponds to the component in $\Lambda^2T^*M \otimes L(F,F)$ of the curvature of γ, compare with 1.3.4. Let us denote this curvature by R^F, so for $s \in \Gamma(F)$ and $\xi, \eta \in \mathfrak{X}(M)$ we have the usual formula

$$R^F(\xi,\eta)(s) = \nabla^F_\xi \nabla^F_\eta s - \nabla^F_\eta \nabla^F_\xi s - \nabla^F_{[\xi,\eta]} s.$$

In abstract index notation, $R := R^F$ has the form $R^{A}_{A'}{}^{B}_{B'}{}_{C'}{}^{D'}$, where the first two pairs of indices describe the form part. According to the description of the highest weight component in the end of Step (D), we can continue as follows: First, we take the complete symmetrization $R^{A}_{(A'}{}^{B}_{B'}{}_{C')}{}^{D'}$ in the three lower primed indices. By skew symmetry in the first two pairs of indices, this expression is automatically skew in A and B, so we can compute ρ_1 as the tracefree part of this. Expanding the symmetrization, one computes

$$R^{A}_{(A'}{}^{B}_{B'}{}_{D')}{}^{D'} = \tfrac{1}{3} R^{A}_{(A'}{}^{B}_{B')}{}_{D'}{}^{D'} + \tfrac{2}{3} R^{[A}_{D'}{}^{B]}_{(A'}{}_{B')}{}^{D'}.$$

On the other hand, the imbedding $S^2\mathbb{R}^{q*} \to S^3\mathbb{R}^{q*} \otimes \mathbb{R}^q$ maps $X_{A'B'}$ to

$$\tfrac{1}{q+2}(X_{A'B'}\delta_{C'}{}^{D'} + X_{C'A'}\delta_{B'}{}^{D'} + X_{B'C'}\delta_{A'}{}^{D'}) = \tfrac{3}{q+2} X_{(A'B'}\delta_{C')}{}^{D'}.$$

Consequently, we obtain

$$(\rho_1)^{A}_{A'}{}^{B}_{B'}{}_{C'}{}^{D'} = R^{A}_{(A'}{}^{B}_{B'}{}_{C')}{}^{D'} - \tfrac{1}{q+2} R^{A}_{(A'}{}^{B}_{B'|I'|}{}^{I'}\delta_{C')}{}^{D'} - \tfrac{2}{q+2} R^{[A}_{I'}{}^{B]}_{(A'}{}_{B'}{}^{I'}\delta_{C')}{}^{D'}.$$

In the last remaining case $p = q = 2$, we automatically have $\tau_1 = \tau_2 = 0$, so the above procedure leads to a principal connection γ on $\mathcal{G}_0$ with vanishing torsion. Looking at the corresponding linear connections ∇^E and ∇^F we can compute the two curvatures ρ_1 and ρ_2 from their curvatures as above. Vanishing of one of these two curvatures is usually referred to as *semi–flatness*.

Step (F): The description of the basic tractor bundles is simple in this case. Let $\mathcal{T}$ be the *standard tractor bundle*, i.e. the bundle corresponding to the standard representation of G on $\mathbb{R}^{p+q}$. By definition, $P \subset G$ is the stabilizer of $\mathbb{R}^p \subset \mathbb{R}^{p+q}$, and the associated bundle to the representation of P on $\mathbb{R}^p$ is E. This means that $\mathcal{T}$ contains E as a smooth subbundle. On the other hand, the quotient representation of P on $\mathbb{R}^{p+q}/\mathbb{R}^p$ is the trivial extension of the representation $\mathbb{R}^q$ of the subgroup $G_0 \subset P$. This corresponds to the bundle F, so we get a short exact sequence $0 \to E \to \mathcal{T} \to F \to 0$ of natural vector bundles. We will also indicated this short exact sequence (and more general filtrations later on) by writing $\mathcal{T} = E \,\oplus\!\!\!\!+\, F$. Of course, this sequence does not admit a natural splitting since the representation of P on $\mathbb{R}^{p+q}$ is indecomposable. On the other hand, we can view $E \subset \mathcal{T}$ as a filtration of the vector bundle $\mathcal{T}$, and then the associated graded bundle $\mathrm{gr}(\mathcal{T})$ is naturally isomorphic to $E \oplus F$.

Since any irreducible representation of $\mathfrak{sl}(p+q,\mathbb{R})$ is isomorphic to a subrepresentation of some tensor power of the standard representation, any tractor bundle corresponding to an irreducible representation can be found in some tensor power of the standard tractor bundle. The filtration $E \subset \mathcal{T}$ gives rise to a filtration of any such tractor bundle, which may, however, be more complicated. As an example, consider the tractor bundles $\Lambda^k\mathcal{T}$ for $k = 2,\dots,p+q-1$ which correspond to the other fundamental representations. Then the number of nontrivial subbundles involved in the filtration is the minimum of p, q, and k, and the associated graded bundle has the form $\mathrm{gr}(\Lambda^k\mathcal{T}) = \bigoplus_{i+j=k} \Lambda^i E \otimes \Lambda^j F$.

4.1.4. An alternative interpretation in the case $p = q = 2$. We have observed in example (2) of 2.2.6 that $\mathfrak{sl}(4,\mathbb{C})$ is isomorphic to $\mathfrak{so}(6,\mathbb{C})$, and we can construct an analog of this isomorphism for the real form $\mathfrak{g} := \mathfrak{sl}(4,\mathbb{R})$. Consider the second exterior power $\Lambda^2\mathbb{R}^4$ of the standard representation. Since $\Lambda^4\mathbb{R}^4$ is a trivial $\mathfrak{g}$–representation, the wedge product induces a $\mathfrak{g}$–invariant nondegenerate symmetric bilinear form on the six–dimensional space $\Lambda^2\mathbb{R}^4$. Taking the standard basis $e_1,\dots,e_4$, we see that the elements $e_1\wedge e_i$ for $i = 2,3,4$ span a three–dimensional isotropic subspace. Hence, this bilinear form must have signature $(3,3)$ and we obtain a homomorphism $\mathfrak{g}\to\mathfrak{so}(3,3)$. This must be injective since $\mathfrak{g}$ is simple and thus an isomorphism for dimensional reasons. The parabolic subalgebra $\mathfrak{p}\le\mathfrak{g}$ is exactly the stabilizer of the line through $e_1\wedge e_2$ (compare with 3.2.10), which is a null vector.

Passing to the group level, we obtain a homomorphism from $G = SL(4,\mathbb{R})$ to $SO(3,3)$, which maps the parabolic subgroup P to the stabilizer of the null line through $e_1\wedge e_2$. Connectedness of G and the fact that we obtain an isomorphism on the Lie algebra level implies that this homomorphism maps G onto the connected component $SO_0(3,3)$ of the identity. The kernel of this homomorphism must be contained in the center $\pm\mathbb{I}$ of G, so evidently it coincides with this center and the homomorphism is a two–fold covering. It is easy to check that the standard representation of G and its dual realize real forms of the two spin representations of $\mathfrak{so}(3,3)$, so our homomorphism actually identifies G with the spin group $Spin(3,3)$. Hence, we are in the situation of conformal structures of split signature $(2,2)$ on four–dimensional manifolds as discussed (in higher dimensions) in 4.1.2.

We can easily show explicitly that G_0 is the conformal spin group $CSpin(2,2)$. By definition, $\mathfrak{g}_{-1}$ is the space $M_2(\mathbb{R})$ of real 2×2–matrices. The determinant defines a quadratic form on $M_2(\mathbb{R})$ and since there clearly are two–dimensional subspaces in $M_2(\mathbb{R})$ which consist entirely of matrices with zero determinant, the inner product inducing the determinant must have split signature $(2,2)$. From 4.1.3 we know that $G_0\cong S(GL(2,\mathbb{R})\times GL(2,\mathbb{R}))$ and given $A,B\in GL(2,\mathbb{R})$ with $\det(A)\det(B) = 1$, the adjoint action on $X\in M_2\mathbb{R}$ is given by BXA^{-1}. Using the relation between the determinants of A and B, we get $\det(BXA^{-1}) = \det(B)^2\det(X)$, so the adjoint action defines a homomorphism $G_0\to CSO(2,2)$. Evidently, the kernel of this homomorphism is $\{(\mathbb{I},\mathbb{I}),(-\mathbb{I},-\mathbb{I})\}$, and the quotient of G by this subgroup is connected. By dimensional reasons, we obtain a two–fold covering $G_0\to CSO_0(2,2)$ of the connected component of the identity of the conformal group. The two obvious two–dimensional representations of G_0 are easily seen to realize the two spin representations, so $G_0\cong CSpin(2,2)$.

In the conformal picture the interpretation of basic curvature components is fairly simple. The principal G_0–bundle $p_0 : \mathcal{G}_0\to M$ is a two–fold covering of the conformal frame bundle. Taking the Levi–Civita connection of any metric in the conformal class, we obtain a torsion–free connection on $\mathcal{G}_0$. Then the two basic curvatures ρ_1 and ρ_2 can be computed directly from the curvatures of the induced connections on the two spin bundles using the formulae in 4.1.3. There is an alternative interpretation of these curvatures, which fits better to conformal structures in general dimensions.

The curvature of a Levi–Civita connection on M is an element of $\Lambda^2T^*M\otimes\mathfrak{so}(TM)$. As in the identification of G_0 with $CSpin(2,2)$ above, one concludes that $\mathfrak{so}(TM)\cong\mathfrak{sl}(E)\oplus\mathfrak{sl}(F)$, where we denote by E and F the two real spin bundles. On the other hand, as we have seen in 4.1.3, the isomorphism $TM \cong E^* \otimes F$ gives rise to a decomposition $\Lambda^2 T^*M = (\Lambda^2 E \otimes S^2 F^*) \oplus (S^2 E \otimes \Lambda^2 F^*)$. One verifies directly that this is exactly the decomposition of two forms into self–dual and anti–self–dual two–forms; see the discussion of conformal structures in 4.1.2. Now $\mathfrak{so}(TM) \cong \Lambda^2 T^*M$ via the metric, so the two decompositions above are just two isomorphic pictures of the same decomposition. Moreover, it is well known that the curvature of a pseudo–Riemannian manifold actually has values $S^2(\Lambda^2 T^*M) \subset \Lambda^2 T^*M \otimes \Lambda^2 T^*M$. Consequently, the curvature actually has values in

$$(\Lambda^2 E \otimes S^2 F^* \otimes \mathfrak{sl}(F)) \oplus (S^2 E \otimes \Lambda^2 F^* \otimes \mathfrak{sl}(E)).$$

Recall from 1.6.8 that the *Weyl curvature* is the totally tracefree part of the Riemann curvature. This is also contained in $S^2(\Lambda^2 T^*M)$ so it splits accordingly, and from the construction it follows that the two components are the curvatures ρ_1 and ρ_2. Hence, the two basic curvatures are exactly the self–dual and the anti–self–dual part of the Weyl curvature, and the two possible semi–flatness conditions are self–duality and anti–self–duality of a conformal structure.

4.1.5. Classical projective structures. Historically, these were among the first examples of parabolic geometries that have been studied. These structures form one of the two basic examples of geometries which are not determined by the underlying infinitesimal flag structure. Therefore, we can only follow the general scheme described in 4.1.1 in the beginning, but will have to deviate from it later on.

Step (A): Fix $n \geq 2$ and consider the Lie algebra $\mathfrak{g} = \mathfrak{sl}(n+1, \mathbb{R})$. The grading $\mathfrak{g} = \mathfrak{g}_{-1} \oplus \mathfrak{g}_0 \oplus \mathfrak{g}_1$ is the extremal case $p = 1$ of the one discussed in Example 3.1.2 (2). Hence, we view elements of $\mathfrak{g}$ as block matrices $\begin{pmatrix} -\operatorname{tr}(A) & Z \\ X & A \end{pmatrix}$ with $X \in \mathbb{R}^n$, $Z \in \mathbb{R}^{n*}$ and $A \in \mathfrak{gl}(n, \mathbb{R})$. The entry X represents $\mathfrak{g}_{-1}$, Z represents $\mathfrak{g}_1$, while the block–diagonal part determined by A is $\mathfrak{g}_0$. The Dynkin diagram describing this grading is simply the A_n–diagram with the leftmost node crossed.

There are two interesting choices for a Lie group G with Lie algebra $\mathfrak{g}$. One is the obvious choice $G = SL(n+1, \mathbb{R})$. In this case, the good choice for P is to take the connected component of the identity in the subgroup determined by the grading. Hence, P is the subgroup of all matrices of the form $\begin{pmatrix} \frac{1}{\det(C)} & W \\ 0 & C \end{pmatrix}$, where $C \in GL(n, \mathbb{R})$ has positive determinant. This is the stabilizer of the ray spanned by the first basis vector. Since G evidently acts transitively on the set of rays in $\mathbb{R}^{n+1}$, we conclude that G/P is the space of all such rays and hence diffeomorphic to S^n. The subgroup $G_0 \subset P$ is characterized by $W = 0$, so $G_0 \cong GL^+(n, \mathbb{R})$, the group of invertible $n \times n$–matrices with positive determinant.

The second interesting choice of groups is $G = PGL(n+1, \mathbb{R})$, the quotient of $GL(n+1, \mathbb{R})$ by the closed normal subgroup consisting of all multiples of the identity. Here, we take P to be the maximal parabolic subgroup for the given grading. While G does not act on $\mathbb{R}^{n+1}$, it does act on the space $\mathbb{R}P^n$ of lines through the origin in $\mathbb{R}^{n+1}$. The subgroup P by definition is the stabilizer of the line generated by the first vector in the standard basis. Hence, $G/P \cong \mathbb{R}P^n$ and we exactly recover the situation discussed in 1.1.3. The subgroup $G_0 \subset P$ is given by the classes of block diagonal matrices, and since any such class has a unique representative of the form $\begin{pmatrix} 1 & 0 \\ 0 & A \end{pmatrix}$ we see that $G_0 \cong GL(n, \mathbb{R})$ in this case.

Below, we will treat both choices of groups simultaneously, by working in $GL(n+1,\mathbb{R})$ and taking into acount that we either work module scalar factors or restrict to matrices of determinant one. In this picture, P consists of the classes of matrices whose first column is a multiple of the first unit vector.

Steps (B) and (C): Consider a block diagonal matrix $\left(\begin{smallmatrix} c & 0 \\ 0 & C \end{smallmatrix}\right) \in GL(n+1,\mathbb{R})$. Then the adjoint action on $\mathfrak{g}_{-1}$ is given by $X \mapsto CXc^{-1}$. Notice that this is unchanged if we replace the matrix by a nonzero multiple. Conversely, if $CXc^{-1} = X$ for all $X \in \mathbb{R}^n$, then the matrix is a multiple of the identity. Hence, we conclude that for the case $G = PGL(n+1,\mathbb{R})$ the adjoint action induces an isomorphism $G_0 \to GL(\mathfrak{g}_{-1})$. Likewise, if $c = \det(C) = 1$, then we simply get the standard action of $SL(n,\mathbb{R})$ on $\mathbb{R}^n$. Finally, for $\lambda > 0$ we can form the $(n+1)$st root $\mu := \lambda^{1/(n+1)}$, and the matrix $\left(\begin{smallmatrix} 1/\mu^n & 0 \\ 0 & \mu \mathrm{id} \end{smallmatrix}\right)$ acts on $\mathfrak{g}_{-1}$ by multiplication by λ. Hence, for $G = SL(n+1,\mathbb{R})$, the adjoint action induces an isomorphism $G_0 \to GL^+(n,\mathbb{R})$.

Hence, determining the basic irreducible representations and the corresponding vector bundles is very easy. The standard representation on $\mathfrak{g}_{-1} = \mathbb{R}^n$ corresponds to the tangent bundle, and its dual on $\mathfrak{g}_1$ corresponds to the cotangent bundle. Any irreducible representation of $SL(n,\mathbb{R})$ can be realized within a tensor product of copies of these representations (see 2.2.13), and correspondingly irreducible bundles can be realized in tensor bundles. On the other hand, there are some natural line bundles. One series of such bundles comes from the one–dimensional representations given by $C \mapsto |\det(C)|^\alpha$ for $\alpha \in \mathbb{R}$, so these are density bundles. Finally, for $GL(n,\mathbb{R})$ there is one additional representation corresponding to $A \mapsto \mathrm{sgn}(\det(A))$. The corresponding bundle with fiber $\mathbb{Z}_2$ is a two–fold covering, usually called the orientation covering. Finally, one may use tensor products of the bundles obtained so far. Tensor products of tensor bundles with a density bundle are often referred to as *weighted tensor bundles*.

At this point, we have to deviate from the standard route. This is due to the fact that $H^1(\mathfrak{g}_{-1},\mathfrak{g})$ has a nontrivial component in homogeneity one. However, we can also see this directly from the developments so far. In the $|1|$–graded case an infinitesimal flag structure (which is automatically regular) on a manifold M is a reduction of structure group of the frame bundle $\mathcal{P}M$ corresponding to $\mathrm{Ad} : G_0 \to GL(\mathfrak{g}_{-1})$. From above, we know that this is either an isomorphism, and hence contains no information at all, or it is the inclusion of the connected component of the identity, and hence we only obtain an orientation on M. In any case, we need more data to describe a Cartan geometry. From Proposition 3.3.7 we know that $H^1(\mathfrak{g}_{-1},\mathfrak{g})$ has trivial components in homogeneities higher than one (since we have assumed $n \geq 2$). According to Theorem 3.1.16, parabolic geometries of type (G,P) are equivalent to normal P–frame bundles of degree one of type (G,P), so we have to understand these.

***P*–frame bundles of degree one**: We have already sketched in 1.1.3 some facts on projective equivalence of connections.

DEFINITION 4.1.5. (1) Let M be a smooth manifold of dimension $n \geq 2$ and let ∇ and $\hat{\nabla}$ be two linear connections on the tangent bundle TM. Then ∇ and $\hat{\nabla}$ are called *projectively equivalent* if and only if there is a one–form $\Upsilon \in \Omega^1(M)$ such that for all vector fields $\xi, \eta \in \mathfrak{X}(M)$ we have

$$\hat{\nabla}_\xi \eta = \nabla_\xi \eta + \Upsilon(\eta)\xi + \Upsilon(\xi)\eta.$$

(2) A *projective structure* on M is a projective equivalence class of linear connections on TM.

Note that if ∇ and $\hat{\nabla}$ are projectively equivalent, then their difference is symmetric, so they have the same torsion. In particular, it makes sense to talk about a torsion–free projective structure on M.

PROPOSITION 4.1.5. *Let $G = PGL(n+1, \mathbb{R})$ and $P \subset G$ the subgroup described above. Then there is an equivalence of categories between P–frame bundles of degree one of type G/P and projective structures. Under this equivalence, normal P–frame bundles exactly correspond to torsion–free projective structures.*

For $G = SL(n+1, \mathbb{R})$ and P the subgroup described above, one obtains an analogous statement for oriented projective structures.

PROOF. The general definition of of a P–frame bundle of degree one from 3.1.15 simplifies considerably in the $|1|$–graded case. The only ingredients are a principal P–bundle $p : \mathcal{G} \to M$ and a one–form $\theta_{-1} \in \Omega^1(\mathcal{G}, \mathfrak{g}/\mathfrak{g}_1)$. In $T\mathcal{G}$ we have two natural subbundles, the vertical subbundle $T^0\mathcal{G}$ of $p : \mathcal{G} \to M$ and the subbundle $T^1\mathcal{G} \subset T^0\mathcal{G}$ spanned by the fundamental vector fields generated by elements of $\mathfrak{g}_1 \subset \mathfrak{p}$. The defining properties of the frame form can be expressed in terms of θ_{-1} as:

- $(r^g)^*\theta_{-1} = \underline{\mathrm{Ad}}(g^{-1}) \circ \theta_{-1}$ for $g \in P$, with $\underline{\mathrm{Ad}}$ denoting the action on $\mathfrak{g}/\mathfrak{g}_1$ induced by Ad.
- For $u \in \mathcal{G}$ the kernel of $\theta_{-1}(u)$ is $T^1_u\mathcal{G}$.
- θ_{-1} maps $T^0\mathcal{G}_0$ to $\mathfrak{p}/\mathfrak{g}_1$, and for the fundamental vector field ζ_A generated by $A \in \mathfrak{p}$, we get $\theta_{-1}(\zeta_A) = A + \mathfrak{g}_1 \in \mathfrak{p}/\mathfrak{g}_1$.

As we have observed at the end of 3.1.15, one may pass from a P–frame bundle to an underlying infinitesimal flag structure. One defines $\mathcal{G}_0 := \mathcal{G}/P_+$ and observes that projecting the values of θ_{-1} to $\mathfrak{g}/\mathfrak{p}$, one obtains a one–form that descends to $\underline{\theta} \in \Omega^1(\mathcal{G}_0, \mathfrak{g}/\mathfrak{p})$. Now $\underline{\theta}$ is G_0–equivariant and strictly horizontal, so this is a first order G_0–structure. From the description of G_0 above, we see that, depending on the choice of G, this gives either the full or the oriented frame bundle.

Now take a point $u_0 \in \mathcal{G}_0$ and a tangent vector $\xi \in T_{u_0}\mathcal{G}_0$. For a point $u \in \mathcal{G}$ over u_0, choose a lift $\tilde{\xi} \in T_u\mathcal{G}$ of ξ. Since $\tilde{\xi}$ is unique up to elements of $T^1_u\mathcal{G}$, the $\mathfrak{g}_0$–component of $\theta_{-1}(\tilde{\xi})$ is indepent of the choice of the lift, so we obtain a well–defined element $\gamma^u(\xi) \in \mathfrak{g}_0$. For $A \in \mathfrak{g}_0$, the fundamental vector field on $\mathcal{G}$ generated by A lifts the one on $\mathcal{G}_0$, which immediately implies that γ^u reproduces the generators of fundamental vector fields.

Next, choose a local smooth section $\sigma : U \to \mathcal{G}$ and let $\underline{\sigma} : U \to \mathcal{G}_0$ be the induced section of $\mathcal{G}_0$. Any point in $\mathcal{G}_0|_U$ can then be uniquely written as $\underline{\sigma}(x) \cdot g_0$ for some $x \in U$ and $g_0 \in G_0$. Define $\gamma_\sigma \in \Omega^1(\mathcal{G}_0|_U, \mathfrak{g}_0)$ by

$$\gamma_\sigma(\underline{\sigma}(x) \cdot g_0) := \gamma^{\sigma(x) \cdot g_0}.$$

By construction, this reproduces the generators of fundamental vector fields. For $h \in G_0$, a tangent vector $\xi \in T_{\underline{\sigma}(x)\cdot g_0}\mathcal{G}_0$, and a lift $\tilde{\xi} \in T_{\sigma(x)\cdot g_0}\mathcal{G}$ we see that $Tr^h \cdot \tilde{\xi}$ is a lift of $Tr^h\xi$. Using this we see that

$$((r^h)^*\gamma_\sigma)(\underline{\sigma}(x) \cdot g_0)(\xi) = \gamma_\sigma(\underline{\sigma}(x) \cdot g_0 h)(Tr^h \cdot \xi)$$

is the $\mathfrak{g}_0$–component of

$$\theta_{-1}(\sigma(x) \cdot g_0 h)(Tr^h \cdot \tilde{\xi}) = \mathrm{Ad}(h^{-1})(\theta_{-1}(\sigma(x) \cdot g_0)(\tilde{\xi})).$$

Since $\mathrm{Ad}(h^{-1})$ preserves the grading of $\mathfrak{g}$, this equals

$$\mathrm{Ad}(h^{-1})(\gamma_\sigma(\underline{\sigma}(x)\cdot g_0)(\xi)).$$

Hence, γ_σ is G_0–equivariant, so it defines a principal connection on $\mathcal{G}_0|_U$.

Now let us determine how this depends on σ. Given σ, a general section $\hat\sigma$ of $\mathcal{G}|_U$ has the form $\hat\sigma(x) = \sigma(x)\cdot g_0(x)\exp(Z(x))$ for smooth functions $g_0 : U \to G_0$ and $Z : U \to \mathfrak{g}_1$; compare with Theorem 3.1.3. Let us first assume that $Z = 0$, i.e. $\hat\sigma(x) = \sigma(x)\cdot g_0(x)$. Then by definition

$$\gamma_{\hat\sigma}(\underline{\hat\sigma}(x)) = \gamma^{\hat\sigma(x)} = \gamma^{\sigma(x)\cdot g_0(x)} = \gamma_\sigma(\underline{\sigma}(x)\cdot g_0(x)) = \gamma_\sigma(\underline{\hat\sigma}(x)).$$

By equivariancy, this implies $\gamma_{\hat\sigma} = \gamma_\sigma$, so we may restrict to the case $\hat\sigma(x) = \sigma(x)\cdot\exp(Z(x))$, and thus $\underline{\hat\sigma} = \underline{\sigma}$. For $\xi \in T_{\underline{\sigma}(x)}\mathcal{G}_0$ and a lift $\tilde\xi \in T_{\sigma(x)}\mathcal{G}$ we get the lift $Tr^{\exp(Z(x))}\cdot\tilde\xi \in T_{\hat\sigma(x)}\mathcal{G}$. By construction, $\gamma_{\hat\sigma}(\underline{\sigma}(x))(\xi)$ is the $\mathfrak{g}_0$–component of

$$\theta_{-1}(\hat\sigma(x))(Tr^{\exp(Z(x))}\cdot\xi) = \mathrm{Ad}(\exp(-Z))(\theta_{-1}(\sigma(x))(\xi)).$$

This is the sum of the $\mathfrak{g}_0$–component of $\theta_{-1}(\sigma(x))(\xi)$ (which equals $\gamma_\sigma(\xi)$) and the bracket of $-Z$ with the $\mathfrak{g}_{-1}$–component of $\theta_{-1}(\sigma(x))(\xi)$, which simply represents the tangent vector ξ. Now one immediately computes that for $Z \in \mathfrak{g}_1$ and $X \in \mathfrak{g}_{-1}$, the action on the bracket $[-Z, X] \in \mathfrak{g}_0$ maps $Y \in \mathfrak{g}_{-1}$ to $XZY + ZXY$. (In the first term we evaluate Z on Y and multiply X by the result, and in the second term it is the other way round.) But this means that looking at the linear connection ∇ on TM associated to γ_σ then the connections associated to $\gamma_{\hat\sigma}$ run exactly through the projective equivalence class of ∇. Since principal (or linear connections) can be patched together smoothly, we conclude that a P–frame bundle of degree one induces a projective structure. From this construction it is also clear that if we have a morphism of P–frame bundles, then the base map is compatible (in the obvious sense) with the projective structures.

Conversely, assume that we have given a projective equivalence class $[\nabla]$ of linear connections on a manifold M. Let $\mathcal{P}M$ be the (oriented) linear frame bundle of M, and let $\theta \in \Omega^1(\mathcal{P}M, \mathfrak{g}_{-1})$ be the soldering form (where we identify $\mathbb{R}^n$ with $\mathfrak{g}_{-1}$). Given a point $y \in \mathcal{P}M$ over $x \in M$, define $\mathcal{G}_y$ as the values in y of the principal connections associated to the linear connections in the projective class. We can view the disjoint union $\mathcal{G}$ of all $\mathcal{G}_y$ as a subset of the bundle $T^*\mathcal{P}M \otimes \mathfrak{g}_0$ of $\mathfrak{g}_0$–valued forms, where we identify $\mathfrak{gl}(n,\mathbb{R})$ with $\mathfrak{g}_0$.

We can view $\mathcal{P}M$ as a principal bundle with structure group G_0, and we want to extend the principal right action to a P–action on $\mathcal{G}$. Given $y \in \mathcal{P}M$, a connection form $\gamma(y)$ and $g = g_0\exp(Z) \in P$, we define $\gamma(y)\cdot g$ to be the connection form at $y\cdot g_0$ defined by

$$\xi \mapsto \gamma(y\cdot g_0)(\xi) + [Z, \theta(\xi)].$$

Explicitly, the first summand equals $\mathrm{Ad}(g_0^{-1})(\gamma(y)(Tr^{g_0^{-1}}\cdot\xi))$. Since θ vanishes on vertical vectors, we immediately conclude that this map reproduces the generators of fundamental vector fields. Hence, we obtain a connection form. Moreover, as above, we conclude that varying Z these forms exactly run through the connections in the projective class. Acting with another element $g_0'\exp(Z')$ we obtain

$$\xi \mapsto \gamma(y\cdot g_0g_0')(\xi) + \mathrm{Ad}(g_0^{-1})([Z, \theta(Tr^{(g_0')^{-1}}\cdot\xi)]) + [Z', \theta(\xi)].$$

Using equivariancy of θ and the fact that the adjoint action is by Lie algebra homomorphisms, the second summand can be rewritten as $[\mathrm{Ad}(g_0^{-1})(Z), \theta(\xi)]$. Since

$$g_0 \exp(Z) g_0' \exp(Z') = g_0 g_0' (g_0')^{-1} \exp(Z) g_0' \exp(Z') = g_0 g_0' \exp(\mathrm{Ad}((g_0')^{-1})(Z) + Z'),$$

we have really defined an action of P on $\mathcal{G}$, which is free and transitive on each fiber. A local smooth section of $\mathcal{P}M$ together with the choice of a connection in the projective class gives rise to a local smooth section of $\mathcal{G}$. Hence, $\mathcal{G}$ is a smooth principal P–bundle. We can define $\theta_{-1} \in \Omega^1(\mathcal{G}, \mathfrak{g}_{-1} \oplus \mathfrak{g}_0)$ tautologically by

$$\theta_{-1}(\gamma(y))(\xi) = (\theta \oplus \gamma(y))(T\pi \cdot \xi),$$

where $\pi : \mathcal{G} \to \mathcal{P}M$ is the projection. One easily verifies directly that this makes $\mathcal{G}$ into a P–frame bundle of degree one, and clearly this construction is functorial. Hence, we have established the equivalence, and it remains to relate normality to torsion freeness.

If γ is the principal connection on $\mathcal{P}M$ corresponding to a linear connection ∇ and θ is the soldering form, then the torsion is induced by

$$d\theta(\xi, \eta) + \gamma(\xi)(\theta(\eta)) - \gamma(\eta)(\theta(\xi));$$

see 1.3.5. Viewing θ as having values in $\mathfrak{g}_{-1}$ and γ as having values in $\mathfrak{g}_0$, this can be rewritten as

$$d\theta(\xi, \eta) + [\gamma(\xi), \theta(\eta)] + [\theta(\xi), \gamma(\eta)].$$

In the equivalent picture of P–frame bundles of degree one, θ corresponds to the $\mathfrak{g}_{-1}$–component of the frame form θ_{-1}, while γ corresponds to its $\mathfrak{g}_0$–component. By the compatibility conditions on frame forms from 3.1.15 we can equivalently view γ as the $\mathfrak{g}_0$–component of θ_0 (or any extension $\tilde{\theta}_0$ of θ_0 to a form defined on all of $T\mathcal{G}$). But then the above expression just computes the $\mathfrak{g}_{-1}$–component of

$$d\theta_{-1}(\xi, \eta) + [\tilde{\theta}_0(\xi), \theta_{-1}(\eta)] + [\theta_{-1}(\xi), \tilde{\theta}_0(\eta)],$$

which by definition computes the torsion of the P–frame bundle. We will see immediately below that ∂^* is injective on the homogeneous part of degree one, so normality of the P–frame bundle is equivalent to vanishing of the torsion of the connections in the projective class. □

Step (D): We have already determined the Hasse diagram in 3.2.17. There is only one element of length two, which corresponds to acting first with the second simple reflection and then with the first one. The set Φ_w corresponding to this element of length two is $\{\alpha_1, \alpha_1 + \alpha_2\}$. Notice that the weight vectors with these weights are exactly the highest and the second highest weight vectors of $\mathfrak{g}_{-1}$. According to Theorem 3.3.5, the other thing we have to determine is the image of the maximal root $\alpha_1 + \cdots + \alpha_n$ under w. For $n = 2$, this is $-\alpha_1$, while for $n > 2$ it is $\alpha_2 + \cdots + \alpha_n$. The latter weight evidently is the highest weight of $\mathfrak{g}_0$. Hence, for $n > 2$, Theorem 3.3.5 shows that the cohomology $H^2(\mathfrak{g}_-, \mathfrak{g})$ is dual to the highest weight component in $\Lambda^2\mathfrak{g}_{-1} \otimes \mathfrak{g}_0$. In particular, this cohomology is concentrated in degree 2. Since any nonzero map $\Lambda^3\mathfrak{g}_{-1} \to \mathfrak{g}$ is homogenous of degree ≥ 2, we see that $\mathrm{im}(\partial^*) \subset L(\Lambda^2\mathfrak{g}_{-1}, \mathfrak{g})$ is contained in homogeneous degree ≥ 2. Since there is no cohomology in degree one, we conclude that $\partial^* : L(\Lambda^2\mathfrak{g}_{-1}, \mathfrak{g}) \to L(\mathfrak{g}_{-1}, \mathfrak{g})$ is injective on maps homogeneous of degree one, which was the last open point in the proof of the proposition above.

Likewise, in the case $n = 2$, the cohomology $H^2(\mathfrak{g}_-, \mathfrak{g})$ is dual to the highest weight component in $\Lambda^2\mathfrak{g}_{-1}\otimes\mathfrak{g}_{-1}$. In particular, this cohomology is concentrated in homogeneous degree 3, and the necessary fact on injectivity of ∂^* follows as before.

Step (E): As we have noted in the proof of Theorem 4.1.1, the proof remains valid if one starts with a connection γ on an infintesimal flag structure (E, θ) such that $(E\times P+, \theta\oplus\gamma)$ is a P–frame bundle of degree one. Hence, we may apply the results of Theorem 4.1.1 interpreting "distinguished connection" as "connection from the projective class". Now we can easily compute the Rho tensor in our setting. We use abstract indices and denote by $R_{ij}{}^k{}_\ell$ the curvature of a fixed connection ∇ in the projective class, with the first two indices being the form indices. By definition,

$$\partial(\mathsf{P})(\xi, \eta) = \{\xi, \mathsf{P}(\eta)\} - \{\eta, \mathsf{P}(\xi)\}.$$

To express this in abstract indices, we have to recall that we identify $\mathfrak{g}_0$ with $L(\mathfrak{g}_{-1}, \mathfrak{g}_{-1})$ via the adjoint action. Now for elements $X, Y \in \mathfrak{g}_{-1}$ and $Z \in \mathfrak{g}_1 \cong \mathfrak{g}^*_{-1}$ we get $[[X, Z], Y] = Z(X)Y + Z(Y)X$. For P we use the abstract index notion fixed by $\mathsf{P}(\xi)_j = \mathsf{P}_{ij}\xi^i$. Then we get

$$\{\xi, \mathsf{P}(\eta)\}^k{}_\ell = \xi^i\mathsf{P}_{ji}\eta^j\delta^k{}_\ell + \xi^k\mathsf{P}_{j\ell}\eta^j,$$

which in turn immediately implies

$$(\partial(\mathsf{P}))_{ij}{}^k{}_\ell = \delta^k_i\mathsf{P}_{j\ell} - \delta^k_j\mathsf{P}_{i\ell} - \mathsf{P}_{ij}\delta^k_\ell + \mathsf{P}_{ji}\delta^k_\ell.$$

To obtain the formula for $\mathrm{gr}_0(\partial^*)$, we observe that in the $|1|$–graded case the formula for ∂^* from 3.1.12 simplifies to

$$\phi_1 \wedge \phi_2 \otimes s \mapsto -\phi_2 \otimes \{\phi_1, s\} + \phi_1 \otimes \{\phi_2, s\}.$$

Since the bracket between $\mathrm{End}_0(TM)$ and T^*M is (up to a nonzero factor) just evaluation of the dual map, we see that, up to a factor, ∂^* is just given by contracting the upper index into one of the form indices, and we use the second form index. Then

$$(\partial(\mathsf{P}))_{ik}{}^k{}_j = -n\mathsf{P}_{ij} + \mathsf{P}_{ji},$$

and the defining property is that this equals $R_{ij} := -R_{ik}{}^k{}_\ell$ (which gives the usual sign to the Ricci type contraction). Symmetrizing we obtain $-(n-1)\mathsf{P}_{(ij)} = R_{(ij)}$ and alternating we get $-(n+1)\mathsf{P}_{[ij]} = R_{[ij]}$. This implies

$$\mathsf{P}_{ij} = \tfrac{-1}{(n-1)}R_{(ij)} - \tfrac{1}{(n+1)}R_{[ij]} = \tfrac{-1}{(n-1)(n+1)}(nR_{ij} + R_{ji}).$$

In view of part (1) of Theorem 4.1.1, this gives a complete description of the normal projective Cartan connection, and the description of harmonic curvatures can be deduced directly from part (2) of that theorem. There is never cohomology in homogeneity one, and thus no harmonic curvature in this homogeneity. This reflects the fact that we deal with torsion–free projective classes. If $n > 2$, the $H^2(\mathfrak{g}_{-1}, \mathfrak{g})$ is concentrated in homogeneity 2 and the Weyl curvature $R + \partial(\mathsf{P})$ of any connection in the projective class is a complete obstruction to local flatness. For $n = 2$, cohomology sits in homogeneity 3, Weyl curvature vanishes identically, so $R = \partial(\mathsf{P})$ and the complete obstruction to local flatness is the Cotton–York tensor $d^\nabla\mathsf{P}$ of any connection ∇ in the projective class.

Step (F): To have the full supply of tractor bundles, we consider the case $G = SL(n+1,\mathbb{R})$ of oriented projective structures. By definition, $P \subset G$ is the stabilizer of the ray generated by the first basis vector in the standard representation $V := \mathbb{R}^{n+1}$. Hence, this vector spans a one–dimensional invariant subspace $V^1 \subset V$. The associated bundle corresponding to the representation V is the *standard tractor bundle* $\mathcal{T}$. Associated to V^1, we get a line subbundle $\mathcal{T}^1 \subset \mathcal{T}$. Now a matrix $\begin{pmatrix} \det(C)^{-1} & w \\ 0 & C \end{pmatrix}$ in P acts on V^1 by multiplication by $\det(C)^{-1}$. On the other hand, it acts on $\mathfrak{g}_{-1}$ by $X \mapsto CX\det(C)$, which has determinant $\det(C)^{n+1}$. Hence, we conclude that, using the convention that 1–densities can be integrated, the line subbundle $\mathcal{T}^1$ is isomorphic to the bundle of $\frac{1}{n+1}$–densities. It is common to denote the latter bundle by $\mathcal{E}(-1)$; see [**BEG94**]. Consequently, in this convention $\mathcal{E}(w)$ denotes the bundle of $-\frac{w}{n+1}$–densities for each $w \in \mathbb{R}$. From the block decomposition defining the grading of $\mathfrak{g}$ it is clear, that $\mathfrak{g}_{-1} \cong L(V^1, V/V^1)$ as a P–module. Therefore, we see that $\mathcal{T}/\mathcal{T}^1 \otimes \mathcal{E}(1) \cong TM$ and hence $\mathcal{T}/\mathcal{T}^1 = TM(-1) := TM \otimes \mathcal{E}(-1)$. The description of $\mathcal{T}$ can be conveniently phrased as a composition series $\mathcal{T} \cong \mathcal{E}(-1) \oplus\!\!\!\!+ \, TM(-1)$.

From this composition series it is easy to see the composition series of other tractor bundles. For example, taking into account that an inclusion dualizes to a quotient map, the composition series for the *standard cotractor bundle* is $\mathcal{T}^* \cong T^*M(1) \oplus\!\!\!\!+ \, \mathcal{E}(1)$. It turns out (see [**BEG94**]) that $\mathcal{T}^*$ can be identified with the first jet prolongation of $\mathcal{E}(1)$. The *adjoint tractor bundle* $\mathcal{A} = \mathfrak{sl}(\mathcal{T})$ can be realized as the tracefree part in $\mathcal{T}^* \otimes \mathcal{T}$, and the resulting composition series is $\mathcal{A} \cong T^*M \oplus\!\!\!\!+ \, (T^*M \otimes TM) \oplus\!\!\!\!+ \, TM$. This of course corresponds to the P–invariant composition series $\mathfrak{g} \cong \mathfrak{g}_1 \oplus\!\!\!\!+ \, \mathfrak{g}_0 \oplus\!\!\!\!+ \, \mathfrak{g}_{-1}$.

For other tensor products of tractor bundles one can proceed similarly. For example, using that $S^2\mathcal{E}(-1) = \mathcal{E}(-2)$ we get $S^2\mathcal{T} \cong \mathcal{E}(-2) \oplus\!\!\!\!+ \, TM(-2) \oplus\!\!\!\!+ \, S^2TM(-2)$, while $\Lambda^2\mathcal{T} \cong TM(-2) \oplus\!\!\!\!+ \, \Lambda^2TM(-2)$.

4.1.6. Projective structures and geodesics. In 4.1.5 above, we have defined a projective structure as an equivalence class of linear connections on the tangent bundle. There is, however, a more geometric interpretation of projective equivalence, that we will briefly discuss here. The basis for this interpretation is the following:

PROPOSITION 4.1.6. *Consider two linear connections ∇ and $\hat{\nabla}$ on the tangent bundle of a smooth manifold M of dimension $n \geq 2$ which have the same torsion. Then ∇ and $\hat{\nabla}$ are projectively equivalent if and only if they have the same geodesics up to parametrization.*

PROOF. The first step is to show that a smooth curve $c : I \to M$ is a geodesic for ∇ up to reparametrization if and only if $\nabla_{c'}c'$ is always proportional to c'. Indeed, by the chain rule one immediately verifies that a reparametrization of a geodesic has this property. Conversely, the same computation shows how the factor of proportionality can be used to set up an ODE whose solution gives a reparametrization to a geodesic.

Now suppose that $\hat{\nabla}_\xi\eta = \nabla_\xi\eta + \Upsilon(\xi)\eta + \Upsilon(\eta)\xi$ for some $\Upsilon \in \Omega^1(M)$. If $c : I \to M$ is a geodesic for ∇, then along c we get $\hat{\nabla}_{c'}c' = 2\Upsilon(c')c'$. This implies that ∇ and $\hat{\nabla}$ have the same geodesics up to parametrization.

Conversely, suppose that ∇ and $\hat{\nabla}$ have the same torsion and the same geodesics up to reparametrization. Consider the difference tensor

$$A(\xi, \eta) := \hat{\nabla}_\xi \eta - \nabla_\xi \eta.$$

Since the connections have the same torsion, this tensor is symmetric, and since any tangent vector occurs as the derivative of some geodesic, we conclude that $A(\xi, \xi)$ must be a multiple of ξ for each ξ. Define $a : TM \to \mathbb{R}$ by $A(\xi, \xi) = a(\xi)\xi$. Then evidently a is homogeneous of degree one and $a(0) = 0$. Using that A is bilinear and symmetric, we obtain

$$A(\xi + \eta, \xi + \eta) = A(\xi, \xi) + 2A(\xi, \eta) + A(\eta, \eta),$$

and inserting we get

$$2A(\xi, \eta) = (a(\xi + \eta) - a(\xi))\xi + (a(\xi + \eta) - a(\eta))\eta. \tag{4.4}$$

Now expand the equation $A(\xi, t\eta) = tA(\xi, \eta)$. Assuming that ξ and η are linearly independent and $t \neq 0$, this implies $a(\xi + t\eta) - a(t\eta) = a(\xi + \eta) - a(\eta)$. Taking the limit $t \to 0$ in the left–hand side, we conclude that a is additive and hence defines a one–form. Then (4.4) reads as

$$2A(\xi, \eta) = a(\eta)\xi + a(\xi)\eta,$$

and we have established projective equivalence. □

This result means that a projective structure can be equivalently described by the family of unparametrized curves defined by the geodesics. It is a bit awkward to formulate the definition precisely in these terms, so instead we will concentrate on the resulting description of morphisms. We will take up the issue of geometries given by families of curves again in the discussion of (generalized) path geometries; see 4.4.3. The relation between projective structures and these path geometries is a nice example of the general concept of correspondence spaces.

COROLLARY 4.1.6. *Let $(M, [\nabla])$ and $(\tilde{M}, [\tilde{\nabla}])$ be projective structures on smooth manifolds, and let $f : M \to \tilde{M}$ be a local diffeomorphism. Then f is a morphism of the projective structures, i.e. $f^*\tilde{\nabla}$ is projectively equivalent to ∇ if and only if for any geodesic $c : I \to M$ of ∇ the composition $f \circ c$ is a geodesic of $\tilde{\nabla}$ up to parametrization.*

PROOF. Of course, $f \circ c$ is a geodesic of $\tilde{\nabla}$ (up to parametrization) if and only if c is a geodesic of $f^*\tilde{\nabla}$ (up to parametrization). Hence, the result follows immediately from the proposition above. □

4.1.7. Some background on quaternions. Next, we consider the real form $\mathfrak{sl}(n+1, \mathbb{H})$ of $\mathfrak{sl}(2n+2, \mathbb{C})$. Let us first recall some background on quaternions. The quaternionic multiplication is compatible with the Euclidean norm on $\mathbb{H} = \mathbb{R}^4$, so $|pq| = |p||q|$ for all $p, q \in \mathbb{H}$. For $q \in \mathbb{H}$, one has the conjugate quaternion $\bar{q}$. The basic properties of the conjugation are $\overline{pq} = \bar{q}\bar{p}$ and $q\bar{q} = \bar{q}q = |q|^2 1$, so $q^{-1} = \frac{1}{|q^2|}\bar{q}$. A quaternion is called *real* if $\bar{q} = q$ and *purely imaginary* if $\bar{q} = -q$. The real quaternions are exactly the real multiples of 1, and the purely imaginary ones form a complementary three–dimensional subspace $\mathrm{im}(\mathbb{H}) \subset \mathbb{H}$. Hence, any q can be split as $\mathrm{re}(q) + \mathrm{im}(q)$ into real and imaginary part. If q is purely imaginary, then $q^2 = -q\bar{q} = -|q|^2 1$, so the square roots of -1 form a two–sphere in $\mathbb{H}$. The standard inner product on $\mathbb{R}^4$ can be written in quaternionic terms as $\langle p, q\rangle = \mathrm{re}(p\bar{q})$. In particular, $\mathbb{H}$ is the orthgonal direct sum of $\mathrm{re}(\mathbb{H})$ and $\mathrm{im}(\mathbb{H})$.

An important difference between $\mathbb{H}$ and the fields $\mathbb{R}$ and $\mathbb{C}$ is that $\mathbb{H}$ has a large group of automorphisms. First note that $q \mapsto \bar{q}$ can be viewed as an isomorphism from $\mathbb{H}$ to the opposite algebra. Consider the group $Sp(1) := \{q \in \mathbb{H} : |q| = 1\}$ of unit quaterions. For $q \in Sp(1)$ consider the map $\mathbb{H} \to \mathbb{H}$ defined by $p \mapsto qp\bar{q}$. Since $\bar{q} = q^{-1}$, this evidently defines an automorphism of the real algebra $\mathbb{H}$ and we have constructed a homomorphism $Sp(1) \to \mathrm{Aut}(\mathbb{H})$. The kernel of this homomorphism consists of ± 1.

It is also evident that any of the maps $p \mapsto qp\bar{q}$ is orthogonal, and maps $\mathrm{im}(\mathbb{H})$ to itself. From this one easily concludes that restricting the map $p \mapsto qp\bar{q}$ to $\mathrm{im}(\mathbb{H})$ leads to a surjective homomorphism $Sp(1) \to SO(\mathrm{im}(\mathbb{H}))$, which identifies $Sp(1)$ with $Spin(3)$. On the other hand, suppose that $\phi : \mathbb{H} \to \mathbb{H}$ is an automorphism. Then $\phi(1) = 1$, so ϕ has to map the orthogonal complement $\mathrm{im}(\mathbb{H})$ to itself. Since the unit sphere in $\mathrm{im}(\mathbb{H})$ consists exactly of the square roots of -1, ϕ must preserve this unit sphere and hence is an orthogonal map. For two orthogonal purely imaginary unit quaternions p and q we have $0 \neq pq \in \mathrm{im}(\mathbb{H})$ and this element is orthogonal to both p and q. Hence, $\{p, q, pq\}$ is an orthonormal basis of $\mathrm{im}(\mathbb{H})$. This turns out to lead exactly to the positively oriented orthonormal bases of $\mathrm{im}(\mathbb{H})$, and we conclude that $\phi|_{\mathrm{im}(\mathbb{H})}$ must be orientation preserving. Hence, $\mathrm{Aut}(\mathbb{H}) = SO(\mathrm{im}(\mathbb{H}))$ and the homomorphism $Sp(1) \to \mathrm{Aut}(\mathbb{H})$ is surjective and thus a two–fold covering.

There is a natural notion of a quaternionic vector space, with the standard example being $\mathbb{H}^n$. To get the usual convention on matrix multiplication one has to view $\mathbb{H}^n$ as a right vector space, so we will in general consider right quaternionic vector spaces. The large automorphism group of $\mathbb{H}$ has the consequence that for quaternionic vector spaces V and W, there is an interesting larger class of maps than just the quaternionic linear ones. Namely, one may look at real linear maps $f : V \to W$ such that there is an automorphism ϕ of $\mathbb{H}$ with $f(vq) = f(v)\phi(q)$ for all $v \in V$ and all $q \in \mathbb{H}$.

To define such kind of maps, one actually does not need the spaces to be quaternionic vector spaces, and the quaternionic vector space structure is not invariant under invertible maps of this type.

DEFINITION 4.1.7. (1) A *prequaternionic vector space* is a real vector space V together with the three–dimensional subspace $\mathcal{Q} = \mathcal{Q}^V \subset L(V, V)$, which admits a basis of the form $\{I, J, I \circ J\}$ where $I, J : V \to V$ are linear maps such that $I^2 = J^2 = -\mathrm{id}$ and $I \circ J = -J \circ I$.

(2) A *morphism* between two prequaternionic vector spaces, say $(V, \mathcal{Q}^V)$ and $(W, \mathcal{Q}^W)$, is a pair (f, ϕ) of linear maps $f : V \to W$, $\phi : \mathcal{Q}^V \to \mathcal{Q}^W$ such that for all $v \in V$ and $A \in \mathcal{Q}^V$ we have $f(A(v)) = \phi(A)(f(v))$.

The main point of this definition is that the basis $\{I, J, I \circ J\}$ is not part of the structure, but an additional choice. Fixing such a basis clearly makes V into a quaternionic vector space, so in particular, V has real dimension $4n$. To see that the concept of morphisms introduced here coincides with our original motivation, assume that V and W are actually quaternionic vector spaces, i.e. we have chosen appropriate bases for $\mathcal{Q}^V$ and $\mathcal{Q}^W$. For a morphism (f, ϕ), we can then view ϕ as a linear map $\mathrm{im}(\mathbb{H}) \to \mathrm{im}(\mathbb{H})$ and the condition on a morphism reads as $f(vq) = f(v)\phi(q)$ for all $q \in \mathrm{im}(\mathbb{H})$. Extending ϕ to a linear map $\mathbb{H} \to \mathbb{H}$ by

$\phi(1) = \phi(1)$, we get $f(vq) = f(v)\phi(q)$ for all $v \in V$ and $q \in \mathbb{H}$. If f is nonzero, then this immediately implies that the extension $\phi : \mathbb{H} \to \mathbb{H}$ must be an automorphism of $\mathbb{H}$. From above we conclude that there is an element $q_0 \in Sp(1)$ (which is unique up to sign) such that $\phi(q) = q_0 q \bar{q}_0$.

4.1.8. Almost quaternionic structures. The Satake diagram of $\mathfrak{sl}(n+1, \mathbb{H})$ has $2n + 1$ dots, alternatingly black and white, starting and ending with a black dot; see example (5) of 2.3.9. By 3.2.9, the parabolic subalgebras giving rise to $|1|$–gradings are given by choosing one of the noncompact simple roots, which correspond to the white dots. We only discuss the structures corresponding to the second simple root, which are by far the most important ones. Hence, these are similar to almost Grassmannian structures of type $(2, 2n)$ as discussed in 4.1.3. We will first discuss the case $n > 1$, the case $n = 1$ will be discussed separately below.

Step (A): Put $\mathfrak{g} := \mathfrak{sl}(n + 1, \mathbb{H})$, the Lie algebra of quaternionic $(n + 1) \times (n + 1)$–matrices with vanishing real trace. From 3.2.10 we know that the parabolic subalgebra $\mathfrak{p} \subset \mathfrak{g}$ corresponding to the second simple root is the stabilizer of the quaternionic line spanned by the first element of the standard basis. Hence, we have a presentation as block matrices of the form $\left(\begin{smallmatrix} a & Z \\ X & A \end{smallmatrix}\right)$ with blocks of size 1 and n, where $X, Z^t \in \mathbb{H}^n$, $a \in \mathbb{H}$ and $A \in M_n(\mathbb{H})$, and $\mathrm{re}(a) + \mathrm{re}(\mathrm{tr}(A)) = 0$. The entries a and A span $\mathfrak{g}_0$, X spans $\mathfrak{g}_{-1} \cong \mathbb{H}^n$ and Z spans $\mathfrak{g}_1 \cong L_{\mathbb{H}}(\mathbb{H}^n, \mathbb{H})$. Hence, this is the quaterionic analog of the algebraic background used to describe classical projective structures in 4.1.5.

As the group G we choose the group $PGL(n+1, \mathbb{H})$, the quotient of all invertible quaternionic linear endomorphisms of $\mathbb{H}^{n+1}$ by the closed normal subgroup of all real multiples of the identity. Then we define $P \subset G$ to be the (quotient of the) stabilizer of the quaternionic line spanned by the first basis vector. Hence, the homogeneous space G/P can be identified with the space $\mathbb{H}P^n$ of quaternionic lines in $\mathbb{H}^{n+1}$, the *quaternionic projective space*. The subgroup $G_0 \subset P$ consists of the classes in P of all block diagonal matrices, i.e. the quotient of

$$\left\{ \begin{pmatrix} q & 0 \\ 0 & \Phi \end{pmatrix} : 0 \neq q \in \mathbb{H}, \Phi \in GL(n, \mathbb{H}) \right\},$$

by real multiples of the identity. Let us write (q, Φ) for the class in G_0 of the block diagonal matrix with the given entries. Then the adjoint action of G_0 on $\mathfrak{g}_{-1} \cong \mathbb{H}^n$ is given by $\Phi(X)q^{-1} = \Phi(Xq^{-1})$. Note that (q, Φ) and $(aq, a\Phi)$ with $a \in \mathbb{R}$ act in the same way, so this really descends to an isomorphism from G_0 to a subgroup of $GL(\mathfrak{g}_{-1})$.

Step (B): We can view $\mathfrak{g} = \mathfrak{sl}(n + 1, \mathbb{H})$ as a Lie subalgebra of its complexification $\mathfrak{g}_{\mathbb{C}} = \mathfrak{sl}(2n + 2, \mathbb{C})$, the space of tracefree complex linear endomorphisms of $\mathbb{H}^{n+1}$. From 2.2.13 we know that the fundamental representations of $\mathfrak{g}_{\mathbb{C}}$ are the complex exterior powers of the standard representation. As in 4.1.3 we obtain the fundamental (complex) representations of $\mathfrak{g}_0$ as the subspaces of $\mathfrak{g}_1$–invariant elements. From above we have $\mathfrak{g}_0 = \mathfrak{s}(\mathfrak{gl}(1, \mathbb{H}) \oplus \mathfrak{gl}(n, \mathbb{H}))$. Using this and comparing with 4.1.3, we see that in terms of the standard representations $\mathbb{H}$ of the first factor and $\mathbb{H}^n$ of the second factor, the fundamental representations of $\mathfrak{g}_0$ are $\mathbb{H}$, $\Lambda^2_{\mathbb{C}}\mathbb{H} \cong \Lambda^{2n}_{\mathbb{C}}\mathbb{H}^{n*}$, $\Lambda^{2n-1}_{\mathbb{C}}\mathbb{H}^{n*}, \dots, \mathbb{H}^{n*}$.

From Example 2.3.15 (3) we know that the even fundamental representations of $\mathfrak{sl}(n, \mathbb{H})$ are complexifications of real representations, while the odd fundamental representations have invariant quaternionic structures. Of course, we have the corresponding real subspaces in $\Lambda^{2i}_{\mathbb{C}}\mathbb{H}^n$. For the second fundamental representation we observe that the presentation of $\mathfrak{gl}(n,\mathbb{H})$ as complex block matrices of the form $\left(\begin{smallmatrix} A & B \\ -\bar{B} & \bar{A}\end{smallmatrix}\right)$ from 2.1.7 shows that the action on $\Lambda^{2n}_{\mathbb{C}}\mathbb{H}^n$ is given by multiplication by $2\,\mathrm{re}(\mathrm{tr}(A))$, so this is evidently a real representation.

Compared to 4.1.3, we have an additional problem, since for the choice of the group G made above, not all fundamental representations of $\mathfrak{g}_0$ integrate to G_0. We can use representatives such that the matrix $\left(\begin{smallmatrix} q & 0 \\ 0 & \Phi\end{smallmatrix}\right)$ has real determinant ± 1, which reduces the ambiguity to $(q,\Phi)=(-q,-\Phi)$. This shows that the even fundamental representations integrate, while the odd ones do not. In particular, the second fundamental representation is the one–dimensional representation given by multiplication by $|q|^2$. The higher even fundamental representations are given by the even exterior powers of $\mathbb{H}^n$, on which the sign ambiguity evidently poses no problem. Notice that the representations which integrate to G_0 are exactly those which admit an invariant real structure.

Notice that $\mathfrak{g}_{-1}\subset\mathfrak{g}$ is a real representation of $\mathfrak{g}_0$ and the highest weight of the complexification clearly is the negative of the second simple root. Denoting by λ_i the fundamental weights of $\mathfrak{g}$, this implies that $\mathfrak{g}_{-1}$ is a real form of the complex irreducible representation of highest weight $\lambda_1-2\lambda_2+\lambda_3$. Similarly, $\mathfrak{g}_1$ is a real form of the complex irreducible representation with highest weigh $\lambda_1+\lambda_{2n+1}$, which is the highest root in $\mathfrak{g}_{\mathbb{C}}$. The two summands in the semisimple part of $\mathfrak{g}_0$ are real forms of the representations with highest weights $2\lambda_1-\lambda_2$ (the first simple root), respectively, $-\lambda_2+\lambda_3+\lambda_{2n+1}$ (the sum of all but the first two simple roots).

Step (C): We claim that $\mathrm{Ad}: G_0\to GL(\mathfrak{g}_{-1})$ defines an isomorphism between G_0 and the group of automorphisms of the prequaternionic vector space $\mathfrak{g}_{-1}\cong\mathbb{H}^n$. For a quaternionic linear map Φ and $0\neq p,q\in\mathbb{H}$, we get $\Phi(Xp)q=\Phi(X)qq^{-1}pq$, which shows that G_0 acts by automorphisms. (We may as well replace the last two copies of q by $\frac{q}{|q|}\in Sp(1)$.) Conversely, given $f:\mathbb{H}^n\to\mathbb{H}^n$ and $q_0\in Sp(1)$ such that $f(Xp)=f(X)q_0p\bar{q}_0$ for all $p\in\mathbb{H}$, we define $\Phi(X):=f(X)q_0$. Then this is evidently quaternionically linear and the adjoint action of (q_0,Φ) induces f.

From this description, the geometric interpretation of an infintesimal flag structure of type (G,P) follows easily. We have to consider smooth manifolds of dimension $\dim(\mathfrak{g}_{-1})=4n$, and an infinitesimal flag structure is a first order G_0–structure. We claim that such a structure is equivalent to a rank three subbundle $\mathcal{Q}=\mathcal{Q}^M\subset L(TM,TM)$ such that locally around each point of M we can find a local smooth frame $\{I,J,I\circ J\}$ for $\mathcal{Q}$, where I and J are such that $I\circ I=J\circ J=-\mathrm{id}$ and $I\circ J=-J\circ I$. Having given such a subbundle $\mathcal{Q}$, the fiber $\mathcal{Q}_x$ over $x\in M$ defines a prequaternionic structure on the tangent space T_xM. We consider $\mathfrak{g}_{-1}$ as the modelling vector space for M and define a subbundle $\mathcal{G}_0$ in the linear frame bundle $\mathcal{P}M$ by letting $(\mathcal{G}_0)_x$ be the space of all isomorphisms $\mathfrak{g}_{-1}\to T_xM$ of prequaternionic vector spaces. Using a local frame $\{I,J,I\circ J\}$ as above one easily constructs local smooth sections of $\mathcal{G}_0$, and since G_0 is the group of automorphisms of the almost quaternionic vector space $\mathfrak{g}_{-1}$ we conclude that $\mathcal{G}_0\to M$ is a G_0–principal bundle.

Conversely, given a reduction $\mathcal{G}_0\to M$ to the structure group G_0 we know that $TM\cong\mathcal{G}_0\times_{G_0}\mathfrak{g}_{-1}$, where G_0 acts on $\mathfrak{g}_{-1}$ via the adjoint action. The purely imaginary quaternions form one of the two simple summands of the semisimple part of $\mathfrak{g}_0$, and we define $\mathcal{Q}$ as the corresponding associated bundle to $\mathcal{G}_0$. Via the adjoint action of $\mathfrak{g}_0$ on $\mathfrak{g}_{-1}$ we can view this summand as a Lie subalgebra of $L(\mathfrak{g}_{-1},\mathfrak{g}_{-1})$, so the associated bundle $\mathcal{Q}$ can be viewed as a subbundle of $L(TM,TM)$. For a local trivialization of $\mathcal{G}_0$, the standard basis $\{i,j,k\}$ of $\mathrm{im}(\mathbb{H})$ gives rise to a local smooth frame $\{I,J,I\circ J\}$ as requried above.

A choice of a subbundle $\mathcal{Q}\subset L(TM,TM)$ as above is usually called an *almost quaternionic structure* on the manifold M. A morphism between almost quaternionic manifolds is a smooth map such that each tangent map is a morphism of prequaternionic vector spaces. Choosing a local frame $\{I,J,I\circ J\}$ of $\mathcal{Q}$ as above, one can locally make the tangent spaces into a smooth family of quaternionic vector spaces. Since there are many possible choices of such frames, this structure is far from being unique. One should also notice that in general it is not possible to make all tangent spaces of M into a smooth family of quaternionic vector spaces. Indeed, if this is possible, then the multiplications by i and j lead to a global smooth frame $\{I,J,I\circ J\}$ of $\mathcal{Q}$, which is usually called a *hypercomplex structure* . In particular, the bundle $\mathcal{Q}$ has to be trivial, which is not true in general. Since preferred smooth frames as above are exactly positively oriented orthonormal frames, we see that an almost quaternionic structures admits a compatible almost hypercomplex structure if and only if the bundle $\mathcal{Q}$ is trivial.

If $(V,\mathcal{Q}^V)$ is a prequaternionic vector space, then choosing a basis $\{I,J,I\circ J\}$ for $\mathcal{Q}^V$ as above, we see that for any A, the composition $A\circ A$ is a negative multiple of the identity. Moreover, the Euclidean norm with respect to this basis can be expressed as $A\circ A=-|A|^2\mathrm{id}$. From above, we see that identifying $\mathcal{Q}^V$ with $\mathrm{im}(\mathbb{H})$, automorphisms of the prequaternionic vector space $(V,\mathcal{Q}^V)$ act as conjugations by unit quaternions. Hence, the Euclidean inner product on $\mathcal{Q}^V$, as well as the orientiation obtained by using any basis of the above form, is an invariant of the prequaternionic vector space. Passing to almost quaternionic structures, this immediately implies that for any structure $(M,\mathcal{Q})$ of this type, the subbundle $\mathcal{Q}\subset L(TM,TM)$ naturally carries an orientation and a Euclidean inner product. The Euclidean norm can be characterized by $A\circ A=-|A|^2\mathrm{id}$.

Steps (D) and (E): By Proposition 3.3.6, the complexification of $H^2(\mathfrak{g}_-,\mathfrak{g})$ is $H^2(\mathfrak{g}_-^{\mathbb{C}},\mathfrak{g}^{\mathbb{C}})$, and the latter cohomology has been computed in 4.1.3. In particular, the basic harmonic curvature components of an almost quaternionic manifold $(M,\mathcal{Q})$ are one torsion τ and one curvature ρ. To give a more detailed description of these invariants, we only have to interpret the representation theoretic results from 4.1.3 in the language of prequaternionic vector spaces.

To formulate the results, we need a few more notions related to prequaternionic vector spaces. First, if $(V,\mathcal{Q}^V)$ is a prequaternionic vector space, then we may look at the space $L(V,V)$ of endomorphisms of V. In there, we have the distinguished three–dimensional subspace $\mathcal{Q}^V=:\mathcal{Q}$. Further, we define $L_{\mathcal{Q}}(V,V)$ as the space of those linear maps which commute with any element of $\mathcal{Q}$. Of course, it suffices to check this for the elements of a distinguished basis $\{I,J,K=I\circ J\}$ of $\mathcal{Q}$. The space $L_{\mathcal{Q}}(V,V)$ further splits as $\mathbb{R}\oplus L^0_{\mathcal{Q}}(V,V)$, where the first summand consists of real multiples of the identity and the second consists of maps with vanishing real trace.

In particular, for a choice $\{I,J,K\}$ of distinguished basis of $\mathcal{Q}$ and an arbitrary linear map $f:V\to V$ we can define

$$\pi_{\mathcal{Q}}(f):=\tfrac{1}{4}(f-I\circ f\circ I-J\circ f\circ J-K\circ f\circ K).$$

One immediately sees that this lies in $L_{\mathcal{Q}}(V,V)$ and hence $\pi_{\mathcal{Q}}$ defines a projection onto this subspace. Passing to another distinguished basis $\{\hat{I},\hat{J},\hat{K}\}$, the matrix describing the base change is in $SO(\mathcal{Q})$. Using this, one easily verifies that $\pi_{\mathcal{Q}}$ is independent of the choice of basis. Note that for $V=\mathfrak{g}_{-1}$ with its natural quaternionic structure, the adjoint action identifies the semisimple part of $\mathfrak{g}_0$ with $\mathcal{Q}^{\mathfrak{g}_{-1}}\oplus L^0_{\mathcal{Q}}(\mathfrak{g}_{-1},\mathfrak{g}_{-1})$, and the direct sum decomposition coincides with the decomposition into simple ideals.

Next, we consider bilinear maps defined on a prequaternionic vector space $(V,\mathcal{Q}^V)$. Let $|\ |$ be the natural norm on $\mathcal{Q}^V$. In analogy with the complex case, we say that for any vector space W, a bilinear map $f:V\times V\to W$ is of type $(1,1)$ if and only if $f(A(x),A(y))=f(x,y)$ for all $A\in\mathcal{Q}^V$ such that $|A|=1$, and all $x,y\in V$. As before, one easily verifies that it suffices to verify this condition for the elements of a distinguished basis $\{I,J,K=I\circ J\}$ for $\mathcal{Q}^V$. Fixing a choice of distinguished basis, for an arbitrary bilinear map $f:V\times V\to W$ we define $\pi_{1,1}(f):V\times V\to W$ by

$$\pi_{1,1}(f)(x,y):=\tfrac{1}{4}(f(x,y)+f(I(x),I(y))+f(J(x),J(y))+f(IJ(x),IJ(y)).$$

Evidently, $\pi_{1,1}(f)$ is bilinear and of type $(1,1)$. If $\{\hat{I},\hat{J},\hat{K}\}$ is another distinguished basis of $\mathcal{Q}^V$, then the base change is represented by a matrix in $SO(3)$. Using this, one verifies directly that $\pi_{1,1}(f)$ is actually independent of the choice of distinguished basis. Hence, we see that it defines a natural projection onto the subspace of maps of type $(1,1)$.

For the special case $W=\mathbb{R}$, we denote by $\Lambda^{1,1}V^*$ (respectively $S^{1,1}V^*$) the space of all skew symmetric (respectively symmetric) bilinear maps $V\times V\to\mathbb{R}$ of type $(1,1)$. For arbitrary W, the space of skew symmetric bilinear maps $V\times V\to W$ of type $(1,1)$ is then naturally isomorphic to $\Lambda^{1,1}V^*\otimes W$, and similarly in the symmetric case. Since $\pi_{1,1}$ defines a natural projection onto the subspace $\Lambda^{1,1}V^*\otimes W\subset\Lambda^2V^*\otimes W$, its kernel is a natural complement.

Using the obvious analog of the notation introduced so far for bundles of prequaternionic vector spaces, we can now formulate the description of the torsion and curvature of an almost quaternionic structure:

PROPOSITION 4.1.8. *Let $(M,\mathcal{Q})$ be an almost quaternionic structure on a manifold of dimension $4n$ with $n>1$.*

(1) *The torsion $\tau\in\Omega^2(M,TM)$ is a section of the subbundle $\ker(\pi_{1,1})\otimes T^*M\subset\Lambda^2T^*M\otimes TM$. This bundle is naturally isomorphic to $S^{1,1}T^*M\otimes\mathcal{Q}\otimes TM$ and the irreducible subbundle in which the values of τ lie is the intersection of the kernels of the two evident natural bundle maps*

$$S^{1,1}T^*M\otimes\mathcal{Q}\otimes TM\to S^{1,1}T^*M\otimes TM,$$
$$S^{1,1}T^*M\otimes\mathcal{Q}\otimes TM\to T^*M\otimes\mathcal{Q}.$$

(2) *The curvature ρ is a section of the subbundle*

$$\Lambda^{1,1}T^*M\otimes L^0_{\mathcal{Q}}(TM,TM)\subset\Lambda^2T^*M\otimes L(TM,TM).$$

It lies in the kernel of the natural Ricci type contraction.

PROOF. For a prequaternionic vector space $(V,\mathcal{Q}^V)$, let us first determine the dimensions of $\Lambda^{1,1}V^*$ and $S^{1,1}V^*$. Suppose that the real dimension of V is $4n$, and choose a distinguished basis $\{I,J,K\}$ of $\mathcal{Q}^V$ and $v_1,\dots,v_n\in V$ in such a way that the $\{v_\ell, I(v_\ell), J(v_\ell), IJ(v_\ell) : \ell = 1, \dots, n\}$ is a real basis of V. To specify an element f of $\Lambda^{1,1}V^*$ or $S^{1,1}V^*$, one may choose $f(v_i, v_j)$, $f(v_i, I(v_j))$, $f(v_i, J(v_j))$ and $f(v_i, IJ(v_j))$ arbitrarily for $i < j$. In the symmetric case, one may, in addition, choose $f(v_i, v_i)$ arbitrarily, while $f(v_i, A(v_i)) = 0$ for all $A \in \mathcal{Q}^V$. This then fixes f completely, which shows that $\dim_{\mathbb{R}}(S^{1,1}V^*) = 4\frac{n(n-1)}{2} + n = n(2n-1)$. In the skew symmetric case, $f(v_i, v_i) = 0$ but one may arbitrarily choose $f(v_i, I(v_i))$, $f(v_i, J(v_i))$, and $f(v_i, IJ(v_i))$, and then f is fixed. Hence, we conclude that $\dim_{\mathbb{R}}(\Lambda^{1,1}V^*) = n(2n+1)$.

(1) The natural decomposition $\Lambda^2 V^* = \Lambda^{1,1}V^* \oplus \ker(\pi_{1,1})$ reflects the decomposition of $\Lambda^2(\mathbb{R}^2 \boxtimes \mathbb{R}^{2n*})$ from 4.1.3. From the dimensions determined above, we see that $\Lambda^{1,1}V^* \subset \Lambda^2 V^*$ corresponds to $\Lambda^2\mathbb{R}^2 \boxtimes S^2\mathbb{R}^{2n*}$, while $\ker(\pi_{1,1})$ corresponds to $S^2\mathbb{R}^2 \boxtimes \Lambda^2\mathbb{R}^{2n*}$. Applied to $\mathfrak{g}_{-1}$ and translated to geometry, the description of the torsion in 4.1.3 immediately implies that τ is a section of $\ker(\pi_{1,1}) \otimes TM \subset \Lambda^2 T^*M \otimes TM$.

For the second claim, we need an alternative description of $\ker(\pi_{1,1}) \subset \Lambda^2 V^*$. Take a map $f \in S^{1,1}V^*$ and an element $A \in \mathcal{Q}^V$, and consider the mapping $(x, y) \mapsto f(A(x), y)$. This actually is skew symmetric since $A^2 = A \circ A = -|A|^2\mathrm{id}$, and hence

$$|A|^2 f(A(x), y) = f(A^2(x), A(y)) = -|A|^2 f(x, A(y)) = -|A|^2 f(A(y), x).$$

This shows that we obtain a map $S^{1,1}V^* \otimes \mathcal{Q}^V \to \Lambda^2 V^*$, which is evidently injective. The image of this mapping forms a subspace of dimension $3n(2n-1)$, which makes it a candidate for a complement to $\Lambda^{1,1}V^*$. To see that this image coincides with $\ker(\pi_{1,1})$, it suffices to consider the case that $|A| = 1$, i.e. $A^2 = -\mathrm{id}$. But then we can find $B \in \mathcal{Q}^V$ such that $\{A, B, A \circ B\}$ is a distinguished basis. Using this basis to compute $\pi_{1,1}$, we immediately see that $(x, y) \mapsto f(A(x), y)$ lies in $\ker(\pi_{1,1})$. In particular, $\ker(\pi_{1,1}) \cong S^{1,1}V^* \otimes \mathcal{Q}^V$. Applying this to $\mathfrak{g}_{-1}$ and passing to associated bundles we see that $\ker(\pi_{1,1}) \otimes TM \cong S^{1,1}T^*M \otimes \mathcal{Q} \otimes TM$.

To obtain the description of the irreducible subbundle in which τ has its values, we again have to return to the description of 4.1.3. There we saw that the harmonic part in

$$(S^2\mathbb{R}^2 \boxtimes \Lambda^2\mathbb{R}^{2n*}) \otimes (\mathbb{R}^{2*} \boxtimes \mathbb{R}^{2n})$$

is the intersection of the kernels of the two natural contractions defined on that space, which have values in $\mathbb{R}^2 \boxtimes (\Lambda^2\mathbb{R}^{2n*} \otimes \mathbb{R}^{2n})$, respectively, in $(S^2\mathbb{R}^2 \otimes \mathbb{R}^{2*}) \boxtimes \mathbb{R}^{2n*}$.

In the picture of a prequaternionic vector space $(V, \mathcal{Q}^V)$, the counterparts are easy to identify by looking at dimensions. From above, we know that we can start from $S^{1,1}V^* \otimes \mathcal{Q}^V \otimes V$, so there are two obvious constractions. On the one hand, we can map $f \otimes A \otimes v$ to $f \otimes A(v)$ to obtain a contraction with values in $S^{1,1}V^* \otimes V$, which is evidently surjective. This space has dimension $4n^2(2n-1)$, so it corresponds to $\mathbb{R}^2 \boxtimes (\Lambda^2\mathbb{R}^{2n*} \otimes \mathbb{R}^{2n})$. On the other hand, we can use the evident contraction $S^{1,1}V^* \otimes V \to V^*$, to obtain a contraction with values in $V^* \otimes \mathcal{Q}^V$, which must be surjective by irreducibility of V^*. Since $V^* \otimes \mathcal{Q}^V$ has dimension $12n$, it corresponds to $(S^2\mathbb{R}^2 \otimes \mathbb{R}^{2*}) \boxtimes \mathbb{R}^{2n*}$. Applying this to $V = \mathfrak{g}_{-1}$ and passing to associated bundles, the last claim of (1) follows.

(2) Again, we have to adapt the results of 4.1.3 to the prequaternionic setting. From part (1) we know that

$$\Lambda^2\mathbb{R}^2 \boxtimes S^2\mathbb{R}^{2n*} \subset \Lambda^2(\mathbb{R}^2 \boxtimes \mathbb{R}^{2n*})$$

corresponds to $\Lambda^{1,1}V^* \subset \Lambda^2 V^*$. On the other hand, we have seen above how to decompose $L(V,V)$ into four summands. This corresponds to the decomposition

$$(\mathbb{R}^2 \boxtimes \mathbb{R}^{2n*}) \otimes (\mathbb{R}^{2*} \boxtimes \mathbb{R}^{2n}) \cong (\mathbb{R} \oplus \mathfrak{sl}(\mathbb{R}^2)) \boxtimes (\mathbb{R} \oplus \mathfrak{sl}(\mathbb{R}^{2n})).$$

Looking at the dimensions, we see that the summand $\mathbb{R} \boxtimes \mathfrak{sl}(\mathbb{R}^{2n})$ in the right–hand side corresponds to $L^0_Q(V,V)$. From 4.1.3 we know that the cohomology component corresponding to the curvature ρ is contained in $\Lambda^2\mathbb{R}^2 \boxtimes S^2\mathbb{R}^{2n*} \otimes \mathfrak{sl}(\mathbb{R}^{2n})$. Applying this to $\mathfrak{g}_{-1}$ and passing to associated bundles, we see that ρ is a section of the irreducible subbundle of

$$\Lambda^{1,1}T^*M \otimes L^0_Q(TM,TM) \subset \Lambda^2 T^*M \otimes L(TM,TM)$$

corresponding to the maximal possible weight. In particular, this implies that it lies in the kernel of the Ricci type contraction. □

Step (F): In order to have all tractor bundles available, let us deal with the choice $G = SL(n+1,\mathbb{H})$. Then all G reperesentations live in tensor products of standard representation on $\mathbb{R}^{4n+4} = \mathbb{H}^{n+1}$ and this representation space is quaternionic, i.e. it enjoys a G–invariant quaternionic structure. Furthermore, there is the P–invariant quaternionic one–dimensional subspace spanned by the highest weight vector. This quaternionic structure and filtration (in particular the natural endomorphisms I, J and K) are inherited on the standard tractor bundle $\mathcal{T} = \mathcal{G} \times_P \mathbb{H}^{n+1}$ and each reduction of the structure group to G_0 leads to the grading

$$\mathcal{T} = \operatorname{gr}\mathcal{T} = \mathcal{T}_0 \oplus \mathcal{T}^1$$

by quaternionic subbundles.

Compared to the case of projective structures, the situation is similar but a bit more complicated here, since the bundle $\mathcal{T}_0$ is not just a weighted tangent bundle. Indeed, $\mathcal{T}_0$ has a canonical quaternionic structure, while no such structure is present on a weighted tangent bundle, which only carries a prequaternionic structure. From the form of the Lie algebra it is evident, that TM can be identified with the bundle $L_{\mathbb{H}}(\mathcal{T}^1, \mathcal{T}_0)$ of quaternionic linear maps.

4.1.9. Four–dimensional conformal Riemannian structures. We next discuss the case $n = 1$ of almost quaternionic structures, i.e. the real form $\mathfrak{g} := \mathfrak{sl}(2,\mathbb{H})$ of $\mathfrak{sl}(4,\mathbb{C})$. By definition $\mathfrak{g}$ is the set of quaternionic 2×2–matrices $\left(\begin{smallmatrix} a & z \\ x & b \end{smallmatrix}\right)$ with $a, b, x, z \in \mathbb{H}$ such that $a + b$ has vanishing real part. The $|1|$–grading on $\mathfrak{g}$ is as in 4.1.8, but now $\mathfrak{g}_{\pm 1} \cong \mathbb{H}$ corresponding to the entries z, respectively, x and $\mathfrak{g}_0 = \mathfrak{s}(\mathbb{H} \oplus \mathbb{H})$. In this case, there are two interesting choices for a group with Lie algebra $\mathfrak{g}$, namely $G = PSL(2,\mathbb{H})$ and $\tilde{G} = SL(2,\mathbb{H})$. The parabolic subgroups $P \subset G$ and $\tilde{P} \subset \tilde{G}$ are the stabilizers of the quaternionic line spanned by the first vector in the standard basis of $\mathbb{H}^2$, and thus $G/P = \tilde{G}/\tilde{P} \cong \mathbb{H}P^1$ is the quaternionic projective line. As in 4.1.8, the subgroup $\tilde{G}_0 \subset \tilde{P}$ consists of all block diagonal matrices contained in $\tilde{G}$, i.e., all matrices of the form $\left(\begin{smallmatrix} p & 0 \\ 0 & q \end{smallmatrix}\right)$ with $p, q \in \mathbb{H}$ such that $|pq| = 1$. The subgroup $G_0 \subset P$ is the quotient of $\tilde{G}_0$ by its center $\{\pm \mathrm{id}\}$. We will denote elements of $\tilde{G}_0$ by pairs (p,q) and also use this notation for the corresponding class in G_0. Mapping (p,q) to $(|q|, \frac{p}{|p|}, \frac{q}{|q|})$ induces isomorphism from $\tilde{G}_0$ to $\mathbb{R}^+ \times Sp(1) \times Sp(1)$ and from G_0 to the quotient of $\mathbb{R}^+ \times Sp(1) \times Sp(1)$ by the subgroup consisting of $(1,1,1)$ and $(1,-1,-1)$.

The action of $\tilde{G}_0$ on $\mathfrak{g}_{-1} \cong \mathbb{H}$ is given by $(p,q) \cdot x = qxp^{-1}$ and this visibly descends to G_0. In the picture of $\tilde{G}_0$ from above, this corresponds to $(r,p,q) \cdot x = rqx\bar{p}$. It is well known that assigning to $(p,q) \in Sp(1) \times Sp(1)$ the map $x \mapsto qx\bar{p}$ defines a two fold covering $Sp(1)\times Sp(1) \to SO(\mathbb{H})$ and thus an isomorphism $Sp(1)\times Sp(1) \to Spin(\mathbb{H})$, where the inner product on $\mathbb{H}$ is the standard positive definite one. Hence, we conclude that the adjoint action of $\tilde{G}_0$ and G_0 on $\mathfrak{g}_{-1}$ identifies G_0 with the conformal group $CSO(\mathfrak{g}_{-1})$ and $\tilde{G}_0$ with the conformal spin group $CSpin(\mathfrak{g}_{-1})$. This immediately implies that a regular infinitesimal flag structure of type (G,P) is equivalent to a conformal Riemannian structure on a 4–manifold, while for type $(\tilde{G},\tilde{P})$ we get conformal Riemannian spin structures.

From 4.1.3 we see that the harmonic curvature is given by two curvatures, and as in 4.1.4 these are the self–dual and the anti–self–dual parts of the Weyl curvature.

4.1.10. Four–dimensional conformal Lorentzian structures. We have seen in 3.2.10 that among the real Lie algebras $\mathfrak{su}(p,q)$, only $\mathfrak{su}(p,p)$ admits a $|1|$–grading, and this grading is unique. It corresponds to the $(p-1)$st of the $2p-1$ simple roots and thus the "middle" node of the Satake diagram. We discuss the corresponding structures only in the case $p=2$, for which there is a conformal interpretation.

From example (2) of 2.3.4 we know that $\mathfrak{g} := \mathfrak{su}(2,2)$ consists of all block matrices $\begin{pmatrix} A & Z \\ X & -A^* \end{pmatrix}$ with $A \in \mathfrak{gl}(2,\mathbb{C})$ such that $\operatorname{tr}(A)$ is real and $X, Z \in \mathfrak{u}(2)$. Choosing the group $G = SU(2,2)$ with Lie algebra $\mathfrak{g}$, and $P \subset G$ as the subgroup of block upper triangular matrices contained in G, then G_0 consits of all block diagonal matrices contained in G. We can identify G_0 with $\{\Phi \in GL(2,\mathbb{C}) : \det(\Phi) \in \mathbb{R}\}$ via

$$\Phi \mapsto \begin{pmatrix} \Phi & 0 \\ 0 & (\Phi^*)^{-1} \end{pmatrix},$$

and in this picture the adjoint action on $\mathfrak{g}_{-1}$ is given by $\Phi \cdot X = (\Phi^*)^{-1} X \Phi^{-1}$. The determinant defines a quadratic form on the real vector space $\mathfrak{u}(2)$ of skew Hermitian 2×2–matrices. Looking at the obvious basis for $\mathfrak{u}(2)$ one sees that the corresponding real inner product has Lorentzian signature $(3,1)$.

Now as in 4.1.9 and 4.1.4 this implies that normal parabolic geometries of type (G,P) are equivalent to 4–dimensional conformal structures in Lorentzian signature $(3,1)$. The interpretation of the two harmonic curvatures as the self–dual and the anti–self–dual part of the Weyl curvature also works as before.

4.1.11. Almost Lagrangean structures. Next we discuss $|1|$–gradings on Lie algebra of type C_n. From 3.2.3 we know that on algebras of type C_n only the last root gives rise to a $|1|$–grading in the complex case. Since $C_2 \cong B_2$, that grading has already been discussed in connection with three–dimensional conformal structures in 4.1.2, so we assume $n > 2$. Under this assumption, we see from the list of Satake diagrams in Appendix B, that this grading exists for two real forms, but we will restrict to the geometry corresponding to the split real form $\mathfrak{sp}(2n,\mathbb{R})$.

Step (A): From 2.2.13 we know that the fundamental representation of $\mathfrak{sp}(2n,\mathbb{R})$ corresponding to the last simple root can be realized as a subspace of $\Lambda^n\mathbb{R}^{2n}$, namely as the kernel of the natural map to $\Lambda^{n-2}\mathbb{R}^{2n}$ induced by the symplectic form on $\mathbb{R}^{2n}$. Hence, we may realize the corresponding parabolic subalgebra $\mathfrak{p} \subset \mathfrak{g} := \mathfrak{sp}(2n,\mathbb{R})$ as the stabilizer of the highest weight line in that representation. In the matrix

4.1. STRUCTURES CORRESPONDING TO |1|–GRADINGS

realization for $\mathfrak{g}$ from example (4) of 2.2.6, the highest weight line is spanned by the wedge product of the first n basis vectors, and we get

$$\mathfrak{p} = \left\{ \begin{pmatrix} -A^t & B \\ 0 & A \end{pmatrix} \right\} \subset \mathfrak{g} = \left\{ \begin{pmatrix} -A^t & B \\ C & A \end{pmatrix} \right\}$$

with $n \times n$–matrices A, B and C such that B and C are symmetric. Hence, $\mathfrak{p}$ is the stabilizer of the isotropic n–dimensional subspace spanned by the first n basis vectors. Choosing the group $G := Sp(2n, \mathbb{R})$ we may therefore use the stabilizer P of that subspace as a parabolic subgroup for the given grading. This means that P consists of all those matrices in G which are block upper triangular with two blocks of size n. From this description it is clear that the homogeneous model G/P is the space of all isotropic n–dimensional subspaces in the symplectic vector space $\mathbb{R}^{2n}$. Such maximal isotropic subspaces are usually called *Lagrangean subspaces*, whence the variety of all these subspaces is referred to as the *Lagrange-Grassmann manifold*. This is also the reason why the corresponding geometries are called almost Lagrangean.

The Levi subgroup $G_0 \subset P$ is formed by the block diagonal matrices contained in G. This preserves the subspaces spanned by the first, respectively, last n basis vectors, and it is easy to see that the action of G_0 on either of these subspaces induces an isomorphism $G_0 \cong GL(n, \mathbb{R})$. We fix the identification determined by the subspace spanned by the last n basis vectors, which is also in accordance with the matrix presentation for the Lie algebras above. The reason for this choice is that then we obtain $\mathfrak{g}_{-1} = S^2 \mathbb{R}^n$, and $\mathfrak{g}_1 = S^2 \mathbb{R}^{n*}$, which fits with these two spaces corresponding to the tangent respectively cotangent bundle.

Step (B): Since the semisimple part of $\mathfrak{g}_0$ is $\mathfrak{sl}(n, \mathbb{R})$, the discussion of all G_0–representations is just a simplified repetition of the Step (B) in 4.1.3. Indeed, we are again dealing with a split real form, so there is no difference between the complexified real and complex representations. The fundamental weights $\lambda_1, \ldots \lambda_{n-1}$ of $\mathfrak{g}$ lead to the exterior powers $\Lambda^k \mathbb{R}^n$, $k = 1, \ldots, n-1$, as the $\mathfrak{p}_+$–irreducible subspaces of the same highest weight. The last fundamental weight gives rise to the one dimensional representation on which the grading element acts by the scalar $\frac{n}{2}$. As we have seen in 3.2.12, the grading element acts on the weight $\lambda = a_1 \lambda_1 + \cdots + a_n \lambda_n$ by the scalar computed with the help of the last row of the inverse Cartan matrix (determined by the position of the cross in the Satake diagram), i.e. in our case by $\sum_{i=1}^n \frac{i}{2} a_i$. The grading of the Lie algebra has got irreducible components with highest weights

$$\begin{array}{c}
0 \quad 0 \quad\quad 2 \;\; -2 \\
\circ\!\!-\!\!\circ\cdots\!\!-\!\!\circ\!\!\Leftarrow\!\!\times
\end{array} \oplus \begin{array}{c}
1 \quad 0 \quad\quad 1 \;\; -1 \\
\circ\!\!-\!\!\circ\cdots\!\!-\!\!\circ\!\!\Leftarrow\!\!\times \\
0 \quad 0 \quad\quad 0 \quad 0 \\
\circ\!\!-\!\!\circ\cdots\!\!-\!\!\circ\!\!\Leftarrow\!\!\times
\end{array} \oplus \begin{array}{c}
2 \quad 0 \quad\quad 0 \quad 0 \\
\circ\!\!-\!\!\circ\cdots\!\!-\!\!\circ\!\!\Leftarrow\!\!\times
\end{array}$$

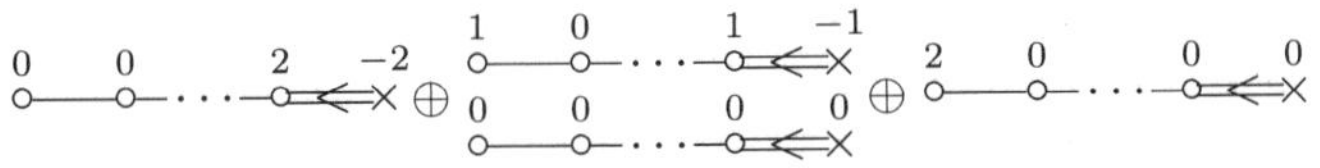

Step (C): As we have seen in Step (A), we get $G_0 \cong GL(n, \mathbb{R})$ and $\mathfrak{g}_{-1} \cong S^2 \mathbb{R}^n$ as a representation of G_0. From this it follows easily that a first order structure with structure group G_0 on a manifold M of dimension $\frac{n(n+1)}{2}$ is given by an auxiliary rank n–vector bundle $E \to M$ and an isomorphism $\Phi : S^2 E \to TM$. Given the G_0–structure, we simply obtain E as the associated bundle with respect to the standard representation. Conversely, given E and an isomorphism $\Phi : S^2 E \to TM$, let $\mathcal{G}_0$ be the full linear frame bundle on E. Together with Φ^{-1}, a point in the fiber of $\mathcal{G}_0$ over x induces a linear isomorphism $T_x M \to S^2 E_x \to S^2 \mathbb{R}^n \cong \mathfrak{g}_{-1}$, which we can use to define a soldering form. In the homogeneous model, the bundle E is the dual of the tautological bundle.

Steps (D) and (E): There is only one element of length two in the subset $W^{\mathfrak{p}} \subset W$ in the Weyl group, namely $s_n \circ s_{n-1}$. One easily verfies that the corresponding cohomology component sits in homogeneity -1. Now $\Lambda^2(S^2\mathbb{R}^{n*})$ is an irreducible representation of $\mathfrak{sl}(n,\mathbb{R})$, so our cohomology component is the highest weight subspace in the space $\Lambda^2(S^2\mathbb{R}^{n*}) \otimes S^2\mathbb{R}^n$, which corresponds to $\Lambda^2T^*M \otimes TM$. This implies that the harmonic curvature is concentrated in one torsion component and this part of the torsion will be shared by all the distinguished connections γ; cf. Theorem 4.1.1. To describe this highest weight bit, one observes that, up to multiples, there is a unique contraction

$$\Lambda^2(S^2\mathbb{R}^{n*}) \otimes S^2\mathbb{R}^n \to \otimes^3\mathbb{R}^{n*} \otimes \mathbb{R}^n.$$

The highest weight bit is simply the kernel of this contraction.

This is easily translated to geometry. To use an abstract index notation, we denote the auxiliary bundle E by $\mathcal{E}^A$ and its dual E^* by $\mathcal{E}_A$. Then the tangent bundle is isomorphic to S^2E, so it has two symmetric upper indices. Likewise, the cotangent bundle has two symmetric lower indices. Since $\mathcal{G}_0 \to M$ is the full linear frame bundle of E, a principal connection on $\mathcal{G}_0$ is equivalent to a linear connection on E. Via the isomorphism $S^2E \cong TM$, such a connection induces a linear connection on TM. In abstract index notation, the torsion of such a connection then has the form $T = T_{ABCD}{}^{EF}$ with symmetry in each of the pairs AB, CD and EF of indices, and such that $T_{CDAB}{}^{EF} = -T_{ABCD}{}^{EF}$. Evidently, all contractions of one of the upper indices into one of the lower indices agree up to sign. The basic invariant of an almost Lagrangean structure is the part of the torsion, which is contained in the kernel of this trace.

An explicit formula for this tracefree part can be computed along the lines of what we did for almost Grassmannian structures in 4.1.4. Then one may continue to compute one of the preferred connections and determine the corresponding Rho tensor, but we do not go into detail on these issues.

Step (F): Since all representations of $Sp(2n,\mathbb{R})$ live in the tensor products of the first fundamental representation, i.e. the standard representation on $\mathbb{R}^{2n}$, all tractor bundles can be obtained from tensor products of the standard tractor bundle. The standard tractor bundle $\mathcal{T} = \mathcal{G} \times_P \mathbb{R}^{2n}$ comes equipped with a canonical symplectic structure inherited from the skew symmetric bilinear form on $\mathbb{R}^{2n}$ used to define $\mathfrak{sp}(2n,\mathbb{R})$. In view of our conventions described in Step (A) it is clear that $\mathcal{T}$ contains a natural subbundle $\mathcal{T}^1 \cong E^*$, and the quotient $\mathcal{T}/\mathcal{T}^1$ is isomorphic to E. This gives rise to natural filtrations on tensor powers of the standard tractor bundles and hence on general tractor bundles.

4.1.12. Almost spinorial structures. From 3.2.3 we know that on complex Lie algebras of type D_n, we always have the $|1|$–grading, which leads to conformal structures as discussed in 4.1.2. For $n = 4$, this is the unique $|1|$–grading up to automorphism, but for $n \geq 5$, there is a second, non–isomorphic $|1|$–grading. This corresponds to either of the last two simple roots. From the list of Satake diagrams in Table B.4 in Appendix B we conclude that there is a corresponding real $|1|$–grading on the real forms of type $\mathfrak{so}(n,n)$ and $\mathfrak{so}^*(4\ell)$. We will restrict our discussion to the split real form $\mathfrak{so}(n,n)$ for $n \geq 5$. The interpretation of the geometric structures to be discussed below also makes sense for $n = 4$, but there the result is equivalent to split signature conformal structures in dimension 6.

Steps (A) and (B): We put $\mathfrak{g} := \mathfrak{so}(n,n)$ for $n \geq 5$. Following the general principles from 3.2.5, we can realize the parabolic subalgebra $\mathfrak{p} \subset \mathfrak{g}$ as the stabilizer of a highest weight line in any irreducible representation whose highest weight is a multiple of the last fundamental weight. One possibility would be to take a spin representation, but it is easier to use the representation on self–dual n–forms. In the matrix presentation from example (2) of 2.2.6 (in which the real matrices form the split real form $\mathfrak{g}$ of $\mathfrak{so}(2n, \mathbb{C})$) it is clear that the resulting highest weight line is spanned by the wedge product of the first n basis vectors. Hence, in terms of block matrices with two blocks of size n each we get

$$\mathfrak{p} = \left\{ \begin{pmatrix} -A^t & B \\ 0 & A \end{pmatrix} \right\} \subset \mathfrak{g} = \left\{ \begin{pmatrix} -A^t & B \\ C & A \end{pmatrix} \right\},$$

where the matrices B and C are skew symmetric. This indicates already that these geometries are very similar to almost Lagrangean structures.

As the group G with Lie algebra $\mathfrak{g}$, we choose $SO_0(n,n)$, and then we can take the stabilizer of the isotropic n–dimensional subspace spanned by the first n basis vectors as the parabolic subgroup P for the given grading. Hence, these are block upper triangular matrices and the Levi subgroup G_0 corresponds to block diagonal matrices. The homogeneous model G/P can be realized as the space of all self–dual isotropic n–dimensional subspaces in $\mathbb{R}^{(n,n)}$. However, using the spin group rather than G, one may also view G/P as the orbit of the highest weight line in the projectivized spin representation. This orbit consists of the lines spanned by the so–called pure spinors, which also motivates the name "almost spinorial structures" for the corresponding geometries.

The action of G_0 on the space spanned by the last n basis vectors induces an isomorphism $G_0 \cong GL(n, \mathbb{R})$, which again shows the similarity to the almost Lagrangean case. In view of this similarity, we keep this presentation short. The convention is chosen in such a way that $\mathfrak{g}_{-1} \cong \Lambda^2\mathbb{R}^n$ and $\mathfrak{g}_1 \cong \Lambda^2\mathbb{R}^{n*}$ as representations of G_0. As before, all the basic representations of G_0 can be obtained from tensor powers of the standard representation and its dual and from one–dimensional representations.

Step (C): First order structures with structure group $G_0 \subset GL(\mathfrak{g}_{-1})$ make sense on manifolds of dimension $\frac{n(n-1)}{2}$. We have seen above that $\mathfrak{g}_{-1} \cong \Lambda^2\mathbb{R}^n$ as a G_0–module, which leads to a description analogous to almost Lagrangean structures. Similarly to that case, one proves that these structures are given by an auxiliary vector bundle E of rank n together with an isomorphism $\Phi : \Lambda^2 E \to TM$. On the homogeneous model, E is again the tautological bundle.

Let us briefly describe why these structures specialize to conformal structures for $n = 4$. The decomposable elements in $\Lambda^2 E$ define a cone in each tangent space. By the Plücker relations, an element is decomposable if and only if the wedge product with itself is trivial. But if $n = 4$, then this wedge product has values in the line bundle $\Lambda^4 E$, so any trivialization of this line bundle gives a metric on $\Lambda^2 E$ having the given cone as its null cone. Alternatively, for $n = 4$ decomposability is equivalent to non–invertibilty of an element viewed as a skew symmetric matrix. The determinant on antisymmetric matrices allows a square root called the Pfaffian, and this is a quadratic form for $n = 4$.

Steps (D) and (E): There is only one element of length two in the subset $W^{\mathfrak{p}} \subset W$, namely $s_n \circ s_{n-2}$. The corresponding cohomology component sits in homogeneity -1 (this uses $n > 4$) and it is the highest weight subspace in $\Lambda^2(\Lambda^2\mathbb{R}^{n*}) \otimes \Lambda^2\mathbb{R}^n$. This highest weight subspaces can be characterized as the kernel of the unique (up to multiples) nonzero contraction

$$\Lambda^2(\Lambda^2\mathbb{R}^{n*}) \otimes \Lambda^2\mathbb{R}^n \to \otimes^3\mathbb{R}^{n*} \otimes \mathbb{R}^n.$$

Geometrically, this means that the full harmonic curvature will be given as the tracefree part of the torsion of any connection on the G_0–structure. The connections on TM coming from the G_0–structure are exactly those, which are induced from linear connections on E via the isomorphism $\Lambda^2 E \cong TM$.

Step (F): The standard tractor bundle $\mathcal{T} = \mathcal{G} \times_P \mathbb{R}^{2n}$ comes equipped by the canonical split signature scalar product inherited from the defining one on $\mathbb{R}^{2n}$. Moreover, it comes with a natural subbundle $\mathcal{T}^1 \subset \mathcal{T}$ which is isomorphic to E^* and such that $\mathcal{T}/\mathcal{T}^1 \cong E$. This gives rise to filtrations on tensor products, and hence on more general tractor bundles. Enlarging G to the spin group $Spin(n,n)$ also the spin representations give rise to tractor bundles, but we do not go into details here.

4.2. Parabolic contact structures

These are parabolic geometries for which the underlying geometric structure is a contact structure with some additional structure on the contact subbundle. The most important example is provided by CR–structures, for which the additional structure is a complex structure on the contact subbundle. We start by recalling some background on contact structures.

4.2.1. Contact structures and contact connections. Recall from linear algebra that nondegenerate skew symmetric bilinear forms exist only on vector spaces of even dimension, and there they are uniquely determined up to isomorphism. Looking at an even dimensional smooth manifold M, a smooth family of skew symmetric bilinear forms on the tangent spaces of M is just a two–form τ on M. A form $\tau \in \Omega^2(M)$ such that $\tau_x : T_xM \times T_xM \to \mathbb{R}$ is nondegenerate for each $x \in M$, is called an *almost symplectic form* on M, and τ is called a *symplectic form* if, in addition, $d\tau = 0$. In this case (M, τ) is called a *symplectic* manifold. Note that the nondegeneracy condition can also be expresses as the fact that $\tau \wedge \dots \wedge \tau$ (with half the dimension many factors) is a volume form on M. Since there is an obvious symplectic form on $\mathbb{R}^{2n}$, they exist locally on any manifold of even dimension. The question of global existence of symplectic structures is surprisingly difficult with lots of recent progress, for example, via the Seiberg–Witten equations; see e.g. [**Tau94**].

The standard example of a symplectic structure is provided by the cotangent bundle T^*N of an arbitrary smooth manifold N. This carries a tautological one form $\alpha \in \Omega^1(T^*N)$ defined as follows. Let $\pi : T^*N \to N$ be the projection and $T\pi : TT^*N \to TN$ its tangent map. Then for $\phi \in T_x^*N$ and $\xi \in T_\phi T^*N$ one defines $\alpha(\phi)(\xi) := \phi(T\pi \cdot \xi)$. Choosing local coordinates q^i on M and using the induced coordinates (q^i, p_i) on T^*N one gets $\alpha = \sum_i p_i dq^i$ and hence $d\alpha = \sum dp_i \wedge dq^i$. This is immediately seen to be nondegenerate and thus defines a symplectic structure. Putting $N = \mathbb{R}^n$, we obtain the (constant) standard linear symplectic structure on

$\mathbb{R}^{2n}$, which usually is written in terms of local coordinates as

$$((x_1,\dots,x_{2n}),(y_1,\dots,y_{2n}))\mapsto \sum_{i=1}^n(x_iy_{n+i}-x_{n+i}y_i).$$

A basic result in symplectic geometry is the Darboux theorem, which states that for any symplectic form τ on M and around each point $x\in M$, there exist local coordinates (q^i,p_i) for M such that $\tau=\sum dp_i\wedge dq^i$. In particular, symplectic structures do not have any local invariants.

Looking for an odd–dimensional analog of this concept, one is led to the notion of a *contact form* $\alpha\in\Omega^1(M)$ on a smooth manifold M of dimension $2n+1$, i.e. a form such that $\alpha\wedge(d\alpha)^n$ is a volume form on M. Again there are simple examples of such forms on $\mathbb{R}^{2n+1}$ and hence locally on any manifold of odd dimension. Namely, using coordinates (t,q_i,p_i), one defines $\alpha=dt+\sum p_i dq^i$, which is immediately seen to be a contact form. There is a contact version of the Darboux theorem which says that given any contact form, one may locally choose coordinates in which it is given by the above expression. So contact forms do not have local invariants either.

If α is a contact form on M, then by definition $\alpha(x)\neq 0$ for all $x\in M$, so the pointwise kernels of α form a codimension one subbundle H of the tangent bundle TM, called the *contact subbundle.* Again by construction, the restriction of $d\alpha$ to $H\times H$ is a nondegenerate skew symmetric bilinear form. The quotient bundle $Q:=TM/H$ is a real line bundle, which is trivialized by α.

Defining $T^{-2}M=TM$ and $T^{-1}M:=H$, we obtain a filtration of the tangent bundle of M with associated graded $\operatorname{gr}(TM)=Q\oplus H$. The condition on compatibility with the Lie bracket from Definition 3.1.7 is vacuous in this case, so this makes M into a filtered manifold, and we can look at the Levi bracket $\mathcal{L}:H\times H\to Q$ induced by this filtration. For sections ξ and η of $H=\ker(\alpha)$ the definition of the exterior derivative implies that $d\alpha(\xi,\eta)=-\alpha([\xi,\eta])$, so we see that, viewing α as a trivialization of Q, we have $\alpha\circ\mathcal{L}=-d\alpha$. Hence, $\mathcal{L}$ is nondegenerate, and the symbol algebra in each point is $\mathbb{R}\oplus\mathbb{R}^{2n}$ with the bracket $\mathbb{R}^{2n}\times\mathbb{R}^{2n}\to\mathbb{R}$ being nondegenerate. This graded Lie algebra is called the real *Heisenberg algebra* $\mathfrak{h}_{2n+1}$.

Now one defines a *contact structure* on a smooth manifold M of dimension $2n+1$ as a smooth subbundle $H\subset TM$ of rank $2n$ such that, putting $Q=TM/H$, the Levi bracket $\mathcal{L}:H\times H\to Q$ is nondegenerate in each point. More elegantly, one can say that a contact manifold is a filtered manifold for which each symbol algebra is a Heisenberg algebra.

As we have observed in 3.1.7, the last statement implies that for any contact structure $H\subset TM$, the associated graded to the tangent bundle, $\operatorname{gr}(TM)=Q\oplus H$ has a natural frame bundle with structure group $\operatorname{Aut}_{\mathrm{gr}}(\mathfrak{h}_{2n+1})$, the group of automorphisms of the graded Lie algebra $\mathfrak{h}_{2n+1}$. The fiber of this bundle over $x\in M$ is just the space of isomorphisms $\mathfrak{h}_{2n+1}\to(Q_x\oplus H_x,\mathcal{L})$ of graded Lie algebras. Let us first determine the group $\operatorname{Aut}_{\mathrm{gr}}(\mathfrak{h}_{2n+1})$.

LEMMA 4.2.1. *Any automorphism of $\mathfrak{h}_{2n+1}$ is uniquely determined by its restriction $\mathbb{R}^{2n}\to\mathbb{R}^{2n}$. Viewed as a subgroup of $GL(2n,\mathbb{R})$, the group $\operatorname{Aut}_{\mathrm{gr}}(\mathfrak{h}_{2n+1})$ is generated by $Sp(2n,\mathbb{R})$, multiples of the identity, and the diagonal matrix $\mathbb{I}_{n,n}$ with the first n entries equal to 1 and the other n equal to -1.*

The subgroup of those automorphisms, which, in addition, preserve an orientation on $\mathbb{R}\subset\mathfrak{h}_{2n+1}$ is the conformal symplectic group $CSp(2n,\mathbb{R})$ generated by $Sp(2n,\mathbb{R})$ and multiples of the identity.

PROOF. As before let $[\ ,\ ] : \mathbb{R}^{2n} \times \mathbb{R}^{2n} \to \mathbb{R}$ be the bracket in $\mathfrak{h}_{2n+1}$ given by the standard symplectic form on $\mathbb{R}^n$. Surjectivity of this bracket immediately implies the first statement. For $A \in Sp(2n, \mathbb{R})$ we by definition have $[Ax, Ay] = [x, y]$, so this gives rise to an automorphism. Likewise, for $a \in \mathbb{R}\setminus 0$, we get $[ax, ay] = a^2[x, y]$, so multiples of the identity are in $\mathrm{Aut}_{\mathrm{gr}}(\mathfrak{h}_{2n+1})$, too. Finally, by definition of $[\ ,\ ]$, we see that $[\mathbb{I}_{n,n}x, \mathbb{I}_{n,n}y] = -[x, y]$.

Conversely, suppose that $A \in GL(2n, \mathbb{R})$ and $a \in \mathbb{R} \setminus 0$ are such that $[Ax, Ay] = a[x, y]$. If $a < 0$, then replace A by $\mathbb{I}_{n,n}A$ to get an element for which $a > 0$. Dividing then by $\sqrt{a}$, we obtain an element which preserves $[\ ,\ ]$ and thus lies in $Sp(2n, \mathbb{R})$. The last statement is then obvious. □

The first statement implies that any isomorphism $\mathfrak{h}_{2n+1} \to (H_x \oplus Q_x, \mathcal{L}_x)$ is uniquely determined by the component mapping $\mathbb{R}^{2n}$ to H_x. Hence, the natural frame bundle for $H \oplus Q$ can be viewed as a subbundle of the frame bundle of H. If we choose an orientation on Q (for example the one induced by a contact form), then the structure group of the natural frame bundle is reduced to $CSp(2n, \mathbb{R})$.

We collect some important properties of contact structures in the following:

PROPOSITION 4.2.1. *Let $H \subset TM$ be a contact structure on a smooth manifold M of dimension $2n+1$ with quotient bundle $Q = TM/H$. Let $p : E \to M$ be the natural frame bundle for $H \oplus Q$ with structure group $\mathrm{Aut}_{\mathrm{gr}}(\mathfrak{h}_{2n+1})$. Let $\mathcal{L} : \Lambda^2 H \to Q$ be the Levi bracket and let $\Lambda^2_0 H \subset \Lambda^2 H$ be the kernel of $\mathcal{L}$.*

(1) Locally, there exists a contact form α which has H as its contact subbundle, and this form is unique up to multiplication by a nowhere vanishing function. In particular, contact structures have no local invariants. There exists a global contact form α for H if and only if the quotient bundle Q is orientable and hence trivial.

(2) Any principal connection on E is completely determined by the induced linear connection on the vector bundle H. A linear connections on H arises in this way if and only if the induced connection on $\Lambda^2 H$ preserves the subbundle $\Lambda^2_0 H$.

(3) If $\alpha \in \Omega^1(M)$ is a contact form with contact subbundle H, then there is a unique vector field r on M such that $\alpha(r) = 1$ and $i_r d\alpha = 0$. In particular, α induces an isomorphism $TM \cong H \oplus \mathbb{R}$.

(4) Given α as in (3), there is a linear connection ∇ on TM such that ∇ preserves the subbundle H, $\nabla\alpha = 0$, $\nabla d\alpha = 0$, and $\nabla r = 0$, and such that the restriction to H is induced by a principal connection on E as in (2).

PROOF. (1) The bundle $Q = TM/H$ is locally trivial, and a local trivialization α of this bundle can be viewed as a local one–form on M whose kernel in each point is the fiber of H. In this picture, $d\alpha|_{\Lambda^2 H} = -\alpha \circ \mathcal{L}$, which implies that α is a contact form. Conversely, a local contact form with contact subbundle H factors to a local trivialization of Q, so we get a bijective correspondence. From this, (1) follows immediately.

(2) Since E can be viewed as a subbundle of the frame bundle of H, a principal connection on E is uniquely determined by the induced linear connection on H. On the other hand, by definition, a linear map $A \in GL(2n, \mathbb{R})$ extends to an automorphism of $\mathfrak{h}_{2n+1}$ if and only if the induced map on $\Lambda^2\mathbb{R}^{2n*}$ preserves the line generated by $[\ ,\ ]$. By duality this is equivalent to the fact that the induced map on $\Lambda^2\mathbb{R}^{2n}$ preserves the kernel of $[\ ,\ ]$. From this, the description of the induced connections follows immediately.

(3) Since α is nowhere vanishing, we can locally find a vector field ξ such that $\alpha(\xi)$ is nowhere vanishing, and mutliplying by an appropriate function we may assume $\alpha(\xi) = 1$. Then we can look at the restriction of $i_\xi d\alpha$ to H, which defines a section of H^*. By nondegeneracy, there is a section η of H such that $i_\xi d\alpha = i_\eta d\alpha$, and $r := \xi - \eta$ has the required properties. If $\hat{r}$ has the same properties, then $\alpha(\hat{r} - r) = 0$, so $\hat{r} - r \in \Gamma(H)$. But then $i_{\hat{r}-r} d\alpha = 0$ implies $\hat{r} = r$ by nondegeneracy of $d\alpha$ on H, and uniqueness follows. The isomorphism $TM \cong H \oplus \mathbb{R}$ is then given by $\xi \mapsto (\xi - \alpha(\xi)r, \alpha(\xi))$.

(4) The choice of contact form α reduces the structure group of the frame bundle E from part (2) further to $Sp(2n, \mathbb{R})$. Explicitly, the fiber over $x \in M$ of this reduction is given by all linear isomorphisms $\phi : \mathbb{R}^{2n} \to H_x$ for which $d\alpha(\phi(v), \phi(w)) = [v, w]$. This bundle admits a principal connection, and we take the induced linear connection on H and extend it by the trival connection to a connection on $H \oplus \mathbb{R}$. Via the isomorphism from (3), this gives a linear connection on TM which preserves the subbundle H. By construction of the frame bundle, this linear connection satisfies $\nabla d\alpha = 0$ and by the trival extension we have $\nabla r = 0$. Since H is preserved, this easily implies $\nabla \alpha = 0$. □

The connections described in part (2) are called *contact connections* for the contact structure H. Given a choice α of contact form, the vector field r from (3) is called the *Reeb vector field* for α. Linear connections as in (4) are called contact connections adapted to the contact form α. There are evident analogs of (2) and (4) for partial connections (see 1.3.7) which leads to the notion of partial contact connections.

There is a contact analog of the canonical symplectic structure on a cotangent bundle. Namely, let M be a smooth manifold of dimension $n + 1$ and let $\mathcal{P}T^*M$ be the *projectivized cotangent bundle.* This is the $\mathbb{R}P^n$–bundle over M whose fiber over x is the space of all lines in T_x^*M. If ℓ is such a line, we define a hyperplane $H_\ell \subset T_\ell \mathcal{P}T^*M$ as the space of those tangent vectors, whose projection to T_xM is annihilated by ℓ. This can be viewed as the image of the kernel of the tautological one–form α on T^*M under the tangent map of the projection from $T^*M \setminus M$ (the complement of the zero section) to $\mathcal{P}T^*M$. Choosing a local section σ of this projection, the subbundle H is realized as the kernel of $\sigma^*\alpha$. Now in a point ϕ of T^*M, the restriction of $d\alpha(\phi)$ to $\ker(\alpha(\phi))$ is degenerate with null space given by those vertical tangent vectors which are multiples of the foot point ϕ. This immediately implies that $\sigma^*\alpha$ defines a local contact form for H. By part (1) of the proposition, any contact structure in dimension $2n + 1$ is locally isomorphic to $\mathcal{P}T^*M$.

4.2.2. Generalities on parabolic contact structures. Recall from 3.2.4 that a contact grading is a $|2|$–grading such that $\mathfrak{g}_{-2}$ is one–dimensional and the Lie bracket $\mathfrak{g}_{-1} \times \mathfrak{g}_{-1} \to \mathfrak{g}_{-2}$ is nondegenerate. This exactly means that the Lie algebra $\mathfrak{g}_- = \mathfrak{g}_{-2} \oplus \mathfrak{g}_{-1}$ is a Heisenberg algebra. In 3.2.4 and 3.2.10 we have seen that contact gradings exist only on simple Lie algebras and obtained a complete classification of both complex and real contact gradings.

Given a real contact grading and corresponding groups $P \subset G$, the first ingredient for an infinitesimal flag structure of type (G, P) is a filtration $TM = T^{-2}M \supset T^{-1}M$, where $\dim(M) = \dim(\mathfrak{g}_-)$ and $H := T^{-1}M$ has corank one. Regularity of the infinitesimal flag structure in particular requires that each symbol algebra of this filtration is isomorphic to $\mathfrak{g}_-$, i.e. that H defines a contact structure on M. The subgroup $G_0 \subset P$ acts on $\mathfrak{g}_-$ by Lie algebra automorphism, so it can be viewed as a subgroup of $GL(\mathfrak{g}_{-1})$, and the additional ingredient of a regular infinitesimal flag structure is a reduction of structure group of H to this subgroup.

If $H^1(\mathfrak{g}_-, \mathfrak{g})$ is concentrated in homogeneous degrees ≤ 0, then regular normal parabolic geometries are equivalent to regular infinitesimal flag structures and hence to contact structures with an additional reduction of structure group to $\mathrm{Ad}(G_0) \subset \mathrm{Aut}_{gr}(\mathfrak{g}_-)$. There is only one parabolic contact structure for which this condition is not satisfied, namely contact projective structures. These will be discussed separately in 4.2.6 below.

Next, let us move towards a description of harmonic curvature components. We will be less detailed here than in the case of $|1|$–gradings. A general approach to the description of harmonic curvature of arbitrary parabolic geometry will be developed using Weyl structures in Chapter 5. First we can prove a general fact on the cohomology for contact gradings, which heavily restricts the possibilities for harmonic curvatures of torsion type.

LEMMA 4.2.2. *Consider a contact grading* $\mathfrak{g} = \mathfrak{g}_{-2} \oplus \cdots \oplus \mathfrak{g}_{-2}$, *and the Kostant Laplacian* $\square$ *on* $\Lambda^2\mathfrak{g}_-^* \otimes \mathfrak{g}$. *Then*

$$\ker(\square) \cap (\Lambda^2\mathfrak{g}_-^* \otimes \mathfrak{g}_-) \subset (\Lambda_0^2\mathfrak{g}_{-1})^* \otimes \mathfrak{g}_-.$$

PROOF. Since the statement of the lemma is invariant under complexification, we may assume that we deal with a complex contact grading. The key point here is that Kostant's version of the Bott–Borel–Weyl theorem asserts that the highest weight of any irreducible component of $\ker(\square)$ occurs with multiplity one in $\Lambda^*\mathfrak{g}_-^* \otimes \mathfrak{g}$; see part (2) of Theorem 3.3.5.

Now as a $\mathfrak{g}_0$–module $\Lambda^2\mathfrak{g}_-^*$ decomposes as $\mathfrak{g}_{-2}^* \otimes \mathfrak{g}_{-1}^* \oplus \Lambda^2\mathfrak{g}_{-1}^*$. Via the decomposition $\Lambda^2\mathfrak{g}_{-1} = \Lambda_0^2\mathfrak{g}_{-1} \oplus \mathfrak{g}_{-2}$ induced by the bracket, the second summand decomposes further as the sum of $(\Lambda_0^2\mathfrak{g}_{-1})^*$ with $\mathfrak{g}_{-2}^* \cong \mathfrak{g}_2$. Tensoring with $\mathfrak{g}$ we again obtain a decomposition into three summands. The last of these is (isomorphic to) $\mathfrak{g}_2 \otimes \mathfrak{g}$. Since the same module is also contained in $\mathfrak{g}_-^* \otimes \mathfrak{g}$, none of its weights can occur with multiplicity one inside $\Lambda^*\mathfrak{g}_-^* \otimes \mathfrak{g}$.

For the first summand, we can apply similar arguments in a more restricted situation. Since $\mathfrak{g}_{-2}$ is one–dimensional, the representation $\mathfrak{g}_{-2}^* \otimes \mathfrak{g}_{-2}$ is trivial, so $\mathfrak{g}_{-2}^* \otimes \mathfrak{g}_{-1}^* \otimes \mathfrak{g}_{-2} \cong \mathfrak{g}_1$, which also sits in $\mathfrak{g} = \Lambda^0\mathfrak{g}_-^* \otimes \mathfrak{g}$. Finally, consider the bracket $\mathfrak{g}_2 \otimes \mathfrak{g}_{-1} \to \mathfrak{g}_1$. Recall that the Killing form B induces dualities between $\mathfrak{g}_{-1}$ and $\mathfrak{g}_1$ and between $\mathfrak{g}_{-2}$ and $\mathfrak{g}_2$. Now for $X, Y \in \mathfrak{g}_{-1}$ and $0 \neq \beta \in \mathfrak{g}_2$ we get $B([\beta, X], Y) = B(\beta, [X, Y])$. Nondegeneracy of the bracket on $\mathfrak{g}_{-1}$ shows that $\mathrm{ad}(\beta) : \mathfrak{g}_{-1} \to \mathfrak{g}_1$ is injective and thus an isomorphism, so $\mathfrak{g}_2 \otimes \mathfrak{g}_{-1} \cong \mathfrak{g}_1$. Hence, we conclude that

$$\mathfrak{g}_{-2}^* \otimes \mathfrak{g}_{-1}^* \otimes \mathfrak{g}_{-1} \cong \mathfrak{g}_{-1}^* \otimes \mathfrak{g}_1,$$

which also sits in $\mathfrak{g}_-^* \otimes \mathfrak{g}$.

Altogether we see that for any weight of $\Lambda^2\mathfrak{g}_-^* \otimes \mathfrak{g}_-$ that occurs with multiplicity one in $\Lambda^*\mathfrak{g}_-^* \otimes \mathfrak{g}_-$ the weight space has to be contained in $(\Lambda_0^2\mathfrak{g}_{-1})^* \otimes \mathfrak{g}_-$, so the same is true for the $\mathfrak{g}_0$–submodule generated by this weight space. $\square$

Let us analyze the consequences of this proposition for the possible locations of harmonic curvature components. Components of $\ker(\square)$ contained in homogeneity zero are irrelevant for harmonic curvature. For homogeneity one, there is only one possibility, namely components of $(\Lambda^2_0\mathfrak{g}_{-1})^* \otimes \mathfrak{g}_{-1}$. For homogeneity two, the lemma does not leave any room in torsion types. Therefore, the only possibility is curvatures coming form $\Lambda^2\mathfrak{g}^*_{-1} \otimes \mathfrak{g}_0$, and from the proof we again see that these actually have to be contained in $(\Lambda^2_0\mathfrak{g}_{-1})^* \otimes \mathfrak{g}_0$.

To describe the harmonic curvatures in homogeneity one and two, we have to interpret the procedure for constructing a normal Cartan connection from an infinitesimal flag structure similarly as in the proof of Theorem 4.1.1. Suppose that (E, θ) is a regular infinitesimal flag structure of some type (G, P) corresponding to a contact grading. The idea of the prolongation procedure in Section 3.1 was to first make some choices in order to construct a Cartan connection $\tilde{\omega}$ on $\mathcal{G} := E \times P_+$ and then modify this to a normal Cartan connection ω. To obtain $\tilde{\omega}$, one has to choose a principal connection γ on E, as well as a projection π from TM onto the subbundle H. Having made these choices, one constructs from θ a Cartan connection $\tilde{\theta}$ on E, which then can be trivially extended to a regular Cartan connection $\tilde{\omega}$ on $\mathcal{G}$. Now we can interpret this in terms of linear connections.

Since both H and $Q = TM/H$ are associated bundles to E, the principal connection γ induces linear connections ∇^H on H and ∇^Q on Q. By construction, these have the property that $\mathcal{L} : H \times H \to Q$ is parallel, which can be used as a characterization of ∇^Q. Since E is a subbundle of the natural frame bundle of $Q \oplus H$, we see from 4.2.1 that ∇^H is a contact connection. Conversely, a contact connection ∇^H which is compatible with the additional structure induced by E can be used to define γ, and then ∇^Q is obtained from $\nabla\mathcal{L} = 0$. A choice of a projection $\pi : TM \to H$ induces an isomorphism $TM \cong H \oplus Q$ via $\xi \mapsto (\pi(\xi), \xi + H)$. Using this isomorphism, the connections ∇^H and ∇^Q induce a linear connection ∇ on TM. Using this, we can now formulate the basic result on the interpretation of harmonic curvature components of parabolic contact structures.

THEOREM 4.2.2. *Let $(E \to M, \theta)$ be an infinitesimal flag structure of type (G, P) corresponding to a contact grading such that $H^1(\mathfrak{g}_-, \mathfrak{g})$ is concentrated in homogeneities ≤ 0. Let $H \subset TM$ be the contact subbundle, $Q = TM/H$ the quotient bundle, and $q : TM \to Q$ the natural quotient map. Let κ_H be the harmonic curvature of a regular normal parabolic geometry of type (G, P) with underlying infinitesimal flag structure (E, θ). For a principal connection γ and a projection $\pi : TM \to H$ let ∇^H, ∇^Q and ∇ be the induced linear connections on H, Q, and TM.*

(1) For any choice of γ and π, the component $(\kappa_H)_1$ in homogeneity 1 is represented by the component in $\ker(\square) \subset \Lambda^2 H^* \otimes H$ *of the tensor $\tau \in \Gamma(\Lambda^2 H^* \otimes H)$ defined by*

$$\tau(\xi, \eta) := \nabla_\xi \eta - \nabla_\eta \xi - \pi([\xi, \eta]), \tag{4.5}$$

for $\xi, \eta \in \Gamma(H)$. The tensor τ depends only on ∇^H in H-directions.

(2) For any choice of γ, there is a unique π such that component $H \otimes Q \to Q$ of the torsion of ∇ vanishes. The projection π is characterized by

$$\mathcal{L}(\pi(\eta), \xi) = \nabla^Q_\xi q(\eta) - q([\xi, \eta]) \tag{4.6}$$

for all $\xi \in \Gamma(H)$ and all $\eta \in \mathfrak{X}(M)$. In particular, if ∇ is compatible with a contact form α, then this is equivalent to $\pi(\eta) = \eta - \alpha(\eta) r$, where r is the Reeb vector field.

(3) Suppose that γ and π are chosen in such a way that the homogeneous component of degree one of the torsion of ∇ has values in $\ker(\square)$ *(which in particular*

implies that π is the projection from (2)). Then the component $(\kappa_H)_2$ in homogeneity 2 is represented by the component in $\ker(\Box) \subset \Lambda^2_0 H^* \otimes (E \times_{G_0} \mathfrak{g}_0)$ *of the curvature R of ∇.*

PROOF. (1) As indicated above, we put $\mathcal{G} = E \times P_+$, and use γ and π to define $\tilde{\omega}$. Denoting by $i : E \to \mathcal{G}$ the inclusion and by $\tilde{\kappa}$ the curvature of $\tilde{\omega}$, the pullback $i^*\tilde{\kappa}$ is given by the torsion and the curvature of ∇. Now normalizing $\tilde{\omega}$, we see that the homogeneous components of degree one $\mathrm{gr}_1(\tilde{\kappa})$ of $\tilde{\kappa}$ and $\mathrm{gr}_1(\kappa)$ differ by an element in the image of ∂. Normality of ω implies that $\mathrm{gr}_1(\kappa)$ lies in $\ker(\Box)$. As in the proof of Theorem 4.1.1, the harmonic curvature component $(\kappa_H)_1$ is represented by $\mathrm{gr}_1(\kappa)$, which coincides with the $\ker(\Box)$–component of the $\mathrm{gr}_1(\tilde{\kappa})$. It follows directly from the definition of torsion that the component $H \times H \to H$ of the torsion of ∇ is given by (4.5). The last claim is obvious from the formula.

(3) Now if we manage to choose γ and π in such a way that $\mathrm{gr}_1(\tilde{\kappa})$ actually is a section of $\ker(\Box)$, then the homogeneous component in degree two of $\tilde{\kappa}$ differs from the one of κ only by elements in the image of ∂. Hence, the $\ker(\Box)$–components in homogeneity two coincide and by the lemma we know that $\ker(\Box) \subset L(\Lambda^2 TM, E \times_{G_0} \mathfrak{g}_0)$. Then the description follows immediately.

(2) First look at the right–hand side of (4.6) for fixed η and variable ξ. Since $q(\xi) = 0$, this is linear over smooth functions, and hence defines (still for fixed η) a bundle map $H \to Q$. By nondegeneracy, there is a unique section $\pi(\eta)$ such that (4.6) holds. But by the Leibniz rule, the right–hand side is also linear over smooth functions in η, so we actually get a bundle map $\pi : TM \to H$ in this way. Finally, if $\eta \in \Gamma(H)$, then $q(\eta) = 0$ and $q([\xi, \eta]) = \mathcal{L}(\xi, \eta)$ and hence $\pi(\eta) = \eta$. Therefore, (4.6) uniquely defines a projection $\pi : TM \to H$.

To get the component $Q \otimes H \to Q$ of the torsion, we have to proceed as follows. For a vector field η consider $\eta - \pi(\eta)$ (which represents the section $q(\eta)$ of Q), take a section ξ of H and consider

$$q\left(\nabla_\xi(\eta - \pi(\eta)) - \nabla_{\eta - \pi(\eta)}\xi - [\xi, \eta - \pi(\eta)]\right) = q(\nabla_\xi \eta) - q([\xi, \eta]) + \mathcal{L}(\xi, \pi(\eta)),$$

where we have used that ∇ preserves the subbundle H. But by construction of ∇ we have $q(\nabla_\xi \eta) = \nabla^Q_\xi q(\eta)$, so the characterization of π follows.

In the case that ∇ is compatible with a contact form α, we can rewrite the characterizing equation as

$$-d\alpha(\pi(\eta), \xi) = \xi \cdot \alpha(\eta) - \alpha([\xi, \eta]).$$

Since $\alpha(\xi) = 0$, the right–hand side equals $d\alpha(\xi, \eta)$, which shows that $\eta - \pi(\eta)$ must be a multiple of the Reeb field. But then $\alpha(\pi(\eta)) = 0$ implies that the factor must equal $\alpha(\eta)$. □

This result is not sufficient to describe all harmonic curvatures in homogeneity one and two. While it tells us how to choose π for given γ, it does not show how to choose γ in order to satisfy the condition in (3). This depends on the concrete choice of structure, and we will indicate it in some cases below. Finally, let us remark that a complete description of the harmonic curvature will be obtained using Weyl structures in Chapter 5.

4.2.3. Lagrangean contact structures. We start the discussion of the individual parabolic contact structures with the A_n–series. From 3.2.10 we know that in this series the algebras $\mathfrak{sl}(n, \mathbb{R})$ and $\mathfrak{su}(p, q)$ with $p, q > 0$ admit contact gradings.

Lagrangean contact structures correspond to the split real form $\mathfrak{sl}(n,\mathbb{R})$. The name of this structures goes back to M. Takeuchi; see [**Tak94**]. It is derived from the fact that maximal isotropic subspaces in symplectic vector spaces are called *Lagrangean subspaces*. In contact geometry, maximal isotropic subbundles of the contact bundle are often called *Legendrean*, so the name Legendrean contact strucutures would also be appropriate.

For $n \geq 1$ consider the Lie algebra $\mathfrak{g} := \mathfrak{sl}(n+2,\mathbb{R})$. The contact grading on this algebra comes from decomposing into blocks of size 1, n, and 1, so

$$\mathfrak{g} = \left\{ \begin{pmatrix} a & Z & \gamma \\ X & A & W \\ \beta & Y & b \end{pmatrix} : a, b, \beta, \gamma \in \mathbb{R}; X, W \in \mathbb{R}^n; Z, Y \in \mathbb{R}^{n*}; a + b + \operatorname{tr}(A) = 0 \right\}.$$

The grading components are indicated by

$$\begin{pmatrix} \mathfrak{g}_0 & \mathfrak{g}_1^E & \mathfrak{g}_2 \\ \mathfrak{g}_{-1}^E & \mathfrak{g}_0 & \mathfrak{g}_1^F \\ \mathfrak{g}_{-2} & \mathfrak{g}_{-1}^F & \mathfrak{g}_0 \end{pmatrix},$$

where we have indicated the splittings $\mathfrak{g}_{\pm 1} = \mathfrak{g}_{\pm 1}^E \oplus \mathfrak{g}_{\pm 1}^F$ for later use. The following facts are easily seen from this block form. The trace form (and hence any invariant form on $\mathfrak{g}$) induces a duality between $\mathfrak{g}_{-1}^E$ and $\mathfrak{g}_1^E$ and between $\mathfrak{g}_{-1}^F$ and $\mathfrak{g}_1^F$. The splittings of $\mathfrak{g}_{\pm 1}$ are invariant under the adjoint action of $\mathfrak{g}_0$ which induces a surjection $\mathfrak{g}_0 \to \mathfrak{gl}(\mathfrak{g}_{-1}^E)$ with one–dimensional kernel. The Lie bracket $\mathfrak{g}_{-1} \times \mathfrak{g}_{-1} \to \mathfrak{g}_{-2}$ is trivial on $\mathfrak{g}_{-1}^E \times \mathfrak{g}_{-1}^E$ as well as on $\mathfrak{g}_{-1}^F \times \mathfrak{g}_{-1}^F$ and its restriction to $\mathfrak{g}_{-1}^E \times \mathfrak{g}_{-1}^F$ induces an isomorphism $\mathfrak{g}_{-1}^F \cong L(\mathfrak{g}_{-1}^E, \mathfrak{g}_{-2})$. In particular, this bracket is nondegenerate, so we really have found a contact grading. Viewing this bracket as a symplectic form on $\mathfrak{g}_{-1}$, the subspaces $\mathfrak{g}_{-1}^E$ and $\mathfrak{g}_{-1}^F$ of $\mathfrak{g}_{-1}$ are Lagrangean.

As a group G with Lie algebra $\mathfrak{g}$ we take $PGL(n+2,\mathbb{R})$. We can either realize this group as the quotient of $GL(n+2,\mathbb{R})$ by scalar multiples of the identity or by taking the subgroup of matrices whose determinant has absolute value one and identifying each matrix with its negative. In any case, we will work with representative matrices. For the parabolic subgroup $P \subset G$ we take the subgroup of matrices which are block upper triangular with blocks of sizes 1, n, and 1. The resulting Levi subgroup $G_0 \subset P$ consists of the block diagonal matrices with these block sizes.

The homogeneous model G/P is the flag manifold $F_{1,n+1}(\mathbb{R}^{n+2})$ of lines in hyperplanes in $\mathbb{R}^{n+2}$. Mapping such a flag to its line makes $F_{1,n+1}$ into a fiber bundle over $\mathbb{R}P^{n+1}$. The fiber of this bundle is the space of all hyperplanes containing a fixed line, which can be identified with hyperlanes in the quotient by that line. Since a hyperplane in a vector space is equivalent to a line in its dual, the fiber is $\mathbb{R}P^{n*}$. Evidently, the vertical bundle of this fibration corresponds to $\mathfrak{g}_{-1}^F$. Likewise, projecting to the hyperplane shows that $F_{1,n+1}$ is a fiber bundle over $\mathbb{R}P^{(n+1)*}$ with fiber $\mathbb{R}P^n$, and the vertical bundle of this fibration corresponds to $\mathfrak{g}_{-1}^E$. To complete the interpretation of the structure, one shows that the projection to $\mathbb{R}P^{n+1}$ actually identifies $F_{1,n+1}$ with the projectivized cotangent bundle $\mathcal{P}T^*\mathbb{R}P^{n+1}$. The subspace spanned by the two vertical bundles (which are transversal) thereby gets identified with the tautological subbundle, so it defines the canonical contact structure on $\mathcal{P}T^*\mathbb{R}P^{n+1}$.

By Proposition 3.3.7, $H^1(\mathfrak{g}_-, \mathfrak{g})$ is concentrated in homogeneous degrees ≤ 0, so we only have to understand regular infinitesimal flag structures of type (G, P). Let us denote elements of $\mathfrak{g}_-$ as triples (β, X, Y). Likewise, we denote a block diagonal matrix $\left(\begin{smallmatrix} c & 0 & 0 \\ 0 & C & 0 \\ 0 & 0 & e \end{smallmatrix}\right)$ with $c, e \in \mathbb{R} \setminus 0$, $C \in GL(n, \mathbb{R})$ as (c, C, e). In this language, the adjoint action is given by $(c, C, e) \cdot (\beta, X, Y) = (\frac{e}{c}\beta, c^{-1}CX, eYC^{-1})$. Observe that this is unchanged if we replace (c, C, e) by a nonzero multiple. Taking the representative $(1, c^{-1}C, \frac{e}{c})$ we see that the second component represents the action on $\mathfrak{g}^E_{-1}$ and the last component the one on $\mathfrak{g}_{-2}$, and this completely determines the element of G_0. In particular, the action (as expected) preserves the bracket and the decomposition of $\mathfrak{g}_{-1}$. Conversely, suppose that we take an automorphism of the graded Lie algebra $\mathfrak{g}_-$, which preserves the decomposition of $\mathfrak{g}_{-1}$. If $C \in GL(\mathfrak{g}^E_{-1})$ denotes the restriction of this automorphism, then compatibility with the bracket implies that the automorphism must be given by $(\beta, X, Y) \mapsto (e\beta, CX, eYC^{-1})$ for some nonzero number e. Since this is the action of $(1, C, e)$, we conclude that the adjoint action identifies G_0 with the subgroup of those autmorphisms of the graded Lie algebra $\mathfrak{g}_-$, which, in addition, preserve the decomposition $\mathfrak{g}_{-1} = \mathfrak{g}^E_{-1} \oplus \mathfrak{g}^F_{-1}$.

From the discussion in 4.2.2 we know that an infinitesimal flag structure of type (G, P) on a smooth manifold M of dimension $2n+1$ is given by a contact structure $H \subset TM$ together with a redcution of structure group corresponding to $G_0 \subset \mathrm{Aut}_{\mathrm{gr}}(\mathfrak{g}_-)$. From the description of G_0 above it is clear that such a reduction is equivalent to a decomposition $H = E \oplus F$ of the contact subbundle as a direct sum of two Legendrean subbundles. This means that each of the subbundles has rank n, and the restriction of $\mathcal{L}$ to $E \times E$ and $F \times F$ vanishes identically. Note that this implies that $\mathcal{L}$ identifies F with the bundle $L(E, Q)$ of linear maps. A contact structure with an additonal decomposition $H = E \oplus F$ into the direct sum of two Legendrean subbundles is called a *Lagrangean contact structure.*

To get an overview of the basic completely reducible natural bundles for these structures we have to look at representations of G_0. Now this has a two–dimensional center, so there is a two–parameter family of one–dimensional representations and correspondingly a two–parameter family of natural real line bundles. We do not go into detail of how these are best parametrized, but just observe that Q, $\Lambda^n E$ and $\Lambda^n F \cong \Lambda^n E^* \otimes Q$ are typical examples. The semisimple part of G_0 is $SL(n, \mathbb{R})$ with the standard representation $\mathfrak{g}^E_{-1}$ corresponding to the bundle E. Hence, all natural bundles can be obtained from tensor bundles of E and natural line bundles.

Let us next compute the cohomology group $H^2(\mathfrak{g}_-, \mathfrak{g})$. This is completely different for $n = 1$ and $n > 1$, and we consider the case $n = 1$ first. If $n = 1$, then we actually deal with the Borel subalgebra in A_2, i.e. the Dynkin diagram ×—×. There are two elements of length two in the Weyl group, namely the two possible compositions of the two simple reflections. The corresponding sets Φ_w are $\{\alpha_1, \alpha_1 + \alpha_2\}$, respectively, $\{\alpha_2, \alpha_1 + \alpha_2\}$. On the other hand, the highest root $\alpha_1 + \alpha_2$ is mapped by these two Weyl group elements to $-\alpha_1$, respectively, $-\alpha_2$. By Theorem 3.3.5 (and dualization to get from cohomology of $\mathfrak{p}_+$ to cohomology of $\mathfrak{g}_-$), we conclude that the two irreducible components in the cohomology are represented by the one–dimensional representations consisting of maps $\mathfrak{g}_{-2} \times \mathfrak{g}^E_{-1} \to \mathfrak{g}^E_1$, respectively, $\mathfrak{g}_{-2} \times \mathfrak{g}^F_{-1} \to \mathfrak{g}^F_1$. The corresponding harmonic curvatures are represented by Cotton–York type tensors mapping $Q \otimes E$ to E^* and $Q \otimes F \to F^*$. These can be determined more explicity using Weyl structures.

For $n > 1$, the Hasse diagram contains three elements of length two. Denoting by $\alpha_1, \dots, \alpha_{n+1}$ the simple roots and by σ_i the simple reflection corresponding to

α_i, these three elements are given as $\sigma_1 \circ \sigma_2$, $\sigma_{n+1} \circ \sigma_n$, and $\sigma_1 \circ \sigma_{n+1} = \sigma_{n+1} \circ \sigma_1$. The corresponding sets Φ_w evidently are $\{\alpha_1, \alpha_1 + \alpha_2\}$, $\{\alpha_{n+1}, \alpha_n + \alpha_{n+1}\}$, and $\{\alpha_1, \alpha_{n+1}\}$. The images of the highest root $\alpha_1 + \cdots + \alpha_{n+1}$ under these three elements are $\alpha_2 + \cdots + \alpha_{n+1}$, $\alpha_1 + \cdots + \alpha_n$, and $\alpha_2 + \cdots + \alpha_n$, respectively. Again using Theorem 3.3.5 and dualization, we conclude that the irreducible components of $\ker(\square)$ are the highest weight parts in the sets of maps $\Lambda^2 \mathfrak{g}_{-1}^E \to \mathfrak{g}_{-1}^F$, $\Lambda^2 \mathfrak{g}_{-1}^F \to \mathfrak{g}_{-1}^E$, and $\mathfrak{g}_{-1}^E \otimes \mathfrak{g}_{-1}^F \to \mathfrak{g}_0$, respectively. The first two components are torsions in homogeneity 1 and the last one is a curvature in homogeneity 2, so we can use Theorem 4.2.2 to analyze the corresponding harmonic curvature components.

For the two torsions, the interpretation is simple. Suppose that ∇^H is the contact connection induced by a principal connection on the regular infinitesimal flag structure determined by a Lagrangean contact structure $H = E \oplus F \subset TM$. Then of course $\nabla^H = \nabla^E \oplus \nabla^F$ for connections on the subbundles, so, in particular, the subbundles are preserved. Now from 4.2.2 we know that that we have to look at components of

$$\tau(\xi, \eta) := \nabla_\xi^H \eta - \nabla_\eta^H \xi - \pi([\xi, \eta])$$

for $\xi, \eta \in \Gamma(H)$ and a certain projection π from TM onto the subbundle H. To get the first torsion component, we have to take $\xi, \eta \in \Gamma(E)$ and project the result to F. But then the covariant derivatives produce sections of E, and since E is Legendrean the bracket $[\xi, \eta]$ is a section of H, so we can leave out π. Hence, we end up with mapping $\xi, \eta \in \Gamma(H)$ to the F–component of $-[\xi, \eta] \in \Gamma(H)$. Since ξ and η actually have trivial F–components, this is bilinear over smooth functions, and hence defines a tensor $\tau_E \in \Gamma(\Lambda^2 E^* \otimes F)$.

To understand the highest weight component, recall that $F \cong E^* \otimes Q$ via $\mathcal{L}$. Thus, $\Lambda^2 E^* \otimes F \cong \Lambda^2 E^* \otimes E^* \otimes Q$ and the highest weight part in there is the kernel of the alternation map to $\Lambda^3 E^* \otimes Q$. Now viewed as a trilinear map on E with values in Q, the torsion τ_E maps (ξ, η, ζ) to $\mathcal{L}$ applied to the F–component of $[\xi, \eta]$ and ζ. Replacing the F–component by $[\xi, \eta]$ does not change the value of $\mathcal{L}$, so we are left with

$$(\xi, \eta, \zeta) \mapsto \mathcal{L}([\xi, \eta], \zeta) = q([[\xi, \eta], \zeta]).$$

But this has trivial alternation by the Jacobi identity. Hence, we obtain

PROPOSITION 4.2.3. *The two harmonic curvature components in homogeneity one of the regular normal parabolic geometry determined by a Lagrangean contact structure $H = E \oplus F \subset TM$ are represented by the torsions $\tau_E \in \Gamma(\Lambda^2 E^* \otimes F)$ and $\tau_F \in \Gamma(\Lambda^2 F^* \otimes E)$, induced by projecting the negative of the Lie bracket of two sections of one subbundle to the other subbundle. In particular, τ_E vanishes identically if and only if the subbundle $E \subset TM$ is integrable and likewise for τ_F. Vanishing of both τ_E and τ_F is equivalent to torsion freeness of the normal parabolic geometry.*

PROOF. Apart from the last claim, everything has been proved already above. For the last claim, recall that the lowest nontrivial homogeneous component of the curvature of a regular normal parabolic geometry is harmonic; see Theorem 3.1.12. Vanishing of τ_E and τ_F implies that this lowest nonzero component has homogeneity at least two. By Lemma 4.2.2 the harmonic part of homogeneity two cannot produce any torsions. Since the same is true for arbitrary maps of homogeneity at least three, the result follows. $\square$

To interpret the remaining harmonic curvature component, one has to choose a contact connection ∇^H and a projection π as above, and then modify it in such a way that the homogeneity one component of the torsion is contained in $\ker(\square)$. Then one looks at the appropriate part of the curvature of the resulting connection. We will give a detailed description of connections adapted to parabolic contact structure in this sense (as well as in stronger senses) in Section 5.2. At this point, we only give a short sketch how such a connection can be constructed.

Observe that a contact connection ∇^H comes from a principal connection on the infintesimal flag strucutre if and only if it is of the form $\nabla^E \oplus \nabla^F$. Starting with an arbitrary choice of such a connection, the tensor τ from above is a section of $\Lambda^2 H^* \otimes H$. Decomposing this bundle, we obtain

$$(\Lambda^2 E^* \oplus (E \otimes F)^*_0 \oplus Q^* \oplus \Lambda^2 F^*) \otimes (E \oplus F).$$

Here we have denoted by $(E \otimes F)_0$ the kernel of $\mathcal{L} : E \otimes F \to Q$ and identified a complementary subbundle with Q. Note that by definition of τ, the only part of this that depends on π is the part in $Q^* \otimes (E \oplus F)$. Now take the components in $(\Lambda^2 E^* \oplus (E \otimes F)^*_0) \otimes E$, interepret them as a section of $H^* \otimes E^* \otimes E$ and subtract this from ∇^E. Likewise, take the components in $((E \otimes F)^*_0 \oplus \Lambda^2 F^*) \otimes F$, view them as a section of $H^* \otimes F^* \otimes F$ and subtract this from ∇^F. Finally, we can use the component in $Q^* \otimes (E \oplus F)$ to change the projection π.

The resulting pair of connnection and projection by construction has the property that the nonzero components of the tensor τ only lie in $\Lambda^2 E^* \otimes F$ and in $\Lambda^2 F^* \otimes E$, and we know from above that these parts automatically lie in $\ker(\square)$. We claim that this is already an appropriate connection, i.e. the part of the torsion which maps $Q \otimes H$ to Q has to vanish automatically. To see this, observe that we are dealing with the lowest homogeneous component of the curvature of a regular Cartan connection, so by the Bianchi identity, it is contained in the kernel of ∂; see Theorem 3.1.12. Let us denote the homogeneity one part of the torsion by ψ and for sections $\xi, \eta, \zeta \in \Gamma(H)$ expand the equation $0 = \partial\psi(\xi, \eta, \zeta)$. Using that $\{\ ,\ \}$ coincides with $\mathcal{L}$, this gives

$$\begin{aligned} 0 &= \mathcal{L}(\xi, \psi(\eta, \zeta)) - \mathcal{L}(\eta, \psi(\xi, \zeta)) + \mathcal{L}(\zeta, \psi(\xi, \eta)) \\ &\quad - \psi(\mathcal{L}(\xi, \eta), \zeta) + \psi(\mathcal{L}(\xi, \zeta), \eta) - \psi(\mathcal{L}(\eta, \zeta), \xi). \end{aligned} \tag{4.7}$$

Now assume that $\xi, \eta \in \Gamma(E)$ and $\zeta \in \Gamma(F)$. Then in the first two terms, ψ already gives zero while in the third term ψ has values in F, so this does not contribute either. In the fourth term we get $\mathcal{L}(\xi, \eta) = 0$ since both are sections of E, so (4.7) reduces to $\psi(\mathcal{L}(\xi, \zeta), \eta) = \psi(\mathcal{L}(\eta, \zeta), \xi)$. But now given $\beta \in \Gamma(Q)$ and $\eta \in \Gamma(E)$ we can choose $\xi \in \Gamma(E)$ and $\zeta \in \Gamma(F)$ such that $\mathcal{L}(\xi, \zeta) = \beta$ and $\mathcal{L}(\eta, \zeta) = 0$, and we get

$$\psi(\beta, \eta) = \psi(\mathcal{L}(\xi, \zeta), \eta) = \psi(\mathcal{L}(\eta, \zeta), \xi) = 0.$$

Thus, ψ vanishes on $Q \otimes E$ and likewise one shows that it vanishes on $Q \otimes F$.

4.2.4. Partially integrable almost CR–structures. This is certainly the most important example of a parabolic contact structure, which has often been studied independently. The constructions of canonical Cartan connections by N. Tanaka in [**Tan62**] and by S.S. Chern and J. Moser in [**ChMo76**] (for the subclass of integrable CR–structures) are among the best known results of this kind and were a strong motivation for the development of the general theory.

We have partly discussed this example in 3.1.7, so we will go through the basics rather quickly. For $p+q=n\geq 1$ we consider the real form $\mathfrak{su}(p+1,q+1)$ of $\mathfrak{sl}(n+2,\mathbb{C})$. We choose the Hermitian form on $\mathbb{C}^{n+2}$ which is given by

$$\langle(z_0,\dots,z_{n+1}),(w_0,\dots,w_{n+1})\rangle = z_0\overline{w_{n+1}}+z_{n+1}\overline{w_0}+\sum_{j=1}^{p} z_j\overline{w_j}-\sum_{j=p+1}^{n} z_j\overline{w_j}.$$

Denoting by $\mathbb{I}=\mathbb{I}_{p,q}$ the $n\times n$–diagonal matrix with the first p entries equal to 1 and the remaining entries equal to -1, we can represent the Lie algebra in block form with blocks of sizes 1, n, and 1, similarly to 4.2.3 as

$$\mathfrak{g}=\left\{\begin{pmatrix} a & Z & iz\\ X & A & -\mathbb{I}Z^*\\ ix & -X^*\mathbb{I} & -\bar{a}\end{pmatrix} : \begin{matrix} A\in\mathfrak{u}(n), a\in\mathbb{C}, X\in\mathbb{C}^n, Z\in\mathbb{C}^{n*},\\ x,z\in\mathbb{R};\quad a+\operatorname{tr}(A)-\bar{a}=0\end{matrix}\right\}.$$

The grading components are as for Lagrangean contact structures in 4.2.3 above. Rather than the splitting of $\mathfrak{g}_{\pm1}$ into two irreducible pieces we have a complex structure on these subspaces. After complexification, the splitting into two components is recovered as the splitting of $\mathfrak{g}_{\pm1}\otimes\mathbb{C}$ into holomorphic and anti–holomorphic parts. This will also be crucial for the interpretation of cohomologies. The bracket $\mathfrak{g}_{-1}\times\mathfrak{g}_{-1}\to\mathfrak{g}_{-2}$ is given by $[X,Y]=Y^*\mathbb{I}X-X^*\mathbb{I}Y$, so this is twice the imaginary part of the standard Hermitian inner product of signature (p,q). Note that this is compatible with the complex structure in the sense that $[iX,iY]=[X,Y]$.

As a group with Lie algebra $\mathfrak{g}$, we take $G=PSU(p+1,q+1)$. The parabolic subgroup P is then the stabilizer of the isotropic complex line generated by the first basis vector. (This automatically stabilizes also its orthocomplement, which is a hyperplane containing the given line.) The subgroup G_0 again is given by block diagonal matrices, i.e. we have matrices $\left(\begin{smallmatrix} c & 0 & 0\\ 0 & C & 0\\ 0 & 0 & 1/\bar{c}\end{smallmatrix}\right)$ with $c\in\mathbb{C}\setminus 0$ and $C\in U(n)$ such that $c\det(C)/\bar{c}=1$. We have to identify matrices which are multiples of each other, which leaves the freedom of multiplying by an $(n+2)$nd root of unity. Using a notation similar to 4.2.3, the adjoint action is immediately computed to be given by $(c,C)\cdot(ix,X)=(|c|^{-2}ix,c^{-1}CX)$, which is complex linear on $\mathfrak{g}_{-1}$ and orientation preserving on $\mathfrak{g}_{-2}$. Notice that there is a p–dimensional subspace in $\mathfrak{g}_{-1}$ on which $X\mapsto[X,iX]$ is nonzero with all values of the same sign and a q–dimensional subspace for which the same is true for the opposite sign. Hence, if $p\neq q$, then preserving the bracket and the complex structure on $\mathfrak{g}_{-1}$ implies that the orientation on $\mathfrak{g}_{-2}$ is preserved. For $p=q$, this is an additional condition.

Conversely, assume that $A:\mathfrak{g}_{-1}\to\mathfrak{g}_{-1}$ is a complex linear isomorphism such that $[AX,AY]=\lambda[X,Y]$ for some $\lambda>0$. Since the standard Hermitian form of signature (p,q) is obtained as $1/2(i[X,iY]+[X,Y])$, we conclude that A has the same compatibility with that Hermitian form. In particular, $|\det(A)|^2=\lambda^n$. Now choose $c\in\mathbb{C}$ such that $|c|^2c^{-n-2}=\det(A)$. Then we get $|\det(A)|^2=|c|^{-2n}=\lambda^n$, and since $\lambda>0$ this implies $\lambda=|c|^{-2}$. Hence, cA has the property that $[cAX,cAY]=[X,Y]$ and hence $cA\in U(n)$. But then A is realized by the adjoint action of (c,cA) and $c\det(cA)/\bar{c}=c^{n+2}|c|^{-2}\det(A)=1$ as required. Note that in this procedure c is only unique up to multiplication with an $(n+2)$nd root of unity.

From the discussion in 4.2.2 we conclude that a regular infinitesimal flag structure (and hence a regular normal parabolic geometry) of type (G,P) on a smooth manifold M of dimension $2n+1$ is equivalent to a contact structure $H\subset TM$ together with a complex structure J on H such that $\mathcal{L}(J\xi,J\eta)=\mathcal{L}(\xi,\eta)$ for all

$\xi, \eta \in \Gamma(H)$. If this last condition is satisfied, then identifying the fiber Q_x of Q over $x \in M$ with $\mathbb{R}$, the map $\mathcal{L}$ is the imaginary part of a Hermitian form, and one requires that this form has signature (p, q). If $p = q$, one in addition has to choose an orientation on Q (which requires Q to be trivial). Since as a complex vector bundle H is canonically oriented, this is equivalent to choosing an orientation on M. For $p \neq q$, this orientation is automatically chosen by deciding between signature (p, q) and signature (q, p).

Let us rephrase this in the language of CR geometry. Given a real smooth manifolds M of dimension $2n + 1$, a rank n complex subbundle (H, J) in TM is called an *almost CR–structure of hypersurface type*. Correspondingly, there is the notion of a (local) *CR–diffeomorphism*, which requires the tangent map to preserve the CR–subbundle H and the restriction to H being complex linear. The almost CR–structure is called *nondegenerate* if H defines a contact structure on M. Next, we have the condition that $\mathcal{L}(J\xi, J\eta) = \mathcal{L}(\xi, \eta)$ for all $\xi, \eta \in H$. This condition is not used very often in CR geometry, since it is implied by the integrability condition to be discussed below. One usual terminology (see e.g. [**Miz93**]) for this condition is *partial integrability*. Then $\mathcal{L}$ becomes the imaginary part of a Hermitian form and choosing an orientation on Q the signature of this form is called the signature of (M, H, J). Hence, we conclude that regular normal parabolic geometries of type $(PSU(p+1, q+1), P)$ are equivalent to oriented nondegenerate partially integrable hypersurface type almost CR–structures of signature (p, q).

To understand the terminology "partial integrability" and its relation to the integrability condition, it is best to pass to the complexified setting. Since H is a complex vector bundle, the image $H \otimes \mathbb{C}$ in the complexified tangent bundle $TM \otimes \mathbb{C}$ splits into holomorphic and anti–holomorphic part as $H \otimes \mathbb{C} = H^{1,0} \oplus H^{0,1}$. Typical sections of $H^{0,1}$ are of the form $\xi + iJ\xi$ for $\xi \in \Gamma(H)$. Applying the complex bilinear extension of $\mathcal{L}$ to two such sections, we obtain

$$\mathcal{L}(\xi + iJ\xi, \eta + iJ\eta) = (\mathcal{L}(\xi, \eta) - \mathcal{L}(J\xi, J\eta)) + i(\mathcal{L}(J\xi, \eta) + \mathcal{L}(\xi, J\eta)).$$

Evidently, partial integrability is equivalent to vanishing of this expression and hence to the fact that the bracket of two sections of $H^{0,1}$ is a section of $H \otimes \mathbb{C}$. Alternatively, this can be phrased as follows. Consider the complex linear extension $q_{\mathbb{C}} : TM \otimes \mathbb{C} \to Q \otimes \mathbb{C}$, and the tensorial map $H^{0,1} \times H^{0,1} \to \mathbb{C}$ induced by an imaginary multiple of $q_{\mathbb{C}}([\xi, \bar{\eta}])$ for sections $\xi, \eta \in \Gamma(H^{0,1})$. Partial integrability is equivalent to this form being Hermitian, thus defining (with appropriate normalization) the classical *Levi form*. The signature of M is then the signature of this form.

A partially integrable almost CR–manifold is called *integrable* or a *CR–manifold* if the bundle $H^{0,1}$ is involutive. In the real picture, this is expressed by vanishing of the *Nijenhuis tensor* $N : \Lambda^2 H \to H$, which is induced by

$$(\xi, \eta) \mapsto [\xi, \eta] - [J\xi, J\eta] + J([J\xi, \eta] + [\xi, J\eta]).$$

Note that N is of type $(0, 2)$ i.e. conjugate linear in both arguments.

The most important examples for CR–structures come from complex analyis. Let $(\mathcal{M}, J)$ be a complex manifold of complex dimension $n+1$, and let $M \subset \mathcal{M}$ be a smooth real hypersurface. For $x \in M$ define $H_x := T_xM \cap J(T_xM)$, the maximal complex subspace of $T_xM \subset T_x\mathcal{M}$. These subspaces must have complex dimension n and they fit together to define a complex subbundle $H \subset TM$. By definiton $H^{0,1} = TM \otimes \mathbb{C} \cap T^{0,1}\mathcal{M}$ and as the interesection of two involutive subbundles, this is automatically involutive. Generically, the subbundle H will be maximally nondegenerate, and then (M, H, J) is automatically a CR–structure. Note further that a biholomorphism of $\mathcal{M}$ which maps M to itself automatically restricts to a CR–diffeomorphism on M.

This picture can be nicely used to understand the homogeneous model G/P. The subgroup P is the stabilizer of a null line and G acts transitively on the space of all such lines. Hence, G/P can be identified with the projectivized null cone, which is a smooth real hypersurface in $\mathbb{C}P^{n+1}$. Since G acts by biholomorphisms on $\mathbb{C}P^{n+1}$, it acts by CR–diffeomorphisms on G/P. It is easy to verify directly the the CR–structure on G/P is nondegenerate of signature (p, q). Alternatively, one may describe the CR–structure in an elementary way along the lines of Example 1.1.6: It is easy to describe the manifold G/P more explicitly. Writing $\mathbb{C}^{n+2} = V' \oplus V''$ such that the Hermitian form is positive definite on V' and negative definite on V'', the null cone can be written as $\{(z', z'') : |z'| = |z''|\}$ for the Euclidean norm on both factors. Factoring by complex multiples we may first assume that $|z'| = |z''| = 1$, and then the remaining freedom is multiplication by elements of $U(1)$. Hence, G/P is obtained by factoring $S^{2p+1} \times S^{2q+1}$ by the diagonal action of $U(1)$. Note that in the special case $q = 0$, we obtain S^{2p+1} as described in Example 1.1.6.

The further discussion of this geometry is closely parallel to the case of Lagrangean contact structures in 4.2.3. As for the basic completely reducible bundles, we have real and complex line bundles as well as tensor bundles of H (taking into account the complex structure to form subbundles). Concerning harmonic curvature components, the cohomologies here have the same complexifications as the ones in 4.2.3. To interpret these we have to recall that on the complexification, the splitting of the subbundles corresponds to holomorphic and anti–holomorphic parts, so this admits an interpretation in terms of complex linearity and anti–linearity. For $n = 1$, the complexified cohomology corresponds to sections of the complex line bundles $Q^* \otimes \mathbb{C} \otimes (H^{1,0})^* \otimes (H^{1,0})^*$ and $Q^* \otimes \mathbb{C} \otimes (H^{0,1})^* \otimes (H^{0,1})^*$. The sum of these bundles is the complexification of the bundle whose sections are bilinear maps $Q \times H \to H^*$, which are complex linear in the second variable. Hence, there is only one basic curvature invariant for three–dimensional CR–structures.

If $n > 1$, then the two components in homogeneity one are the complexification of a single complex representation. Namely, they correspond to bilinear maps $\Lambda^2 H^{1,0} \to H^{0,1}$ and $\Lambda^2 H^{0,1} \to H^{1,0}$, which is the complexification of maps $\Lambda^2 H \to H$ which are conjugate linear in both variables. So there is just one torsion in this case, which corresponds to bilinear maps of type $(0, 2)$. The other cohomology component corresponds to the complexification of the bundle $\Lambda^{1,1} H^* \otimes \mathfrak{su}(H)$ of forms of type $(1, 1)$, so the last basic curvature has its values there.

The development in 4.2.3 already suggests that the torsion in this case should be the obstruction to involutivity of the bundles $H^{0,1}$ and $H^{1,0}$ (which are conjugate to each other) and hence to integrability of the almost CR–structure. This is indeed true and can be verified nicely as follows. If ∇^H is a contact connection induced by a principal connection on the infinitesimal flag structure, then J is parallel for the induced connection. By 4.2.2 we have to consider the $(0, 2)$–component of the tensor

$$\tau(\xi, \eta) = \nabla^H_\xi \eta - \nabla^H_\eta \xi - \pi([\xi, \eta]),$$

for some chosen projection $\pi : TM \to H$. Up to a nonzero factor, the $(0,2)$–component is given by

$$\tau(\xi,\eta) - \tau(J\xi, J\eta) + J(\tau(J\xi,\eta) + \tau(\xi, J\eta)).$$

Expanding the corresponding expression, one immediately concludes that π can be left out by partial integrability. Moreover, all terms involving ∇^H cancel since J is parallel, and the bracket terms add up to the negative of the Nijenhuis tensor. In particular, we conclude that the category of CR–structures is equivalent to the category of torsion–free normal parabolic geometries of type (G,P).

4.2.5. Lie contact structures. We next switch to the contact structures associated to the orthogonal Lie algebras. Hence, we have to consider real forms of $\mathfrak{so}(n,\mathbb{C})$ for $n \geq 7$. Our convention will be that for $p+q=n\geq 3$, we consider $\mathfrak{g} := \mathfrak{so}(p+2,q+2)$. Let V be the real vector space $\mathbb{R}^{n+4}$. Let $\mathbb{J}$ be the matrix $\left(\begin{smallmatrix} 0 & 1 \\ 1 & 0\end{smallmatrix}\right)$ and let $\mathbb{I}_{p,q}$ be the diagonal matrix with the first p entries equal to 1 and the others equal to -1. On V, we use the inner product associated to the matrix $\begin{pmatrix} 0 & 0 & \mathbb{J} \\ 0 & \mathbb{I}_{p,q} & 0 \\ \mathbb{J} & 0 & 0\end{pmatrix}$. This means that the first two and the last two basis vectors are null, and the only non–trivial inner products among these vectors are between the first and the last and between the second and the last but one. The orthocomplement of these four vectors carries a standard inner product of signature (p,q). In particular, the whole inner product has the right signature $(p+2,q+2)$. A direct computation shows that with respect to this inner product, the Lie algebra $\mathfrak{so}(V)$ has the following form with blocks of size 2, n, and 2, where we write $\mathbb{I}$ for $\mathbb{I}_{p,q}$,

$$\mathfrak{g} = \left\{ \begin{pmatrix} A & Z & z\mathbb{I}_{1,1} \\ X & E & -\mathbb{I}Z^t\mathbb{J} \\ x\mathbb{I}_{1,1} & -\mathbb{J}X^t\mathbb{I} & -\mathbb{J}A^t\mathbb{J} \end{pmatrix} : E \in \mathfrak{so}(p,q), x, z \in \mathbb{R} \right\}.$$

Here A, X, and Z are arbitrary matrices of size 2×2, $n\times 2$, and $2\times n$, respectively, and we have used that a 2×2–matrix B such that $\mathbb{J}B = -B^t\mathbb{J}$ must be a real multiple of $\mathbb{I}_{1,1} = \left(\begin{smallmatrix} 1 & 0 \\ 0 & -1\end{smallmatrix}\right)$. The grading is by blocks as usual, so, in particular, $\mathfrak{g}_{\pm 2}$ has real dimension one. The bracket $[\ ,\ ] : \mathfrak{g}_{-1}\times\mathfrak{g}_{-1} \to \mathfrak{g}_{-2}$ is given by $[X,Y] := \mathbb{J}Y^t\mathbb{I}X - \mathbb{J}X^t\mathbb{I}Y$. Denoting the columns of an element $X \in \mathfrak{g}_{-1}$ by X_1 and X_2, and by $\langle\ ,\ \rangle$ the standard inner product of signature (p,q) on $\mathbb{R}^{p+q}$, we obtain

$$[X,Y] = (\langle X_1, Y_2\rangle - \langle X_2, Y_1\rangle)\mathbb{I}_{1,1}.$$

This is evidently nondegenerate, so we have obtained a contact grading. For an element $A \in GL(p+q,\mathbb{R})$ the composition with X is obtained by applying A to the columns of X, which immediately shows that $[AX,AY] = [X,Y]$ for $A \in O(p,q)$. This uniquely characterizes the bracket as a bilinear form:

LEMMA 4.2.5. *Up to real multiples, there is a unique nondegenerate skew symmetric bilinear form on $L(\mathbb{R}^2,\mathbb{R}^{p+q})$ which is invariant under composition by elements of $O(p,q)$.*

PROOF. The space $\Lambda^2(\mathbb{R}^{2*}\otimes\mathbb{R}^{p+q})$ decomposes as

$$(\Lambda^2\mathbb{R}^{2*}\otimes S^2\mathbb{R}^{p+q}) \oplus (S^2\mathbb{R}^{2*}\otimes\Lambda^2\mathbb{R}^{p+q}).$$

This decomposition is valid over $GL(2,\mathbb{R})\times GL(p+q,\mathbb{R})$, but we have to analyze it over $O(p,q)$, with $\mathbb{R}^2$ being viewed as a trivial representation. Then $\Lambda^2\mathbb{R}^{2*}$ is a trivial one–dimensional representation which we can forget. Hence, the first summand is isomorphic to $S^2\mathbb{R}^{p+q} \cong \mathbb{R} \oplus S^2_0\mathbb{R}^{p+q}$. In the second summand, $S^2\mathbb{R}^{2*}$ is a trivial three–dimensional representation, so this summand is isomorphic to the sum of three copies of the irreducible representation $\Lambda^2\mathbb{R}^{p+q}$. In total we conclude that $\Lambda^2(\mathbb{R}^{2*} \otimes \mathbb{R}^{p+q})$ contains a unique one–dimensional subspace on which $O(p,q)$ acts trivially, which completes the proof. □

For the group with Lie algebra $\mathfrak{g}$ we choose $G := O(p+2, q+2)$. For the parabolic subgroup P we take the stabilizer of the isotropic plane spanned by the first two basis vectors, i.e. the subgroup of matrices which are block upper triangular with blocks of size 2, n, and 2. The Levi subgroup $G_0 \subset P$ then consists of all block diagonal matrices with these block sizes contained in $O(p+2, q+2)$. A short computation shows that these are exactly the matrices of the form

$$\begin{pmatrix} B & 0 & 0 \\ 0 & C & 0 \\ 0 & 0 & \mathbb{J}(B^t)^{-1}\mathbb{J} \end{pmatrix} : B \in GL(2,\mathbb{R}), C \in O(p,q).$$

We will denote such a matrix as a pair (B, C) and elements of $\mathfrak{g}_-$ as pairs (x, X). For a 2×2–matrix B one easily computes that $\mathbb{J}B^t\mathbb{J}\mathbb{I}_{1,1}B = \det(B)\mathbb{I}_{1,1}$, and using this one verfies that the adjoint action is given by $(B, C)\cdot(x, X) = (\det(B^{-1})x, CXB^{-1})$. Using the same identity once more, one checks that this is really compatible with the bracket and by definition it is compatible with the identification of $\mathfrak{g}_{-1}$ with $L(\mathbb{R}^2, \mathbb{R}^{p+q})$, including the inner product on $\mathbb{R}^{p+q}$.

DEFINITION 4.2.5. For $p + q = n \geq 3$ a *Lie contact structure* of signature (p, q) on a smooth manifold M of dimension $2n + 1$ is given by a contact structure $H \subset TM$, two auxilliary vector bundles $E \to M$ of rank 2 and $F \to M$ of rank n, a bundle metric on F of signature (p, q), and an isomorphism $H \cong L(E, F)$ such that for each $x \in M$ the Levi bracket $\mathcal{L}_x$ is invariant under the resulting action of the orthogonal group $O(F_x)$ on H_x.

A morphism of Lie contact structures from (M, H, E, F) to $(\tilde{M}, \tilde{H}, \tilde{E}, \tilde{F})$ is a local contact diffeomorphism $f : M \to \tilde{M}$ such that for each $x \in M$ the restriction $T_xf : H_x \to \tilde{H}_{f(x)}$ comes, via the fixed isomorphisms from a linear isomorphisms $E_x \to \tilde{E}_{f(x)}$ and an orthogonal isomorphism $F_x \to \tilde{F}_x$.

PROPOSITION 4.2.5. *For $G = O(p+2, q+2)$ and $P \subset G$ the stabilizer of a null plane, the category of regular normal parabolic geometries of type (G, P) is equivalent to the category of Lie contact structures of signature (p, q).*

PROOF. By Proposition 3.3.7, the Lie algebra cohomology $H^1(\mathfrak{g}_{-1}, \mathfrak{g}_0)$ is concentrated in homogeneous degrees ≤ 0, so it suffices to prove that regular infinitesimal flag structures of type (G, P) are equivalent to Lie contact structures of signature (p, q).

Consider a regular infinitesimal flag structure $(H := T^{-1}M, \mathcal{G}_0 \to M, \theta)$ of type (G, P). Then $\mathcal{G}_0$ is a principal bundle with structure group $G_0 \cong GL(2,\mathbb{R}) \times O(p,q)$. Define $E := \mathcal{G}_0 \times_{G_0} \mathbb{R}^2$ and $F := \mathcal{G}_0 \times_{G_0} \mathbb{R}^{p+q}$, where we use the defining representations of the two components of G_0. Then F carries a natural bundle metric of signature (p, q). Next, we have the component $\theta_{-1} \in \Gamma(L(T^{-1}\mathcal{G}_0, \mathfrak{g}_{-1}))$ of the frame form θ. By definition, $T^{-1}\mathcal{G}_0$ is the preimage of H, and $\mathfrak{g}_{-1} = L(\mathbb{R}^2, \mathbb{R}^{p+q})$. Via θ_{-1}, we get an isomorphism

$$H \cong \mathcal{G}_0 \times_{G_0} L(\mathbb{R}^2, \mathbb{R}^{p+q}) \cong L(E, F);$$

see 3.1.6. On the other hand, the component $\theta_{-2} \in \Omega^2(\mathcal{G}_0, \mathfrak{g}_{-2})$ identifies $Q = TM/H$ with $\mathcal{G}_0 \times_{G_0} \mathfrak{g}_{-2}$. Regularity of the infinitesimal flag structure now exactly means that via these identifications, the Levi bracket $\mathcal{L} : H \times H \to Q$ is induced by $[\ ,\ ] : \mathfrak{g}_{-1} \times \mathfrak{g}_{-1} \to \mathfrak{g}_{-2}$. This immediately implies that H is a contact structure and $\mathcal{L}$ is invariant under the action of $O(F)$, so we have obtained a Lie contact structure. In the same way, a morphism of regular infinitesimal flag structures induces a morphism of Lie contact structures.

Conversely, assume we have given a Lie contact structure (H, E, F) on M. Define $p : \mathcal{G}_0 \to M$ to be the fibered product of the linear frame bundle of E and the orthonormal frame bundle of F. Then this is a principal bundle with structure group $GL(2, \mathbb{R}) \times O(p, q) \cong G_0$. A point in $\mathcal{G}_0$ over $x \in M$ by definition is a pair (ϕ, ψ) of a linear isomorphism $\phi : \mathbb{R}^2 \to E_x$ and an orthogonal isomorphism $\psi : \mathbb{R}^{p+q} \to F_x$. For a tangent vector $\xi \in T^{-1}_{(\phi,\psi)}\mathcal{G}_0$ we have $Tp \cdot \xi \in H_xM$, so via the fixed isomorphism $H \cong L(E, F)$, we can view $Tp \cdot \xi$ as a linear map $E_x \to F_x$. Now define

$$\theta_{-1}(\xi) := \psi^{-1} \circ (Tp \cdot \xi) \circ \phi \in L(\mathbb{R}^2, \mathbb{R}^{p+q}) = \mathfrak{g}_{-1}.$$

This evidently defines a smooth section θ_{-1} of the bundle $L(T^{-1}\mathcal{G}_0, \mathfrak{g}_{-1})$, and since ϕ and ψ are isomorphisms, the kernel of θ_{-1} in each point is the vertical subbundle of $\mathcal{G}_0 \to M$. The principal right action of $(B, C) \in G_0 = GL(2, \mathbb{R}) \times O(p, q)$ maps (ϕ, ψ) to $(\phi \circ B, \psi \circ C)$. Since acting on ξ by the derivative of the principal action leaves the image under Tp unchanged, we conclude that

$$(r^{(B,C)})^*\theta_{-1}(\xi) = C^{-1} \circ \psi^{-1} \circ (Tp \cdot \xi) \circ \phi \circ B = Ad((B.C)^{-1})(\theta_{-1}(\xi)),$$

so we obtain the right equivariancy condition. Fixing the point (ϕ, ψ), the form θ_{-1} induces an isomorphism $H_x \to \mathfrak{g}_{-1}$ which intertwines the action of $O(F_x)$ on H_x with the action of $O(p, q)$ on $\mathfrak{g}_{-1} = L(\mathbb{R}^2, \mathbb{R}^{p+q})$. Pulling back $\mathcal{L}_x$ via the inverse β of this isomorphism, we obtain a nondegenerate skew symmetric bilinear map $\mathfrak{g}_{-1} \times \mathfrak{g}_{-1} \to Q_x$ which is invariant under the action of $O(p, q)$. By the lemma, there is a unique linear isomorphism $\gamma : Q_x \to \mathfrak{g}_{-2}$ such that

$$\gamma \circ \mathcal{L}_x \circ (\beta \times \beta) = [\ ,\ ] : \mathfrak{g}_{-1} \times \mathfrak{g}_{-1} \to \mathfrak{g}_{-2}.$$

Now for $\xi \in T_{(\phi,\psi)}\mathcal{G}_0$ we define $\theta_{-2}(\xi) := \gamma(Tp \cdot \xi + H_x) \in \mathfrak{g}_{-2}$. Clearly, this defines $\theta_{-2} \in \Omega^1(\mathcal{G}_0, \mathfrak{g}_{-2})$, and the kernel of this form in each point u is $T^{-1}_u\mathcal{G}_0$. Now take $(B, C) \in G_0$ and pass from (ϕ, ψ) to $(\phi \circ B, \psi \circ C)$. By construction, this replaces β by $\beta \circ \mathrm{Ad}(B, C)$. Since $\gamma \circ \mathcal{L}_X \circ (\beta \times \beta) = [\ ,\ ]$ is Ad–equivariant, we see that we have to replace γ by $\mathrm{Ad}((B, C)^{-1}) \circ \gamma$. As above, this shows that $(r^{(B,C)})^*\theta_{-2} = \mathrm{Ad}((B, C)^{-1}) \circ \theta_{-2}$, so $(\mathcal{G}_0 \to M, \theta)$ is an infinitesimal flag structure of type (G, P). The construction of θ_{-2} was done in such a way that under the isomorphism $H \oplus Q \cong \mathcal{G}_0 \times_{G_0} \mathfrak{g}_-$, the Levi bracket corresponds to $[\ ,\ ]$. This implies regularity of the infinitesimal flag structure.

A morphism of Lie contact structures by definition gives rise to bundle maps between the auxilliary bundles, with the F–part being orthogonal and both covering the given local diffeomorphism. Using the induced maps on the frame bundles, one obtains a principal bundle map between the corresponding infinitesimal flag structures. From the above construction of the frame form θ one easily deduces that this principal bundle map is compatible with the frame forms, and thus a morphism of infinitesimal flag structures. □

Since $G_0 \cong GL(2,\mathbb{R}) \times O(p,q)$, all irreducible bundles in this case can be obtained from natural line bundles (with $Q = TM/H$ as a basic example) and tensor bundles of E, E^*, and F. Tensor bundles constructed from H can be decomposed into sums of these basic bundles similarly as discussed for almost Grassmannian structures of type $(2,n)$ in 4.1.3, but the fact that F is equipped with a bundle metric leads to finer decompositions. It has to be emphasized, however, that the bundles E and F, as well as the metric on F, have almost no intrinsic meaning. For example, a change of the bundle metric on F can be absorbed into a change of the isomorphism $H \to L(E,F)$. The main purpose of these data is, on the one hand, to express a compatibility condition between the Levi bracket and the tensor product decomposition. On the other hand, the finer decompositions of tensor bundles constructed from H are uninfluenced by the freedom in the choices one has.

The interpretation of harmonic curvature components is relatively easy in this case, since they are all contained in homogeneity one. In high dimensions, this is very similar to the discussion for almost Grassmanian structures in 4.1.4 (apart from the fact that the highest root is different). In lower dimensions, one has to discuss the cases of even and odd n (which correspond to D_n or B_n series) separately, but the final outcome is always the same (apart from a small speciality for $n = 4$ to be discussed below). The cohomology $H^2(\mathfrak{g}_-,\mathfrak{g})$ consists of two irreducible components, which are both contained in homogeneity one. In the language of bundles, they can be described as follows. Via the isomorphism $H \cong E^* \otimes F$, we get the decomposition

$$\Lambda^2 H^* \otimes H \cong (\Lambda^2 E \otimes E^* \otimes S^2 F^* \otimes F) \oplus (S^2 E \otimes E^* \otimes \Lambda^2 F^* \otimes F).$$

The two harmonic curvature components are the highest weight components of the two summands. In the first summand, $\Lambda^2 E \otimes E^*$ is already irreducible and via the bundle metric we can identify F with F^*, and the highest weight component is then given by the totally symmetric and tracefree part $S^3_0 F^*$. For the other summand, the highest weight part in $S^2 E \otimes E^*$ is the tracefree part, while in $\Lambda^2 F^* \otimes F$, we can identify F with F^* and then have to take the tracefree part of the kernel of the alternation. If $n = 4$, then the bundle F has rank four and the middle dimensional exterior power $\Lambda^2 F^*$ splits into self–dual and anti–self–dual parts. Accordingly, we get a splitting of the corresponding harmonic curvature into two components. Hence, for $n = 4$ there are three basic harmonic curvature quantities.

Following the general procedure from 4.2.2, we can determine the two harmonic curvature components from any contact connnection ∇^H, which is induced from a principal connection on the infinitesimal flag structure. The latter condition clearly means the ∇^H is induced from a linear connection ∇^E on E and a linear connection ∇^F on F, which is compatible with the bundle metric. Having chosen such connections and a projection π from TM onto the subbundle H, we can form the tensor $\tau \in \Gamma(\Lambda^2 H^* \otimes H)$ given by

$$\tau(\xi,\eta) = \nabla^H_\xi \eta - \nabla^H_\eta \xi - \pi([\xi,\eta]),$$

and the components of τ in $\ker(\square)$ represent the harmonic curvatures by Theorem 4.2.2. These components can be determined explicitly along similar lines as for almost Grassmannian structures in 4.1.3.

Let us finally remark that for even n, there is another real form of $\mathfrak{so}(n+2,\mathbb{C})$ for which the contact grading makes sense, namely $\mathfrak{so}^*(\frac{n}{2}+1)$. From the Satake diagram of this algebra we see that the semisimple part of the subalgebra $\mathfrak{g}_0$ for this grading is given by $\mathfrak{sp}(1) \oplus \mathfrak{so}^*(\frac{n}{2})$. In 2.3.11 we have seen that $\mathfrak{so}^*(\frac{n}{2})$ is the Lie algebra of maps compatible with a quaternionic Hermitian form with skew symmetric real part and symmetric imaginary part on $\mathbb{H}^n$, while $\mathfrak{sp}(1)$ can be viewed as acting on $\mathbb{H}^n$ by quaternionic scalar multiplications. Thus, this grading is related to Lie contact structures similarly as the almost quaternionic structures studied in 4.1.8 are related to the almost Grassmannian structures from 4.1.3. This analogy actually goes further, since one may easily show that a parabolic contact structure corresponding to $\mathfrak{so}^*(\frac{n}{2}+1)$ is given by a contact structure H on a manifold of dimension $4n+1$ together with a prequaternionic structure on the contact subbundle H for which the Levi–bracket is Hermitian. This means that there is a rank three subbundle $\mathcal{Q} \subset L(H,H)$, which can be locally spanned by smooth sections I, J, K, such that $I \circ I = J \circ J = -\mathrm{id}$, $I \circ J = -J \circ I = K$, and such that the Levi bracket $\mathcal{L}$ is Hermitian with respect to I, J and K. To our knowledge, the study of these structures has just begun recently; see [**Ž09**].

4.2.6. Contact projective structures. This is the last remaining example of a parabolic contact structure associated to a classical Lie algebra. At the same time, it provides the second basic example of a parabolic geometry which is not determined by the underlying infinitesimal flag structure.

For $n \geq 1$, consider $\mathfrak{g} := \mathfrak{sp}(2n+2,\mathbb{R})$. To obtain a description similar to the other examples, it is better not to use the standard symplectic form on $\mathbb{R}^{2n+2}$, but the form defined by

$$((x_0,\dots,x_{2n+1}),(y_0,\dots,y_{2n+1})) \mapsto x_0y_{2n+1} - y_0x_{2n+1} + \sum_{i=1}^{n}(x_iy_{n+i} - x_{n+i}y_i).$$

For this choice, the Lie algebra $\mathfrak{g}$ gets the following form with blocks of size 1, n, n, and 1,

$$\mathfrak{g} = \left\{ \begin{pmatrix} a & Z & W & z \\ X & A & B & W^t \\ Y & C & -A^t & -Z^t \\ x & Y^t & -X^t & -a \end{pmatrix} : B^t = B, C^t = C \right\}.$$

The grading is by the usual block form, but taking the two middle blocks as one, i.e. $\mathfrak{g}_{\pm 2} \cong \mathbb{R}$ via the entries x and z, $\mathfrak{g}_{-1} \cong \mathbb{R}^{2n}$ via the entries X and Y, and $\mathfrak{g}_1 \cong \mathbb{R}^{2n*}$ via the entries Z and W. Finally, $\mathfrak{g}_0$ is the block diagonal part given by the a entry and the central block, which evidently is $\mathfrak{sp}(2n,\mathbb{R})$. The bracket $[\ ,\ ] : \mathfrak{g}_{-1} \times \mathfrak{g}_{-1} \to \mathfrak{g}_{-2}$ is (with obvious notation) given by

$$\left[\begin{pmatrix} X_1 \\ Y_1 \end{pmatrix}, \begin{pmatrix} X_2 \\ Y_2 \end{pmatrix}\right] = Y_1^tX_2 - X_1^tY_2 - Y_2^tX_1 + X_2^tY_1,$$

which, up to a factor -2, is just the standard symplectic form on $\mathbb{R}^{2n}$. Hence, we have found a contact grading.

As a group with Lie algbra $\mathfrak{g}$, we take $G := PSp(2n+2,\mathbb{R})$ the quotient of $Sp(2n+2,\mathbb{R})$ by its center $\{\pm\mathrm{id}\}$. As usual, we will work with representative matrices taking into account that they are only defined up to sign. As the parabolic subgroup we use the stabilizer of the line generated by the first basis vector. This immediately shows that the homogeneous model G/P is the projective space $\mathbb{R}P^{2n+1}$ as discussed in 1.1.4. If we use $G = Sp(2n+2,\mathbb{R})$ and the connected component of the identity of the stabilizer for the parabolic subgroup P (which means that the entry in the top left corner is positive), then we obtain the sphere S^{2n+1}. The subgroup $G_0 \subset P$ is given by the classes of block diagonal matrices in G. The middle block of such a matrix simply represents the standard action of $Sp(2n, \mathbb{R})$ on $\mathfrak{g}_{-1} = \mathbb{R}^{2n}$, while the rest just gives multiples of the identity. Hence, Ad identifies G_0 with $CSp(\mathfrak{g}_{-1})$. In view of 4.2.2 we conclude that a regular infinitesimal flag structure of type (G, P) is equivalent to a contact structure H together with an orientation of the quotient bundle $Q = TM/H$. Hence, it is clear, that parabolic geometries of type (G, P) cannot be determined by the underlying ininfinitesimal flag structure. We also see that all irreducible bundles come from line bundles and tensor bundles of H.

By part (1) of Proposition 3.3.7, $H^1(\mathfrak{g}_-, \mathfrak{g})$ is concentrated in homogeneous degrees ≤ 1 so by Theorem 3.1.16, regular normal parabolic geometries of type (G, P) are equivalent to regular normal P–frame bundles of degree one. Interpreting these is parallel to the case of projective structures discussed in 4.1.5. Any such P–frame bundle $(\mathcal{G}_1 \to M, \theta)$ has an underlying regular infinitesimal flag structure of type (G, P) and we have seen above that this is equivalent to a contact structure $H \subset TM$ plus an orientation of the quotient bundle $Q = TM/H$. Now $\mathcal{G}_1 \to M$ by definition is a principal bundle with structure group P/P_+^2. The frame form θ consists of two components, namely

$$\theta_{-1} \in \Gamma(L(T^{-1}\mathcal{G}_1, \mathfrak{g}^{-1}/\mathfrak{g}^1)) \qquad \theta_{-2} \in \Omega^1(\mathcal{G}_1, \mathfrak{g}/\mathfrak{p}).$$

Here, $T^{-1}\mathcal{G}_1$ is the preimage of H. The underlying infinitesimal flag structure is $\mathcal{G}_0 = \mathcal{G}_1/(P_+/P_+^2)$ with the frame form induced by θ. Now we follow the discussion in 4.1.5. Suppose that $u_0 \in \mathcal{G}_0$ is a point. Then for $u \in \mathcal{G}_1$ over u_0, we can view the component in $\mathfrak{g}^0/\mathfrak{g}^1 = \mathfrak{g}_0$ of θ_{-1} as a map $T_{u_0}^{-1}\mathcal{G}_0 \to \mathfrak{g}_0$, which reproduces the generators of fundamental vector fields. Choosing a local section of $\mathcal{G}_1 \to M$, one obtains a local partial principal connection on $\mathcal{G}_0$; see 1.3.7. Equivalently, we can view this as a local partial contact connection ∇^σ on H. This means that ∇^σ is an operator $\Gamma(H) \times \Gamma(H) \to \Gamma(H)$ which is linear over smooth functions in the first variable and satisfies a Leibniz rule in the second variable. Morover, the induced partial connection on $\Lambda^2 H$ preserves the subbundle $\Lambda_0^2 H$.

The component θ_{-2} of the frame form can be interpreted as follows. As above, we fix u_0 and u and suppose they lie over $x \in M$. For a tangent vector $\xi \in T_x M$ we can choose a lift $\tilde{\xi} \in T_u\mathcal{G}_1$, and since this is unique up to elements of $T^0\mathcal{G}_1 = \ker(\theta_{-2}(u))$ the value of $\theta_{-2}(\tilde{\xi}) \in \mathfrak{g}/\mathfrak{p}$ depends only on ξ. Recall that we have a natural linear isomorphism $\mathfrak{g}_- \to \mathfrak{g}/\mathfrak{p}$. Then define $\pi^u(\xi) := T_u p \cdot \eta \in H_x \subset T_x M$, where $\eta \in T_u\mathcal{G}_1$ has the property that $\theta_{-2}(\eta)$ has vanishing $\mathfrak{g}_{-2}$–component, while the $\mathfrak{g}_{-1}$–component coincides with the one of $\theta_{-2}(\tilde{\xi})$. This is immediately seen to be well defined and by definition $\pi(\xi) = \xi$ for $\xi \in H_x$. Using equivariancy of θ_{-2} one shows similarly as for θ_{-1} that a local section σ of $\mathcal{G}_1$ leads to a locally defined projection π^σ from TM onto the subbundle H.

Next, one has to compute how a change of the section σ changes the data $(\nabla^\sigma, \pi^\sigma)$. The possible change of a section over $U \subset M$ is detemined by functions $g_0 : U \to G_0$ and $Z : U \to \mathfrak{g}_1$, via $\hat{\sigma}(x) = \sigma(x) \cdot g_0(x) \exp(Z(x))P_+^2$. As in 4.1.5 the data remain unchanged if $Z = 0$, so we may restrict to the case that $\hat{\sigma}(x) = \sigma(x) \cdot \exp(Z(x))P_+^2$. Using equivariancy of θ, it follows as in 4.1.5 that the change is determined by $-\operatorname{ad}(Z(x))$. Keeping in mind that we view $\mathfrak{g}_0$ as a subalgebra of $\mathfrak{gl}(\mathfrak{g}_{-1})$ via the adjoint action, we compute as follows: Take (with obvious notation) $(Z,W) \in \mathfrak{g}_1$ and $\binom{X_1}{Y_1}, \binom{X_2}{Y_2} \in \mathfrak{g}_{-1}$. Then one computes that $-\left[\left[(Z,W),\binom{X_1}{Y_1}\right],\binom{X_2}{Y_2}\right]$ is given by

$$(ZX_2+WY_2)\begin{pmatrix}X_1\\Y_1\end{pmatrix}+(ZX_1+WY_1)\begin{pmatrix}X_2\\Y_2\end{pmatrix}+(X_1^tY_2-Y_1^tX_2)\begin{pmatrix}W^t\\-Z^t\end{pmatrix}.$$

Taking the traceform on $\mathfrak{g}$ to describe the duality between $\mathfrak{g}_{-1}$ and $\mathfrak{g}_1$, the expression (ZX_2+WY_2) is simply the pairing between (Z,W) and $\binom{X_2}{Y_2}$, so the first two terms are easy to interpret and look exactly as in the projective case. To interpret the last term, we observe that taking its bracket with $\binom{X}{Y}$ gives

$$(ZX+WY)2(Y_1^tX_2-X_1^tY_2)=(ZX+WY)\left[\begin{pmatrix}X_1\\Y_1\end{pmatrix},\begin{pmatrix}X_2\\Y_2\end{pmatrix}\right].$$

Using this we see that changing σ to $\hat{\sigma}$, the partial connection changes as

$$\nabla^{\hat{\sigma}}_\xi\eta=\nabla^{\sigma}_\xi\eta+\Upsilon(\xi)\eta+\Upsilon(\eta)\xi+\Upsilon^{\#}(\mathcal{L}(\xi,\eta)) \tag{4.8}$$

for $\xi,\eta\in\Gamma(H)$. Here Υ is the map Z interpreted as a section of H^* and $\Upsilon^{\#}:Q\to H$ is characterized by $\mathcal{L}(\Upsilon^{\#}(\beta),\zeta)=\Upsilon(\zeta)\beta$ for $\beta\in Q$ and $\zeta\in H$.

For the projection, the interpretation is easier. For $(Z,W)\in\mathfrak{g}_1$ and $x\in\mathfrak{g}_{-2}$ we obtain $-[(z,W),x]=x\binom{W^t}{-Z^t}$. This shows that

$$\pi^{\hat{\sigma}}(\xi)=\pi^{\sigma}(\xi)+2\Upsilon^{\#}(q(\xi)),$$

for $\xi\in\mathfrak{X}(M)$, with $q:TM\to TM/H=Q$ the natural quotient map.

This relation has two nice consequences. On the one hand, consider the tensor τ from 4.2.2. For the data associated to σ this is given by

$$\tau(\xi,\eta)=\nabla^{\sigma}_\xi\eta-\nabla^{\sigma}_\eta\xi-\pi^{\sigma}([\xi,\eta]).$$

Changing to $\hat{\sigma}$, the terms in (4.8) involving Υ (rather than $\Upsilon^{\#}$) are symmetric in ξ and η and hence do not contribute to the change of τ. Hence, the full contribution to the change of τ from (4.8) is $\Upsilon^{\#}(\mathcal{L}(\xi,\eta)-\mathcal{L}(\eta,\xi))=2\Upsilon^{\#}(\mathcal{L}(\xi,\eta))$, which exactly cancels with the contribution from the projection term. Thus, for all choices of sections, we obtain the same torsion tensor τ.

For the other part of homogeneity one in the torsion, we have to use the induced connection on Q, so we have to compute its change first. This connection is characterized by the fact that $\mathcal{L}$ is parallel, so

$$\nabla^{\hat{\sigma}}_\xi\mathcal{L}(\eta,\zeta)=\mathcal{L}(\nabla^{\hat{\sigma}}_\xi\eta,\zeta)+\mathcal{L}(\eta,\nabla^{\hat{\sigma}}_\xi\zeta),$$

for $\xi,\eta,\zeta\in\Gamma(H)$. Expanding the right–hand side, collecting terms and using that $\mathcal{L}$ is parallel for ∇^{σ}, one obtains $\nabla^{\sigma}_\xi\mathcal{L}(\eta,\zeta)+2\Upsilon(\xi)\mathcal{L}(\eta,\zeta)$. In the proof of Theorem 4.2.2, we have seen that the homogeneity one part of the torsion for the choice associated to σ is given by

$$\nabla^{\sigma}_\xi q(\eta)-q([\xi,\eta])+\mathcal{L}(\xi,\pi^{\sigma}(\eta)),$$

for $\xi\in\Gamma(H)$ and $\eta\in\mathfrak{X}(M)$. Passing to $\hat{\sigma}$, the middle term remains unchanged, while the changes caused by the first and last term evidently cancel. Hence, the whole homogeneity one part of the torsion is independent of the choice of σ.

DEFINITION 4.2.6. Let (M,H) be a contact manifold, and let $\nabla:\Gamma(H)\times\Gamma(H)\to\Gamma(H)$ be a partial contact connection.

(1) The *contact torsion* $\tau : \Lambda^2 H \to H$ of ∇ is the tensor induced by

$$\tau(\xi, \eta) = \nabla_\xi \eta - \nabla_\eta \xi - \pi([\xi, \eta]),$$

where $\pi : TM \to H$ is the projection associated to ∇ in part (2) of Theorem 4.2.2.

(2) A partial contact connection $\hat{\nabla}$ on H, is said to be *contact projectively equivalent* to ∇ if and only if there is a smooth section $\Upsilon \in \Gamma(H^*)$ such that

$$\hat{\nabla}_\xi \eta = \nabla_\xi \eta + \Upsilon(\xi)\eta + \Upsilon(\eta)\xi + \Upsilon^\#(\mathcal{L}(\xi, \eta)),$$

where $\Upsilon^\# : Q \to H$ is characterized by $\mathcal{L}(\Upsilon^\#(\beta), \zeta) = \Upsilon(\zeta)\beta$ for $\beta \in Q$ and $\zeta \in H$.

The upshot of the above discussion was that a regular P–frame bundle of degree one over M induces a contact structure H on M as well as a contact projective equivalence class of partial contact connections. Parallel to 4.1.5, one establishes the converse. On the other hand, we have seen that contact projectively equivalent partial connections have the same contact torsion, so it make sense to talk about the contact torsion of a projective class.

Thus, it remains to discuss the normality condition, which needs some basic information on $H^2(\mathfrak{g}_-, \mathfrak{g})$. There is only one element w of length two in the Hasse diagram, namely acting with the reflection corresponding to the second simple root and then with the one corresponding to the first simple root, which is crossed. Applying this to the highest root $2\alpha_1 + \cdots + 2\alpha_{n-1} + \alpha_n$ we get $2\alpha_2 + \cdots + 2\alpha_{n-1} + \alpha_n$ (respectively α_2 if $n = 2$). The corresponding root space lies in $\mathfrak{g}_0$. It is still a positive root though, so the root spaces corresponding to elements of Φ_w must be contained in $\mathfrak{g}_1$. This shows that the irreducible representation $H^2(\mathfrak{g}_-, \mathfrak{g})$ sits in $\Lambda^2 \mathfrak{g}_{-1}^* \otimes \mathfrak{g}_0$. In particular, there is no cohomology in homogeneity one. Using this, one verifies (again parallel to 4.1.5 and using 4.2.2) that a regular P–frame bundle of degree one is normal if and only if it corresponds to a contact projective class with vanishing contact torsion. The harmonic curvature can then be read off as in part (3) of Theorem 4.2.2 using any connection extending a member of the contact projective equivalence class.

4.2.7. Contact projective structures and geodesics. Similarly, as discussed for projective structures in 4.1.6, there is an interpretation of contact projective structures in terms of unparametrized geodesics. We discuss this only briefly, more details can be found in [**Fox05a**] and [**Fox05b**]. Suppose first that (M, H) is a contact structure, ∇ is a contact connection on TM and $c : I \to M$ is a geodesic for ∇. Let α be a contact form for H with Reeb vector field r, and put $f(t) := \alpha(c'(t))$. Then $c'(t) - f(t)r \in H$ for all t, and since ∇ is a contact connection we get

$$0 = \alpha(\nabla_{c'}(c' - fr)) = -\alpha(\nabla_{c'} fr) = -c' \cdot f - f\alpha(\nabla_{c'} r).$$

This is a linear first order ODE on f, so if f vanishes in one point, then it vanishes identically. Hence, we conclude that any geodesic for a contact connection that is tangent to H in one point is tangent to H everywhere. We call these geodesics the *contact geodesics* of ∇. Evidently, they depend only on the partial connection underlying ∇.

Now given a partial contact connection, consider the set of all partial contact connections which have the same contact torsion and the same geodesics up to parametrization. Here the contact torsion is determined with respect to the projection π associated to the partial contact connection according to part (2) of

Theorem 4.2.2. Similarly, as in 4.1.6, one shows that this recovers the contact projective equivalence class as defined in 4.2.6 above. The family of contact geodesics can then be viewed as a smooth family of unparametrized curves, with exactly one curve through each point in each direction in the contact subbundle. Among such families there are those, which are the geodesics of a (partial) contact connection with vanishing contact torsion.

Suppose that we have contact manifolds (M, H) and $(\tilde{M}, \tilde{H})$ endowed with such families of paths and a contact diffeomorphism $f : M \to \tilde{M}$, which is compatible with these families. Then the families determine contact projective equivalence classes of partial contact connections on M and $\tilde{M}$. Take a representative of the class on $\tilde{M}$, and pull it back to M using f. The result is a partial contact connection with vanishing contact torsion, which by construction has the distinguished paths as contact geodesics. Thus, it lies in the contact projective equivalence class, and f is a morphism of contact projective structures. Hence, we see that the families of paths provide an equivalent description of projective contact structures.

At this point there occurs a subtlety which is not present for classical projective structures. Any linear connection on the tangent bundle of a manifold can be changed into a torsion–free connection. Since the necessary change is given by a skew symmetric tensor, this does not change the set of geodesics. Thus, the family of geodesics of an arbitrary linear connection can always be described by a torsion–free projective structure and hence a regular normal parabolic geometry.

This is no longer true in the contact case. The deformation tensor between two partial contact connections is a section of $H^* \otimes \mathfrak{csp}(H) \subset H^* \otimes H^* \otimes H$. Such a change does not affect the contact geodesics, if at the same time it is contained in $\Lambda^2 H^* \otimes H$, and then it also directly describes the change of torsion. The intersection $(H^* \otimes \mathfrak{csp}(H)) \cap (\Lambda^2 H^* \otimes H)$ turns out to be too small to remove arbitrary torsions. Only those families of contact paths, which can be described by a partial contact connection with vanishing contact torsion, admit an equivalent description as a regular normal parabolic geometry.

It turns out, however, that also contact projective structures with nonvanishing contact torsion admit a canonical regular Cartan connection of type (G, P). This was shown by D.J.F. Fox in [**Fox05a**], where he generalized the normalization condition on the curvature of a Cartan connection. In the case of vanishing contact torsion, the original normalization condition is recovered. Hence, this extends the approach via Cartan geometries to arbitrary families of contact geodesics.

4.2.8. Exotic parabolic contact structures. In this section we briefly indicate what the parabolic contact structures associated to exceptional Lie algebras look like. We also use this to demonstrate that a rough picture of the nature of a parabolic geometry can be obtained with very little effort. To our knowledge, the details have not been worked out yet for any of these geometries. It should, however, be remarked that there are relations between contact gradings on simple Lie algebras and Jordan algebras (see [**Kan73**]), which should be useful for a more detailed study of these geometries.

We will not discuss harmonic curvature components here. Indeed, one shows that in all cases, there is only one irreducible component in $H^2(\mathfrak{g}_-, \mathfrak{g})$ and that component sits in homogeneity one. Hence, the harmonic curvatures can always be read off as appropriate components of the tensor τ associated to any contact connection induced by from a principal connection on the infinitesimal flag structure.

The principles along which we discuss the geometries are fairly easy: The list of complex contact gradings can be found in 3.2.4 and from the Satake diagram of each real form one immediately sees whether the contact grading exists on that real form or not. In that table one also finds the type of $\mathfrak{g}_0$. Using this, one computes the dimension, in which the geometry exists as $1/2(\dim(\mathfrak{g}) - \dim(\mathfrak{g}_0))$. For the exceptional algebras, $\mathfrak{g}_0$ always has one–dimensional center and the Satake diagram of the semisimple part is obtained by removing the crossed node and all edges connecting to it; see [**Kane93**]. The single crossed node represents the simple root for which the corresponding root space is contained in $\mathfrak{g}_1$. But this immediately implies that the negative of this root is the highest weight of $\mathfrak{g}_{-1}$. From the definition of the Dynkin diagram, one may immediately write out this weight, thus finding the nature of the reduction of structure group of a contact structure which is equivalent to the parabolic contact structure.

The parabolic contact structure associated to G_2. Here there is just one noncompact noncomplex real form, namely the split real form. From the table in 3.2.4 we see that the semisimple part of $\mathfrak{g}_0$ is $\mathfrak{sl}(2,\mathbb{R})$, so $\mathfrak{g}_0 \cong \mathfrak{gl}(2,\mathbb{R})$. Since G_2 has dimension 14, the associated parabolic contact structure exists on manifolds of dimension 5. The highest weight of $\mathfrak{g}_{-1}$ is 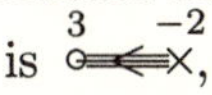

, so up to the action of the center, this is the third symmetric power of the standard representation of $\mathfrak{g}_0$. The weights of $S^3\mathbb{R}^2$ are 3, 1, -1, and -3, all with multiplicity one, which shows that the weights of $\Lambda^2(S^3\mathbb{R}^2)$ are 4, 2, -2, and -4 with multiplicity one and 0 with multiplicity two. This shows that $\Lambda^2(S^3\mathbb{R}^2) \cong S^4\mathbb{R}^2 \oplus \mathbb{R}$, so up to scale there is a unique skew symmetric bilinear form on $S^3\mathbb{R}^2$ which is invariant under $\mathfrak{sl}(2,\mathbb{R})$.

Thus, we conclude that a parabolic contact structure associated to G_2 on a smooth manifold M of dimension 5 is given by a contact structure $H \subset TM$ and an auxilliary rank two vector bundle $E \to M$, together with an isomorphism $S^3E \to H$, such that the Levi bracket $\mathcal{L}$ is invariant under the action of $\mathfrak{sl}(E)$. The contact connections on H coming from a principal connection on the associated infinitesimal flag structure are exactly those induced by connections on E.

The parabolic contact structure associated to F_4. Again there is only the split real form to consider. The subalgebra $\mathfrak{g}_0$ has semisimple part $\mathfrak{sp}(6,\mathbb{R})$, which shows that the geometry exists in dimension 15. The highest weight of $\mathfrak{g}_{-1}$ is 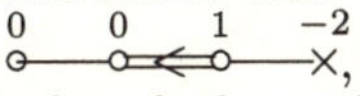

, so this is the tracefree part $\Lambda^3_0\mathbb{R}^6 \subset \Lambda^3\mathbb{R}^6$. The wedge product $\Lambda^3\mathbb{R}^6 \times \Lambda^3\mathbb{R}^6 \to \Lambda^6\mathbb{R}^6$ defines a nondegenerate skew symmetric bilinear form, which is invariant under $\mathfrak{sp}(6,\mathbb{R})$.

Hence, the parabolic contact structure associated to F_4 on a contact manifold (M,H) of dimension 15 is given by an auxilliary vector bundle E of rank 6 which is endowed with a nondegenerate skew symmetric bilinear form and an isomorphism $\Lambda^3_0E \to H$, such that $\mathcal{L}$ is invariant under $\mathfrak{sp}(E)$. From this it is also clear what the associated contact connections look like.

The parabolic contact structure associated to E_6. Here there are three different real forms to consider. For the semisimple part of $\mathfrak{g}_0$ we obtain $\mathfrak{sl}(6,\mathbb{R})$, $\mathfrak{su}(3,3)$, and $\mathfrak{su}(5,1)$, respectively. Hence, these types of geometries exist in dimension 21. The highest weight of $\mathfrak{g}_{-1}$ is given by

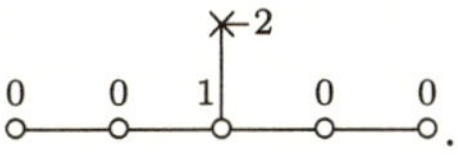

.

For the split form, we therefore get $\mathfrak{g}_{-1} \cong \Lambda^3 \mathbb{R}^6$, and the description is completely parallel to the F_4 case above but without a symplectic form on the auxilliary bundle. For the two $\mathfrak{su}$–algebras, $\mathfrak{g}_{-1}$ is a real subrepresentation in $\Lambda^3 \mathbb{C}^6$. Existence of this real subrepresentation comes from the fact that on one hand, the wedge product identifies $\Lambda^3 \mathbb{C}^6$ with its dual. On the other hand, we get an induced Hermitian form on this space, which leads to an identification with the conjugate dual. Together with the above, this defines a invariant conjugation, whose fixed points form the real subrepresentation. Hence, in these two cases, one has an auxiliary complex vector bundle E of rank 6 with a Hermitian bundle metric of signature $(3,3)$, respectively, $(5,1)$ and an identification of the contact subbundle with the appropriate real subspace in $\Lambda^3_{\mathbb{C}} E$.

The parabolic contact structure associated to E_7. In this case, we do not know an explicit interpretation of the reduction of structure group one obtains. Again there are three different real forms to consider for which the semisimple parts of $\mathfrak{g}_0$ are isomorphic to $\mathfrak{so}(6,6)$, $\mathfrak{so}(4,8)$, and $\mathfrak{so}^*(12)$. In particular, the geometries exist in dimension 33. For the highest weight of the complexification of $\mathfrak{g}_{-1}$, one obtains

$$\begin{array}{c} \overset{0}{\circ} \\ \underset{0}{\circ}\!\!-\!\!\underset{0}{\circ}\!\!-\!\!\underset{0}{\circ}\!\!-\!\!\underset{0}{\circ}\!\!-\!\!\underset{1}{\circ}\!\!-\!\!\underset{-2}{\times}, \end{array}$$

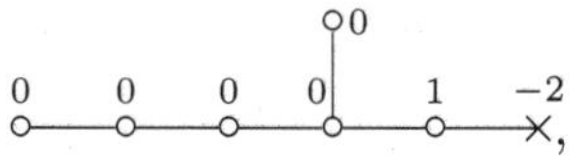

so this corresponds to one of the spin representations. As before it turns out that on this representation there is a unique skew symmetric bilinear form. Hence, the geometry is given by a reduction of the contact subbundle to the appropriate group G_0, which is included in $GL(32,\mathbb{R})$ via a real subrepresentation of the appropriate basic spin representation.

The parabolic contact structure associated to E_8. Here things get really involved, since the semisimple part of $\mathfrak{g}_0$ is a real form of E_7, and two of these real forms actually occur. The geometry exists in dimension 57 and is given by a reduction of structure group of the contact subbundle to the appropriate real form of E_7, which has a unique representation in dimension 56.

4.3. Examples of general parabolic geometries

4.3.1. Geometries determined by filtrations. A general regular infinitesimal flag structure has two ingredients, the filtration of the tangent bundle and the reduction of structure group. In the case of $|1|$–gradings discussed in section 4.1, the filtration was trivial and all the geometry was given by the reduction of structure group. Now we consider the other extreme, in which the geometry is determined by the filtration only. The reduction of structure group corresponds to the homomorphism $\mathrm{Ad}: G_0 \to \mathrm{Aut}_{\mathrm{gr}}(\mathfrak{g}_-)$, where $\mathrm{Aut}_{\mathrm{gr}}(\mathfrak{g}_-)$ denotes the group of automorphisms of the graded Lie algebra $\mathfrak{g}_-$. If this is an isomorphism, then the whole geometry is determined by the filtration. While it is not evident a priori, whether this condition is satisfied, there is a simple cohomological criterion, which shows that it is even true generically. Recall that for a semisimple Lie algebra $\mathfrak{g}$, the group $\mathrm{Aut}(\mathfrak{g})$ is a Lie group with Lie algebra $\mathfrak{der}(\mathfrak{g})$, the Lie algebra of derivations of $\mathfrak{g}$. Since any derivation of $\mathfrak{g}$ is inner by Corollary 2.1.6, this is isomorphic to $\mathfrak{g}$. Hence, $G := \mathrm{Aut}(\mathfrak{g})$ has Lie algebra $\mathfrak{g}$ and the adjoint action is given by applying automorphisms. Consequently, given a $|k|$–grading on $\mathfrak{g}$ the maximal parabolic subgroup $P \subset \mathrm{Aut}(\mathfrak{g})$ for the given grading is the group $\mathrm{Aut}_f(\mathfrak{g})$ of all automorphisms of $\mathfrak{g}$ which preserve the filtration. For the Levi subgroup $G_0 \subset P$ we then obtain the group $\mathrm{Aut}_{\mathrm{gr}}(\mathfrak{g})$ of automorphisms of the graded Lie algebra $\mathfrak{g}$.

PROPOSITION 4.3.1. *Let $\mathfrak{g}$ be a $|k|$-graded semisimple Lie algebra such that no simple ideal is contained in $\mathfrak{g}_0$.*

(1) If the cohomology group $H^1(\mathfrak{g}_-, \mathfrak{g})$ is contained in homogeneous degrees < 0, then putting $G = \mathrm{Aut}(\mathfrak{g})$, the adjoint action induces an isomorphism $G_0 \to \mathrm{Aut}_{\mathrm{gr}}(\mathfrak{g}_-)$. Regular normal parabolic geometries of type (G, P) are then equivalent to filtered manifolds such that the bundle of symbol algebras is locally trivial with fiber $\mathfrak{g}_-$.

(2) The cohomological condition in (1) is satisfied unless one of the $|k_i|$-graded simple ideals of the complexification of $\mathfrak{g}$ belongs to one of the following 4 classes:

(i) *$|1|$-gradings as described in 3.2.3.*
(ii) *Complex contact gradings as described in 3.2.4.*
(iii) *Type A_ℓ with $\ell \geq 3$ and gradings corresponding to $\Sigma \subset \Delta_+$ such that $\Sigma = \{\alpha_i, \alpha_j\}$ with $i < j$ and $i = 1$ or $j = n$.*
(iv) *Type C_ℓ with $\ell \geq 2$ and the grading corresponding to $\Sigma = \{\alpha_1, \alpha_\ell\} \subset \Delta_+$.*

PROOF. (1) By definition, $G_0 = \mathrm{Aut}_{\mathrm{gr}}(\mathfrak{g})$ and the relevant map $\mathrm{Ad} : G_0 \to \mathrm{Aut}_{\mathrm{gr}}(\mathfrak{g}_-)$ is given by restriction. Now the Killing form on $\mathfrak{g}$ is invariant under each automorphism and it induces a duality between $\mathfrak{g}_-$ and $\mathfrak{p}_+$. This implies that for $\phi \in G_0$, the restriction $\phi|_{\mathfrak{p}_+}$ must be the dual map of $(\phi|_{\mathfrak{g}_-})^{-1}$. On the other hand, since no simple ideal of $\mathfrak{g}$ is contained in $\mathfrak{g}_0$, the map $\mathrm{ad} : \mathfrak{g}_0 \to \mathfrak{gl}(\mathfrak{g}_{-1})$ is injective by part (5) of Proposition 3.1.2. For $A \in \mathfrak{g}_0$ and $X \in \mathfrak{g}_{-1}$, we must have $[\phi(A), X] = \phi([A, \phi^{-1}(X)])$. Thus, $\phi|_{\mathfrak{g}_0}$ is also determined by $\phi|_{\mathfrak{g}_-}$ and $\mathrm{Ad} : G_0 \to \mathrm{Aut}_{\mathrm{gr}}(\mathfrak{g}_-)$ is always injective.

The computations done above also indicate how to prove surjectivity. Given $\psi \in \mathrm{Aut}_{\mathrm{gr}}(\mathfrak{g}_-)$, there is only one reasonable way to extend it to $\mathfrak{g}$. We define $\psi : \mathfrak{p}_+ \to \mathfrak{p}_+$ as the dual map (with respect to the Killing form) of $\psi^{-1} : \mathfrak{g}_- \to \mathfrak{g}_-$. For $A \in \mathfrak{g}_0$, consider the map $\Phi : \mathfrak{g}_- \to \mathfrak{g}_-$ defined by $\Phi(X) := \psi([A, \psi^{-1}(X)])$. We claim that there is a unique element $\psi(A) \in \mathfrak{g}_0$ such that $\Phi(X) = [\psi(A), X]$ for all $X \in \mathfrak{g}_-$. To show this, we have to invoke the cohomological condition. An element $f \in C^1(\mathfrak{g}_-, \mathfrak{g})$ by definition is a linear map $\mathfrak{g}_- \to \mathfrak{g}$ and if one requires f to be homogeneous of degree zero, then it has values in $\mathfrak{g}_-$. The cocycle equation is

$$0 = \partial f(X, Y) = [X, f(Y)] - [Y, f(X)] - f([X, Y]),$$

or equivalently $f([X, Y]) = [f(X), Y] + [X, f(Y)]$. Hence, cocycles of homogeneity zero are exactly derivations of the graded algebra $\mathfrak{g}_-$. Now we compute

$$\begin{aligned}\Phi([X, Y]) &= \psi([A, \psi^{-1}([X, Y])]) \\ &= \psi([[A, \psi^{-1}(X)], \psi^{-1}(Y)] + [\psi^{-1}(X), [A, \psi^{-1}(Y)]]) = [\Phi(X), Y] + [X, \Phi(Y)],\end{aligned}$$

so Φ is indeed a derivation. Vanishing of cohomology says that Φ is given by the adjoint action of an element of $\mathfrak{g}_0$ and uniqueness follows from the injectivity of the adjoint action of $\mathfrak{g}_0$ on $\mathfrak{g}_-$.

Now we have extended ψ to a linear endomorphism of $\mathfrak{g}$, which is evidently invertible, so it remains to check compatibility of ψ with the Lie bracket. Compatibility with the bracket of two elements of $\mathfrak{g}_-$ holds by definition. For the bracket $\mathfrak{g}_0 \times \mathfrak{g}_- \to \mathfrak{g}_-$ we get

$$[\psi(A), \psi(X)] = \Phi(\psi(X)) = \psi([A, X]).$$

Using this and the Jacobi identity, we get for $A, C \in \mathfrak{g}_0$ and $X \in \mathfrak{g}_-$

$$\begin{aligned}[][\psi(A),\psi(C)],\psi(X)] =&[\psi([A,X]),\psi(C)] + [\psi(A),\psi([C,X])] \\ =&\psi([[A,C],X]) = [\psi([A,C]),\psi(X)],\end{aligned}$$

which implies compatibility with the bracket $\mathfrak{g}_0 \times \mathfrak{g}_0 \to \mathfrak{g}_0$. Thus, we have verified compatibility with all brackets on $\mathfrak{g}_{\leq 0} \times \mathfrak{g}_{\leq 0}$.

By definition, $\psi|_{\mathfrak{p}_+}$ is characterized by $B(\psi(Z),\psi(X)) = B(Z,X)$ for $Z \in \mathfrak{p}_+$ and $X \in \mathfrak{g}_-$, where B is the Killing form. Now take $Z \in \mathfrak{p}_+$ and $Y \in \mathfrak{g}_{\leq 0}$ such that $[Z,Y] \in \mathfrak{p}_+$. Then for $X \in \mathfrak{g}_-$ we obtain

$$\begin{aligned}B([\psi(Z),\psi(Y)],\psi(X)) =&B(\psi(Z),[\psi(Y),\psi(X)]) \\ =&B(Z,[Y,X]) = B(\psi([Z,Y]),\psi(X)).\end{aligned}$$

In particular, for $A \in \mathfrak{g}_0$ we have now verified that $\mathrm{ad}(\psi(A)) = \psi \circ \mathrm{ad}(A) \circ \psi^{-1}$, which implies that $B(\psi(A),\psi(C)) = B(A,C)$ holds for $A, C \in \mathfrak{g}_0$. Having this at hand, the last argument continues to hold for $Z \in \mathfrak{p}_+$ and $Y \in \mathfrak{g}_-$ such that $[Z,Y] \in \mathfrak{g}_0$.

For $Z \in \mathfrak{p}$ and $X \in \mathfrak{g}_{-1}$ we therefore always have $\psi([Z,X]) = [\psi(Z),\psi(X)]$. Taking and additional element $Y \in \mathfrak{g}_{-1}$, and using $[Z,[X,Y]] = [[Z,X],Y] + [X,[Z,Y]]$, we conclude that $\psi([Z,[X,Y]]) = [\psi(Z),\psi([X,Y])]$. Since elements of the form $[X,Y]$ span $\mathfrak{g}_{-2}$, this together with the above implies compatibility with all brackets defined on $\mathfrak{p} \times \mathfrak{g}_{-2}$. Continuing inductively, we conclude that all brackets on $\mathfrak{p} \times \mathfrak{g}_-$ are compatible with ψ. Together with the above, it remains to verify that ψ is compatible with the bracket on $\mathfrak{p}_+ \times \mathfrak{p}_+$. This now follows immediately from the compatibility with the Killing form.

Having shown that $\mathrm{Ad} : G_0 \to \mathrm{Aut}_{\mathrm{gr}}(\mathfrak{g}_-)$ is an isomorphism, it is clear that regular infinitesimal flag structures of type (G,P) are equivalent to filtered manifolds such that the bundle of symbol algebras is locally trivial with model $\mathfrak{g}_-$. By Theorem 3.1.14, the description of regular normal parabolic geometries follows.

(2) Suppose that $\mathfrak{g} = \mathfrak{g}' \oplus \mathfrak{g}''$ is a sum of a $|k'|$–graded Lie algebra and a $|k''|$–graded Lie algebra. Then from 3.3.6 and 3.3.7 we know that $H^1(\mathfrak{g}_-,\mathfrak{g})$ is the sum of

$$H^1(\mathfrak{g}'_-,\mathfrak{g}') \boxtimes H^0(\mathfrak{g}''_-,\mathbb{K}) \oplus H^0(\mathfrak{g}'_-,\mathfrak{g}') \boxtimes H^1(\mathfrak{g}''_-,\mathbb{K}),$$

and the same terms with $\mathfrak{g}'$ and $\mathfrak{g}''$ exchanged. Moreover, we have seen there that $H^0(\mathfrak{g}''_-,\mathbb{K})$, $H^1(\mathfrak{g}''_-,\mathbb{K})$, and $H^0(\mathfrak{g}'_-,\mathfrak{g}')$ are concentrated in homogeneities 0, 1, and $-k'$, respectively. Hence, $H^1(\mathfrak{g}_-,\mathfrak{g})$ contains a component of nonnegative homogeneity if and only if either $\mathfrak{g}'$ or $\mathfrak{g}''$ has this property or one of the two summands is $|1|$–graded. Inductively, this immediately implies that we may restrict to the case that $\mathfrak{g}$ itself is complex and simple.

Now we can deduce the result as an application of Kostant's version of the Bott–Borel–Weil Theorem 3.3.5. The elements of length one in the Hasse diagram are the simple reflections for roots in the subset $\Sigma \subset \Delta_+$ determined by the grading. Let us denote by σ_i the reflection corresponding to the simple root $\alpha_i \in \Sigma$. The corresponding subset Φ_{σ_i} is $\{\alpha_i\}$. On the other hand, the highest weight of $\mathfrak{g}$ is the highest root β. The explicit representative for the component in $H^1(\mathfrak{p}_+,\mathfrak{g})$ determined by σ_i from Theorem 3.3.5 has weight $-\alpha_i+\sigma_i(\beta)$. Dualizing, we see that the corresponding component of $H^1(\mathfrak{g}_-,\mathfrak{g})$ is contained in homogeneity $\mathrm{ht}_\Sigma(\alpha_i - \sigma_i(\beta)) = 1 - \mathrm{ht}_\Sigma(\sigma_i(\beta))$. Hence, we have to determine for which choices of $\mathfrak{g}$ and

Σ we get $\text{ht}_\Sigma(\sigma_i(\beta)) \leq 1$. Since $|1|$–gradings are in the list of exceptions, we may assume $\text{ht}_\Sigma(\beta) = k > 1$.

Let us first assume that $\mathfrak{g}$ is not of type A_ℓ or C_ℓ. Then the adjoint representation is one of the fundamental representations, so β is the fundamental weight corresponding to a simple root, say α_{i_0}. In particular, $\sigma_i(\beta) = \beta$ unless $i = i_0$, so we must have $\alpha_{i_0} \in \Sigma$. On the other hand, $\sigma_{i_0}(\beta) = \beta - \alpha_{i_0}$ since β is fundamental, so $\text{ht}_\Sigma(\sigma_{i_0}(\beta)) = \text{ht}_\Sigma(\beta) - 1$. But $\Sigma = \{\alpha_{i_0}\}$ corresponds to the contact grading and hence $k = 2$, while enlarging this subset would increase the Σ–height of β. Thus, $k > 1$ and $\text{ht}_\Sigma(\sigma_{i_0}(\beta)) \leq 1$ in these cases is only possible for the contact grading corresponding to $\Sigma = \{\alpha_{i_0}\}$.

For type A_ℓ we have $\beta = \alpha_1 + \cdots + \alpha_\ell$ and $\sigma_i(\beta)$ equals β for $i \neq 1, \ell$ and $\beta - \alpha_i$ for $i = 1$ or $i = \ell$. Consequently, $\text{ht}_\Sigma(\sigma_i(\beta))$ equals k for $i \neq 1, \ell$ and $k - 1$ for $i = 1$ or $i = \ell$. Since k in this case equals the number of elements of Σ, we see that we get $\text{ht}_\Sigma(\sigma_i(\beta)) = 1$ if and only if $\Sigma = \{\alpha_1, \alpha_i\}$ or $\Sigma = \{\alpha_i, \alpha_\ell\}$ and $\Sigma = \{\alpha_1, \alpha_\ell\}$ corresponds to the contact grading.

Finally, in the C_ℓ–case, $\beta = 2\alpha_1 + \cdots + 2\alpha_{\ell-1} + \alpha_\ell$ is twice the fundamental weight corresponding to the simple root α_1. Hence, $\sigma_i(\beta) = \beta$ for $i > 1$ and $\sigma_1(\beta) = \beta - 2\alpha_1$, so this decreases the Σ–height by 2. Hence, we have to determine subsets $\Sigma \subset \Delta_+$ which contain α_1 and have the property that $\text{ht}_\Sigma(\beta) \leq 3$. These are exactly $\{\alpha_1\}$, which corresponds to the contact grading, and $\{\alpha_1, \alpha_\ell\}$. □

From the description of contact gradings in 3.2.4 it follows that the subset of $\mathfrak{p}_+$ corresponding to the contact grading consists of those simple roots which are not orthogonal to the highest root. Hence, there are only two roots with this property in the A_n case, while for the other simple Lie algebras, there is only one such simple root. Hence, part (2) of the proposition shows that for most of the possible gradings, parabolic geometries are determined by the filtration alone.

4.3.2. Generic distributions. In general it is not evident how to obtain examples of filtrations with prescribed symbol algebras. In some cases, however, having the "right" symbol algebra turns out to be a generic condition, which leads to several simplifications. We discuss a few examples of this situtation in more detail next.

Let us start with some general observations. Suppose that $TM = T^{-k}M \supset \cdots \supset T^{-1}M$ is a filtration of the tangent bundle with symbol algebra $(\text{gr}(TM), \mathcal{L})$. Then by definition, for $\xi \in \Gamma(T^iM)$ and $\eta \in \Gamma(T^jM)$ we have $[\xi, \eta] \in \Gamma(T^{i+j}M)$. If we assume (as it is always the case for the filtrations underlying parabolic geometries), that the symbol algebra is generated by $\text{gr}_{-1}(TM)$, then we conclude that iterated Lie brackets of sections of $T^{-1}M$ span TM. This says that the distibution $H := T^{-1}M \subset TM$ is *bracket generating.* Moreover, in this case, H already determines the complete filtration. Namely, the subbundle $T^{-2}M$ is spanned by sections of $T^{-1}M$ and single Lie brackets of such sections. Inductively, $T^{-i-1}M$ is spanned by sections of $T^{-i}M$ and their brackets with sections of $T^{-1}M$. In Tanaka's work, $T^{-1}M$ is viewed as a differential system, and the above fact is phrased as stating that $T^{-i-1}M$ is the ith *weak derived system* of $T^{-1}M$. In this situation, one calls the sequence $(\text{rank}(T^iM))$ (starting with $\text{rank}(H)$) the *small growth vector* of the distribution $H = T^{-1}M$. Since this is the only type of growth vector we will deal with, we will often omit the adjective "small" and just talk about the growth vector.

Growth vector $(3,6)$ **and** $(n, n(n+1)/2)$. This is the simplest example of a generic distribution. It is associated to the Lie algebra $\mathfrak{g} = \mathfrak{so}(n+1,n)$ and the grading associated to the last simple root α_n for $n \geq 3$. This leads to a $|2|$–grading, and since the adjoint representation is the second fundamental representation, we always obtain a geometry determined by the filtration by part (2) of Proposition 4.3.1. From the list of roots of $\mathfrak{g}$ in example (3) of 2.2.6 we see that $\mathfrak{g}_{-1}$ is spanned by the root spaces corresponding to the roots α_n, $\alpha_n + \alpha_{n-1}, \dots, \alpha_n + \dots + \alpha_1$ and hence has dimension n. The semisimple part of $\mathfrak{g}_0$ is of type A_{n-1} and hence $\dim(\mathfrak{g}_0) = n^2$. Since $\dim(\mathfrak{g}) = n(2n+1)$ we conclude that $\dim(\mathfrak{g}_{\pm 2}) = \frac{n(n-1)}{2}$, so the geometry exists in dimension $\frac{n(n+1)}{2}$. Finally, from the general theory we know that $\mathfrak{g}_-$ is generated by $\mathfrak{g}_{-1}$, so since $\dim(\mathfrak{g}_{-2}) = \dim(\Lambda^2\mathfrak{g}_{-1})$, we see that $[\ ,\ ] : \Lambda^2\mathfrak{g}_{-1} \to \mathfrak{g}_{-2}$ must be a linear isomorphism.

Now suppose that M is a smooth manifold of dimension $\frac{n(n+1)}{2}$ and $H \subset TM$ is a distribution of rank n. Putting $Q = TM/H$, we get the Levi bracket $\mathcal{L} : \Lambda^2 H \to Q$, so this is a bundle map between two bundles of rank $\frac{n(n-1)}{2}$. Hence, the growth vector of H is $(n, \frac{n(n+1)}{2})$ if and only if $\mathcal{L}$ is an isomorphism of vector bundles. Otherwise put, this means that H is as non–integrable as possible. This condition is generic, i.e. preserved under sufficiently small deformations.

Suppose that $\mathcal{L} : \Lambda^2 H \to Q$ is an isomorphism of vector bundles, $x \in M$ is a point and $\phi : \mathfrak{g}_{-1} \to H_x$ is a linear isomorphism. Then the composition

$$\mathfrak{g}_{-2} \xrightarrow{([\ ,\ ])^{-1}} \Lambda^2\mathfrak{g}_{-1} \xrightarrow{\Lambda^2\phi} \Lambda^2 H_x \xrightarrow{\mathcal{L}} Q_x$$

defines a (uniquely determined) linear isomorphism $\psi : \mathfrak{g}_{-2} \to Q_x$ such that $\psi([X,Y]) = \mathcal{L}(\phi(X), \phi(Y))$ for all $X, Y \in \mathfrak{g}_{-1}$. This shows that the bundle of symbol algebras is locally trivial and modelled on $\mathfrak{g}_- = \mathfrak{g}_{-2} \oplus \mathfrak{g}_{-1}$. Conversely, this property of course implies that $\mathcal{L} : \Lambda^2 H \to Q$ is an isomorphism. Since the Dynkin diagram B_n has no automorphisms, the group $\mathrm{Aut}(\mathfrak{so}(n+1,n))$ coincides with the group of inner automorphisms. This group can be realized as the quotient of any connected Lie group with Lie algebra $\mathfrak{so}(n+1,n)$ by its center, and since we are in odd dimensions, the connected component $SO_0(n+1,n)$ has trivial center. The appropriate parabolic subgroup $P \subset G$ can be identified with the stabilizer of an isotropic subspace of dimension n in $\mathbb{R}^{2n+1}$; see 3.2.5 and 3.2.11. Using Proposition 4.3.1 we conclude that regular normal parabolic geometries of type $(SO_0(n+1,n), P)$ are equivalent to distributions of growth vector $(n, \frac{n(n+1)}{2})$. The simplest special case is $n = 3$, where we obtain growth vector $(3,6)$, so we deal with generic rank three distributions on manifolds of dimension six. In this case, a canonical Cartan connection was also constructed by R. Bryant in his thesis; see [**Br79, Br06**].

From Proposition 4.3.1 we know that the Levi subgroup $G_0 \subset P$ coincides with the automorphism group of the graded Lie algebra $\mathfrak{g}_- = \mathfrak{g}_{-2} \oplus \mathfrak{g}_{-1}$. Since $\mathfrak{g}_-$ is generated by $\mathfrak{g}_{-1}$, such an automorphism is determined by its restriction to $\mathfrak{g}_{-1}$. On the other hand, since the bracket induces an isomorphism $\Lambda^2\mathfrak{g}_{-1} \to \mathfrak{g}_{-2}$, we see that any invertible linear map on $\mathfrak{g}_{-1}$ extends to an automorphism of $\mathfrak{g}_-$. Hence, we conclude that the restriction of the adjoint action identifies G_0 with $GL(\mathfrak{g}_{-1}) \cong GL(n, \mathbb{R})$.

Using this, we can immediately describe the regular infinitesimal flag structure determined by such a generic distribution. The linear frame bundle of the distribution H is a principal bundle $p_0 : \mathcal{G}_0 \to M$ with structure group $GL(n,\mathbb{R}) \cong G_0$. This carries a canonical soldering form θ_{-1} defined on $(Tp_0)^{-1}(H) \subset T\mathcal{G}_0$, and we can view this form as having values in $\mathfrak{g}_{-1}$. Explicitly, we can interpret a point $u \in \mathcal{G}_0$ as a linear isomorphism $u : \mathfrak{g}_{-1} \to H_x$, where $x = p_0(u)$. For $\xi \in T_u\mathcal{G}_0$ such that $T_up_0 \cdot \xi \in H_x \subset T_xM$, we then have $\theta_{-1}(\xi) := u^{-1}(T_up_0 \cdot \xi)$. Genericity of the distribution H then implies that given $u : \mathfrak{g}_{-1} \to H_x$, there is a unique map $\tilde{u} : \mathfrak{g}_{-2} \to Q_x = T_xM/H_x$ such that $\tilde{u}([X,Y]) = \mathcal{L}_x(u(X), u(Y))$. Using this, we define $\theta_{-2} \in \Omega^1(\mathcal{G}_0, \mathfrak{g}_{-2})$ by $\theta_{-2}(\xi) := \tilde{u}^{-1}(T_up_0 \cdot \xi + H_x)$ for $\xi \in T_u\mathcal{G}_0$. It is an easy exercise to verify that $(\theta_{-1}, \theta_{-2})$ makes $p_0 : \mathcal{G}_0 \to M$ into a regular inifinitesimal flag structure.

To understand the basic obstructions to local flatness (i.e. local isomorphism to $SO_0(n+1,n)/P$, we can compute the cohomology group $H^2(\mathfrak{g}_-, \mathfrak{g})$. The recipes from 3.2.18 show that there is just one irreducible component in the cohomology. The nature of the cohomology group depends on n. For $n > 3$, the cohomology is the highest weight part in $\mathfrak{g}_{-1}^* \otimes \mathfrak{g}_{-2}^* \otimes \mathfrak{g}_{-2}$. Geometrically, this means that the complete obstruction to local flatness is given by a torsion, which can be interpreted as a bundle map $H \times Q \to Q$, where $H \subset TM$ is the distribution and $Q = TM/H$. We shall see in Chapter 5, how such torsions can be described in terms of compatible connections.

For $n = 3$, the cohomology $H^2(\mathfrak{g}_-, \mathfrak{g})$ is the highest weight component in $\mathfrak{g}_{-1}^* \otimes \mathfrak{g}_{-2}^* \otimes \mathfrak{g}_0$. From above, we know that $\mathfrak{g}_0 = \mathfrak{gl}(\mathfrak{g}_{-1})$. Hence, for $n = 3$, the basic obstruction to local flatness is a curvature, which can be interpreted as a bundle map $H \times Q \to L(H, H)$. General tools to describe such curvatures will be developed in Chapter 5, an explicit description of the curvature can be found in [**Br06**].

Growth vector (2,3,5). This is a classical example whose study goes back to E. Cartan's famous "five variables paper" [**Car10**] from 1910. In this fundamental paper, Cartan classified subbundles in the tangent bundle of five–dimensional manifolds, showing that in some cases such subbundles may have local invariants. For the generic type of such distributions, Cartan constructed a canonical Cartan connection related to the exceptional Lie algebra of type G_2 and determined the basic curvature quantity. This paper has motivated many further developments, in particular in the study of differential systems.

The complex exceptional Lie algebra of type G_2 has two simple roots α_1 and α_2 and we number them in such a way that the highest root becomes $3\alpha_1 + 2\alpha_2$. From Table B.2 in Appendix B, we can see that the positive roots are

$$\{\alpha_1, \alpha_2, \alpha_1 + \alpha_2, 2\alpha_1 + \alpha_2, 3\alpha_1 + \alpha_2, 3\alpha_1 + 2\alpha_2\}.$$

If we consider the parabolic subalgebra corresponding to $\Sigma = \{\alpha_1\}$, then we obtain a $|3|$–grading with $\dim(\mathfrak{g}_{\pm 1}) = \dim(\mathfrak{g}_{\pm 3}) = 2$ and $\dim(\mathfrak{g}_{\pm 2}) = 1$. Since $\mathfrak{g}_-$ is generated by $\mathfrak{g}_{-1}$, the Lie bracket on $\mathfrak{g}_-$ has to induce isomorphisms $\Lambda^2\mathfrak{g}_{-1} \to \mathfrak{g}_{-2}$ and $\mathfrak{g}_{-1} \otimes \mathfrak{g}_{-2} \to \mathfrak{g}_{-3}$. Conversely, it is clear that, together with the dimensions of the grading components, these two properties characterize the Lie algebra structure of $\mathfrak{g}_-$.

The split real form of the exceptional Lie algebra of type G_2 has an analogous grading and we define G to be the automorphism group of this Lie algebra and

$P \subset G$ as the subgroup of filtration preserving automorphisms. Then the P–invariant subspace $\mathfrak{g}^{-1}/\mathfrak{p} \subset \mathfrak{g}/\mathfrak{p}$ corresponds to a distribution of rank two on the five–dimensional manifold G/P. Moreover, since $\mathfrak{g}_{-1}$ generates $\mathfrak{g}_-$ we see that the growth vector of this distribution is $(2, 3, 5)$. It turns out that the space G/P is diffeomorphic to $(S^3 \times S^2)/\mathbb{Z}_2$, and the rank two distribution corresponding to $\mathfrak{g}^{-1}/\mathfrak{p}$ admits a nice explicit description in this picture; see [**Sag06b**].

Conversely, consider a manifold M of dimension 5 and a distribution $\mathcal{H} \subset TM$ of rank two. Given a local frame $\{\xi, \eta\}$ for $\mathcal{H}$, we see that the bracket of any two sections of $\mathcal{H}$ can be locally written as a linear combination of ξ, η, and $[\xi, \eta]$ with smooth coefficients. Likewise, brackets of three such sections can be written as linear combinations of ξ, η, $[\xi, \eta]$, $[\xi, [\xi, \eta]]$ and $[\eta, [\xi, \eta]]$ with smooth coefficients. Hence, we see that $(2, 3, 5)$ is the quickest possible growth for rank two distributions, so distributions with this growth vector are as non–integrable as possible. Since we have seen above that there exists a distribution with this growth vector such distributions are generic.

Given a distribution H with growth vector $(2, 3, 5)$, brackets of two sections of H together with H span a rank three subbundle of TM. Defining this rank three subbundle to be $T^{-2}M$ and putting $T^{-1}M := H$, we obtain a filtration $TM = T^{-3}M \supset T^{-2}M \supset T^{-1}M$. From the growth vector, we can see that $\operatorname{gr}_{-2}(TM)$ has rank one, while $\operatorname{gr}_{-3}(TM)$ has rank two, and that the Levi bracket $\mathcal{L}$ has to induce isomorphisms $\Lambda^2 T^{-1}M \to \operatorname{gr}_{-2}(TM)$ as well as $T^{-1}M \otimes \operatorname{gr}_{-2}(TM) \to \operatorname{gr}_{-3}(TM)$. We have observed above, that this implies that $(\operatorname{gr}(T_xM), \mathcal{L}_x)$ is isomorphic to $\mathfrak{g}_-$ for each $x \in M$. Using local frames, we see that the bundle of symbol algebras is locally trivial and modelled on $\mathfrak{g}_-$. From Proposition 4.3.1 we conclude that regular normal parabolic geometries of type (G, P) are equivalent to rank two distributions on five–dimensional manifolds with (small) growth vector $(2, 3, 5)$.

Similarly as before, we can describe the infinitesimal flag structure determined by a generic rank two distribution H on a manifold M of dimension five explicitly. A moment of thought shows that $G_0 = \operatorname{Aut}_{\operatorname{gr}}(\mathfrak{g}_-) = GL(\mathfrak{g}_{-1})$. Hence, as for growth vector $(n, \frac{n(n+1)}{2})$ one concludes that the regular infinitesimal flag structure of type (G, P) describing the distribution is defined on the linear frame bundle of $H \subset TM$. Computing the cohomology $H^2(\mathfrak{g}_-, \mathfrak{g})$ shows that the basic obstruction to local flatness in this case is a curvature, which can be interpreted as a bundle map $H \times \operatorname{gr}_{-3}(TM) \to L(H, H)$. This basic curvature quantity has already been found by Cartan in [**Car10**].

4.3.3. Quaternionic contact structures. These structures lie on the borderline between generic distributions and general filtrations with prescribed symbol algebra. They have been first introduced (for definite signature) by O. Biquard in his work on conformal infinities of quaternion–Kähler metrics; see [**Bi00**] and the overview article [**Bi02**]. It was only realized later, that these structures provide an example of a parabolic geometry.

As in 4.1.7, we will always consider $\mathbb{H}^n$ as a right vector space over $\mathbb{H}$. A bilinear form $\langle\ ,\ \rangle : \mathbb{H}^n \times \mathbb{H}^n \to \mathbb{H}$ is called *quaternionic–Hermitian* if $\langle v, wq\rangle = \langle v, w\rangle q$ and $\langle w, v\rangle = \overline{\langle v, w\rangle}$, where we use the conjugation on quaternions as discussed in 4.1.7. For nonnegative integers p and q such that $p + q = n$, one defines the standard quaternionic Hermitian form of signature (p, q) on $\mathbb{H}^n$ by

$$\langle(v_1, \dots, v_n), (w_1, \dots, w_n)\rangle := \bar{v}_1 w_1 + \dots + \bar{v}_p w_p - \bar{v}_{p+1} w_{p+1} - \dots - \bar{v}_n w_n.$$

It turns out that any nondegenerate quaternionic Hermitian form on an n–dimensional quaternionic vector space is isomorphic to one of these, so such forms are determined by their signature.

Given a Hermitian form $\langle\ ,\ \rangle$, we can look at its imaginary part, which has values in the three–dimensional space of purely imaginary quaternions. By definition, this imagnary part can be written as $(v,w) \mapsto \frac{1}{2}(\langle v,w\rangle - \langle w,v\rangle)$, so it is skew symmetric. Now one defines the *quaternionic Heisenberg algebra* of signature (p,q) (with $n = p+q$) as $\mathbb{H}^n \oplus \mathrm{im}(\mathbb{H})$ endowed with the bracket

$$[(v,p),(w,q)] := (0, \tfrac{1}{2}(\langle v,w\rangle - \langle w,v\rangle)).$$

Since triple brackets are zero by definition, this satisfies the Jacobi identity, hence making $\mathbb{H}^n \oplus \mathrm{im}(\mathbb{H})$ into a two–step nilpotent graded Lie algebra. For our purposes, the grading is best written as $\mathfrak{n} = \mathfrak{n}_{-2} \oplus \mathfrak{n}_{-1}$ with $\mathfrak{n}_{-2} = \mathrm{im}(\mathbb{H})$ and $\mathfrak{n}_{-1} = \mathbb{H}^n$. In particular, $\dim(\mathfrak{n}) = 4n+3$.

DEFINITION 4.3.3. Let M be a smooth manifold of dimension $4n+3$ for some $n \geq 1$, and let p, q be nonnegative integers such that $p+q = n$. A *quaternionic contact structure* on M of signature (p,q) is a subbundle $H \subset TM$ of rank $4n$ such that the bundle of symbol algebras of the filtration $H := T^{-1}M \subset TM = T^{-2}M$ is locally trivial and modelled on a quaternionic Heisenberg algebra of signature (p,q).

This is not the basic definition used (for signature $(n,0)$) in [**Bi00**], but it is noted there as an equivalent definition. We will discuss this further below.

To get a connection to parabolic geometries, consider (for fixed p and q) the Lie algebra $\mathfrak{g} = \mathfrak{sp}(p+1,q+1)$ of linear maps on $\mathbb{H}^{p+q+2}$ which are skew Hermitian with respect to a quaternionic Hermitian form of signature $(p+1,q+1)$; see 2.1.7. Let $\mathfrak{p} \subset \mathfrak{g}$ be the stabilizer of an isotropic quaternionic line in $\mathbb{H}^{p+q+2}$. Then $\mathfrak{g}$ is an algebra of type C_{p+q+2}, the Satake diagram can be found in Table B.4 in Appendix B. Using Theorem 3.2.9 and arguments similar to the ones used in 3.2.10 for the algebras of type $\mathfrak{su}$, respectively, $\mathfrak{sl}(n,\mathbb{H})$, one shows that the parabolic subalgebras of $\mathfrak{g}$ are exactly the stabilizers of flags of isotropic quaternionic subspaces. In particular, $\mathfrak{p}$ is the standard parabolic subalgebra corresponding to the second simple root in the Satake diagram.

An explicit realization of $\mathfrak{g} = \mathfrak{sp}(p+1,q+1)$ can be obtained analogously to the realizations of $\mathfrak{so}(n+1,1)$ in 1.6.3 and of $\mathfrak{su}(n+1,1)$ in Example 3.1.2 (3). This shows that $\mathfrak{g}_{-1} \cong \mathbb{H}^{p+q}$ and $\mathfrak{g}_{-2} \cong \mathrm{im}(\mathbb{H})$ and under this identification the bracket $[\ ,\ ] : \mathfrak{g}_{-1} \times \mathfrak{g}_{-1} \to \mathfrak{g}_{-2}$ is the imaginary part of a nondegenerate quaternionic Hermitian form of signature (p,q). This means that $\mathfrak{g}_-$ is a quaternionic Heisenberg algebra. Putting $G := \mathrm{Aut}(\mathfrak{g})$ and $P := \mathrm{Aut}_f(\mathfrak{g})$, Proposition 4.3.1 implies that regular normal parabolic geometries are equivalent to quaternionic contact structures of signature (p,q). The subalgebra $\mathfrak{g}_0 \subset \mathfrak{g}$ turns out to be isomorphic to $\mathbb{H} \oplus \mathfrak{sp}(p,q)$.

Since $\mathfrak{g}_-$ is generated by $\mathfrak{g}_{-1}$, the group $G_0 \cong \mathrm{Aut}_{\mathrm{gr}}(\mathfrak{g}_-)$ can be realized as a subgroup of $GL(\mathfrak{g}_{-1})$. The computation is similar as in the complex case which was done in 4.2.4. It turns out that G_0 is generated by the scalar multiplications by nonzero quaternions and by the elements of $Sp(p,q)$ acting on $\mathfrak{g}_{-1} \cong \mathbb{H}^{p+q}$. This shows that $\mathfrak{g}_{-1}$ carries a G_0–invariant prequaternionic structure as discussed in 4.1.7. Passing to associated bundles, it follows that, given a quaternionic contact structure $H \subset TM$, the subbundle H inherits a natural almost quaternionic structure. Explicitly, this means that there is a canonical rank three subbundle

$\mathcal{Q} \subset L(H, H)$, which can be locally spanned by I, J and $I \circ J$, where $I^2 = J^2 = -\mathrm{id}$ and $J \circ I = -I \circ J$. Moreover, the actions of elements of G_0 leave the real part of the Hermitian form on $\mathfrak{g}_{-1}$ invariant up to scale, so there is a canonical conformal class of real inner products on H, which are Hermitian with respect to all the complex structures in $\mathcal{Q}$. These data provide Biquard's original definition of a quaternionic contact structure.

The homogeneous model of quaternionic contact structures can be equivalently described as the quotient of $Sp(p+1, q+1)$ modulo the stabilizer of an isotropic quaternionic line, compare with 3.2.6 and 3.2.5. Hence, this is a real hypersurface $\mathcal{N}$ in the quaternionic projective space $\mathbb{H}P^{p+q+1}$. A point $x \in \mathcal{N}$ corresponds to an isotropic quaternionic line $\ell \subset \mathbb{H}^{p+q+2}$. Choosing $q \in \ell$, we obtain an identification $T_x\mathcal{N} \cong q^\perp/\ell$, where $q^\perp$ denotes the real orthocomplement of q. In particular, the quaternionic orthocomplement of q gives rise to a codimension three subspace in $T_x\mathcal{N}$, which is immediately seen to be independent of the choice of q. Passing from q to qa for $a \in \mathbb{H}$, the quaternionic vector space structure on this subspace changes by conjugation by a. Hence, we can see the ingredients of the quaternionic contact structure coming up.

While for $n > 1$, quaternionic contact structures in dimension $4n + 3$ behave uniformly, there are some special features in dimension 7. Consider the imaginary part of a Hermitian form (which is automatically definite in this dimension) as a map $\Lambda^2\mathbb{R}^4 \to \mathbb{R}^3$. The space of all such maps has dimension 18 and it carries a natural action of the group $GL(4, \mathbb{R}) \times GL(3, \mathbb{R})$, which has dimension 25. Now an element in the stabilizer of the bracket can be interpreted as an endomorphism of $\mathbb{R}^4 \oplus \mathbb{R}^3$ and as such it is by definition an automorphism of the quaternionic Heisenberg algebra. Conversely, such an automorphism defines an element in the stabilizer. Now by Proposition 4.3.1 we know that the automorphism group is G_0. In this dimension, $\mathfrak{g}_0 = \mathbb{H} \oplus \mathfrak{sp}(1)$, so this has dimension 7. This shows that the orbit of the bracket in $L(\Lambda^2\mathbb{R}^4, \mathbb{R}^3)$ under $GL(4, \mathbb{R}) \times GL(3, \mathbb{R})$ has dimension 18 and is therefore open.

This shows that any sufficiently small deformation of the bracket of the quaternionic Heisenberg algebra leads to an isomorphic Lie algebra. Hence, the distributions defining quaternionic contact structures in dimension 7 are generic, i.e. stable under small deformations. It turns out (see [**Mon02**, 7.12]) that the group $GL(4, \mathbb{R}) \times GL(3, \mathbb{R})$ has only two open orbits on $L(\Lambda^2\mathbb{R}^4, \mathbb{R}^3)$, so apart from quaternionic contact structures there is only one other generic type of rank four distributions in dimension seven. These correspond to split quaternionic contact structures to be discussed below. In particular, the genericity of the distributions implies that there are lots of examples.

Dimension seven is also special from the point of view of the basic curvature quantities. For any value of n, the cohomology $H^2(\mathfrak{g}_-, \mathfrak{g})$ has two irreducible components, one of which is contained in $\Lambda^2\mathfrak{g}^*_{-1} \otimes \mathfrak{g}_0$. This component corresponds to a curvature, which can be interpreted as a bundle map $\Lambda^2 H \to L(H, H)$. In dimension seven, the second cohomology component is contained in $\mathfrak{g}^*_{-1} \otimes \mathfrak{g}^*_{-2} \otimes \mathfrak{g}_{-2}$. Hence, it gives rise to a torsion, which can be interpreted as a bundle map $H \times TM/H \to TM/H$, and there are two independent basic curvature quantities in dimension seven. In higher dimensions, the second cohomology component is contained in $\Lambda^2\mathfrak{g}^*_{-1} \otimes \mathfrak{g}_{-2}$. Hence, it has homogeneity zero, and the corresponding component of the harmonic curvature has to vanish for regular normal parabolic geometries by definition of regularity. Nontrivial harmonic curvature in this component would change the symbol algebras of the filtration. It turns out that in most respects torsion–free quaternionic contact structures in dimension seven behave similarly to higher–dimensional quaternionic contact structures. Structures with torsion behave differently, for example in twistor theory; see 4.5.5. The different location of the second cohomology also expresses the fact that the seven–dimensional quaternionic Heisenberg algebra is rigid in the algebraic sense (as we have seen above) while higher dimensional quaternionic Heisenberg algebras are non–rigid.

As we have noticed already, in general it is not so easy to obtain examples for filtered manifolds with prescribed symbol algebras. In the case of quaternionic contact structures, however, there are general results ensuring the existence of many examples. The work of O. Biquard and C. Le Brun (on quaternion Kähler metrics, see [**Bi00**]), even shows that there exist quaternionic contact structures on compact manifolds, which admit an infinite–dimensional family of nontrivial deformations.

4.3.4. Split quaternionic contact structures. The quaternions can be characterized as the unique four–dimensional real algebra, which admits a definite quadratic form N, which is compatible with the multiplication, i.e. satisfies $N(pq) = N(p)N(q)$ for all $p, q \in \mathbb{H}$. If one allows the quadratic form to be indefinite, then there is a second algebra of real dimension four with a multiplicative quadratic form. It turns out that an indefinite multiplicative quadratic form has to be of split signature $(2, 2)$. This second algebra is well known. It is simply the algebra $M_2(\mathbb{R})$ of real (2×2)–matrices with the determinant as the multiplicative quadratic form. (This is the only size of matrices for which the determinant is quadratic.) In this context, we will refer to $M_2(\mathbb{R})$ as the algebra of *split quaternions* and denote it by $\mathbb{H}_s$ to emphasize the analogy to the quaternions. One may also realize $\mathbb{H}_s$ using a basis $\{1, i, j, k\}$ for which $i^2 = j^2 = 1$, $N(i) = N(j) = -1$ and $k = ij = -ji$, whence $k^2 = -1$, and $N(k) = 1$.

There is also a notion of conjugation on $\mathbb{H}_s$, with the crucial property being that $a\bar{a} = N(a)1$ for all $a \in \mathbb{H}_s$. Since N is indefinite, this does not imply that any element of $\mathbb{H}_s$ is invertible. In the picture of matrices, the conjugation is given by forming the matrix of cofactors, which for (2×2)–matrices just amounts to permutation of the entries and sign changes. Having the conjugation at hand, one may consider Hermitian forms on the right $\mathbb{H}_s$–vector space $\mathbb{H}_s^n$. It turns out that for each n, there is a unique such form up to isomorphism. Using the imaginary part of the standard form to define a bracket $\mathbb{H}_s^n \times \mathbb{H}_s^n \to \mathrm{im}(\mathbb{H}_s)$, one arrives at the definition of the *split quaternionic Heisenberg algebras.*

DEFINITION 4.3.4. A *split quaternionic contact structure* on a smooth manifold M of dimension $4n + 3$ is a subbundle $H \subset TM$ of rank $4n$ such that the bundle of symbol algebras of the filtration $H =: T^{-1}M \subset T^{-2}M := TM$ is locally trivial and modelled on a split quaternionic Heisenberg algebra.

Parallel to the quaternionic case, we can now look at the special unitary algebras of the split quaternions, and the stabilizer of an isotropic split quaternionic line in there. Using matrix realizations similar to the real, complex, and quaternionic cases, dealt with in 1.6.3, 3.1.2, and 4.3.3, respectively, one concludes that one obtains a grading with the negative part forming a split quaternionic Heisenberg algebra. The split quaternionic special unitary algebras admit a more classical description, however.

To see this, we first note that for a (2×2)–matrix $A \in \mathbb{H}_s$ and $\mathbb{I} := \left(\begin{smallmatrix} 0 & -1 \\ 1 & 0 \end{smallmatrix}\right)$, we get $\overline{A} = -\mathbb{I}A^t\mathbb{I}$. Now we can view $(n \times n)$–matrices with entries from $\mathbb{H}_s$ as real $(2n \times 2n)$–matrices. In particular, we define $\mathbb{J}$ to be the matrix of this type which has $\mathbb{I}$'s on the main diagonal and zeros everywhere else. Then for a matrix $A = (A_{ij})$ we get $A\mathbb{J} = (A_{ij}\mathbb{I})$ and $\mathbb{J}^tA = (\mathbb{I}A_{ji})$. Now A is skew symmetric with respect to $\mathbb{J}$ if and only if $A_{ij}\mathbb{I} = -\mathbb{I}A_{ji}$. This is equivalent to $A_{ij} = -\overline{A_{ji}}$ and hence to the fact that A is skew Hermitian. But now $\mathbb{J}$ evidently defines a symplectic form on $\mathbb{R}^{2n}$, which shows that the unitary algebras of $\mathbb{H}_s$ can be identified with the symplectic algebras $\mathfrak{sp}(2n, \mathbb{R})$. The corresponding parabolic subalgebra is given as the stabilizer of an isotropic plane in $\mathbb{R}^{2n}$.

Having this description at hand, Proposition 4.3.1 shows that split quaternionic contact structures in dimension $4n + 3$ are equivalent to regular normal parabolic geometries of type $(Sp(2n+4, \mathbb{R}), P)$, where P is the stabilizer of an isotropic plane in $\mathbb{R}^{2n+4}$. In particular, the homogeneous model of these split quaternionic contact structures is the Grassmannian of isotropic planes in $\mathbb{R}^{2n+4}$.

The basic facts about split quaternionic contact structures are closely parallel to quaternionic contact structures. In dimension seven, one may see from the dimensions of the Lie algebras involved that the bracket on the split quaternionic Heisenberg algebra must have open orbit in $L(\Lambda^2\mathbb{R}^4, \mathbb{R}^3)$, so this must be the second type of generic rank four distributions. In particular, there are many examples of seven–dimensional split quaternionic contact structures. In this dimension, there are two basic curvature quantities, one curvature and one torsion, whose form is exactly as in the quaternionic contact case. In higher dimensions there is a single curvature quantity (again of the same form as in the quaternionic contact case) which provides the complete obstruction to local flatness.

4.3.5. Rigid geometries. It can happen that for a semisimple Lie algebra $\mathfrak{g}$ and a parabolic subalgebra $\mathfrak{p}$, the second cohomology group $H^2(\mathfrak{g}_-, \mathfrak{g})$ is concentrated in nonpositive homogeneities. The full information when this happens can be found in the article [**Ya93**]. If this is the case, then any regular normal parabolic geometry has vanishing harmonic curvature, and hence is locally flat by Theorem 3.1.12. While this means that the theory of regular normal geometries is rather vacuous, it leads to interesting rigidity results for the underlying structures.

An interesting example for this phenomenon is provided by *octonionic contact structures* and split octonionic contact structures. These structures are related to exceptional Lie algebras of type F_4. Consider the complex simple Lie algebra of that type and the parabolic subalgebra corresponding to ×——o⇐o——o. By Proposition 3.2.2, the subalgebra $\mathfrak{g}_0$ determined by this grading is the sum of a one–dimensional center and a simple Lie algebra of type B_3. Since F_4 has dimension 52 and B_3 has dimension 21, we conclude that $\dim(\mathfrak{g}_-) = 15$ in this case. The expression for the highest root in Table B.2 in Appendix B shows that the corresponding grading is a $|2|$–grading. From the list of roots in that table one easily concludes that $\dim(\mathfrak{g}_{-1}) = 8$ and $\dim(\mathfrak{g}_{-2}) = 7$. Hence, passing to any real form of this situation, we will arrive at a geometry which exists on manifolds of dimension 15 and involves a subbundle of dimension 8 in the tangent bundle.

Using the recipes from 3.2.18 to determine the Hasse diagram for this parabolic, Kostant's version of the Bott–Borel–Weil theorem implies that both $H^1(\mathfrak{g}_-, \mathfrak{g})$ and $H^2(\mathfrak{g}_-, \mathfrak{g})$ are concentrated in negative homogeneous degrees. By Proposition 4.3.1, the geometries in question are given by a rank 8 subbundle $H \subset TM$ such that sections of H together with brackets of two such sections span TM and such that the bundle of symbol algebras is locally trivial and modelled on $\mathfrak{g}_-$. As we have seen above, Theorem 3.1.12 implies that any such geometry is locally flat and hence locally isomorphic to the homogeneous model.

From Table B.4 in Appendix B, we see that the split real form FI and unique noncompact, nonsplit real form FII of F_4 admit such a grading. The semisimple part of $\mathfrak{g}_0$ will be $\mathfrak{so}(3,4)$, respectively, $\mathfrak{so}(7)$ in the two cases; see Remark 3.2.9. The nilpotent Lie algebra $\mathfrak{g}_-$ obtained in these two cases are related to the octonions, respectively, to the split octonions. In our discussion of quaternionic and split quaternionic contact structures, we have met the notion of a composition algebra. This is a finite–dimensional real alternative algebra (i.e. any two elements generate an associative subalgebra) endowed with a quadratic form which is compatible with the multiplication. The quaternions and the split quaternionins are the two real composition algebras of dimension four. There are only two real composition algebras of dimension larger than four, and they both have dimension eight. For one of them, the quadratic form is definite, while for the other it has split signature $(4,4)$. These two algebras are called the *octonions* $\mathbb{O}$ and the *split octonions* $\mathbb{O}_s$, respectively.

On both $\mathbb{O}$ and $\mathbb{O}_s$, there is a natural conjugation map, and $(a,b) \mapsto a\bar{b}$ can be viewed as the standard Hermitian form on $\mathbb{O}$, respectively, $\mathbb{O}_s$. Using the imaginary parts of these, one obtains the octonionic, respectively, the split–octonionic Heisenberg algebra. It turns out that these are exactly the algebras $\mathfrak{g}_-$ for the two real forms described above. Due to the non–associativity of the (split) octonions, the notion of (split) octonionic vector spaces does not make sense, so there are no higher–dimensional (split) octonionic Heisenberg algebras.

With a bit of trickery, one may, however, define projective planes over $\mathbb{O}$ and $\mathbb{O}_s$. This is done via passing to skew Hermitian 3×3–matrices with entries in one of the algebras. Starting from $\mathbb{O}$, the result is the exceptional Jordan algebra of dimension 27. The projective plane $\mathbb{O}P^2$, respectively, $\mathbb{O}_sP^2$ is then defined as the subspace of these skew–Hermitian 3×3–matrices consisting of projections of (octonionic) rank one. It turns out that these spaces are homogenous under the appropriate real form of F_4 (which admits an interpretation as automorphisms of the Jordan algebra of skew–Hermitian 3×3–matrices, with isotropy group the appropriate parabolic subgroup). More information on these issues (in the case of $\mathbb{O}$) can be found in [**Ba02**].

Parallel to the quaternionic case, one defines a *(split) octonionic contact strcuture* on a smooth manifold M of dimension 15 as a subbundle $H \subset TM$ such that sections of H together with Lie brackets of two such sections span TM and such that the bundle of symbol algebras is locally trivial and modelled on a (split) octonionic Heisenberg algebra. In view of the above discussion, we obtain the following result, which is stated in [**Bi00**] with the observation that this follows form Yamaguchi's results in [**Ya93**] atributed to R. Bryant.

PROPOSITION 4.3.5. *Any octonionic (respectively split octonionic) contact structure is locally isomorphic to $\mathbb{O}P^2$ (respectively $\mathbb{O}_sP^2$).*

4.3.6. Parabolic geometries in dimensions one and two. So far, we have met many different examples of geometric structures in this chapter, each of them in a distinguished dimension (or series of dimensions). Now we will systematically determine all possible parabolic geometries in the lowest dimensions $1 \leq m \leq 5$. In fact, we have met most of these geometries already.

The dimension of the geometry for a given $(\mathfrak{g}, \mathfrak{p})$ is computed easily using the data in the tables in Appendix B. Indeed, the geometry may exist only in dimension $m = \dim \mathfrak{g}_-$, and the known dimension $\mathfrak{g}$ (as listed in Table B.1) is given by

$$\dim \mathfrak{g} = \dim \mathfrak{g}_0 + 2m.$$

The real dimension of a real form of a complex Lie algebra equals the complex dimension of the complex algebra, so again we only need the information from Table B.1. For the underlying real algebra of a complex one the real dimension is of course twice the complex dimension. The list of real forms can be found in Table B.4. Furthermore, the same table describes the semisimple parts of $\mathfrak{g}_0$ once we describe the choice of $\mathfrak{p}$ by Satake diagrams with crossed nodes. The corresponding rules exhausting all possible choices were discussed in Remark 3.2.9. In particular, the real dimension of the center is equal to the number of crosses while the Satake diagram of the semisimple part of $\mathfrak{g}_0$ is obtained by removing in the Satake diagram for $\mathfrak{g}$ all crossed nodes, all edges connecting to these nodes, and all arrows joining two crossed nodes.

Dimension one. Adding any type of node (crossed or uncrossed) to a crossed Satake diagram or changing an uncrossed node into a crossed node clearly increases the corresponding dimension m. Thus, the only possibility to search for dimension one homogeneous spaces is the diagram for A_1 with the only node being crossed. The homogeneous model is $SL(2,\mathbb{R})/B = \mathbb{R}P^1$, i.e. a circle. Here B denotes the Borel subgroup of G. The other real form $SU(2)$ of $SL(2,\mathbb{C})$ is compact and hence does not admit any parabolic subgroups.

The first cohomology $H^1(\mathfrak{p}, \mathfrak{g})$ has highest weight $-4\lambda_1$ and hence its dual $H^1(\mathfrak{g}_-, \mathfrak{g})$ is contained in homogeneity two. In analogy with the developments in 3.1.15 and 3.1.16 this implies that normal parabolic geometry of type (G, B) is determined by the underlying P–frame bundle of degree two, but since we are dealing with a $|1|$–grading here, this is already the full Cartan geometry. So there is no normalization procedure there and the whole geometry is given by the choice of the Cartan connection. (It turns out that the underlying P–frame bundle of degree one is the full second order frame bundle of the one–dimensional manifold M.)

On the other hand, since we are in dimension one, there cannot be any curvature. Thus, all Cartan geometries of type G/B are locally isomorphic to the projective line. Since the projective line comes equipped with the space of distinguished projective parametrizations, the Cartan connection on the one-dimensional M carries locally this property over from the homogeneous model. The best way to describe such a geometry is following the description of locally flat geometries in Remark 1.5.2 (3) by an atlas for M with charts that have values in open subsets of $\mathbb{R}P^1$ and chart changes which are restrictions of projective transformations. The Cartan connection can, however, also be described as a linear connection plus a choice of Rho tensor, which turns out to be a fruitful point of view; see [**BE90**].

Dimension two. To get an overview, we next list all possible crossed Satake diagrams with two nodes together with the dimension of $\mathfrak{g}/\mathfrak{p}$. The types of diagrams we have to consider are $A_1 \times A_1$, A_2, $B_2 \cong C_2$, and G_2. For the A–types, we obtain the following five possibilites:

$(\mathfrak{g}, \mathfrak{p})$	× ×	× × (with arc)	×—o	×—×	×—× (with arc)
$\dim \mathfrak{g}/\mathfrak{p}$	2	2	2	3	3

For diagrams of types B and G, we get 7 further possibilities, namely

$(\mathfrak{g}, \mathfrak{p})$	×⇒o	×⇒•	o⇒×	×⇒×	×⇛o	o⇛×	×⇛×
$\dim \mathfrak{g}/\mathfrak{p}$	3	3	3	4	5	5	6

Hence, there are only three possible types of parabolic geometries on two–dimensional manifolds. They all belong to classical objects studied in the literature, however. This is most obvious for the last of these three diagrams, which simply gives rise to projective structures in dimension 2 as discussed in 4.1.5. Recall that this geometry has the curvature obstruction concentrated in homogeneity three, so the only obstruction to a local flatness is a Cotton–York type tensor.

Next consider $\mathfrak{g} = \mathfrak{sl}(2, \mathbb{C})$, viewed as the real algebra with $\mathfrak{p} = \mathfrak{b}$, the Borel subalgebra (the second case in the first table above). The first crucial observation to be made here is that, as a real Lie algebra, $\mathfrak{sl}(2, \mathbb{C})$ is isomorphic to $\mathfrak{so}(3, 1)$. This isomorphism is induced via the 4–dimensional invariant subspace $\Lambda^{1,1}\mathbb{C}^2$ in the real representation $\Lambda^2_{\mathbb{R}}\mathbb{C}^2$ which has dimension six. The Lorentzian inner product is induced by the wedge product using the fact that $\Lambda^4_{\mathbb{R}}\mathbb{C}^2 \cong \mathbb{R}$. This representation integrates to the group $G = SL(2, \mathbb{C})$ which thereby, in view of simple connectedness, gets identified with $Spin(3, 1)$. Likewise, the subgroup $G_0 \cong \mathbb{C}^*$ can also be viewed as a two–fold covering of $CSO(2, \mathbb{R})$. Thus, a Cartan connection of type G/B over M will induce a 2–dimensional conformal structure.

On the other hand, on any Riemann surface M, the above noticed isomorphisms of groups automatically gives rise to an almost complex structure (which must be integrable by dimensional reasons). So we have encountered the two–dimensional conformal geometry, which we excluded from our considerations in 4.1.2. A Cartan geometry of type G/B, however, is a much stronger structure than that. Clearly, the algebra is $|1|$–graded and the first cohomology $H^1(\mathfrak{g}_-, \mathfrak{g})$ is concentrated in homogeneity two. Therefore, exactly as for one–dimensional projective structures, the Cartan connection itself represents the defining ingredient of the structure in question. These geometries, which are modelled on S^2, viewed as the projectivized light cone in $\mathbb{R}^{(3,1)}$, are called *Möbius structures* in the literature. A detailed treament of such structures and links to Einstein–Weyl geometries is presented in [**Cal98**].

The first entry in our table of A–type diagrams leads to a closely related geometry. Here we get $\mathfrak{g} = \mathfrak{sl}(2, \mathbb{R}) \times \mathfrak{sl}(2, \mathbb{R})$ and $\mathfrak{p}$ is the product of the Borel subalgebras of the components. Now $\mathfrak{g}$ is isomorphic to $\mathfrak{so}(2, 2)$. On the group level, we can realize this by considering the action of $G = SL(2, \mathbb{R}) \times SL(2, \mathbb{R})$ on $M_2(\mathbb{R})$ defined by $(A, B) \cdot X := AXB^{-1}$, which leaves the quadratic form induced by the determinant invariant. This identifies G with $Spin(2, 2)$, with the two standard representations corresponding to the two spinor representations. Likewise, $G_0 = \mathbb{R}^* \times \mathbb{R}^*$ is realized as a two–fold covering of $SO_0(1, 1)$. Note that this is a split–quaternionic (see 4.3.4) version of the two–fold covering $Sp(1) \times Sp(1) \to SO(4)$ constructed via quaternions.

Thus again, every geometry modeled over the product $\mathbb{R}P^1 \times \mathbb{R}P^1$ carries a product structure (which must be integrable by dimension reasons) and we deal with the pseudo–Riemannian conformal structure of signature $(1,1)$ defined by the cone determined by the product structure. The Cartan geometry is then given by the choice of the Cartan connection.

Let us briefly analyze how our general procedures of 4.1.1 work in these two special cases. We have started with the choice of appropriate connections on M. In our two cases, there are torsion–free connections compatible with either the almost complex structure or the almost product structure on the manifold, so these have to be the distinguished ones. Clearly, the space of all such connections will be parametrized by one–forms (it is easy to verify this directly, cf. 1.6.4). Next, we can solve the equation $\partial^*(R + \partial \mathsf{P}) = 0$ from Lemma 4.1.1, but the solution is not uniquely determined. In fact, the freedom in the choice of P is given fiber–wise by the homogeneity two cohomology component $H^1(\mathfrak{g}_{-1}, \mathfrak{p})_2 \neq 0$. As we have seen in Theorem 4.1.1, the knowledge of a torsion–free connection γ and the corresponding tensor P determines the normal Cartan connection ω, so P cannot be uniquely determined. Fixing one choice of P for a connection γ, we obtain a Cartan connection ω and hence a parabolic geometry.

Describing the geometry in that way, all the conclusions of Theorem 4.1.1 remain valid. The standard computation shows that the entire cohomology $H^2(\mathfrak{g}_{-1}, \mathfrak{g})$ sits in homogeneity 3. Therefore, the complete obstruction to local isomorhisms to the homogeneous Möbius structure on S^2, respectively, to the product of homogeneous projective structures on $\mathbb{R}P^1$ is given by the Cotton–York tensor $d^\gamma \mathsf{P}$.

4.3.7. The dimensions three through five. Some of the possible types of three–dimensional parabolic geometries are already in tables in 4.3.6. In particular, the last two missing types of A–type diagrams (the two rightmost entries) correspond to Lagrangean contact structures, respectively, CR–structures in the lowest possible dimension three. We have discussed these examples already in 4.2.3 and 4.2.4, respectively. It should be remarked that historically three–dimensional CR–structures were among the early examples of geometries which were studied using Cartan connections; see [**Car32**]. In these low–dimensional cases, the harmonic curvature is concentrated in homogeneity three and so a version of Cotton–York tensor appears as the only curvature obstruction to local flatness.

The first two colums in the table of digrams of type B and G in 4.3.6 correspond to conformal pseudo–Riemannian structures of the two signatures $(3,0)$ and $(2,1)$ which are possible in dimension three. We have discussed them in detail in Section 1.6 and in 4.1.2. The complete obstruction to the local flatness is the Cotton–York tensor of any of the Levi–Civita connections of the metrics in the conformal class, and this is the historical source for the name of this type of tensors.

The third diagram in that table is better understood by interpreting it as an algebra of type C_2. Then it corresponds to contact projective structures in the lowest possible dimension three. These have been discussed in 4.2.6.

This exhausts the three–dimensional geometries which showed up in the tables in 4.3.6, i.e. which correspond to Satake diagrams with two nodes. However, there are also three–dimensional geometries coming from diagrams with three nodes. In the following table, we list all available simple real algebras $\mathfrak{g}$ of rank 3, together with their possible gradings and the dimensions of $\mathfrak{g}/\mathfrak{p}$. The choice of the parabolic subalgebra is indicated above the colums by the positions of crosses over the Satake diagram:

	xoo	oxo	oox	xxo	xox	oxx	xxx
$\mathfrak{sl}(4,\mathbb{R})$	3	4	3	5	5	5	6
$\mathfrak{sl}(2,\mathbb{H})$	—	4	—	—	—	—	—
$\mathfrak{su}(1,3)$	—	—	—	—	5	—	—
$\mathfrak{su}(2,2)$	—	4	—	—	5	—	6
$\mathfrak{so}(4,3)$	5	7	6	8	8	8	9
$\mathfrak{so}(5,2)$	5	7	—	8	—	—	—
$\mathfrak{so}(6,1)$	5	—	—	8	—	—	—
$\mathfrak{sp}(6,\mathbb{R})$	5	7	6	8	8	8	9
$\mathfrak{sp}(2,1)$	—	7	—	—	—	—	—

In particular, only two (isomorphic) pairs lead to three–dimensional quotients $\mathfrak{g}/\mathfrak{p}$, namely the first and third entry in the first row. These are three–dimensional projective structures as discussed in 4.1.5.

Apart from the geometries corresponding to simple algebras (which we all have listed), there are also the ones corresponding to semisimple algebras. They are modeled on products of homogeneous spaces of dimensions one and two (two different cases with models $\mathbb{C}P^2 \times \mathbb{R}P^1$ and $\mathbb{R}P^1 \times \mathbb{R}P^1 \times \mathbb{R}P^1$). They are never determined by some underlying structure, so the choice of Cartan connection is an essential ingredient of the geometry. While the almost product structures may lead to interesting curvature components, we will not go into detail here.

Dimension four. The first example here comes from the table in 4.3.6, namely an algebra of type $B_2 \cong C_2$ with the grading coming from the Borel subalgebra. For the interpretation it is better to view it as C_2, so we are looking at the Borel subalgebra in $\mathfrak{sp}(4,\mathbb{R})$. This is the lowest dimensional example of so called *contact path geometries* which we will not discuss in this book. Contact path geometries can be viewed as generalizations of contact projective structures with vanishing contact torsion, in a similar spirit as we will exhibit path geometries as a generalization of projective structures in 4.4.3 below. These geometries are $|3|$–graded, since the highest root of C_2 has the form $2\alpha_1 + \alpha_2$. In dimension four, we get $\dim(\mathfrak{g}_{-1}) = 2$, while the other two components are one–dimensional, so this is the smallest possible $|3|$–graded algebra. Four–dimensional contact path geometries have been studied classically (although not under this name) due to their relation to the geometry of a single ODE of third order modulo contact equivalence. This geometry has been studied in [**Wü1905**] and worked out within the Cartan theory in [**Ch40**] in the context of 3rd order ODEs. An account on contact path geometries in higher dimensions can be found in [**Fox05b**].

Next, we go through the list of parabolic subalgebras in simple Lie algebras of rank three, such that $\mathfrak{g}/\mathfrak{p}$ is 4–dimensional. According to the table above, there are only three examples, all of type A_3 with the middle node crossed. These are pseudo–Riemannian conformal structures of the three signatures which are possible in dimension four. We discussed them thoroughly in Section 1.6, and also in 4.1.2, 4.1.4, 4.1.10, and 4.1.9.

The two tables above also show that among the simple Lie algebras of rank bigger than three, only the A_ℓ series may lead to further four–dimensional geometries. In fact, the only possibility is the algebra $\mathfrak{sl}(5,\mathbb{R})$ with the grading corresponding to either the first or the last node. These are projective structures in dimension four; see 4.1.5.

In addition to these real simple algebras with simple complexification, one can also obtain four–dimensional geometries from semismiple Lie algebras or from the underlying real algebras of complex simple Lie algebras. There are examples in this dimension, in which these geometries are determined by an underlying structure. Most notably, there are the cases with homogeneous models $\mathbb{R}P^2 \times \mathbb{R}P^2$ as a homogeneous space of $SL(3, \mathbb{R}) \times SL(3, \mathbb{R})$, respectively, of $\mathbb{C}P^2$ as a homogeneous space of $SL(3, \mathbb{C})$. We will not go into detail of these examples here. One can expect that they behave similarly as the examples of CR–structures of CR–dimension and codimension two which are discussed in 4.3.9 and 4.3.10 below.

Dimension five. From the second table in 4.3.6 we get two geometries in dimension five corresponding to the two maximal parabolic subgebras in a simple Lie algebra $\mathfrak{g}$ of type G_2. The first of these was discussed in 4.3.2 leading to the geometry of generic rank two distributions on manifolds of dimension five. The study of this example has a long history, it has been among the first applications of Cartan's approach in the early twentieth century; see [**Car10**]. The second maximal parabolic subalgebra gives rise to the (unique) parabolic contact structure associated to a simple Lie algebra of type G_2, which we discussed in 4.2.8.

Let us next collect the gradings from the table of rank three Lie algebras above, which lead to five–dimensional geometries. Let us go through the table column by column. In the first column, we find the conformal pseudo–Riemannian structures in the three signatures $(5, 0)$, $(4, 1)$, and $(3, 2)$, which are possible in dimension five. We have discussed them all in section 1.6 and also in 4.1.2. The other 5–dimensional geometry in the first column are five–dimensional projective contact structures as treated in 4.2.6.

The next examples arise from the algebras of type A_3 in the fifth column. The crosses in the diagram are exactly at the position of the nonzero coefficients in the expession of the highest weight $\lambda_1 + \lambda_3$ of the adjoint representation. Thus, all three geometries are parabolic contact structures. In the first row, we find Lagrangean contact structures, while the next two rows correspond to partially integrable almost CR–structures of the two possible non–isomorphic signatures $(2, 0)$ and $(1, 1)$.

The final five–dimensional example in the above table is provided by $\mathfrak{sl}(4, \mathbb{R})$ with the grading corresponding to either the first two or the last two roots. These are generalized path geometries in dimension five, which we will discuss in 4.4.3 in connection with correspondence spaces.

Now, we come to diagrams with four nodes. Adding a node or cross to an existing Satake diagram with crosses always increases the dimension. Thus, already the table with diagrams with three nodes shows that we may restrict ourselves to the types A and to check the remaining types D, E, and F. But the lowest available dimension with D_4 and one cross is already six (the conformal Riemannian 6–dimensional structures), while the other option with one cross is of dimension 9. The F diagrams with one cross produce dimensions 15 and 20. The A diagram with four nodes provides examples of dimensions 4 and 6. Thus the only remaining case for us are the five–dimensional projective structures coresponding to the diagram with five nodes and the cross over the first or the last one.

Of course, again examples coming from products of homogeneous spaces of lower dimensions are available. With increasing number of nodes and crosses, the dimensions expand very quickly. However, there are some more geometries of very special interest in quite low dimensions. We shall see a particularly nice example next.

4.3.8. Codimension two CR–structures on six–dimensional manifolds. It is a well–known phenomenon that submanifolds may inherit geometric structures from an ambient space. The best known example is the induced Riemannian metric on a submanifold in a Riemannian manifold. In the realm of parabolic geometries, well–known examples are provided by generic hypersurfaces in the projective space $\mathbb{RP}^{n+1}$, which inherit a conformal structure (related to the second fundamental form), and by generic hypersurfaces in $\mathbb{C}^{n+1}$ (or in some complex manifold) which inherit a CR–structure; see 4.2.4. There we have also discussed the more general partially integrable almost CR–structures (of hypersurface type). These geometries are obtained on codimension one submanifolds in ambient manifolds with an almost complex structure.

The first steps towards the induced CR–structure on a real hypersurface have an analog for submanifolds of higher codimension. For a submanifold M in a manifold $\tilde{M}$ endowed with an almost complex structure $\tilde{J}$ and each point $x \in M$, there is the maximal complex subspace $H_x = T_xM \cap \tilde{J}(T_xM) \subset T_xM$. Assuming that these spaces are of constant complex dimension n (which is automatically satisfied in the case of a hypersurface), they form a smooth subbundle $H \subset TM$. The almost complex structure $\tilde{J}$ restricts to an almost complex structure J on H. Next $Q := TM/H$ is a real vector bundle on M, and by construction, the rank of Q equals the real codimension k of M in $\tilde{M}$. Viewing $H \subset TM$ as a filtration of the tangent bundle, the associated graded then is $\mathrm{gr}(TM) = Q \oplus H$. This is the usual setup for the CR geometry in complex analysis. In the language of (abstract) CR geometry, a complex subbundle $H \subset TM$ of complex rank n on a manifold M of real dimension $2n + k$ is called an *almost CR–structure of CR–dimension n and codimension k.*

Having $H \subset TM$, we get the Levi bracket $\mathcal{L} : H \times H \to Q$. Similarly to the case of hypersurface type CR–structures in 4.2.4, we can next impose non–degeneracy and integrability conditions. In the case of a submanifold in a complex manifold, integrability automatically follows from integrability of the ambient complex structure. In the abstract setting, the partial integrability condition from 4.2.4 continues to make sense. We call an abstract almost CR–structure *partially integrable* if $\mathcal{L}(J\xi, J\eta) = \mathcal{L}(\xi, \eta)$ for all $\xi, \eta \in HM$, i.e. if $\mathcal{L}$ is totally real. If this is the case, then for each $x \in M$, there is a Hermitian form on the complex vector space $H_x \cong \mathbb{C}^n$ with values in the complex vector space $Q_x \otimes \mathbb{C} \cong \mathbb{C}^k$ whose imaginary part is $\mathcal{L}_x$.

Now suppose that $h : \mathbb{C}^n \times \mathbb{C}^n \to \mathbb{C}^k$ is an arbitrary vector–valued Hermitian form. Then there is a natural codimension k submanifold in $\mathbb{C}^{n+k}$ whose Levi bracket in each point is isomorphic (in the obvious sense) to the imaginary part of h; see [**Fo92**]. This is the *quadric* associated to h. Let us view $\mathbb{C}^{n+k}$ as $\mathbb{C}^n \times \mathbb{C}^k$ with coordinates z on $\mathbb{C}^n$ and $w = u + iv$ on $\mathbb{C}^k$. Then we define the quadric as

$$\mathcal{Q}_h := \{(z, u + iv) : v = h(z, z)\} \subset \mathbb{C}^{n+k}.$$

This is a smooth submanifold, since it can be interpreted as the graph of a smooth function. Now the tangent spaces of $\mathcal{Q}_h$ are given by

$$T_{(z,u+ih(z,z))}\mathcal{Q}_h = \{(w, r + 2i\,\mathrm{re}(h(z, w))) : w \in \mathbb{C}^n, r \in \mathbb{R}\}.$$

Using this, one easily verifies that $H_{(z,u+ih(z,z))} = \{(w, 2ih(w,z)) : w \in \mathbb{C}^n\}$ and that the Levi bracket is induced by a nonzero multiple of the imaginary part of h. For $k = 1$ and h nondegenerate one easily verifies that the quadric $\mathcal{Q}_h$ is locally isomorphic to the homogeneous model of partially integrable hypersurface type CR–structures as discussed in 4.2.4.

Next, we need an appropriate nondegeneracy condition. Of course, we will require the evident condition that $\mathcal{L}(\xi, \eta) = 0$ for all η implies $\xi = 0$. This is much to weak, however, since it does not even ensure that $\mathcal{L}$ is onto. We will call the structure *nondegenerate* at a point $x \in M$ if, in addition, for each nonzero $\psi \in Q_x^*M$ the skew symmetric bilinear map $\mathcal{L}_x^\psi : H_xM \times H_xM \to \mathbb{R}$ defined by $\mathcal{L}_x^\psi(\xi,\eta) = \psi(\mathcal{L}_x(\xi,\eta))$ is nonzero. Identifying Q_x with $\mathbb{R}^k$, the latter condition just means that the components of the vector valued Levi form are linearly independent. An almost CR–structure (H, J) on M is called nondegenerate if it is nondegenerate at each point $x \in M$.

From the definition it is evident that nondegeneracy is an open condition. For codimension one, it reduces to the nondegeneracy condition from 4.2.4 and ensures that the Levi bracket is characterized by the signature of the corresponding Hermitian form. Hence, in codimension one, nondegeneracy ensures that the isomorphism type of the Levi bracket is locally constant. This is no more true in higher codimension. Indeed, for general n and k, $\mathbb{C}^k$–valued Hermitian forms on $\mathbb{C}^n$ admit continuous invariants; see e.g. [**GaMi98**]. Correspondingly, one may construct continuous families of non-isomorphic intersections of k quadrics; see e.g. [**ES94, ES99**].

The situation simplifies drastically, if one looks at the case of CR–dimension $n = 2$ and codimension $k = 2$ with the basic examples provided by submanifolds of real dimension six in complex manifolds of complex dimension four. Assuming non–degeneracy and partial integrability, there are only three possible types of points, which we shall call *hyperbolic*, *exceptional*, and *elliptic*, according to parts (1), (2), and (3) in the following lemma.

LEMMA 4.3.8. *Let M be a six–dimensional smooth manifold endowed with a nondegenerate, partially integrable almost CR–structure (H, J) of CR–dimension two and codimension two. Then at each point $x \in M$ exactly one of the following possibilities happens:*

1. *There are two distinct points $[\psi_1]$ and $[\psi_2]$ in the projectivization $\mathcal{P}(Q_x^*)$ such that $\mathcal{L}^\psi : \Lambda^2 H_x \to \mathbb{R}$ is degenerate if and only if $\psi \in [\psi_1]$ or $\psi \in [\psi_2]$. In this case $H_x = H_x^{[\psi_1]} \oplus H_x^{[\psi_2]}$, where each of the null spaces $H_x^{\psi_i}$ for $\mathcal{L}^{\psi_i}$ is a complex line.*
2. *There is one point $[\psi_0] \in P(Q_x^*)$ such that $\mathcal{L}^\psi$ is degenerate only if $\psi \in [\psi_0]$.*
3. *The forms $\mathcal{L}^\psi$ are nondegenerate for all nonzero elements $\psi \in Q_x^*$.*

PROOF. Assume first that $\mathcal{L}^{\psi_1}$ and $\mathcal{L}^{\psi_2}$ are degenerate for two linearly independent forms $\psi_i \in Q_x^*$. The subsets $H_x^{[\psi_i]}$ are complex lines since the Levi form is totally real. If $\xi \in H_x^{[\psi_1]} \cap H_x^{[\psi_2]}$, then $\mathcal{L}^\psi(\xi,\eta) = 0$ for all $\eta \in T_xM$ and $\psi \in Q_x^*$. Thus, $\xi = 0$ by nondegeneracy of $\mathcal{L}$, and $H_x = H_x^{[\psi_1]} \oplus H_x^{[\psi_2]}$.

If $\psi = a\psi_1 + b\psi_2$ is another form for which $\mathcal{L}^\psi$ is degenerate and both $a, b \neq 0$, then any two of the lines $H_x^{[\psi]}$, $H_x^{[\psi_1]}$, and $H_x^{[\psi_2]}$ are complementary. Thus, for each $\xi \in H_x^{[\psi_1]}$, there is a unique $\phi(\xi) \in H^{[\psi_2]}$ such that $\xi + \phi(\xi) \in H_x^{[\psi]}$, and the map ϕ defined in this way is a linear isomorphism. According to our choices,

$0 = \mathcal{L}^{\psi}(\xi + \phi(\xi), \eta) = a\mathcal{L}^{\psi_1}(\phi(\xi), \eta) + b\mathcal{L}^{\psi_2}(\xi, \eta)$ for all $\eta \in H_x$. But inserting $\eta \in H_x^{[\psi_1]}$ leads to $\mathcal{L}^{\psi_2}(\xi, \eta) = 0$ for all $\xi, \eta \in H_x^{[\psi_1]}$ and this implies $\mathcal{L}^{\psi_2} = 0$, which is not possible. Thus, there are at most two different forms as in (1). This proves all the claims. □

The conditions in (1) and (3) are evidently open, so hyperbolic and elliptic points always form open subsets. We shall see below that such point are related to highly interesting parabolic geometries. The exceptional points are not of that character and they may form boundaries between the two stable types. In the literature, these unstable points are called parabolic, but we do not want to use this terminology here since they are the only ones which do not define a parabolic geometry.

DEFINITION 4.3.8. A nondegenerate, partially integrable almost CR–structure of CR–dimension and codimension two is called *hyperbolic*, respectively, *elliptic* if all its points are hyperbolic, respectively, elliptic.

REMARK 4.3.8. We already know from above that appropriate quadrics provide examples of partially integrable almost CR–structures with prescribed Levi bracket. For the three types of points from the lemma, the appropriate quadrics (with coordinates z_j and $w_j = u_j + iv_j$ for $j = 1, 2$) are:

$$v^1 = h^1(z, \bar{z}) = z_1\bar{z}_1, \qquad v^2 = h^2(z, \bar{z}) = z_2\bar{z}_2, \tag{4.9}$$

$$v^1 = h^1(z, \bar{z}) = z_1\bar{z}_1, \qquad v^2 = h^2(z, \bar{z}) = \operatorname{Re} z_1\bar{z}_2, \tag{4.10}$$

$$v^1 = h^1(z, \bar{z}) = \operatorname{Re} z_1\bar{z}_2, \quad v^2 = h^2(z, \bar{z}) = \operatorname{Im} z_1\bar{z}_2. \tag{4.11}$$

The first of these quadrics evidently is the product of two 3–dimensional quadrics in $\mathbb{C}^2$, which makes it easy to describe its symmetries. For the elliptic quadric (4.11) this is much more complicated and it took quite some time until the relation to $SL(3, \mathbb{C})$ which we will derive below was found; cf. [**SchSl00**] and [**EIS99**].

It turns out that a general embedded CR–manifold M of CR–dimension two and codimension two can be approximated by a quadric in each point. There are complex coordinates $(z_1, z_2, w_1 = u_1 + iv_1, w_2 = u_2 + iv_2)$, such that M is locally described by the two equations

$$v_j = f_j(z_1, z_2, \bar{z}_1, \bar{z}_2, u_1, u_2), \quad j = 1, 2$$

with $f_j(0) = 0$ and $df_j(0) = 0$. This is the usual approach addopted in complex analysis. Geometrically this means that we have chosen $x \in M$ to be the origin and fixed the coordinates so that T_0M is given by $v = 0$. A bit of second order normalization of the coordinates than allows to kill all second derivatives of $f = (f_1, f_2)$ up to the Hermitian part $h(z, \bar{z}) = \frac{1}{2}\sum \frac{\partial^2 f}{\partial z_i \partial \bar{z}_j}|_0 z_i\bar{z}_j$ and so we see that M is locally given by equations

$$v_1 = h_1(z, \bar{z}) + O(3), \quad v_2 = h_2(z, \bar{z}) + O(3)$$

with two Hermitian forms h_1 and h_2. In the same coordinates, a straightforward computation (using the partial integrability) reveals that the Levi bracket $\mathcal{L} : \Lambda^2 H \to TM/H$ is a nonzero multiple of the imaginary part of the vector–valued Hermitian form $h = (h_1, h_2)$.

4.3.9. The hyperbolic case. In Remark 4.3.8, we have seen that the product of two quadrics in $\mathbb{C}^2$ carries a hyperbolic CR–structure of CR–dimension and codimension two. More generally, any product of two three–dimensional hypersurface type CR–structures has this property. Since by 4.2.4 three–dimensional hypersurface type CR–structures are equivalent to parabolic geometries of type $(PSU(2,1), B)$, where B is the Borel subgroup, it is natural to expect that the hyperbolic codimension two CR–structures are related to parabolic geometries of type $(PSU(2,1) \times PSU(2,1), B \times B)$. This is actually easy to prove.

PROPOSITION 4.3.9. *Oriented hyperbolic partially integrable almost CR–structures of CR–dimension and codimension 2 on 6–dimensional manifolds are equivalent to regular normal parbolic geometries of type* $(PSU(2,1) \times PSU(2,1), B \times B)$, *where* $B \subset PSU(2,1)$ *is the stabilizer of an isotropic complex line.*

PROOF. In view of Proposition 3.3.7 and Theorem 3.1.14 we only have to prove that the structures are equivalent to regular infinitesimal flag structures of the given type. Now put $G := PSU(2,1) \times PSU(2,1)$ and $P := B \times B$, let $\mathfrak{g}$ and $\mathfrak{p}$ be the corresponding Lie algebras. Let us denote $\mathfrak{g} = \mathfrak{g}^+ \oplus \mathfrak{g}^-$ the splitting into the two factors. Then the resulting $|2|$–grading is of the form $\mathfrak{g}_i = \mathfrak{g}_i^- \oplus \mathfrak{g}_i^+$ and likewise for the filtration, and these decompositions are invariant under G_0, respectively, P. In particular, $\mathfrak{g}_{-1}$ is the sum of two complex lines and $\mathfrak{g}_{-2}$ is the sum of two real lines. The Lie bracket $[\ ,\ ] : \mathfrak{g}_{-1} \times \mathfrak{g}_{-1} \to \mathfrak{g}_{-2}$ vanishes on $\mathfrak{g}^+_{-1} \times \mathfrak{g}^-_{-1}$ and the components $\Lambda^2 \mathfrak{g}^{\pm}_{-1} \to \mathfrak{g}^{\pm}_{-2}$ are the imaginary parts of nondegenerate Hermitian forms. From the description in 4.2.4 we conclude that the adjoint actions of elements of G_0 are exactly those automorphisms of the graded Lie algebra $\mathfrak{g}_-$, which preserve the decompositions $\mathfrak{g}_i = \mathfrak{g}_i^+ \oplus \mathfrak{g}_i^-$ as well as the complex structures on $\mathfrak{g}^{\pm}_{-1}$.

From the description of regular infinitesimal flag structures in Observation 3.1.7 we thus conclude that a regular infinitesimal flag structure of type (G, P) on a smooth manifold M of dimension 6 is given by:

- two transversal complex line subbundles H^+ and H^- in TM,
- a decomposition of $Q = TM/H$, where $H := H^+ \oplus H^-$, into a direct sum $Q^+ \oplus Q^-$ of two line bundles,
- such that $\mathcal{L} : \Lambda^2 H \to Q$ is totally real, vanishes on $H^+ \times H^-$, and maps $\Lambda^2 H^{\pm}$ to $Q^{\pm}$ in a nondegenerate way.

Indeed, the last condition is evidently equivalent to the fact that for each $x \in M$, one can choose complex linear isomorphisms $H^{\pm}_x \to \mathfrak{g}^{\pm}_{-1}$ and real linear isomorphisms $Q^{\pm}_x \to \mathfrak{g}^{\pm}_{-2}$, which intertwine $\mathcal{L}$ with the bracket on $\mathfrak{g}_-$. Such a family of isomorphisms is then unique up to the action of an element of G_0.

These data evidently define a hyperbolic partially integrable almost CR–structure on M. The two functionals ψ_i have to be chosen in such a way that their kernels are Q^+, respectively, Q^-. To obtain an orientation, we observe that the bundles $H^{\pm}$ are canonically oriented by their complex structure. Nondegeneracy of $\mathcal{L} : \Lambda^2 H^+ \to Q^+$ implies that $\mathcal{L}(\xi, J\xi) \neq 0$ for any $\xi \in H^+$. Any other element in the same fiber of H^+ is a complex multiple of ξ, and for such a multiple, we obtain $\mathcal{L}(a\xi + bJ\xi, J(a\xi + bJ\xi)) = (a^2 + b^2)\mathcal{L}(\xi, J\xi)$. Hence, these values determine an orientation of Q^+, and in the same way Q^- is canonically oriented. Altogether, this gives us an orientation of TM.

Conversely, assume that M is a 6–dimensional oriented manifold endowed with a hyperbolic partially integrable almost CR–structure of CR–dimension and codimension two. By Lemma 4.3.8, we get at each point $x \in M$ two distinguished real lines $[\psi_1], [\psi_2] \subset Q_x^*$. Their kernels form a decomposition of Q_x, and as above each of the two summands is oriented by the values of the Levi bracket. We can choose the sucession of the two subbundles in such a way that together with the orientation of H provided by the complex structure, $Q_x = Q_x^+ \oplus Q_x^-$ induces the given orientation of M. Again by Lemma 4.3.8 we get a decomposition $H_x = H_x^+ \oplus H_x^-$ as a direct sum of two complex lines. Here we use the convention that if $Q_x^+ = \ker(\psi_1)$, then H_x^+ is the complex line on which $\mathcal{L}^{\psi_1}$ vanishes identically.

Since $Q_x^\pm$ and $H_x^\pm$ are given as solutions to smooth equations, they fit together to define local smooth (real respectively complex) line subbundles. Hence, we get a decomposition $H = H^+ \oplus H^-$ and $Q = Q^+ \oplus Q^-$ as required. The Levi bracket $\mathcal{L}$ is totally real by partial integrability. Further, for $\xi \in H^+$ and $\eta \in H^-$, we know that ξ inserts trivially into $\mathcal{L}^{\psi_1}$ and η inserts trivially into $\mathcal{L}^{\psi_2}$, which shows that $\mathcal{L}(\xi, \eta) = 0$. Nondegeneracy of $\mathcal{L}$ then implies that $\mathcal{L}(\xi, J\xi) \neq 0$ for $\xi \neq 0$ and this has values in Q^+ by construction. Since the same holds for η, we see that $\mathcal{L}$ has all required properties, and the proof is complete. □

From the description of the Lie algebra in the proof, it is evident that $\mathfrak{g}/\mathfrak{p} = \mathfrak{g}^+/\mathfrak{p}^+ \oplus \mathfrak{g}^-/\mathfrak{p}^-$. Passing to associated bundles, this shows that we get a decomposition $TM = T^+M \oplus T^-M$ on M such that $T^\pm M \cap H = H^\pm$ and $Q^\pm = q(T^\pm M)$, where $q : TM \to Q$ is the natural quotient map. Hence, there is a natural almost product structure on M. This can be easily constructed directly. Indeed, consider a local nonvanishing section $\xi \in \Gamma(H^+)$. Then $\{\xi, J\xi\}$ is a local frame for H^+ and we have seen above that $\mathcal{L}(\xi, J\xi)$ is nonvanishing. Hence, ξ, $J\xi$, and $[\xi, J\xi]$ span a rank three subbundle of TM, and one easily computes that this bundle is independent of the choice of ξ. We shall see below that it coincides with T^+M.

Having associated a canonical Cartan connection to every partially integrable hyperbolic CR–manifold of dimension and codimension two, we can now proceed to understand the geometry of these structures. We first have to analyze the harmonic curvature components. Kostant's theorem and its consequences described in 3.3.5 and 3.3.6 allow us to compute the cohomology $H^2(\mathfrak{g}_-, \mathfrak{g})$. It turns out that this splits into eight components, which are listed in the table below. As in the proof above, we use the decomposition $\mathfrak{g} = \mathfrak{g}^+ \oplus \mathfrak{g}^-$. The first column of the table gives the homogeneity of the component. In the second column, we give the location of the component within the space of maps $\Lambda^2\mathfrak{g}_- \to \mathfrak{g}$. If the target space is complex, then the last column lists the complex linearity properties with respect to the complex source space(s).

homog.	location	complex linearity
1	$\mathfrak{g}^+_{-2} \times \mathfrak{g}^+_{-1} \to \mathfrak{g}^-_{-2}$	
1	$\mathfrak{g}^-_{-2} \times \mathfrak{g}^-_{-1} \to \mathfrak{g}^+_{-2}$	
1	$\mathfrak{g}^-_{-1} \times \mathfrak{g}^+_{-1} \to \mathfrak{g}^-_{-1}$	conjugate linear in both arguments
1	$\mathfrak{g}^-_{-1} \times \mathfrak{g}^+_{-1} \to \mathfrak{g}^+_{-1}$	sesquilinear

homog.	location	complex linearity
1	$\mathfrak{g}^+_{-1} \times \mathfrak{g}^-_{-1} \to \mathfrak{g}^+_{-1}$	conjugate linear in both arguments
1	$\mathfrak{g}^+_{-1} \times \mathfrak{g}^-_{-1} \to \mathfrak{g}^-_{-1}$	sesquilinear
4	$\mathfrak{g}^-_{-2} \times \mathfrak{g}^-_{-1} \to \mathfrak{g}^-_1$	complex linear
4	$\mathfrak{g}^+_{-2} \times \mathfrak{g}^+_{-1} \to \mathfrak{g}^+_1$	complex linear

Fortunately, all the homogeneity one components of the curvature allow a quite nice expression in terms of Lie brackets of vector fields. In order to understand this, let us recall the technique from 3.1.8 of extending tangent vectors $\xi, \eta \in T_xM$ to locally defined vector fields $\tilde{\xi}$, $\tilde{\eta}$ in such a way that

$$[\tilde{\xi}, \tilde{\eta}](x) = T_u p \cdot \omega_u^{-1}([X, Y] - \kappa(u)(X, Y)). \tag{4.12}$$

Here u is a fixed frame over x in the Cartan bundle with the Cartan connection ω and curvature function κ, $\xi = T_u p.\omega^{-1}(X)$, and $\eta = T_u p.\omega^{-1}(Y)$.

Since the whole homogeneity one component of the curvature κ is harmonic, there cannot be other components of the homogeneity one torsion but those listed in the table. Using this, (4.12) immediately provides nontrivial information. For two sections of H^+, we can form the Lie bracket and then project the result to TM/T^+M. Since $\mathcal{L}(H^+, H^+) \subset Q^+$, the result lies in H^-, and the construction defines a tensorial map $H^+ \times H^+ \to H^-$. To apply (4.12) to this problem, assume that $\xi, \eta \in H^+_x$. Then $X, Y \in \mathfrak{g}^+_{-1}$ and hence $[X, Y] \in \mathfrak{g}^+_{-2}$. From the table, we see that there is no nonzero cohomology component which involves maps defined on $\mathfrak{g}^+_{-1} \times \mathfrak{g}^+_{-1}$, so there can be no curvature in homogeneity one defined on that space. Hence, $\kappa(u)(X, Y) \in \mathfrak{p}$ and this does not contribute, so we conclude that $[\tilde{\xi}, \tilde{\eta}](x) \in T^+_x M$. From the compatibility of the extension map with data coming from ω observed in 3.1.8 we know that $\tilde{\xi}, \tilde{\eta} \in \Gamma(H^+)$, so the algebraic bracket $H^+ \times H^+ \to H^-$ constructed above actually vanishes. This can be done in the same way with the roles of H^+ and H^- exchanged. In particular, this shows that our description of $T^{\pm}M$ above was correct.

Consequently, forming the Lie bracket on $\Gamma(T^+M) \times \Gamma(H^+)$ and projecting to $\Gamma(Q^-)$ and likewise with + and − exchanged, the result factors to define bundle maps

$$Q^+ \times H^+ \to Q^- \qquad Q^- \times H^- \to Q^+.$$

Using (4.12), we see that these are the only homogeneity one obstructions to integrability of the almost product structure.

It is quite easy to guess the geometric relevance of the conjugate linear parts of the torsion. Exactly as in the case of the partially integrable CR–structures of codimension one, these corrrespond to the Nijenhuis tensor, which obstructs the integrability of the almost complex structure on H. We can see this also with the help of the complexification. The complexification $\omega_{\mathbb{C}} : T_{\mathbb{C}}\mathcal{G} \to \mathfrak{g}_{\mathbb{C}}$ of the canonical Cartan connection ω is a complex linear isomorphism on each complex tangent space. The Lie bracket of real vector fields extends to the complex ones. Now the right–hand side of (4.12) can also be written as $T_u p \cdot [\omega^{-1}(X), \omega^{-1}(Y)]$. The analog holds in the complex setting, so for each choice of $u \in \mathcal{G}$ and $X, Y \in (\mathfrak{g}_-)_{\mathbb{C}}$, we can choose projectable complex vector fields $\tilde{\xi}$, $\tilde{\eta}$ such that $[\tilde{\xi}, \tilde{\eta}](x) = T_u p.[\omega_{\mathbb{C}}^{-1}(X), \omega_{\mathbb{C}}^{-1}(Y)](u)$. Furthermore, the expansion of $\omega_{\mathbb{C}}([\omega_{\mathbb{C}}^{-1}(X), \omega_{\mathbb{C}}^{-1}(Y)])$

into the real and imaginary parts shows that the latter expression yields exactly the complexification $\kappa_{\mathbb{C}}$ of the curvature. Thus, we may proceed exactly as above in order to link the component in the homgeneity one part $\kappa_{\mathbb{C}}^{(1)}$, which acts on two sections of the holomorphic part of $\mathfrak{g}_{-1} \otimes \mathbb{C}$ and has values in antiholomorphic part, to the obstruction against the integrability of the holomorphic part of $H_{\mathbb{C}}$. Of course, since the Nijenhuis tensor is skew symmetric and conjugate linear in both variables, it vanishes on $H^{\pm} \times H^{\pm}$ and the resulting map $H^{+} \times H^{-} \to H$ splits into two components according to the decomposition $H = H^{+} \oplus H^{-}$ of the values.

The sesquilinear components in the homogeneity one curvature allow a similar interpretation. Having the splitting $H = H^{+} \oplus H^{-}$ at hand, we can define a second complex structure $\tilde{J}$ on H by putting $\tilde{J}|_{H^{\pm}} := \pm J|_{H^{\pm}}$. Since the only nonzero components of the Levi bracket are the ones defined on $\Lambda^2 H^{+}$ and $\Lambda^2 H^{-}$, we see that $\mathcal{L}$ is totally real with respect to $\tilde{J}$. Hence, we can form the Nijenhuis tensor of $\tilde{J}$, which is conjugate linear with respect to $\tilde{J}$ in both arguments, so the only nonzero components map $H^{+} \times H^{-}$ to H^{+}, respectively, H^{-}. With respect to J, they are sesquilinear. As above one verifies that they represent (up to nonzero factors) the remaining two harmonic curvature components of homogeneity one. Of course, it is easy to express the parts of the Nijenhuis tensor for $\tilde{J}$ in terms of J. For example, the part with values in H^{-} is represented by ,

$$S_{-}(\xi,\eta) := \pi_{H^{-}}([\xi,\eta] + [J\xi, J\eta] - J[J\xi,\eta] + J[\xi, J\eta]). \tag{4.13}$$

Similarly, we obtain $S_{+} \in (H^{-})^{*} \otimes (H^{+})^{*} \otimes H^{+}$.

THEOREM 4.3.9. *Let (M, H, J) be a nondegenerate partially integrable oriented hyperbolic CR–geometry of dimension and codimension two.*

(1) The canonical almost product structure $TM = T^{+}M \oplus T^{-}M$ is integrable if and only if the two algebraic brackets $Q^{+} \times H^{+} \to Q^{-}$ and $Q^{-} \times H^{-} \to Q^{+}$ vanish.

(2) The almost complex structure J on H is integrable if and only if the conjugate linear part of the torsion vanishes (i.e. the two components of the Nijenhuis tensor on H have to vanish). This part of the torsion vanishes automatically for embedded 6–dimensional real surfaces in $\mathbb{C}^4$ locally around a hyperbolic point in M.

(3) The two sesquilinear components of the homogeneity one part of the harmonic curvature vanish if and only if the partially integrable almost CR–structure $(H, \tilde{J})$ on M, where $\tilde{J}|_{H^{\pm}} = \pm J|_{H^{\pm}}$, is integrable. This is equivalent to vanishing of the tensor S_{-} defined by (4.13) *and its analog S_{+}.*

(4) The CR–structure is locally isomorphic to a product of two codimension one CR–geometries of dimension three if and only if the corresponding Cartan geometry is torsion free. This is equivalent to vanishing of all the six obstructions described in (1)–(3).

PROOF. Let us sketch the proof, more details are available in [**SchSl00**]. In order to prove (1), we have to check all components of the obstructions against integrability of the two distributions $T^{\pm}M$. We have checked the componets of homogeneity one already and there are only two further possibilites, which both are of homogeneity two. One of these involves two arguments from either Q^{+} or Q^{-} and hence vanishes because $\dim Q^{\pm} = 1$. Vanishing of the algebraic brackets from (1) implies that we get well–defined tensorial maps $Q^{+} \times H^{+} \to H^{-}$ and $Q^{-} \times H^{-} \to H^{+}$ induced by the Lie bracket. A technical but straighforward discussion of the possible terms in the Bianchi identity and the co-closedness of κ imply that these last obstructions vanish as well. This completes the proof of (1).

The proofs of (2) and (3) follow directly from the interpretation of the obstructions in terms of the Nijenhuis tensors.

If the Cartan connection ω comes from a product of two Cartan geometries of the right type on 3–dimensional manifolds, then clearly all curvature components mixing the arguments from different components of the product structure have to vanish. Thus, all torsion components of the curvature have to vanish; see the above table of harmonic components of the curvature. On the other hand, looking at the same table again, we see that vanishing of the six torsion components implies that all the curvature is concentrated in homogeneities at least four. This allows to identify a compatible integrable product structure on $\mathcal{G}$. At the same time, the curvature in homogeneity four does not mix the arguments in the product structure and so a careful check exploiting the Bianchi identity proves the claim. □

REMARK 4.3.9 (The hyperbolic normal form, [**SchSl00**]). The function theory approach leads to a normalized osculation of 6–dimensional surfaces $M \subset \mathbb{C}^4$ by quadrics. In our case of hyperbolic points in M, the normal form reads as

$$v^1 = |z_1|^2 + n_{11\bar{2}} z_1^2 \bar{z}_2 + n_{22\bar{2}} z_2^2 \bar{z}_2 + \bar{n}_{11\bar{2}} \bar{z}_1^2 z_2 + \bar{n}_{22\bar{2}} \bar{z}_1^2 z_2 + O(4),$$
$$v^2 = |z_2|^2 + n_{11\bar{1}} z_1^2 \bar{z}_1 + n_{22\bar{1}} z_2^2 \bar{z}_2 + \bar{n}_{11\bar{1}} \bar{z}_1^2 z_1 + \bar{n}_{22\bar{1}} \bar{z}_1^2 z_1 + O(4).$$

A straighforward computation of the algebraic brackets in terms of vector fields provides the following link between the coefficients of the normal form and the torsion of the hyperbolic geometry:

- Integrability of the almost product structure is equivalent to vanishing of the coefficients $n_{11\bar{1}}$ and $n_{22\bar{2}}$.
- Vanishing of the tensors $S_\pm$ is equivalent to vanishing of the coefficients $n_{11\bar{2}}$ and $n_{22\bar{1}}$.

In particular, the surface allows locally an embedding in the form of the product of two real hypersurfaces in $\mathbb{C}^2$ if and only if the normal form does not involve any terms of third order at each point.

4.3.10. The elliptic case. There is a close similarity between the elliptic and hyperbolic cases, but some inital work is needed to uncover this similarity. The quadrics (4.9) and (4.11) in 4.3.8 suggest that the almost product structure in the elliptic case should be replaced by an almost complex structure. This suggests that rather than the product of two copies of the real form $\mathfrak{su}(2,1)$ of $\mathfrak{sl}(3,\mathbb{C})$, we should consider $\mathfrak{sl}(3,\mathbb{C})$ itself, but viewed as a real Lie algebra. Otherwise put, we hope to get a relation to real regular normal parabolic geometries of type (G,P), where $G = SL(3,\mathbb{C})$ and $P \subset SL(3,\mathbb{C})$ is the Borel subgroup. Evidently, this has the same complexification as $(PSU(2,1) \times PSU(2,1), B \times B)$, which is the basis for the relation between the two types.

Now by Proposition 3.3.7 and Theorem 3.1.14, regular normal parabolic geometries of type (G,P) are equivalent to regular infinitesimal flag structures, and these are easy to describe explicitly. The grading on $\mathfrak{g}$ is simply the complex version of the grading defining Lagrangean contact structures as in 4.2.3 in the simplest case $n = 1$. This means that $\mathfrak{g}_{-1} = \mathfrak{g}_{-1}^+ \oplus \mathfrak{g}_{-1}^-$ (where we have replaced the superscripts E and F by $+$ and $-$ for compatibility with the hyperbolic case) for one–dimensional complex subspaces $\mathfrak{g}_{-1}^\pm$. Further $\mathfrak{g}_{-2}$ is of complex dimension one, and the bracket $[\ ,\ ] : \Lambda^2\mathfrak{g}_{-1} \to \mathfrak{g}_{-2}$ is nondegenerate and complex bilinear. This implies that it vanishes on $\mathfrak{g}^{\pm}_{-1} \times \mathfrak{g}^{\pm}_{-1}$ and induces a nondegenerate complex bilinear pairing $\mathfrak{g}^{+}_{-1} \times \mathfrak{g}^{-}_{-1} \to \mathfrak{g}_{-2}$.

Since the bracket is complex bilinear rather than totally real, such structures do not directly give rise to partially integrable almost CR–structures. An equivalence to such structures can still be established as follows.

PROPOSITION 4.3.10. *Let $(M, H, \tilde{J})$ be an oriented elliptic partially integrable CR–structure of CR–dimension and codimension two.*

(1) There is a unique almost complex structure J^Q on Q, which is compatible with the orientation of M and has the property that for each $x \in M$ there are nonzero vectors $\eta \in T_xM$ for which $\mathcal{L}(\_, \eta)$ is complex linear respectively conjugate linear with respect to $\tilde{J}$ and J^Q.

(2) The vectors η from (1) span two complementary smooth complex line bundles $H^{\mp} \subset H$, which are both isotropic with respect to the Levi bracket $\mathcal{L}$.

(3) If we define a new almost complex structure J on H such that $J = \tilde{J}$ on H^+ and $J = -\tilde{J}$ on H^-, then the Levi bracket $\mathcal{L}$ is complex bilinear with respect to J and J^Q.

(4) The construction in (3) establishes an equivalence with regular infinitesimal flag structures of type (G, P), so our structures are equivalent to regular normal parabolic geometries of that type.

PROOF. The simplest way to start the discussion is to complexify the tangent space so that the situation gets similar to the hyperbolic case. The Levi bracket is totally real and thus the imaginary part of the Hermitian form $\mathcal{H} : H \times H \to Q^{\mathbb{C}} = Q \otimes \mathbb{C}$, $\mathcal{H}(\xi, \eta) = -\mathcal{L}(\tilde{J}\xi, \eta) + i\mathcal{L}(\xi, \eta)$. Next, similarly to Lemma 4.3.8, for $x \in M$ and each complex linear map $\psi : Q^{\mathbb{C}}_x \to \mathbb{C}$, we consider the mapping $\mathcal{H}^{\psi} = \psi \circ \mathcal{H} : H_x \times H_x \to \mathbb{C}$. This is complex linear in the first argument and conjugate linear in the second one. As before, the existence of a nonzero nullspace of $\mathcal{H}^{\psi}$ depends only on the class $[\psi]$ in the complex projectivization $\mathcal{P}(Q^{\mathbb{C}}_x)$. Choosing a complex basis ξ_1, ξ_2 of H_x and a real basis of Q_x, the Hermitian form $\mathcal{H}$ is given by two Hermitian matrices $\mathcal{H}_1$ and $\mathcal{H}_2$ which have to be linearly independent due to nondegeneracy of $\mathcal{L}$. The induced homogeneous coordinates $(\lambda : \mu)$ on the projectivization $\mathcal{P}(Q^{\mathbb{C}})^*$ allow to express degeneracy of $\mathcal{H}^{[\psi]}$ as the solution of the quadratic homogeneous equation $\det(\lambda\mathcal{H}_1 + \mu\mathcal{H}_2) = 0$.

If this equation had a real root $(\lambda : \mu)$, then the restriction of ψ to the imaginary part $iQ_x \subset Q^{\mathbb{C}}_x$ would provide a real linear form ϕ, for which $\mathcal{L}^{\phi}$ is degenerate and our geometry would be either hyperbolic or exceptional. Thus, there are just two complex conjugate points $[\psi]$ and $[\bar{\psi}] \in \mathcal{P}(Q^{\mathbb{C}}_x)^*$, for which the forms $\mathcal{H}^{\psi}$ are degenerate. At the same time, the restriction of such a map ψ to $Q_x \subset Q^{\mathbb{C}}_x$ is injective, thus inducing a complex structure J^Q_x on Q_x. Since $\bar{\psi}$ leads to the negative complex structure, we can fix ψ by requiring that J^Q_x, together with the orientation on H_x coming from $\tilde{J}_x$, induces the given orientation on T_xM.

Since $\mathcal{H}^{\psi}$ is degenerate, there is a nonzero element $\eta \in H_x$ such that $\psi \circ \mathcal{H}(\_, \eta) = 0$, and inserting the definition of $\mathcal{H}$ this gives $\psi(\mathcal{L}(\tilde{J}\xi, \eta)) = i\psi(\mathcal{L}(\xi, \eta))$ for all $\xi \in H$. By definition of J^Q, this is equivalent to $\xi \mapsto \mathcal{L}(\xi, \eta)$ being complex linear. As in the hyperbolic case, the vectors η for which this holds form a complex line H^-_x in H_x. Skew symmetry of $\mathcal{L}$ then implies that $\mathcal{L}(\_, \eta)$ vanishes on any complex multiple of η which shows that H_x^- is isotropic for $\mathcal{L}$. Starting from $\bar{\psi}$ we obtain a complementary complex line H_x^+ which is also isotropic for $\mathcal{L}$.

As in the hyperbolic case, all our constructions so far were described as solutions of smooth equations. Hence, they fit together smoothly to define an almost complex structure J^Q on Q and a decomposition $H = H^+ \oplus H^-$ into a sum of complex line bundles which are isotropic for $\mathcal{L}$. This completes the proofs of (1) and (2). Since $H^\pm$ are isotropic, the only nonzero component of $\mathcal{L}$ maps $H^+ \times H^-$ to Q, and by construction this is complex linear in the first variable and conjugate linear in the second variable. Swapping the complex structure on H^- therefore makes $\mathcal{L}$ complex bilinear, which proves (3).

From the description of regular infinitesimal flag structures of type (G, P) above, we see that the construction of (3) defines such a structure on M. Conversely, if $(H = H^+ \oplus H^-, J, J^Q)$ is a regular infinitesimal flag structure on M, then we define $\tilde{J}$ on M by $\tilde{J}|_{H^\pm} = \pm J$. Then $H \subset TM$ is a rank four subbundle and $\tilde{J}$ makes H into a complex rank two bundle. Since the subbundles $H^\pm$ are isotropic for $\mathcal{L}$, the only nonzero component maps $H^+ \times H^-$ to Q. Since this is complex bilinear for J, it is totally real for $\tilde{J}$, so we have defined a partially integrable almost CR–structure. It is immediately verified that this structure is elliptic and hence, in particular, nondegenerate. Since the construction is evidently compatible with morphisms, this completes the proof. □

The computation of the real cohomology $H^2(\mathfrak{g}_-, \mathfrak{g})$ proceeds via complexification, so the result looks very similar as in the hyperbolic case. The resulting eight components together with their homogeneity, their location within linear maps $\Lambda^2\mathfrak{g}_- \to \mathfrak{g}$ and their complex linearity properties with respect to J are listed in the following table.

homog.	location	complex linearity properties w.r.t. J
1	$\mathfrak{g}_{-2} \times \mathfrak{g}_{-1}^- \to \mathfrak{g}_{-2}$	conjugate linear in both arguments
1	$\mathfrak{g}_{-2} \times \mathfrak{g}_{-1}^+ \to \mathfrak{g}_{-2}$	conjugate linear in both arguments
1	$\mathfrak{g}_{-1}^- \times \mathfrak{g}_{-1}^- \to \mathfrak{g}_{-1}^+$	totally real
1	$\mathfrak{g}_{-1}^+ \times \mathfrak{g}_{-1}^+ \to \mathfrak{g}_{-1}^-$	totally real
1	$\mathfrak{g}_{-1}^+ \times \mathfrak{g}_{-1}^- \to \mathfrak{g}_{-1}^-$	sesquilinear
1	$\mathfrak{g}_{-1}^- \times \mathfrak{g}_{-1}^+ \to \mathfrak{g}_{-1}^+$	sesquilinear
4	$\mathfrak{g}_{-2} \times \mathfrak{g}_{-1}^- \to \mathfrak{g}_1^-$	complex linear in both arguments
4	$\mathfrak{g}_{-2} \times \mathfrak{g}_{-1}^+ \to \mathfrak{g}_1^+$	complex linear in both arguments

Of course, in the terminology of 3.1.17, the first two lines form the coholomogy of type $H^{(0,2)}$, the next four lines are of type $H^{(1,1)}$, while the last two lines are of type $H^{(2,0)}$. We will only sketch the interpretation of the harmonic curvatures of homogeneity one; details can be found in [**ČSc02**].

For some parts the interpretation is very easy. Since the subbundles $H^\pm$ are isotropic for $\mathcal{L}$, the bracket of two sections of either of the two subbundles is a section of H, and hence may be projected to the other subbundle. This defines tensorial maps $\Lambda^2 H^+ \to H^-$ and $\Lambda^2 H^- \to H^+$. Using the extension map from Propostion 3.1.8 one easily shows that these brackets (up to nonzero factors) represent the two harmonic curvature components corresponding to totally real maps.

Second, we can again consider the Nijenhuis tensor $\tilde{N}$ of $\tilde{J}$. Since this is of type $(0,2)$ and skew symmetric, it vanishes on $H^\pm \times H^\pm$ so the only nonzero component maps $H^+ \times H^-$ to $H = H^+ \oplus H^-$. According to the latter splitting, we obtain two componenents. The conjugate linearity in both factors with respect to $\tilde{J}$ shows that, in terms of J, both these components become sesquilinear. Using the extension map from Propostion 3.1.8, one easily shows that they represent (up to nonzero factors) the two harmonic curvature components corresponding to sesquilinear maps. In particular, vanishing of these two components is equivalent to integrability of the almost CR–structure $(H, \tilde{J})$.

The interpretation of the curvatures corresponding to $H^{(0,2)}$ is a bit more involved. First, note that via the Cartan geometry $(\mathcal{G}, \omega)$ on M, many features of the Lie algebra $\mathfrak{g}$ are carried over to M and $\mathcal{G}$. In particular, the complex structure on $\mathfrak{g}$ induces almost complex structures $J^{\mathcal{G}}$ on $\mathcal{G}$ and J on M. Of course, J restricts to the almost complex structure J from the lemma on H and induces J^Q on Q, but the extension to TM is not so obvious. In analogy to the hyperbolic case, one should expect that the two harmonic curvatures corresponding to $H^{(0,2)}$ are the obstructions against integrability of the almost complex structure J on M.

Choose an almost complex structure $\hat{J}$ on TM which agrees with J on H and projects to J^Q on Q. This can be done by choosing a subbundle complementary to H and identifying it with Q via $q : TM \to Q$. Then for $\xi \in \mathfrak{X}(M)$ and $\eta \in \Gamma(H^\pm)$ consider the expression

$$(\xi, \eta) \mapsto q([\hat{J}\xi, \eta]) - J^Q q([\xi, \eta]).$$

Since $q(\eta) = 0$ and $q(\hat{J}\xi) = J^Q q(\xi)$, this is bilinear over smooth functions. Moreover, if $\xi \in \Gamma(H)$, then $\hat{J}\xi = J\xi$ and the expression vanishes by complex bilinearity of $\mathcal{L}$. Thus, it is given by $S^\pm(q(\xi), \eta)$ for tensorial maps $S^\pm : Q \times H^\pm \to Q$. Using once more that $J^Q q(\xi) = q(\hat{J}\xi)$ we conclude that $S^\pm$ is conjugate linear in the first variable.

Now suppose that $\hat{J} = J$, the almost complex structure coming from the canonical Cartan connection. Using the technique of extensions of vectors to vector fields exactly as in the hyperbolic case, we see that $S^\pm(\xi, \eta)$ is the projection of

$$-\omega(u)^{-1}\Big(\kappa^{(1)}(iX_{-2}, Y) + [iX, Y]_{-2} - i\kappa^{(1)}(X_{-2}, Y) - i[X, Y]_{-2}\Big).$$

Here X and Y represent ξ and η in the frame $u \in \mathcal{G}$, and the subscripts indicate the components with respect to the grading of $\mathfrak{g}$. Since the Lie bracket on $\mathfrak{g}$ is complex bilinear, the two bracket terms cancel. The list of all cohomologies in homogeneity one shows that the restriction of the homogeneity one component $\kappa^{(1)}$ to $\mathfrak{g}_{-2} \times \mathfrak{g}_{-1}$ is conjugate linear in the second argument. This shows that the almost complex structure J on TM has the property that $S^\pm$ is also conjugate linear in the second variable.

This pins down J uniquely. For another almost complex structure $\hat{J}$ as above and any ξ we must have $\hat{J}\xi = J\xi + \xi'$ for some $\xi' \in \Gamma(H)$. But then S changes by $\mathcal{L}(\xi', \eta)$, which is complex linear in η and nonzero as long as ξ' is nonzero.

By definition of J and since $S^{\pm}$ is conjugate linear in both arguments, it is also induced by the map

$$(\xi, \eta) \mapsto \frac{1}{2} q\left([J\xi, \eta] - J[\xi, \eta] + J[J\xi, J\eta] + [\xi, J\eta]\right).$$

This shows that $S^{\pm}(\xi, \eta)$ is a nonzero multiple of $q(N(J\xi, \eta))$, where N denotes the Nijenhuis tensor of J, so it essentially is a component of the Nijenhuis tensor. Using the Bianchi identity one shows, similarly as in the hyperbolic case, that vanishing of this components already implies vanishing of N and hence integrability of J.

Similarly to the hyperbolic case, one may also give a nice description of torsion–free geometries. Here the tedious verifications using the Bianchi identity can be nicely replaced by general arugments. We have characterized complex parabolic geometries among real ones in Proposition 3.1.17. There we have also mentioned that using the BGG machinery one can also give a characterization in terms of harmonic curvature rather than the full Cartan curvature. The BGG techniques (which we will discuss in detail in volume two) show that in our situation the fact that the harmonic curvature is of type $(2, 0)$ already implies that the full Cartan curvature is of type $(2, 0)$, and hence the geometry is a complex (holomorphic) parabolic geometry; see [**Čap05**]. Summarizing our results we get.

THEOREM 4.3.10. *Let $(M, H, \tilde{J})$ be an oriented elliptic partially integrable almost CR geometry of dimension and codimension two. Then there is a canonical regular normal parabolic geometry $(\mathcal{G}, \omega)$ of type $(SL(3, \mathbb{C}), B)$ on M. This gives rise to a decomposition $H = H^+ \oplus H^-$ and a canonical almost complex structure J on M as described above. Furthermore:*

(1) *The almost complex structure J is integrable if and only if the tensors $S^{\pm}$ vanish. This is equivalent to vanishing of the $(0, 2)$ part of the harmonic curvature.*
(2) *The subbundles $H^{\pm} \subset TM$ are integrable if and only if the corresponding totally real components in the homogeneity one part of the harmonic curvature vanish identically.*
(3) *The original almost complex structure $\tilde{J}$ on H is integrable if and only if the two sesquilinear components in the homogeneity one part of the harmonic curvature vanish identically.*
(4) *The Cartan geometry $(\mathcal{G}, \omega)$ is holomorphic if and only if the homogenity one part of the harmonic curvature vanishes identically.*

REMARK 4.3.10. (1) Similarly to the hyperbolic CR geometry of dimension and codimension two, elliptic points in embedded 6–dimensional surfaces $M \subset \mathbb{C}^4$ provide the most important examples of such geometries. Of course, in the embedded case the almost complex structure $\tilde{J}$ is automatically integrable, so the obstructions in part (3) of the theorem vanish automatically.

Again the function theory approach leads to normal forms of such surfaces and we can directly relate the individual coefficients of the third order terms to the four components of the torsion. The CR geometry will be holomorphic if and only if all third order terms of the normal form vanish at all points. See [**SchSl00**] for more details in this direction.

Our theorem has surprising consequences for embeddability. Namely, by part (4), a torsion–free geometry is related to a complex parabolic geometry. While holomorphicity is lost by passing back to the almost complex structure $\tilde{J}$ describing the CR–structure, we certainly can conclude that the orginial almost CR–structure is real analytic. Using this, it is easy to show that it is embeddable; see [**ČSc02**].

(2) In 4.3.2 we have studied various types of generic distributions. In particular, we have seen that there is just one generic type of distributions of rank three in dimension six. Now one can similarly look at distributions of rank four in dimension six. If $H \subset TM$ is such a distribution with quotient bundle $Q = TM/H$, then the Levi bracket is a map $\mathcal{L} : \Lambda^2 H \to Q$. Since the wedge product is symmetric on two–forms, the induced map $\Lambda^4 H \to Q \otimes Q$ has values in $S^2 Q$. Since $\Lambda^4 H$ has rank one, we can consider this map as a quadratic form on Q^* defined up to scale. Nondegeneracy of this form clearly is a generic property (provided that there are examples for which it is satisfied). In the nondegenerate case, we can look at the signature of the quadratic form. Since it is defined on a rank two bundle, there are just two possibilities, namely definite and indefinite forms.

Now it is easy to verify that the rank four distribution underlying a hyperbolic (respectively elliptic) partially integrable almost CR–structure of CR–dimension and codimension two is always of this generic type with indefinite (respectively definite) nondegenerate quadratic forms. Hence, we conclude that there are exactly two types of generic rank four distributions in dimension six, which are called of elliptic and of hyperbolic type.

The simplest examples of these distributions are provided by a product of two three–dimensional contact manifolds, respectively, by the real manifold underlying a three–dimensional complex contact manifold. Such examples are always locally isomorphic by the Darboux theorem.

One can use the ideas developed for codimension two CR–structures to study the geometry of generic rank four distributions in dimension six. This is worked out in [**ČE03**]. In particular, it is shown that one can construct a canonical almost product structure, respectively, almost complex structure from such a distribution. Then analogs of the tensors $S^{\pm}$ provide complete obstructions to local isomorphism with the model spaces described above. This has surprising consequences. For example, it implies that the complex structure of a three–dimensional complex contact manifold can be recovered from the contact subbundle (viewed as a real bundle).

4.4. Correspondence spaces and twistor spaces

In this section we study the constructions of correspondence spaces and twistor spaces, which were introduced for general Cartan geometries in 1.5.13 and 1.5.14 in the special case of parabolic geometries. The input needed for these general constructions are two nested closed subgroups $K \subset H \subset G$, and one obtains relations between Cartan geometries of type (G, H) and of type (G, K).

For parabolic geometries, there are two basic families of examples for which these constructions are of interest. On the one hand, both H and K can be taken to be parabolic subgroups of G. In this case, one obtains relations between parabolic geometries of different types. It turns out, that these constructions always preserve normality, so they are nicely compatible with interpretations in terms of underlying structures. Typically, the geometries corresponding to the smaller parabolic subgroup are more complicated in nature than the ones corresponding to the larger subgroup. Hence, already the construction of correspondence spaces provides a systematic way of passing from simple structures to more complicated structures. The construction of twistor spaces reproduces some of the classical twistor correspondences in this case. The most prominent example is twistor theory for (anti–) self–dual split signature conformal structures in dimension four.

In the second case of interest, only the larger subgroup H is parabolic, but $K \subset H$ is a more general subgroup. In this setting, we can directly use the machinery for Cartan connections developed in Section 1.5 to obtain canonical geometric structures on the total spaces of certain natural fiber bundles over manifolds endowed with a parabolic geometry. This for example leads to twistor theory for conformal four manifolds in Riemannian signature.

4.4.1. The basic setup for nested parabolic subalgebras. Let G be a semisimple Lie group with Lie algebra $\mathfrak{g}$, and consider two nested parabolic subalgebras $\mathfrak{q} \subset \mathfrak{p} \subset \mathfrak{g}$. Suppose that we have chosen a parabolic subgroup P corresponding to $\mathfrak{p} \subset \mathfrak{g}$. Then we define $Q \subset P$ to be the normalizer of $\mathfrak{q}$ in P, i.e. $Q = \{g \in P : \mathrm{Ad}(g)(\mathfrak{q}) \subset \mathfrak{q}\}$. By definition, this is a closed subgroup of P. Moreover, by Corollary 3.2.1, the normalizer $N_G(\mathfrak{q})$ of $\mathfrak{q}$ in G has Lie algebra $\mathfrak{q}$, so since $Q = N_G(\mathfrak{q}) \cap P$, we see that Q has Lie algebra $\mathfrak{q}$, too.

We will always deal with standard parabolic subalgebras corresponding to choices of a Cartan subalgebra and positive (restricted) roots. This means that $\mathfrak{p}$ and $\mathfrak{q}$ correspond to sets $\Sigma_{\mathfrak{p}} \subset \Sigma_{\mathfrak{q}}$ of simple (restricted) roots. In the language of Dynkin diagrams, respectively, Satake diagrams with crossed nodes from 3.2.2 and 3.2.10, this means that the diagram for $\mathfrak{q}$ is obtained from the one of $\mathfrak{p}$ by changing some uncrossed nodes into crossed nodes (taking into account which nodes may be crossed in the real case). We will write the decompositions of $\mathfrak{g}$ as $\mathfrak{g} = \mathfrak{p}_- \oplus \mathfrak{p}_0 \oplus \mathfrak{p}_+$ and $\mathfrak{g} = \mathfrak{q}_- \oplus \mathfrak{q}_0 \oplus \mathfrak{q}_+$, and we will write $P_0 \subset P$ and $Q_0 \subset Q$ for the subgroups of elements whose adjoint actions preserve the gradings corresponding to $\mathfrak{p}$ and $\mathfrak{q}$, respectively.

LEMMA 4.4.1. *Let G be a semisimple Lie group with Lie algebra $\mathfrak{g}$ and let $\mathfrak{q} \subset \mathfrak{p} \subset \mathfrak{g}$ be nested parabolic subalgebras.*

(1) $\mathfrak{p}_\pm \subset \mathfrak{q}_\pm$, $\mathfrak{q}_0 \subset \mathfrak{p}_0$ and $\mathfrak{z}(\mathfrak{p}_0) \subset \mathfrak{z}(\mathfrak{q}_0)$.

(2) Let $P \subset G$ be a parabolic subgroup corresponding to $\mathfrak{p}$ and put $Q := N_P(\mathfrak{q})$. Denoting by P_0^{ss} the semisimple part of P_0, the subgroup $P_0^{ss} \cap Q \subset P_0^{ss}$ is parabolic. If the homogeneous space P/Q is connected, then it is naturally diffeomorphic to the generalized flag variety $P_0^{ss}/(P_0^{ss} \cap Q)$.

PROOF. We may assume, without loss of generality, that we are dealing with standard parabolics corresponding to subsets $\Sigma_{\mathfrak{p}} \subset \Sigma_{\mathfrak{q}}$ of simple (restricted) roots. Then from the description of the grading components in terms of Σ–heights (1) is evident.

(2) Suppose that $\mathfrak{g}$ is complex. Then from 3.2.2 we know what the root decomposition of $\mathfrak{p}_0^{ss}$ looks like. In particular, $\mathfrak{p}_0^{ss} \cap \mathfrak{q}$ is simply the standard parabolic subalgebra of $\mathfrak{p}_0^{ss}$ corresponding to the set $\Sigma_{\mathfrak{q}} \setminus \Sigma_{\mathfrak{p}}$ of simple roots. By definition of Q, any element of $P_0^{ss} \cap Q$ normalizes the subalgebra $\mathfrak{p}_0^{ss} \cap \mathfrak{q} \subset \mathfrak{p}_0^{ss}$. On the other hand, the Lie algebra of $P_0^{ss} \cap Q$ is $\mathfrak{p}_0^{ss} \cap \mathfrak{q}$, so it is a parabolic subgroup and $P_0^{ss}/(P_0^{ss} \cap Q)$ is a generalized flag variety. In the real case, one may use the same argument after complexification.

Now consider the action of P_0^{ss} on P/Q. By part (1) we have $\mathfrak{p}_+ \subset \mathfrak{q}_+$ and hence $P_+ \subset Q$, which shows that P_0 acts transitively on P/Q. Since $\mathfrak{z}(\mathfrak{p}_0) \subset \mathfrak{q}$, we see that the orbit of eQ under P_0^{ss} is open in P/Q. Hence, we obtain an open embedding of $P_0^{ss}/(P_0^{ss}\cap Q)$ into P/Q. But by Proposition 3.2.6 the generalized flag variety $P_0^{ss}/(P_0^{ss}\cap Q)$ is compact, so the image is closed and the result follows. $\square$

Given nested parabolic subgroups $Q \subset P \subset G$, we can apply the machinery of correspondence spaces as introduced in 1.5.13. Starting from a parabolic geometry $(p : \mathcal{G} \to N, \omega)$ of type (G, P), we form the correspondence space $\mathcal{C}N = \mathcal{G}/Q$. We can identify $\mathcal{C}N$ with the associated bundle $\mathcal{G} \times_P (P/Q)$ whose fiber is a generalized flag manifold or a disjoint union of copies of a generalized flag manifold. By Proposition 1.5.13, the obvious projection $\pi : \mathcal{G} \to \mathcal{C}N$ is a principal Q bundle on which ω defines a Cartan connection. Hence, $\mathcal{C}N$ carries a natural parabolic geometry of type (G, Q). Now the fact that there is a uniform normalization condition for parabolic geometries pays off:

PROPOSITION 4.4.1. *Let $(p : \mathcal{G} \to N, \omega)$ be a parabolic geometry of type (G, P) with correspondence space $\mathcal{C}N$. Then the canonical geometry of type (G, Q) on $\mathcal{C}N$ is normal if and only if $(p : \mathcal{G} \to N, \omega)$ is normal.*

PROOF. By Proposition 1.5.13 the curvature function $\kappa^{\mathcal{C}N}$ is induced from κ^N via the natural surjection $\mathfrak{g}/\mathfrak{q} \to \mathfrak{g}/\mathfrak{p}$ or the corresponding inclusion $(\mathfrak{g}/\mathfrak{p})^* \hookrightarrow (\mathfrak{g}/\mathfrak{q})^*$. Otherwise put, both curvature functions are the same when viewed as functions $\mathcal{G} \to \Lambda^2\mathfrak{g}^* \otimes \mathfrak{g}$. We may identify the latter space with $\Lambda^2\mathfrak{g} \otimes \mathfrak{g}$ via the Killing form. In this picture, we know that κ^N has values in $\Lambda^2\mathfrak{p}_+ \otimes \mathfrak{g} \subset \Lambda^2\mathfrak{q}_+ \otimes \mathfrak{g}$, which is the natural target space for $\kappa^{\mathcal{C}N}$. But from the definition in 3.1.11 we see that the Kostant codifferentials for the parabolic subalgebras $\mathfrak{p}$ and $\mathfrak{q}$ are obtained as the restriction to $\Lambda^2\mathfrak{p}_+\otimes\mathfrak{g}$, respectively, to $\Lambda^2\mathfrak{q}_+\otimes\mathfrak{g}$ of a map $\Lambda^2\mathfrak{g}\otimes\mathfrak{g} \to \mathfrak{g}\otimes\mathfrak{g}$. Hence, on $\Lambda^2\mathfrak{p}_+ \otimes \mathfrak{g}$ the codifferentials for the two parabolic subalgebras coincide. $\square$

In contrast to normality, regularity and torsion freeness are *not* preserved in the process of forming a correspondence space. For torsion freeness this is evident, since there may be elements $X, Y \in \mathfrak{g}$ such that $\kappa(X, Y)$ has values in $\mathfrak{p} \setminus \mathfrak{q}$. Similarly, it may happen that there are elements X and Y which have the same homogeneity with respect to $\mathfrak{p}$ and $\mathfrak{q}$, but such that $\kappa(X, Y)$ has strictly lower homogeneity with respect to $\mathfrak{q}$ than with respect to $\mathfrak{p}$. We shall see in examples below that this already occurs in fairly simple cases.

Finally, we can also directly use the characterization of correspondence spaces from 1.5.14. Let us start with a general parabolic geometry $(\mathcal{G} \to M, \omega)$ of type (G, Q). Since $Q \subset P$, the subspace $\mathfrak{p} \subset \mathfrak{g}$ is Q–invariant and hence descends to a Q–invariant subspace $\mathfrak{p}/\mathfrak{q} \subset \mathfrak{g}/\mathfrak{q}$. Passing to associated bundles, this gives rise to a distribtuion $VM \subset TM$. In the case of a correspondence space $\mathcal{C}N$, this is the vertical bundle of $\mathcal{C}N \to N$ and hence globally integrable. In the general case, local integrability of VM is equivalent to $T(VM, VM) \subset VM$ by Lemma 1.5.14, where $T \in \Omega^2(M, TM)$ is the torsion of ω.

If this is the case, then one may form a *local twistor space* for M, i.e. a local leaf space N for the foliation defined by VM. Theorem 1.5.14 then characterizes when, for sufficiently small local twistor spaces, one gets an induced geometry of type (G, P) on N such that M is locally isomorphic to the correspondence space $\mathcal{C}N$. We extend this result a bit by proving a result for larger leaf spaces.

COROLLARY 4.4.1 (of Theorem 1.5.14). *Suppose that P/Q is connected, and $(p : \mathcal{G} \to M, \omega)$ is a parabolic geometry of type (G, Q) with Cartan curvature κ such* *that $i_\xi \kappa = 0$ for all $\xi \in \Gamma(VM)$. Let $\psi : U \to N$ be a local twistor space such that there exists a smooth section $s : N \to U$ of ψ. Then there is a unique parabolic geometry of type (G, P) on N such that an open neighborhood of $s(N) \subset U$ is isomorphic (as a parabolic geometry) to an open subset in the correspondence space $\mathcal{C}N$. This geometry is normal provided that $(p : \mathcal{G} \to M, \omega)$ is normal.*

PROOF. The construction in the proof of Theorem 1.5.14 is based on a local smooth section σ of $\psi \circ p : p^{-1}(U) \to N$. Locally around $x \in N$ we can apply this construction for σ being the composition of a local smooth section of $p : p^{-1}(U) \to U$ with the fixed section s of $\psi : U \to N$. This gives us an open neighborhood V_x of $s(x)$ in M and a diffeomorphism Φ_x from a Q–invariant open subset in $\psi(V_x) \times P$ to $p^{-1}(V_x) \subset \mathcal{G}$, such that $\Phi_x^* \omega$ extends by equivariancy to a Cartan connection $\tilde{\omega}_x$ on $\psi(V_x) \times P$. The map Φ_x defines an isomorphism between V_x and an open subspace in the correspondence space $\mathcal{C}(\psi(V_x))$. If $x, y \in N$ are such that $V_x \cap V_y \neq \emptyset$, then the restriction of $\Phi_y^{-1} \circ \Phi_x$ to $\psi(V_x \cap V_y)$ defines an isomorphism between open subsets of the correspondence spaces $\mathcal{C}\psi(V_x)$ and $\mathcal{C}\psi(V_y)$. Since P/Q is connected, part (4) of Proposition 1.5.13 implies that this is actually an isomorphism of the two induced parabolic geometries of type (G, P) on $\psi(V_x \cap V_y)$.

Hence, we can use the bundles $\psi(V_x) \times P$ to glue a global principal P–bundle over N, and the forms $\tilde{\omega}_x$ fit together to define a Cartan connection $\tilde{\omega}$ on this principal bundle. Thus, we have obtained a parabolic geometry of type (G, P) on N, and the statement on normality follows immediately from the proposition above. By construction, the open neighborhood $\bigcup_{x \in N} V_x$ of $s(N)$ in M is isomorphic to an open subset of the correspondence space $\mathcal{C}N$. □

There is a much stronger version of this characterization in which the full curvature of the Cartan connection is replaced by the harmonic curvature. This result is one of the striking applications of BGG sequences, so we will discuss it in detail in the second volume of this book. Here we just state the result, a proof can be found in [**Čap05**].

THEOREM 4.4.1. *Suppose that $(\mathcal{G} \to M, \omega)$ is a parabolic geometry of type (G, Q) such that the harmonic curvature κ_H of ω has the properties that all components involving entries from VM vanish identically. Then the Cartan curvature κ of ω statifies $i_\xi \kappa = 0$ for all $\xi \in \Gamma(VM)$, and the corollary above applies.*

4.4.2. Example: Lagrangean contact structures from projective structures. Consider the Lie algebra $\mathfrak{g} := \mathfrak{sl}(n+2, \mathbb{R})$ and the parabolic subalgebra $\mathfrak{q} \subset \mathfrak{p} \subset \mathfrak{g}$ corresponding to the crossed Satake diagrams ×—∘—···—∘—×, respectively, ×—∘—···—∘—∘. We can realize $\mathfrak{p}$ as the stabilizer of a line ℓ in $\mathbb{R}^{n+2}$, while elements of $\mathfrak{q}$ in addition have to stabilize a hyperplane. From 4.2.3 we know that the decomposition of $\mathfrak{g}$ correspoding to $\mathfrak{q}$ has the form

$$\begin{pmatrix} \mathfrak{q}_0 & \mathfrak{q}_1^E & \mathfrak{q}_2 \\ \mathfrak{q}_{-1}^E & \mathfrak{q}_0 & \mathfrak{q}_1^F \\ \mathfrak{q}_{-2} & \mathfrak{q}_{-1}^F & \mathfrak{q}_0 \end{pmatrix},$$

and the decomposition corresponding to $\mathfrak{p}$ is characterized by $\mathfrak{p}_{\pm 1} = \mathfrak{q}_{\pm 1}^E \oplus \mathfrak{q}_{\pm 2}$.

As a group with Lie algebra $\mathfrak{g}$, we take $G := PGL(n+2, \mathbb{R})$ and as the parabolic subgroup $P \subset G$ we use the stabilizer of the point in $\mathbb{R}P^{n+1}$ corresponding to ℓ. From 4.1.5 we know that elements of P correspond to matrices of the form

$\begin{pmatrix} 1/\det(C) & W \\ 0 & C \end{pmatrix}$ and one easily verifies that $Q \subset P$ consists of those matrixes for which the last row of C has zeros in all but the last column. The action of P on $\mathfrak{g}/\mathfrak{p} \cong \mathbb{R}^{n+1}$ is given by $X \mapsto \det(C)^{-1}CX$. Hence, we conclude that Q is the stabilizer in P of a hyperplane in $\mathfrak{g}/\mathfrak{p}$, or equivalently the stabilizer of a line in $(\mathfrak{g}/\mathfrak{p})^*$. Since P acts transitively on $(\mathfrak{g}/\mathfrak{p})^*$, we can therefore naturally identify the quotient P/Q with the projective space $\mathcal{P}(\mathfrak{g}/\mathfrak{p})^*$.

From 4.1.5 we know that normal parabolic geometries of type (G, P) are equivalent to classical projective structures on manifolds of dimension $n+1$. On the other hand, by 4.2.3 regular normal parabolic geometries of type (G, Q) are equivalent to Lagrangean contact structures. Such a geometry is given by a contact distribution H together with a decomposition $H = E \oplus F$ into the direct sum of two Legendrean subbundles. In 4.2.3 we have also described the harmonic curvature of a Lagrangean contact structure, which consists of two torsions τ_E and τ_F, and one curvature, which we denote by ρ.

PROPOSITION 4.4.2. *Let N be a smooth manifold of dimension $n+1$ endowed with a projective equivalence class $[\nabla]$ of torsion–free linear connections on TN.*

(1) The projectivized cotangent bundle $\mathcal{P}T^*N$ carries a canonical Lagrangean contact structure $E \oplus F \subset T\mathcal{P}T^*N$. Here F is the vertical subbundle of the projection $\mathcal{P}T^*N \to N$ and $E \oplus F =: H \subset T\mathcal{P}T^*N$ is the natural contact structure as introduced in the end of 4.2.1.

(2) Let τ_E, τ_F and ρ be the harmonic curvatures of the Lagrangean contact structure from (1). Then τ_F and ρ are identically zero, while $\tau_E = 0$ if and only if $(N, [\nabla])$ is locally projectively flat.

PROOF. (1) Let $(p : \mathcal{G} \to N, \omega)$ be the regular normal parabolic geometry of type (G, P) determined by $[\nabla]$. Then we can form the correspondence space $\mathcal{C}N$ for $\mathfrak{q} \subset \mathfrak{p}$. The associated bundle $\mathcal{G} \times_P (\mathfrak{g}/\mathfrak{p})^*$ is the cotangent bundle T^*N. Since we have verified above that P/Q can be identified with the projectivization of $(\mathfrak{g}/\mathfrak{p})^*$ we see that $\mathcal{C}N = \mathcal{G} \times_P (P/Q) \cong \mathcal{P}T^*N$.

Denoting by $\pi : \mathcal{G} \to \mathcal{C}N$ the natural projection, we know from Proposition 4.4.1 that the natural parabolic geometry $(\pi : \mathcal{G} \to \mathcal{C}N, \omega)$ of type (G, Q) is normal. From 4.1.5 we know that the normal Cartan geometries corresponding to projective structures are automatically torsion free, so the curvature function has values in $\mathfrak{p} \subset \mathfrak{q}_{-1} \oplus \mathfrak{q}$. In particular, this implies that $(\pi : \mathcal{G} \to \mathcal{C}N, \omega)$ is regular and hence it is the canonical parabolic geometry associated to a Lagrangean contact structure on $\mathcal{C}N$.

To obtain an explicit description of this Lagrangean contact structure, observe that via the Cartan connection ω, we get $TN \cong \mathcal{G} \times_P (\mathfrak{g}/\mathfrak{p})$ and $T\mathcal{C}N \cong \mathcal{G} \times_Q (\mathfrak{g}/\mathfrak{q})$. In this picture, the tangent map of the projection $\mathcal{C}N \to N$ corresponds to the canonical projection $\mathfrak{g}/\mathfrak{q} \to \mathfrak{g}/\mathfrak{p}$. Via the linear isomorphism $\mathfrak{q}_- \to \mathfrak{g}/\mathfrak{q}$ the subspaces $\mathfrak{q}_{-1}^E$ and $\mathfrak{q}_{-1}^F$ give rise to Q–invariant subspace of $\mathfrak{g}/\mathfrak{q}$, which induce the subbundles E and F. Now $\mathfrak{q}_{-1}^F$ exactly corresponds to $\mathfrak{p}/\mathfrak{q} \subset \mathfrak{g}/\mathfrak{q}$. By part (3) of Proposition 1.5.13, the subbundle $F \subset H$ is the vertical subbundle of $\mathcal{C}N \to N$.

On the other hand, the image of $\mathfrak{q}_{-1}^E$ in $\mathfrak{g}/\mathfrak{p}$ is exactly the hyperplane stabilized by $Q \subset P$. Otherwise put, this is the kernel of the line in $(\mathfrak{g}/\mathfrak{p})^*$ stabilized by Q. This shows that in a point of $\mathcal{G} \times_P (P/Q) \cong \mathcal{P}T^*MN$, the subbundle H is the preimage of the kernel in TM of the line in T^*M determined by the foot point. But this is the canonical contact structure from 4.2.1.

(2) By parts (2) and (3) of Proposition 1.5.13 and the description of F above, the Cartan curvature $\kappa = \kappa^{\mathcal{C}N}$ has the property that $i_\xi \kappa = 0$ for all $\xi \in F$. Since the harmonic curvature components τ_F and ρ are defined on $\Lambda^2 F$, respectively, on $E \otimes F$, they have to vanish identically. But then vanishing of τ_E is equivalent to vanishing of the harmonic curvature, and hence to vanishing of κ by Theorem 3.1.12. Since κ also is the curvature function of $(p : \mathcal{G} \to N, \omega)$, the result follows. □

The second subbundle $E \subset H$ can also be described explicitly. Any connection ∇ in the projective class induces a principal connection on the linear frame bundle of N. Since $\mathcal{P}T^*N$ is associated to this principal bundle, it inherits a general connection. This general connection depends on the choice of ∇, but it turns out that the intersection of the horizontal distribution with the tautological subbundle H depends only on the projective class. This intersection is exactly the subbundle E. Further, the projective structure can be recovered from this subbundle. This reflects the fact that correspondence spaces form a full subcategory by part (4) of Proposition 1.5.13. More details on this can be found in section 4 of [**Čap05**].

Of course, one may ask in general whether a geometric structure on a manifold N can be interpreted as a refinement of the canonical contact structure on $\mathcal{P}T^*N$. We will study this problem for conformal structures, where the answer is given by an analog of the Fefferman construction, later in this chapter.

Consider a general Lagrangean contact structure $E \oplus F = H \subset TM$ on a smooth manifold M of dimension $2n+1$. From above we know that the subbundle $F \subset TM$ corresponds to $\mathfrak{p}/\mathfrak{q} \subset \mathfrak{g}/\mathfrak{q}$. We have already seen in 4.2.3 that the harmonic torsion τ_F is eactly the obstruction to integrability of the bundle F. Hence, existence of local twistor spaces corresponding to $\mathfrak{p} \subset \mathfrak{q}$ is equivalent to vanishing of τ_F. Suppose that this is the case, and let $\psi : U \to N$ be a local twistor space. Then for each $x \in U$, the map $T_x\psi : T_xM \to T_{\psi(x)}N$ is surjective with kernel F_x. Hence, $T_x\psi(E_x)$ is a hyperplane in $T_{\psi(x)}N$ which can be interpreted as a point $\tilde{\psi}(x) \in \mathcal{P}T^*_xN$. It is easy to show that, possibly shrinking U, one obtains a diffemorphism $\tilde{\psi}$ from U onto an open subset in $\mathcal{P}T^*N$. By construction, this is a contact diffeomorphism for the canonical contact structure on $\mathcal{P}T^*N$.

Hence, we have obtained a twistorial intepretation of Lagrangean contact structures in dimension $2n+1$ which are *semi-torsion-free* in the sense that one of the two harmonic torsion components vanishes. Locally, they are given by the projectivized cotangent bundle of a smooth manifold of dimension $n+1$ with its canonical contact structure plus a Legendrean subbundle E which is transversal to the vertical subbundle. Theorem 4.4.1 implies that such a bundle E is induced by a projective structure if and only if the harmonic curvature ρ vanishes.

4.4.3. Example: Generalized path geometries. Consider $\mathfrak{g} = \mathfrak{sl}(n+2, \mathbb{R})$ with $n \geq 2$ and let $\mathfrak{q}$ be the stabilizer of a flag consisting of a line lying in a plane in $\mathbb{R}^{n+2}$. (For $n = 1$, the situation here is identical to the one discussed in 4.4.2.) We can view $\mathfrak{q}$ as $\mathfrak{p} \cap \tilde{\mathfrak{p}}$, where $\mathfrak{p}$ is the stabilizer of the line and $\tilde{\mathfrak{p}}$ is the stabilizer of the plane. The grading of $\mathfrak{g}$ induced by $\mathfrak{q}$ can be visualized by the following block form with blocks of sizes 1, 1, and n:

$$\begin{pmatrix} \mathfrak{q}_0 & \mathfrak{q}_1^E & \mathfrak{q}_2 \\ \mathfrak{q}_{-1}^E & \mathfrak{q}_0 & \mathfrak{q}_1^V \\ \mathfrak{q}_{-2} & \mathfrak{q}_{-1}^V & \mathfrak{q}_0 \end{pmatrix}.$$

Hence, $\mathfrak{q}_{\pm 2}$ has dimension n in this case, while $\mathfrak{q}_{\pm 1}$ splits into the direct sum of an n–dimensional component and a 1–dimensional component. The gradings of $\mathfrak{g}$ corresponding to $\mathfrak{p}$ and $\tilde{\mathfrak{p}}$ are both $|1|$–gradings, and they are determined by $\mathfrak{p}_{\pm 1} = \mathfrak{q}^E_{\pm 1} \oplus \mathfrak{q}_{\pm 2}$ respectively $\tilde{\mathfrak{p}}_{\pm 1} = \mathfrak{q}^V_{\pm 1} \oplus \mathfrak{q}_{\pm 2}$. The Lie bracket $\mathfrak{q}_{-1} \times \mathfrak{q}_{-1} \to \mathfrak{q}_{-2}$ has to vanish on $\mathfrak{q}^E_{-1} \times \mathfrak{q}^E_{-1}$ by skew symmetry. One immediately computes that it also vanishes on $\mathfrak{q}^V_{-1} \times \mathfrak{q}^V_{-1}$ and induces a linear isomorphism $\mathfrak{q}^E_{-1} \otimes \mathfrak{q}^V_{-1} \to \mathfrak{q}_{-2}$. Motivated by these bracket relations, we make the following definition.

DEFINITION 4.4.3. Let M be a smooth manifold of dimension $2n+1 \geq 5$. Then a *generalized path geometry* on M is given by two subbundles $E, V \subset TM$ of rank 1 and n, respectively, such that:

(i) $E \cap V = 0$.
(ii) The Lie bracket of two sections of V is a section of $E \oplus V$.
(iii) For sections $\xi \in \Gamma(E)$ and $\eta \in \Gamma(V)$ and a point $x \in M$, the equation $[\xi, \eta](x) \in E_x \oplus V_x$ implies that $\xi(x) = 0$ or $\eta(x) = 0$.

The reason for the name "generalized path geometry" will become clear later.

As a group G with Lie algebra $\mathfrak{g}$, let us take $PGL(n+1, \mathbb{R})$, and for the parabolic subgroups we take the appropriate stabilizers in G, so we get $Q = P \cap \tilde{P}$. From 4.1.5 we know that normal parabolic geometries of type (G, P) are equivalent to classical projective structures on n–dimensional manifolds. In 4.1.3 we have seen that normal parabolic geometries of type $(G, \tilde{P})$ are equivalent to almost Grassmannian structures. We have not discussed parabolic geometries of type (G, Q) so far, since they are best described in a twistorial picture.

PROPOSITION 4.4.3. *Regular normal parabolic geometries of type (G, Q) are equivalent to generalized path geometries on smooth manifolds of dimension $2n+1$.*

PROOF. By definition, the subgroup $Q \subset G$ is represented by matrices in $GL(n+2, \mathbb{R})$ which are block upper triangular with block sizes 1, 1, and n. One easily verifies directly that $Q_0 \subset Q$ is the subgroup of block diagonal matrices. If we denote the entries of such a matrix by $a, b \in \mathbb{R}$ and $A \in GL(n, \mathbb{R})$, then the adjoint actions on $\mathfrak{q}^E_{-1}$, $\mathfrak{q}^V_{-1}$, and $\mathfrak{q}_{-2}$ are given by $a^{-1}b$, $b^{-1}A$, and $a^{-1}A$, respectively. Taking a representative matrix with $b = 1$, we conclude that the adjoint action identifies Q_0 with the group of all automorphisms of the Lie algebra $\mathfrak{q}_-$, which in addition preserve the decomposition $\mathfrak{q}_{-1} = \mathfrak{q}^E_{-1} \oplus \mathfrak{q}^V_{-1}$.

By Theorem 3.1.14, regular normal parabolic geometries of type (G, Q) are equivalent to regular infinitesimal flag structures of that type. Such a structure on a manifold M of dimension $2n+1$ first needs a subbundle $H \subset TM$ of rank $n+1$ such that the bundle of symbol algebras is locally trivial and modelled on $\mathfrak{q}_-$. From the above description of Q_0 we see that a reduction of structure group of $\mathrm{gr}(TM)$ to $Q_0 \subset \mathrm{Aut}_{\mathrm{gr}}(\mathfrak{q}_-)$ is equivalent to a decomposition $H = E \oplus V$, where E has rank 1, and V has rank n. This decomposition has to be compatible with the isomorphism from each symbol algebra to $\mathfrak{q}_-$. But the bracket on $\mathfrak{q}_-$ is completely determined by the facts that it vanishes on $\mathfrak{q}^V_{-1} \times \mathfrak{q}^V_{-1}$ and induces a linear isomorphism $\mathfrak{q}^E_{-1} \otimes \mathfrak{q}^V_{-1} \to \mathfrak{q}_{-2}$. These two conditions exactly correspond to conditions (ii) and (iii) in the definition of a generalized path geometry. □

We can also analyze the harmonic curvature components of generalized path geometries following the standard recipes. In the case $n = 2$, the complete Hasse diagram for this parabolic is computed in 3.2.16. The structure of the first two steps looks the same in higher dimensions, so, in particular, there are always three irreducible components in $H^2(\mathfrak{q}_-,\mathfrak{g})$. From the sets Φ_w in the diagrams in 3.2.16 one can see where representative bilinear maps for these three components are defined. To compute where they have their values, one has to apply the Weyl group elements to the highest weight of the adjoint representation, and there is a slight but important difference between $n=2$ and $n>2$. The first two components are always represented by maps

$$\mathfrak{q}^E_{-1}\times\mathfrak{q}_{-2}\to\mathfrak{q}^V_{-1},$$
$$\mathfrak{q}^V_{-1}\times\mathfrak{q}_{-2}\to\mathfrak{q}_0.$$

Geometrically, they correspond to a torsion $\tau_E : E\times TM/H\to V$, respectively, a curvature $\rho : V\times TM/H\to \operatorname{End}(V)$, where $H=E\oplus V$. For $n=2$, the last component is represented by maps $\Lambda^2\mathfrak{g}^V_{-1}\to\mathfrak{g}^E_{-1}$, so it corresponds to another torsion $\tau_V : \Lambda^2V\to E$. However, for $n>2$, the component is represented by maps $\Lambda^2\mathfrak{g}^V_{-1}\to\mathfrak{g}_{-2}$. This has homogeneity zero, and hence cannot correspond to a harmonic curvature component of a regular normal parabolic geometry.

At this stage, it is not clear at all how to obtain examples of generalized path geometries. The simplest source of such examples is provided by correspondence spaces of projective structures.

THEOREM 4.4.3. *Let $(N,[\nabla])$ be a classical projective structure on a smooth manifold N of dimension $n+1$. Then the projectivized tangent bundle $M:=\mathcal{P}TN$ carries a canonical generalized path geometry. Explicitly, the bundle $H\subset T\mathcal{P}(TN)$ is the tautological bundle, i.e. $\xi\in H_\ell$ if and only if $T\pi\cdot\xi$ lies in the line $\ell\subset T_{\pi(\ell)}N$. The subbundle $V\subset H$ is the vertical subbundle of $\pi:\mathcal{P}(TN)\to N$, and the line bundle $E\subset TM$ is determined by the horizontal lifts of the connections in the projective class.*

The curvature ρ of this generalized path geometry vanishes identically. If $n=2$, then also the torsion τ_V vanishes identically. On the other hand, vanishing of τ_E is equivalent to $(N,[\nabla])$ being locally projectively flat.

PROOF. Let $(p:\mathcal{G}\to N,\omega)$ be the normal parabolic geometry of type (G,P) associated to the projective structure. Similarly, as in the proof of Proposition 4.4.2, one shows the subgroup $Q\subset P$ is the stabilizer of the line corresponding to $\mathfrak{q}^E_{-1}$ in $\mathfrak{g}/\mathfrak{p}$. Since P acts transitively on the projectivization $\mathcal{P}(\mathfrak{g}/\mathfrak{p})$, we get $P/Q\cong\mathcal{P}(\mathfrak{g}/\mathfrak{p})$. Therefore, the correspondence space $\mathcal{C}N$ for $Q\subset P$ is naturally identified with $\mathcal{P}TN$. Torsion freeness of ω then implies that the curvature function has values in $\mathfrak{p}=\mathfrak{q}^V_{-1}\oplus\mathfrak{q}$. This shows that the parabolic geometry $(\mathcal{G}\to\mathcal{C}N,\omega)$ is regular and it is normal by Proposition 4.4.1. Hence, it is the canonical parabolic geometry associated to a generalized path geometry.

The facts that the subbundles V and $H=E\oplus V$ of $T\mathcal{P}TN$ are the vertical subbundle, respectively, the tautological subbundle are verified exactly as in Proposition 4.4.2. Any connection ∇ in the projective class gives rise to a horizontal lift of tangent vectors from N to TN; see 1.3.1. By linearity, the horizontal subspaces descend to a horizontal distribution on $\mathcal{P}(TN)$, thus defining a general connection on this fiber bundle; see 1.3.2. Passing to a projectively equivalent connection $\hat{\nabla}$, the horizontal lift of a vector $\eta\in T_xM$ at the point $\xi\in T_xM$ changes by a linear combination of ξ and η. In particular, if η and ξ are collinear, then the only change is by a multiple of ξ, which is killed by the projection to $\mathcal{P}(TN)$. In particular, in a point $\ell \in \mathcal{P}(TN)$ the horizontal lifts of elements of ℓ are independent of the choice of the connection from the projective class. Hence, one obtains a line subbundle of the tautological bundle, which is complementary to the vertical subbundle. From the description in 4.1.5 we conclude that the horizontal subspaces in TN for the connections in the projective class are just the images of subspaces of the form $\omega_u^{-1}(\mathfrak{p}_{-1})$. Using this, one immediately verifies that the line subbundle from above exactly corresponds to E.

The harmonic curvature components ρ and τ_V (for $n = 2$) take one or two entries from V, so they have to vanish by part (3) of Proposition 1.5.13. But then vanishing of τ_E is equivalent to vanishing of the harmonic curvature and hence to local projective flatness. □

4.4.4. Twistor spaces for generalized path geometries. The construction of correspondence space associated to projective structures in Theorem 4.4.3 suggests a way to obtain more examples for generalized path geometries. Suppose that N is an arbitrary smooth manifold, and consider the projectivized tangent bundle $M := \mathcal{P}TN$ as above. Let $V \subset H \subset TM$ be the vertical subbundle, respecitively, the tautological subbundle. These two bundles exist independently of any choice of a projective structure. Choose a linear connection ∇ on TN. Then by Theorem 4.4.3 we obtain a line subbundle $E^\nabla \subset H$ (which depends only on the projective class of ∇) that is complementary to V and defines a generalized path geometry on M.

Now suppose that $E \subset H$ is any line subbundle which is complementary to V. Then any section $\xi \in \Gamma(E)$ can be uniquely written as $\xi = \xi_1 + \xi_2$ with $\xi_1 \in \Gamma(E^\nabla)$ and $\xi_2 \in \Gamma(V)$. By construction for any point $x \in M$ we have $\xi_1(x) = 0$ if and only if $\xi(x) = 0$. Now for a section $\eta \in \Gamma(V)$ we obtain $[\xi_2, \eta] \in \Gamma(V)$ and hence $[\xi, \eta](x) \in H_x$ if and only if $[\xi_1, \eta](x) \in H_x$. This shows that the subbundles $E, V \subset TM$ satisfy the conditions of Definition 4.4.3, and thus give rise to a generalized path geometry on $M = \mathcal{P}TN$. Such a geometry on $\mathcal{P}TN$ is classically called a *path geometry* on N.

Here the word "path" has to be understood as "unparametrized curve" or "immersed 1–dimensional submanifold". A subbundle $E \subset H \subset T\mathcal{P}TN$ as above defines an integrable distribution. The fact that E is transverse to V implies that the restriction of the projection $\pi : \mathcal{P}TN \to N$ to an integral submanifold of E is an immersion. Projecting the leaves of the foliation defined by E therefore gives rise to a family of paths in N. For a point $\ell \in \mathcal{P}TN$ defined by a line $\ell \subset T_xN$, let $c \subset N$ be the projection of the leaf through ℓ. Since $E \subset H$, we see that $T_xc = \ell$. Hence, we conclude that for each point $x \in M$ and each direction $\ell \subset T_xM$, there is a unique path in the family which passes through x in direction ℓ.

Consider an immersed submanifold c in a smooth manifold N. Then this canonically lifts to an immersed submanifold $\hat{c}$ in $\mathcal{P}TN$, whose points are the lines given by the tangent spaces of c. A local regular parametrization of c canonically lifts to a regular parametrization of $\hat{c}$. Since $\hat{c}$ is a lift of c, its tangent spaces lie in H and are transverse to V. Now a family of paths in N with exactly one path through each point in each direction therefore lifts to a one–dimensional foliation of $\mathcal{P}TN$. We call the family of paths smooth if this foliation is smooth. By construction, the subbundle $E \subset T\mathcal{P}TN$ defined by such a foliation is contained in H and transverse to V, and we get

OBSERVATION 4.4.4. A path geometry on N is equivalent to a smooth family of one–dimensional immersed submanifolds in N with exactly one submanifold through each point in each direction. Such a path geometry gives rise to a generalized path geometry on $\mathcal{P}TN$.

The description of path geometries as generalized path geometries (i.e. via the bundles E and V) has the immediate advantage of generalizing to open subsets. It is not necessary to have one path through each point in each direction. It is sufficient to have pathes through each point in an open set of directions (which may depend on the point) in order to get a generalized path geometry and hence a parabolic geometry. If $n \neq 2$, this locally exhausts all generalized path geometries:

PROPOSITION 4.4.4. *Let M be a smooth manifold of dimension $2n+1$ endowed with a generalized path geometry defined by subbundles $E, V \subset TM$. If $n > 2$, then the subbundle $V \subset TM$ is automatically involutive. In the notation of 4.4.3, this means that there are always local twistor spaces for $\mathfrak{p} \supset \mathfrak{q}$. If $\psi : U \to N$ is a sufficiently small local twistor space of this type, then ψ canonically lifts to an open embedding $\tilde{\psi} : U \to \mathcal{P}TN$ such that $T\tilde{\psi}$ maps V, respectively, $E \oplus V$ to the vertical subbundle, respectively, the tautological subbundle of $\mathcal{P}TN \to N$. In particular, M is locally isomorphic to a path geometry on N.*

PROOF. Let $(p : \mathcal{G} \to M, \omega)$ be the parabolic geometry of type (G, Q) determined by the given generalized path geometry. Let $T \in \Omega^2(M, TM)$ be the torsion of ω. By Lemma 1.5.14 we have to prove that T maps $V \times V$ to V to conclude integrability of V. By Theorem 3.1.12, the lowest nonzero homogeneous component of the Cartan curvature κ of ω is harmonic. Looking at the description of the harmonic curvature in 4.4.3, we conclude that, for $n > 2$, κ is homogeneous of degree ≥ 2. Since $V \subset T^{-1}M$, we see that the restriction of T to $V \times V$ has to vanish.

Now let $\psi : U \to N$ be a local leaf space for the foliation determined by the involutive subbundle $V \subset TM$. For each $x \in U$, the tangent map $T_x\psi$ by definition induces a linear isomorphism $T_xM/V_x \to T_{\psi(x)}N$. Hence, $E_x \in T_xM$ is mapped to a line in $T_{\psi(x)}N$, which determines a point $\tilde{\psi}(x) \in \mathcal{P}TN$. Choosing a local nonvanishing section σ of E, we can write $\tilde{\psi} = q \circ T\psi \circ \sigma$, where $q : TN \setminus N \to \mathcal{P}TN$ is the natural surjection. Hence, $\tilde{\psi}$ is smooth. From the construction it is evident that $T\tilde{\psi}$ maps V to the vertical subbundle and E to the tautological subbundle of $\mathcal{P}TN \to N$. Hence, we can complete the proof by showing that $\tilde{\psi}$ has invertible tangent maps, since then it locally is an open embedding.

To do this, fix a point $x_0 \in U$. Since ψ is a surjective submersion, we can choose local coordinates $(u^0, \dots, u^{2n})$ around x_0 such that $\psi(u^0, \dots, u^{2n}) = (u^0, \dots, u^n)$. We will write ∂_i for the coordinate vector field $\frac{\partial}{\partial u^i}$. Then by construction the subbundle V is, locally around x_0, spanned by $\{\partial_{n+1}, \dots, \partial_{2n}\}$. Further, we take a locally nonvanishing smooth section $\xi = \xi^i \partial_i$ of the subbundle $E \subset TM$. We may choose the coordinates in such a way that $\xi(x_0) = \partial_0$. Evidently, $[\partial_j, \xi] = \frac{\partial \xi^i}{\partial u^j}\partial_i$. The assumption on the Lie bracket between sections of E and V thus implies that that matrix $(\frac{\partial \xi^i}{\partial u^j}(x_0))$ with $i = 1, \dots, n$ and $j = n+1, \dots, 2n$ is invertible.

By construction, we can use $(u^0, \dots, u^n)$ as local coordinates around $\psi(x_0)$ on N. Further, $T_{x_0}\psi(E_{x_0}) \subset T_{\psi(x_0)}N$ is the line spanned by ∂_0, so the one–form du^0 restricts to a nonzero functional on this line. Now consider the set of all lines in tangent spaces with foot point contained in our chart, to which du^0 has nontrivial restriction. This is an open neighborhood of $T_{x_0}\psi(E_{x_0})$ in $\mathcal{P}TN$, on which we get local coordinates $(u^0,\dots,u^n,a^1,\dots,a^n)$ which are characterized by $du^i|_\ell = a^i(\ell)du^0|_\ell$. (This is an analog of the standard inhomogenous coordinates on $\mathbb{R}P^n$.)

We can easily compute the mapping $\tilde{\psi}$ in these coordinates. Namely, by construction $T\tilde{\psi}\cdot\xi = \xi^i\partial_i$, where now the indices run only from 0 to n. But this means that, in our coordinates,

$$\tilde{\psi}(u^0,\dots,u^{2n}) = (u^0,\dots,u^n,\xi^1/\xi^0,\dots,\xi^n/\xi^0).$$

In the point x_0, we have $\xi_0(x_0) = 1$ and $\xi^i(x_0) = 0$ for $i > 0$. This implies that the partial derivative of ξ^i/ξ^0 in direction of u^j in x_0 equals $\frac{\partial\xi^i}{\partial u^j}(x_0)$. Hence, from above we conclude that $T_{x_0}\tilde{\psi}$ has the block form $\left(\begin{smallmatrix}\mathbb{I} & B\\ 0 & A\end{smallmatrix}\right)$, where $\mathbb{I}$ is the unit matrix of size $n+1$ and A is an invertible $n\times n$–matrix. Hence, $T_{x_0}\tilde{\psi}$ is invertible. □

Identifying a generalized path geometry on M locally with a path geometry on $\mathcal{P}TN$, we can next ask when M is locally isomorphic to a correspondence space. Otherwise put, this is the question when a path geometry on $\mathcal{P}TN$ comes from a projective structure $[\nabla]$ on N. In view of our description of the harmonic curvature, this immediately follows from Theorem 4.4.1.

COROLLARY 4.4.4. *The paths of a path geometry on a manifold N can be realized as the unparametrized geodesics of a linear connection ∇ on TN if and only if the harmonic curvature ρ of the canonical Cartan connection determined by the path geometry vanishes identically.*

4.4.5. Generalized path geometries from Grassmannian structures. In our discussion of generalized path geometries in 4.4.3 we have realized the relevant parabolic subalgebra $\mathfrak{q}\subset\mathfrak{g} = \mathfrak{sl}(n+2,\mathbb{R})$ as the intersection $\mathfrak{p}\cap\tilde{\mathfrak{p}}$, where $\mathfrak{p}$ and $\tilde{\mathfrak{p}}$ are the stabilizers of a line, respectively, a plane in the standard representation $\mathbb{R}^{n+2}$ of $\mathfrak{g}$. Besides the inclusion $\mathfrak{q}\hookrightarrow\mathfrak{p}$, which gives rise to the correspondence and twistor spaces studied in 4.4.3 and 4.4.4 above, we also have the inclusion $\mathfrak{q}\hookrightarrow\tilde{\mathfrak{p}}$. Here it is better to choose the group $G = SL(n+2,\mathbb{R})$, to take the subgroups P and $\tilde{P}$ to be the stabilizers of the appropriate subspaces, and to put $Q := P\cap\tilde{P}$. This causes only minimal changes compared to the discussion in 4.4.3, so regular normal parabolic geometries of type (G,Q) are essentially generalized path geometries.

On the other hand, by 4.1.3 normal parabolic geometries of type $(G,\tilde{P})$ are almost Grassmannian structures on manifolds of dimension $2n$. Such a structure on a smooth manifold $\tilde{N}$ comes with two auxiliary bundles $\tilde{E},\tilde{F}\to\tilde{N}$ of rank 2 and n together with an isomorphism $\Phi : \tilde{E}^*\otimes\tilde{F}\to T\tilde{N}$. The harmonic curvature of such a structure consists of a torsion τ_2, which is a section of $S^2\tilde{E}\otimes\tilde{E}^*\otimes\Lambda^2\tilde{F}^*\otimes\tilde{F}$ and a curvature ρ_1, which is a section of $\Lambda^2\tilde{E}\otimes S^2\tilde{F}^*\otimes\mathfrak{sl}(\tilde{F})$.

PROPOSITION 4.4.5. *Let $\Phi : \tilde{E}^*\otimes\tilde{F}\to T\tilde{N}$ be an almost Grassmannian structure on a manifold $\tilde{N}$ of dimension $2n$. Then the correspondence space $\mathcal{C}\tilde{N}$ for $\mathfrak{q}\subset\tilde{\mathfrak{p}}$ is the projectivization $\mathcal{P}(\tilde{E})$ of the rank two bundle $\tilde{E}\to\tilde{N}$.*

The normal parabolic geometry of type (G,Q) on $\mathcal{P}(\tilde{E})$ is regular, if and only if the given structure on $\tilde{N}$ is torsion free, i.e. Grassmannian rather than almost Grassmannian. Hence, in the Grassmannian case, $\mathcal{P}(\tilde{E})$ naturally carries a generalized path geometry. The torsion T of this geometry vanishes identically, while the curvature ρ is a natural lift of the Grassmannian curvature ρ_1.

Explicitly, the generalized path geometry on $\mathcal{P}(\tilde{E})$ is given as follows. The subbundle $E \subset T\mathcal{P}(\tilde{E})$ is the vertical subbundle of $\pi : \mathcal{P}(\tilde{E}) \to \tilde{N}$. For a line $\ell \subset \tilde{E}_x$, the subbundle $H_\ell \subset T\mathcal{P}(\tilde{E})$ consits of all tangent vectors ξ such that $T_\ell\pi\cdot\xi \in T_x\tilde{N} \cong L(\tilde{E}_x, \tilde{F}_x)$ vanishes on the line $\ell \subset \tilde{E}_x$. The subbundle $V \subset H$ comes from the horizontal lifts of torsion–free connections preserving the Grassmannian structure.

PROOF. This is similar to the proofs of Proposition 4.4.2 and Theorem 4.4.3, so we sktech the proof briefly, emphasizing mainly the differences. The group $\tilde{P}_0$ is isomorphic to $S(GL(2,\mathbb{R}) \times GL(n,\mathbb{R}))$ and the bundle $\tilde{E} \to \tilde{N}$ corresponds to the standard representation of the first factor. The subgroup $Q \subset \tilde{P}$ can be characterized as the stabilizer of the line through the first basis vector in this representation, which leads to the description of the correspondence space. It is also clear then that the line bundle E on $\mathcal{C}\tilde{N}$ is the vertical subbundle of $\mathcal{C}\tilde{N} \to \tilde{N}$.

If one starts from a torsion–free parabolic geometry of type $(G, \tilde{P})$, then the fact that $\mathfrak{q}_{-2} \cap \tilde{\mathfrak{p}} = \{0\}$ implies regularity of the parabolic geometry of type (G,Q) on $\mathcal{P}(\tilde{E})$. Conversely, suppose that the parabolic geometry of type (G,Q) on $\mathcal{P}(\tilde{E})$ is regular. Since $\mathfrak{q}^E_{-1} \subset \tilde{\mathfrak{p}}$, vanishing of the torsion τ_E follows directly from part (3) of Proposition 1.5.13. Hence, the curvature function has to be homogeneous of degree ≥ 3, and the only nonzero part of homogeneous degree 3 maps $\mathfrak{q}^V_{-1} \otimes \mathfrak{q}_{-2}$ to $\mathfrak{q}_0$. Since homogeneous components of degree ≥ 4 automatically have values in $\mathfrak{q} \subset \mathfrak{p}$, the whole curvature function has to have values in $\mathfrak{p}$. Hence, the original geometry on $\tilde{N}$ must be torsion free.

Remaining in the torsion–free case, description of the subbundle H follows immediately from the fact that $\mathfrak{q}^V_{-1} \subset \tilde{\mathfrak{p}}$ are those maps which annihilate the line in $\tilde{E}$ corresponding to $\mathfrak{q}^E_{-1}$. For the subbundle $V \subset H$ one observes that a torsion–free connection preserving the Grassmannian structure induces linear connections on $\tilde{E}$ and $\tilde{F}$, and then proceeds as in the proofs of Proposition 4.4.2 and Theorem 4.4.3. □

Concerning twistor spaces, the situation with the inclusion $\mathfrak{q} \hookrightarrow \tilde{\mathfrak{p}}$ is rather easy. For a regular normal parabolic geometry $(p : \mathcal{G} \to M, \omega)$ of type (G,Q), the subbundle in TM corresponding to $\tilde{\mathfrak{p}}/\mathfrak{q} \subset \mathfrak{g}/\mathfrak{q}$ is just the line bundle $E \subset TM$. As a line bundle, E is always integrable, so we always have local twistor spaces. In the case of a path geometry (i.e. $M = \mathcal{P}TN$), the integral submanifolds of E project to the paths defining the geometry. Hence, a local leaf space for E locally parametrizes the paths defining the geometry. If $\tilde{N}$ is such a local leaf space, then the question of whether M is locally isomorphic to the correspondence space $\mathcal{C}\tilde{N}$ can be naturally phrased as the question of whether the generalized path geometry on M descends to a Grassmannian structure on $\tilde{N}$. In the case of path geometries, the question is thus under which conditions the spaces which locally parametrize the paths of the geometry inherit a canonical Grassmannian structure.

By Theorem 1.5.14 sufficiently small parameter spaces inherit such a structure if and only if $i_\xi\kappa = 0$ for all sections $\xi \in \Gamma(E)$. By Theorem 4.4.1, this is equivalent to vanishing of the torsion T of the generalized path geometry.

A nice application of generalized path geometries is the geometric theory of systems of second order ODEs. Let us start with a smooth manifold $\underline{N}$ of dimension $n-1$, for example, an open subset of $\mathbb{R}^{n-1}$. Then a system of second order ODEs on smooth functions on $\underline{N}$ can be rephrased as a family of paths in an open subset N of

$\underline{N} \times \mathbb{R}$, with the paths representing the graphs of the solutions of the initial system. If this initial system is of order two, then there will be a unique path through each point in each direction (in some open set of directions), so we obtain a path geometry on N. Of course, any invariant of this path geometry is an invariant of the corresponding system of ODEs (with respect to point transformations), so in particular, it is independent of choices of coordinates, etc.

In this language, the homogeneous model which is given by projective lines in $\mathbb{R}P^n$ corresponds to the trivial system $y'' = 0$. In particular, the torsion τ_E and the curvature ρ associated to the path geometry in 4.4.3 are complete obstructions to equivalence of a given system of second order ODEs to $y'' = 0$. Otherwise put, vanishing of these two invariants is equivalent to the fact that the solutions of the given system of ODEs are straight lines in appropriate coordinates.

The correspondence space $\mathcal{C}N$ of a projective structure $(N, [\nabla])$ as in Theorem 4.4.3 represents a geodesic equation. The characterization of these correspondence spaces as discussed in the end of 4.4.4 thus leads to a characterization of systems of second order ODEs which (up to switching from the solution to its graph) can be written as geodesic equations of torsion–free connections; see [**Čap05**].

Local twistor spaces in the "other direction" (i.e. corresponding to $\mathfrak{q} \hookrightarrow \tilde{\mathfrak{p}}$) represent local spaces of solutions of the given system of ODEs. Hence, the characterization of correspondence spaces amounts to characterizing the existence of a canonical Grassmannian structure on local spaces of solutions of a given system. As we have noted above, this is equivalent to vanishing of the torsion τ_E of the corresponding path geometry. By a result of D. Grossman (see [**Gro00**]) for generic systems of that type, the curvature of the Grassmannian structure separates points in the local space of solutions. These curvature quantities thus lift to enough constants of the motion on $\mathcal{P}TN$ to explicitly solve the given system. This is a very remarkable result since it asserts explicit solvability for the system provided that it is not too symmetric.

4.4.6. Twistor correspondences. Twistor correspondences can be obtained by combining the constructions of correspondence and twistor spaces. On the level of the homogeneous model, the basic situation giving rise to such a correspondence is given by a semisimple Lie group G together with two parabolic subgroups P and $\tilde{P}$ which contain the same Borel subgroup. Then $Q := P \cap \tilde{P}$ is a parabolic subgroup, and one has the natural projections $\pi : G/Q \to G/P$ and $\tilde{\pi} : G/Q \to G/\tilde{P}$. They are smooth fiber bundles with fibers the generalized flag manifolds P/Q, respectively, $\tilde{P}/Q$; see Lemma 4.4.1.

On the other hand, the map $\pi \times \tilde{\pi} : G/Q \to G/P \times G/\tilde{P}$ is immediately seen to be an embedding, so G/Q defines a *correspondence* between G/P and $G/\tilde{P}$. Correspondences of this type are the subject of the book [**BE89**]. The most classical example is given by $G = SL(4, \mathbb{C}) \cong Spin(6, \mathbb{C})$, P the stabilizer of a line, and $\tilde{P}$ the stabilizer of a plane in $\mathbb{C}^4$, which contains the line stabilized by P. Then $G/P = \mathbb{C}P^3$ and $G/\tilde{P}$ is the Grassmannian $\mathrm{Gr}_2(\mathbb{C}^4)$ of two–planes in $\mathbb{C}^4$. The intersection $Q = P \cap \tilde{P}$ is the stabilizer of the flag defined by the two subspaces, and G/Q is the flag manifold $F_{1,2}(\mathbb{C}^4)$ of lines contained in planes in $\mathbb{C}^4$. The canonical projections onto $\mathbb{C}P^3$ and $\mathrm{Gr}_2(\mathbb{C}^4)$ are given by forgetting one of the two subspaces. The product to these two projections embeds $F_{1,2}(\mathbb{C}^4)$ as the incidence subspace into $\mathbb{C}P^3 \times \mathrm{Gr}_2(\mathbb{C}^4)$. This shows explicitly that we have obtained a correspondence.

Geometrically, one views $\mathrm{Gr}_2(\mathbb{C}^4)$ as compactified, complexified Minkowski space, i.e. the homogeneous model of four–dimensional complex conformal geometry. Via the twistor correspondence, this is related to the projective geometry of $\mathbb{C}P^3$.

Passing to the case of curved geometries, the situation loses its symmetry. Let us start with a parabolic of type $(G, \tilde{P})$ on M, and form the correspondence space $\tilde{\pi} : \mathcal{C}M \to M$ with respect to $Q \subset \tilde{P}$. From Proposition 1.5.13 we know that M and $\mathcal{C}M$ have the same curvature function κ. In terms of this curvature function, we can immediately see the condition for existence of a local twistor space with respect to $P \supset Q$; see Lemma 1.5.14. One may form local (i.e. on open subsets of $\mathcal{C}M$) twistor spaces if and only if $\kappa(u)(\mathfrak{p}, \mathfrak{p}) \subset \mathfrak{p} + \tilde{\mathfrak{p}}$ for all u.

This condition is not sufficient to get a local twistor correspondence. For this, we need twistor spaces for open subsets of the form $\tilde{\pi}^{-1}(U) \subset \mathcal{C}M$, where $U \subset M$ is a sufficiently small open subset. One may construct such twistor spaces by gluing together twistor spaces defined on small open subsets of $\mathcal{C}M$, but the problem is to prove that the resulting spaces are Hausdorff. In the most classical case of four–dimensional conformal structures (complex or real split signature), the Hausdorff property is proved using geodesic convexity. For the other known examples of twistor correspondences, e.g. the ones for Grassmannian structures, one has to use analogous conditions. Both of these cases will be discussed in 4.4.7 below.

Since these types of questions are outside the main line of this book, we will not discuss them, but simply assume that we have given a global twistor correspondence. Hence, we assume that the local intebrability conditions are satisfied and that apart from the canonical projection $\tilde{\pi} : \mathcal{C}M \to M$, we also have a projection $\pi : \mathcal{C}M \to N$, whose fibers are the integral submanifolds of the distribution on $\mathcal{C}M$ induced by $\mathfrak{g}/\mathfrak{p} \subset \mathfrak{g}/\mathfrak{q}$. As in the homogeneous case, this defines a correspondence, at least on the local level. The kernels of $T\pi$ and $T\tilde{\pi}$ by construction have zero intersection in each point of $\mathcal{C}M$, so $\pi \times \tilde{\pi} : \mathcal{C}M \to N \times M$ is an immersion. For many applications, one still has to restrict this correspondence to appropriate open subsets of M. If $U \subset M$ is open, then we consider $\mathcal{C}U = \tilde{\pi}^{-1}(U)$ and $N_U := \pi(\mathcal{C}U)$.

Having a correspondence consisting of $\tilde{\pi} : \mathcal{C}M \to M$ and of $\pi : \mathcal{C}M \to N$, we can view each of the two spaces M and N as parametrizing a family of distinguished immersed submanifolds in the other space. For $y \in N$ consider the fiber $\pi^{-1}(y)$, which is a smooth submanifold in $\mathcal{C}N$. By construction, $\tilde{\pi}$ restricts to an immersion on $\pi^{-1}(y)$, so $A_y := \tilde{\pi}(\pi^{-1}(y)) \subset M$ is an immersed submanifold in M, of dimension $\dim(P/Q)$. In the same way, a point $x \in M$ gives rise to the immersed submanifold $B_x := \pi(\tilde{\pi}^{-1}(x))$ in N of dimension $\dim(\tilde{P}/Q)$.

The main application of such correspondences is the Penrose transform; see [**BE89**] for the homogeneous version. Following the fundamental article [**EPW81**], this transform is formulated in terms of sheaf cohomology as follows. Take some sheaf $\mathcal{V}$ on N, and consider the pullback sheaf $\pi^{-1}\mathcal{V}$ on $\mathcal{C}M$. Then the projection $\tilde{\pi} : \mathcal{C}M \to M$ is a locally trivial fiber bundle, so one can use (variants of) a Serre spectral sequence to compute the sheaf cohomology of $\pi^{-1}\mathcal{V}$ in terms of data on M.

To use this in practice, one has to restrict to open subsets $U \subset M$, which are chosen in such a way that the fibers of π are contractible in a nice way compatible with π. This ensures that the sheaf cohomology of $\mathcal{V}$ is isomorphic to the sheaf cohomology of $\pi^{-1}\mathcal{V}$; see [**BE89**, Chapter 7]. On the other hand, the spectral sequence leads to maps between sheafs on M (or on U). For appropriate choices of $\mathcal{V}$, it turns out that the sheaves in this process are sections of natural bundles and the operators are natural differential operators. In this way, the transform interepretes cohomologies of complexes of natural operators (which in many cases of interest boil down to kernels and images of such operators) in terms of sheaf cohomology groups on N.

The main examples of the Penrose transform are actually set up in the holomorphic category. Our setup works in this category without problems, so we assume that M is equipped with a holomorphic geometry which satisfies the necessary local integrability conditions. Further, we assume that the local twistor spaces (which naturally are complex manifolds) fit together to define a global holomorphic correspondence consisting of $\tilde{\pi} : \mathcal{C}M \to M$ and $\pi : \mathcal{C}M \to N$. For the sheaf $\mathcal{V}$ one uses holomorphic sections of a holomorphic vector bundle $E \to N$. If this bundle is natural (i.e. depends only on the structure of a complex manifold on N), then the pullback bundle $\pi^*E \to \mathcal{C}M$ is a natural bundle on $\mathcal{C}M$, since it was constructed from the given geometric structure. Restricting to an appropriate subset U, the cohomologies of the sheafs of holomorphic sections of E and of π^*E are isomorphic. Next, one can restrict π^*E to each of the fibers of $\tilde{\pi} : \mathcal{C}U \to U$, and each of these fibers is isomorphic to the generalized flag manifold $\tilde{P}/Q$.

Now it turns out that each of these restrictions is a homogeneous vector bundle, and one may use a relative Bernstein–Gelfand–Gelfand resolution to resolve the sheaf of local holomorphic sections of this bundle. This has the advantage that the spaces showing up in the resolution are local holomorphic sections of bundles induced by irreducible representations. The first step in the resulting spectral sequence is given by sheaf cohomology of U with coefficients in direct images of these sheafs. These direct images can be computed by a slight generalization of the Bott–Borel–Weil theorem (Theorem 3.3.8). In particular, they are nonzero in at most one degree, so the spectral sequence degenerates in a controlled way to a sequence of differential operators. More information on the homogeneous version of the transform can be found in [**BE89**]; examples of curved versions of the transform are described e.g. in [**Pe76, Ea90, BE90, LeB90**].

4.4.7. Example: Grassmannian structures. Generalizing the case of anti–self–dual conformal structures in dimension four (see below), this example of a twistor correspondence was developed (in the holomorphic category) in [**BaE91**]. Since we have discussed both sides of this correspondence individually in 4.4.3–4.4.5, we already have many necessary results at our disposal. Also, the notation in our discussion of twistor correspondences in 4.4.6 above was chosen in such a way, that in this example we can keep most of the notations from 4.4.3–4.4.5. In particular, this applies to the group $G = SL(n+2, \mathbb{R})$ and the parabolic subgroups P, $\tilde{P}$, and $Q = P \cap \tilde{P}$.

Let us start with a smooth manifold M of dimension $2n$ endowed with a Grassmannian structure $\Phi : \tilde{E}^* \otimes \tilde{F} \xrightarrow{\cong} TM$, where $\tilde{E}$ and $\tilde{F}$ are auxilliary bundles of rank two and n, respectively. For the first part, we assume $n > 2$, the case $n = 2$ will be discussed below. We have seen in Proposition 4.4.5 that the correspondence space $\mathcal{C}M$ corresponding to $Q \subset \tilde{P}$ is the projectivization $\mathcal{P}E$ of the auxiliary bundle E and the fiber of $\tilde{\pi} : \mathcal{C}M \to M$ is $\mathbb{R}P^1$. Integrability of the Grassmannian structure was necessary there to obtain a regular Cartan geometry and hence a generalized path geometry on $\mathcal{C}M$. In the same proposition, we have seen that this generalized path geometry is torsion free, while its curvature ρ vanishes if and only if the Grassmannian structure on M is locally flat.

Proposition 4.4.4 then implies, that there are always local twistor spaces corresponding to $P \supset Q$. Locally, we therefore may always view $\mathcal{C}M$ as a path geometry on a twistor space N. However, by the last part of 4.4.4 this path geometry is not induced by a projective structure on N, unless the Grassmannian structure on M is locally flat. (In the latter case, we are locally in the situation of the homogeneous double fibration of G/Q over G/P and $G/\tilde{P}$ as discussed in 4.4.6.)

Things can be easily made more explicit using the description of the correspondence as one of the spaces parametrizing immersed submanifolds in the other space. On the one side, this is fairly evident. The path geometry on N is given by a family of one–dimensional immersed submanifolds, which are the projections of the integral submanifolds of the distribution $\mathfrak{q}^E_{-1} = \tilde{\mathfrak{p}}/\mathfrak{q} \subset \mathfrak{g}/\mathfrak{q}$. Hence, locally the paths are exactly the projections of the fibers of $\tilde{\pi} : \mathcal{C}M \to M$, and M locally is the space of paths in the path geometry.

The interpretation on the other side of the correspondence is a bit more complicated. The fibers of the projections $\pi : \mathcal{C}M \to N$ are leaves of the integrable distribution $\mathcal{V}$ on $\mathcal{C}M$ induced by $\mathfrak{q}^V_{-1} = \mathfrak{p}/\mathfrak{q} \subset \mathfrak{g}/\mathfrak{q}$. Fix a point $\ell \in \mathcal{C}M$ which lies over $x \in M$, so ℓ is a line in $\tilde{E}_x$. Then we have noted in Proposition 4.4.5 that the image of $\mathcal{V}_\ell \subset T_\ell\mathcal{C}M$ in $\mathcal{T}_xM \cong L(\tilde{E}_x, \tilde{F}_x)$ is given by the n–dimensional subspace of all maps which vanish on $\ell \subset \tilde{E}_x$. Fixing ℓ, the projection of a local integral submanifold for $\mathcal{V}$ therefore provides a distinguished n–dimensional immersed submanifold $A_\ell \subset M$ (defined locally around x), whose tangent space in x is formed by the maps vanishing on ℓ. (This is a generalization of the concept of distinguished curves which we will study in Section 5.3 below.)

Through each point $x \in M$, this construction provides us with a distinguished family of (local) immersed n–dimensional submanifolds which are parametrized by $\mathcal{P}(\tilde{E}_x) \cong \mathbb{R}P^1$. The tangent spaces of these distinguished submanifolds in some point y form the natural family of n–dimensional generators of the canonical cone in $T_yM = L(\tilde{E}_y, \tilde{F}_y)$ formed by the rank one maps. These distinguished immersed submanifolds exist only for Grassmannian (as opposed to almost Grassmannian) structures), since only for those the distribution on $\mathcal{C}M$ defined by $\mathfrak{p}/\mathfrak{q}$ is involutive; see 4.4.8 below.

This also indicates, how properties like geodesic convexity can be used to ensure existence of large enough twistor spaces. Like all parabolic geometries, Grassmannian structures come with a family of distinguished curves, which we will study in Section 5.3. Now from the construction of these distinguished curves we shall see, that the distinguished n–dimensional immersed submanifolds can be viewed as the unions of the images of some of the distinguished curves. Now the local twistor space N (locally) parametrizes the integral submanifolds. Hence, the question of whether one can form a twistor space which is Hausdorff boils down to questions about the local behavior of these integral submanifolds. Since they are swept out by distinguished curves, appropriate conditions on distinguished curves can be used to control this behavior.

The case $n = 2$. Let us briefly discuss this case which leads to twistor theory for anti–self–dual (complex or split signature real) conformal structures in dimension 4. This was the first version of twistor theory, originally introduced in [**Pe76**]. On the way, we will also describe the anonalogs of the results from 4.4.5 in this special case. We have discussed the equivalence between almost Grassmannian structures for $n = 2$ and split signature conformal structures in 4.1.4. We have also noted there that the two basic harmonic curvature quantities, which are both of homogeneity two, are the self–dual and the anti–self–dual part of the Weyl curvature.

The structure of the harmonic curvature for generalized path geometries in dimension five has been discussed in 4.4.3. There are three basic curvature quantities represented by maps

$$\mathfrak{q}_{-1}^{V} \times \mathfrak{q}_{-2} \to \mathfrak{q}_0,$$
$$\mathfrak{q}_{-1}^{E} \times \mathfrak{q}_{-2} \to \mathfrak{q}_{-1}^{V},$$
$$\Lambda^2 \mathfrak{q}_{-1}^{V} \to \mathfrak{q}_{-1}^{E},$$

and corresponding to the curvature ρ and the torsions τ_E and τ_V. Following the general principles it is easy to see that the torsion τ_V is exactly the obstruction against integrability of the subbundle V in the tangent bundle and hence against forming local twistor spaces corresponding to $\mathfrak{p} \supset \mathfrak{q}$.

Suppose that M is a holomorphic or real split signature conformal structure. By 4.4.3, the two bundles $\tilde{E}$ and $\tilde{F}$ can be interpreted as the two bundles of half spinors, so the correspondence space $\mathcal{C}M$ (corresponding to $\mathfrak{q} \subset \tilde{\mathfrak{p}}$) is a projectivized spinor bundle. To have local twistor spaces corresponding to $\mathfrak{p} \supset \mathfrak{q}$ one has to require that the torsion τ_V of $\mathcal{C}M$ vanishes. Using that the curvatures of M and $\mathcal{C}M$ are the same, one proves (either directly or as a simple consequence of Theorem 4.4.1) that this is equivalent to vanishing of the self–dual part of the Weyl curvature of the original conformal structure.

The distinguished cones in the tangent spaces of M in this case simply are the cones of isotropic vectors. The two families of planes contained in these cones (corresponding to maps vanishing on a given line in $\tilde{E}_x$, respectively, maps having values in a given line in $\tilde{F}_x$) are known in twistor theory as alpha–planes and beta–planes. Hence anti–self–duality ensures that one finds local immersed submanifolds whose tangent spaces are alpha–planes, and the twistor space locally parametrizes these integral submanifolds. The alpha–planes are swept out by (null–)geodesics (which depend only on the conformal class up to parametrization), and one may directly use geodesic convexity to prove existence of Hausdorff local twistor spaces.

4.4.8. Cone structures and generalized conformal structures. In the twistor correspondence for Grassmannian structures discussed in 4.4.7 above, cones in the individual tangent spaces play a central role. This has motivated the study of a family of geometric structures, which, in particular, contains parabolic geometries corresponding to $|1|$–gradings. This approach has been put forward by S. Gindikin and A. Goncharov; see [**Gi90, Gon87**]. We first outline this interpretation of almost Grassmannian structures and then briefly sketch the generalizations.

Recall that a *cone* in a vector space W is a subset which is stable under multiplication by positive scalars. Fixing a cone $C \subset W$, one gets a closed subgroup $GL_C(W) \subset GL(W)$ by looking at those maps which preserve the cone C. Now let M be a smooth manifold of dimension $\dim(W)$. Then a *cone–structure* modelled on C on M is given by choosing, for each $x \in M$, a cone $C_x \subset T_xM$ in such a way that there are local trivializations of TM which map each T_xM to W and each C_x to C. Evidently, such a structure is equivalent to a reduction of structure group to the subgroup $GL_C(W) \subset GL(W)$. There is an evident flat model for a cone structure, namely W with the cone structure $W \times C \subset W \times W \cong TW$.

For real almost Grassmannian structures, one has to take $W := L(\mathbb{R}^p, \mathbb{R}^q)$ and $C \subset W$ to be the cone formed by all linear maps of rank one. It then turns out that $GL_C(V) = S(GL(p, \mathbb{R}) \times GL(q, \mathbb{R}))$, so this is exactly the subgroup G_0 used in 4.1.3. Hence, we see that the corresponding cone–structures are exactly the almost Grassmannian structures.

Now the cone $C \subset W$ has two distinguished families of generators. Namely, for each hyperplane $A \subset \mathbb{R}^p$, we get a q–dimensional linear subspace contained in C, which consists of the maps whose kernel is A. Thus, we obtain a decomposition of C into a family Σ_1 of q–dimensional subspaces parametrized by $\mathbb{R}P^{p*}$. Likewise, given a line $\ell \subset \mathbb{R}^q$, we obtain a p–dimensional subspace contained in C, which consists of the maps with image contained in ℓ. This gives us a decomposition of C into a family Σ_2 of p–dimensional subspaces parametrized by $\mathbb{R}P^q$. Clearly, choosing one subspace from each of the two families, the two spaces always intersect in a line. Moreover, the spaces in the two families are exactly the maximal linear subspaces contained in the cone C.

The immediate advantage of this point of view is that it gives rise to two integrability conditions. Namely, we call a cone structure Σ_i–integrable if for each $x \in M$ each maximal subspace $A \subset T_xM$ contained in Σ_i, there is a unique local immersed submanifold S with $x \in S$ and $T_xS = A$ such that all tangent spaces of S lie in Σ_i. If one of these integrability conditions is satisfied, then one can construct local twistor spaces as local parameter spaces for the integral submanifolds. In the case of Grassmannian structures integrability of each the two families of generators of the cones is equivalent to vanishing of the corresponding component of the harmonic torsion, respectively, curvature; see [**Gon87**].

In [**Gi90**], the author proposes a general approch to cone structures and integrability. As a specific example, he discusses cone structure related to Hermitian symmetric spaces. The modelling cone used for these structures is the cone formed by the orbit of the highest weight line in the representation inducing the tangent bundle. In this way, such cone structures can be viewed as curved analogs of Hermitian symmetric spaces. The resulting geometries are called *generalized conformal structures.* They are exactly the underlying structures of parabolic geometries corresponding to $|1|$–gradings as discussed in Section 4.1.

In [**Gon87**] the author considers a certain type of geometric structures which he calls *Frobenius structures.* These are given by associating to each point x of a manifold M of dimension n a family of k–dimensional subspaces in the tangent space T_xM which satisfies a smoothness condition. He then develops a theory of prolongations for such structures and of obstructions against local isomorphisms. In these questions, cohomology groups play a central role. This theory is then applied to study generalized conformal structures. It is shown that the cohomology groups in the sense of Frobenius structures that show up in these cases coincide with the Lie algebra cohomology of the nilradical of the parabolic with coefficients in the adjoint representation. These cohomology groups are then computed using Kostant's version of the Bott–Borel–Weil theorem. In this way, Goncharov arrives at an independent proof for the existence of canonical Cartan connections as well as a description of their fundamental curvature quantities.

4.4.9. Twistor spaces II. Since we have developed the basic techniques for correspondence spaces and twistor spaces in the realm of general Cartan geometries, we can apply them to arbitrary subgroups in a parabolic subgroup. This leads to a second version of twistor theory, which probably is even better known than the one we have discussed so far. The prototypical example is twistor theory for self–dual conformal Riemannian manifolds in dimension 4, which generalizes to quaternionic structures.

We describe the version for quaternionic structures, so we consider the group $G := PSL(n+1,\mathbb{H})$ and let P be the stabilizer of the quaternionic line in $\mathbb{H}P^n$, which is generated by the first vector in the standard basis of $\mathbb{H}^{n+1}$; see 4.1.8. Hence, we are dealing with a different real form of the complexification of the example studied in 4.4.7 above. The parabolic subalgebra defining P corresponds to the second simple root and in 4.4.7 we obtained the correspondence space by looking at the parabolic associated to the first two simple roots. There is no parabolic subalgebra associated to these two roots in the real form $\mathfrak{sl}(n+1,\mathbb{H})$ of $\mathfrak{sl}(2n+2,\mathbb{C})$, since the first simple root is compact. Still one can imitate the characterization of the smaller parabolic subgroup as the stabilizer of a line in an appropriate representation; compare also with 3.2.8 and 3.2.11.

According to the general principles, the representation to consider is an irreducible representation of P, whose highest weight is a multiple of the first fundamental weight. The natural candidate is the three–dimensional representation of P on the space $\mathcal{Q}^{\mathfrak{g}_{-1}} \subset L_{\mathbb{R}}(\mathfrak{g}_{-1},\mathfrak{g}_{-1})$, which defines the almost quaternionic structure on $\mathfrak{g}_{-1} \subset \mathfrak{g}$. Using matrix representations as in 4.1.8, elements of P correspond to matrices of the form $\begin{pmatrix} q & v^t \\ 0 & \Phi \end{pmatrix}$ with $q \in \mathbb{H}$, $\Phi \in GL(n,\mathbb{H})$ and $v \in \mathbb{H}^n$ such that $|q|^4 \det_{\mathbb{R}}(\Phi) = 1$. Via the adjoint action, the subspace $\mathcal{Q}^{\mathfrak{g}_{-1}}$ can be identified with elements of $\mathfrak{g}$ of the form $\begin{pmatrix} p & 0 \\ 0 & 0 \end{pmatrix}$ with $p \in \operatorname{im}(\mathbb{H})$. Correspondingly, the representation of P on $\mathcal{Q}^{\mathfrak{g}_{-1}}$ is induced by the adjoint action and hence is given by $p \mapsto qpq^{-1}$ (which visibly factors to P).

From 4.1.8 we see that this action defines a homomorphism $P \to SO(3)$ whose kernel consists of the matrices with $q \in \mathbb{R}$. It turns out that it is better to consider the stabilizer of a ray in $\mathcal{Q}^{\mathfrak{g}_{-1}}$ rather than a line, and since we have an invariant inner product, we can as well take the stabilizer of a single vector. For this vector let us choose the matrix with $p = i$. Viewing $\mathbb{C}$ as a subspace of $\mathbb{H}$ (via i), we can write elements of $\mathbb{H}$ uniquely as $q = z + wj$ with $z, w \in \mathbb{C}$ and the product is fixed by the rule $jw = \overline{w}j$. The standard conjugation on $\mathbb{H}$ takes the form $\overline{z+wj} = \overline{z} - j\overline{w} = \overline{z} - wj$. Since we already know that elements of $\mathbb{R}$ act trivially on $\mathcal{Q}^{\mathfrak{g}_{-1}}$, we may assume that $|q| = 1$ and hence $q^{-1} = \overline{q}$. Using this, we obtain

$$(z+wj)i(\overline{z} - wj) = i(|z|^2 - |w|^2) - 2izwj,$$

which immediately implies that the stabilizer of i is the subgroup $H \subset P$ corresponding to matrices $\begin{pmatrix} q & v^t \\ 0 & \Phi \end{pmatrix}$ with $q \in \mathbb{C} \subset \mathbb{H}$.

DEFINITION 4.4.9. *Let $(M, \mathcal{Q})$ be a manifold endowed with an almost quaternionic structure, and let $(p : \mathcal{G} \to M, \omega)$ be the corresponding regular normal parabolic geometry of type (G, P). Then we define the twistor space Z of M by $Z := \mathcal{G}/H$.*

Hence, the twistor space of a quaternionic structure is just the correspondence space for the subgroup $H \subset P \subset G$, which is not parabolic.

PROPOSITION 4.4.9. *Let $(M, \mathcal{Q})$ be an almost quaternionic manifold with twistor space Z. Then Z is the total space of a natural fiber bundle over M with fiber S^2. The points in Z lying over $x \in M$ parametrize the complex structures on T_xM, which are contained in $Q_x \subset L(T_xM, T_xM)$. The twistor space Z naturally carries an almost complex structure.*

PROOF. From Proposition 1.5.13 we know that Z is the total space of a fiber bundle over M with typical fiber P/H. This space can be identified with the orbit of a unit vector in $\mathbb{R}^3$ under $SO(3)$ and hence is isomorphic to S^2. For $q \in \operatorname{im}(\mathbb{H})$ we have $q^2 = -q\bar{q} = -|q|^2 1$, so unit vectors in $\mathcal{Q}^{\mathfrak{g}_{-1}}$ are exactly those linear maps, which define an almost complex structure on $\mathfrak{g}_{-1}$. Otherwise put, P/H can be viewed as the space of almost complex structures contained in the almost quaternionic structure on $\mathfrak{g}_{-1}$. Passing to associated bundles, we immediately obtain the claimed description of Z in terms of almost complex structures.

Again by Proposition 1.5.13, we know that $(\mathcal{G} \to Z, \omega)$ is a Cartan geometry of type (G, H). Hence, to construct a natural almost complex structure on Z, it suffices to construct an H–invariant complex structure on $\mathfrak{g}/\mathfrak{h}$. Writing quaternions in the form $z + wj$ for $z, w \in \mathbb{C}$ as above, a complementary subspace to $\mathfrak{h}$ in $\mathfrak{g}$ is formed by all matrices of the form $\left(\begin{smallmatrix} wj & 0 \\ X & 0 \end{smallmatrix}\right)$ with $w \in \mathbb{C}$ and $X \in \mathbb{H}^n$. For $v \in \mathbb{H}^n$, $0 \neq q \in \mathbb{C}$ and $\Phi \in GL(n, \mathbb{H})$, we obtain

$$\operatorname{Ad}\begin{pmatrix} q & v^t \\ 0 & \Phi \end{pmatrix}\begin{pmatrix} wj & 0 \\ X & 0 \end{pmatrix} = \begin{pmatrix} (qwj + v^tX)q^{-1} & \dots \\ \Phi X q^{-1} & \dots \end{pmatrix}.$$

Writing $v^tX \in \mathbb{H}$ as $\alpha + \beta j$ for $\alpha, \beta \in \mathbb{C}$, the action of H on $\mathfrak{g}/\mathfrak{h}$ is therefore given by

$$\begin{pmatrix} q & v^t \\ 0 & \Phi \end{pmatrix} \cdot \begin{pmatrix} wj \\ X \end{pmatrix} = \begin{pmatrix} q\bar{q}^{-1}(w + \beta)j \\ \Phi X q^{-1} \end{pmatrix}.$$

Now $v^tXi = (\alpha + \beta j)i = i\alpha - i\beta j$, and $q \in \mathbb{C}$ commutes with i. Hence, we conclude that this action is complex linear for the complex structure on $\mathfrak{g}/\mathfrak{h}$ defined by

$$\begin{pmatrix} wj \\ X \end{pmatrix} \mapsto \begin{pmatrix} iwj \\ -Xi \end{pmatrix}.$$

□

Notice that via the adjoint action the matrix $\left(\begin{smallmatrix} i & 0 \\ 0 & 0 \end{smallmatrix}\right)$ acts on $\mathbb{H}^n$ by $X \mapsto -Xi$. Hence, the complex structure on $\mathbb{H}^n$ used in the end of the proof ist exactly the one whose stabilizer is the subgroup H.

4.4.10. Integrability of the natural almost complex structure. A crucial point in twistor theory is to characterize integrability of the natural almost complex structure on the twistor space Z. We can derive this characterization from a general result on Cartan geometries admitting a natural almost complex structure. This result is a generalization of the well–known fact that integrability of an almost complex structure is equivalent to the fact that for one (or equivalently any) compatible linear connection, the torsion has vanishing $(0, 2)$–part.

To formulate this generalization, assume that we have given a Lie group G, a closed subgroup $H \subset G$ and an H–invariant complex structure on $\mathfrak{g}/\mathfrak{h}$, which we simply denote by $X \mapsto iX$. Then for any Cartan geometry $(p : \mathcal{G} \to N, \omega)$ of type (G, H) the identification $TN \cong \mathcal{G} \times_H (\mathfrak{g}/\mathfrak{h})$ gives rise to a natural almost complex structure J on N. Viewing the Cartan curvature as a two form on N with values in $\mathcal{G}\times_H\mathfrak{g}$, we can apply the natural projection $\mathcal{G}\times_H\mathfrak{g}\to TN$, to obtain the torsion $\tau\in\Omega^2(N,TN)$ of ω. Using the almost complex structure J, we can split τ into (p,q)–types as $\tau_{2,0}+\tau_{1,1}+\tau_{0,2}$. Explicitly,

$$\tau_{0,2}(\xi,\eta)=\tfrac{1}{4}\big(\tau(\xi,\eta)-\tau(J\xi,J\eta)+J(\tau(J\xi,\eta)+\tau(\xi,J\eta))\big).$$

On the other hand, we need an algebraic observation. Since $\mathfrak{h}$ is not an ideal in $\mathfrak{g}$, the Lie bracket does not descend to $\mathfrak{g}/\mathfrak{h}$. However, since we have the H–invariant almost complex structure on $\mathfrak{g}/\mathfrak{h}$, we can construct a map $\Lambda^2(\mathfrak{g}/\mathfrak{h})\to\mathfrak{g}/\mathfrak{h}$, which can be viewed as the $(0,2)$–component of the map induced by the Lie bracket of $\mathfrak{g}$:

Let $\pi:\mathfrak{g}\to\mathfrak{g}/\mathfrak{h}$ be the natural projection. For two elements $a,b\in\mathfrak{g}/\mathfrak{h}$ choose $X,\hat{X},Y,\hat{Y}\in\mathfrak{g}$ such that $\pi(X)=a$, $\pi(\hat{X})=ia$, $\pi(Y)=b$ and $\pi(\hat{Y})=ib$. Then consider

$$\tfrac{1}{4}\big(\pi([X,Y]-[\hat{X},\hat{Y}])+i\pi([\hat{X},Y]+[X,\hat{Y}])\big)\in\mathfrak{g}/\mathfrak{h}.$$

H–invariance of the almost complex structure on $\mathfrak{g}/\mathfrak{h}$ implies that for all $A\in\mathfrak{h}$, we get $\pi([A,\hat{Y}])=i\pi([A,Y])$. Since X and $\hat{X}$ are unique up to addition of elements of $\mathfrak{h}$, this implies that the above expression depends only on a and not on the choices of X and $\hat{X}$. In the same way, it depends only on b, so we obtain a well–defined map

$$\mathcal{S}:\mathfrak{g}/\mathfrak{h}\times\mathfrak{g}/\mathfrak{h}\to\mathfrak{g}/\mathfrak{h}.$$

By construction $\mathcal{S}(b,a)=-\mathcal{S}(a,b)$. Moreover, for chosen elements $X,\hat{X}$ associated to a, the pair $(\hat{X},-X)$ is associated to ia. This immediately shows that $\mathcal{S}$ is conjugate linear in the first variable, so by skew symmetry it is of type $(0,2)$. Finally, for $h\in H$ we can choose $\mathrm{Ad}(h)(X)$ and $\mathrm{Ad}(h)(\hat{X})$ as natural representatives associated to $\underline{\mathrm{Ad}}(h)(a)$, and likewise for b. This immediately implies that $\mathcal{S}$ is H–equivariant. Thus, for any Cartan geometry $(p:\mathcal{G}\to N,\omega)$, we get an induced natural tensor field $S\in\Omega^2(N,TN)$, which by construction is of type $(0,2)$.

LEMMA 4.4.10. *Let $(p:\mathcal{G}\to N,\omega)$ be a Cartan geometry of type (G,H). Then the Nijenhuis tensor $\mathcal{N}$ of the almost complex structure J on N is given by $\mathcal{N}=-4(\tau_{0,2}-S)$. In particular, J is integrable if and only if $\tau_{0,2}=S$.*

PROOF. Given a local vector field $\underline{\xi}$ on N, choose projectable local vector fields ξ and $\hat{\xi}$ on $\mathcal{G}$, which project onto $\underline{\xi}$ and $J\underline{\xi}$, respectively. Similarly, for $\underline{\eta}\in\mathfrak{X}(N)$, choose η and $\hat{\eta}$. Denoting by $\mathcal{N}$ the Nijenhuis tensor of J, we by definition get

$$\mathcal{N}(\underline{\xi},\underline{\eta})=Tp\cdot([\xi,\eta]-[\hat{\xi},\hat{\eta}])+JTp\cdot([\hat{\xi},\eta]+[\xi,\hat{\eta}]).$$

Therefore, the function $\mathcal{G}\to\mathfrak{g}/\mathfrak{h}$ representing $\mathcal{N}(\underline{\xi},\underline{\eta})$ is given by

$$\pi(\omega([\xi,\eta])-\omega([\hat{\xi},\hat{\eta}]))+i\pi(\omega([\hat{\xi},\eta])+\omega([\xi,\hat{\eta}])). \tag{4.14}$$

Now we get

$$\pi(\xi\cdot\omega(\hat{\eta}))=\xi\cdot\pi(\omega(\hat{\eta}))=i(\xi\cdot\pi(\omega(\eta))).$$

Using this identity and its analogs for other combinations, one immediately verifies that (4.14) can also be written as

$$-\pi(d\omega(\xi,\eta)-d\omega(\hat{\xi},\hat{\eta}))-i\pi(d\omega(\hat{\xi},\eta)+d\omega(\xi,\hat{\eta})).$$

On the other hand, the expression

$$-\pi([\omega(\xi),\omega(\eta)]-[\omega(\hat{\xi}),\omega(\hat{\eta})])-i\pi([\omega(\hat{\xi}),\omega(\eta)]+[\omega(\xi),\omega(\hat{\eta})])$$

by defintion represents $-4S(\underline{\xi},\underline{\eta})$. Adding this to the above, by definition we obtain the function representing $-4\tau_{0,2}(\underline{\xi},\underline{\eta})$, which shows that $\mathcal{N}=-4(\tau_{0,2}-S)$. □

It is easy to apply this characterization to the case of twistor spaces for almost quaternionic structures:

THEOREM 4.4.10. *Let $(M,\mathcal{Q})$ be an almost quaternionic structure with twistor space Z. Then the canonical almost complex structure on Z is integrable if and only if the almost quaternionic structure $(M,\mathcal{Q})$ is torsion free and hence quaternionic.*

PROOF. Using the notation of 4.4.9, the Lie algebra $\mathfrak{h}\subset\mathfrak{g}=\mathfrak{sl}(n+1)(\mathbb{H})$ has a natural complement formed by matrices of the form $\left(\begin{smallmatrix} wj & 0\\ X & 0\end{smallmatrix}\right)$ with $w\in\mathbb{C}$ and $X\in\mathbb{H}^n$. The commutator of $\left(\begin{smallmatrix} wj & 0\\ X & 0\end{smallmatrix}\right)$ and $\left(\begin{smallmatrix} w'j & 0\\ X' & 0\end{smallmatrix}\right)$ is immediately seen to be congruent to $\left(\begin{smallmatrix} 0 & 0\\ Xw'j-X'wj & 0\end{smallmatrix}\right)$ modulo $\mathfrak{h}$. Since the complex structure on $\mathfrak{g}/\mathfrak{h}$ is given by $\left(\begin{smallmatrix} wj & 0\\ X & 0\end{smallmatrix}\right)\mapsto\left(\begin{smallmatrix} iwj & 0\\ -Xi & 0\end{smallmatrix}\right)$, the commutator map evidently is of type $(1,1)$, so S vanishes identically.

If $(M,\mathcal{Q})$ is a quaternionic structure, then the associated canonical Cartan connection ω is torsion free (see 4.1.8). The lowest homogeneous component of the curvature of ω then has values in the lower right $n\times n$–block in $\mathfrak{g}_0$, which is contained in $\mathfrak{h}$. Thus, ω is also torsion free when viewed as a Cartan connetion on $\mathcal{G}\to Z$ rather than $\mathcal{G}\to M$, and integrability of the almost complex structure on Z follows from the lemma.

To prove the converse assertion, we have to describe the almost complex structure on Z in more detail. Let $\pi:Z\to M$ be the natural projection. Its tangent map corresponds to the natural projection $\mathfrak{g}/\mathfrak{h}\to\mathfrak{g}/\mathfrak{p}$. Now we have defined the subgroup $H\subset P$ as the stabilizer of $i\in\mathcal{Q}^{\mathfrak{g}_{-1}}$, and we have used the natural action of this element to define the almost complex structre on $\mathfrak{g}/\mathfrak{h}$. This implies that the tangent map $T_y\pi:T_yZ\to T_xM$ is complex linear if we endow T_xM with the complex structure in $\mathcal{Q}_x$, which is determined by the point $y\in\pi^{-1}(x)$.

Let us denote by $\tilde\tau\in\Omega^2(Z,TZ)$ the torsion of ω, viewed as a Cartan connection on $\mathcal{G}\to Z$. Since ω comes from the bundle $\mathcal{G}\to M$, we see that for $y\in Z$ with $\pi(y)=x$, the map $\tilde\tau(y):\Lambda^2T_yZ\to T_yZ$ descends to Λ^2T_xM. Of course, $T_y\pi\circ\tilde\tau(y)$ descends to the value $\tau(x)$ of the torsion of ω, viewed as a Cartan connection on $\mathcal{G}\to M$.

The upshot of the last two observations is that if $\tilde\tau_{0,2}(y)=0$, then the map $\tau(x):\Lambda^2T_xM\to T_xM$ must have trivial $(0,2)$–component with respect to each of the complex structures contained in $\mathcal{Q}_x$. Now $\tau(x)$ is the torsion of the canonical normal Cartan connection associated to the almost quaternionic manifold $(M,\mathcal{Q})$, so it has values in an irreducible subbundle of $\Lambda^2T^*M\otimes TM$, which is described in Proposition 4.1.8. Consider the corresponding subrepresentation of G_0 in $\Lambda^2\mathfrak{g}_{-1}^*\otimes\mathfrak{g}_{-1}$. Having trivial $(0,2)$–component with respect to each of the complex structures in $\mathcal{Q}^{\mathfrak{g}_{-1}}$ is by definition preserved under automorphisms of the prequaternionic vector space $\mathfrak{g}_{-1}$. Hence, by irreducibility, this property has to be satisfied either for all or no nonzero elements in this subrepresentation.

Viewing $\mathfrak{g}_{-1}$ as $\mathbb{H}^n$ and taking the standard basis $\{e_\ell\}$, consider the unique element $f\in S^{1,1}\mathfrak{g}_{-1}^*$ such that $f(e_1,e_1)=1$, and $f(e_i,e_j)=0$ if either $i\neq 1$ or $j\neq 1$; see 4.1.8 for the notation. Let $\{I,J,K=I\circ J\}$ be any distinguished basis for $\mathcal{Q}^{\mathfrak{g}_{-1}}$. By the representation theory version of part (1) of Propositon 4.1.8, $f\otimes I\otimes Ie_2-f\otimes J\otimes Je_2$ defines an element of the right irreducible component.

The corresponding map is $(x, y) \mapsto f(Ix, y)Ie_2 - f(Jx, y)Je_2$, and one easily verifies directly that this has nontrivial $(0, 2)$–part with respect to I. $\square$

4.4.11. Twistor theory for conformal Riemannian 4–manifolds. To conclude the discussion of correspondence spaces and twistor spaces, we briefly discuss the case $n = 1$ of the twistor theory from 4.4.9 and 4.4.10 above. Hence, we have $G = PSL(2, \mathbb{H})$ for which $G_0 \cong CSO(4)$ and hence we obtain oriented conformal Riemannian structures in dimension four; see 4.1.9. As in 4.4.9, the twistor space Z associated to a manifold M is obtained as a space of almost complex structures. More precisely, one obtains a fiber bundle $\pi : Z \to M$ such that the fiber $\pi^{-1}(x)$ over $x \in M$ can be identified with a set of complex structures on the tangent space T_xM.

From the description of the isomorphism of G_0 with $CSO(4)$ in 4.1.9 one easily concludes that the complex structures contained in the distinguished three–dimensional subspace of $L(T_xM, T_xM)$ are exactly those, which are orthogonal with respect to the conformal class of inner products and induce the right orientation on T_xM. This means that in order that a complex structure J on T_xM lies in $\pi^{-1}(x) \subset Z$, we must on the one hand have $g_x(J_x\xi, J_x\eta) = g_x(\xi, \eta)$ for a metric g in the conformal class and all $\xi, \eta \in T_xM$. On the other hand, the identifaction $T_xM \cong \mathbb{C}^2$ induced by J must lead to the given orientation on T_xM. Note that these conditions are evidently conformally invariant.

As in Proposition 4.4.9, one obtains a natural almost complex structure on the twistor space Z. The discussion of integrability is parallel to 4.4.10, but there are changes due to the different structure of curvature. Recall from 4.1.9 that in this case the harmonic curvature consists of two irreducible components, which both map $\Lambda^2 TM$ to $L(TM, TM)$. They correspond to the self–dual and anti–self–dual part of the Weyl curvature, and they are distinguished by the fact that their values lie in the two simple ideals of $\mathfrak{so}(TM)$. Now the subalgebra $\mathfrak{h}$ used to define Z as in 4.4.9 contains one of these ideals (corresponding to the lower right entry in $\mathfrak{sl}(2, \mathbb{H})$). This block corresponds to the anti–self–dual part of the curvature. This means that for an anti–self–dual conformal structure on M, the curvature function of the associated normal Cartan connection is torsion free, even when viewed as a Cartan connection on $\mathcal{G} \to Z$. As in 4.4.10, this implies integrability of the natural almost complex structure on the twistor space.

Conversely, one has to show that if the self–dual part of the Weyl curvature of an oriented conformal 4–manifold is nontrivial, then the associated canonical Cartan connection has nontrivial torsion of type $(0, 2)$, when viewed as a Cartan connection on $\mathcal{G} \to Z$. In the quaternionic picture, the self–dual part of the Weyl curvature can be viewed as a two form W_+ on M with values in the bundle $\mathcal{Q} \subset L(TM, TM)$. From the description of the almost complex structure on Z in 4.4.9 one easily concludes that vanishing of the torsion of type $(0, 2)$ as a Cartan connection on $\mathcal{G} \to Z$ is equivalent to the following condition: For any $x \in M$, any $A \in \mathcal{Q}_x$ with $\|A\| = 1$, and all tangent vectors $\xi, \eta \in T_xM$, the element

$$\phi(\xi, \eta) - \phi(A\xi, A\eta) + A \circ \phi(A\xi, \eta) + A \circ \phi(\xi, A\eta) \in \mathcal{Q}_x$$

is a real multiple of A. As in 4.4.10 this condition is (in the representation theory picture) immediately seen to be invariant under the action of G_0. By irreducibility of the representation in which self–dual Weyl curvatures have their values, this condition either has to be satisfied for all nonzero elements or for no nonzero element in this representation. Similarly, as in 4.4.10, one then easily checks that it is not satisfied for one simple element.

Collecting our results, we obtain the basic result on four–dimensional twistor theory in Riemannian signature as developed in [**AHS78**]:

THEOREM 4.4.11. *Let $(M, [g])$ be an oriented conformal 4–manifold, and let $Z \to M$ be the space of orthogonal complex structures on the tangent spaces of M, which induce the given orientation. Then Z carries a canonical almost complex structure J, which is integrable if and only if the conformal structure on M is anti–self–dual.*

4.5. Analogs of the Fefferman construction

These constructions are another way to associate to a parabolic geometry of some type a parabolic geometry of different type on the total space of a natural fiber bundle. The basic input here is an inclusion $G \hookrightarrow \tilde{G}$ of semisimple Lie groups with certain properties, together with appropriate choices of parabolic subgroups $P \subset G$ and $\tilde{P} \subset \tilde{G}$. Starting from a geometry of type (G, P) one then obtains a geometry of type $(\tilde{G}, \tilde{P})$ on the total space of a natural fiber bundle, which is called the Fefferman space. In contrast to the construction of twistor spaces discussed in Section 4.4, normality is a difficult problem here and has to be discussed case by case. For most examples, this will need a bit of input from the theory of BGG–sequences, that we will develop in volume two.

There are cases in which structures of type $(\tilde{G}, \tilde{P})$ are more complicated than structures of type (G, P), so one may use the construction to obtain examples of complicated structures from examples of simple structures. For other cases (e.g. for the classical Fefferman construction for CR–structures) the geometries of type $(\tilde{G}, \tilde{P})$ are simpler in nature than the ones of type (G, P). In these cases, one may use the construction to reduce some aspects of the study of the complicated structures to the study of simpler structures. In addition, Fefferman spaces form a highly interesting subclass of geometries of type $(\tilde{G}, \tilde{P})$ in these cases.

4.5.1. The general setup. The first ingredient for an analog of the Fefferman construction is a homomorphism $i : G \to \tilde{G}$ between two semisimple Lie groups. By semisimplicity, we may always assume that i is infinitesimally injective, i.e. $i' : \mathfrak{g} \to \tilde{\mathfrak{g}}$ is injective, since otherwise we can factor G by the normal subgroup corresponding to the kernel of i'. Second, we need a parabolic subgroup $\tilde{P} \subset \tilde{G}$, which has the property that the G–orbit of $o = e\tilde{P} \in \tilde{G}/\tilde{P}$ is open. This can be equivalently characterized infinitesimally as surjectivity of the map $\mathfrak{g} \to \tilde{\mathfrak{g}}/\tilde{\mathfrak{p}}$ induced by $i' : \mathfrak{g} \to \tilde{\mathfrak{g}}$. Having chosen $\tilde{P}$, the subgroup $Q := i^{-1}(\tilde{P})$ is a closed subgroup of G, which is usually not parabolic. The homomorphism i then induces a smooth map $G/Q \to \tilde{G}/\tilde{P}$, whose image is the G–orbit of o. As the final ingredient, we need a parabolic subgroup $P \subset G$ which contains Q.

Having all these ingredients at hand, the Fefferman type construction is carried out in two steps. Starting from a parabolic geometry $(p : \mathcal{G} \to M, \omega)$ of type (G, P), one first forms the correspondence space $\tilde{M}$ for the subgroup $Q \subset P$. In this context, $\tilde{M}$ is referred to as the *Fefferman space* of M. By definition $\tilde{M} = \mathcal{G}/Q = \mathcal{G} \times_P (P/Q)$, so it is the total space of a natural fiber bundle with fiber a homogeneous space. In the second step, we use an extension functor as introduced in 1.5.15, based on $i|_Q : Q \to \tilde{P}$ and $i' : \mathfrak{g} \to \tilde{\mathfrak{g}}$; see Example 1.5.16

for a discussion of this class of examples. We first extend the structure group of the principal Q–bundle $\mathcal{G} \to \tilde{M}$ by putting $\tilde{\mathcal{G}} := \mathcal{G} \times_Q \tilde{P}$, where Q acts on $\tilde{P}$ via $i : Q \to \tilde{P}$. Then there is a natural map $j : \mathcal{G} \to \tilde{\mathcal{G}}$, which is equivariant over i. From 1.5.15 we know that there is a unique Cartan connection $\tilde{\omega} \in \Omega^1(\tilde{\mathcal{G}}, \tilde{\mathfrak{g}})$ such that $j^*\tilde{\omega} = i' \circ \omega$ and that mapping $(\mathcal{G} \to \tilde{M}, \omega)$ to $(\tilde{\mathcal{G}} \to \tilde{M}, \tilde{\omega})$ defines a functor. Together with the evident functorial properties of the construction of correspondence spaces this implies the following result.

THEOREM 4.5.1. *Let $i : G \to \tilde{G}$ be an infinitesimally injective homomorphism between semisimple Lie groups with derivative $i' : \mathfrak{g} \to \tilde{\mathfrak{g}}$, let $\tilde{P} \subset \tilde{G}$ be a parabolic subgroup such that the G–orbit of $e\tilde{P} \in \tilde{G}/\tilde{P}$ is open, and let $P \subset G$ be a parabolic subgroup which contains $Q := i^{-1}(\tilde{P})$.*

Then mapping $(\mathcal{G} \to M, \omega)$ to $(\tilde{\mathcal{G}} \to \tilde{M}, \tilde{\omega})$ as defined above defines a functor from parabolic geometries of type (G, P) to parabolic geometries of type $(\tilde{G}, \tilde{P})$.

There are interesting examples of the construction, in which the first step is actually trivial. This means that there is an appropriate homomorphism $i : G \to \tilde{G}$ and a parabolic subgroup $\tilde{P} \subset \tilde{G}$ such that $i^{-1}(\tilde{P}) \subset G$ is already a parabolic subgroup of G. In this case, we can choose $P = Q$, and the whole construction reduces to a simple special case of extension functors.

It should be mentioned that in many cases the underlying structure corresponding to a parabolic geometry of type $(\tilde{G}, \tilde{P})$ on the Fefferman space $\tilde{M}$ can be obtained directly, without invoking the machinery of extension functors as follows. We have required above that the derivative $i' : \mathfrak{g} \to \tilde{\mathfrak{g}}$ induces a surjection $\mathfrak{g} \to \tilde{\mathfrak{g}}/\tilde{\mathfrak{p}}$. Since $Q = i^{-1}(\tilde{P})$, we see that the kernel of this surjection is the Lie algebra $\mathfrak{q}$ of Q, and we obtain a linear isomorphism $\mathfrak{g}/\mathfrak{q} \to \tilde{\mathfrak{g}}/\tilde{\mathfrak{p}}$. By construction, this map is equivariant over the homomorphism $i : G \to \tilde{G}$, i.e. $i'(\underline{\mathrm{Ad}}(g)(X + \mathfrak{q})) = \underline{\mathrm{Ad}}(i(g))(i'(X)) + \tilde{\mathfrak{p}}$ for all $g \in G$ and $X \in \mathfrak{g}$. Here we denote by $\underline{\mathrm{Ad}}$ the representations on the quotients induced by the adjoint representations. Suppose that we have some representation $\tilde{W}$ constructed functorially from $\tilde{\mathfrak{g}}/\tilde{\mathfrak{p}}$ and a $\tilde{P}$–invariant element $\tilde{w} \in \tilde{W}$. Then we can apply the same functor to $\mathfrak{g}/\mathfrak{q}$ to obtain a representation W of Q and a linear isomorphism $W \to \tilde{W}$, which is equivariant over i. Hence, the preimage of $\tilde{w}$ under this isomorphism is a Q–invariant element of W. Let us discuss this point of view for the classical Fefferman construction.

EXAMPLE 4.5.1 (The classical Fefferman construction). Fix a signature (p, q), put $G := SU(p+1, q+1)$, $n = p+q$, let $V = \mathbb{C}^{n+2}$ be the standard representation of G and $\langle\ ,\ \rangle$ the G–invariant Hermitian form on V. Forgetting about the complex structure, we can view V as $\mathbb{R}^{2n+4}$ and the real part of the Hermitian form as a real inner product of signature $(2p+2, 2q+2)$, which is invariant under the action of any element of G. Hence, we obtain an inclusion i of G into $\tilde{G} := SO(2p+2, 2q+2)$. Next fix a real line ℓ in V which is isotropic for the real inner product and let $\tilde{P}$ be the stabilizer of ℓ. Then $\tilde{G}/\tilde{P}$ is the space of all isotropic real lines in V. Elementary linear algebra shows that G acts transitively on the space of nonzero isotropic vectors in V and hence also on $\tilde{G}/\tilde{P}$. Defining $Q = i^{-1}(\tilde{P}) = G \cap \tilde{P}$ we conclude that i induces a diffeomorphism $G/Q \to \tilde{G}/\tilde{P}$.

Finally, we define $P \subset G$ to be the stabilizer of the complex line generated by ℓ. This is a parabolic subgroup of G (see 3.2.6) and G/P is the space of complex isotropic lines in V. Since a complex linear map which stabilizes ℓ also stabilizes $\ell \otimes \mathbb{C}$, we see that $Q \subset P$ as required. The group P evidently acts on the complex line $\ell\otimes\mathbb{C}$ and the stabilizer of the real line ℓ is just $Q \subset P$. Hence, we see that P/Q can be identified with the space of real lines in a complex line, so $P/Q \cong \mathbb{R}P^1 \cong S^1$.

In 4.2.4 we have described regular normal parabolic geometries of type $(\underline{G}, \underline{P})$, where $\underline{G} := PSU(p+1, q+1)$ and $\underline{P}$ is the stabilizer of an isotropic line. They are equivalent to partially integrable almost CR–structures of hypersurface type which are nondegenerate of signature (p, q). Hence, we know what is going on if we replace G by the quotient $\underline{G}$ of G by its center. In 4.2.4 we have actually described the group $\underline{G}_0$ as a quotient of G_0. We can identify G_0 with pairs (c, C) with $c \in \mathbb{C}$ and $C \in U(p,q)$ such that $\frac{c}{\bar{c}}\det(C) = 1$, and $\underline{G}_0$ is the quotient by the $(p+q+2)$nd roots of unity, which are embedded as $(\zeta, \zeta\mathrm{id})$. For forming infinitesimal flag structures, we have to view $\underline{G}_0$ as a subgroup of $GL(\mathfrak{g}_{-1}) \cong GL(p+q, \mathbb{C})$ via $(c, C) \mapsto c^{-1}C$. We have also seen in 4.2.4 that the action on $\mathfrak{g}_{-2} \cong \mathbb{R}$ is given by multiplication by $|c|^{-2}$. This means that on the action of $\underline{G_0}$ on the one–dimensional complex vector space $\mathfrak{g}_{-2} \otimes \Lambda^{p+q}_{\mathbb{C}}\mathfrak{g}_{-1}$ is induced by multiplication by $|c|^{-2}c^{-n}\det(C) = c^{-n-2}$.

A principal G_0–bundle evidently has an underlying principal $\underline{G}_0$–bundle, and an infinitesimal flag structure of type (G, P) has an underlying infinitesimal flag structure of type $(\underline{G}, \underline{P})$. Regularity is clearly preserved in this process. Hence, we can view a regular infinitesimal flag structure of type (G, P) on a manifold M as a partially integrable almost CR–structure plus some additional data. The additional data can be described as a complex line bundle $\mathcal{E} \to M$ such that the $(n+2)$nd tensor power $\mathcal{E}^{\otimes^{n+2}}$ is isomorphic to $Q \otimes \Lambda^n_{\mathbb{C}}H$, where $H \subset TM$ is the CR–subbundle and $Q = TM/H$:

Having given a regular infinitesimal flag structure of type (G, P), the bundle $\mathcal{E}$ is obtained as the associated bundle corresponding to the action of G_0 on $\mathbb{C}$ given by multiplication by c^{-1}. The $(n+2)$nd tensor power of this bundle is induced by multiplication by c^{-n-2}, so from above we see that this is the associated bundle corresponding to $\mathfrak{g}_{-2}\otimes\Lambda^n_{\mathbb{C}}\mathfrak{g}_{-1}$ and hence $Q\otimes\Lambda^n_{\mathbb{C}}H$ as required. Conversely, given a partially integrable almost CR–structure plus an appropriate root $\mathcal{E}$ of this bundle, we take a regular infinitesimal flag structure of type $(\underline{G}, \underline{P})$ inducing the almost CR–structure. The chosen root naturally gives rise to a $(n+2)$–fold covering of the bundle, and one easily verifies that this can be made into a regular infinitesimal flag structure of type (G, P). It should be remarked the for hypersurfaces in $\mathbb{C}^{n+1}$, the trivialization of the tangent bundle $T\mathbb{C}^{n+1}$ naturally induces a trivialziation of $Q \otimes \Lambda^n_{\mathbb{C}}H$, so there also is a canonical $(n+2)$nd root of this bundle.

Hence, we conclude that a regular normal parabolic geometry $(p : \mathcal{G} \to M, \omega)$ of type (G, P) is equivalent to a partially integrable almost CR–structure of hypersurface type together with an appropriate complex line bundle $\mathcal{E} \to M$. From our constructions it is clear that $Q \subset P$ is the stabilizer of a real line in the one–dimensional complex representation of P which induces the bundle $\mathcal{E}$. Hence, the Fefferman space $\tilde{M} = \mathcal{G}/Q \cong \mathcal{G} \times_P (P/Q)$ can be viewed as the space of real lines in the complex line bundle $\mathcal{E}$. It is the total space of a natural circle bundle over M.

On the other hand, from Example 4.1.2 we know that parabolic geometries of type $(\tilde{G}, \tilde{P})$ are oriented conformal pseudo–Riemannian structures of signature $(2p+1, 2q+1)$. This comes from the fact that there is a $\tilde{P}$–invariant conformal class of inner products of that signature on $\tilde{\mathfrak{g}}/\tilde{\mathfrak{p}}$. Under the linear isomorphism $\mathfrak{g}/\mathfrak{q} \to \tilde{\mathfrak{g}}/\tilde{\mathfrak{p}}$ induced by the inclusion $i' : \mathfrak{su}(p+1, q+1) \to \mathfrak{so}(2p+2, 2q+2)$ this gives rise to a Q–invariant conformal class of inner products of signature $(2p+1, 2q+1)$ on

$\mathfrak{g}/\mathfrak{q}$. Viewing ω as a Cartan connection on the bundle $\mathcal{G} \to \tilde{M}$, this gives rise to a canonical conformal class of that signature on $\tilde{M}$, which implies the following result. For $q = 0$ and embedded CR–structures this is due to Ch. Fefferman (see [**Fe76**]), for abstract CR–structures and $q = 0$ it is due to [**BDS77**]. The fact that the construction is still possible in the partially integrable case was, to our knowledge, first observed in [**Čap02**], reporting on the work that appeared in [**ČGo08**].

PROPOSITION 4.5.1. *Let M be a smooth manifold endowed with a partially integrable almost CR–structure of hypersurface type, which is nondegenerate of signature (p, q), and with a complex line bundle $\mathcal{E}$ as above. Then there is a natural conformal class of signature $(2p+1, 2q+1)$ on the total space $\tilde{M}$ of a certain natural circle bundle over M.*

4.5.2. The Cartan geometry interpretation. While the structure underlying a regular parabolic geometry of type $(\tilde{G}, \tilde{P})$ may be obtained without applying the theory of extension functors for Cartan geometries discussed in 4.5.1, the Cartan geometry interpretation offers some immediate advantages. Again, we will discuss this in the example of the classical Fefferman construction. Consider the inclusion $G := SU(p+1, q+1) \hookrightarrow SO(2p+2, 2q+2) = \tilde{G}$ obtained by looking at the underlying real vector space of $\mathbb{C}^{p+q+2}$ and the real part of the Hermitian inner product. Now we can improve the construction by showing that this inclusion lifts to a homomorphism to the associated spin group.

LEMMA 4.5.2. *The inlusion $SU(p+1, q+1) \hookrightarrow SO(2p+2, 2q+2)$ lifts to a homomorphism to the spin group $Spin(2p+2, 2q+2)$.*

PROOF. In definite signature, the fact that the inclusion of $SU(n)$ into $SO(2n)$ lifts into $Spin(2n)$ follows immediately from the fact that $SU(n)$ is simply connected. In the case of indefinite signature that we need here, things are a bit more complicated. We assume $p \geq q$ and $p > 0$. The maximal compact subgroup of $SU(p+1, q+1)$ is $S(U(p+1) \times U(q+1))$ and by the Iwasawa decomposition (see 2.3.5), the inclusion of the maximal compact subgroup induces an isomorphism in the fundamental group, so $\pi_1(SU(p+1, q+1)) \cong \mathbb{Z}$. Since $SU(p+1, q+1)$ is connected, the inclusion has values in the connected component $SO_0(2p+2, 2q+2)$ of the identity. The maximal compact subgroup in there is $SO(2p+2) \times SO(2q+2)$ which implies that $\pi_1(SO(2p+2, 2q+2))$ is isomorphic to $\mathbb{Z}_2 \times \mathbb{Z}_2$ for $q > 0$ and to $\mathbb{Z}_2 \times \mathbb{Z}$ for $q = 0$.

In indefinite signature, the spin group is again a two–fold covering of the special orthogonal group, so it is no longer simply connected; see the book [**Bau81**] for details on this. More precisely, $\pi_1(Spin(2p+2, 2q+2))$ is isomorphic to $\mathbb{Z}_2$ for $q > 0$ and to $\mathbb{Z}$ for $q = 0$, with the image in $\pi_1(SO(2p+2, 2q+2))$ generated by $(1, 1)$ in both cases. From the construction above we see that the homomorphism of fundamental groups induced by the inclusion $SU(p+1, q+1) \hookrightarrow SO(2p+2, 2q+2)$ is induced by $n \mapsto (n, -n)$, where in the right–hand side either the first component or both components have to be taken modulo 2. In any case, this lies in the image of $\pi_1(Spin(2p+2, 2q+2))$ and the claim follows from general results on coverings. □

In view of this result, we can also use $Spin(2p+2, 2q+2)$ for the group $\tilde{G}$ and the stabilizer of a null line in the standard representation for the parabolic subgroup $\tilde{P} \subset \tilde{G}$, and still obtain an extension functor from Cartan geometries of type (G, Q) to parabolic geometries of type $(\tilde{G}, \tilde{P})$. From 4.1.2 we know that these geometries (for which the regularity condition is vacuous) have an underlying pseudo–Riemannian spin structure. Hence, we conclude

COROLLARY 4.5.2. *The Fefferman space $\tilde{M}$ associated to nondegenerate partially integrable almost CR–structure together with an appropriate choice $\mathcal{E}$ of a complex line bundle on M as described in Example 4.5.1 carries a canonical spin structure.*

We have proved in Theorem 4.5.1 that a Fefferman type construction defines a functor between parabolic geometries of different types. The next natural question is whether this functor is compatible with regularity and/or normality. Once we prove that regular normal geometries are mapped to regular ones, we get a regular inifinitesimal flag structure of type $(\tilde{G}, \tilde{P})$ on the Fefferman space, and hence the underlying structures corresponding to regular normal parabolic geometries of that type. Moreover, we know that the canonical regular normal Cartan connection associated to this structure can be constructed from the extended Cartan connection $\tilde{\omega}$ via the normalization procedure discussed in 3.1.13.

In contrast to the case of correspondence spaces, there seems to be no general answer to the question of regularity and normality for the analogs of the Fefferman construction. The known normality results do not reduce to a purely algebraic comparison of the normalization conditions, but one always has to use the fact that one deals with the curvature of a regular normal Cartan connection. As we shall immediately see, regularity and normality are not preserved in general.

The starting point for all these considerations is of course to describe the relation between the curvatures of the two Cartan connections. This relation can be deduced from the general theory of correspondence spaces and of extension functors. First, we know from 1.5.13 that a correspondence space has the same curvature function as the initial geometry. On the other hand, the effect of extension functors on the curvature is described in Proposition 1.5.16. We have assumed that $i' : \mathfrak{g} \to \tilde{\mathfrak{g}}$ is injective and induces a linear isomorphism $\mathfrak{g}/\mathfrak{q} \to \tilde{\mathfrak{g}}/\tilde{\mathfrak{p}}$. Taking the inverse of this isomorphism and composing with the obvious surjection $\mathfrak{g}/\mathfrak{q} \to \mathfrak{g}/\mathfrak{p}$, we obtain a surjective linear map $\pi : \tilde{\mathfrak{g}}/\tilde{\mathfrak{p}} \to \mathfrak{g}/\mathfrak{p}$. In the notation of 1.5.16, the linear map α is the Lie algebra homomorphism i', so the map Φ_α defined there vanishes identically in our situation. Hence, we obtain

PROPOSITION 4.5.2. *Consider a parabolic geometry $(p : \mathcal{G} \to M, \omega)$ of type (G, P) and let $(\tilde{p} : \tilde{\mathcal{G}} \to \tilde{M}, \tilde{\omega})$ be the result of the Fefferman type construction. Let $\kappa : \mathcal{G} \to L(\Lambda^2(\mathfrak{g}/\mathfrak{p}), \mathfrak{g})$ and $\tilde{\kappa} : \tilde{\mathcal{G}} \to L(\Lambda^2(\tilde{\mathfrak{g}}/\tilde{\mathfrak{p}}), \tilde{\mathfrak{g}})$ be the curvature functions of the two geometries, and let $j : \mathcal{G} \to \tilde{\mathcal{G}} = \mathcal{G} \times_i \tilde{P}$ be the natural map.*

Then for each $u \in \mathcal{G}$, we have $\tilde{\kappa}(j(u)) = i' \circ \kappa(u) \circ \Lambda^2\pi$, and this completely determines $\tilde{\kappa}$.

At the current stage it is not evident how important it is to know whether $\tilde{\omega}$ is normal or not. It turns out, however, that this condition is crucial for deeper applications of Fefferman type constructions. For the classical Fefferman construction, several deep consequences of normality of $\tilde{\omega}$ are deduced in [**ČGo08**]. We will discuss these results in volume two with the necessary background on BGG–sequences at our disposal.

The description of the curvature of $\tilde{\omega}$ in the proposition above easily implies that normality is not preserved in the classical Fefferman construction in general.

OBSERVATION 4.5.2. In the setting of the classical Fefferman construction, let $(p : \mathcal{G} \to M, \omega)$ be a regular normal parabolic geometry of type (G, P) and let $(\tilde{p} : \tilde{\mathcal{G}} \to \tilde{M}, \tilde{\omega})$ be the resulting geometry of type $(\tilde{G}, \tilde{P})$.

If $\tilde{\omega}$ is normal, then the original Cartan connection ω is torsion free. This implies that the partially integrable almost CR–structure on M is actually integrable and hence a CR–structure.

PROOF. A normal Cartan connection of type $(\tilde{G}, \tilde{P})$ is automatically torsion free; see 4.1.2. Hence, normality of $\tilde{\omega}$ implies that $\tilde{\kappa}$ has values in $\tilde{\mathfrak{p}}$, which by the proposition implies that κ must have values in $(i')^{-1}(\tilde{\mathfrak{p}}) = \mathfrak{q} \subset \mathfrak{p}$. But this implies that ω is torsion free. The interpretation in terms of the underlying almost CR–structure then follows from 4.2.4. □

In [**ČGo08**] it is shown that if ω is regular and normal, then $\tilde{\omega}$ is normal if and only if ω is torsion free. This needs some facts on the curvature of regular normal Cartan connections which are deduced using BGG sequences, so we will discuss it in volume two.

4.5.3. Lie contact structures induced by conformal structures, I. This is an example for which we can solve the problem of normality of the induced Cartan geometry using only tools which are currently at our disposal. It is also fairly transparent from a geometric point of view, so we discuss it in a bit more detail here. The example was initially worked out by the Japanese school in the case of definite signature, phrased as a construction of a canonical Lie contact structure on the unit sphere bundle of a Riemannian manifold, which depends only on the conformal class of the metric; see [**SaYa88, SaYa93**]. While this was based on the canonical Cartan connections constructed by Tanaka, the analogy to Fefferman's construction was only realized much later.

In the end of 4.2.1 we have noted the existence of a canonical contact structure on the projectivized cotangent bundle $\mathcal{P}(T^*M)$ of any smooth manifold M. Here we will need a slight variant of this, namely we consider the bundle of rays $\mathcal{R}(T^*M)$ in T^*M. This is obtained by identifying two nonzero elements of T^*M if they are positive multiples of each other. If $\dim(M) = n$, then $\mathcal{R}(T^*M)$ is a fiber bundle over M with fiber S^{n-1} and a two–fold covering of $\mathcal{P}(T^*M)$. For any choice of a Riemannian metric on M, the bundle $\mathcal{R}(T^*M)$ can be identified with the unit sphere bundle of T^*M or (again using the metric) of TM. The canonical contact structure on this bundle is obtained in the same way as for $\mathcal{P}(T^*M)$. The contact hyperplane in the point corresponding to a ray r in T^*_xM is formed by those tangent vectors whose projection to T^*_xM is annihilated by r. Otherwise put, this is just the pullback of the canonical contact structure on $\mathcal{P}(T^*M)$ via the covering, and hence it is a contact structure.

Now contact structures do not have any local invariants, but it is natural to ask whether a choice of some local geometric structure on M induces a refinement of this contact structure to some local geometric structure on $\mathcal{R}(T^*M)$ or $\mathcal{P}(T^*M)$. Here we show that a conformal structure on M induces a Lie contact structure on $\mathcal{R}(T^*M)$ and the two structures are related in the way described in 4.5.1 and 4.5.2 above.

Recall from 4.2.5 that a Lie contact structure of signature (p, q) (which is then defined on a manifold of dimension $2(p + q) + 1$) is equivalent to a regular normal parabolic geometry of type $(\tilde{G}, \tilde{P})$, with $\tilde{G} = PO(p + 2, q + 2)$. We will often work in the orthogonal group keeping in mind that the matrices are only determined up to sign. Let $\tilde{V} = \mathbb{R}^{p+q+4}$ be the standard representation of this orthogonal group and let $\langle\ ,\ \rangle$ be the invariant inner product of signature $(p+2, q+2)$. Then $\tilde{P} \subset \tilde{G}$ comes from the stabilizer of a null plane $\tilde{W} \subset \tilde{V}$. On the other hand, from 4.1.2 we know that conformal structures of signature $(p+1, q)$ are equivalent to normal parabolic geometries of type (G, P), where $G = O(p+2, q+1)$ and $P \subset G$ is the stabilizer of a null line in the standard representation.

To construct an inclusion $G \hookrightarrow \tilde{G}$, we fix a vector $\tilde{v} \in \tilde{V}$ such that $\langle \tilde{v}, \tilde{v} \rangle = -1$, and let $V := \tilde{v}^\perp$ be its orthocomplement. Then $\langle\ ,\ \rangle$ restricts to an inner product of signature $(p+2, q+1)$ on V. Choosing G to be the orthogonal group of this restriction, we obtain an inclusion $G \hookrightarrow O(\tilde{V})$ as those maps which leave $\tilde{v}$ fixed. Projecting to the quotient $\tilde{G}$ we obtain a homomorphism $i : G \to \tilde{G}$, and a moment of thought shows that i is injective.

Now there is a point here to be careful about, which frequently shows up in analogs of the Fefferman construction. Replacing $\tilde{v}$ by another negative unit vector leads to a subgroup of $\tilde{G}$, which is conjugate to G. Likewise, replacing $\tilde{W}$ by another isotropic plane leads to a subgroup of $\tilde{G}$, which is conjugate to $\tilde{P}$. The reason for this is that any two negative unit vectors as well as any two isotropic planes can be mapped to each other by an orthogonal tranformation. However, we have to be careful about the relative position of $\tilde{v}$ and $\tilde{W}$. Namely, we may either choose $\tilde{v}$ to be orthogonal to $\tilde{W}$ (and hence $\tilde{W} \subset V$) or not, and this is invariant under orthogonal maps. In particular, if $\tilde{W} \subset V$, then the same is true for the image of $\tilde{W}$ under each element of G. This immediately implies that for such a choice the G–orbit of $o = e\tilde{P}$ in $\tilde{G}/\tilde{P}$ cannot be open. Hence, we have to require that $\tilde{v}$ is *not* orthogonal to $\tilde{W}$, which means that $\tilde{W}$ is transversal to V. Notice that this is automatically satisfied in the case $q = 0$, which corresponds to definite conformal structures.

Taking this setup, the image of $\tilde{W}$ under each element of G is transversal to V. If $E \subset \tilde{V}$ is an isotropic plane such that $E \cap V$ has dimension one, then choose a basis $\{w, v\}$ for E such that $v \in V$. Rescaling w, we may assume that it has the form $v_1 + \tilde{v}$ with $v_1 \in V$, and the equations $\langle w, v \rangle = \langle w, w \rangle = 0$ are equivalent to $\langle v_1, v \rangle = 0$ and $\langle v_1, v_1 \rangle = 1$. Of course, any two pairs (v, v_1) of elements of V such that $\langle v, v \rangle = \langle v, v_1 \rangle = 0$ and $\langle v_1, v_1 \rangle = 1$ can be mapped to each other by an element of G. In particular, we can find an element in G which maps $\tilde{W}$ to E. Hence, the G–orbit of $o = e\tilde{P}$ in the space $\tilde{G}/\tilde{P}$ of isotropic planes in $\tilde{V}$ consists of all those planes which are transversal to V and hence is open.

Since $\tilde{W}$ is not contained in the codimension one subspace V, the intersection $\ell := \tilde{W} \cap V$ is an isotropic line in V. The stabilizer of ℓ is a parabolic subgroup $P \subset G$, and elements of $Q := i^{-1}(\tilde{P})$ stabilize both V and $\tilde{W}$, whence $Q \subset P$. Therefore, the theory developed in 4.5.1 and 4.5.2 applies to the pairs (G, P) and $(\tilde{G}, \tilde{P})$ and the homomorphism $i : G \to \tilde{G}$.

PROPOSITION 4.5.3. *Let $(p : \mathcal{G} \to M, \omega)$ be the regular normal parabolic geometry induced by a conformal structure of signature $(p+1, q)$ on M. Then the Fefferman space $\tilde{M} := \mathcal{G}/Q$ can be naturally identified with the open subspace $\mathcal{R}_+(T^*M) \subset \mathcal{R}(T^*M)$ of those rays in T^*M, on which the conformal structure is positive definite.*

The conformal structure on M induces a Lie contact structure of signature (p,q) on $\mathcal{R}_+(T^*M)$. The underlying contact structure is the restriction of the natural contact structure on $\mathcal{R}(T^*M)$.

PROOF. The first step is to characterize the subgroup $Q = i^{-1}(\tilde{P}) \subset P$. Projecting the plane $\tilde{W}$ orthogonally into V, we obtain a plane $W \subset V$ which by construction is stabilized by any element of Q. From above, we see that on W the inner product is degenerate with one positive eigenvalue, and the degenerate line is ℓ. The orthogonal projection defines a linear isomorphism $\phi : \tilde{W} \to W$. Mapping w to $\langle \phi^{-1}(w), \tilde{v}\rangle$ defines a nonzero linear functional on W with kernel ℓ. In particular, we obtain an orientation on W/ℓ, which is again preserved by the induced action of any element of Q. Now it is easy to verify directly that $Q \subset P$ is exactly the subgroup of those elements which stabilize W and have the property that the induced action on W/ℓ is orientation preserving.

Now for $A \in \mathfrak{g} = \mathfrak{so}(V)$, we have $A(\ell) \subset \ell^\perp$. Projecting to the quotient we get an induced linear map $\ell \to \ell^\perp/\ell$. Hence, we obtain a linear map $\mathfrak{g} \to L(\ell, \ell^\perp/\ell)$, whose kernel by definition is the stabilizer $\mathfrak{p}$ of ℓ. The subgroup $P \subset G$ evidently acts on ℓ and $\ell^\perp$ and thus also on $L(\ell, \ell^\perp/\ell)$, and our map clearly is P–equivariant. For dimensional reasons, it must induce an isomorphism $\mathfrak{g}/\mathfrak{p} \to L(\ell, \ell^\perp/\ell)$ of P–modules. This is just a basis independent version of the matrix representation of $\mathfrak{g}$ and $\mathfrak{p}$ used in 1.6.3.

Via the isomorphism $\mathfrak{g}/\mathfrak{p} \cong L(\ell, \ell^\perp/\ell)$, the subspace $W/\ell \subset \ell^\perp/\ell$ gives rise to an oriented line in $\mathfrak{g}/\mathfrak{p}$, and $Q \subset P$ can be characterized as the stabilizer of this oriented line. The inner product $\langle\ ,\ \rangle$ induces an inner product of signature $(p+1, q)$ on $\ell^\perp/\ell$, which, via the above isomorphism, gives rise to a P–invariant conformal class of inner products of that signature on $\mathfrak{g}/\mathfrak{p}$. We have noted above that the induced inner product on W/ℓ is positive definite, so any inner product from that conformal class is positive on this distinguished line. Using the conformal class of inner products, we may replace the oriented line in $\mathfrak{g}/\mathfrak{p}$ by the annihilator of its orthocomplement, which is a line in $\mathfrak{p}_+ = (\mathfrak{g}/\mathfrak{p})^*$ that also inherits a natural orientation. The duality also gives rise to a conformal class of inner products on $\mathfrak{p}_+$ and all these products are positive on the distinguished line.

The action of P on $\mathfrak{p}_+$ is simply the standard action of $CO(p+1,q)$ on (the dual of) $\mathbb{R}^{p+q+1}$, so it has exactly three nonzero orbits corresponding to the sign of the inner product. Since Q is the stabilizer of an oriented positive line, we can identify the homogeneous space P/Q with the open subset $\mathcal{R}_+(\mathfrak{p}_+) \subset \mathcal{R}(\mathfrak{p}_+)$ of positive rays in $\mathfrak{p}_+$. Since $\mathcal{G} \times \mathfrak{p}_+ \cong T^*M$, we conclude that $\tilde{M} = \mathcal{G}/Q = \mathcal{G} \times_P (P/Q)$ is the open subspace $\mathcal{R}_+(T^*M) \subset \mathcal{R}(T^*M)$ of rays on which the conformal structure is positive. In particular, in the case $q = 0$, when we start with a Riemannian conformal structure, the space $\tilde{M}$ is the full space $\mathcal{R}(T^*M)$.

Next, we need a bit of information on the filtration $\{\tilde{\mathfrak{g}}^i\}$ of $\tilde{\mathfrak{g}}$ coming form the parabolic subalgebra $\tilde{\mathfrak{p}}$. This comes from the filtration $\tilde{V} = \tilde{V}^{-1} \supset \tilde{V}^0 \supset \tilde{V}^1$, where $\tilde{V}^1 = \tilde{W}$ and $\tilde{V}^0 = \tilde{W}^\perp$. In particular, the filtration component $\tilde{\mathfrak{g}}^{-1} \subset \tilde{\mathfrak{g}}$ consists of all $\tilde{A} \in \tilde{\mathfrak{g}}$ such that $\tilde{A}(\tilde{W}) \subset \tilde{W}^\perp$. Now let $\{v_0, \tilde{w}\}$ be a basis for $\tilde{W}$ such that $v_0 \in \ell = \tilde{W} \cap V$. Since the plane $\tilde{W}$ is isotropic, we have $\langle v_0, v_0\rangle = \langle v_0, \tilde{w}\rangle = \langle \tilde{w}, \tilde{w}\rangle = 0$. Applying $\tilde{A} \in \tilde{\mathfrak{g}}$ we obtain $\langle \tilde{A}(v_0), v_0\rangle = \langle \tilde{A}(\tilde{w}), \tilde{w}\rangle = 0$ and $\langle \tilde{A}(v_0), \tilde{w}\rangle = -\langle v_0, \tilde{A}(\tilde{w})\rangle$. Hence, $\tilde{A} \in \tilde{\mathfrak{g}}^{-1}$ if and only if $\langle \tilde{A}(v_0), \tilde{w}\rangle = 0$. This equation is certainly satisfied if $A(v_0)$ is a multiple of v_0. Since $\mathfrak{p}$ by definition is the stabilizer of the line generated by v_0, we see that $i'(\mathfrak{p}) \subset \tilde{\mathfrak{g}}^{-1}$.

Now we use the Cartan geometry interpretation of the Fefferman construction as discussed in 4.5.2. Form the extension $\tilde{\mathcal{G}} = \mathcal{G} \times_Q \tilde{P} \to \tilde{M}$, and let $\tilde{\omega} \in \Omega^1(\tilde{\mathcal{G}}, \tilde{\mathfrak{g}})$ be the Cartan connection obtained as in 4.5.2. Let $j : \mathcal{G} \to \tilde{\mathcal{G}}$ be the natural inclusion. Since ω is the normal Cartan connection associated to a conformal structure, it is torsion free, i.e. its curvature function κ has values in $\mathfrak{p}$. By Proposition 4.5.2, this implies that the restriction of the curvature function $\tilde{\kappa}$ of $\tilde{\omega}$ to $j(\mathcal{G})$ has values in $i'(\mathfrak{p}) \subset \tilde{\mathfrak{g}}^{-1}$. Since $\tilde{\mathfrak{g}}^{-1} \subset \tilde{\mathfrak{g}}$ is a $\tilde{P}$–invariant subspace, we conclude that $\tilde{\kappa}$ has values in $\tilde{\mathfrak{g}}^{-1}$ on all of $\tilde{\mathcal{G}}$. Since we are dealing with a $|2|$–grading on $\tilde{\mathfrak{g}}$, this implies that $\tilde{\omega}$ is regular, and hence induces a Lie contact structure on $\tilde{M}$.

To complete the proof, we have to describe the contact subbundle $\tilde{H} \subset T\tilde{M}$ explicitly. In the picture of the extended bundle, we have $\tilde{H} = \tilde{\mathcal{G}} \times_{\tilde{P}} (\tilde{\mathfrak{g}}^{-1}/\tilde{\mathfrak{p}})$. Under the isomorphism $\mathfrak{g}/\mathfrak{q} \to \tilde{\mathfrak{g}}/\tilde{\mathfrak{p}}$ induced by i', the subspace $\tilde{\mathfrak{g}}^{-1}/\tilde{\mathfrak{p}}$ corresponds to a Q–invariant codimension one subspace of $\mathfrak{g}/\mathfrak{q}$. Hence, we have to determine when an element $A \in \mathfrak{g}$ has the property that $i'(A) \in \tilde{\mathfrak{g}}^{-1}$. From above, we see that this holds if and only if $i'(A)(\ell) \subset \tilde{W}^{\perp}$. Since $A(V) \subset V$, this is equivalent to $A(\ell) \subset W^{\perp}$. But this exactly means that $A + \mathfrak{p} \in \mathfrak{g}/\mathfrak{p}$ lies in the kernel of the distinguished line in $\mathfrak{p}_+$ constructed above.

In the picture of associated bundles, we have $T\tilde{M} = \mathcal{G} \times_Q (\mathfrak{g}/\mathfrak{q})$ and $TM = \mathcal{G} \times_P (\mathfrak{g}/\mathfrak{p})$, and the natural projection $\mathfrak{g}/\mathfrak{q} \to \mathfrak{g}/\mathfrak{p}$ corresponds to the tangent map of the projection $\tilde{M} = \mathcal{R}_+(T^*M) \to M$. Hence, we conclude that the contact subbundle in a point of $\tilde{M}$ is formed by those tangent vectors whose projection to TM lies in the kernel of the line defining the foot point. Hence, this is exactly the restriction of the canonical contact structure on $\mathcal{R}(T^*M)$. □

REMARK 4.5.3. In addition to the contact subbundle $H \subset T\tilde{M}$, the Lie contact structure consists of two auxiliary bundles E and F such that $H \cong L(E, F)$. Here E has rank two and F has rank $p+q$ and is endowed with a bundle metric of signature (p, q). These bundles are given as $E = \tilde{\mathcal{G}} \times_{\tilde{P}} \tilde{W}$, respectively, $F = \tilde{\mathcal{G}} \times_{\tilde{P}} (\tilde{W}^{\perp}/\tilde{W})$. Analyzing $\tilde{W}$ and $\tilde{W}^{\perp}/\tilde{W}$ as representations of Q, one can explicity describe the bundles E and F in terms of conformally natural vector bundles on M.

4.5.4. Lie contact structures induced by conformal structures, II. We now turn to the question of normality of the extended Cartan connection. While we can prove normality without additional tools in this case, the proof is fairly subtle. In contrast to the case of correspondence spaces (see 4.4.1), the result does not follow from a purely algebraic comparison of the normalization conditions. Apart from normality of the curvature of the canonical conformal Cartan connection the proof also needs the fact that the lowest homogeneous component is harmonic and therefore totally tracefree. We continue to work in the setting of the last subsection.

THEOREM 4.5.4. *Let $(p : \mathcal{G} \to M, \omega)$ be the normal parabolic geometry of type (G, P) corresponding to a conformal structure of signature $(p+1, q)$ on M. Let $\tilde{M} = \mathcal{G}/Q = \mathcal{R}_+(T^*M)$ be the Fefferman space of M, and consider the extended parabolic geometry $(\tilde{\mathcal{G}} \to \tilde{M}, \tilde{\omega})$ of type $(\tilde{G}, \tilde{P})$. Then the Cartan connection $\tilde{\omega}$ is automatically regular and normal and hence is the canonical Cartan connection associted to the Lie contact structure on $\tilde{M}$ from Theorem 4.5.3.*

PROOF. We have already observed in the proof of Theorem 4.5.3 that $\tilde{\omega}$ is regular, so it remains to prove normality. To do this, it is easiest to use an explicit description of the Lie algebras in questions in terms of matrices. In the terminology of 4.5.3, we choose a basis $\{e_1, \dots, e_{p+q+4}\}$ for $\tilde{V}$ such that $\langle e_1, e_j\rangle = \delta_{j,p+q+4}$, $\langle e_2, e_j\rangle = \delta_{j,p+q+3}$, $\langle e_i, e_j\rangle = \delta_{i,j}$ for $i = 3, \dots, p+2$ and $\langle e_i, e_j\rangle = -\delta_{i,j}$ for $i = p+3, \dots, p+q+2$. Putting $J = \left(\begin{smallmatrix} 0 & 1 \\ 1 & 0 \end{smallmatrix}\right)$ and denoting by $\mathbb{I} = \mathbb{I}_{p,q}$ the $(p+q)\times(p+q)$ diagonal matrix whose first p diagonal entires are equal to 1 and who last q diagonal entries are -1, this inner product corresponds to the block matrix $\left(\begin{smallmatrix} 0 & 0 & J \\ 0 & \mathbb{I} & 0 \\ J & 0 & 0 \end{smallmatrix}\right)$. Now one directly computes that with respect to this matrix, elements of $\tilde{\mathfrak{g}} = \mathfrak{so}(\tilde{V})$ have the following block form with block sizes 2, n, and 2:

$$\tilde{\mathfrak{g}} = \left\{ \begin{pmatrix} A & B & c\mathbb{I}_{1,1} \\ D & E & -\mathbb{I}B^tJ \\ f\mathbb{I}_{1,1} & -JD^t\mathbb{I} & -JA^tJ \end{pmatrix} : E \in \mathfrak{so}(p,q) \right\}.$$

The advantage of this choice of basis is of course that we can use the plane generated by e_1 and e_2 for $\tilde{W}$. In particular, $\tilde{\mathfrak{p}} \subset \tilde{\mathfrak{g}}$ corresponds exactly to those matrices for which $D = 0$ and $f = 0$. Moreover, the $|2|$–grading on $\tilde{\mathfrak{g}}$ directly corresponds to the distance of blocks from the diagonal, i.e. $\tilde{\mathfrak{g}}_{-2}$ corresponds to the block $f\mathbb{I}_{1,1}$, $\tilde{\mathfrak{g}}_{-1}$ to the D–block, and so on.

Next, the vector $\tilde{v} := \frac{1}{2}(e_2 - e_{p+q+3})$ evidently satisfies $\langle \tilde{v}, \tilde{v}\rangle = -1$ and is not orthogonal to $\tilde{W}$. For this choice of $\tilde{v}$, the subalgebra $\mathfrak{g} \subset \tilde{\mathfrak{g}}$ consists of those matrices in $\tilde{\mathfrak{g}}$ for which the second column equals the $(p+q+3)$rd column. More explicitly, this means that A is of the form $\left(\begin{smallmatrix} a & c \\ f & 0 \end{smallmatrix}\right)$, and denoting by Y the right column of D, the lower row of B has to be equal to $-Y^t\mathbb{I}$. The intersection of $\tilde{W}$ with $\tilde{v}^\perp$ is then exactly the isotropic line spanned by e_1. In particular, $\mathfrak{p}$ consists of those matrices in $\mathfrak{g}$ for which $f = 0$ and the first column of D vanishes. More generally, the $|1|$–grading on $\mathfrak{g}$ comes from the block decomposition into blocks of size 1, $p+q+2$ and 1, so we obtain a similar presentation as in 1.6.3.

The relation between the curvature $\tilde{\kappa}$ of $\tilde{\omega}$ an the curvature κ of ω is described in Proposition 4.5.2. Let $j : \mathcal{G} \to \tilde{\mathcal{G}}$ be the inclusion and let us interpret the values of the curvature functions as bilinear maps on $\tilde{\mathfrak{g}}$, respectively, on $\mathfrak{g}$ which vanish if one of their entries is from $\tilde{\mathfrak{p}}$, respectively, from $\mathfrak{p}$. Since we view $\mathfrak{g}$ as a subalgebra of $\tilde{\mathfrak{g}}$, the result of Proposition 4.5.2 simply reads as $\tilde{\kappa}(j(u))(X+\tilde{\mathfrak{p}}, Y+\tilde{\mathfrak{p}}) = \kappa(u)(X+\mathfrak{p}, Y+\mathfrak{p})$ for all $X, Y \in \mathfrak{g} \subset \tilde{\mathfrak{g}}$. Since the inclusion of $\mathfrak{g}$ induces a linear isomorphism $\mathfrak{g}/\mathfrak{q} \to \tilde{\mathfrak{g}}/\tilde{\mathfrak{p}}$, this determines $\tilde{\kappa}(j(u))$. By equivariancy of $\tilde{\kappa}$, it is determined by its restriction to $j(\mathcal{G}) \subset \tilde{\mathcal{G}}$. Since ∂^* is also P–equivariant, the same argument shows that to prove normality of $\tilde{\omega}$, it suffices to verify that $\partial^*\tilde{\kappa}(j(u)) = 0$ for all $u \in \mathcal{G}$.

To compute $\partial^*\tilde{\kappa}(j(u))$, we use Lemma 3.1.11, so we have to choose elements of $\tilde{\mathfrak{g}}$ which project onto a basis of $\tilde{\mathfrak{g}}/\tilde{\mathfrak{p}}$. Again, since $\mathfrak{g}/\mathfrak{q} \cong \tilde{\mathfrak{g}}/\tilde{\mathfrak{p}}$, we choose these elements to be in $\mathfrak{g}$. More specifically, we let $X_0 \in \mathfrak{g}$ be the element with A–block $\left(\begin{smallmatrix} 0 & 0 \\ 1 & 0 \end{smallmatrix}\right)$ and correspondingly $f = 1$, while all other blocks are trivial. Next, for $i = 1, \dots, p+q$ we denote by X_i and X_{p+q+i} the elements of $\mathfrak{g}$ which have all entries equal to zero except for one entry in the D–block which equals one, namely the element in the ith row of the first column for X_i and the one in the ith row of the second column for X_{p+q+i}. From the descriptions of the various subalgebras above it is evident that $\{X_0 + \tilde{\mathfrak{p}}, \dots, X_{2(p+q)} + \tilde{\mathfrak{p}}\}$ is a basis of $\tilde{\mathfrak{g}}/\tilde{\mathfrak{p}}$, that $X_i \in \mathfrak{p}$ for $i > p+q$, and that $\{X_0 + \mathfrak{p}, \dots, X_{p+q} + \mathfrak{p}\}$ is a basis of $\mathfrak{g}/\mathfrak{p}$.

Next, we determine the dual bases $\{\tilde{Z}_0, \dots, \tilde{Z}_{2(p+q)}\}$ of $\tilde{\mathfrak{p}}_+$ and $\{Z_0, \dots, Z_{p+q}\}$ of $\mathfrak{p}_+$. Since there is no harm in replacing ∂^* by a nonzero multiple, we may use the trace–form of $\tilde{\mathfrak{g}}$ rather than the Killing forms to induce the dualities for both algebras. Consider the matrix which has a 1 in the ith column of the first row of the B–block and all other entries equal to zero. Then this lies both in $\tilde{\mathfrak{p}}_+$ and in $\mathfrak{p}_+$, it pairs to 1 with X_i under the trace–form and to zero with all other X_j. Hence, this element equals $\tilde{Z}_i$ and Z_i, so in particular, we have $\tilde{Z}_i = Z_i$ for $i = 1, \dots, p+q$. The elements $\tilde{Z}_i$ for $i = p+q+1, \dots, 2(p+q)$ can be described similarly, but we will not need them. Hence, it remains to determine $\tilde{Z}_0$ and Z_0. Since $\tilde{Z}_0 \in \tilde{\mathfrak{p}}_+$ it may have nonzero entries only in the B–block and the block $c\mathbb{I}_{1,1}$. Since $\operatorname{tr}(\tilde{Z}_0 X_i) = 0$ for $i = 1, \dots, 2(p+q)$, also the B–block must be zero, while $\operatorname{tr}(\tilde{Z}_0 X_0) = 1$ implies that $c = 1/2$. Likewise, one determines Z_0 using that it has to lie in $\mathfrak{g} \subset \tilde{\mathfrak{g}}$ and have nonzero entries only in the first row and in the last colum and above the main diagonal. One obtains that the only nontrivial blocks in Z_0 are the A–block, which equals $\left(\begin{smallmatrix} 0 & 1/4 \\ 0 & 0 \end{smallmatrix}\right)$, and correspondingly the one for $c = 1/4$.

According to Lemma 3.1.11, we get

$$\partial^* \tilde{\kappa}(j(u))(X) = 2 \sum_{i=0}^{2(p+q)} [\tilde{Z}_i, \tilde{\kappa}(j(u))(X, X_i)] - \sum_{i=0}^{2(p+q)} \tilde{\kappa}(j(u))([\tilde{Z}_i, X], X_i)$$

for all $X \in \tilde{\mathfrak{g}}$. As before, it suffices to take $X \in \mathfrak{g}$. Now we have $\tilde{\kappa}(j(u))(X, X_i) = \kappa(u)(X, X_i)$ and $\tilde{\kappa}(j(u))([\tilde{Z}_i, X], X_i) = \kappa(u)(Y, X_i)$, where $Y \in \mathfrak{g}$ has the property that $Y + \tilde{\mathfrak{p}} = [\tilde{Z}_i, X] + \tilde{\mathfrak{p}}$. In both cases, the expression vanishes if $i > p + q$, since then $X_i \in \mathfrak{p}$. Hence, in both summands, we only have to sum up to $i = p + q$. Now for $i = 1, \dots, p + q$, we have $\tilde{Z}_i = Z_i \in \mathfrak{p}_+$ and since $X \in \mathfrak{g}$ this implies $[\tilde{Z}_i, X] \in \mathfrak{p}$. From the definitions, one immediately concludes that $[\tilde{Z}_0, Z_i] = 0$, and hence $0 = \operatorname{tr}(X[\tilde{Z}_0, Z_i]) = -\operatorname{tr}([\tilde{Z}_0, X] Z_i)$ for all X and $i = 1, \dots, p + q$. But this implies that $[\tilde{Z}_0, X]$ is congruent to a multiple of X_0 mod $\mathfrak{p}$, and we see that the second summand in the above formula does not contribute at all. Hence, we are left with

$$\partial^* (\tilde{\kappa}(j(u)))(X) = 2[\tilde{Z}_0 - Z_0, \kappa(u)(X, X_0)] + 2 \sum_{i=0}^{p+q} [Z_i, \kappa(u)(X, X_i)].$$

On $\mathfrak{g}$, we are dealing with a $|1|$–grading, so $[\mathfrak{g}, \mathfrak{p}_+] \subset \mathfrak{p}$. Hence, the second summand expresses the full normalization condition for $(\mathfrak{g}, \mathfrak{p})$, and therefore vanishes by normality of ω. Finally, for $\tilde{Z}_0 - Z_0$ the only nonzero blocks are the A–block, which equals $\left(\begin{smallmatrix} 0 & -1/4 \\ 0 & 0 \end{smallmatrix}\right)$, and correspondingly the one for $c = 1/4$. This matrix has the property that it vanishes on $V = \tilde{v}^\perp$ and maps $\tilde{v}$ to a multiple of e_1. On the other hand, $\kappa(u)(X, X_0) \in \mathfrak{g}$, so it vanishes on $\tilde{v}$ and has values in V. Hence, $(\tilde{Z}_0 - Z_0) \circ \kappa(u)(X, X_0)$ vanishes identically and $\kappa(u)(X, X_0) \circ (\tilde{Z}_0 - Z_0)$ vanishes on V. But from Corollary 1.6.8 we know that $\kappa(u)(X, X_0) \in \mathfrak{p}$ and its $\mathfrak{g}_0$–component is totally tracefree. In particular, it acts trivially on the preferred null line spanned by e_1, and we also get $\kappa(u)(X, X_0) \circ (\tilde{Z}_0 - Z_0) = 0$. □

4.5.5. Twistor theory for quaternionic contact structures. For all the other analogs of the Fefferman construction that we are aware of, stronger tools are needed in order to settle the question of normality of the induced Cartan connections. In the rest of this chapter, we will therefore only briefly outline several other examples of analogs of the Fefferman construction. The first of these examples has been mainly worked out by O. Biquard; see [**Bi00**]. This was done only in terms of the underlying structures without any reference to Cartan connections. In the picture of underlying structures, the construction is similar to the twistor construction for quaternionic structures as described in 4.4.9 and 4.4.10. In terms of Cartan connections, it is closely parallel to the classical Fefferman construction.

We have discussed quaternionic contact structures in 4.3.3. Fixing a signature (p, q) and putting $n = p+q$, these geometries exist on manifolds of dimension $4n+3$. (In Biquard's work, only the definite case $q = 0$ is considered.) A quaternionic contact structure of signature (p, q) on a manifold M of that dimension is given by a corank three subbundle $H \subset TM$, such that the bundle of symbol algebras is locally trivial and modelled on a quaternionic Heisenberg algebra of signature (p, q). This means that one can locally identify H with $M \times \mathbb{H}^n$ and TM/H with $M \times \text{im}(\mathbb{H})$ such that the Levi bracket becomes the imaginary part of a quaternionic Hermitian form of signature (p, q).

In 4.3.3 we have seen that quaternionic contact structures of signature (p, q) are equivalent to regular normal parabolic geometries of a certain type (G, P) to be discussed in more detail below. We have also seen there that the subbundle $H \subset TM$ inherits an almost quaternionic structure, i.e. a preferred rank three subbundle $\mathcal{Q} \subset L(H, H)$, which locally can be spanned by I, J, and $I \circ J$ for two anti-commuting almost complex structures I and J on H. To define the twistor space for such a structure, Biquard used an analog of the construction for quaternionic structures we have described in 4.4.9. The subbundle $\mathcal{Q}$ carries a natural norm, which is characterized by the fact that an element of $\mathcal{Q}$ is a unit vector if and only if it defines a complex structure on the corresponding fiber of H. Then the *twistor space* $\tilde{M}$ of the quaternionic contact structure (M, H) is defined as the unit sphere subbundle of $\mathcal{Q}$. By construction, $\tilde{M}$ is the total space of an S^2-bundle over M, so in particular, its dimension equals $4(p + q) + 5$.

Next, Biquard directly defines a corank one subbundle $\tilde{H} \subset T\tilde{M}$ and an almost complex structure $\tilde{J}$ on $\tilde{H}$. Basically, these can be described as follows. For a point $x \in M$, elements of $\mathcal{Q}_x$ act on H_x as scalar multiplications by purely imaginary quaternions. The induced action on $T_xM/H_x \cong \text{im}(\mathbb{H})$ is given by the commutator with the given imaginary quaternion. Hence, any nonzero element in $\mathcal{Q}_x$ determines a two-dimensional subspace in T_xM/H_x, namely the orthocomplement of the kernel of this commutator map. The elements of $\tilde{H}$ in that point are then those tangent vectors, which under the composition of projections $T\tilde{M} \to TM \to TM/H$ land in this codimension one subspace. The complex structure on $\tilde{H}$ is defined directly using the footpoint acting as a complex structure on various spaces.

Next, one shows that $\tilde{H}$ defines a contact structure on M, so $\tilde{J}$ defines a natural almost CR-structures on the twistor space $\tilde{M}$ of (M, H). It turns out that the signature of this CR-structure is $(2p + 1, 2q + 1)$. Using a notion of preferred connetions for quaternionic contact structures, which we will discuss further in Chapter 5, Biquard then proves directly that for definite quaternionic contact structures of dimension at least eleven, this is an integrable and hence a CR-structure. The case of dimension seven is more involved and was sorted out later by Biquard's student D. Duchemin; see [**Du06**]. It turns out that the almost CR-structure on $\tilde{M}$ in this case is integrable if and only if the quaternionic contact structure is torsion free; compare with 4.3.3.

We will next describe a construction of a natural CR-structure on the same space via an analog of the Fefferman construction, which works for general signatures. While we expect that the result coincides with the structures constructed by

Biquard and Duchemin in the positive definite case, to our knowledge this has not been proved until now.

The basic construction is closely parallel to the classical Fefferman construction as discussed in 4.5.1 and 4.5.2. We first have to describe the groups G and P in more detail. The Lie algebra $\mathfrak{g}$ of G is $\mathfrak{sp}(p+1,q+1)$ and G is the automorphism group $\text{Aut}(\mathfrak{g})$. Since the Dynkin diagram of $\mathfrak{sp}(p+1,q+1)$ is of type C, it has no automorphisms, so $\text{Aut}(\mathfrak{g}) = \text{Int}(\mathfrak{g})$, the group of inner automorphisms. Hence, we can realize G as the quotient of $Sp(p+1,q+1)$ by its center, which consists of $\pm\text{id}$ only. The parabolic subgroup $P \subset G$ is the stabilizer of an isotropic quaternionic line in $\mathbb{H}^{p+q+2}$. Now we have to interpret quaternionic vector spaces as special complex vector spaces and quaternionic Hermitian forms as special complex Hermitian forms. So let us view $\mathbb{H}^{p+q+2}$ as $\mathbb{C}^{2p+2q+4}$ with an additional almost complex structure j which anti–commutes with i and a quaternionic Hermitian form of signature $(p+1,q+1)$ as an extension of a complex Hermitian form of signature $(2p+2,2q+2)$ for which j is orthogonal. This gives rise to an inclusion $Sp(p+1,q+1) \hookrightarrow U(2p+2,2q+2)$, and since $Sp(p+1,q+1)$ is simple, the image must be contained in $SU(2p+2,2q+2)$. Projecting to the quotient $\tilde{G} := PSU(2p+2,2q+2)$, the result factorizes to a homomorphism $i : G \to \tilde{G}$, which is easily seen to be injective.

Choose a complex isotropic line $\ell \subset \mathbb{C}^{2p+2q+4}$ and let $\tilde{P} \subset \tilde{G}$ be the stabilizer of ℓ. Elementary linear algebra shows that $Sp(p+1,q+1)$ acts transitively on the space of nonzero null vectors in $\mathbb{C}^{2p+2q+4}$. Hence, G acts transitively on the projectivized null–cone which can be identified with $\tilde{G}/\tilde{P}$. The subgroup $G \cap \tilde{P}$ of course is the stabilizer of ℓ in G. Now a quaternionic linear map which stabilizes ℓ also stabilizes the quaternionic line $\ell_{\mathbb{H}}$ generated by ℓ. Hence, using $\ell_{\mathbb{H}}$ to define the parabolic subgroup $P \subset G$, we see that $G \cap \tilde{P} \subset P$, and we can apply the theory developed in 4.5.1 and 4.5.2. With the tools we have available at this stage, we can easily prove that in the torsion free case, we get a partially integrable almost CR–structure. To prove integrability, a small bit of input from BGG–sequences is needed.

PROPOSITION 4.5.5. *Let (M,H) be a quaternionic contact structure (with vanishing harmonic torsion if* $\dim(M) = 7$*) of signature (p,q) with twistor space $\tilde{M}$. Then $\tilde{M}$ inherits a natural integrable CR–structure of signature* $(2p+1,2q+1)$.

PROOF. Let $(p : \mathcal{G} \to M, \omega)$ be the parabolic geometry of type (G,P) determined by the quaternionic contact structure. We first claim that $\mathcal{G}/(G \cap \tilde{P})$ can be identified with the twistor space $\tilde{M}$ of M. We have noted in 4.3.3 that the group $G_0 \cong P/P_+$ can be identified with the subgroup of $GL(\mathfrak{g}_{-1})$ generated by elements of $Sp(p,q)$ and quaternionic scalar multiplications. The subbundle $\mathcal{Q} \subset L(H,H)$ is the associated bundle to $\mathcal{G}$ corresponding to the natural representation of P (which factors through G_0) on the space of purely imaginary quaternions acting on $\mathfrak{g}_{-1}$. This is given by conjugating imaginary quaternions by the part of G_0 which acts by scalar multiplications on $\mathfrak{g}_{-1}$. The stabilizer of the imaginary quaternion i under this action is is simply given by $\mathbb{C} \setminus \{0\} \subset \mathbb{H} \setminus \{0\}$ and hence coincides with the stabilizer of a complex line in the natural representation of $\mathbb{H} \setminus \{0\}$ on $\mathbb{H}$.

Thus, we can view $G \cap \tilde{P} \subset P$ as the stabilizer of a unit quaternion in $\text{im}(\mathbb{H})$ and since P acts transitively on the unit vectors in $\text{im}(\mathbb{H})$, we conclude that $P/(G \cap \tilde{P})$ can be viewed as the space of this unit vectors. Passing to associated bundles, we obtain an identification of $\mathcal{G}/(G \cap \tilde{P}) \cong \mathcal{G} \times_P P/(G \cap \tilde{P})$ with the space of unit vectors in $\mathcal{Q}$, which is the twistor space $\tilde{M}$. By Theorem 4.5.1, we obtain a canonical Cartan geometry $(\tilde{\mathcal{G}} \to \tilde{M}, \tilde{\omega})$ of type $(\tilde{G}, \tilde{P})$ on $\tilde{M}$.

To proceed, we need some facts about the compatibility of the inclusion $\mathfrak{g} = \mathfrak{sp}(p+1, q+1) \hookrightarrow \mathfrak{su}(2p+2, 2q+2) = \tilde{\mathfrak{g}}$ with the filtrations of the two algebras. The filtration component $\tilde{\mathfrak{g}}^{-1} \subset \tilde{\mathfrak{g}}$ consists of those maps which map the distinguished complex line ℓ into its complex orthocomplement. Likewise, $\mathfrak{g}^{-1} \subset \mathfrak{g}$ consists of those maps which map the quaternionic line $\ell_{\mathbb{H}}$ to its quaternionic orthocomplement, which shows that $\mathfrak{g}^{-1} \subset \tilde{\mathfrak{g}}^{-1}$. On the other hand, $\mathfrak{p}$ is not contained in $\tilde{\mathfrak{p}}$. However, the quotient $\mathfrak{p}/\mathfrak{p}_+ \cong \mathfrak{g}_0$ is the direct sum of a one–dimensional center and two simple ideals isomorphic to $\mathfrak{sp}(1)$ and $\mathfrak{sp}(p,q)$, respectively. The preimage of $\mathfrak{sp}(p,q)$ in $\mathfrak{p}$ clearly is an ideal in $\mathfrak{p}$. Since this acts trivially on $\ell_{\mathbb{H}}$, it is contained in $\tilde{\mathfrak{p}}$.

If we can prove that $(\tilde{\mathcal{G}} \to \tilde{M}, \tilde{\omega})$ is regular, then we obtain an underlying partially integrable almost CR–structure on $\tilde{M}$. Since we have assumed that M has vanishing harmonic torsion, the discussion of harmonic curvature components in 4.3.3 shows that the only nonzero part of the harmonic curvature has to be contained in the part corresponding to the irreducible cohomology component which lies in $\Lambda^2\mathfrak{g}^*_{-1} \otimes \mathfrak{g}_0$. From Theorem 3.1.12 we conclude that the curvature κ of ω has to be homogeneous of degree ≥ 2 and the homogenous component of degree two of the curvature function has to have values in this irreducible cohomology component. But since homogeneous components of degree ≥ 3 automatically have values in $\mathfrak{g}^{-1} \subset \tilde{\mathfrak{g}}^{-1}$, we conclude that the whole curvature function has values in there. Proposition 4.5.2 then implies that, along $j(\mathcal{G})$, also the curvature function $\tilde{\kappa}$ of $\tilde{\omega}$ has values in $\tilde{\mathfrak{g}}^{-1}$, so this has to hold everywhere by equivariancy of $\tilde{\kappa}$. Since we are dealing with a $|2|$–grading on $\tilde{\mathfrak{g}}$, having values in $\tilde{\mathfrak{g}}^{-1}$ implies regularity.

To prove that our CR–structure actually is integrable, some input from BGG–sequences in the form of Corollary 3.2 of [**Čap05**] is needed. First, $(\Lambda^2\mathfrak{g}_- \otimes \mathfrak{p}) \cap \ker(\partial^*)$ is a P–submodule in $\ker(\partial^*)$ whose intersection with the harmonic part evidently contains the cohomology component we are interested in. Hence, we can use Corollary 3.2 of [**Čap05**] to conclude that the curvature function κ has values in this submodule, so ω is torsion free. Now computing the cohomology group $H^2(\mathfrak{g}_-, \mathfrak{g})$ following the recipes from 3.2.18, one sees that the irreducible component in homogeneity two is actually contained in $\Lambda^2\mathfrak{g}^*_{-1} \otimes \mathfrak{sp}(p,q) \subset \Lambda^2\mathfrak{g}^*_{-1} \otimes \mathfrak{g}_0$. Denoting by $\mathfrak{a} \subset \mathfrak{p}$ the preimage of $\mathfrak{sp}(p,q) \subset \mathfrak{p}/\mathfrak{p}_+$, we see that $\Lambda^2\mathfrak{g}^*_- \otimes \mathfrak{a} \subset \Lambda^2\mathfrak{g}^*_- \otimes \mathfrak{g}$ is a P–submodule. Intersecting this with $\ker(\partial^*)$ we again obtain a P–submodule, whose intersection with the harmonic part evidently contains the irreducible cohomology component of homogeneity two. Applying Corollary 3.2 of [**Čap05**] once more (using torsion freeness), we now see that κ has values in $\Lambda^2\mathfrak{g}^*_- \otimes \mathfrak{a}$.

As we have noted above, $\mathfrak{a} \subset \tilde{\mathfrak{p}}$, so along $j(\mathcal{G})$ (and hence everywhere by equivariancy) $\tilde{\kappa}$ has values in $\Lambda^2\tilde{\mathfrak{g}}^*_- \otimes \tilde{\mathfrak{p}}$, so $\tilde{\omega}$ is torsion free. This also implies that $\tilde{\kappa}$ is homogeneous of degree ≥ 2, so normalizing $\tilde{\omega}$ to obtain the normal Cartan connections associated to our almost CR–structure, we only have to add parts of homogeneity at least two; see Propositon 3.1.13. Hence, the curvature of the normal Cartan connection differs from $\tilde{\kappa}$ only in homogeneity ≥ 2. But the Nijenhuis tensor of the almost CR–structure can be computed from the the homogeneity one component of the curvature of the normal Cartan connection; see 4.2.4. Hence, the Nijenhuis tensor has to vanish and integrability follows. □

4.5.6. Projective structures induced by contact projective structures. This is the first analog of the Fefferman construction we discuss for which the Fefferman space coincides with the initial space. Thus, we just associate to one parabolic geometry on a manifold M a parabolic geometry of different type on M via an extension functor. This example is based on the evident inclusion $i : G := Sp(2n+2,\mathbb{R}) \to SL(2n+2,\mathbb{R}) =: \tilde{G}$. The basic parabolic subgroups in $\tilde{G}$ are the stabilizers of k–dimensional subspaces in $\mathbb{R}^{2n+2}$. Correspondingly, the basic homogeneous spaces are the Grassmannians of all k–dimensional subspaces in $\mathbb{R}^{2n+2}$. In general, the size of the orbits of G on the Grassmannians depends on the properties of the restriction of the symplectic form to the subspace, but there is a case in which everything is very easy.

Namely, defining $\tilde{P} \subset \tilde{G}$ to be the stabilizer of a line, we get $\tilde{G}/\tilde{P} = \mathbb{R}P^{2n+1}$, and of course the subgroup $G \subset \tilde{G}$ acts transitively on $\mathbb{R}P^{2n+1}$. Moreover, the subgroup $G \cap \tilde{P} \subset G$ itself is a parabolic subgroup, so we take $P = Q = G \cap \tilde{P}$. This means that the inclusion $i : G \to \tilde{G}$ induces a diffeomorphism $G/P \cong \tilde{G}/\tilde{P}$. Passing to the curved case, the fact that $P = Q$ implies that $\tilde{M} = M$. Therefore, our construction leads from a parabolic geometry $(p : \mathcal{G} \to M, \omega)$ of type (G, P) to a parabolic geometry $(\tilde{p} : \tilde{\mathcal{G}} \to M, \tilde{\omega})$ of type $(\tilde{G}, \tilde{P})$ on the same manifold M.

From 4.2.6 we know that regular normal parabolic geometries of type (G, P) are equivalent to contact projective structures with vanishing contact torsion. On the other hand, by 4.1.5 parabolic geometries of type $(\tilde{G}, \tilde{P})$ are related to projective structures. Hence, the analog of the Fefferman construction leads from a contact projective structure on a contact manifold (M, H) to a projective structure on M. Interpreting the two types of structures in terms of geodesics (see 4.2.7, respectively, 4.1.6), the construction can be interpreted as extending a contact projective structure to a projective structure on the same manifold. This will be discussed in Section 5.3 below, when we have the theory of distinguished curves at our disposal.

The basic features of the construction are easy to prove, also the proof of normality is relatively easy, although as usual, a bit of input from BGG–sequences is needed.

PROPOSITION 4.5.6. *Let $(M, H, [\nabla])$ be a contact projective structure with vanishing contact torsion. Then this induces a canonical classical projective structure on M. Denoting by $(p : \mathcal{G} \to M, \omega)$ the parabolic geometry determined by the contact projective structure, the result $(\tilde{\mathcal{G}} \to M, \tilde{\omega})$ of the Fefferman type construction is automatically normal.*

PROOF. Since the parabolic subgroup $\tilde{P} \subset \tilde{G}$ corresponds to a $|1|$–grading, the extended Cartan connection $\tilde{\omega}$ is automatically regular, and thus gives rise to a projective structure on M. As we have seen in 4.2.6, the cohomology group $H^2(\mathfrak{g}_-, \mathfrak{g})$ is irreducible in this case and contained in $\Lambda^2\mathfrak{g}_{-1}^* \otimes \mathfrak{g}_0$. Hence, the curvature κ of ω is homogenous of degree ≥ 2 and the homogeneous component of degree two of the curvature function has values in $\mathfrak{g}_0$. Since $\mathfrak{g}_{-2}$ has dimension one, homogeneous components of degree > 2 automatically have values in $\mathfrak{p}$, so we conclude that ω is torsion free.

Now in our situation, the inclusion $i' : \mathfrak{g} \hookrightarrow \tilde{\mathfrak{g}}$ of the Lie algebras induces a linear isomorphism $\mathfrak{g}/\mathfrak{p} \to \tilde{\mathfrak{g}}/\tilde{\mathfrak{p}}$, which is equivariant over the embedding $P \hookrightarrow \tilde{P}$. Hence, i' induces an inclusion $\iota : \Lambda^2(\mathfrak{g}/\mathfrak{p})^* \otimes \mathfrak{g} \hookrightarrow \Lambda^2(\tilde{\mathfrak{g}}/\tilde{\mathfrak{p}})^* \otimes \tilde{\mathfrak{g}}$, which by construction is equivariant over $P \hookrightarrow \tilde{P}$. In particular, for the codifferential $\tilde{\partial}^* : \Lambda^2(\tilde{\mathfrak{g}}/\tilde{\mathfrak{p}})^* \otimes \tilde{\mathfrak{g}} \to$

$(\tilde{\mathfrak{g}}/\tilde{\mathfrak{p}})^* \otimes \tilde{\mathfrak{g}}$, the composition $\tilde{\partial}^* \circ \iota$ is equivariant over this inclusion. Restricting this to $\ker(\partial^*)$, where $\partial^* : \Lambda^2(\mathfrak{g}/\mathfrak{p})^* \otimes \mathfrak{g} \to (\mathfrak{g}/\mathfrak{p})^* \otimes \mathfrak{g}$ is the codifferential for $(\mathfrak{g}, \mathfrak{p})$, we still get an equivariant map, so its kernel is a P–submodule. Now it is easy to verify directly that the harmonic part is contained in this submodule. But then Corollary 3.2 of [**Čap05**] implies that the whole curvature function κ has values in this submodule. By Proposition 4.5.2 this implies that the curvature $\tilde{\kappa}$ of $\tilde{\omega}$ has, along $j(\mathcal{G})$ and hence everywhere by equivariancy, values in $\ker(\tilde{\partial}^*)$. But this exactly means that $\tilde{\omega}$ is normal. □

REMARK 4.5.6. The construction of the canonical projective structure induced by a projective contact structure is orginally due to D. Fox; see [**Fox05a**]. His approach was based on the so–called ambient description of projective structures, which is equivalent to the Cartan and/or tractor description; see 5.3.10. In his article, Fox develops an ambient description of contact projective structures, and relates the two structures in this picture. The analogy of this construction to the Fefferman construction was only pointed out later and worked out in [**ČŽ08**].

As we have mentioned in 4.2.6 and 4.2.7, Fox found a canonical Cartan connection also for contact projective structures with nonzero contact torsion, and his construction also works in this more general setting. The approach via the extension functor corresponding to the inclusion $i : G \to \tilde{G}$ and the parabolic subgroups $\tilde{P}$ and $P = G \cap \tilde{G}$ can also be used for these more general geometries. It produces a parabolic geometry of type $(\tilde{G}, \tilde{P})$ and hence a classical projective structure on M, since there is no problem with regularity for a structure corresponding to a $|1|$–grading. However, the contact torsion is nonzero if and only if the canonical Cartan connection associated to the contact projective structure has nontrivial tosion; see [**Fox05a**]. This easily implies that the extended Cartan connection $\tilde{\omega}$ has nonzero torsion, too, and hence cannot be normal. So while the construction continues to work on the level of underlying structures, the nice compatibility with the canonical Cartan connections is lost. See [**ČŽ08**] for more details on this.

4.5.7. Conformal structures induced by generic distributions. These are two more examples of analogs of the Fefferman construction for which the Fefferman space coincides with the original manifold. They were discovered quite recently and while the original constructions were based on Cartan connections, the analogy to the Fefferman construction was only pointed out later. In his article [**Nu05**], P. Nurowski studied Cartan's construction of a canonical Cartan connection associated to a generic rank two distribution on a manifold of dimension five via the equivalence method. Via the trivialization of the tangent bundle of the Cartan bundle by the Cartan connection, any bilinear form on the Lie algebra can be pulled back to a tensor field on the total space of the Cartan bundle. Nurowski showed that for an appropriate choice of a degenerate bilinear form, one obtains a degenerate Riemannian metric on the total space of the Cartan bundle. This is done in such a way that the degenerate directions are exactly the vertical ones, and using this, one can show that the metric descends to a well–defined conformal class on the underlying manifold. After Nurowski's work had appeared, R. Bryant realized that similar ideas can be applied in the case of generic rank three distributions in dimension 6, which he had studied in his thesis [**Br79**]. The construction is described in [**Br06**].

From our current point of view, both these constructions are very similiar to the construction of a projective structure from a contact projective structure in 4.5.6 above. The parabolic geometries equivalent to the two kinds of generic distributions are described in 4.3.2. For rank two distributions in dimension five, the basic Lie algebra $\mathfrak{g}$ is the split real form of the exceptional Lie algebra of type G_2. This can be most easily realized as the Lie algebra of derivations of the (non–associative) algebra of split octonions; see e.g. [**SF00**] or [**Sag06b**]. The corresponding group G is the automorphism group of the split octonions. The split octonions are an eight–dimensional unital algebra and they carry an inner product of split signature $(4,4)$, which is preserved by any automorphism. Moreover, any automorphism preserves the unit element, and hence its orthocomplement, the space of purely imaginary split octonions. This gives rise to a seven–dimensional representation of G which by construction is injective. Since it has an invariant inner product of signature $(3,4)$, it can be viewed as an inclusion $i : G \hookrightarrow SO(3,4) =: \tilde{G}$.

Since we are dealing with a split real form here, the stabilizer of a highest weight line in this representation is a parabolic subgroup $P \subset G$; see 3.2.10 and 3.2.11. From the classification of parabolics it follows immediately that this is the parabolic subgroup which was used in 4.3.2. Now the highest weight line is isotropic, so its stabilizer in $\tilde{G}$ is a parabolic subgroup $\tilde{P} \subset \tilde{G}$. The homogeneous space $\tilde{G}/\tilde{P}$ is the space of all isotropic lines and hence has dimension five. Since we know from 4.3.2 that G/P has the same dimension, the G–orbit of $e\tilde{P}$ in $\tilde{G}/\tilde{P}$ must be open. Since both G/P and $\tilde{G}/\tilde{P}$ are compact (see 3.2.11) we conclude that $i : G \hookrightarrow \tilde{G}$ induces a diffeomorphism $G/P \to \tilde{G}/\tilde{P}$, so the theory developed in 4.5.1 and 4.5.2 applies. The parabolic subgroup $\tilde{P} \subset \tilde{G}$ corresponds to a $|1|$–grading, so parabolic geometries of type $(\tilde{G}, \tilde{P})$ are automatically regular and determine an underlying conformal structure of (split) signature $(2,3)$; see 4.1.2. Hence, Theorem 4.5.1 directly implies existence of a canonical conformal structure of signature $(2,3)$ induced by a generic rank two distribution on a five–dimensional manifold. Using ideas from the theory of Weyl structures, one can derive explicit formulae for metrics in the conformal class based on a notion of generalized contact forms; see [**ČSa07**]. Finally, similar arguments as in in the proof of Proposition 4.5.6 show that the extended Cartan connection is automatically normal; see [**Sag08**].

The case of generic rank three distributions in dimension six is closely parallel. Compared to the discussion in 4.3.2, we have to replace the group $SO_0(4,3)$ used to describe the geometry by its two–fold covering $Spin(4,3) =: G$. At least locally, such an extension is always possible and uniquely determined, so it causes no problems. Doing this, we have the spin representation at our disposal, which for this signature can be chosen to be real, of dimension eight and canonically endowed with an inner product of split signature $(4,4)$. Hence, the spin representation defines an inclusion $i : G \to \tilde{G} := SO(4,4)$. From 4.3.2 we know that the grading on $\mathfrak{g}$ we are dealing with corresponds to the last simple root. From Theorem 3.2.12 and 3.2.15 we conclude that a parabolic subgroup $P \subset G$ for this grading is given as the stabilizer of a highest weight line in the spin representation. This highest weight line is isotropic, so its stabilizer $\tilde{P}$ in $\tilde{G}$ is a parabolic subgroup, and $\tilde{G}/\tilde{P}$ can be identified with the space of null lines in $\mathbb{R}^{(4,4)}$.

Using the fact that G/P and $\tilde{G}/\tilde{P}$ have the same dimension and are both compact, one concludes similarly as before that the spin representation $G \hookrightarrow \tilde{G}$ induces a diffeomorphism $G/P \cong \tilde{G}/\tilde{P}$. Hence, we may apply the theory developed in 4.5.1 to conclude that a generic rank three distribution H on a smooth manifold M of dimension six induces a canonical conformal structure of split signature $(3,3)$ on M. One can also prove in this case, that the extended Cartan connection $\tilde{\omega}$ is automatically normal. More information on this case can be found in [**Br06**] and [**Arm07c**].

REMARK 4.5.7. (1) The representations of split G_2 in $SO(3,4)$ and of $Spin(4,3)$ in $SO(4,4)$ have analogs for compact real forms. These are given by a homomorphism from the compact real form of G_2 to $SO(7)$ and the spin representation $Spin(7) \to SO(8)$. Remarkably, these two homomorphisms correspond to two special Riemannian holonomies.

(2) Bryant's example of a conformal structure associated to a generic rank three distribution in dimension six generalizes to higher dimensions in an unexpected way. From 4.3.2 we know that replacing $SO_0(4,3)$ by $SO_0(n+1,n)$ and taking P to be the parabolic subgroup given by the last simple root, one obtains generic rank n distributions in dimension $\frac{n(n+1)}{2}$. The parabolic subgroup P can be realized as the stabilizer of an isotropic subspace of (maximal) dimension n in $\mathbb{R}^{(n+1,n)}$. Now we can simply include $\mathbb{R}^{(n+1,n)}$ into $\mathbb{R}^{(n+1,n+1)}$ thus defining an inclusion $G := SO_0(n+1,n) \hookrightarrow SO_0(n+1,n+1) =: \tilde{G}$. Choosing an isotropic subspace of (maximal) dimension $n+1$ in $\mathbb{R}^{(n+1,n+1)}$ and letting $\tilde{P}$ be its stabilizer, one easily verifies that $G \cap \tilde{P} = P$ is a parabolic subgroup determined for the grading on $\mathfrak{so}(n+1,n)$ we are dealing with. One further shows that the G–orbit of $e\tilde{P}$ in $\tilde{G}/\tilde{P}$ is open, so the theory from 4.5.1 and 4.5.2 applies.

From 3.2.12 and 3.2.15 we can see that the parabolic subalgebra $\tilde{\mathfrak{p}} \subset \tilde{\mathfrak{g}}$ corresponds to either the last but one or the last (depending on whether our $(n+1)$–dimensional isotropic subspace is self–dual or anti–self–dual) simple root. From 4.1.12 we see that normal parabolic geometries of type $(\tilde{G}, \tilde{P})$ are equivalent to almost spinorial structures, so this is the structure on a manifold of dimension $\frac{n(n+1)}{2}$, which we canonically get from a generic distribution of rank n. The compatibility of the construction of the extended Cartan geometry with normality in this case is subtle. The problem is that for both geometries in question the harmonic curvatures are of torsion type. Hence non–flat geometries always have nontrivial torsion, which makes it difficult to apply BGG–sequences to the problem. This example is studied in [**DS**].

The reason why in the special case $n = 3$ a different structure is obtained is triality. In this case, the group $\tilde{G}$ can be taken to be $Spin(4,4)$. Triality tells us that the group of outer automorphisms of $\tilde{G}$ which is the permutation group $\mathfrak{S}_3$ can be realized as permuting the two spin representations and the standard representation. Correspondingly, the three parabolic subalgebras coming from these representations are all conjugate by an outer automorphism and almost spinorial structures in this dimension are equivalent to conformal structures.

(3) The examples discussed in this subsection together with the one from 4.5.6 exhaust all analogs of the Fefferman construction for which the Fefferman space coincides with the original space. This is proved in [**DS**] based on deep results on generalized flag manifolds due to A. Onishchik; see §15 of [**On94**]. Onishchik proved that for a complex generalized flag manifold G/P, the group of biholomorphisms of the complex manifold G/P coincides with G, except in (the complex versions

of) the three cases considered here. In these cases, the group of biholomorphisms coincides with $\tilde{G}$.

CHAPTER 5

Distinguished connections and curves

In Chapter 3 we have discussed how to associate to a parabolic geometry $(p : \mathcal{G} \to M, \omega)$ of some fixed type (G, P) an infinitesimal flag structure $(p_0 : \mathcal{G}_0 \to M, \theta)$ of the same type. In Chapter 4 we have explicitly described this underlying infinitesimal flag structure in a variety of examples. In view of the equivalences of categories proved in 3.1.14, 3.1.16, and 3.1.18, this provides an equivalent description of regular normal parabolic geometries in almost all cases. In the two exceptional cases (see 4.1.5 and 4.2.6), one may specify a regular normal parabolic geometry by choosing additional data on the level of the infinitesimal flag structure. In any case, the bundle $p_0 : \mathcal{G}_0 \to M$ is easy to construct from the underlying geometric structure.

We have also noticed in Chapter 3 that for topological reasons the principal P_+–bundle $\pi : \mathcal{G} \to \mathcal{G}_0$ is always trivial. In this chapter, we will systematically use this fact to give (in the regular normal case) equivalent descriptions of the Cartan connection ω in terms of objects defined on $\mathcal{G}_0$. The motivating example for these general developments comes from conformal structures, where the bundle $p_0 : \mathcal{G}_0 \to M$ is simply the conformal frame bundle. In 1.6.4 the canonical principal bundle $p : \mathcal{G} \to M$ is constructed in terms of Weyl connections on M, i.e. torsion–free linear connections on the tangent bundle TM, which are compatible with the given conformal class. The value of the canonical normal Cartan connection in a point is then determined by the soldering form, the connection form of the Weyl connection, and the Rho tensor as described in 1.6.7.

These classical constructions for conformal structures admit close analogs for all parabolic geometries. The description which generalizes most easily is the interpretation of Weyl connections as equivariant sections of the bundle $\pi : \mathcal{G} \to \mathcal{G}_0$ from Proposition 1.6.4. Such sections exist for all smooth parabolic geometries, and in view of the conformal case we call them *Weyl structures.* Similarly, as in the conformal case, one may view a general Weyl structure as consisting of a soldering form, a Weyl connection, and a Rho tensor. Technically, the main point about Weyl structures is that they form an affine space modelled on $\Omega^1(M)$ and, although it is involved, the affine structure can be described explicitly. Using this affine structure we obtain a generalization of the interpretation of conformal Weyl structures in terms of scales (see 1.6.5) to general parabolic geometries.

There is an obvious abstract version of the data (soldering form, Weyl connection, and Rho tensor) associated to a Weyl structure. In Section 5.2 we show that among these abstract objects, the ones obtained via Weyl structures from a regular normal parabolic geometry can be characterized by an analog of the normalization conditions for Cartan connections. This normalization condition can be expressed in terms of torsion and curvature quantities that are naturally associated to the abstract objects.

This then allows us to directly construct the soldering form, Weyl connection, and Rho tensor associated to a Weyl structure from the underlying geometry. We obtain an explicit description of tractor bundles, which are equivalent to the Cartan bundle and Cartan connection as discussed in 3.1.19–3.1.22. Hence, this is an alternative way to describe objects that are equivalent to the Cartan geometry. We work out these ideas in detail for structures corresponding to $|1|$–gradings and for parabolic contact structures. In the latter case, we also obtain a version of Webster–Tanaka connections for arbitrary parabolic contact structures. In Section 5.2 we discuss several applications of the concepts developed so far, for example, to questions of affine holonomy theory and of Einstein rescalings of pseudo–Riemannian metrics.

The last part of the chapter is devoted to the study of distinguished curves, as introduced in 1.5.17 for general Cartan geometries, in the special case of parabolic geometries. We discuss the various classes of distinguished curves. Any such curve is determined by some finite jet in one point, and we derive general results on which order of the jet is needed for each class of curves. Particular emphasis is put on cases in which distinguished curves are uniquely determined up to parametrization by their direction in one point. The basic example of this situation is provided by chains in hypersurface type CR–structures. We closed the chapter with some results on the geometry of chains.

5.1. Weyl structures and scales

The basic theme of this section is describing the Cartan connection associated to a parabolic geometry of some fixed type in terms of data on the underlying infinitesimal flag structure. The model case is the family of Weyl connections on a manifold endowed with a conformal structure, so we use the term *Weyl structures* for the general concept. Weyl structures form an affine space modelled on the space of one–forms on the underlying manifold. Understanding the behavior of the various objects associated to a Weyl structure under these affine changes is a key to many results.

For conformal structures, the Levi–Civita connections of the metrics in the conformal class provide a nice subclass of Weyl connections. Replacing the bundle of metrics in the conformal class by appropriate line bundles called *bundles of scales*, this concept generalizes to all parabolic geometries. This leads to the concept of *closed* and of *exact* Weyl structures. A more subtle class of local Weyl structures is provided by the so–called *normal* Weyl structures, which are closely related to normal coordinates and provide the best approximations to the Cartan connection in a point.

Fixing a Weyl structure, one may identify any natural bundle with a bundle associated to a completely reducible representation. Usually this can be nicely phrased as an identification with the associated graded bundle (with respect to a natural filtration). Specialized to tractor bundles, this leads to an explicit description of tractor calculus in terms of a Weyl structure as well as formulae for the effect of a change of Weyl structures. This topic will also be taken up in Section 5.2 below, where we will have more explicit descriptions of Weyl structures at hand.

5.1.1. Weyl structures. Let $\mathfrak{g} = \mathfrak{g}_{-k} \oplus \cdots \oplus \mathfrak{g}_k$ be a $|k|$–graded semisimple Lie algebra, G a Lie group with Lie algebra $\mathfrak{g}$, let $P \subset G$ be a parabolic subgroup for the given grading and $G_0 \subset P$ the Levi subgroup; see 3.1.3. Let $(p : \mathcal{G} \to M, \omega)$

be a parabolic geometry of type (G,P), and consider the underlying principal G_0–bundle $p_0:\mathcal{G}_0\to M$ introduced in 3.1.5. By definition, $\mathcal{G}_0=\mathcal{G}/P_+$, so there is a natural projection $\pi:\mathcal{G}\to\mathcal{G}_0$, which is a principal bundle with structure group P_+.

DEFINITION 5.1.1. A (local) *Weyl structure* for the parabolic geometry $(p:\mathcal{G}\to M,\omega)$ is a (local) smooth G_0–equivariant section $\sigma:\mathcal{G}_0\to\mathcal{G}$ of the projection $\pi:\mathcal{G}\to\mathcal{G}_0$.

As a first basic result we want to prove that global Weyl structures always exist and that they form an affine space. From 3.1.5 we know that the associated graded $\mathrm{gr}(TM)$ of the tangent bundle TM is the associated bundle $\mathcal{G}_0\times_{G_0}(\mathfrak{g}_{-k}\oplus\cdots\oplus\mathfrak{g}_{-1})$. The action of G_0 on each $\mathfrak{g}_i$ is induced by the restriction of the adjoint action of G to the subgroup G_0. By part (2) of Proposition 3.1.2, for each $i\geq 0$ the G_0–module $\mathfrak{g}_i$ is dual to $\mathfrak{g}_{-i}$, so we conclude that the associated graded $\mathrm{gr}(T^*M)$ of the cotangent bundle is isomorphic to $\mathcal{G}_0\times_{G_0}(\mathfrak{g}_1\oplus\cdots\oplus\mathfrak{g}_k)$. The dualities are induced by the Killing form of $\mathfrak{g}$. In particular, smooth sections of $\mathrm{gr}(T^*M)$ can be identified with smooth functions $f=(f_1,\dots,f_k):\mathcal{G}_0\to\mathfrak{g}_1\oplus\cdots\oplus\mathfrak{g}_k$ such that $f(u\cdot g)=\mathrm{Ad}(g^{-1})(f(u))$ for all $u\in\mathcal{G}_0$ and $g\in G_0$.

PROPOSITION 5.1.1. *For any parabolic geometry $(p:\mathcal{G}\to M,\omega)$, there exists a global Weyl structure $\sigma:\mathcal{G}_0\to\mathcal{G}$.*

Fixing one Weyl structure σ, there is a bijective correspondence between the set of all Weyl structures and the space $\Gamma(\mathrm{gr}(T^*M))$ of smooth sections of the associated graded of the cotangent bundle. Explicitly, this correspondence is given by mapping $\Upsilon\in\Gamma(\mathrm{gr}(T^*M))$ with corresponding functions $\Upsilon_i:\mathcal{G}_0\to\mathfrak{g}_i$ for $i=1,\dots,k$ to the Weyl structure $\hat\sigma(u):=\sigma(u)\exp(\Upsilon_1(u))\cdots\exp(\Upsilon_k(u))$.

PROOF. By topological dimension theory, any fiber bundle over the smooth manifold M admits an atlas with finitely many (usually disconnected) charts; see [**GHV72**]. Hence, there is a finite open covering $\{U_1,\dots,U_N\}$ of M such that both $\mathcal{G}$ and $\mathcal{G}_0$ are trivial over each U_i. Via a chosen trivialization, the inclusion $G_0\hookrightarrow P$ induces a smooth G_0–equivariant section $\sigma_i:p_0^{-1}(U_i)\to p^{-1}(U_i)$ for each $i=1,\dots,N$. Moreover, we can find open sets $V_1,\dots,V_N$ such that $\bar V_i\subset U_i$ for all i, and such that $\{V_1,\dots,V_N\}$ still is a covering of M.

By Theorem 3.1.3, the exponential map restricts to a diffeomorphism $\mathfrak{p}_+\to P_+$, where $\mathfrak{p}_+=\mathfrak{g}_1\oplus\cdots\oplus\mathfrak{g}_k$. Thus, there is a smooth map $\Psi:p_0^{-1}(U_1\cap U_2)\to\mathfrak{p}_+$ such that $\sigma_2(u)=\sigma_1(u)\exp(\Psi(u))$ for all $u\in p_0^{-1}(U_1\cap U_2)$. Equivariance of σ_1 and σ_2 immediately implies that $\Psi(u\cdot g)=\mathrm{Ad}(g^{-1})(\Psi(u))$ for all $g\in G_0$. Now let $f:M\to[0,1]$ be a smooth function with support contained in U_2, which is identically one on V_2 and define $\sigma:p_0^{-1}(U_1\cup V_2)\to p^{-1}(U_1\cup V_2)$ by $\sigma(u)=\sigma_1(u)\exp(f(p_0(u))\Psi(u))$ for $u\in U_1$ and by $\sigma(u)=\sigma_2(u)$ for $u\in V_2$. Then obviously these two definitions agree on $U_1\cap V_2$, so σ is smooth. Moreover, from equivariancy of the σ_i and of Ψ one immediately concludes that σ is equivariant. In the same way, one next extends the section to $U_1\cup V_2\cup V_3$ and by induction one reaches a globally defined smooth equivariant section.

Fixing a global equivariant section σ, any other section $\hat\sigma$ of $\pi:\mathcal{G}\to\mathcal{G}_0$ can be written uniquely as $\hat\sigma(u)=\sigma(u)\Phi(u)$ for some a smooth function $\Phi:\mathcal{G}_0\to P_+$. The section $\hat\sigma$ is G_0–equivariant if and only if $\Phi(u\cdot g)=g^{-1}\Phi(u)g$ for all $u\in\mathcal{G}_0$ and all $g\in G_0$. By Theorem 3.1.3, the map $(Z_1,\dots,Z_k)\mapsto\exp(Z_1)\cdots\exp(Z_k)$ is a diffeomorphism $\mathfrak{p}_+\to P_+$, so we may uniquely write Φ

as $\Phi(u) = \exp(\Upsilon_1(u)) \cdots \exp(\Upsilon_k(u))$ for smooth functions $\Upsilon_i : \mathcal{G}_0 \to \mathfrak{g}_i$. Since

$$g^{-1} \exp(\Upsilon_1(u)) \cdots \exp(\Upsilon_k(u))g = g^{-1} \exp(\Upsilon_1(u))g \cdots g^{-1} \exp(\Upsilon_k(u))g$$
$$= \exp(\mathrm{Ad}(g^{-1})(\Upsilon_1(u))) \cdots \exp(\mathrm{Ad}(g^{-1})(\Upsilon_k(u))),$$

uniqueness of the representation as a product implies that $\Phi(u \cdot g) = g^{-1}\Phi(u)g$ is equivalent to $\Upsilon_i(u \cdot g) = \mathrm{Ad}(g^{-1})(\Upsilon_i(u))$ for all $i = 1, \dots, k$, and the result follows. □

Fixing one Weyl structure σ, there is another bijection between the set of all Weyl structures and the space $\Gamma(\mathrm{gr}(T^*M))$. Namely, for $\Upsilon \in \Gamma(\mathrm{gr}(T^*M))$ corresponding to the functions $\Upsilon_i : \mathcal{G}_0 \to \mathfrak{g}_i$ one may also put

$$\widehat{\sigma}(u) = \sigma(u) \exp(\Upsilon_1(u) + \cdots + \Upsilon_k(u)).$$

The proof that this indeed defines a bijection is completely parallel to the proof in the proposition. The advantage of the convention used there is that it makes it easier to separate homogeneous degrees.

5.1.2. Weyl connections, soldering form and Rho tensor. Given a Weyl structure $\sigma : \mathcal{G}_0 \to \mathcal{G}$ for a parabolic geometry $(p : \mathcal{G} \to M, \omega)$, we can consider the pullback $\sigma^*\omega \in \Omega^1(\mathcal{G}_0, \mathfrak{g})$ of the Cartan connection. Equivariancy of σ reads as $r^g \circ \sigma = \sigma \circ r^g$ for all $g \in G_0$. Using this, we obtain

$$(r^g)^*(\sigma^*\omega) = \sigma^*((r^g)^*\omega) = \mathrm{Ad}(g^{-1}) \circ (\sigma^*\omega),$$

so the one–form $\sigma^*\omega$ is G_0–equivariant. As a G_0–module, the Lie algebra $\mathfrak{g}$ decomposes as $\mathfrak{g}_{-k} \oplus \cdots \oplus \mathfrak{g}_k$, so decomposing $\sigma^*\omega = \sigma^*\omega_{-k} + \cdots + \sigma^*\omega_k$ accordingly, each of the components is G_0–equivariant.

PROPOSITION 5.1.2. *Let $\sigma : \mathcal{G}_0 \to \mathcal{G}$ be a Weyl structure on a parabolic geometry $(p : \mathcal{G} \to M, \omega)$. Then we have:*

(1) *The component $\sigma^*\omega_0 \in \Omega^1(\mathcal{G}_0, \mathfrak{g}_0)$ defines a principal connection on the bundle $p_0 : \mathcal{G}_0 \to M$.*

(2) *The components $\sigma^*\omega_{-k}, \dots, \sigma^*\omega_{-1}$ can be interpreted as defining an element of $\Omega^1(M, \mathrm{gr}(TM))$. This form determines an isomorphism*

$$TM \to \mathrm{gr}(TM) = T^{-k}M/T^{-k+1}M \oplus \cdots \oplus T^{-1}M$$

which is a splitting of the filtration of TM. This means that for each $i = -k, \dots, -1$, the subbundle T^iM is mapped to $\bigoplus_{j \geq i} \mathrm{gr}_j(TM)$ and the component in $\mathrm{gr}_i(TM)$ is given by the canonical surjection $T^iM \to T^iM/T^{i+1}M$.

(3) *The components $\sigma^*\omega_1, \dots, \sigma^*\omega_k$ can be interpreted as a one–form $\mathsf{P} \in \Omega^1(M, \mathrm{gr}(T^*M))$.*

PROOF. For an element $A \in \mathfrak{g}_0$, the corresponding fundamental vector field on $\mathcal{G}_0$ is by definition given by $\zeta_A^{\mathcal{G}_0}(u) = \frac{d}{dt}|_{t=0} u \cdot \exp(tA)$ for all $u \in \mathcal{G}_0$. Equivariancy of σ immediately implies that $\sigma(u \cdot \exp(tA)) = \sigma(u) \cdot \exp(tA)$, which in turn implies that $T_u\sigma \cdot \zeta_A^{\mathcal{G}_0}(u) = \zeta_A^{\mathcal{G}}(\sigma(u))$, the fundamental vector field on $\mathcal{G}$. By definition, the Cartan connection ω reproduces the generators of fundamental vector fields, so we conclude that $\sigma^*\omega_0$ reproduces the generators of fundamental vector fields. Since we have already observed that $\sigma^*\omega_0$ is G_0–equivariant, this proves (1).

On the other hand, the components $\sigma^*\omega_i$ for $i \neq 0$ vanish upon insertion of a fundamental field, so they are horizontal. Since we have already observed that

$\sigma^*\omega_i$ is G_0–equivariant, Corollary 1.2.7 shows that it determines an element of $\Omega^1(M, \mathcal{G}_0 \times_{G_0} \mathfrak{g}_i)$. For $i > 0$ we have $\mathcal{G}_0 \times_{G_0} \mathfrak{g}_i \cong \text{gr}_i(T^*M)$ and the proof of (3) is complete.

For $i < 0$ we get $\mathcal{G}_0 \times \mathfrak{g}_i = \text{gr}_i(TM)$, so we can interpret $\sigma^*\omega_-$ as an element of $\Omega^1(M, \text{gr}(TM))$. To verify the remaining claims in (2) consider $\omega(\sigma(u))$ for a point $u \in \mathcal{G}_0$. This is a linear isomorphism $T_{\sigma(u)}\mathcal{G} \to \mathfrak{g}$ which reproduces the generators of fundamental vector fields, so it descends to a linear isomorphism $T_{\sigma(u)}\mathcal{G}/\ker(T_u p) \cong \mathfrak{g}/\mathfrak{p}$. The map $T_u\sigma$ induces a linear isomorphism $T_u\mathcal{G}_0/\ker(T_u p_0) \to T_{\sigma(u)}\mathcal{G}/\ker(T_u p)$. Hence, we conclude from the construction that the element of $\Omega^1(M, \text{gr}(TM))$ defined by the negative components of $\sigma^*\omega$ restricts to an injection on each tangent space. Since the bundles TM and $\text{gr}(TM)$ have the same rank, we must obtain an isomorphism $TM \to \text{gr}(TM)$.

From 3.1.5 we know that the identification of $\text{gr}_i(TM)$ with $\mathcal{G}_0 \times_{G_0} \mathfrak{g}_i$ is obtained in terms of the frame form $\theta = (\theta_{-k}, \dots, \theta_{-1})$ as follows. For $u \in \mathcal{G}_0$ and $\xi \in T^i_u\mathcal{G}_0$, the pair $(u, \theta_i(u)(\xi)) \in \mathcal{G}_0 \times \mathfrak{g}_i$ represents the class of $T_u p_0 \cdot \xi$ in $T^iM/T^{i+1}M$. But recall that the component θ_i of the frame form was obtained by choosing any lift of ξ to a tangent vector on $\mathcal{G}$ and take the $\mathfrak{g}_i$–component of the value of ω on that lift. This immediately shows that $\theta_i(u)(\xi) = \sigma^*\omega_i(u)(\xi)$ for all $\xi \in T^i_u\mathcal{G}_0$, which implies the last claim in (2). □

DEFINITION 5.1.2. Let $\sigma : \mathcal{G}_0 \to \mathcal{G}$ be a Weyl structure for a parabolic geometry $(p : \mathcal{G} \to M, \omega)$.

(1) The principal connection $\sigma^*\omega_0$ on the bundle $\mathcal{G}_0 \to M$ is called the *Weyl connection* associated to the Weyl structure σ.

(2) The $\text{gr}(TM)$–valued one–form on M determined by the negative components of $\sigma^*\omega$ is called the *soldering form* associated to the Weyl structure σ.

(3) The one–form $\mathsf{P} \in \Omega^1(M, \text{gr}(T^*M))$ induced by the positive components of $\sigma^*\omega$ is called the *Rho tensor* associated to the Weyl structure σ.

5.1.3. Weyl connections and soldering forms on natural bundles. Let $\sigma : \mathcal{G}_0 \to \mathcal{G}$ be a Weyl structure for a parabolic geometry $(p : \mathcal{G} \to M, \omega)$ of type (G, P). Then the corresponding Weyl connection $\sigma^*\omega_0$ is a principal connection on the bundle $\mathcal{G}_0$. As discussed in 1.3.4, this principal connection gives rise to an induced connection on any fiber bundle associated to the principal bundle $\mathcal{G}_0$. In the case of an associated vector bundle, the resulting connection is automatically linear. All these induced connections will be referred to as the Weyl connections corresponding to the Weyl structure σ.

Assume that $\ell : P \times S \to S$ is a smooth left action of the group P on a smooth manifold S. Since G_0 by definition is a subgroup of P, we can restrict this action to a smooth left action $\underline{\ell} : G_0 \times S \to S$. Via the action ℓ, we can form the associated bundle $\mathcal{G} \times_P S \to M$, while $\underline{\ell}$ gives rise to the fiber bundle $\mathcal{G}_0 \times_{G_0} S \to M$. From above we know that we have a Weyl connection on the associated bundle $\mathcal{G}_0 \times_{G_0} S \to M$.

PROPOSITION 5.1.3. *Let $(p : \mathcal{G} \to M, \omega)$ be a parabolic geometry of some fixed type (G, P), and let S be a smooth manifold endowed with a smooth left P–action. Then choosing a Weyl structure $\sigma : \mathcal{G}_0 \to \mathcal{G}$ induces an isomorphism $\mathcal{G} \times_P S \cong \mathcal{G}_0 \times_{G_0} S$ and thus gives rise to a connection on the natural bundle $\mathcal{G} \times_P S$. In the case of a natural vector bundle this connection is automatically linear.*

PROOF. Consider the smooth map $\sigma \times \mathrm{id}_S : \mathcal{G}_0 \times S \to \mathcal{G} \times S$. Composing with $q : \mathcal{G} \times S \to \mathcal{G} \times_P S$ we obtain a smooth map, which by equivariancy of σ factors to a smooth map $\mathcal{G}_0 \times_{G_0} S \to \mathcal{G} \times_P S$. By construction, this is a fiber bundle map covering the identity on M and it restricts to diffeomorphisms on the fibers. Hence, it is an isomorphism of fiber bundles. We have already observed above that σ gives rise to a connection on the bundle $\mathcal{G}_0 \times_{G_0} S$, which is linear in the case of a natural vector bundle. □

For a large class of natural vector bundles, we can interpret this result in a more conceptual way, which shows that the soldering form on the tangent bundle generalizes to this class of vector bundles.

COROLLARY 5.1.3. *Let V be a finite-dimensional representation of P, which is completely reducible as a representation of G_0.*

(1) There is a P–invariant filtration $V = V^0 \supset V^1 \supset \cdots \supset V^N \supset \{0\}$ such that for each i, the action of $\mathfrak{p}_+$ maps V^i to V^{i+1}.

(2) For a manifold M endowed with a parabolic geometry, let $\mathcal{V}M \to M$ be the natural vector bundle induced by V. Consider the filtration $\{\mathcal{V}^iM\}$ of $\mathcal{V}M$ by smooth subbundles induced by the filtration of V from (1). Then we can naturally identify $\mathrm{gr}(\mathcal{V}M)$ with $\mathcal{G}_0 \times_{G_0} V$, so any Weyl structure induces an isomorphism between $\mathcal{V}M$ and its associated graded bundle, which defines a splitting of the filtration.

PROOF. (1) We have proved in Proposition 3.2.12 that one obtains a $\mathfrak{p}$–invariant filtration on V by putting $V^N := \{v \in \mathbb{V} : Z \cdot v = 0 \quad \forall Z \in \mathfrak{p}_+\}$ and then inductively $V^{i-1} := \{v \in V : Z \cdot v \in V^i \quad \forall Z \in \mathfrak{p}_+\}$. We have also seen there, that each of the subquotients V^i/V^{i+1} is completely reducible as a representation of $\mathfrak{g}_0$. Now clearly the filtration is also invariant under the action of the group P and the subquotients are completely reducible as G_0–modules. Since $\mathfrak{p}_+ \cdot V^i \subset V^{i+1}$ by construction, this completes the proof of (1).

(2) The associated graded $\mathrm{gr}(V)$ to the filtered vector space V carries a natural representation of P; see 3.1.1. Since $\mathfrak{p}_+ \cdot V^i \subset V^{i+1}$, we see that $\mathfrak{p}_+$ and thus P_+ acts trivially on $\mathrm{gr}(V)$. On the other hand, since the G_0–action on V is completely reducible, we can find for each $i \geq 0$ a G_0–invariant subspace $V_i \subset V^i$ such that $V^i = V_i \oplus V^{i+1}$. In particular, $V_i \cong V^i/V^{i+1} = \mathrm{gr}_i(V)$ and as a G_0–module $V = V_0 \oplus \cdots \oplus V_N$, so V is naturally isomorphic to $\mathrm{gr}(V)$ as a G_0–module.

Now let $\pi : \mathcal{G} \to \mathcal{G}_0 = \mathcal{G}/P_+$ be the natural projection. Since P_+ acts trivially on $\mathrm{gr}(V)$ and $P/P_+ \cong G_0$, the map $\pi \times \mathrm{id} : \mathcal{G} \times \mathrm{gr}(V) \to \mathcal{G}_0 \times \mathrm{gr}(V)$ factors to a diffeomorphism $\mathcal{G} \times_P \mathrm{gr}(V) \to \mathcal{G}_0 \times_{G_0} \mathrm{gr}(V)$, and the latter bundle can be identified with $\mathcal{G}_0 \times_{G_0} V$. Hence, we can naturally interpret $\mathcal{G}_0 \times_{G_0} V$ as the associated graded vector bundle $\mathrm{gr}(\mathcal{G} \times_P V)$. From the construction it is clear, that this is a splitting of the filtration. □

REMARK 5.1.3. (1) From 3.2.12 we know that on the level of the Lie algebra $\mathfrak{g}_0$, complete reducibility of a complex representation is equivalent to the center $\mathfrak{z}(\mathfrak{g}_0)$ acting diagonalizably. Hence, the assumption that a representation of P is completely reducible as a representation of G_0 is a very weak condition.

(2) Denoting the largest filtration component by V^0 is just one possible convention. In some cases (for example for the adjoint tractor bundle) other conventions are more natural and we will use these.

(3) The filtration from part (1) of the corollary can be described on the level of bundles. The infinitesimal action of $\mathfrak{p}_+$ defines a map $\mathfrak{p}_+ \times V \to V$, which

is a P–homomorphism. Since $\mathcal{G} \times_P \mathfrak{p}_+ = T^*M$, we get an induced bundle map $T^*M \times \mathcal{V}M \to \mathcal{V}M$, which makes $\mathcal{V}M$ into a bundle of modules over the bundle T^*M of Lie algebras. We will write this bundle map as $(\phi, s) \mapsto \phi \bullet s$. The filtration of the bundle $\mathcal{V}M$ is then given by $\mathcal{V}_x^N M = \{v \in \mathcal{V}_x M : \phi \bullet v = 0 \quad \forall \phi \in T_x^*M\}$ and $\mathcal{V}_x^{i-1}M = \{v \in \mathcal{V}_x M : \phi \bullet v \in \mathcal{V}_x^i M \quad \forall \phi \in T_x^*M\}$.

(4) Choosing a Weyl structure σ induces an isomorphism $\mathcal{V}M \to \operatorname{gr}(\mathcal{V}M)$, which can be used to transfer the Weyl connection on $\operatorname{gr}(\mathcal{V}M)$ to $\mathcal{V}M$. Of course, this generalizes the isomorphism $TM \cong \operatorname{gr}(TM)$ given by the soldering form associated to σ, see 5.1.2.

EXAMPLE 5.1.3. Consider the adjoint tractor bundle $\mathcal{A}M = \mathcal{G} \times_P \mathfrak{g}$ and its associated graded $\operatorname{gr}(\mathcal{A}M) = \operatorname{gr}_{-k}(\mathcal{A}M) \oplus \cdots \oplus \operatorname{gr}_k(\mathcal{A}M)$. Each fiber of $\operatorname{gr}(\mathcal{A}M)$ is a graded Lie algebra isomorphic to $\mathfrak{g}$, so in each fiber $\operatorname{gr}(\mathcal{A}_x M)$ there is a unique element $E(x)$ such that $\{E(x), \_\}$ is multiplication by j on the grading component $\operatorname{gr}_j(\mathcal{A}_x M)$; see Proposition 3.1.2. These elements fit together to define a smooth section $E \in \Gamma(\operatorname{gr}_0(\mathcal{A}M))$ called the *grading section* of $\operatorname{gr}(\mathcal{A}M)$. Indeed, viewing $\operatorname{gr}(\mathcal{A}M)$ as $\mathcal{G}_0 \times_{G_0} \mathfrak{g}$, the section E corresponds to the constant function mapping all of $\mathcal{G}_0$ to the grading element of $\mathfrak{g}$.

Choosing a Weyl structure $\sigma : \mathcal{G}_0 \to \mathcal{G}$ we obtain an isomorphism $\mathcal{A}M \to \operatorname{gr}(\mathcal{A}M)$ and thus a section $E^\sigma \in \Gamma(\mathcal{A}M)$ corresponding to the grading section E. By construction, E^σ is a section of the filtration component $\mathcal{A}^0M$, which maps to E under the natural projection $\mathcal{A}^0M \to \operatorname{gr}_0(\mathcal{A}M)$. Otherwise put, the choice of the Weyl structure σ gives us a lift $E^\sigma \in \Gamma(\mathcal{A}^0M)$ of the grading section. We shall see later, that $\sigma \mapsto E^\sigma$ defines a bijection between the set of Weyl structures and the set of all such lifts.

5.1.4. Bundles of scales. Our next aim is to prove that a small part of the data associated to a Weyl structure σ in 5.1.2 is already sufficient to uniquely pin down σ. More precisely, we want to show that there is a class of natural line bundles, such that the induced Weyl connection on one of this line bundles determines σ uniquely. A natural line bundle associated to $\mathcal{G}_0$ is given by a homomorphism $\lambda : G_0 \to \mathbb{R}$, which defines a one–dimensional representation of G_0 on $\mathbb{R}$. The bundle is oriented (and hence can be trivialized) if and only if the representation has values in $\mathbb{R}_+$. The (oriented) line bundle $L^\lambda = \mathcal{G}_0 \times_\lambda \mathbb{R}$ can be equivalently described by its (oriented) frame bundle $\mathcal{L}^\lambda$, which is a principal bundle with structure group $\mathbb{R}$ (respectively $\mathbb{R}_+$) on M. The frame bundle can be most easily described as the orbit space $\mathcal{G}_0/\ker(\lambda)$ of the normal subgroup $\ker(\lambda) \subset G_0$.

A homomorphism $\lambda : G_0 \to \mathbb{R}$ gives rise to a Lie algebra homomorphism $\lambda' : \mathfrak{g}_0 \to \mathbb{R}$. This homomorphism vanishes on the semisimple part of $\mathfrak{g}_0$, so it is just a linear functional on the center $\mathfrak{z}(\mathfrak{g}_0)$. To describe the homomorphisms which are appropriate for our purposes, recall from 3.1.2 that the Killing form B of $\mathfrak{g}$ restricts to a nondegenerate bilinear form on $\mathfrak{g}_0$. The splitting of $\mathfrak{g}_0$ into center and semisimple part is orthogonal with respect to B, so the restriction of B to $\mathfrak{z}(\mathfrak{g}_0)$ still is nondegenerate. Given a homomorphism $\lambda : G_0 \to \mathbb{R}$ we therefore get a unique element $E_\lambda \in \mathfrak{z}(\mathfrak{g}_0)$ such that $\lambda'(A) = B(E_\lambda, A)$ for all $A \in \mathfrak{g}_0$.

DEFINITION 5.1.4. (1) An element $F \in \mathfrak{z}(\mathfrak{g}_0)$ is called a *scaling element* if and only if the restriction to $\mathfrak{p}_+$ of the adjoint action $\operatorname{ad}_F : \mathfrak{g} \to \mathfrak{g}$ is injective.

(2) A *bundle of scales* for parabolic geometries of type (G, P) is a natural principal $\mathbb{R}$–bundle $\mathcal{L}^\lambda$ associated to a homomorphism $\lambda : G_0 \to \mathbb{R}$, such that the

corresponding element $E_\lambda \in \mathfrak{z}(\mathfrak{g}_0)$ is a scaling element. If λ has values in $\mathbb{R}_+$, then we obtain an *oriented bundle of scales.*

(3) Having chosen a bundle $\mathcal{L}^\lambda$ of scales, a *(local) scale* for a parabolic geometry $(p : \mathcal{G} \to M, \omega)$ of type (G, P) is a (local) smooth section of the principal $\mathbb{R}_+$–bundle $\mathcal{L}^\lambda \to M$.

PROPOSITION 5.1.4. *(1) For any type of parabolic geometry there exist natural oriented bundles of scales.*

(2) A bundle of scales admits global smooth sections, if and only if it is oriented.

PROOF. (1) The grading element $E \in \mathfrak{z}(\mathfrak{g}_0)$ (see 3.1.2) acts on $\mathfrak{g}_i$ by multiplication with i, so it is a scaling element. For $g \in G_0$ the adjoint action $\mathrm{Ad}(g) : \mathfrak{g} \to \mathfrak{g}$ by definition respects the grading of $\mathfrak{g}$, and we denote by $\mathrm{Ad}_j(g)$ the restriction of $\mathrm{Ad}(g)$ to $\mathfrak{g}_j$. Now define $\lambda : G_0 \to \mathbb{R}$ by

$$\lambda(g) := \prod_{j=1}^{k} |\det(\mathrm{Ad}_j(g))|^{2j} .$$

This clearly is a homomorphism with derivative $\lambda' : \mathfrak{g}_0 \to \mathbb{R}$ given by $\lambda'(A) = \sum_{j=1}^k 2j \operatorname{tr}(\mathrm{ad}_j(A))$, where $\mathrm{ad}_j(A)$ denotes the restriction of $\mathrm{ad}(A)$ to $\mathfrak{g}_j$. By definition, $B(E, A) = \operatorname{tr}(\mathrm{ad}(E) \circ \mathrm{ad}(A))$, and this composition acts by $j\,\mathrm{ad}_j(A)$ on each component $\mathfrak{g}_j$. Hence, $B(E, A) = \sum_{j=-k}^k j \operatorname{tr}(\mathrm{ad}_j(A))$, and since from Proposition 3.1.2 we know that $\mathrm{ad}_{-j}(A)$ is the dual map of $\mathrm{ad}_j(A)$, we conclude that $\lambda'(A) = B(E, A)$. Thus the one–dimensional representation λ gives rise to a natural bundle of scales.

(2) This is just due to the fact that orientable real line bundles and thus principal $\mathbb{R}^+$–bundles are automatically globally trivial and hence admit global smooth sections. □

REMARK 5.1.4. For maximal parabolic subalgebras, i.e. those corresponding to Satake diagrams with just one crossed root, the center of $\mathfrak{g}_0$ has dimension one and is generated by the grading element E. Hence, any nonzero element of $\mathfrak{z}(\mathfrak{g}_0)$ is a scaling element in this case, and the bundle of scales constructed in the proposition is the unique oriented bundle of scales up to forming roots or tensor powers (which poses no problem for trivial bundles).

In all applications we know of it is sufficient to restrict to bundles of scales corresponding to multiples of the grading element (however, some nontrivial multiples maybe particularly convenient). We have chosen to present the theory in a more general version, since this causes no additional difficulties.

5.1.5. The effect of a change of Weyl structures on soldering forms. The key step towards the basic results on Weyl structures is understanding the effect of a change of Weyl structure on the various data associated to it. We start by analyzing the effect of the change on soldering forms, and we will only deal with the case of natural vector bundles. Considering two Weyl structures σ and $\hat{\sigma}$ we will always denote quantities corresponding to σ by unhatted symbols and quantities corresponding to $\hat{\sigma}$ by hatted symbols. For a representation V of P, which is completely reducible as a representation of G_0, and the corresponding natural vector bundle $\mathcal{V}M$, we will denote the isomorphism $\mathcal{V}M \to \mathrm{gr}(\mathcal{V}M)$ induced by the Weyl structure σ by $\nu \mapsto (\nu)_\sigma := (\nu_0, \dots, \nu_N)$ (if the filtration indices start with 0) and the isomorphism corresponding to $\hat{\sigma}$ by $\nu \mapsto (\nu)_{\hat{\sigma}} = (\hat{\nu}_0, \dots, \hat{\nu}_N)$.

To formulate the results efficiently, we introduce some notation for multi–indices. We will consider multi–indices consisting of k components (in the case of a $|k|$–grading), and write $\underline{i} = (i_1, \dots, i_k)$ with $i_1, \dots, i_k \geq 0$. For such a multi–index $\underline{i}$ we put $\underline{i}! := i_1! \cdots i_k!$, $\|\underline{i}\| := i_1 + 2i_2 + \cdots + ki_k$, and $(-1)^{\underline{i}} := (-1)^{i_1 + \cdots + i_k}$.

As we have noted in Remark 5.1.3, the restriction to $\mathfrak{p}_+$ of the infinitesimal action of $\mathfrak{p}$ on V induces a bundle map $\operatorname{gr}(T^*M) \times \operatorname{gr}(\mathcal{V}M) \to \operatorname{gr}(\mathcal{V}M)$. This bundle map is homogeneous of degree zero, and we will denote it by $(\phi, \nu) \mapsto \rho(\phi)(\nu)$, or by $\phi \bullet \nu$ if the action is clear from the context. Using this, we now formulate:

PROPOSITION 5.1.5. *Let σ and $\hat{\sigma}$ be two Weyl structures related by*

$$\hat{\sigma}(u) = \sigma(u) \exp(\Upsilon_1(u)) \cdots \exp(\Upsilon_k(u)),$$

with corresponding section $\Upsilon = (\Upsilon_1, \dots, \Upsilon_k)$ of $\operatorname{gr}(T^*M)$. For a representation V of P which is completely reducible as a representation of G_0 let ρ denote the corresponding action of $\operatorname{gr}(T^*M)$ on $\operatorname{gr}(\mathcal{V}M)$. Then the isomorphisms $\mathcal{V}M \to \operatorname{gr}(\mathcal{V}M)$ corresponding to σ and $\hat{\sigma}$ are related by

$$\hat{\nu}_\ell = \sum_{\|\underline{i}\| + j = \ell} \frac{(-1)^{\underline{i}}}{\underline{i}!} \rho(\Upsilon_k)^{i_k} \circ \cdots \circ \rho(\Upsilon_1)^{i_1}(\nu_j).$$

PROOF. We view $\mathcal{V}M$ as $\mathcal{G} \times_P V$ and $\operatorname{gr}(\mathcal{V}M)$ as $\mathcal{G}_0 \times_{G_0} V$. By definition, the fact that $(\nu)_\sigma = (\nu_0, \dots, \nu_N)$ means that given $u \in \mathcal{G}_0$ over the same point as ν, and the unique elements $v_i \in V_i$ for $i = 0, \dots, N$ such that $\nu_i = [\![u, v_i]\!]$, we have $\nu = [\![\sigma(u), v]\!]$, where $v = v_0 + \cdots + v_N$. Again by definition, this means that

$$\nu = [\![\hat{\sigma}(u), \exp(-\Upsilon_k(u)) \cdots \exp(-\Upsilon_1(u)) \cdot v]\!],$$
$$(\nu)_{\hat{\sigma}} = [\![u, \exp(-\Upsilon_k(u)) \cdots \exp(-\Upsilon_1(u)) \cdot v]\!].$$

Hence, we get $(\nu)_{\hat{\sigma}} = \exp(-\Upsilon_k(u)) \cdots \exp(-\Upsilon_1(u)) \cdot v$, from which the claimed formula follows by collecting pieces of fixed homogeneity. □

The low grading components can be easily spelled out explicitly. If the filtration index starts with 0, then we get

$$\hat{v}_0 = v_0,$$
$$\hat{v}_1 = v_1 - \Upsilon_1 \bullet v_0,$$
$$\hat{v}_2 = v_2 - \Upsilon_1 \bullet v_1 - \Upsilon_2 \bullet v_0 + \tfrac{1}{2} \Upsilon_1 \bullet \Upsilon_1 \bullet v_0,$$

and so on. Of course, the first line simply expresses the fact that $\operatorname{gr}_0(\mathcal{V}M)$ is naturally a quotient of $\mathcal{V}M$.

We may apply the result of the proposition, in particular, to the isomorphism $\operatorname{gr}(TM) \to TM$ induced by a choice of a Weyl structure. In this case, the action of $\operatorname{gr}(T^*M)$ on $\operatorname{gr}(TM)$ comes from the adjoint action, so this is given by the algebraic bracket $\{\ ,\ \}$.

We can also use the proposition to obtain a first alternative interpretation of Weyl structures. Recall from Example 5.1.3 that for a given Weyl structure σ we obtain a section $E^\sigma \in \Gamma(\mathcal{A}^0 M)$ such that $\operatorname{gr}_0(E^\sigma) = E \in \Gamma(\operatorname{gr}_0(\mathcal{A}M))$, the natural grading section.

COROLLARY 5.1.5. *The map $\sigma \mapsto E^\sigma$ defines a bijection between the set of all Weyl structures and the set of all sections $s \in \Gamma(\mathcal{A}^0 M)$ such that $\operatorname{gr}_0(s) = E \in \Gamma(\operatorname{gr}_0(\mathcal{A}M))$.*

PROOF. The natural action of $\mathrm{gr}(T^*M)$ on $\mathrm{gr}(AM)$ comes from the adjoint action, so we denote it by ad. By definition, E^σ is the image of E under the isomorphism $\mathrm{gr}(\mathcal{A}M) \to \mathcal{A}M$ induced by σ. For a lift $s \in \Gamma(\mathcal{A}^0M)$ of E we must have $(s)_\sigma = (0, \dots, 0, E, s_1, \dots, s_n)$, so fixing σ we obtain a bijection between the set of all such lifts and $\Gamma(\mathrm{gr}_1(\mathcal{A}M) \oplus \cdots \oplus \mathrm{gr}_k(\mathcal{A}M))$. For an arbitrary Weyl structure $\hat{\sigma}$ we obtain a section $\Upsilon \in \Gamma(\mathrm{gr}(T^*M))$ such that $\sigma(u) = \hat{\sigma}(u)\exp(\Upsilon_1(u)) \cdots \exp(\Upsilon_k(u))$. By the proposition we obtain $(E^{\hat{\sigma}})_\sigma = (0, \dots, s_0, \dots, s_k)$ with

$$s_j = \sum_{\|\underline{i}\|=j} \frac{(-1)^{\underline{i}}}{\underline{i}!} \mathrm{ad}(\Upsilon_k)^{i_k} \circ \dots \circ \mathrm{ad}(\Upsilon_1)^{i_1}(E).$$

In particular, $s_0 = E$ and $s_1 = -\{\Upsilon_1, E\} = \Upsilon_1$. Next, $s_2 = -\{\Upsilon_2, E\} + \frac{1}{2}\{\Upsilon_1, \{\Upsilon_1, E\}\} = 2\Upsilon_2 - \frac{1}{2}\{\Upsilon_1, \Upsilon_1\}$. Inductively, we conclude that each s_j is given by the sum of $j\Upsilon_j$ plus some expression in the Υ_i for $i < j$. This shows that we can compute each Υ_j from $(E^{\hat{\sigma}})_\sigma$ and conversely, any choice for s_j with $s_0 = E$ can be realized by choosing an appropriate section $\Upsilon = (\Upsilon_1, \dots, \Upsilon_k)$. □

REMARK 5.1.5. The modern treatment of tractor bundles was initially based on using Proposition 5.1.5 as a definition. In [**BEG94**], the standard tractor bundles for conformal and projective structures were defined as being given by a direct sum for each choice of a metric in the conformal class, respectively, a connection in the projective class with the appropriate behavior under changes of these choices. This approach was extended to more general geometries and bundles in [**ČGo02**] and [**ČGo00**]. The idea to look at grading sections and use them to parametrize Weyl structures is taken from [**CDS05**].

5.1.6. The effect of a change of Weyl structure on Weyl connections. We derive the complete formula only in the case of a completely reducible representation V of P, so that $\mathcal{V}M = \mathcal{G} \times_P V = \mathcal{G}_0 \times_{G_0} V$. The infinitesimal action of $\mathfrak{g}_0$ on V induces a bundle map $\mathrm{gr}_0(\mathcal{A}M) \times \mathcal{V}M \to \mathcal{V}M$, which we denote by $\bullet$.

PROPOSITION 5.1.6. *Let σ and $\hat{\sigma}$ be two Weyl structures related by*

$$\hat{\sigma}(u) = \sigma(u)\exp(\Upsilon_1(u)) \cdots \exp(\Upsilon_k(u)),$$

with corresponding section $\Upsilon = (\Upsilon_1, \dots, \Upsilon_k)$ of $\mathrm{gr}(T^*M)$. For a smooth section ν of a bundle $\mathcal{V}M$ associated to a completely reducible representation of P, the Weyl connections ∇ and $\hat{\nabla}$ are related by

$$\hat{\nabla}_\xi \nu = \nabla_\xi \nu + \sum_{\|\underline{i}\|+j=0} \frac{(-1)^{\underline{i}}}{\underline{i}!} \Big(\mathrm{ad}(\Upsilon_k)^{i_k} \circ \cdots \circ \mathrm{ad}(\Upsilon_1)^{i_1}(\xi_j) \Big) \bullet \nu,$$

where $(\xi)_\sigma = (\xi_{-k}, \dots, \xi_{-1})$.

PROOF. Let us also denote by ν the G_0–equivariant function $\mathcal{G}_0 \to V$ representing the section ν. By definition of the Weyl connection, the value of the function representing $\nabla_\xi \nu$ in $u \in \mathcal{G}_0$ is given by $\tilde{\xi} \cdot \nu - \omega_0(T_u\sigma \cdot \tilde{\xi}) \bullet \nu$, where $\tilde{\xi}$ is any lift of ξ. Denoting by r the principal right action of P_+ on $\mathcal{G} \to \mathcal{G}_0$ and putting $\Phi(u) := \exp(\Upsilon_1(u)) \cdots \exp(\Upsilon_k(u))$, we have $\hat{\sigma} = r \circ (\sigma, \Phi)$. Putting $r^g = r(\ , g)$ and $r_u = r(u, \ )$ we get

$$T_u\hat{\sigma} \cdot \tilde{\xi} = T_{\sigma(u)} r^{\Phi(u)} \cdot T_u\sigma \cdot \tilde{\xi} + T_{\Phi(u)} r_{\sigma(u)} \cdot T_u\Phi \cdot \tilde{\xi}.$$

The second summand is tangent to the fibers of $\mathcal{G} \to \mathcal{G}_0$, and we obtain

$$\omega_0(T_u\hat{\sigma} \cdot \tilde{\xi}) = \omega_0(T_{\sigma(u)}r^{\Phi(u)} \cdot T_u\sigma \cdot \xi).$$

By equivariancy of ω, this equals the $\mathfrak{g}_0$–component of $\mathrm{Ad}(\Phi(u)^{-1})(\omega(T_u\sigma \cdot \xi))$. By definition $\Phi(u)^{-1} = \exp(-\Upsilon_k(u)) \cdots \exp(-\Upsilon_1(u))$. Taking into account the fact that for $i < 0$ the function $\omega_i(T_u\sigma \cdot \xi)$ represents ξ_i, the result follows by expanding $\mathrm{Ad}(\exp(-\Upsilon_j)) = e^{-\operatorname{ad}(\Upsilon_j)}$ and collecting the terms of degree zero. □

The result can be easily expressed explicitly in the case of short gradings. In the case of a $|1|$–grading, we have $\Upsilon = \Upsilon_1$, while $\xi = \xi_{-1}$ independently of σ, and we simply obtain $\hat{\nabla}_\xi \nu = \nabla_\xi \nu - \{\Upsilon, \xi\} \bullet \nu$. For a $|2|$–grading, one gets

$$\hat{\nabla}_\xi \nu = \nabla_\xi \nu + \left(\tfrac{1}{2}\{\Upsilon_1, \{\Upsilon_1, \xi_{-2}\}\} - \{\Upsilon_2, \xi_{-2}\} - \{\Upsilon_1, \xi_{-1}\}\right) \bullet \nu.$$

The change of Weyl connection on a natural vector bundle associated to a representation V of P, which is completely reducible as a representation of G_0, can then be computed by combining the Propositions 5.1.6 and 5.1.5. For a section ν of $\mathcal{V}M$ one has to compute the change from $(\nabla_\xi \nu_0, \dots, \nabla_\xi \nu_N)$ to $(\hat{\nabla}_\xi \hat{\nu}_0, \dots, \hat{\nabla}_\xi \hat{\nu}_N)$ using the two formulae. One has to take into account that the actions ρ and ad are induced by equivariant maps on the standard fibers and thus are parallel for any Weyl connection. Since the result becomes rather messy, we do not write out the general formula here.

An important application of this result is that we obtain a description of Weyl structures in terms of connections on a bundle of scales which generalizes the results for conformal structures from 1.6.5.

COROLLARY 5.1.6. *Let L^λ be a bundle of scales. Then mapping a Weyl structure to the induced linear connection on L^λ induces a bijection between the set of all Weyl structures and the set of all linear connections on L^λ.*

PROOF. Let $\lambda' : \mathfrak{g}_0 \to \mathbb{R}$ be the functional giving rise to the bundle L^λ of scales, and let $E_\lambda \in \mathfrak{z}(\mathfrak{g}_0)$ be the associated scaling element, so $\lambda'(A) = B(A, E_\lambda)$ for the Killing form B. For $i > 0$, $X \in \mathfrak{g}_{-i}$ and $Z \in \mathfrak{g}_i$ we have

$$\lambda'([Z, X]) = B([Z, X], E_\lambda) = -B(X, [Z, E_\lambda]).$$

By assumption, the adjoint action of E_λ on $\mathfrak{p}_+$ is injective, which implies that $(X, Z) \mapsto \lambda'([Z, X])$ defines a duality between $\mathfrak{g}_{-i}$ and $\mathfrak{g}_i$. Passing to the induced bundle maps, we conclude that $(\xi_{-i}, \Upsilon_i) \mapsto \{\Upsilon_i, \xi_i\} \bullet s$ defines a duality provided that $s \in L^\lambda$ is nonzero.

Assume that we have given two Weyl structures σ and $\hat{\sigma}$ with the change from σ to $\hat{\sigma}$ corresponding to Υ. For a section ξ of $T^{-1}M$, the formula in the proposition simplifies to $\hat{\nabla}_\xi s = \nabla_\xi s - \{\Upsilon_1, \xi\} \bullet s$. Hence, we conclude that $\hat{\nabla}_\xi s = \nabla_\xi s$ for all $s \in \Gamma(L^\lambda)$ and all $\xi \in \Gamma(T^{-1}M)$ if and only if $\Upsilon_1 = 0$. If this is the case, then the formula for $\xi \in \Gamma(T^{-2}M)$ simplifies to $\hat{\nabla}_\xi s = \nabla_\xi s - \{\Upsilon_2, \mathrm{gr}_{-2}(\xi)\} \bullet s$. As above, we conclude that the derivatives coincide for all $s \in \Gamma(L^\lambda)$ and all $\xi \in \Gamma(T^{-2}M)$ if and only if $\Upsilon_1 = 0$ and $\Upsilon_2 = 0$. Inductively, we see that $\hat{\nabla} = \nabla$ implies $\sigma = \hat{\sigma}$.

Conversely, fix σ and consider any linear connection $\hat{\nabla}$ on L^λ. Then $\hat{\nabla}_\xi s = \nabla_\xi s + \tau(\xi)s$ for some $\tau \in \Omega^1(M)$. Restricting τ to $T^{-1}M$ we can find a unique section Υ_1 of $\mathrm{gr}_1(T^*M)$ such that $\tau(\xi)s = -\{\Upsilon_1, \xi\} \bullet s$. Repeating the argument with σ replaced by $u \mapsto \sigma(u)\exp(\Upsilon_1(u))$, we see that $\hat{\nabla}_\xi s = \nabla_\xi s$ for all $\xi \in$

$\Gamma(T^{-1}M)$. Hence, the corresponding one–form τ vanishes on $T^{-1}M$, so there is a section $\Upsilon_2 \in \Gamma(\mathrm{gr}_2(T^*M))$ such that $\tau(\xi)s = -\{\Upsilon_2, \xi_{-2}\} \bullet s$ for all $\xi \in \Gamma(T^{-2}M)$. Iterating this, we see that $\hat{\nabla}$ is induced by an appropriate Weyl structure. □

REMARK 5.1.6. One can extend the result of the corollary to obtain an interpretation of the Cartan bundle $\mathcal{G}$. A linear connection on the bundle of scales L^λ is equivalent to a principal connection on its frame bundle $\mathcal{L}^\lambda$. Now there is a natural fiber bundle $\mathcal{QL}^\lambda \to M$ whose smooth sections are in bijective correspondence with principal connections on $\mathcal{L}^\lambda$. Consider the projection $\pi : \mathcal{G} \to \mathcal{G}_0$ such that $p = p_0 \circ \pi$ for the bundle projection $p_0 : \mathcal{G}_0 \to M$. Forming the pullback $p_0^*\mathcal{QL}^\lambda \to \mathcal{G}_0$, there is a tautological bijective correspondence between smooth sections of $\mathcal{QL}^\lambda \to M$ and G_0–equivariant smooth sections of $p_0^*\mathcal{QL}^\lambda \to \mathcal{G}_0$. Using this, one verifies that $\mathcal{G} \cong p_0^*\mathcal{QL}^\lambda$ as a bundle over $\mathcal{G}_0$. In this picture, the principal right action of P on $\mathcal{G}$ is expressed by the formula for the change of connection in Proposition 5.1.6; see [**ČSl03**, 3.8] for more details.

5.1.7. Closed and exact Weyl structures. The bijective correspondence between Weyl structures and linear connections on a fixed bundle L^λ of scales suggests two subclasses of Weyl structures.

DEFINITION 5.1.7. Let L^λ be a fixed bundle of scales.

(1) A Weyl structure $\sigma : \mathcal{G}_0 \to \mathcal{G}$ is called *closed* if and only if the induced connection ∇ on L^λ is flat.

(2) A Weyl structure $\sigma : \mathcal{G}_0 \to \mathcal{G}$ is called *exact* if and only if the induced connection ∇ on L^λ comes from a global trivialization of L^λ.

By definition, exact Weyl structures are automatically closed by definition. By part (2) of Proposition 5.1.4, the frame bundle $\mathcal{L}^\lambda$ of a bundle L^λ of scales admits global smooth sections, if and only L^λ is oriented. Since global smooth sections of $\mathcal{L}^\lambda$ are exactly global nonzero sections of L^λ, global exact Weyl structures exist precisely in the oriented case. For conformal structures, the obvious choice for $\mathcal{L}^\lambda$ is the bundle of metrics in the conformal class (see 1.6.5), so exact Weyl structures are in bijective correspondence with these metrics. The reason for the names "closed" and "exact" becomes apparent when looking at the affine structure on the space of Weyl structures.

PROPOSITION 5.1.7. *Let L^λ be a fixed bundle of scales. Then the space of all Weyl structures for $(p : \mathcal{G} \to M, \omega)$ is an affine space modelled on the vector space $\Omega^1(M)$ of one–forms on M. If they exist, then the spaces of closed (respectively exact) Weyl structures are affine subspaces modelled on closed (respectively exact) one–forms. Denoting by λ' also the bundle map induced by the infinitesimal representation, the one form $\tau^\lambda_{\sigma,\hat{\sigma}}$ describing the change from σ to $\hat{\sigma}(u) = \sigma(u)\exp(\Upsilon_1(u)) \cdots \exp(\Upsilon_k(u))$ is given by*

$$\tau^\lambda_{\sigma,\hat{\sigma}}(\xi) = \sum_{\|\underline{i}\|+j=0} \frac{(-1)^{\underline{i}}}{\underline{i}!} \lambda'(\mathrm{ad}(\Upsilon_k)^{i_k} \circ \cdots \circ \mathrm{ad}(\Upsilon_1)^{i_1}(\xi_j)).$$

PROOF. The affine structure on the space of Weyl structures is simply obtained by pulling back the affine structure on the space of linear connections on L^λ. The formula for $\tau^\lambda_{\sigma,\hat{\sigma}}$ directly follows from Proposition 5.1.6. Changing a linear connection on a line bundle by a one–form τ, the change of curvature is given by $d\tau$. Hence, starting from a closed Weyl structure σ, the Weyl structure $\hat{\sigma}$ is closed if and only if $d\tau^\lambda_{\sigma,\hat\sigma}=0$. If the structure σ is exact, then it is given by a global non–vanishing section $\phi\in\Gamma(L^\lambda)$. Any other such section can be written as $\hat\phi=e^f\phi$ for a smooth function $f:M\to\mathbb{R}$. A smooth section ψ of L^λ then can be written as $\psi=h\phi=he^{-f}\hat\phi$, and by definition we have $\nabla\psi=dh\otimes\phi$ and $\hat\nabla\psi=d(he^{-f})\otimes\hat\phi$, which immediately implies $\hat\nabla\psi=\nabla\psi-df\otimes\psi$. □

REMARK 5.1.7. (1) Let us specialize to the case of $|1|$–gradings for illustration. There the formula for τ from the proposition simply reads as $\tau^\lambda_{\sigma,\hat\sigma}=-\lambda'([\Upsilon,\xi])$. As we have noted in Remark 5.1.4, any scaling element is a multiple of the grading element E in the $|1|$–graded case. Denoting by B the Killing form, this means that there is a nonzero number a such that

$$-\lambda'([\Upsilon,\xi])=-aB(E,[\Upsilon,\xi])=-aB(\Upsilon,\xi),$$

and hence $\tau^\lambda_{\sigma,\hat\sigma}=-a\Upsilon$.

(2) There is another useful point of view for exact Weyl structures which generalizes the case of conformal structures discussed in 1.6.5. The frame bundle $\mathcal{L}^\lambda$ of L^λ can be naturally identified with the quotient $\mathcal{G}_0/\ker(\lambda)$. By Lemma 1.2.6, global smooth sections of $\mathcal{G}_0/\ker(\lambda)\to M$ are in bijective correspondence with reductions of the principal bundle $\mathcal{G}_0\to M$ to the structure group $\ker(\lambda)\subset G_0$. Hence, we can interpret exact Weyl structures as reductions of the structure group of $\mathcal{G}_0$ to $\ker(\lambda)$. For conformal structures we exactly recover the reductions to the structure group $O(n)\subset CO(n)$ given by the choices of metrics in the conformal class.

5.1.8. The change of the Rho tensor. Let us next clarify the dependence of the Rho tensor on the choice of Weyl structure.

PROPOSITION 5.1.8. *Let σ and $\hat\sigma$ be two Weyl structures related by*

$$\hat\sigma(u)=\sigma(u)\exp(\Upsilon_1(u))\cdots\exp(\Upsilon_k(u)),$$

with corresponding section $\Upsilon=(\Upsilon_1,\dots,\Upsilon_k)$ of $\operatorname{gr}(T^*M)$. Then the Rho tensors $\mathsf{P}=(\mathsf{P}_1,\dots,\mathsf{P}_k)$ and $\hat{\mathsf{P}}=(\hat{\mathsf{P}}_1,\dots,\hat{\mathsf{P}}_k)$ associated to σ and $\hat\sigma$ are related by

$$\begin{aligned}\hat{\mathsf{P}}_i(\xi)=&\sum_{\|\underline{j}\|+\ell=i}\frac{(-1)^{\underline{j}}}{\underline{j}!}\operatorname{ad}(\Upsilon_k)^{j_k}\circ\cdots\circ\operatorname{ad}(\Upsilon_1)^{j_1}(\xi_\ell)\\&+\sum_{m=1}^k\sum_{\substack{\|\underline{j}\|+m=i\\ j_1=\dots=j_{m-1}=0}}\frac{(-1)^{\underline{j}}}{\underline{j}!(j_m+1)}\operatorname{ad}(\Upsilon_k)^{j_k}\circ\cdots\circ\operatorname{ad}(\Upsilon_m)^{j_m}(\nabla_\xi\Upsilon_m)\\&+\sum_{\|\underline{j}\|+\ell=i}\frac{(-1)^{\underline{j}}}{\underline{j}!}\operatorname{ad}(\Upsilon_k)^{j_k}\circ\cdots\circ\operatorname{ad}(\Upsilon_1)^{j_1}(\mathsf{P}_\ell(\xi)).\end{aligned}$$

where $(\xi)_\sigma=(\xi_{-k},\dots,\xi_{-1})$.

PROOF. Fix $u\in\mathcal{G}_0$ and $\xi\in T_u\mathcal{G}_0$. By definition, $\hat{\mathsf{P}}_i(\xi)$ corresponds to the $\mathfrak{g}_i$–component of $\omega(\hat\sigma(u))(T_u\hat\sigma\cdot\xi)$. Since this depends only on the projection of ξ to M we may assume that ξ is horizontal with respect to σ. Putting $\Phi(u)=\exp(\Upsilon_1(u))\cdots\exp(\Upsilon_k(u))$ we know from the proof of Proposition 5.1.6 that

$$T_u\hat\sigma\cdot\xi=T_{\sigma(u)}r^{\Phi(u)}\cdot T_u\sigma\cdot\xi+T_{\Phi(u)}r_{\sigma(u)}\cdot T_u\Phi\cdot\xi.$$

Denoting by λ left translations in the group P_+, it follows from the definition of the principal right action that $r_{\sigma(u)} = r_{\hat{\sigma}(u)} \circ \lambda_{\Phi(u)^{-1}}$. Differentiating this, we obtain

$$T_{\Phi(u)}r_{\sigma(u)} \circ T_u\Phi = T_e r_{\hat{\sigma}(u)} \circ T_{\Phi(u)}\lambda_{\Phi(u)^{-1}} \circ T_u\Phi.$$

Now $T_e r_{\hat{\sigma}(u)}$ is the fundamental vector field map in $\hat{\sigma}(u)$ while the composition of the other two factors by definition is $\delta\Phi(u)$, where $\delta\Phi \in \Omega^1(\mathcal{G}_0, \mathfrak{p}_+)$ is the left logarithmic derivative of $\Phi : \mathcal{G}_0 \to P_+$; see 1.2.4. Hence, we conclude that

$$\omega(T_u\hat{\sigma} \cdot \xi) = \mathrm{Ad}(\Phi(u)^{-1})(\omega(T_u\sigma \cdot \xi)) + \delta\Phi(u)(\xi).$$

Using $\omega_0(T_u\sigma \cdot \xi) = 0$, inserting $\mathrm{Ad}(\Phi(u)^{-1}) = e^{\mathrm{ad}(-\Upsilon_k(u))} \circ \cdots \circ e^{\mathrm{ad}(-\Upsilon_1(u))}$, and collecting the terms of the right degrees, we see that $\mathrm{Ad}(\Phi(u)^{-1})(\omega(T_u\sigma \cdot \xi))$ represents the first line and the last line in the claimed formula for $\hat{\mathsf{P}}_i$.

We have noted in 1.2.4 that the left logarithmic derivative satisfies a Leibniz rule of the form $\delta(fg)(x) = \delta g(x) + \mathrm{Ad}(g(x)^{-1})\delta f(x)$. Iteratively, this implies that

$$\delta\Phi = \sum_{j=1}^{k} e^{\mathrm{ad}(-\Upsilon_k)} \circ \cdots \circ e^{\mathrm{ad}(-\Upsilon_{j+1})} \circ \delta(\exp \circ \Upsilon_j).$$

Finally, $\delta(\exp \circ \Upsilon_j)(u)(\xi) = \delta(\exp)(\Upsilon_j(u))(T_u\Upsilon_j \cdot \xi)$. Since ξ was chosen to be horizontal with respect to σ, the function $T_u\Upsilon_j \cdot \xi$ represents the covariant derivative of Υ_j in direction of the vector field on M underlying ξ. The formula for the right logarithmic derivative of the exponential mapping from [**KMS**, Lemma 4.27] can be easily adapted to the left logarithmic derivative, showing that $\delta(\exp)(X) = \sum_{r=0}^{\infty} \frac{(-1)^r}{(r+1)!} \mathrm{ad}(X)^r$. Inserting this, the result follows. □

REMARK 5.1.8. Together with Propositions 5.1.5 and 5.1.6 we now have a complete description of how a change of Weyl structure affects the associated data. For many applications, one does not need the complete changes but the linearized changes are sufficient. More precisely, suppose that we have given a Weyl structure σ and a smooth section $\Upsilon = (\Upsilon_1, \dots, \Upsilon_k)$ of $\mathrm{gr}(T^*M)$. Then for $t \in \mathbb{R}$ we can consider the Weyl structure σ_t defined by

$$\sigma_t(u) = \sigma(u)\exp(t\Upsilon_1(u)) \cdots \exp(t\Upsilon_k(u)),$$

to obtain a smooth family $\{\sigma_t : t \in \mathbb{R}\}$ of Weyl structures. This leads to smooth families of soldering forms, Weyl connections, and Rho tensors, and the linearized changes are by definition the derivatives at $t = 0$ of these curves, which we indicate by the symbol δ. From Propositions 5.1.5, 5.1.6, and 5.1.8 one immediately reads off that (in the setting and notation of the respective proposition), they are given by

$$\delta v_\ell = -\sum_{i=1}^{k} \Upsilon_i \bullet v_{\ell-i},$$

$$\delta\nabla_\xi s = -\sum_{i=1}^{k} \{\Upsilon_i, \xi_{-i}\} \bullet s,$$

$$\delta\mathsf{P}_\ell(\xi) = \nabla_\xi\Upsilon_\ell - \sum_{i=\ell+1}^{k} \{\Upsilon_i, \xi_{\ell-i}\} - \sum_{i=1}^{\ell-1} \{\Upsilon_i, \mathsf{P}_{\ell-i}(\xi)\}.$$

5.1.9. The Rho–corrected derivative. In 5.1.3 we have defined Weyl connections on arbitrary natural bundles by combining the principal connection $\sigma^*\omega_0$ and the identification of a natural bundle with an associated bundle to $\mathcal{G}_0$ provided by σ. For bundles associated to actions of P coming from actions of G_0 this is the best one can do, but for general bundles there is an alternative possibility, which was suggested in [**Sl97**] and formally introduced in [**CDS05**].

Given a Weyl structure $\sigma : \mathcal{G}_0 \to \mathcal{G}$, we can consider the image $\sigma(\mathcal{G}_0) \subset \mathcal{G}$ and use $\omega^{-1}(\mathfrak{g}_-)$ as a horizontal distribution along this subset. Then we can extend this horizontal distribution equivariantly to all of $\mathcal{G}$, hence defining a principal connection on $\mathcal{G}$. Alternatively, this can be formulated as follows.

DEFINITION 5.1.9. Let $\sigma : \mathcal{G}_0 \to \mathcal{G}$ be a Weyl structure.

(1) The *principal connection on $\mathcal{G}$ associated to σ* is defined by the connection form $\gamma^\sigma \in \Omega^1(\mathcal{G}, \mathfrak{p})$ defined by

$$\gamma^\sigma(\sigma(u) \cdot g)(\xi) := \omega_{\mathfrak{p}}(\sigma(u))(Tr^{g^{-1}} \cdot \xi)$$

for $u \in \mathcal{G}_0$ and $g \in P_+$, where $\omega_{\mathfrak{p}}$ is the $\mathfrak{p}$–component of $\omega \in \Omega^1(\mathcal{G}, \mathfrak{g}_- \oplus \mathfrak{p})$.

(2) For a natural vector bundle the covariant derivative induced by the principal connection γ^σ is called the *Rho–corrected derivative* ∇^{P} associated to the Weyl structure σ.

Using the facts that σ is G_0 equivariant and that ω reproduces the generators of fundamental vector fields, one immediately verifies that γ^σ indeed defines a principal connection on $\mathcal{G}$. It is easy to relate the Rho–corrected derivative to the Weyl connection. This relation also explains the choice of the name "Rho–corrected derivative".

PROPOSITION 5.1.9. *Let $\sigma : \mathcal{G}_0 \to \mathcal{G}$ be a Weyl structure with Weyl connection ∇ and Rho–corrected derivative ∇^{P}. For a natural vector bundle $\mathcal{V}M = \mathcal{G} \times_P V$ let $\bullet : T^*M \otimes \mathcal{V}M \to \mathcal{V}M$ be the action induced by the $\mathfrak{p}_+$–action on V.*

(1) For any section $s \in \Gamma(\mathcal{V}M)$ and any vector field $\xi \in \mathfrak{X}(M)$ we have

$$(\nabla^{\mathsf{P}}_\xi s)_i = \nabla_\xi s_i + \sum_{j=1}^{k} \mathsf{P}_j(\xi) \bullet s_{i-j},$$

where $(s)_\sigma = (s_0, \dots, s_N)$ and $(\nabla^{\mathsf{P}}_\xi s)_\sigma = ((\nabla^{\mathsf{P}}_\xi s)_0, \dots, (\nabla^{\mathsf{P}}_\xi s)_N)$. Otherwise put, $\nabla^{\mathsf{P}}_\xi s = \nabla_\xi s + \mathsf{P}(\xi) \bullet s$.

(2) For $\hat{\sigma}(u) = \sigma(u)\exp(\Upsilon_1(u)) \cdots \exp(\Upsilon_k(u))$ with associated Rho–corrected derivative $\hat{\nabla}^{\mathsf{P}}$ and $s \in \Gamma(\mathcal{V}M)$ we have

$$\hat{\nabla}^{\mathsf{P}}_\xi s = \nabla^{\mathsf{P}}_\xi s + \sum_{\|\underline{i}\|+j \geq 0} \frac{(-1)^{\underline{i}}}{\underline{i}!} \Big(\operatorname{ad}(\Upsilon_k)^{i_k} \circ \dots \circ \operatorname{ad}(\Upsilon_1)^{i_1}(\xi_j) \Big) \bullet s.$$

PROOF. (1) Let us also denote by $s : \mathcal{G} \to V$ the equivariant function corresponding to $s \in \Gamma(\mathcal{V}M)$. By definition, the section $s_i \in \Gamma(\operatorname{gr}_i(\mathcal{V}M))$ is represented by the V_i–component $(s \circ \sigma)_i$ of the function $s \circ \sigma : \mathcal{G}_0 \to V$. Let us denote by $\xi^h \in \mathfrak{X}(\mathcal{G}_0)$ the horizontal lift of $\xi \in \mathfrak{X}(M)$ with respect to the Weyl connection of σ. Then the section $\nabla_\xi s_i$ is represented by $\xi^h \cdot (s \circ \sigma)_i = ((T\sigma \cdot \xi^h) \cdot s)_i$. On the other hand, choosing a lift $\tilde{\xi} \in \mathfrak{X}(\mathcal{G})$ of ξ, the section $\nabla^{\mathsf{P}}_\xi s \in \Gamma(\mathcal{V}M)$ is represented by the function $\tilde{\xi} \cdot s - \omega_{\mathfrak{p}}(\tilde{\xi}) \bullet s$. The component $(\nabla^{\mathsf{P}}_\xi s)_i$ is then obtained by composing with σ and looking at the component in V_i. Along the image σ, we can use $T\sigma \cdot \xi^h$

as $\tilde{\xi}$. Since ξ^h is the horizontal lift of ξ, we have $\omega_0(T\sigma \cdot \xi^h) = 0$ and by definition the $\mathfrak{p}_+$–component of $\omega(T\sigma \cdot \xi^h)$ represents $\mathsf{P}(\xi)$. Equivariancy of s then implies $\nabla^{\mathsf{P}}_\xi s = \nabla_\xi s + \mathsf{P}(\xi) \bullet s$ along the image of σ, and the formula follows by collecting terms in V_i.

(2) Let $\tilde{\xi} \in \mathfrak{X}(\mathcal{G})$ be horizontal lift of $\xi \in \mathfrak{X}(M)$ with respect to γ^σ. Then $\nabla^{\mathsf{P}}_\xi s$ and $\hat{\nabla}^{\mathsf{P}}_\xi s$ are represented by $\tilde{\xi} \cdot s$ and $\tilde{\xi} \cdot s + \gamma^{\hat{\sigma}}(\tilde{\xi}) \bullet s$, respectively. By definition $\omega(\sigma(u))(\tilde{\xi})$ has values in $\mathfrak{g}_-$. Putting $\Phi(u) = \exp(\Upsilon_1(u)) \cdots \exp(\Upsilon_k(u)) \in P_+$, by definition we get

$$\gamma^{\hat{\sigma}}(\sigma(u))(\tilde{\xi}) = \omega_{\mathfrak{p}}(\hat{\sigma}(u))(Tr^{\Phi(u)} \cdot \tilde{\xi}(\sigma(u))),$$

and this is the $\mathfrak{p}$–component of $\operatorname{Ad}(\Phi(u)^{-1})(\omega(\sigma(u))(\tilde{\xi}))$. Now $\omega(\sigma(u))(\tilde{\xi})$ has values in $\mathfrak{g}_-$, and its components represent the sections $\xi_j \in \Gamma(\operatorname{gr}_j(TM))$. The result then follows from the fact that $\operatorname{Ad}(\Phi(u)^{-1}) = e^{\operatorname{ad}(-\Upsilon_k(u))} \circ \cdots \circ e^{\operatorname{ad}(-\Upsilon_1(u))}$ by sorting out the components in $\mathfrak{p}$. □

Note that part (1) shows that on completely reducible bundles the operations ∇ and ∇^{P} coincide, and part (2) gives the same transformation law in this case. The advantage of the Rho–corrected derivative is that the transformation law in (2) holds for arbitrary natural bundles. Part (2) also shows that the linearized change (as discussed in Remark 5.1.8) of the Rho–corrected derivative on an arbitrary natural vector bundle is given by

$$\delta\nabla^{\mathsf{P}}_\xi s = -\sum_{i+j\geq 0} \{\Upsilon_i, \xi_j\} \bullet s.$$

5.1.10. Fundamental derivatives and tractor connections. Using the Rho–corrected derivative, we can now express two basic invariant differential operators in terms of the data associated to a chosen Weyl structure. Recall from 1.5.8 that for any natural vector bundle $\mathcal{V}M = \mathcal{G} \times_P V$ the fundamental derivative is an invariant differential operator $\Gamma(\mathcal{A}M) \times \Gamma(\mathcal{V}M) \to \Gamma(\mathcal{V}M)$, where $\mathcal{A}M = \mathcal{G} \times_P \mathfrak{g}$ denotes the adjoint tractor bundle of M. The fundamental derivative is written as $(s,t) \mapsto D_s t$ for $s \in \Gamma(\mathcal{A}M)$ and $t \in \Gamma(\mathcal{V}M)$ to emphasize the analogy to a covariant derivative. By definition, the function $\mathcal{G} \to V$ representing $D_s t$ is obtained by using the vector field $\xi(u) = \omega_u^{-1}(s(u))$ induced by the function representing s to differentiate the function representing t.

On the other hand, we know from 1.5.7 that in the case that $\mathcal{V}M$ is a tractor bundle, i.e. V is the restriction to P of a representation of G (or more generally a $(\mathfrak{g}, P)$–module), then there is a natural linear connection $\nabla^{\mathcal{V}}$ on $\mathcal{V}M$ called the tractor connection. The easiest way to express this operation is via the fundamental derivative; see 1.5.8. Since V is a $(\mathfrak{g}, P)$–module, the action of $\mathfrak{g}$ on V induces a bundle map $\bullet : \mathcal{A}M \times \mathcal{V}M \to \mathcal{V}M$, which extends the actions denoted by the same symbol before. Denoting by $\Pi : \mathcal{A}M \to TM$ the natural projection, we then have $\nabla_{\Pi(s)} t = D_s t + s \bullet t$ for all $s \in \Gamma(\mathcal{A}M)$ and $t \in \Gamma(\mathcal{V}M)$.

PROPOSITION 5.1.10. *Let $\sigma : \mathcal{G}_0 \to \mathcal{G}$ be a Weyl structure, $\mathcal{V}M = \mathcal{G} \times_P V$ a natural vector bundle and $\mathcal{A}M$ the adjoint tractor bundle.*

(1) The fundamental derivative is given by

$$(D_s t)_i = (\nabla^{\mathsf{P}}_{\Pi(s)} t)_i - \sum_{j=0}^{k} s_j \bullet t_{i-j} = \nabla_{\Pi(s)} t_i - s_0 \bullet t_i + \sum_{j=1}^{k} (\mathsf{P}_j(\Pi(s)) - s_j) \bullet t_{i-j},$$

where $s \in \Gamma(\mathcal{A}M)$ *with* $(s)_\sigma = (s_{-k}, \dots, s_k)$ *and* $t \in \Gamma(\mathcal{V}M)$ *with* $(t)_\sigma = (t_0, \dots, t_N)$.

(2) If $\mathcal{V}M$ *is a tractor bundle, then the tractor connection* $\nabla^{\mathcal{V}}$ *is given by*

$$(\nabla^{\mathcal{V}}_\xi t)_i = (\nabla^{\mathsf{P}}_\xi t)_i + \sum_{j=1}^{k} \xi_{-j} \bullet t_{i+j} = \nabla_\xi t_i + \sum_{j=1}^{k} \mathsf{P}_j(\xi) \bullet t_{i-j} + \sum_{j=1}^{k} \xi_{-j} \bullet t_{i+j}.$$

PROOF. (1) We can write $s \circ \sigma : \mathcal{G}_0 \to \mathfrak{g}$ as the sum of $(s_{-k}, \dots, s_{-1}, 0, \dots, 0)$ and $(0, \dots, 0, s_0, \dots, s_k)$. Along the image of σ, the first summand by definition represents the horizontal lift of $\Pi(s)$ with respect to the principal connection γ^σ on $\mathcal{G}$, so the derivative of the function corresponding to t in direction of this field represents $\nabla^{\mathsf{P}}_{\Pi(s)} t$. Along the image of σ, the second summand represents the fundamental vector field with generator $(s_0, \dots, s_k) \in \mathfrak{p}$. Differentiating the function corresponding to t in this direction gives $-(s_0, \dots, s_k) \bullet t$ by equivariancy and the formula in terms of ∇^{P} follows. The formula in terms of the Weyl connection then immediately follows using Proposition 5.1.9.

(2) The horizontal lift of ξ to $\mathcal{G}$ with respect to the connection γ^σ is, along the image of σ, given by $(\xi_{-k}, \dots, \xi_{-1}, 0, \dots, 0)$, where $(\xi)_\sigma = (\xi_{-k}, \dots, \xi_{-1})$. Interpreting this as a section s of $\mathcal{A}M$, the formula in terms of ∇^{P} follows immediately from $\nabla_{\Pi(s)} t = D_s t + s \bullet t$ and part (1). The formula in terms of the Weyl connection then follows from Proposition 5.1.9. □

REMARK 5.1.10. It is a very instructive exercise for the reader to verify invariance of the fundamental derivative and the tractor connection, i.e. independence of the choice of the Weyl structure σ from the formulae in Propositions 5.1.5, 5.1.6, and 5.1.8. As we shall see in the discussion of invariant differential operators in the second volume, it suffices to consider the linearized transformation laws from Remark 5.1.8. For simple cases, e.g. the fundamental derivative on an irreducible bundle, this is almost trivial, but in general the computations quickly become fairly involved.

5.1.11. Example. On the one hand, we want to show in this example that it is easy to convert the general formulae in terms of Weyl structures that we have derived so far into a very explicit form for a given structure. On the other hand, we want to explain that knowing the soldering form, the Weyl connection and the Rho tensor associated to one Weyl structure is sufficient to recover the full Cartan geometry. By definition, the soldering form and the Rho tensor are just one–forms on M with values in $\mathrm{gr}(TM)$, respectively, $\mathrm{gr}(T^*M)$, while the Weyl connection is a principal connection on $\mathcal{G}_0$. Thus, they all can be interpreted in terms of underlying structures and the Cartan bundle is not needed.

Consider the example of an almost Grassmannian structure of type (p, q) on a smooth manifold M of dimension pq with $p, q \geq 2$. As described in 4.1.3, such a geometry is given by two auxiliary bundles E and F over M of rank p and q, respectively, together with a trivialization of $\Lambda^p E \otimes \Lambda^q F$ and an isomorphism $\Phi : E^* \otimes F \to TM$ of vector bundles. The fibered product of the linear frame bundles of E and F has structure group $GL(p, \mathbb{R}) \times GL(q, \mathbb{R})$ and via the trivialization of $\Lambda^p E \otimes \Lambda^q F$ one obtains the subbundle $\mathcal{G}_0 \to M$ with structure group $G_0 = S(GL(p, \mathbb{R}) \times GL(q, \mathbb{R}))$. Hence, a principal connection on $\mathcal{G}_0$ is equivalent to a pair (∇^E, ∇^F) of linear connections on E and F, such that the induced connection on $\Lambda^p E \otimes \Lambda^q F$ preserves the fixed trivialization. Via the isomorphism Φ, these induce a linear connection ∇ on TM.

Since our geometry corresponds to a $|1|$–grading, there is no filtration on TM and the soldering form is just id_{TM} viewed as a TM–valued one–form, while the Rho tensor is a one–form $\mathsf{P} \in \Omega^1(M, T^*M)$. We will obtain a characterization of the Weyl connection and a formula for the Rho tensor in terms of the curvature of a Weyl connection in 5.2.3 below. There we will also discuss other examples in more detail.

Passing to a different Weyl structure with corresponding change $\Upsilon \in \Omega^1(M)$, consider the new Weyl connection $\hat{\nabla}$. For $\xi, \eta \in \mathfrak{X}(M)$ we know from 5.1.6 that $\hat{\nabla}_\xi \eta = \nabla_\xi \eta - \{\{\Upsilon, \xi\}, \eta\}$. To express this more explicitly, we use the abstract index notation from 4.1.4. In this notation $E = \mathcal{E}^A$ and $F = \mathcal{E}^{A'}$, so $TM = \mathcal{E}^{A'}_A$ and $T^*M = \mathcal{E}^A_{A'}$. Computing the brackets in the Lie algebra $\mathfrak{sl}(p+q, \mathbb{R})$, one immediately verifies that

$$\{\{\Upsilon, \xi\}, \eta\}^{A'}_A = -\xi^{A'}_B \Upsilon^B_{B'} \eta^{B'}_A - \xi^{B'}_A \Upsilon^B_{B'} \eta^{A'}_B .$$

Hence, the change of Weyl connection caused by Υ is given by

$$\hat{\nabla}^A_{A'} \eta^{B'}_B = \nabla^A_{A'} \eta^{B'}_B + \delta^{B'}_{A'} \Upsilon^A_{C'} \eta^{C'}_B + \delta^A_B \Upsilon^C_{A'} \eta^{B'}_C .$$

Let us also describe the two basic tractor bundles in terms of a Weyl structure. These correspond to the standard representation of $G = SL(p+q, \mathbb{R})$ and its dual. As a representation of $G_0 \subset G$, the standard representation $\mathbb{R}^{p+q}$ evidently splits as the direct sum $\mathbb{R}^p \oplus \mathbb{R}^q$ of the standard representations of the two factors of G_0. Likewise, the dual $\mathbb{R}^{(p+q)*}$ splits as $\mathbb{R}^{p*} \oplus \mathbb{R}^{q*}$ when restricted to G_0. Let $\mathcal{T}$ be the standard tractor bundle and $\mathcal{T}^*$ its dual. By 5.1.3, the choice of a Weyl structure gives rise to isomorphisms $\mathcal{T} \cong E \oplus F$ and $\mathcal{T}^* = F^* \oplus E^*$. (We have used the convention that the P–invariant summand comes first.) In abstract index notation, elements of $E \oplus F$ are naturally denoted as $(v^A, w^{A'})$ while for the dual we get $(\phi_{A'}, \psi_A)$. Looking at the action of $\mathfrak{sl}(p+q, \mathbb{R})$ on its standard representation, one immediately reads off that for $\Upsilon = \Upsilon^A_{A'} \in \Omega^1(M)$ and $\xi = \xi^{A'}_A \in \mathfrak{X}(M)$ we get

$$\Upsilon^A_{A'} \bullet (v^B, w^{B'}) = (\Upsilon^A_{A'} w^{A'}, 0), \qquad \xi^{A'}_A \bullet (v^B, w^{B'}) = (0, \xi^{A'}_A v^A),$$
$$\Upsilon^A_{A'} \bullet (\phi_{B'}, \psi_B) = (-\Upsilon^A_{A'} \psi_A, 0), \qquad \xi^{A'}_A \bullet (\phi_{B'}, \psi_B) = (0, -\xi^{A'}_A \phi_{A'}).$$

Proposition 5.1.5 now shows that under a change of Weyl structure, we get

$$\widehat{(v^A, w^{A'})} = (v^A - \Upsilon^A_{B'} w^{B'}, w^{A'}), \qquad \widehat{(\phi_{A'}, \psi_A)} = (\phi_{A'} + \Upsilon^B_{A'} \psi_B, \psi_A).$$

On the other hand, we can easily compute the action of the tractor connection from part (2) of Proposition 5.1.10. In abstract index notation the Rho tensor has the form $\mathsf{P}^{AB}_{A'B'}$, and we use the convention that $\mathsf{P}(\xi)^{A'}_A = \xi^{B'}_B \mathsf{P}^{BA}_{B'A'}$. Then we obtain

$$(\nabla^{\mathcal{T}})^A_{A'}(v^B, w^{B'}) = (\nabla^A_{A'} v^B + \mathsf{P}^{AB}_{A'C'} w^{C'}, \nabla^A_{A'} w^{B'} + \delta^{B'}_{A'} v^A),$$
$$(\nabla^{\mathcal{T}^*})^A_{A'}(\phi_{B'}, \psi_B) = (\nabla^A_{A'} \phi_{B'} - \mathsf{P}^{AC}_{A'B'} \psi_C, \nabla^A_{A'} \psi_B - \delta^A_B \phi_{A'}).$$

Knowing one Weyl connection and its Rho tensor, we have obtained a complete description of the standard tractor bundle and its tractor connection, and the Cartan bundle and the Cartan connection can be recovered; see 3.1.21. Indeed, one may turn this into a definition of the standard tractor bundle. Given a Weyl structure, one considers appropriate pairs and then factors by an equivalence relation involving the change of Weyl structure. Then one can use the formula above to define a linear connection on the resulting bundle, and verify directly that the connection is independent of the chosen Weyl connection. This approach is taken essentially in

[**BEG94**] for conformal and projective structures and explicitly in [**GoGr05**] for CR–structures.

5.1.12. Normal Weyl structures. We finish this section by discussing a special class of Weyl structures, which are intimately tied to the Cartan connection ω. They are closely related to the construction of normal coordinates, which partly has been discussed in 3.1.8 already. In contrast to the closed and exact Weyl structures discussed in 5.1.7, normal Weyl structures are independent of the choice of a bundle of scales, but in general they exist only locally. They can be nicely characterized in terms of the Rho tensor.

Choosing a Weyl structure $\sigma : \mathcal{G}_0 \to \mathcal{G}$ we obtain two linear connections on TM, the Weyl connection ∇ and the Rho–corrected derivative ∇^{P}. For either of these two connections we can consider geodesics on M.

LEMMA 5.1.12. *For a Weyl structure $\sigma : \mathcal{G}_0 \to \mathcal{G}$ with Rho tensor P and a point $x \in M$ the following conditions are equivalent:*

(1) *For each geodesic $c : I \to M$ of ∇ with $c(0) = x$ we have $\mathsf{P}(c(t))(c'(t)) = 0$ for all $t \in I$.*
(2) *For each geodesic $c : I \to M$ of ∇^{P} with $c(0) = x$ we have $\mathsf{P}(c(t))(c'(t)) = 0$ for all $t \in I$.*

If these conditions are satisfied, then ∇ and ∇^{P} have the same geodesics emanating in x.

PROOF. For a smooth curve $c : I \to M$ we have by Proposition 5.1.9,

$$\nabla^{\mathsf{P}}_{c'(t)} c'(t) = \nabla_{c'(t)} c'(t) + \mathsf{P}(c(t))(c'(t)) \bullet c'(t).$$

Hence, if c is a geodesic for ∇ emanating in x, and (1) is satisfied, then c is also a geodesic for ∇^{P}. In the other direction, the same argument applies. □

DEFINITION 5.1.12. A Weyl structure $\sigma : \mathcal{G}_0 \to \mathcal{G}$ is called *normal at a point* $x \in M$ if it satisfies the equivalent conditions of the lemma.

THEOREM 5.1.12. *Let $(p : \mathcal{G} \to M, \omega)$ be a parabolic geometry with underlying bundle $p_0 : \mathcal{G}_0 \to M$.*

(1) For each point $x \in M$ there exists an open neighborhood U of x in M and a local Weyl structure $\sigma : p_0^{-1}(U) \to p^{-1}(U)$ which is normal at x.

(2) Locally around x, any Weyl structure which is normal at x is uniquely determined by its value in one point $u_0 \in p_0^{-1}(x)$.

(3) For any Weyl structure which is normal at x, the Rho tensor has the property that for each $k \in \mathbb{N}$ and all vector fields $\xi_0, \dots, \xi_k \in \mathfrak{X}(M)$ the symmetrization over all ξ_i of $(\nabla_{\xi_k} \cdots \nabla_{\xi_1} \mathsf{P})(\xi_0)$ vanishes at x.

PROOF. (1) Let $\pi : \mathcal{G} \to \mathcal{G}_0$ be the natural projection. Fix a point $u \in \mathcal{G}$ with $p(u) = x$ and put $u_0 := \pi(u)$. Consider the map $\phi(X) := \mathrm{Fl}_1^{\omega^{-1}(X)}(u)$ which is well defined on a neighborhood V of 0 in $\mathfrak{g}_-$. The tangent map of $p \circ \phi$ in 0 is given by $X \mapsto T_u p \cdot \omega^{-1}(X)$ so it is a linear isomorphism. Possibly shrinking V, we may assume $p \circ \phi$ is a diffeomorphism from V onto an open neighborhood U of x in M. Now the map $\pi(\phi(X)) \mapsto \phi(X)$ uniquely extends to a G_0–equivariant smooth section $\sigma : p_0^{-1}(U) \to p^{-1}(U)$ of π, thus defining a Weyl structure on U.

For any $X \in \mathfrak{g}_-$ and sufficiently small $t \in \mathbb{R}$, we have the curve $c_{\mathcal{G}}(t) = \mathrm{Fl}_t^{\omega^{-1}(X)}(u) = \phi(tX)$. By construction, $\omega(c_{\mathcal{G}}'(t)) = X \in \mathfrak{g}_-$, so this tangent vector is horizontal for the principal connection γ^σ associated to σ. The projection $c := p \circ c_{\mathcal{G}}$ is a smooth curve in M with $c(0) = x$, and by construction $c'_{\mathcal{G}}(t)$ is the horizontal lift in $c_{\mathcal{G}}(t)$ of the tangent vector $c'(t)$. Hence, the function $\omega(c'_{\mathcal{G}}(t))$ (equivariantly extended) represents the vector field $c'(t)$, and since this is constant along $c_{\mathcal{G}}$ we conclude that $\nabla^{\mathsf{P}}_{c'(t)} c'(t) = 0$ for all t. Thus, c is a geodesic for ∇^{P} emanating in x, and we can obtain all such geodesics in this way.

Likewise, we can consider the curve $c_{\mathcal{G}_0} = \pi \circ c_{\mathcal{G}}$ in $\mathcal{G}_0$. By construction $c'_{\mathcal{G}_0}(t)$ lifts $c'(t)$ and $c_{\mathcal{G}} = \sigma \circ c_{\mathcal{G}_0}$, so $T\sigma \cdot c'_{\mathcal{G}_0}(t) = c'_{\mathcal{G}}(t)$. This implies that $\mathsf{P}(c'(t)) = \omega_{\mathfrak{p}_+}(c'_{\mathcal{G}}(t)) = 0$, so the Weyl structure σ is normal at x.

(2) Let σ be a locally defined Weyl structure which is normal at x, and let $\hat{\sigma}$ be the local normal Weyl structure as constructed in (1) starting from $u = \sigma(u_0)$. Let Υ be the $\mathfrak{p}_+$–valued function describing the change from σ to $\hat{\sigma}$. For some $X \in \mathfrak{g}_-$ consider the curve $c_{\mathcal{G}_0}(t)$ constructed in (1). Then $t \mapsto \Upsilon(c_{\mathcal{G}_0}(t))$ is a smooth curve in $\mathfrak{p}_+$. The derivative of this curve can be written as $(c'_{\mathcal{G}_0}(t) \cdot \Upsilon)(c_{\mathcal{G}_0}(t))$. Since $c'_{\mathcal{G}_0}(t)$ is horizontal for $\gamma^{\hat{\sigma}}$, this derivative represents $(\hat{\nabla}_{c'(t)}\Upsilon)(c(t))$, where c is the projection of $c_{\mathcal{G}_0}$ to M. By Propositions 5.1.5 and 5.1.6, we can compute $(\hat{\nabla}_{c'(t)}\Upsilon)(c(t))$ algebraically from Υ and $(\nabla_{c'(t)}\Upsilon)(c(t))$.

From (1) we know that the projection c of $c_{\mathcal{G}_0}$ is a geodesic for $\hat{\nabla}$ emanating in x, so by normality of $\hat{\sigma}$ we must have $\hat{\mathsf{P}}(c(t))(c'(t)) = 0$. By Proposition 5.1.8, this can be interpreted as an equation involving $\Upsilon(c(t))$, $(\nabla_{c'(t)}\Upsilon)(c(t))$, $\mathsf{P}(c(t))(c'(t))$, and the components of the tangent vector $c'(t)$ with respect to σ. Passing to the picture of equivariant functions on $\mathcal{G}_0$, we conclude that the curve $t \mapsto \Upsilon(c_{\mathcal{G}_0}(t))$ satisfies an equation of the form $F(\Upsilon(c_{\mathcal{G}_0}(t)), \frac{d}{dt}\Upsilon(c_{\mathcal{G}_0}(t)), t) = 0$ for a smooth function $F : \mathfrak{p}_+ \times \mathfrak{p}_+ \times I \to \mathfrak{p}_+$. Looking at the formulae in Propositions 5.1.5, 5.1.6, and 5.1.8, we see that in terms of the splitting $\mathfrak{p}_+ = \mathfrak{g}_1 \oplus \cdots \oplus \mathfrak{g}_k$ we get $F_i(Z, W, t) = W_i + \phi(Z, W_1, \dots, W_{i-1}, t)$ for some function ϕ. This implies that the second partial derivative of F is injective and thus a linear isomorphism in each point. By the implicit function theorem, the implicit equation $F(\Upsilon(c_{\mathcal{G}_0}(t)), \frac{d}{dt}\Upsilon(c_{\mathcal{G}_0}(t)), t) = 0$ is locally equivalent to an equation of the form $\frac{d}{dt}\Upsilon(c_{\mathcal{G}_0}(t)) = \Phi(\Upsilon(c_{\mathcal{G}_0}(t)), t)$. Since this is a first order ODE, Υ is uniquely determined by $\Upsilon(0)$. But by construction $\hat{\sigma}(u_0) = \sigma(u_0)$ and thus $\hat{\sigma} = \sigma$ and the claim follows.

(3) By part (2) we may, without loss of generality, assume that we deal with a Weyl structure σ as constructed in (1). By assumption, we have $\mathsf{P}(c(t))(c'(t)) = 0$ for each geodesic c emanating in x. In particular, all derivatives of this curve at $t = 0$ have to vanish. Now the function on $\mathcal{G}_0$ representing $\mathsf{P}(c(t))(c'(t))$ is simply given by $\mathsf{P}(c_{\mathcal{G}_0}(t))(X)$, where $X = \omega(c'_{\mathcal{G}}(t))$. Since $c'_{\mathcal{G}_0}$ is horizontal for the Weyl connection, we conclude that the iterated derivatives of this curves at $t = 0$ simply represent iterated covariant derivatives of P in direction ξ evaluated at ξ, where $\xi = c'(0)$. The general claim then follows by polarization. □

REMARK 5.1.12. Using Proposition 5.1.8 one shows that the condition on the Rho tensor in (3) and the value of a normal Weyl structure σ in one point over x characterizes the infinite jet of σ in that point; see [**ČSl03**].

EXAMPLE 5.1.12. As an example of normal Weyl structures we describe a specific Weyl structure on the homogeneous model of any parabolic geometry, called the *very flat Weyl structure*. Recall from 3.1.13 that given a parabolic pair (G, P), there also is the opposite parabolic subgroup $P^{op} \subset G$ and the Carnot group $G_- \subset P^{op}$, which is a simply connected Lie group with Lie algebra $\mathfrak{g}_-$. Now we want to determine the normal Weyl structure for $p : G \to G/P$ determined by the points $e \in G$ and $eP_+ \in G/P_+$ over $o := eP \in G/P$.

In this case, for $X \in \mathfrak{g}$ the vector field $\omega^{-1}(X)$ is the left invariant vector field L_X, so its flow through e is simply $\exp(X)$. Now from 3.1.13 we know that $\exp : \mathfrak{g}_- \to G_-$ is a diffeomorphism and $G_- \cap P = \{e\}$ so $X \mapsto \exp(X)P$ defines a diffeomorphism from $\mathfrak{g}_-$ onto an (actually dense) open subset of G/P. Moreover, the restriction of both the bundles $G \to G/P$ and $G/P_+ \to G/P$ to this open subset is canonically trivialized by the restrictions of the group multiplication to maps $G_- \times P \to G$, respectively, $G_- \times G_0 \to G$. Now the construction in the proof of Theorem 5.1.12 shows that our Weyl structure is characterized by $\sigma(\exp(X)P_+) = \exp(X)$ for all $X \in \mathfrak{g}_-$. Under our trivializations $p^{-1}(G_-) \cong G_- \times P$ and $\pi^{-1}(G_-) = G_- \times G_0$, this section is simply given by the inclusion $G_0 \to P$. Otherwise put, in the trivialization of $\pi^{-1}(G_-)$, σ is given by the group multiplication viewed as a map $G_- \times G_0 \to G$.

This map can be viewed as the composition of the isomorphism $G_- \times G_0 \cong P^{op}$ and the inclusion of this subgroup into G. The pullback of the Maurer–Cartan form of G along the inclusion of P^{op} clearly is the Maurer–Cartan form of P^{op}. The induced principal connection is defined by the $\mathfrak{g}_0$–component of this Maurer–Cartan form. In particular, on $\{e\} \times G_0$, this is the Maurer–Cartan form of G_0. Via the diffeomorphism $G_- \times G_0 \to P^{op}$ defined by the group multiplication, this is extended to the whole bundle, so we exactly obtain the flat connection determined by this trivialization. Hence, the induced linear connection on any associated bundle is flat. On the other hand, the soldering form is induced by the Maurer–Cartan form of G_-, i.e. by the trivialization of TG_- by left translations. Finally, for vectors tangent to $P^{op} \subset G$ the Maurer–Cartan form of G of course has values in $\mathfrak{p}^{op} = \mathfrak{g}_- \oplus \mathfrak{g}_0$, so the Rho tensor of our Weyl structure vanishes identically. Finally, note that since any bundle of scales is associated to $G/P_+ \to G/P$, our trivialization of $\pi^{-1}(G_-) \to G_-$ gives rise to a section of the bundle of scales which is trivial for the induced Weyl connection. Hence, the very flat Weyl structure is always an exact Weyl structure.

The very flat Weyl structure has remarkable properties. First, it is at the same time normal in $o = eP$ and exact. Second, since its Rho–tensor vanishes identically, it is not only normal in the single point o but in *all* points in which it is defined. Since locally flat geometries are locally isomorphic to G/P, we can find Weyl structures with these properties on all locally flat geometries. On general geometries, Weyl structures with these properties do not exist.

5.2. Characterization of Weyl structures

As we have seen in 5.1.2, a Weyl structure determines a soldering form, a Weyl connection and a Rho tensor. These can be interpreted as a principal connection on $\mathcal{G}_0$ and as one forms with values in $\operatorname{gr}(TM)$, respectively, $\operatorname{gr}(T^*M)$, so discussing such objects does not need the Cartan bundle at all. In particular, there is a simple abstract version of these data, which can be described in terms of the underlying structure. Our aim in this section is to characterize the data obtained by Weyl structures among their abstract analogs. This will also clarify the interpretation of the Cartan curvature and the harmonic curvature in terms of a Weyl structure, which we have not discussed so far.

This characterization of the data associated to Weyl structures can actually be viewed as an alternative approach to the description of parabolic geometries. Knowing them for one Weyl structure leads to an explicit description of tractor bundles and tractor connections, which are equivalent to the Cartan bundle and Cartan connection; see 5.1.11 for an example. Since one also knows the behavior of the data under a change of Weyl structure, there is a way to check whether some expression is actually independent of the choice of Weyl structure and thus intrinsic to the parabolic geometry.

We will work out the characterization results in detail for structures corresponding to $|1|$–gradings as well as for parabolic contact structures. After having these results at hand, we discuss several geometric applications. In particular, we discuss in 5.2.8 and 5.2.19 how most non–metric affine holonomy groups can be realized using Weyl connections on generalized flag manifolds.

5.2.1. Weyl forms.

DEFINITION 5.2.1. Let $(p_0 : \mathcal{G}_0 \to M, \theta = (\theta_{-k}, \dots, \theta_{-1}))$ be a regular infinitesimal flag structure of some fixed type (G, P); see 3.1.6 and 3.1.7.

(1) A *Weyl form* on $\mathcal{G}_0$ is a G_0–equivariant one form $\tau \in \Omega^1(\mathcal{G}_0, \mathfrak{g})$ such that:

- For $A \in \mathfrak{g}_0$ with corresponding fundamental vector field $\zeta_A \in \mathfrak{X}(\mathcal{G}_0)$ we have $\tau(\zeta_A) = A$.
- For $i = -k, \dots, -1$ and $\xi \in T^i\mathcal{G}_0$ we have $\tau(\xi) \in \mathfrak{g}^i$ and the class of $\tau(\xi)$ in $\mathfrak{g}^i/\mathfrak{g}^{i+1}$ coincides with $\theta_i(\xi)$.

(2) The *curvature* $K_\tau \in \Omega^2(\mathcal{G}_0, \mathfrak{g})$ of a Weyl form τ is defined by

$$K_\tau(\xi, \eta) := d\tau(\xi, \eta) + [\tau(\xi), \tau(\eta)].$$

If $(p : \mathcal{G} \to M, \omega)$ is a regular parabolic geometry with underlying infinitesimal flag structure $(p_0 : \mathcal{G}_0 \to M, \theta)$ and $\sigma : \mathcal{G}_0 \to \mathcal{G}$ is a Weyl structure, then evidently $\sigma^*\omega$ is a Weyl form. Our aim is to characterize the Weyl forms which are obtained in this way.

Similarly to the case of Weyl structures, one may split general Weyl forms into pieces. The $\mathfrak{g}_0$ component τ_0 of a Weyl form τ by definition reproduces the generators of fundamental vector fields, so it defines a principal connection on $p_0 : \mathcal{G}_0 \to M$. As before, we call this the *Weyl connection* associated to the Weyl form τ. The $\mathfrak{g}_-$–component τ_- is horizontal and G_0–equivariant, and thus may be viewed as an element of $\Omega^1(M, \operatorname{gr}(TM))$. The second condition in the definition of a Weyl form implies that τ_- maps $T^iM \subset TM$ to $\operatorname{gr}_i(TM) \oplus \cdots \oplus \operatorname{gr}_{-1}(TM)$ and the first component coincides with the natural projection $T^iM \to \operatorname{gr}_i(TM)$. In particular, this implies that τ_- induces a splitting $TM \cong \operatorname{gr}(TM)$ of the filtration of TM. We call τ_- the *soldering form* associated to τ. Finally, the $\mathfrak{p}_+$–component τ_+ of τ is horizontal and G_0–equivariant and hence can be viewed as an element of $\Omega^1(M, \operatorname{gr}(T^*M))$, which is called the *Rho tensor* associated to τ.

The defining properties of a Weyl form immediately imply that the curvature $K_\tau \in \Omega^2(\mathcal{G}_0, \mathfrak{g})$ of any Weyl form τ is horizontal and G_0–equivariant. Thus, it can be interpreted as a two–form k_τ on M with values in the associated bundle $\mathcal{G}_0 \times_{G_0} \mathfrak{g}$. This associated bundle splits as $\operatorname{gr}(TM) \oplus \operatorname{End}_0(\operatorname{gr}(TM)) \oplus \operatorname{gr}(T^*M)$, for an appropriate subbundle $\operatorname{End}_0(\operatorname{gr}(TM))$ in the bundle of endomorphisms of $\operatorname{gr}(TM)$, compare with 3.1.9.

LEMMA 5.2.1. *Let $(p_0 : \mathcal{G}_0 \to M, \theta)$ be a regular infinitesimal flag structure and let $\tau \in \Omega^1(\mathcal{G}_0, \mathfrak{g})$ be a Weyl form with curvature $k_\tau \in \Omega^2(M, \mathcal{G}_0 \times_{G_0} \mathfrak{g})$. Then k_τ is homogeneous of positive degree, i.e. for $\xi \in T^iM$ and $\eta \in T^jM$ the components of k_τ in $\mathrm{gr}_\ell(TM)$ for $\ell \leq i+j$ vanish.*

PROOF. Choose local vector fields $\xi \in \Gamma(T^i\mathcal{G}_0)$ and $\eta \in \Gamma(T^j\mathcal{G}_0)$. Regularity of the infinitesimal flag structure, in particular, implies that $[\xi, \eta] \in \Gamma(T^{i+j}\mathcal{G}_0)$; see 3.1.7 and 3.1.8. By definition of a Weyl form, we get $\tau([\xi,\eta]) \in \mathfrak{g}^{i+j}$ and together with $\tau(\xi) \in \mathfrak{g}^i$ and $\tau(\eta) \in \mathfrak{g}^j$ this shows that $K_\tau(\xi,\eta) \in \mathfrak{g}^{i+j}$. Again by definition of a Weyl form, the $\mathfrak{g}_{i+j}$–component of $\tau([\xi,\eta])$ equals $\theta_{i+j}([\xi,\eta])$, and similarly for the lowest nonzero components of $\tau(\xi)$ and $\tau(\eta)$. Hence, the component of $K_\tau(\xi,\eta)$ in $\mathfrak{g}_{i+j}$ is given by $-\theta_{i+j}([\xi,\eta])+[\theta_i(\xi),\theta_j(\eta)]$, which vanishes by Proposition 3.1.7. □

5.2.2. Normal Weyl forms. Let $(p : \mathcal{G}_0 \to M, \theta)$ be a regular infinitesimal flag structure of type (G,P) endowed with a Weyl form $\tau \in \Omega^1(\mathcal{G}_0, \mathfrak{g})$. Then $\tau_- + \tau_0 \in \Omega^1(\mathcal{G}_0, \mathfrak{g}_- \oplus \mathfrak{g}_0)$ defines a Cartan connection on the G_0–principal bundle $\mathcal{G}_0$. In particular, the soldering form τ_- induces an identification $TM \cong \mathcal{G}_0 \times_{G_0} \mathfrak{g}_-$, which dualizes to an identification $T^*M \cong \mathcal{G}_0 \times_{G_0} \mathfrak{p}_+$. Since the Kostant codifferential ∂^* from 3.1.11 is G_0–equivariant, it induces bundle maps on the bundles of forms with values in $\mathcal{G}_0 \times_{G_0} \mathfrak{g}$. In particular, we have $\partial^* : \Omega^2(M, \mathcal{G}_0 \times_{G_0} \mathfrak{g}) \to \Omega^1(M, \mathcal{G}_0 \times_{G_0} \mathfrak{g})$.

DEFINITION 5.2.2. A Weyl form τ is called normal if and only if its curvature $k_\tau \in \Omega^2(M, \mathcal{G}_0 \times_{G_0} \mathfrak{g})$ satisfies $\partial^* k_\tau = 0$.

Now we are ready to characterize the Weyl forms obtained from Weyl structures for those parabolic geometries which are equivalent to regular infinitesimal flag structures.

THEOREM 5.2.2. *Let $(p : \mathcal{G} \to M, \omega)$ be a regular normal parabolic geometry of type (G,P) and let $(p_0 : \mathcal{G}_0 \to M, \theta)$ be the underlying regular infinitesimal flag structure. If $H^1(\mathfrak{g}_-, \mathfrak{g})^1 = 0$, then a Weyl form $\tau \in \Omega^1(\mathcal{G}_0, \mathfrak{g})$ is of the form $\sigma^*\omega$ for some Weyl structure $\sigma : \mathcal{G}_0 \to \mathcal{G}$ if and only if τ is normal.*

PROOF. If $\tau = \sigma^*\omega$ then the definition of the curvature immediately implies that $K_\tau = \sigma^*K \in \Omega^2(\mathcal{G}_0, \mathfrak{g})$, where $K \in \Omega^2(\mathcal{G}, \mathfrak{g})$ is the curvature of ω. Hence, k_τ is exactly the image of the Cartan curvature $\kappa \in \Omega^2(M, \mathcal{A}M)$ under the map induced by the isomorphism between $\Lambda^2T^*M \otimes \mathcal{A}M$ and its associated graded coming from the Weyl structure σ. We have observed in 3.1.11 that ∂^* is compatible with the grading on $\Lambda^k\mathfrak{p}_+ \otimes \mathfrak{g}$, which implies that the induced bundle map is unchanged when passing to the associated graded. Hence, $\partial^*\kappa = 0$ implies that any Weyl form of the form $\sigma^*\omega$ is normal.

Conversely, assume that $\tau \in \Omega^1(\mathcal{G}_0, \mathfrak{g})$ is a normal Weyl form, and choose any Weyl structure $\sigma : \mathcal{G}_0 \to \mathcal{G}$. For a point $u \in \mathcal{G}_0$ any tangent vector at $\sigma(u) \in \mathcal{G}$ can be uniquely written as $T_u\sigma \cdot \xi + \zeta_A(\sigma(u))$ for some $\xi \in T_u\mathcal{G}_0$ and $A \in \mathfrak{p}_+$. Now we define $\tilde{\omega}(\sigma(u)) : T_{\sigma(u)}\mathcal{G} \to \mathfrak{g}$ by

$$\tilde{\omega}(\sigma(u))(T_u\sigma \cdot \xi + \zeta_A(\sigma(u))) := \tau(\xi) + A.$$

Since $\tau_-(u) + \tau_0(u) : T_u\mathcal{G}_0 \to \mathfrak{g}_- \oplus \mathfrak{g}_0$ is injective, we see that $\tilde{\omega}(\sigma(u))$ is injective and thus a linear isomorphism. Moreover, by construction $\tilde{\omega}(\sigma(u))$ reproduces the generators of fundamental vector fields.

We claim that this uniquely extends to a Cartan connection $\tilde{\omega} \in \Omega^1(\mathcal{G}, \mathfrak{g})$ by equivariancy. This means that we have to define

$$\tilde{\omega}(\sigma(u) \cdot g)(Tr^g \cdot \eta) := \mathrm{Ad}(g^{-1})(\omega(\sigma(u))(\eta))$$

for all $u \in \mathcal{G}_0$ and $g \in P$. To see that this is well defined, suppose that $\sigma(u) \cdot g = \sigma(u') \cdot g'$. Writing $g = g_0 \exp(Z)$ and likewise for g', we see that $\sigma(u \cdot g_0) \cdot \exp(Z) = \sigma(u' \cdot g_0') \cdot \exp(Z')$, which implies $u \cdot g_0 = u' \cdot g_0'$ and $Z = Z'$. Hence it suffices to show that for $g_0 \in G_0$ we have $\tilde{\omega}(\sigma(u \cdot g_0))(Tr^{g_0} \cdot \eta) := \mathrm{Ad}(g_0^{-1})(\tilde{\omega}(\sigma(u))(\eta))$. For vertical vectors this follows immediately from equivariancy of the fundamental vector field map. For tangent vectors of the form $T_u\sigma \cdot \xi$ it follows easily from G_0–equivariancy of σ and τ.

Next we claim that the Cartan connection $\tilde{\omega}$ induces the same infinitesimal flag structure as ω. To compute the value of the frame form induced by $\tilde{\omega}$ on $\xi \in T^i_u\mathcal{G}_0$, we have to choose a point in $\mathcal{G}$ over u and a lift of ξ, so we may take $T_u\sigma \cdot \xi \in T_{\sigma(u)}\mathcal{G}$. Then $\tilde{\omega}(T_u\sigma \cdot \xi) = \tau(\xi) \in \mathfrak{g}^i$, which implies that ω and $\tilde{\omega}$ induce the same filtration. Since the $\mathfrak{g}_i$–component of $\tau(\xi)$ by definition coincides with $\theta_i(\xi)$, we further conclude that $\tilde{\omega}$ induces the same frame form as ω and hence the same underlying infinitesimal flag structure.

Finally, consider the curvature $\tilde{K}$ of the Cartan connection $\tilde{\omega}$. By construction, we have $\tau = \sigma^*\tilde{\omega}$ and thus $K_\tau = \sigma^*\tilde{K}$. Passing to the curvature functions and using normality of τ, we conclude that the function representing $\partial^*\tilde{\kappa}$ vanishes along the image of σ and hence everywhere by equivariancy.

Since ω and $\tilde{\omega}$ are two normal Cartan connections which induce the same underlying infinitesimal flag structure, we may invoke Proposition 3.1.14. This shows that there is a principal bundle isomorphism $\Psi : \mathcal{G} \to \mathcal{G}$, which induces the identity on $\mathcal{G}_0$ such that $\Psi^*\omega = \tilde{\omega}$. Then $\hat{\sigma} := \Psi \circ \sigma$ is a smooth G_0–equivariant section of $\mathcal{G}_0 \to \mathcal{G}$ and thus a Weyl structure, and $\hat{\sigma}^*\omega = \sigma^*\Psi^*\omega = \sigma^*\tilde{\omega} = \tau$. □

REMARK 5.2.2. Let us briefly discuss what happens in the two basic cases for which $H^1(\mathfrak{g}_-, \mathfrak{g})^1 \neq 0$. First we discuss the case of projective structures; see 4.1.5. In this case a regular infinitesimal flag structure on M can be simply identified with the full linear frame bundle of M endowed with the tautological soldering form $\theta = \theta_{-1}$. Since we are dealing with a $|1|$–grading, we have $\tau_- = \theta$, so a Weyl form consists just of the principal connection τ_0, and the Rho tensor τ_+. Equivalently, we may view τ_0 as a linear connection on TM. Now fix a projective class of linear connections on TM and let $(p : \mathcal{G} \to M, \omega)$ be the corresponding normal parabolic geometry. Then a Weyl form τ is of the form $\sigma^*\omega$ if and only if τ is normal and τ_0 belongs to the projective class. The proof that $\sigma^*\omega$ is normal is as in the proof of Theorem 5.2.2 above. In 4.1.5 we have observed that the connections $\sigma^*\omega_0$ are exactly the elements of the projective class. Hence, the necessity of the condition follows.

Conversely, if τ is normal and τ_0 belongs to the projective class, then we can find a Weyl structure σ such that $\sigma^*\omega_0 = \tau_0$. Now as in the proof of the theorem we construct a Cartan connection $\tilde{\omega}$ from σ and τ and see that it is normal and $\tilde{\omega} - \omega$ is homogeneous of degree ≥ 2. Then the first part of the proof of Proposition 3.1.14 shows that $\tilde{\omega} = \Psi^*\omega$ for an automorphism Ψ of $\mathcal{G}$ which induces the identity on $\mathcal{G}_0$.

For contact projective structures (see 4.2.6) the situation is very similar. The regular infinitesimal flag structure $(\mathcal{G}_0 \to M, \theta)$ is simply the full frame bundle of the bundle $\mathrm{gr}(TM)$ of Lie algebras. For a Weyl form τ, the restriction of τ_0 to $T^{-1}\mathcal{G}_0$ can be viewed as a partial contact connection on $T^{-1}M$. Now fix a projective contact structure on M and let $(p : \mathcal{G} \to M, \omega)$ be the corresponding regular normal parabolic geometry. We claim that τ is of the form $\sigma^*\omega_0$ if and only if τ is normal and the restriction of τ_0 to $T^{-1}\mathcal{G}_0$ belongs to the projective class. The necessity of the condition is proved as in the projective case.

Suppose conversely that the condition is satisfied. Then we can find a Weyl structure σ such that $\sigma^*\omega_0|_{T^{-1}\mathcal{G}_0} = \tau_0|_{T^{-1}\mathcal{G}_0}$, and we construct a normal Cartan connection $\tilde{\omega}$ from σ and τ as before. To see that $\tilde{\omega} - \omega$ is homogeneous of degree ≥ 2, one has to show that $\sigma^*\omega_{-1} = \tau_{-1}$, or otherwise put that the isomorphisms $TM \to \mathrm{gr}(TM)$ induced by τ and $\sigma^*\omega$ coincide. But vanishing of the torsion component $T^{-1}M \otimes TM/T^{-1}M \to T^{-1}M$ (which is a consequence of normality, see 4.2.6) shows that this isomorphism is determined by the corresponding partial connections. Hence, Proposition 3.1.14 gives an automorphism Ψ on $\mathcal{G}$ and as in the proof of the theorem, this leads to a Weyl structure $\hat{\sigma}$ such that $\hat{\sigma}^*\omega = \tau$.

5.2.3. Example: Structures corresponding to $|1|$–gradings. To motivate further development, we give a complete description of the Weyl forms corresponding to Weyl structures in the case of $|1|$–gradings. In this case, an infinitesimal flag structure $(p_0 : \mathcal{G}_0 \to M, \theta)$ is simply a reduction of the linear frame bundle of M to the structure group G_0 with respect to the homomorphism $G_0 \to GL(\mathfrak{g}_{-1})$ induced by the adjoint action; see Example 3.1.6 and 4.1.1. In particular, for any Weyl form τ, the component τ_- coincides with the tautological soldering form θ, and τ_0 is a principal connection compatible with the first order structure as discussed in 1.3.6. Now the curvature K_τ can be split according to its values into components. Since these components are horizontal and equivariant, they represent elements $T \in \Omega^2(M, TM)$, $W \in \Omega^2(M, \mathrm{End}_0(TM))$ and $Y \in \Omega^2(M, T^*M)$. Since the decomposition $k_\tau = T+W+Y$ coincides with the decomposition into homogeneous components, we see that $\partial^* k_\tau = 0$ if and only if $\partial^* T = 0$ and $\partial^* W = 0$, while $\partial^* Y = 0$ is automatically satisfied by homogeneity.

The quantities T, W, and Y are easy to interpret for a given Weyl form τ. Since $\mathfrak{g}_-$ is abelian, T by construction corresponds to the element of $\Omega^2(\mathcal{G}_0, \mathfrak{g}_-)$ given by

$$d\tau_-(\xi, \eta) + [\tau_0(\xi), \tau_-(\eta)] + [\tau_-(\xi), \tau_0(\eta)].$$

Since τ_- is the soldering form of the first order G_0–structure $\mathcal{G}_0$, we see from 1.6.1 that T is the torsion of the linear connection on TM induced by the principal connection τ_0. From 1.6.1 we also know that the possible values of this torsions (for arbitrary τ_0) form an affine space modelled on $\mathrm{im}(\partial)$. In particular, the condition that $\partial^* T = 0$ fixes T uniquely. By Theorem 4.1.1 we then even have $\square(T) = 0$ and T coincides with the homogeneous component of degree one of the harmonic curvature.

If τ is a Weyl form such that $\partial^* T = 0$, then by part (2) of Theorem 4.1.1 we know that we can compute the homogeneous component of degree two of the harmonic curvature from the curvature R of the principal connection τ_0. By construction the form W is induced by the element of $\Omega^2(\mathcal{G}_0, \mathfrak{g}_+)$ given by

$$d\tau_0(\xi, \eta) + [\tau_0(\xi), \tau_0(\eta)] + [\tau_-(\xi), \tau_+(\eta)] + [\tau_+(\xi), \tau_-(\eta)].$$

By 1.6.1 the first two summands represent R. The remaining two summands represent the element of $\Omega^2(M, \mathrm{End}_0(TM))$ given by

$$(\xi, \eta) \mapsto \{\xi, \mathsf{P}(\eta)\} + \{\mathsf{P}(\xi), \eta\} = \partial\mathsf{P}(\xi, \eta),$$

where $\{\ ,\ \} : TM \times T^*M \to \mathrm{End}_0(TM)$ is the algebraic bracket, ∂ is induced by the Lie algebra differential, and P is the Rho tensor, which is induced by τ_+. Hence, we conclude that $W = R + \partial\mathsf{P}$. The equation $\partial^* W = 0$ reads as $\partial^*\partial\mathsf{P} = -\partial^* R$, and since $\partial^*\mathsf{P}$ vanishes automatically, we get $\Box\mathsf{P} = -\partial^* R$. By the Hodge decomposition, $\Box$ is invertible on $\mathrm{im}(\partial^*)$, which implies that this equation always has a unique solution.

Hence, we see that the only condition for normality of a Weyl form τ is $\partial^* T = 0$. If this is satisfied, then normality is equivalent to the fact that P is the unique solution of $\Box\mathsf{P} = -\partial^* R$. Moreover, since $W - R \in \mathrm{im}(\partial)$ we see that the components of W and R in $\ker(\Box)$ coincide, so we can compute the corresponding harmonic curvature component either from W or from R.

Finally, the form Y is induced by the element of $\Omega^2(\mathcal{G}_0, \mathfrak{g}_1)$ given by

$$(\xi, \eta) \mapsto d\tau_+(\xi, \eta) + [\tau_0(\xi), \tau_+(\eta)] + [\tau_+(\xi), \tau_0(\eta)].$$

Choosing ξ and η to be horizontal and inserting the definition of the exterior derivative we see that Y is the covariant exterior derivative of P. In particular, by Theorem 4.1.1 the component of Y in $\ker(\Box)$ represents the homogeneous component of degree 3 of the harmonic curvature. Since $W = R + \partial\mathsf{P}$, the equation $\partial^* W = 0$ implies that P coincides with the Rho tensor as defined in 4.1.1. This implies that also the Weyl curvature W and the Cotton–York tensor Y coincide with their counterparts defined in 4.1.1. In particular, we can use all the computations made in the examples in Section 4.1.

Collecting our results we get:

THEOREM 5.2.3. *Let $(p_0 : \mathcal{G}_0 \to M, \theta)$ be an infinitesimal flag structure of type (G, P) corresponding to a $|1|$–grading $\mathfrak{g} = \mathfrak{g}_{-1} \oplus \mathfrak{g}_0 \oplus \mathfrak{g}_1$ such that $H^1(\mathfrak{g}_-, \mathfrak{g})^1 = 0$. Let $(p : \mathcal{G} \to M, \omega)$ be the unique normal parabolic geometry of type (G, P) extending this infinitesimal flag structure, κ its curvature and κ_H its harmonic curvature.*

(1) The Weyl connections associated to Weyl structures $\sigma : \mathcal{G}_0 \to \mathcal{G}$ are exactly the principal connections on $\mathcal{G}_0$ with ∂^*–closed torsion. All of these connections have the same torsion T, which coincides with the homogeneous component of degree one of κ_H.

(2) For such a Weyl connection, let R be the curvature. Then the Rho tensor coincides with the one defined in 4.1.1 and is uniquely determined by $\Box\mathsf{P} = -\partial^* R$. The remaining homogeneous components of the harmonic curvature are given by the components in $\ker(\Box)$ of R, respectively, of the covariant exterior derivative Y of P.

(3) The Cartan curvature κ of ω is represented by

$$(T, W, Y) \in \Omega^2(M, TM \oplus \mathrm{End}_0(TM) \oplus T^*M),$$

where T and Y are defined in (1) and (2) and $W := R + \partial\mathsf{P}$.

For projective structures, we easily get the analogous result using the description of Weyl forms coming from Weyl structures in Remark 5.2.2. The Weyl connections associated to Weyl structures are exactly the torsion–free connections in the projective class. The Rho tensor is determined by $\Box\mathsf{P} = -\partial^* R$. The harmonic curvature has just one nonzero homogeneous component which is represented by $W = R + \partial \mathsf{P}$. Finally, the Cartan curvature is represented by $(0, W, Y)$, where Y is the covariant exterior derivative of P.

We will next discuss a few applications of the ideas developed so far in the $|1|$–graded case and return to the general theory in 5.2.9 below.

5.2.4. Conformal standard tractors. As a first application of our results, we give an explicit description of the conformal standard tractor bundle, i.e. the tractor bundle induced by the standard representation. In view of Theorem 3.1.22, this gives an equivalent description of the Cartan bundle and the Cartan connection associated to the conformal structure.

Let $(M, [g])$ be a conformal manifold of dimension $n \geq 3$ and arbitrary signature. By Theorem 5.2.3, the Weyl connections on M are exactly the torsion–free connections preserving the conformal class, i.e. the classical Weyl connections as introduced in 1.6.4. In particular, a choice of metric in the conformal class determines an exact Weyl structure, and the corresponding Weyl connection is the Levi–Civita connection of the metric. Let us use abstract index notation as introduced in 1.6.6 and 1.6.10, and denote by $\mathbf{g}_{ij}$ and $\mathbf{g}^{ij}$ the conformal metric and its inverse, which are canonical sections of $\mathcal{E}_{ij}[2]$ and $\mathcal{E}^{ij}[-2]$, respectively.

Assume that ∇ is the Weyl connection corresponding to a Weyl structure σ. Let us first compute the change of ∇ caused by changing from σ to $\hat{\sigma}$ given by $\hat{\sigma}(u) = \sigma(u) \exp(\Upsilon(u))$. Viewing Υ as a one–form Υ_a on M, we see from formula (1.32) in 1.6.6 that for $\xi, \eta \in \mathfrak{X}(M)$, the field $\{\Upsilon, \xi\} \bullet \eta$ is given by

$$-\xi^i \Upsilon_j \eta^j + \mathbf{g}^{ij} \Upsilon_j \mathbf{g}_{k\ell} \xi^k \eta^\ell - \xi^j \Upsilon_j \eta^i.$$

From now on, we will use $\mathbf{g}_{ij}$ and $\mathbf{g}^{ij}$ to raise and lower indices without further mention. Using the specialization of Proposition 5.1.6 for structures corresponding to $|1|$–gradings, we conclude the Weyl connection $\hat{\nabla}$ corresponding to $\hat{\sigma}$ on TM is given by

$$\hat{\nabla}_i \eta^j = \nabla_i \eta^j + \Upsilon_i \eta^j - \Upsilon^j \eta_i + \Upsilon^k \eta_k \delta_i{}^j. \tag{5.1}$$

For a Weyl connection ∇, let $R = R_{ij}{}^k{}_\ell$ be the curvature tensor. Then by Lemma 1.6.6, the Rho tensor is given by

$$\mathsf{P}_{ij} = \tfrac{-1}{n-2} \left(R_{ki}{}^k{}_j + \tfrac{1}{n}(R_{kj}{}^k{}_i - R_{ki}{}^k{}_j) - \tfrac{1}{2(n-1)} \mathbf{g}^{ab} R_{ka}{}^k{}_b \mathbf{g}_{ij} \right), \tag{5.2}$$

with the convention that $\mathsf{P}(\xi)_i = \xi^j \mathsf{P}_{ji}$.

Next, we can give an explicit description of the standard tractor bundle. For simplicity, let us consider oriented conformal structures, so from 4.1.2 we know that we may choose $G := SO(p+1, q+1)$ and $P \subset G$ to be the connected component of the identity of the stabilizer of an isotropic line in the standard representation $\mathbb{R}^{(p+1,q+1)}$ of G. Denoting by $\mathcal{G}$ the corresponding Cartan bundle, the standard tractor bundle $\mathcal{T}M$ is the associated bundle $\mathcal{G} \times_P \mathbb{R}^{(p+1,q+1)}$. By construction, $\mathcal{T}M$ carries a natural bundle metric of signature $(p+1, q+1)$, and from 1.5.7 we know that ω induces a canonical linear connection $\nabla^{\mathcal{T}}$ on $\mathcal{T}M$. Now we can describe all of these objects explicitly.

PROPOSITION 5.2.4. *Any choice of a Weyl connection ∇ gives rise to an identification $\mathcal{T}M \cong \mathcal{E}[1] \oplus \mathcal{E}_a[1] \oplus \mathcal{E}[-1]$, and we denote elements and sections of the right–hand side by triples of the form (σ, μ_i, ρ). Changing the Weyl connection to*

$\hat{\nabla}$ *according to formula* (5.1) *with respect to* $\Upsilon \in \Omega^1(M)$, *this identification changes as*

$$(5.3) \qquad \widehat{(\sigma, \mu_i, \rho)} = (\sigma, \mu_i + \sigma\Upsilon_i, \rho - \Upsilon^j\mu_j - \tfrac{1}{2}\Upsilon^j\Upsilon_j\sigma).$$

In the identification corresponding to ∇, *the canonical bundle metric on* $\mathcal{T}M$ *is given by*

$$(5.4) \qquad \langle(\sigma, \mu_i, \rho), (\sigma', \mu'_j, \rho')\rangle = \sigma\rho' + \mathbf{g}^{ij}\mu_i\mu'_j + \rho\sigma',$$

while the canonical tractor connection is given by

$$(5.5) \qquad \nabla^{\mathcal{T}}_j(\sigma, \mu_j, \rho) = (\nabla_i\sigma - \mu_i, \nabla_i\mu_j - \sigma\mathsf{P}_{ji} + \rho\boldsymbol{g}_{ij}, \nabla_i\rho + \mathsf{P}_{ij}\mu^j).$$

PROOF. By definition, $\mathcal{T}M$ is the associated bundle corresponding to the standard representation. From 4.1.2, we know that the composition series of $\mathcal{T}M$ is given by $\mathcal{T}M = \mathcal{E}[-1] \,+\!\!\!\!\supset\, \mathcal{E}_a[1] \,+\!\!\!\!\supset\, \mathcal{E}[1]$. In particular, the associated graded is given by

$$\mathrm{gr}(\mathcal{T}M) = \mathrm{gr}_0(\mathcal{T}M) \oplus \mathrm{gr}_1(\mathcal{T}M) \oplus \mathrm{gr}_2(\mathcal{T}M) = \mathcal{E}[1] \oplus \mathcal{E}_a[1] \oplus \mathcal{E}[-1].$$

Here the numbering the grading components follows the conventions of 5.1.5. There we have seen that the choice of a Weyl structure gives rise to an identification $\mathcal{T}M \cong \mathrm{gr}(\mathcal{T}M)$ and how this identification changes under a change of Weyl structure. This change is given in terms of the action of T^*M on $\mathrm{gr}(\mathcal{T}M)$, and for later use, we compute the action of TM as well.

We use the matrix representation of $\mathfrak{g}$ from 1.6.3. Then the component of $\mathrm{gr}_2(\mathcal{T}M)$ corresponds to vectors with only the top component nonzero, $\mathrm{gr}_1(\mathcal{T}M)$ corresponds to the middle $(p+q)$–components, and $\mathrm{gr}_0(\mathcal{T}M)$ to the bottom component. Hence, (σ, μ_i, ρ) corresponds to a column vector with σ in the bottom and ρ in the top, and the formula (5.4) for the tractor metric follows immediately. To compute the relevant actions, we just have to note that

$$\begin{pmatrix} 0 & Z & 0 \\ X & 0 & -Z^t\mathbb{I}_{p,q} \\ 0 & -X^t\mathbb{I}_{p,q} & \end{pmatrix} \begin{pmatrix} r \\ Y \\ s \end{pmatrix} = \begin{pmatrix} ZY \\ rX - sZ^t\mathbb{I}_{p,q} \\ -X^t\mathbb{I}_{p,q}Y \end{pmatrix}$$

directly implies

$$(5.6) \qquad \xi^i \bullet (\sigma, \mu_j, \rho) = (-\xi^k\mu_k, \rho\xi_j, 0),$$

$$(5.7) \qquad \phi_i \bullet (\sigma, \mu_j, \rho) = (0, -\sigma\phi_j, \mathbf{g}^{ij}\phi_i\mu_j).$$

Using (5.7) for $\phi_i = \Upsilon_i$, formula (5.3) for the change of identification follows directly from the Proposition 5.1.5 specialized to $|1|$–gradings.

By part (2) of Proposition 5.1.10, we can compute the action of $\nabla^{\mathcal{T}}_\xi$ by the sum of the componentwise Weyl connection and the actions of ξ^i and $\xi^k\mathsf{P}_{kj}$. Inserting (5.6) and (5.7) this immediately leads to (5.5). □

REMARK 5.2.4. (1) Note that in view of Theorem 3.1.22 the data described in the proposition provide an equivalent description of the conformal Cartan bundle and the canonical Cartan connection. The first ad–hoc definitions of tractor bundles in this spirit are due to Tracy Thomas in the 1920s; see for example [**Tho26, Tho31**]. These ideas were rediscovered and formulated in modern language in [**BEG94**], which initiated a lot of research in this direction. In comparing the formulae to the latter reference, one has to take into account that our Rho tensor differs in sign from the one used there.

(2) It is a great exercise in abstract index calculus to verify directly that (5.5) gives rise to a well–defined connection on $\mathcal{T}M$, i.e. that the result behaves well under a change of Weyl connections. Doing this exercise should also convince the reader that it is preferable to replace direct computations by general arguments.

5.2.5. Parallel standard tractors. We next analyze the geometric meaning of existence of a parallel section of the standard tractor bundle $\mathcal{T}M$ of a conformal manifold $(M, [g])$. This should just be viewed as a teaser for volume two, in which similar questions will be studied in a conceptual and systematic way.

PROPOSITION 5.2.5. *Let $(M, [g])$ be an oriented manifold of dimension $n \geq 3$ endowed with a pseudo–Riemannian conformal structure of arbitrary signature.*

(1) The space of those sections of the standard tractor bundle $\mathcal{T}M$, which are parallel for the canonical tractor connection is in bijective correspondence with the space of all $\sigma \in \Gamma(\mathcal{E}[1])$ such that for some (or equivalently any) Weyl–connection ∇ with Rho–tensor P we have

$$\nabla_{(i}\nabla_{j)_0}\sigma - \mathsf{P}_{(ji)_0}\sigma = 0. \tag{5.8}$$

Here the subscript denotes the tracefree symmetric part.

(2) For any section $\sigma \in \Gamma(\mathcal{E}[1])$ which satisfies (5.8)*, $\{x \in M : \sigma(x) \neq 0\}$ is a dense open subset of M.*

(3) Let $\sigma \in \Gamma(\mathcal{E}[1])$ be nowhere vanishing. Then $g_{ij} := \sigma^{-2}\mathbf{g}_{ij}$ is a metric in the conformal class, and σ satisfies (5.8) *if and only if the metric g is Einstein.*

PROOF. Suppose first that $s \in \Gamma(\mathcal{T}M)$ is a parallel section. For a Weyl connection ∇, we get $s = (\sigma, \mu_i, \rho)$ in the splitting of $\mathcal{T}M$ determined by ∇ as in Proposition 5.2.4. Then formula (5.5) from that Proposition shows that $\nabla_i^{\mathcal{T}} s = 0$ implies $\nabla_i\sigma = \mu_i$ and $\rho\mathbf{g}_{ij} = -\nabla_i\mu_j + \sigma\mathsf{P}_{ji}$, so we conclude that σ satisfies (5.8) for the given Weyl connection. Changing to another Weyl connection, the component σ remains unchanged, so we see that $\sigma \in \Gamma(\mathcal{E}[1])$ depends only on s and satisfies (5.8) for any Weyl connection. Moreover, we see that for any Weyl connection ∇, with Rho–tensor P_{ij}, the section is given by

$$s = (\sigma, \nabla_i\sigma, -\tfrac{1}{n}(\nabla^j\nabla_j\sigma - \mathsf{P}^j{}_j\sigma))$$

in the splitting corresponding to ∇. In particular, this shows that if $x \in M$ is a point such that $\sigma(x) = 0$, $\nabla_i\sigma(x) = 0$ and $\Delta\sigma(x) = \nabla^j\nabla_j\sigma(x) = 0$, then $s(x) = 0$. Since s is parallel, this implies that s vanishes identically. Hence, for nonzero s, the two–jet of σ in each point $x \in M$ must be nontrivial, so in particular, $\{x : \sigma(x) \neq 0\} \subset M$ is open and dense.

Conversely, let us start from $\sigma \in \Gamma(\mathcal{E}[1])$ and some Weyl connection ∇ with curvature $R_{ij}{}^k{}_\ell$. Formula (5.2) in 5.2.4 for the Rho tensor immediately implies that

$$\mathsf{P}_{ij} - \mathsf{P}_{ji} = \tfrac{1}{n}(R_{kj}{}^k{}_i + R_{ik}{}^k{}_j) = \tfrac{1}{n}R_{ij}{}^k{}_k,$$

where in the second step we have used the Bianchi identity. From 4.1.2 we know that $\mathcal{E}[1]$ is the bundle of $(-\frac{1}{n})$–densities, so the curvature of the connection on this bundle induced by ∇ is given by $-\frac{1}{n}R_{ij}{}^k{}_k$. This shows that the expression $\nabla_i\nabla_j\sigma - \mathsf{P}_{ji}\sigma$ is automatically symmetric. Hence, putting $\mu_i := \nabla_i\sigma$ and $\rho := -\frac{1}{n}(\nabla^j\nabla_j\sigma - \mathsf{P}^j{}_j\sigma)$, the equation (5.8) on σ is equivalent to

$$\mathbf{g}_{ij}\rho = -\nabla_i\nabla_j\sigma + \mathsf{P}_{ji}\sigma. \tag{5.9}$$

Otherwise put, this is equivalent to the fact that the tractor $s := (\sigma, \mu_i, \rho)$ in the splitting corresponding to ∇ has the property that $\nabla_i^{\mathcal{T}} s = (0, 0, \nabla_i \rho + \mathsf{P}_{ij}\mu^j)$. Now we can apply ∇_k to (5.9), and then contract with $\mathbf{g}^{jk}$ to compute $\nabla_i \rho$. On the right-hand side, we can swap the derivatives ∇_k and ∇_i at the expense of a curvature. After the contraction, the term without curvature gets the form $-\nabla_i \Delta\sigma$, which we can rewrite as $\nabla_i(-n\rho + \mathsf{P}^j{}_j\sigma)$. The curvature terms can all be expressed in terms of the Rho tensor and its trace, and one can use the differential Bianchi identity to compute $\nabla^j \mathsf{P}_{ji}$. Putting all that together, one verifies that $\nabla_i\rho + \mathsf{P}_{ij}\mu^j = 0$ is always satisfied automatically, so the tractor s defined above is actually parallel. This completes the proofs of (1) and (2).

(3) For a section $\sigma \in \mathcal{E}[1]$ without zeros, σ^{-2} can be interpreted as a section of $\mathcal{E}[-2]$. Since the conformal metric $\mathbf{g}_{ij}$ is a section $S^2T^*M[2]$, we see that $\sigma^{-2}\mathbf{g}_{ij}$ is a section of S^2T^*M, and by construction this is a metric in the conformal class. The Levi–Civita connection ∇ of this metric is the Weyl connection for the exact Weyl structure defined by σ. In particular, $\nabla_i\sigma = 0$ for the induced connection on $\mathcal{E}[1]$. Hence, (5.8) reduces to $\mathsf{P}_{(ij)_0}\sigma = 0$, and since σ is nowhere vanishing this is equivalent to $\mathsf{P}_{(ij)_0} = 0$. For a Levi–Civita connection $R_{ki}{}^k{}_j$ is symmetric, so formula (5.2) for the Rho tensor from 5.2.4 shows that $\mathsf{P}_{(ij)_0} = -\frac{1}{m-2}R_{k(i}{}^k{}_{j)_0}$. Hence, $\mathsf{P}_{(ij)_0} = 0$ is equivalent to the Ricci curvature being pure trace and hence to our metric being Einstein. □

REMARK 5.2.5. (1) Conformal structures which admit a parallel standard tractor are referred to as *almost Einstein* conformal structures in the literature; see [**Go05**]. By parts (2) and (3) of the proposition, for an almost Einstein structure $[g]$ on M, there is an open dense subset $U \subset M$ and an Einstein metric on U which is contained in the restriction of the conformal class.

(2) The question of existence of parallel sections of the standard tractor bundle is closely related to the holonomy of canonical tractor connection $\nabla^{\mathcal{T}}$.

For a linear connection on a vector bundle $E \to M$, one always has the corresponding horizontal distribution $\mathcal{H} \subset TE$; compare with 1.3.1. This gives rise to the concept of parallel transport along smooth curves. Given a smooth curve in M from a point x to a point y, one looks for smooth lifts to curves in E which are tangent to the horizontal distribution. If such lifts exist, then they are uniquely determined by their value in one point, so they may be viewed as providing a map $E_x \to E_y$ between the fibers. It turns out that this map is always linear. In particular, one may fix a point $x \in M$ and then look at closed curves through x. The parallel transports along such curves are linear endomorphisms of the fiber E_x, and it is easy to see that they form a closed subgroup of $GL(E_x) \cong GL(N, \mathbb{R})$, $N = \dim(E_x)$, which is called the *holonomy group* of the connection. If M is connected, then changing the point x leads to a conjugate subgroup of $GL(N, \mathbb{R})$.

One may play the same game for a principal connection on a principal bundle with structure group H. In this case, it is easy to see that the parallel transports commute with the principal right action. Identifying a fiber with the structure group H, the parallel transports are thus given by left translations with elements of H and hence the holonomy group can be naturally viewed as a subgroup of H. Passing to the induced bundle determined by a representation $H \to GL(V)$ and the induced linear connection on that bundle is compatible with holonomy, i.e. the holonomy group of the linear connection is (conjugate to) the image of the holonomy group of the principal connection under the representation.

The relation between parallel sections and the holonomy group is easy to understand. If s is a parallel section of a vector bundle E, then for any smooth curve c from x to y, the curve $t \mapsto s(c(t))$ by construction is the horizontal lift of c with initial value $s(x)$. In particular, if c is a closed curve through x, then $s(x)$ is transported along c to itself, so any element of the holonomy group has to fix the element $s(x)$. Conversely, suppose that we have an element $v \in E_x$ which is preserved by all elements of the holonomy group. Then for points y sufficiently close to x, we can choose a smooth curve from x to y for which the parallel transport is defined, and then define $s(y)$ to be the element obtained by parallel transporting v. Then condition on the holonomy group ensures that the result is independent of the choice of the curve, so this defines a local parallel section.

Now let us apply this idea to the standard tractor connection $\nabla^{\mathcal{T}}$ of a conformal structure $(M, [g])$. If $(\mathcal{G} \to M, \omega)$ is the normal parabolic geometry of type (G, P) determined by the conformal structure, then the standard tractor connection $\nabla^{\mathcal{T}}$ is induced by the principal connection on $\mathcal{G} \times_P G$ induced by ω, compare with 1.5.7. Hence, the holonomy group of $\nabla^{\mathcal{T}}$ is a subgroup of $G = SO(p+1, q+1)$, which is called the *conformal holonomy* of $(M, [g])$. As we have seen above, existence of a local parallel section of $\mathcal{T}M$ is equivalent to the fact that the conformal holonomy acts trivially on some vector in the standard representation $\mathbb{R}^{(p+1,q+1)}$ of G. Since all the canonical tractor connections are induced in this way, similar arguments apply to other tractor bundles.

The question of which groups may occur as conformal holonomy groups as well as the question of the geometric interpretation of conditions on the conformal holonomy have been intensively studied during the last few years; see e.g. [**Leit04, Leis06, Arm07b, Nu08**]. For the case of definite signature, a complete classification of the possible conformal holonomy groups has been obtained in [**Arm07a**].

5.2.6. Standard tractors and cone description for projective structures. The basic description of the standard tractor bundle $\mathcal{T}M$ and its dual $\mathcal{T}^*M$ for an oriented projective structure $(M, [\nabla])$ is simpler than in the conformal case. Here the Weyl connections are just the connections in the projective class. The change from such a connection ∇ to another connection $\hat{\nabla}$ in the class is described by a one–form $\Upsilon \in \Omega^1(M)$; see Definition 4.1.5. In abstract index notation, this reads as

$$\hat{\nabla}_i \eta^j = \nabla_i \eta^j + \Upsilon_i \eta^j + \Upsilon_k \eta^k \delta_i{}^j. \tag{5.10}$$

Using the notation introduced in 4.1.5, the composition series for the standard tractor bundle is $\mathcal{T}M = \mathcal{E}(-1) \rlap{+}\supset TM(-1)$ and hence we get $\mathcal{T}^*M = T^*M(1) \rlap{+}\supset \mathcal{E}(1)$. Hence, any choice of a connection in the projective class gives rise to identifications $\mathcal{T}M \cong TM(-1) \oplus \mathcal{E}(-1)$ and $\mathcal{T}^*M \cong \mathcal{E}(1) \oplus T^*M(1)$. Following [**BEG94**], we denote elements in the latter sum as (μ_i, σ) and elements in the former sum as $\binom{\eta^i}{\rho}$. Then the dual pairing between the two bundles is simply given by $((\mu_i, \sigma), \binom{\eta^i}{\rho}) \mapsto \eta^i \mu_i + \sigma\rho$.

The actions of TM and T^*M on $\operatorname{gr}(\mathcal{T}M)$ and $\operatorname{gr}(\mathcal{T}^*M)$ are given by

$$\xi^i \bullet \begin{pmatrix} \eta^j \\ \rho \end{pmatrix} = \begin{pmatrix} \rho\xi^j \\ 0 \end{pmatrix}, \qquad \xi^i \bullet (\mu_j, \sigma) = (0, -\xi^k \mu_k),$$

$$\phi_i \bullet \begin{pmatrix} \eta^j \\ \rho \end{pmatrix} = \begin{pmatrix} 0 \\ \phi_k \eta^k \end{pmatrix}, \qquad \phi_i \bullet (\mu_j, \sigma) = (-\sigma\phi_j, 0).$$

Hence, changing from ∇ to $\hat{\nabla}$ according to (5.10) our identifications change as

$$\widehat{\begin{pmatrix} \eta^i \\ \rho \end{pmatrix}} = \begin{pmatrix} \eta^i \\ \rho - \Upsilon_j \eta^j \end{pmatrix} \qquad \widehat{(\mu_i, \sigma)} = (\mu_i + \sigma \Upsilon_i, \sigma).$$

From 4.1.5, we know that the Rho tensor corresponding to ∇ is given in terms of the curvature $R_{ij}{}^k{}_\ell$ of ∇ by

$$\mathsf{P}_{ij} = \tfrac{-1}{(n+1)(n-1)}(nR_{ki}{}^k{}_j + R_{kj}{}^k{}_i).$$

Using this, the tractor connections are given by

$$\nabla_i^{\mathcal{T}} \begin{pmatrix} \eta^j \\ \rho \end{pmatrix} = \begin{pmatrix} \nabla_i \eta^j + \rho \delta_i{}^j \\ \nabla_i \rho + \mathsf{P}_{ik} \eta^k \end{pmatrix},$$

$$\nabla_i^{\mathcal{T}^*}(\mu_j, \sigma) = (\nabla_i \mu_j - \sigma \mathsf{P}_{ij}, \nabla_i \sigma - \mu_i).$$

These formulae coincide with the ones in [**BEG94**, Section 3.2] in view of the fact that the Rho tensor used here is the negative of the one used in that reference.

Similarly, as discussed in Remark 5.2.5 for conformal structures, it is natural to study the holonomy group of the standard tractor connection of a projective structure, which is called the *projective holonomy* of the structure. For an oriented n–dimensional projective structure, this is a subgroup of $G = SL(n+1, \mathbb{R})$. Existence of a parallel section of the standard tractor bundle $\mathcal{T}M$ is again equivalent to the fact the the projective holonomy fixes a vector in the standard representation $\mathbb{R}^{n+1}$ of G.

In the case of projective structures, there is a nice way to equivalently describe the tractor bundle and the tractor connection in more traditional terms. This is called the *cone description* or the *ambient description* of projective structures. The basic ideas again go back to the work of Tracy Thomas, but the details have not been available in the literature for a long time. The complete construction is presented in [**Fox05a**] where the analogous construction for contact projective structures was introduced; see also 5.3.10 below.

The cone description is easy to understand from the Cartan picture. Consider an oriented projective structure $[\nabla]$ on a smooth manifold M. From 4.1.5 we know that there is an equivalent Cartan geometry $(p : \mathcal{G} \to M, \omega)$ of type (G, P), where $G = SL(n+1, \mathbb{R})$ and $P \subset G$ is the connected component of the identity of the stabilizer in G of a line in $\mathbb{R}^{n+1}$. In the matrix representation from 4.1.5, this means that P consists of all matrices of the form $\left(\begin{smallmatrix} \det(C)^{-1} & W \\ 0 & C \end{smallmatrix}\right)$ with $C \in GL(n, \mathbb{R})$ such that $\det(C) > 0$ and $W \in \mathbb{R}^{n*}$. Consider the closed normal subgroup $Q \subset P$ consisting of all elements with $C \in SL(n, \mathbb{R})$.

Defining $M_\# := \mathcal{G}/Q$, we see that the obvious projection $\pi : M_\# \to M$ is a bundle with fiber $P/Q \cong \mathbb{R}_+$, while the natural projection $\mathcal{G} \to M_\#$ is a principal bundle with structure group Q. In particular, the tangent bundle $TM_\#$ can be realized as $TM_\# = \mathcal{G} \times_Q (\mathfrak{g}/\mathfrak{q})$, where Q acts on $\mathfrak{g}/\mathfrak{q}$ via the adjoint action. But now for $C \in SL(n, \mathbb{R})$, $a \in \mathbb{R}$ and $X \in \mathbb{R}^n$, we get

$$\begin{pmatrix} 1 & W \\ 0 & C \end{pmatrix} \begin{pmatrix} a & 0 \\ X & 0 \end{pmatrix} \begin{pmatrix} 1 & -WC^{-1} \\ 0 & C^{-1} \end{pmatrix} = \begin{pmatrix} a + WX & * \\ CX & * \end{pmatrix}.$$

On the one hand, this shows that the elements with $X = 0$ form a trivial Q–submodule in $\mathfrak{g}/\mathfrak{q}$, which gives rise to a natural trivial line subbundle in $TM_\#$. Evidently, this is the vertical subbundle $\ker(T\pi)$ of $\pi : M_\# \to M$. On the other hand, $\mathfrak{g}/\mathfrak{q}$ is isomorphic, as a representation of Q, to the restriction of the defining representation $\mathbb{R}^{n+1}$ of G. In particular, this shows that $TM_\# \cong \mathcal{G} \times_Q \mathbb{R}^{n+1}$, so from the general theory developed in 1.5.7 we know that the Cartan connection ω on $\mathcal{G}$ induces a linear connection $\nabla^\#$ on $TM_\#$, called the *cone connection.*

At first glance, it is not obvious, why constructing a natural affine connection associated to a projective structure would be a big simplification. One should keep in mind, however, that it is very easy to construct invariants of such a connection, using the curvature and its iterated covariant derivatives, and any such expression will automatically be an invariant of the projective structure. Also, it is rather easy to construct invariant differential operators using the cone connection.

Next, we can easily describe $M_\#$ explicitly. The subgroup $Q \subset P$ by construction is the stabilizer of a nonzero point in the one–dimensional representation of P defined by the action on the P–invariant line in the standard representation $\mathbb{R}^{n+1}$ of G. The P–orbit of this point is given by all positive multiples, so we can view P/Q as the space of positive elements in this representation. The line bundle over M corresponding to this representation is the density bundle $\mathcal{E}(-1)$. Hence, $M_\# = \mathcal{G} \times_P (P/Q)$ can be either viewed as the space $\mathcal{E}_+(-1)$ of positive elements in $\mathcal{E}(-1)$ or as the oriented frame bundle of this line bundle. In particular, $\pi : M_\# \to M$ is a principal bundle with structure group $\mathbb{R}_+$, so the vertical subbundle is canonically trivialized by the Euler vector field $X \in \mathfrak{X}(M_\#)$, the fundamental vector field corresponding to $1 \in \mathbb{R}$. Now clearly $\mathcal{E}_+(-1) \to M$ can be used as a bundle of scales, so any (local) smooth section of this bundle defines a (local) Weyl structure for $(\mathcal{G} \to M, \omega)$.

To formulate the properties of the ambient connection, we need one more observation. Suppose that $s : M \to M_\#$ is a local section of $\pi : M_\# \to M$. Then for a vector field $\eta \in \mathfrak{X}(M)$, we can choose a vector field $\tilde{\eta} \in \mathfrak{X}(M_\#)$ which is s–related to η, i.e. such that $\tilde{\eta}(s(x)) = T_x s \cdot \eta(x)$ for all $x \in M$. Since $s(M) \subset M_\#$ is a smooth submanifold, the value of $\nabla^\# \tilde{\eta}$ along $s(M)$ in directions tangent to $s(M)$ depends only on η and not on the choice of $\tilde{\eta}$. Using this, one immediately concludes that there is a well–defined linear connection ∇^s on TM such that for all $\xi, \eta \in \mathfrak{X}(M)$ and any $\tilde{\eta} \in \mathfrak{X}(M_\#)$ which is s–related to η we have

$$\nabla^s_\xi \eta(x) = T_{s(x)}\pi \cdot (\nabla^\#_{T_x s \cdot \xi} \tilde{\eta}(s(x))). \tag{5.11}$$

PROPOSITION 5.2.6. *Let M be an oriented smooth manifold of dimension $n \geq 2$, and let $\pi : M_\# \to M$ be the oriented frame bundle of the density bundle $\mathcal{E}(-1) \to M$. For $t \in \mathbb{R}_+$ let $\rho^t : M_\# \to M_\#$ be the principal right action of t. Let $X \in \mathfrak{X}(M_\#)$ be the Euler vector field, and let $[\nabla]$ be a projective equivalence class of torsion–free linear connections on TM. Then the associated cone connection $\nabla^\#$ on $TM_\# \to M_\#$ has the following properties:*

- (i) *For each $t \in \mathbb{R}_+$, ρ^t preserves $\nabla^\#$.*
- (ii) *$\nabla^\#_\xi X = \xi$ for all $\xi \in \mathfrak{X}(M_\#)$.*
- (iii) *$\nabla^\#$ is torsion free.*
- (iv) *The Ricci type contraction of the curvature $R^\#$ of $\nabla^\#$ vanishes identically.*
- (v) *For any smooth section $s : M \to M^\#$ of $\pi : M_\# \to M$, the linear connection ∇^s defined by (5.11) lies in the projective class $[\nabla]$.*

PROOF. We have seen that $TM_\# \cong \mathcal{G} \times_Q \mathbb{R}^{n+1}$, so vector fields on $M_\#$ correspond to smooth Q–equivariant functions $\mathcal{G} \to \mathbb{R}^{n+1}$. If $\eta \in \mathfrak{X}(M_\#)$ corresponds to $f : \mathcal{G} \to \mathbb{R}^{n+1}$ and $\xi \in \mathfrak{X}(M_\#)$ is another vector field, then from 1.5.7 we know that the function corresponding to $\nabla^{\#}_{\xi}\eta$ is given by $\tilde{\xi}\cdot f+\omega(\tilde{\xi})\circ f$ where $\tilde{\xi}\in\mathfrak{X}(\mathcal{G})$ is any lift of ξ and in the second summand $\omega(\tilde{\xi})$ acts algebraically on the values of f.

The Euler vector field X by construction corresponds to the constant map $\mathcal{G}\to\mathbb{R}^{n+1}$ with value the first unit vector $e_1\in\mathbb{R}^{n+1}$. Now again by construction, for $\xi\in\mathfrak{X}(M_{\#})$ and a lift $\tilde{\xi}\in\mathfrak{X}(\mathcal{G})$, the function $\mathcal{G}\to\mathbb{R}^{n+1}$ corresponding to ξ is given by $\omega(\xi)(e_1)$, so (ii) holds.

Next, we know from 1.5.7 that the curvature and torsion of $\nabla^{\#}$ are induced by the action the curvature of the Cartan connection ω on $\mathbb{R}^{n+1}$. From 4.1.5 we know that, viewed as a Cartan connection on $\mathcal{G}\to M$, ω has vanishing torsion, and the homogeneity two part of its curvature is the totally tracefree part of the curvature of any connection in the projective class. In particular, the curvature of ω has values in $\mathfrak{q}$, and hence is torsion free as a Cartan connection on $\mathcal{G}\to M_{\#}$. Hence, the linear connection $\nabla^{\#}$ is torsion free, and its curvature $R^{\#}$ satisfies $i_XR^{\#}=0$. Expanding the latter equation we obtain

$$\begin{aligned}0=&\nabla^{\#}_X\nabla^{\#}_{\xi}\eta-\nabla^{\#}_{\xi}\nabla^{\#}_X\eta-\nabla^{\#}_{[X,\xi]}\eta\\=&\nabla^{\#}_X\nabla^{\#}_{\xi}\eta-\nabla^{\#}_{\xi}\nabla^{\#}_{\eta}X-\nabla^{\#}_{\xi}[X,\eta]-\nabla^{\#}_{[X,\xi]}\eta,\end{aligned}$$

where we have used torsion freeness. Using (ii) twice, to rewrite the second summand as $-\nabla^{\#}_{\nabla^{\#}_{\xi}\eta}X$, we can use torsion freeness once more to get

$$0=[X,\nabla^{\#}_{\xi}\eta]-\nabla^{\#}_{\xi}[X,\eta]-\nabla^{\#}_{[X,\xi]}\eta.$$

But this exactly says that X is an infinitesimal automorphism of $\nabla^{\#}$, so its flow preserves $\nabla^{\#}$, and (i) follows.

To compute the Ricci type contraction of $R^{\#}$, we can use a local frame $\{X_i\}$ for $TM_{\#}$ consisting of $X_1=X$ and the pullback of a local frame of TM. Denoting the dual frame for $T^*M_{\#}$ by ϕ^i, we see that $\{\phi^2,\dots,\phi^{n+1}\}$ descends to the dual of the above frame for TM. Then the Ricci type contraction is given by

$$(\xi,\eta)\mapsto\textstyle\sum_i\phi^i(R^{\#}(\xi,X_i)(\eta)).$$

Since $i_XR^{\#}=0$, the summand for $i=1$ vanishes, and the rest descends to an expression on TM which vanishes since $\partial^*\kappa=0$, so (iv) holds.

To verify (v), fix a smooth section $s:M\to M_{\#}$. For a point $x\in M$ choose a local smooth section $\tilde{\tau}$ of $\mathcal{G}\to M_{\#}$ defined in an open neighborhood $U_{\#}$ of $s(x)$. Then we can consider the pullback $\tilde{\tau}^*\omega$ which is a local $\mathfrak{g}$–valued one–form on $M_{\#}$. For any smooth function $\phi:U_{\#}\to Q$ also $\tau(y)=\tilde{\tau}(y)\cdot\phi(y)$ is a smooth section $U_{\#}\to\mathcal{G}$, and $\tau^*\omega(y)-\operatorname{Ad}(\phi(y))^{-1}\circ\tilde{\tau}^*\omega(y)$ has values in $\mathfrak{q}$. In particular, if we assume that $\phi(y)=\exp(Z(y))$ for a smooth function $Z:U_{\#}\to\mathfrak{g}_1$, and for some vector field $\xi_{\#}\in\mathfrak{X}(M_{\#})$ the first column of $\tilde{\tau}^*\omega(\xi_{\#})$ has the form $\binom{a(y)}{X(y)}$, then the first column of $\tau^*\omega(\xi_{\#})$ has the form $\binom{a(y)-Z(y)X(y)}{X(y)}$. For $y\in s(M)$ we have the codimension one subspace formed by the image of Ts, and by construction, the $\mathfrak{g}_{-1}$–component of $\tilde{\tau}^*\omega(y)$ restricts to a linear isomorphism on this subspace. Otherwise put, elements of this subspace are characterized by the fact that $a(y)=\psi(y)(X(y))$ for some linear functional $\psi(y)$ on $\mathfrak{g}_{-1}$. This functional can be written as $X(y)\mapsto Z(y)X(y)$ for an element $Z(y)\in\mathfrak{g}_1$. Since the whole construction is smooth, this defines a smooth function $Z:U_{\#}\to\mathfrak{g}_{-1}$. Using this function to modify $\tilde{\tau}$ to τ we see that for any $x \in M$ with $s(x) \in U_{\#}$ and $\xi \in T_xM$, the first column of $\tau^*\omega(T_xs \cdot \xi)$ has the form $\binom{0}{X}$ for some $X \in \mathfrak{g}_{-1}$.

Putting $U := s^{-1}(U_{\#})$, $\tau \circ s : U \to \mathcal{G}$ is a local smooth section of $\mathcal{G} \to M$. This gives rise to a G_0–equivariant smooth section $\sigma : p_0^{-1}(U) \to p^{-1}(U)$ of $\tilde{\pi} : \mathcal{G} \to \mathcal{G}_0$ which is characterized by $\sigma(\tilde{\pi}(\tau \circ s(u))) = \tau(s(u))$, and hence a local Weyl structure. We want to show that the local Weyl connection determined by σ coincides with ∇^s, which completes the proof.

Let ξ be a vector field on M. Consider $T(\tau \circ s) \cdot \xi$ along $\tau(s(U))$, and extend it to a vector field $\hat{\xi} \in \mathfrak{X}(\mathcal{G})$, which is projectable to a local vector field $\tilde{\xi} \in \mathfrak{X}(M_{\#})$. Then by construction $\tilde{\xi}$ is s–related to ξ. Applying the same construction to another vector field η on M, we also see that the local function $f : \mathcal{G} \to \mathfrak{g}$ defined by $f(u)\omega(\hat{\eta}(u))$ induces both the P–equivariant function $\mathcal{G} \to \mathfrak{g}/\mathfrak{p}$ corresponding to η and the Q–equivariant function $\mathcal{G} \to \mathfrak{g}/\mathfrak{q}$ corresponding to $\tilde{\eta}$. Now by construction, the Q–equivariant function $\mathcal{G} \to \mathfrak{g}/\mathfrak{q}$ corresponding to $\nabla^{\#}_{\tilde{\xi}}\tilde{\eta}$ is given by $\hat{\xi} \cdot f(u) + \omega(\hat{\xi})(f(u)) + \mathfrak{q}$. Applying $T\pi$ simply corresponds to taking the image in $\mathfrak{g}/\mathfrak{p}$, so $\nabla^s_{\xi}\eta$ corresponds to the function $\hat{\xi} \cdot f(u) + \omega(\hat{\xi})(f(u)) + \mathfrak{p}$. By construction, both $\omega(\hat{\xi})$ and $f(u)$ have the form $\left(\begin{smallmatrix} 0 & * \\ X & * \end{smallmatrix}\right)$. Using this, it is evident that $\omega(\hat{\xi})(f(u)) + \mathfrak{p} = \omega_0(\xi)(f(u)+\mathfrak{p})$, and since the horizontal lift of ξ with respect to the Weyl connection determined by σ is given by $\hat{\xi}(u) - \zeta_{\omega_0(\hat{\xi}(u))}(u)$, this completes the proof. □

REMARK 5.2.6. There is no direct analog of this construction for the other geometries corresponding to $|1|$–graded Lie algebras. While there is always a subgroup $Q \subset P$ such that G/Q is the space of nonzero elements in a scale bundle, the representation $\mathfrak{g}/\mathfrak{q}$ does not extend to $\mathfrak{g}$ in general. We will see in 5.3.10 below that a direct analog of the ambient description of projective structures exists for contact projective structures.

Consider for example the case of conformal structures. Viewing elements of P as block upper triangular matrices of the form

$$\begin{pmatrix} \lambda & -\lambda w^t \mathbb{I}_{p,q} C & -\frac{\lambda}{2}\langle w^t, w^t\rangle \\ 0 & C & w \\ 0 & 0 & \lambda^{-1} \end{pmatrix}$$

(see Proposition 1.6.3), the subgroup Q corresponds to those matrices for which $\lambda = 1$. This easily implies that $\mathfrak{g}/\mathfrak{q}$ can be identified (as a Q–module) with the subspace in $\mathbb{R}^{n+2}$ spanned by the first $n+1$ vectors in the standard basis. Since this subspace is not invariant under $\mathfrak{g}$, we cannot obtain an ambient connection in this way. This idea still leads to an interesting construction in the case of conformal structures containing an Einstein metric. For such structures, Proposition 5.2.5 implies existence of a parallel standard tractor. Hence, one obtains a linear connection on the orthocomplement of this parallel standard tractor. In a similar way as discussed for projective structures above, this can be used to construct a linear connection on the tangent space of $\mathcal{G}/Q$. This construction is also referred to as the *cone construction* for Einstein conformal classes; see [**Arm07a**].

There is an ambient description for general conformal structures, which, however, is much more complicated. This was introduced in [**FG85**], the details of the construction have been worked out in [**FG07**]. To obtain that description, one has to artificially enlarge the total space of a density bundle (usually one takes the ray subbundle of S^2T^*M which defines the conformal class) by one dimension. One then shows that on this enlarged space there exists a Ricci–flat pseudo–Riemannian metric (formally along the density bundle up to some order) which induces the given conformal class in a certain sense.

5.2.7. From the cone description to the Cartan description. We next want to prove that the ambient description of projective structures from 5.2.6 is actually equivalent to the Cartan description. This will also prove that the cone connection is uniquely determined by the properties (i)–(v) in Proposition 5.2.6. This equivalence can be nicely formulated via abstract tractor bundles.

Let M be an oriented smooth n–dimensional manifold and let $\pi : M_\# \to M$ be the space of nonzero elements in the bundle of $\frac{1}{n+1}$–densities of M. (From 4.1.5 we know that for any projective structure on M, the bundle $\mathcal{E}(-1)$ will be isomorphic to the bundle of $\frac{1}{n+1}$–densities.) Then we can construct a canonical volume form on $M_\#$ as follows: Since M is oriented, for a point $z \in M_\#$ one may interpret the 1–density z^{n+1} as a nonzero element in $\Lambda^n T^*_{\pi(z)}M$. Hence, we can define a tautological n–form $\alpha \in \Omega^n(M_\#)$ by $\alpha(z)(\xi_1, \dots, \xi_n) = z^{n+1}(T_z p \cdot \xi_1, \dots, T_z p \cdot \xi_n)$, and we claim that $\nu := d\alpha \in \Omega^{n+1}(M_\#)$ is a volume form. By construction, $\pi : M_\# \to M$ is a principal fiber bundle with structure group $\mathbb{R}_+$, and we denote by ρ^t the principal right action of $t \in \mathbb{R}_+$. Then by construction $(\rho^t)^*\alpha = t^{n+1}\alpha$. Denoting by X the Euler vector field on $M_\#$, the flow of X is given by $\mathrm{Fl}^X_t = \rho^{e^t}$. For the Lie derivative of α along X, we thus get $\mathcal{L}_X\alpha = (n+1)\alpha$, and since $i_X\alpha = 0$, this shows that $i_X\nu = (n+1)\alpha$, which proves the claim. Note that $(\rho_t)^*\nu = t^{n+1}\nu$ by construction.

Next, for $t \in \mathbb{R}_+$ and $\xi \in T_z M_\#$ put $\xi \cdot t := t^{-1}T_z\rho^t \cdot \xi$. This evidently defines a smooth right action on $TM_\#$, which by construction lifts the principal right action on $M_\#$. Consequently, the action is free and the orbit space $\mathcal{T} := TM_\#/\mathbb{R}_+$ is a smooth manifold and a vector bundle over $M_\#/\mathbb{R}_+ = M$. (A vector bundle atlas for $\mathcal{T}$ can be easily constructed from a principal bundle atlas for $\pi : M_\# \to M$ and the induced atlas for TM.)

From the definition of $\mathcal{T}$ we conclude that smooth sections of $\mathcal{T} \to M$ are in bijective correspondence with vector fields $\xi \in \mathfrak{X}(M_\#)$, which are homogeneous of degree -1 in the sense that $\xi(z \cdot t) = t^{-1}T_z\rho^t \cdot \xi(z)$. This can be equivalently characterized infinitesimally as $[X, \xi] = -\xi$.

PROPOSITION 5.2.7. *(1) The vertical bundle of $\pi : M_\# \to M$ descends to a line subbundle $\mathcal{T}^1 \subset \mathcal{T}$ such that $\mathcal{T}^1 \cong \mathcal{E}(-1)$ and $\mathcal{T}/\mathcal{T}^1 \cong TM \otimes \mathcal{E}(-1)$. The canonical volume form ν on $M_\#$ descends to a canonical nonvanishing section of $\Lambda^{n+1}\mathcal{T}$.*

(2) Let $\nabla^\#$ be a linear connection on $TM_\#$ which preserves the volume form ν and satisfies conditions (i)–(iii) from Proposition 5.2.6. Then $\nabla^\#$ descends to a tractor connection $\nabla^{\mathcal{T}}$ on $\mathcal{T}$, making it into an abstract standard tractor bundle of projective type. In particular, we obtain an induced (torsion free) projective structure on M.

(3) The curvature $R^\#$ of $\nabla^\#$ descends to the tractor curvature of $\nabla^{\mathcal{T}}$. The tractor connection $\nabla^{\mathcal{T}}$ is normal if and only if the Ricci–type contraction of $R^\#$ vanishes.

PROOF. (1) Since $\mathbb{R}_+$ is commutative, we have $\rho^t \circ \rho^s = \rho^s \circ \rho^t$, and differentiating this, we get $X(z \cdot t) = T_z\rho^t \cdot X(z)$. Since X spans the vertical bundle of $\pi : M_\# \to M$, this shows that we get an induced subbundle $\mathcal{T}^1 \subset \mathcal{T}$. Sections of this bundle are isomorphic to vector fields of the form fX, where $f : M_\# \to \mathbb{R}$ is a smooth function such that $f(z \cdot t) = t^{-1} f(z)$. But then $s(\pi(z)) := f(z)z$ (scalar multiplication in $\mathcal{E}(-1)$) is a well-defined section of $\mathcal{E}(-1)$. Conversely, given a section $s \in \Gamma(\mathcal{E}(-1))$ this equation defines a function f with the right equivariancy property. This shows that $\mathcal{T}^1 \cong \mathcal{E}(-1)$.

In the same way, smooth functions $f : M_\# \to \mathbb{R}$ such that $f(z \cdot t) = t^w f(z)$ for some fixed $w \in \mathbb{R}$ can be identified with sections of $\mathcal{E}(w)$. Given a function f corresponding to a section of $\mathcal{E}(1)$ and a vector field ξ which is homogeneous of degree -1, the product $f\xi$ is homogeneous of degree zero and hence projectable to M. The kernel of this projection is given by the vertical fields. This induces an isomorphism from $\mathcal{T}/\mathcal{T}^1$ to the bundle of linear maps from $\mathcal{E}(1)$ to TM, i.e. to $TM \otimes \mathcal{E}(-1)$.

For sections $s_1, \dots, s_{n+1} \in \Gamma(\mathcal{T})$, we can consider the corresponding vector fields $\xi_1, \dots, \xi_{n+1} \in \mathfrak{X}(M_\#)$, which are homogeneous of degree -1. As we have observed above, the volume form $\nu \in \Omega^{n+1}(M_\#)$ satisfies $(\rho_t)^*\nu = t^{n+1}\nu$. Hence, the function $\nu(\xi_1, \dots, \xi_{n+1}) : M_\# \to \mathbb{R}$ is homogeneous of degree zero, so it descends to a smooth function on M. Assigning this function to $s_1, \dots, s_{n+1}$ defines a section of $\Lambda^{n+1}\mathcal{T}^*$, which is nowhere vanishing by construction.

(2) If $\eta \in \mathfrak{X}(M_\#)$ is homogeneous of degree -1, then torsion freeness of $\nabla^\#$ implies that $\nabla^\#_X \eta = \nabla^\#_\eta X + [X, \eta] = 0$. Consequently, for a vector field $\xi \in \mathfrak{X}(M)$ and a lift $\tilde{\xi} \in \mathfrak{X}(M_\#)$ of ξ, the vector field $\nabla^\#_{\tilde{\xi}} \eta$ depends only on ξ and not on the choice of the lift. Further, any lift $\tilde{\xi}$ is homogeneous of degree zero, so $[X, \tilde{\xi}] = 0$. By assumption, for each $t \in \mathbb{R}_+$ the principal right action ρ^t by t preserves $\nabla^\#$. In particular, this implies that $\nabla^\#$ is compatible with homogeneities, so for $\tilde{\xi}$ and η as above, the vector field $\nabla^\#_{\tilde{\xi}} \eta$ is homogeneous of degree -1. Thus, it again corresponds to a section of $\mathcal{T}$, and if η corresponds to $s \in \Gamma(\mathcal{T})$, then we denote that section by $\nabla^{\mathcal{T}}_\xi s$. It is straightforward to verify that this defines a linear connection on $\mathcal{T}$.

Now consider $G = SL(n+1, \mathbb{R})$ and let $P \subset G$ be the stabilizer of a line in $\mathbb{R}^{n+1}$. Then via the line subbundle $\mathcal{T}^1 \subset \mathcal{T}$ and the nonvanishing section of $\Lambda^{n+1}\mathcal{T}$ constructed in (1), we get a natural frame bundle $\mathcal{G}$ for $\mathcal{T}$ with structure group P. As discussed in 3.1.21, this gives rise to an abstract adjoint tractor bundle via $\mathcal{A} = \mathcal{G} \times_P \mathfrak{g}$. A linear connection on $\mathcal{T}$ then is a $\mathfrak{g}$–connection as defined in 3.1.21 if and only if it preserves the distinguished section of $\Lambda^{n+1}\mathcal{T}$. Since this is evidently the case for $\nabla^{\mathcal{T}}$, we only have to verify the nondegeneracy condition from 3.1.21 to complete the proof of (2). As we have noted in (1), sections of the subbundle $\mathcal{T}^1$ are represented by vector fields of the form fX, where $f : M_\# \to \mathbb{R}$ is a smooth function which is homogeneous of degree -1. But then we get $\nabla^\#_{\tilde{\xi}} fX = (\tilde{\xi} \cdot f)X + f\tilde{\xi}$, so if f is nonzero in a point (and hence along the corresponding fiber), the class of this in $\mathcal{T}/\mathcal{T}^1 \cong TM \otimes \mathcal{E}(-1)$ is simply $\xi \otimes f$.

(3) It follows directly from the construction that $R^\#$ descends to the curvature R of the tractor connection $\nabla^{\mathcal{T}}$. The computations done in the proof of Proposition 5.2.6 read backwards show that $i_X R^\# = 0$. This easily implies that the Ricci–type contraction of $R^\#$ descends to $\partial^* R$. □

REMARK 5.2.7. The passage from the cone description of projective structures to the Cartan description via abstract tractor bundles was motivated by a similar construction in conformal geometry, which was worked out in [**ČGo03**]. There the alternative description is provided by an ambient metric, which is a weakening of the Fefferman–Graham ambient metric as introduced in [**FG85**]; see also Remark 5.2.6.

5.2.8. Geometries corresponding to $|1|$–gradings and affine holonomies. We have briefly discussed holonomy in part (2) of Remark 5.2.5. While the concept of holonomy is defined for connections on general bundles, the case of linear connections on the tangent bundle is of particular interest. It turns out that, under a few additional restrictions, the possible holonomy groups are not arbitrary but confined to a rather short list. Describing the possible holonomy groups was a long term project in differential geometry, starting from the basic work of Marcel Berger in the 1950s.

We have seen in 5.2.5 that for a linear connection on a vector bundle $E \to M$, the holonomy groups are subgroups of $GL(E_x)$, where E_x is the fiber over $x \in M$. Likewise, the holonomy groups of a principal connection are subgroups of the structure group of the principal bundle. Now it turns out that in both cases the holonomy group is a closed subgroup and hence a Lie subgroup. Consequently, there is the Lie algebra of the holonomy group, which is called the *holonomy Lie algebra* of the connection in a point. If the base of the bundle is connected, then the holonomy group and the holonomy Lie algebra is essentially independent of the given point. Indeed, the parallel transport along any path between the two points induces an isomorphism between the two groups. Hence, the holonomy group can be viewed as a subgroup of the general linear group of the standard fiber, respectively, the structure group of the principal bundle defined up to conjugation. Likewise, the holonomy Lie algebra can be viewed as a Lie subalgebra of the corresponding Lie algebra defined up to the adjoint action of a group element. In particular, the holonomy Lie algebra can be viewed as a Lie algebra endowed with a fixed representation defined up to isomorphism.

Now let us specialize to the case of the tangent bundle TM of a smooth manifold M. Then the holonomy group is a closed subgroup of $GL(n, \mathbb{R})$ and the holonomy Lie algebra is a Lie subalgebra of $\mathfrak{gl}(n, \mathbb{R})$, where $n = \dim(M)$. It turns out that arbitrary holonomy Lie algebras can occur if one allows connections with torsion. Likewise, locally symmetric connections (those, which have parallel curvature) form a special class, which can be classified by other means. Hence, in holonomy theory one is usually only interested in torsion free connections, which are not locally symmetric.

The first step in determining possible holonomy groups or Lie algebras is to look at the case of Levi–Civita connections of Riemannian metrics. In that case, a theorem of de Rham ensures that a decomposition of a holonomy representation into a direct sum of two invariant subspaces is induced by a local direct product decomposition of the manifold into corresponding factors. Hence, it suffices to study the case that the holonomy representation is irreducible, i.e. that $\mathfrak{g} \subset \mathfrak{gl}(n, \mathbb{R})$ does not admit a nontrivial invariant subspace in $\mathbb{R}^n$. In his classical work [**Be55**], M. Berger not only determined a list of possible holonomy Lie algebras of Levi–Civita connections but also extended this to a list of possible irreducible holonomy representations of arbitrary affine connections (with the above restrictions). (Later on, a few small corrections to the list were found.) Obtaining the list is a purely algebraic task, the main step is to analyze the consequences of the Bianchi identities.

Berger's original claim was that his list is complete up to possibly finitely many exceptions.

For some of the holonomy algebras in Berger's list, there is a simple geometric interpretation, which easily leads to examples. In general, however, proving existence of examples (and in particular of compact or complete examples) of such holonomies turned out to be a difficult task. A variety of different methods for constructing connections with certain holonomies was developed, some of them leading to interesting relations to other parts of mathematics. Step by step, this lead to extensions of Berger's list, which became known as *exotic holonomies*, and it turned out that there even is an infinite family of holonomy Lie algebras, which is not in Berger's list. The whole program was completed by S. Merkulov and L. Schwachhöfer in [**MS99**], where the last remaining cases were sorted out, so the classification of affine holonomies was complete.

The list of possible non–Riemannian holonomy Lie algebras has surprisingly close relations to the classification of certain types of parabolic subalgebras in simple Lie algebras. We will now use the theory of Weyl structures for geometries corresponding to $|1|$–gradings to prove existence of a class of holonomy representations. Suppose that $\mathfrak{g} = \mathfrak{g}_{-1} \oplus \mathfrak{g}_0 \oplus \mathfrak{g}_1$ is a $|1|$–graded simple Lie algebra. Then from 3.2.3 and 3.2.10 we know that the adjoint action restricts to an irreducible representation of $\mathfrak{g}_0$ on $\mathfrak{g}_{-1}$. Further, we know that $\mathfrak{g}_0 = \mathbb{R}E \oplus \mathfrak{g}_0^{ss}$, where E is the grading element and $\mathfrak{g}_0^{ss}$ is the semisimple part of $\mathfrak{g}_0$. We want to prove that both the natural representation of $\mathfrak{g}_0$ on $\mathfrak{g}_{-1}$ and its restriction to $\mathfrak{g}_0^{ss}$ are holonomy representations. Except for the full symplectic algebra (which is easy to realize), this exhausts all of Berger's original non–metric list, as well as some exotic holonomies related to the exceptional algebras of type E_6 and E_7.

LEMMA 5.2.8. *Let $\mathfrak{g} = \mathfrak{g}_{-1} \oplus \mathfrak{g}_0 \oplus \mathfrak{g}_1$ be a $|1|$–graded simple Lie algebra of rank larger than one. Then the mapping $(\mathsf{P}, X \wedge Y) \mapsto \partial\mathsf{P}(X,Y)$ induces surjections $S^2\mathfrak{g}_1 \otimes \Lambda^2\mathfrak{g}_{-1} \to \mathfrak{g}_0^{ss}$ as well as $\mathfrak{g}_1 \otimes \mathfrak{g}_1 \otimes \Lambda^2\mathfrak{g}_{-1} \to \mathfrak{g}_0$.*

PROOF. Note that both maps under consideration are evidently $\mathfrak{g}_0$–equivariant, so it suffices to show that their image meets each irreducible component of the target space. By the assumption on the rank, we know that $\dim(\mathfrak{g}_{-1}) > 1$, so we can choose linearly independent elements $X, Y \in \mathfrak{g}_{-1}$ and a linear map $\mathsf{P} : \mathfrak{g}_{-1} \to \mathfrak{g}_1$ such that $\mathsf{P}(Y) = 0$ and $B(\mathsf{P}(X), Y) \neq 0$, where B denotes the Killing form. But then $\partial\mathsf{P}(X,Y) = -[Y, \mathsf{P}(X)]$ and hence

$$B(\partial\mathsf{P}(X,Y), E) = B(\mathsf{P}(X), [Y, E]) = B(\mathsf{P}(X), Y) \neq 0.$$

Since the decomposition $\mathfrak{g}_0 = \mathbb{R}E \oplus \mathfrak{g}_0^{ss}$ is orthogonal for B, this shows that the image of the second map is not contained in $\mathfrak{g}_0^{ss}$, so it suffices to prove surjectivity of the first map.

Complexifying if necessary, we may assume that $\mathfrak{g}$ is a complex $|1|$–graded semisimple Lie algebra (with each of the simple ideals being $|1|$–graded), and then we can use the root decomposition. In each of the simple ideals of $\mathfrak{g}$, there is a unique simple root α such that the root space $\mathfrak{g}_\alpha$ is contained in $\mathfrak{g}_1$. From the description of the Dynkin diagram of $\mathfrak{g}_0$ in Proposition 3.2.2, we see that for any simple ideal of $\mathfrak{g}_0^{ss}$ there is a simple root β, which is not orthogonal to one of the roots α, such that the root space $\mathfrak{g}_\beta$ is contained in the given simple ideal. Since β is not orthogonal to α, $\alpha + \beta$ is a root and by construction $\mathfrak{g}_{\alpha+\beta} \in \mathfrak{g}_1$. Now choose a basis $\{X_\gamma\}$ of $\mathfrak{g}_{-1}$ such that each X_γ lies in the root space $\mathfrak{g}_{-\gamma}$ and let $\{Z_\gamma\}$ be the dual basis of $\mathfrak{g}_1$. Then consider $\mathsf{P} = Z_\alpha \vee Z_\alpha + Z_{\alpha+\beta} \vee Z_{\alpha+\beta} \in S^2\mathfrak{g}_1$. Then

$$\partial\mathsf{P}(X_\alpha, X_{\alpha+\beta}) = [X_\alpha, Z_{\alpha+\beta}] - [X_{\alpha+\beta}, Z_\alpha],$$

which is the sum of a nonzero element of $\mathfrak{g}_\beta$ and a nonzero element of $\mathfrak{g}_{-\beta}$. Hence, the image of our map meets the simple ideal containing $\mathfrak{g}_\beta$. □

THEOREM 5.2.8. *Let $\mathfrak{g} = \mathfrak{g}_{-1} \oplus \mathfrak{g}_0 \oplus \mathfrak{g}_1$ be a simple $|1|$–graded Lie algebra of rank larger than one. Then the representations $\mathfrak{g}_0 \hookrightarrow \mathfrak{gl}(\mathfrak{g}_{-1})$ and $\mathfrak{g}_0^{ss} \hookrightarrow \mathfrak{gl}(\mathfrak{g}_{-1})$ can be realized as holonomy representations of torsion free linear connections on the tangent bundle of a compact manifold, which are not locally symmetric.*

PROOF. Let G be a Lie group with Lie algebra $\mathfrak{g}$, $P \subset G$ a parabolic subgroup corresponding to $\mathfrak{p} = \mathfrak{g}_0 \oplus \mathfrak{g}_1$, and consider the generalized flag manifold G/P. Putting $P_+ := \exp(\mathfrak{g}_1) \subset P$ as usual, we know that all Weyl connections for the canonical parabolic geometry of type (G, P) on G/P are induced from principal connections on the principal G_0–bundle $G/P_+ \to G/P$, and they are all torsion free by Theorem 5.2.3. Since we are dealing with a $|1|$–graded simple Lie algebra here, the center of $\mathfrak{g}_0$ is generated by the grading element E. Hence, for any functional λ defining a bundle of scales, the kernel of λ coincides with the semisimple part $\mathfrak{g}_0^{ss}$. From 5.1.7 we thus see that in the special case of an exact Weyl structure, we get a further reduction to a principal bundle whose structure group has Lie algebra $\mathfrak{g}_0^{ss}$. In particular, the holonomy Lie algebra of any exact Weyl connection is contained in $\mathfrak{g}_0^{ss}$, while the holonomy Lie algebras of arbitrary Weyl connections are contained in $\mathfrak{g}_0$ (compare with 5.2.5).

To start our construction, we need a specific exact Weyl structure on G/P. First of all, we know that there always exist global exact Weyl structures, and we choose one of those. On the other hand, from Example 5.1.12 we know that there is the very flat Weyl structure defined on the open subset $G_- \subset G/P$. This is also an exact Weyl structure, its Rho tensor vanishes identically, and the corresponding Weyl connection on any induced vector bundle is flat. Now on the open subset G_-, the change from the restriction of the globally defined Weyl structure to the very flat Weyl structure is described by an exact one–form $\Upsilon = df$. Multiplying f by a function which has support in a ball around zero and is identically one on a smaller ball, the product can be smoothly extended by zero to all of G/P. Using the corresponding exact one–form to modify our initial global Weyl structure, we obtain a globally defined exact Weyl structure σ, which coincides with the very flat Weyl structure locally around $o = eP$.

Now we modify the Weyl structure σ using a one–form Υ, call the result $\hat{\sigma}$ and use notation as before. The following computations are done on the open neighborhood of o on which ∇ is flat. There we have $R = 0$ and $\mathsf{P} = 0$, and hence $\hat{R} = \partial(\hat{\mathsf{P}})$, $\hat{\mathsf{P}}(\xi) = \nabla_\xi\Upsilon + \frac{1}{2}\{\Upsilon, \{\Upsilon, \xi\}\}$ (see 5.1.8) and $\hat{\nabla}_\xi s = \nabla_\xi s - \{\Upsilon, \xi\} \bullet s$ (see 5.1.6) for any $\xi \in \mathfrak{X}(G/P)$ and any section s of an associated bundle. In particular, assume that for some $k \geq 0$, we have $j_o^k\Upsilon = 0$, where j_o^k denotes the k–jet in o. Then $j_o^{k-1}\hat{\mathsf{P}} = 0$, $j_o^k\hat{\mathsf{P}} = j_0^k(\nabla\Upsilon)$ and, for any s we get $j_o^k(\hat{\nabla}s) = j_o^k(\nabla s)$. Iterating this, we conclude that $\hat{\nabla}^k\hat{\mathsf{P}}(o) = \nabla^{k+1}\Upsilon(o)$. Finally, since ∂ is a bundle map on a vector bundle associated to $G/P_+ \to G/P$, which comes from a G_0–equivariant map between the inducing representations, it is parallel for any of the Weyl connections. Using this, we conclude that $\nabla^k\hat{R}(o) = (\mathrm{id} \otimes \partial)(\nabla^{k+1}\Upsilon(o))$.

Now it is well known that all values of $\nabla^k\hat{R}(o)$ lie in the holonomy Lie algebra. Hence, we can complete the proof by showing that we can find a one–form Υ such that $j^k_o\Upsilon = 0$ and the values of $(\mathrm{id}\otimes\partial)(\nabla^{k+1}\Upsilon(o))$ span all of $\mathfrak{g}_0$, respectively, the same statement for $\mathfrak{g}_0^{ss}$ with an exact one form Υ. Now of course we can find a one–form (respectively an exact one–form) Υ with $j^k_o\Upsilon = 0$, such that $\nabla^{k+1}\Upsilon(o)$ is an arbitrarily prescribed element of $S^{k+1}T^*_o(G/P)\otimes T^*_o(G/P)$ (respectively of $S^{k+2}T^*_o(G/P)$). Passing to the inducing representations and using the lemma, we see that it suffices to construct elements of $S^{k+1}\mathfrak{g}_1\otimes\mathfrak{g}_1$, respectively, of $S^{k+2}\mathfrak{g}_1$, which are surjective when viewed as maps from $S^k\mathfrak{g}_{-1}$ to $\mathfrak{g}_1\otimes\mathfrak{g}_1$, respectively, to $S^2\mathfrak{g}_1$.

For the first case, we can use $k = 4$ and, using a basis $\{e_i\}$ of $\mathfrak{g}_1$ with dual basis $\{e^i\}$ of $\mathfrak{g}_{-1}$, consider the element $\sum_{i,j} e_i^3e_j^2\otimes e_i\in S^5\mathfrak{g}_1\otimes\mathfrak{g}_1$. Evaluating on $(e_i)^4$, this gives $e_i\otimes e_i$, while evaluating on $(e^i)^3e^j$, we get $e_j\otimes e_i$. For the second case, we can also use $k = 4$ and consider the element $\sum_{i<j} e_i^4e_j^2\in S^6\mathfrak{g}_1$. Putting $n = \dim(\mathfrak{g}_{-1})$, we can get e_i^2 for $i < n$ as the value on $(e^i)^2(e^{i+1})^2$ and e_n^2 as the value on $(e^{n-1})^4$. For $i < j$, we can obtain e_ie_j as the value on $(e^i)^3e^j$. □

5.2.9. Weyl forms for general parabolic geometries. Let us return to the general theory of Weyl forms. To generalize the results of 5.2.3 to arbitrary parabolic geometries, we first define appropriate replacements for the torsion and curvature quantities used there. In 5.2.2, we have already observed that for a Weyl form $\tau\in\Omega^1(\mathcal{G}_0,\mathfrak{g})$, the components $\tau_- + \tau_0\in\Omega^1(\mathcal{G}_0,\mathfrak{g}_-\oplus\mathfrak{g}_0)$ define a Cartan connection on $\mathcal{G}_0$. Hence, it is a natural idea to look at the curvature of this Cartan connection. Since the splitting $\mathfrak{g}_-\oplus\mathfrak{g}_0$ is $\mathfrak{g}_0$–equivariant, this curvature splits accordingly into a torsion part and the curvature of the principal connection τ_0. Since τ_- identifies TM with $\mathrm{gr}(TM)$, the torsion part can be naturally interpreted as a form $T\in\Omega^2(M,\mathrm{gr}(TM))$. We call T the *torsion* of the Weyl form τ. The curvature of the Weyl connection τ_0 can be naturally viewed as $R\in\Omega^2(M,\mathrm{End}_0(\mathrm{gr}(TM)))$.

The appropriate definition for the analog of the tensor Y is a bit more difficult to guess, since there is the possibility to include various terms that automatically vanish in the $|1|$–graded case. Recall that we view the Rho tensor associated to τ as $\mathsf{P}\in\Omega^1(M,\mathrm{gr}(T^*M))$. The Weyl connection ∇ on $\mathrm{gr}(T^*M)$ extends to the covariant exterior derivative d^∇ on $\mathrm{gr}(T^*M)$–valued forms. Second, using the identification $TM\cong\mathrm{gr}(TM)$ provided by the soldering form τ_-, we can carry over the algebraic bracket to $\{\ ,\ \} : TM\times TM\to TM$. Finally, we have the algebraic bracket on T^*M, which we also denote by $\{\ ,\ \}$. Using these ingredients, we now define $Y\in\Omega^2(M,\mathrm{gr}(T^*M))$ by

$$Y(\xi,\eta) := d^\nabla\mathsf{P}(\xi,\eta) + \mathsf{P}(\{\xi,\eta\}) + \{\mathsf{P}(\xi),\mathsf{P}(\eta)\}.$$

Motivated by the conformal case (compare with 1.6.8), we call Y the *Cotton–York tensor* of the Weyl form τ.

We can now easily describe the relation between the curvature k_τ of τ and the triple (T,R,Y). Using the identification $TM\cong\mathrm{gr}(TM)$ provided by τ_-, we can apply the map ∂ to P, and view the result as

$$\partial\mathsf{P}\in\Omega^2(M,\mathrm{gr}(TM)\oplus\mathrm{End}_0(\mathrm{gr}(TM))\oplus\mathrm{gr}(T^*M)).$$

Using this, we can now generalize a part of Theorem 5.2.3:

THEOREM 5.2.9. *Let τ be a Weyl form on a regular infinitesimal flag structure $(p_0 : \mathcal{G}_0 \to M, \theta)$. Then*

$$k_\tau = (T, R, Y) + \partial(\mathsf{P}) \in \Omega^2(M, \mathrm{gr}(TM) \oplus \mathrm{End}_0(\mathrm{gr}(TM)) \oplus \mathrm{gr}(T^*M)).$$

In particular, the harmonic part of k_τ coincides with the component in $\ker(\Box)$ of (T, R, Y), so for normal Weyl forms, this component represents the harmonic curvature κ_H.

PROOF. Take vector fields $\xi, \eta \in \mathfrak{X}(M)$ and let $\xi^h, \eta^h \in \mathfrak{X}(\mathcal{G}_0)$ be their horizontal lifts with respect to τ_0. Then the section $k_\tau(\xi, \eta) \in \Gamma(\mathcal{G}_0 \times_{G_0} \mathfrak{g})$ is by definition represented by

$$d\tau(\xi^h, \eta^h) + [\tau(\xi^h), \tau(\eta^h)].$$

Since ξ^h is horizontal we have $\tau(\xi^h) = \tau_-(\xi^h) + \tau_+(\xi^h)$ and likewise for η. Splitting the bracket term accordingly, we see that by definition

$$d\tau(\xi^h, \eta^h) + [\tau_-(\xi^h), \tau_-(\eta^h)]$$

represents $(T(\xi, \eta), R(\xi, \eta), d^\nabla \mathsf{P}(\xi, \eta))$. The term $[\tau_+(\xi), \tau_+(\eta)]$ of course represents $\{\mathsf{P}(\xi), \mathsf{P}(\eta)\}$. Finally, the expression

$$[\tau_+(\xi^h), \tau_-(\eta^h)] + [\tau_-(\xi^h), \tau_+(\eta^h)]$$

evidently represents

$$\{\mathsf{P}(\xi), \eta\} - \{\xi, \mathsf{P}(\eta)\} = \partial \mathsf{P}(\xi, \eta) + \mathsf{P}(\{\xi, \eta\}),$$

and the result follows by definition of Y. □

From this result, in particular, we see how to describe the Cartan curvature and the harmonic curvature of a regular normal parabolic geometry in terms of a Weyl structure.

5.2.10. Constructing normal Weyl forms. The part of Theorem 5.2.3 that we have not generalized so far is the characterization of normal Weyl forms. For general parabolic geometries, this is significantly more complicated than in the $|1|$–graded case. The main reason is than in the $|1|$–graded case the tensors T, R, and Y coincide with the homogeneous components of (T, R, Y) and $\partial \mathsf{P}$ is concentrated in one homogeneity.

For general geometries, it is still possible to split the forms τ, T, R, and Y according to their values. However, the conceptually more important splitting is according to homogeneities. The objects we consider here are one–forms and two–forms with values in $\mathrm{gr}(TM) \oplus \mathrm{End}_0(\mathrm{gr}(TM)) \oplus \mathrm{gr}(T^*M)$. Hence, these values naturally split into components of degree $-k, \dots, k$. For the one form τ and $r \geq 0$, we mean by the "component of homogeneity $\leq r$" of τ, the restrictions of the forms τ_i to $T^{i-r}M$ for $i = -k, \dots, r-1$. Note, in particular, that the component of homogeneity ≤ 0 of any Weyl form τ coincides with the frame form of the infinitesimal flag structure, so this is always known in advance. For the two–forms T, R, and Y, we use an analogous terminology. For example, the component of T of homogeneity $\leq r$ consists of the restrictions of the forms $T_\ell \in \Omega^2(M, \mathrm{gr}_\ell(TM))$ to $T^iM \times T^jM$ where $i + j + r = \ell$ and $\ell = -k, \dots, -1$. Note that the components of R of homogeneity ≤ 1 and of Y of homogeneity ≤ 2 vanish by definition.

The main technical result now is to describe the dependence of the homogeneous components of T, R, and Y on the homogeneous components of τ. This exhibits an important difference between Y and the other two components.

LEMMA 5.2.10. *Let $(p : \mathcal{G}_0 \to M, \theta)$ be a regular infinitesimal flag structure, and let τ be a Weyl form.*

(1) The torsion T of τ is given by

$$T_\ell(\xi,\eta) = \nabla_\xi(\tau_\ell(\eta)) - \nabla_\eta(\tau_\ell(\xi)) - \tau_\ell([\xi,\eta]) + \sum_{a,b<0,a+b=\ell} \{\tau_a(\xi), \tau_b(\eta)\}$$

for $\xi, \eta \in \mathfrak{X}(M)$ and $\ell = -k, \dots, -1$, where we view each τ_i as an element of $\Omega^1(M, \mathrm{gr}_i(TM))$. In particular, the component of homogeneity ≤ 0 of T vanishes and for $r > 0$, the component of homogeneity $\leq r$ of T depends only on the component of homogeneity $\leq r$ of τ.

(2) For each $r \geq 2$, the component of homogeneity $\leq r$ of R depends only on the component of homogeneity $\leq r$ of τ.

(3) For each $r \geq 3$, the component of homogeneity $\leq r$ of Y depends only on the component of homogeneity $\leq r-1$ of τ.

PROOF. (1) As in the proof of Theorem 5.2.9, consider vector fields $\xi, \eta \in \mathfrak{X}(M)$ and let $\xi^h, \eta^h \in \mathfrak{X}(\mathcal{G}_0)$ be their horizontal lifts with respect to τ_0. In that proof we have seen that $T(\xi,\eta)$ is represented by

$$d\tau_-(\xi^h, \eta^h) + [\tau_-(\xi^h), \tau_-(\eta^h)].$$

Inserting the definition of the exterior derivative and taking into account that differentiating by a horizontal lift represents a covariant derivative, the formulae for the components of T follow by splitting the values into components.

Now suppose that $i, j < 0$ and $\ell < i + j$. By definition of a Weyl form, τ_ℓ vanishes on $T^{\ell+1}M$ and thus, in particular, on T^iM, T^jM and $T^{i+j}M$. Hence we see that the first three terms in the formula form T_ℓ vanish on $T^iM \times T^jM$. Moreover, if $a+b = \ell$, then either $i > a$ or $j > b$, so the last term cannot contribute either. Hence, the component of homogeneity ≤ -1 of T vanishes automatically.

If $\ell = i+j$, then $i > \ell$ and $j > \ell$ and for $\xi \in T^iM$ and $\eta \in T^jM$ the formula for $T_\ell(\xi,\eta)$ reduces to $-\tau_\ell([\xi,\eta]) + \{\tau_i(\xi), \tau_j(\eta)\}$. Since $[\xi,\eta]$ is a section of $T^\ell M$, we may replace in this expression each τ by the frame form θ and the vanishing follows from Proposition 3.1.7. Thus, the component of homogeneity ≤ 0 of T vanishes automatically.

Finally, suppose that $\ell = i + j + r$ for some $r > 0$. Then $\ell - i = r + j < r$, $\ell - j = r + i < r$, and $\ell - (i+j) = r$, so the first three summands in the formula for T_ℓ depend only on the components of homogeneity $\leq r$ of τ. For the last term, we must have $a+b = \ell$, but only summands with $a \leq i$ and $b \leq j$ can produce a nonzero contribution on $T^iM \times T^jM$. From $a+b = i+j+r$ we obtain $a - i = j - b + r \leq r$ and likewise $b - j \leq r$.

(2) Since the curvature $R(\xi,\eta)$ is represented by $d\tau_0(\xi^h, \eta^h)$, we see that the restriction of R to $T^iM \times T^jM$ depends only on the restriction of τ_0 to $T^{i+j}M$ and the result follows.

(3) From the definition of Y in 5.2.9 we know that it only depends on the Rho tensor, which is represented by τ_+. Expanding the definition, we see that

$$Y_\ell(\xi,\eta) = \nabla_\xi \mathsf{P}_\ell(\eta) - \nabla_\eta \mathsf{P}_\ell(\xi) - \mathsf{P}_\ell([\xi,\eta] - \{\xi,\eta\}) + \sum_{a,b>0;a+b=\ell} \{\mathsf{P}_a(\xi), \mathsf{P}_b(\eta)\}.$$

Assume that for some $r \geq 2$, $i, j < 0$ we have $\ell = i + j + r > 0$. Then for sections ξ of T^iM and η of T^jM, the expression $\{\xi,\eta\} \in T^{i+j}M$ depends only on the restriction of τ_- to $T^{i+j}M$, and this concerns only the components $\tau_{i+j}, \dots, \tau_{-1}$.

Hence, $\{\xi, \eta\}$ only depends on the component of homogeneity $-(i+j)-1<r$ of τ. Via the covariant derivatives, we have a dependence on the restrictions of τ_0 to T^iM, respectively, T^jM, but this again only gives homogeneities $<r$. In the first two terms and the last sum in the formula for Y above, we only use restrictions of P_a to T^iM and T^jM for $a \leq \ell$, so we meet at most homogeneities $\ell - i < r$, respectively, $\ell - j < r$. Finally, by regularity $[\xi, \eta] - \{\xi, \eta\}$ is a section of $T^{i+j-1}M$ so in the last remaining term we only need the restriction of P_ℓ to $T^{i+j-1}M$, which depends only on the component of homogeneity $r-1$ of τ. □

Now we describe in principle how to use these results in order to determine normal Weyl forms. The case of parabolic contact structures will be discussed in more detail below. First make any choice for $\tilde{\tau}_- \in \Omega^1(M, \mathrm{gr}(TM))$, which only amounts to choosing a splitting $TM \cong \mathrm{gr}(TM)$ of the filtration of TM. Further, choose a principal connection $\tilde{\tau}_0 \in \Omega^1(\mathcal{G}_0, \mathfrak{g}_0)$. This gives rise to linear connections on all $\mathrm{gr}_i(TM)$. For a start, we put $\tilde{\tau}_+ = 0$. The strategy is now to work homogeneity by homogeneity, on the one hand correcting the choices and at the same time computing parts of the Rho tensor $\mathsf{P} \in \Omega^1(M, \mathrm{gr}(T^*M))$, which are then used as $\tilde{\tau}_+$. Recall from Theorem 5.2.9 that $k_{\tilde{\tau}} = (T, R, Y) + \partial\mathsf{P}$, so the basic equation describing normality is $0 = \partial^*(T, R, Y) + \partial^*\partial\mathsf{P}$. Recall further that ∂^* is compatible with homogeneities; see 3.1.12. Hence, this equation can be analyzed homogeneity by homogeneity.

Since ∂ is also compatible with homogeneities, and the components of homogeneity ≤ 2 of P evidently vanish, we conclude that the same is true for $\partial^*\partial\mathsf{P}$. For homogeneity ≤ 1 we are therefore left with $\partial^* T^{(1)} = 0$. Here and below we use upper indices in braces to indicate homogeneous components. From the lemma we see that this depends only on the component of homogeneity ≤ 1 of $\tilde{\tau}$, i.e. the restriction of each $\tilde{\tau}_i$ to $T^{i-1}M$ and the restriction of $\tilde{\tau}_0$ to $T^{-1}M$. The general theory tells us that we can rearrange these data in such a way that $\partial^* T^{(1)} = 0$. Let us denote the Weyl form obtained in this way again by $\tilde{\tau}$. Then $\partial^*(\kappa_{\tilde{\tau}})$ is by construction homogeneous of degree ≥ 2, and similarly to the proof of Theorem 5.2.2, one shows that we can normalize $\tilde{\tau}$ by changing the part of homogeneity ≥ 2. Hence, the homogeneous components of degree one of $\tilde{\tau}$ already coincide with the components of some normal Weyl form.

Next, we look at the component of homogeneity 2. The normalization equation has the form $\partial^*(T^{(2)} + R^{(2)}) + \partial^*\partial\mathsf{P}^{(2)} = 0$. In this step, we first have to correct the components of homogeneity 2 of $\tilde{\tau}$ that we have chosen so far, and then compute $\mathsf{P}^{(2)}$. For the given choice of $\tilde{\tau}$ we can compute $\partial^*(T^{(2)} + R^{(2)})$ and this depends only on the components of homogeneity ≤ 2 of $\tilde{\tau}$ by the lemma. Now we can rewrite the normalization equation as

$$\partial^*(T^{(2)} + R^{(2)}) + \square\mathsf{P}^{(2)} - \partial\partial^*\mathsf{P}^{(2)} = 0, \tag{5.12}$$

where $\square$ is induced by the Kostant Laplacian; see 3.1.11. Since $\partial^* T^{(1)} = 0$, the map $\partial^*(T^{(2)} + R^{(2)})$ induces bundle maps $\mathrm{gr}_i(TM) \to \mathrm{gr}_{i+2}(TM)$ for $i \leq -3$, $\mathrm{gr}_{-2}(TM) \to \mathrm{End}_0(TM)$, and $\mathrm{gr}_{-1}(TM) \to \mathrm{gr}_1(T^*M)$.

We claim that (5.12) admits a solution if and only if there is a section $\alpha_2 \in \Gamma(\mathrm{gr}_2(T^*M))$ such that for the map induced by $\partial^*(T^{(2)} + R^{(2)}) + \partial\alpha_2$ only the component $\mathrm{gr}_{-1}(TM) \to \mathrm{gr}_1(T^*M)$ is nonzero. By Proposition 3.3.4, the Kostant Laplacian acts by a scalar on each isotypical component, so it is a linear combination of projections to isotypical components and thus commutes with any G_0–equivariant map. Moreover, from Proposition 3.3.7 we know that $\square$ acts invertibly on bundle maps $\mathrm{gr}(TM) \to \mathrm{gr}(TM) \oplus \mathrm{End}_0(\mathrm{gr}(TM)) \oplus \mathrm{gr}(T^*M)$, which are homogeneous of degree ≥ 2. In particular, if Φ is such a bundle map, then Φ has values in $\mathrm{gr}(T^*M)$ if and only if $\square\Phi$ has values in $\mathrm{gr}(T^*M)$.

If (5.12) has a solution, then $\mathsf{P}^{(2)}$ has values in $\mathrm{gr}(T^*M)$, so the same is true for $\square\mathsf{P}^{(2)}$, and then $\alpha_2 := -\partial^*\mathsf{P}^{(2)}$ has the required property. Conversely, given α_2 such that $\partial^*(T^{(2)} + R^{(2)}) + \partial\alpha_2$ has values in $\mathrm{gr}(T^*M)$ we define $\mathsf{P}^{(2)} := -\square^{-1}\left(\partial^*(T^{(2)} + R^{(2)}) + \partial\alpha_2\right)$, and this has values in $\mathrm{gr}(T^*M)$, too. This also implies that

$$\partial^*\partial\mathsf{P}^{(2)} + \partial\partial^*\mathsf{P}^{(2)} = \square\mathsf{P}^{(2)} = -\partial^*(T^{(2)} + R^{(2)}) - \partial\alpha_2.$$

Since $\mathrm{im}(\partial^*)$ and $\mathrm{im}(\partial)$ are complementary, we get $\partial\alpha_2 = -\partial\partial^*\mathsf{P}^{(2)}$, and hence (5.12) is satisfied.

In fact, we can do even better, since there is only one possibility for α_2, which can be computed in advance. Let $\phi : \mathrm{gr}_{-2}(TM) \to \mathrm{End}_0(\mathrm{gr}(TM))$ be the restriction of $\partial^*(T^{(2)} + R^{(2)})$, and let B denote the bilinear form induced by the Killing form of $\mathfrak{g}$. Then $\mathrm{gr}_2(T^*M)$ is dual to $\mathrm{gr}_{-2}(T^*M)$ via B, so $\alpha_2(\xi) = B(\xi, \alpha_2)$ for $\xi \in \Gamma(\mathrm{gr}_{-2}(TM))$. By assumption, ϕ must coincide with $-\partial\alpha_2$ on $\mathrm{gr}_{-2}(TM)$, so we must have $\phi(\xi) = -\{\xi, \alpha_2\}$. Now if $E \in \Gamma(\mathrm{End}_0(\mathrm{gr}(TM)))$ is the canonical grading section (see 5.1.5), then we obtain

$$B(E, \phi(\xi)) = -B(E, \{\xi, \alpha_2\}) = -B(\{E, \xi\}, \alpha_2) = 2B(\xi, \alpha_2).$$

Thus, $\alpha_2(\xi) := \frac{1}{2}B(E, \phi(\xi))$ is the only possible solution.

Now we can collect what has to be done for homogeneity 2: We can compute the candidate for α_2 by the above formula, and then we have to modify the component of homogeneity 2 of $\tilde{\tau}_{\leq 0}$ in such a way that $\partial^*(T^{(2)} + R^{(2)}) + \partial\alpha_2$ has values in $\mathrm{gr}(T^*M)$ only. (In the $|1|$–graded case, this step is vacuous.) Then we can compute $\mathsf{P}^{(2)} := -\square^{-1}\left(\partial^*(T^{(2)} + R^{(2)}) + \partial\alpha_2\right)$ and the resulting form, which we again call $\tilde{\tau}$, coincides with a normal Weyl form up to components of homogeneity ≥ 3. In particular, the whole procedure is finished here in the $|1|$–graded case, and we recover Theorem 5.2.3.

Now the last crucial fact is that by part (3) of the lemma, $Y^{(3)}$ depends only on the components of homogeneity ≤ 2 of $\tilde{\tau}$ that we have fixed already. Hence, in the normalization equation

$$0 = \partial^*(T^{(3)} + R^{(3)}) + \partial^*Y^{(3)} + \partial^*\partial\mathsf{P}^{(3)}$$

the second term is already known, and the first one depends only on the components of homogeneity ≤ 3 of $\tilde{\tau}_{\leq 0}$. We can proceed as before, first normalizing the component of $\tilde{\tau}_{\leq 0}$ in such a way that there is a section α_3 of $\mathrm{gr}_3(T^*M)$ such that $\partial^*(T^{(3)} + R^{(3)}) + \partial\alpha_3$ has values in $\mathrm{gr}(T^*M)$. (This step becomes vacuous in the $|2|$–graded case.) Similarly as before, denoting by $\phi : \mathrm{gr}_{-3}(TM) \to \mathrm{End}_0(\mathrm{gr}(TM))$ the map induced by $\partial^*(T^{(3)} + R^{(3)})$ we must have $\alpha_3(\xi) = \frac{1}{3}B(E, \phi(\xi))$ for all $\xi \in \Gamma(\mathrm{gr}_{-3}(TM))$. Having accomplished these normalizations, we define

$$\mathsf{P}^{(3)} := -\square^{-1}\left(\partial^*(T^{(3)} + R^{(3)}) + \partial^*Y^{(3)} + \partial\alpha_3\right).$$

From this stage, we can continue in exactly the same way. For the case of a $|k|$–grading, we get conditions on $\tilde{\tau}_{\leq 0}$ for the components of homogeneity up to

k. From that point on, $\tau_{\leq 0}$ remains fixed, and one only computes the remaining homogeneous components of P.

REMARK 5.2.10. In the remainder of this section, we will discuss the case of parabolic contact structures in more detail. The program of completely determining some Weyl structure and the associated data has only been carried out in very few more general cases. For generic rank two distributions in dimension five (which correspond to a $|3|$–grading), this has been done in [**Sag08**]. In [**DS**], the essential parts of a Weyl structure (everything but the Rho tensor) for generic rank ℓ–distributions in dimension $\frac{1}{2}\ell(\ell+1)$ are determined. Distinguished connections for quaternionic contact structures play an important role in O. Biquard's study of these geometries and of their twistor theory as discussed in 4.5.5. For quaternionic contact structures, there is a distinguished conformal class of metrics on the subbundle $H \subset TM$ which defines the geometry. The bundle of these metrics can be used as a bundle of scales in this case. Biquard associates to each choice of such a metric a connection on the tangent bundle. This has all the structural properties to be expected from the Weyl connection of the corresponding Weyl structure. We expect, however, that Biquard's normalization condition on the torsion is rather an analog of the condition for Webster–Tanaka connections (see 5.2.12 below), than the normalization conditions for Weyl connections. The Weyl connection should be computable from the Biquard connection and its torsion and curvature similarly to the results in 5.2.13 below.

5.2.11. The case of parabolic contact structures. We next specialize the methods developed in 5.2.10 to the case of parabolic contact structures and indicate simplifications, some of which have analogs in more general cases. For a parabolic contact structure, the filtration of TM consists only of the subbundle $H := T^{-1}M$ which defines a contact structure on M. In particular, $Q := TM/H$ is a real line bundle, and we denote by $q : TM \to Q$ the canonical bundle map. The dual bundle Q^* of Q can be naturally viewed as the annihilator of H in T^*M and hence is isomorphic to $\text{gr}_2(T^*M)$. Assuming that M is orientable, also Q^* is orientable and hence trivial. The first specific feature of parabolic contact structures is that Q^* can be used as a bundle of scales, since it corresponds to the scaling element $\frac{1}{2}E$, where E is the grading element. In particular, an exact Weyl structure in this case is determined by a global smooth section of Q^*, which may be viewed as a one–form $\theta \in \Omega^1(M)$ such that $H = \ker(\theta)$, i.e. a contact form on M. In the further discussion of parabolic contact structures we will therefore try to describe the exact Weyl structure corresponding to a contact form θ. Recall from 4.2.1 that the *Reeb field* associated to a contact form θ is the unique vector field $r \in \mathfrak{X}(M)$ such that $\theta(r) = 1$ and $i_r d\theta = 0$.

Choosing a splitting $TM \cong \text{gr}(TM)$ of the filtration is equivalent to choosing a bundle map $\pi : TM \to H$, which is a projection onto H. Then $\ker(\pi)$ is a complementary subbundle to H and the projection $q : TM \to Q$ restricts to an isomorphism between this kernel and Q. From 4.2.1 we know that any principal connection on the infinitesimal flag structure $\mathcal{G}_0$ of a parabolic contact geometry induces a contact connection ∇^H on H. This means that the connection on $\Lambda^2 H$ induced by ∇^H respects the decomposition $\Lambda^2 H = \Lambda^2_0 H \oplus Q$, where the first summand is the kernel of $\mathcal{L} : \Lambda^2 H \to Q$. Via this decomposition, we obtain an induced linear connection ∇^Q on Q, which is characterized by

$$\nabla^Q_\xi \mathcal{L}(\eta,\zeta) = \mathcal{L}(\nabla^H_\xi \eta,\zeta) + \mathcal{L}(\eta,\nabla^H_\xi \zeta), \tag{5.13}$$

for $\xi \in \mathfrak{X}(M)$, and $\eta,\zeta \in \Gamma(H)$. Of course, ∇^H also has to respect the structure on H given by the choice of G_0. For example, in the case of partially integrable almost CR–structures discussed in 4.2.4, this is the almost complex structure $J : H \to H$, so we must have $\nabla^H J = 0$. Likewise, for Lagrangean contact structures as discussed in 4.2.3, ∇^H must respect the decomposition $H = E \oplus F$ into a sum of Legendrean subbundles, so it actually is induced from connections ∇^E and ∇^F on the two factors.

THEOREM 5.2.11. *Consider a parabolic contact structure on a smooth manifold M with contact subbundle $H \subset TM$ and quotient bundle $Q := TM/H$, and let $q : TM \to Q$ be the natural bundle map. Let θ be a contact form for H with Reeb field r. Then we have:*

(1) The isomorphism $TM \cong H \oplus (TM/H)$ determined by the exact Weyl structure corresponding to θ is given by $\zeta \mapsto (\pi(\zeta), q(\zeta))$, where $\pi(\zeta) = \zeta - \theta(\zeta)r$.

(2) The Weyl connection determined by the exact Weyl structure corresponding to θ has the property that for the induced connection ∇^H on H, the bundle map $\Lambda^2 H \to H$ induced by

$$(\xi,\eta) \mapsto \nabla^H_\xi \eta - \nabla^H_\eta \xi - \pi([\xi,\eta])$$

has values in $\ker(\square)$. *This uniquely determines the restriction of the Weyl connection to directions in H.*

PROOF. For the normal Weyl form corresponding to the exact Weyl structure determined by θ, ∇^Q has to be the flat connection induced by θ. Hence, for $\xi,\eta \in \mathfrak{X}(M)$, the section $\nabla^Q_\xi q(\eta)$ must correspond to $\xi \cdot \theta(\eta)$. For $\eta,\zeta \in \Gamma(H)$, in particular, we have $\mathcal{L}(\eta,\zeta) = q([\eta,\zeta])$ and $\theta([\eta,\zeta]) = -d\theta(\eta,\zeta)$. Hence, the defining equation for ∇^Q above can be rephrased in the form that viewing $d\theta$ as a section of $\Lambda^2 H^*$, we have $\nabla d\theta = 0$ for the induced connection.

Next, the homogeneity one component $T^{(1)}$ of the torsion splits into two parts, one mapping $\Lambda^2 H$ to H and the other mapping $H \otimes Q$ to Q. The explicit formula for these components follows immediately from part (1) of Lemma 5.2.10. For the first component we have to consider $\xi,\eta \in \Gamma(H)$. Then $\tau_{-1}(\xi) = \xi$ and similarly for η, but $\tau_{-1}([\xi,\eta])$ corresponds to $\pi([\xi,\eta]) \in \Gamma(H)$. For the second component take $\xi \in \Gamma(H)$ and $\eta \in \mathfrak{X}(M)$. Then $\tau_{-2}(\xi) = 0$, while $\tau_{-2}(\eta)$ represents $q(\eta)$, and $\tau_{-1}(\eta)$ represents $\pi(\eta)$. Similarly, $\tau_{-2}([\xi,\eta])$ represents $q([\xi,\eta])$. Hence, for the two components we obtain

$$T_{-1}(\xi,\eta) = \nabla^H_\xi \eta - \nabla^H_\eta \xi - \pi([\xi,\eta]) \qquad \xi,\eta \in \Gamma(H), \tag{5.14}$$

$$T_{-2}(\xi,q(\eta)) = \nabla^Q_\xi q(\eta) - q([\xi,\eta]) + \mathcal{L}(\xi,\pi(\eta)) \qquad \xi \in \Gamma(H), \eta \in \mathfrak{X}(M). \tag{5.15}$$

Now we know that for a normal Weyl form, $T^{(1)}$ coincides with the homogeneous component of degree one of the Cartan curvature. In particular, it must have values in $\ker(\square)$ by Theorem 3.1.12, which proves the first part of (2). In Lemma 4.2.2 we have seen that the component of $\ker(\square)$ in homogeneity one is contained in $\Lambda^2 \mathfrak{g}^*_{-1} \otimes \mathfrak{g}_{-1}$, so (5.15) has to vanish identically. Rewriting this in terms of the contact form θ we obtain

$$0 = \xi \cdot \theta(\eta) - \theta([\xi,\eta]) - d\theta(\xi,\pi(\eta)) = d\theta(\xi,\eta - \pi(\eta)),$$

for all ξ such that $\theta(\xi) = 0$. Since H has codimension one, it follows that this equality must be satisfied for all $\xi \in \mathfrak{X}(M)$. But this exactly means that $\eta - \pi(\eta)$ must be a multiple of the Reeb field, which immediately implies the formula for π in (1).

Having described the projection $\pi : TM \to HM$, the remaining information about the component of the Weyl form in homogeneity one is contained in the restriction of ∇^H to $\Gamma(H) \times \Gamma(H)$. We know by the general theory that this restriction is uniquely determined by the requirement that $T_{-1} : \Lambda^2 H \to H$ lies in $\ker(\Box)$. □

5.2.12. Webster–Tanaka connections. Rather than proceeding directly to the determination of a Weyl connection, we will take a slight detour here and generalize one of the standard tools of CR–geometry to all parabolic contact structures. In 5.2.11, we have associated to a contact form θ on a parabolic contact structure (and the corresponding isomorphism $TM = H \oplus Q$) a unique partial connection on H, or equivalently, a partial principal connection on the infinitesimal flag structure $p_0 : \mathcal{G}_0 \to M$ which describes the geometry. This was characterized by the fact that the homogeneous component of degree one of the associated torsion is harmonic.

The Weyl connection associated to the contact form θ is a canonical extension of this partial connection to a connection. There is, however, a second natural extension to such a connection, which is determined by a normalization condition that involves only the torsion. This was introduced (in the setting of integrable CR–structures) in [**Tan75**] and [**We78**], and is known as the Webster–Tanaka connection. This idea extends to general parabolic contact structures and we have kept this name in the more general setting.

To formulate the normalization condition, we need a bit of algebraic background. Consider a contact grading $\mathfrak{g} = \mathfrak{g}_{-2} \oplus \cdots \oplus \mathfrak{g}_2$. Then $\mathfrak{g}_-$ is a Heisenberg algebra, so by Lemma 4.2.1 any automorphism of the graded Lie algebra $\mathfrak{g}_-$ is determined by its restriction to $\mathfrak{g}_{-1}$ and the group of all these automorphisms is isomorphic to $CSp(2n, \mathbb{R})$, where $2n = \dim(\mathfrak{g}_{-1})$. On the Lie algebra level, this means that each derivation of $\mathfrak{g}_-$ which is homogeneous of degree zero is determined by its restriction to $\mathfrak{g}_{-1}$, and the algebra of all such derivations is isomorphic to $\mathfrak{csp}(2n, \mathbb{R})$. Now we have the Lie algebra differential $\partial : L(\mathfrak{g}_-, \mathfrak{g}) \to L(\Lambda^2\mathfrak{g}_-, \mathfrak{g})$, and we may restrict it to maps which are homogeneous of degree zero. Then by definition (see also the proof of Proposition 4.3.1) $\ker(\partial) \subset L(\mathfrak{g}_-, \mathfrak{g}_-)_0$ is the space of derivations of $\mathfrak{g}_-$, which are homogeneous of degree zero.

On the other hand, $\partial : \mathfrak{g}_0 \to L(\mathfrak{g}_-, \mathfrak{g}_-)_0$ is injective, so we may view $\mathfrak{g}_0$ as a subspace of either $L(\mathfrak{g}_-, \mathfrak{g}_-)_0$ or $L(\mathfrak{g}_{-1}, \mathfrak{g}_{-1})$. From Theorem 3.3.1 we further know that $\ker(\Box)$ is a natural complement to $\operatorname{im}(\partial) \cong \mathfrak{g}_0$ in $\ker(\partial) \subset L(\mathfrak{g}_-, \mathfrak{g}_-)_0$. Notice that $\ker(\Box)$ is easily computable explicitly using Kostant's version of the Bott–Borel–Weyl theorem (Theorem 3.3.5) and the algorithms for determining the Hasse diagram from 3.2.18.

Now all this has a geometric counterpart. Let M be a manifold endowed with a parabolic contact structure of type (G, P) where G has Lie algebra $\mathfrak{g}$ and P is a parabolic subgroup for the given contact grading. Then we have the corresponding regular infinitesimal flag structure $p : \mathcal{G}_0 \to M$, and passing to associated bundles we have $\ker(\partial) = \operatorname{im}(\partial) \oplus \ker(\Box) \subset L(\operatorname{gr}(TM), \operatorname{gr}(TM))$, and each map in this space is determined by its restriction in $L(H, H)$.

Now if we extend the unique partial connection compatible with θ to a principal connection on $\mathcal{G}_0$, then we can form the component of homogeneity two in the torsion. This is the restriction of T_{-1} to a tensor $T_{-1} : Q \times H \to H$, so we can view it as a section of $Q^* \otimes L(H, H)$. Note that the Reeb field r associated to θ by definition satisfies $\pi(r) = 0$ and $[r, \xi] \in \Gamma(H)$ for all $\xi \in \Gamma(H)$. Hence, this torsion component is determined by $T_{-1}(q(r), \xi) = \nabla_r \xi - [r, \xi]$.

PROPOSITION 5.2.12. *Let $p_0 : \mathcal{G}_0 \to M$ be a regular infinitesimal flag structure of type (G, P), where P corresponds to a contact grading on $\mathfrak{g}$. Let $H \subset TM$ be the corresponding underlying contact structure and let θ be a contact form for H.*

Then there exists a unique principal connection on $\mathcal{G}_0$ which is compatible with θ and whose torsion T has the following properties:

- *Its homogeneous component of degree one is harmonic.*
- *Its homogeneous component of degree two is a section of $Q^* \otimes \ker(\Box) \subset Q^* \otimes L(H, H)$.*

PROOF. Let $\nabla : \Gamma(H) \times \Gamma(H) \to \Gamma(H)$ be the partial connection compatible with θ as determined in 5.2.11. Choose any extension of the corresponding partial principal connection on $\mathcal{G}_0$ to a true principal connection, and let $\nabla : \mathfrak{X}(M) \times \Gamma(H) \to \Gamma(H)$ be the induced linear connection on H. This is a contact connection, and hence induces a linear connection on Q. Now any other extension of this form is given by

$$\tilde{\nabla}_\zeta \xi = \nabla_\zeta \xi + A(q(\zeta))(\xi) \tag{5.16}$$

for some section A of $Q^* \otimes \mathrm{End}_0(H)$, where $\mathrm{End}_0(H) = \mathcal{G}_0 \times_{G_0} \mathfrak{g}_0 \subset L(H, H)$.

On the other hand, consider

$$\hat{\nabla}_\zeta \xi := \theta(\zeta)[r, \xi] + \nabla_{\pi(\zeta)} \xi,$$

for $\zeta \in \mathfrak{X}(M)$ and $\xi \in \Gamma(H)$. Here r is the Reeb field associated to θ and $\pi(\zeta) = \zeta - \theta(\zeta) r$. By definition of the Reeb field, we have $[r, \xi] \in \Gamma(H)$, so $\hat{\nabla}_\zeta \xi \in \Gamma(H)$. Further, the expression is evidently linear over smooth functions in ζ and since $\zeta = \theta(\zeta) r + \pi(\zeta)$, it satisfies a Leibniz rule in ζ. Hence, $\hat{\nabla}$ defines an extension of ∇ to a linear connection on H.

Expanding the equation $0 = dd\theta(r, \xi, \eta)$ for $\xi, \eta \in \Gamma(H)$ and using the definition of r, we get

$$0 = r \cdot d\theta(\xi, \eta) - d\theta([r, \xi], \eta) - d\theta(\xi, [r, \eta]) = (\hat{\nabla}_r d\theta)(\xi, \eta).$$

This means that $\hat{\nabla}$ is a contact connection on H, which is compatible with θ. Consequently, we get

$$\hat{\nabla}_\zeta \xi - \nabla_\zeta \xi = B(q(\zeta))(\xi)$$

for some section B of $Q^* \otimes \mathfrak{csp}(H) \subset Q^* \otimes L(H, H)$. From Lemma 4.2.1 we know that $CSp(\mathfrak{g}_{-1})$ is the automorphism group of the Heisenberg algebra $\mathfrak{g}_-$, viewed as a graded Lie algebra. Hence, $\mathfrak{csp}(\mathfrak{g}_{-1})$ is the Lie algebra of derivations of $\mathfrak{g}_-$, which are homogeneous of degree zero. This is exactly the kernel of $\partial : L(\mathfrak{g}_-, \mathfrak{g}_-)_0 \to L(\Lambda^2 \mathfrak{g}_-, \mathfrak{g}_-)_0$; compare with the proof of Proposition 4.3.1. The bundle $\mathrm{End}_0(H)$ corresponds to the image of $\partial : \mathfrak{g}_0 \to L(\mathfrak{g}_-, \mathfrak{g}_-)_0$, which is injective. By Proposition 3.1.11, the subspace $\ker(\Box) \subset L(\mathfrak{g}_-, \mathfrak{g}_-)_0$ is complementary to $\mathrm{im}(\partial)$ within $\ker(\partial)$.

Translating this back to geometric terms, we see that $\ker(\Box) \subset \mathfrak{csp}(H)$ is complementary to the subspace $\mathrm{End}_0(H)$. Hence, we see we can uniquely choose

$A \in \Gamma(Q^* \otimes \mathrm{End}_0(H))$ in such a way that the connection $\tilde{\nabla}$ defined by (5.16) has the property that

$$\hat{\nabla}_\zeta \xi = \tilde{\nabla}_\zeta \xi - C(q(\zeta))(\xi)$$

for a section C of $Q^* \otimes \ker(\Box)$. But this exactly means that $\tilde{\nabla}_r \xi - [r, \xi] = -C(q(r))(\xi)$, so $\tilde{\nabla}$ satisfies the claimed torsion conditions.

Conversely, if a principal connection on $\mathcal{G}_0$ satisfies the two torsion conditions, then the underlying partial connection has harmonic torsion in homogeneity one, so it coincides with the one constructed in 5.2.11. Hence, the induced linear connection on H can be written in the form of (5.16) above. Then the remaining torsion condition is evidently equivalent to the fact that the difference to $\hat{\nabla}$ is a section of $Q^* \otimes \ker(\Box)$, which completes the proof. □

5.2.13. From Webster–Tanaka connections to Weyl connections. To determine the Weyl connection, we next need a bit of information on the homogeneity two components of the curvature and torsion of a Webster–Tanaka connection. For the torsion, we simply have T_{-1} which is a section of $Q^* \otimes \ker(\Box) \subset Q^* \otimes H^* \otimes H$ as determined above. The homogeneity two part of the curvature is the section R of $\Lambda^2 H^* \otimes \mathrm{End}_0(H)$ determined by the usual formula. Now depending on the structure in question, this splits into several components, but some components are available for all parabolic contact structures. Namely, the bracket $\Lambda^2 \mathfrak{g}_1 \to \mathfrak{g}_2$ induces $\{\ ,\ \} : \Lambda^2 H^* \to Q^*$, and applying this to R, we obtain a tensor $Ric^W \in \Gamma(Q^* \otimes \mathrm{End}_0(H))$ called the *Webster–Ricci curvature.*

Next, we have to analyze the normality condition in homogeneity 2, and we do this for a general extension of the partial connection determined in 5.2.11. According to 5.2.10, what we need to compute is the restriction of $\partial^*(T^{(2)} + R^{(2)})$ to $\mathrm{gr}_{-2}(TM)$, so in our situation we have to evaluate it on the Reeb field r only. Here $T^{(2)}$ represents the part $Q \otimes H \to H$ of the torsion, and $R^{(2)}$ represents the part $\Lambda^2 H \to \mathrm{End}_0(H)$ of the curvature, so we can always see from the entries which quantity is needed. Now we can derive a formula for the relevant component of $\partial^*(T^{(2)} + R^{(2)})$ and use the result to compute Weyl connections from Webster–Tanaka connections.

THEOREM 5.2.13. *Consider a parabolic contact structure on a smooth manifold M with contact subbundle $H \subset TM$.*

(1) For a given Weyl form, let $T^{(2)} \in \Gamma(Q^* \otimes H^* \otimes H)$ and $R^{(2)} \in \Gamma(\Lambda^2 H^* \otimes \mathrm{End}_0(H))$ be the homogeneity two components of torsion and curvature. Then in terms of the maps $\partial^* : H^* \otimes H \to \mathrm{End}_0(H)$ and $\{\ ,\ \} : \Lambda^2 H^* \to Q^*$ the restriction to Q of $\partial^*(T^{(2)} + R^{(2)})$ is given by

$$-(\mathrm{id} \otimes \partial^*)(T^{(2)}) - (\{\ ,\ \} \otimes \mathrm{id})(R^{(2)}) : Q \to \mathrm{End}_0(H).$$

(2) Let ∇ be the Webster–Tanaka connection associated to a contact form θ on M, and let $Ric^W \in \Gamma(Q^* \otimes \mathrm{End}_0(H))$ be its Webster–Ricci curvature. Further, let $s \in \mathbb{R}$ be the number characterized by $(\{\ ,\ \} \otimes \mathrm{id})(\mathcal{L}) = s \cdot \mathrm{id}_Q$, where we view $\mathcal{L}$ as a section of $\Lambda^2 H^* \otimes Q$.

Then the Weyl connection associated to θ is given by $\nabla_\zeta \xi + A(q(\zeta))(\xi)$, where $A \in \Gamma(Q^* \otimes H^* \otimes H^*)$ is characterized by $(\mathrm{id} \otimes (\Box + s \cdot \mathrm{id}))(A) = Ric^W$, with $\Box : \mathrm{End}_0(H) \to \mathrm{End}_0(H)$ induced by the Kostant Laplacian on $\mathfrak{g}_0$.

PROOF. (1) Recall from 3.1.12 that for a decomposable element of $\Lambda^2T^*M \otimes \mathcal{A}M$ the map ∂^* acts by

$$\partial^*(\phi_1 \wedge \phi_2 \otimes s) = -\phi_2 \otimes \{\phi_1, s\} + \phi_1 \otimes \{\phi_2, s\} - \{\phi_1, \phi_2\} \otimes s.$$

Now $T^{(2)}$ can be written as a sum of terms of this type with $\phi_1 \in \Gamma(Q^*)$, $\phi_2 \in \Gamma(H^*)$ and $s \in \Gamma(H)$, while $R^{(2)}$ can be written as a sum of terms with $\phi_1, \phi_2 \in \Gamma(H^*)$ and $s \in \Gamma(\mathrm{End}_0(H))$. Since we only want to compute the restriction of the result to Q, we only have to look at those terms for which the first factor in the tensor product lies in Q^*. For $R^{(2)}$ these are exactly the terms of the form $\{\phi_1, \phi_2\} \otimes s$. For $T^{(2)}$ these bracket terms always vanish identically, and we only have to consider the terms $\phi_1 \otimes \{\phi_2, s\}$. However, by definition $\partial^*(\phi_2 \otimes s) = \{\phi_2, s\}$, so the result follows.

(2) By construction, the Webster–Tanaka connection ∇ associated to θ extends the partial connection associated to a normal Weyl form. Since the Weyl connection $\hat{\nabla}$ associated to θ has the same property, there must be a section A of the bundle $Q^* \otimes \mathrm{End}_0(H)$ such that

$$\hat{\nabla}_\zeta \xi = \nabla_\zeta \xi + A(q(\zeta))(\xi)$$

holds for all $\zeta \in \mathfrak{X}(M)$ and $\xi \in \Gamma(H)$. Since ∇ and $\hat{\nabla}$ both leave θ parallel, $A(q(\zeta))$ must induce the zero map on Q.

Let us denote by $T^{(2)}$ and $R^{(2)}$ the relevant torsion and curvature quantities of ∇, and by $\hat{T}^{(2)}$ and $\hat{R}^{(2)}$ the ones for $\hat{\nabla}$. Then from the definitions we immediately get

$$\hat{T}^{(2)}(q(\zeta), \xi) = T^{(2)}(q(\zeta), \xi) + A(q(\zeta))(\xi),$$
$$\hat{R}^{(2)}(\xi, \eta) = R^{(2)}(\xi, \eta) - A(\mathcal{L}(\xi, \eta)).$$

Since $A(q(\zeta))$ is a section of $\mathrm{End}_0(H) \subset H^* \otimes H$ and induces the zero map on Q, we may actually write the first equation as $i_{q(\zeta)}\hat{T}^{(2)} = i_{q(\zeta)}T^{(2)} + \partial(A(q(\zeta)))$, where ∂ is induced by $\partial : \mathfrak{g}_0 \to \mathfrak{g}_{-1}^* \otimes \mathfrak{g}_{-1}$. By definition of the Webster–Tanaka connection, $T^{(2)}$ is a section of $Q^* \otimes \ker(\square)$ so, in particular, $(\mathrm{id} \otimes \partial^*)(T^{(2)}) = 0$. Hence, from part (1) we conclude that the restriction of $\partial^*(\hat{T}^{(2)} + \hat{R}^{(2)})$ to Q is given by

$$-(\mathrm{id} \otimes \square)(A) - sA + (\{\ ,\ \} \otimes \mathrm{id})(R^{(2)}) = -(\mathrm{id} \otimes (\square + s \cdot \mathrm{id}))(A) + Ric^W.$$

By construction, both Ric^W and A have values which are orthogonal to the grading element E (since they act trivial on Q). In the notation of 5.2.10 this implies that $\alpha_2 = 0$ and hence we know from there that $\partial^*(\hat{T}^{(2)} + \hat{R}^{(2)})$ has to restrict to zero on Q. □

Note that in each example the condition characterizing the difference between the Webster–Tanaka and the Weyl connection in part (2) of the theorem becomes very simple. The Kostant Laplacian $\square$ acts by a multiple of the identity on each irreducible component of $\mathfrak{g}_0$, and there are very few such components.

5.2.14. Example: Distinguished connections for Lagrangean contact structures. We next show how the constructions of 5.2.11–5.2.13 above can be made explicit for specific structures. We do this in detail for Lagrangean contact structures, other examples are briefly discussed below. Suppose that $H = E \oplus F \subset TM$ is a Lagrangean contact structure on a smooth manifold M of dimension $2n+1$. As we have noted in 5.2.11 above, the (partial) contact connections on H which come from the infinitesimal flag structure are exactly those which respect the two Lagrangean subbundles, or equivalently are of the form $\nabla^E \oplus \nabla^F$. Note that by definition, the Reeb field r associated to a contact form θ has the property that $[r,\xi] \in \Gamma(H)$ for all $\xi \in \Gamma(H)$. In particular, the splitting $H = E \oplus F$ defining the Lagrangean contact structure allows us to further split $[r,\xi] = [r,\xi]_E + [r,\xi]_F$.

We also need a bit of information about the Webster–Ricci curvature. By definition, this is a section of $Q^* \otimes \mathrm{End}_0(TM)$. Since the Webster–Tanaka connection leaves θ parallel, its curvature has to act trivially on $Q = TM/H$. Now recall from 4.2.3 that the Lie algebra $\mathfrak{g} = \mathfrak{sl}(n+2,\mathbb{R})$ has the form

$$\mathfrak{g} = \left\{ \begin{pmatrix} a & Z & \gamma \\ X & A & W \\ \beta & Y & b \end{pmatrix} : a,b,\beta,\gamma \in \mathbb{R}; X,W \in \mathbb{R}^n; Y,Z \in \mathbb{R}^{n*}; a+b+\mathrm{tr}(A) = 0 \right\}$$

with the evident grading coming from the distance to the main diagonal. The subalgebra of $\mathfrak{g}_0$ which acts trivially on $\mathfrak{g}_{-2}$ is formed by block diagonal matrices with entries $(a, A - \frac{2a}{n}\mathbb{I}, a)$ for $a \in \mathbb{R}$ and $A \in \mathfrak{sl}(n,\mathbb{R})$. For $X, Y \in \mathfrak{g}_{-1}$, the action of this element is given by $X \mapsto AX - \frac{n+2}{n}aX$ and $Y \mapsto -YA + \frac{n+2}{n}aY$. Hence, we see that acting on $\mathfrak{g}^E_{-1}$ induces an isomorphism between that subalgebra and $\mathfrak{gl}(\mathfrak{g}^E_{-1})$, for which A can be recovered as the tracefree part of an endomorphism of $\mathfrak{g}^E_{-1}$, while a can be recovered as $\frac{-1}{n+2}$ times the trace of the endomorphism.

Consequently, we can split the Webster–Ricci curvature into a tracefree part Ric^W_0 and the *Webster scalar curvature* $R^W \in \Gamma(Q^*)$, which is defined as the trace of the action of Ric^W on the bundle E. Using these observations, we can now describe all the distinguished connections for Lagrangean contact structures.

PROPOSITION 5.2.14. *Let $(M, H = E \oplus F)$ be a Lagrangean contact structure, and let $\theta \in \Omega^1(M)$ be a contact form.*

(1) The partial connections ∇^E and ∇^F which induce the distinguished partial connection corresponding to the exact Weyl structure determined by θ are given by

$$\begin{aligned} d\theta(\nabla^E_{\eta_1}\xi, \eta_2) &= d\theta([\eta_1,\xi],\eta_2), \\ d\theta(\nabla^E_{\xi_1}\xi_2, \eta) &= \xi_1 \cdot d\theta(\xi_2,\eta) + d\theta(\xi_2, [\xi_1,\eta]), \\ d\theta(\nabla^F_{\xi_1}\eta, \xi_2) &= d\theta([\xi_1,\eta],\xi_2), \\ d\theta(\nabla^F_{\eta_1}\eta_2, \xi) &= \eta_1 \cdot d\theta(\eta_2,\xi) + d\theta(\eta_2, [\eta_1,\xi]), \end{aligned} \tag{5.17}$$

for $\xi, \xi_1, \xi_2 \in \Gamma(E)$ and $\eta, \eta_1, \eta_2 \in \Gamma(F)$.

(2) The Webster–Tanaka connection associated to θ is characterized by $\nabla^{WT} r = 0$ and

$$\begin{aligned} \nabla^{WT}_\zeta \xi &= \theta(\zeta)[r,\xi]_E + \nabla^E_{\pi(\zeta)}\xi, \\ \nabla^{WT}_\zeta \eta &= \theta(\zeta)[r,\eta]_F + \nabla^F_{\pi(\zeta)}\eta, \end{aligned} \tag{5.18}$$

for $\xi \in \Gamma(E)$, $\eta \in \Gamma(F)$, and $\zeta \in \mathfrak{X}(M)$, where ∇^E and ∇^F are the partial connections from part (1).

(3) Let Ric^W be the Webster–Ricci curvature of the Webster–Tanaka connection from (2) and let R^W be its Webster scalar curvature. The the Weyl connection associated to θ is characterized by $\nabla r = 0$ and

$$\begin{aligned} \nabla_\zeta \xi &= \nabla^{WT}_\zeta \xi - \tfrac{1}{2(n+1)} Ric^W(q(\zeta))(\xi) + \tfrac{1}{n(3n+2)} R^W(q(\zeta))\xi, \\ \nabla_\zeta \eta &= \nabla^{WT}_\zeta \eta - \tfrac{1}{2(n+1)} Ric^W(q(\zeta))(\eta) - \tfrac{1}{n(3n+2)} R^W(q(\zeta))\eta, \end{aligned} \tag{5.19}$$

for $\xi \in \Gamma(E)$*,* $\eta \in \Gamma(F)$*, and* $\zeta \in \mathfrak{X}(M)$*, where* ∇^{WT} *is the Webster–Tanaka connection from (2).*

PROOF. (1) Observe first that the right–hand sides of all claimed equations contain only known data. Further, on the left–hand side we always have inserted the result of a differentiation with values in one of the two subbundles with a general element in the other subbundle. By nondegeneracy of $d\theta$, we conclude that the equations always completely determine the values of ∇^E, respectively, ∇^F in E–directions, respectively, F–directions. Hence, the four equations together determine the partial connections ∇^E and ∇^F, and hence the partial connection ∇^H.

One may also verify directly (although this is formally not necessary), that the right–hand side of each equation has the right behaviour under multiplications of the individual fields by smooth functions in order to guarantee that one really obtains partial connections. For example, consider the right–hand side of the first equation in (5.17). Since $\eta_1, \eta_2 \in \Gamma(F)$, we get $d\theta(\eta_1, \eta_2) = 0$, which implies that the expression is linear over $C^\infty(M, \mathbb{R})$ in η_1. On the other hand, for $f \in C^\infty(M, \mathbb{R})$ we get

$$d\theta([\eta_1, f\xi], \eta_2) = f d\theta([\eta_1, \xi], \eta_2) + (\eta_1 \cdot f) d\theta(\xi, \eta_2).$$

Hence, we get $\nabla^E_{\eta_1} f\xi = f\nabla^E_{\eta_1}\xi + (\eta_1 \cdot f)\xi$. For the other equations, these properties are verified similarly.

The lowest possibly nonzero homogeneous component of the Cartan curvature must be harmonic by Theorem 3.1.12. We have described the harmonic curvature for Lagrangean contact structures in 4.2.3. If $\dim(M) = 3$, then the geometry is automatically torsion free, so there is no harmonic curvature in homogeneity one. If $\dim(M) > 3$, then the harmonic curvature in homogeneity one has two irreducible components, one of which is given by a section of $\Lambda^2 E^* \otimes F$ while the other is given by a section of $\Lambda^2 F^* \otimes E$. (These two torsions are exactly the obstructions against integrability of the Legendrean subbundles E and F.) In particular, this means that $T_{-1}(\xi, \eta) = 0$ for $\xi \in \Gamma(E)$ and $\eta \in \Gamma(F)$. Expanding the formula (5.14) for T_{-1} from 5.2.11 and splitting into components in $\Gamma(E)$ and $\Gamma(F)$, we see that this is equivalent to

$$\begin{aligned} \nabla^F_\xi \eta - \pi_F([\xi, \eta]) &= 0, \\ -\nabla^E_\eta \xi - \pi_E([\xi, \eta]) &= 0. \end{aligned} \tag{5.20}$$

Here we have split $\pi([\xi, \eta]) \in \Gamma(H)$ into its components in $\Gamma(E)$ and $\Gamma(F)$. Now we just have to observe that

$$[\xi, \eta] = d\theta(\xi, \eta) r + \pi_E([\xi, \eta]) + \pi_F([\xi, \eta]).$$

By definition, multiples of r insert trivially into $d\theta$, and $d\theta$ also vanishes if both its entries are either from E or F. Hence, we conclude that for $\eta_2 \in \Gamma(F)$ we get $d\theta([\xi, \eta], \eta_2) = d\theta(\pi_E([\xi, \eta]), \eta_2)$, while for $\xi_2 \in \Gamma(E)$, we get $d\theta([\xi, \eta], \xi_2) = d\theta(\pi_F([\xi, \eta]), \xi_2)$. Hence, the first and third line of (5.17) follow directly from (5.20).

Next, we know that $\nabla d\theta = 0$. Expanding this, we obtain

$$d\theta(\nabla^E_{\xi_1} \xi_2, \eta) = \xi_1 \cdot d\theta(\xi_2, \eta) - d\theta(\xi_2, \nabla^F_{\xi_1} \eta)$$

for $\xi_1, \xi_2 \in \Gamma(E)$ and $\eta \in \Gamma(F)$. Using this, the second line in (5.17) follows immediately from the third line. Likewise, the last line in (5.17) is deduced from the first line.

(2) As in the proof of Proposition 5.2.12, one immediately verifies that the two lines in (5.18) define linear connections on E and F, so together they define a linear connection on H, which is induced by a principal connection on $\mathcal{G}_0$. It also follows directly from the construction, that the torsion component in homogeneity two maps $Q \otimes E$ to F and $Q \otimes F$ to E. Finally, (5.18) is the unique extension of the partial connection determined by ∇^E and ∇^F with that property. Now it is easy to see that $\ker(\Box) \subset L(H, H)$ is contained in those maps which map E to F and F to E, which completes the proof.

(3) We compute the Weyl connection using Theorem 5.2.13. Therefore, we have to compute the number $s \in \mathbb{R}$ that occurs in this theorem, as well as the action of the Kostant Laplacian on $\mathfrak{g}_0$. For the duality between $\mathfrak{g}_-$ and $\mathfrak{p}_+$, we use the trace form of $\mathfrak{g}$. Take the standard bases $\{e^i\}$ of $\mathfrak{g}^E_{-1}$, $\{e_i\}$ of $\mathfrak{g}^F_{-1}$, λ_i of $\mathfrak{g}^E_1$ and λ^i of $\mathfrak{g}^F_1$, which are dual with respect to the trace form. Also, the elements $\phi \in \mathfrak{g}_{-2}$ and $\psi \in \mathfrak{g}_2$ which have their unique nonzero matrix entry equal to one are dual with respect to the trace form. The Lie bracket $[\ ,\ ] : \mathfrak{g}_{-1} \times \mathfrak{g}_{-1} \to \mathfrak{g}_{-2}$ maps $(X_1, Y_1) \times (X_2, Y_2)$ to $(Y_1 X_2 - Y_2 X_1)\phi$. Viewed as an element of $\Lambda^2 \mathfrak{g}_1 \otimes \mathfrak{g}_{-2}$, this bracket is therefore given by $\sum_i (\lambda^i \otimes \lambda_i - \lambda_i \otimes \lambda^i) \otimes \phi$. Applying $[\ ,\ ] \otimes \mathrm{id}$ to this element, we get $-2n\psi \otimes \phi$, so the number s from Theorem 5.2.13 equals $-2n$.

Since the Kostant Laplacian acts by a scalar on each irreducible component, it preserves the subalgebra of $\mathfrak{g}_0$ which acts trivially on $\mathfrak{g}_{-2}$. As we have seen above, there are only two irreducible components in this subalgebra, one isomorphic to $\mathbb{R}$ and one to $\mathfrak{sl}(n, \mathbb{R})$. We can compute the eigenvalue of $\Box$ on each component by inserting some fixed element. The matrix corresponding to $a = 1$ and $A = 0$, acts on $\mathfrak{g}^E_{-1}$ as $-\frac{n+2}{n}\mathrm{id}$ and on $\mathfrak{g}^F_{-1}$ as $\frac{n+2}{n}\mathrm{id}$. Applying ∂ simply means viewing this element as an endomorphism of $\mathfrak{g}_-$, i.e. as sitting in $\mathfrak{p}_+ \otimes \mathfrak{g}_-$. This is clearly given by $-\frac{n+2}{n} \sum_i \lambda_i \otimes e^i + \frac{n+2}{n} \sum_i \lambda^i \otimes e_i$. To compute ∂^*, we have to apply the bracket to these elements, which immediately shows that the eigenvalue of $\Box$ on this component is $-(n+2)$. For the other component, one can simply take the matrix for which $a = 0$ and A is the highest weight vector of $\mathfrak{sl}(n, \mathbb{R})$, i.e. the matrix with a 1 in the top right corner and zeros everywhere else. This corresponds to $\lambda_n \otimes e^1 - \lambda^1 \otimes e_n$, and applying the bracket we see that $\Box$ acts by multiplication by -2 on that component.

According to Theorem 5.2.13, we have to consider the map $\Box + s \cdot \mathrm{id}$, which acts by $-2(n+1)$ on the tracefree part and by $-3n-2$ on the trace part. Decomposing the deformation tensor A from Theorem 5.2.13 as $A_0 + \alpha \mathrm{id}$ when acting on E, we must have $A_0 = \frac{-1}{2(n+1)} Ric^W_0$ and $\alpha = \frac{-1}{n(3n+2)} R^W$. From this, the formula for the Weyl connection on E follows immediately. The formula on F then can be deduced from compatibility of ∇, ∇^{WT} and Ric^W with $d\theta$. □

5.2.15. Tractor calculus for Lagrangean contact structures. Having at hand the Weyl connection determined by a contact form θ, one can compute its torsion T, curvature R, and Cotton–York tensor Y. As described in 5.2.10, knowing these quantities, one can compute, homogeneity by homogeneity, the Rho tensor associated to the exact Weyl structure determined by θ. Compared to the $|1|$–graded cases we have studied before, the Rho tensor now has several independent components. In terms of the Rho tensor, we can then give a complete description of the tractor bundles associated to a Lagrangean contact structure.

To formulate the results, we use an abstract index notation, which is particularly useful to compare our results to the tractor calculus for CR–structures introduced in [**GoGr05**]. We denote the bundle E by $\mathcal{E}^\alpha$ and the bundle F by $\mathcal{E}^{\bar{\alpha}}$. Further, using $G = SL(n+2,\mathbb{R})$ as our basic group, we have the standard representation which gives rise to the standard tractor bundle $\mathcal{T}$. The P–invariant one–dimensional subspace in the standard representation gives rise to a line subbundle $\mathcal{T}^1 \subset \mathcal{T}$, which we denote by $\mathcal{E}(-1,0)$. The P–invariant $(n+1)$–dimensional subspace gives us a subbundle $\mathcal{T}^0 \subset \mathcal{T}$, and the quotient $\mathcal{T}/\mathcal{T}^0$ is a line bundle, which we denote by $\mathcal{E}(0,1)$. For $k,\ell \in \mathbb{Z}$ we then define $\mathcal{E}(k,\ell)$ via tensor products and duals of these bundles. From the matrix representation it is evident that $\mathcal{T}^0/\mathcal{T}^1 \cong \mathcal{E}^\alpha(-1,0)$, and that $Q = TM/H \cong \mathcal{E}(-1,0)^* \otimes \mathcal{E}(0,1) = \mathcal{E}(1,1)$.

The Levi bracket can then be viewed as an invertible section $\mathcal{L}_{\alpha\bar{\beta}}$ of $\mathcal{E}_{\alpha\bar{\beta}}(1,1)$, and we denote the inverse as $\mathcal{L}^{\alpha\bar{\beta}} \in \Gamma(\mathcal{E}^{\alpha\bar{\beta}}(-1,-1))$. We will use these two sections to raise and lower indices (at the expense of a weight).

In these terms, we can now express the components of the Rho tensor. We choose the notation to simplify comparison to [**GoGr05**]. The homogeneity two part of the Rho tensor is a section of $H^* \otimes H^*$, so it splits into four components, which we denote by $A_{\alpha\beta}$, $A_{\bar{\alpha}\bar{\beta}}$, $\mathsf{P}_{\alpha\bar{\beta}}$, and $\mathsf{P}_{\bar{\alpha}\beta}$. As usual, we use the convention that the form index comes first, so for example, $\mathsf{P}_{\bar{\alpha}\beta}$ represents the component of the Rho tensor which maps F to E^*. In homogeneity three, we have a part mapping H to Q^* and a part mapping Q to H^*. The first part is represented by sections $T_\alpha \in \Gamma(\mathcal{E}_\alpha(-1,-1))$ and $T_{\bar{\alpha}} \in \Gamma(\mathcal{E}_{\bar{\alpha}}(-1,-1))$ while the second part is represented by sections S_α and $S_{\bar{\alpha}}$ of the same bundles. Finally, the homogeneity four component of the Rho tensor is represented by a single section $S \in \Gamma(\mathcal{E}(-2,-2))$.

Having the ingredients at hand, we can now describe the standard tractor bundle and its dual.

PROPOSITION 5.2.15. *Let $(M, H = E \oplus F)$ be a Lagrangean contact structure, $\mathcal{T} \to M$ the standard tractor bundle and $\mathcal{T}^* \to M$ its dual.*

(1) Any choice of Weyl structure (and in particular any choice of contact form) gives rise to isomorphisms

$$\mathcal{T} \cong \mathcal{E}(0,1) \oplus \mathcal{E}^\alpha(-1,0) \oplus \mathcal{E}(-1,0),$$
$$\mathcal{T}^* \cong \mathcal{E}(1,0) \oplus \mathcal{E}^{\bar{\alpha}}(0,-1) \oplus \mathcal{E}(0,-1).$$

In terms of this splitting, the dual pairing is given by

$$((\sigma,\mu^\alpha,\rho),(\tau,\nu^{\bar{\alpha}},\omega)) \mapsto \sigma\omega + \rho\tau + \mu^\alpha\nu_\alpha.$$

Changing the Weyl structure by $\Upsilon \in \Gamma(\operatorname{gr}(T^*M))$ with components $\Upsilon_\alpha \in \Gamma(\mathcal{E}_\alpha)$, $\Upsilon_{\bar{\alpha}} \in \Gamma(\mathcal{E}_{\bar{\alpha}})$, and $\Upsilon \in \Gamma(\mathcal{E}(-1,-1))$ these identifications change as

$$\widehat{(\sigma,\mu^\alpha,\rho)} = (\sigma, \mu^\alpha - \sigma\Upsilon^\alpha, \rho - \Upsilon_\alpha\mu^\alpha + \sigma(\tfrac{1}{2}\Upsilon_\alpha\Upsilon^\alpha - \Upsilon)),$$
$$\widehat{(\tau,\nu^{\bar{\alpha}},\omega)} = (\tau, \nu^{\bar{\alpha}} + \tau\Upsilon^{\bar{\alpha}}, \omega + \Upsilon_{\bar{\alpha}}\nu^{\bar{\alpha}} + \tau(\Upsilon + \tfrac{1}{2}\Upsilon_\alpha\Upsilon^\alpha)).$$

(2) Denoting all Weyl connections by ∇ and using the splittings from (1), the normal tractor connection on $\mathcal{T}$ in directions of H is given by

$$\nabla^{\mathcal{T}}_\alpha \begin{pmatrix} \sigma \\ \mu^\beta \\ \rho \end{pmatrix} = \begin{pmatrix} \nabla_\alpha\sigma \\ \nabla_\alpha\mu^\beta + \rho\delta^\beta_\alpha + \sigma\mathsf{P}_\alpha{}^\beta \\ \nabla_\alpha\rho + A_{\alpha\beta}\mu^\beta + \sigma T_\alpha \end{pmatrix},$$

$$\nabla^{\mathcal{T}}_{\bar{\alpha}} \begin{pmatrix} \sigma \\ \mu^\beta \\ \rho \end{pmatrix} = \begin{pmatrix} \nabla_{\bar{\alpha}}\sigma + \mu_{\bar{\alpha}} \\ \nabla_{\bar{\alpha}}\mu^\beta + \sigma A_{\bar{\alpha}}{}^\beta \\ \nabla_{\bar{\alpha}}\rho + \mathsf{P}_{\bar{\alpha}\beta}\mu^\beta + \sigma T_{\bar{\alpha}} \end{pmatrix}.$$

For a section $s \in \Gamma(Q) \cong \mathcal{E}(1,1)$, the covariant derivative in direction s is given by

$$\nabla^{\mathcal{T}}_s \begin{pmatrix} \sigma \\ \mu^\beta \\ \rho \end{pmatrix} = \begin{pmatrix} \nabla_s \sigma + s\rho \\ \nabla_s \mu^\beta + \sigma s S^\beta \\ \nabla_s \rho + s S_\beta \mu^\beta + s\sigma S \end{pmatrix}.$$

PROOF. We have already observed the form of the composition series for $\mathcal{T}$ above. From this the composition series for $\mathcal{T}^*$ follows immediately taking into account that $\mathcal{E}_\alpha(1,0) \cong \mathcal{E}^{\bar{\alpha}}(0,-1)$ via $\mathcal{L}^{\alpha\bar{\beta}}$. Then part (1) follows from Proposition 5.1.5 and straightforward computations, similar to the ones in 5.1.11. Likewise, part (2) is deduced via straightforward computations from Proposition 5.1.10. □

REMARK 5.2.16. (1) From the formulae for the tractor connection on $\mathcal{T}$, the ones for $\mathcal{T}^*$ can be easily deduced by duality.

(2) The formulae in the proposition compare nicely to the ones in [**GoGr05**]. One can see that some of the curvature terms showing up in the formulae there simply express the difference between Webster–Tanaka and Weyl connections, so the formulae in terms of Weyl connections are simpler. While the tractor calculus in [**GoGr05**] only works for integrable CR–structures, our results hold for arbitrary Lagrangean contact structures. In particular, the analogous results in the CR case work assuming only partial integrability of the CR–structure. Of course, in the non–integrable case, the expressions for the components of the Rho tensor (which we have not derived explicitly) will become more complicated than the formulae in [**GoGr05**].

5.2.17. Preferred connections for other parabolic contact structures. For the other parabolic contact structures, the Weyl connections associated to a choice of contact form can be computed similarly as for Lagrangean contact structures. Let us briefly outline the cases of partially integrable almost CR–structures, for which the results are known best, and the one of projective contact structures, which is slightly special but turns out to be easier in the end.

In the CR case, there are two basic possibilities. Either one starts by complexifying the tangent bundle. Then the situation becomes very closely parallel to the Lagrangean contact case and one may directly follow the developments there. This leads to formulae for the complex linear and the conjugate linear parts of derivatives which is also the usual approach to Webster–Tanaka connections in CR–geometry. Alternatively, one may proceed as follows. Consider a contact form θ on a partially integrable almost CR–structure (M, H, J). Then $(\xi, \eta) \mapsto d\theta(\xi, J\eta)$ defines a nondegenerate symmetric bilinear form on H. The partial connection ∇ associated to θ by definition satisfies $\nabla d\theta = 0$ and $\nabla J = 0$, so it preserves this bilinear form. Hence, one can imitate the procedure for obtaining an explicit formula for the Levi–Civita connection on a Riemannian manifold. First expand the expression $0 = \nabla d\theta(\ ,J\ )$ for three sections $\xi, \eta, \zeta \in \Gamma(H)$. Then permute the entries cyclically and subtract one of the resulting terms from the other two to obtain

$$\begin{aligned} 0 =& \xi \cdot d\theta(\eta, J\zeta) - \zeta \cdot d\theta(\xi, J\eta) + \eta \cdot d\theta(\zeta, J\xi) \\ &+ d\theta(\nabla_\xi \eta + \nabla_\eta \xi, J\zeta) + d\theta(\nabla_\xi \zeta - \nabla_\zeta \xi, J\eta) + d\theta(\nabla_\eta \zeta - \nabla_\zeta \eta, J\xi). \end{aligned} \tag{5.21}$$

Now from the definition of torsion, we know that $\nabla_\xi \zeta - \nabla_\zeta \xi$ can be computed from

$$\pi_{-1}([\xi, \zeta]) = [\xi, \zeta] - d\theta(\xi, \zeta) r$$

and the Nijenhuis tensor $\mathcal{N}$ of the CR–structure. Using this, we can directly rewrite the last two summands in (5.21) in terms of known quantities. Rewriting $\nabla_\xi\eta+\nabla_\eta\xi$ as $2\nabla_\xi\eta-(\nabla_\xi\eta-\nabla_\eta\xi)$, we then end up with an explicit formula for $2d\theta(\nabla_\xi\eta, J\zeta)$, which completely determines the partial connection on H associated to θ. Notice that this shows that the partial connection ∇ on H is determined by $\nabla d\theta=0$, $\nabla J=0$ and the torsion conditions. The other two usual defining conditions for a Webster–Tanaka connection, $\nabla\theta=0$ and $\nabla r=0$, simply say that the connection on the tangent bundle preserves H and how this restriction to H is extended to all of TM.

To determine the full Webster–Tanaka connection ∇^{WT} associated to θ, we have to understand $\ker(\square)\subset L(H,H)$. For Lagrangean contact structures, these were maps exchanging the two subbundles, so in the CR case we see directly from the complexified picture that $\ker(\square)\subset L(H,H)$ consists of conjugate linear maps. Using $\nabla^{WT}J=0$, the equation that $\xi\mapsto T(r,\xi)$ for $\xi\in\mathfrak{X}(M)$ has vanishing complex linear part immediately gives

$$\nabla^{WT}_r\xi=\tfrac{1}{2}([r,\xi]-J[r,J\xi]),$$

which together with the partial connection completely determines ∇^{WT}. Note that via the conjugate linear isomorphism $\mathfrak{g}_{-1}\cong\mathfrak{g}^*_{-1}\otimes\mathfrak{g}_{-2}$, and one can identify $\ker(\square)$ with $S^2_{\mathbb{C}}\mathfrak{g}^*_{-1}\otimes\mathfrak{g}_{-2}$. The fact that the torsion is described by an element of $S^2_{\mathbb{C}}H^*\otimes Q$ is one of the usual ways to formulate the torsion condition on the Webster–Tanaka connection, compare with [**GoGr05**]. Other common versions of stating the condition involve the complex linear part of the connection, boiling down to the fact that no torsion in homogeneity two is visible in this complex linear part.

The Webster–Ricci curvature in our situation has values in the space of skew Hermitian endomorphisms of H. Hence, it splits into a part Ric^W_0 with vanishing complex trace and a purely imaginary multiple of the identity, which defines the *Webster scalar curvature* R^W. Using these curvature quantities, one computes the Weyl connection associated to ∇ similarly as in the proof of Proposition 5.2.14. Having this at hand, one can explicitly describe the standard tractor bundle for partially integrable almost CR–structures, and so on.

As already indicated above, the situation for contact projective structures is slightly special. From the description in 4.2.6 it is clear that in this case the distinguished partial connections are exactly the ones in the projective equivalence class which defines the geometry. Let us verify that in a projective equivalence class, there is exactly one partial connection that is compatible with a given contact form. Recall from 4.2.6 that two partial connections are projectively equivalent if and only if there is a smooth section $\Upsilon\in\Gamma(H^*)$ such that

$$\hat\nabla_\xi\eta=\nabla_\xi\eta+\Upsilon(\xi)\eta+\Upsilon(\eta)\xi+\Upsilon^\#(\mathcal{L}(\xi,\eta)),\tag{5.22}$$

for all $\xi,\eta\in\Gamma(H)$. Here $\Upsilon^\#:TM/H\to H$ is characterized by $\mathcal{L}(\Upsilon^\#(\beta),\zeta)=\Upsilon(\zeta)\beta$. Now suppose that θ is a contact form. For each partial connection ∇ in the projective class, we can consider $\nabla d\theta\in\Gamma(H^*\otimes\Lambda^2H^*)$. Since any (partial) contact connection leaves the subbundle $\Lambda^2_0H^*$ invariant, we see that

$$(\nabla_\xi d\theta)(\eta,\zeta)=\alpha(\xi)d\theta(\eta,\mathcal{E})$$

for some $\alpha\in\Gamma(H^*)$. Replacing ∇ by a projectively equivalent connection $\hat\nabla$ with the change corresponding to $\Upsilon\in\Gamma(H^*)$, we conclude from (5.22) and the definition of $\Upsilon^\#$ that

$$(\hat{\nabla}_\xi d\theta) = (\nabla_\xi d\theta) - 2\Upsilon(\xi)d\theta(\eta,\zeta).$$

This shows that the projective class contains a unique connection which is compatible with θ as claimed.

Obtaining the Webster–Tanaka connection and Weyl connection is particularly simple for projective contact structures. The unique irreducible component in $H^1(\mathfrak{g}_-,\mathfrak{g})$ is contained in homogeneity 1, so there is no harmonic piece in $\ker(\partial) \subset L_0(\mathfrak{g}_-,\mathfrak{g}_-)$. Since we also know that projective contact structures have no harmonic torsion in homogeneity one, the Webster–Tanaka connections for projective contact structures must have vanishing torsion in homogeneity one and two. Hence, the Webster–Tanaka connection corresponding to θ is given by

$$\nabla^{WT}_\zeta \xi = \theta(\zeta)[r,\xi] + \nabla_{\pi(\zeta)}\xi,$$

where in the right–hand side we use the unique partial connection in the projective class which is compatible with θ. Finally, the Webster–Ricci curvature has values in an irreducible bundle, so there is no further splitting into components. Hence, by Proposition 5.2.13 the Weyl connection associated to θ is obtained from the Webster–Tanaka connection by adding an appropriate multiple of the Webster–Ricci curvature.

5.2.18. Special symplectic connections. To conclude this section, we give a brief outline of another application of Weyl connections for parabolic contact structures. This comes from the beautiful article [**CahS04**], in which the authors describe a construction of connections compatible with a symplectic structure. They show that, among other interesting examples, this construction locally produces all connections with special symplectic holonomy, which leads to several striking consequences. While the construction in [**CahS04**] is based on generalized flag manifolds, it does not use Weyl structures. The relation to the latter was first observed in [**PŽ08**]. A full discussion of the results of [**CahS04**] would take us too far from the main line of development, so we only sketch the basic ideas and the relation to Weyl structures.

The first step is to define the class of so–called special symplectic subalgebras. Consider a real simple Lie algebra $\mathfrak{g}$ endowed with a contact grading $\mathfrak{g} = \mathfrak{g}_{-2}\oplus\cdots\oplus\mathfrak{g}_2$. In the classification of complex contact gradings in 3.2.4, we have seen that, for complex $\mathfrak{g}$, one can choose a Cartan subalgebra and positive roots in such a way that $\mathfrak{g}_2$ is the root space for the highest root, $\mathfrak{g}_{-2}$ is the root space for the lowest root, and $[\mathfrak{g}_{-2},\mathfrak{g}_2] \subset \mathfrak{g}_0$ consists of all complex multiples of the grading element E. The classification in the real case proceeds via complexification, so we see that $\mathfrak{s} := \mathfrak{g}_{-2} \oplus \mathbb{R}\cdot E \oplus \mathfrak{g}_2$ is a subalgebra of $\mathfrak{g}$, which is isomorphic to $\mathfrak{sl}(2,\mathbb{R})$. On the other hand, the restriction of the Killing form B to $\mathfrak{g}_0$ is always nondegenerate, and $B(E,E) > 0$, so we conclude that $\mathfrak{h} := E^\perp \subset \mathfrak{g}_0$ forms a complementary subspace to $\mathbb{R}\cdot E$. Invariance of the Killing form and the fact that E lies in the center of $\mathfrak{g}_0$ immediately imply that $\mathfrak{h}$ is a subalgebra of $\mathfrak{g}_0$, which is automatically reductive.

It also follows easily that $[\mathfrak{h},\mathfrak{s}] = \{0\}$, so $\mathfrak{g}_{-2}\oplus\mathfrak{g}_0\oplus\mathfrak{g}_2 \cong \mathfrak{s}\oplus\mathfrak{h}$ as a Lie algebra. By the grading property, the subspace $\mathfrak{g}_{-1}\oplus\mathfrak{g}_1 \subset \mathfrak{g}$, which evidently is complementary to $\mathfrak{s}\oplus\mathfrak{h}$, is invariant under the adjoint action of $\mathfrak{s}\oplus\mathfrak{h}$. We can determine $\mathfrak{g}_{-1}\oplus\mathfrak{g}_1$ as a representation of $\mathfrak{s}\oplus\mathfrak{h}$ from the fact that the only eigenvalues of $E \in \mathfrak{s}$ are ± 1. It must be of the form $\mathbb{R}^2 \boxtimes V$, i.e. an exterior tensor product of the standard representation of $\mathfrak{s} \cong \mathfrak{sl}(2,\mathbb{R})$ and some representation V of $\mathfrak{h}$. Fixing an appropriate nonzero vector in $\mathbb{R}^2$ leads to an identification of V with $\mathfrak{g}_1$. Since $\mathfrak{h}$ acts trivially on $\mathfrak{g}_2$, the bracket $[\ ,\ ] : \mathfrak{g}_1 \times \mathfrak{g}_1 \to \mathfrak{g}_2$ gives rise to an $\mathfrak{h}$–invariant symplectic form on V, so $\mathfrak{h} \subset \mathfrak{sp}(V)$.

The subalgebras of the symplectic Lie algebras obtained in this way are called *special symplectic Lie subalgebras* in [**CahS04**], and the corresponding groups are called *special symplectic Lie subgroups.* The complete list of these, which is given in Table 1 of [**CahS04**], can be easily obtained from the classification of contact gradings in 3.2.4 and 3.2.10. There also is an "abstract" definition of special symplectic Lie subalgebras. They can be characterized as subalgebras $\mathfrak{h} \subset \mathfrak{sp}(V)$ which are endowed with an $\mathfrak{h}$–invariant nondegenerate bilinear form and a bilinear map $\circ : S^2(V) \to \mathfrak{h}$, which satisfy certain compatibility conditions. If $\mathfrak{h}$ is obtained from a contact grading as above, these are obtained as the restriction of the Killing form, respectively, from the $\mathfrak{h}$–component of the bracket $[\ ,\ ] : \mathfrak{g}_{-1} \otimes \mathfrak{g}_1 \to \mathfrak{g}_0$. Conversely, given the abstract data, one uses them to construct a Lie bracket on $\mathbb{R}^2 \otimes V \oplus (\mathfrak{sl}(2, \mathbb{R}) \oplus \mathfrak{h})$ and proves that the result is simple and inherits a contact grading.

Given a special symplectic Lie algebra $\mathfrak{h} \subset \mathfrak{sp}(V)$, the authors define (using the operation $\circ$ from above) an $\mathfrak{h}$–equivariant injection $\mathfrak{h} \hookrightarrow \Lambda^2 V^* \otimes \mathfrak{h}$. They show that the image $\mathcal{R}_{\mathfrak{h}}$ of this map is contained in the space of formal curvatures, i.e. its elements satisfy the first Bianchi identity. Moreover, the contraction defining Ricci curvature restricts to an injection on $\mathcal{R}_{\mathfrak{h}}$. In most cases, the subspace $\mathcal{R}_{\mathfrak{h}}$ exhausts all possible formal curvatures corresponding to $\mathfrak{h}$.

Now let (M, τ) be a symplectic manifold; see 4.2.1. A *symplectic connection* on M is a linear connection ∇ on the tangent bundle TM such that $\nabla\tau = 0$. This implies that the holonomy of ∇ (see 5.2.8) is contained in the symplectic group $Sp(2n, \mathbb{R})$ where $2n = \dim(M)$. A symplectic connection ∇ is called *special* if it is torsion free and its curvature is contained in $\mathcal{R}_{\mathfrak{h}}$ for some special symplectic Lie subalgebra $\mathfrak{h} \subset \mathfrak{sp}(2n, \mathbb{R})$.

The Ambrose–Singer theorem then implies that the holonomy of a special symplectic connection lies in the corresponding special symplectic subgroup $H \subset Sp(2n, \mathbb{R})$. Using the classification of these groups, the definition of special symplectic connections can be turned into something much more concrete. On the one hand, recall from 5.2.8 that there is a complete classification of the possible irreducibly acting holonomy groups of torsion–free affine connections. These groups can be collected into three categories. First, there are pseudo–Riemannian holonomy groups, which admit an invariant nondegenerate symmetric bilinear form. Second, there are symplectic holonomy groups, which admit an invariant nondegenerate skew symmetric bilinear forms. The last type is given by those holonomy groups, which do not admit any nondegenerate invariant bilinear forms. (Almost all of the latter are covered by the holonomy groups related to $|1|$–gradings as discussed in 5.2.8.) On the other hand, the classifications show that the possible symplectic holonomy groups (apart from the full symplectic group), are exactly the special symplectic subgroups corresponding to simple Lie algebras of types different from A_ℓ and C_ℓ. Hence, for these cases, special symplectic connections are exactly connections with special symplectic holonomy.

Special symplectic connections of the two remaining types A_ℓ and C_ℓ turn out to be highly interesting as well. For the real forms $\mathfrak{su}(p, q)$ of type A_ℓ, they are the Levi–Civita connections of so–called Bochner–Kähler metrics (see [**Br01**]), while for the real form $\mathfrak{sl}(\ell+1,\mathbb{R})$ one obtains Bochner–bi–Lagrangean connections. Finally, for type C_ℓ one obtains so–called symplectic connections of Ricci type, which have also been studied intensively; see e.g. [**BC99**].

5.2.19. The relation to parabolic contact structures. After introducing and studying the concepts of special symplectic Lie subalgebras, the authors of [**CahS04**] construct a family of examples of special symplectic connections. This starts from the generalized flag manifold G/P corresponding to a given contact grading on $\mathfrak{g}$. We will describe a Weyl connection, which is the basic ingredient in the construction of [**CahS04**], using a notation similar to this reference to simplify comparison. For an element $a \in \mathfrak{g}$, left translation by $\exp(ta)$ defines a one–parameter group of automorphisms of the canonical parabolic contact structure on G/P. Consequently, the right invariant vector field $R_a \in \mathfrak{X}(G)$ generated by a is an infinitesimal automorphism. This projects to a vector field $a^* \in \mathfrak{X}(G/P)$. Now define $\mathcal{C}_a := \{x \in G/P : a^*(x) \notin T^{-1}_x(G/P)\}$, i.e. the set of those points in which a^* is transverse to the contact subbundle. Evidently, this is an open subset of G/P, and we will only be interested in elements a such that $\mathcal{C}_a \neq \emptyset$. This is satisfied generically.

Denoting by $p : G \to G/P$ the canonical map, we can consider $p : p^{-1}(\mathcal{C}_a) \to \mathcal{C}_a$ with the restriction of the Maurer–Cartan form ω. By definition, $\omega(R_a(g)) = \mathrm{Ad}(g^{-1})(a)$ for all $g \in G$. Let us fix nonzero elements $\phi \in \mathfrak{g}_{-2}$ and $\psi \in \mathfrak{g}_2$ such that $B(\phi,\psi) = 1$. Then we get $p^{-1}(\mathcal{C}_a) = \{g \in G : \psi(\mathrm{Ad}(g^{-1})(a)) \neq 0\}$. Restricting a^* to $\mathcal{C}_a$, we get a vector field which is everywhere transversal to the contact distribution and thus defines a nowhere vanishing section of $\mathrm{gr}_{-2}(T\mathcal{C}_a)$. Consequently, there is a unique exact Weyl structure for the parabolic geometry $(p^{-1}(\mathcal{C}_a) \to \mathcal{C}_a, \omega)$, such that this section is parallel for the induced connection. As we have seen in 5.1.7, such an exact Weyl structure provides us with a reduction of the Cartan bundle to the structure group $\ker(\lambda) \subset G_0$, where $\lambda : G_0 \to \mathbb{R}_+$ is the representation inducing the bundle of scales. Since our bundle of scales corresponds to (a multiple of) the grading element, we can identify $\ker(\lambda)$ with the special symplectic subgroup H.

PROPOSITION 5.2.19. *The image of the reduction to the structure group H of the principal bundle $p^{-1}(\mathcal{C}_a) \to \mathcal{C}_a$ coming from the exact Weyl structure described above is given by*

$$\Gamma_a = \{g \in G : \mathrm{Ad}(g^{-1})(a) - \phi \in \mathfrak{p}'\} \subset G,$$

where $\mathfrak{p}' := \mathfrak{h} \oplus \mathfrak{g}_1 \oplus \mathfrak{g}_2 \subset \mathfrak{p}$ with $\mathfrak{h} \subset \mathfrak{g}_0$ denoting the special symplectic subalgebra coming from the contact grading. Moreover, this Weyl structure is invariant under the one–parameter group of automorphisms induced by a.

PROOF. First note that for $g \in \Gamma_a$ by construction we get $\psi(\mathrm{Ad}(g^{-1})(a)) = \psi(\phi) = 1$, which shows that $\Gamma_a \subset p^{-1}(\mathcal{C}_a)$. Second, for an element $h \in H$ the adjoint action $\mathrm{Ad}(h)$ preserves $\mathfrak{p}'$ and is trivial on $\mathfrak{g}_{-2}$, which immediately implies that Γ_a is invariant under right translations by elements of H. We claim that $p|_{\Gamma_a} : \Gamma_a \to \mathcal{C}_a$ is surjective with the fibers being the H–orbits.

For $g \in p^{-1}(\mathcal{C}_a)$ and $b \in P$, we have $\omega(R_a(gb)) = \mathrm{Ad}(b^{-1})(\omega(R_a(g)))$. First, we take $b = \exp(tE)$, where E is the grading element. This acts by multiplication by e^{-2t} on $\mathfrak{g}_{-2}$, so we can uniquely choose t in such a way that

$$\psi(\mathrm{Ad}(b^{-1})(\omega(R_a(g)))) = 1.$$

Assuming that $\omega(R_a(g))$ already has this property, we can take $b = \exp(Z)$ for $Z \in \mathfrak{g}_1$. Then acting by $\mathrm{Ad}(b^{-1})$ on $\omega(R_a(g))$ leaves the $\mathfrak{g}_{-2}$–component unchanged, and since that component is nonzero, we can uniquely choose Z in such a way that $\mathrm{Ad}(b^{-1})(\omega(R_a(g)))$ has vanishing $\mathfrak{g}_{-1}$–component. Assuming again that $\omega(R_a(g))$ already has both properties, we can finally consider $b = \exp(t\psi)$. Since $[\phi, \psi]$ is a nonzero multiple of the grading element E, we can uniquely choose t in such a way that $\mathrm{Ad}(b^{-1})(\omega(R_a(g))) \in \Gamma_a$. This also shows that $g \in \Gamma_a$, $b \in P$ and $gb \in \Gamma_a$ implies $b \in H$, thus completing the proof of the claim.

Finally, we can start with a local smooth section of $p^{-1}(\mathcal{C}_a) \to \mathcal{C}_a$ and apply the above construction depending smoothly on the base point, to obtain smooth local sections of $p|_{\Gamma_a} : \Gamma_a \to \mathcal{C}_a$. Hence, $\Gamma_a \to \mathcal{C}_a$ is a principal fiber bundle with structure group H and the inclusion $\Gamma_a \hookrightarrow p^{-1}(\mathcal{C}_a)$ is a reduction of structure group to $H \subset P$.

Now let us denote by $\pi : G \to G/P_+$ and by $p_0 : G/P_+ \to G/P$ the natural projections. Then we see that π restricts to a reduction $j : \Gamma_a \to p_0^{-1}(\mathcal{C}_a)$ to the structure group $H \subset G_0$. Now it is easy to see that any element $g \in G_0$ can be written as hg' with $h \in H$ and g' in the center of G_0. This immediately shows that the unique map $\sigma : G/P_+ \to G$ characterized by $\sigma(j(g)) = g$ for all $g \in \Gamma_a$ is G_0–equivariant and hence defines a Weyl structure for the parabolic geometry $(p^{-1}(\mathcal{C}_a) \to \mathcal{C}_a, \omega)$. By construction of Γ_a, the equivariant function $p^{-1}(\mathcal{C}_a) \to \mathfrak{g}_-$ corresponding to the vector field $a^* \in \mathfrak{X}(\mathcal{C}_a)$ restricts to the constant function ϕ on Γ_a. But this exactly means that the vector field a^* is parallel for the Weyl connection corresponding to σ, so the induced section of $\mathrm{gr}_{-2}(TM)$ must be parallel, too. But the latter condition pins down the Weyl structure uniquely.

It remains to prove that the Weyl structure σ is invariant under the one–parameter subgroup of automorphisms induced by a. For $g \in G$ and $t \in \mathbb{R}$, we get

$$\omega(R_a(\exp(ta)g)) = \mathrm{Ad}(g^{-1}) \circ Ad(\exp(-ta))(a) = \mathrm{Ad}(g^{-1})(a).$$

This shows that both $p^{-1}(\mathcal{C}_a)$ and Γ_a are invariant under left translations by elements of the one–parameter subgroup of G generated by a. As we observed above, any element of $p_0^{-1}(\mathcal{C}_a)$ can be written as $j(g)g'$ with $g \in \mathcal{C}_a$ and g' in the center of G_0. But then $\exp(ta) \cdot j(g)g' = j(\exp(ta)g)g'$, and hence $\sigma(\exp(ta) \cdot j(g)g') = \exp(ta)gg' = \exp(ta)\sigma(j(g)g')$. □

To prove their first main result, the authors of [**CahS04**] directly define the set $\Gamma_a \subset G$ exhibited above and consider the restriction of the components in $\mathfrak{g}_-$ and $\mathfrak{g}_0$ of the Maurer–Cartan form of G to this subset, so these are exactly the soldering form and the Weyl connection of the Weyl structure determined by a. Then they consider open subsets of $\mathcal{C}_a$, which are small enough to form local leaf spaces for the foliation determined by the nowhere–vanishing vector field a^*. Of course, such a leaf space is just the quotient by the one–parameter group of automorphisms induced by a^*, so from the last part of the proposition it follows that all data associated to our Weyl structure descend to the quotient. In particular, the contact structure descends to a symplectic structure on the leaf space, and Γ_a descends to a reductions of the symplectic frame bundle to the structure group H. The authors then show that the Weyl connection descends to a special symplectic connection on this leaf space.

The results in [**CahS04**] actually go much further than that. Namely, it is proved that locally the connections constructed from parabolic contact structures as above exhaust all possible special symplectic connections. This has strong consequences like automatic analyticity, finite dimensionality of the moduli space of symplectic connections, and the fact that special symplectic connections always admit nontrivial infinitesimal automorphisms. Finally, the authors use these results to obtain a complete classification of special symplectic connections on simply connected compact manifolds.

5.3. Canonical curves

In this section, we specialize the concept of canonical curves for Cartan geometries as discussed in 1.5.18 to the case of parabolic geometries. We will restrict our attention to the simplest classes of curves, namely those coming from exponential curves in the homogeneous model, which still leads to a very rich and diverse theory. We will mainly address the question how many distinguished curves emanate from a given point in a given direction. Otherwise put, we study how many derivatives in a point are needed to pin down a distinguished curve uniquely or uniquely up to reparametrization. As we shall see, this heavily depends on the direction in question, so understanding the possible types of geometrically different directions will be an important task, too. Via the concept of development from 1.5.17 these local questions need to be studied for the homogeneous model only, and we will reduce them to purely Lie algebraic considerations. The basic source for this section is [**ČSŽ03**].

5.3.1. The basic setup. Let us fix some type (G, P) of parabolic geometry. Given a geometry $(p : \mathcal{G} \to M, \omega)$ of this type, Cartan's space $\mathcal{S}M$ is defined as $\mathcal{G} \times_P (G/P)$, the associated bundle with fiber the homogeneous model of the geometry. In 1.5.17 we have shown that the Cartan connection ω induces a connection on the fiber bundle $\mathcal{S}M \to M$. Using this connection, we defined the *development of curves*. To a point $x \in M$, a curve $c : I \to M$ defined on some open interval with $0 \in I$ and $c(0) = x$, and a frame $u \in \mathcal{G}_x$, this associates a curve $\operatorname{dev}_c : I' \to G/P$ defined on a subinterval $I' \subset I$ with $\operatorname{dev}_c(0) = o = eP \in G/P$. By Theorem 1.5.17, the map $c \mapsto \operatorname{dev}_c$ induces a bijection between the space of germs in the point x of smooth curves in M and the space of germs in o of smooth curves in G/P. For each $r \in \mathbb{N}$, this relation is compatible with the notion of rth order contact, so two curves have the same r–jet in x if and only if their developments have the same r–jet in o.

The definition of canonical curves in 1.5.18 was based on the notion of development. The starting point was a family $\mathcal{C}$ of curves in G/P, which is admissible in the sense that for $\gamma \in \mathcal{C}$, t_0 in the domain of γ, and $g \in G$ such that $\gamma(t_0) = g^{-1} \cdot o$, also $t \mapsto g \cdot \gamma(t + t_0)$ lies in $\mathcal{C}$. Consider the bundle $\mathcal{S}M = \mathcal{G} \times_P (G/P)$ and let us denote the natural projection $\mathcal{G} \times (G/P) \to \mathcal{S}M$ by $(u, y) \mapsto [\![u, y]\!]$. Given M and x as above, canonical curves of type $\mathcal{C}$ through x are then defined to be the curves whose development in x can be expressed in the form $t \mapsto [\![u, \gamma(t)]\!]$ for some $u \in \mathcal{G}_x$ and some $\gamma \in \mathcal{C}$. Changing the element $u \in \mathcal{G}_x$ amounts to replacing $\gamma(t)$ by $g \cdot \gamma(t)$ for some $g \in P$, and the latter curve lies in $\mathcal{C}$ by admissibility.

The definition of canonical curves and the properties of the development map hence imply that understanding local properties of the family of canonical curves of type $\mathcal{C}$ through x in M is equivalent to understanding local properties of the family $\mathcal{C}$ of curves through o in G/P.

We will mainly look at the case that the initial family $\mathcal{C}$ is a family of exponential curves. Let $\mathfrak{g} = \mathfrak{g}_- \oplus \mathfrak{p} = \mathfrak{g}_- \oplus \mathfrak{g}_0 \oplus \mathfrak{p}_+$ be the decomposition of the Lie algebra $\mathfrak{g}$ of G from the grading corresponding to the parabolic subgroup $P \subset G$. Take a G_0–invariant subset $A_0 \subset \mathfrak{g}_-$, and define $A := \{\mathrm{Ad}(\exp(Z))(X) : X \in A_0, Z \in \mathfrak{p}_+\} \subset \mathfrak{g}$. By Theorem 3.1.3, any element of P can be written as $\exp(Z)g_0$ for some $g_0 \in G_0$ and $Z \in \mathfrak{p}_+$, so the subset $A \subset \mathfrak{g}$ is P–invariant. Denoting by $p : G \to G/P$ the canonical projection, it follows immediately that $\mathcal{C}_A := \{t \mapsto p(\exp(tX)) : X \in A\}$ is an admissible family of curves through o in G/P; see 1.5.18. The tangent vectors of curves from $\mathcal{C}_A$ in o are described by the P–invariant subset $A + \mathfrak{p} \subset \mathfrak{g}/\mathfrak{p}$. Note that under the identification of $\mathfrak{g}/\mathfrak{p}$ with $\mathfrak{g}_-$, this subset may be much larger than the initial subset A_0. For a parabolic geometry $(p : \mathcal{G} \to M, \omega)$ and a point $x \in M$, the set of possible tangent vectors of canonical curves of type $\mathcal{C}_A$ is the subset of T_xM determined by $A + \mathfrak{p}$ via $TM \cong \mathcal{G} \times_P (\mathfrak{g}/\mathfrak{p})$.

The families of canonical curves induced by families of exponential curves also have the advantage that there is a general description of canonical curves of type $\mathcal{C}_A$ which does not need the development and depends only on A_0.

PROPOSITION 5.3.1. *Let $A_0 \subset \mathfrak{g}_-$ be a G_0–invariant subset and put*

$$A := \{\mathrm{Ad}(\exp(Z))(X) : X \in A_0, Z \in \mathfrak{p}_+\} \subset \mathfrak{g}.$$

Consider a parabolic geometry $(p : \mathcal{G} \to M, \omega)$ of type (G, P). Then a curve $c : I \to M$ is canonical of type $\mathcal{C}_A$ if and only if it locally coincides up to a constant shift of parameter with the projection of a flow line in $\mathcal{G}$ of a constant vector field of the form $\omega^{-1}(X)$ with $X \in A_0$.

PROOF. In Corollary 1.5.18 we have seen in the realm of general Cartan geometries that c is a canonical curve of type $\mathcal{C}_A$ if and only if it locally coincides up to a constant shift of parameter with the projection of a flow line in $\mathcal{G}$ of a vector field of the form $\omega^{-1}(Y)$ for some $Y \in A$. By assumption, $Y = \mathrm{Ad}(\exp(Z))(X)$ for some $Z \in \mathfrak{p}_+$ and some $X \in A_0$. But then for each $u \in \mathcal{G}$ we get

$$\omega^{-1}(Y)(u) = Tr^{\exp(-Z)} \cdot \omega^{-1}(X)(u \cdot \exp(Z)),$$

so $\omega^{-1}(Y)$ and $\omega^{-1}(X)$ are $(r^{\exp(Z)})$–related. Hence, their flows are $(r^{\exp(Z)})$–related, so for each flow line of $\omega^{-1}(Y)$, we can find a flow line of $\omega^{-1}(X)$ which has the same projection to M and vice versa. □

From this result, we immediately obtain a relation to normal Weyl structures as introduced in 5.1.12.

COROLLARY 5.3.1. *In the setup of the previous proposition, a curve c with $c(0) = x \in M$ is canonical of type $\mathcal{C}_A$ if and only if there is a Weyl structure σ, which is normal in x, and an element $X \in A_0$ with $c'(0) = [\![\sigma(x), X]\!]$, such that c is a geodesic for the Weyl connection corresponding to σ.*

EXAMPLE 5.3.1. Let us illustrate the above description of canonical curves on the best known examples of homogeneous models.

(1) Let us first consider conformal structures in Riemannian signature. Here the homogeneous model G/P is the Möbious sphere; see 1.6.2 and 1.6.3. Since $G_0 = CO(n)$ and $\mathfrak{g}/\mathfrak{p} \cong \mathbb{R}^n$ as a G_0–module, there is just one nontrivial G_0–invariant subset A_0, namely $\mathbb{R}^n \setminus \{0\}$, and we get only one type of canonical curves. We can immediately compute their form explicitly using the formulae from 1.6.3. There we have used the flat coordinates coming from the exponential mapping

(which equal the coordinates from the very flat Weyl structure discussed in Example 5.1.12). We have computed there the action of $\mathrm{Ad}(\exp(Z))$ in these coordinates. For $X \in \mathfrak{g}_{-1}$ and $Z \in \mathfrak{g}_1$, the element $\mathrm{Ad}(\exp(Z))(X)P \in G/P$ is given by

$$\frac{X + \frac{1}{2}\langle X, X\rangle Z^t}{1 + \langle Z^t, X\rangle + \frac{1}{4}\langle Z, Z\rangle^2 \langle X, X\rangle}. \tag{5.23}$$

To get the expression for the canonical curve, we simply have to replace X by tX in this formula. Of course, for $Z = 0$ we simply get the line through X with its natural parametrization, and if Z^t is a multiple of X, then we get a reparametrization of this line. In general, the curve always lies in the plane spanned by X and Z^t. We saw in 1.6.3 that the above mapping is actually obtained from an inversion on the unit circle, followed by translations by $\frac{1}{2}Z^t$ and another inversion on the unit circle. This implies that our curve is a circle whose midpoint is given by a certain multiple of the projection of Z^t onto the line perpendicular to X. In particular, we see that we can obtain any circle through zero and tangent to the line spanned by X in this way.

It was this description that led to the names conformal circles and generalized geodesics given to our canonical curves in the older literature on conformal Riemannian manifolds. Notice that we need the two–jet in 0 to fix the canonical curve, even if we are not interested in its parametrization.

(2) For conformal structures in indefinite signature, the situation changes drastically. Namely, we obtain three different nontrivial G_0–invariant subsets of vectors, corresponding to $\langle X, X\rangle$ being positive, negative or zero. The formula from (1) remains correct in general signatures (if we interpret the transpose as being taken with respect to the indefinite inner product), and for $\langle X, X\rangle \neq 0$ the behavior is similar as above. However, for $\langle X, X\rangle = 0$ the formula shows that for any choice of Z, we only get a reparametrization of the line spanned by X.

Thus, we see that the canonical curves in null directions are determined by their first jets as unparametrized curves. This corresponds to the well–known fact that the null–geodesics of pseudo–Riemannian metrics are conformally invariant as unparametrized curves.

(3) Let us finally look at projective structures, for which the homogeneous model G/P is the projective space $\mathbb{R}P^n$. As in the first example, there are no distinguished directions here and the available transformations in the very flat coordinates are just the projective transformations fixing the origin. Thus, we end up in a situation similar to the null directions in (2), and the canonical curves for projective geometries are just the geodesics of the connections in the projective class, with their distinguished projective parametrizations.

5.3.2. Reduction to algebra and a fundamental estimate. For $Y \in \mathfrak{g}$ let us denote by $c^Y : \mathbb{R} \to G/P$ the curve $t \mapsto p(\exp(tY))$, so the canonical curves of type $\mathcal{C}_A$ through $o \in G/P$ are the curves c^Y for $Y \in A$. Recall from 1.2.4 that for G–valued functions one has the left logarithmic derivative δf. Here we will be interested in curves $c : \mathbb{R} \to G$. Via the usual trivialization of the tangent bundle of $\mathbb{R}$, we can consider the left logarithmic derivative of c simply as a smooth function $\delta c : \mathbb{R} \to \mathfrak{g}$. Explicitly, this is given by $\delta c(t) = T_{c(t)}\lambda_{c(t)^{-1}} \cdot c'(t)$. Then we can form the iterated derivatives $(\delta c)^{(i)} \in C^\infty(\mathbb{R}, \mathfrak{g})$.

LEMMA 5.3.2. *Let G be a Lie group with Lie algebra $\mathfrak{g}$, $P \subset G$ a closed subgroup with Lie algebra $\mathfrak{p}$, and let $p : G \to G/P$ be the canonical projection. Let c_1, c_2 :*

$\mathbb{R} \to G$ be smooth curves with $c_1(0) = c_2(0) = e$, and let $u : \mathbb{R} \to G$ be the curve defined by $u(t) = c_2(t)^{-1}c_1(t)$.

(1) For each $r \in \mathbb{N}$ with $r \geq 1$, the following conditions are equivalent:

(i) *c_1 and c_2 have the same r–jet in 0.*
(ii) *The curve u has the same r–jet in 0 as the constant curve e.*
(iii) *The left logarithmic derivatives δc_1 and δc_2 have the same $(r-1)$–jet in 0.*

(2) For each $r \in \mathbb{N}$ with $r \geq 1$, the curves $p \circ c_1$ and $p \circ c_2$ have the same r–jet in 0 if and only if $(\delta u)^{(i)}(0) \in \mathfrak{p}$ for all $i = 0, \dots, r-1$.

PROOF. (1) We can use $\exp : \mathfrak{g} \to G$ as a local parametrization for G. Hence, we write $c_i(t) = \exp(\phi_i(t))$ for $i = 1, 2$ and smooth $\mathfrak{g}$–valued functions ϕ_1, ϕ_2 defined locally around zero such that $\phi_1(0) = \phi_2(0) = 0$.

(i) $\implies$ (iii) By definition, (i) is equivalent to ϕ_1 and ϕ_2 having the same r–jet in 0. We have noted in the proof of Proposition 5.1.8 already that the left logarithmic derivative of the exponential mapping is given by $\delta(\exp)(X) = \sum_{j=0}^{\infty} \frac{(-1)^j}{(j+1)!} \operatorname{ad}(X)^j$. Clearly, this implies that

$$\delta(\exp \circ \phi)(t) = \delta(\exp)(\phi(t)) \cdot \phi'(t) = \sum_{j=0}^{\infty} \tfrac{(-1)^j}{(j+1)!} \operatorname{ad}(\phi(t))^j(\phi'(t)). \tag{5.24}$$

This shows that $\delta(c_i(t))$ is given by a universal formula in terms of $\phi_i(t)$ and $\phi_i'(t)$, which shows that (i) implies (iii).

(iii) $\implies$ (ii) Observe first that equation (5.24) shows that $\delta(\exp \circ \phi)(t)$ is the sum of $\phi'(t)$ and a sum of iterated Lie brackets involving several copies of $\phi(t)$ and one copy of $\phi'(t)$. Inductively, this shows that $\delta(\exp \circ \phi)^{(s)}$ is given by $\phi^{(s+1)}(t)$ plus a sum of iterated Lie brackets with entries equal to $\phi^{(\ell)}(t)$ for $0 \leq \ell \leq s+1$ and the total number of derivatives in each term is $s+1$. But this shows that if $\delta(\exp \circ \phi)$ has vanishing $(s-1)$–jet in 0, then ϕ has vanishing s–jet in 0 and the s–jet of $\exp \circ \phi$ coincides with the s–jet of the constant curve e.

Now the compatibility of the left logarithmic derivative with pointwise products and inverses observed in the proof of Theorem 1.2.4 implies that $\delta u(t) = \delta c_1(t) - \operatorname{Ad}(u(t)^{-1})\delta c_2(t)$. By assumption, $u(0) = e$ so we obtain $\delta u(0) = \delta c_1(0) - \delta c_2(0) = 0$, so the one–jet of u in 0 coincides with the one–jet of the constant map e. Hence, the one–jet in 0 of $\delta u(t)$ coincides with the one–jet of $\delta c_1(t) - \delta c_2(t)$. Inductively, we conclude that (iii) implies that δu has vanishing $(r-1)$–jet in 0, and from above we see that this implies (ii).

(ii) $\implies$ (i) is obvious since $c_1(t) = c_2(t)u(t)$ and the right–hand side has the same r–jet in zero as $c_2(t)e = c_2(t)$.

(2) Let σ be a smooth section of $p : G \to G/P$ defined on an open neighborhood U of $o = eP \in G/P$. Then $(x, b) \mapsto \sigma(x)b$ defines a local diffeomorphism $U \times P \to p^{-1}(U)$, so locally around zero, we may write our curves uniquely as $c_i(t) = \sigma(p(c_i(t))) \cdot b_i(t)$ for smooth P–valued curves b_1, and b_2 defined locally around zero. This means that, locally around zero, we get

$$u(t) = b_2(t)^{-1}\sigma(p(c_2(t)))^{-1}\sigma(p(c_1(t)))b_1(t).$$

Now if $p \circ c_1$ and $p \circ c_2$ have the same r–jet in 0, then the same is true for $\sigma \circ p \circ c_1$ and $\sigma \circ p \circ c_2$. By part (1), this implies that $\sigma(p(c_2(t)))^{-1}\sigma(p(c_1(t)))$ has the same

r–jet in 0 as the constant curve e. Thus, the r–jet of u in 0 coincides with the r–jet of $b_2(t)^{-1}b_1(t)$, which implies that $(\delta u)^{(i)}(0)$ lies in $\mathfrak{p}$ for $i = 0, \dots, r-1$.

Conversely, assume that $(\delta u)^{(i)}(0)$ lies in $\mathfrak{p}$ for $i = 0, \dots, r-1$. Solving an ODE, we can find a smooth curve b in P (defined locally around zero) such that $(\delta b)^{(i)}(0) = (\delta u)^{(i)}(0)$ for $i = 0, \dots, r-1$. This easily implies that also for the curves $t \mapsto u(t)^{-1}$ and $t \mapsto b(t)^{-1}$, the left logarithmic derivatives have the same $(r-1)$–jet in 0. By part (1), the curve $t \mapsto u(t)b(t)^{-1}$ has the same r–jet in 0 as the constant curve e. But by construction, $c_2(t) = c_1(t)u(t)$, so

$$p(c_2(t)) = p(c_1(t)u(t)) = p(c_1(t)u(t)b(t)^{-1}),$$

and the latter curve has the same r–jet in 0 as $p(c_1(t))$. □

Using this, we can now formulate an efficient algebraic criterion to check how many derivatives in a point are needed to pin down a canonical curve uniquely.

DEFINITION 5.3.2. Let G be a semisimple Lie group, $P \subset G$ a parabolic subgroup, $\mathfrak{g} = \mathfrak{g}_- \oplus \mathfrak{g}_0 \oplus \mathfrak{p}_+$ the corresponding grading of the Lie algebra of G, $G_0 \subset P$ the Levi subgroup, and put $P_+ := \exp(\mathfrak{p}_+) \subset P$. For $r \in \mathbb{N}$ we say that a G_0–invariant subset $A_0 \subset \mathfrak{g}_-$ is *r–determined* if and only if for all $X_1, X_2 \in A_0$ and all $g \in P_+$ the condition that $\operatorname{ad}(X_1)^\ell(X_1 - \operatorname{Ad}(g)(X_2)) \in \mathfrak{p}$ for all $0 \le \ell \le r$ implies that $\operatorname{ad}(X_1)^\ell(X_1 - \operatorname{Ad}(g)(X_2)) \in \mathfrak{p}$ for all $\ell \in \mathbb{N}$.

THEOREM 5.3.2. *Let (G, P) be a type of parabolic geometry corresponding to $\mathfrak{g} = \mathfrak{g}_- \oplus \mathfrak{g}_0 \oplus \mathfrak{p}_+$, let $A_0 \subset \mathfrak{g}_-$ be a G_0–invariant subset, and put $A := \operatorname{Ad}(P)(A_0) \subset \mathfrak{g}$ as before. If for some $r \in \mathbb{N}$, the subset A_0 is r–determined, then any canonical curve of type $\mathcal{C}_A$ in a parabolic geometry of type (G, P), which is defined on a connected open interval, is uniquely determined by its $(r+1)$–jet in a single point.*

PROOF. Consider two elements $Y_1, Y_2 \in A$ and the associated curves $c_i(t) = \exp(tY_i)$ in G for $i = 1, 2$, and put $u(t) = c_2(t)^{-1}c_1(t)$. From formula (5.24) in the proof of the lemma, we see that

$$\delta c_i(t) = \sum_{j=0}^{\infty} \tfrac{(-1)^j}{(j+1)!} \operatorname{ad}(tY_i)^p(Y_i) = Y_i,$$

and as observed in that proof, this implies that $\delta u(t) = Y_1 - \operatorname{Ad}(u(t)^{-1})Y_2$ for all $t \in \mathbb{R}$. Now we claim that

$$(\delta u)^{(r)}(t) = (-1)^r \operatorname{ad}(Y_1)^r(\delta u(t)). \tag{5.25}$$

To prove this, we have to compute the derivative of $t \mapsto \operatorname{Ad}(u(t)^{-1})$, which clearly is given by $T_{u(t)^{-1}} \operatorname{Ad} \cdot T_{u(t)}\nu \cdot u'(t)$, where ν denotes the inversion mapping. Now $\operatorname{Ad}' = T_e \operatorname{Ad} = \operatorname{ad}$ and from $\operatorname{Ad} \circ \lambda_g = \operatorname{Ad}(g) \circ \operatorname{Ad}$ we get $T_g \operatorname{Ad} = \operatorname{Ad}(g) \circ \operatorname{ad} \circ T_g\lambda_{g^{-1}}$. Likewise, $T_e\nu = -\mathrm{id}$ and $\nu \circ \lambda_g = \rho^{g^{-1}} \circ \nu$ imply that $T_g\nu = -T_e\rho^{g^{-1}} \circ T_g\lambda_{g^{-1}}$. Since $u'(t) = T_e\lambda_{u(t)} \cdot \delta u(t)$ and $T_{u(t)^{-1}}\lambda_{u(t)} \circ T_e\rho^{u(t)^{-1}} = \operatorname{Ad}(u(t))$, we finally end up with

$$\begin{aligned}\tfrac{d}{dt} \operatorname{Ad}(u(t)^{-1}) &= \big(\operatorname{Ad}(u(t)^{-1}) \circ \operatorname{ad} \circ T_{u(t)^{-1}}\lambda_{u(t)} \circ (-T_e\rho^{u(t)^{-1}})\big)(\delta u(t)) \\ &= -(\operatorname{Ad}(u(t)^{-1}) \circ \operatorname{ad} \circ \operatorname{Ad}(u(t)))(\delta u(t))\end{aligned}$$

and hence

$$\begin{aligned}(\delta u)'(t) &= \operatorname{Ad}(u(t)^{-1})[\operatorname{Ad}(u(t))(\delta u(t)), Y_2] \\ &= -\,[\operatorname{Ad}(u(t)^{-1})(Y_2), \delta u(t)].\end{aligned} \tag{5.26}$$

Since $\mathrm{Ad}(u(t)^{-1})(Y_2) = Y_1 - \delta u(t)$, this proves (5.25) for $r = 1$. For general r, (5.25) then immediately follows by induction.

Now suppose that the curves $p \circ c_1$ and $p \circ c_2$ have the same $(r+1)$–jet in 0. By the lemma and our claim, this implies

$$(\delta u)^{(\ell)}(0) = (-1)^\ell \,\mathrm{ad}(Y_1)^\ell (Y_1 - Y_2) \in \mathfrak{p} \text{ for } \ell = 0, \dots, r.$$

Since $Y_i \in A$ and A_0 is G_0–invariant, there are elements $X_i \in A_0$ and $g_i \in P_+$ such that $Y_i = \mathrm{Ad}(g_i)(X_i)$ for $i = 1, 2$. Putting $g = g_1^{-1} g_2 \in P_+$, we get

$$\mathrm{ad}(Y_1)^\ell (Y_1 - Y_2) = \mathrm{Ad}(g_1)(\mathrm{ad}(X_1)^\ell (X_1 - \mathrm{Ad}(g)(X_2)))$$

for all $\ell \in \mathbb{N}$. Since $g_1 \in P_+$, we see that $\mathrm{ad}(X_1)^\ell (X_1 - \mathrm{Ad}(g)(X_2)) \in \mathfrak{p}$ for all $0 \leq \ell \leq r$. If A_0 is r–determined, then this holds for all $\ell \in \mathbb{N}$ and hence $\mathrm{ad}(Y_1)^\ell (Y_1 - Y_2) \in \mathfrak{p}$ for all $\ell \in \mathbb{N}$. Again by the lemma, $p \circ c_1$ and $p \circ c_2$ have the same infinite jet in 0, and since both curves are analytic by construction, they coincide locally.

Together with the observations on developments from 5.3.1, this shows that two canonical curves of type $\mathcal{C}_A$ in any manifold with a parabolic geometry of type (G, P), which are defined on an open interval $I \subset \mathbb{R}$ have the same germ in a point $t \in I$ if and only if they have the same $(r+1)$–jet in t. Now the subset of I, in which two smooth curves have the same germ, is evidently open. But also the subset, on which two curves have different $(r+1)$–jet is evidently open. If I is connected, this implies that the two canonical curves coincide on I if they have the same $(r+1)$–jet and hence the same germ in one point $t_0 \in I$. □

Using this, we get a simple first fundamental estimate for the jet needed to pin down a canonical curve.

COROLLARY 5.3.2. *Let (G, P) be a type of parabolic geometry corresponding to a $|k|$–grading of the Lie algebra $\mathfrak{g}$ of G. Suppose that $A_0 \subset (\mathfrak{g}_{-k} \oplus \cdots \oplus \mathfrak{g}_{-j})$ for some $j = 1, \dots, k$ is G_0–invariant, and put $A = \mathrm{Ad}(P)(A_0)$. If r is such that $rj > k$, then any canonical curve of type $\mathcal{C}_A$ in a parabolic geometry of type (G, P), which is defined on a connected open interval, is uniquely determined by its $(r+1)$–jet in a single point.*

PROOF. It suffices to show that A_0 is r–determined, so suppose that $X_1, X_2 \in A_0$ and $g \in P_+$. By assumption $C := X_1 - \mathrm{Ad}(g)(X_2) \in \mathfrak{p}$ and $X_1 \in \mathfrak{g}_{-k} \oplus \cdots \oplus \mathfrak{g}_{-j}$. Hence, by the grading property, $\mathrm{ad}(X_1)^\ell (C) \in \mathfrak{g}_{-k} \oplus \cdots \oplus \mathfrak{g}_{k-\ell j}$. In particular, $\mathrm{ad}(X_1)^r (C) \in \mathfrak{g}_-$, so if this also lies in $\mathfrak{p}$, then $\mathrm{ad}(X_1)^r (C) = 0$ and hence $\mathrm{ad}(X_1)^\ell (C) = 0$ for all $\ell > r$. □

Example 5.3.1 shows that this bound is not sharp in general. There we always had $j = k = 1$, so the corollary shows that any canonical curve is determined by its three–jet in a point. In Example 5.3.1, two–jets were always sufficient, and we will give a general proof of this fact below.

5.3.3. Improvements of the fundamental estimate. The bound on the number of derivatives needed to pin down a canonical curve proved in Corollary 5.3.2 only used the grading of $\mathfrak{g}$ and no finer information on the algebraic structure. Going deeper into this algebraic structure, we can improve this result for subsets contained in one grading component. Example 5.3.1 shows that the resulting estimates are sharp in some situations.

THEOREM 5.3.3. *Let (G,P) be a type of parabolic geometry corresponding to a $|k|$–grading on $\mathfrak{g}$. For some $j = 1,\dots,k$ let $A_0 \subset \mathfrak{g}_{-j}$ be a G_0–invariant subset and put $A = \mathrm{Ad}(P)(A_0)$.*

Then any canonical curve of type $\mathcal{C}_A$ in a parabolic geometry of type (G,P) defined on a connected interval is uniquely determined by its r–jet in a single point provided that $rj > k$.

We shall need the following lemma in the quite technical proof of our theorem.

LEMMA 5.3.3. *Let X and Z be elements of an arbitrary Lie algebra such that for some $n > 0$ we have $\mathrm{ad}(X)^{n+1}(Z) = 0$. Then for each $\ell > n$, there is a linear map ϕ such that $\mathrm{ad}(X)^\ell \circ \mathrm{ad}(Z) = \phi \circ \mathrm{ad}(X)^{\ell-n}$.*

PROOF. The Jacobi identity says that $\mathrm{ad}(X)\circ\mathrm{ad}(Z) = \mathrm{ad}(\mathrm{ad}(X)(Z))+\mathrm{ad}(Z)\circ \mathrm{ad}(X)$. Inductively, this implies that for each $\ell > 0$, the map $\mathrm{ad}(X)^\ell \circ \mathrm{ad}(Z)$ can be written as a linear combination of the maps $\mathrm{ad}(\mathrm{ad}(X)^i(Z)) \circ \mathrm{ad}(X)^{\ell-i}$ for $i = 0,\dots,\ell$. In particular, if $\mathrm{ad}(X)^{n+1}(Z) = 0$ and $\ell > n$, then each of these summands can be written as the composition of some map with $\mathrm{ad}(X)^{\ell-n}$. □

PROOF OF THE THEOREM. In view of Theorem 5.3.2, we only have to show that if $rj > k$, then A_0 is $(r-1)$–determined. According to Theorem 3.1.3, any element $g \in P_+$ can be written as $g = \exp(Z_1)\cdots\exp(Z_k)$ for some $Z_i \in \mathfrak{g}_i$. Hence, for $X \in \mathfrak{g}$, we get

$$\mathrm{Ad}(g)(X) - X = \sum_{i_1,\dots,i_k} \tfrac{1}{i_1!\cdots i_k!}\, \mathrm{ad}(Z_1)^{i_1} \circ \cdots \circ \mathrm{ad}(Z_k)^{i_k}(X), \tag{5.27}$$

where at least one of the i_ℓ is nonzero. By the grading property, if $X \in \mathfrak{g}_{-j}$, then a nonzero contribution can only come from summands for which $i_1 + 2i_2 + \cdots + ki_k \leq k+j$.

Now consider $X_1, X_2 \in A_0 \subset \mathfrak{g}_{-j}$, and compute $\mathrm{Ad}(g)(X_2)$ according to (5.27). Of course, the $\mathfrak{g}_{-j}$ component of this is X_2, so $X_1 - \mathrm{Ad}(g)(X_2) \in \mathfrak{p}$ implies $X_2 = X_1 =: X$. Further, the $\mathfrak{g}_{-j+1}$–component of $\mathrm{Ad}(g)(X)$ equals $[Z_1, X]$, so if $j > 1$, then $X - \mathrm{Ad}(g)(X) \in \mathfrak{p}$ implies $Z_1 = 0$. In the same way, we get $Z_2 = \cdots = Z_{j-1} = 0$, so the sum in (5.27) is actually only over $i_j,\dots,i_k$. Now for $\ell = 1,\dots,k$ we write W'_ℓ for the sum of all terms in (5.27), for which $i_{\ell+1} = \cdots = i_k = 0$. Hence, we know that if $j > 1$, then $W'_1 = \cdots = W'_{j-1} = 0$, while $W'_k = \mathrm{Ad}(g)(X) - X$. Now if $rj > k$, then the following claim in the case $n = r-1$ implies that A_0 is $(r-1)$–determined. (Notice that for $n = r-1$, the claim covers the case $m = k$ by assumption.)

Claim: If $\mathrm{ad}(X)^i(X - \mathrm{Ad}(g)(X)) \in \mathfrak{p}$ for $i = 1,\dots,n$ then for each $m < (n+1)j$, we have $\mathrm{ad}(X)^{n+1}(Z_m) = 0$ and $\mathrm{ad}(X)^\ell(W'_m) \in \mathfrak{p}$ for each $\ell > n$.

We prove this by induction on n. First, the component of $[X, X - \mathrm{Ad}(g)(X)]$ in $\mathfrak{g}_-$ evidently is $[X,[Z_j,X]] + \cdots + [X,[Z_{2j-1},X]]$. Since the summands lie in different grading components, $\mathrm{ad}(X)(X - \mathrm{Ad}(g)(X)) \in \mathfrak{p}$ implies that $\mathrm{ad}(X)^2(Z_m) = 0$ for $m < 2j$. We already know that $Z_m = 0$ for $m < j$, and hence $W'_m = 0$ for $m < j$, so $W'_j = \sum_i \frac{1}{i!}\mathrm{ad}(Z_j)^i(X)$. Now if $\ell > 1$, then

$$\mathrm{ad}(X)^\ell(W'_j) = \sum_i \tfrac{1}{i!}\,\mathrm{ad}(X)^\ell\,\mathrm{ad}(Z_j)^i(X).$$

Summands with $i > \ell$ automatically lie in $\mathfrak{p}$. From the lemma we know that $\mathrm{ad}(X)^\ell \circ \mathrm{ad}(Z_j)$ can be written as the composition of some map with $\mathrm{ad}(X)^{\ell-1}$.

Inductively, this shows that for $i \leq \ell$, we get

$$\operatorname{ad}(X)^\ell \operatorname{ad}(Z_j)^i(X) = \psi \circ \operatorname{ad}(X)^{\ell-i+1} \operatorname{ad}(Z_j)(X) = -\psi \circ \operatorname{ad}(X)^{\ell-i+2}(Z_j)$$

for some linear map ψ, and this vanishes since $\ell - i + 2 \geq 2$. Still for $n = 1$, assume inductively, that for some $j < m < 2j$ we have proved that $\operatorname{ad}(X)^\ell(W'_{m-1}) \in \mathfrak{p}$ for all $\ell \geq 2$. To prove that $\operatorname{ad}(X)^\ell(W'_m) \in \mathfrak{p}$ it suffices to prove the same statement for $W'_m - W'_{m-1}$, which by definition can be written as

$$\sum_{i_j,\dots,i_m} \tfrac{1}{i_j!\cdots i_m!} \operatorname{ad}(Z_j)^{i_j} \circ \cdots \circ \operatorname{ad}(Z_m)^{i_m}(X),$$

with $i_m > 0$. Applying $\operatorname{ad}(X)^\ell$, we only have to consider terms for which $ji_j + \cdots + mi_m < (\ell+1)j$, whence $i_j + \cdots + i_m \leq \ell$. Now as above, we can write

$$\operatorname{ad}(X)^\ell \circ \operatorname{ad}(Z_j)^{i_j} \circ \cdots \circ \operatorname{ad}(Z_m)^{i_m} = \psi \circ \operatorname{ad}(X)^{\ell - i_j - \cdots - i_m + 1} \circ \operatorname{ad}(Z_m)$$

for some linear map ψ. Applying this to X, we obtain $\operatorname{ad}(X)^{\ell - i_j - \cdots - i_m + 2}(Z_m) = 0$. This completes the proof of the claim for $n = 1$.

Now assume that $n > 1$ and the claim has been proved for all $s < n$. This means that for $s = 1, \dots, n-1$ and $m < (s+1)j$ we have $\operatorname{ad}(X)^{s+1}(Z_m) = 0$ and $\operatorname{ad}(X)^\ell(W'_m) = 0$ for all $\ell > s$. In particular, we know that $\operatorname{ad}(X)^n(W'_{nj-1}) \in \mathfrak{p}$. Hence, $\operatorname{ad}(X)^n(X - \operatorname{Ad}(g)(X)) \in \mathfrak{p}$ implies that $\operatorname{ad}(X)^n(X - \operatorname{Ad}(g)(X) - W'_{nj-1}) \in \mathfrak{p}$. Now modulo $\mathfrak{g}^{nj}$ (which is automatically mapped to $\mathfrak{p}$ by $\operatorname{ad}(X)^n$), the element $X - \operatorname{Ad}(g)(X) - W'_{n-1}$ is congruent to

$$[Z_{nj}, X] + \cdots + [Z_{(n+1)j-1}, X].$$

Since the individual summands lie in different grading components, we conclude that $\operatorname{ad}(X)^{n+1}(Z_m) = 0$ for $m < (n+1)j$. To prove that $\operatorname{ad}(X)^\ell(W'_{nj}) \in \mathfrak{p}$ for all $\ell > n$, it suffices to prove the same fact for $W'_{nj} - W'_{nj-1}$. But this can be written as

$$\sum_{i_j,\dots,i_{nj}} \tfrac{1}{i_j!\cdots i_{nj}!} \operatorname{ad}(Z_j)^{i_j} \circ \cdots \circ \operatorname{ad}(Z_{nj})^{i_{nj}}(X),$$

with the sum over $i_{nj} > 0$ and $ji_j + (j+1)i_{j+1} + \cdots + nji_{nj} \leq k + j$. Summands for which $ji_j + (j+1)i_{j+1} + \cdots + nji_{nj} > \ell j$ are automatically mapped to $\mathfrak{p}$ by $\operatorname{ad}(X)^\ell$, so we may ignore them. Now let $\alpha(m) \in \mathbb{N}$ be the unique number such that $\alpha(m)j \leq m < (\alpha(m)+1)j$. Then we conclude that $\alpha(j)i_j + \alpha(j+1)i_{j+1} + \cdots + \alpha(nj)i_{nj} \leq \ell$. Applying the lemma inductively, we conclude that

$$\begin{aligned}&\operatorname{ad}(X)^\ell \circ \operatorname{ad}(Z_j)^{i_j} \circ \cdots \circ \operatorname{ad}(Z_{nj})^{i_{nj}} \\ &\quad = \psi \circ \operatorname{ad}(X)^{\ell - \alpha(j)i_j - \cdots - \alpha(nj-1)i_{nj} - n(i_{nj}-1)} \circ \operatorname{ad}(Z_{n_j}),\end{aligned}$$

for some linear map ψ. Applying this to X, we get zero as before, since

$$\ell - \alpha(j)i_j - \cdots - \alpha(nj-1)i_{nj} - n(i_{nj} - 1) + 1 > n.$$

Having proved this, one concludes inductively in the same way that $\operatorname{ad}(X)^\ell(W'_m) = 0$ for $\ell > n$ and $m < (n+1)j$. □

5.3.4. Parametrizations of homogeneous curves. As we saw already in Example 5.3.1, there may be canonical curves which only differ by their parametrizations. Correspondingly, any canonical curve comes with a distinguished class of parametrizations. These can be studied by a direct computation in the spirit of 5.3.2, as worked out in [**ČSŽ03**]. This leads to the conclusion, that the possibilities for reparametrizations are very limited. Since the reason for that is much more general (and thus simpler), we take a small detour at this point, following [**EaSl04**] and [**Do05**].

Let us for a while consider an arbitrary homogeneous space, i.e. a manifold M with smooth transitive left action $\lambda : G \times M \to M$ of a Lie group G. Then there is the infinitesimal action of the Lie algebra $\mathfrak{g}$ of G, given by the derivative $\lambda' : \mathfrak{g} \to \mathcal{X}(M)$, $\lambda'(X)(x) = \frac{d}{dt}(\exp(-tX)x)|_{t=0}$. Given a connected unparametrized curve, i.e. an immersed 1–dimensional submanifold $C \subset M$, we define its *symmetry algebra* as

$$\mathfrak{s} = \{X \in \mathfrak{g};\ \lambda'(X)(x) \text{ is tangent to } C \text{ for all } x \in C\}.$$

By definition, $\mathfrak{s}$ is a Lie subalgebra of $\mathfrak{g}$. Moreover, since C is one–dimensional, it is itself a homogeneous space (under the restriction of the action on M) if and only if for any $x \in C$ there is an element in $\lambda'(\mathfrak{s})$ which is nonzero in x.

A canonical curve in a parabolic homogeneous spaces $p : G \to G/P$ is given as $c(t) = p(u \cdot \exp(tX)) = \exp(\mathrm{Ad}(u)(tX)) \cdot o$, so its image C is a homogeneous curve with X in its symmetry algebra. Notice that in the cases of interest, X lies in $\mathfrak{g}_-$ and hence is nilpotent in the sense that $\mathrm{ad}(X) : \mathfrak{g} \to \mathfrak{g}$ is a nilpotent map. Since $\mathrm{ad}(\mathrm{Ad}(u)(X)) = \mathrm{Ad}(u) \circ \mathrm{ad}(X) \circ \mathrm{Ad}(u)^{-1}$, also $\mathrm{Ad}(u)(X)$ is a nilpotent element of $\mathfrak{g}$, so the symmetry algebras of canonical curves always contain nilpotent elements. The image $\lambda'(\mathfrak{s}) \subset \mathfrak{X}(C)$ is a nontrivial finite–dimensional Lie subalgebra.

Fortunately, the finite–dimensional Lie algebras of vector fields on one–dimensional manifolds were already classified by Sophus Lie. For the convenience of the reader, we present an elementary proof of the local version of this result, which we will need in the sequel.

PROPOSITION 5.3.4. *Suppose $\mathfrak{s}$ is a finite-dimensional Lie subalgebra of the Lie algebra of vector fields on a neighbourhood of $0 \in \mathbb{R}$. Suppose $\mathfrak{s}$ contains a vector field that does not vanish at 0. Then there is a neighbourhood U of 0 and a change of coordinates such that one of the following three possibilities holds on U*

$$\mathfrak{s} = \mathrm{span}\left\{\tfrac{\partial}{\partial x}\right\}, \qquad \mathfrak{s} = \mathrm{span}\left\{\tfrac{\partial}{\partial x}, x\tfrac{\partial}{\partial x}\right\}, \qquad \mathfrak{s} = \mathrm{span}\left\{\tfrac{\partial}{\partial x}, x\tfrac{\partial}{\partial x}, x^2\tfrac{\partial}{\partial x}\right\}. \tag{5.28}$$

In particular, the dimension of $\mathfrak{s}$ is at most three.

PROOF. If $\dim(\mathfrak{s}) = 1$, then $\mathfrak{s}$ is spanned by a vector field which does not vanish at 0, so after a change of coordinates $\mathfrak{s} = \mathrm{span}\{\frac{\partial}{\partial x}\}$ and we are done. Next, if $\dim(\mathfrak{s}) = 2$, then $\mathfrak{s} = \mathrm{span}\{\frac{\partial}{\partial x}, g(x)\frac{\partial}{\partial x}\}$ for some nonconstant smooth function $g(x)$. Now $\left[\frac{\partial}{\partial x}, g(x)\frac{\partial}{\partial x}\right] = g'(x)\frac{\partial}{\partial x}$, so closure under the Lie bracket implies $g'(x) = \mu + \lambda g(x)$. Solving this differential equation we get

$$g(x) = \begin{cases} Ce^{\lambda x} + D & \text{if } \lambda \neq 0, \\ \mu x + C & \text{if } \lambda = 0. \end{cases}$$

This describes all two-dimensional subalgebras up to coordinate changes. However, for the local change of coordinates $y = \frac{1-e^{-\lambda x}}{\lambda}$ around zero, we get $e^{\lambda x}\frac{\partial}{\partial x} = \frac{\partial}{\partial y}$ and

$\frac{\partial}{\partial x} = (1-\lambda y)\frac{\partial}{\partial y}$, so

$$\operatorname{span}\left\{\tfrac{\partial}{\partial x}, e^{\lambda x}\tfrac{\partial}{\partial x}\right\} \cong \operatorname{span}\left\{\tfrac{\partial}{\partial y}, y\tfrac{\partial}{\partial y}\right\}.$$

In particular, we have proved the claimed classification of the algebras of dimension at most two.

So suppose $\dim(\mathfrak{s}) = k+1 \geq 3$ and choose a basis $\frac{\partial}{\partial x}, g_1(x)\frac{\partial}{\partial x}, \ldots, g_k(x)\frac{\partial}{\partial x}$ of $\mathfrak{s}$. From closure under Lie bracket by $\frac{\partial}{\partial x}$, we immediately deduce a system of ordinary differential equations with constant coefficients

$$g_i'(x) = \mu_i + \sum_{j=1}^{k} \lambda_{ij} g_j(x), \quad \text{for } i = 1, \ldots, k.$$

We may conclude that the functions $g_i(x)$ and, therefore, all vector fields in $\mathfrak{s}$ are real-analytic.

Since $\dim(\mathfrak{s}) \geq 3$, there is a vector field $g(x)\frac{\partial}{\partial x} \in \mathfrak{s}$ which vanishes to second order in zero, so

$$g(x) = x^N + aX^{N+1} + \cdots \quad \text{for some } N \geq 2.$$

Because $\mathfrak{s}$ is finite-dimensional, we may choose $g(x)$ in such a way that N is maximal. But then the vector field

$$\begin{aligned}\left[\left[\tfrac{\partial}{\partial x}, g(x)\tfrac{\partial}{\partial x}\right], g(x)\tfrac{\partial}{\partial x}\right] &= \left[g'(x)\tfrac{\partial}{\partial x}, g(x)\tfrac{\partial}{\partial x}\right] \\ &= \left((g'(x))^2 - g(x)g''(x)\right)\tfrac{\partial}{\partial x} = \left(Nx^{2(N-1)} + \cdots\right)\tfrac{\partial}{\partial x}\end{aligned}$$

lies in $\mathfrak{s}$. This contradicts maximality of N unless $N = 2$.

Therefore, $\dim(\mathfrak{s}) = 3$ and

$$\mathfrak{s} = \operatorname{span}\left\{\tfrac{\partial}{\partial x}, g(x)\tfrac{\partial}{\partial x}, g'(x)\tfrac{\partial}{\partial x}\right\}, \tag{5.29}$$

where

$$g(x) = x^2 + ax^3 + \cdots. \tag{5.30}$$

But then $\mathfrak{s}$ contains the vector field

$$\begin{aligned}\left[g'(x)\tfrac{\partial}{\partial x}, g(x)\tfrac{\partial}{\partial x}\right] - 2g(x)\tfrac{\partial}{\partial x} &= \left((g'(x))^2 - g(x)g''(x) - 2g(x)\right)\tfrac{\partial}{\partial x} \\ &= \left(2ax^3 + \cdots\right)\tfrac{\partial}{\partial x},\end{aligned}$$

which again contradicts maximality of N unless $a = 0$. Now, in order for (5.29) to be closed under Lie bracket we must have

$$g''(x)\tfrac{\partial}{\partial x} = \left[\tfrac{\partial}{\partial x}, g'(x)\tfrac{\partial}{\partial x}\right] \in \operatorname{span}\left\{\tfrac{\partial}{\partial x}, g(x)\tfrac{\partial}{\partial x}, g'(x)\tfrac{\partial}{\partial x}\right\}$$

and to be, in addition, consistent with $a = 0$ in (5.30), we conclude that $g''(x) = 2 + \nu g(x)$, for some constant ν. This differential equation, with initial conditions imposed by (5.30), has the solutions

$$g(x) = \begin{cases} (2/\lambda^2)(\cos(\lambda x) - 1) & \text{if } \nu < 0, \\ x^2 & \text{if } \nu = 0, \\ (2/\lambda^2)(\cosh(\lambda x) - 1) & \text{if } \nu > 0. \end{cases}$$

It is easy to check that (5.29) is, indeed, closed under the Lie bracket in these cases.

Now clearly there are no more than three different Lie algebras (5.29) with the above choices for g. Although these algebras are different globally, locally we may choose the coordinate change $y = \tan((\lambda x)/2)$, which gives

$$\tfrac{\partial}{\partial x} = \frac{\lambda}{2}(1+y^2)\tfrac{\partial}{\partial y},\ \sin(\lambda x)\tfrac{\partial}{\partial x} = \lambda y \tfrac{\partial}{\partial y},\ \cos(\lambda x)\tfrac{\partial}{\partial x} = \frac{\lambda}{2}(1-y^2)\tfrac{\partial}{\partial y}$$

whence

$$\operatorname{span}\left\{\tfrac{\partial}{\partial x}, \sin(\lambda x)\tfrac{\partial}{\partial x}, \cos(\lambda x)\tfrac{\partial}{\partial x}\right\} \cong \operatorname{span}\left\{\tfrac{\partial}{\partial y}, y\tfrac{\partial}{\partial y}, y^2\tfrac{\partial}{\partial y}\right\}.$$

Similarly, $y = \tanh((\lambda x)/2)$ gives

$$\tfrac{\partial}{\partial x} = \frac{\lambda}{2}(1-y^2)\tfrac{\partial}{\partial y},\ \sinh(\lambda x)\tfrac{\partial}{\partial x} = \lambda y \tfrac{\partial}{\partial y},\ \cosh(\lambda x)\tfrac{\partial}{\partial x} = \frac{\lambda}{2}(1+y^2)\tfrac{\partial}{\partial y},$$

whence

$$\operatorname{span}\left\{\tfrac{\partial}{\partial x}, \sinh(\lambda x)\tfrac{\partial}{\partial x}, \cosh(\lambda x)\tfrac{\partial}{\partial x}\right\} \cong \operatorname{span}\left\{\tfrac{\partial}{\partial y}, y\tfrac{\partial}{\partial y}, y^2\tfrac{\partial}{\partial y}\right\}$$

and the proposition is proved. □

Let us apply this result to understand the possible reparametrizations of homogeneous curves. Guided by the example of canonical curves in parabolic homogeneous spaces, we are interested in parametrizations $t \mapsto \exp(tX)\cdot o$ of a curve C for nilpotent elements X, which lie in the symmetry algebra $\mathfrak{s}$ of C. These will be called *preferred parametrizations* of C. Notice that we have not included the (evidently possible) constant shifts of the parameter by fixing the value at $t = 0$ to be o.

THEOREM 5.3.4. *Let C be an unparametrized homogeneous curve with a preferred parameter t. The freedom among the preferred parametrizations is one of the following types:*

(1) *affine, $t \mapsto at$ for $a \neq 0$,*
(2) *projective, $t \mapsto \frac{at}{bt+1}$ for $a \neq 0$ and b arbitrary.*

Moreover, the freedom is affine if and only if the dimension of the image of the symmetry algebra $\mathfrak{s}$ in $\mathfrak{X}(C)$ has dimension one or two.

PROOF. Since the generator of a fixed preferred parametrization X is a nilpotent element of $\mathfrak{g}$, it is nilpotent in $\mathfrak{s}$, and hence also $\lambda'(X) \in \lambda'(\mathfrak{s}) \subset \mathfrak{X}(C)$ is nilpotent. By inspection, we may find the nilpotent elements in each of the local forms of the symmetry algebras in the proposition above.

$$\begin{aligned} a\tfrac{\partial}{\partial x} &\in \operatorname{span}\left\{\tfrac{\partial}{\partial x}\right\}, \\ a\tfrac{\partial}{\partial x} &\in \dim\left\{\tfrac{\partial}{\partial x}, x\tfrac{\partial}{\partial x}\right\}, \\ (p-qx)^2\tfrac{\partial}{\partial x} &\in \dim\left\{\tfrac{\partial}{\partial x}, x\tfrac{\partial}{\partial x}, x^2\tfrac{\partial}{\partial x}\right\}. \end{aligned}$$

In the first two cases, $a\frac{\partial}{\partial x} = \frac{\partial}{\partial t}$ if and only if $x = at$, which gives affine freedom, whilst in the third case

$$(p-qx)^2\frac{\partial}{\partial x} = \frac{\partial}{\partial t} \iff x = \frac{p^2 t}{1+pqt},$$

which gives projective freedom. □

5.3.5. Reparametrization of canonical curves. Having understood the symmetry algebras of homogeneous curves, we may easily derive the results for canonical curves.

THEOREM 5.3.5. *Let c be a canonical curve of any type on a manifold M with a fixed parabolic geometry. Then for the reparametrizations of c which are again canonical curves of the same type we have the following two possibilities:*

(1) *affine,* $t \mapsto at$ *for* $a \neq 0$,
(2) *projective,* $t \mapsto \frac{at}{bt+1}$ *for* $a \neq 0$ *and b arbitrary.*

PROOF. Clearly, it is enough to prove the result for canonical curves on the homogeneous models. Then the parametrizations of a canonical curve have the form $t \mapsto \exp(tY) \cdot o$ where Y is P-conjugate to an element of $\mathfrak{g}_-$. Certainly, there is at least an affine freedom in such parametrizations, because Y can be replaced by aY. But the possible Y are, in particular, nilpotent elements of $\mathfrak{g}$. Therefore, the parametrizations of the image C of c as a distinguished curve are preferred parametrizations of C as a homogeneous curve. Theorem 5.3.4 then implies that, if there is any additional freedom, it must be projective. But just one projective transformation, together with the affine transformations, generates the full projective freedom and the proof is complete. □

Before we discuss various types of canonical curves in the individual parabolic geometries, we shall draw some more conclusions from the general setup of homogeneous curves in homogeneous spaces.

Let us consider a fixed element $X \in \mathfrak{g}_-$ and look for Y in the symmetry algebra $\mathfrak{s} \subset \mathfrak{g}$ of the curve $C \subset G/P$ parametrized by $c(t) = \exp(tX)P$. By definition (see 5.3.4), this amounts to the requirement that

$$\lambda'(Y)(\exp(sX)) = \tfrac{d}{dt}|_0 \exp(-tY) \cdot \exp(sX)$$

is tangent to C for all s in a neighborhood of the origin. Moving the tangent vector to the origin, we get

$$\begin{aligned} T\lambda_{\exp(-sX)}(\lambda'(Y)(\exp(sX))) &= \tfrac{d}{dt}|_0 \big(\exp(-sX)\exp(-tY)\exp(sX)\big)P \\ &= -\operatorname{Ad}(\exp(-sX))(Y) + \mathfrak{p} \in \mathfrak{g}/\mathfrak{p} = T_o(G/P). \end{aligned}$$

Thus, the condition for $Y \in \mathfrak{s}$ is simply

$$\operatorname{Ad}(\exp(-sX))(Y) = Y - s[X,Y] + \frac{1}{2}s^2[X,[X,Y]] - \cdots \in \langle X \rangle \oplus \mathfrak{p}, \tag{5.31}$$

where $\langle X \rangle$ is the line spanned by X. This provides an efficient iterative procedure to determine $\mathfrak{s}$.

LEMMA 5.3.5. *Let* $\mathfrak{a}_i$ *be the non–increasing sequence of subspaces of* $\mathfrak{p}$ *defined by*

$$\mathfrak{a}_0 = \mathfrak{p}, \qquad \mathfrak{a}_{i+1} = \{Y \in \mathfrak{a}_i : [X,Y] \subset \langle X \rangle \oplus \mathfrak{a}_i\}.$$

Then there is the smallest number r for which the sequence stabilizes, i.e. $\mathfrak{a}_r = \mathfrak{a}_{r+1}$, *and* $\mathfrak{s} = \langle X \rangle \oplus \mathfrak{a}_r$ *is the symmetry algebra of the canonical curve generated by X.*

PROOF. The sequence has to stabilize because $\mathfrak{p}$ is finite–dimensional. Since $\mathfrak{a}_0$ is a Lie subalgebra of $\mathfrak{g}$ it follows by induction that each $\mathfrak{a}_i$ is a subalgebra of $\mathfrak{g}$, whence $\langle X \rangle \oplus \mathfrak{a}_r$ is a subalgebra of $\mathfrak{g}$ by construction.

To prove that this equals $\mathfrak{s}$, assume first that $Y \in \mathfrak{s}$. If (5.31) is satisfied for s close to zero, then $\operatorname{ad}(X)^k(Y) \in \langle X \rangle \oplus \mathfrak{p}$ for all $k \geq 0$. For $k = 0$, this means that $Y = \alpha_0 X + Y_0$ for some $Y_0 \in \mathfrak{a}_0 = \mathfrak{p}$. But then $\text{ad}(X)^k(Y) = \text{ad}(X)^k(Y_0)$ for all $k > 1$, so Y_0 also satisfies (5.31), and it suffices to show that $Y_0 \in \mathfrak{a}_r$. Now (5.31) for Y_0 and $k = 1$ says that $Y_0 \in \mathfrak{a}_1$, so $[X, Y_0] = \alpha_1 X + Y_1$ with $Y_1 \in \mathfrak{a}_0$. Then $[X, [X, Y_0]] = [X, Y_1] \in \langle X \rangle \oplus \mathfrak{a}_0$ shows that $Y_1 \in \mathfrak{a}_1$. Thus, $Y_0 \in \mathfrak{a}_2$, and inductively we arrive at $Y_0 \in \mathfrak{a}_i$ for all i.

Conversely, for $Y \in \mathfrak{a}_r$ we evidently have $\text{ad}(X)^k(Y) \in \langle X \rangle \oplus \mathfrak{p}$ for all k, so Y satisfies (5.31). Thus $\mathfrak{a}_r \subset \mathfrak{s}$, which completes the proof. □

REMARK 5.3.5. As a homogeneous space, the curve C is locally described by the pair $(\mathfrak{s}, \mathfrak{a}_r)$, or its effective quotient $(\mathfrak{s}/\mathfrak{m}, \mathfrak{a}_r/\mathfrak{m})$. Here $\mathfrak{m}$ is the maximal ideal of $\mathfrak{s}$ contained in $\mathfrak{a}_r$. By Theorems 5.3.4 and 5.3.5, the reparametrization freedom on C is projective if and only if $\dim(\mathfrak{s}/\mathfrak{m}) = 3$, while the dimension is one or two for the affine case. Since we have noted above that $\text{ad}(X)$ will be nilpotent in $\mathfrak{s}$ and hence in $\mathfrak{s}/\mathfrak{m}$ we see that if $\text{ad}(X)^2 \neq 0$ on $\mathfrak{s}/\mathfrak{m}$, then $\mathfrak{s}$ must be three–dimensional.

Notice also that the algebra $\mathfrak{a}_r$ describes the vector fields in the symmetry algebra, which vanish at the origin. By the arguments from the proof, we may conclude that the smallest number r from the lemma provides the smallest order of the contact element by which the unparametrized homogeneous curve C is determined.

The lemma is also in the background of the following result proved in [**Do05**].

PROPOSITION 5.3.5. *(1) Let $X \in \mathfrak{g}$ be an element not contained in $\mathfrak{p}$, and suppose that there is $Z \in \mathfrak{p}$ such that $H := [Z, X] \in \mathfrak{p}$ and $[X, [X, Z]] = 2X$. Then there is an element $Y \in \mathfrak{p}$ completing X and H to an $\mathfrak{sl}(2, \mathbb{R})$–triple. Moreover, the curve $\exp(tX) \cdot o \subset G/P$ admits a projective family of preferred parametrizations.*

(2) Let $X \in \mathfrak{g}_{-i}$ be a homogeneous element of $\mathfrak{g}_-$. Then there exists an $\mathfrak{sl}(2, \mathbb{R})$–triple (X, H, Y) satisfying the conditions in (1) with $H \in \mathfrak{g}_0$ and $Y \in \mathfrak{g}_i$. In particular, the canonical curves of type $\mathcal{C}_A$ coming from a subset $A_0 \subset \mathfrak{g}_-$ which is contained in one grading component, always admit projective reparametrizations.

PROOF. (1) By assumption $[H, X] = 2X$, so the span $\mathfrak{b} = \langle H, X \rangle$ is a solvable Lie subalgebra of $\mathfrak{g}$, and X spans the derived algebra $[\mathfrak{b}, \mathfrak{b}]$. Thus, Lie's theorem (see 2.1.1) applied to the restriction of the adjoint action of $\mathfrak{b}$ on $\mathfrak{g}$ implies that X is a nilpotent element of $\mathfrak{g}$.

We want to modify Z to an element $Y \in \mathfrak{p}$ without changing the bracket $[Z, X]$, so we consider the kernel $\mathfrak{n}$ of $\text{ad}(X)$ and, in particular, the subspace $\mathfrak{n}_0 = \mathfrak{n} \cap \mathfrak{p}$. Clearly, $[\text{ad}(X), \text{ad}(H)] = -2\,\text{ad}(X)$, so both $\mathfrak{n}$ and $\mathfrak{n}_0$ are $\text{ad}(H)$ invariant.

Claim. The restrictions of the linear mapping $\text{ad}(H)+2\text{id}$ to $\mathfrak{n}$ and $\mathfrak{n}_0$ are invertible.

To verify this, let us consider $V_n = \text{ad}(X)^n(\mathfrak{g})$ for $n \geq 0$. By construction $[\text{ad}(Z), \text{ad}(X)] = \text{ad}(H)$ and by a straightforward induction using $[\text{ad}(H), \text{ad}(X)] = 2\,\text{ad}(X)$, this implies that

$$[\text{ad}(Z), \text{ad}(X)^n] = n(\text{ad}(H) - (n-1)\text{id})\,\text{ad}(X)^{n-1}.$$

Consequently, for each $W \in V_{n-1}$ the element $n[H, W] - n(n-1)W$ is congruent to $[Z, [X, W]]$ modulo V_n. Moreover, $\mathfrak{n}$ is invariant with respect to $\text{ad}(H)$, and so

$$(\text{ad}(H) - (n-1)\text{id})(\mathfrak{n} \cap V_{n-1}) \subset \mathfrak{n} \cap V_n.$$

Now X is nilpotent, and therefore $V_n = 0$ for sufficiently large n. For such a number n, we conclude that $(\text{ad}(H) - (n-1)\text{id}) \circ \cdots \circ (\text{ad}(H) - \text{id}) \circ \text{ad}(H)$ is nilpotent on $\mathfrak{n}$, so all eigenvalues of $\text{ad}(H)|_{\mathfrak{n}}$ are nonnegative integers. In particular, this implies the claim.

Next, observe that $[H,Z]+2Z \in \mathfrak{n}_0$ by assumption. By the claim, there is an element $Y' \in \mathfrak{n}_0$ such that $[H,Z]+2Z = [H,Y']+2Y'$. But now $Y = Z - Y'$ is in $\mathfrak{p}$, $[Y,X] = [Z,X] = H$, and $[H,Y] = [H,Z-Y'] = 2(Y'-Z) = -2Y$. This proves the first part of (1).

To discuss the preferred parametrizations of the homogeneous curve $\exp tX \cdot o$, observe that both H and Y belong to the symmetry algebra $\mathfrak{s}$; see the condition (5.31). Since the ideal $\mathfrak{m}$ is contained in $\mathfrak{a}_r$, we see that it cannot contain X. Hence, its intersection with the subalgebra spanned by X, H and Y must be zero since otherwise it would be a proper ideal in $\mathfrak{sl}(2,\mathbb{R})$. This shows that $\dim(\mathfrak{s}/\mathfrak{m}) \geq 3$ and the rest of (1) follows from Theorem 5.3.4.

(2) Since $X \in \mathfrak{g}_{-i}$, it is a nilpotent element in $\mathfrak{g}$. There is the well-known Jacobson–Morozov theorem (see subsection X.2 in [**Kn96**]), which says that every nilpotent element X in a semisimple Lie algebra over a field $\mathbb{K}$ of characteristic zero can be completed to an $\mathfrak{sl}(2,\mathbb{K})$ triple (X,H',Z'). Given the corresponding embedding of $\mathfrak{sl}(2,\mathbb{R})$, we will adjust it to a triple (X,H,Z) for which $H \in \mathfrak{g}_0$ and $Z \in \mathfrak{g}_i$.

Let us write $H' = \sum_{j=-k}^{k} H'_j$ and $Z' = \sum_{j=-k}^{k} Z'_j$ for the decompositions into homogeneous components. By homogeneity of X, we obtain $[H'_0, X] = 2X$ and $[X, Z'_i] = H'_0$. Thus, putting $H = H'_0$ and $Z = Z'_i$ we arrive at the assumptions for (1). Thus, the projective reparametrization freedom has been proved. A direct inspection of the construction in the proof of (1) shows that the computed adjustment Y of Z'_i also lies in $\mathfrak{g}_i$. □

In order to distinguish the projective parameters we need the second order jet of the curve. Thus, taking together the above proposition and Theorem 5.3.3, we obtain

COROLLARY 5.3.5. *Suppose M is equipped with a parabolic geometry of type (G,P) with $|k|$-graded Lie algebra $\mathfrak{g}$ and suppose $A_0 \subset \mathfrak{g}_{-k}$ is a G_0-invariant subset. Then the order determining the canonical curves of type $A = \mathrm{Ad}(P)(A_0)$ is two.*

5.3.6. Examples corresponding to $|1|$-gradings. By Theorem 5.3.3, the parametrized canonical curves of all types in geometries corresponding to $|1|$-gradings are determined by their two-jet in one point. Moreover, they all admit projective reparametrizations by Proposition 5.3.5, and so this estimate is sharp; see Corollary 5.3.5. Still, there is an interesting diversity in the behavior of the various types of canonical curves in the individual geometries.

Conformal geometry. As a warm up, let us reconsider the canonical curves for n-dimensional conformal structures, $n \geq 3$, which we have discussed already in 5.3.1. At this point, we also establish the procedure to be followed in the next examples.

Step (A): First we determine the geometrically different types of directions. In the positive definite case, the G_0-action on $\mathfrak{g}_{-1} = \mathbb{R}^n$ is just the standard representation of $CO(n,\mathbb{R})$. Hence, every nonzero vector $X \in \mathfrak{g}_{-1}$ generates the entire set A_0 of nonzero elements in $\mathbb{R}^n = \mathfrak{g}/\mathfrak{p}$ and all directions are equivalent. In indefinite signature, we have three different G_0-orbits in $\mathfrak{g}/\mathfrak{p}$ corresponding to positive, negative, and isotropic vectors, respectively.

Step (B): Next, we discuss how many parametrized canonical curves emanate from a point with given initial tangent vector, and deduce conditions on jets which make sure that two curves of the form $c(t) = \exp(Z) \cdot \exp(tY)P \in$

G/P with $Z \in \mathfrak{p}$, $Y \in A_0$ coincide. Without loss of generality, we may fix any $X \in A_0$, fix the curve $c_1(t) = \exp(tX)P \in G/P$ and look for conditions on $c_2(t) = \exp(t\operatorname{Ad}(\exp(Z))(Y))P$ with $Z \in \mathfrak{p}_+$ and $Y \in A_0$. This was done by means of the logarithmic derivative δu from Lemma 5.3.2 in the proof of Theorem 5.3.2. In particular, formula (5.25) is the most useful tool. Indeed, in the $|1|$–graded cases

$$\operatorname{Ad}(\exp(Z))(Y) = Y + [Z,Y] + \frac{1}{2}[Z,[Z,Y]] \tag{5.32}$$

and so the condition that c_1 and c_2 share the same tangent vector at the origin is $\delta u(0) = X - \operatorname{Ad}(\exp(Z))(Y) \in \mathfrak{p}$, i.e. $Y = X$.

Next, $\delta u'(0) = -[X, \delta u(0)] \in \mathfrak{p}$ means $[X,[Z,X]] \in \mathfrak{p}$ and so this has to vanish. At the same time we know that, if $[X,[Z,X]]$ is a nonzero multiple of X, then there is a $Z' \in \mathfrak{g}_1$ such that the corresponding curve corresponds to a reparametrization. But then the curve corresponding to Z must be a reparametrization of $c_1(t)$, too. This very nicely corresponds to the intuitive expectation that $[X,[Z,X]]$ describes the change of the acceleration of the curve at the origin.

From the description of the brackets in formula (1.29) in 1.6.3, we see that for $X \in \mathfrak{g}_{-1} \cong \mathbb{R}^n$ and $Z \in \mathfrak{g}_1 \cong \mathbb{R}^{n*}$ we obtain $[X,[Z,X]] = 2XZX - \langle X,X\rangle Z^t$, where the transpose is taken with respect to the indefinite inner product. This clearly demonstrates the difference between isotropic and non–isotropic directions. Namely, if $\langle X,X\rangle = 0$, then the curve $c_2(t)$ coincides with $c_1(t)$ if and only if $ZX = 0$, while the remaining values of Z lead only to reparametrizations of the same curve. In all other directions, $[X,[Z,X]] = 0$ is only possible if Z^t is a multiple of X.

Step (C): Finally, we determine the size of the family of curves emanating in the given direction. It suffices again to analyze the linear mapping $Z \mapsto 2XZX - \langle X,X\rangle Z^t$. We have noted already that this is injective if $\langle X,X\rangle \neq 0$. Hence, for a nonzero element Z, we always have $c_2 \neq c_1$ as parametrized curves, while a one–dimensional subspace of $\mathfrak{g}_1$ leads to reparametrizations of c_1. Summarizing, there is a $(n-1)$–parameter family of unparametrized canonical curves in every non–isotropic direction. Each of them carries a distinguished projective family of parametrizations. There is just one unparametrized canonical curve in every isotropic direction.

By definition, these curves $c(t)$ in null directions coincide with the isotropic geodesics of arbitrary Weyl connections as unparametrized curves; see formula (5.1) in 5.2.4. Of course, the null geodesics of the normal Weyl connections also provide the distinguished parameters.

Almost Grassmannian structures. The discussion is very similar to the previous example. Let us fix two positive integers p, q and consider the corresponding group $SL(p+q,\mathbb{R})$ and its Lie algebra as discussed in 4.1.3.

Step (A): The adjoint action of $G_0 = S(GL(p,\mathbb{R}) \times GL(q,\mathbb{R}))$ on $\mathfrak{g}_{-1} = L(\mathbb{R}^p,\mathbb{R}^q)$ is given by the standard action on linear maps, so its orbits are classified by the ranks of matrices $X \in \mathfrak{g}_{-1}$. Thus, we have got as many geometrically different directions in the tangent bundle as ranks of matrices of size p times q.

Step (B): The computation is the same as in the conformal case, except this time the bracket $[X,[Z,X]]$ is given by

$$\left[\begin{pmatrix} 0 & 0 \\ X & 0 \end{pmatrix}, \left[\begin{pmatrix} 0 & Z \\ 0 & 0 \end{pmatrix}, \begin{pmatrix} 0 & 0 \\ X & 0 \end{pmatrix}\right]\right] = \begin{pmatrix} 0 & 0 \\ 2XZX & 0 \end{pmatrix}.$$

Hence, we have to understand the mapping $\psi_X : \mathfrak{g}_1 \to \mathfrak{g}_{-1}$, $Z \mapsto \frac{1}{2}(\mathrm{ad}_X)^2 Z = XZX$. Clearly, this corresponds to the composition of linear mappings $\mathbb{R}^p \to \mathbb{R}^q \to \mathbb{R}^p \to \mathbb{R}^q$ and therefore

$$\mathrm{ad}(X)^2(\mathfrak{g}_1) \cong L(\mathbb{R}^p/\ker(X), \mathrm{im}(X)).$$

If the rank of X is r, then the dimension of both domain and target of the above space of mappings is r. Thus, we obtain an r^2–parameter family of different parametrized canonical curves with the same tangent vector.

Step (C): The curves described by the target of the mapping ψ_X will produce the same unparametrized curve if and only if the image is in the one–dimensional subspace spanned by X. Thus, there is a $(r^2 - 1)$–parameter family of unparametrized curves emanating in directions corresponding to a matrix X of rank r.

In particular, there is always the subclass of curves in the direction of decomposable tensors, i.e. matrices X of rank 1. In this case, the unparametrized canonical curve is determined by the direction itself.

This is consistent with the results for four–dimensional split signature conformal structures (compare with 4.1.4). Null directions correpond to elements of rank one and give rise to a unique unparametrized curve. Non–isotropic directions correspond to metrices of rank 2 and generate a 3–parameter family of unparametrized canonical curves.

Almost quaternionic geometries. These geometries were discussed in detail in 4.1.8. They correspond to the Lie algebra $\mathfrak{sl}(n+1, \mathbb{H})$ which has the same complexification as the algebra for Grassmannian structures with $p = 2$, so the discussion is parallel to the Grassmannian case. We are concerned with the action of $G_0 = S(\mathbb{H}^* \times GL(n, \mathbb{H}))$ on $\mathfrak{g}_{-1} \cong \mathbb{H}^n$, so there is only one nontrivial G_0–orbit and all directions are geometrically equivalent.

As before $[X, [Z, X]] = XZX$, but now ZX has to be interpreted as the result of the $\mathbb{H}$–valued pairing between $\mathbb{H}^n$ and its dual. In particular, $\mathrm{ad}(X)^2(\mathfrak{g}_1)$ is the quaternionic line spanned by X, and hence has real dimension 4, so we obtain a 3–parameter family of unparametrized canonical curves in any given direction. Again this is consistent with the results for four–dimensional conformal structures in definite signature, which are a special case by 4.1.9.

The formula for the change of covariant derivative of Weyl connections in 5.1.6, specialized to the case that one differentiates a vector field ξ in direction ξ, reads as

$$\hat{\nabla}_\xi \xi = \nabla_\xi \xi + \{\{\Upsilon, \xi\}, \xi\}.$$

We have just observed that $[[Z, X], X]$ always lies in the quaternionic line spanned by X, which of course implies that $\{\{\Upsilon, \xi\}, \xi\}(x)$ always lies in the quaternionic line spanned by $\xi(x)$ (which is well–defined in a prequaternionic vector space). In particular, if a curve c has the property that $\nabla_{c'(t)} c'(t)$ lies in the quaternionic line generated by $c'(t)$ for one Weyl connection, then the same holds for all Weyl connections. These curves were studied in the literature under the name *quaternionic curves*. In this context, it is interesting that the quaternionic curves on an almost quaternionic manifold M are exactly the geodesics of all Weyl connections on M (as parametrized curves). This is almost obvious from the above formula, since we just have to check, that there are enough one–forms to make any quaternionic curve a geodesic by deforming a given Weyl connection. Some consequences of this observation were discussed in [**HS06**]. In particular, in dimensions ≥ 8, a diffeomorphism is a morphism of almost quaternionic manifolds if and only if it preserves the class of unparametrized quaternionic curves, i.e. the class of all unparametrized geodesics of all Weyl connections.

Lagrangean and spinorial geometries. These two geometries discussed in 4.1.11 and 4.1.12 are analogous to the Grassmannian case up to some minor adjustments. In the Lagrangian case, the tangent space is identified with symmetric matrices and so the above reasoning applies with $p = q$. If the rank of X is $r = 1, \dots, p$, then the dimension of the image of $\mathrm{ad}(X)^2(\mathfrak{g}_{-1})$ equals the dimension of the space of all symmetric matrices of size r. Indeed, we have to consider only the restriction of the self–adjoint maps $\mathbb{R}^p \to \mathbb{R}^p$ to the image of X. Thus, there are $\frac{1}{2}r(r+1)$ different parametrized curves sharing the same tangent vector X. Again, in direction of a decomposable tensor X, there is a unique unparametrized canonical curve.

Dealing with spinorial geometries, we have to replace symmetric by skew symmetric matrices. Since there are no skew symmetric matrices of odd rank, the available ranks r for the directions X are all even numbers $r \leq p$. For the same reason as above, the image of $\mathrm{ad}(X)^2(\mathfrak{g}_1)$ corresponds to all skew symmetric mappings on a space of dimension r, thus giving $\frac{1}{2}(r-1)r$ parameters. For the lowest possible rank 2, we again obtain a uniquely determined canonical unparametrized curve in the given direction.

5.3.7. Chains in parabolic contact geometries. The diversity of the behavior of canonical curves increases quickly with the length of the grading. Before we illustrate this explicitly by the simple example of Lagrangean contact structures, we introduce a particularly nice class of canonical curves, which are available for all parabolic contact structures. They behave similarly to the null–geodesics in conformal pseudo–Riemannian geometry, and have been studied intensively in the case of CR–structures under the name "chains". We use the same name and concept in general.

DEFINITION 5.3.7. For all parabolic contact geometries, the canonical curves corresponding to the G_0–invariant subset $A_0 = \mathfrak{g}_{-2} \subset \mathfrak{g}_-$ are called *chains*.

THEOREM 5.3.7. *Let M be an arbitrary manifold equipped with a parabolic contact geometry. Then there is exactly one unparametrized chain in each direction transverse to the contact distribution. Each chain comes equipped with a distinguished projective family of parametrizations.*

PROOF. We shall see that for the choice $A_0 = \mathfrak{g}_{-2}$, the procedure from 5.3.6 becomes very similar to the $|1|$–graded case. In particular, Proposition 5.3.5 applies and the preferred parameters are always projective. Now consider a contact grading on $\mathfrak{g}$ and corresponding groups $G_0 \subset P \subset G$, and let use determine the subset $A \subset \mathfrak{g}/\mathfrak{p}$ induced by A_0. By Theorem 3.1.3, any element of P_+ can be uniquely written in the form $\exp(Z_1)\exp(Z_2)$ with $Z_i \in \mathfrak{g}_i$ for $i = 1, 2$. Since we are dealing with a $|2|$–grading, $\exp(Z_2)$ acts trivially on $\mathfrak{g}/\mathfrak{p}$. On the other hand, the action of $\exp(Z_1)$ maps $X \in \mathfrak{g}_{-2}$ to $X + [Z, X]$. Similarly, as in the proof of Lemma 4.2.2, nondegeneracy of the bracket $\mathfrak{g}_1 \times \mathfrak{g}_1 \to \mathfrak{g}_2$ implies that for a nonzero element $X \in \mathfrak{g}_{-2}$, the map $\mathrm{ad}(X) : \mathfrak{g}_1 \to \mathfrak{g}_{-1}$ is a linear isomorphism. Hence, A is the whole complement of the P–invariant subspace $\mathfrak{g}^{-1}/\mathfrak{p}$, so chains are available in all directions transverse to the contact distribution.

Next, by Corollary 5.3.5 we know that the order of the jet determining each chain as a parametrized curve is two. Let us fix $X \in \mathfrak{g}_{-2}$, $Z = Z_1 + Z_2 \in \mathfrak{g}_1 \oplus \mathfrak{g}_2$, and $Y \in \mathfrak{g}_{-2}$. We are going to compare the curves $c_1(t) = \exp(tX)P$ and $c_2(t) = \exp(t \operatorname{Ad}(\exp(Z))(Y))P$ in G/P exactly as in 5.3.6. Due to our special choice, formula (5.32) gets only slightly more complicated than the one for $|1|$–graded geometries:

$$\operatorname{Ad}(\exp(Z))(Y) = Y + [Z_1, Y] + [Z_2, Y] + \frac{1}{2}[Z_1 + Z_2, [Z_1 + Z_2, Y]] \mod \mathfrak{p}_+.$$

The condition $\delta u(0) = X - \operatorname{Ad}(\exp(Z))(Y) \in \mathfrak{p}$ evidently implies $Y = X$. Moreover, the $\mathfrak{g}_{-1}$–component of $X - \operatorname{Ad}(\exp(Z))(Y)$ equals $-[Z_1, Y]$. From above we know that $\operatorname{ad}(Y) : \mathfrak{g}_1 \to \mathfrak{g}_{-1}$ is injective, so we also get $Z_1 = 0$ and hence $Z \in \mathfrak{g}_2$. Then the above expression for $\operatorname{Ad}(\exp(Z))(X)$ gets the same form as (5.32), and we may continue exactly as in 5.3.6.

The final condition is $\delta u'(0) = -[X, \delta u(0)] \in \mathfrak{p}$, which implies $[X, [Z, X]] \in \mathfrak{p}$. Since we already know that $Z \in \mathfrak{g}_2$, this bracket lies in $\mathfrak{g}_{-2}$ and hence is always a multiple of X. Thus, the curves c_1 and c_2 coincide only for $Z = 0$, while for $Z \neq 0$, we get two parametrizations of the same curve. □

Recall that the Weyl connections of the normal Weyl structures introduced in 5.1.12 form a distinguished class of affine connections on our manifold M. In the normal coordinates $\phi : \mathfrak{g}_{-2} \oplus \mathfrak{g}_{-1} \to M$ on M induced by any of these connections, the line defined by $\mathfrak{g}_{-2} \times \{0\}$ is a chain.

5.3.8. Canonical curves for Lagrangean contact structures. As an illustration of general features, we shall now treat in detail the example of Lagrangean contact geometries. We shall proceed step by step as in 5.3.6.

Step (A): Let us consider a manifold M of dimension $2n+1 \geq 3$ with contact distribution $H = E \oplus F$, where both E and F are Lagrangean subspaces in H of dimension n. The splitting of H is conveniently expressed by the involutive automorphism $\mathbb{J}$ on H, which acts as the identity on E and as minus the identity on F. Then $\mathcal{L} \circ (\mathrm{id}_H \times \mathbb{J})$ clearly is a nondegenerate symmetric bilinear bundle map $H \times H \to TM/H$. We shall call vectors in H *isotropic*, if they are isotropic with respect to this bilinear form, and *non–isotropic* otherwise. Correspondingly, we will talk about isotropic and non–isotropic directions in H. Of course, all vectors in E and F are isotropic.

We shall use the same notation for the elements in the graded Lie algebra

$$\begin{pmatrix} a & Z & \gamma \\ X & A & W \\ \beta & Y & b \end{pmatrix} \in \mathfrak{g} = \mathfrak{sl}(n+2, \mathbb{R})$$

as in 5.2.14. In particular, the elements of $\mathfrak{g}_-$ are the lower triangular matrices and we shall write X, Y, β, and so on, for the elements of $\mathfrak{g}$ completed by zeros at all other positions in the matrix. At the same time, the one–dimensional values β and γ will also be used as scalar multiples, but this will always be clear from the context. The concatenation of symbols means either matrix multiplications or tensor product if multiplication is not possible.

A straightforward check shows that the following is the complete list of orbits of the adjoint action of G_0 on $\mathfrak{g}_-$. For each orbit, we also indicate for which directions in the tangent spaces representatives of the corresponding type are available.

(1) $X \in \mathfrak{g}_{-1}$ corresponding to the directions in the distinguished subspace E in the contact distribution $H \subset TM$.
(2) $Y \in \mathfrak{g}_{-1}$ corresponding to the directions in the subspace F in the contact distribution $H \subset TM$.
(3) $X+Y \in \mathfrak{g}_{-1}$ with $YX = 0$ and $X \neq 0$, $Y \neq 0$, corresponding to isotropic directions in H, which are neither in E nor in F.
(4) $X+Y \in \mathfrak{g}_{-1}$ with $YX \neq 0$ corresponding to non–isotropic directions in $H \subset TM$.
(5) $\beta \in \mathfrak{g}_{-2}$, available for all directions complementary to H. This corresponds to the chains from 5.3.7.
(6) $\beta + X \in \mathfrak{g}_-$ with $X \neq 0$, available for all directions complementary to H.
(7) $\beta + Y \in \mathfrak{g}_-$ with $Y \neq 0$, available for all directions complementary to H.
(8) $\beta + X + Y \in \mathfrak{g}_-$ with $Y \neq 0$, $X \neq 0$ but $YX = 0$, available for all directions complementary to H.
(9) $\beta + X + Y \in \mathfrak{g}_-$ with $YX \neq 0$, available for all directions complementary to H.

The first four orbits determine geometrically distinct directions, while for all other orbits representatives are available for any direction outside the contact distribution. The first five orbits give rise to different types of canonical curves, but they all correspond to G_0–invariant subsets in one grading component, so the strong estimates on the orders from Theorem 5.3.3 apply.

Next, it is easy to see that the $\underline{\mathrm{Ad}}(P)$–orbit of $\mathfrak{g}_{-2}$ contains the G_0–orbits listed in (6), (7), and (8). Indeed, these three orbits are generated by elements $\beta + X + Y$ such that $YX = 0$. Now assuming that $YX = 0$, we compute

$$\mathrm{Ad}\left(\exp\begin{pmatrix} 0 & -Y & 0 \\ 0 & 0 & X \\ 0 & 0 & 0 \end{pmatrix}\right)\begin{pmatrix} 0 & 0 & 0 \\ 0 & 0 & 0 \\ 1 & 0 & 0 \end{pmatrix} = \begin{pmatrix} 0 & 0 & 0 \\ X & 0 & 0 \\ 1 & Y & 0 \end{pmatrix},$$

since the matrix in the exponential has zero square. Hence, taking A_0 to be one of the orbits from (6), (7), and (8), we always obtain the same set A, and hence the chains as for orbit (5).

The last orbit (9) in our list is the first example of a subset A_0 which is not contained in one of the grading components. As we shall see, the resulting canonical curves show completely different behavior than chains.

Steps (B) and (C): Next we check the jets of the curves of the individual types emanating in a fixed direction. Let us first formulate the result.

PROPOSITION 5.3.8. *Let M be a manifold of dimension $2n+1$ endowed with a Lagrangean contact structure $E \oplus F = H \subset TM$. Then apart from the chains, there are the following types of (parametrized) canonical curves with a fixed tangent vector $\xi \in TM$.*

(a) If ξ is in E or F, then the canonical curve is uniquely determined by its two–jet in a point. There is just one unparametrized canonical curve in direction $\langle\xi\rangle$, which carries a natural projective family of parametrizations and remains being tangent to the subbundle in question.

(b) If ξ is an isotropic vector not contained in E or F, then the canonical curve is determined uniquely by its two–jet in a point. There is a one–parameter family of unparametrized canonical curves in direction $\langle\xi\rangle$. Each of these canonical curves

carries a projective family of parametrizations, and its tangent direction remains isotropic along the curve.

(c) If $\xi \in H$ is not isotropic, then the canonical curve is determined by its three–jet in a point. There is a $2n$–parameter family of unparametrized canonical curves in direction $\langle\xi\rangle$. All these curves carry a natural projective family of parametrizations.

(d) If $\xi \notin H$, then there is an additional $(2n+1)$–parameter family of unparametrized canonical curves in direction $\langle\xi\rangle$, all of which are distinct from the chain in direction $\langle\xi\rangle$. They carry only an affine family of parametrizations. These canonical curves are determined by their three–jet in a point.

PROOF. It follows from the discussion of the orbits above, that we have exactly these four classes of distinguished curves and the chains. Proposition 5.3.5 shows that for the types (a)–(c), we will always have a projective family of preferred parametrizations. By Theorem 5.3.3, curves of these three types are determined by their three–jet in one point. For type (d) we only have the rough estimate from Corollary 5.3.2, which shows that they are determined by their 4–jet in a point. Let us also notice that the description of the canonical curves as projections of flow lines of constant vector fields on the Cartan bundle $\mathcal{G}$ (see Proposition 5.3.1) implies directly our claims that the type of the tangent vector is preserved along the curve.

Let us fix X, Y, and β, and put $\Xi = X + Y + \beta \in \mathfrak{g}_-$. Let Ψ be another element with components $\bar{X}$, $\bar{Y}$ and $\bar{\beta}$, put $\Phi = Z + W + \gamma \in \mathfrak{p}_+$, and consider the curves

$$c_1(t) = \exp(t\Xi)P, \qquad c_2(t) = \exp(t\,\mathrm{Ad}(\exp(\Phi))(\Psi))P$$

in the homogeneous model G/P. The condition for first order contact of the curves is easy to express. The only component of $\delta u(0) = \Xi - \mathrm{Ad}(\exp(\Phi)(\Psi))$ in $\mathfrak{g}_-$ is given by

$$\beta - \bar{\beta} + X - \bar{X} - \bar{\beta}W' + Y - \bar{Y} - \bar{\beta}Z', \tag{5.33}$$

where we add a prime to the name of an element of $\mathfrak{g}_{\pm 1}$ to indicate its bracket with the matrix in $\mathfrak{g}_{\mp 2}$ whose unique nonzero entry equals 1. For initial directions in H, i.e. for the orbits (1)–(4) of the above list, we have $\beta = 0$, so $\delta u(0) \in \mathfrak{p}$ implies $\bar{\beta} = 0$ and then $\bar{X} = X$ and $\bar{Y} = Y$. In the remaining two cases, (5) and (9) in our list, $\delta u(0) \in \mathfrak{p}$ is equivalent to $\bar{\beta} = \beta$ and the restrictions $\beta Z' = Y - \bar{Y}$, $\beta W' = X - \bar{X}$ on Φ.

Curves of type (a). These correspond to the orbits (1) and (2) in the above list. Assume $\beta = 0$ and $Y = 0$, and recall the simple check for higher order jets deduced in (5.25) in the proof of Theorem 5.3.2. The condition that two curves c_1 and c_2 have the same two–jet in zero is $[X, \mathrm{Ad}(\exp(\Phi)\Psi)] \in \mathfrak{p}$. Since we already know that $\bar{X} = X$, $\bar{\beta} = 0$, and $\bar{Y} = 0$, this reduces to $[X,[Z,X]] = 0$. Using this, one verifies that, modulo elements in $\mathfrak{p}_+$,

$$[X, \mathrm{Ad}(\exp(\Phi)\Psi)] = \frac{1}{2}[X,[Z+W+\gamma,[Z,X]+\gamma X']] = \frac{1}{2}[X,[Z,[Z,X]]] = 0.$$

Therefore, $\mathrm{ad}(X)^2(\mathrm{Ad}(\exp(\Phi)\Psi)) \in \mathfrak{p}$ is automatically satisfied. Hence, our curves have the same three–jet in 0, so they coincide by Theorem 5.3.3.

Next, recalling that $\delta u'(0) = [X,[Z,X]] = 2(ZX)X \mod \mathfrak{p}$, we see again that there is the projective reparametrization freedom; see Proposition 5.3.5. But since all available values of $\delta u'(0)$ coincide with one of reparametrizations of $c_1(t)$, and the parametrized curves are determined by their two–jet, there cannot be any other unparametrized curve C of type (1) in the same direction as $c_1(t)$. Of course the curves corresponding to the orbit (2) can be treated in the same way, which completes the proof of part (a).

Curves of type (b). These correspond to the orbit (3) in our list, so we suppose that both X and Y are nonzero but $YX = 0$ and $\beta = 0$. Proceeding as above, we get

$$\delta u'(0) = [X,[Z,X]] + [Y,[Z,X]] + [X,[W,Y]] + [Y,[W,Y]] \mod \mathfrak{p}.$$

Computing the brackets explicitly, we arrive at

$$2(ZX)X - (YX)W' = 0, \qquad 2(YW)Y - (YX)Z' = 0. \tag{5.34}$$

Since $YX = 0$ by assumption, this means that $ZX = 0$ and $YW = 0$, and hence all four terms in the above expression for $\delta u'(0)$ vanish individually. Now we compute modulo $\mathfrak{p}_+$ as follows:

$$\begin{aligned}\delta u'(0) =&[X+Y, \gamma X' + \gamma Y' + \tfrac{1}{2}[Z+W,[Z,X]+[W,Y]]]\\ =&\gamma([X,Y'] + [Y,X']) - \tfrac{1}{2}[[X+Y,[Z,X]+[W,Y]],Z+W]\\ &- \tfrac{1}{2}[[Z,X]+[W,Y],[X+Y,Z+W]]\\ &\text{(notice the last term in the first line vanishes by assumption)}\\ =&\gamma([X,Y'] + [Y,X']) - \tfrac{1}{2}[[Z,X]+[W,Y],[X,Z]+[Y,W]]\\ =&\gamma([X,Y'] + [Y,X']).\end{aligned}$$

Finally, still assuming that the second jets coincide,

$$\delta u''(0) = [X+Y, \gamma[X,Y'] + \gamma[Y,X']] = -\gamma(YX)(X+Y) \text{ mod } \mathfrak{p}_+.$$

As above, we conclude that curves of type (b) are determined by their two–jet in one point.

Let us look again at the number of different curves in the same direction up to reparametrization. As computed above, the freedom in $\delta u'(0)$ is

$$2(ZX)X - (YX)W' + 2(YW)Y - (YX)Z' = 2(ZX)X + 2(YW)Y.$$

Now, every common multiple of X and Y is available from the reparametrized curves and the two–jet in a point determines the curve. Thus, for $ZX = WY$ we can only get the reparametrizations. However, if $ZX \neq YW$, then we arrive at a new curve in the same direction $\langle \xi \rangle$. This proves the claim about the number of parameters of the family.

Curves of type (c). We have to repeat the computations for type (b) with the assumption $YX \neq 0$. If we multiply the two vector equations (5.34) by Y and X, then they imply that $4(ZX)(YX) = (ZX)(YX)$ so $ZX = 0$. But then the original equations imply $W = 0$ and then $Z = 0$. Therefore, the curves c_1 and c_2 of type (c) have the same two–jet in zero if and only if $W = 0$ and $Z = 0$.

Then, $\delta u'(0)$ is equal to $[X+Y, \gamma X' + \gamma Y'] = \gamma[X,Y'] + \gamma[Y,X']$, up to elements in $\mathfrak{p}_+$. Finally, if the second jets coincide, then $\delta u''(0) = -\gamma(YX)(X+Y)$, up to elements in $\mathfrak{p}$. Thus, the only possibility to share the three–jet as well is $\Phi = 0$ and there clearly are different curves with the same second jet (because they cannot allow more then projective freedom in the reparametrization).

We have seen on the way, that $\mathfrak{g}_1$ parametrizes all possible values of $\delta u'(0)$ and, moreover, the different values of $\delta u''(0)$ are parametrized by $\mathfrak{g}_2$. Thus, we get a

$2n$–parameter family of different unparametrized curves of type (c) in each given direction. This completes the discussion of curves tangent to H.

Curves of type (d). Following the same track as above would be tricky, since we do not have a general result on the freedom in the reparametrization. We only know that the parametrized curves must be determined by their 4–jets according to Corollary 5.3.2. Fortunately, it turns out that the general procedure for treating the unparametrized homogeneous curves in Lemma 5.3.5 provides a nice way to completely sort out this case.

Thus, our next goal will be to compute iteratively the vector spaces $\mathfrak{a}_0 = \mathfrak{p}$, $\mathfrak{a}_{i+1} = \{\Phi \in \mathfrak{a}_i;\ [\Xi, \Phi] \in \langle \Xi \rangle + \mathfrak{a}_i\}$; see Lemma 5.3.5. Let us start with the general bracket formula for $[\Xi, \Phi]$ in our algebra $\mathfrak{sl}(n+2, \mathbb{R})$:

$$\left[\begin{pmatrix} 0 & 0 & 0 \\ X & 0 & 0 \\ 1 & Y & 0 \end{pmatrix}, \begin{pmatrix} a & Z & \gamma \\ 0 & A & W \\ 0 & 0 & b \end{pmatrix}\right] = \begin{pmatrix} -ZX - \gamma & -\gamma Y & 0 \\ aX - AX - W & XZ - WY & \gamma X \\ a - b & Z + YA - bY & \gamma + YW \end{pmatrix}.$$

Without loss of generality, we only consider the case $\beta = 1$, and $YX \neq 0$ will be assumed in the sequel. Now the condition on Φ to be in $\mathfrak{a}_1$ is that the lower triangular part of the bracket has to equal $(a-b)\Xi$. This is equivalent to

$$W = (b - A)X, \qquad Z = Y(a - A). \tag{5.35}$$

These equations form the full description of $\mathfrak{a}_1 \subset \mathfrak{p}$.

Next, we have to consider the same general bracket with $\Phi \in \mathfrak{a}_1$, i.e. subject to the equations (5.35). The elements of $\mathfrak{a}_2$ are then given by the requirement that the result again satisfies the conditions (5.35) (with a, b, A, W, and Z denoting the appropriate entries of the matrix). This means that the matrix

$$\begin{pmatrix} -aYX + YAX - \gamma & -\gamma Y & 0 \\ * & (a-b)XY - XYA + AXY & \gamma X \\ * & * & \gamma + bYX - YAX \end{pmatrix}$$

has to be in $\mathfrak{a}_1$ (notice that we do not need entries indicated by stars). The conditions (5.35) now read as

$$\begin{aligned} \gamma X &= \gamma X + b(YX)X - (YAX)X - (a-b)X(YX) + X(YAX) - AX(YX) \\ -\gamma Y &= -aY(YX) + Y(YAX) - \gamma Y - (a-b)(YX)Y + (YX)YA - (YAX)Y \end{aligned}$$

and observing all cancellations (and the assumption $YX \neq 0$) we arrive at

$$YA = (2a - b)Y, \qquad AX = (2b - a)X.$$

Multiplying the first equation by X from left and the other by Y from right, we deduce the consequence $3a - 3b = 0$ and so the complete description of $\mathfrak{a}_2$ is given by the equations

$$YA = aY, \qquad AX = aX, \qquad b = a. \tag{5.36}$$

Substituting (5.36) into (5.35), we obtain

$$\mathfrak{a}_2 = \left\{ \begin{pmatrix} a & 0 & \gamma \\ 0 & A & 0 \\ 0 & 0 & a \end{pmatrix};\ AX = aX,\ YA = aY,\ \mathrm{tr}(A) + 2a = 0 \right\}.$$

Finally, repeating the same procedure again we compute $[\Xi, \Phi]$ with $\Phi \in \mathfrak{a}_2$ as described above to obtain

$$\begin{pmatrix} -\gamma & -\gamma Y & 0 \\ * & 0 & \gamma X \\ * & * & \gamma \end{pmatrix}.$$

But this is in $\mathfrak{a}_2$ only if $\gamma = 0$ and all elements in $\mathfrak{a}_2$ with vanishing γ are

$$\mathfrak{a}_3 = \left\{ \begin{pmatrix} a & 0 & 0 \\ 0 & A & 0 \\ 0 & 0 & a \end{pmatrix} ;\ AX = aX,\ YA = aY,\ \operatorname{tr}(A) + 2a = 0 \right\},$$

and clearly this is the step where the iterations stabilize.

Let us notice, that in the lowest dimensional case $n = 1$, the computation simplifies, but is essentially covered by the above. By the trace condition $A = -a-b$, so in the first step the conditions are $W = (2b+a)X$, $Z = (2a+b)Y$. The conditions for $\mathfrak{a}_2$ already imply $a = b = A = 0$, but γ still remains free. Of course $\mathfrak{a}_3 = 0$ in this case.

We have computed that the entire symmetry algebra of our curve $c_1(t)$ is $\mathfrak{s} = \langle \Xi \rangle \oplus \mathfrak{a}_3$. But we know that $\mathfrak{a}_3$ is a subalgebra of $\mathfrak{s}$ and from the explicit description of $\mathfrak{a}_3$ we see that all its elements have trivial bracket with Ξ. Hence, $\mathfrak{a}_3 = \mathfrak{m}$, the maximal ideal in $\mathfrak{s}$ contained in $\mathfrak{a}_3$ (compare with Remark 5.3.5), and we conclude that the image of $\mathfrak{s}$ in $\mathfrak{X}(C)$ is 1–dimensional, so it must be isomorphic to the algebra $\langle \frac{\partial}{\partial x} \rangle$ from 5.3.4. Thus, we only get an affine freedom for reparametrizations. At the same time, our symmetry algebra stabilized in the third step and so we need the three-jet of the curve to determine it.

Finally, we want to discuss how many canonical curves of type (d) emanate in a given direction. This question has also been addressed in Remark 5.3.5. In general, the difference of the dimensions of $\mathfrak{p}$ and $\mathfrak{a}_3$ describes the freedom in all unparametrized curves of given type (without any restriction on the direction or second order jets). In our case, $\dim(\mathfrak{p}) = n^2+1+2n+1$ and $\dim(\mathfrak{a}_3) = n^2+1-2n$, and so the difference is $4n + 1$. At the same time, there is a $2n$–parameter family of different directions complementary to the contact distribution. Thus, there is a $(2n + 1)$–parameter family of distinct unparametrized canonical curves of type (d) in a given direction. Since the curves allow only affine reparametrizations, the orders of jet needed to determine the parametrized and unparametrized curves coincide. □

REMARK 5.3.8. The description of canonical curves for partially integrable almost CR–structures of hypersurface type is very similar to the Lagrangean contact case. In fact, all of the above computations go through with some minor modifications following from the comparison of our description of the algebra $\mathfrak{g} = \mathfrak{su}(p+1, q+1)$ in 4.2.4 with the one of $\mathfrak{sl}(n+2, \mathbb{R})$ above. Thus, we have to deal with complex vectors $X \in \mathbb{C}^{p+q}$, and substitute $Y = -X^*$ and $W = -Z^*$, where the conjugate is taken with respect to the appropriate Hermitian form of signature (p, q). The entries β and γ become purely imaginary, $b = -\bar{a} \in \mathbb{C}$ and $A \in \mathfrak{u}(n)$. The results are as follows:

In the strictly pseudoconvex case (i.e. for definite signature $(n, 0)$), all nontrivial tangent vectors in the contact distribution H are of the same geometric type, and all canonical curves tangent to H behave as curves of type (c) in the proposition. At the same time, the chains as well as analogs of the curves of type (d) are available in all directions not contained in H.

For indefinite signature, there are three orbits of vectors in $\mathfrak{g}_{-1}$ corresponding to the sign of $\langle X, X\rangle$. For null–vectors $\xi \in H$, the canonical curves behave exactly as those of type (a) in the proposition, for other directions one obtains the behavior of curves of type (c). Again, chains and analogs of the curves of type (d) allowing only affine reparametrizations, are available in all directions complementary to H.

The curves of type (a) in CR–geometries of hypersurface type were first described by L. Koch; see [**Koch88, Koch93**]. It is remarkable that she has also introduced some of the general ideas on canonical curves in our approach in the special setting of CR–geometry in an unpublished preprint from about the same time.

5.3.9. Further remarks on canonical curves. While the behavior of the distinguished curves for Lagrangean contact structures summarized in Proposition 5.3.8 looks fairly involved, the situation is still rather nice. In particular, for each direction, there are canonical curves, which correspond to a G_0–invariant subset $A_0 \subset \mathfrak{g}_-$ contained in one grading component. A direction in H corresponds to exactly one of the types (a)–(c), while for transverse directions we can use the chains. (It is not clear, whether the curves of type (d) are actually needed in applications.)

This is not just a consequence of the fact that we are dealing with a parabolic contact structure here. Indeed, the same type of behavior occurs whenever the map $\mathrm{ad}(X) : \mathfrak{g}_1 \oplus \dots \oplus \mathfrak{g}_{i-1} \to \mathfrak{g}_{-i+1} \oplus \dots \oplus \mathfrak{g}_{-1}$ is onto for any $i = 2, \dots, k$ and any nonzero element $X \in \mathfrak{g}_{-i}$. For example, consider quaternionic contact structures as discussed in 4.3.3. These correspond to a $|2|$–grading, so we only have to check $\mathrm{ad}(X) : \mathfrak{g}_1 \to \mathfrak{g}_{-1}$ for a nonzero element $X \in \mathfrak{g}_{-2}$. But here $\mathfrak{g}_{-2} \cong \mathrm{im}(\mathbb{H})$ and $\mathfrak{g}_{\pm 1} \cong \mathbb{H}^n$ and the bracket is given by (quaternionic) scalar multiplication, so it is surjective for nonzero X. This also shows that there are no distinguished transverse directions in this case.

There are also examples of longer gradings with this nice behavior. Generic rank two distributions in dimension five as discussed in 4.3.2 correspond to a three grading with $\dim(\mathfrak{g}_{\pm 2}) = 1$ and $\dim(\mathfrak{g}_{\pm 1}) = \dim(\mathfrak{g}_{\pm 3}) = 2$, and it is easy to see that the bracket condition is satisfied.

It may also happen, however, that there are directions in which no canonical curves corresponding to a subset of one grading component emanate. For these geometries, one has to deal with canonical curves analogous to those of type (d) in Proposition 5.3.8. To give an example of such a structure, consider split quaternionic contact structures as discussed in 4.3.4. These are similar to quaternionic contact structures, but based on the algebra $\mathbb{H}_s$ of split quaternions, which can be viewed as the algebra $M_2(\mathbb{R})$ of two times two matrices. This also corresponds to a $|2|$–grading with $\mathfrak{g}_{-2} \cong \mathrm{im}(\mathbb{H}_s) = \mathfrak{sl}(2,\mathbb{R})$ and $\mathfrak{g}_1 \cong \mathbb{H}_s^n$, and the bracket is given by (split quaternionic) scalar multiplication. Now for a direction transverse to the subbundle H, one may look at the image in the quotient TM/H. In there, there is a natural cone corresponding to the rank one elements in $\mathrm{im}(\mathbb{H}_s)$. If the image lies outside of this cone, then there will be a canonical curve corresponding to some subset $A_0 \subset \mathfrak{g}_{-2}$ in the given direction. However, among the directions whose image is in the cone, there are two geometrically distinct subclasses. Choosing a representative vector on the Cartan bundle and looking at its image under the Cartan connection, the $\mathfrak{g}_{-2}$–component, let us call it ψ, will be nonzero but non–invertible. The question of whether the $\mathfrak{g}_{-1}$–component of our element lies in the image of $\mathrm{ad}(\psi)$ does not depend on the choice of lift, so it expresses a geometric property of the original direction. Evidently, there will be a canonical curve corresponding to some subset $A_0 \subset \mathfrak{g}_{-2}$ in the given direction if and only if the $\mathfrak{g}_{-1}$–component does indeed lie in the image of $\mathrm{ad}(\psi)$.

5.3.10. The ambient description of contact projective structures. In 4.5.6 we have seen that there is a generalized Fefferman construction, which associates to a contact projective structure $(M, H, [\nabla])$ a projective structure on M. Our next aim is to interpret this construction in terms of canonical curves, in particular, using the chains of the contact projective structure. Before doing this, we briefly discuss the ambient description (or cone description) of contact projective structures, which was introduced by D.J.F. Fox in [**Fox05a**]. This article also contained the first description of a projective structure associated to a contact projective structure, and we shall prove that this structure coincides with the one constructed in 4.5.6.

The cone description of contact projective structures is similar to the description of projective structures discussed in 5.2.6. As in 4.5.6, we use the group $G := Sp(2n+2, \mathbb{R})$ and the stabilizer $P \subset G$ of an oriented line in $\mathbb{R}^{2n+2}$ to describe contact projective structures, so we deal with structures admitting global contact forms. We have described the Lie algebra $\mathfrak{g}$ in 4.2.6 already, namely

$$\mathfrak{g} = \left\{ \begin{pmatrix} a & Z & W & z \\ X & A & B & W^t \\ Y & C & -A^t & -Z^t \\ x & Y^t & -X^t & -a \end{pmatrix} : B^t = B, C^t = C \right\},$$

with blocks of size 1, n, n, and 1, $\mathfrak{g}_{-2}$ corresponding to the entry x and $\mathfrak{g}_{-1}$ corresponding to the entries X and Y. On the group level, P consists of matrices which are block upper triangular, and the block diagonal part has the form (c, Φ, c^{-1}) with $c > 0$ and $\Phi \in Sp(2n, \mathbb{R})$ corresponding to the central 2×2–block. The elements with $c = 1$ form a closed normal subgroup $Q \subset P$.

Given a contact projective structure $(M, H, [\nabla])$ with corresponding parabolic geometry $(p : \mathcal{G} \to M, \omega)$ of type (G, P), we now define $M_\# := \mathcal{G}/Q$. The obvious projection $\pi : M_\# \to M$ is a principal bundle with structure group $P/Q \cong \mathbb{R}^+$. On the other hand, $\mathcal{G} \to M_\#$ clearly is a principal Q–bundle and ω defines a Cartan connection on this bundle. Via this Cartan connection, we get an isomorphism $TM_\# \cong \mathcal{G} \times_Q (\mathfrak{g}/\mathfrak{q})$, where $\mathfrak{q} \subset \mathfrak{g}$ is the Lie algebra of Q. As in 5.2.6 one easily verifies directly that, as a representation of Q, $\mathfrak{g}/\mathfrak{q}$ is isomorphic to the restriction of the standard representation $\mathbb{R}^{2n+2}$.

Since Q is contained in the symplectic group of $\mathbb{R}^{2n+2}$, this shows that we obtain a natural almost symplectic structure (i.e. a nondegenerate two–form $\tau \in \Omega^2(M_\#)$) on $M_\#$. Further, viewing $TM_\#$ as a tractor bundle, we obtain a natural linear connection $\nabla^\#$ on $TM_\#$ induced by ω. From the construction of this connection one immediately deduces (using again that $\mathfrak{g}$ is the symplectic algebra of $\mathbb{R}^{2n+2}$) that $\nabla^\#$ is compatible with the natural almost symplectic structure on $M_\#$. The connection $\nabla^\#$ is called the *cone connection* or the *ambient connection* associated to the contact projective structure.

The manifold $M_\#$ can be identified either with the frame bundle or with the subsets of positive elements in an oriented natural line bundle over M. The appropriate line bundle by construction is the natural line subbundle contained in the standard tractor bundle of the contact projective structure. From the Lie algebra description, it follows immediately that this is a square root of the bundle of contact forms. Hence, we can use it as a bundle of scales. The vertical bundle of $M_\# \to M$ is canonically trivialized by the Euler vector field. As in 5.2.6 one shows that a (local) smooth section s of $M_\# \to M$ can be used to pull back $\nabla^\#$ to a (locally defined) linear connection ∇^s on TM.

PROPOSITION 5.3.10. *Let (M,H) be an oriented smooth contact manifold of dimension $2n+1 \geq 3$, and let $\pi : M_\# \to M$ be the principal $\mathbb{R}_+$-bundle described above. For $t \in \mathbb{R}_+$ let $\rho^t : M_\# \to M_\#$ be the principal right action of t. Let $X \in \mathfrak{X}(M_\#)$ be the Euler vector field, and let $[\nabla]$ be a projective equivalence class of partial connections on H with vanishing contact torsion. Then the cone connection $\nabla^\#$ on $TM_\# \to M_\#$ has the following properties:*

(i) *For each $t \in \mathbb{R}_+$, ρ^t preserves the connection $\nabla^\#$.*
(ii) *$\nabla^\#$ is compatible with the almost symplectic structure on $M_\#$.*
(iii) *$\nabla^\#_\xi X = \xi$ for all $\xi \in \mathfrak{X}(M_\#)$.*
(iv) *$\nabla^\#$ is torsion free, so the natural almost symplectic structure on $M_\#$ is symplectic.*
(v) *The Ricci type contraction of the curvature $R^\#$ of $\nabla^\#$ vanishes identically.*
(vi) *For any smooth section $s : M \to M_\#$ of $\pi : M_\# \to M$, the restriction of linear connection ∇^s to a partial connection on H lies in the projective class $[\nabla]$.*

PROOF. We have already observed (ii). Having proved that $\nabla^\#$ is torsion free, we can write the exterior derivative of any differential form on $M_\#$ as the alternation of the covariant derivative with respect to $\nabla^\#$. Then the fact that natural almost symplectic structure τ is actually symplectic follows directly from (ii). The proof that $\nabla^\#$ is torsion free, as well as the proofs of all other parts is closely parallel to the proof of Proposition 5.2.6. Notice that in the construction of a Weyl structure from a section s for the proof of part (vi), one only has to deal with the underlying partial connection (i.e. with derivatives in contact directions). Therefore, even though we are dealing with a $|2|$–grading here, the proof remains similar to the one of Proposition 5.2.6. □

In [**Fox05a**] the author obtains a cone description without assuming vanishing of the contact torsion. While it turns out that the natural almost symplectic structure on $M_\#$ is always symplectic (see below), torsion freeness of the cone connection $\nabla^\#$ is equivalent to vanishing of the contact torsion of the contact projective structure (and to torsion freeness of the associated canonical Cartan connection).

Also, the passage from the ambient picture to the Cartan picture can be done for projective contact structures in a way closely parallel to the case of projective structures discussed in 5.2.7. First, $M_\#$ is available as a square root of the bundle of contact forms without any reference to a projective contact structure. It turns out that the same is true for the canonical almost symplectic structure. Namely, by construction for $z \in M_\#$ with $\pi(z) = x \in M$, we can interpret z^2 as the restriction of a contact form to T_xM. Hence, there is a tautological one–form $\alpha \in \Omega^1(M_\#)$ given by projecting down tangent vectors and then applying the contact form determined by the foot point. By construction, $(\rho^t)^*\alpha = t^2\alpha$ which implies that for the Euler vector field X we get $i_X d\alpha = \mathcal{L}_X\alpha = 2\alpha$. This easily implies that $\tau = d\alpha$ is a nondegenerate two form and that the pullback of the contact distribution in $TM_\#$ is exactly the annihilator of X under the symplectic form. One easily verifies directly that this coincides with the canonical almost symplectic form constructed above, which also shows that the latter is always symplectic.

As in 5.2.7, one then defines a right action of $\mathbb{R}_+$ on $TM_\#$ by $\xi \cdot t := t^{-1}T\rho^t \cdot \xi$. This is free and the orbit space is a vector bundle $\mathcal{T}$ of rank $2n+2$ over $M_\#/\mathbb{R}_+ = M$. In particular, sections of $\mathcal{T} \to M$ can be identified with vector fields on $M_\#$, which are homogeneous of degree -1. The vertical subbundle descends to a line subbundle $\mathcal{T}^1 \subset \mathcal{T}$ which is a square root of the bundle of contact forms on M. The natural symplectic form descends to a section of $\Lambda^2\mathcal{T}^*$, which is nondegenerate in each point. Next, one defines $\mathcal{T}^0$ to be the annihilator of $\mathcal{T}^1$ with respect to this form to obtain a filtration $\mathcal{T} = \mathcal{T}^{-1} \supset \mathcal{T}^0 \supset \mathcal{T}^1$.

Next suppose that $\nabla^\#$ is a linear connection on $TM_\#$, which satisfies conditions (i)–(iv) from the proposition above. Then this descends to a linear connection on $\mathcal{T}$ for which the section of $\Lambda^2\mathcal{T}^*$ constructed above is parallel. As in the proof of Proposition 5.2.7 one verifies that this actually is a torsion–free tractor connection and hence induces a contact projective structure on M with vanishing contact torsion. Finally, normality of the induced tractor connection is equivalent to vanishing of the Ricci type contraction of the curvature $R^\#$ of $\nabla^\#$. Again, this construction can also be done without assuming torsion freeness of $\nabla^\#$. The construction of a canonical Cartan connection in [**Fox05a**] is done in that way, starting from a general version of the cone connection.

5.3.11. The induced projective structure and its relation to chains. In the setting of the cone connection $\nabla^\#$ on $TM_\#$ as discussed above, the construction of a projective structure induced by a contact projective structure is very easy. Recall that $M_\#$ can be viewed as the space of positive elements in a square root of the bundle of contact forms. Hence, for a point $z \in M_\#$ with $\pi(z) = x \in M$, we can interpret z^2 as the value of a contact form at x. Now for a contact form θ on M, $\theta \wedge d\theta^n$ defines a volume form. Replacing θ by $f^2\theta$ for a positive smooth function f, we get $d(f^2\theta) = f^2 d\theta + 2f df \wedge \theta$, which shows that the associated volume form changes by multiplication by f^{2n+2}. But this implies that a positive square root of the bundle of contact forms can be identified with the bundle of $\frac{1}{2n+2}$–densities. Since $\dim(M) = 2n+1$, this is exactly the bundle to consider for the ambient description of a projective structure.

Hence, given a contact projective structure on M, we can form the ambient connection $\nabla^\#$ on $TM_\#$, then forget about the contact and symplectic structures, and just interpret $\nabla^\#$ as a linear connection on the tangent bundle of the "right" density bundle. Proposition 5.2.7 then shows that we obtain an induced projective structure on M. This is the subordinate projective structure as defined by Fox. This also works in the presence of contact torsion, but in that case, one uses the symmetrization of the cone connection $\nabla^\#$ rather than the cone connection itself. In the case of vanishing contact torsion, there is a nice description of the induced projective structure, which was first obtained in [**ČŽ08**]:

THEOREM 5.3.11. *Let $(M, H, [\nabla])$ be a contact projective structure with vanishing contact torsion.*

(1) The associated projective structure obtained from the ambient connection coincides with the projective structure obtained from the generalized Fefferman construction described in 4.5.6.

(2) The paths of the associated projective structure are exactly the paths of the contact projective structure (in contact directions) and the chains of the contact projective structure (in directions transverse to the contact distribution).

PROOF. (1) Consider the inclusion $G = Sp(2n+2,\mathbb{R}) \hookrightarrow SL(2n+2,\mathbb{R}) = \tilde{G}$ which maps the subgroups $Q \subset P \subset G$ to their counterparts $\tilde{Q} \subset \tilde{P} \subset \tilde{G}$ described in 5.2.6. On the level of Lie algebras, we have the inclusion $i : \mathfrak{g} \hookrightarrow \tilde{\mathfrak{g}}$, which maps $\mathfrak{p}$ to $\tilde{\mathfrak{p}}$ and $\mathfrak{q}$ to $\tilde{\mathfrak{q}}$ and induces linear isomorphisms $\mathfrak{g}/\mathfrak{p} \to \tilde{\mathfrak{g}}/\tilde{\mathfrak{p}}$ and $\mathfrak{g}/\mathfrak{q} \to \tilde{\mathfrak{g}}/\tilde{\mathfrak{q}}$. We shall also denote the two latter isomorphisms by i. Let $(\mathcal{G} \to M, \omega)$ be the regular normal parabolic geometry corresponding to the projective contact structure, and put $M_{\#} = \mathcal{G}/Q$. To obtain the Fefferman construction, one defines $\tilde{\mathcal{G}} := \mathcal{G} \times_P \tilde{P}$, so there is a natural map $j : \mathcal{G} \to \tilde{\mathcal{G}}$. On the extended bundle, there is a unique Cartan connection $\tilde{\omega}$ with values in $\tilde{\mathfrak{g}}$ such that $j^*\tilde{\omega} = i \circ \omega$. From the construction it is clear that we may naturally identify $\mathcal{G}/Q = M_{\#}$ with $\tilde{\mathcal{G}}/\tilde{Q}$.

To complete the proof of (1), it suffices to show that the ambient connection on $TM_{\#}$ induced by $(\mathcal{G} \to M, \omega)$ as in 5.3.10 coincides with the one induced by $(\tilde{\mathcal{G}} \to M, \tilde{\omega})$ as in 5.2.6. To see this, consider vector fields $\underline{\xi}$ and $\underline{\eta}$ on $M_{\#}$. Let $f : \mathcal{G} \to \mathfrak{g}/\mathfrak{q}$ be the Q–equivariant function and let $\tilde{f} : \tilde{\mathcal{G}} \to \tilde{\mathfrak{g}}/\tilde{\mathfrak{q}}$ be the $\tilde{Q}$–equivariant function representing η. By definition, for $u \in \mathcal{G}$ and lifts $\eta \in T_u\mathcal{G}$ and $\tilde{\eta} \in T_{j(u)}\tilde{\mathcal{G}}$ of η we have $f(u) = \omega(\eta) + \mathfrak{q}$ and $\tilde{f}(j(u)) = \tilde{\omega}(\tilde{\eta}) + \tilde{\mathfrak{q}}$. Using $\tilde{\eta} = T_u j \cdot \eta$, we see that $\tilde{f}(j(u)) = i(f(u))$. Hence, we get $\tilde{f} \circ j = i \circ f$ and this completely determines $\tilde{f}$ by equivariancy.

Likewise, let us choose a local lift $\xi \in \mathfrak{X}(\mathcal{G})$ of ξ, then consider $Tj \circ \xi$ along $j(\mathcal{G}) \subset \tilde{\mathcal{G}}$, and extend this to a lift $\tilde{\xi} \in \mathfrak{X}(\tilde{\mathcal{G}})$ of ξ. Now by construction, the covariant derivatives of η in direction ξ with respect to the two ambient connections in question are represented by $\xi \cdot f + \omega(\xi) \circ f$, respectively, $\tilde{\xi} \cdot \tilde{f} + \tilde{\omega}(\tilde{\xi}) \circ f$. But by construction, composing the second function with j, we obtain i composed with the first function, which completes the proof of (1).

(2) Consider first the ambient connection $\nabla^{\#}$ associated to a projective structure. For a point $y \in M_{\#}$ consider a geodesic c for $\nabla^{\#}$ with $c(0) = y$ such that $c'(0)$ does not lie in the vertical subbundle of $M_{\#} \to M$. Then locally around y, the curve c projects to a smooth curve $\underline{c}$ in M, and we can choose a local smooth section s of $M_{\#} \to M$ such that $s \circ \underline{c} = c$ holds locally around zero. But then from the definition of the connection ∇^s in formula (5.11) in 5.2.6 it is evident that $\underline{c}$ is a geodesic for the connection ∇^s, which by construction lies in the projective class. Since we can obtain curves through each point in each direction in that way, we see that the paths determined by the projective structure locally are exactly the projections of geodesics of $\nabla^{\#}$, which are transverse to the vertical bundle.

Now assume that $\nabla^{\#}$ comes from a contact projective structure, and let τ be the canonical symplectic form on $M_{\#}$. For a geodesic c of $\nabla^{\#}$ and the Euler vector field X, compatibility of $\nabla^{\#}$ with τ gives

$$\tfrac{d}{dt}\tau(c'(t), X(c(t))) = \tau(\nabla^{\#}_{c'(t)}c'(t), X(c(t))) + \tau(c'(t), \nabla^{\#}_{c'(t)}X(c(t))) = 0.$$

Using $\nabla^{\#}_{\xi}X = \xi$, we conclude that if $\tau(c'(0), X) = 0$, the $\tau(c'(t), X) = 0$ for all t. This exactly means that if the initial direction of the projection $\underline{c}$ of c is tangent to the contact distribution, the $\underline{c}$ remains to be tangent to the contact distribution for all times. But then the same argument as above can be applied, showing that the paths associated to the contact projective structure are exactly the projections of such geodesics.

In particular, this shows that the paths of the contact projective structure are also paths for the associated projective structure, so it remains to prove the claim about chains. From the explicit description of the Lie algebra $\mathfrak{g}$ in 5.3.10 we see that (in contrast to $\mathfrak{g}_{-1}$) the subspace $\mathfrak{g}_{-2} \subset \mathfrak{g}$ is contained in $\tilde{\mathfrak{g}}_{-1}$. Hence, for a nonzero element $\psi \in \mathfrak{g}_{-2}$ and a point $u \in \mathcal{G}$, the flow line of $\omega^{-1}(\psi)$ starting in u is mapped by j to the flow line of $\tilde{\omega}^{-1}(\psi)$ starting in $j(u)$. By definition, the former flow line projects to a chain of the contact projective structure, while the latter projects to a path of the associated projective structure. Hence, the chains of the contact projective structure are among the paths of the associated projective structure, and since together with the paths in contact directions they exhaust all directions, the proof is complete. □

Notice, in particular, that this implies that, viewed as unparametrized curves, the chains of a contact projective structure can be locally realized as geodesics of a linear connection. As we shall see below, this is in sharp contrast to the situation for other parabolic contact structures.

As a consequence, we can also prove that a contact projective structure is completely determined by its chains.

COROLLARY 5.3.11. *Let $(M^i, H^i, [\nabla^i])$ for $i = 1, 2$ be two contact projective structures of the same dimension, and assume that $f : M^1 \to M^2$ is a contactomorphism, which maps chains to chains (as unparametrized curves). Then f is an isomorphism of projective contact structures.*

PROOF. In 5.3.10 we have seen that the ambient space and its canonical symplectic structure depends only on the contact structure. Hence, f lifts to a symplectomorphism $f_\# : M^1_\# \to M^2_\#$, which is also compatible with the natural $\mathbb{R}_+$–actions and the Euler vector fields on the two ambient manifolds. In 5.2.7 we have constructed the standard tractor bundle from the tangent bundle of the ambient manifold using only these data, and the Cartan bundle of the induced projective structure can then be obtained as a frame bundle of the tractor bundle. The upshot of this is that denoting by $\tilde{\mathcal{G}}^i$ the Cartan bundle of the induced projective structure on M^i, the contactomorphism f naturally lifts to an isomorphism $F : \tilde{\mathcal{G}}^1 \to \tilde{\mathcal{G}}^2$ of principal bundles.

Now we claim that F is compatible with the Cartan connections, i.e. that $F^*\tilde{\omega}^2 = \tilde{\omega}^1$. To prove this, recall from 4.4.3 that given a projective structure, we obtain the projectivized tangent bundle as a correspondence space, which then naturally carries a path geometry. Applying this in our setting, we can view $\tilde{\mathcal{G}}^i$ as a bundle over $\mathcal{P}(TM^i)$ for $i = 1, 2$. In the projectivized tangent bundle of a contact manifold, we can consider lines which are transverse to the contact distribution, and of course they form a dense open subset. Let us denote these subsets by $\mathcal{P}_0(TM^i)$, and consider the restrictions of the principal bundles $\tilde{\mathcal{G}}^i$ to these subsets. Since f is a contactomorphism, the induced map $\mathcal{P}(TM^1) \to \mathcal{P}(TM^2)$ of course maps $\mathcal{P}_0(TM^1)$ to $\mathcal{P}_0(TM^2)$, so F restricts to these subsets. But on $\mathcal{P}_0(TM^i)$, the paths determined by the path geometry are the paths of the projective structure which are transverse to the contact distribution, so by the theorem these are exactly the chains. Hence, $F^*\tilde{\omega}^2 = \tilde{\omega}^1$ holds over $\mathcal{P}_0(TM^1)$ and since this is a dense open subset, it holds everywhere.

This shows that f maps all the paths of the associated projective structure on M^1 to paths of the associated projective structure on M^2. In contact directions, these are just the paths of the contact projective structures by the theorem, and the result follows from 4.2.7. □

Notice that, as promised in 1.1.4, this shows that morphisms of contact projective structures can be characterized as diffeomorphisms, which preserve both a contact structure and a projective structure.

5.3.12. The path geometry of chains. We conclude our study of chains with results on Lagrangean contact structures and partially integrable almost CR–structures, which were first obtained in [**ČŽ05**] using many of the tools developed in this book. We start the discussion in the setting of general parabolic contact structure.

For any parabolic contact structure on M, the chains introduced in 5.3.7 give rise to a canonical family of unparametrized curves in all directions transverse to the contact distribution. Looking at the projectivized tangent bundle $\mathcal{P}(TM)$, these directions form a dense open subset $\mathcal{P}_0(TM)$. Over this subset, the chains clearly give rise to a generalized path geometry as defined in 4.4.3. Now it is easy to describe this in the language of (non–parabolic) correspondence spaces.

PROPOSITION 5.3.12. *Let $\mathfrak{g} = \mathfrak{g}_{-2} \oplus \cdots \oplus \mathfrak{g}_2$ be a contact grading on a simple Lie algebra and let G be a Lie group with Lie algebra $\mathfrak{g}$. Let $P \subset G$ be a parabolic subgroup corresponding to the contact grading and let $G_0 \subset P$ be the Levi subgroup. Let $Q \subset P$ be the stabilizer of the line in $\mathfrak{g}/\mathfrak{p}$ corresponding to $\mathfrak{g}_{-2} \subset \mathfrak{g}_-$ under the obvious linear isomorphism.*

(1) The subgroup Q contains G_0 and has Lie algebra $\mathfrak{g}_0 \oplus \mathfrak{g}_2$. Moreover, the subspaces $\mathfrak{g}_{-2}$ and $\mathfrak{g}_1$ of $\mathfrak{g}$ descend to Q–invariant subspaces of $\mathfrak{g}/\mathfrak{q}$.

(2) If $(p : \mathcal{G} \to M, \omega)$ is a regular normal parabolic geometry, then the quotient $\mathcal{G}/Q$ is naturally isomorphic to $\tilde{M} := \mathcal{P}_0(TM)$. Viewing ω as a Cartan connection on $\mathcal{G} \to \tilde{M}$, we get $T\tilde{M} = \mathcal{G} \times_Q (\mathfrak{g}/\mathfrak{q})$. In this identification, the line bundle defining the path geometry of chains corresponds to the Q–invariant subspace of $\mathfrak{g}/\mathfrak{q}$ induced by $\mathfrak{g}_{-2}$.

PROOF. (1) We have described the action of P on the line in $\mathfrak{g}/\mathfrak{p}$ determined by $\mathfrak{g}_{-2}$ in the proof of Theorem 5.3.7. From the computations made there, we see that an element $g \in P$, which we write as $g_0 \exp(Z_1) \exp(Z_2)$ with $g_0 \in \mathfrak{g}_0$ and $Z_i \in \mathfrak{g}_i$ for $i = 1, 2$, lies in Q if and only if $Z_1 = 0$. From this, all statements in (1) follow immediately.

(2) We have also seen in the proof of Theorem 5.3.7 that the P–orbit of the line $\mathfrak{g}_{-2}$ consists of all lines in $\mathfrak{g}_-$ which are not contained in $\mathfrak{g}_{-1}$. Hence, the homogeneous space P/Q can be identified with the lines in $\mathfrak{g}_-$ which are transversal to $\mathfrak{g}_{-1}$. Since $TM \cong \mathcal{G} \times_P (\mathfrak{g}/\mathfrak{p})$, we conclude that $\mathcal{G}/Q = \mathcal{G} \times_P (P/Q)$ can be naturally identified with $\mathcal{P}_0(TM) = \tilde{M}$. Since ω clearly is a Cartan connection on $\mathcal{G} \to \tilde{M}$, we get the claimed description of $T\tilde{M}$. In this picture, the tangent map of the natural projection $\tilde{M} \to M$ corresponds to the natural quotient map $\mathfrak{g}/\mathfrak{q} \to \mathfrak{g}/\mathfrak{p}$. Hence, the vertical subbundle of this projection is represented by the Q–invariant subspace in $\mathfrak{g}/\mathfrak{q}$ induced by $\mathfrak{g}_1$. From the construction it also follows immediately

that the tautological subbundle corresponds to the Q–submodule induced by $\mathfrak{g}_{-2} \oplus \mathfrak{g}_1$. Hence, $\mathfrak{g}_{-2}$ gives rise to a line bundle which is complementary to the vertical subbundle inside the tautological subbundle, and hence defines a generalized path geometry; see 4.4.3. Finally, since the chains are by definition the projections of the flow lines of constant vector fields corresponding to elements of $\mathfrak{g}_{-2}$, we obtain the path geometry defined by the chains in this way. □

From 4.4.3 we know that a generalized path geometry will give rise to a regular normal parabolic geometry. If our original contact manifold has dimension $2n+1$, then this geometry is associated to the $|2|$–grading of $\tilde{\mathfrak{g}} = \mathfrak{sl}(2n+2, \mathbb{R})$ corresponding to the first two simple roots. The appropriate group is $\tilde{G} = PGL(2n+2, \mathbb{R})$ with the maximal parabolic subgroup $\tilde{P} \subset \tilde{G}$ corresponding to this grading. Our aim is to obtain a direct description of this parabolic geometry.

Let us first consider the homogeneous model G/P of our parabolic contact structure. Then the proposition simply says that $\mathcal{P}_0(T(G/P)) = G/Q$, so our path geometry is actually homogeneous under the group G. Hence, also the associated regular normal parabolic geometry $(\tilde{\mathcal{G}} \to G/Q, \tilde{\omega})$ will be homogeneous under G, and we can apply the theory of homogeneous Cartan geometries from 1.5.15. Any homogeneous Cartan geometry $(\tilde{\mathcal{G}} \to G/Q, \tilde{\omega})$ of type $(\tilde{G}, \tilde{P})$ is given by a pair (i, α) consisting of a homomorphism $i : Q \to \tilde{P}$ and a linear map $\alpha : \mathfrak{g} \to \tilde{\mathfrak{g}}$, which satisfy certain compatibility conditions, as follows. The bundle $\tilde{\mathcal{G}} = G \times_i \tilde{P}$ is the associated bundle with respect to the left action of Q on $\tilde{P}$ defined by $q \cdot h = i(q)h$. Denoting by $j : G \to \tilde{\mathcal{G}}$ the obvious map, the Cartan connection is then uniquely determined by the fact that $j^*\tilde{\omega} = \alpha \circ \omega$. The curvature of this geometry is described in Proposition 1.5.16. The compatibility conditions on i and α imply that

$$(X, Y) \mapsto [\alpha(X), \alpha(Y)] - \alpha([X, Y])$$

factors to a Q–equivariant bilinear map $\mathfrak{g}/\mathfrak{q} \times \mathfrak{g}/\mathfrak{q} \to \tilde{\mathfrak{g}}$. Identifying $\mathfrak{g}/\mathfrak{q}$ with $\tilde{\mathfrak{g}}/\tilde{\mathfrak{p}}$, we obtain the bilinear map Φ_α describing the Cartan curvature.

Knowing that there is a regular normal parabolic geometry inducing the given path geometry on G/Q, we conclude that it must be possible to find a pair (i, α) for which the map Φ_α satisfies $\partial^* \Phi_\alpha = 0$. Uniqueness of the resulting Cartan geometry together with Proposition 1.5.15 then implies that any other pair $(\hat{\imath}, \hat{\alpha})$ with this property has the form $\hat{\imath}(g) = \tilde{g} i(g) \tilde{g}^{-1}$, $\hat{\alpha} = \mathrm{Ad}(\tilde{g}) \circ \alpha$ for some $\tilde{g} \in \tilde{P}$.

As we have seen in 1.5.15, the pair (i, α) gives rise to an extension functor mapping Cartan geometries of type (G, Q) to Cartan geometries of type $(\tilde{G}, \tilde{P})$, and any equivalent pair $(\hat{\imath}, \hat{\alpha})$ as above will lead to a naturally isomorphic extension functor. For any parabolic contact structure of type (G, P) this extension functor provides a parabolic geometry of type $(\tilde{G}, \tilde{P})$ which induces the path geometry determined by the chains of the parabolic contact structure. Hence, the obvious question is for which geometries we actually obtain the canonical regular normal parabolic geometry induced by the path geometry of chains. For this, one only has to check when the geometry produced by the extension functor is normal. We will next carry out this program for Lagrangean contact structures.

5.3.13. Chains in the homogeneous model for Lagrangean contact structures. As we have seen in 4.2.3, Lagrangean contact structures in dimension $2n+1$ are obtained from the group $G := PGL(n+2, \mathbb{R})$, which we view as being obtained from the set of all matrices with determinant ± 1 by identifying each matrix with its negative. As there, we will always work with representative matrices. The parabolic subgroup $P \subset G$ consists of the classes of matrices which are block upper triangular with blocks of sizes 1, n, and 1, and its Levi subgroup G_0 corresponds to block diagonal matrices. We have described the appropriate presentation of the Lie algebra $\mathfrak{g}$ in 4.2.3, namely

$$\mathfrak{g} = \left\{ \begin{pmatrix} a & Z & \gamma \\ X & A & W \\ \beta & Y & b \end{pmatrix} : a, b, \beta, \gamma \in \mathbb{R}; X, W \in \mathbb{R}^n; Z, Y \in \mathbb{R}^{n*}; a + b + \operatorname{tr}(A) = 0 \right\}.$$

From this we can read off that the subgroup $Q \subset P$ as introduced in 5.3.12 comes from matrices of the form $\begin{pmatrix} c & 0 & \psi \\ 0 & C & 0 \\ 0 & 0 & d \end{pmatrix}$ where $cd \det(C) = 1$. Likewise,the subalgebra $\mathfrak{q} \subset \mathfrak{g}$ is given by matrices as above for which the components β, X, Y, Z, and W vanish.

As discussed in 5.3.12 above and in 4.4.3, the path geometry of chains in a Lagrangean contact structure will correspond to the group $\tilde{G} = PGL(2n+2, \mathbb{R})$ with the parabolic subgroup $\tilde{P} \subset \tilde{G}$ induced by matrices which are block upper triangular with block sizes 1, 1, and $2n$. We will actually use a finer decomposition, by decomposing the block of size $2n$ into two blocks of size n. As in 4.4.3, we will use the $\tilde{G}_0$–invariant decomposition of $\tilde{\mathfrak{g}}_{-1}$ into a one–dimensional summand $\tilde{\mathfrak{g}}^E_{-1}$ and a $(2n)$–dimensional summand $\tilde{\mathfrak{g}}^V_{-1}$. According to our finer decomposition, elements of the latter summand will be denoted by $\binom{X_1}{X_2}$ for $X_i \in \mathbb{R}^n$, and similarly for elements of $\tilde{\mathfrak{g}}_{-2}$.

The first thing to do is to define a homomorphism $i : Q \to \tilde{P}$ and a linear map $\alpha : \mathfrak{g} \to \tilde{\mathfrak{g}}$ with certain properties.

LEMMA 5.3.13. *Consider*

(5.37)
$$i \begin{pmatrix} c & 0 & \gamma \\ 0 & C & 0 \\ 0 & 0 & d \end{pmatrix} := \begin{pmatrix} \operatorname{sgn}(\frac{d}{c})\sqrt{|\frac{c}{d}|} & \operatorname{sgn}(\frac{d}{c})\frac{\gamma}{c}\sqrt{|\frac{c}{d}|} & 0 & 0 \\ 0 & \sqrt{|\frac{d}{c}|} & 0 & 0 \\ 0 & 0 & d^{-1}\sqrt{|\frac{d}{c}|}C & 0 \\ 0 & 0 & 0 & d\sqrt{|\frac{c}{d}|}(C^{-1})^t \end{pmatrix},$$

$$(5.38) \qquad \alpha \begin{pmatrix} a & Z & \gamma \\ X & A & W \\ \beta & Y & b \end{pmatrix} := \begin{pmatrix} \frac{a-b}{2} & \gamma & \frac{1}{2}Z & \frac{1}{2}W^t \\ \beta & \frac{b-a}{2} & \frac{1}{2}Y & -\frac{1}{2}X^t \\ X & W & A - \frac{a+b}{2}\mathrm{id} & 0 \\ Y^t & -Z^t & 0 & -A^t + \frac{a+b}{2}\mathrm{id} \end{pmatrix}.$$

(1) The map i induces a well–defined homomorphism $Q \to \tilde{P}$, which together with the linear map $\alpha : \mathfrak{g} \to \tilde{\mathfrak{g}}$ satisfies the conditions of Proposition 1.5.15 for inducing a homogeneous Cartan connection. Explicitly, $\alpha \circ \mathrm{Ad}(g) = \mathrm{Ad}(i(g)) \circ \alpha$ for all $g \in Q$, the restriction of α to $\mathfrak{q} \subset \mathfrak{g}$ coincides with i', and α induces a linear isomorphism $\mathfrak{g}/\mathfrak{q} \to \tilde{\mathfrak{g}}/\tilde{\mathfrak{p}}$.

(2) The map $\Phi_\alpha : \Lambda^2\tilde{\mathfrak{g}}_- \to \tilde{\mathfrak{g}}$ from 5.3.12 has values in the semisimple part of $\tilde{\mathfrak{g}}_0$ and is only nonzero if one of its entries lies in $\tilde{\mathfrak{g}}^V_{-1}$ and the other entry lies in $\tilde{\mathfrak{g}}_{-2}$. For a nonzero element $\beta_0 \in \mathfrak{g}^E_{-1}$, the trilinear map $(\tilde{\mathfrak{g}}^V_{-1})^3 \to \tilde{\mathfrak{g}}^V_{-1}$ defined by $(R, S, T) \mapsto [\Phi_\alpha(R, [S, \beta_0]), T]$ is, up to a nonzero factor, the complete symmetrization of the map $(\binom{R_1}{R_2}, \binom{S_1}{S_2}, \binom{T_1}{T_2}) \mapsto \langle R_1, S_2\rangle \binom{T_1}{-T_2}$. Here we have split

each of the elements of $\tilde{\mathfrak{g}}^V_{-1} \cong \mathbb{R}^{2n}$ into two elements of $\mathbb{R}^n$ and $\langle\ ,\ \rangle$ denotes the standard inner product on $\mathbb{R}^n$.

PROOF. (1) Evidently, the value of i remains unchanged if we replace the matrix by its negative, so we obtain a well–defined map $Q \to \tilde{P}$. An easy direct computation then shows that this induced map is a homomorphism. The map α evidently maps $\mathfrak{q}$ to $\tilde{\mathfrak{p}}$, and it injects $\mathfrak{g}_- \oplus \mathfrak{g}_1$ into $\tilde{\mathfrak{g}}_-$, so the map $\mathfrak{g}/\mathfrak{q} \to \tilde{\mathfrak{g}}/\tilde{\mathfrak{p}}$ will be injective and thus a linear isomorphism by dimensional reasons. Another easy direct computation shows that α restricts to i' on $\mathfrak{q}$. The verifications of equivariancy of α is a slightly tedious but straightforward computation.

(2) To obtain Φ_α it suffices to compute $[\alpha(\ ), \alpha(\ )] - \alpha([\ ,\ ])$ on two elements of $\mathfrak{g}_- \oplus \mathfrak{g}_1$, which we denote by $(\beta_i, X_i, Y_i, Z_i, W_i)$ for $i = 1, 2$. A direct computation shows that all components of the result vanish except the lower right $(2n) \times (2n)$–block, which equals

$$\begin{pmatrix} -U_{11} - \operatorname{tr}(U_{11})\mathrm{id} & U_{12} + U_{12}^t \\ U_{21} + U_{21}^t & U_{11}^t + \operatorname{tr}(U_{11})\mathrm{id} \end{pmatrix},$$

where the $n \times n$ matrices U_{ij} are given by

$$U_{11} = \tfrac{1}{2}(X_1 Z_2 + W_1 Y_2 - X_2 Z_1 - W_2 Y_1),$$
$$U_{12} = \tfrac{1}{2}(X_1 W_2^t - W_1 X_2^t),$$
$$U_{21} = \tfrac{1}{2}(Y_1^t Z_2 - Z_1^t Y_2).$$

Since the block matrix is evidently tracefree, we see that Φ_α has values in the semisimple part $\tilde{\mathfrak{g}}_0^{ss}$. To get Φ_α as a map defined on $\Lambda^2(\tilde{\mathfrak{g}}/\tilde{\mathfrak{p}})$ one has to interpret the original elements as consisting of $\beta_i \in \tilde{\mathfrak{g}}^E_{-1}$, $\binom{W_i}{-Z_i^t} \in \tilde{\mathfrak{g}}^V_{-1}$ and $\binom{X_i}{Y_i^t} \in \tilde{\mathfrak{g}}_{-2}$. This shows that the only nonzero component of Φ_α is the one taking one entry from $\tilde{\mathfrak{g}}^V_{-1}$ and one entry from $\tilde{\mathfrak{g}}_{-2}$.

To compute the trilinear map $(\tilde{\mathfrak{g}}^V_{-1})^3 \to \tilde{\mathfrak{g}}^V_{-1}$ in the last claim, we have to take the above matrix for $\binom{W_1}{-Z_1^t} = \binom{R_1}{R_2}$, $\binom{X_2}{Y_2^t} = \binom{S_1}{S_2}$, and $X_1 = Y_1 = Z_2 = W_2 = 0$, and then apply this matrix to the vector $\binom{T_1}{T_2}$. Specializing the matrix, we get $U_{11} = R_1 S_2^t + S_1 R_2^t$, $U_{12} = -R_1 S_1^t$, and $U_{21} = R_2 S_2^t$, and from this the last claim follows easily. □

This allows us to determine the path geometry of chains for the homogeneous model of Lagrangean contact structures, which has some unexpected consequences.

THEOREM 5.3.13. *Let $i : Q \to \tilde{P}$ and $\alpha : \mathfrak{g} \to \tilde{\mathfrak{g}}$ be the maps from the lemma. Then the invariant Cartan connection on $G \times_i \tilde{P} \to G/Q$ determined by α is torsion free (and hence regular) and normal. Thus $(G \times_i \tilde{P}, \tilde{\omega}_\alpha)$ is the regular normal Cartan geometry describing the path geometry of chains on G/P. This path geometry is non–flat, so there is no local coordinate system on G/P in which all chains are straight lines. There is not even a local linear connection on $T(G/P)$ which has the chains among its geodesics.*

PROOF. Since the curvature of the homogeneous Cartan geometry determined by (i, α) is induced by the map Φ_α computed in the lemma, torsion freeness follows immediately from the fact that Φ_α has values in $\tilde{\mathfrak{g}}_0 \subset \tilde{\mathfrak{p}}$. To prove normality, we have to show that $\partial^* \Phi_\alpha = 0$. This can be verified by a direct computation, but it is more easily done by using a bit of representation theory. As it stands, Φ_α

is an element of $\tilde{\mathfrak{g}}_{-2}^* \otimes (\tilde{\mathfrak{g}}_{-1}^V)^* \otimes \tilde{\mathfrak{g}}_0^{ss}$, which as a representation of $\tilde{\mathfrak{g}}_0^{ss} \cong \mathfrak{sl}(2n, \mathbb{R})$ is isomorphic to $\mathbb{R}^{2n*} \otimes \mathbb{R}^{2n*} \otimes \mathfrak{sl}(2n, \mathbb{R})$. The last statement in the lemma says that Φ_α actually lies in the intersection of that space with $S^3\mathbb{R}^{2n*} \otimes \mathbb{R}^{2n}$. From the latter representation, there is a unique $\mathfrak{sl}(2n, \mathbb{R})$–equivariant trace with values in $S^2\mathbb{R}^{2n*}$, whose kernel is irreducible. From the explicit description of Φ_α in the lemma, one immediately verifies that it lies in the kernel of this trace.

Now, on the other hand, ∂^* is $\tilde{\mathfrak{g}}_0^{ss}$–equivariant and has values in $\tilde{\mathfrak{g}}_-^* \otimes \tilde{\mathfrak{g}}$. Since Φ_α is homogeneous of degree three, its image under ∂^* must be contained in $\tilde{\mathfrak{g}}_{-2}^* \otimes \tilde{\mathfrak{g}}_1 \oplus \tilde{\mathfrak{g}}_{-1} \otimes \tilde{\mathfrak{g}}_2$. Now, evidently, this contains at most tensor products of two copies of $\mathbb{R}^{2n*}$, so it certainly cannot contain an irreducible component isomorphic to the tracefree part of $S^3\mathbb{R}^{2n*} \otimes \mathbb{R}^{2n}$. Hence, $\partial^*\Phi_\alpha = 0$ and normality follows. Having found the canonical parabolic geometry associated to the path geometry of chains on G/P, we can apply the results from 4.4.3–4.4.5. Since $\Phi_\alpha \neq 0$, this path geometry is non–flat, and in particular, the paths cannot be realized as straight lines in any coordinate system.

We have described the harmonic curvature components of path geometries in 4.4.3. There are three such components for $n = 1$ and two for $n > 1$, but in any case all but one of these components are homogeneous of degree less than three. Since Φ_α is homogeneous of degree three, it must already coincide with the unique harmonic curvature component of homogeneity 3, which was called ρ in 4.4.3. Now the last claim follows from Corollary 4.4.4. □

This result is remarkable in several ways. On the one hand, we see that even in the homogeneous model, the chains form a pretty complicated family of curves. On the other hand, since the automorphism group of the path geometry of chains contains G, we obtain an example of a non–flat path geometry with a large automorphism group. In view of the connection between path geometries and systems of second order ODE's discussed in 4.4.5, we also obtain an example of a nontrivial system of ODE's with large automorphism group. This is the context in which this example was first considered (in the case $n = 1$ and without mentioning the connection to chains) in [**Gro00**]. It also provides an example for the class of torsion–free path geometries which was studied in that reference.

In this context, one may go one step further and pass to the space of chains, which carries a canonical Grassmannian structure; see 4.4.5. Applied in this special case, we obtain an example of a non–flat homogeneous Grassmannian structure with large automorphism group (containing G). This example is particularly interesting, since it is even a Grassmannian symmetric space, i.e. for each point x there is an automorphism σ_x of the structure which has x as a fixed point and satisfies $T_x\sigma_x = -\mathrm{id}$; see [**ZŽ08**].

5.3.14. Chains in Lagrangean contact structures. According to Theorem 1.5.15, the pair (i, α) from Lemma 5.3.13 gives rise to an extension functor mapping Cartan geometries of type (G, Q) to parabolic geometries of type $(\tilde{G}, \tilde{P})$. One simply extends the structure group via the action of Q on $\tilde{P}$ coming from i, i.e. $\tilde{\mathcal{G}} = \mathcal{G} \times_i \tilde{P}$. Denoting by $j : \mathcal{G} \to \tilde{\mathcal{G}}$ the canonical inclusion, there is a unique Cartan connection $\tilde{\omega}_\alpha$ on $\tilde{\mathcal{G}}$ for which $j^*\tilde{\omega}_\alpha = \alpha \circ \omega$. The results in 5.3.13 show that for the homogeneous model (and hence for locally flat Lagrangean contact structures), this extension functor maps the geometry coming from the canonical Cartan connection for the Lagrangean contact structure to the canonical parabolic geometry associated to the path geometry of chains. Hence, it is natural to ask, whether the same holds for a larger class of Lagrangean contact structures.

It is clear how to attack this question, since the curvature of the geometries coming from the extension functor are described in Proposition 1.5.16. To obtain the result, however, we need a rather careful analysis of the Cartan curvature.

THEOREM 5.3.14. *Let* $(p : \mathcal{G} \to M, \omega)$ *be the regular normal parabolic geometry associated to a Lagrangean contact structure on* M. *Put* $\tilde{M} = \mathcal{P}_0(TM) = \mathcal{G}/Q$ *and let* $(\tilde{\mathcal{G}} \to \tilde{M}, \tilde{\omega}_\alpha)$ *be the value of the extension functor induced by the pair* (i, α) *from Lemma 5.3.13. Then the geometry* $(\tilde{\mathcal{G}} \to \tilde{M}, \tilde{\omega}_\alpha)$ *is regular and normal if and only if* ω *is torsion free.*

PROOF. By Proposition 1.5.16, the curvature function $\tilde{\kappa}$ of $(\tilde{\mathcal{G}} \to \tilde{M}, \tilde{\omega}_\alpha)$ is determined by

$$\tilde{\kappa} \circ j = \alpha \circ \kappa \circ \underline{\alpha}^{-1} + \Phi_\alpha, \tag{5.39}$$

where $\underline{\alpha} : \mathfrak{g}/\mathfrak{q} \to \tilde{\mathfrak{g}}/\tilde{\mathfrak{p}}$ is the linear isomorphism induced by α and Φ_α is the map computed in Lemma 5.3.13.

We start by proving the necessity of torsion freeness of ω. This is vacuous if $n = 1$, since in that case any regular normal Cartan connection of type (G, P) is torsion free, so we may assume $n > 1$. If $\tilde{\omega}$ is normal, then the lowest nonzero homogeneous component of $\tilde{\kappa}$ must be harmonic by Theorem 3.1.12. From the description of the harmonic curvature of the parabolic geometries associated to generalized path geometries in 4.4.3, we then conclude that $\tilde{\kappa}$ must be homogeneous of degree at least two, and the homogeneous component of degree two must be contained in the space of maps $\tilde{\mathfrak{g}}^E_{-1} \times \tilde{\mathfrak{g}}_{-2} \to \tilde{\mathfrak{g}}^V_{-1}$. In particular, the restriction of $\tilde{\kappa}$ to $\Lambda^2\tilde{\mathfrak{g}}_{-2}$ has to be homogeneous of degree at least three, which implies that $\tilde{\kappa}$ must have values in $\tilde{\mathfrak{g}}^{-1}$. Now from the definition (5.38) of α in Lemma 5.3.13, we see that α restricts to a linear isomorphism $\mathfrak{g}_{-1} \to \tilde{\mathfrak{g}}_{-2}$. By regularity, κ maps $\Lambda^2\mathfrak{g}_{-1}$ to $\mathfrak{g}^{-1}$, and from Lemma 5.3.13 we know that Φ_α vanishes on $\Lambda^2\tilde{\mathfrak{g}}_{-2}$. Hence, we conclude from (5.39) that κ has to map $\Lambda^2\mathfrak{g}_{-1}$ to $\mathfrak{p}$. From the discussion of harmonic curvature components of Lagrangean contact structures in 4.2.3, we conclude that this implies torsion freeness of ω.

Let us conversely assume that ω is torsion free. Since we already know that Φ_α has values in $\tilde{\mathfrak{g}}_0^{ss}$, (5.39) together with torsion freeness of ω immediately implies that $\tilde{\kappa} \circ j$ has values in $\tilde{\mathfrak{g}}^{-1}$, which implies regularity of $\tilde{\omega}_\alpha$. To proceed, we need a bit of input from BGG–sequences. By torsion freeness, the curvature function κ of ω has values in $\Lambda^2\mathfrak{p}_+ \otimes \mathfrak{p}$. Now the Lie bracket defines a P–equivariant map $\Lambda^2\mathfrak{p}_+ \to \mathfrak{g}_2$, so its kernel is a P–submodule $\Lambda^2_0\mathfrak{p}_+ \subset \Lambda^2\mathfrak{p}_+$. Likewise, the projection $\mathfrak{p} \to \mathfrak{p}/\mathfrak{p}_+ \cong \mathfrak{g}_0$ is P–equivariant, and the semisimple part $\mathfrak{g}_0^{ss} \subset \mathfrak{g}_0$ is a P–submodule. Hence, the elements whose projections to $\mathfrak{g}_0$ lie in $\mathfrak{g}_0^{ss}$ form a P–submodule $\mathfrak{p}_0 \subset \mathfrak{p}$. Now for $n > 1$, the harmonic part of κ has values in the highest weight submodule of $\Lambda^2\mathfrak{g}_1 \otimes \mathfrak{g}_0$, which certainly is contained in the P–submodule $\Lambda^2_0\mathfrak{p}_+ \otimes \mathfrak{p}_0 \subset \Lambda^2\mathfrak{p}_+ \otimes \mathfrak{p}$. For $n = 1$, this is also verified easily, so by Corollary 3.2 of [**Čap05**], torsion freeness implies that all of κ has values in $\Lambda^2_0\mathfrak{p}_+ \otimes \mathfrak{p}_0$.

Since we have already proved that $\partial^*\Phi_\alpha = 0$ in Lemma 5.3.13, it remains to show that for each $u \in \mathcal{G}$, the map $F(u) = \alpha \circ \kappa(u) \circ \underline{\alpha}^{-1}$ lies in $\ker(\partial^*) \subset L(\Lambda^2\tilde{\mathfrak{g}}, \tilde{\mathfrak{g}})$. To verify this, it is better to interpret $F(u)$ as an element of $\Lambda^2\tilde{\mathfrak{p}}_+ \otimes \tilde{\mathfrak{g}}$, and interpreting $\kappa(u)$ in a similar way, we get $F(u) = (\alpha \otimes \Lambda^2\phi)(\kappa(u))$, where $\phi : \mathfrak{p}_+ \to \tilde{\mathfrak{p}}_+$ is dual to the composition of the natural projection $\mathfrak{g}/\mathfrak{q} \to \mathfrak{g}/\mathfrak{p}$ with

$\underline{\alpha}^{-1} : \tilde{\mathfrak{g}}/\tilde{\mathfrak{p}} \to \mathfrak{g}/\mathfrak{q}$. To compute ϕ we can use the trace form on both algebras, which is just a multiple of the Killing form. But then expanding the defining equation, one immediately verifies that

$$\phi\begin{pmatrix} 0 & Z & \gamma \\ 0 & 0 & W \\ 0 & 0 & 0 \end{pmatrix} = \begin{pmatrix} 0 & \gamma & Z & W^t \\ 0 & 0 & 0 & 0 \\ 0 & 0 & 0 & 0 \\ 0 & 0 & 0 & 0 \end{pmatrix}.$$

Now suppose that $\kappa(u) = \sum U_i \wedge V_i \otimes B_i$ for $U_i, V_i \in \mathfrak{p}_+$ and $B_i \in \mathfrak{p}$. Then $F(u) = \sum \phi(U_i) \wedge \phi(V_i) \otimes \alpha(B_i)$, and from 3.1.11 we know that

$$\begin{aligned}\partial^*(F(u)) = \textstyle\sum_i \big(&- \phi(V_i) \otimes [\phi(U_i), \alpha(B_i)] + \phi(U_i) \otimes [\phi(V_i), \alpha(B_i)] \\ &- [\phi(U_i), \phi(V_i)] \otimes \alpha(B_i)\big).\end{aligned} \tag{5.40}$$

From above we see that ϕ has values in $\tilde{\mathfrak{g}}_1^E \oplus \tilde{\mathfrak{g}}_2$, which implies that any two elements in the image of ϕ have vanishing bracket, so for all i we have $[\phi(U_i), \phi(V_i)] = 0$ in (5.40). From our explicit formulae, one next immediately verifies that for $R \in \mathfrak{p}_+$ and $T \in \mathfrak{p}_0$ we have $[\phi(R), \alpha(T)] \in \tilde{\mathfrak{g}}_1^E \oplus \tilde{\mathfrak{g}}_2$, and the first component equals the $\tilde{\mathfrak{g}}_1^E$–component of $\alpha([R,T])$, while the $\tilde{\mathfrak{g}}_2$–component equals twice the $\tilde{\mathfrak{g}}_2$–component of $\alpha([R,T])$. Knowing that $\kappa(u) \in \Lambda^2_0\mathfrak{p}_+ \otimes \mathfrak{p}_0$, we conclude that $\partial^*\kappa(u) = 0$ implies $\partial^* F(u) = 0$, which completes the proof. □

Hence we see that the extension functor associated to the pair (i, α) produces the canonical parabolic geometry associated to the path geometry of chains for a large class of Lagrangean contact structures, which corresponds to the class of CR–structures among partially integrable CR–structures. Hence, for this subclass the geometry of chains is intimately related to the original parabolic contact structure. Analyzing the Cartan curvature, we can draw some first basic consequences.

COROLLARY 5.3.14. *(1) The chains of a torsion–free Lagrangean contact structure on a manifold M cannot be obtained as geodesics of a linear connection on TM.*

(2) The path geometry of chains associated to a torsion–free Lagrangean contact structure is torsion free if and only if the original structure is locally flat.

PROOF. The curvature $\tilde{\kappa}$ of the path geometry of chains is described by equation (5.39) in the proof of the theorem. Now $\tilde{\mathfrak{p}}_+$ contains the $\tilde{P}$–invariant subspace $\tilde{\mathfrak{p}}_+^2 = \mathfrak{g}_2$. This induces a $\tilde{P}$–invariant filtration on $\Lambda^2\tilde{\mathfrak{p}}_+$ which has the form $\Lambda^2\tilde{\mathfrak{g}}_2 \subset \tilde{\mathfrak{p}}_+ \wedge \tilde{\mathfrak{g}}_2 \subset \Lambda^2\tilde{\mathfrak{p}}_+$. From Lemma 5.3.13 we know that the function Φ_α has values in $(\tilde{\mathfrak{p}}_+ \wedge \tilde{\mathfrak{g}}_2) \otimes \tilde{\mathfrak{p}}$. On the other hand, the fact that the map $\phi : \mathfrak{p}_+ \to \tilde{\mathfrak{p}}_+$ constructed in the proof of the theorem has values in $\tilde{\mathfrak{g}}_1^E \oplus \tilde{\mathfrak{g}}_2$ shows that $\Lambda^2\phi$ has values in $\tilde{\mathfrak{p}}_+ \wedge \tilde{\mathfrak{g}}_2$. Hence, from (5.39) we conclude that $\tilde{\kappa} \circ j$ has values in $(\tilde{\mathfrak{p}}_+ \wedge \tilde{\mathfrak{g}}_2) \otimes \tilde{\mathfrak{g}}$ and this then has to hold for all of $\tilde{\kappa}$ by equivariancy.

The quotient $(\tilde{\mathfrak{p}}_+ \wedge \tilde{\mathfrak{g}}_2)/\Lambda^2\tilde{\mathfrak{g}}_2$ is clearly isomorphic to $\tilde{\mathfrak{g}}_1 \wedge \tilde{\mathfrak{g}}_2$ as a representation of $\tilde{P}$, so this decomposes into a direct sum according to $\tilde{\mathfrak{g}}_1 = \tilde{\mathfrak{g}}_1^E \oplus \tilde{\mathfrak{g}}_1^V$. Hence, we obtain P–equivariant projections

$$\tilde{\pi}^E : (\tilde{\mathfrak{p}}_+ \wedge \tilde{\mathfrak{g}}_2) \otimes \tilde{\mathfrak{g}} \to (\tilde{\mathfrak{g}}_1^E \wedge \tilde{\mathfrak{g}}_2) \otimes \tilde{\mathfrak{g}},$$
$$\tilde{\pi}^V : (\tilde{\mathfrak{p}}_+ \wedge \tilde{\mathfrak{g}}_2) \otimes \tilde{\mathfrak{g}} \to (\tilde{\mathfrak{g}}_1^V \wedge \tilde{\mathfrak{g}}_2) \otimes \tilde{\mathfrak{g}}.$$

Now the description of Φ_α in Lemma 5.3.13 shows that $\tilde{\pi}^E(\Phi_\alpha) = 0$ and that $\tilde{\pi}^V(\Phi_\alpha)$ equals Φ_α, so in particular, is nonzero. On the other hand, since the map ϕ has values in $\tilde{\mathfrak{g}}_1^E \oplus \tilde{\mathfrak{g}}_2$, we conclude that $\tilde{\pi}^V((\Lambda^2\phi \otimes \alpha)(\kappa(u))) = 0$ for all u.

(1) The last considerations show that $\tilde{\pi}^V(\tilde{\kappa}) \neq 0$. But if the chains of the Lagrangean contact structure were geodesics of a linear connection on TM, then the associated path geometry would locally drop to a projective geometry on a twistor space; see 4.4.4. In particular, the distribution corresponding to $\tilde{\mathfrak{g}}_1^V$ would be integrable and vectors from this distribution would insert trivially into the Cartan curvature of the path geometry. But this would contradict the fact that $\tilde{\pi}^V(\tilde{\kappa}) \neq 0$.

(2) Lemma 5.3.13 implies that locally flat Lagrangean contact structures give rise to torsion–free path geometries, so let us conversely assume that $\tilde{\omega}$ is torsion free. By Theorem 3.1.11, the lowest nonzero homogeneous component of $\tilde{\kappa}$ has to be harmonic. In view of the description of harmonic curvature quantities for path geometries in 4.4.3, this implies that $\tilde{\kappa}$ is of homogeneity at least three, and the homogeneous component of degree three is a map $\tilde{\mathfrak{g}}_{-1}^V \times \tilde{\mathfrak{g}}_{-2} \to \tilde{\mathfrak{g}}_0$. Moreover, the whole harmonic curvature then has to have values in this space, which is evidently contained in $\ker(\tilde{\pi}_E)$. By Corollary 3.2 of [**Čap05**], the curvature function $\tilde{\kappa}$ has values in this $\tilde{P}$–submodule. Since $\Phi_\alpha \in \ker(\tilde{\pi}^E)$ we conclude that $(\Lambda^2\phi \otimes \alpha)(\kappa(u))$ also has to lie in that space.

From the proof of the theorem, we know that $\phi : \mathfrak{p}_+ \to \tilde{\mathfrak{g}}_1^E \oplus \tilde{\mathfrak{g}}_2$ is a linear isomorphism, and hence $\tilde{\mathfrak{g}}_1^E \wedge \tilde{\mathfrak{g}}_2$ is contained in the image of $\Lambda^2\phi$. In view of this, $\tilde{\pi}^E((\Lambda^2\phi \otimes \alpha)(\kappa(u))) = 0$ implies that all values of $\kappa(u) : \Lambda^2\mathfrak{g}_- \to \mathfrak{p}$ must lie in the kernel of α. Since α is injective, this completes the proof. □

5.3.15. Chain preserving diffeomorphisms. To conclude the discussion of chains in Lagrangean contact structures, we show that a torsion–free Lagrangean contact structure can be essentially reconstructed from its path geometry of chains. This leads to a beautiful proof for the fact that a contactomorphism between Lagrangean contact manifolds which maps chains to chains (as unparametrized curves) has to be a homomorphism or an anti–homomorphism of Lagrangean contact structures. The essential step in that direction is that (as one may expect already from the proof of Corollary 5.3.14) one of the harmonic curvatures is induced by the mapping Φ_α only. Recall from 5.3.13 that Φ_α is induced by the map $\mathfrak{g} \times \mathfrak{g} \to \tilde{\mathfrak{p}}$ defined by $(X, Y) \mapsto [\alpha(X), \alpha(Y)] - \alpha([X, Y])$, which factors to $(\mathfrak{g}/\mathfrak{p}) \times (\mathfrak{g}/\mathfrak{p})$. By construction, the initial map is Q–equivariant, where Q acts on $\mathfrak{g}$ via the adjoint action and on $\tilde{\mathfrak{g}}$ via the composition of the adjoint action of $\tilde{G}$ with $i : Q \to \tilde{P}$.

Recall from 5.3.12 that the subspaces $\mathfrak{g}_1$ and $\mathfrak{g}_{-2}$ descend to Q–invariant subspaces in $\mathfrak{g}/\mathfrak{q}$. Let us denote by $\mathfrak{v} \subset \mathfrak{g}/\mathfrak{q}$ the subspace corresponding to $\mathfrak{g}_1$ and by $\mathfrak{h} \subset \mathfrak{g}/\mathfrak{q}$ the sum of the two subspaces. Then the explicit description of Φ_α from Lemma 5.3.13 in this picture says that Φ_α vanishes unless one of its entries comes from $\mathfrak{v}$, and inserting such an entry, the result factors to $(\mathfrak{g}/\mathfrak{q})/\mathfrak{h}$. Further, Φ_α has values in the semisimple part of $\tilde{\mathfrak{g}}_0$. Putting this together, we may view Φ_α as a Q–equivariant map $\mathfrak{v} \times ((\mathfrak{g}/\mathfrak{q})/\mathfrak{h}) \to \tilde{\mathfrak{g}}_0^{ss}$. Hence, it induces a natural bundle map between associated bundles on Cartan geometries of type (G, Q).

Now take a Lagrangean contact structure and the corresponding parabolic geometry $(\mathcal{G} \to M, \omega)$, put $\tilde{M} = \mathcal{G}/Q$ and take the corresponding Cartan geometry $(\mathcal{G} \to \tilde{M}, \omega)$ of type (G, Q). Clearly, the bundles $\mathcal{G} \times_Q \mathfrak{v}$ and $\mathcal{G} \times_Q \mathfrak{h}$ correspond to the vertical bundle $\tilde{V} \to \tilde{M}$ and the tautological bundle $\tilde{H} = \tilde{V} \oplus \tilde{E}$ for the generalized path geometry on $\tilde{M}$. Defining $\tilde{S} := T\tilde{M}/\tilde{H}$, we conclude that the natural bundle map induced by Φ_α can be actually viewed as a section of $\tilde{V}^* \otimes \tilde{S}^* \otimes L(\tilde{S}, \tilde{S})$. This is the bundle in which the harmonic curvature quantity, which was called ρ in 4.4.3, has its values (taking into account that $L(\tilde{S}, \tilde{S}) \cong L(\tilde{V}, \tilde{V})$).

LEMMA 5.3.15. *For a torsion-free Lagrangean contact structure, the harmonic curvature* ρ *of the associated path geometry of chains (see 4.4.3) coincides (up to a nonzero factor) with the bundle map induced by* Φ_α.

PROOF. We know that the curvature function $\tilde{\kappa}$ of the path geometry is determined by equation (5.39) from the proof of Theorem 5.3.13. Here $\tilde{\kappa} \circ j$ is a function from $\mathcal{G}$ to $\ker(\partial^*) \subset \Lambda^2\tilde{\mathfrak{p}}_+ \otimes \tilde{\mathfrak{g}}$. The function representing the harmonic part is obtained from this by applying the projection $\pi_H : \ker(\partial^*) \to \ker(\partial^*)/\operatorname{im}(\partial^*)$. From the discussion in 4.4.3 we know that (the part of positive homogeneity in) the latter representation splits into the direct sum of two irreducible components, which are the highest weight subspaces in $\tilde{\mathfrak{g}}_1^E \otimes \tilde{\mathfrak{g}}_2 \otimes \mathfrak{g}_{-1}^V$ and in $\tilde{\mathfrak{g}}_1^V \otimes \tilde{\mathfrak{g}}_2 \otimes \tilde{\mathfrak{g}}_0$, respectively. In view of this decomposition, we can write π_H as $\pi_H^E + \pi_H^V$, where the superscript indicates which component in $\mathfrak{g}_1$ is involved. The harmonic curvature ρ is then represented by the function $\pi_H^V \circ \tilde{\kappa} \circ j$.

Now recall from the proof of Theorem 5.3.14 that $\tilde{\kappa}$ has values in $(\tilde{\mathfrak{p}}_+ \wedge \tilde{\mathfrak{g}}_2) \otimes \tilde{\mathfrak{g}}$. In the proof of Corollary 5.3.14, we have constructed a $\tilde{P}$–equivariant projection $\tilde{\pi}^V$ from that space to $(\tilde{\mathfrak{g}}_1^V \wedge \tilde{\mathfrak{g}}_2) \otimes \tilde{\mathfrak{g}}$. Now consider the restriction of π_H^V to $(\tilde{\mathfrak{p}}_+ \wedge \tilde{\mathfrak{g}}_2) \otimes \tilde{\mathfrak{g}}$. The image of π_H^V is an irreducible representation of $\tilde{\mathfrak{g}}_0$, which by Kostant's version of the Bott–Borel–Weil Theorem (Theorem 3.3.5) occurs with multiplicity one in $\Lambda^*\tilde{\mathfrak{p}}_+ \otimes \tilde{\mathfrak{g}}$. In particular, the kernel of $\tilde{\pi}^V$ cannot contain any irreducible component isomorphic to this. Hence, π_H^V has to vanish on $\ker(\tilde{\pi}^V)$ and thus factors through $\tilde{\pi}^V$. But in the proof of Corollary 5.3.14 we have seen that $\tilde{\pi}^V(\tilde{\kappa} \circ j) = \tilde{\pi}^V(\Phi_\alpha)$. Hence, we see that ρ is actually induced by $\pi_H^V \circ \Phi_\alpha$. Finally, in the proof of Theorem 5.3.13, we have seen that Φ_α lies in an irreducible representation of $\tilde{\mathfrak{g}}_0$, which is isomorphic to the highest weight component in $\tilde{\mathfrak{g}}_1^V \otimes \tilde{\mathfrak{g}}_2 \otimes \tilde{\mathfrak{g}}_0$. By Schur's lemma, the restriction of π_H^V to this irreducible representation is an isomorphism, which is unique up to scale. □

Having the harmonic curvature $\rho \in \Gamma(\tilde{V}^* \otimes \tilde{S}^* \otimes L(\tilde{S}, \tilde{S}))$ we may follow the developments in Lemma 5.3.13. Via the isomorphism $\tilde{V} \cong \tilde{E}^* \otimes \tilde{S}$, we can view ρ as a section of $\tilde{E} \otimes \otimes^3\tilde{S}^* \otimes \tilde{S}$, and then Lemma 5.3.13 shows that ρ is completely symmetric and tracefree. Since $\tilde{E}$ is a line bundle, we may also view this as a section of $(S^3\tilde{S}^* \otimes \tilde{S})_0$ determined up to scale. We interpret this map in the underlying Lagrangean contact structure. Let $H \subset TM$ be the contact bundle and $H = E \oplus F$ the decomposition into Legendrean subbundles.

Now fix $x \in M$ and a point $\ell \in \tilde{M}$ over x. For a tangent vector $\xi \in T_xM$, we can choose a lift $\tilde{\xi} \in T_\ell\tilde{M}$, and then project to its class in $\tilde{S}_\ell = T_\ell\tilde{M}/\tilde{H}_\ell$. Since the vertical subbundle of $\tilde{M} \to M$ is $\tilde{V}_\ell \subset \tilde{H}_\ell$, the result depends only on ξ. Moreover, since the projection $T\tilde{M} \to TM$ corresponds to $\mathfrak{g}/\mathfrak{q} \to \mathfrak{g}/\mathfrak{p}$, we conclude that this restricts to a linear isomorphism $H_x \to \tilde{S}_\ell$. Consequently, we can interpret $\rho(\ell)$ as an element of $S^3H_x^* \otimes H_x$ determined up to scale. To describe this map more explicitly, consider the bundle map $\mathbb{J} : H \to H$ from 5.3.8, which is the identity on E and minus the identity on F. There we have also considered the bundle map $H \times H \to TM/H$ defined by $(\xi, \eta) \mapsto \mathcal{L}(\xi, \mathbb{J}\eta)$, where $\mathcal{L}$ denotes the Levi bracket. In each fiber, this is a nondegenerate symmetric bilinear form, for which the subspaces E and F are isotropic. Using this, we can now formulate

THEOREM 5.3.15. *Let* $(M, H = E \oplus F)$ *be a torsion-free Lagrangean contact structure, and let* $\mathbb{J} : H \to H$ *be the map acting as the identity on* E *and as minus the identity on* F. *Then for a point* $x \in M$ *and any point* $\ell \in \tilde{M}$ *over* x, *let*

$\Psi : S^3 H_x \to H_x$ be the map induced by the value $\rho(\ell)$ of the harmonic curvature of the path geometry of chains.

(1) Up to a nonzero multiple, Ψ is the complete symmetrization of the mapping

$$(\xi, \eta, \zeta) \mapsto \mathcal{L}(\xi, \mathbb{J}\eta)\mathbb{J}\zeta.$$

(2) A tangent vector $\xi \in H_x$ lies in $E_x \cup F_x$ if and only $\Psi(\xi, \xi, \xi) = 0$ and there is a nonzero element $\eta \in H_x$ such that $\Psi(\xi, \xi, \eta)$ is a nonzero multiple of ξ.

PROOF. (1) We have to interpret the claimed map, which actually has values in $(T_xM/H_x) \otimes H_x$, as a map with values in H_x defined up to scale. Second, observe that on the Lie algebra $\mathfrak{g}$, the map $\mathfrak{g}_{-1} \times \mathfrak{g}_{-1} \to \mathfrak{g}_{-2}$ corresponding to $\mathcal{L}(\xi, \mathbb{J}(\eta))$ is as follows. Given to elements (X_1, X_2) and (Y_1, Y_2) of $\mathfrak{g}_{-1}$, the resulting matrix has its only nonzero entry in the lower left corner, and this entry is given by $Y_2 X_1 + X_2 Y_1$. On the other hand, $\mathbb{J}$ maps a third element (Z_1, Z_2) to $(Z_1, -Z_2)$. Hence, we see that on the Lie algebra level our map is given by the complete symmetrization of $(X, Y, Z) \mapsto (\langle X_1, Y_2^t \rangle + \langle Y_1, X_2^t \rangle)(Z_1, -Z_2)$. Now part (1) follows directly from the description of Φ_α in Lemma 5.3.13.

(2) then follows by linear algebra: By part (1), for a vector $\xi \in H_x$, we get $\Psi(\xi, \xi, \xi) = \mathcal{L}(\xi, \mathbb{J}\xi)\mathbb{J}\xi$, so this vanishes if and only if $\mathcal{L}(\xi, \mathbb{J}\xi) = 0$. Assuming this, symmetry of $\mathcal{L}(\ , \mathbb{J}\ )$ implies that $\Psi(\xi, \xi, \eta) = 2\mathcal{L}(\xi, \mathbb{J}\eta)\mathbb{J}\xi$, so this is a multiple of $\mathbb{J}\xi$ for any η. Since $\mathcal{L}(\ , \mathbb{J}\ )$ is nondegenerate, we can find a vector η for which this is a multiple of ξ if and only if ξ is an eigenvector for $\mathbb{J}$, i.e. if and only if $\xi \in E_x \cup F_x$. □

Using this, we can now analyze chain preserving contactomorphisms. For two Lagrangean contact structures given by subbundles E_i and F_i for $i = 1, 2$, let us define an *anti–homomorphism* to be a contactomorphism f for which each tangent map $T_x f$ maps $(E_1)_x$ to $(F_2)_{f(x)}$ and $(E_2)_x$ to $(F_1)_{f(x)}$.

COROLLARY 5.3.15. *A contactomorphism between two torsion–free Lagrangean contact structures which maps chains to chains (viewed as unparametrized curves) is either a homomorphism or an anti–homomorphism of the Lagrangean contact structures.*

PROOF. Any contactomorphism $f : M_1 \to M_2$ evidently lifts to a diffeomorphism $\mathcal{P}_0(TM_1) \to \mathcal{P}_0(TM_2)$. The fact that f preserves chains exactly means the this lift is a morphism of the natural generalized path geometries on the two spaces. Thus, it is compatible with the harmonic curvatures of the two geometries. Hence, the restrictions of the tangent maps of f to the contact subbundle are compatible with the maps Ψ, and the result follows immediately from part (2) of the theorem. □

REMARK 5.3.15. The developments in 5.3.12–5.3.15 can be carried out in a completely parallel way for partially integrable CR–structures of hypersurface type; see [**ČŽ05**]. One obtains similar formulae for i and α, and the extension functor produces a normal geometry exactly on the subclass of integrable CR–structures. For this subclass, the CR–structure (including its signature!) can again be recovered from the harmonic curvature of the path geometry of chains. For such structures, chain preserving contactomorphisms are then CR–diffeomorphisms or anti–CR–diffeomorphisms, i.e. the restriction of each tangent map to the contact subbundle is complex linear or conjugate linear.

For contact projective structures, the path geometry of chains is of course obtained by the correspondence space construction from 4.4.3 applied to the induced projective structure discussed in 5.3.11. In particular, one obtains a flat path geometry in the case of the homogeneous model. Describing the contact projective structure in terms of the path geometry is not an issue here, since by Theorem 5.3.11 the associated projective structure is, from the point of view of paths, indeed an extension of the contact projective structure. Still it is true that chain preserving diffeomorphisms preserve the structure in this case.

It is also clear that 5.3.12–5.3.15 must have an analog for any parabolic contact structure, however, only weaker results can be expected. The point here is that for all remaining parabolic contact structures (Lie contact structures as discussed in 4.2.5 and the exotic parabolic contact structure from 4.2.8), all harmonic curvature quantities are of torsion type. Therefore, one has to expect that in all these cases, the close relation between the path geometry of chains and the original parabolic contact structure on the level of the canonical Cartan connections will only work for locally flat structures. In the case of Lie contact structures this has actually been verified in [**Ž09**].

APPENDIX A

Other prolongation procedures

In this appendix, we describe alternative proofs for the equivalence of regular normal parabolic geometries to underlying structures, which were found before the one presented in Section 3.1. We start by discussing Tanaka's approach. This was based on a different description of underlying structures, and we first review this description and its equivalence to regular infinitesimal flag structures. Then we describe Tanaka's construction of a normal Cartan connection. Next, we switch to Morimoto's general results on Cartan connections associated to geometric structures on filtered manifolds. Finally, we describe the prolongation procedure developed by Čap and Schichl.

A.1. Tanaka's description of underlying structures. Starting, in particular, from the example of CR–structures of hypersurface type, N. Tanaka realized that there is a general approach to Cartan connections related to simple graded Lie algebras. Independently of Kostant's work he found the algebraic versions of regularity and normality and showed that resulting conditions on the curvature of a Cartan connection make sure that it is uniquely determined by an underlying geometric structure. The general version of these results can be found in [**Tan79**]. The underlying structures used by Tanaka are different from regular infinitesimal flag structures as defined in 3.1.6 and 3.1.7.

While Tanaka's version of underlying structures in our opinion is geometrically less transparent than infinitesimal flag structures, it has the advantage that it is closely related to classical first order G–structures. Although Tanaka's prolongation procedure which we will discuss below only works under small additional restrictions, we will describe his version of underlying structures and the relation to infinitesimal flag structures in general. We will stick to the real (smooth) version of the theory, but there is also an obvious holomorphic analog available.

The basic idea for Tanaka's description of underlying structures as well as their equivalence to infinitesimal flag structures is easy to explain. To get from a parabolic geometry to the underlying infinitesimal flag structure, one passes from the filtration preserving map $\omega : T\mathcal{G} \to \mathfrak{g}$ to the associated graded object and at the same time to the smallest underlying bundle on which this associated graded object makes sense. To come to Tanaka's description of underlying structures, one passes from the filtration preserving map ω to the equivalence class of all forms which differ from ω by a map of strictly positive homogeneity, and again goes to a simple bundle on which this makes sense.

Let us start with a $|k|$–graded semisimple Lie algebra $\mathfrak{g} = \mathfrak{g}_{-k} \oplus \cdots \oplus \mathfrak{g}_k$ and a Lie group G with Lie algebra $\mathfrak{g}$. Let $P \subset G$ be a parabolic subgroup for the given grading and let $G_0 \subset P$ be the Levi subgroup; see 3.1.3. Consider the filtered vector space $\mathfrak{g}/\mathfrak{p}$ and let $GL_f(\mathfrak{g}/\mathfrak{p})$ be the group of all filtration preserving linear isomorphisms of $\mathfrak{g}/\mathfrak{p}$. This is a closed subgroup of $GL(\mathfrak{g}/\mathfrak{p})$ and thus a Lie group. Any element of $GL_f(\mathfrak{g}/\mathfrak{p})$ induces a linear automorphism of the associated graded $\mathrm{gr}(\mathfrak{g}/\mathfrak{p}) = \mathfrak{g}_-$, so we get a smooth homomorphism $\pi : GL_f(\mathfrak{g}/\mathfrak{p}) \to GL(\mathfrak{g}_-)$, whose kernel we denote by $GL_+(\mathfrak{g}/\mathfrak{p})$. On the other hand, the adjoint action induces a homomorphism $G_0 \to GL(\mathfrak{g}_-)$. Now we define $G_0^\#$ to be the fibered product of G_0 and $GL_f(\mathfrak{g}/\mathfrak{p})$ with respect to this homomorphism, i.e.

$$G_0^\# = \{(g, h) \in G_0 \times GL_f(\mathfrak{g}/\mathfrak{p}) : \mathrm{Ad}(g) = \pi(h)\}.$$

This is a closed subgroup of $G_0 \times GL_f(\mathfrak{g}/\mathfrak{p})$ and thus a Lie group. If G is the adjoint group of $\mathfrak{g}$ (which is the only case considered by Tanaka), then $G_0^\#$ can be identified with the closed subgroup of $GL_f(\mathfrak{g}/\mathfrak{p})$ generated by P and $GL_+(\mathfrak{g}/\mathfrak{p})$. In general, mapping $h \in GL_+(\mathfrak{g}/\mathfrak{p})$ to (e, h) induces a homomorphism $GL_+(\mathfrak{g}/\mathfrak{p}) \to G_0^\#$, and since $GL_+(\mathfrak{g}/\mathfrak{p}) \subset GL_f(\mathfrak{g}/\mathfrak{p})$ is a normal subgroup, the image of this homomorphism is a normal subgroup of $G_0^\#$. Clearly, this image coincides with the kernel of the homomorphism $G_0^\# \to G_0$ induced by the first projection. We can split this projection by mapping $g \in G_0$ to $(g, \mathrm{Ad}(g))$, so $G_0^\#$ is actually the semidirect product of the normal subgroup $GL_+(\mathfrak{g}/\mathfrak{p})$ and the subgroup $\{(g, \mathrm{Ad}(g))\} \cong G_0$.

The second projection defines a smooth homomorphism $\phi : G_0^\# \to GL(\mathfrak{g}/\mathfrak{p})$. The kernel of ϕ is visibly isomorphic to the kernel of $\mathrm{Ad} : G_0 \to GL(\mathfrak{g}_-)$ and thus a discrete subgroup. Hence, it makes sense to consider first order $G_0^\#$–structures on smooth manifolds M of dimension $\dim(\mathfrak{g}/\mathfrak{p})$; see 1.3.6. By definition, such a structure is given by a smooth principal bundle $p : E \to M$ with structure group $G_0^\#$ endowed with a one–form $\tau \in \Omega^1(E, \mathfrak{g}/\mathfrak{p})$ such that $(r^{(g,h)})^*\tau = h \circ \tau$ and which is strictly horizontal, i.e. $\ker(\tau(u)) = V_uE$ for all $u \in E$. As usual for first order G–structures, the one–form τ induces an isomorphism $TM \cong E \times_{G_0^\#} (\mathfrak{g}/\mathfrak{p})$.

By construction, any element of $G_0^\#$ acts on $\mathfrak{g}/\mathfrak{p}$ by a filtration preserving linear automorphism. Hence, the filtration of $\mathfrak{g}/\mathfrak{p}$ induces a filtration $TM = T^{-k}M \supset \dots \supset T^{-1}M$ of the tangent bundle of M by smooth subbundles such that the rank of T^iM coincides with the dimension of $\mathfrak{g}^i/\mathfrak{p}$ for all $i = -k, \dots, -1$. This gives rise to a filtration of TE with $T^iE = (Tp)^{-1}(T^iM)$ for $i = -k, \dots, -1$ and T^0E the vertical bundle. Now Tanaka calls a $G_0^\#$–structure $(p : E \to M, \tau)$ *of type* $\mathfrak{m}$ if and only if identifying $\mathfrak{g}/\mathfrak{p}$ with $\mathfrak{g}_-$, for $\xi \in T^iE$ and $\eta \in T^jE$ the exterior derivative $d\tau(\xi, \eta)$ is congruent to $-[\tau(\xi), \tau(\eta)]$ modulo $\mathfrak{g}^{i+j+1}$. (The terminology comes from the fact that Tanaka uses the notation $\mathfrak{m}$ for what we call $\mathfrak{g}_-$.)

In the language introduced in Section 3.1, Tanaka's way to associate an underlying $G_0^\#$–structure to a parabolic geometry can be described as follows. Let $(p : \mathcal{G} \to M, \omega)$ be a parabolic geometry of type (G, P). Let $P_+^k \subset P$ be the closed subgroup $\exp(\mathfrak{g}_k)$; see 3.1.3. This subgroup lies in the kernel of the homomorphism $\underline{\mathrm{Ad}} : P \to GL(\mathfrak{g}/\mathfrak{p})$ induced by the adjoint action and the resulting homomorphism $P/P_+^k \to GL(\mathfrak{g}/\mathfrak{p})$ (which we again denote by $\underline{\mathrm{Ad}}$) is infinitesimally injective. As described in 3.1.15, $\mathcal{G}_{k-1} = \mathcal{G}/P_+^k$ is a principal bundle over M with structure group P/P_+^k. By part (2) of Proposition 3.1.15, the Cartan connection ω induces a one–form $\omega_{k-1}^{-k} \in \Omega^1(\mathcal{G}_{k-1}, \mathfrak{g}/\mathfrak{p})$ which is strictly horizontal and equivariant for the action $\underline{\mathrm{Ad}}$.

Next, the projection $P \to P/P_+ \cong G_0$ and $\underline{\mathrm{Ad}} : P \to GL_f(\mathfrak{g}/\mathfrak{p})$ evidently induce a homomorphism $P \to G_0^\#$. A moment of thought shows that the kernel of this homomorphism is exactly $P_+^k \subset P$, so we can naturally view P/P_+^k as a subgroup of $G_0^{\#}$. Now $E := \mathcal{G}_{k-1} \times_{P/P_+^k} G_0^{\#}$ is a principal bundle with structure group $G_0^{\#}$ and one easily shows that ω_{k-1}^{-k} can be uniquely extended to a strictly horizontal $G_0^{\#}$–equivariant form $\tau \in \Omega^1(E, \mathfrak{g}/\mathfrak{p})$. Hence, we have constructed an underlying $G_0^{\#}$–structure from a parabolic geometry of type (G, P). Similarly to the proof of Proposition 3.1.7, one shows that a parabolic geometry is regular if and only if its underlying $G_0^{\#}$–structure is of type $\mathfrak{m}$.

A.2. The equivalence to regular infinitesimal flag structures. The relation between $G_0^{\#}$–structures of type $\mathfrak{m}$ and regular infinitesimal flag structures is easy to describe. An analog of the procedure used for parabolic geometries associates to any $G_0^{\#}$–structure an underlying infinitesimal flag structure, which is regular if and only if the $G_0^{\#}$–structure is of type $\mathfrak{m}$. Conversely, an infinitesimal flag structure can be extended to a $G_0^{\#}$–structure by a very simple version of prolongation.

More explicitly, consider a $G_0^{\#}$–structure $(p : E \to M, \omega)$. The subgroup $GL_+(\mathfrak{g}/\mathfrak{p}) \subset G_0^{\#}$ acts freely on E, and we define $E_0 := E/GL_+(\mathfrak{g}/\mathfrak{p})$. The obvious projection $p_0 : E_0 \to M$ is a smooth principal bundle with structure group $G_0^{\#}/GL_+(\mathfrak{g}/\mathfrak{p}) \cong G_0$. The filtration of the tangent bundle of M induces a filtration of TE_0 by smooth subbundles, and we define T^0E_0 to be the vertical bundle. For $i = -k, \dots, -1$, a point $u_0 \in E_0$ and a tangent vector $\xi \in T_{u_0}E_0$ choose a point $u \in E$ lying over u_0 and a tangent vector $\tilde{\xi}$ lying over ξ. By construction $\tau(u)(\tilde{\xi})$ lies in $\mathfrak{g}^i/\mathfrak{p}$ and is independent of the choice of $\tilde{\xi}$ and we define $\theta_i(u)(\xi)$ to be the $\mathfrak{g}_i$–component of $\tau(u)(\tilde{\xi})$.

As in 3.1.5, one shows that θ_i is a well–defined smooth section of the bundle $L(T^iE_0, \mathfrak{g}_i)$, and that $(p_0 : E_0 \to M, (\theta_{-k}, \dots, \theta_{-1}))$ is an infinitesimal flag structure of type (G, P). Any morphism of $G_0^{\#}$–structures induces a morphism of the underlying infinitesimal flag structures, so this actually defines a functor. Using Proposition 3.1.7 one easily shows that regularity of the underlying infinitesimal flag structure is equivalent to the $G_0^{\#}$–structure being of type $\mathfrak{m}$.

From 3.1.13 we know that any infinitesimal flag structure comes from some parabolic geometry, and hence it can be realized as the underlying infinitesimal flag structure of some $G_0^{\#}$–structure. This can also be easily proved directly along the following lines. Given an infinitesimal flag structure $(p_0 : E_0 \to M, \theta)$, one first extends the frame form θ to a G_0–equivariant one–form $\tilde{\theta} \in \Omega^1(E_0, \mathfrak{g}_-)$. To do this, one simply chooses a principal connection γ which gives a splitting $TE_0 = VE_0 \oplus HE_0$ such that $HE_0 \cong p_0^*TM$. Next, one chooses an isomorphism between TM and its associated graded $\operatorname{gr}(TM)$. Pulling this back via p_0, we can identify the filtered vector bundle HE_0 with its associated graded, and since $VE_0 = T^0E_0$, we obtain an isomorphism $TE_0 \cong \operatorname{gr}(TE_0)$, which by construction is G_0–equivariant. Using this, we can then define the $\mathfrak{g}_i$–component of $\tilde{\theta}(\xi)$ to be the value of θ_i on the image of ξ in $\operatorname{gr}_i(TE_0)$. This is an extension of θ as required.

Next, we extend the structure group by putting $E := E_0 \times_{G_0} G_0^{\#}$. Then $p : E \to M$ is a principal fiber bundle with structure group $G_0^{\#}$. We have an evident bundle map $j : E_0 \to E$, which is a homomorphism of principal bundles over the inclusion $G_0 \hookrightarrow G_0^{\#}$. Now for $u \in E_0$, the tangent space $T_{j(u)}E$ is spanned by $T_uj(T_uE_0)$ and the vertical subspace. Hence, there is a unique linear map $\tau(j(u)) : T_{j(u)}E \to \mathfrak{g}_-$ which vanishes on the vertical subspace and such that

$\tau(j(u)) \circ T_u j = \tilde{\theta}(u)$. One easily verifies that this can be uniquely extended to a strictly horizontal $G_0^{\#}$–equivariant one form $\tau \in \Omega^1(E, \mathfrak{g}_-)$, so (E, τ) is a $G_0^{\#}$–structure. By construction, its underlying infinitesimal flag structure is (E_0, θ).

To get the required equivalence of categories it suffices to show that any morphism between the infinitesimal flag structures underlying two $G_0^{\#}$–structures of type $\mathfrak{m}$ uniquely lifts to a morphism of the $G_0^{\#}$–structures. This can be done completely parallel to the proof of Theorem 3.1.14: First, a homomorphism of infinitesimal flag structures locally over the base lifts to a principal bundle homomorphism between the $G_0^{\#}$–structures. The difference between the one–form on the first bundle and the pullback of the other one–form by construction is homogeneous of degrees ≥ 1, so one may view it as a map to the Lie algebra $\mathfrak{gl}_+(\mathfrak{g}/\mathfrak{p})$ of $GL_+(\mathfrak{g}/\mathfrak{p})$. Exponentiating this, one obtains an automorphism of the first bundle that composed with the original map produces a pullback whose difference to the original one–form is homogeneous of degree ≥ 2, and in finitely many steps one arrives at local morphisms of $G_0^{\#}$–structures covering the given morphism of infinitesimal flag structures. Then one verifies that this condition determines the principal bundle map up to multiplication by a smooth function with values in the subgroup of $G_0^{\#}$ consisting of those elements which act trivially on $\mathfrak{g}/\mathfrak{p}$. These elements are of the form (g, e), where $g \in G_0$ is such that $\mathrm{Ad}(g) = \mathrm{id}$ on $\mathfrak{g}_-$. Hence, these multiplications are already visible on the underlying infinitesimal flag structures so we get uniqueness of the local morphisms of $G_0^{\#}$–structures, which thus patch together to a uniquely determined global morphism. Hence, we have proved

PROPOSITION A.2. *Passing to the underlying infinitesimal flag structure determines an equivalence between the category of $G_0^{\#}$–structures of type $\mathfrak{m}$ and the category of regular infinitesimal flag structures.*

A.3. Tanaka's prolongation procedure. Tanaka's description of underlying structures leads to a prolongation procedure in two seemingly different steps. As we have noted in A.1, there is a natural homomorphism $P \to G_0^{\#}$ with kernel P_+^k, which embeds P/P_+^k as a closed subgroup into $G_0^{\#}$. As described in A.1, the starting point of Tanaka's procedure is a first order $G_0^{\#}$–structure $p : E \to M$ with certain properties. In the first part of the prolongation procedure, one constructs a reduction of structure group to the subgroup $P/P_+^k \subset G_0^{\#}$. The result of this step is a principal bundle $\tilde{p} : \tilde{E} \to M$ with structure group P/P_+^k together with a soldering form $\tilde{\theta} \in \Omega^1(\tilde{E}, \mathfrak{g}/\mathfrak{p})$; see 1.3.6. It corresponds to the first order structure $(\mathcal{G}_{k-1}, \omega_{k-1}^{-k})$ obtained as an underlying structure of a parabolic geometry in Proposition 3.1.15. In the second step of the prolongation procedure, this bundle is extended to a principal P–bundle $\mathcal{G} \to M$ and one constructs a Cartan connection $\omega \in \Omega^1(\mathcal{G}, \mathfrak{g})$, which induces $\tilde{\theta}$ in the sense described in 3.1.15.

Let us describe the two steps in more detail. The first step of the procedure is carried out in §3 of [**Tan79**]. One starts by constructing a sequence of subgroups

$$G_0^{\#} = U^0 \supset U^1 \supset \cdots \supset U^{k-2} \supset U^{k-1} = P/P_+^k.$$

As mentioned above, Tanaka assumes that G is the adjoint group of $\mathfrak{g}$. Hence, $\mathrm{Ad} : G_0 \to GL(\mathfrak{g}_-)$ is injective and $G_0^{\#}$ can be viewed as a closed subgroup of $GL_f(\mathfrak{g}/\mathfrak{p})$. Explicitly, this subgroup consists of those maps which are congruent to the adjoint action of an element of P modulo maps of strictly positive homogeneity. In this picture, the subgroup U^ℓ consists of those linear maps which are congruent to the adjoint action of an element of P up to maps which are homogeneous of degree $\geq \ell + 1$. From the length of the filtration of $\mathfrak{g}/\mathfrak{p}$ it then follows that U^{k-1} indeed coincides with $\underline{\mathrm{Ad}}(P) \cong P/P_+^k$.

Since $U^\ell \subset GL(\mathfrak{g}/\mathfrak{p})$, there is the notion of a first order U^ℓ–structure on manifolds of dimension $\dim(G/P)$. Moreover, for such structures, the notion of being of type $\mathfrak{m}$ can be defined completely analogously to the case of U^0–structures discussed in A.1. Given a first order U^ℓ–structure $(p : E^\ell \to M, \theta)$, Tanaka introduces the notion of an ℓ–system. This is given by a collection θ_i of $\mathfrak{g}_i$–valued one forms on E^ℓ for $i = 0, \dots, \ell - 1$, which are used to extend the soldering form. There is a natural notion of equivariancy for such forms. Since the Lie algebra of U^ℓ can be identified with the sum of $\mathfrak{g}_0 \oplus \cdots \oplus \mathfrak{g}_\ell$ and the space of all linear maps on $\mathfrak{g}/\mathfrak{p}$ which are homogeneous of degree $> \ell$, it also makes sense to require that these forms reproduce the generators of fundamental vector fields.

Given such an ℓ–system, one can define the analog of the Cartan curvature. Requiring the homogeneous components of degree $\leq \ell$ of this curvature to have values in the kernel of ∂^*, one obtains the notion of a *normal* ℓ–system. Given a normal ℓ–system, one tries to extend the system further by a form θ_ℓ with values in $\mathfrak{g}_\ell$. This leads to the notion of a pre–$(\ell+1)$–system, and again there is a version of normality for such systems. The key part of the first prolongation step is then to show that a normal ℓ–system on a U^ℓ–structure can be extended to a normal pre–$(\ell+1)$–system, which in turn leads to a reduction to a $U^{\ell+1}$–structure endowed with a normal $(\ell+1)$–system.

The relation of these notions to the procedure developed in Section 3.1 is as follows. For $\ell < k$, consider a P–frame bundle of degree ℓ as defined in 3.1.15. Then one has to extend the structure group $P/P_+^{\ell+1}$ to U^ℓ by adding arbitrary linear maps homogeneous of degree larger than ℓ. The frame form can be uniquely extended equivariantly to a collection of partially defined forms on that larger bundle, and any extension of these partially defined forms to true one–forms gives rise to a U^ℓ–structure endowed with an ℓ–system. Normality of the P–frame bundle is then equivalent to normality of the ℓ–system. A subtle point in Tanaka's procedure is that in many cases one has to take into account identifications of subspaces of $\mathfrak{g}$ and $\mathfrak{gl}(\mathfrak{g}/\mathfrak{p})$ with subquotients of these Lie algebras, which makes it hard to keep track of the natural actions of the groups involved.

At the end of the first step, one arrives at a first order P/P_+^k–structure, say $\mathcal{G}_k \to M$, which admits a normal $(k-1)$–system. This is the input for the second part of the procedure, which is described in §4 of [**Tan79**]. The normal $(k-1)$–system can be extended to a normal pre–k–system, and Tanaka shows that such an extension gives rise to a cocycle of transition functions on $\mathcal{G}_k$ with values in the group P_+^k. This cocycle is used to define a principal P_+^k–bundle $\mathcal{G} \to \mathcal{G}_k$. Next one has to define an action of P on $\mathcal{G}$ and prove that it can be made into a principal bundle over M with structure group P. Finally, one constructs a regular normal Cartan connection on $\mathcal{G}$ from the chosen normal pre–k–system. This completes the proof that any $G_0^\#$–structure of type $\mathfrak{m}$ is induced by a regular normal Cartan connection. In §5 of [**Tan79**], uniqueness of normal Cartan connections is proved. Using the specific features of the construction, Tanaka shows directly that an isomorphism of first order $G_0^\#$–structures of type $\mathfrak{m}$ induces an isomorphism of the resulting normal Cartan geometries.

A.4. Morimoto's prolongation procedure. In his article [**Mo93**], T. Morimoto studied geometric structures on filtered manifolds and proved a general existence theorem for normal Cartan connections in this setting. The key notions in Morimoto's theory are skeletons and towers. Let $V = V^{-k} \supset \cdots \supset V^{-1} \supset V^0 = \{0\}$ be a finite–dimensional filtered vector space. Further, let H be a (possibly infinite–dimensional) Lie group endowed with a decreasing filtration $H = H^0 \supset H^1 \supset \dots$ by closed normal subgroups of finite codimension, which satisfies some technical conditions. Using the induced filtration on the Lie algebra $\mathfrak{h}$ of H, one obtains a filtration on $E := V \oplus \mathfrak{h}$. A *skeleton* on V is then given by an extension of the adjoint representation of H to a representation on E, which has the property that each H^ℓ acts by maps which are homogeneous of degree $\geq \ell$.

The first main result in [**Mo93**] is that for any finite–dimensional filtered vector space V, there is a maximal skeleton $(E(V), H(V))$ into which any skeleton can be embedded. More generally, suppose that there is a skeleton (E, H) on V and for some $k \geq 0$ consider $H^{(k)} := H/H^{k+1}$. Then there is a maximal skeleton $(E(V, H^{(k)}), H(V, H^{(k)}))$ such that $H(V, H^{(k)})/H^{k+1}(V, H^{(k)}) \cong H^{(k)}$. These maximal skeletons are constructed inductively by an algebraic version of prolongation.

Next, suppose that M is a filtered manifold of type V (i.e. each tangent space of M is isomorphic to V as a filtered vector space), and that (E, H) is a skeleton on V. Then a *tower* on M with skeleton (E, H) is a principal H–bundle $p : \mathcal{P} \to M$ endowed with a one–form $\theta \in \Omega^1(\mathcal{P}, E)$ which defines an H–equivariant absolute parallelism that reproduces the generators of fundamental vector fields. Hence, θ has all the defining properties of a Cartan connection, but E is not supposed to be a Lie algebra. There is an obvious notion of morphisms of towers.

Morimoto's second crucial result is that on any filtered manifold of type V, there is a maximal tower with skeleton $(E(V), H(V))$. As before, there is a more general version. Namely, given a tower $(p : \mathcal{P} \to M, \theta)$, one may form $\mathcal{P}^{(k)} := \mathcal{P}/H^{k+1}$. It turns out that $\mathcal{P}^{(k)}$ can be uniquely extended to a maximal tower with skeleton $(E(V, H^{(k)}), H(V, H^{(k)}))$. This is called the universal prolongation of $\mathcal{P}^{(k)}$. Again, the maximal towers are constructed inductively by a formal prolongation procedure.

Next, the kth noncommutative frame bundle $\mathcal{R}^{(k)}M$ of M is obtained by factoring the maximal tower on M by $H^{k+1}(V)$. For $k = 0$, one obtains the frame bundle of the associated graded of TM. A geometric structure of order $k + 1$ is then defined as a principal subbundle $\mathcal{P}^{(k)} \subset \mathcal{R}^{(k)}M$, which can be obtained from factoring some tower as above.

Any tower has an associated structure function, which simply described the exterior derivative of θ via the trivialization of the tangent bundle. Truncating a tower as above, some parts of the structure function descend, thus giving rise to invariants of geometric structures of order $k + 1$. Using the structure function, Morimoto studies the equivalence problem for such geometries. Concerning Cartan connections, the main result of [**Mo93**] is a general existence theorem. By definition, Cartan geometries are special towers (with finite–dimensional structure group).

In his study of Cartan connections, Morimoto considers filtered manifolds as defined in 3.1.7, for which the bundle of symbol algebras is locally trivial with fiber a fixed nilpotent graded Lie algebra $\mathfrak{n} = \mathfrak{n}_{-k} \oplus \cdots \oplus \mathfrak{n}_{-1}$. Then he considers the natural frame bundle for $\text{gr}(TM)$ with structure group $\text{Aut}_{\text{gr}}(\mathfrak{n})$ of automorphisms of the graded Lie algebra $\mathfrak{n}$. The basic question then is when a reduction $\mathcal{P}^{(0)}$

of this frame bundle to the structure group $G_0 \subset \mathrm{Aut}_{\mathrm{gr}}(\mathfrak{n})$ gives rise to a Cartan connection.

From above, we know that the bundle $\mathcal{P}^{(0)}$ admits a maximal prolongation with skeleton $(E(\mathfrak{n}, G_0), G(\mathfrak{n}, G_0))$. Let N be a simply connected Lie group with Lie algebra $\mathfrak{n}$. Via the left trivialization $TN \cong N \times \mathfrak{n}$, this becomes a filtered manifold with the bundle of symbol algebras also isomorphic to $N \times \mathfrak{n}$. In particular, we can view $N \times G_0$ as a reduction of structure group of $\mathrm{gr}(TN)$. Morimoto next proves that this trivial bundle can be canonically extended to a tower with constant structure function. Let $K(\mathfrak{n}, G_0)$ be the structure group of this tower and $\mathfrak{k}(\mathfrak{n}, G_0)$ its Lie algebra. Then $\mathfrak{l}(\mathfrak{n}, G_0) := \mathfrak{n} \oplus \mathfrak{k}(\mathfrak{n}, G_0)$ can be naturally made into a Lie algebra containing $\mathfrak{n}$ as a subalgebra, and $(\mathfrak{l}(\mathfrak{n}, G_0), K(\mathfrak{n}, G_0))$ is a subskeleton of $(E(\mathfrak{n}, G_0), G(\mathfrak{n}, G_0))$.

The final ingredient in Morimoto's procedure is the so–called *condition (C)*. This axiomatically requires the existence of a subspace $W \subset L(\Lambda^2\mathfrak{n}, \mathfrak{l}(\mathfrak{n}, G_0))^1$ (linear maps homogeneous of degree ≥ 1), which is $K(\mathfrak{n}, G_0)$–invariant and complementary to the image of the Lie algebra differential ∂. Assuming this condition, Morimoto proves that any reduction of structure group $\mathcal{P}^{(0)}$ as above can be canonically extended to a principal bundle with structure group $K(\mathfrak{n}, G_0)$ endowed with a canonical Cartan connection with values in $\mathfrak{l}(\mathfrak{n}, G_0)$ whose curvature has values in W. The construction of the bundle and the Cartan connection is done inductively using the techniques of maximal towers developed before. For the case of parabolic subalgebras, the kernel of the Kostant codifferential satisfies condition (C), so Morimoto's procedure applies.

A.5. The procedure of Čap and Schichl. This procedure was derived independently of the results of Tanaka and Morimoto in [**ČS00**]. It uses the description of underlying structures as regular infinitesimal flag structures (using different terminology) and it is tailored to the parabolic case. It avoids the use of infinite prolongations (which show up in Morimoto's procedure as an intermediate step) and directly constructs the Cartan bundle and connection. At the time the procedure was developed, it also had the advantage that it directly led to explicit descriptions of the prolonged bundles. By now, such description can be deduced in the general setting of abstract Cartan geometries using Weyl structures; see Chapter 5.

Let $\mathfrak{g} = \mathfrak{g}_{-k} \oplus \cdots \oplus \mathfrak{g}_k$ be a $|k|$–graded semisimple Lie algebra with no simple ideal contained in $\mathfrak{g}_0$, G a Lie group with Lie algebra $\mathfrak{g}$, and $P \subset G$ a parabolic subgroup for the given grading. Then we have the Levi subgroup $G_0 \subset P$ as well as the normal subgroups $P_+^k \subset \cdots \subset P_+ \subset P$ as introduced in the end of 3.1.3. Given these data, [**ČS00**] is based on the notion of P–frame bundles of degree ℓ from 3.1.15. Compared to the definition there, the degree in [**ČS00**] is shifted by one, but we will use the convention from 3.1.15 for the further exposition. The condition of regularity is phrased as "satisfying the structure equations".

It is then shown that on a regular P–frame bundle (E, θ) of degree ℓ, one has a well–defined torsion map $t_\theta : E \to L(\Lambda^2\mathfrak{g}_-, \mathfrak{g})$, defined analogously to the Cartan curvature, which has nonzero homogeneous components of degrees between 0 and ℓ only. If the homogeneous components of degrees $1, \ldots, \ell$ all have values in the kernel of the codifferential ∂^* (see 3.1.11), then the P–frame bundle is called *harmonic*. For $\ell = 0$ the condition of harmonicity is vacuous, so one recovers regular infinitesimal flag structures. For $\ell = 1$ this notion of harmonicity coincides with the notion of normality used in 3.1.16. Finally, a harmonic P–frame bundle of degree

$\ell = 2k$ by definition is a regular normal parabolic geometry. It is shown that for a P–frame bundle of degree ℓ one may form underlying P–frame bundles of lower degree, and harmonicity is preserved in this procedure.

The main result in [**ČS00**] is that, assuming that $H^1(\mathfrak{g}_-, \mathfrak{g})$ has no component in homogeneity $\ell + 1$, a harmonic P–frame bundle (E, θ) of degree ℓ may be prolonged to a unique (up to isomorphism) harmonic P–frame bundle $(\tilde{E}, \tilde{\theta})$ of degree $\ell + 1$. Iterative application of this result proves Theorems 3.1.14 and 3.1.16. The proof of this main result is actually constructive. For $u \in E$, the value $\theta(u)$ of the frame form is a collection of linear maps $\theta_i(u) : T^i_u E \to \mathfrak{g}_i \oplus \cdots \oplus \mathfrak{g}_{i+\ell}$. These maps are mutually compatible, have prescribed kernels and prescribed behavior when acting on fundamental vector fields. One then defines $\hat{E}_u$ to be the set of all extensions of $\theta(u)$ which add to each $\theta_i(u)$ a map with values in $\mathfrak{g}_{i+\ell+1}$ such that all the conditions are still satisfied. Then $\hat{E} = \bigsqcup_{u\in E} \hat{E}_u$ is a locally trivial bundle over E whose fibers are affine spaces modelled on the space $L(\mathfrak{g}_-, \mathfrak{g})_{\ell+1}$ of linear maps which are homogeneous of degree $\ell + 1$.

Next, from the principal action of $P/P_+^{\ell+1}$ on E and the adjoint action, one constructs a free action of the group $P/P_+^{\ell+2}$ on $\hat{E}$. The bundle $\hat{E}$ inherits a natural filtration of its tangent bundle and carries a family of tautological partially defined differential forms with the properties of a frame form of length $\ell + 1$. Using these, one shows that mimicking the definition of the Cartan curvature, there is a well–defined torsion function $\hat{t} : \hat{E} \to L(\Lambda^2\mathfrak{g}_-, \mathfrak{g})$ with nonzero homogeneous components in degrees $0,\dots,\ell + 1$.

Then one defines $\tilde{E} \subset \hat{E}$ as the subset of those points for which this torsion has values in $\ker(\partial^*)$ and $\tilde{\theta}$ as the restriction of the tautological form. One shows that the affine translation by $\phi \in L(\mathfrak{g}_-, \mathfrak{g})_{\ell+1}$ on a fiber of $\hat{E}$ changes the torsion by $\partial\phi$. Using this, it follows that $\tilde{E} \to E$ is a bundle whose fibers are affine with modelling vector space $\ker(\partial)$. Further, using equivariancy of ∂^*, one proves that $\tilde{E}$ is stable under the action of $P/P_+^{\ell+2}$. Assuming that $H^1(\mathfrak{g}_-, \mathfrak{g})$ has no nonzero component in homogeneity $\ell+1$, it follows that the fibers of $\tilde{E} \to E$ are affine over $\operatorname{im}(\partial)$, which is then used to prove that $P/P_+^{\ell+2}$ acts transitively on each fiber of $\tilde{E} \to M$. Hence, this is a principal bundle and the (relatively easy) verification that $\tilde{\theta}$ is a harmonic frame from of length $\ell + 1$ completes the existence part of the proof.

The uniqueness of the construction follows easily from its tautological nature. If (F, Θ) is any extension of (E, θ) to a harmonic P–frame bundle of degree $\ell + 1$, then we have a natural projection $\pi : F \to E$. For $x \in F$ we can then interpret $\Theta(x)$ as an element in $\hat{E}_{\pi(x)}$, and use this to define a map $F \to \hat{E}$. Harmonicity of F implies that this has values in $\tilde{E}$ and equivariancy of Θ implies that it is a principal bundle map. Compatibility of this map with the frame forms then follows directly from the construction.

APPENDIX B

Tables

These tables are inspired by [**OnVi91**] and by [**On04**], we thank J. Šilhan for help with their preparation.

Table B.1

List of the complex simple Lie algebras and their Dynkin Diagrams.

Type of $\mathfrak{g}$	Dynkin diagram	Matrix algebra $\mathfrak{g}$	$\dim \mathfrak{g}$
A_ℓ $(\ell \geq 1)$	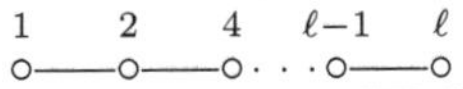	$\mathfrak{sl}(\ell+1, \mathbb{C})$	$\ell^2 + 2\ell$
B_ℓ $(\ell \geq 2)$	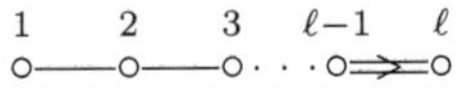	$\mathfrak{so}(2\ell+1, \mathbb{C})$	$2\ell^2 + \ell$
C_ℓ $(\ell \geq 2)$	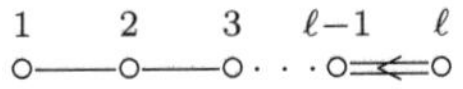	$\mathfrak{sp}(2\ell, \mathbb{C})$	$2\ell^2 + \ell$
D_ℓ $(\ell \geq 3)$	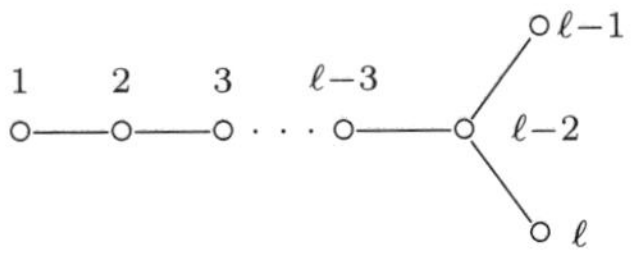	$\mathfrak{so}(2\ell, \mathbb{C})$	$2\ell^2 - \ell$
E_6	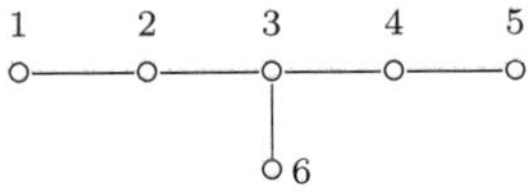		78
E_7	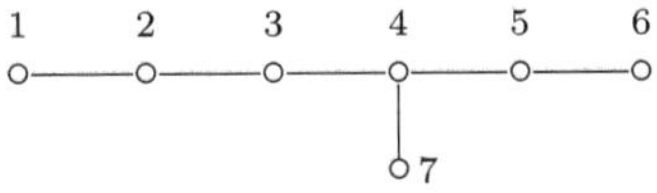		133
E_8	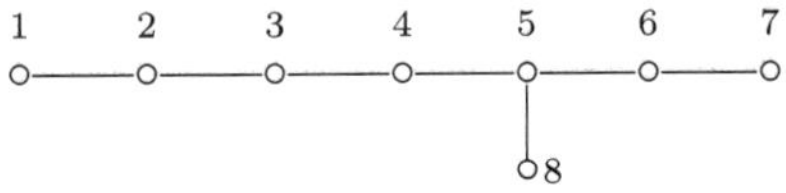		248
F_4	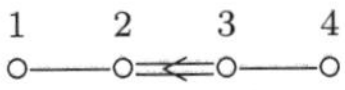		52
G_2			14

Table B.2

For the complex simple Lie algebras, this table lists roots, simple roots α_i, fundamental weights λ_i, and the highest root β. The roots are expressed in terms of auxiliary functionals e_i. For an algebra of rank ℓ, one either uses linearly independent functionals $e_1, \dots, e_\ell$ of functionals $e_1, \dots, e_{\ell+1}$ such that $e_1 + \cdots + e_{\ell+1} = 0$ is the only relation between them.

Type	Roots / Simple roots / Fundamental weights	Highest root
A_ℓ $(\ell \geq 1)$	$e_i - e_j,\ i \neq j\ i,j = 1,\dots,\ell+1$ $\alpha_i = e_i - e_{i+1}$ $\lambda_i = e_1 + \cdots + e_i$	$\beta = e_1 - e_{\ell+1}$ $\beta = \alpha_1 + \alpha_2 + \cdots + \alpha_\ell$ $\beta = \lambda_1 + \lambda_\ell$
B_ℓ $(\ell \geq 2)$	$\pm e_i,\ \pm e_i \pm e_j,\ i \neq j\ i,j = 1,\dots,\ell$ $\alpha_\ell = e_\ell$ $\alpha_i = e_i - e_{i+1}, (i < \ell)$ $\lambda_i = e_1 + \cdots + e_i,\ (i < \ell)$ $\lambda_\ell = \frac{1}{2}(e_1 + \cdots + e_\ell)$	$\beta = e_1 + e_2$ $\beta = \alpha_1 + 2\alpha_2 + \cdots + 2\alpha_\ell$ $\beta = 2\lambda_2,\ \ell = 2$ $\beta = \lambda_2,\ \ell > 2$
C_ℓ $(\ell \geq 2)$	$\pm 2e_i,\ \pm e_i \pm e_j,\ i \neq j\ i,j = 1,\dots,\ell$ $\alpha_\ell = 2e_\ell$ $\alpha_i = e_i - e_{i+1}, (i < \ell)$ $\lambda_i = e_1 + \cdots + e_i$	$\beta = 2e_1$ $\beta = 2\alpha_1 + \cdots + 2\alpha_{\ell-1} + \alpha_\ell$ $\beta = 2\lambda_1$
D_ℓ $(\ell \geq 3)$	$\pm e_i \pm e_j,\ i \neq j\ i,j = 1,\dots,\ell$ $\alpha_\ell = e_{\ell-1} + e_\ell$ $\alpha_i = e_i - e_{i+1}, (i < \ell)$ $\lambda_i = e_1 + \cdots + e_i,\ (i < \ell - 1)$ $\lambda_{\ell-1} = \frac{1}{2}(e_1 + \cdots + e_{\ell-1} - e_\ell)$ $\lambda_\ell = \frac{1}{2}(e_1 + \cdots + e_{\ell-1} + e_\ell)$	$\beta = e_1 + e_2$ $\beta = \alpha_1 + 2\alpha_2 + \cdots + 2\alpha_{\ell-2}$ $+ \alpha_{\ell-1} + \alpha_\ell$ $\beta = \lambda_2 + \lambda_3,\ \ell = 3$ $\beta = \lambda_2,\ \ell > 3$

(to be continued)

Table B.2

(continued)

Type	Roots / Simple roots / Fundamental weights	Highest root
E_6	$\pm 2e_7,\ e_i - e_j,\ e_i + e_j + e_k \pm e_7$ $i, j, k = 1, \ldots, 6$ ― $\alpha_i = e_i - e_{i+1}, (i < 6)$ $\alpha_6 = e_4 + e_5 + e_6 + e_7$ ― $\lambda_i = e_1 + \cdots + e_i$ $+ \min\{i, 6-i\} e_7,\ (i < 6)$ $\lambda_6 = 2e_7$	$\beta = 2e_7$ $\beta = \alpha_1 + 2\alpha_2 + 3\alpha_3 +$ $2\alpha_4 + \alpha_5 + 2\alpha_6$ $\beta = \lambda_6$
E_7	$e_i - e_j,\ e_i + e_j + e_k + e_l\ i, j, k, l = 1, \ldots, 8$ ― $\alpha_i = e_i - e_{i+1}, (i < 7)$ $\alpha_7 = e_5 + e_6 + e_7 + e_8$ ― $\lambda_i = e_1 + \cdots + e_i$ $+ \min\{i, 8-i\} e_8,\ (i < 7)$ $\lambda_7 = 2e_8$	$\beta = -e_7 + e_8$ $\beta = \alpha_1 + 2\alpha_2 + 3\alpha_3 +$ $4\alpha_4 + 3\alpha_5 +$ $2\alpha_6 + 2\alpha_7$ $\beta = \lambda_6$
E_8	$e_i - e_j,\ \pm(e_i + e_j + e_k)\ i, j, k = 1, \ldots, 9$ ― $\alpha_i = e_i - e_{i+1}, (i < 8)$ $\alpha_8 = e_6 + e_7 + e_8$ ― $\lambda_i = e_1 + \cdots + e_i$ $- \min\{i, 15-2i\} e_9,\ (i < 8)$ $\lambda_8 = -3e_9$	$\beta = e_1 - e_9$ $\beta = 2\alpha_1 + 3\alpha_2 + 4\alpha_3 +$ $5\alpha_4 + 6\alpha_5 + 4\alpha_6 +$ $2\alpha_7 + 3\alpha_8$ $\beta = \lambda_1$
F_4	$\pm e_i,\ \pm e_i \pm e_j,\ i \neq j\ i, j = 1, \ldots, 4$ $\frac{1}{2}(\pm e_1 \pm e_2 \pm e_3 \pm e_4)$ ― $\alpha_1 = \frac{1}{2}(e_1 - e_2 - e_3 - e_4)$ $\alpha_2 = e_4,\ \alpha_3 = e_3 - e_4$ $\alpha_4 = e_2 - e_3$ ― $\lambda_1 = e_1,\ \lambda_2 = \frac{1}{2}(3e_1 + e_2 + e_3 + e_4)$ $\lambda_3 = 2e_1 + e_2 + e_3,\ \lambda_4 = e_1 + e_2$	$\beta = e_1 + e_2$ $\beta = 2\alpha_1 + 4\alpha_2 + 3\alpha_3 + 2\alpha_4$ $\beta = \lambda_4$
G_2	$\pm e_i,\ e_i - e_j,\ i \neq j\ i, j = 1, 2, 3$ ― $\alpha_1 = -e_2,\ \alpha_2 = e_2 - e_3$ ― $\lambda_1 = e_1,\ \lambda_2 = e_1 - e_3$	$\beta = e_1 - e_3$ $\beta = 3\alpha_1 + 2\alpha_2$ $\beta = \lambda_2$

Table B.3

Inverses of the Cartan matrices of the complex simple Lie algebras.

Type	Inverse Cartan matrix
A_ℓ $(\ell \geq 1)$	$\frac{1}{\ell+1}\begin{pmatrix} \ell & \ell-1 & \ell-2 & \dots & 2 & 1 \\ \ell-1 & 2(\ell-1) & 2(\ell-2) & \dots & 2\cdot 2 & 2 \\ \ell-2 & 2(\ell-2) & 3(\ell-2) & \dots & 3\cdot 2 & 3 \\ \vdots & \vdots & \vdots & \ddots & \vdots & \vdots \\ 2 & 2\cdot 2 & 3\cdot 2 & \dots & (\ell-1)2 & \ell-1 \\ 1 & 2 & 3 & \dots & \ell-1 & \ell \end{pmatrix}$
B_ℓ $(\ell \geq 2)$	$\frac{1}{2}\begin{pmatrix} 2 & 2 & 2 & \dots & 2 & 1 \\ 2 & 4 & 4 & \dots & 4 & 2 \\ 2 & 4 & 6 & \dots & 6 & 3 \\ \vdots & \vdots & \vdots & \ddots & \vdots & \vdots \\ 2 & 4 & 6 & \dots & 2(\ell-1) & \ell-1 \\ 2 & 4 & 6 & \dots & 2(\ell-1) & \ell \end{pmatrix}$
C_ℓ $(\ell \geq 2)$	$\frac{1}{2}\begin{pmatrix} 2 & 2 & 2 & \dots & 2 & 2 \\ 2 & 4 & 4 & \dots & 4 & 4 \\ 2 & 4 & 6 & \dots & 6 & 6 \\ \vdots & \vdots & \vdots & \ddots & \vdots & \vdots \\ 2 & 4 & 6 & \dots & 2(\ell-1) & 2(\ell-1) \\ 1 & 2 & 3 & \dots & \ell-1 & \ell \end{pmatrix}$
D_ℓ $(\ell \geq 3)$	$\frac{1}{4}\begin{pmatrix} 4 & 4 & 4 & \dots & 4 & 2 & 2 \\ 4 & 8 & 8 & \dots & 8 & 4 & 4 \\ 4 & 8 & 12 & \dots & 12 & 6 & 6 \\ \vdots & \vdots & \vdots & \ddots & \vdots & \vdots & \vdots \\ 4 & 8 & 12 & \dots & 4(\ell-2) & 2(\ell-2) & 2(\ell-2) \\ 2 & 4 & 6 & \dots & 2(\ell-2) & \ell & \ell-2 \\ 2 & 4 & 6 & \dots & 2(\ell-2) & \ell-2 & \ell \end{pmatrix}$

(to be continued)

Table B.3

(continued)

Type	Inverse Cartan matrix
E_6	$\frac{1}{3}\begin{pmatrix} 4 & 5 & 6 & 4 & 2 & 3 \\ 5 & 10 & 12 & 8 & 4 & 6 \\ 6 & 12 & 18 & 12 & 6 & 9 \\ 4 & 8 & 12 & 10 & 5 & 6 \\ 2 & 4 & 6 & 5 & 4 & 3 \\ 3 & 6 & 9 & 6 & 3 & 6 \end{pmatrix}$
E_7	$\frac{1}{2}\begin{pmatrix} 3 & 4 & 5 & 6 & 4 & 2 & 3 \\ 4 & 8 & 10 & 12 & 8 & 4 & 6 \\ 5 & 10 & 15 & 18 & 12 & 6 & 9 \\ 6 & 12 & 18 & 24 & 16 & 8 & 12 \\ 4 & 8 & 12 & 16 & 12 & 8 & 6 \\ 2 & 4 & 6 & 8 & 6 & 4 & 4 \\ 3 & 6 & 9 & 12 & 8 & 4 & 7 \end{pmatrix}$
E_8	$\begin{pmatrix} 2 & 3 & 4 & 5 & 6 & 4 & 2 & 3 \\ 3 & 6 & 8 & 10 & 12 & 8 & 4 & 6 \\ 4 & 8 & 12 & 15 & 18 & 12 & 6 & 9 \\ 5 & 10 & 15 & 20 & 24 & 16 & 8 & 12 \\ 6 & 12 & 18 & 24 & 30 & 20 & 10 & 15 \\ 4 & 8 & 12 & 16 & 20 & 14 & 7 & 10 \\ 2 & 4 & 6 & 8 & 10 & 7 & 4 & 5 \\ 3 & 6 & 9 & 12 & 15 & 10 & 5 & 8 \end{pmatrix}$
F_4	$\frac{1}{2}\begin{pmatrix} 2 & 3 & 4 & 2 \\ 3 & 6 & 8 & 4 \\ 2 & 4 & 6 & 3 \\ 1 & 2 & 3 & 2 \end{pmatrix}$
G_2	$\begin{pmatrix} 2 & 1 \\ 3 & 2 \end{pmatrix}$

Table B.4

Real simple Lie algebras, Satake diagrams, the automorphsims ν and s of the Satake diagram mapping a weight to its dual, respectively, its conjugate, and the indices of irreducible representations. See the end of 2.3.15 for more explanations.

Real form	Satake diagram with a weight	s	ν	Index
$\mathfrak{sl}(l+1,\mathbb{R})$	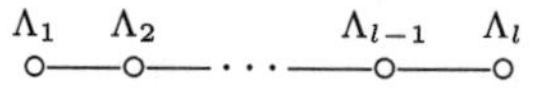	e	$\neq e$	$+1$
$\mathfrak{sl}(m,\mathbb{H})$ $l=2m-1$	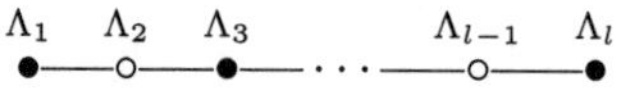	e	$\neq e$	$(-1)^{\sum_{i=1}^{m}\Lambda_{2i-1}}$
$\mathfrak{su}(p,l+1-p)$ $1\leq p\leq\frac{l}{2}$	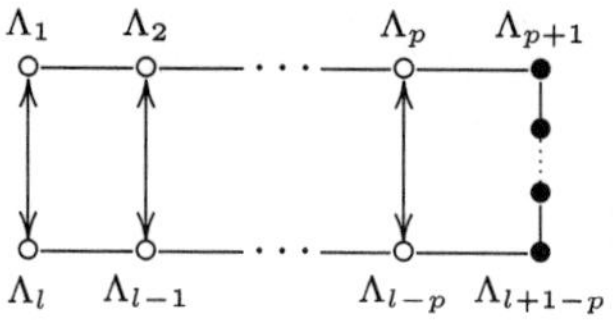			
l even		$\neq e$	$\neq e$	$+1$
$l=2m-1$		$\neq e$	$\neq e$	$(-1)^{(m+p)\Lambda_m}$
$\mathfrak{su}(p,p)$ $l=2p-1$ $p\geq 2$	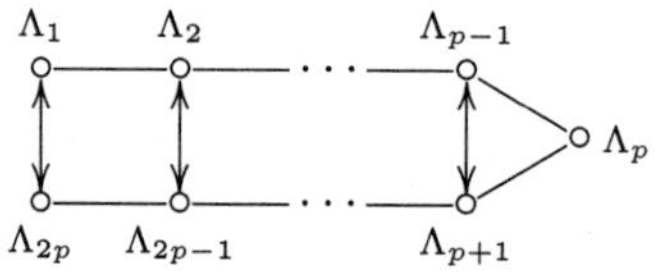	$\neq e$	$\neq e$	$+1$
$\mathfrak{su}(l+1)$	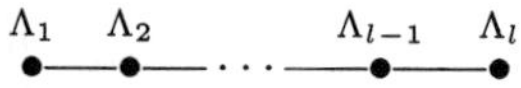			
l even		$\neq e$	$\neq e$	$+1$
$l=2m-1$		$\neq e$	$\neq e$	$(-1)^{m\Lambda_m}$
$\mathfrak{so}(p,2l+1-p)$ $1\leq p\leq l$	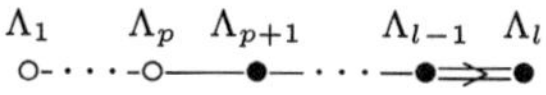			
$p=2k$		e	e	$(-1)^{(k+\frac{l(l+1)}{2})\Lambda_l}$
$p=2k+1$		e	e	$(-1)^{(k+\frac{l(l+3)}{2})\Lambda_l}$
$\mathfrak{so}(2l+1)$	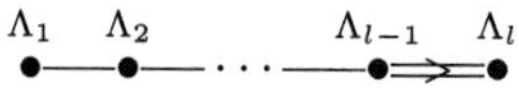	e	e	$(-1)^{\frac{l(l+1)}{2}\Lambda_l}$
$\mathfrak{sp}(2l,\mathbb{R})$	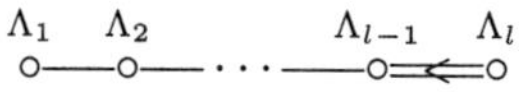	e	e	$+1$
$\mathfrak{sp}(p,l-p)$ $1\leq p\leq\frac{l-1}{2}$	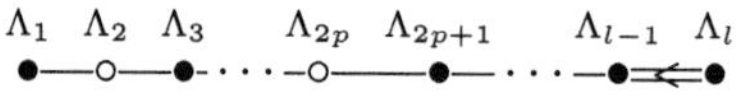	e	e	$(-1)^{\sum_{i=1}^{[\frac{l+1}{2}]}\Lambda_{2i-1}}$
$\mathfrak{sp}(p,p)$ $l=2p$	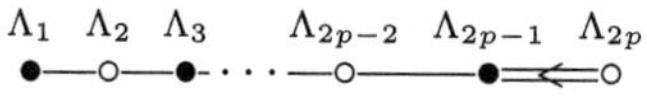	e	e	$(-1)^{\sum_{i=1}^{p}\Lambda_{2i-1}}$
$\mathfrak{sp}(l)$	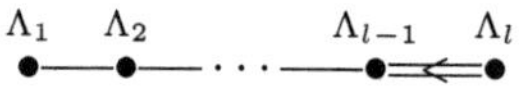	e	e	$(-1)^{\sum_{i=1}^{[\frac{l+1}{2}]}\Lambda_{2i-1}}$

(to be continued)

Table B.4

(continued)

Real form	Satake diagram with a weight	s	ν	Index
$\mathfrak{so}(l,l)$	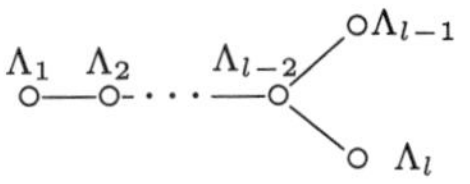			
l even		e	e	$+1$
l odd		e	$\neq e$	$+1$
$\mathfrak{so}(p,2l-p)$ $1 \leq p \leq l-2$	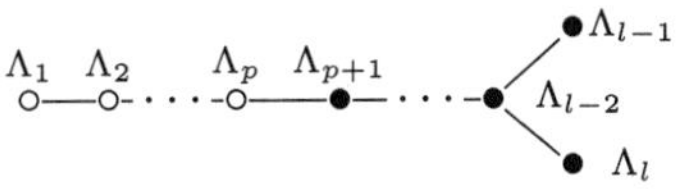			
p, l even		e	e	$(-1)^{\frac{l-p}{2}(\Lambda_{l-1}+\Lambda_l)}$
p, l odd		e	$\neq e$	$(-1)^{\frac{l-p}{2}(\Lambda_{l-1}+\Lambda_l)}$
p even, l odd		$\neq e$	$\neq e$	$+1$
p odd, l even		$\neq e$	e	$+1$
$\mathfrak{so}(l-1,l+1)$	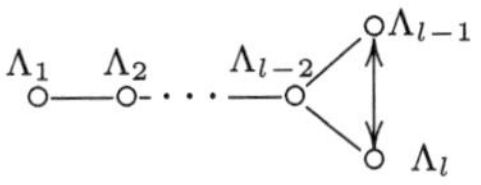			
l even		$\neq e$	e	$+1$
l odd		$\neq e$	$\neq e$	$+1$
$\mathfrak{so}(2l)$	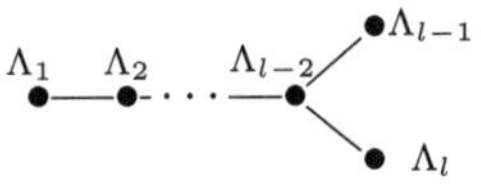			
l even		e	e	$(-1)^{\frac{l}{2}(\Lambda_{l-1}+\Lambda_l)}$
l odd		$\neq e$	$\neq e$	$+1$
$\mathfrak{so}^*(2l)$ $l=2m$	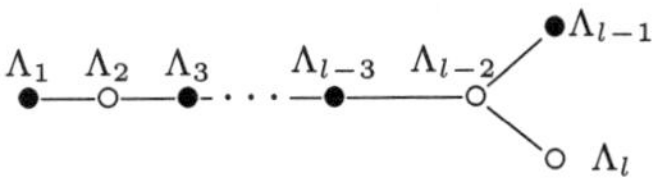	e	e	$(-1)^{\sum_{i=1}^{m}\Lambda_{2i-1}}$
$\mathfrak{so}^*(2l)$ $l=2m+1$	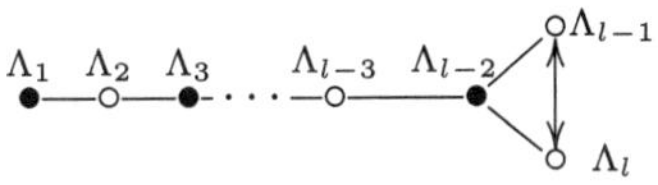	$\neq e$	$\neq e$	$(-1)^{\sum_{i=1}^{m}\Lambda_{2i-1}}$

(to be continued)

Table B.4

(continued)

Real form	Satake diagram with a weight	s	ν	Index
EI	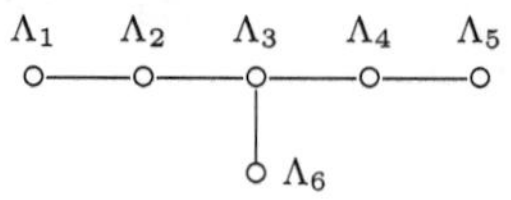	e	$\neq e$	$+1$
EII	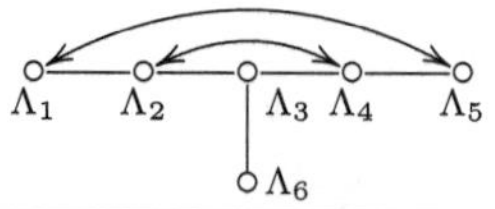	$\neq e$	$\neq e$	$+1$
EIII	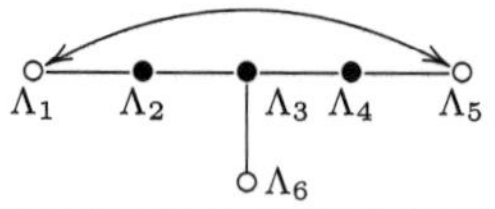	$\neq e$	$\neq e$	$+1$
EIV	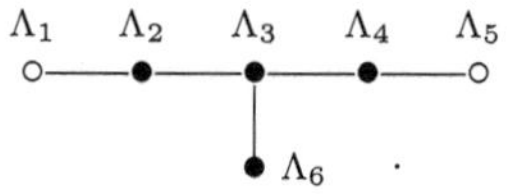	e	$\neq e$	$+1$
compact form of E_6	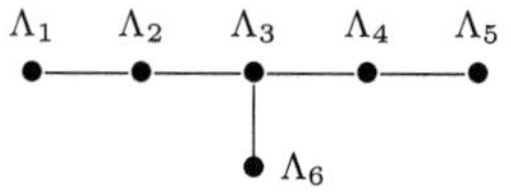	$\neq e$	$\neq e$	$+1$
EV	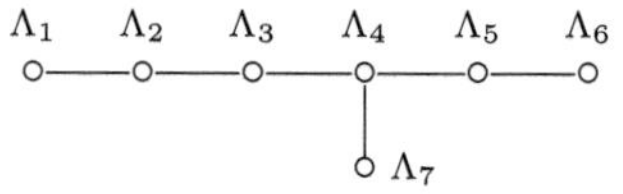	e	e	$+1$
EVI	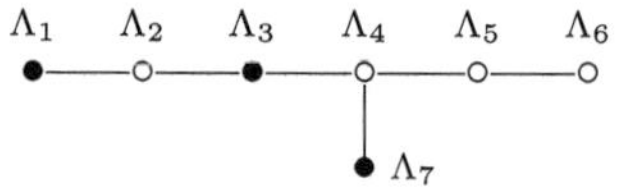	e	e	$(-1)^{\Lambda_1+\Lambda_3+\Lambda_7}$
EVII	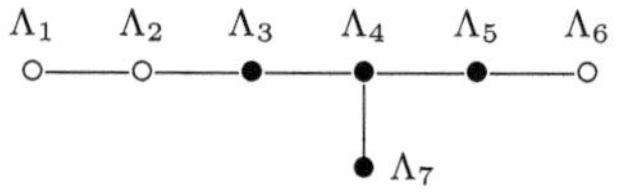	e	e	$+1$
compact form of E_7	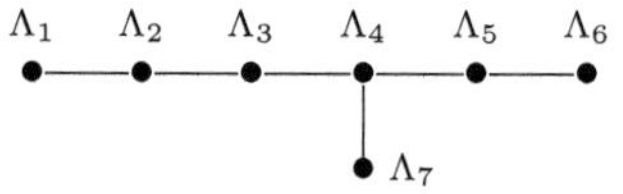	e	e	$(-1)^{\Lambda_1+\Lambda_3+\Lambda_7}$

(to be continued)

Table B.4

(continued)

Real form	Satake diagram with a weight	s	ν	Index
EVIII	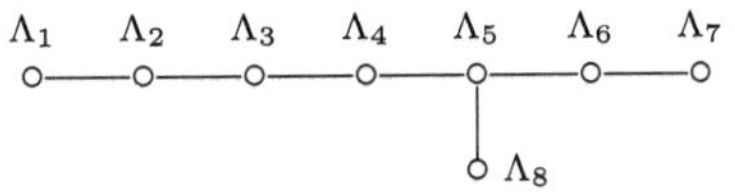	e	e	$+1$
EIX	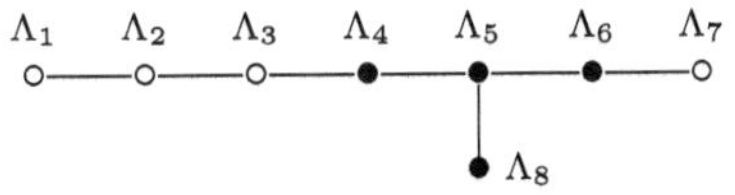	e	e	$+1$
compact form of E_8	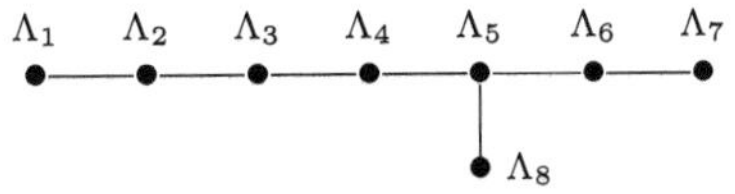	e	e	$+1$
FI	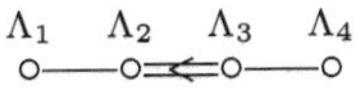	e	e	$+1$
FII	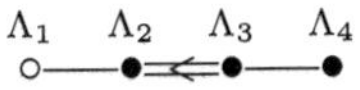	e	e	$+1$
compact form of F_4	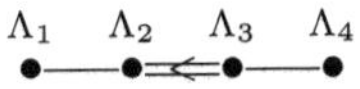	e	e	$+1$
G_2		e	e	$+1$
compact form of G_2		e	e	$+1$

Bibliography

[AMR] D.V. Alekseevsky, P.W. Michor, W. Ruppert, Extensions of Lie algebras, unpublished survey article available as preprint arXiv:math.DG/0005042.

[Ar62] S. Araki, On root systems and an infinitesimal classification of irreducible symmetric spaces, J. Math. Osaka City Univ. **13** (1962) 1–34.

[Arm07a] S. Armstrong, Definite signature conformal holonomy: a complete classification, J. Geom. Phys. **57**, no. 10 (2007) 2024–2048.

[Arm07b] S. Armstrong, Generalised Einstein condition and cone construction for parabolic geometries, preprint arXiv:0705.2390.

[Arm07c] S. Armstrong, Free 3–distributions: holonomy, Fefferman constructions and dual distributions, preprint arXiv:0708.3027.

[AtHi61] M.F. Atiyah, F. Hirzebruch, Vector bundles and homogeneous spaces. Proc. Sympos. Pure Math., Vol. III Amer. Math. Soc. (1961) 7–38.

[AHS78] M.F. Atiyah, N.J. Hitchin, I.M. Singer, Self–duality in four–dimensional Riemannian geometry, Proc. Roy. Soc. London **A 362** (1978) 425–461.

[Ba02] J. Baez, The octonions, Bull. Amer. Math. Soc. **39** no. 2 (2002) 145–205.

[BaE91] T.N. Bailey, M.G. Eastwood, Complex paraconformal manifolds: their differential geometry and twistor theory. Forum Math. **3**, (1991) 61–103.

[BEG94] T.N. Bailey, M.G. Eastwood, A.R. Gover, Thomas's structure bundle for conformal, projective and related structures, Rocky Mountain J. **24** (1994) 1191–1217.

[Ba91] R.J. Baston, R, Almost Hermitian symmetric manifolds, I: Local twistor theory; II: Differential invariants Duke Math. J. **63** (1991) 81–111, 113–138.

[BE89] R.J. Baston, M.G. Eastwood, *The Penrose transform. Its interaction with representation theory*, Oxford Mathematical Monographs, Clarendon Press, Oxford, 1989.

[BE90] R.J. Baston, M.G. Eastwood, Invariant Operators, in T.N. Bailey, R.J. Baston (eds.), *Twistors in Mathematics and Physics*, London Math. Soc. Lecture Notes 156, Cambridge University Press, 1990, 129–163.

[Bau81] H. Baum, *Spin-Strukturen und Dirac-Operatoren über pseudo-Riemannschen Mannigfaltigkeiten.* (German) Teubner-Text zur Mathematik, 41, Teubner-Verlag, Leipzig, 1981.

[Be55] M. Berger, Sur les groupes d'holonomie homogène des variétés à connexion affine et des variétés riemanniennes. Bull. Soc. Math. France **83** (1955) 279–330.

[BGG73] I.N. Bernšteĭn, I.M. Gel'fand, S.I. Gel'fand, Schubert cells, and the cohomology of the spaces G/P, Russian Math. Surveys **28** no. 3 (1973) 1–26.

[BGG75] I.N. Bernšteĭn, I.M. Gel'fand, S.I. Gel'fand, Differential operators on the base affine space and a study of $\mathfrak{g}$-modules. in *Lie groups and their representations* (Proc. Summer School, Bolyai János Math. Soc., Budapest, 1971) Halsted, New York, 1975, 21–64.

[Bi00] O. Biquard, *Métriques d'Einstein asymptotiquement symétriques*, Astérisque **265** (2000).

[Bi02] O. Biquard, Quaternionic contact structures. in *Quaternionic structures in mathematics and physics (Rome, 1999)* (electronic), Univ. Studi Roma "La Sapienza", 1999, 23–30.

[Bor53] A. Borel, Sur la cohomologie des espaces fibrés principaux et des espaces homogènes des groupes de Lie compacts, Ann. of Math. **57** (1953) 115–207.

[BH61] A. Borel, A. Haefliger, La classe d'homologie fondamentale d'un espace analytique, Bull. Soc. Math. France **89** (1961) 461–513.

[Bott57] R. Bott, Homogeneous vector bundles, Ann. of Math. **66** (1957) 203–248.

[Br79] R. Bryant, Some aspects of the local and global theory of Pfaffian systems, PhD Thesis, University of North Carolina at Chapel Hill, 1979.

[Br01] R. Bryant, Bochner–Kähler metrics, J. Amer. Math. Soc. **14** (2001) 623–715.

[Br06] R. Bryant, Conformal geometry and 3–plane fields on 6–manifolds, in "Developments of Cartan Geometry and Related Mathematical Problems", RIMS Symposium Proceedings, vol. **1502** (July, 2006), pp. 1-15, Kyoto University, preprint arXiv:math/0511110

[BC99] F. Bourgeois, M. Cahen, A variational principle for symplectic connections, J. Geom. Phys. **30** (1999) 233–265.

[BDS77] D. Burns, K. Diederich, S. Shnider, Distinguished curves in pseudoconvex boundaries, Duke Math. J. **44** no. 2 (1977) 407–431.

[CahS04] M. Cahen, L.J. Schwachhöfer, Special symplectic connections, preprint arXiv:math/0402221

[Cal98] D.M.J. Calderbank, Möbius structures and two–dimensional Einstein–Weyl geometry, J. Reine Ang. Math. **504** (1998) 37–53.

[CDS05] D.M.J. Calderbank, T. Diemer, V. Souček, Ricci–corrected derivatives and invariant differential operators. Differential Geom. Appl. **23**, no. 2 (2005) 149–175.

[Čap02] A. Čap, Parabolic geometries, CR-tractors, and the Fefferman construction, Differential Geom. Appl. **17** (2002) 123-138.

[Čap05] A. Čap, Correspondence spaces and twistor spaces for parabolic geometries, J. Reine Angew. Math. **582** (2005) 143–172.

[ČE03] A. Čap, M.G. Eastwood, Some special geometry in dimension six, Rend. Circ. Mat. Palermo Suppl. ser. II **71** (2003) 93-98.

[ČGo00] A. Čap, A.R. Gover, Tractor bundles for irreducible parabolic geometries, in *SMF Séminaires et congrès* **4** (2000) 129–154.

[ČGo02] A. Čap, A.R. Gover, Tractor calculi for parabolic geometries, Trans. Amer. Math. Soc. **354**, no. 4 (2002) 1511-1548.

[ČGo03] A. Čap, A.R. Gover, Standard tractors and the conformal ambient metric construction, Ann. Global Anal. Geom. **24**, 3 (2003) 231-259.

[ČGo08] A. Čap, A.R. Gover, CR–tractors and the Fefferman space, Indiana Univ. Math. J. **57** No. 5 (2008) 2519-2570.

[ČSa07] A. Čap, K. Sagerschnig, On Nurowski's conformal structure associated to a generic rank two distribution in dimension five, preprint arXiv:0710.2208.

[ČS00] A. Čap, H. Schichl, Parabolic geometries and canonical Cartan connections, Hokkaido Math. J. **29** No.3 (2000) 453-505.

[ČSc02] A. Čap, G. Schmalz, Partially integrable almost CR manifolds of CR dimension and codimension two, in *Lie Groups Geometric Structures and Differential Equations - One Hundred Years after Sophus Lie*, Adv. Stud. Pure Math. **37**, Math. Soc. of Japan, Tokyo, (2002) 45-79.

[ČSl03] A. Čap, J. Slovák, Weyl structures for parabolic geometries, Math. Scand. **93** (2003) 53–90.

[ČSS97a] A. Čap, J. Slovák, V. Souček, Invariant operators on manifolds with almost Hermitian symmetric structures, I. Invariant differentiation. Acta Math. Univ. Commenianae **66** no. 1 (1997) 33–69.

[ČSS97b] A. Čap, J. Slovák, V. Souček, Invariant operators on manifolds with almost Hermitian symmetric structures, II. Normal Cartan connections. Acta Math. Univ. Comenian. **66** no. 2 (1997) 203–220.

[ČSS00] A. Čap, J. Slovák, V. Souček, Invariant operators on manifolds with almost Hermitian symmetric structures, III. Standard operators. Diff. Geom. Appl. **12** no. 1 (2000) 51–84.

[ČSS01] A. Čap, J. Slovák, V. Souček, Bernstein–Gelfand–Gelfand sequences, Ann. of Math. **154** no. 1 (2001), 97–113.

[ČSŽ03] A. Čap, J. Slovák, V. Žádník, On distinguished curves in parabolic geometries, Transform. Groups **9** no. 2 (2004) 143–166.

[ČŽ05] A. Čap, V. Žádník, On the geometry of chains, preprint arXiv:math.DG/0504469, to appear in J. Differential Geom.

[ČŽ08] A. Čap, V. Žádník, Contact projective structures and chains, preprint arXiv:0810.2721.

[Car10] É. Cartan, Les systèmes de Pfaff a cinq variables et les équations aux derivées partielles du second ordre, Ann. Ec. Normale **27** (1910) 109–192.

[Car14] É. Cartan, *Les groupes réels simples finis et continus*, Ann. Éc. Norm. Sup., 3-ième série, **31** (1914) 263–355.

[Car32] É. Cartan, Sur la géométrie pseudo-conforme des hypersurfaces de l'espace de deux variables complexes, Ann. Mat. Pura Appl., IV. Ser. **11** (1932) 17-90.

[Car37] É. Cartan, *Leçons sur la théorie des espaces à connexion projective.* (Cahiers scientifiques, fasc. XVII) Gauthier–Villars, 1937.

[Car52] É. Cartan, *Les espaces géneralisés*, in *Notice sur les traveaux scientifique*, Ouvres Completes, Partie I, Vol. I (1952) 72–85.

[CSM95] C. Carter, G. Segal, I. Macdonald, *Lectures on Lie Groups and Lie Algebras*, London Mathematical Society Student Texts 32, Cambridge University Press, 1995.

[Cart61] P. Cartier, Remarks on "Lie algebra cohomology and the generalized Borel-Weil theorem", by B. Kostant, Ann. of Math. **74** (1961) 388–390.

[CaSt99] L. Casian, R.J. Stanton, Schubert cells and representation theory. Invent. Math. **137** no. 3 (1999) 461–539.

[Ch40] S.-S. Chern, The geometry of the differential equation $y''' = F(x, y, y', y'')$, Sci. Rep. Nat. Tsing Hua Univ. (A) **4** (1940) 97–111.

[ChMo76] S.S. Chern, J.K. Moser, Real hypersurfaces in complex manifolds, Acta Math. **133** (1974) 219–271; Erratum Acta Math. **150** no. 3-4 (1983) 297.

[Dem76] M. Demazure, A very simple proof of Bott's theorem, Invent. Math. **33**, no. 3 (1976) 271–272.

[Do05] B. Doubrov, Projective parametrization of homogeneous curves, Arch. Math. (Brno) 41 (2005) 129–133.

[DS] B. Doubrov, J. Slovák, Inclusions of parabolic geometries, preprint arXiv:0807.3360, to appear in Pure and Appl. Math. Quart.

[Du06] D. Duchemin, Quaternionic contact structures in dimension 7, Ann. Inst. Fourier **56** (2006) 851–885.

[Ea90] M.G. Eastwood, The Penrose transform, in T.N. Bailey, R.J. Baston (eds.), *Twistors in Mathematics and Physics*, London Math. Soc. Lecture Notes 156, Cambridge University Press, 1990, 87–103.

[Ea92] M.G. Eastwood, A Penrose transform for D_4 and homomorphisms of generalized Verma modules. Bull. London Math. Soc. **24**, no. 4 (1992) 347–350.

[EPW81] M.G. Eastwood, R. Penrose, R.O. Wells, Jr., Cohomology and massless fields, Commun. Math. Phys. **78** (1981) 305–351.

[EaRi87] M.G. Eastwood, J.W. Rice, Conformally invariant differential operators on Minkowski space and their curved analogues, Comm. Math. Phys. **109**, no.2 (1987) 207–228; Erratum Comm. Math. Phys. **144**, no. 1 (1992) 213.

[EaSl04] M.G. Eastwood, J. Slovák, Preferred parameterisations on homogeneous curves, Comment. Math. Univ. Carolin. 45 (2004), no. 4, 597–606.

[ET79] D.B.A. Epstein, W.P. Thurston, Transformation groups and natural bundles. Proc. London Math. Soc. **38** no. 2 (1979) 219–236.

[EIS99] V.V. Ežov, A.V. Isaev, G. Schmalz, Invariants of elliptic and hyperbolic CR-structures of codimension 2, Internat. J. Math. **10**, no. 1 (1999) 1–52.

[ES94] V.V. Ežov, G. Schmalz, Holomorphic automorphisms of quadrics, Math. Z. **216**, no. 3 (1994) 453–470.

[ES99] V.V. Ežov, G. Schmalz, Infinitesimale Starrheit hermitscher Quadriken in allgemeiner Lage, Math. Nachr. **204** (1999) 41–60.

[Fe76] C. Fefferman, Monge–Ampère equations, the Bergman kernel and geometry of pseudoconvex domains, Ann. of Math. **103** (1976) 395–416; Erratum **104** (1976) 393–394.

[Fe79] C. Fefferman, Parabolic invariant theory in complex analysis, Adv. in Math. **31**, no. 2 (1979) 131–262.

[FG85] C. Fefferman, C.R. Graham, Conformal invariants, in *Élie Cartan et les Mathématiques d'Adjourd'hui*, (Astérisque, hors serie) (1985) 95–116.

[FG07] C. Fefferman, C.R. Graham, The ambient metric, 96 pages, preprint arXiv:0710.0919

[Feg79] H.D. Fegan, Conformally invariant first order differential operators. Quart. J. Math. Oxford **27**, no. 107 (1976) 371–378.

[Fo92] F. Forstnerič, Mappings of quadric Cauchy-Riemann manifolds, Math. Ann. **292** (1992) 163–180.

[Fox05a] D.J.F. Fox, Contact projective structures, Indiana Univ. Math. J. **54** no. 6 (2005) 1547–1598.

[Fox05b] D.J.F. Fox, Contact path geometries, preprint arXiv:math.DG/0508343.

[Fu97] W. Fulton, *Young tableaux*, London Mathematical Society Student Texts 35, Cambridge University Press, 1997.

[FH91] W. Fulton, J. Harris, *Representation theory. A first course*, Graduate Texts in Mathematics 129, Springer, 1991.

[GaMi98] T. Garrity, R. Mizner, Vector-valued forms and CR geometry, Adv. Stud. Pure Math., 25, Math. Soc. Japan, Tokyo, 1997, 110–121.

[Gi90] S. Gindikin, Generalized conformal structures, in T.N. Bailey, R.J. Baston (eds.), *Twistors in Mathematics and Physics*, London Math. Soc. Lecture Notes 156, Cambridge University Press, 1990, 36–52.

[Gon87] A.B. Goncharov, Generalized conformal structures on manifolds, Selecta Math. Soviet. **6** (1987) 308–340.

[Go05] A.R. Gover, Laplacian operators and Q–curvature on conformally Einstein manifolds, Math. Ann. **336** no. 2 (2006) 311–334.

[GoHi04] A.R. Gover, K. Hirachi, Conformally invariant powers of the Laplacian — a complete nonexistence theorem. J. Amer. Math. Soc. **17**, no. 2 (2004) 389–405.

[GoGr05] A.R. Gover, C.R. Graham, CR invariant powers of the sub–Laplacian, J. Reine Angew. Math. **583** (2005) 1–27.

[Gr92] C.R. Graham, Conformally invariant powers of the Laplacian. II. Nonexistence. J. London Math. Soc. **46** no. 3 (1992) 566–576.

[GHV72] W. Greub, S. Halperin, R. Vanstone, *Connections, curvature, and cohomology. Vol. I: de Rham cohomology of manifolds and vector bundles.* Pure and Applied Mathematics, Vol. 47. Academic Press, 1972

[GrHa78] P. Griffiths, J. Harris, *Principles of algebraic geometry*, John Wiley & Sons, 1978.

[Gro00] D.A. Grossman, Torsion-free path geometries and integrable second order ODE systems. Selecta Math. **6**, No. 4 (2000) 399–442.

[Ha06] M. Hammerl, Homogeneous Cartan geometries, diploma thesis, University of Vienna, 2006, electronically available via http://www.mat.univie.ac.at/∼cap/theses.html .

[Ha07] M. Hammerl, Homogeneous Cartan geometries, Arch. Math. (Brno) **43**, no. 5 (2007) 431–442.

[HS06] J. Hrdina, J. Slovák, Generalized planar curves and quaternionic geometry, Ann. Global Anal. Geom. **29** no. 4 (2006) 349–360.

[Je00] J. Jeffers, Lost theorems of geometry, Amer. Math. Monthly **107**, no. 9 (2000) 800–812.

[Kac90] V.G. Kac, *Infinite–dimensional Lie algebras*, third edition, Cambridge University Press (1990).

[Kane93] S. Kaneyuki, On the subalgebras $\mathfrak{g}_0$ and $\mathfrak{g}_{ev}$ of semisimple graded Lie algebras, J. Math. Soc. Japan, **45**, No. 1 (1993) 1–19.

[Kan73] I. Kantor, Models of exceptional Lie algebras, Soviet Math. Dokl. **14** (1973) 254–25.

[Kn96] A.W. Knapp, *Lie groups beyond an introduction*, Progress in Mathematics 140, Birkhäuser, 1996.

[Ko72] S. Kobayashi *Transformation groups in differential geometry*, Ergebnisse der Mathematik und ihrer Grenzgebiete, 70. Springer-Verlag, New York-Heidelberg, 1972.

[KN64] S. Kobayashi, T. Naganon, On filtered Lie algebras and geometric structures. I. J. Math. Mech. **13** (1964) 875–907.

[KoNo69] S. Kobabyashi, K. Nomizu, *Foundations of Differential Geometry* Volume II, Interscience Publishers, 1969.

[Koch88] L. Koch, Chains on CR manifolds and Lorenz geometry, Trans. Amer. Math. Soc. **307** (1988) 827–841.

[Koch93] L. Koch, Cains, null-chains, and CR geometry, Tran. Amer. Math. Soc. **338** (1993) 245–261.

[Kol71] I. Kolář, Higher order torsions of spaces with Cartan connection. Cahiers Topologie Géom. Différentielle **12** (1971) 137–146.

[KMS] I. Kolař, P.W. Michor, J. Slovák, *Natural operations in differential geometry*, Springer, 1993.

[Kos61] B. Kostant, Lie algebra cohomology and the generalized Borel-Weil theorem, Ann. of Math. **74** (1961) 329–387.

[Kos63] B. Kostant, Lie algebra cohomology and generalized Schubert cells, Ann. of Math. (2) **77** (1963) 72–144.

[LeB90] C. LeBrun, Twistors, Ambitwistors, and Conformal Gravity, in T.N. Bailey, R.J. Baston (eds.), *Twistors in Mathematics and Physics*, London Math. Soc. Lecture Notes 156, Cambridge University Press, 1990, 71–86.

[Leis06] T. Leistner, Conformal holonomy of C-spaces, Ricci-flat, and Lorentzian manifolds, Differential Geom. Appl. **24**, no. 5 (2006) 458–478.

[Leit04] F. Leitner, Conformal holonomy of bi-invariant metrics, Conform. Geom. Dyn. **12** (2008) 18–31.

[McK06] B. McKay, Complete complex parabolic geometries. Int. Math. Res. Not. (2006) Art. ID 86937, 34 pp.

[MS99] S.A. Merkulov, L.J. Schwachhöfer, Classification of irreducible holonomies of torsion–free affine connections, Ann. Math. **150** (1999) 77–149; Addendum: Ann. Math. **150** (1999) 1177–1179.

[Mi91a] R. Miyaoka, Lie contact structures and normal Cartan connections, Kodai Math. J. **14**, no. 1 (1991) 13–41, MR: 92d:53025.

[Mi91b] R. Miyaoka, Lie contact structures and conformal structures. Kodai Math. J. **14**, no. 1, (1991) 42–71, MR: 92d:53026.

[Miz93] R.I. Mizner, Almost CR structures, f–structures, almost product structures and associated connections, Rocky Mountain J. Math. **23** no. 4 (1993) 1337–1359.

[Mon02] R. Montgomery, *A tour of subriemannian geometries, their geodesics and applications*, Mathematical Surveys and Monographs vol. 91, Amer. Math. Soc. 2002.

[Mo93] T. Morimoto, *Geometric structures on filtered manifolds*, Hokkaido Math. J. **22** (1993) 263–347.

[Ne03] Yu.A. Neretin, A construction of finite–dimensional faithful representation of Lie algebra, Rend. Circ. Mat. Palermo Supp. (2) **71** (2003) 159–161.

[NeNi57] A. Newlander, L. Nierenberg, Complex analytic coordinates in almost complex manifolds, Ann. of Math. **65** (1957) 391–404.

[Nu05] P. Nurowski, Differential equations and conformal structures, J. Geom. Phys. **55**, no. 1 (2005) 19–49.

[Nu08] P. Nurowski, Conformal structures with explicit ambient metrics and conformal G_2 holonomy, in M. Eastwood, W. Miller, Jr. (eds.): *Symmetries and overdetermined systems of partial differential equations*, The IMA values in Mathematics and its Applications 144, Springer 2008, 515–526.

[On93] A.L. Onishchik (Ed.) *Lie groups and Lie algebras I*, Encyclopedia of Mathematical Sciences 20, Springer, 1993.

[On94] A.L. Onishchik, *Topology of transitive transformation groups*, Johann Ambrosius Barth, 1994.

[On04] A.L. Onishchik, *Lectures on real semisimple Lie algebras and their representations*, ESI Lectures in Mathematics and Physics 1, Europ. Math. Soc., Zürich, 2004.

[OnVi91] A.L. Onishchik, E.B. Vinberg (Ed.) *Lie groups and Lie algebras III*, Encyclopedia of Mathematical Sciences 41, Springer, 1991.

[Pa57] R.S. Palais, *A global formulation of the Lie theory of Transformation Groups*, Mem. Amer. Math. Soc. **22** (1957).

[PT77] R.S. Palais, C.L. Terng, Natural bundles have finite order. Topology **19**, no. 3 (1977) 271–277.

[PŽ08] M. Panák, V. Žádník, Remarks on special symplectic connections, Arch. Math. (Brno) **44** no. 5 (2008) 491–510.

[Pe76] R. Penrose, Nonlinear gravitons and curved twistor theory, Gen. Rel. Grav. **7** (1976) 31–52.

[Sag06a] K. Sagerschnig, Parabolic geometries determined by filtrations of the tangent bundle, Rend. Circ. Mat. Palermo Suppl. ser. II, **79** (2006) 175–181.

[Sag06b] K. Sagerschnig, Split octonions and generic rank two distributions in dimension five, Arch. Math. (Brno) **42** (2006) Suppl. 329–339.

[Sag08] K. Sagerschnig, Weyl structures for generic rank two distributions in dimension five, Ph.D. thesis, University of Vienna, 2008, electronically available via http://www.mat.univie.ac.at/~cap/theses.html.

[Sa60] I. Satake,On representations and compactifications of symmetric Riemannian spaces, Ann. of Math. **71** (1960) 77–110.

[SaYa88] H. Sato, K. Yamaguchi, Lie contact manifolds, in *Geometry of manifolds (Matsumoto, 1988)*, Perspect. Math., 8, Academic Press (1989) 191–238.

[SaYa93] H. Sato, K. Yamaguchi, Lie contact manifolds II, Math. Ann. **297**, no.1 (1993) 33–57.

[SchSl00] G. Schmalz, J. Slovák, The geometry of hyperbolic and elliptic CR-manifolds of codimension two, Asian J. Math. **4**, no. 3 (2000) 565–597;

[Sh97] R.W. Sharpe, *Differential Geometry*, Graduate Texts in Mathematics 166, Springer, 1997.

[Ši04] J. Šilhan, A real analog of Kostant's version of the Bott–Borel–Weil theorem, J. Lie Theory **14**, no. 2 (2004) 481–499.

[Sl96] J. Slovák, Principal prolongations and geometries modeled on homogeneous spaces. Arch. Math. (Brno) **32** no. 4 (1996) 325–342.

[Sl97] J. Slovák, Parabolic geometries, Research Lecture Notes, Part of DrSc. Dissertation, Preprint IGA 11/97, electronically available at www.maths.adelaide.edu.au, 70pp.

[SlSo04] J. Slovák, V. Souček, Invariant operators of the first order on manifolds with a given parabolic structure, in J.P. Bourguignon, T. Branson and O. Hijazi (eds.): *Global analysis and harmonic analysis*, Séminaires et Congrés **4**, Soc. Math. de France, Paris, 2000, 251–276.

[SF00] T.A. Springer, F.D. Feldenkamp, *Octonions, Jordan algebras and exceptional groups*, Springer, Berlin, 2000.

[S64] S. Sternberg, *Lectures on differential geometry*, Prentice-Hall, Inc., Englewood Cliffs, N.J. 1964.

[Tak94] M. Takeuchi, Lagrangean contact structures on projective cotangent bundles. Osaka J. Math. **31** (1994) 837–860.

[Tan62] N. Tanaka, On the pseudo-conformal geometry of hypersurfaces of the space of n complex variables, J. Math. Soc. Japan **14** (1962) 397–429.

[Tan75] N. Tanaka, *A differential geometric study on strongly pseudo-convex manifolds.* Lectures in Mathematics, Department of Mathematics, Kyoto University, No. 9. Kinokuniya Book-Store Co., 1975

[Tan79] N. Tanaka, On the equivalence problems associated with simple graded Lie algebras, Hokkaido Math. J. **8**, no. 1 (1979) 23–84.

[Tau94] C.H. Taubes, The Seiberg–Witten invariants and symplectic forms, Math. Res. Lett. **1** no. 6 (1994) 809–822.

[Ti67] J. Tits, *Tabellen zu den einfachen Lie Gruppen und ihren Darstellungen*, Springer-Verlag, Berlin-New York 1967.

[Tho26] T.Y. Thomas, On conformal geometry, Proc. Nat. Acad. Sci. USA **12** (1926) 352–359.

[Tho31] T.Y. Thomas, Conformal tensors, Proc. Nat. Acad. Sci. USA **18** (1931) 103–189.

[Tho34] T.Y. Thomas, *The differential invariants of generalized spaces*, Amer. Math. Soc. (1934) reprinted in 1991.

[Ve68] D.-N. Verma, Structure of certain induced representations of complex semisimple Lie algebras, Bull. Amer. Math. Soc. **74** (1968) 160–166, Erratum same volume p. 628.

[We78] S.M. Webster, Pseudo–hermitian structures on a real hypersurface, J. Differential Geom. **13** (1978) 25–41.

[Wo64] J.A. Wolf, On the classification of hermitian symmetric spaces. J. Math. Mech. **13** (1964) 489–495.

[Wo65] J.A. Wolf, Complex homogeneous contact manifolds and quaternionic symmetric spaces. J. Math. Mech. **14** (1965) 1033–1047.

[Wü1905] K. Wünschmann, *Über Berührungsbedingungen bei Integralkurven von Differentialgleichungen*, Ph.D. thesis, Königliche Universität Greifswald, 1905, 36 pp.

[Ya93] K. Yamaguchi, Differential systems associated with simple graded Lie algebras, Advanced Studies in Pure Math. **22** (1993) 413–494.

[ZŽ08] L. Zalabová, V. Žádník, Remarks on Grassmannian symmetric spaces, Arch. Math. (Brno) **44**, no. 5 (2008) 569–585.

[Ž09] V. Žádník, Lie contact structures and chains, preprint arXiv:0901.4433

Index

Titles in This Series

TITLES IN THIS SERIES

123 **Barbara Fantechi, Lothar Göttsche, Luc Illusie, Steven L. Kleiman, Nitin Nitsure, and Angelo Vistoli,** Fundamental algebraic geometry: Grothendieck's FGA explained, 2005

122 **Antonio Giambruno and Mikhail Zaicev, Editors,** Polynomial identities and asymptotic methods, 2005

121 **Anton Zettl,** Sturm-Liouville theory, 2005

120 **Barry Simon,** Trace ideals and their applications, 2005

119 **Tian Ma and Shouhong Wang,** Geometric theory of incompressible flows with applications to fluid dynamics, 2005

118 **Alexandru Buium,** Arithmetic differential equations, 2005

117 **Volodymyr Nekrashevych,** Self-similar groups, 2005

116 **Alexander Koldobsky,** Fourier analysis in convex geometry, 2005

115 **Carlos Julio Moreno,** Advanced analytic number theory: L-functions, 2005

114 **Gregory F. Lawler,** Conformally invariant processes in the plane, 2005

113 **William G. Dwyer, Philip S. Hirschhorn, Daniel M. Kan, and Jeffrey H. Smith,** Homotopy limit functors on model categories and homotopical categories, 2004

112 **Michael Aschbacher and Stephen D. Smith,** The classification of quasithin groups II. Main theorems: The classification of simple QTKE-groups, 2004

111 **Michael Aschbacher and Stephen D. Smith,** The classification of quasithin groups I. Structure of strongly quasithin K-groups, 2004

110 **Bennett Chow and Dan Knopf,** The Ricci flow: An introduction, 2004

109 **Goro Shimura,** Arithmetic and analytic theories of quadratic forms and Clifford groups, 2004

108 **Michael Farber,** Topology of closed one-forms, 2004

107 **Jens Carsten Jantzen,** Representations of algebraic groups, 2003

106 **Hiroyuki Yoshida,** Absolute CM-periods, 2003

105 **Charalambos D. Aliprantis and Owen Burkinshaw,** Locally solid Riesz spaces with applications to economics, second edition, 2003

104 **Graham Everest, Alf van der Poorten, Igor Shparlinski, and Thomas Ward,** Recurrence sequences, 2003

103 **Octav Cornea, Gregory Lupton, John Oprea, and Daniel Tanré,** Lusternik-Schnirelmann category, 2003

102 **Linda Rass and John Radcliffe,** Spatial deterministic epidemics, 2003

101 **Eli Glasner,** Ergodic theory via joinings, 2003

100 **Peter Duren and Alexander Schuster,** Bergman spaces, 2004

99 **Philip S. Hirschhorn,** Model categories and their localizations, 2003

98 **Victor Guillemin, Viktor Ginzburg, and Yael Karshon,** Moment maps, cobordisms, and Hamiltonian group actions, 2002

97 **V. A. Vassiliev,** Applied Picard-Lefschetz theory, 2002

96 **Martin Markl, Steve Shnider, and Jim Stasheff,** Operads in algebra, topology and physics, 2002

95 **Seiichi Kamada,** Braid and knot theory in dimension four, 2002

94 **Mara D. Neusel and Larry Smith,** Invariant theory of finite groups, 2002

93 **Nikolai K. Nikolski,** Operators, functions, and systems: An easy reading. Volume 2: Model operators and systems, 2002

For a complete list of titles in this series, visit the AMS Bookstore at **www.ams.org/bookstore/**.